Diesel Emissions and Their Control

Other SAE titles of interest:

Diesel Engine Reference Book
Edited by Rodica Baranescu and Bernard J. Challen
(Order No. R-183)

Vehicle and Engine Technology
By Heinz Heisler
(Order No. R-233)

Practical Diesel-Engine Combustion Analysis
By Bertrand Hsu
(Order No. R-327)

The Diesel Engine
By Daniel J. Holt
(Order No. PT-109)

Diesel Emissions and Their Control

W. Addy Majewski
and
Magdi K. Khair

SAE International™
Warrendale, Pa.

For permission and licensing requests, contact:

SAE Permissions
400 Commonwealth Drive
Warrendale, PA 15096-0001 USA
E-mail: permissions@sae.org
Tel: 724-772-4028
Fax: 724-772-4891

Library of Congress Cataloging-in-Publication Data

Majewski, W. Addy
Diesel emissions and their control / W. Addy Majewski and Magdi K. Khair.
p. cm.
Includes bibliographical references and index.
ISBN-13: 978-0-7680-0674-2
ISBN-10: 0-7680-0674-0
1. Motor vehicles—Pollution control devices. 2. Motor vehicles—Motors (Diesel)—Exhaust gas. I. Khair, Magdi K. II. Title.
TL214.P6M35 2006
629.25'28--dc22

2006049659

SAE International
400 Commonwealth Drive
Warrendale, PA 15096-0001 USA
E-mail: CustomerService@sae.org
Tel: 877-606-7323 (inside USA and Canada)
724-776-4970 (outside USA)
Fax: 724-776-1615

ISBN-10 0-7680-0674-0
ISBN-13 978-0-7680-0674-2

SAE Order No. R-303

Printed in the United States of America.

green press INITIATIVE

SAE International is committed to preserving ancient forests and natural resources. We elected to print *R-303 Diesel Emissions And Their Control* on 30% post consumer recycled paper, processed chlorine free. As a result, for this printing, we have saved:

9 Trees (40' tall and 6-8" diameter)
3,751 Gallons of Wastewater
1,509 Kilowatt Hours of Electricity
413 Pounds of Solid Waste
812 Pounds of Greenhouse Gases

SAE International made this paper choice because our printer, Thomson-Shore, Inc., is a member of Green Press Initiative, a nonprofit program dedicated to supporting authors, publishers, and suppliers in their efforts to reduce their use of fiber obtained from endangered forests.

For more information, visit www.greenpressinitiative.org

Preface

The diesel engine is the most efficient powerplant among all known types of internal combustion engines. Heavy trucks, urban buses, and industrial equipment all over the world have been powered almost exclusively by diesel engines. The diesel engine is a major candidate to become the powerplant of the future, at least in the near term. And yet, as the population of diesels grows, so does their relative contribution to air pollution. Today, control of emissions has become the major driving force in the development of diesel engines. The situation with diesel emissions is particularly sensitive because, in contrast to emissions from other types of internal combustion engines, diesel emissions are visible. Commercial trucks, buses, and off-road machinery, powered by conventional and older diesel engines, left a negative impression on the public due to the black smoke—a flagrant intrusion on the environment—they emitted in plain view, as well as the noxious smell. The visibility of emissions has made diesel also an easy target of political campaigns—in some cases justified, in others not.

The fuel crisis of the mid-1970s, whether real or perceived, created an interest in the diesel engine for personal transportation. For a while, during that crisis, it seemed that redemption of diesel's reputation was near. In the United States, public acceptance was manifested by increasing sales of a European import, Volkswagen's diesel Rabbit. Undaunted by its foreign competition, American automaker General Motors responded by offering a dieselized gasoline engine for its larger car models. Other major U.S. car companies either had their own versions of dieselized gasoline powerplants or negotiated with foreign manufacturers to supply them for American-made cars. GM's effort to rush to market with their dieselized version proved to be almost fatal to the U.S. public's emerging acceptance. The engine's lack of reliability and durability, its sluggish performance, and its high cost of maintenance reversed the public's improving view of the diesel engine. Further dieselization activities by domestic developments or foreign supply arrangements were dampened by the return to lower gasoline prices at the end of the oil embargo, a major car downsizing effort, and improved gasoline engine fuel efficiency.

A new period in the evolution of the diesel engine started in the 1990s that was prompted by the introduction of stricter diesel emission standards, first in the United States and then in the European Union (EU) and Japan. In the EU, diesel—due to its superior fuel efficiency—was also embraced as the vehicle to control greenhouse gas emissions from the transportation sector and fight global warming, resulting in massive dieselization of the passenger car market. To meet the regulatory challenges, engine manufacturers had to develop new, cleaner, and even more efficient engines. Revolutionary fuel injection equipment that sported high injection pressures, smaller hole nozzles, shorter ignition delays, new combustion chamber designs, and possessed the capability to shape the rate of combustion, as well as lower the charge air temperature, carried diesel engines to a new level of excellence in performance matched by cleaner exhaust.

In the United States, this success quietly infiltrated the pickup truck market, which is powered by medium-duty engines and is sparking renewed interest in light-duty diesels for personal transportation. If this sounds like a quantum leap forward from the days of sluggish engine performance and the black-smoke-belching image that made the diesel Public Enemy No. 1, it is. This quantum leap forward is currently being helped by unparalleled and very significant contributions from the emission control catalyst industry. Modern postcombustion and auxiliary emission control devices and systems are emerging that have NO_x and particulate matter reduction potentials that were unheard of a few years ago. They are giving diesel engine engineers new freedom to develop good performance and excellent fuel economy, while achieving extremely low emission levels.

. . .

This book is a technical reference on diesel engines and diesel emissions and their control technologies. It is intended for engineers, applied scientists, students, and others working in the field of emissions reduction, engine design and testing, catalytic emission control, and diesel particulate filter technology. No prior knowledge of engine or emission technologies is assumed on the part of the reader and the book can be used as an introductory text. At the same time, it provides practical information that should be useful for experienced readers.

Part I begins with a fundamental description of the diesel engine and combustion process. Later chapters cover the basic subsystems of the diesel engine: air induction, fuel injection, and combustion. The part ends with a discussion on the formation of diesel emissions.

Part II focuses on emission characterization, health and environmental effects, measurement and testing, and regulations. It includes an overview of the current knowledge on the nature of diesel exhaust particulate matter, including particle sizes and numbers. Another chapter discusses emission test methods and equipment for gas phase compounds and particulate matter. Finally, a summary of North American and European emission regulations is provided.

Part III is an overview of diesel fuels and their emission effects. It discusses petroleum diesel fuel, as well as alternative fuels including synthetic fuels, biodiesel, DME, and natural gas.

Part IV discusses the progress in emission control that has been achieved by engine design means, such as combustion modification, injection systems, and the use of electronic control. One entire chapter has been devoted to exhaust gas recirculation technology.

Part V presents diesel exhaust aftertreatment methods. The chapters in this part present in-depth discussions of diesel catalyst technologies (including diesel oxidation catalyst, selective catalytic reduction with ammonia/urea, lean NO_x catalyst, and NO_x adsorber), diesel particulate filters, and other aftertreatment technologies such as nonthermal plasma.

. . .

It is our pleasant duty to end this preamble with words of thanks to individuals and organizations who provided help and support. A number of industry and academia experts took their time to review the manuscript and supplied feedback that made this book much better. The entire manuscript, at different stages of its creation, was reviewed by Mr. Charles T. Hare of SwRI, Dr. Kirby Baumgard of John Deere, and Prof. John Johnson of MTU who provided valuable comments and advice. An excellent review of several chapters on diesel catalyst technologies was provided by Dr. Owen Bailey of OMG. Useful comments were also provided by Dr. Jürgen Gieshoff of OMG, who reviewed the chapter on selective catalytic reduction of NO_x; by Dr. Win Watts of the University of Minnesota, who read the sections on sampling of ambient diesel aerosols; and by Dr. Karl-Heinz Breuer of Deutz, who helped write the section on burner-regenerated particulate filter systems. We are also indebted to a number of people who were always ready to discuss various topics related to the book, providing valuable ideas and answers; among this group we would like to name Mr. Peter Ahlvik of Ecotraffic who also cowrote the chapter on alternative fuels, Dr. Paul Zelenka, Prof. David Kittelson, Dr. Tim Johnson (of Corning), Mr. Phillip Weber, and Mr. Andreas Mayer.

This book is largely based on the material developed and posted on DieselNet.com as the "Technology Guide" papers. We would like to acknowledge the support from all subscribers to the DieselNet service, as well as from its corporate sponsors, including AVL-List, Johnson Matthey, Horiba, Kubota Engine, Engelhard Corporation, and many other companies representing the emission control, emission measurement, and diesel engine industries. Without their support, this project would never have been finished. Finally, we would like to thank Stu Horton, SAE Editorial Consultant (retired). Without his persuasion and encouragement years ago, this project might never have been started.

Last but not least, both authors thank their families, who suffered during lost weekends and holidays. Without their support and understanding this work could not have been completed.

February 2006

Table of Contents

Preface v

Part I **Diesel Engine Basics** 1

Chapter 1 The Case for the Diesel Engine 3
1.1 Introduction 3
1.2 What Is a Diesel Engine? 3
1.3 Types of Diesel Internal Combustion Engines 4
1.4 Energy Conservation and the Diesel Engine 5
1.5 What Are the Advantages of Diesel Engines? 13
1.5.1 Emissions 13
1.5.2 Compression Ratio 14
1.5.3 Friction Losses 15
1.5.4 Diesel Fuel Safety 15
1.5.5 Durability 15
1.5.6 Reliability 15
1.5.7 Ease of Breathing and Lower Pumping Losses 15
References 16

Chapter 2 Engine Fundamental and Design Parameters 17
2.1 Introduction 17
2.2 Heat Engines 17
2.3 Operation of Reciprocating Internal Combustion Diesel Engines 18
2.3.1 Operation of Two-Stroke Engines 18
2.3.2 Scavenging in Two-Stroke Engines 19
2.3.3 Definition of Basic Performance Parameters for Two-Stroke Engines 20
2.3.4 Scavenging Characteristics in Various Two-Stroke Designs 22
2.3.5 Operation of Four-Stroke Engines 22
2.4 Assessment of Engine Performance 26
2.5 The First and Second Laws of Thermodynamics 27
2.6 Mass Conservation in Combustion Thermodynamics 28
2.7 Stoichiometric Ratio in Diesel Combustion 29
2.8 Additional Performance Parameters and Their Definition 29
2.8.1 Power Output 29
2.8.2 Indicated Power 30
2.8.3 Mechanical Efficiency 31
2.8.4 Indicated Mean Effective Pressure 31
2.8.5 Brake Mean Effective Pressure 31
2.8.6 Specific Fuel Consumption 32
2.8.7 Volumetric Efficiency 32
2.8.8 Engine Specific Weight and Volume 32
2.9 Performance Curves 32
References 33

Chapter 3 Air Induction for Diesel Engines 35
3.1 Introduction 35
3.2 Methods of Charge Air Boosting 35
3.3 Types of Superchargers 35
3.3.1 The Roots Blower 36

3.3.2 The Sliding Vane Compressor 37
3.3.3 The Screw Compressor 37
3.3.4 The Rotary Piston Supercharger 38
3.3.5 The Spiral-Type Supercharger 38
3.3.6 The Variable-Displacement Piston Supercharger 38
3.3.7 The Centrifugal Compressor Supercharger 38
3.4 Turbochargers 41
3.5 Types of Turbochargers 44
3.5.1 Waste-Gated Turbochargers 44
3.5.2 Variable-Geometry Turbochargers 44
3.6 Supercharger-Aided Turbochargers 45
3.7 Compressor or Turbine A/R Ratio 46
3.8 Practical Embodiments 47
3.9 Sequential Turbocharging 48
3.10 Turbocompounding 48
3.11 Wave Supercharging 48
References 50

Chapter 4 Diesel Fuel Injection 51
4.1 Introduction 51
4.2 Purposes of the Fuel Injection System 51
4.3 Types of Diesel Fuel Injection Systems 52
4.4 The Pump-Line-Nozzle System 52
4.4.1 The P-L-N In-Line Pump System 55
4.4.2 The P-L-N Distributor/Rotary Pump System 66
4.4.3 The P-L-N Unit Pump System 68
4.5 The Unit Injector System 69
4.6 The Common-Rail Injection System 70
4.7 More on P-L-N Systems 71
4.8 Modernizing the Diesel Fuel Injection System 72
4.9 Modernizing the Pump-Line-Nozzle System 72
4.9.1 Electronic Control of the In-Line Pump System 72
4.9.2 Electronic Control of the Distributor Pump System 76
4.9.3 Electronic Control of the Unit Pump System 77
4.10 Modernizing the Unit Injector System 78
4.11 Electronically Controlled Common-Rail System 79
4.12 Components of an Electronically Controlled Injection System 80
4.12.1 Sensors 82
4.12.2 Electronic Control Unit 82
4.12.3 Actuators 83
4.13 Innovative Injection Systems 84
4.13.1 The Bendix Diesel Engine Control Unit Injector System 84
4.13.2 The Hydraulically Actuated, Electronically Controlled Unit Injector 85
4.13.3 Cummins Accumulator Pump System 90
4.14 Conclusion 91
References 91

Chapter 5 Combustion in Diesel Engines 93
5.1 Introduction 93
5.2 Components of the Combustion Process 93
5.3 Fuel and Air Distribution in the Fuel Spray of a DI Diesel 94
5.3.1 Lean Flame Region 95
5.3.2 Lean Flame-Out Region 95
5.3.3 Spray Core 95
5.3.4 Spray Tail 96
5.3.5 After-Injection or Secondary Injection 96

5.3.6 Fuel Deposited on the Walls 96
5.4 Heat Release Rates in DI Engines 97
5.5 Additional Details of Spray Formation 97
5.5.1 Spray Formation during Injection 98
5.5.2 Spray Atomization 98
5.5.3 Spray Penetration 99
5.5.4 Droplet Size Distribution 99
5.6 The Three Phases of Diesel Combustion 100
5.6.1 Ignition Delay 100
5.6.2 Premixed Combustion 102
5.6.3 Rate-Controlled Combustion 102
References 103

Chapter 6 Emissions Formation in Diesel Engines 105
6.1 Emission Formation in Diesel Engines 105
6.2 Formation of Hydrocarbons in Diesel Engines 105
6.2.1 Air-to-Fuel Ratio 106
6.2.2 Effect of Engine Load 107
6.2.3 Effect of After-Injection and Secondary Injection 107
6.2.4 Effects of Cold Starts and Misfires 107
6.2.5 Effect of Turbocharging 108
6.2.6 Effect of Nozzle Opening Pressure 108
6.2.7 Effect of Injection Timing 108
6.2.8 Effect of Intake Swirl 108
6.2.9 Final Comments on Sensitivity of HC to the Fuel System 109
6.3 Formation of Carbon Monoxide in Diesel Engines 109
6.4 Formation of Nitrogen Oxides in Diesel Engines 110
6.4.1 Fuel-Air Ratio 111
6.4.2 Effect of Nozzle Opening Pressure 111
6.4.3 Effect of Injection Timing 111
6.4.4 Effect of Cetane Number 112
6.4.5 Effect of Swirl 112
6.4.6 Effect of Intake Charge Dilution 112
6.5 Formation of Particulate Matter in Diesel Engines 112
6.5.1 Constituents and Sources of Particulate Matter 112
6.5.2 Carbonaceous Particulate Release and Oxidation 112
6.5.3 Summary of Potential Sources of Particulate Matter 113
6.6 Smoke in Diesel Engines 114
References 116

Part II **Diesel Emissions** 119

Chapter 7 Characterization of Emissions 121
7.1 What Are Diesel Emissions? 121
7.1.1 Diesel Combustion Products 121
7.1.2 Diesel Emission Levels 122
7.2 Gaseous Emissions 123
7.2.1 Nitrogen Oxides 123
7.2.2 Hydrocarbons 123
7.2.3 Carbon Monoxide 124
7.2.4 Sulfur Dioxide 124
7.2.5 Nitrous Oxide 125
7.2.6 Hydrogen 126
7.3 Diesel Particulate Matter 126
7.3.1 What Are Diesel Particulates? 126

7.3.2 Composition of Diesel Particulates ... 127
7.3.3 Solid Fraction ... 128
7.3.4 Soluble Organic Fraction ... 130
7.3.5 Sulfate Particulates ... 133
7.3.6 Particle Size Distribution ... 133
7.4 Diesel Exhaust Particle Size ... 134
7.4.1 Diesel Particle Size Distribution ... 134
7.4.2 Particle Distributions from New Diesel Engines ... 135
7.4.3 Spark-Ignited Engines ... 137
7.4.4 Influence of Particulate Sampling ... 139
7.4.5 Nanoparticles: Composition and Formation ... 140
7.4.6 Engine Nanoparticle Issues ... 143
References ... 143

Chapter 8 Health and Environmental Effects ... 147
8.1 Overview ... 147
8.1.1 Toxic Compounds in Diesel Exhaust ... 147
8.1.2 Inventories and Exposures ... 148
8.1.3 Effects on Health and Environment ... 149
8.1.4 Summary ... 150
8.2 Diesel Emission Inventory ... 150
8.2.1 Diesel Engines and Air Pollution ... 150
8.2.2 United States ... 151
8.2.3 European Union ... 154
8.3 Exposure to Diesel Exhaust ... 155
8.3.1 Introduction ... 155
8.3.2 Exposure Estimation ... 156
8.3.3 Diesel Particulate Matter ... 156
8.3.4 Polynuclear Aromatic Hydrocarbons ... 158
8.3.5 Hydrocarbons and Aldehydes ... 159
8.4 Health Effects of Gas Phase Components ... 159
8.4.1 Introduction ... 159
8.4.2 Carbon Monoxide ... 160
8.4.3 Sulfur Dioxide ... 161
8.4.4 Nitric Oxide ... 162
8.4.5 Nitrogen Dioxide ... 162
8.5 Health Effects of Diesel Particulates ... 163
8.5.1 Introduction ... 163
8.5.2 Noncancer Health Effects ... 164
8.5.3 Carcinogenicity of Diesel Exhaust ... 165
8.5.4 Future Issues ... 167
8.6 Environmental Effects of Emissions ... 167
8.6.1 Pollutants and Air Quality ... 167
8.6.2 Photochemical Smog ... 169
8.6.3 Acid Rain ... 170
8.6.4 Global Warming ... 170
8.6.5 Reduced Atmospheric Visibility ... 172
References ... 172

Chapter 9 Measurement of Emissions ... 175
9.1 Overview ... 175
9.1.1 Introduction ... 175
9.1.2 Emission Test Cycles ... 176
9.1.3 Sampling and Analysis ... 177
9.2 Gas-Phase Measurements ... 179
9.2.1 Techniques of Gas Analysis ... 179

9.2.2 Analysis of Exhaust Pollutants 181
9.3 Particulate Matter Measurements 182
9.3.1 Introduction 182
9.3.2 Gravimetric Method 183
9.3.3 Particle Number and Size Analysis 184
9.3.4 Instrumental Analysis of Other PM Parameters 190
9.3.5 Smoke and Smoke Opacity 192
9.4 Measurement of Ambient Diesel Aerosol 194
9.4.1 Aerosol Sampling and Analysis 194
9.4.2 Measurement in Ambient Air 196
9.4.3 Measurement in Occupational Health Settings 197
9.4.4 Measurement in Mines 199
9.4.5 Gas-Phase Measurements 201
References 202

Chapter 10 Emission Regulations 207
10.1 Historical Background 207
10.2 Types of Emission Regulations 209
10.3 Emission Test Cycles 210
10.3.1 U.S. Heavy-Duty FTP Transient Cycle 211
10.3.2 U.S. FTP-75 Cycle 211
10.3.3 European Stationary Cycle 211
10.3.4 European Transient Cycle 213
10.3.5 ECE 15 + EUDC 213
10.3.6 ISO 8178 214
10.4 Selected Diesel Emission Standards 214
10.4.1 Overview 214
10.4.2 United States 214
10.4.3 European Union 218
10.5 Selected Diesel-Related Occupational Health Regulations 220
10.5.1 United States 220
10.5.2 Germany 222
References 222

Part III **Diesel Fuels** 225

Chapter 11 Petroleum Diesel Fuel 227
11.1 What Is Diesel Fuel? 227
11.1.1 Components of Diesel Fuel 227
11.1.2 Diesel Fuel Properties 227
11.1.3 Typical Specifications 231
11.2 Fuel Properties and Emissions 233
11.2.1 Introduction 233
11.2.2 Heavy-Duty Engines 234
11.2.3 Light-Duty Engines 237
References 237

Chapter 12 Alternative Diesel Fuels 239
12.1 Overview of Alternative Diesel Fuels 239
12.1.1 Environmental Driving Forces 239
12.1.2 Alternative Fuel Options 240
12.1.3 Resource Use and Energy Efficiency 240
12.2 Synthetic Diesel Fuel 241
12.2.1 Introduction 241
12.2.2 Fisher-Tropsch Process 243
12.2.3 FT Fuel Properties 244

12.2.4 Emissions with Synthetic Fuel 244
12.3 Biodiesel 246
12.3.1 Introduction 246
12.3.2 Manufacturing Process 247
12.3.3 Properties and Specifications 248
12.3.4 Biodiesel Emissions 249
12.3.5 Compatibility with Petroleum Diesel Engines 253
12.4 Dimethyl Ether 254
12.4.1 Introduction 254
12.4.2 Manufacture of DME 255
12.4.3 DME as Automotive Fuel 256
12.4.4 DME Engines 257
12.4.5 DME Fuel and Emissions 258
12.5 Natural Gas 259
12.5.1 Utilization of Natural Gas 259
12.5.2 Emission Comparison 260
References 263

Part IV **Emission Control through Engine Design** 267

Chapter 13 Steps Toward the Modern Diesel Engine 269
13.1 Introduction 269
13.2 Reduction of Nitrogen Oxides 269
13.2.1 Injection Timing Retard 269
13.2.2 Intake Charge Cooling 270
13.2.3 Combined Effect of Injection Timing and Intake Charge Cooling 270
13.2.4 Combined Effect of Injection Timing Retard and Charge Air Cooling on Fuel Consumption 271
13.2.5 Effect of Injection Pressure on Fuel Consumption 271
13.3 Reduction of Particulate Matter 272
13.3.1 Reduction of Particulate Matter through Air Management 273
13.3.2 Reduction of Particulate Matter through Combustion System Design 282
13.3.3 Reduction of Particulate Matter through Oil Consumption Control 284
13.3.4 Reduction of Particulate Matter through the Fuel Injection System 286
References 289

Chapter 14 Emerging Technologies and the Modern Diesel Engine 291
14.1 Introduction 291
14.2 A Platform for Future Heavy-Duty Diesels 291
14.3 Platform for Future Light-Duty Passenger Car Diesels 292
14.4 Platform for Future Light-Duty Truck Diesels 292
14.5 Technology Options for Modernizing Diesels 293
14.6 Fuel Injection System Technology Options 294
14.7 Combustion System Technology Options 299
14.8 Air Charging System Technology Options 303
14.9 Air Induction System Technology Options 305
14.10 Variable Valve Actuation Technology Options 307
14.11 Combined Charge Air and Variable Valve Actuation Technology Options 308
14.12 New Combustion Technology Option 311
14.13 Engine Accessories and Subsystem Component Technology Options 311
14.14 Concluding Remarks 314
References 315

Chapter 15 Water Addition to Diesel Combustion 317
15.1 Addition of Water to Diesel Process 317
15.1.1 Methods of Water Addition 317
15.1.2 Impact on Combustion Process 318
15.2 Fumigation of Water into Intake Air 319
15.3 Direct Injection of Water 320
15.4 Fuel Emulsions 321
15.4.1 Types of Emulsions 321
15.4.2 Impact on Combustion and Engine Performance 322
15.4.3 Impact on Emissions 323
15.5 Practical Embodiments 324
15.5.1 Commercial Status 324
15.5.2 PuriNOx Emulsion 325
15.5.3 Wartsila Water Injection System 325
15.5.4 Pielstick Humid Air Motor 326
References 326

Chapter 16 Exhaust Gas Recirculation in Diesel Engines 329
16.1 Introduction 329
16.2 Definition 329
16.3 EGR Configurations 330
16.4 System Components 335
16.5 EGR Cooling and EGR Coolers 336
16.6 EGR Control Strategy 338
16.7 Effects of EGR on Combustion Parameters and Emissions 339
16.8 Effect of EGR on Diesel Engine Wear and Durability 344
16.8.1 Introduction 344
16.8.2 Background 344
16.8.3 Discussion 345
16.9 Alternative In-Cylinder Technologies to EGR for low NOX in Diesel Engines: A Review 347
16.9.1 Effect of Pilot Injection and Injection Rate Shaping 347
16.10 Effect of Intake Charge Air Cooling 348
16.11 Effect of Injection Timing Retard 348
16.12 Issues Related to EGR Control Systems 348
16.12.1 Material Buildup within EGR Systems 349
16.12.2 EGR Contaminants 349
16.12.3 Engine Durability 349
16.12.4 Application/Test Cycle 349
16.12.5 EGR Cooler Design 349
16.12.6 EGR System Packaging 350
16.12.7 EGR Valve Control Considerations 350
16.12.8 Piping 350
16.13 Demonstration of Low NOx and PM Using EGR and a Passive Trap System 350
References 352

Chapter 17 Controls for Modern Diesel Engines 355
17.1 Introduction 355
17.2 Background 355
17.3 Basic Operating Principles of the Control System 356
17.4 Variation in Closed-Loop Control Systems 356
17.5 Variation in Open-Loop Control Systems 356
17.6 The New Principle in Control Systems—A Model-Based Approach 357
17.6.1 EGR System Elements 357
17.6.2 Some Principles of Modeling 358
17.6.3 Discrete and Continuous Models 358
17.6.4 Data Conversion and Diagnostics Models 359

17.7 Challenges in Engine Control Systems 361
17.8 Neural Networks 361
17.9 Conclusion 363
References 363

Part V **Exhaust Gas Aftertreatment** 365

Chapter 18 Emission Control Catalysts 367
18.1 Overview 367
18.1.1 Background 367
18.1.2 Catalytic Reactors 368
18.1.3 Conversion of Pollutants: Basic Terms 369
18.2 Catalyst Fundamentals 370
18.2.1 Introduction 370
18.2.2 Conversion Rate in Heterogeneous Catalyst 372
18.2.3 Catalyst Deactivation 375
18.3 Cellular Monolith Substrates 377
18.3.1 Catalyst Substrates 377
18.3.2 Cellular Properties of Monoliths 378
18.3.3 Substrate Configuration and Performance 379
18.3.4 Catalyst Pressure Drop 379
18.4 Ceramic Catalyst Substrates 380
18.4.1 Introduction 380
18.4.2 Material Characterization Parameters 381
18.4.3 Cordierite 383
18.4.4 Commercial Substrates 383
18.5 Metallic Catalyst Substrates 384
18.5.1 Introduction 384
18.5.2 Substrate Materials 385
18.5.3 Substrate Designs 386
18.6 Catalytic Coating and Materials 389
18.6.1 Catalyst Washcoats 389
18.6.2 Aluminum Oxide 391
18.6.3 Silicon Oxide 392
18.6.4 Cerium Dioxide 392
18.6.5 Titanium Dioxide 392
18.6.6 Zirconium Oxide 392
18.6.7 Zeolites 392
18.7 Catalytic Converters 393
18.7.1 Introduction 393
18.7.2 Packaging Materials 394
18.7.3 Canning Methods 396
18.7.4 Header Design 398
18.7.5 Catalytic Mufflers 399
References 400

Chapter 19 Diesel Catalysts 403
19.1 Overview 403
19.1.1 Catalyst Technologies 403
19.1.2 Diesel Oxidation Catalyst 404
19.1.3 NOx Reduction Catalysts 405
19.2 Diesel Oxidation Catalyst 407
19.2.1 Reactions in Diesel Oxidation Catalyst 407
19.2.2 Emissions Performance 408
19.2.3 Sulfate Formation 413

19.2.4 Catalysts and Configurations 416
19.3 Selective Catalytic Reduction 416
19.3.1 Introduction 416
19.3.2 Reductants and Catalytic Reactions 416
19.3.3 Catalysts 418
19.3.4 Stationary SCR Systems 421
19.3.5 SCR Systems for Mobile Engines 422
19.4 Lean NOx Catalyst 426
19.4.1 Selective Reduction of NOx by Hydrocarbons 426
19.4.2 Catalysts 427
19.4.3 Advanced Catalyst Systems 429
19.4.4 Diesel Engine Experience 432
19.5 NOx Adsorbers 432
19.5.1 NOx Adsorber Concept 432
19.5.2 Performance and Operation 436
19.5.3 Diesel Engine Systems 440
19.6 Commercial Catalyst Technologies 445
19.6.1 Introduction 445
19.6.2 Light-Duty Engines 446
19.6.3 Heavy-Duty Engines 448
19.7 Deactivation of Diesel Catalyst 450
19.7.1 Catalyst Durability 450
19.7.2 Thermal Degradation 451
19.7.3 Poisoning 451
References 453

Chapter 20 Diesel Particulate Filters 459
20.1 Introduction 459
20.2 Principle of Operation 460
20.2.1 Filter Configuration 460
20.2.2 Filtration Mechanisms 460
20.2.3 Regeneration 462
20.3 DPF Performance 462
20.3.1 Particle Mass 462
20.3.2 Particle Numbers 463
20.4 Commercial Status 464
20.5 Diesel Filter Regeneration 465
20.5.1 Oxidation of Diesel Soot 465
20.5.2 Regeneration as an Equilibrium Process 466
20.5.3 Determination of Balance Temperature 469
20.5.4 Regeneration Methods 470
References 471

Chapter 21 Diesel Filter Materials 473
21.1 Overview 473
21.1.1 Material Requirements 473
21.1.2 Types of Materials 474
21.2 Wall-Flow Monoliths 477
21.2.1 Monolith Construction 477
21.2.2 Monolith Characterization Parameters 478
21.2.3 Commercial Materials 482
21.2.4 Pressure Drop 488
21.2.5 Canning 489
21.3 Ceramic Fibers and Cartridges 489
21.3.1 Fiber Materials 489
21.3.2 Filtration Mechanism 490

21.3.3 Filter Cartridges ... 490
21.3.4 Advantages and Disadvantages of Fiber Filters ... 491
21.3.5 Example Design: Nextel Fiber Cartridges ... 491
References ... 493

Chapter 22 Diesel Filter Systems ... 497
22.1 Overview ... 497
22.1.1 Design Considerations ... 497
22.1.2 Classification of Systems ... 497
22.1.3 Regeneration Strategy, Durability and Reliability ... 498
22.1.4 Future Trends ... 499
22.2 Catalyzed Diesel Filters ... 500
22.2.1 Introduction ... 500
22.2.2 Filter Design ... 500
22.2.3 Catalyst Systems ... 501
22.2.4 Emission Performance ... 502
22.2.5 Regeneration and Pressure Drop ... 505
22.2.6 Durability ... 507
22.3 CRT Filter ... 508
22.3.1 Principle of Operation ... 508
22.3.2 Application Limits ... 510
22.3.3 Filter System Design ... 511
22.3.4 Performance ... 512
22.3.5 Experience and Commercial Status ... 513
22.3.6 SCRT System ... 514
22.4 Filters with Fuel Additives ... 514
22.4.1 Fuel Additives ... 514
22.4.2 Catalytic Metals ... 515
22.4.3 Peugeot Filter System ... 517
22.4.4 Fiber Filters with Fuel Additives ... 519
22.5 Electrically Regenerated Filters ... 521
22.5.1 Electric Regeneration ... 521
22.5.2 Off-Board Regenerated Filters ... 522
22.5.3 Shore Power Regenerated Filters ... 523
22.5.4 Filters with On-Board Regeneration ... 525
22.6 Filters with Fuel Burners ... 527
22.6.1 Introduction ... 527
22.6.2 Full-flow Filter Systems ... 528
22.7 Microwave-Regenerated Filters ... 530
22.7.1 Microwave Heating ... 530
22.7.2 Diesel Filter Experience ... 532
References ... 534

Chapter 23 Plasma Exhaust Treatment ... 537
23.1 Introduction ... 537
23.2 Nonthermal Plasma ... 538
23.2.1 What Is Plasma? ... 538
23.2.2 Plasma Generators ... 538
23.2.3 Energy Efficiency ... 539
23.3 Chemical Reactions ... 540
23.3.1 Plasma Chemistry ... 540
23.3.2 NO Reduction by Plasma ... 541
23.3.3 NO Oxidation ... 542
23.3.4 HC Oxidation ... 543

23.4 Plasma Catalysts 543
23.4.1 DeNOx Catalyst 543
23.4.2 SCR (Urea) Catalyst 544
23.5 Particulate Matter Control by Plasma 544
23.5.1 PM Oxidation by Plasma 544
23.5.2 Single-Stage Packed-Bed Plasma Reactor 545
23.5.3 Two-Stage Reactor: NTP-DPF Configuration 546
23.6 Future Issues 546
References 547

Index 549

About the Authors 563

Part I

Diesel Engine Basics

Chapter 1:
The Case for the Diesel Engine

Chapter 2:
Engine Fundamental and Design Parameters

Chapter 3:
Air Induction for Diesel Engines

Chapter 4:
Diesel Fuel Injection

Chapter 5:
Combustion in Diesel Engines

Chapter 6:
Emissions Formation in Diesel Engines

Chapter 1

The Case for the Diesel Engine

1.1 Introduction

Amid gloomy predictions from some groups regarding the future of diesel engines, evidence seems to suggest that impressive gains continue to be scored. The diesel share of the market is growing, not only in Europe where diesels enjoy an excellent reputation, but also in the United States where diesels are much maligned. Recent reports from regulatory agencies in the United States, such as the Environmental Protection Agency (EPA), indicate that in spite of very restrictive emission regulations, diesel sales in the U.S. light-duty truck segment are growing at the rate of 4% per year [1]. Expectations are for sales to accelerate in the next few years with much of the growth occurring in the light-duty market segment.

Rudolph Diesel (Figure 1.1), who is best known for his invention of the engine that bears his name, was born in France in 1858. He received Patent No. 608,845 for his invention of a highly efficient, slow burning, compression ignition, internal combustion engine. His invention came while the steam engine was the dominant power source for large industries. The prototype of Dr. Diesel's engine ran on its own power on August 10, 1893, after 13 years of intensive development. Two years later, on the last day of 1896, Dr. Diesel demonstrated another model with a spectacular theoretical mechanical efficiency of 75.6%, while the then-popular steam engine had an anemic 10% mechanical efficiency [2].

Development of Diesel's invention needed more time and work to attain the commercial success that it was destined for. Many engineers and developers joined in the work to improve the market viability of Dr. Diesel's idea. He, on the other hand, became somewhat threatened by this process and was troubled by criticisms of his role in creating the engine, and in 1913, he vanished from a ship on her voyage to England.

Even in its infancy, Diesel's invention had begun to sow the seeds of discord and envy among its rivals. The engine was clearly superior to other forms of energy conversion machines.

Figure 1.1 *Dr. R. Diesel (1858–1913).*

It was certainly much more efficient than Watt's steam engine and was a major threat to England's coal industry. This sort of rivalry appears to still manifest itself today although those who are against Dr. Diesel's concept may advocate other fuels, combustion cycles, and other energy conversion concepts.

Of all the internal combustion engines known today, the diesel engine is the most efficient. This means that it extracts the greatest amount of energy from a specific amount of fuel [3]. In this chapter, reasons why the diesel engine is superior are highlighted. Differences between the diesel combustion concept and other concepts are explained and analyzed. However, no matter what else may be praiseworthy about the diesel engine, its fuel economy and its durability will always be the foundations of its strength.

1.2 What Is a Diesel Engine?

The diesel engine is an embodiment of the internal combustion engine. It was invented by Dr. R. Diesel for the purpose of efficiently producing mechanical power from the chemical energy stored in fuel. Diesel engines use the conventional cylinder and piston arrangement common to other internal combustion engines such as the gasoline engine. For the most part, there is a great deal of similarity between diesel and gasoline engine components.

Conceptually, diesels achieve their high performance and excellent fuel economy by compressing air to high pressures, then injecting a small amount of fuel into this highly compressed air. Temperatures created when air is compressed cause the small amount of highly atomized injected fuel to evaporate. Mixing with the hot surrounding air in the combustion chamber, the evaporated fuel reaches its auto-ignition temperature and burns, thus releasing the energy that is stored in that fuel [4].

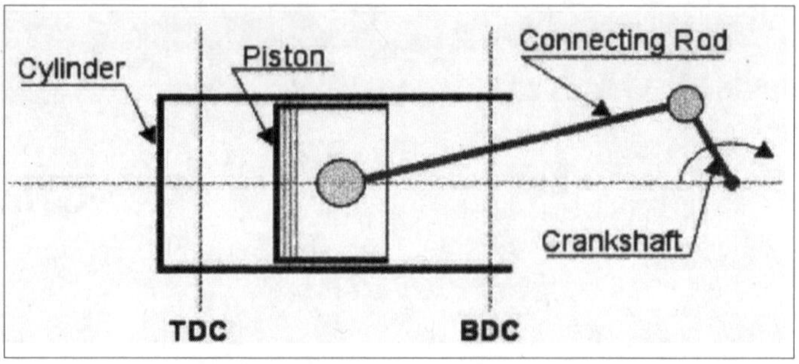

Figure 1.2 *Diagrammatic sketch of cylinder/piston internal combustion engine.*

1.3 Types of Diesel Internal Combustion Engines

It is widely accepted that cylinder- and piston-type internal combustion engines are divided into two main categories: the two- and the four-stroke concepts. By definition, two-stroke engines complete a combustion cycle (intake, compression, expansion, and exhaust) with two piston strokes or one complete rotation of the crankshaft. Air is admitted just before the compression begins, and the burned gases are exhausted near the end of the power stroke. A two-stroke engine is usually smaller in size than a four-stroke engine having the same power output, and tends to have higher specific power (power output for a given engine displacement) than its four-stroke counterpart. Two-stroke engines are generally less fuel efficient than four-stroke engines. The main reasons for the fuel inefficiency in two-stroke engines are poor scavenging and, in some designs, low volumetric efficiency. Although two-stroke engines have twice as many power strokes as four-stroke engines, they do not produce twice the power of four-stroke engines. The main reason for this is their shorter power stroke, which generally occupies 90 to 95 crankshaft degrees versus 120 to 140 for the four-stroke design. Two-stroke engines are popular in applications where small one- or two-cylinder engines are needed [5]. However, Burmeister and Wain in Copenhagen and Sulzer in Winterthur have built extremely large two-stroke engines. These are typically 900-mm bore and 1800-mm stroke turning at 60 to 100 rpm, producing as high as 4000 hp per cylinder. Their thermal efficiency is greater than 50%—without the constraints of emissions limits.

By contrast, four-stroke engines require four piston strokes, or two crankshaft rotations, to complete a combustion cycle. In the first stroke, the *intake stroke,* the piston moves from its position at top-dead-center (TDC) toward the bottom-dead-center (BDC). A diagrammatic sketch of the cylinder/piston internal combustion engine arrangement is given in Figure 1.2. During most of the intake stroke, filtered air is inducted into the cylinder. In the second stroke, air that was inducted into the cylinder is compressed by the piston moving back to TDC from its starting position at BDC. This second stroke is known as the *compression stroke,* where air in the cylinder heats up to a temperature usually above the auto-ignition temperature of the fuel that is injected into the cylinder near TDC. As the fuel burns, heat energy is released, raising the pressure inside a greatly reduced volume near TDC. This energy release produces pressure that is applied to the top surface of the piston, thus pushing it back toward its BDC position. This stroke is known as the *expansion stroke,* because it is through that expansion that power is imparted to the piston and causes it to move to BDC. The expansion stroke is also known as the *power stroke* for obvious reasons. It is also referred to by some as the *work stroke* because the expanding gases produce work by applying their pressure to the top of piston.

The last of the four strokes is the *exhaust stroke,* where combustion by-products are exhausted into the exhaust system for evacuation into the atmosphere. Figure 1.3 is a schematic representation of the four-stroke combustion cycle as applied to a diesel engine. In general, today's four-stroke diesel engines are equipped with devices that enhance air charging and allow additional fuel to be injected proportional to the additional air inducted, a step that improves specific power output.

There are two basic types of four-stroke diesel engines: direct- and the indirect-injected engines. In indirect-injected (IDI) engines, fuel is injected into a prechamber connected via a relatively narrow passage to the main combustion chamber. Prechamber or IDI engines are sometimes referred to as divided chamber engines. A glowplug is usually installed in the prechamber to assist in cold starting. With a glowplug, the fuel spray comes in contact with the red-hot tip of the plug and starts combustion in that limited volume. Gaseous products flow through the throat connecting the two sections of that

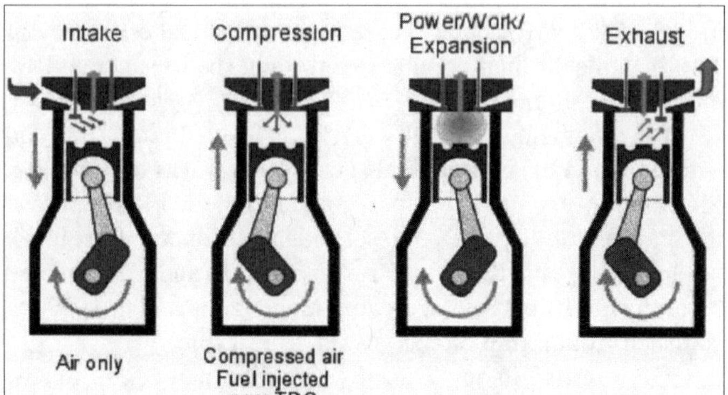

Figure 1.3 *The four-stroke diesel operation.*

divided chamber and continue their oxidation process into the main chamber. Combustion in the prechamber is so intense that resulting gases and partially combusted by-products pass through the throat area and create substantial turbulence in the main chamber. This turbulence is credited with mixing air in the main chamber with the remaining unburned fuel as well as the hot by-products from the prechamber. A similar action is also experienced while air flows in the opposite direction, from the main chamber through the throat and into the prechamber, during the compression stroke. The kinetic energy of the air flowing into the prechamber creates the turbulence that is credited with mixing fuel and air for proper combustion. For this reason, injection pressure in IDI applications does not have to be too high because mixing depends on the kinetic energy of the air much more than on fuel atomization.

Figure 1.4 shows three different prechamber designs. They represent various shapes, lengths, and widths of the throat area, but they all share a few common features. They all sport a glowplug to assist in cold starting and they all have a throat connecting both chambers, which has an adverse effect on the engine's efficiency. The flow of air during the compression stroke into the prechamber volume is invariably hindered by the presence of the throat, which increases pumping losses, thus lowering the overall efficiency of this concept. Another feature common to all prechamber engine designs is the relatively high surface area exposed to the engine coolant, leading to increased cooling losses.

Prechamber diesel engines have several other disadvantages such as lower thermal efficiency, flame impingement on the piston crown while exiting the throat, and in some designs hot gases from the throat may impinge on the valves, causing overheating of some critical components and rapid deterioration in lube oil life.

On the other hand, prechamber engines have lower noise characteristics that are related directly to the lower rate of pressure rise as well as lower maximum combustion pressure and temperature. All of these attributes have historically made them ideal for passenger car applications where creature comfort is a high priority. Recent development in fuel injection systems, however, as well as emphasis on fuel economy have made it possible for direct-injected (DI) designs to enter this market segment at the expense of the IDI engines.

A simple definition of DI combustion is that the fuel is injected directly into the combustion chamber. In DI engines, the combustion chamber is usually in the form of a bowl in the piston crown (see Figure 1.5). Both air and fuel energies contribute to the mixture preparation and consequently to combustion efficiency. Therefore, significant effort is expended in carefully designing the combustion system and matching it with the fuel spray characteristics to not only optimize engine performance, but also minimize harmful exhaust emissions. Because mixture preparation in this combustion system has more dependency on the kinetic energy of the fuel spray, injection pressure and injector nozzle hole geometry (hole diameter and length) are more important in DI than IDI engines.

The absence of a divided chamber (as in IDI systems), as well as a restriction like that of the throat in prechamber designs, leads to reduced pumping losses. In addition, without a prechamber the surface area exposed to the coolant is much reduced from that of the IDI system, which leads to lower heat loss to the coolant. These features contribute to DIs having as much as 10% to 15% better fuel economy than IDIs. For this reason most new diesel engine designs are now exclusively open-chamber direct-injected engines.

1.4 Energy Conservation and the Diesel Engine

Having briefly discussed one of the major advantages of diesel engines—their fuel economy—one must consider how this parameter competes among other forms of energy conversion machines. A study by the Chrysler Corporation titled "The Basis of Engine Efficiency" was published during the China Automotive Technology Conference/Workshop in August

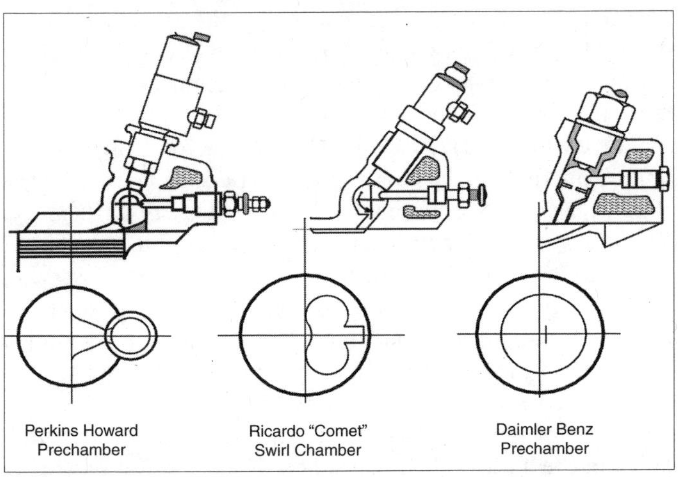

Figure 1.4 *Schematic representation of three IDI designs. (Reprinted from Ref. 6.)*

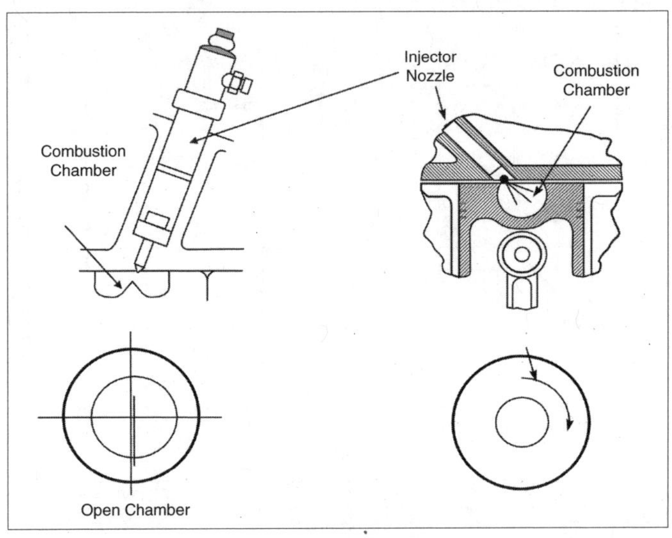

Figure 1.5 *Schematics of two DI combustion systems.*

1995. Figure 1.6 was adapted from the published material and gives a comparative analysis of the thermal efficiencies of six different powerplants.

The idea of freeing the public from total dependence on the conventional reciprocating internal combustion engine has long fascinated scientists and engineers. To recall the recent history in the United States, the gas turbine engine was made popular thanks to a pilot program by the Chrysler Corporation during the 1960s and 1970s. Attracted by its quiet, vibration-free operation, the fact that it had fewer moving parts than its reciprocating rival, and its fuel versatility, the gas turbine engine's development was substantially funded by government-sponsored grants. Unfortunately, two major factors led to its downfall, and its development as the main prime mover for the automobile of the future came to a halt in the mid-1970s. Its fuel economy was not a great improvement over that of the reciprocating gasoline engine, a very serious negative in view of the ongoing oil embargo by several oil producers in the Middle East during that same period. The second problem was related to its high cost as scientists realized that for the gas turbine engine to compete, new turbine materials had to be invented and engineered that would allow higher temperatures to be used at the inlet of the power turbine (shown as T_5 in Figure 1.7). The cost associated with the use of high-temperature materials inevitably increased overall engine costs, thus dashing its chances to compete against the well-entrenched reciprocating engine. Other drawbacks associated with gas turbine engines included:

- Slow acceleration
- Low efficiency of the regenerator seals
- Corrosion of recuperators (when used)
- Fire danger in cases of false starts.

Since the buying public often demands performance, the gas turbine's slow acceleration was considered a major problem.

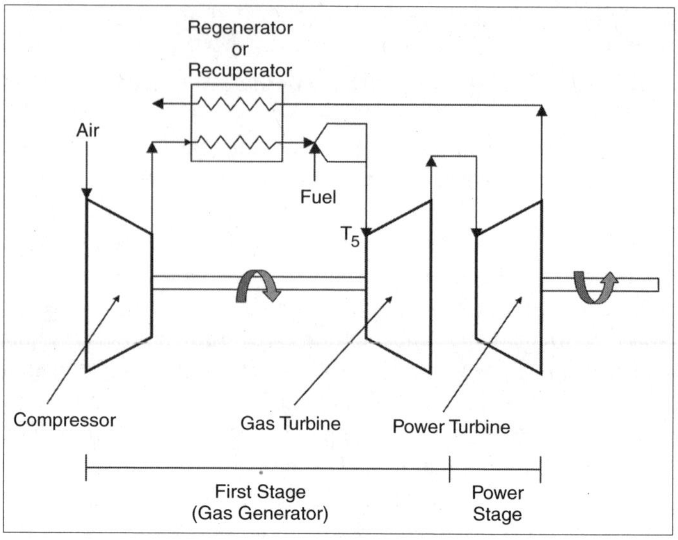

Figure 1.7 *Schematic representation of a two-shaft gas turbine engine.*

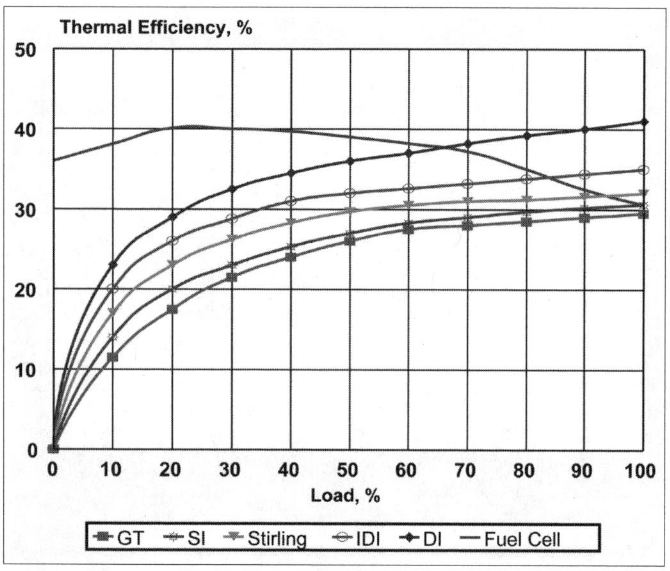

Figure 1.6 *Energy conversion efficiencies of several types of powerplants.*

One of the reasons for the engine's sluggish acceleration was attributed to the high accessory load and parasitic losses incurred by the gas generator shaft, because it had to drive most of the engine's accessories. Efforts were then launched to design a new accessory drive system powered by the second-stage turbine (power turbine). An incremental efficiency improvement was noted, yet the overall efficiency still depended on the ability to provide greater expansion through the first- and second-stage turbines. This depended mainly on how high T_5 was allowed to be. Replacing the metal turbine with a high-temperature ceramic allowed T_5 to increase to temperatures in excess of 2500°F (1370°C) [7]. The ceramic technology that was prevalent when these developments were taking place failed to give this powerplant the fuel efficiency edge it needed to enter its commercial phase.

It is clear from Figure 1.6 that the gas turbine engine lags far behind other forms of energy conversion concepts in terms of thermal efficiency. Recent activity in hybrid drivetrains has revived interest in gas turbines, yet its appetite for fuel, the costs, the need for new high-temperature and high-strength materials, and the critical tolerances of its turbomachinery continue to challenge its few advocates in the automotive industry.

From Figure 1.6, we find that the reciprocating spark-ignited internal combustion engine provides a small improvement over the gas turbine engine in terms of thermal efficiency. The spark-ignited reciprocating (SI) engine has proven its reliability and sociability in more than 100 years of serving mankind. Innovation evolved to meet various challenges as they materialized. Advancements in engines and vehicles included increased engine efficiency, improved drivetrains, and lowered transportation costs through improved fuel efficiency by means of lighter weight vehicles, decreased aerodynamic drag, and reduced rolling resistance. These innovations made it easy for the public to embrace the SI engine and propel it to a dominant place in the industrialized world. Yet, this concept is still limited by a relatively low compression ratio, a fact that directly affects its thermal efficiency, which is described by the following relationship:

$$\eta_{th} = \left[1 - \frac{1}{(C_r)^{\gamma-1}}\right] \times 100 \qquad (1.1)$$

where η_{th} is thermal efficiency, C_r is compression ratio, and γ is the ratio of specific heat at constant pressure to the specific heat at constant volume. Throttling the conventional SI engine is a means to control its power, but has direct adverse implication on the engine's volumetric and overall efficiency. The invention and development of the three-way catalyst has provided the SI engine with new capabilities in meeting extremely low emissions. Yet, the pressure of preserving natural resources, specifically fossil fuel supplies, is still driving engineers toward developing more efficient powerplants.

Even though Dr. Robert Stirling was a clergyman rather than a scientist or engineer, he is credited with inventing the closed-cycle hot-air engine that carries his name in 1816 [8]. The working fluid in this cycle was air that recirculated between a hot and a cold zone. Heat was added to the working fluid through a heat exchanger, causing air to expand, thus imparting pressure on the hot side of the piston. When pressure is applied to the piston, it moves, thus causing the evacuation of the colder air, which migrates to the hot side of the piston. The piston sliding motion is used to transfer work to mechanical energy to power various drivetrains. Figure 1.8 shows a diagram that helps explain the operation of the Stirling cycle and highlights its major components.

The fact that the combustion process was isolated from the cyclic changes in the working fluid pressure and temperature changes made it easier to deal with controlling emissions. Another advantage of this concept is the ability to provide heat to the working fluid from a variety of fuels. The only limitation to the fuel is its packaging, delivery, and service complexity. Combustion in the Stirling engine is continuous and, therefore, free from the intermittent and explosive nature of that of the internal combustion engine. This characteristic provides a smooth and quiet operation.

From the late 1930s until the 1950s, a Dutch company (Phillips) intensified research on this concept and established that using hydrogen or helium instead of air led to greater thermal efficiency. Yet, it was most difficult if not impossible to prevent the working fluid from leaking out of the closed components of the engine. In addition, all of the heat generated, then imparted to the working fluid, had to be dissipated into a cooling medium to protect the physical integrity of the engine's components since there is no exhaust gas exiting the working fluid area.

It is worth noting that in the closed-cycle Stirling engine case, ideal thermal efficiency can be expressed as follows:

$$E = \left(1 - \frac{T_{min}}{T_{max}}\right) \times 100 \qquad (1.2)$$

where E is the ideal thermal efficiency, T_{min} is the minimum cycle absolute temperature, and T_{max} is the maximum cycle absolute temperature. It follows then that for the highest thermal efficiency it is beneficial to minimize T_{min} and maximize T_{max}. Efforts to reduce T_{min} led to increasing the radiator size, which made it cumbersome for packaging in over-the-road applications. Similarly, efforts to increase T_{max} usually led to expanding the surface area of the heater head and/or use of a material that would withstand extremely high temperatures without losing its mechanical integrity. Other, more practical limitations exist that make it difficult to use Stirling engines in transportation. One such limitation is its extremely long starting time, because a lot of heat must be added to the hot side before the engine can develop any power. Another limitation is that after shutting the engine down, heat from the heater head escapes into the environment without producing any useful work. Therefore, the Stirling engine has failed to step into the tough arena where internal combustion engines still reign supreme.

Reference to Figure 1.6 is made once again, where the next two powerplants having higher energy conversion efficiency are the diesel indirect-injected and the direct-injected engines. Because the diesel engine is the main subject of this chapter, we will refer to them later and describe their attributes, both good and bad. While still referring to Figure 1.6, let us focus our attention on the fuel cell as an efficient conversion machine. Its energy conversion graph seems to stand alone from the rest of the powerplants discussed thus far, and it exhibits a rather high efficiency at part load conditions, where all others appear very deficient.

From a historical perspective, Sir William Grove is credited with creating the first fuel cell in 1839, however, he could not produce enough power to compete with other sources of energy available in his time [9]. In the 1930s, approximately 100 years later, Francis Bacon made significant engineering advances in fuel cell technology. By 1959, after 27 years of research and development, he was able to produce a 5-kW fuel cell system that powered a fork-lift truck. Since this modest beginning, fuel cells have been further developed and used in various military applications. They are used to provide power for life-support systems aboard space shuttles, to power homes and businesses, and now they are being considered as propulsion systems for clean transportation.

Fuel cells are electrochemical devices that use hydrogen and oxygen to produce electricity. Their by-product is water and they reject heat as a result of the chemical conversion process. Unlike batteries, which must be periodically recharged, fuel cells can produce power as long as they are supplied with fuel and oxidant.

At the heart of the fuel cell is a solid electrolyte consisting of proton-conducting plastic foil (proton-exchange membrane), as shown in Figure 1.9. This foil is coated with a platinum catalyst and an electrode made from gas-permeable graphite paper. Graphite bipolar plates, in which fine gas channels are milled, are positioned on both sides of the catalyst. Hydrogen is fed through the channels on one side of the membrane (solid electrolyte), and air flows through the channels on the other side

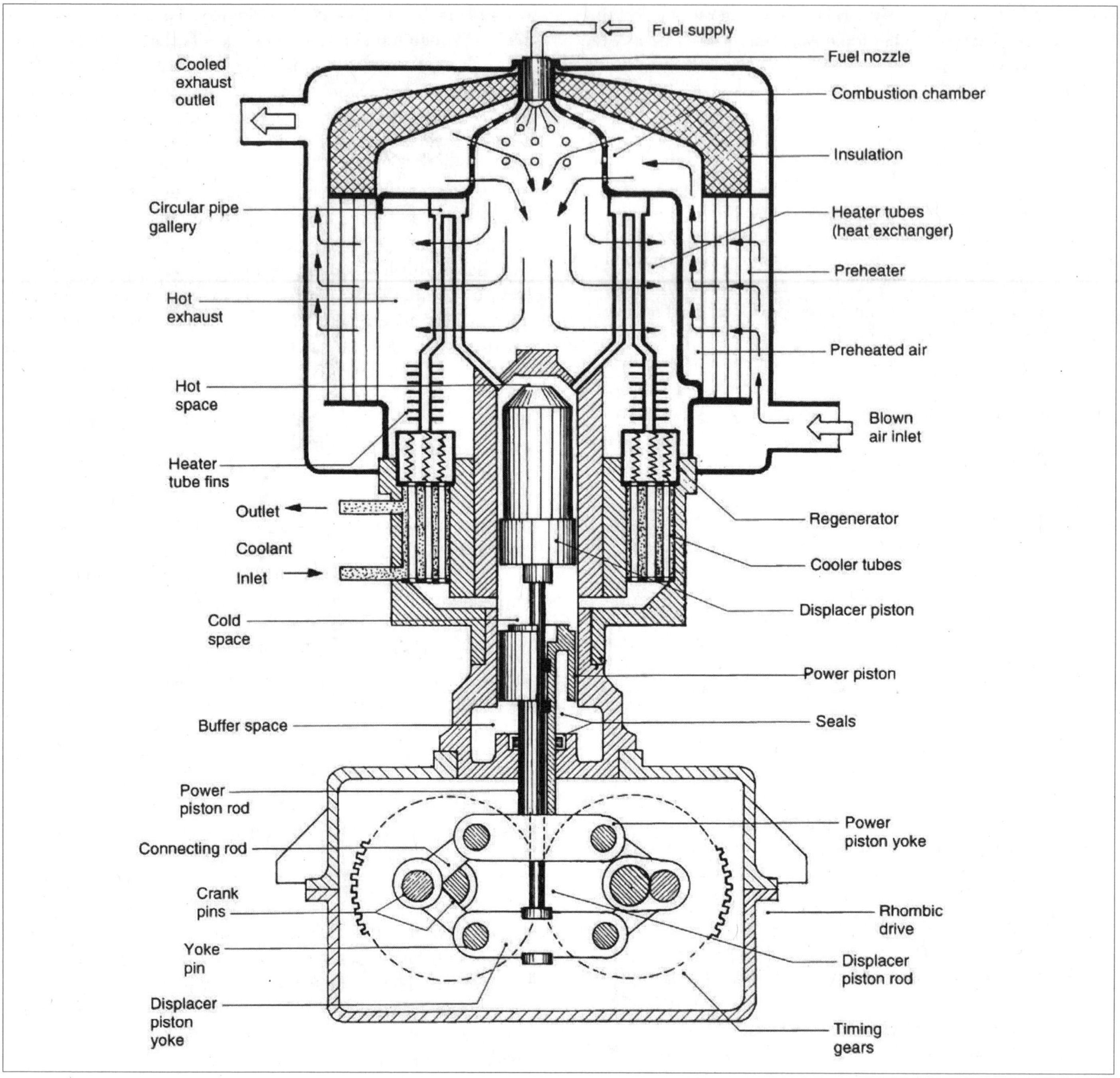

Figure 1.8 *Schematic of a Stirling engine. (From* Advanced Engine Technology *by Heinz Heisler. Reprinted by permission of Butterworth-Heinemann.)*

of the electrolyte. The hydrogen side is designated the anode (negative polarity), since the platinum catalyst essentially ionizes the hydrogen molecules at the anode into negatively charged electrons and hydrogen ions (protons), which migrate through the electrolyte toward the cathode. Oxygen flows through the second electrode, where it combines with hydrogen to form water vapor, which eventually exits from the powerplant. The result is an electrical voltage between the negative terminal (the anode) and the positive terminal (the cathode). Stacks of electrodes sandwiching proton-exchange membranes are used to generate the power required for a given application.

An alternative to the proton-exchange membrane technology is the solid oxide cell, used mostly for high-power applications such as industrial and large-scale power generating plants. Solid oxide fuel cells (SOFCs) use ceramics instead of either liquid or dry electrolytes. Operating temperatures may

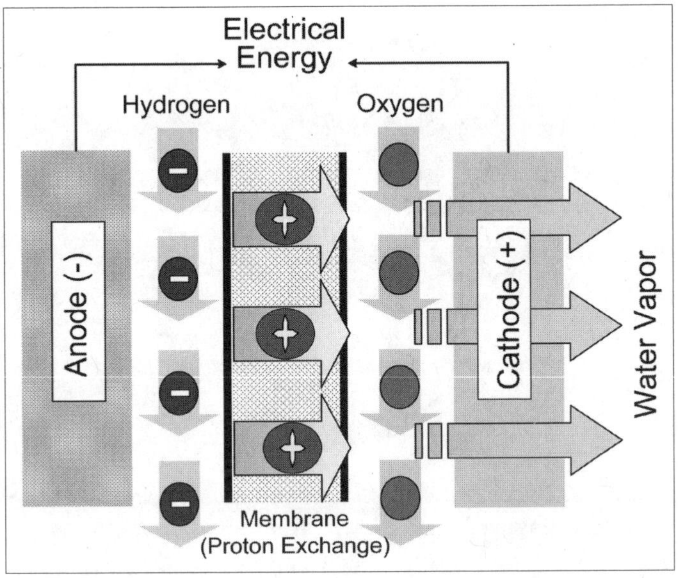

Figure 1.9 *Functional schematic of the fuel cell.*

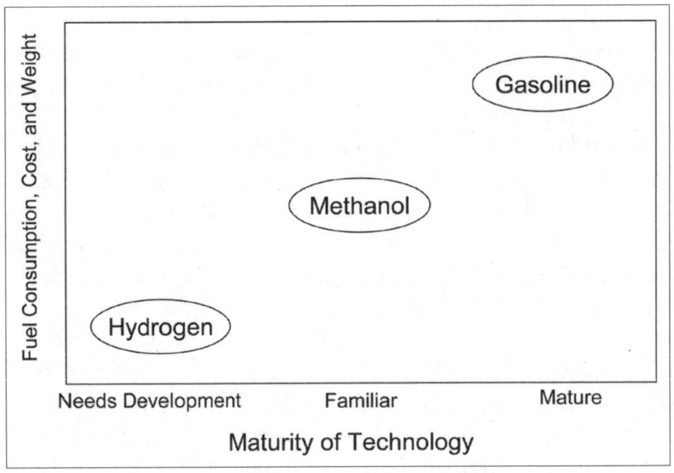

Figure 1.10 *Options and implications for fueling the fuel cell (Redrawn from Ref. 10.)*

reach 1800°F (980°C) and thermal efficiencies nearing 60% are quite possible. Many observers feel that SOFCs are also a viable alternative for automotive applications.

Beyond its relatively simple concept, the fuel cell still faces many challenges before it can be considered a commercially viable powerplant for automotive use. An important issue to be resolved is which fuel will most likely be the energy source for future fuel cells. Hydrogen appears to be an obvious choice and, theoretically at least, there is an abundance of water to produce hydrogen fuel by electrolysis. However, the process of deriving hydrogen from water by electrolysis is energy intensive and seems to defeat the purpose of economical conversion of energy for transportation. Some feel that the most environmentally and economically sound way to provide hydrogen fuel is to use hydroelectric or solar power in the electrolysis process, and only then will it be environmentally responsible and possibly economically viable. Therefore, as reflected in Figure 1.10, producing hydrogen as the energy source for future fuel cell–powered vehicles will require further development [10].

Obtaining hydrogen from methanol is the next best choice, but will require onboard fuel refining using a device known as a reformer. Among the advantages of using methanol, the prospect of low CO_2 emissions (estimated at 30% less than today's conventional internal combustion engines) is the most attractive. This advantage is followed closely by the fact that methanol is produced from natural gas, which is available in large quantities in many areas of the world. In addition, methanol can be made from renewable sources such as organic waste or timber residues. Even though a methanol distribution infrastructure is currently nonexistent, it could be distributed through other existing networks such as filling stations as we know them today.

The least desirable choice of fuel for the fuel cell is gasoline. As with methanol, an onboard refining process is also necessary, but it requires a more complicated reformer than that of methanol. To balance this major negative, the existing gasoline infrastructure should be considered one of the more significant advantages for this fuel. However, gasoline fuel as we know it may have to change drastically to encourage the formation of large quantities of hydrogen from a reasonably sized reformer.

Although there is a great deal of excitement regarding the outlook for fuel cells, this sentiment should be moderated by the challenges that lie ahead. Fuel selection, reformer functionality and efficiency, fuel handling and infrastructure, safety, environmental concerns, toxicity, ground water contamination, fuel efficiency, performance, and consumer acceptance are just a few of the topics that present both challenges and promises for this concept. Yet, there is a lot of optimism and even enthusiasm in the engineering community regarding some recent fuel cell–powered prototypes. While the debate about the most likely fuel to power fuel cell cars continues, methanol was favored by Nissan for its fuel cell vehicle.

Several other powerplants have seen limited commercialization either in a niche market or for a short duration until market and economic conditions changed. For instance, the rotary engine (Wankel), which is shown in Figure 1.11, was produced in limited volume and installed in a sports-type car made by Mazda of Japan [11]. The engine was invented by a German, Felix Wankel, and was further developed by the German automaker NSU. This concept uses a wide triangular disk having rounded flanks. A recess in the middle of the side of each flank represents the combustion chamber. This triangular rotor drives an eccentric shaft around which the rotor orbits (see Figure 1.11). The rotor moves in an orbital fashion inside a housing whose shape resembles a figure eight. At each apex of the triangular rotor, a sliding hard carbon seal mechanism is placed to reduce gas leakage between the two chambers on either side of the seal during the combustion phases. The combustion cycle is very similar to that of the reciprocating piston cycle described earlier. However, the rotating action of the triangular rotor/disk completes all phases of the combustion cycle without the excessive unbalanced forces experienced in the reciprocating engine.

Figure 1.11 shows the induction phase of the rotary engine where the charge is drawn into the engine through the intake port as the rotor moves in a counterclockwise direction. The preceding charge is being compressed as the volume of the space between the flank of the rotor and the engine housing decreases. The seal at apex P (Figure 1.11) is responsible for preventing leakage from the charge under compression to the freshly inducted charge side of the rotor. At the proper time, the spark plug is energized, and combustion follows shortly after while the rotor continues its counterclockwise rotation. The compression phase of the combustion cycle is shown in Figure 1.12. Both the expansion and exhaust phases of the combustion cycle are represented schematically in Figure 1.13.

The Wankel engine, whether in its spark-ignited or compression-ignited version, suffers from basic design problems. The most severe of these problems is that the apex seal, made of hard carbon fiber material, experienced excessive wear resulting from moving across the ports' sharp edges. Another serious problem is engine housing distortion, because a small area of the engine is cooled by the incoming charge, leaving the rest of the engine without adequate cooling. This fact led to an excessive temperature differential and housing distortion that affected oil consumption and cooling heat loss. Another major limitation of

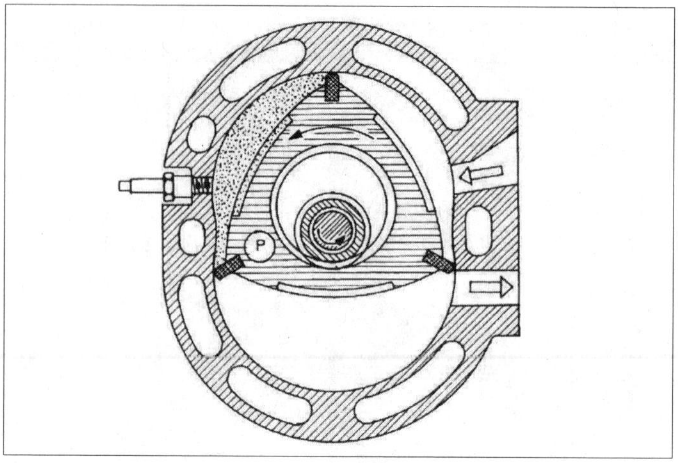

Figure 1.12 Compression phase of the Wankel engine. (From Advanced Engine Technology *by Heinz Heisler. Reprinted by permission of Butterworth-Heinemann.)*

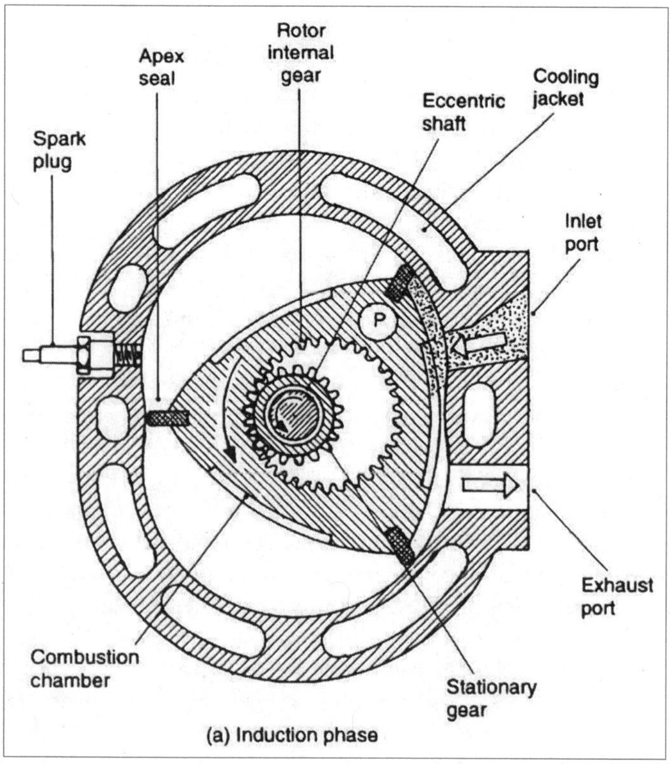

Figure 1.11 Construction of the Wankel engine—induction phase. (From Advanced Engine Technology *by Heinz Heisler. Reprinted by permission of Butterworth-Heinemann.)*

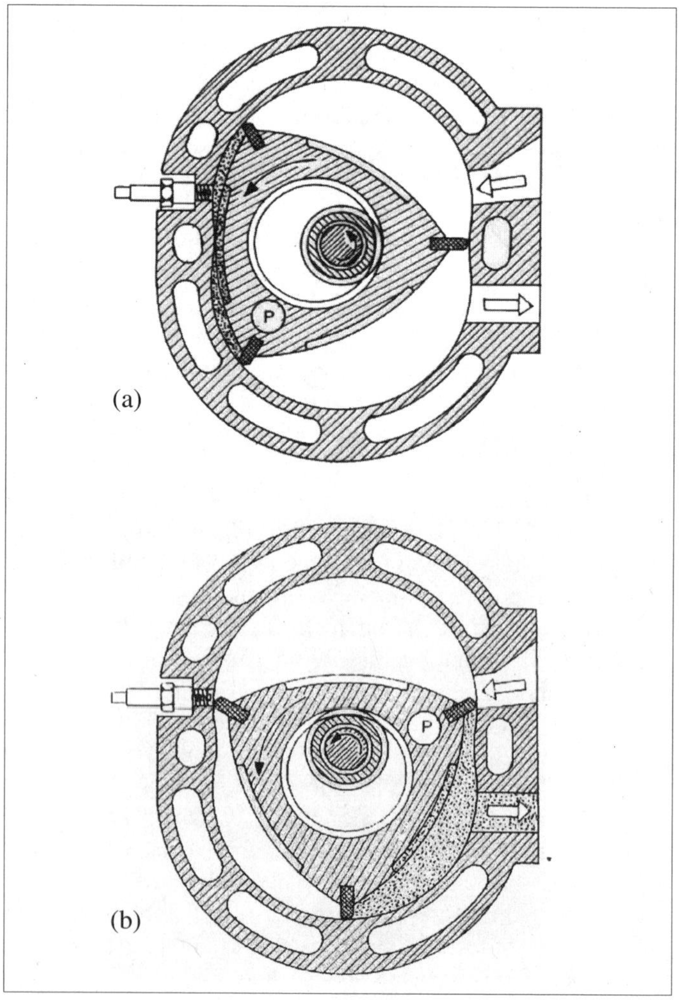

Figure 1.13 (a) Power and (b) exhaust phases of the Wankel engine. (From Advanced Engine Technology *by Heinz Heisler. Reprinted by permission of Butterworth-Heinemann.)*

this engine is its low compression ratio, which is limited by the rotor's geometric eccentricity. This problem was compounded by the apex seals that prevented high compression, leading eventually to low thermal efficiencies. However, the engine is compact, simple, and capable of achieving high speeds and, therefore, capable of developing high engine outputs.

Another alternative to the internal combustion engine is electric energy, which is a subject that has received much attention in recent years. Electrically powered vehicles are attractive because of the general perception that they emit extremely low or nearly no emissions, and certainly no CO_2 emissions at the vehicle itself. Indeed if such vehicles are to reduce the global warming phenomenon, their electric energy should not be produced from fossil fuel. In fact, it is quite appropriate to think of the concept of well-to-wheels emissions if true zero CO_2 emissions is our goal. In spite of the promise of a purer environment, electric cars built by General Motors Corporation in the 1990s did not sell, and the company ended up halting their production. Yet, even if these cars were sold in significant numbers, it is doubtful that their impact on the CO_2 emissions would have been felt. The reason is that electric energy production is predominantly based on fossil fuels, and is projected to remain this way for some time to come. According to projections by the U.S. Department of Energy (DOE), electricity production in the United States will involve fossil fuel well into the 21st century. Figure 1.14 shows DOE's projection of how electricity will be generated in the United States in 2010 [12].

From Figure 1.14, we can see that about two-thirds of the electricity produced in the United States in 2010 will come from coal, oil, and natural gas. Only 20% will come from nuclear energy and 15% from renewable energy. As such, electric vehicles may not be totally vindicated when it comes to CO_2 emission, because fossil fuel combustion is still required to produce the electricity used by these vehicles. The possible advantage is the location at which these emissions may be generated. A powerplant burning fossil fuels and producing

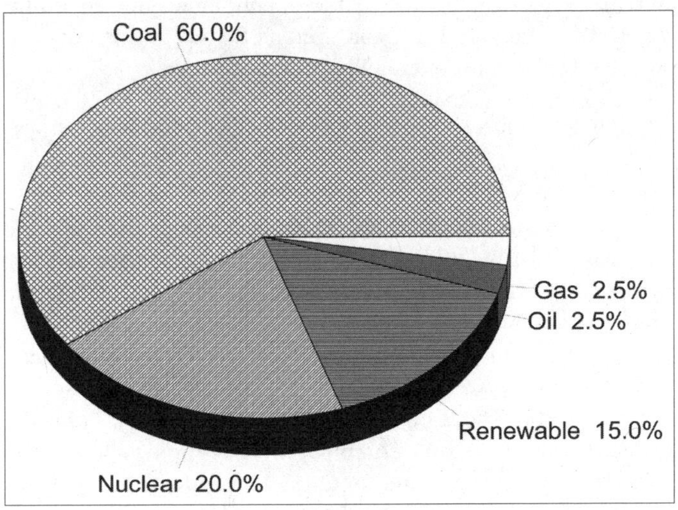

Figure 1.14 *Distribution of primary energy sources for U.S. electricity in 2010. (Redrawn from Ref. 12.)*

electricity could be located at some distance from the residential and commercial neighborhoods where electric vehicles are being used. The location of the powerplant away from densely populated areas may serve as a natural barrier for keeping emissions away from that population. However, wind direction and other weather conditions give rise to emission transport phenomena that could jeopardize human health.

The next plausible choice for generating electricity is nuclear power. Disasters such as the accidents at Three Mile Island in Pennsylvania and at Chernobyl in the Soviet Union have given nuclear power an unfavorable public image. In addition, the cost of building nuclear powerplants and the time required to complete their construction are major undertakings and may even be prohibitive for today's economies. Nevertheless, if a satisfactory source for electric power is found, power from electrical generators can be transmitted through high-tension wires to onboard batteries for storage and use in vehicular applications. Developing a battery with an energy density high enough to be equivalent to a 300- to 400-mile tank range has been difficult. Although the lead/acid battery has a relatively high power density (about 80 W/kg), it falls far short of the mark in its energy density and cannot provide the power required for long-range vehicular applications. Other battery candidates include nickel/cadmium and nickel/iron, which have greater power density than lead/acid batteries and a slightly improved energy density at room temperature. Both sodium/sulfur and lithium aluminum/iron sulfide batteries operate at high temperatures; 300 to 350°C for the former and 400 to 500°C for the latter. While these two types of batteries hold promise for best performance and range, they present some operational difficulties if used in cold climates. Their active elements would have to be unfrozen before they could be used [12]. Other factors affecting battery applications for traction include the number of times a battery can be charged before it must be discarded, the depth of charge and its effect on battery performance, the effect of driving habits on battery life and performance, and the effect of driving speed on battery performance and range.

Because successful replacement of the internal combustion engine as prime mover depends in large part on customer acceptance, electric vehicles must compete in the same arena. In spite of its many technical accomplishments, the electric vehicle must overcome its battery limitations, which include battery energy as well as power density, performance and life, time to charge the batteries, range, depth of charge versus performance, battery cost associated with the need to discard and replace over the vehicle service life, reliability, and cost. While it is conceivable that these issues can be resolved, the success of the electric vehicle will depend on our ability to invent the "superbattery." Meanwhile, it is worth mentioning that General Motors introduced an electrically powered vehicle in the mid-1990s. The company halted its production soon after that introduction, citing poor sales as the main reason.

A potential short-term solution to the limitations of electric batteries is a hybrid combustion engine/electric powerplant. In this case, the combustion engine is used to supplement the battery if its energy or power density drops below those of the driver demand. If the combustion engine in a hybrid vehicle uses fossil fuels, one might argue that we have made the full circle back to the very engine we were trying to replace. Whereas there is some truth to this assertion, it must be noted that combustion engines in hybrids work in concert with the battery. In this respect, over-the-road emissions can be significantly reduced relative to the amount of energy contributed by batteries when excess power is required. Of course, once developed, fuel cells can fulfill the role of the combustion engine in hybrids and if hydrogen is their fuel, then both energy efficiency and clean air can be accomplished. Although this goal may seem too difficult to reach in the foreseeable future, it seems that the engineering community is edging closer to this feat. DaimlerChrysler's Necar 4 is a hybrid fuel cell/electric vehicle that features a 55-kW asynchronous electric motor and two fuel cell stacks, each stack comprising 160 fuel cells. Together, the stacks generate a total output of 70 kW.

Meanwhile, the highly publicized government-supported Partnership for a New Generation of Vehicles (PNGV) has selected a powertrain of the internal combustion engine/electric hybrid type as the most likely to meet future fuel efficiency and emission standards. Not surprisingly, the internal combustion engine selected by the nation's top scientists and researchers was a compression ignition, direct-injected (diesel) engine. At present, the nickel metal hydride battery appears to be the choice for hybrid demonstrations offered by car companies.

In hybrid powertrains, the heat engine, electric motor, or both provide propulsion power. Generally, two configurations are considered. The first is referred to as a parallel configuration in which the heat engine is the primary power source, while the electric motor is used as a power assist. This assistance can include a variety of functions from operating most onboard systems to providing supplemental power for acceleration, hill climbing, or any combination of these functions. The power split between the heat engine and the electric motor is an important parameter affecting the total powertrain emission level. For instance, if the heat engine can be downsized, there will be ample opportunity for minimizing fuel consumption and emission levels. Figure 1.15a is a schematic representation of the parallel hybrid configuration.

The second approach is the series hybrid arrangement in which the electric motor is used as the primary power source. A small heat engine is used as a generator to recharge the batteries and extend the vehicle range. This configuration is shown schematically in Figure 1.15b.

For either hybrid arrangement, parallel or series, there is always the option of having heat engines and electric motors operate in a dual mode, each supplying power as conditions demand it [13]. For instance, a vehicle could operate solely on electric power while being driven in the city, where it is desirable to minimize or avoid emissions altogether. The heat engine power would be used at highway cruising speeds to conserve electric energy. Both the heat engine and electric motor would be used during hard accelerations and other high-power demand conditions. In all of these cases, braking energy could be captured and effectively converted from mechanical energy to electrical energy through a generator to help charge the batteries. Sophisticated electronic controls are required for the optimum apportionment of the power between the heat engine and electric motor.

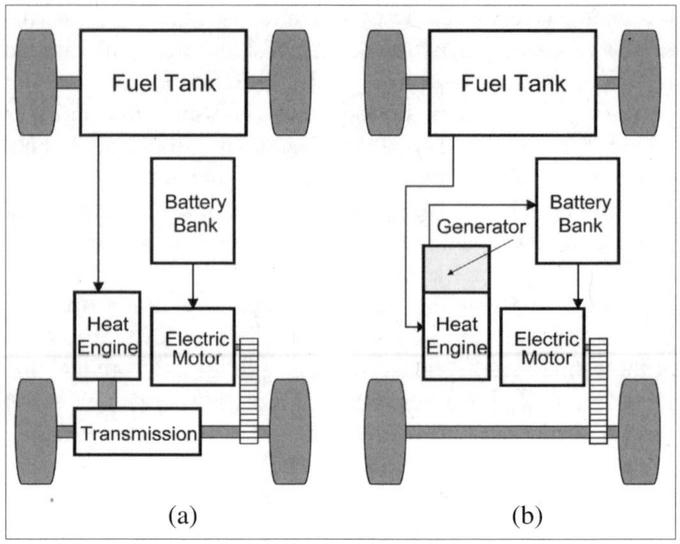

Figure 1.15 *Hybrid powertrain configurations: (a) parallel and (b) series.*

Among the notable hybrid electric accomplishments to date is Ford's P2000 LSR (Prodigy) equipped with a four-cylinder, 1.2-L, direct-injection diesel engine. This open-chamber diesel engine supplies the part load power required by the drivetrain and is assisted by an electric motor for acceleration and high-power demands. Braking energy is used to recharge the nickel metal hydride storage battery. The combined diesel/electric powertrain takes advantage of the vehicle's reduced weight to give an average of more than 60 mpg. General Motors' entry into the hybrid electric field (Precept) has a three-cylinder, 1.3-L, turbocharged diesel engine that drives a 6.5-hp, permanent magnet, DC brushless motor/generator. General Motors claims that this vehicle, which uses other technologies developed for its electrical car, the EV1, can average 80 mpg when powered by the diesel/electric parallel hybrid powertrain.

Even though the choice for the PNGV, Ford, and General Motors hybrids was a diesel reciprocating engine, Honda decided that it was better off with a spark-ignited gasoline engine for its first production hybrid. The engine is a three-cylinder, 1.0-L engine that is rated for 73 hp at 5700 rpm. Its electric motor is a 10-kW, 3000-rpm DC brushless permanent magnet motor. Like its Ford and General Motor counterparts, the Honda hybrid uses a nickel metal hydride battery pack for energy storage. Honda claims that its hybrid version can deliver 70 mpg and has a range of between 600 and 700 miles per tank of gasoline. Another hybrid heat engine/electric offered by Toyota, the Prius, is equipped with a combination heat engine and electric motor that is generally similar to the Honda hybrid. The Prius is currently sold in the United States and its fuel economy is advertised at 52 mpg in the city and 45 mpg on the highway.

With oil prices rising at the same time as the sales of light trucks, minivans, and sport utility vehicles, which account for more than 50% of auto sales in the United States, research has shifted toward hybrid technology applications in this market segment. However, whether or not hybrids will ultimately play a significant role in the transportation sector, one thing remains clear: Fuel economy will be emphasized more by environmentalists, consumers, and legislators as time goes by.

The fact that more hybrid demonstrations sport a direct-injection diesel engine may suggest that diesels have an advantage over other powerplants. The balance of this chapter is therefore devoted to further review the actual and potential advantages of the diesel engine.

1.5 What Are the Advantages of Diesel Engines?

After exploring the subject of fuel economy, it may be appropriate to review other advantages of diesel engines and perhaps expound more on the reasons for the superior fuel economy of diesels. Not only should one focus on the subject of superior performance and fuel economy, but also on environmental responsibility and accountability. In short, future powerplants and drivetrains must minimize energy consumption as well as emission of pollutants. In doing so, they should also meet customer demands for reliability, durability, transportation passenger and cargo capacities needs and wants, as well as a host of economic, safety, and social considerations. Although it is difficult to accurately assess the cost/benefit ratio of any new technology, certain boundaries must be placed before one can properly evaluate the alternatives. For instance, future designs will have to meet emission regulations and limits, otherwise the product cannot be legally sold in those regulated markets. In addition, future products may have to meet certain mandated fuel economy levels unless market forces are able to dictate these levels to the automotive industry.

1.5.1 Emissions

Although diesel engines have been much maligned in many circles, they are indeed very efficient powerplants as described earlier in this chapter. Besides their superior fuel economy advantage, they emit extremely low concentrations of unburned hydrocarbon as well as carbon monoxide emissions [14]. The reason for these extremely low hydrocarbon (HC) and carbon monoxide (CO) emissions is that diesels operate in very lean regimes where λ (relative air/fuel ratio) is greater than 1.0 [15].

Among the other exhaust gases resulting from the combustion of fossil fuels is carbon dioxide (CO_2). This exhaust species is generally referred to as a greenhouse gas and considered responsible for global warming. In spite of this belief, CO_2 remains an unregulated emission species, but at recent environmental world meetings, concerns over the global effects of emissions were discussed and commitments were made to collectively work at reducing CO_2 emissions. Interestingly, CO_2 emissions produced by natural phenomena far exceed those which are man-made. Figure 1.16 shows the apportionment of CO_2 emission from natural sources. Reference [14] indicates that total CO_2 emission from natural sources is estimated at 770×10^9 tons/annum. By contrast, the total man-made CO_2 emissions is estimated at 26×10^9 tons/annum, or less than 4% of the total CO_2 inventory.

Even though man-made CO_2 emissions may represent a small percentage of the global CO_2 problem, it is important that action be taken to minimize its impact on the environment. Figure 1.17 gives the CO_2 allocation of each man-made source. Power generation, heating, and industrial activities are responsible for about 70% of man-made CO_2, while transportation-related CO_2 emissions represent about 16% of all manmade CO_2.

With the worldwide predominance of gasoline-powered personal transportation, conversion to modern diesel-powered vehicles could reduce transportation-related CO_2 by about 25%

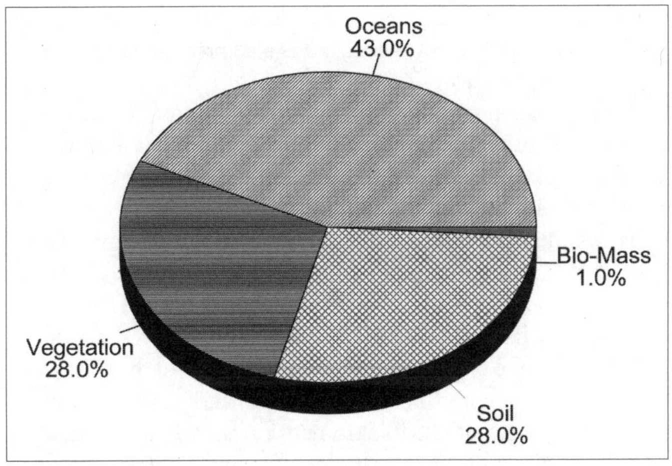

Figure 1.16 *Sources of natural CO_2 emission. (Adapted from Ref. 14.)*

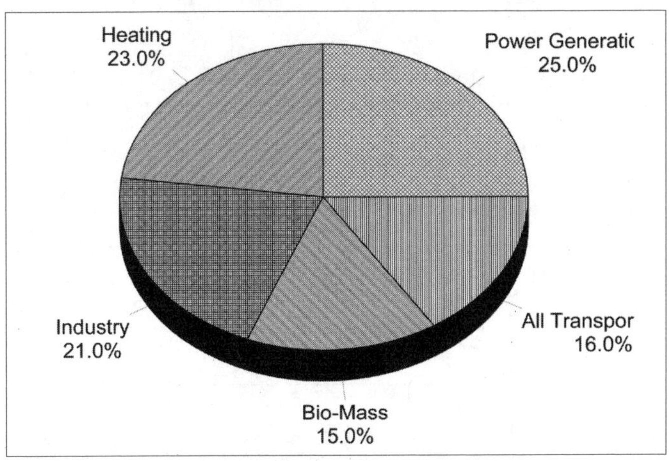

Figure 1.17 *Allocation of CO_2 emissions from man-made sources. (Adapted from Ref. 14.)*

from current levels. Figure 1.18 represents the results obtained by the German Federal Environmental Agency on 99 gasoline- and 36 diesel-powered vehicles in 1991 [14]. From that study, it was concluded that diesels had an average 19% advantage over gasoline engines in terms of CO_2 emissions. It was also reported by Schindler [14] that the diesel engines involved in that study were mostly indirect-injected engines, which are lower in fuel efficiency than their direct-injected counterparts, by about 10% to 15%. Hence a 25% reduction in transportation-related CO_2 emissions by encouraging dieselization of gasoline-powered vehicles is thought to be quite feasible.

While diesel engines are known for their high engine-out emission of nitrogen oxides (NO_x), it is perhaps worth further examination to put matters in their proper perspective. A sample of about 15 heavy-duty gasoline engines calibrated to meet the 1991 U.S. federal emission standards was tested according to EPA specifications. The mathematical average of their emissions is given in Table 1.1, where the corresponding results of a statistically representative sample of diesel engines of the same capacity are included. In addition, a line of emission data obtained by testing the same heavy-duty gasoline engines when equipped with a three-way catalyst is also provided in Table 1.1 for comparative purposes.

Note that in diesel engines both HC and CO emissions are a small fraction of their gasoline engine-out counterparts. Even diesel engine-out NO_x emissions, in the example of Table 1.1, are almost 1.0 g/bhp-h (grams/brake horsepower-hour) less than corresponding gasoline engine emissions. However, thanks to the efficiency of three-way catalyst conversion, the same gasoline engines emit extremely low HC, CO, and NO_x emissions at the tailpipe. Of course, conditions in the exhaust have to be conducive to the optimum operation of the three-way catalyst. With accurate control of the fuel and air, modern gasoline engines operate at stoichiometric ratio where the catalyst performs at its highest conversion efficiency. Unfortunately, in this instance, the diesel exhaust is extremely lean, and reducing NO_x in an oxygen-rich environment is a very challenging task. The catalyst industry is developing solutions for the diesel NO_x problem, but so far there is a lot of debate regarding the most plausible method of dealing with this problem.

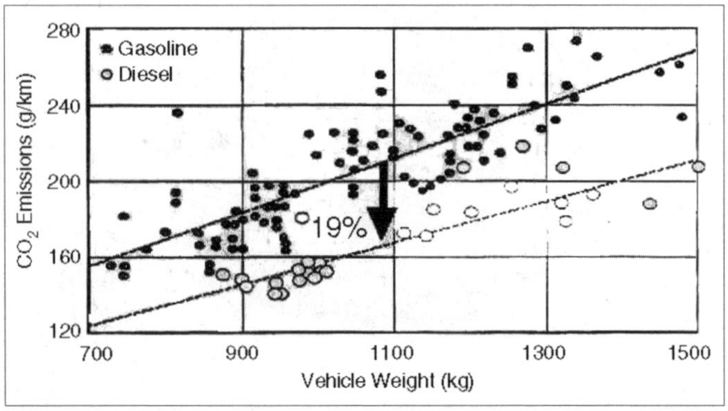

Figure 1.18 *Comparison of gasoline- and diesel-powered CO_2 emissions. (Adapted from Ref. 14.)*

TABLE 1.1
EMISSIONS FROM DIESEL AND GASOLINE ENGINES

Test Condition	Emissions (g/bhp-h)			
	HC	CO	NO_x	PM
Diesel engine-out[a]	0.15	1.50	3.40	0.07
Gasoline engine-out	0.81	30.22	4.30	—
Gasoline with three-way catalyst	0.07	2.30	0.04	—

Note: Results are composites for EPA heavy-duty transient FTP cycle.

a. Engine-out results are the mathematical average of 15 gasoline engines, 9 diesel engines, and 3 heavy-duty gasoline engines equipped with three-way catalysts. All engines had about a 7.0-L capacity.

Another problematic pollutant associated with diesel engines is particulate matter. The casual observer is made aware of this pollutant in the form of black smoke or soot emitted from either the tailpipes of many diesel-equipped passenger cars or the stacks of diesel-powered, heavy-duty vehicles. Emission of soot is also accompanied by other matter suspended in the exhaust, such as unburned lube oil, unburned fuel, trace metals, and sulfur by-products. Emission of soot in particulate matter results from the nature of the heterogeneous combustion process or diffusion-type combustion that is prevalent in diesel engines. Fuel and air mixture preparation in modern diesel engines has greatly reduced this problem. In addition, the development of diesel particulate filters promises to eliminate it almost entirely. Particulate matter from gasoline engines is not yet regulated.

1.5.2 Compression Ratio

In itself, a higher compression ratio is not necessarily an advantage. However, the compression ratio acquires its importance in diesels because of how it relates to thermal efficiency. In Eq. (1.1), mentioned earlier in this chapter, thermal efficiency was defined as follows:

$$\eta_{th} = \left[1 - \frac{1}{(C_r)^{\gamma-1}}\right] \times 100 \qquad (1.1)$$

where η_{th} is the thermal efficiency, C_r is the compression ratio, and γ is the ratio of the specific heat at constant pressure (c_p) and the specific heat at constant volume (c_v). The compression ratio is defined as:

$$C_r = \frac{V_s + V_c}{V_c} \qquad (1.3)$$

where V_s is the swept volume and V_c is the clearance volume. From Eq. (1.1), it is clear that the higher the compression ratio, the higher the thermal efficiency. However, thermal efficiency cannot be exchanged for the total efficiency of the engine, since it is only one of many efficiencies that make up the overall energy conversion efficiency of the engine. Other efficiencies include mechanical, volumetric, fuel conversion, and thermodynamic. Simply put, the overall efficiency may be calculated on the basis of the work output of an engine relative to its fuel input.

1.5.3 Friction Losses

The power output of an engine is proportional to the product of its torque and speed. Because diesels produce more torque at lower engine speeds than gasoline engines, their power is generally produced at lower engine speed. With lower engine speeds, friction losses are reduced, which improves the brake power of diesel engines and adds to their service life.

1.5.4 Diesel Fuel Safety

Diesel fuel is usually more difficult to ignite than gasoline and, therefore, is safer to handle, even in cases of spills resulting from road accidents. Diesel fuel is also less volatile than gasoline (does not evaporate as readily), which reduces the chances of releasing excessive hydrocarbons into the environment during its handling and transfer.

1.5.5 Durability

The compression ratio in diesel engines is approximately twice as high as in gasoline engines. It follows that the expansion ratio during which heat is released is also greater in diesel than gasoline engines. Greater expansion ratios provide longer time for the heat released to impart pressure on the piston and produce power. For this reason, diesel engines tend to have lower speeds than their gasoline counterparts. They also produce more torque and power at lower engine speeds than gasoline engines. Not only is the heat released earlier in the combustion cycle, but also in a much more sudden manner, which necessitates more robust engine construction. For this reason, diesel engines are known for their long useful life, which may be three to four times that of gasoline engines. In heavy-duty applications, it is not unusual to have diesel engines last close to a million miles before requiring a total overhaul. In passenger car applications, diesel engines normally outlast the service life of the vehicles they power.

In general, durability is more of a concern to the trucking industry than to passenger car owners. Durability is usually defined by the distance traveled prior to engine overhaul. Overhauling a heavy-duty diesel engine is usually required when its components show signs of wear, its fuel and/or oil consumption is exceptionally high, it smokes badly, and its performance is drastically lacking. In addition to these traditional aspects, new emission regulations proposed for the 2007 to 2010 time frame require manufacturers to guarantee emission control equipment to 435,000 miles. The implication here is that engine components that affect emissions formation and control need to be durable enough to perform their intended function for a very long time. Therefore, it is not unusual for heavy-duty engine and truck customers to demand longevity and many call for durability that extends to even twice the warranty period proposed by the EPA.

1.5.6 Reliability

As members of industrialized societies have become increasingly dependent on their personal cars or public transportation for their mobility, the reliability of these vehicles has become increasingly important. For consumers throughout the world, a steady and consistent supply of goods is of paramount importance. For these reasons, the reliability of the diesel engine which powers vehicles for a wide range of applications is central to its reputation.

Although it is difficult to assign a uniform definition to reliability, different people define the term using whichever scales are of significance to them. For instance, to the owner-operator, reliability can be defined by how few times he was stranded on the highway because of an engine failure and how quickly that failure was repaired, at a minimum expense. To the fleet manager, a criterion such as *mean time between failures* (MTBF) may be the best way reliability could be expressed. For this reason, engine manufacturers build reliability measures into the design, development, field test, and early engine production phases. Extensive *failure modes and effects analysis* (FMEA) is conducted to ensure that customers receive an extremely reliable product.

In general, the number of complaints from engine breakdowns is much smaller in the case of diesel than in gasoline engines. In the past, this was mainly due to the fact that diesel engines did not use spark plugs, distributors, and ignition systems, which tend to have more breakdowns than rugged mechanical parts and components. However, today's diesel engine is using sophisticated electronics and an increasing number of actuators and electrical components. Yet, despite the demanding applications and tough environments in which they work, diesel engines have maintained their superior reliability reputation and record.

1.5.7 Ease of Breathing and Lower Pumping Losses

Power output in diesel engines is controlled by varying the fuel rate, rather than by throttling inlet air as in gasoline engines. By not resorting to throttling, air induction into the engine is not interfered with, thus preserving good engine breathing and volumetric efficiency as well as reducing pumping losses. In other words, the work done by the engine to induct enough air for efficient combustion is minimized.

Lest the reader may think that diesels are without fault, one must include the other side of the coin where diesel engines fall short and where improvements should be made to meet both the social and legislative demands of today's environment. Figure 1.19 is a tabulated list of the diesel's advantages and disadvantages that serves as the summary and conclusion for this chapter.

ADVANTAGES	DISADVANTAGES
FUEL ECONOMY	NOISE *
DURABILITY	WEIGHT
LOW HC *	HIGH NOx *
LOW CO *	HIGH PM *
HIGH TORQUE	LOW SPEED
RELIABILITY	LOW AIR UTILIZATION
LOW FUEL COST	HIGH ENGINE COST
LOW MAINTENANCE COST	LOW EXHAUST TEMPERATURE **

* Regulated
** Significant for exhaust aftertreatment & turbo efficiency.

Figure 1.19 *Advantages and disadvantages of diesel engines.*

Even though modern diesel engines have made big strides toward noise abatement, more work is still required to quiet the engine at idle, especially in DI diesel engine designs. This powerplant is known for poor specific power output, that is, low power per unit weight. Its specific power output is generally 50% to 65% of that of gasoline engines. This feature is highly undesirable, especially for passenger car applications where emphasis on engine size and weight may be more important than in heavy-duty applications. This trend is changing.

The engine's high NO_x emission and the NO_x/PM trade-off challenge remain as the most important technical issues for the diesel engine today. Engine design changes, mainly focusing on in-cylinder emission controls, are projected to fall short of meeting future emission limits. Research has now progressed in new areas of technology including fuels and postcombustion emission control devices (exhaust aftertreatment) to assist in meeting future emission standards in the United States, Europe, and Japan.

As mentioned earlier, diesel engines develop high torque at relatively low engine speeds. Whereas this may be desirable for reducing friction loss and producing good torque at low speeds for commercial applications, it forces the use of more robust and expensive transmissions. Another aspect of diesel engine combustion is its low air utilization, meaning that in spite of inducting large amounts of air, it uses only a small fraction for combustion.

In spite of the many design changes and evolutions diesel engines have experienced in the last two decades, high engine cost remains an important issue. The main reason for this high cost is believed to be its highly precise fuel injection system. Manufactured from high-strength tool steel, the diesel fuel injection system is responsible for the fine atomization of the fuel as well as some of the mixture preparation that controls to a great degree the combustion efficiency. Increased use of electronics has made it possible for modern injection systems to perform a variety of control functions at a reasonable cost. Introducing novel means for creating the required high injection pressures has improved engine performance and fuel efficiency to reduce the impact of the high cost of diesel fuel injection equipment.

Finally, with new emphasis placed on postcombustion emission control devices, the low exhaust temperatures characteristic of diesel engines are viewed as a mixed blessing. They represent good efficiency through minimized exhaust heat loss, but they also present a challenge to exhaust aftertreatment systems, which often require higher temperatures for their efficient operation.

References

1. Presentation at Exploring Future Diesel Engine Lube Oils Symposium, February 1, 2000, Scottsdale, AZ.
2. National Inventors Hall of Fame, "Internal Combustion Engine, Rudolf Diesel, Patent No. 608,845." Retrieved from www.invent.org/hall_of_fame/42.html.
3. Compton's Encyclopedia Online, "Diesel Engine," Retrieved from www.optonline.com/comptons/ceo/01330.
4. Heywood, J.B., "Engine Types and Their Operation," Chap. 1 in *Internal Combustion Engine Fundamentals,* McGraw-Hill, New York, 1988.
5. Blair, G.P., *Design and Simulation of Two-Stroke Engines,* SAE International, Warrendale, PA, 1991.
6. Jones, J.H., W.L. Kingsbury, H.H. Lyon, P.R. Mutty, and K.W. Thurston, "Development of a 5.7 Litre V-8 Automotive Diesel Engine," SAE 780412, SAE International, Warrendale, PA, 1978.
7. Amann, C.A., "The Automotive Engine—A Future Perspective," SAE Paper No. 891666, SAE International, Warrendale, PA, 1989.
8. Heisler, H., *Advanced Engine Technology,* Butterworth-Heinemann, New York, 1995.
9. Energy Partners, "Fuel Cells, the Cleaner Energy Alternative," *S A Automotive,* July 30, 1993.
10. Jost, K., "Fuel-Cell Concepts and Technology," *Automotive Engineering International,* SAE International, Warrendale, PA, March 2000.
11. Taylor, C.F., *The Internal Combustion Engine in Theory and Practice,* Vol. 2, rev. ed., The MIT Press, Cambridge, MA, 1985.
12. Amann, C.A., "The Passenger Car and the Greenhouse Effect", SAE Paper No. 902099, SAE International, Warrendale, PA, 1990.
13. Burke, A.F., and M. Miller, "Assessment of the Greenhouse Gas Emission Reduction Potential of Ultra-Clean Hybrid-Electric Vehicles," Institute of Transportation Studies, University of California, Davis, December 1997.
14. Schindler, K.-P., "Why Do We Need the Diesel?" SAE Paper No. 972684, SAE International, Warrendale, PA, 1997.
15. Heywood, J.B., *Internal Combustion Engine Fundamentals,* McGraw-Hill, New York, 1988.

Chapter 2

Engine Fundamental and Design Parameters

2.1 Introduction

New emission regulations and performance requirements imposed on modern diesel engines are strong motivators for the development of new technology. Traditional design methods and parameters are continuously being upgraded and new ones adopted to meet the competitive market pressures. However, to those who are involved in modern engine design, a good knowledge of the many aspects that not only control engine performance, but also its emissions profile, is imperative. This knowledge must evolve from simple design principles to sophisticated interactions between various engine systems as well as postcombustion exhaust emissions control devices.

In this chapter a review of the basic engine components, design parameters, geometric properties, and performance issues are presented. These factors are discussed with a view toward maximizing performance and fuel economy and minimizing exhaust pollutants. Although many of the principles discussed in this chapter may have application to other powerplants, the main emphasis is on diesel engines.

2.2 Heat Engines

Heat engines can generally be considered energy conversion machines. In essence, chemical energy supplied (generally in the form of fossil fuel) to a heat engine is combusted by mixing with oxygen at a certain temperature, producing heat that is eventually converted to useful work. Heat engines can be classified as (1) external combustion engines or (2) internal combustion engines.

Examples of the external combustion engine include the Stirling engine (Figure 1.8, Chapter 1), in which heat is added to the working fluid at high temperature and rejected at low temperature. A net work is produced by the working fluid. Heat added to the working fluid can be generated from practically any heat source, such as burning fossil fuels, wood, or any other organic material. The steam engine is another example of external combustion systems. Heat added from an external source elevates the water temperature until it is converted into steam that provides pressure and eventually the net work. Steam engines powered cars in the United States between 1900 and 1916; however, they had all but disappeared by 1924. Reasons for their demise were mainly the size and number of the major components required for their operation such as furnace, boiler, valving, and their complicated controls [1].

In internal combustion engines, the same combustible mixture that produces heat is responsible for producing useful work. In Chapter 1, two categories of internal combustion engines were introduced: rotating and reciprocating engines. A very well-known example of the rotating engine is the Wankel engine (Figures 1.11, 1.12, and 1.13, Chapter 1), which is also known as the rotary engine. The second category, the reciprocating engine, can be divided into two types: the two-stroke and the four-stroke engine. In internal combustion engines, the working fluid is the combustible mixture of air and fuel. This fluid is responsible for releasing heat energy that is stored in the fuel and producing useful work. Worthwhile goals common to the design and development of all heat engines include maximizing work (power output), minimizing energy consumption, and reducing pollutants that may be formed in the process of producing work. Figure 1.2 in Chapter 1 shows the main components of a reciprocating engine, where TDC refers to top-dead-center and BDC refers to bottom-dead-center. Intake and exhaust valves are omitted for simplicity; however, it is worth noting that in some two-stroke engine designs, inlet and exhaust ports are used rather than the valves that are commonly used in four-stroke engines.

Both internal combustion reciprocating engine categories (two- and four-stroke) may incorporate either a spark-ignited (SI) combustion system or a compression-ignited (CI) system. Spark-ignited engines use a combustion cycle known as the Otto cycle, whereas compression-ignited engines use the

Diesel cycle. Spark-ignited systems are characterized by a homogeneous or mostly well-mixed charge of fuel and air. In this medium, combustion is initiated by a spark, and the flame propagates along a front from the spark location to the opposite side of the combustion chamber. Diesel engines achieve their high performance and excellent fuel economy by compressing air to high pressures and temperatures and then injecting a small amount of fuel into this highly compressed air. The charge of air and fuel in these engines is described as heterogeneous, meaning that the fuel remains separate from the air for a finite time in the combustion chamber. Heat generated due to the compression of air in the cylinder causes the small amount of highly atomized injected fuel to evaporate. Mixing with the hot surrounding air in the combustion chamber, the evaporated fuel reaches its auto-ignition temperature and burns, thus releasing the energy that is stored in that fuel. The auto-ignition temperature of a fuel depends on its chemistry. Unlike the SI system, combustion in CI engines occurs at many points where the air-to-fuel (A/F) ratio and temperature can sustain this process.

2.3 Operation of Reciprocating Internal Combustion Diesel Engines

As mentioned earlier, reciprocating internal combustion diesel engines can be classified as two-stroke or four-stroke designs. In this chapter, the operation of each classification is described and some of the parameters affecting performance are detailed.

2.3.1 Operation of Two-Stroke Engines

By definition, two-stroke engines require two strokes to complete their combustion cycle. Figure 2.1 gives the details of the two-stroke combustion cycle. With the transfer and exhaust ports open, air under slight pressure in the crankcase flows into the cylinder. The rising piston eventually covers the transfer ports, thus trapping the inducted air in the cylinder. Further upward motion toward top-dead-center compresses the air where fuel is injected at the appropriate timing. Heat absorbed from the surrounding hot compressed air causes the fuel to evaporate and mix with the air. Once the auto-ignition temperature is reached, combustion begins and causes the working fluid (combustible mixture) to expand, thus applying pressure on the surface of the piston and producing useful work at the engine output shaft. Meanwhile, fresh air flows into the crankcase to be compressed by the descending piston on its way to bottom-dead-center. While descending, the piston uncovers the exhaust port, starting the scavenging of the cylinder and causing a slight increase in crankcase pressure. This increase in crankcase pressure causes the fresh air induction into the cylinder through the transfer port and resumes the cycle once again.

The two-stroke engine represented by the schematic in Figure 2.1 is not the only design possibility for a two-stroke engine. Several other designs exist with various mechanical arrangements, but their principle of operation is essentially the same. Some two-stroke engines have inlet and exhaust ports placed at one end of their cylinders as shown in Figure 2.2 [1].

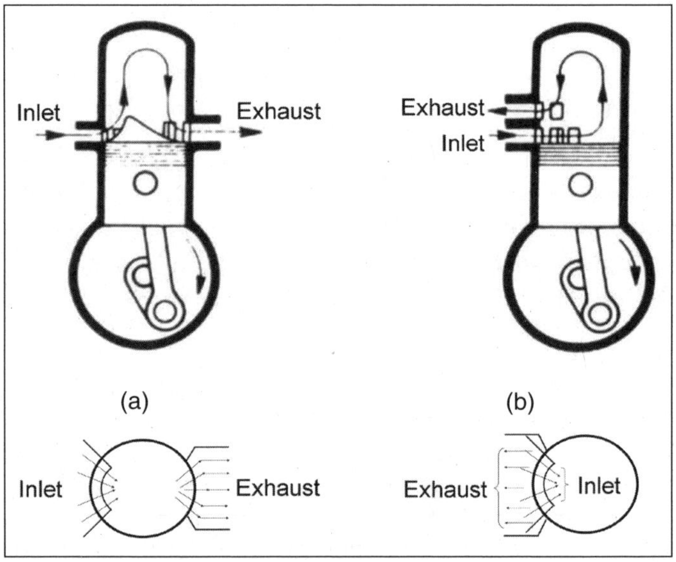

Figure 2.2 *Schematics of two-stroke engines showing (a) an opposite side port arrangement and (b) a same side port arrangement. (Reprinted with permission from* The Internal Combustion Engine in Theory and Practice *[rev. ed.], Vol. 1, by Charles Fayette Taylor. ©1985 The MIT Press.)*

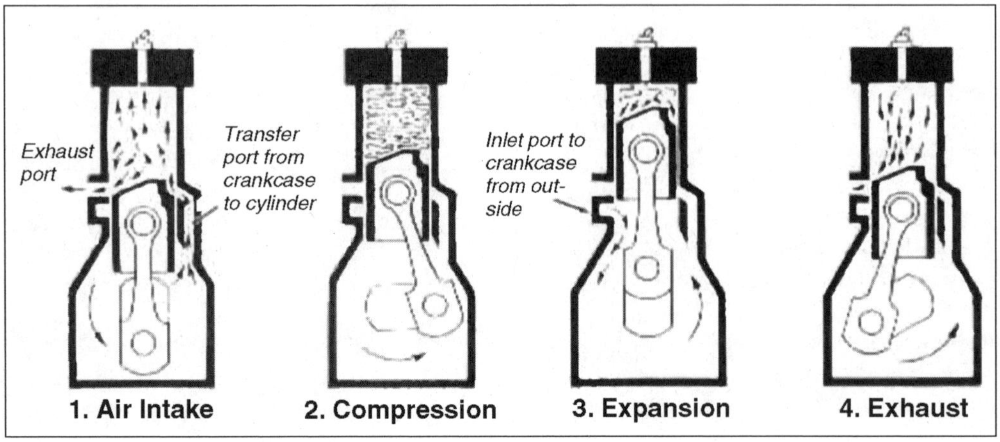

Figure 2.1 *Details of the two-stroke combustion cycle.*

2.3.2 Scavenging in Two-Stroke Engines

The process of purging exhaust gases from a previous cycle and filling the cylinder with fresh air for a new cycle is referred to as *scavenging*. The main method for scavenging two-stroke engines is to use the pressure of the inducted fresh air to purge or displace the burned gases from the previous cycle. Generally, the greater the oncoming air pressure, the more complete the scavenging process. Therefore, better scavenging in two-stroke engines is achieved, in part, by raising the pressure of the fresh air being inducted into the cylinder. This process is accomplished by using various devices such as blowers, compressors, or pumps [2].

Scavenging in two-stroke engines is performed mainly by one of three methods:

- Cross scavenging
- Loop scavenging
- Uniflow scavenging.

There are other variations to these three methods, but these remain the principal implementations of scavenging in two-stroke engines. Figures 2.3a–c illustrate the differences between the three methods. The uniflow-scavenging system may use inlet ports and exhaust valves, as shown in Figure 2.3c, or inlet and exhaust ports in opposed piston engines such as shown in Figure 2.4.

As the exhaust ports open (EPO), a rapid drop in cylinder pressure is observed. Evacuation of the exhaust gas through the exhaust ports is known as the blowdown process. A blowdown angle is defined as the angle between EPO and the angle at which the cylinder and exhaust system pressures equalize. The inlet ports open (IPO) late into the blowdown process and allow fresh air to enter the cylinder. This step can begin only when the fresh air charge pressure exceeds that in the cylinder. Meanwhile, the piston reaches its bottom-dead-center (BDC) position and reverses its direction, thus moving toward its top-dead-center (TDC). The point at which the piston reverses its direction is shown as BC (bottom center) in Figure 2.5. The next event in the cycle is the inlet port closing (IPC), which takes place as the piston moves toward its TDC and covers the inlet ports. Further piston movement toward its TDC causes the exhaust ports to close (EPC), thus trapping the fresh charge and any residual exhaust in the cylinder. Thereafter, a steady but fast pressure rise is observed in the cylinder. This pressure rise is also accompanied by a proportional increase in temperature. The crank angle during which both inlet and exhaust ports remain open is referred to as the scavenging angle.

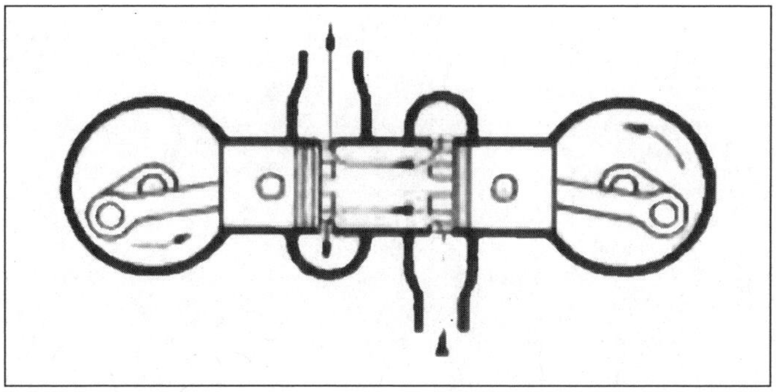

Figure 2.4. *Opposed piston, two-stroke engine design. (Reprinted with permission from* The Internal Combustion Engine in Theory and Practice *[rev. ed.], Vol. 1, by Charles Fayette Taylor. ©1985 The MIT Press.)*

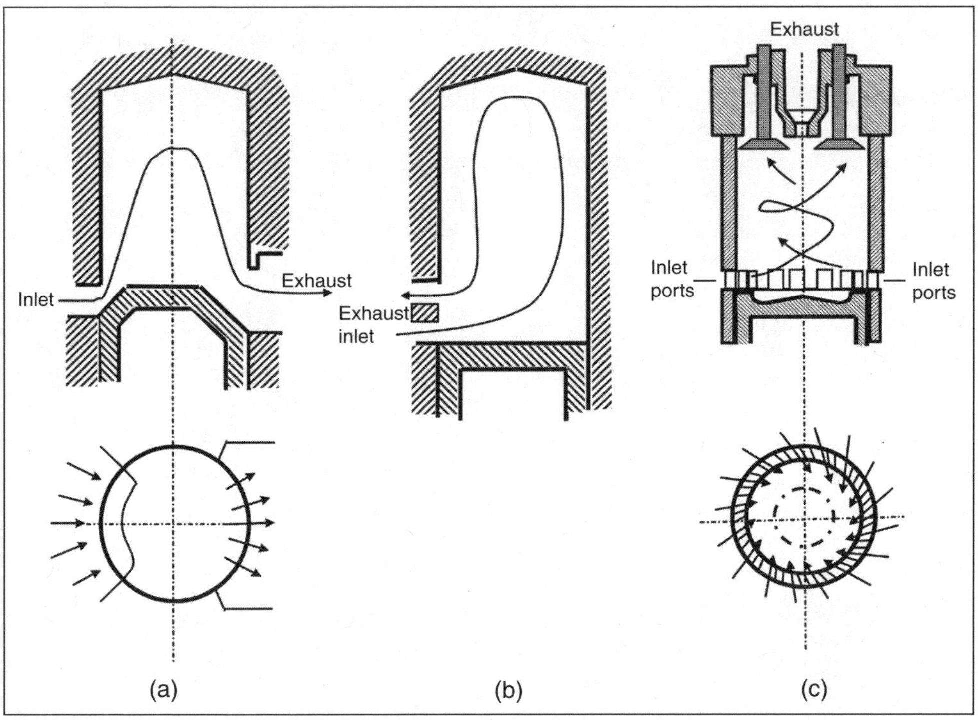

Figure 2.3 *(a) Cross-scavenging method, (b) loop-scavenging method, and (c) uniflow-scavenging method. (Modified from Ref. 2, Figs. 6.23a–c.)*

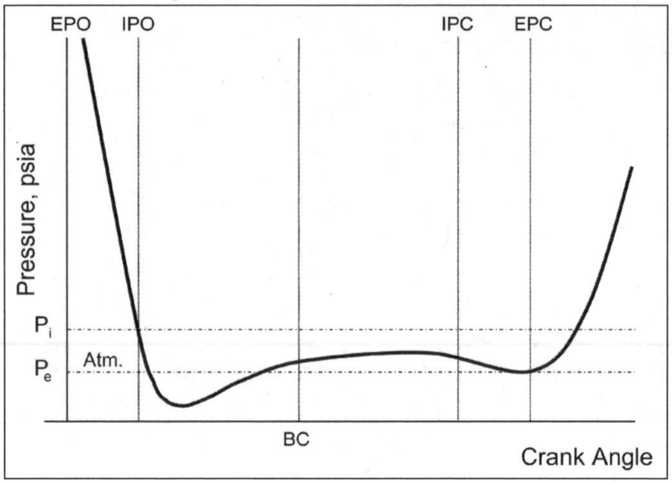

Figure 2.5 Indicator diagram of a two-stroke cycle.

The description given to the scavenging process so far was somewhat ideal and comprised distinct steps. However, the actual scavenging in two-stroke engines is far less than ideal. In fact, during scavenging not only does the fresh charge exchange heat with the residual gases, it also mixes with them and changes its chemical composition in the process. The final chemical makeup of the mixture at the end of the scavenging process plays an important role in the combustion quality as well as its resultant emissions. An inherent loss in two-stroke engines results when some fresh charge escapes through the exhaust ports during scavenging. This phenomenon is often referred to as *short-circuiting,* which leads to lower volumetric efficiency.

A two-stroke is usually smaller in size than a four-stroke engine, yet it has the same power output and tends to have higher specific power (power output for a given engine displacement) than its four-stroke counterpart. Two-stroke engines are generally less fuel efficient than four-stroke engines. The main reason for this relative fuel inefficiency in two-stroke engines is poor scavenging and relatively low volumetric efficiency.

2.3.3 Definition of Basic Performance Parameters for Two-Stroke Engines

Having described the two-stroke combustion cycle, we can now define several parameters that help us assess engine performance.

2.3.3.1 Compression Ratio

The compression ratio is generally defined as the ratio of total (maximum) to minimum cylinder volume. According to this general definition, the compression ratio is the product of dividing the cylinder volume when the piston is at BDC by the cylinder volume when the piston is at TDC. This definition complies with the traditional or conventional understanding that applies to reciprocating internal combustion engines.

It is worth noting, however, that the traditional definition may not reflect the actual compression ratio in two-stroke combustion systems. In fact, compression is almost nonexistent until the piston covers both inlet and exhaust ports in two-stroke engines. Therefore, a more accurate definition of compression ratio in two-stroke engines should limit the total cylinder volume to that in effect at the time of exhaust port closure when the working fluid is trapped within the cylinder. In spite of this technicality, this definition is preferred over other alternatives. Therefore, the compression ratio can be defined as follows:

$$\text{Compression ratio}(C_r) = \frac{\text{Maximum cylinder volume}}{\text{Minimum cylinder volume}} \qquad (2.1)$$

The compression ratio is an important parameter by which engine efficiency can be determined. Although the overall efficiency of the engine may depend on a number of other parameters such as mechanical efficiency, volumetric efficiency, and pumping efficiency to name a few, the compression ratio has a direct impact on thermal efficiency. Equation (2.2) shows the mathematical relationship between compression ratio and thermal efficiency:

$$\eta_{th} = \left[1 - \frac{1}{(C_r)^{\gamma - 1}}\right] \times 100 \qquad (2.2)$$

where η_{th} is thermal efficiency, C_r is compression ratio, and γ is the ratio of specific heat at constant pressure to the specific heat at constant volume. It follows then that the greater the C_r, the higher the thermal efficiency of an engine.

2.3.3.2 Swept Volume

For any cylinder in an engine having a bore B_{cyl} and a piston stroke S_P, the swept volume is defined by the product of the area of that cylinder ($\pi B^2_{cyl}/4$) and the piston stroke (S_P). In other words:

$$\text{Swept volume}(V_s) = \left(\frac{\pi B^2_{cyl}}{4}\right) \times S_p \qquad (2.3)$$

2.3.3.3 Clearance Volume

Clearance volume (V_c) is the volume above the piston when it is at its TDC position. Both the swept and clearance volumes are used to define the compression ratio. Figure 2.6 is a helpful illustration of the swept as well as the clearance volumes and Eq. (2.4) is used to determine the geometric compression ratio of an engine:

$$C_r = \frac{Vs + Vc}{Vc} \qquad (2.4)$$

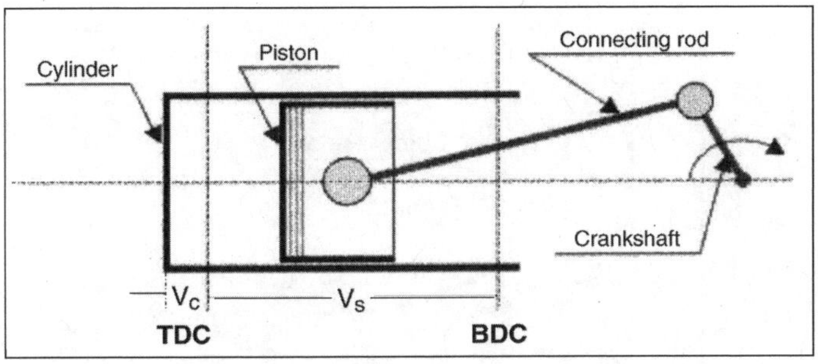

***Figure 2.6** Relationship between swept and clearance volumes.*

2.3.3.4 Delivery Ratio

The delivery ratio is the ratio of the mass of fresh air delivered during the scavenge duration and the mass of air required to fill the swept volume at ambient conditions. The mass of air required to fill the swept volume at ambient conditions serves as reference and is conventionally calculated on the basis of the prevailing ambient pressure and temperatures conditions.

$$\text{Delivery ratio } (D_r) = \frac{\text{Actual air mass delivered during scavenge}}{\text{Reference air mass to fill swept volume}} \quad (2.5)$$

The delivery ratio defines how well an engine is able to fill the cylinder with air from the prevailing ambient conditions. Obviously the more air introduced into the cylinder, the more oxygen available to the combustion process. Even though the majority of the air inducted into the cylinder may not engage during the combustion process, still the opposite action of starving a diesel engine of air may have devastating consequences on combustion efficiency. The evidence of a diesel engine starved for air is usually seen in the form of very visible black smoke.

2.3.3.5 Scavenge Ratio

The scavenge ratio is the ratio of the mass of fresh air supplied during the scavenge duration and the mass of air required to fill the total cylinder volume at ambient conditions in naturally aspirated engines. The mass of air required to fill the total cylinder volume at ambient conditions serves as a reference and is conventionally calculated on the basis of the prevailing ambient pressure and temperatures conditions. The scavenge ratio should not be confused with the delivery ratio, the difference being that in the scavenge ratio the reference is total cylinder volume, whereas in the case of the delivery ratio the reference is just the swept volume.

$$\text{Scavenge ratio } (S_r) = \frac{\text{Actual air mass delivered during scavenge}}{\text{Reference air mass to fill cylinder volume}} \quad (2.6)$$

Naturally, a successful scavenge is required before we can introduce the maximum amount of air in the cylinder. Therefore, good scavenging is a prerequisite for a good delivery as well as good scavenge ratios.

2.3.3.6 Scavenge Efficiency

The working fluid in the cylinder at the point of exhaust port closure may consist of trapped fresh air ($m_{fa,}$), residual air from a previous cycle that had not engaged in the combustion process (m_{ra}), and exhaust that had not been scavenged (m_{ex}). The scavenge efficiency is the ratio of the trapped fresh air (m_{fa}) to the total mass trapped in the cylinder ($m_{fa} + m_{ra} + m_{ex}$):

$$\text{Scavenge efficiency } (S_e) = \frac{m_{fa}}{m_{fa} + m_{ra} + m_{ex}} \quad (2.7)$$

2.3.3.7 Purity of Charge

It is important to differentiate between the terms *purity of charge* and *scavenge efficiency* because they are often confused with each other. Charge purity is the ratio of the air trapped in the cylinder at the start of combustion to the total mass of the working fluid. The air trapped in the cylinder prior to combustion consists of the charge of fresh air trapped (m_{fa}) plus any residual fresh air that had not engaged in prior combustion (m_{ra}):

$$\text{Purity} (P_u) = \frac{m_{fa}}{m_{fa} + m_{ra}} \quad (2.8)$$

From a combustion efficiency standpoint, it is important to utilize as much of the fresh charge inducted into the cylinder as possible. In other words, we would hope to have an extremely small m_{ra} if any. Ideally, the highest purity of charge is achieved when all of the charge is consumed during the combustion process.

2.3.3.8 Trapping Efficiency

The trapping efficiency (T_e) is defined as the ratio of the charge retained in the cylinder to the charge supplied. This definition applies at the point of port closure at which the charge is trapped within the volume of the cylinder. The higher the T_e, the higher the fresh charge trapped together with its oxygen content being made available to the combustion to follow.

$$\text{Trapping efficiency } (T_e) = \frac{\text{Mass of fresh air at port closing}}{\text{Mass of air supplied to cylinder}} \quad (2.9)$$

Other design parameters could be defined as having direct implications on engine performance and emissions characteristics. However, the preceding are the more common parameters and are most often used.

2.3.4 Scavenging Characteristics in Various Two-Stroke Designs

It may be useful at this point to apply some of the definitions to various scavenging methods and examine how well these concepts may meet design objectives. Figure 2.7 shows scavenging behavior in large two-stroke diesel engines for three scavenging configurations.

The line showing perfect displacement indicates that the fresh charge has displaced all of the exhaust products from the prior combustion cycle, and the entire cylinder volume is being occupied by a fresh charge. This line also implies that there was no mixing between the fresh charge and the products of combustion that were scavenged. On the other end of the spectrum, complete mixing may take place when the scavenging process allows enough interaction between the fresh charge and the products of combustion. Obviously, one should aim for perfect displacement as the design objective. Examination of the results shown in Figure 2.7 indicates that uniflow scavenging is the most effective followed by loop scavenging and then cross scavenging.

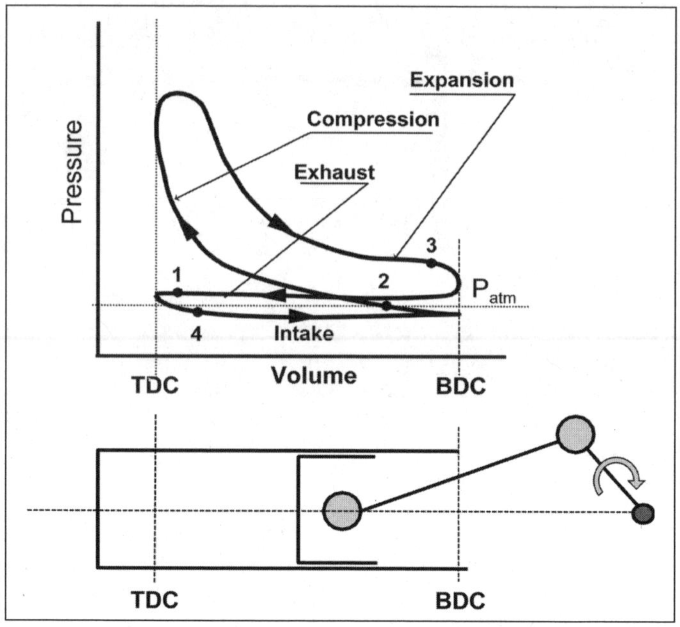

Figure 2.8 *Pressure–volume diagram for four-stroke naturally aspirated engines.*

2.3.5 Operation of Four-Stroke Engines

It takes four strokes to complete the combustion cycle in four-stroke engines. Figure 1.3 in Chapter 1 is a schematic representation of the four-stroke combustion cycle as applied to a diesel engine. See the text in Chapter 1 that describes Figure 1.3 for a description of the combustion cycle for four-stroke diesel engines.

Changes in pressure, volume, temperature, and mixture composition occur during the four strokes. The pressure–volume diagram is often used to describe changes in pressure and volume in the cylinder during a complete cycle. Figure 2.8 illustrates pressure and volume changes for a naturally aspirated diesel engine.

In Figure 2.8, intake and exhaust valve events are marked by points 1 through 4, where point 1 is the point at which the intake valve opens, point 2 is where the intake valve closes, point 3 is where the exhaust valve opens, and point 4 is where the exhaust valve closes. It is important to note that both intake and exhaust valves remain open during the time between points 1 and 4 as well as its equivalent crank angle duration. This period is referred to as *valve overlap* and plays an important role in engine performance and its emission characteristics.

The intake valve closing occurs a few degrees beyond BDC to improve cylinder filling and, therefore, the volumetric

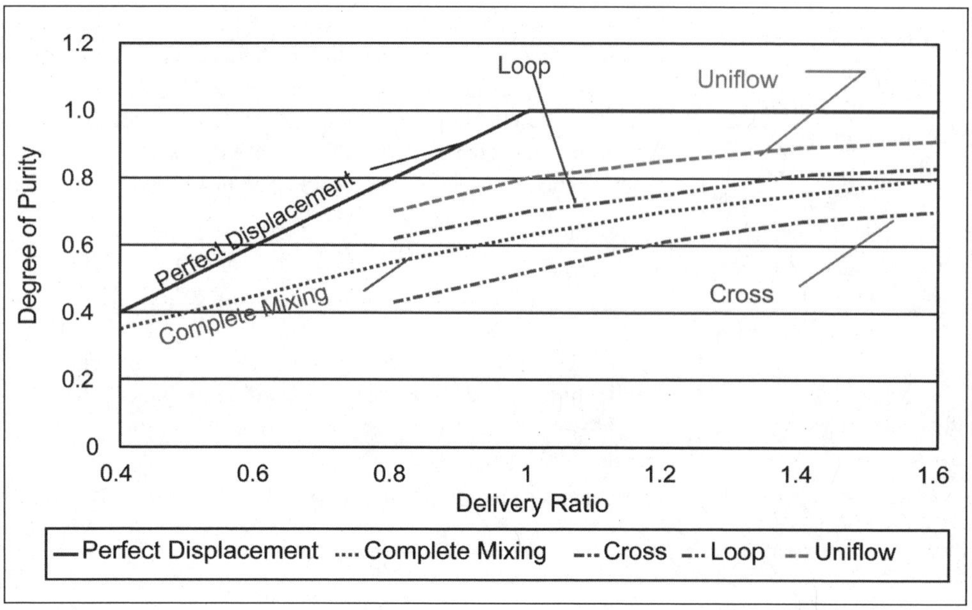

Figure 2.7 *Relationship between purity of charge and delivery ratio for various scavenging configurations.*

efficiency of the engine. Effective and rapid compression of the working fluid (air) begins after the intake valve closes as the piston travels from BDC to TDC. In naturally aspirated engines, pressure inside the cylinder during the intake stroke is below atmospheric pressure. Restrictions through the air intake filter, air inlet piping, intake manifold, intake port, and intake valve contribute to pressure loss and help reduce cylinder pressure to below atmospheric. Shortly following combustion, the expansion stroke begins and is marked by a number of chemical reactions and heat transfer processes while the piston travels from TDC to BDC. At point 3, the exhaust valve opens, thus allowing some of the combustion products to go through a blowdown process as a result of the pressure differential between the cylinder and the exhaust system. The remainder of the exhaust gases are expelled from the cylinder by virtue of the piston motion from BDC to TDC during the exhaust stroke.

Another way to illustrate the four-stroke cycle is through the pressure–crank angle diagram shown in Figure 2.9. In addition to the details outlined in the description of the pressure–volume diagram, the pressure–crank angle diagram of Figure 2.9 highlights the point at which fuel is injected (I) as well as ignition delay. During this delay fuel injected into the cylinder evaporates using heat from the working fluid that had been compressed. The result of the heat transfer from the compressed air to the fuel is a reduction in the rate of pressure rise, as illustrated in Figure 2.9. Following the start of combustion, the rate of pressure rise increases dramatically and the combustion pressure peaks a few crank angle degrees past TDC. Factors controlling the rate of pressure rise include the ignition delay, fuel quality, and the rate of injection. In many designs, the engine noise, vibration, and harshness characteristics are often tied to the rate of pressure rise in the cylinder.

Together with the rise in cylinder pressure, cylinder temperature also increases and reaches its peak. The maximum combustion temperature depends on several factors including fuel rate, fuel injection timing, fuel quality (especially its calorific value and cetane number), initial cylinder pressure at intake valve closing, and charge temperature.

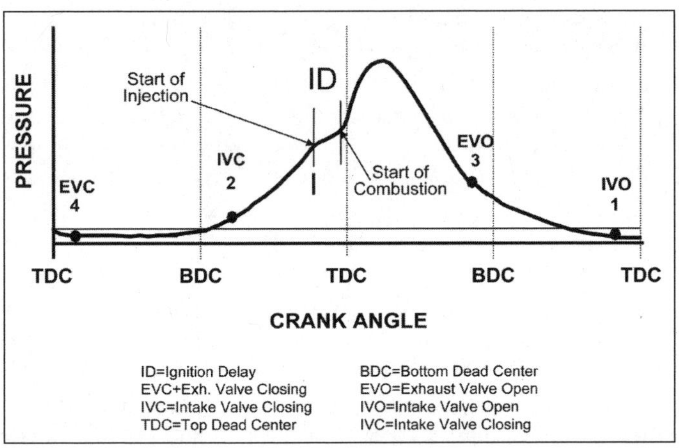

Figure 2.9 *Pressure–crank angle diagram for a four-stroke diesel engine.*

So far, intake and exhaust valve motion has been treated generally in relation to the specifics of the pressure–volume or pressure–crank angle diagrams. Earlier, valve overlap was cited as a parameter that has a major influence on engine performance and its emission characteristics. A more detailed look into the opening and closing of these two valves reveals more insight into their effect on volumetric efficiency and total charge composition. Starting with the exhaust valve, its opening takes place at a point near the end of the power stroke. The timing of EVO should not be too early lest useful work be lost, yet it is beneficial to effect the valve opening while cylinder pressure is at a level capable of clearing exhaust products through the exhaust valve area during the blowdown time. To ensure complete clearing of the exhaust gases through the exhaust valve opening it is usually kept open a few degrees past TDC.

It is worth noting that both valves do not reach their fully open position instantaneously, but have a finite length of time during which they move from a fully closed to a fully open position. To make the best use of both valves, they must be in their fully open position at the time when they are exposed to the maximum pressure differential causing the working fluid to flow across them. To ensure this aspect of the cylinder filling and emptying processes, the intake and exhaust valves are usually opened before the start of the intake and exhaust strokes, respectively. They are also held open a few degrees past their respective strokes for the same purpose. As the exhaust valve is opened, exhaust gases are expelled by their own kinetic energy, thus reducing the work required by the piston during the exhaust stroke to expel the remaining exhaust gases. Ensuring the expulsion of all exhaust products from the cylinder facilitates the induction of a fresh and generous charge to sustain combustion of the tiny quantity of fuel injected into the cylinder.

The intake valve opens toward the end of the exhaust stroke while exhaust gases are exiting through the exhaust valve at high velocity. Their high exit velocity creates a pressure drop in the cylinder that helps draw the fresh charge and maximizes induction. If the intake valve opening is too early, some of the exhaust products may actually flow back through its opening into the intake port and then on to the intake manifold. This portion will be inducted once again into the cylinder during the intake stroke and contaminates the fresh charge with exhaust. The proper function of both valves may require timing their opening and closing in a way that may interfere with the piston position. In such cases two solutions may be used: valve recess in the cylinder head (as shown in Figure 2.10) or valve cutouts in the piston crown. Therefore, it is extremely important to carefully design the timing of intake and exhaust valve opening, closing, lift, flow area, and intake and exhaust valve overlap.

As mentioned in Chapter 1, the two basic types of four-stroke diesel engines are direct- and the indirect-injected engines. The design and operation of both types was described in Chapter 1 and basic differences in their construction were illustrated in Figures 1.4 and 1.5. Recent interest in energy conservation has caused most, if not all, new four-stroke diesel engine designs to adopt the direct-injection concept for its superior fuel economy and its future promise of low emissions.

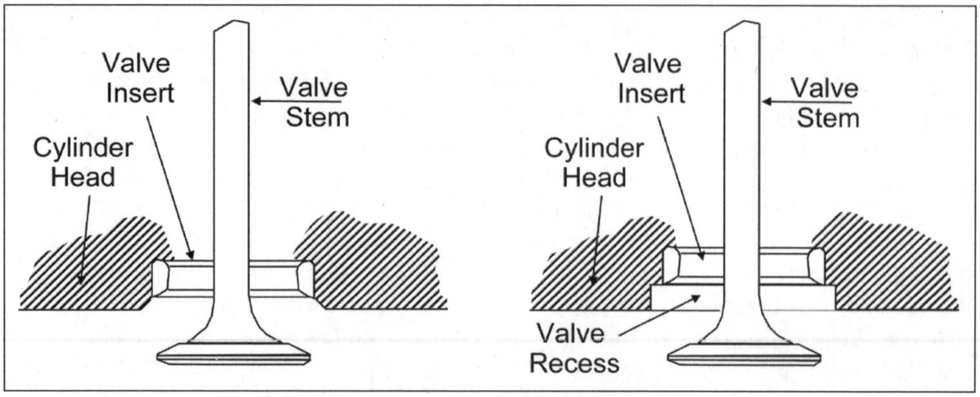

Figure 2.10 *Valve (a) without recess and (b) with recess.*

In spite of the promising future for direct-injected (DI) diesel engines, the indirect-injected (IDI) engine clearly dominated the passenger car market. This dominance was due to its relatively quiet and smooth operation when compared with older DI diesels. In addition, IDIs have higher power density than DIs and are characterized by lower nitric oxide emissions than their DI counterparts. IDI engines are more popular elsewhere in the world than in the United States. The main reason for their popularity is their superior fuel economy (15% to 20%) over their gasoline competitors. Their fuel economy advantage over gasoline engines is more prevalent in urban driving conditions where the spark-ignited engine is normally heavily throttled. Modern, port-injected gasoline engines may have fuel evaporation difficulties during cold weather operation, and that is another area where IDI engines may be more superior. This IDI dominance over the DI engine in the passenger car market has recently been challenged. In the mid-1980s, Ford Motor introduced its first production light-duty, high-speed DI diesel engine (HSDI) to the European market. The new breed of HSDI engines is even more fuel frugal than its IDI predecessors and has already demonstrated an average of between 10% and 15% lower fuel consumption than IDI engines. When compared to gasoline engines, the fuel economy advantage may approach 30% to 40%. New engine designs and materials used in producing new DI engines for passenger car applications are proving that HSDIs can be made to operate with very acceptable noise, vibration, and harshness (NVH) characteristics [3].

Although IDI engines have lower combustion noise, a less costly fuel injection system, higher power density, and in general lower emissions levels, HSDIs have better starting abilities in cold weather, less heat rejection to the water jacket, more EGR tolerance, greater durability, and lower overall carbon dioxide emissions [3].

So far we have treated reciprocating engines as either spark- or compression-ignited engines. We have even subdivided compression-ignition engines into two categories. By so doing, we obviously run the risk of oversimplifying engine categorization. Therefore, to put this potential misunderstanding to rest, we need to mention here that engines can be categorized in many other ways, including the following:

- *Mobility*: engines can serve both stationary as well as mobile applications. Stationary engines range from very small power outputs (perhaps below 2 hp, such as in Hatz single-cylinder engines) to as high as 21,000 hp, such as in the Harland and Wolff uniflow-scavenged two-stroke engine. Stationary engines may be used for power generation in either standby or continuous applications. They are also used for driving machinery such as large compressors that specialize in moving bulk.
- *Fuel*: So far we have mainly addressed engines that use either gasoline or diesel fuels. Yet, fuels are not necessarily limited to these two. Other fuel types include compressed natural gas (CNG), liquid propane gas (LPG), methanol, ethanol, biodiesel, broad-cut, kerosene, and dimethyl-ether. The less common fuels may be used in special applications that may, for instance, be located near an abundant supply of that fuel.
- *Application*: Engines can be installed in vehicles serving off-road or on-highway applications. Off-road engines may cover multiple applications ranging from earthmovers, back hoes, and other construction equipment to agricultural machinery such as combines, tractors, mowers, and garden tools. On-highway engines may power heavy trucks as well as passenger cars and mopeds. Other applications that do not quite fit the description of on-highway or off-road vehicles may include aircraft, marine, locomotive, and portable units.
- *Configuration*: Engines are designed to fit a variety of installations. For many years in-line engine designs (Figure 2.11a) were most popular for vehicular applications. Yet, in the mid-1970s front-wheel drives became popular to provide more room in the new downsized car designs. Transverse engine installations became common practice and led to rash of conversions to vee engine configurations (Figure 2.11b) from the in-line designs. The vee configuration compressed the length of the engine and allowed it to be packaged under the hood of passenger cars. Other configurations include the horizontally opposed piston (Figure 2.12), radial (Figure 2.13a) for aircraft, and delta (Figure 2.13b) configurations.
- *Valve /port design*: The position of intake and exhaust valves or ports can be used to differentiate among various engine

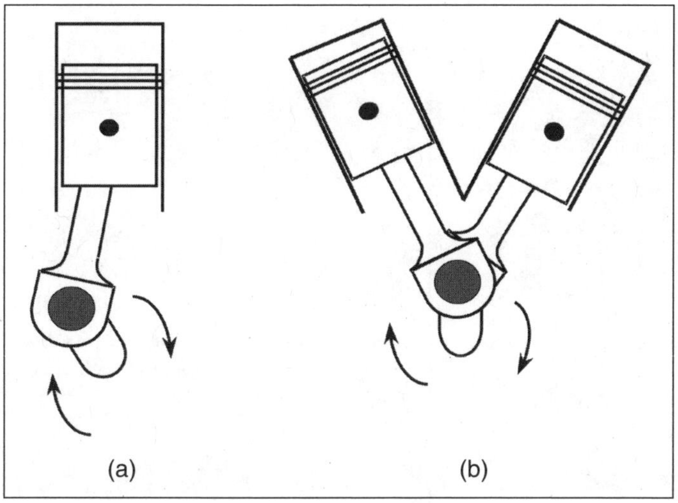

Figure 2.11 *(a) In-line and (b) vee engine configurations.*

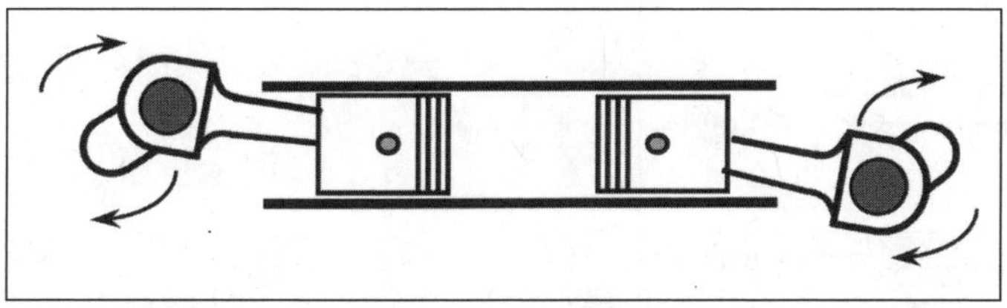

Figure 2.12 *Horizontally opposed piston engine.*

designs. For instance, an overhead cam design facilitates the actuation of unit injector fuel injection systems. Actuation of the unit injector is often accomplished by inserting an additional cam between the intake and exhaust cams. Figure 2.14 shows a variety of cam arrangements as applied in practical engine embodiments.

- *Induction*: Induction systems come in two basic designs: naturally aspirated or turbocharged. Hence, engines that do not have devices to boost their charge are referred to as *naturally aspirated engines* and those that have charge boost devices, such as turbochargers or superchargers, are called *turbocharged engines.* Figure 2.15 illustrates a turbocharged engine.
- *Charge air cooling*: Turbocharging or supercharging of engines leads to increasing the charge air temperature and reducing its density. To improve charging of the combustion chamber, designers resorted to cooling the charging after it had been boosted (increasing its pressure); hence, the term *aftercooling* is used to describe a system through which air passes through a charge cooler. In the aftercooler heat is exchanged to another medium, normally water or air, prior to its introduction into the cylinder. Engines are often described as turbocharged and aftercooled (TA). Some designers refer to this heat exchanger as an *intercooler* because it is placed between the turbocharger and the intake manifold of an engine or port of a cylinder.
- *Engine cooling*: Combustion produces heat that must be dissipated away from the engine's hottest regions to preserve its mechanical integrity. The process of carrying heat away from the engine's critical component is achieved through its cooling system. While most engines use a water jacket around the hot engine parts to transfer the heat from the engine to the environment via a heat exchanger (radiator), some engines use air cooling. The liners in these engines have fins on their outer surface that exchange heat by means of air blowing across those fins.
- *Power modulation*: Increasing or decreasing the engine's power output generally requires controlling its air-to-fuel ratio. In gasoline, spark-ignited engine power modulation is achieved through throttling the engine, thus reducing its air flow and enriching its fuel and air mixture. Some engineers refer to this method as *qualitative control* because it involves

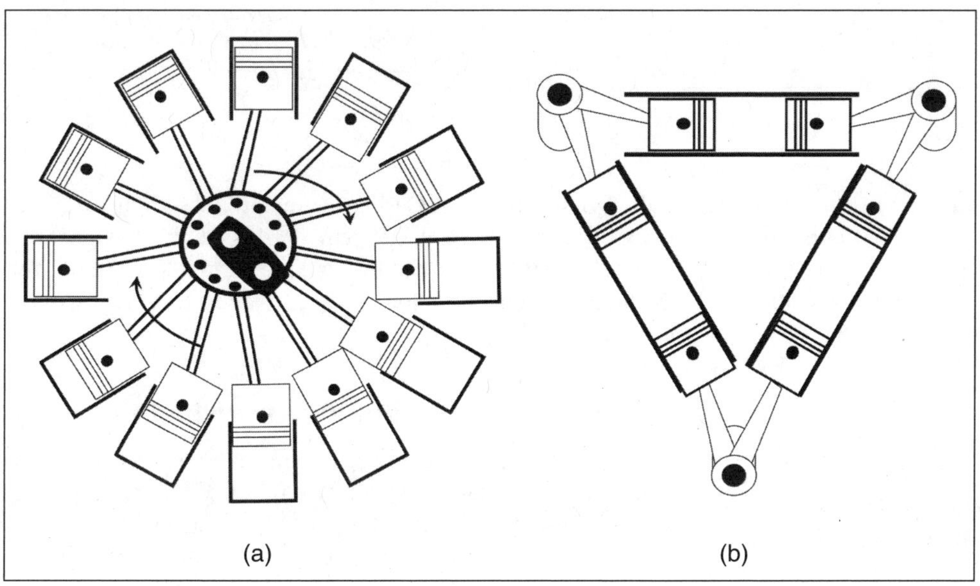

Figure 2.13 *(a) Radial and (b) delta engine configurations*

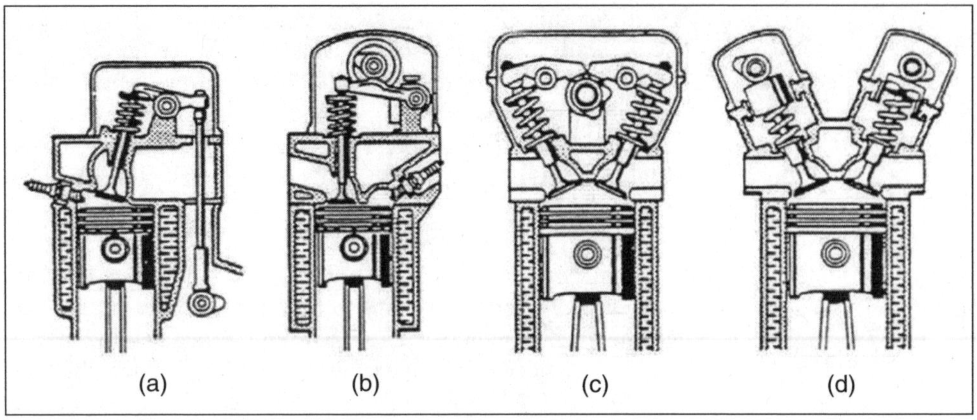

Figure 2.14 *Various cam arrangements. (Reprinted from Ref. 4 with permission from Robert Bosch GmbH.)*

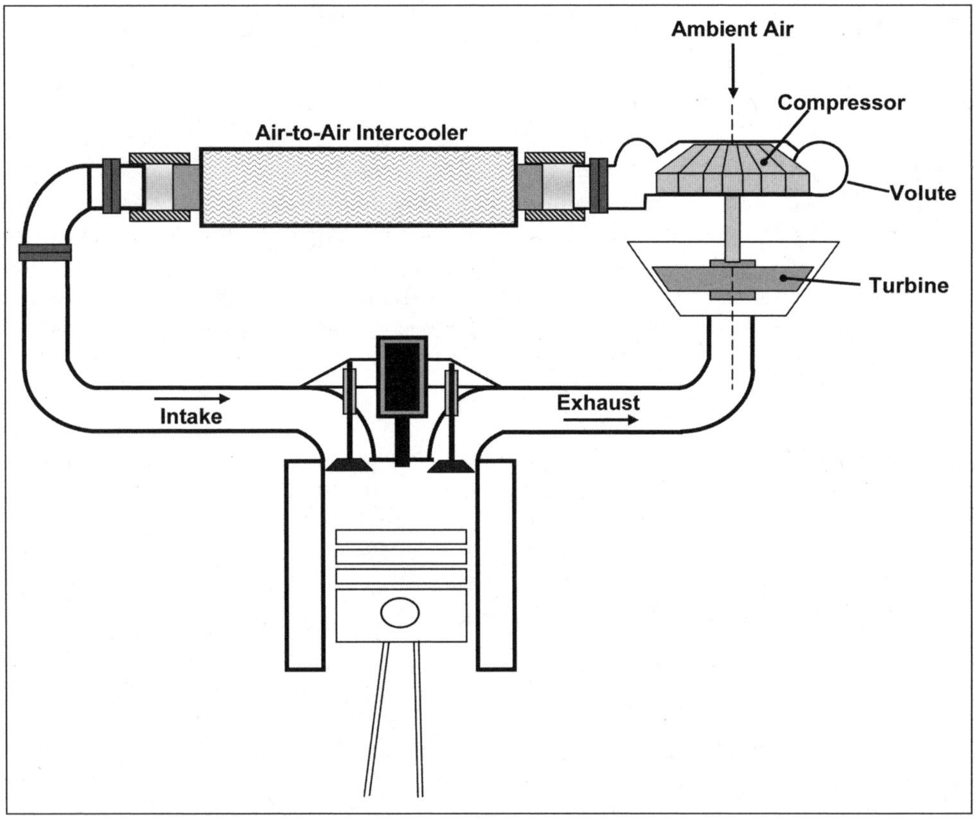

Figure 2.15 *Illustration of a turbocharged engine.*

changing the quality of the fuel-to-air mixture from lean to rich or vice versa. On the other hand, compression ignition engines have an overall lean mixture at all operating conditions. They are referred to as *quantitatively controlled* engines because power is controlled by varying the amount of fuel in an engine where air flow is mostly unchanged for a fixed speed.

Several other designations could be considered categories of engines, but they are a bit more technically involved and used by those who are well versed with engine designs. These designations may include the type of combustion chamber used in an engine, such as a M.A.N chamber for a system having a design specific to a German company (see Figure 1.5), or Quadram for a system designed by Perkins. Other designations may refer to the method of mixture preparation, and so on. Yet, the preceding classification or categorization covers most of the designs available.

2.4 Assessment of Engine Performance

Regardless of categories, design features, importance of application, displacement, or any other descriptive feature, engines must perform a certain task. It is reasonable to expect engines to give maximum performance at minimum cost. In other words, what return should customers expect on their capital investment (price paid for the engine and its installation) and their operating costs (costs of running and maintaining the engine). In addition and in view of current and future environmental concerns, it is not enough just to maximize performance and minimize cost; we need to do so while also preserving the environment.

It is perhaps time that we investigate engine performance through applying thermodynamic principles. This material is presented in a simplified manner to facilitate its understanding, especially for those not familiar with this science. The fundamental reason for resorting to thermodynamics is that it is an elegant way to treat the balance of energy as well as the mass of the working fluid in a controlled volume. The controlled volume in this case is the cylinder, into and out of which energy and mass flow. Figure 2.16 is an illustration of the control volume in an engine where both chemical energy in the form of fuel as well as mass in the form of fuel and air are introduced.

Fuel is usually metered into the control volume where air is also inducted. Both fuel and air are introduced into the cylinder in amounts (mass) that support efficient combustion and produce the desired power output. In summary, mass and energy are transferred into and out of the control volume. Applying the laws of thermodynamics helps in quantifying the relationship between the energy delivered to that received by the control volume.

2.5 The First and Second Laws of Thermodynamics

The first law of thermodynamics simply states that energy cannot be created nor destroyed but can only be converted from one form to another. For instance, a mass of hydrocarbon fuel contains chemical energy that is converted to work or mechanical energy within the cylinder (control volume) of an internal combustion engine. Theoretically, if this process was ideal and no losses were incurred, this energy conversion would be 100% efficient. Yet, in reality, converting energy from one form to another involves many losses resulting in an overall loss in efficiency. This fact is what the second law of thermodynamics expresses as it states that the useful work from a combustion system should be less than the energy input [5]. In general, the ratio between useful work and the chemical energy added to the control volume represents the brake thermal efficiency of the system. Considering the cylinder and piston arrangement shown in Figure 2.16, the combustible mixture of fuel and air is burned in the control volume producing heat that results in the expansion of that volume, causing the piston to move. Motion of the piston creates friction against the cylinder walls, leading to friction heat loss.

Another source of loss results from the temperature associated with the heat generated by the combustion process itself. As the combustion temperature increases, the cylinder material approaches its limitation in mechanical strength. Therefore, cylinders or control volumes are cooled, by water or air, to move heat away from the material, thus preserving its mechanical strength. Heat transferred away from the control volume material is another loss added to the balance between the energy received by that control volume and the energy it delivers back. Another major source of loss in this energy conversion system is exhaust gases flowing out of the control volume. Exhaust gases exit the control volume with heat energy delivered to ambient without any benefit. They also exit with a great deal of pressure as well as speed. Therefore, considering the system on hand we can think of a control volume where fuel and air are supplied and in return piston motion (work) is delivered, but the work delivered is much less than the value of energy supplied to the control volume. The difference between the energy supplied to the control volume and the energy that it delivers is the sum of losses including, but not limited to, cooling and exhaust losses. A helpful illustration of this balance of energies is given in Figure 2.17.

Whereas the description given so far for the energy exchange through a control volume is generally correct, it may not be very complete. A better accounting of all the energies entering and leaving the control volume will have to include other types of energies not yet accounted for. For instance, any mass entering the control volume brings with it several forms of energy [6]:

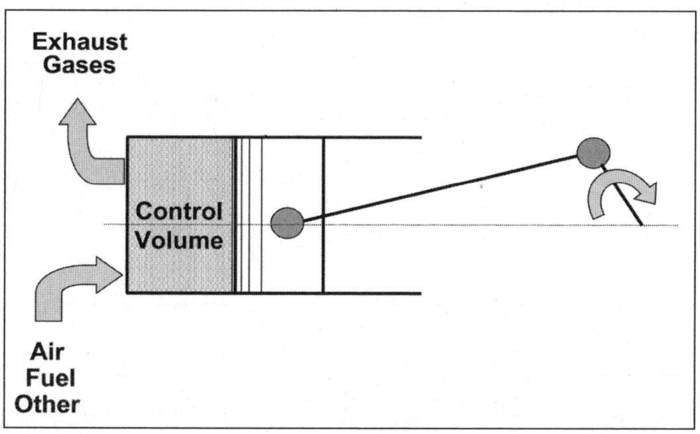

Figure 2.16 *Schematic of control volume (mass conservation).*

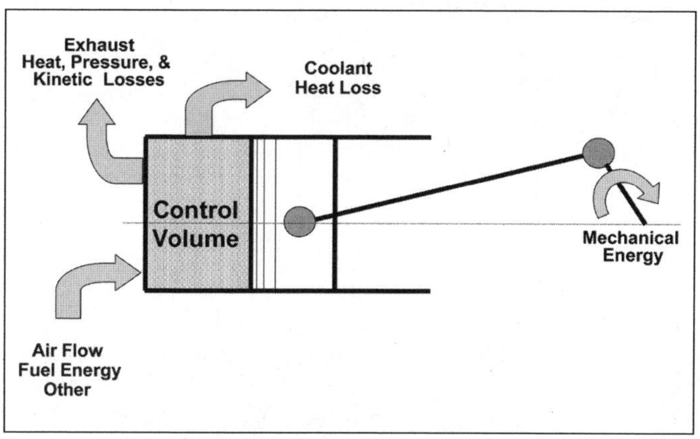

Figure 2.17 *Energy balance in a control volume.*

- Internal energy; mainly due to its temperature, which is generally very small
- Kinetic energy; mainly due to injection characteristics, which usually leads to important interactions between the fuel and air within the control volume
- Potential energy; generally associated with pressure admitting mass into the control volume
- Flow energy; principally associated with the interrelationship between the control volume and its pressure.

Revisiting the first law of thermodynamic and considering the various forms of energy we are now acquainted with, one could think of the energy balance in a control volume as follows:

$$\text{Net output} = \text{Energy supplied to the control volume} - \text{Total energy loss} \tag{2.10}$$

In other words, a careful accounting of the energies supplied to the control volume and those delivered by the same volume can assist in assessing the system conversion efficiency. Caution must be exercised when defining the control volume. So far, we have considered one cylinder in an engine as the control volume. However, an entire engine or the total vehicle can be viewed as that control volume. Therefore, defining the control volume and its boundaries is extremely important in the energy conservation equation.

2.6 Mass Conservation in Combustion Thermodynamics

To complete our discussion regarding the control volume, consideration should be given to the mass entering and leaving it. As illustrated earlier in Figure 2.16, fuel and air are the two principal constituents entering the control volume. In view of modern diesel engine technology, this may be a limited treatment of the masses entering the control volume since the need to control emission may dictate recirculating some exhaust products back into the cylinder. In addition, lube oil contribution to exhaust emissions is being scrutinized and efforts are made to limit its consumption within the cylinder. Therefore, a more accurate representation of the mass balance in a control volume may be described by the following relationship:

$$\text{Exhaust mass} = \text{Mass of air} + \text{Mass of recirculated exhaust} + \text{Mass of fuel} + \text{Mass of lube oil} \tag{2.11}$$

or

$$M_{exh} = M_{air} + M_{egr} + M_{fuel} + M_{lo} \tag{2.12}$$

In practical embodiments of reciprocating internal combustion engines, accessories are used to perform various functions in support of engine operation. For example, lubricating the engine requires an oil pump driven by the engine itself, thus subtracting a portion of the work produced by that engine. Therefore, if the control volume is the cylinder, then the conversion efficiency of such a system will be higher than if the control volume were an engine. In fact, in diesel engines power is consumed when driving its fuel injection system. Other drives required for engine operation include the camshaft, alternator, coolant pump, and superchargers.

Even though combustion in diesel engines is not the subject of this chapter, an alternative treatment of the mass balance in a control volume, the cylinder in this case, can be helpful. A simplified expression for mass conservation can be limited to reactant species and reaction products such as fuel (C_nH_m) and air (mostly nitrogen and oxygen) reacting with each other in the proper environment and producing exhaust constituents as follows:

$$\begin{aligned} &C_nH_m + aO_2 + 3.76aN_2 \\ &\rightarrow bCO_2 + cCO + dH_2O + eOH_2 + fH_2 + \\ &g(HC) + hNO + iHCHO + jNH_3 + kN_2 \end{aligned} \tag{2.13}$$

In Eq. (2.13), C_nH_m represents hydrocarbon fuel reacting with air consisting of oxygen and nitrogen in a volumetric ratio of 20.99% to 79.01%, respectively. It follows that for every mole of oxygen provided, 3.76 moles of nitrogen would be present in the reaction. The majority of the exhaust gaseous species are CO_2, H_2O, N_2, and in diesels excess O_2 as well. In fact, these species constitute about 99% of the exhaust from an engine, leaving only 1% unaccounted for but made of mostly undesirable species.

The molecular weight of typical diesel fuel is 191 when determined by UOP Method 375-86 [6]. Because the molecular weight of C is 12.0111 and that of hydrogen is 1.00797, then the hydrocarbon designation of diesel fuel can be determined as follows:

$$\alpha(12) + \beta(1) = 191 \tag{2.14}$$

From actual fuel analysis each kilogram of diesel fuel contains 0.8687 kg of carbon or:

$$\frac{0.8687\,\text{kg of Carbon}}{12.0111} = 0.07232\,\text{kmol of Carbon} \tag{2.15}$$

Similarly,

$$\frac{0.1263\,\text{kg of Hydrogen}}{1.00797} = 0.12530\,\text{kmol of Hydrogen} \tag{2.16}$$

From Eqs. (2.15) and (2.16), the hydrogen to carbon ratio is as follows:

$$(H/C) \text{ is } \frac{0.12530}{0.037232} = \frac{\beta}{\alpha} \tag{2.17}$$

By solving Eqs. (2.14) and (2.17), we can define diesel fuel as $C_{13.883} H_{24.053}$.

2.7 Stoichiometric Ratio in Diesel Combustion

Having determined the composition of diesel fuel, it is relatively simple to calculate its stoichiometric ratio. By definition, the stoichiometric ratio is the ratio of air to fuel that when combusted would yield nothing but CO_2, H_2O, and N_2. It is sometimes referred to as the chemically correct ratio. Applying this definition to the diesel fuel from the previous section yields the following equation:

$$\begin{aligned} &C_{13.883}H_{24.053} + 94.744\left[0.21\ O_2 + 0.79\ N_2\right] \\ &\rightarrow 13.883\ CO_2 + 12.026\ H_2O + 74.848\ N_2 \end{aligned} \tag{2.18}$$

From Eq. (2.18), the molar A/F ratio is 94.744 kmol air/kmol fuel. On a mass basis, the A/F ratio can be calculated as follows:

$$\begin{aligned} &\left[\frac{94.744\,\text{kmol of air}}{\text{kmol of fuel}}\right] \times \left[\frac{28.97\,\text{kg of air}}{\text{kmol of air}}\right] \times \\ &\left[\frac{\text{kmol of fuel}}{191\ \text{kg of fuel}}\right] = \frac{14.37\ \text{kg of air}}{\text{kg of fuel}} \end{aligned} \tag{2.19}$$

2.8 Additional Performance Parameters and Their Definition

After months and years designing an engine, it must be put to the test of verifying whether or not all design assumptions and choices can indeed perform the intended output in an efficient manner. To fully appreciate what an engine can do, the engineer must be well versed in the science of testing engines including laboratory tools designed for this purpose, evaluation and performance parameters, test methods, interpretation of results, statistical tools, equipment calibration, and maintenance of test and analysis equipment. Of importance to the end user are the following:

- Engine acquisition cost
- Engine operating characteristics (output and speed)
- Engine operating cost (fuel consumption and maintenance)
- Engine reliability and its durability
- Engine exhaust emission profile and the cost of making it environmentally friendly.

Of course, the importance of any of these items will depend on the specific application of the engine. Therefore, priorities have to be established before acquiring an engine to reflect the purpose for which the engine will be used. Nevertheless, certain common methods are used to evaluate engine performance and emission profiles. In this section, engine performance parameters, test tools, and test methodologies will be described and in some cases quantified.

2.8.1 Power Output

Of primary importance to the designer, engine development engineer, end user, and others in the industry is the power output of an engine. The maximum power output is defined by a maximum torque at a given engine speed. The engine maximum power output, sometimes referred to as rated power, is defined as follows:

$$\text{Power output} = \frac{T \times N}{\text{Constant}} \tag{2.20}$$

where T is engine torque in lb-ft or N·m, and N is engine speed in rotations per minute. The appearance of torque in Eq. (2.20) leads us to seek a more basic definition of power by considering the fundamental purpose for using an engine. An engine produces work that we can use in various applications, and power is the rate at which this work is produced. The most common method used to measure power is a device designed to brake the engine that is called a *dynamometer*. There are many types of dynamometers, but they have certain features in common. Most dynamometers are designed with a stator that does not rotate and is coupled electromagnetically to a rotor as shown in Figure 2.18. The rotor is the second major feature of a dynamometer and is concentric with the stator. It is also coupled to the engine and rotates around the same axis as the stator. In electromagnetic dynamometers, the rotor is driven by the engine, and an electric field in the stator tries to oppose its motion. The electromagnetic force (F) exerted is measured by a load cell that is placed at a distance (b) from the center of the dynamometer as shown in Figure 2.18. The product of the force (F) and distance (b) as expressed in Eq. (2.21) defines *torque*:

$$T = F \times b \tag{2.21}$$

In one revolution the rotor travels a distance $2\pi r$ against resistance force f; thus its work for this one revolution is:

$$\text{Work} = 2\pi rf \tag{2.22}$$

The rotor turning moment (rf) must be balanced exactly by the external turning moment; that is, the product (Fb) and this relationship can be expressed as follows:

$$rf = Fb \tag{2.23}$$

Substituting from Eq. (2.23) into Eq. (2.22), we can then express work as:

$$\text{Work} = 2\pi Fb \tag{2.24}$$

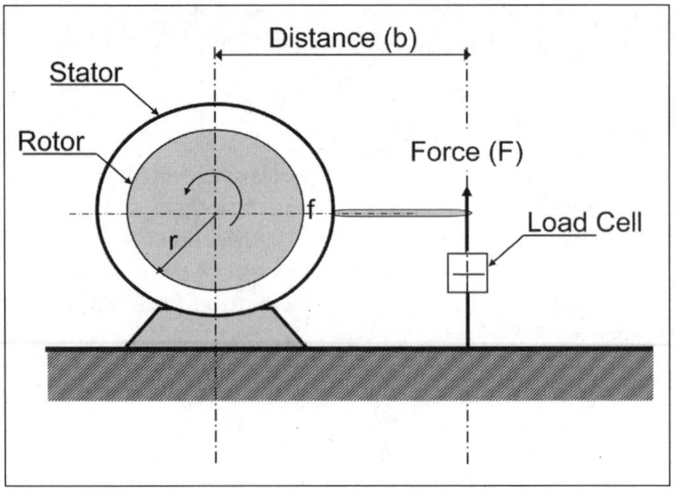

Figure 2.18 Schematic diagram of an engine dynamometer.

Using the relationship of Eq. (2.21) to express work per unit time gives the following:

$$\text{Work per minute} = 2\pi\text{TN} \tag{2.25}$$

Because power is defined as the rate of doing work, we can then write the following:

$$\text{Work per minute} = \text{Power} = 2\pi\text{TN} \tag{2.26}$$

Horsepower (hp) is a unit of power equal to 33,000 lb-ft per minute (550 lb-ft per second) in the English system, and the kilowatt (kW) is its equivalent in the metric system. The kilowatt is 550 × 0.746 = 738 lb-ft per second. Having defined these units, we can then express horsepower as follows:

$$\text{hp} = \frac{2\pi\text{TN}}{33{,}000} = \frac{\text{T} \times \text{N}}{5252} \tag{2.27}$$

From Eq. (2.27), a relationship is clearly established between power output, torque, and engine speed. From this relationship we conclude that since hp is a function of both torque and speed, then it is possible to design an engine to achieve power through high torque or through high speed. Engines designed for high-torque output are usually large and built to withstand high internal forces. Their maximum speeds are quite low, ranging from a few hundred revolutions per minute (rpm) to perhaps 1800 to 2100 rpm. Stationary engines are on the low end of the speed spectrum, while on-highway trucks and midrange engines occupy the high end of the speed range. High-torque, low-speed engines are very suitable for heavy-truck applications enabling them to move extremely high loads away from loading docks or traffic lights at very low engine speeds. Meanwhile small, high-speed engines, such as those powering many passenger cars, produce their high power at very high engine speeds (greater than 5000 rpm). An extreme example of these types of applications is the race car engine with speeds far exceeding 9000 rpm.

2.8.2 Indicated Power

In Section 2.8.1, we investigated brake power. It is so named because of the method used to quantify it. Brake power is a good measure for the useful power produced by an engine. Indicated power is that which is produced by the direct application of the gas pressure to the surface of the piston. By integrating cylinder pressure during a complete combustion cycle, one can obtain a value for the cycle pressure that can be used to calculate indicated power.

From Figure 2.19 the net pressure applied to the surface of the piston is (area A – area B). Area A is bordered by the clockwise arrows and is regarded as positive pressure on the piston. This is a conventional way to express this rather than a purely scientific way since, during the compression stroke, it is the piston that is doing work on the gas. A similar consideration is given to area B, where the piston is found once again to be pushing exhaust gases while it moves from BDC to TDC. Area B is often considered a measure for pumping losses. In addition to area B, friction losses are also subtracted from indicated pressure to obtain brake power. In other words:

$$\text{Indicated power} = \text{Brake power} + \text{Pumping power} + \text{Friction power} \tag{2.28}$$

In turbocharged engines, pumping losses are relatively small so they can be neglected. The result is the following relationship:

$$\text{Indicated power} = \text{Brake power} + \text{Friction power} \tag{2.29}$$

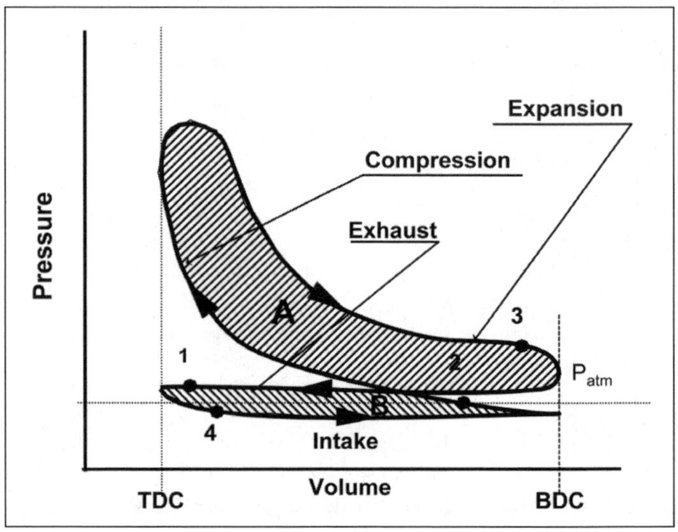

Figure 2.19 Pressure–volume diagram.

2.8.3 Mechanical Efficiency

The term *friction power* in Eq. (2.29) includes power required to expel exhaust gases, to induct fresh air, and to overcome piston ring/liner friction, bearing friction, and drive engine accessories. It also accounts for parasitic losses elsewhere within the engine. It follows that mechanical efficiency can be defined as the ratio of useful power (brake power) over available power (indicated power). In mathematical form it can be expressed as follows:

$$\text{Mechanical efficiency } (\eta_{\text{mech}}) = \frac{\text{Brake power}}{\text{Indicated power}} \tag{2.30}$$

$$\eta_{\text{mech}} = \frac{P_b}{P_i} = 1 - \frac{P_f}{P_i} \tag{2.31}$$

where P_i is indicated power, P_b is brake power, and P_f is friction power. Normally friction power losses increase with engine speed. An estimate of friction power can be obtained from normal engine brake power and the fuel consumption basic relationship. An example is given in Figure 2.20 where regular engine dynamometer test information about brake horsepower and fuel consumption can be used at constant engine speed to derive an estimate of friction power. By plotting brake power on the x-axis and fuel consumption on the y-axis, data from 0 to 300 hp are represented by an almost straight line except for the high-output points. As this line is extended to the left and intersects the x-axis, it gives us an estimate of the friction power as shown in Figure 2.20. At the point where that line intersects the vertical line at zero power output, the fuel consumption required to overcome engine friction at that speed can be read (about 15 lb/hr in this case). The line describing engine brake power versus fuel consumption, its extension to the left, shown as a dashed line, and the estimate of friction fuel consumption were attributed to an Englishman by the name of Willans; hence, the name given to this characteristic line.

Even though diesels are not usually throttled, operating them at high speed and requiring a greater volume of air to be inducted into the cylinder through the narrow passage of an intake valve causes pumping losses to increase and volumetric efficiency (see definition in Section 2.8.7) to degrade. For this as well as other reasons, in modern engines the valve area was increased by adopting two intake valves per cylinder instead of just one.

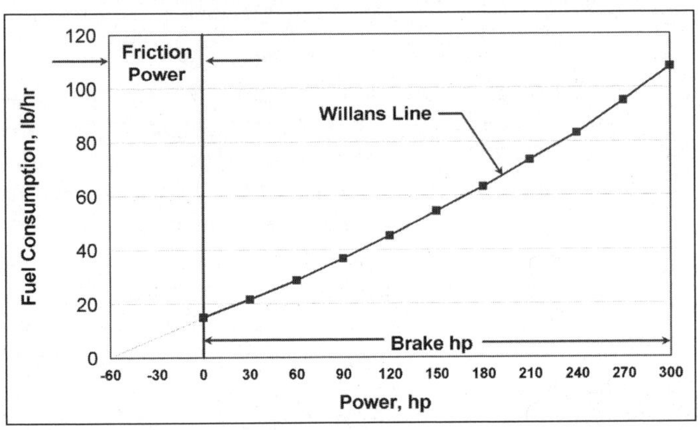

Figure 2.20 The Willans line.

2.8.4 Indicated Mean Effective Pressure

Area A in Figure 2.19 was described as the area indicating the work exerted by the working fluid on the surface of the piston. Pressure resulting from the combustion process, as a function of the cylinder volume, is applied to the piston surface to produce power. The indicated mean effective pressure (imep) is the work exerted by the gas on the piston per unit swept volume. The following is an expression of imep:

$$\text{imep} = \oint \frac{P_i\,dV}{Vd} \tag{2.32}$$

The indicated mean effective pressure is illustrated graphically in Figure 2.21.

2.8.5 Brake Mean Effective Pressure

Whereas imep is a form of bulk pressure exerted on the surface of the piston, brake mean effective pressure may be described as the portion of that imep that produces useful power. It represents the net work after subtracting friction, pumping, and other parasitic losses, the largest of these losses being that of friction as expressed by in Eq. (2.29). There is a relation of proportionality between brake mean effective pressure (bmep) and engine torque. Rather than using torque, which depends on engine size, bmep is preferred because it is a normalized parameter as given in Eq. (2.33):

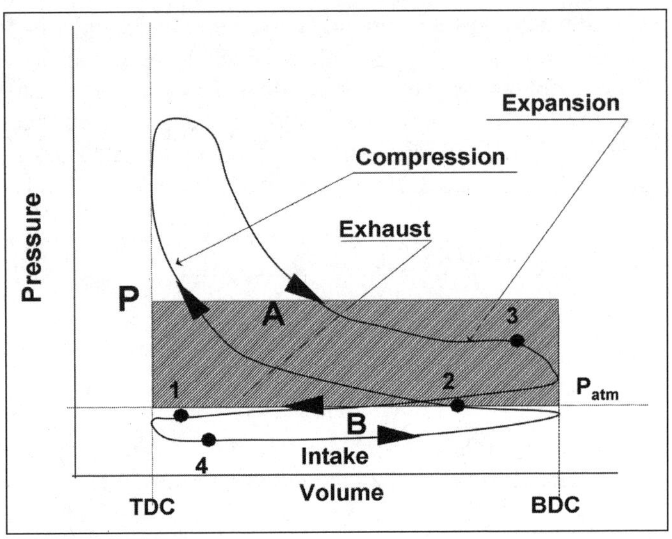

Figure 2.21 Graphical representation of indicated mean effective pressure.

$$\text{bmep} = \oint \frac{P_b dV}{Vd} \quad (2.33)$$

Other useful relationships for bmep include those given in Eq. (2.31) and (2.32). Numerically, bmep can be calculated using its proportional relationship to torque as shown in Eqs. (2.34) and (2.35) for two- and four-stroke engines, respectively [7]:

$$\text{bmep} = \frac{75.4 \times T}{\text{CID}} \text{ psi} \quad (2.34)$$

$$\text{bmep} = \frac{150.8 \times T}{\text{CID}} \text{ psi} \quad (2.35)$$

where CID is the engine displacement in cubic inches. It is customary to compare engines on the basis of their bmep using their peak torque value. However, a more accurate comparison of engines should consider the entire bmep characteristic of the engine versus its speed range.

2.8.6 *Specific Fuel Consumption*

Fuel consumption in mass per unit time is normally recorded during engine testing. One way to evaluate engine efficiency is via the ratio of fuel consumption to useful power. The result of this mathematical treatment produces the term *brake specific fuel consumption* (bsfc). In essence, bsfc is a measure of how much fuel an engine consumes in the process of producing an output of 1 hp:

$$\text{bsfc} = \frac{\text{Fuel consumption per unit time}}{\text{Brake power output}} \quad \frac{\text{lb}}{\text{bhp-h}} \quad (2.36)$$

Single-cylinder engines are popular in engine research work. Because power is produced by one cylinder only, friction and parasitic losses are disproportionately high relative to the power output of these engines. In these cases, rather than using the term bsfc engineers compute indicated specific fuel consumption (isfc). The term isfc is calculated on the basis of indicated power output as given in Eq. (2.37):

$$\text{isfc} = \frac{\text{Fuel consumption per unit time}}{\text{Indicated power output}} \quad \frac{\text{lb}}{\text{ihp-h}} \quad (2.37)$$

2.8.7 *Volumetric Efficiency*

Volumetric efficiency is a measure of the breathing quality of an engine. It is the ratio of the mass of air actually inducted by the engine during its intake stroke to the theoretical mass of air that could be inducted given the displacement of that engine. Equation (2.38) puts this relationship into a mathematical form:

$$\text{Volumetric efficiency } (\eta_{vol}) = \frac{\text{Actual mass of air inducted}}{\text{Theoretical mass of air to fill displacement}} \quad (2.38)$$

Reference [7] describes this definition as a misnomer because it is a ratio of two masses rather than two volumes. Yet, it is the conventional way of calculating volumetric efficiency since it takes air density into consideration.

2.8.8 *Engine Specific Weight and Volume*

Total engine weight and volume are very important for packaging and cost considerations. For instance, underhood space for many on-highway as well as off-road applications is very limited. Therefore, engines having the right power output and torque characteristics may have different volumes where the smallest volume would be of greatest advantage. Not only is engine volume very important, but also its weight. The lighter the weight, the greater its advantage for a given application because additional fuel costs are associated with transporting additional weight. The terms *specific engine weight* and *specific engine volume* are used to compare engines of similar power outputs, but having different sizes and weights. Equations (2.39) and (2.40) give the calculation for specific volume and specific weight, respectively:

$$\text{Specific volume} = \frac{\text{Engine volume}}{\text{Rated power}} \quad (2.39)$$

$$\text{Specific weight} = \frac{\text{Engine weight}}{\text{Rated power}} \quad (2.40)$$

2.9 Performance Curves

Engine testing is usually conducted with a wide-open throttle and variable speeds or at constant speed but with a variable throttle positions (term commonly used in spite of the fact that diesels do not have a throttle). If tests are conducted at variable engine speeds starting at maximum speed and proceeding to low idle speed, they may be referred to as lug tests. Wide-open throttle and variable-speed tests are conducted to determine the engine's maximum power output and best fuel consumption at different speeds (see Figure 2.22). Part load tests at constant speed are used to determine the specific fuel consumption characteristic as the load varies. They are also used to determine the smoke-limited output of an engine at various speeds for older engine designs.

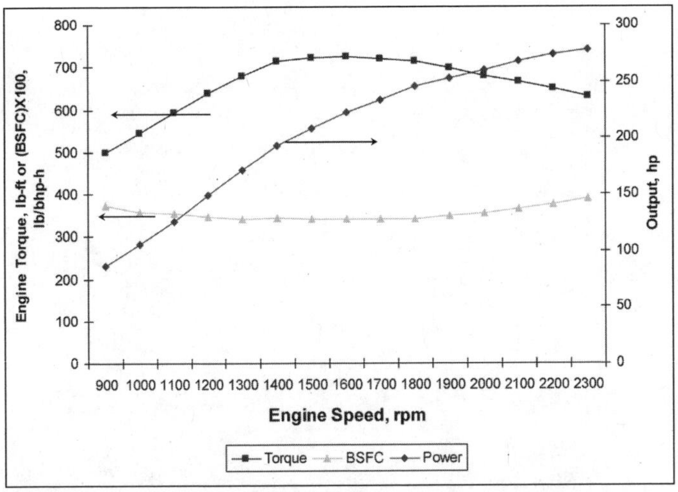

Figure 2.22 *Engine performance—variable-speed, wide-open throttle curves.*

References

1. Taylor, C. F., *The Internal Combustion Engine in Theory and Practice,* Vol. 1, rev. ed., The MIT Press, Cambridge, MA, 1985.
2. Heywood, J. B., *Internal Combustion Engine Fundamentals,* McGraw-Hill, New York, 1988.
3. Foulkes, D. M., "Developing Light-Duty Diesel Engines for Low Emissions and High Fuel Economy," Internal Report, Ford Motor Company, Detroit, MI, 1996.
4. *Automotive Handbook,* 2nd ed., R. Bosch GmbH, Germany, 1986.
5. Henein, N., "Engine Fundamentals," Lecture Notes to Ford Tractor Operations, Ford Motor Company, Detroit, MI, October 16–November 20, 1985.
6. Van Gerpen, J., "The Origins of Fuel Economy," Diesel Engine Technology Engineering Academy, SAE Engineering Academies, Warrendale, PA, 2000.
7. Obert, E. G., *Internal Combustion Engines,* International Text Book, Scranton, PA, 1968.

Chapter 3

Air Induction for Diesel Engines

3.1 Introduction

Diesel engines have traditionally suffered from low specific power output, that is, the maximum output per unit weight. Specific power output can also be compared on the basis of engine displacement. The principal reason for this phenomenon is that power output is limited by how much fuel one can efficiently burn in each combustion chamber [1]. For efficient combustion, an adequate supply of oxygen must be provided. Not only is the amount of air inducted important for efficient combustion, but good air and fuel mixing must also be accomplished. Only then can we achieve the chemically correct ratio that ideally leads to complete conversion of the hydrocarbon fuel to carbon dioxide and water.

The air induction system normally consists of an air filter, inlet ducts, intake manifold, and intake ports; intake valves may also be considered part of the induction system. Once an engine is designed, its capacity to induct air is essentially fixed. The maximum amount of air inducted depends on intake filter restriction, on pressure drop through its inlet duct, and on the intake manifold, intake ports, and valves. However, if the inducted air can be compressed to a higher density than ambient before introducing it into the engine, power output can be augmented simply by adding more fuel in proportion to the additional mass of air inducted. This added fuel and air would mix, combust, and produce more output.

The process used to increase air density before inducting it in the engine is referred to as *boosting* of the charge air (the more common descriptive term used is *supercharging*). Charge air boosting is usually accomplished through various pumping or compression devices. This chapter presents a review of these devices and their general features. Their effects on engine performance and emissions profile are also highlighted.

3.2 Methods of Charge Air Boosting

Three basic methods are used for charge air boosting. The first method uses a portion of the engine power to provide the mechanical drive of an air pump, compressor, or blower to raise the charge air pressure and increase its density. In general, these mechanically driven devices are called *superchargers*. The second method comes in the form of exhaust heat recovery in that it uses excess exhaust heat to power an air compressor, which, in turn, raises the charge air pressure and increases its density. This method is referred to as *turbocharging,* because it uses a turbine wheel placed in the path of exhaust gases exiting from the exhaust manifold to drive the air compressor. The third method of boosting charge air pressure and density can be described as a combination supercharger/turbocharger. The pulsating exhaust pressure waves that are exiting from the exhaust manifold are used to impart their energy to the incoming air within the channels of a mechanically driven device, as described later in this chapter. This type of device is referred to as a *wave supercharger*.

3.3 Types of Superchargers

Superchargers are generally positive-displacement devices and can be classified into several types. Figure 3.1 shows the variety of superchargers used in charge air (or mixture) boosting.

The first six devices listed in Figure 3.1 are positive displacement, whereas the device listed at the bottom, the centrifugal compressor, is classified as an aerodynamic or continuous-flow device. Positive-displacement devices deliver a specific volume of air per revolution. Because the volumetric efficiency is almost constant, airflow is usually proportional to

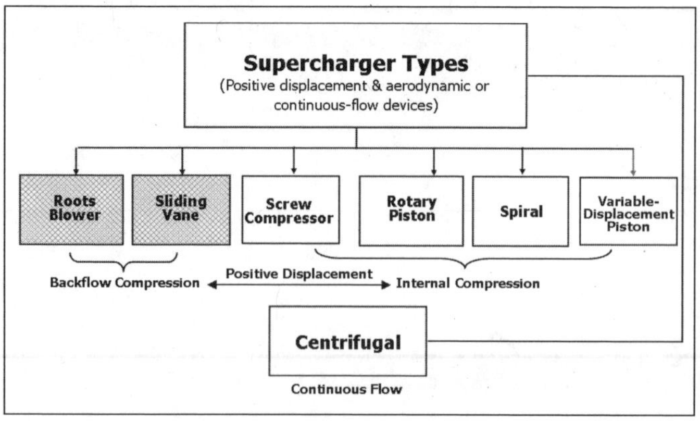

Figure 3.1 *Types of superchargers.*

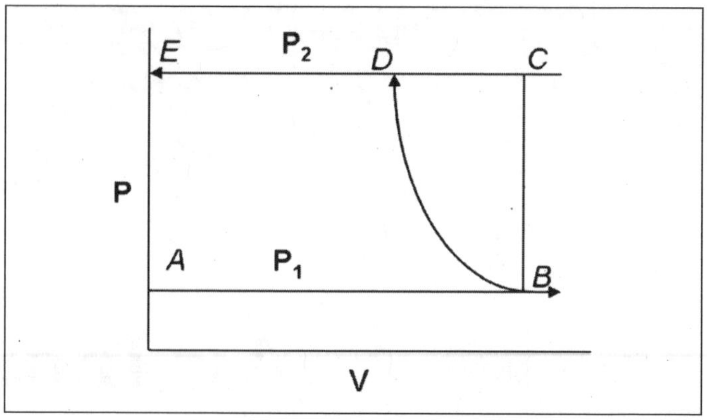

Figure 3.3 *Roots blower pressure vs. volume relationship.*

the supercharger or engine speed. Each of these devices has its advantages as well as its disadvantages, but each supercharger is best suited for a specific application. Generally, boost pressure is proportional to the square of the supercharger speed for most of these devices.

3.3.1 The Roots Blower

The Roots blower may have two or three rotors, as shown in Figure 3.2 [1]. Rotors are straight, but can also be helical for noise suppression. As the rotors turn, air enters in the displacement volume between the rotors and the housing (line *AB* in Figure 3.3). It is then carried across to the discharge port without compression. When the discharge port opens, the relatively hot air volume is instantaneously delivered from the blower and the pressure rises to P_2 as indicated by line *BC* in Figure 3.3. In practical packages, leakage between the rotors as well as backflow from the receiver to the inlet side of the blower take place, thus reducing its overall compression process. Therefore, the delivery of air from the blower may be better represented by line *BD* in Figure 3.3. In fact, Roots blowers are used in applications where the pressure ratio is rather low (1.0 to 1.3) [2]. More losses would be experienced at higher pressure ratios where the use of Roots-type blowers would be questionable.

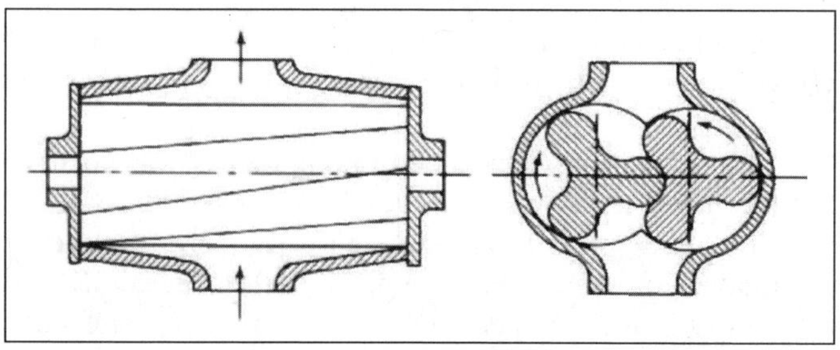

Figure 3.2 *Roots blower supercharger. (From* Internal Combustion Engine Fundamentals *by John B. Heywood. ©1988. Reproduced with permission of The McGraw-Hill Companies.)*

However, Roots blowers are popular because of their potential high-speed operation, high mechanical efficiency, simplicity, and cost. A typical performance map for a Roots blower appears in Figure 3.4. Its performance is characterized by its low pressure ratio, and its flow rate at constant speed is a function of the pressure ratio. However, Roots blowers are generally noisy and large in size. One of the more famous applications of a Roots blower in the United States is the two-stroke Detroit Diesel engine that powers the majority of the public transit buses.

Roots blowers may have good volumetric efficiency (about 90%) as well as reasonable mechanical efficiency (85%). However, their isentropic efficiency (measure of heat loss per unit mass of air) is barely 65% and contributes strongly to the blower's low overall efficiency (product of the various efficiencies) of about 55 percent [3].

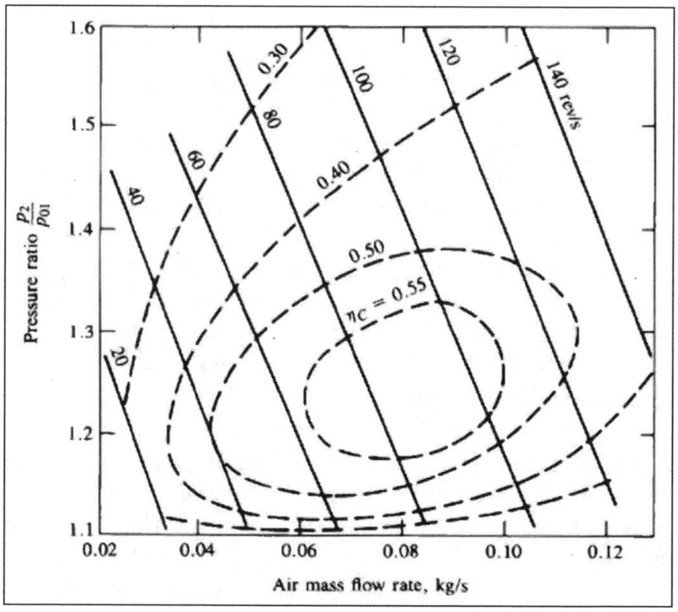

Figure 3.4 *Typical performance map for a Roots blower. (From* Internal Combustion Engine Fundamentals *by John B. Heywood. ©1988. Reproduced with permission of The McGraw-Hill Companies.)*

3.3.2 The Sliding Vane Compressor

In the sliding vane compressor, slots in the rotor house thin silicon carbide vanes that move in a radial direction (see Figure 3.5) [1]. Their motion is dictated by centrifugal force that results from the high-speed rotation of the rotor. The rotor itself is mounted eccentrically in the housing. Therefore, the high-speed rotation of the rotor causes the vanes to move toward the housing, thus trapping air between two vanes, the housing, and the rotor.

Air enters through the inlet port of the sliding vane compressor and immediately occupies a crescent-like volume as shown in Figure 3.5. As the rotor continues its rotation, which is controlled by its eccentricity relative to the housing, trapped air undergoes a volume increase followed by compression (decreasing volume) at the point of delivery. The capacity of the sliding vane compressor depends on several factors. Among them is the amount of eccentricity, the volume of air inducted (which depends on the size of the housing and rotor), the number of vanes, and its speed. As in the case of the Roots blower, actual compressor performance suffers from leakage between the blades and the housing, especially at low speeds where little centrifugal force is experienced, thus reducing the sealing between the vanes and the housing.

Heating results from the friction of the rotors against the housing. Unless this heat is dissipated through cooling, it is transferred to the air, thus decreasing its density and increasing its volume. This development eventually reduces the compressor efficiency and adds to the engine cooling system load. A performance map for the sliding vane compressor is shown in Figure 3.6 [1]. It is worth noting that the compressor isentropic efficiency is rather low.

The overall efficiency of the sliding vane compressor is only 40%. This low performance is due to a combination of low volumetric efficiency (85%), mechanical efficiency (about 65%), and an isentropic efficiency of just 60% [3].

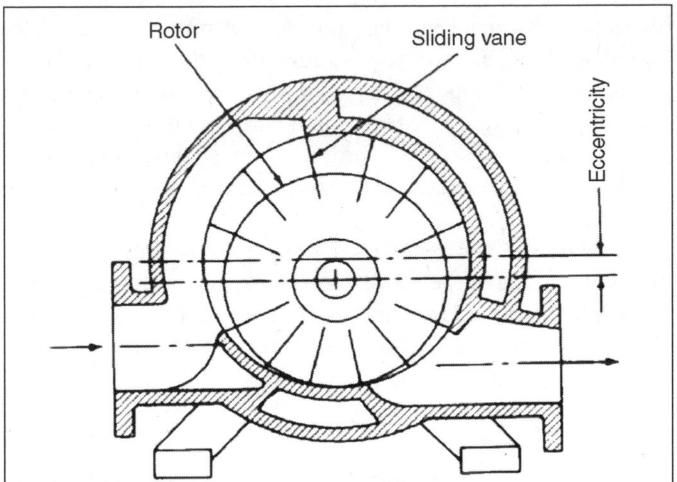

Figure 3.5 Sliding vane compressor. (From Internal Combustion Engine Fundamentals *by John B. Heywood. ©1988. Reproduced with permission of The McGraw-Hill Companies.)*

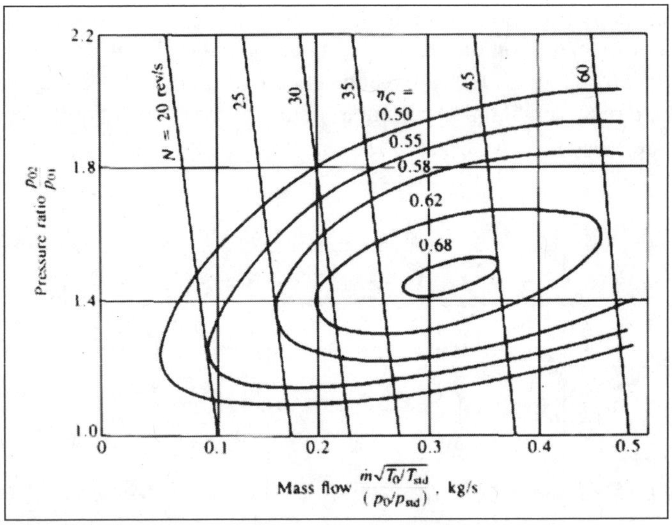

Figure 3.6 Sliding vane compressor performance map. (From Internal Combustion Engine Fundamentals *by John B. Heywood. ©1988. Reproduced with permission of The McGraw-Hill Companies.)*

3.3.3 The Screw Compressor

At first glance, the screw-type compressor (see Figure 3.7) may have a strong resemblance to the Roots blower. They do indeed share similar features such as a housing and a rotor. The rotor in the screw compressor is precision machined to maintain very tight tolerances between the rotor and the housing.

As the rotor turns, air is inducted through ports arranged around the cylindrical housing and occupies the volume between two consecutive screws and the housing. This air is delivered through a discharge port as shown in Figure 3.7. The delivery pressure is a function of the rotor speed and the discharge port flow area. Screw compressors have a rotational speed that ranges from 3000 to 30,000 rpm, and they generate substantial heat from friction between the rotor and the housing. Measures are usually taken to dissipate that heat to maintain the compressor's mechanical integrity. Screw compressors enjoy high volumetric

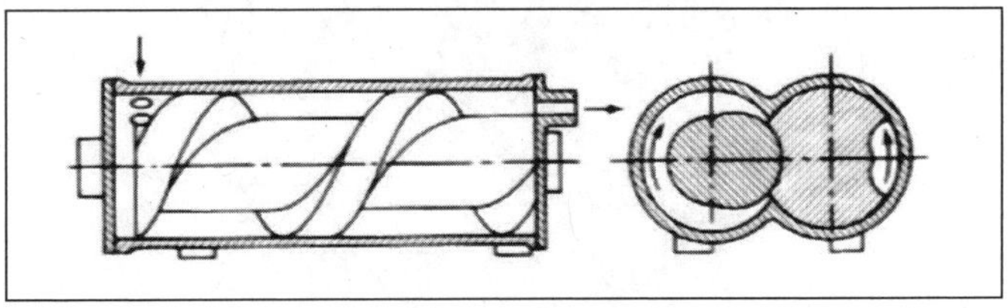

Figure 3.7 Screw-type compressor. (From Internal Combustion Engine Fundamentals *by John B. Heywood. ©1988. Reproduced with permission of The McGraw-Hill Companies.)*

efficiencies as long as their clearances are kept extremely small. Performance of the screw compressor depends on the rotor speed as long as leakage can be kept to a minimum. This leads to a relatively flat performance characteristic over a wide speed range as seen in Figure 3.8.

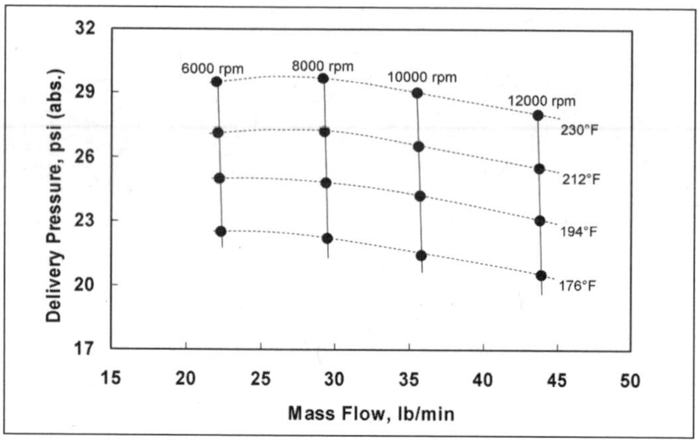

Figure 3.8 *Performance map of a screw compressor.*

3.3.4 The Rotary Piston Supercharger

The rotary piston supercharger is an internal-shaft rotary piston device in which the shaft rotates eccentrically inside a cylindrical housing, denoted as item 1 in Figure 3.9. In the example shown in Figure 3.9, the outer rotor has three teeth (2). The driven inner rotor (3) has two teeth (lobes) and rotates eccentrically within the outer rotor (2). The rotors have a ratio of 3:2 and do not contact each other while rotating. The level of internal compression is determined by the timing of the outer rotor clearing the outlet edge (4) and allowing compressed air to escape into the delivery duct. The compressed volume of air is

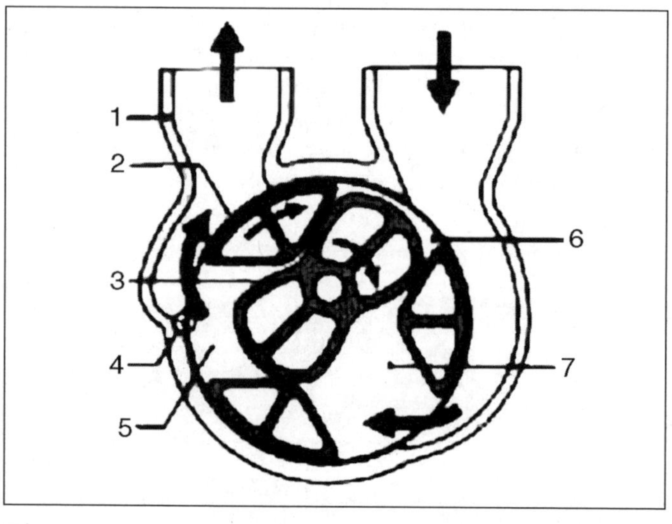

Figure 3.9 *Rotary piston supercharger. (Reprinted from Ref. 4 with permission from Robert Bosch GmbH.)*

represented by area (5), while area (6) represents the evacuated volume. Area 7 represents the fill volume that is exposed to the inlet duct as the rotors continue to rotate. A pair of cylindrical gears synchronize the motion of the rotors and prevent them from contacting each other.

The isentropic efficiency of a rotary piston supercharger is about 65%, whereas its volumetric efficiency is about 90%. These superchargers are usually belt driven and can reach speeds of 15,000 rpm. They are capable of delivering a charge air boost of up to 12 psi.

3.3.5 The Spiral-Type Supercharger

The spirals are arranged in a flat-sided casing having a shaft that rotates eccentrically. Sandwiched between the fixed spirals are moving displacer walls attached to a disk that is connected to an eccentric pin roller bearing (see Figure 3.10). As the drive shaft rotates, the displacer performs an oscillating circular motion of double eccentricity. Air entering the blower moves from one working chamber to the next performing filling, transporting, and expelling of the air at a hub, then delivering it through its discharge. The rotation of the eccentric and the rotor is through a toothed belt as shown in Figure 3.10 [5].

The overall efficiency of this supercharger is about 55%. Its isentropic efficiency is 68% and its volumetric efficiency is close to 90%. The speed range for this type of supercharger is 0 to 13,000 rpm, and it can deliver up to 12 psi of boost pressure. The spiral-type supercharger is often referred to as the scroll-type supercharger. The casing of the supercharger is die cast aluminum and the displacer is made of die cast magnesium.

3.3.6 The Variable-Displacement Piston Supercharger

This supercharger was developed in the mid-1990s and has a design that incorporates four radially arranged rectangular pistons (see Figure 3.11). The pistons are driven within their respective chambers in a circular motion referred to as *nutation.* The chambers are free to move laterally to allow for the pistons' side movement. The chambers' movement opens and closes inlet and outlet passages at the top of the cylinders. The design also incorporates a sliding pin mechanism that allows the eccentricity to be adjusted, thus permitting the piston stroke to be varied.

The overall efficiency of this supercharger, about 65%, is greater than that of the scroll-type supercharger. Its isentropic, mechanical, and volumetric efficiencies are about 80%, 80%, and 90%, respectively. The speed range for this type of supercharger is a relatively low (0 to 5000 rpm), and it can deliver up to 15 psi of boost pressure.

3.3.7 The Centrifugal Compressor Supercharger

The centrifugal compressor supercharger is normally used with an exhaust-driven turbine [1]. However, it is also used independent of exhaust-driven applications. In one configuration

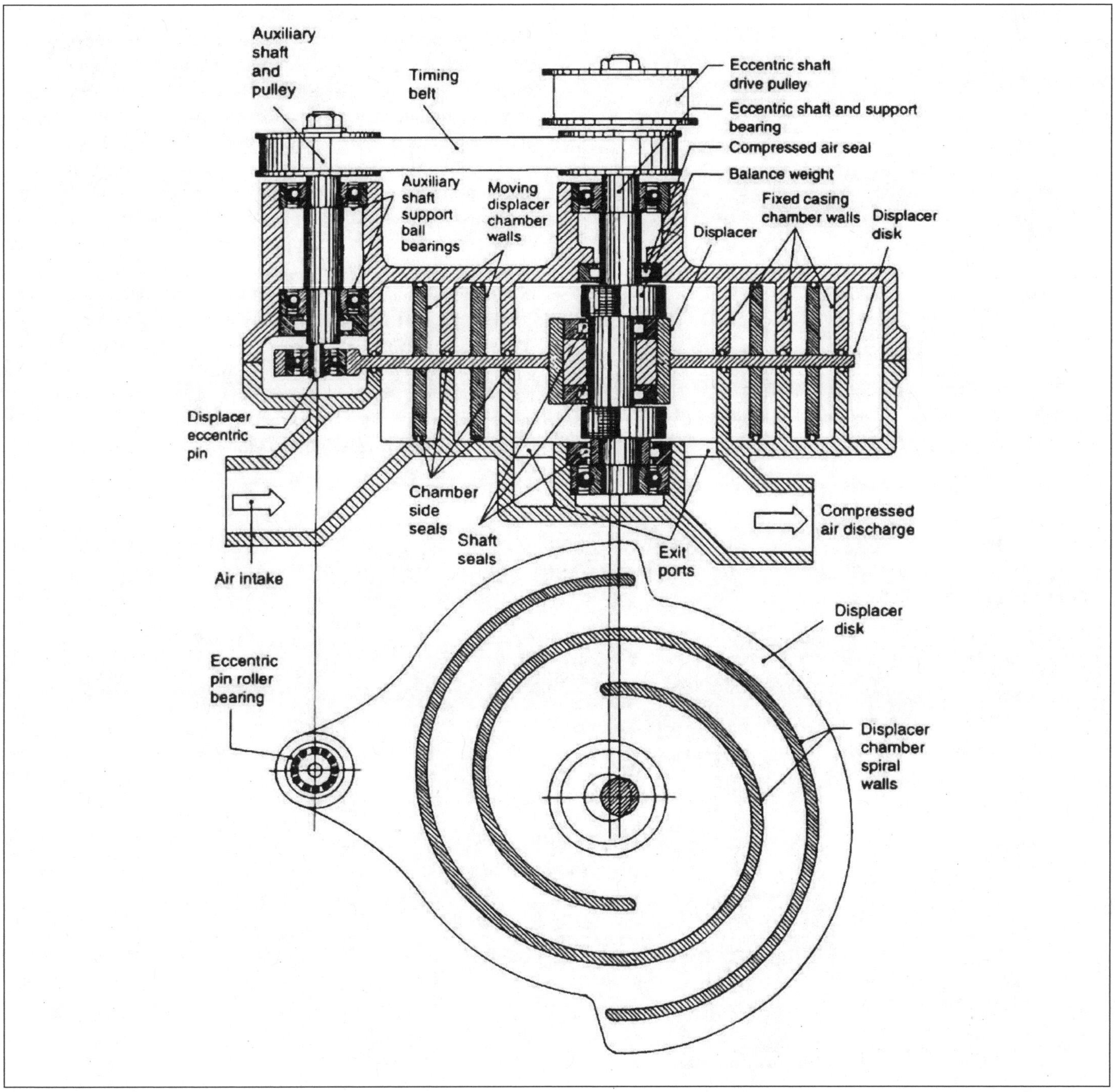

Figure 3.10 *Details of a spiral (scroll)-type supercharger. (From* Advanced Engine Technology *by Heinz Heisler. Reprinted by permission of Butterworth-Heinemann.)*

it can be driven by a high-speed electric motor, and in another it can be driven hydraulically.

It consists mainly of a single-stage radial compressor through which air is pulled from the ambient. The air accelerates to a high velocity and flows radially outward via a stationary diffuser stage toward a shell-shaped housing called the *volute* as shown in Figure 3.12. The volute is shaped in such a way that its diameter, as well as its cross-sectional flow area, is always increasing. For this reason, the air velocity decreases, and its kinetic energy is converted to potential energy as manifested by increasing pressure. The centrifugal compressor is ideal for providing high mass flow rates at the relatively low pressure ratio (less than 3.5) that is normally the case in internal combustion engine applications.

Figure 3.13 is a performance map for a centrifugal compressor. This map relates the compressor's airflow characteristics

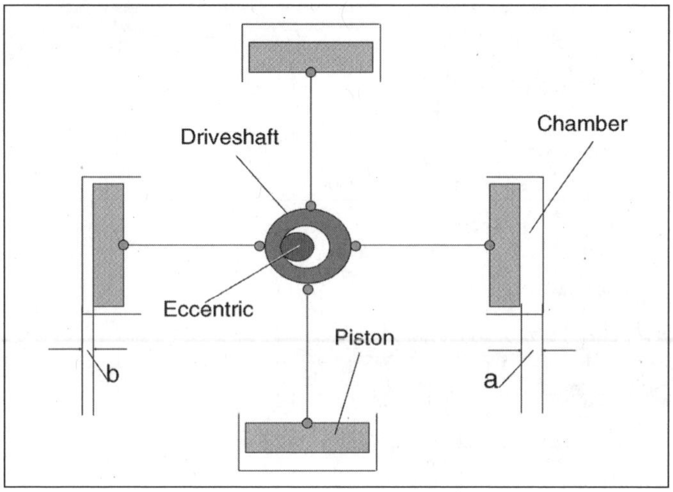

Figure 3.11 *Conceptual drawing of a variable-displacement piston supercharger.*

with its pressure ratio. The topographical contour lines indicate islands of constant efficiency. The area to the left of the operating range of the compressor is an area of unstable operation. The *surge line* is the line that separates the stable from the unstable operating regime on the left side of the compressor. Performance at constant compressor speeds is also indicated in Figure 3.13. On the other hand, performance is limited on the right side of the compressor map by *choking,* which occurs at each of the compressor speeds. Choking, of course, is the inability to flow any more air through the compressor blades or the diffuser channels. Normally, we want the compressor to operate at its highest possible mechanical efficiency at any speed. A line that would connect the most efficient points at each operating speed is referred to as the most efficient operating line. However, in most cases we may settle for a point on a constant-speed line that may not be the most efficient for specific design reasons. Therefore, the resultant line that attempts to describe the optimum operation of a compressor is referred to as the *operating line*.

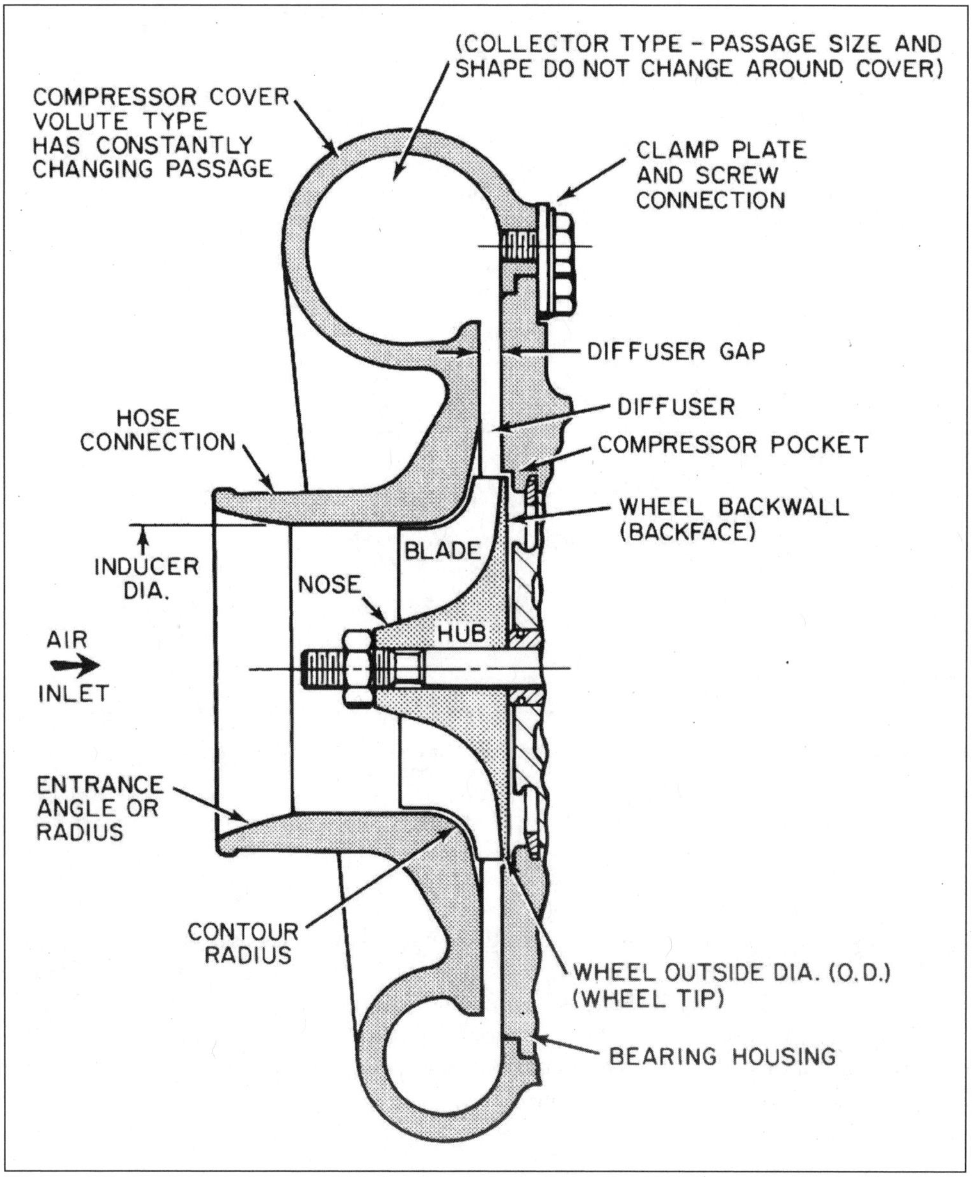

Figure 3.12 *Centrifugal compressor. (Reprinted with permission from Ref. 6.)*

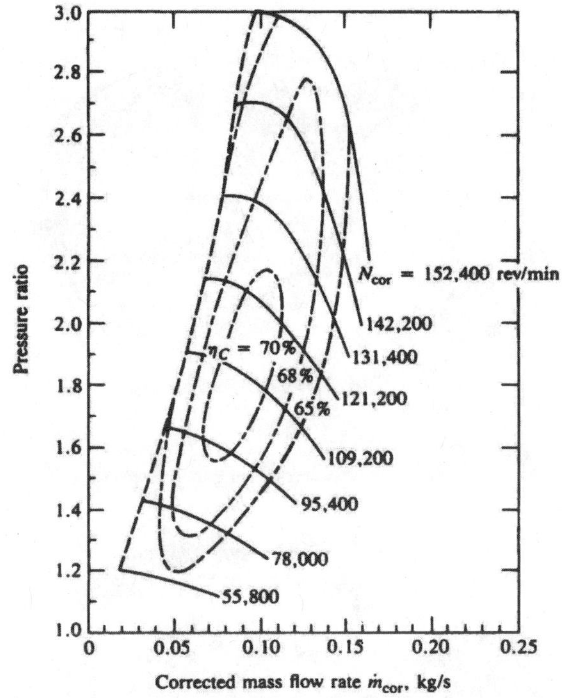

Figure 3.13 Performance map for a centrifugal compressor. (From Internal Combustion Engine Fundamentals *by John B. Heywood. ©1988. Reproduced with permission of The McGraw-Hill Companies.)*

Figure 3.14 is a photograph of an electrically driven centrifugal supercharger. The motor is a brushless, permanent magnet, high-speed motor capable of up to 40,000 rpm using short bursts of relatively high current to achieve good response time. (This concept will be further discussed in Chapter 14.)

Figure 3.14 *An electrically driven centrifugal compressor.*

3.4 Turbochargers

Like superchargers, turbochargers are devices that increase the amount of air inducted into an engine. As described earlier, increasing the air inducted into an engine allows adding a commensurate amount of fuel, thus giving the potential of producing more power than the naturally aspirated version of the same engine. The main difference between turbochargers and superchargers is the fact that the former utilize engine exhaust and a turbine wheel to provide its rotary motion, while superchargers are generally mechanically-, electrically-, or hydraulically-driven.

In turbochargers, the compressor and the turbine wheel are mounted on the same shaft (see Figure 3.15). The turbine wheel uses exhaust gas energy to drive the compressor. The latter draws ambient air, increases its pressure, and sends it on its way to the engine cylinders.

Exhaust turbochargers are used in many diesel engine applications including on- and off-highway, marine, stationary, passenger car, locomotive, and small power generation engines. Modern turbochargers can reach speeds in excess of 150,000 rpm (see Figure 3.13). This speed is not dependent on engine speed, but rather on the power balance between exhaust energy imparted to the turbine and the power required to drive the compressor. A compressor performance map is shown in Figure 3.16 for several of the more common turbocharger applications. The compressor performance is to the right of the surge line. These lines define the stable compressor performance, whereas to the left of the surge line there is no performance activity because this is a region of unstable performance.

In radial compressors, filtered air enters through the center of the compressor cover and proceeds through the inducer into the compressor wheel. As air proceeds through the compressor wheel, it makes a 90-degree turn, thus changing its flow from an axial to a radial direction. Air exiting the compressor enters

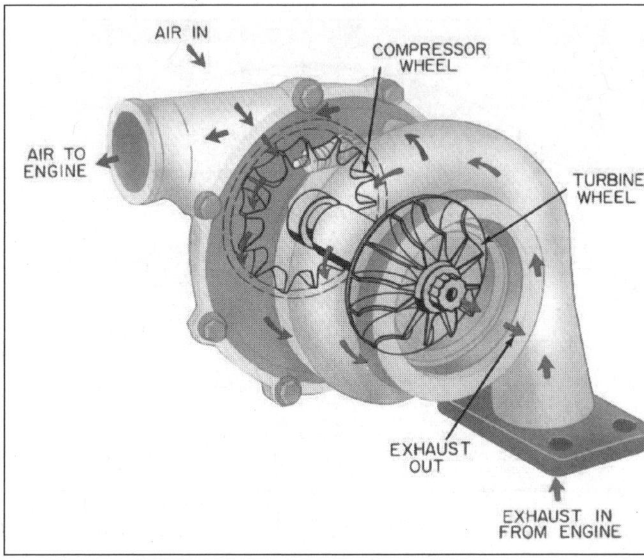

Figure 3.15 *A turbocharger. (Reprinted with permission from Ref. 6.)*

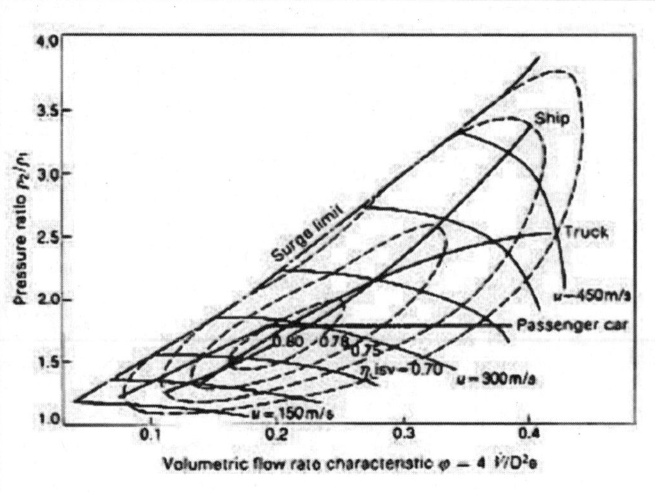

Figure 3.16 Compressor map showing example engine performance lines. (Reprinted from Ref. 4 with permission from Robert Bosch GmbH.)

Figure 3.18 *Volute covers. (Reprinted with permission from Ref. 6.)*

stationary diffuser passages arranged on a diffuser plate such as that shown in Figure 3.17. From the diffuser passages, the air is then discharged into a collector called the *volute*.

Figure 3.18 shows two volutes and their corresponding compressor wheels. The difference between these volutes lies in the way in which they connect to the piping at the exit from the volutes. It is worth noting that the cross-sectional area of the collector passage (volute) steadily increases until the air is discharged in the pipe at the exit of the volute. This gradual cross-sectional area increase helps convert kinetic energy into potential energy, or air velocity into pressure. Together with increased pressure, the compressed air experiences an increase in its temperature and therefore a reduction in its density. Although many applications use turbocharged air, a majority of these also use aftercoolers for the purpose of increasing air density.

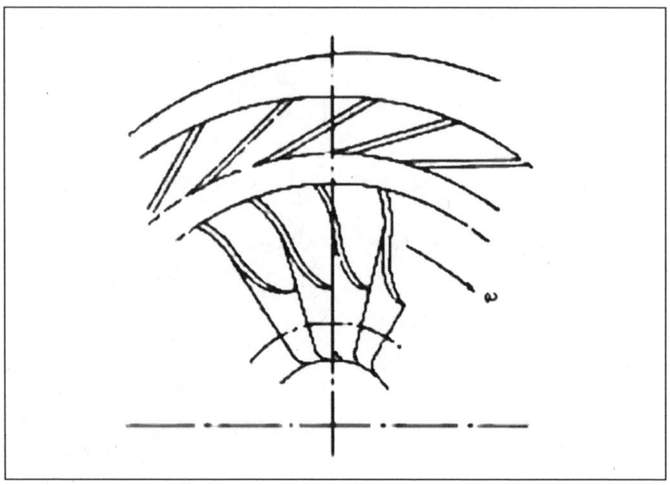

Figure 3.17 *Segment of a radial compressor and diffuser plate. (From* Internal Combustion Engine Fundamentals *by John B. Heywood. ©1988. Reproduced with permission of The McGraw-Hill Companies.)*

Compressor wheel blades are either straight (radial) from the center of the wheel's hub or backward leaning (i.e., away from the direction of rotation). The blade form is designed to maximize efficiency by minimizing turbulent flow within the channels formed by two consecutive blades. The compressor wheel is made of aluminum, and its blades have extremely close tolerances with the housing to minimize backflow. Researchers at Engelhard have coated the inside of the compressor housing with an abradable coating in an attempt to eliminate tip clearances (radial clearance between the blade and the housing) altogether. Reference 7 and Figure 3.19 describe the observed improvement in fuel economy resulting from applying this coating to the inside of the compressor housing. Not only was fuel economy improved, but there were also indications that compressor efficiency improvements were achieved leading to lower carbonaceous particulate emissions during transient operation.

There is probably much more to be said about centrifugal compressors, but for the generalist the explanations given are adequate for the general understanding of the basic concept.

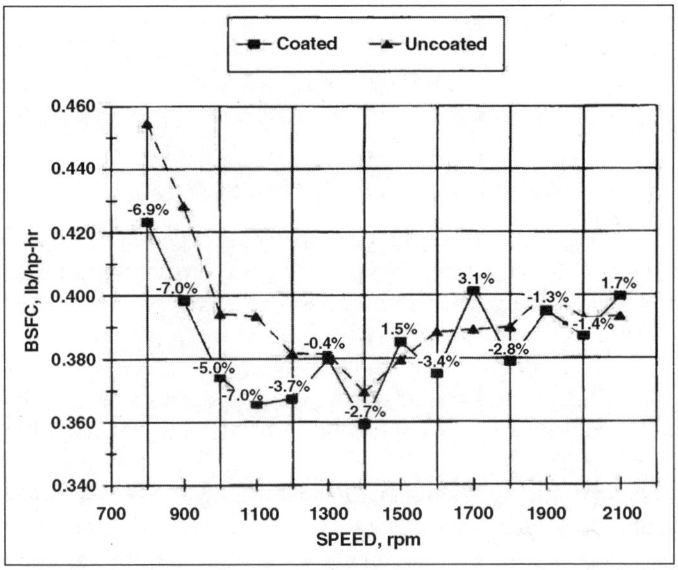

Figure 3.19 *Influence of an abradable coating on turbocharger fuel economy. (Reprinted from Ref. 7.)*

The other important component of the exhaust turbocharger is the turbine. The turbine wheel is made of a nickel-base alloy and rotates around the common shaft connecting it with the compressor wheel. The sole purpose for the turbine section is to use exhaust gases during the blowdown process (discharge of exhaust gases when the exhaust valve opens) to drive the compressor wheel. The turbine rotates inside a housing that bolts onto the exhaust manifold via a flange that is normally rectangular in shape.

The connection between the exhaust manifold and the turbine housing plays a very important role in the dynamic operation of the turbocharger. By enlarging the volume of that connection as well as the manifold, pulsations from the exhaust valves discharging into the manifold are dampened and converted into pressure that is nearly constant. For this reason, this method of connecting turbochargers to exhaust manifolds is referred to as *constant pressure*. While this method ensures the continuous flow of exhaust over the turbine blades, the fact that it is a higher pressure than that of the inlet disrupts the positive flow of air through the engine and reduces its overall volumetric efficiency.

A second method of connecting the exhaust manifold to the turbine housing makes use of the pulsating nature of exhaust emanating from each of the exhaust valves and is called the *pulse* method. A common design couples the front cylinders in a single manifold that leads to one side of the connecting flange, and the rear cylinders in another single manifold leading to the other side of the same connecting flange (see Figure 3.20a). This method is preferred for applications where engine response is important (most of the on-highway applications), since pulse energy dissipates through the turbine blades and keeps the exhaust pressure low.

The third and last method of connecting the exhaust manifold to the turbine housing is called the *pulse converter*. Figure 3.20b describes one embodiment of the pulse converter method showing a venturi box that converts exhaust kinetic energy to potential energy or pressure that is a close reminder of the constant pressure system. In general, constant pressure systems are common in two-stroke engine applications and pulse conservation systems are popular in four-stroke, on-highway applications [2].

It is worth noting that turbine housings serve as safety shields since they are designed with a generous wall thickness to contain the turbine blades in case a turbine bursts. Turbine housings are also referred to as *hot housings* by those in the trade. Naturally, using superchargers or turbochargers improves volumetric efficiency and allows higher output through increased fuel rates, thus producing higher cylinder pressures. With increased cylinder pressures comes increased mechanical loadings and the need for more robust design of the engine components. The usual process is to reduce the compression ratio by one point (from 18 to 17:1 for instance) when turbocharging a naturally aspirated engine.

Even though good attributes have been assigned to turbochargers, they have some undesirable characteristics. An example of these is the phenomenon called *turbocharger lag*. Because a turbocharger is driven by the pulse power from the exhaust blowdown, it is necessary to form generous amounts of

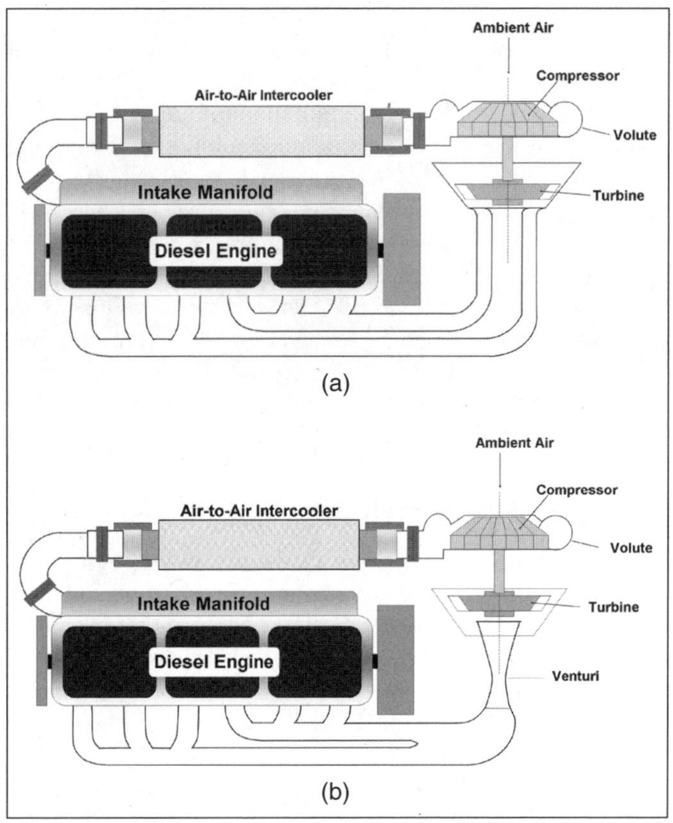

Figure 3.20 *(a) Pulse conservation and (b) pulse converter systems.*

exhaust before we can expect the turbocharger to rotate at speeds that would be beneficial to charging the engine. For instance, when a driver demands more power by stepping on the accelerator pedal, more fuel is injected into the cylinder. However, this acceleration fuel needs to combust and produce exhaust first before the turbine can accelerate and drive the compressor to deliver more air into the cylinder. The time lapse between introducing additional fuel into the cylinder and the delivery of more charge air by the compressor is generally referred to as *turbocharger lag*. Combustion during turbocharger lag is characterized by a momentary dip in the air-to-fuel (A/F) ratio, causing the release of a visible puff of smoke.

Another problem encountered with turbocharger design and its matching to a particular engine is its poor performance at low engine speeds and loads. In this engine operating regime, exhaust pulse power is rather low given the fact that the engine is using low airflow and fuel rate. In these cases, engine designers are forced into turbocharger designs that incorporate smaller turbine wheel moments of inertia to allow them to achieve faster accelerations and rotational speeds even when exposed to low exhaust pulse power. Such designs may be favored for low engine speeds and loads, but can be detrimental to high-speed, high-load operation. If this is the case in a given application, serious consideration is given to other turbocharger design modifications to avoid developing extremely high charging pressure during high engine output operating regimes. In addition to the potential problems associated with high charging pressures, serious combustion quality problems may be faced due to distortion of the fuel spray pattern in the combustion chamber.

Diesel engines perform best when their inlet and exhaust systems are not restricted. Placing a turbine wheel in the path of exhaust gases restricts their flow and may cause a drop in volumetric efficiency. However, in most designs if not all, this drop in volumetric efficiency is more than compensated for through the added charge air inducted into the engine thanks to turbocharging. This is especially evident when engines/vehicles operate in high-altitude environments where atmospheric ambient pressure is lower than near sea level. In these cases, naturally aspirated engine performance suffers tremendous deficits through lower *smoke-limited* power outputs (opacity-limited output) that may even force the engine to be derated. Thus turbocharging can be used as a form of altitude compensation that maintains rated output power and avoids the need to derate the engine.

3.5 Types of Turbochargers

Most turbochargers share the common single-flow turbine-shaft-radial compressor design. However, with the proliferation of applications requiring turbocharging for various reasons, including emissions control, new designs and types of turbochargers have become necessary. Some of these designs are not new concepts, but instead involve modifications to the base *fixed-geometry turbocharger* described thus far. The main premise for the modern use of turbochargers is that they achieve nearly constant charge air pressure over a wide range of operating conditions. Achieving this goal requires controlling the turbocharger operation in a flexible manner. This process can be achieved through regulation at the exhaust gas end or the fresh air end. Regulation at the exhaust gas end can be achieved principally through waste gating and variable-geometry turbocharging.

3.5.1 Waste-Gated Turbochargers

Waste-gated turbochargers are generally actuated pneumatically, based on an inlet system pressure signal. Figure 3.21 is a schematic representation of a waste-gated turbocharger that illustrates a mechanism that bypasses exhaust around and away from the turbine [4]. Regulation of this bypass is achieved via a spring-loaded diaphragm actuating a poppet valve. This entire assembly constitutes the body of the waste gate (see item 3 in Figure 3.21).

In practical embodiments, a pressure signal from the intake manifold is applied to the waste-gate diaphragm. The force generated by the pressure application depresses the spring at the prescribed design point, thus opening the bypass (poppet) valve, which routes exhaust around the turbine and dissipates pulse energy directly into the ambient via the exhaust pipe. This process allows the design of turbochargers that can easily accelerate when exposed to low exhaust energy, thus developing adequate inlet pressure (intake manifold pressure), and also avoiding excessive inlet pressure when exposed to high exhaust flow when the engine operates in the high-output regime.

3.5.2 Variable-Geometry Turbochargers

In waste-gated turbochargers some of the exhaust pulsating energy is not used to rotate the turbine, as described earlier. Dissipating this energy directly into the atmosphere rather than making use of it may be viewed as waste that can reduce the potential overall powerplant efficiency. An alternative to this method is the variable-geometry turbocharger, which is shown schematically in Figure 3.22 [4].

Figure 3.22 shows the hot housing of a variable-geometry turbocharger (VGT), variable-angle vanes, and a saw-toothed adjusting ring (items 1, 2, and 3 in Figure 3.22, respectively). The variable-angle vanes are sandwiched between the exhaust

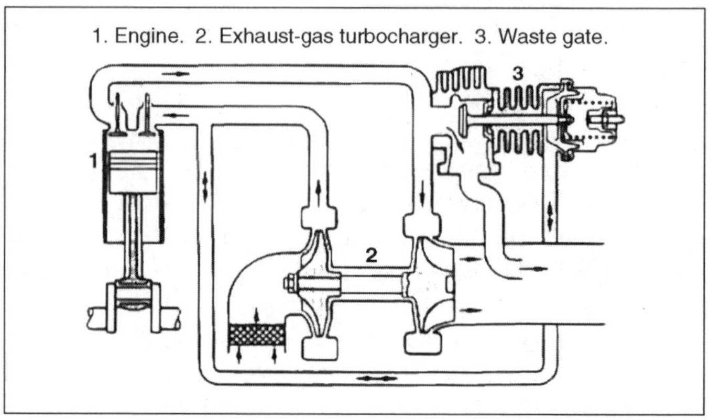

Figure 3.21 *A waste-gated turbocharger. (Reprinted from Ref. 4 with permission from Robert Bosch GmbH.)*

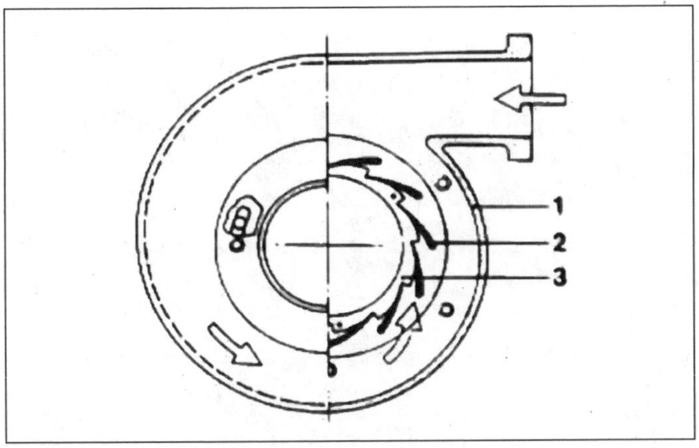

Figure 3.22 *A variable-geometry turbocharger. (Reprinted from Ref. 4 with permission from Robert Bosch GmbH.)*

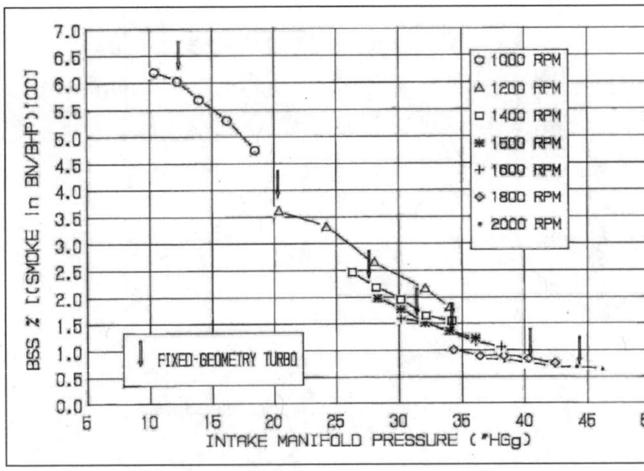

Figure 3.23 *Relationship between intake manifold pressure and smoke emission.*

manifold and the turbine. At low-energy exhaust conditions, the variable-angle vanes reduce the flow area between two consecutive vanes, thus converting the exhaust potential energy (pressure) into kinetic energy (speed). Faster exhaust impacts the turbine blades, causing them to rotate faster and providing the compressor with more energy to move more charge air into the cylinder. Conversely, at high engine output conditions, the vanes are positioned such that little conversion of potential energy is experienced. This condition prevents the compressor from achieving too high a rotating speed, and thus leads to reduced (correct) additional charge air being inducted into the cylinder.

Variable-geometry turbochargers offer a great deal of flexibility in controlling engine performance as well as reducing emissions, especially in modern engines where the use of electronics is more prevalent. For instance, sensing a driver demand for fast acceleration, the electronic control module could be programmed to position the variable-angle vanes in a VGT to reduce the exhaust flow area and increase turbine speed, thus providing additional charge air. Additional charge air could also be scheduled to occur prior to the arrival of the bulk of acceleration fuel into the combustion chamber. This control methodology helps eliminate the momentary smoke puff that usually occurs during sudden accelerations.

In addition to this control methodology, one can think of other advantages of variable-geometry turbocharging. Among them is controlling the pressure between the exhaust and intake manifolds. By positioning the variable-angle vanes to reduce exhaust flow area, exhaust manifold pressure increases while intake manifold pressure may not increase by the same proportion. Emissions control engineers use this feature to achieve more exhaust gas recirculation, thus lowering NO_x emissions. Experiments conducted at Ford Tractor Operations in the mid-1980s provided a comparison between the performance of a fixed-geometry and a variable-geometry turbocharger. Figure 3.23 illustrates the relationship between intake manifold pressure and brake-specific smoke for a VGT at constant engine speeds. The performance of the fixed-geometry turbocharger is also indicated.

From Figure 3.23 several conclusions can be derived:

- Using a VGT can substantially lower brake-specific smoke, which is defined as smoke (measured in Bosch No.) divided by the engine power output and multiplied by 100.
- Using a VGT is more effective at low engine speeds, which corresponds to a low intake manifold pressure, compared to fixed-geometry turbocharging.
- Using a VGT at high engine output conditions, versus a well-matched fixed-geometry turbocharger, may not be justified.

Gains in torque and power output as well as fuel economy corresponding to those achieved in smoke control have also been demonstrated. Figure 3.24 is an example of the superior performance achieved with VGTs at speeds below peak torque.

In some cases, regulating charge air is achieved on the cold side (compressor). The system is often referred to as the *air gas regulator*. This method of bypassing air after it has been compressed is not as popular as the exhaust bypass described in conjunction with the waste-gated turbocharger. The reason for this is the loss of energy used to compress the air that is bypassed rather than used in the engine.

3.6 Supercharger-Aided Turbochargers

Recent developments in turbocharging and supercharging technology produced interesting combinations of supercharger-aided turbocharging systems. In Figure 3.14, a high-speed electric motor driving a centrifugal compressor is shown. The same company (Turbodyne Systems, Inc.) further developed this concept and placed an electrically driven, high-speed motor on the shaft connecting the turbine and the compressor [8]. This high-speed motor is used intermittently and only to supplement the turbocharger. An example of its use is during acceleration to minimize or eliminate the smoke puff produced during turbocharger lag. Figure 3.25 is a schematic representation of the

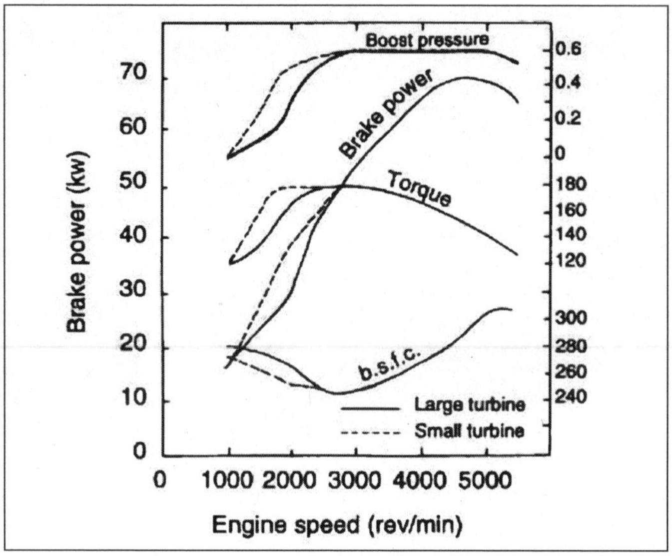

Figure 3.24 Effect of VGT on torque, power, and BSFC at below beak torque speeds for a small diesel engine.

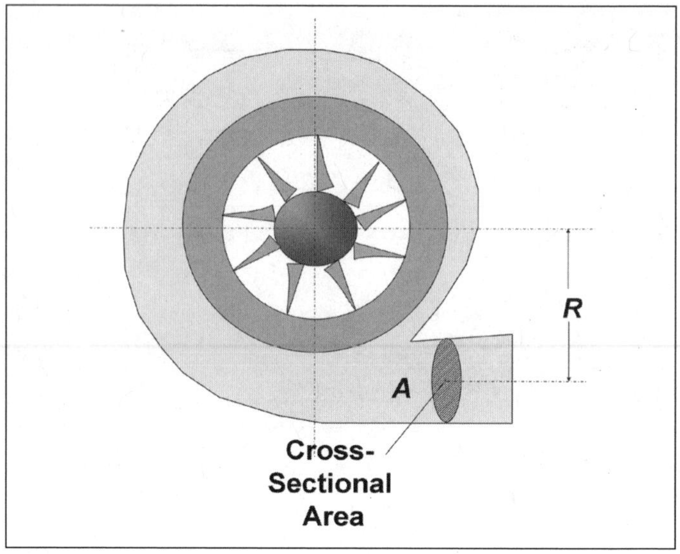

Figure 3.26 Definition of A/R ratio.

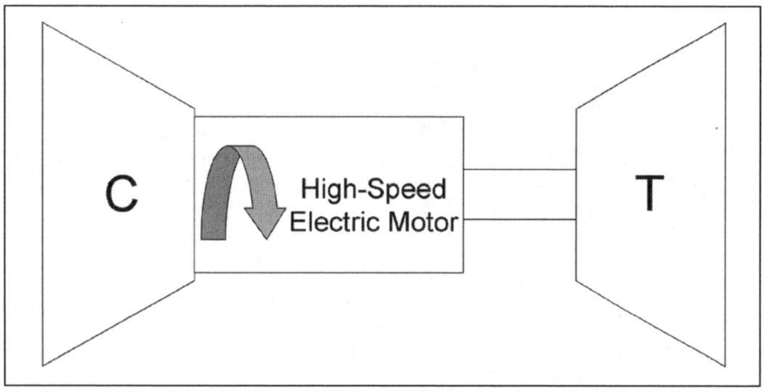

Figure 3.25 A supercharger-aided turbocharger.

supercharger-aided turbocharger that carries the trade name of *Dynacharger*.

3.7 Compressor or Turbine A/R Ratio

As explained at the beginning of this chapter, the main purpose for using supercharging or turbocharging is to improve power output through increased air capacity. It is reasonable then to have, as part of the design objectives, a goal of increasing the air charge delivered by the centrifugal machinery. Several factors are considered to fulfill this task and some, such as tip clearance, have already been discussed. Another important design parameter is the *A/R ratio,* defined as the smallest cross-sectional area of the passage leading to the turbine wheel divided by the distance between the center of area A and the center of the turbine or compressor wheel (R). This basic relationship is illustrated in Figure 3.26 and applies to both turbine and compressor.

A small A/R ratio increases the turbine rotational speed and leads to higher compressor speeds. This action usually leads to greater air inducted by the compressor. The opposite is also true. The larger the A/R ratio, the slower the rotational speed of both turbine and compressor wheels. Therefore, for greater boost a smaller A/R ratio is required, and for lower boost a greater A/R ratio is used.

For good engine performance, especially during fast or sudden accelerations, a fast delivery of charge air is required. This helps reduce or even eliminate the diesel's characteristic puff of smoke that results in part due to the turbocharger lag. A smaller A/R ratio is able to spin sooner and accelerate faster, thus it delivers air boost sooner and more rapidly. Of course, as more exhaust energy is imparted to the turbine wheel, such as during high-speed and high-torque engine conditions, the turbocharger speed will dramatically increase, causing the incoming air to distort the injector spray pattern, reducing combustion efficiency. Faster turbocharger spin can also be accomplished by using ceramic turbines due to their lower rotating mass.

Turbocharger performance is not only important during acceleration, but also during steady-state operation. In Figures 3.23 and 3.24, the effect of inducting additional air at engine speeds below peak torque speed is very evident, especially at full load conditions. In fact, optimizing the air inducted into the engine at speeds below that of peak torque is one of the main objectives for turbocharger matching. Figure 3.27 provides more proof of the beneficial effect of additional air inducted below peak torque speed. It shows how smoke emissions were reduced as the engine was operated at constant speeds and varying loads. Two speeds were used in this demonstration (1200 and 1400 rpm), while engine load was increased in several steps. The turbocharger having the smaller A/R ratio helped reduce smoke emissions at these two speeds by a substantial margin.

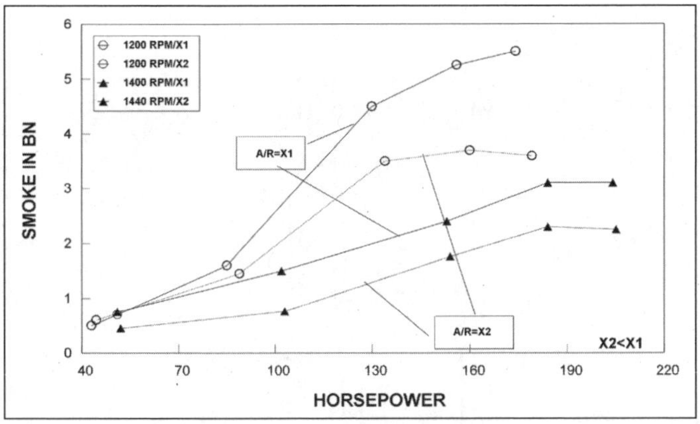

Figure 3.27 *Effect of A/R ratio on smoke characteristics.*

Of course, one would expect these improvements to decrease as engine speeds and load increased.

3.8 Practical Embodiments

Applications involving superchargers and turbochargers are diverse and range from simple to complex configurations. Figure 3.28a illustrates a schematic for a turbocharged engine, and Figure 3.28b is the corresponding schematic for a mechanical, engine-driven supercharger. Both figures indicate that whether a supercharger or a turbocharger is used, cooling of the charge air did not occur. Many engines fall under this category of supercharged, uncooled engines. Although this type of engine was popular prior to emphasizing fuel economy and exhaust emissions, most modern engines are turbocharged and cooled as shown in Figure 3.28c. The heat exchanger that cools combustion air as it exits the compressor is referred to as an *intercooler* or *aftercooler*. The term *intercooler* refers to the fact that this heat exchanger performs its task *between* the compressor and the intake manifold of the engine. The term *aftercooler* refers to the charge air being cooled *after* being compressed in the compressor. The need to cool the charge air stems from the fact that, while converting the kinetic energy of the charge air to potential energy, its temperature increases, thus reducing its density. When its density is reduced by heating, charge air occupies a larger volume in the combustion chamber, leading to a lower overall mass of charge air in the cylinder and, hence, the need to cool the charge air to recover its density and increase its mass in the cylinder.

So far we have dealt with simple systems involving supercharging or turbocharging. However, in some applications supercharging may be complex and may include combinations of superchargers and turbochargers. Figure 3.29, for instance, illustrates a combination of a mechanically driven supercharger and a turbocharger. This arrangement has the potential to develop a high charge pressure and/or to develop a higher charge pressure more rapidly to overcome the adverse effects of turbocharger lag. For instance, the mechanical drive shown in Figure 3.29 may incorporate an electrically controlled clutch that is activated when the driver commands an acceleration.

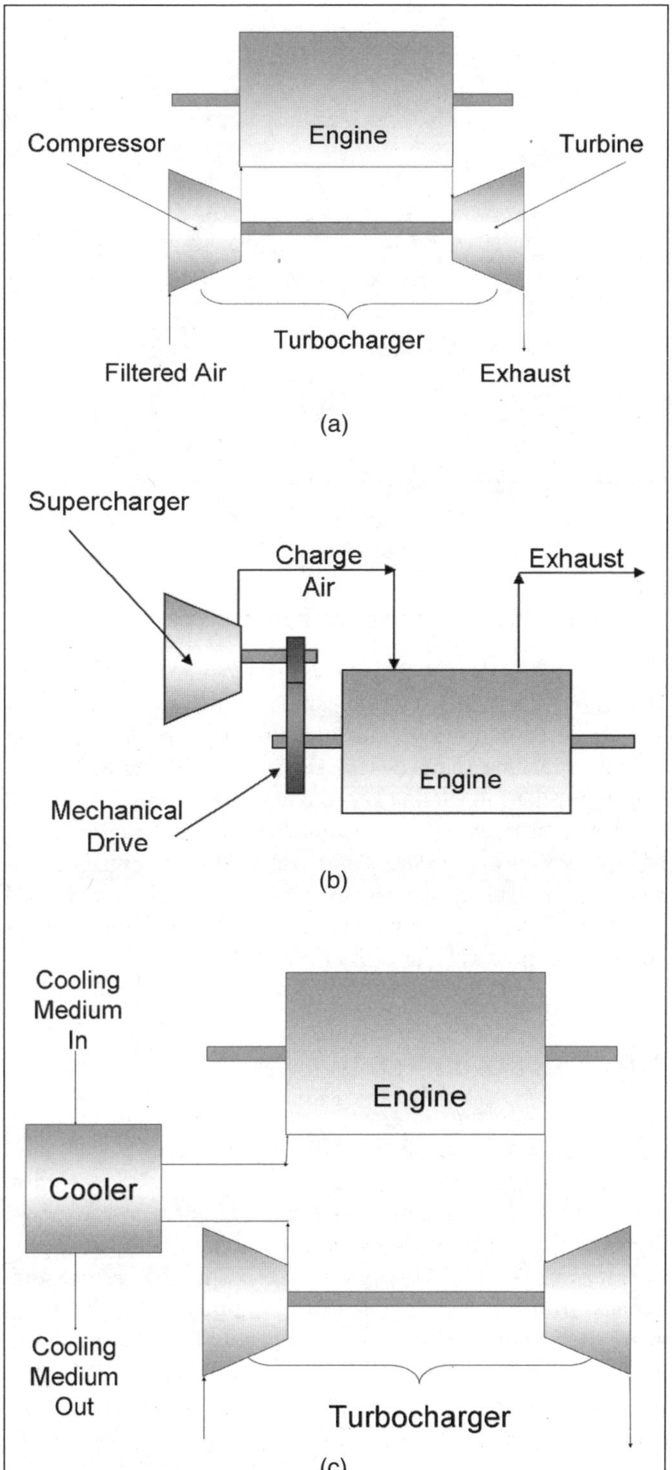

Figure 3.28 *(a) Uncooled turbocharger, (b) uncooled supercharger, and (c) cooled turbocharger.*

Once activated the clutch engages the mechanical drive, which begins to develop charge air pressure immediately. Once the acceleration is complete, the supercharger drive is de-clutched and the operation continues with the stock turbocharger in full control.

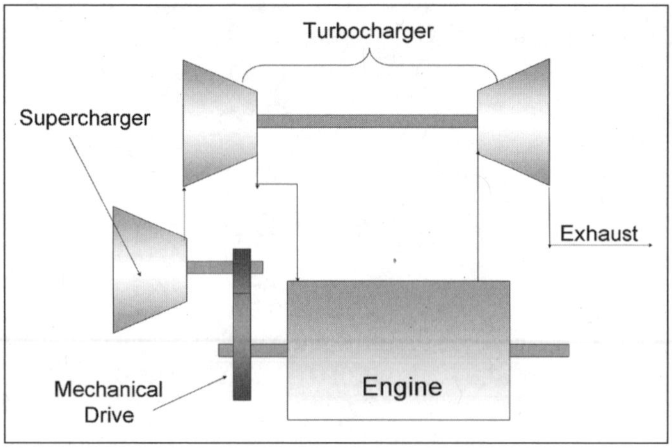

Figure 3.29 *Supercharger-assisted turbocharger.*

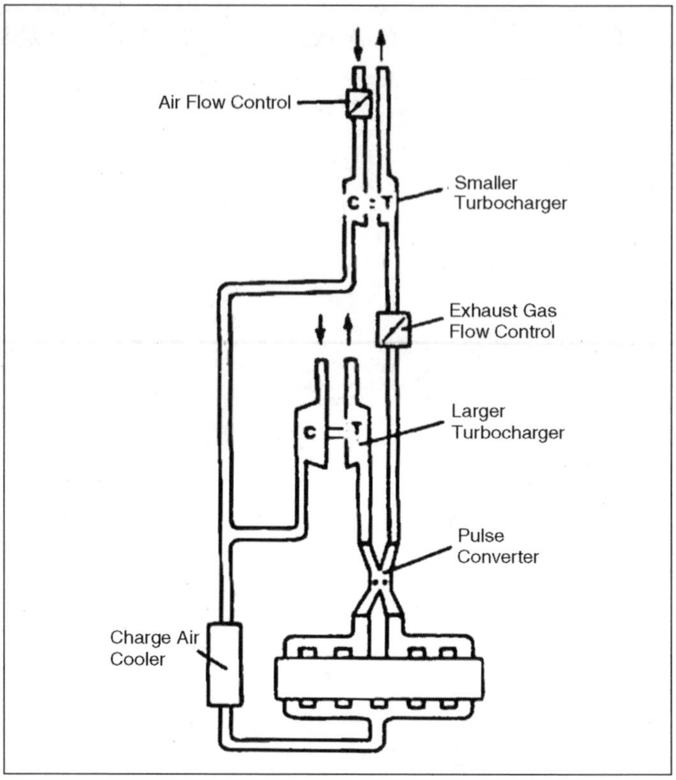

Figure 3.30 *Sequential turbocharging. (Courtesy MTU Detroit Diesel.)*

3.9 Sequential Turbocharging

Another system worth noting is *sequential turbocharging,* which involves two turbochargers of different size to meet the engine demand in various operating regimes [9]. A small turbocharger that requires less energy to spin and accelerate is used for the low engine speed and load regime. The second and larger turbocharger is designed to meet the engine needs when operating at high speed and load conditions. Fresh filtered airflow is controlled by a valve, and exhaust flow is controlled by a second valve, as shown in Figure 3.30, to divert charge air and exhaust flow as dictated by the needs of the engine according to its operating regime.

3.10 Turbocompounding

A schematic representation is given in Figure 3.31 for turbocompounding systems. This configuration is generally reserved for large engine installations such as power generation or bulk fluid pumping stations. The reason for this is the expense involved in gearing down from high turbine rotational speeds to crankshaft speeds. This system is based on the premise that exhaust from the first turbine (T1 in Figure 3.31) has enough energy left to turn a second turbine wheel (T2 in Figure 3.31). This excess energy is high enough to give T2 power that is adequate to lend the crankshaft some assistance. This additional power provided to the crankshaft by T2 is considered part of the exhaust heat recovery, and it results in real fuel economy benefits as shown in Figure 3.32. For the same power output, BSFC for a turbocompounded engine is lower than that for a turbocharged-intercooled engine [10].

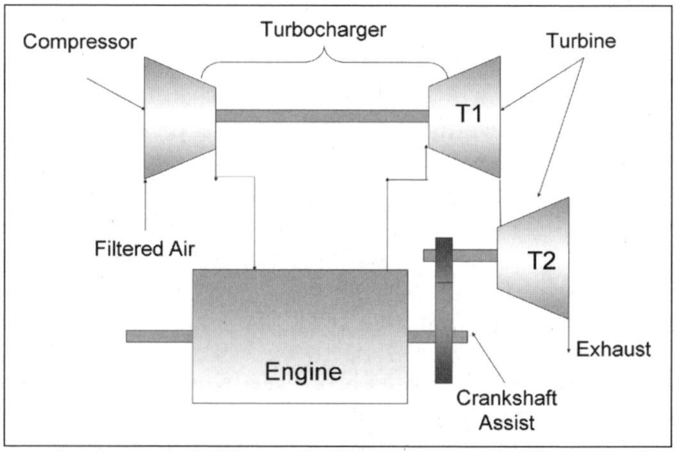

Figure 3.31 *Turbocompounding.*

3.11 Wave Supercharging

One of the better known wave superchargers is the *Comprex*. The Comprex is considered by many in the industry as a combination supercharger/turbocharger device. Because it is somewhat difficult to understand the operation of the wave supercharger, Figure 3.33 is presented to assist in describing its principle of operation.

Filtered air proceeds into one of many channels of the wave supercharger. The filtered air travels through that channel where it is met by a pressure wave from the exhaust pulse leaving the exhaust manifold. Both filtered air and exhaust pulse travel toward each other and eventually meet within volume A shown in Figure 3.33. When the exhaust pulse meets with the filtered

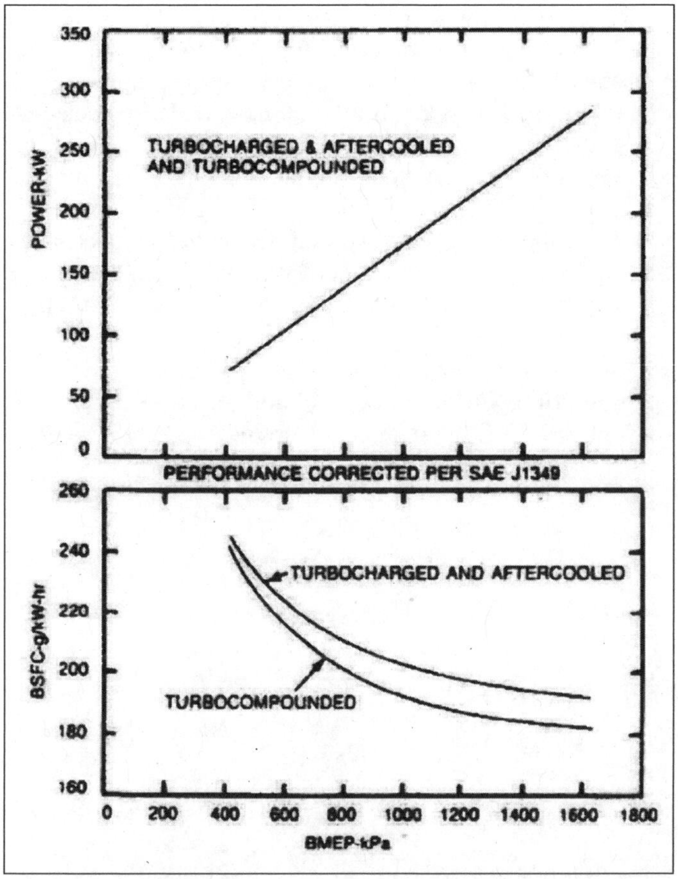

Figure 3.32 *Effect of turbocompounding on fuel consumption.*

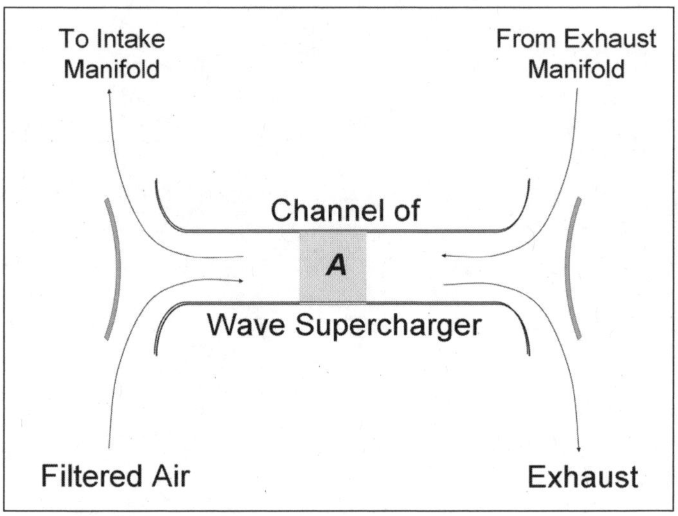

Figure 3.33 *Wave supercharger operation.*

air, it imparts some of its energy to the filtered air thus increasing its pressure. In the process, the exhaust pressure pulse loses a portion of its own energy. Having imparted some of its energy to the filtered air, the exhaust pulse exits through the exhaust port as shown in Figure 3.33. In the meantime, filtered air, having received a portion of the exhaust energy, gains potential energy that accelerates it toward the intake manifold or port. This entire process takes place over an extremely short period since the total number of these channels is synchronized to expose a new fresh charge to the exhaust pulse as described. This synchronization or indexing is achieved by adjusting the drive ratio of the entire wave compressor, which is arranged in a cylindrical configuration and is belt driven as shown in Figure 3.34.

In Figure 3.34, several components are shown to facilitate the understanding of the wave supercharger. For instance, item 1 represents the combustion chamber, and the 2 indicates the multichannel wave supercharger with its channels arranged cylindrically. The entire wave supercharger assembly is driven by a belt/pulley drive designated as 3. Items 4 and 5 represent the exhaust and intake pipes, respectively. Item 6 represents the inlet of the fresh filtered air, whereas 7 is the exhaust gas exit to ambient.

It is interesting to note that the contact time between the fresh air and the exhaust in the channel of the wave supercharger is only a fraction of a second. During that time, the kinetic energy of the exhaust gas is converted to potential energy of the charge air. If the contact time is lengthened, there may be more chemical interaction as well as physical mixing between the exhaust on one side and the fresh air on the other. The exchange of mass between exhaust gas and inlet fresh air leads to introducing exhaust along with the fresh air and the loss of some fresh air to the exhaust. This process acts as exhaust gas recirculation to help reduce nitric oxide emissions.

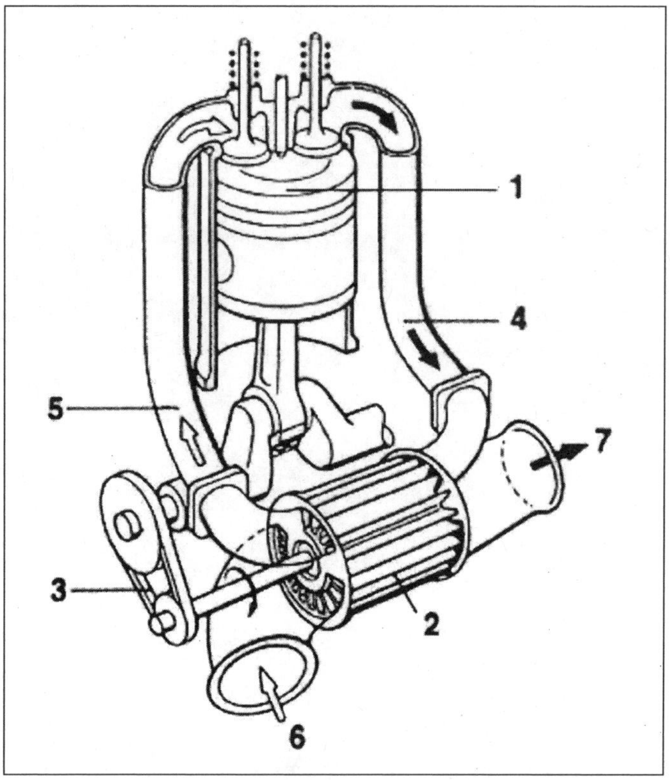

Figure 3.34 *Wave supercharger. (Reprinted from Ref. 4 with permission from Robert Bosch GmbH.)*

References

1. Heywood, J.B., *Internal Combustion Engine Fundamentals,* McGraw-Hill, New York, 1988.
2. Obert, E.F., *Internal Combustion Engines,* International Text Book, Scranton, PA, 1968.
3. Ronzi, W., B. Ericson, and J.M. McNaughton, "Supercharger Bench Testing," Internal Report, General Motors Institute and Eaton Corporation, Detroit, MI, 1995.
4. "Superchargers," chapter in *Automotive Handbook,* 2nd ed., Bosch GmbH, Germany, 1986, pp. 355–359.
5. Heisler, H., *Advanced Engine Technology,* Butterworth-Heinemann, New York, 1995.
6. "Introduction to Turbochargers," Schwitzer Turbochargers, Indianapolis, IN, 1991.
7. Sharp, C.A., M.K. Khair, and A. Gorel, "The Effect of a Turbocharger Clearance Control Coating on the Performance and Emissions of a 2-Stroke Diesel Engine," SAE Paper 1999-01-3665, SAE International, Warrendale, PA, 1999.
8. "Performance for Your Engine and the Environment," Turbodyne Systems, Carpinteria, CA, 1995.
9. Kech, J., et al., "Model-Based Sequential Turbocharging Optimization for Series 8000 M70/M90 Engines," SAE Technical Paper 2002-01-0378, SAE International, Warrendale, PA, 2002.
10. Wilson, D.E., "The Design of a Low Specific Fuel Consumption Turbocompound Engine," SAE Paper No. 860072, SAE International, Warrendale, PA, 1986.

Chapter 4

Diesel Fuel Injection

4.1 Introduction

The performance of diesel engines is heavily influenced by the design of their injection system. In fact, the more notable advances achieved in diesel engines resulted directly from superior fuel injection system designs. Although the main purpose of the system is to deliver fuel to the cylinders of a diesel engine, it is *how* that fuel is delivered that makes the difference in engine performance, emissions, and noise characteristics.

Unlike its spark-ignited engine counterpart, the diesel injection system delivers fuel under extremely high injection pressures. This aspect implies that the system component designs and materials should be selected to withstand higher stresses, besides being able to perform for extended durations, matching the engine's excellent durability records. Greater manufacturing precision and tight tolerances are also required if the system is to function efficiently. In addition to expensive materials and manufacturing costs, diesel injection systems are characterized by complex control requirements. All of these features add up to a system whose cost may represent as much as 30% of the total cost of the engine.

In this chapter we review the basic diesel fuel injection system and its components, the various types of injection systems, and how they meter and deliver fuel into the cylinders of a diesel engine. The mechanical function of these systems is described with emphasis on their mixture formation role. Various injection timing and metering schemes are also described. The need for greater flexibility in control is presented and several modern electronically controlled systems are reviewed.

4.2 Purposes of the Fuel Injection System

As stated in the preceding section, the main purpose of the fuel injection system is to deliver fuel into the cylinders of an engine. Although this may be the general purpose of the system, some of the specifics may be described by the following functions: (1) timely introduction of the fuel into the cylinder, that is, injection timing and its control; and (2) delivery of an accurate amount of fuel to meet power requirements, that is, injection metering and its control.

To settle for just these two general functions of the fuel injection system would be premature, because it is not enough to deliver an accurately metered amount of fuel at the proper time to achieve good combustion. How well this delivered fuel mixes with the inducted air is also of major importance. Therefore, fuel atomization, that is, how small the fuel particle size is and how dispersed it is in the cylinder, should be a primary design objective for a successful diesel fuel injection system. Yet, even a properly atomized, accurately metered, and well-timed fuel delivery system may not provide the desired combustion efficiency. Another important parameter in this respect is the quality of the mixture (fuel and air). Although good atomization goes a long way toward improving combustion efficiency, how well the available air is used is also an indicator of high combustion efficiency. To achieve this objective, better air utilization must be achieved. This parameter can be accomplished through a combination of fuel penetration into the dense air that is compressed in the cylinder and by equally dividing the total fuel among the cylinders of an engine. The primary purposes of the diesel fuel injection system are graphically represented in Figure 4.1.

To accomplish the functions described earlier is no small task, and to do it effectively requires the collective experience of many experts specialized in the disciplines of materials,

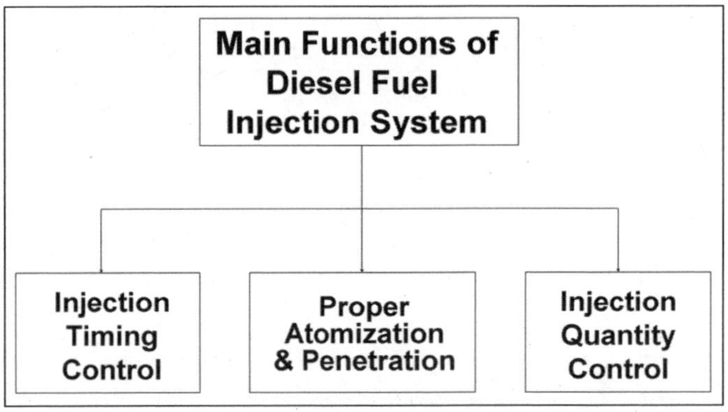

Figure 4.1 *Diesel fuel injection system functions.*

mechanical design, hydraulics, fluid dynamics, and combustion systems. As more limits were imposed on emissions from diesel engines, and pressures were felt from customers for better performance, the need for additional fuel injection system functions became apparent. Indeed, the modern diesel injection system must perform the tasks of metering, timing, and atomizing as well as contribute to good air and fuel mixing. In addition, experience gained during the last 25 years has shown that scheduling of the metered fuel into the cylinder is very important to the performance, emissions, and noise characteristics of the combustion process. A general term describing the scheduling of injected fuel is *rate shaping*. Yet, modern systems are not only concerned with how much fuel is delivered per crank angle degree, but also with the number of injections per combustion cycle. Figure 4.2 summarizes the additional functions that may be required of the modern diesel injection system.

Combustion kinetics dictate the right injection rate for a given engine speed and load condition. What may be preferred for low-speed and low-load conditions may be diametrically opposed to the desired injection characteristics at high-speed and high-load conditions. A good illustration of this aspect is given in Figure 4.3, which is a compilation of optimized injections for various engine conditions.

As suggested by Figure 4.3, a modern fuel injection system with built-in design flexibility may be used to produce the various injection rates and/or multiple injections for best results. The term *best results* may mean different goals for different operating regimes. For instance, for the best noise control and to avoid the formation of white smoke during a cold start, pilot injection may be preferred. However, to achieve the best fuel economy in the midsection below the full-load torque curve, a sharp beginning and end for injection may be preferred. Yet under low-load and part-load conditions, split injections may be preferred, whereas at high-speed and high-load conditions, a ramped injection might be preferred. We must emphasize, however, that this may very well be the right prescription for a certain combustion system for a specific engine, but may not be the preference for a significantly different type of combustion system.

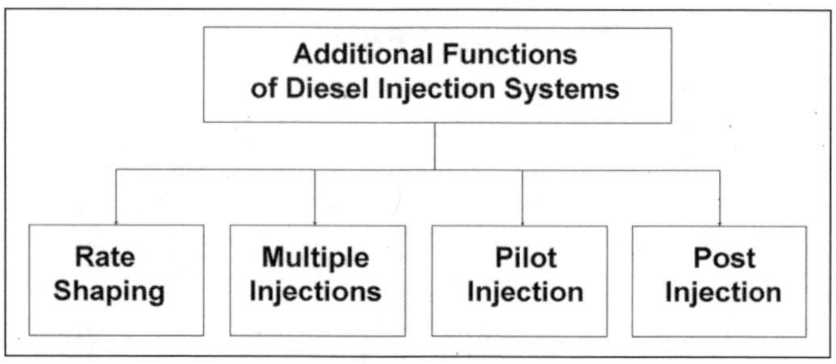

Figure 4.2 Additional functions for modern diesel fuel injection systems.

4.3 Types of Diesel Fuel Injection Systems

Diesel fuel injection systems can be placed into the following categories:

- Pump-line-nozzle (see Figure 4.4)
- Unit injector (see Figure 4.5)
- Common-rail injection system (see Figure 4.6).

For each of the preceding categories, there are a number of subcategories that will be detailed through the course of this chapter. However, we can make some generalized comments at this stage to further clarify how quickly these three very general categories can branch out into many unique and distinct designs. For instance, the pump-line-nozzle (P-L-N) system includes in-line (Figure 4.4), distributor or rotary, as well as unit pump designs. Not only could we differentiate them as such, but we can also distinguish among them on the basis of whether they are mechanically or electronically controlled. Similarly, unit injectors can be mechanically or electronically controlled. Unit injector systems can also be designed to deliver extremely high injection pressures, in which case they may feature mechanical intensifiers in the form of plungers having two different diameters. They may also be actuated by brute force imparted to the top of the injector by a large cam such as the Cummins pressure-time controlled (PT) system. The common-rail category enjoys a similar variety of design details, as explored later in this chapter.

4.4 The Pump-Line-Nozzle System

The pump-line-nozzle injection system is so-called for producing high fuel pressure in a pumping element, transferring the fuel pressure pulse through a high-pressure injection

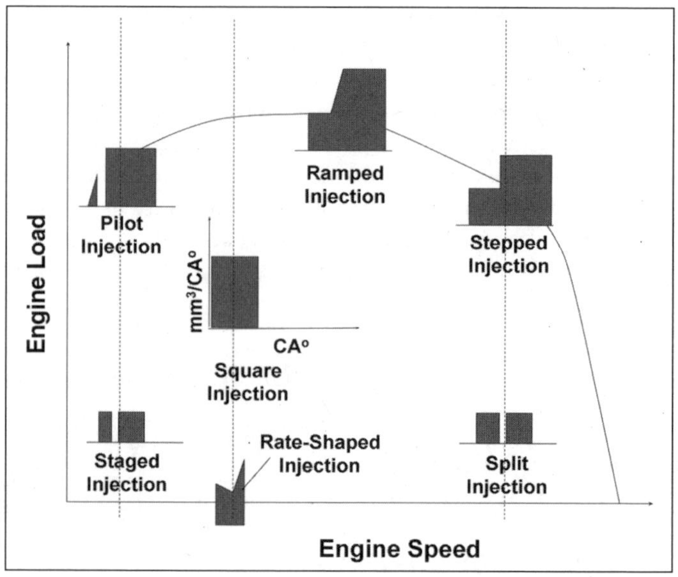

Figure 4.3 Example of preferred injection scheduling.

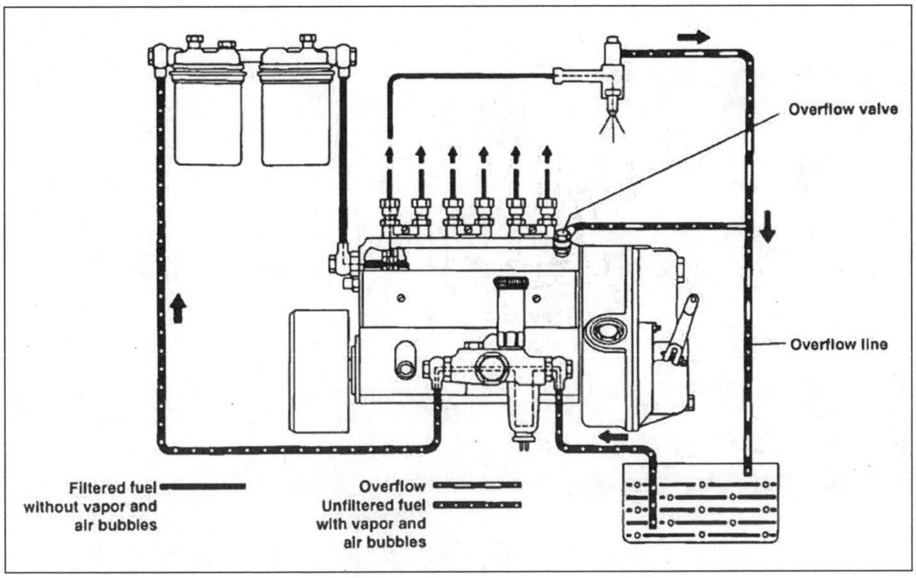

Figure 4.4 *Pump-line-nozzle system. (Reprinted from Ref. 1, Fig. 65, courtesy of Robert Bosch GmbH.)*

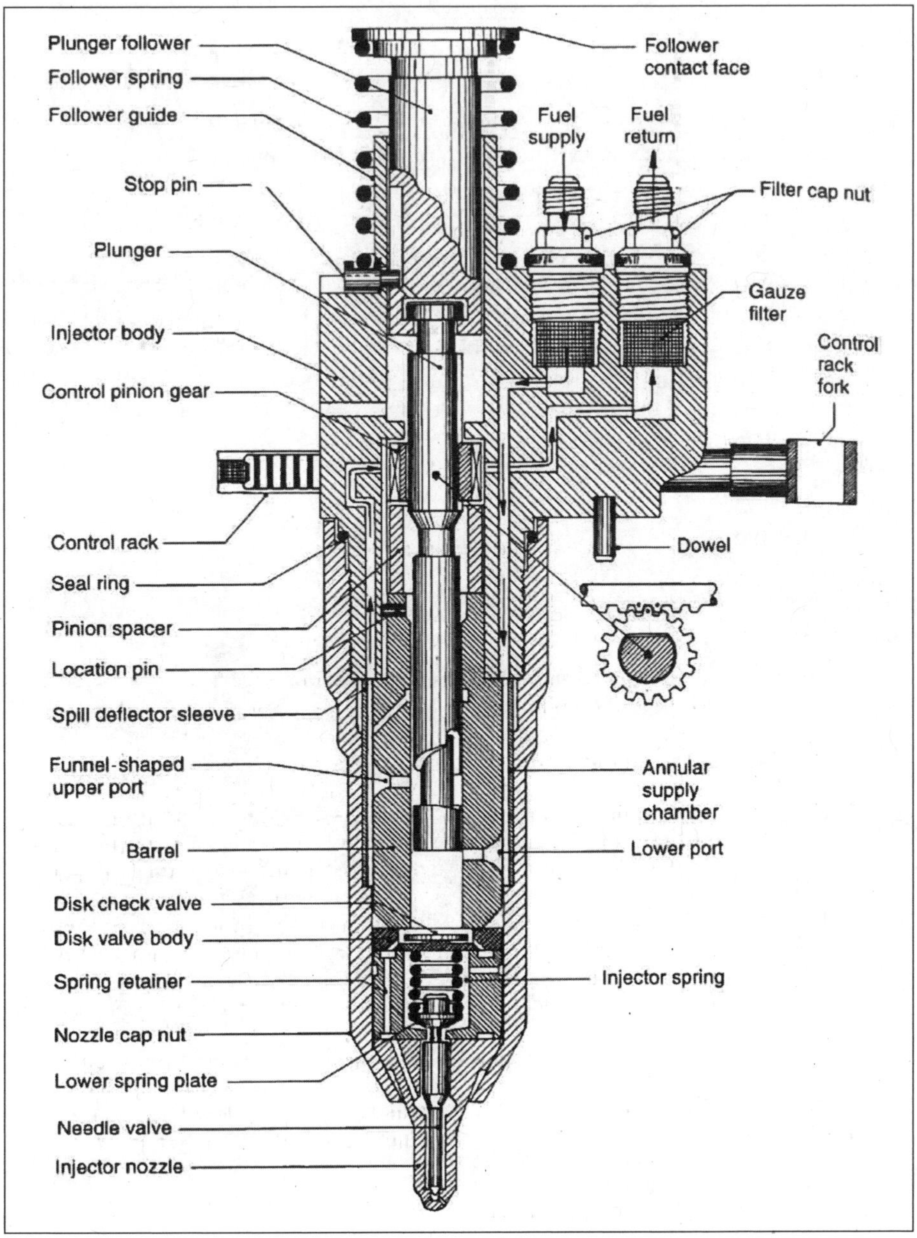

Figure 4.5 *Cross section of a unit injector. (From* Advanced Engine Technology *by Heinz Heisler [2]. Reprinted by permission of Butterworth-Heinemann.)*

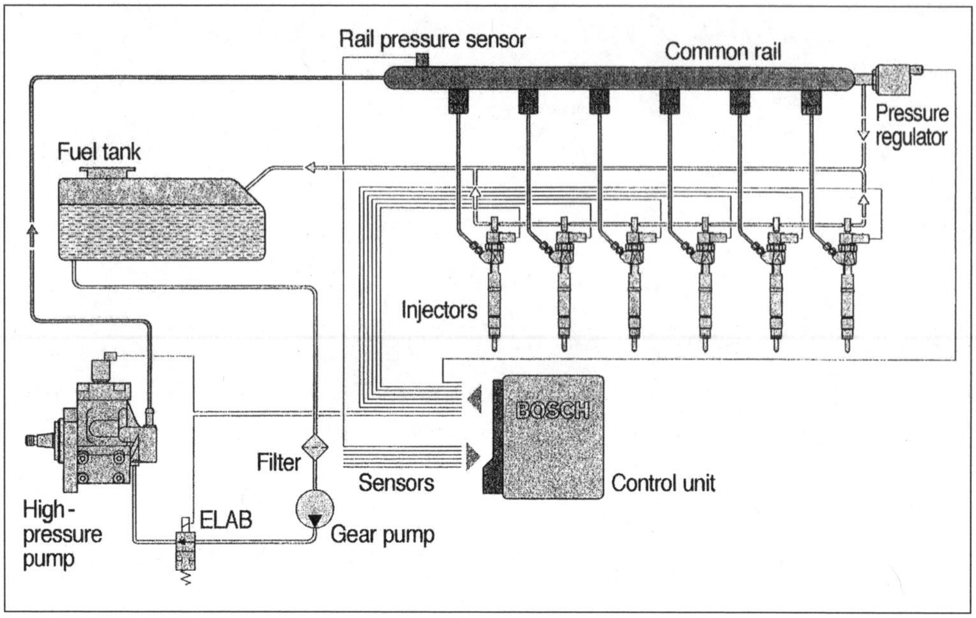

Figure 4.6 *Bosch electronic common-rail diesel injection system. (Reprinted from Ref. 3, Fig. 40, courtesy of Robert Bosch GmbH.)*

line, and then spraying this fuel into the cylinder via the nozzle of an injector. In-line pumps such as that shown in Figure 4.7, distributor or rotary pumps (see Figure 4.8), as well as unit pumps (see Figure 4.9) have those features in common, and are therefore classified as P-L-N systems.

It may seem redundant to some readers that several P-L-N systems should exist. Yet there is economic and/or technical justification for each of these systems. For instance, in-line pumps serve multicylinder engines and, therefore, house as many pumping elements as there are cylinders in an engine. The pump is usually gear driven by the camshaft and is positioned in a central location relative to the engine assembly. Engine and fuel system designers strive to locate the pump such that the injection lines are equal in length between the injection pump and the entry to the injectors. With highly pulsating systems and pressure waves traveling through narrow pipes, line dynamics can be difficult to manage and may cause erratic injection behavior at the nozzle. In their desire to avoid line dynamics complications, designers strive to keep the total line length as short as possible. In many cases this objective may not be possible to accomplish. One such case is large marine and stationary power plants where the sheer size of the engine would prohibit the feasibility of short injection lines. Examples of this type of application are given in Figures 4.10a and b, which show the DDC/MTU Series 2000 and MTU/DDC Series 4000 engines, respectively [6]. In such cases unit pump systems are used to

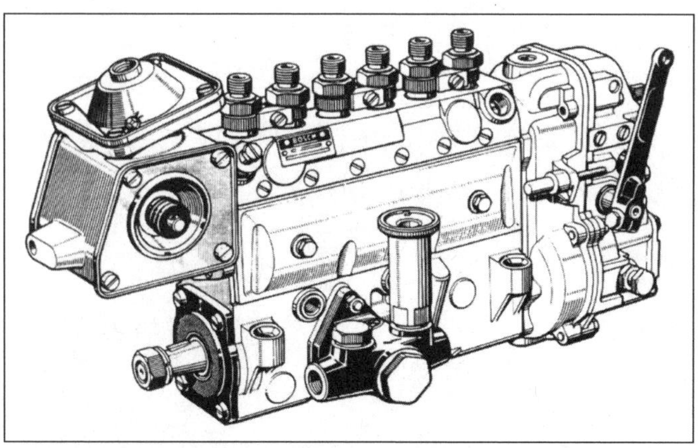

Figure 4.7 *Features of a Bosch high-pressure in-line injection pump. (Reprinted from Ref. 4, page 15, Fig. 21, courtesy of Robert Bosch GmbH.)*

1 Pressure-control valve, 2 Governor assembly, 3 Overflow restriction, 4 distributor head with high-pressure pump, 5 Vane-type fuel-supply, 6 Timing device, 7 Cam plate, 8 Electromagnetic shutoff valve.

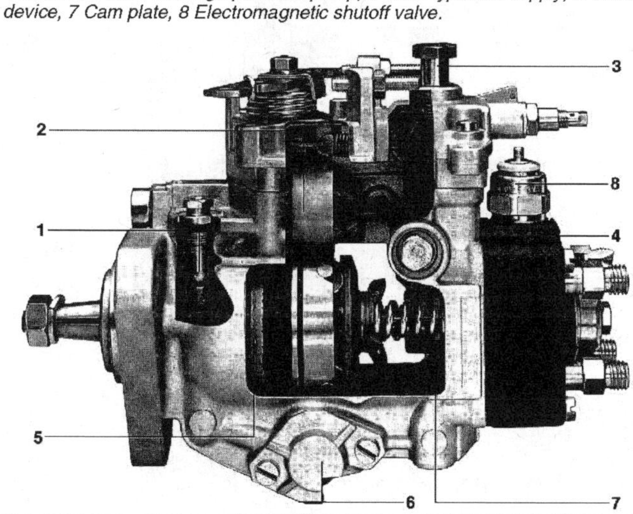

Figure 4.8 *Details of a Bosch distributor injection pump. (Reprinted from Ref. 5, page 151, Fig. 3, courtesy of Robert Bosch GmbH.)*

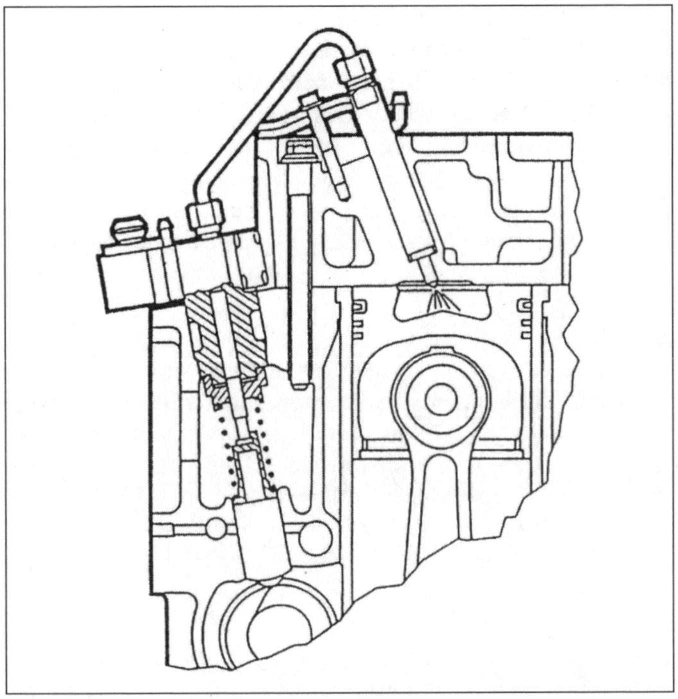

Figure 4.9 Bosch unit pump diesel fuel injection system. (Reprinted from Ref. 3, courtesy of Robert Bosch GmbH.)

maintain short injection lines between the pump and injector. Each unit pump is installed on the engine in proximity to the cylinder it serves and is driven by the camshaft of the engine.

High-pressure pumping elements consisting of plunger and barrel combinations are made of high-strength tool steel, and extremely close tolerances are kept between the sliding/rotating parts. Such high-precision machining is required throughout the mechanical components of the injection system to maintain accurate metering and injection timing within a 1-degree crankangle while maintaining extremely high pressures. The cost of such fuel systems is rather high and is difficult to justify in small passenger car size engines. A solution to this problem is the distributor pump, where one central pumping element is used to produce the high injection pressure. This high-pressure fuel is then introduced into a commutator head or distributor assembly that diverts it to the proper injector and cylinder according to the engine firing order. Reducing the number of pumping elements for a multicylinder diesel engine application to only one reduces the cost of the expensive high-precision machined parts of the pumping element and, hence, makes its cost more reasonable for the small car market.

4.4.1 The P-L-N In-Line Pump System

The in-line pump injection system is one of the oldest and proven systems for diesel engines. Reference is made to Figure 4.4 for the description of its components. To start with, the system is divided into low-pressure and high-pressure subgroups. The *low-pressure side* of the system includes the following components:

(a)

(b)

Figure 4.10 Two large displacement engines that use a unit pump injection system: (a) DDC/MTU Series 2000 and (b) MTU/DDC Series 4000. (Courtesy of MTU Detroit Diesel.)

- Fuel tank
- Fuel supply pump (lift or priming pump)
- Fuel filter
- Fuel deaerator.

The *fuel tank* is a reservoir that holds the fuel supply and helps maintain its temperature at a level below its flash point. The fuel tank is also where heated overflow from the fuel gallery as well as the leak off fuel from each injector is returned. The expanded surface area of the tank, along with total system design, generally transfers heat away from the heated/returned fuel and keeps its temperature at about 5°C above ambient. The *fuel supply pump,* often referred to as the *lift pump,* is responsible for drawing fuel from the tank and delivering it to the high-pressure pump. Although this may be its primary function, it must also be capable of flowing a large quantity of fuel to maintain constant temperature along its path through the gallery of the high-pressure pump. This feature is quite important since the high-pressure pump is a volumetric metering device, and variations in fuel density from one cylinder to another (as a function of its temperature) would affect the mass of the metered fuel for each cylinder. Fuel system designers strive to avoid uneven distribution of fuel from one cylinder to another, because it is good engineering practice to have the cylinders of an engine contribute equally to the total power output of the engine. The lift pump delivers fuel to the gallery of the high-pressure pump at about 15 psi (1 bar). It is driven by the camshaft of the high-pressure pump through an eccentric that is machined on that camshaft. The top of the lift pump is a hand primer that is screwed into its housing. By unscrewing this top, it becomes a hand-operated pump to prime the fuel system and facilitate start-ups after repair or when other maintenance work is performed on the injection system.

Trouble-free operation of a diesel injection system is possible only with filtered fuel. *Fuel filters* are designed to retain very fine particles and prevent them from getting into the very tight tolerances of the fuel injection system. Thus they help reduce damage and premature wear from contaminants. Filter elements are to be replaced at regular intervals specified by the fuel system or engine manufacturer. The interval for fuel filter changes is usually 10,000 to 15,000 miles, or about 400 hours of service. Filters can be of the box-type or replacement element design. Box-type filters can be completely replaced as needed and do not require cleaning. Replacement element-type filters have to be thoroughly cleaned when elements are replaced, and care must be taken to avoid any dirt residue that could migrate to the intricate parts of the fuel injection system. Figure 4.11 illustrates both the box-type (a) and replacement element-type filters (b).

Fuel filters use different materials for filtering diesel fuel. The most common material is pleated paper, yet packed cotton thread, wood chips, a mixture of packed cotton thread and wood fibers, fiberglass, and wound cotton or synthetic yarn have all been used to make fuel filters. The degree of filtration required depends on the specific application. In general, when two filters are used in series, the primary filter retains particles down to about 30 µm, while the secondary filter is capable of retaining particles greater than 8 to 12 µm.

In addition to keeping solid particles out of the fuel supply and injection equipment, water in fuel must also be avoided. Water freezes in cold temperature conditions, and if ice gets into the small fuel injection system passages it will risk blocking those passages, thus cutting off the fuel supply to the rest of the fuel injection system. Therefore, fuel filters are often coupled with water separators such as shown in Figure 4.12.

Fuel overflow and leak-off fuel returning to the tank carry along air and fuel vapor. The presence of gaseous substances in the fuel can cause difficulties in starting as well as normal engine operation in high-temperature environments. Therefore, bleeder valves and *deaerators* are used to rid the fuel supply from vapors and air to ensure trouble-free engine operation. The high-pressure side of the in-line pump system includes the following components:

- In-line, high-pressure injection pump
- High pressure lines
- Nozzle holder (injector).

The main function of the *high-pressure fuel injection* system is to meter fuel according to the power requirement

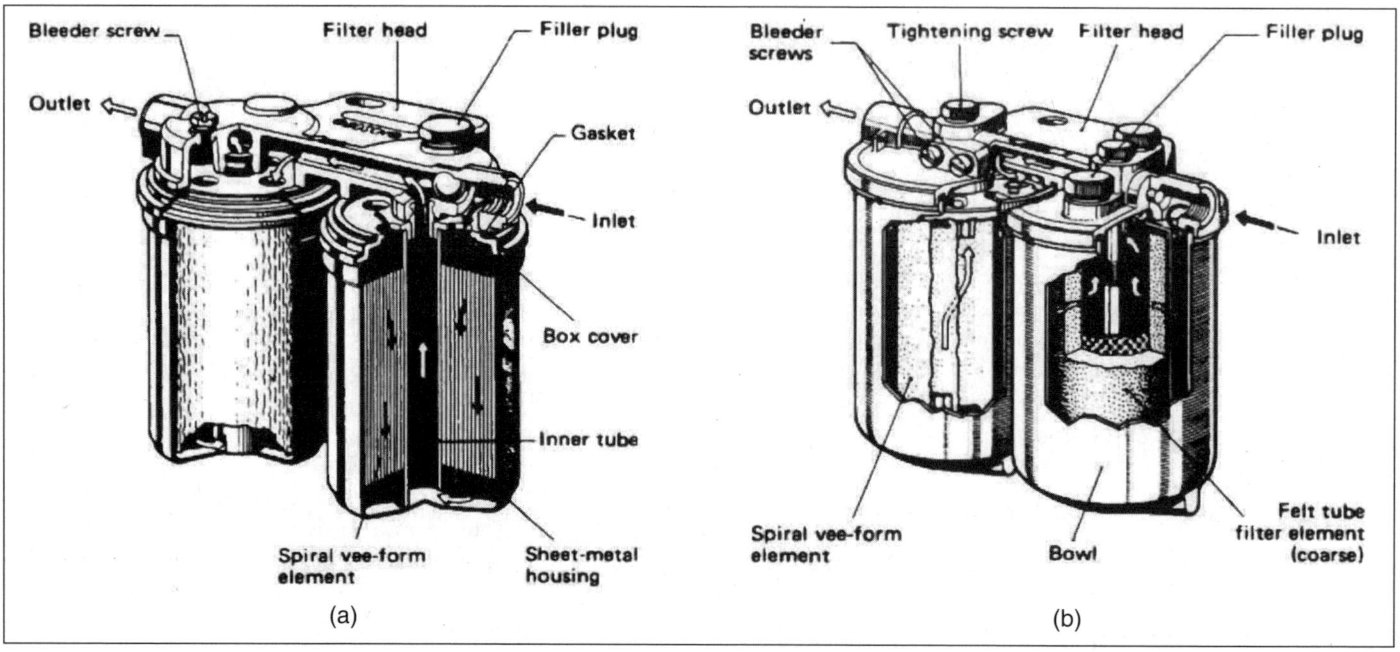

Figure 4.11 *Two types of fuel filters: (a) box-type filter and (b) replacement element-type filter. (Reprinted from Ref. 4, page 39, Figures 61 and 62, courtesy of Robert Bosch GmbH.)*

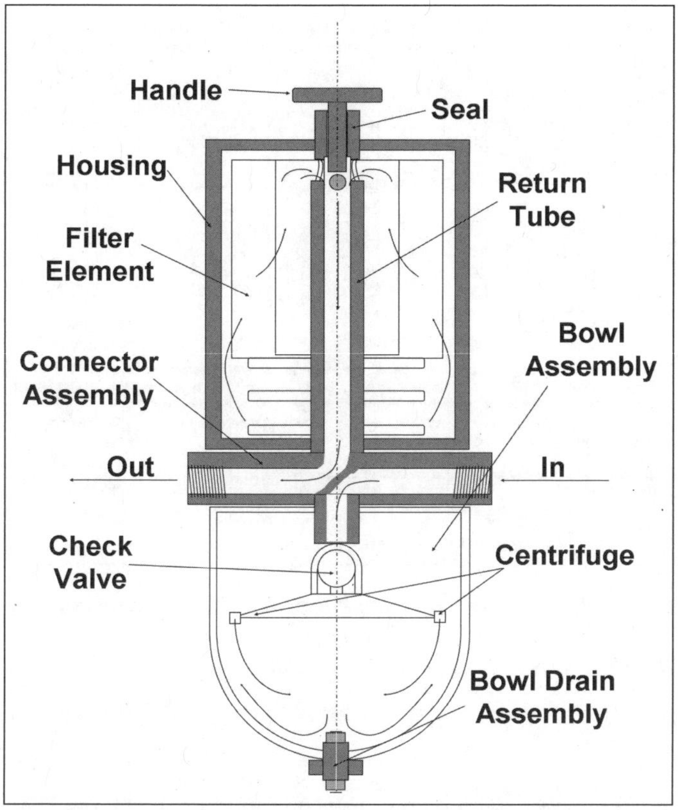

Figure 4.12 Fuel filter equipped with water separator.

and deliver that fuel at the appropriate time into the cylinder. The high-pressure pump is an assembly that houses several subassemblies important to the overall function of the system. The primary purpose of the high-pressure pump is to produce the high fuel pressure that will ultimately help the atomization process when the fuel is sprayed from the injector nozzle into a cylinder. The high-pressure pump performs this function through the pumping elements consisting of a number of plunger/barrel assemblies corresponding to the number of cylinders in a given engine. Fuel delivered by the supply pump flows through a fuel gallery along the length of the high-pressure pump and is introduced into the plunger/barrel assembly via one or two fill ports (see Figure 4.13). The plungers are given a vertical motion by means of the pump camshaft through the cam profile, with the return vertical motion made possible by return springs (see Figure 4.14).

Fuel from the gallery enters into the cavity on top of the plunger through the two spill ports as shown in Figure 4.13. The upward motion of the plunger resulting from the camshaft rotation, cam profile, and the cam follower (see Figure 4.14) continues to push the fuel atop the plunger upward toward a delivery valve. As the plunger continues its upward movement, it eventually blocks both fill ports and thus traps the fuel and forces it to unseat the delivery valve, flow into the high-pressure pipe, and proceed on to the injector. Each plunger is machined with a slot and a helix near its top and further upward movement of the plunger causes a direct fluid communication between the high-pressure fuel and the low-pressure supply side through the slot, the bottom of the helix, and via the fill port that switches its role and becomes a spill port. This action diffuses the high pressure and helps end the injection. The filling, delivery, and spilling of the fuel injection cycle is described graphically in Figure 4.13 and more comprehensively in Figure 4.15.

Not only do the plungers move in a vertical direction, but they also rotate around their vertical axis to change their effective stroke. They achieve this rotating motion by means of a control rod (control rack), which is shown in Figure 4.16[11]. The control rack is attached to the driver's foot via a series of linkages. Thus, when the driver demands more power, these linkages transmit this need to the pump metering system via a distinct control rack motion. The control rack is attached to the pump plunger via gear segments as shown in Figure 4.14[9]. Therefore, the control rack linear movement is transmitted to the plunger in the form of a circumferential movement that changes the relative position of the plunger helix to the pump spill port. This action is shown graphically in Figure 4.16[11].

The clearance between the plunger and barrel in each pumping element is on the order of a few 1/1000 mm in order to seal

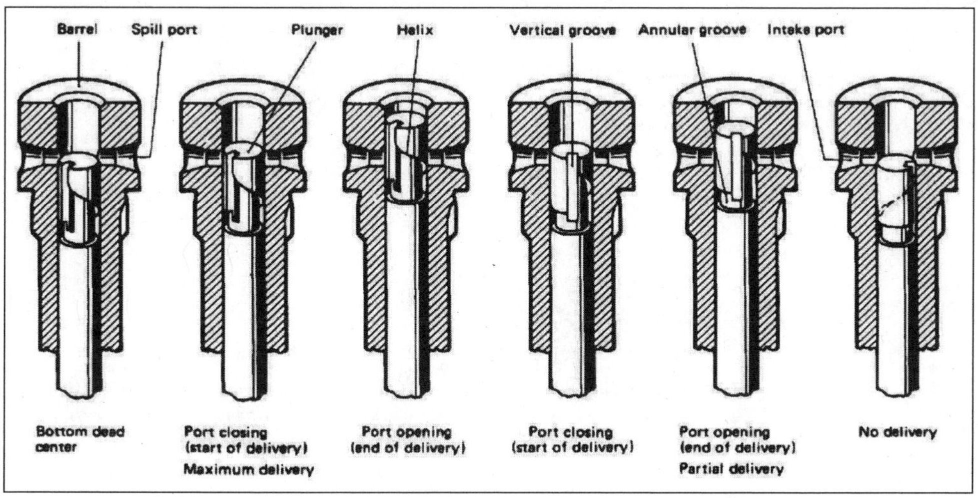

Figure 4.13 Plunger/barrel assembly showing plunger position relative to fill ports for various delivery conditions. (Reprinted from Ref. 1, Fig. 8, courtesy of Robert Bosch GmbH.)

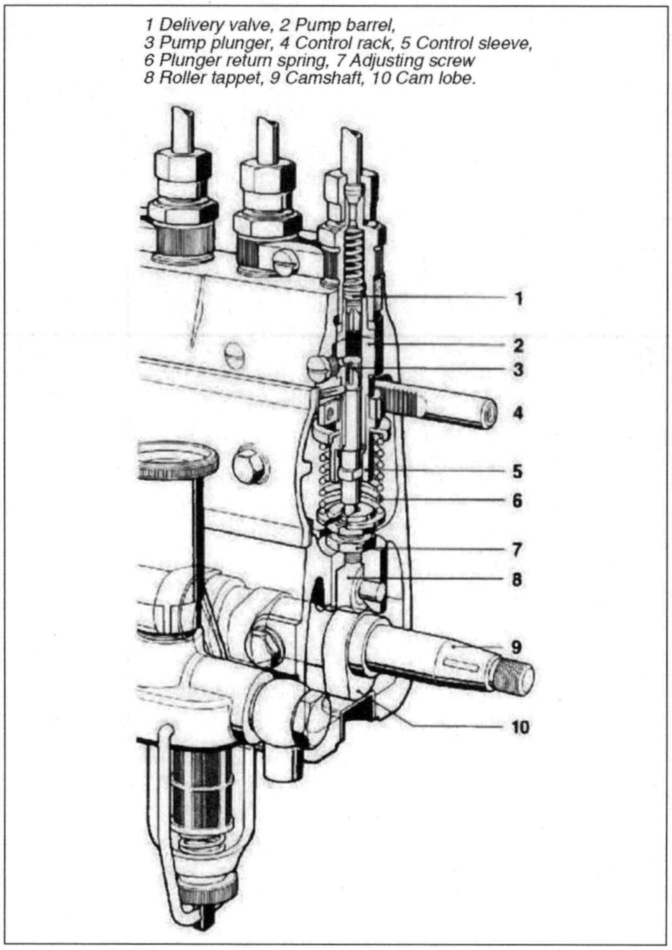

Figure 4.14 Cross-sectional view of an in-line pump type PE...A. (Reprinted from Ref. 5, page 61, Fig. 11, courtesy of Robert Bosch GmbH.)

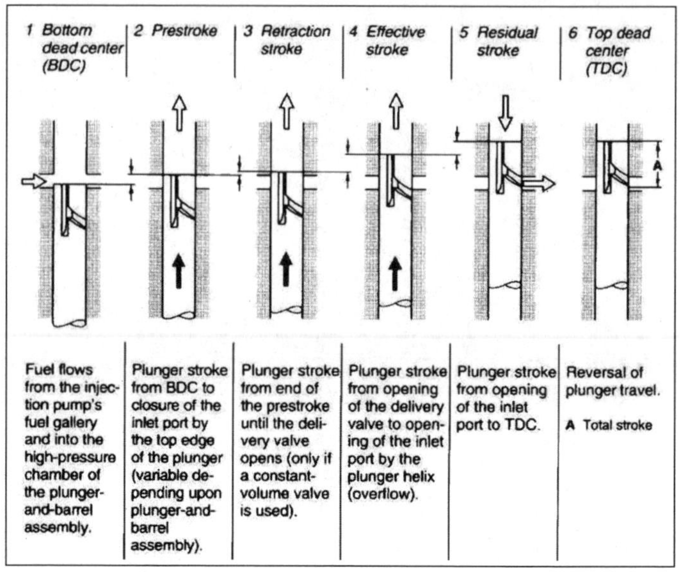

Figure 4.15 Plunger-stroke phases in a complete injection cycle. (Reprinted from Ref. 3, Fig. 13, courtesy of Robert Bosch GmbH.)

even under very high pressures and low pump (engine) speeds. Therefore, each pumping element has a matched plunger and barrel set and should be changed as an assembly.

The primary purpose for the *high-pressure pipes* or *lines* is to connect the pumping elements to their respective injectors. They are equal in length to avoid cylinder-to-cylinder timing and fuel delivery problems. They are made of steel tubing and designed to have gentle bends. Tubes have two popular outer diameter sizes, either 6 or 8 mm, and their internal diameter is either 2 or 2.5 mm depending on the fuel pressure. The sealing conical edges of the high-pressure lines are brazed or swaged and carefully cleaned to avoid introducing contaminants into the system.

The two basic types of *injector nozzles* are throttling pintle and hole-type nozzles. Injector nozzles play an important role in mixture preparation, injection timing, injection rate and shape, and sealing at the end of injection. In essence, they are the last link between the injection system and the engine. Therefore, their design details are closely matched to the combustion system to which they cater.

As described in Chapters 1 and 2, diesel engines can be either indirect or direct injected. While it is true that most of the modern diesels can be classified as direct-injected engines, a

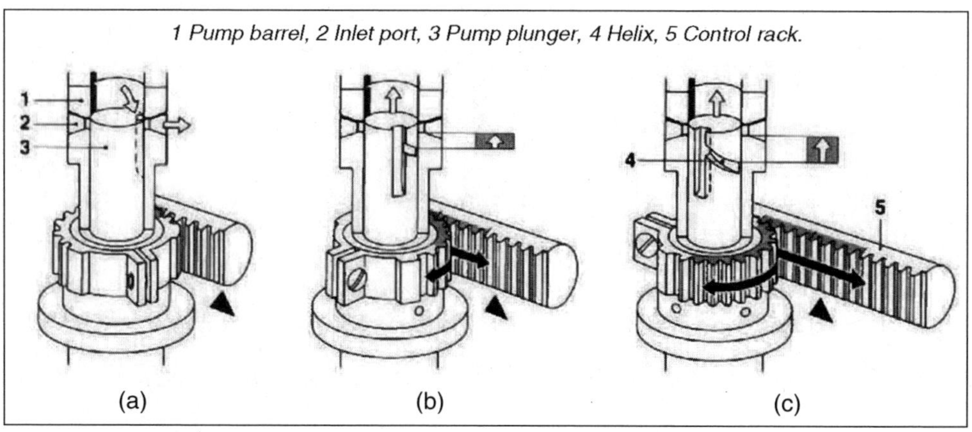

Figure 4.16 Fuel delivery control: (a) zero delivery, (b) partial delivery, and (c) maximum delivery. (Reprinted from Ref. 4, page 10, Fig. 10, courtesy of Robert Bosch GmbH.)

large number of indirect-injected engines still exist. The *throttling pintle nozzle* type is the popular nozzle used in indirect-injected engines. A detailed diagram of this type of nozzle is given in Figure 4.17.

One important characteristic of the throttling pintle nozzle is the extension of its needle and its effect on the rate of injection control. At small strokes, that is, small needle lifts, that extension remains in the middle of the spray, thus reducing the area available to the fuel flow. In fact, that flow area is in the shape of a ring with the needle extension in its middle and bordered by the circumference of the hole in the nozzle itself. Yet, when longer needle lifts are used, the needle extension lifts entirely out of the hole and uncovers the total area of the nozzle hole, making it available to the fuel flow and spray formation. This operation leads to a slow injection start followed by a sharp rate of increase once the needle extension clears the nozzle hole.

Specific designs may control injection characteristics through nozzle hole size, needle lift, system pressure, needle return spring rate and initial setting, and needle extension dimension. In general, throttling pintle nozzles are associated with injection systems having low to moderate injection pressures that are typical of indirect-injected diesel combustion systems. Close tolerances between the needle extension and the nozzle hole, coupled with exposure of the injector tip to high combustion temperatures, lead to coking of the throttling gap. To overcome this problem, a variation on the pintle nozzle design, called the flat-pintle injection nozzle is used. This design directs the fuel flow to a specific area of the nozzle and creates a self-cleaning action, thus keeping the nozzle free from coking and maintaining good flow pattern.

The second type of injector nozzle is the hole-type, which is predominantly used in direct-injected engines. Figure 4.18 is a schematic showing the details of a hole-type injection nozzle. Hole-type injectors are indexed to the combustion chamber to accurately position the fuel sprays relative to the combustion chamber. Mis-indexing of the injector nozzle, or errors in machining the nozzle holes and their appropriate angle of inclination relative to the combustion bowl, can lead to inefficient combustion. Hole geometry (diameter, length), angle of inclination, and the spray angle (included angle of the spray cone, as shown in Figure 4.18) are extremely important for good mixture preparation and combustion efficiency.

The needle in the hole-type injector usually has a 6-mm diameter that reduces down to 4 mm. Fuel supplied to the injector arrives to the pressure chamber (see Figure 4.18) and applies its pressure to the projected area between the two different diameters of the needle. This action results in an upward force that compresses a retaining spring, shown in Figure 4.19,

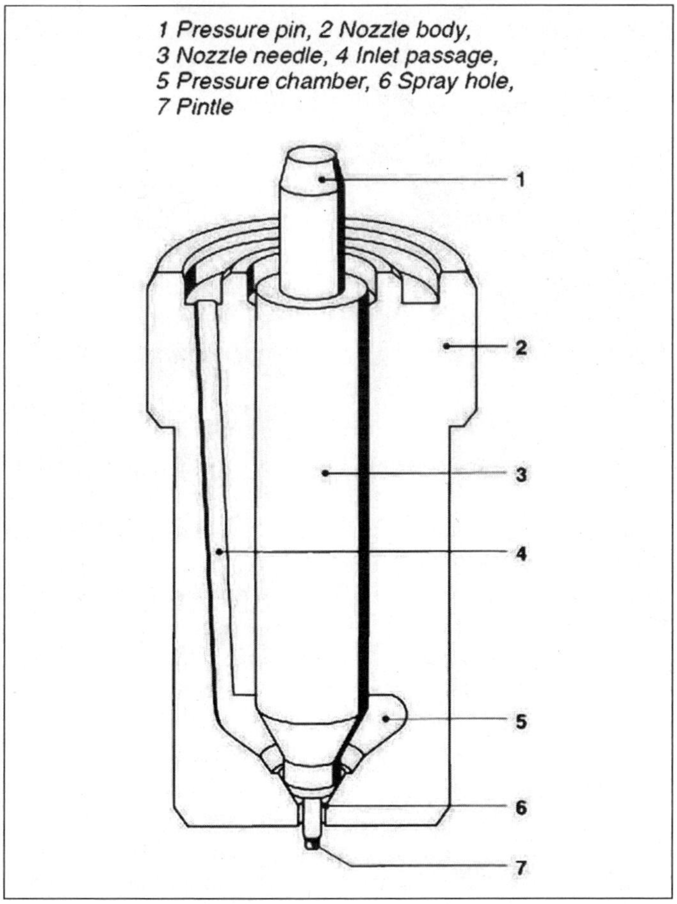

Figure 4.17 Details of a throttling pintle nozzle. (Reprinted from Ref. 5, page 193, Fig. 1, courtesy of Robert Bosch GmbH.)

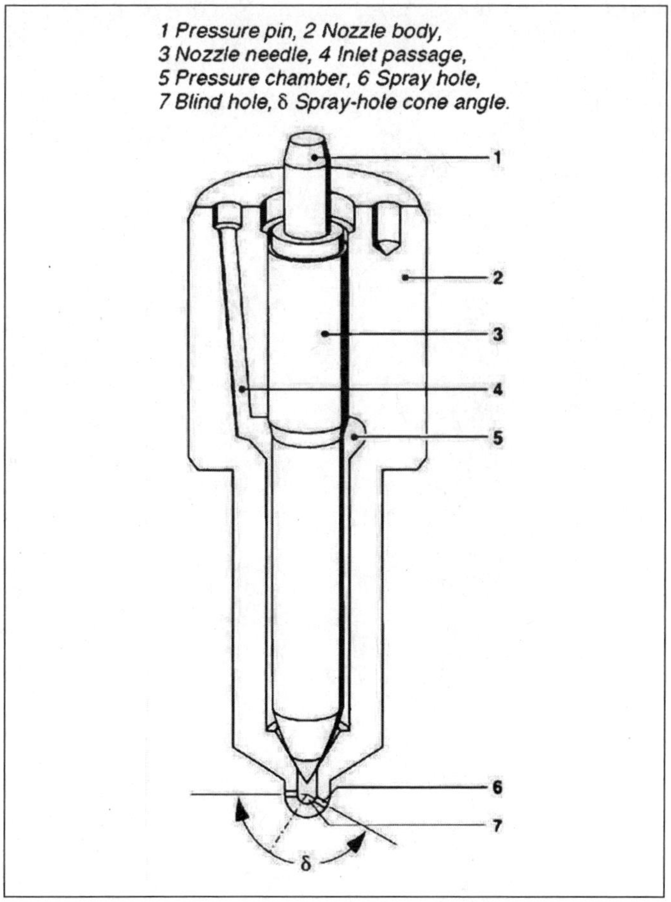

Figure 4.18 Hole-type injector nozzle. (Reprinted from Ref. 5, page 195, Fig. 3, courtesy of Robert Bosch GmbH.)

and causes the needle tip to lift off of its seat, thus exposing the nozzle holes to high-pressure fuel. Fuel pressure measured at the point of initial needle lift is referred to as *nozzle opening pressure* (NOP). Hole-type injectors have a NOP generally ranging from 200 to 250 bars. Flow of the high-pressure fuel through those tiny holes (0.2 mm in diameter) leads to its atomization. Proper control of fuel pressure, nozzle opening pressure, hole size, fuel temperature, spray angle, and entrained air motion are some of the parameters used to define fuel penetration into the combustion chamber. Thus, not only is fuel atomization necessary for good combustion quality, but equally important is its penetration and its reach for the available oxygen.

Because of the tight tolerance between the needle and its nozzle, it requires lubrication. This function is provided by the fuel, which invariably finds its way through these small clearances between the needle and the nozzle body. Fuel also serves as the heat transfer medium through which friction heat generated by the relative motion of the needle and its nozzle can be dissipated. This tiny amount of fuel that is allowed to leak through these clearances is eventually returned to the fuel tank through a *leak-off connection and line*, as shown in Figure 4.19.

Figure 4.20 shows various types of nozzle tip shapes for both the throttling pintle and hole-type nozzles. Nozzle 1 in Figure 4.20 shows the tip of a standard throttling pintle nozzle; views 2a and 2b give two views of the flat-cut throttling pintle nozzle with its self-cleaning feature. Nozzles 3, 4, and 5 in Figure 4.20 represent three different designs for the hole-type nozzle: nozzle 3, conical sac tip; nozzle 4, cylindrical sac tip; and nozzle 5, seat-hole or valve-covering-orifice (VCO) tip.

The term *sac* refers to the volume between the needle tip and the inside of the nozzle tip. It is designed to provide damping of the needle motion as it moves as a projectile toward its seat once the high fuel pressure is removed. The mass of the needle and its dependence on the return spring rate represent a classical mass-spring dynamic problem that greatly influences combustion in diesel engines. Sac volume retains fuel that could evaporate into the combustion chamber, especially toward the end of the expansion stroke. This evaporated fuel proceeds into the exhaust system through the exhaust valve opening, resulting in higher unburned hydrocarbon emissions. The conical sac as well as the VCO designs are two ways to reduce the potential hydrocarbon emissions problem through reducing or even eliminating the sac volume altogether. The VCO injector tip design gained popularity in the early to mid-1980s, but almost disappeared when it failed to show significant improvement in hydrocarbon emission control over the conical sac tip. In addition, the continuous hammering of the needle on its seat that was weakened by injection holes drilled in the sealing line led

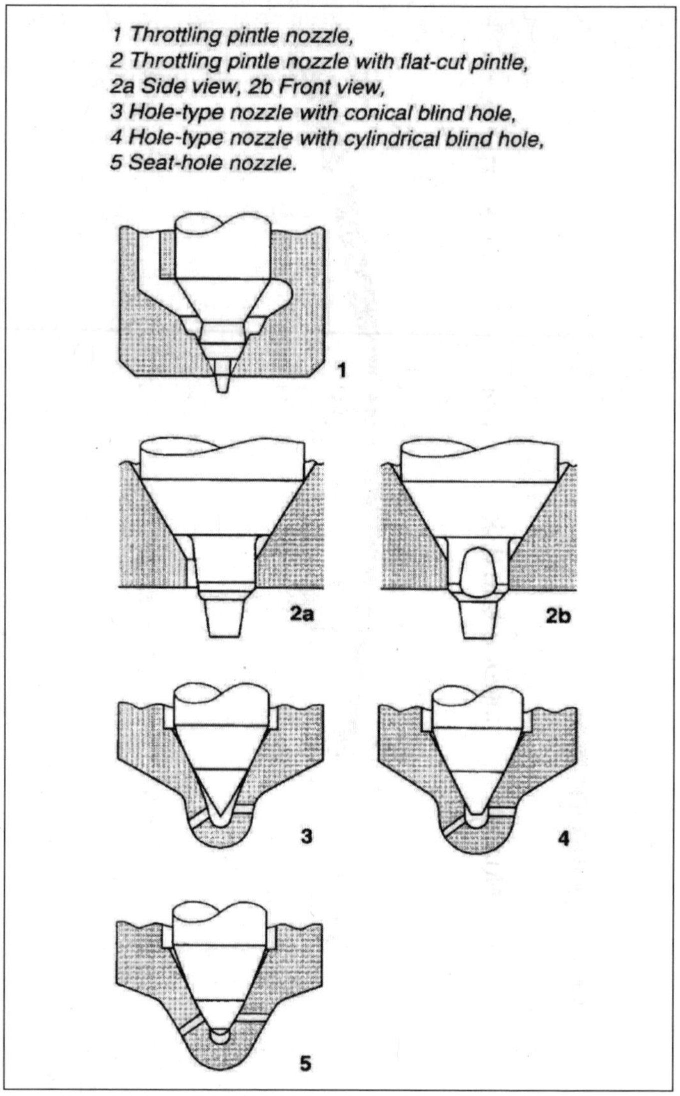

Figure 4.20 *Details of various types of nozzle tips. (Reprinted from Ref. 5, page 194, Fig. 2, courtesy of Robert Bosch GmbH.)*

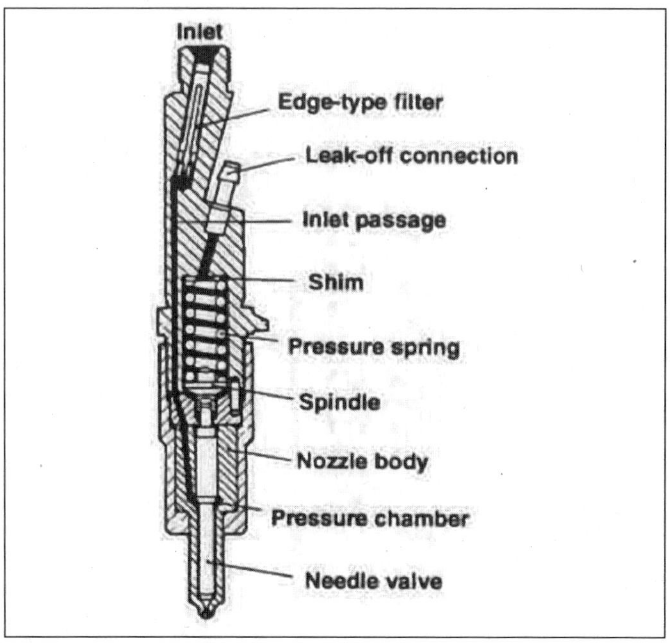

Figure 4.19 *Hole-type injector holder. (Reprinted from Ref. 1, Fig. 42, courtesy of Robert Bosch GmbH.)*

to the nozzle tip breaking off. However, new designs have been made and have been successfully commercialized.

There is much more to be said about fuel injectors, their materials, designs, and other unique features than has been described here. There are also other developments aimed at injection rate shaping such as the dual spring injector shown schematically in Figure 4.21. The collapse of one spring permits the slow start of injection, and when followed by the collapse of the second spring, the injection rate increases thus controlling combustion quality as well as combustion-generated noise characteristics. As we continue to address and consider the features of the injection system, we will have additional opportunities to revisit the subject of fuel injectors. However, it is worthwhile considering other features of the high-pressure injection pump at this time.

Other injection pump design features relate to its control functions and include the following:

- Pump governor
- Timing device
- Fuel delivery valve
- Boost control device (BCD)
- Torque control capsule.

Engine applications may require maintaining a certain speed regardless of load. One example of this is a farm tractor towing an implement for seeding a field. This tractor must maintain a fixed speed if the distance between two consecutive plants is expected to be the same throughout that field. This way farmers can maximize their yield from a given acreage. That controlled distance between seeds is to be maintained even if the tractor is moving up or down a slope. It should also be maintained if the tractor suddenly encounters an obstacle along the way. Injection *pump governors* are designed to control fueling in response to road load conditions so as to perform the required task, whether it is constant engine speed within a speed range or constant torque under varying speed conditions.

To hold the engine speed constant, the control rod (rack) in an in-line pump system must be moved to control the injected fuel quantity. In on-highway applications, this is accomplished by the driver depressing or releasing the accelerator pedal between idle and maximum speed. However, idle speed as well as maximum speed are both controlled via an automatic governor. The responsibility of that idle governor is to prevent stalling while the engine is idling, while the maximum speed governor would prevent the engine from overspeeding or exceeding the engine design limit. These governors have historically been either mechanical or pneumatic. An example of a mechanical governor appeared earlier in Figure 4.7. The mechanical governor is generally equipped with spring-loaded flyweights that are speed sensitive. As speed increases above the set threshold, the flyweights extend in response to their centrifugal force and against the spring force. This movement is used to control the rack movement, thus increasing or decreasing speed accordingly. Figure 4.22 is an illustration of a minimum/maximum speed (min/max) mechanical governor, and Figure 4.23 is a schematic diagram explaining how it functions.

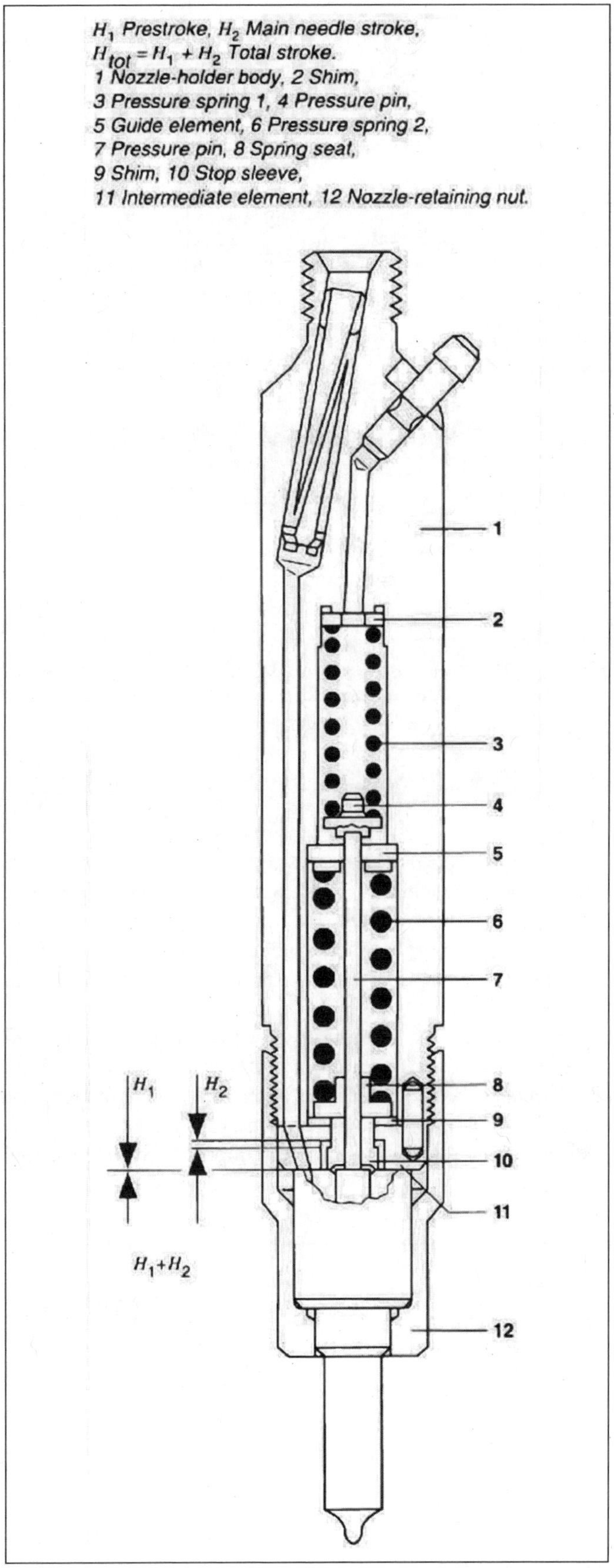

Figure 4.21 *Hole-type injector equipped with dual springs for staged injection. (Reprinted from Ref. 5, page 197, Fig. 6, courtesy of Robert Bosch GmbH.)*

1 Control rack, 2 Link fork, 3 Play-compensating spring, 4 Adjusting nut, 5 Governor springs, 6 Flyweight, 7 Bell crank, 8 Sliding bolt, 9 Slider, 10 Guide pin, 11 Control lever, 12 Fulcrum lever, 13 Guide block, 14 Linkage lever.

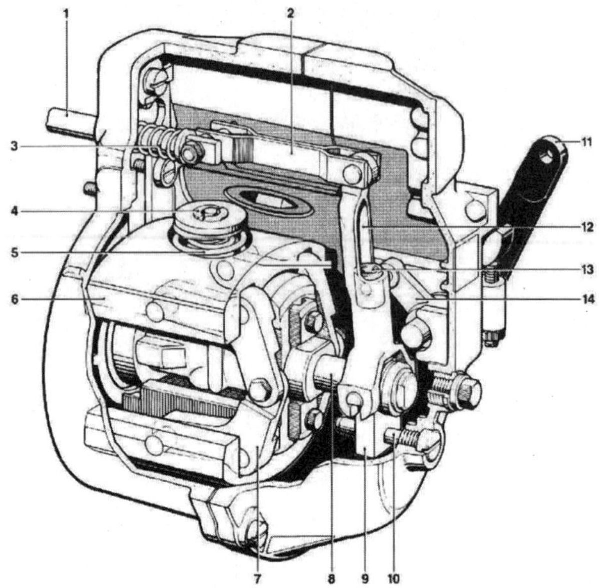

Figure 4.22 *Minimum/maximum speed mechanically controlled governor. (Reprinted from Ref. 5, page 93, Fig. 2, courtesy of Robert Bosch GmbH.)*

In addition to min-max speed governors for vehicle applications, there are constant-speed governors that operate machinery, such as power-generating sets strictly at steady speeds, and also variable-speed governors. Variable-speed governors are automatically controlled over the entire speed range. (Please see page 61, left col., 3rd paragraph). The min-max governor characteristic curve is given in Figure 4.24.

Other types of governors deserve to be addressed for completeness. However, the purpose of this chapter is not to detail all aspects of the fuel injection system, but to give the reader a generalized review of the types, function, and capabilities of the fuel injection equipment and, more importantly, how the fuel injection characteristics help achieve best engine performance, lowest undesirable emissions, and best fuel economy. For a more thorough review of the various types of governors and their operation, the reader is directed to Ref. [5] by R. Bosch titled *Diesel Fuel Injection.*

During the injection process, the needle is opened by a pressure wave started by the plunger in the pumping element and propagated through the high-pressure line at the speed of sound. The injection event in each combustion cycle occurs at the same time in terms of crankangle degrees BTDC. Yet, as engine speed increases, injection lag increases (occupies a greater number of crankangle degrees). To counter this phenomenon, a *timing device* is used. Because most favorable combustion happens at a fixed position of the piston relative to its TDC position, a timing device is often used to reposition the injection pump camshaft relative to the plunger follower. This new camshaft position is such that the fuel pressure wave is advanced by a proportional number of crankangle degrees to ensure that the injection takes place at the same time (crankangle) BTDC.

It is important to note that the creation of high injection pressure is via that pressure pulse or wave that is formed when the plunger imparts its force onto the column of fuel accumulated on top of the plunger. That pressure wave and its intensity depend on the high-pressure pump cam profile, which controls the plunger linear speed. It also depends on the amount of fuel accumulated on top of the plunger, which is controlled by the volume of the cavity receiving fuel from the fuel gallery. Therefore, one often hears fuel injection specialists speak of a *fast cam,* which means that the cam profile is steep enough to create a large pulse (or wave) that would propagate through the high-pressure pipe and on to the nozzle, early in the cycle. From this discussion, it follows that the desire to have high injection pressure can be accomplished through a faster cam, a larger plunger, a smaller injector hole area, or combinations of any or all of these.

The end of injection begins by spilling the last portion of the fuel that had been trapped on top of the plunger, as previously explained. By diverting the high-pressure pulse away from the injector, the needle is catapulted toward its seat by means of the return spring. It is important to observe that the

Stop (shutoff) position.
1 Shutoff stop,
2 Control lever,
3 Full-load stop,
4 Guide block,
5 Fulcrum lever,
6 Link fork,
7 Control rack,
8 Pump plunger,
9 Control-rod stop (spring-loaded),
10 Slider,
11 Guide pin,
12 Sliding bolt,
13 Bell crank,
14 Governor hub,
15 Adjusting nut,
16 Governor spring,
17 Flyweight,
18 Camshaft..

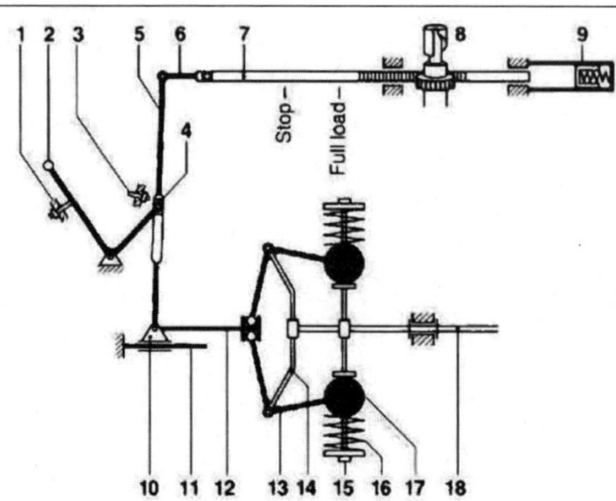

Figure 4.23 *Schematic describing the operation of a min-max speed governor. (Reprinted from Ref. 5, page 93, Fig. 3, courtesy of Robert Bosch GmbH.)*

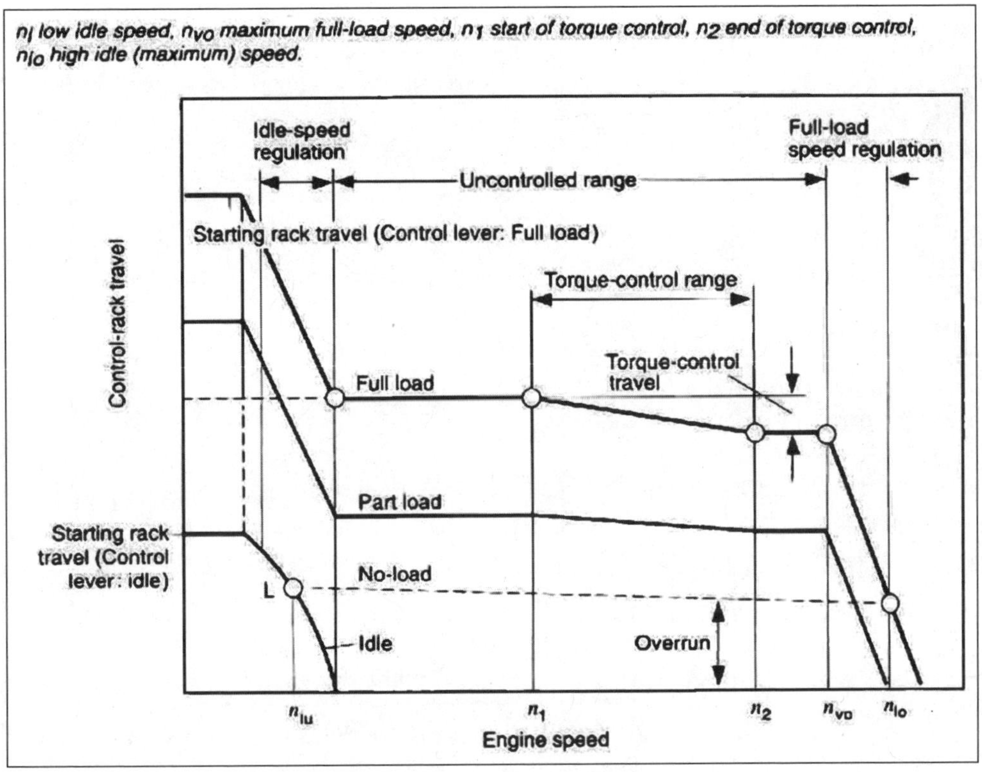

Figure 4.24 Minimum/maximum governor characteristic curves. (Reprinted from Ref. 5, page 95, Fig. 7, courtesy of Robert Bosch GmbH.)

residual pressure in the high-pressure line must be kept below the threshold that may cause a reflected pulse to unseat the needle again, thus creating the potential for after-injection (see Figure 4.25). This phenomenon leads to injecting uncontrolled fuel during the latter portion of the expansion stroke, thus causing higher hydrocarbon, carbon monoxide, smoke, and particulate matter emissions. In addition to after-injection, the nature of a pressure wave traveling back and forth within the high-pressure pipe may lead to wave superpositioning, which may in turn lead to a secondary injection as shown in Figure 4.25. It is worth noting that the traces given in Figure 4.25 are referred to as *needle lift traces.* They are obtained by a proximity probe inserted within the injector spring cage. The transducer wires are nestled through the leak-off connection where the signal can be captured.

Delivery valves are used to help avoid both after-injections as well as secondary injections. As soon as the plunger helix comes into registry with the spill port and starts to dissipate the high pressure, the delivery valve reseats and adds to the volume of the high-pressure tubing the equivalent of the volume of the retraction piston. The increase in the pipe volume helps maintain its residual pressure below the dangerous threshold and avoid unwanted injections. The complete valve assembly appears in Figure 4.26. Other details of the constant-volume delivery valve are shown in Figure 4.27.

As injection pressures increased to meet the rigorous demands of emission controls, injection systems adopted the constant-pressure delivery valve design (see Figure 4.28). The main characteristic of this valve is that it maintains the residual pipe pressure at constant pressure at any speed and load condition. This constant pressure is always kept below the point that could cause irregular injection quality.

In diesel engines, and especially turbocharged diesels, low-speed, high-load operation is usually characterized by heavy smoke emission (see Figure 4.29). The torque curve represented by line b-c-d in the schematic drawing in Figure 4.29 is referred to as the *natural lug curve.* Line a-b′-c-d is referred to as the *modified lug curve.*

The shaded area in Figure 4.29 between the two lug curves (natural and modified) represents the point in engine operation when it would produce heavy smoke. In other words, if the engine were to run with a positive-displacement pump, such

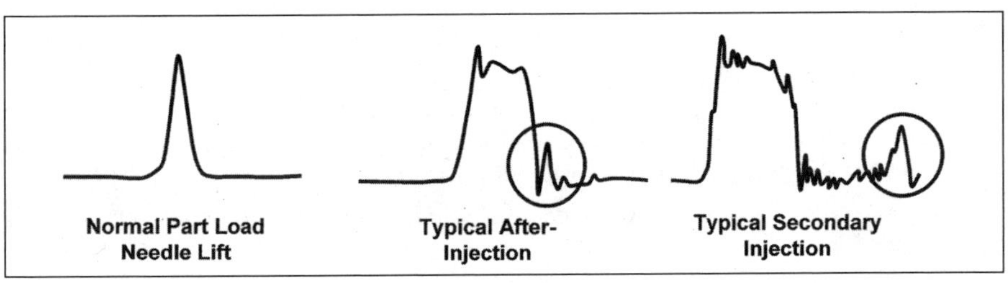

Figure 4.25 Needle lift traces illustrating after-injection and secondary injection.

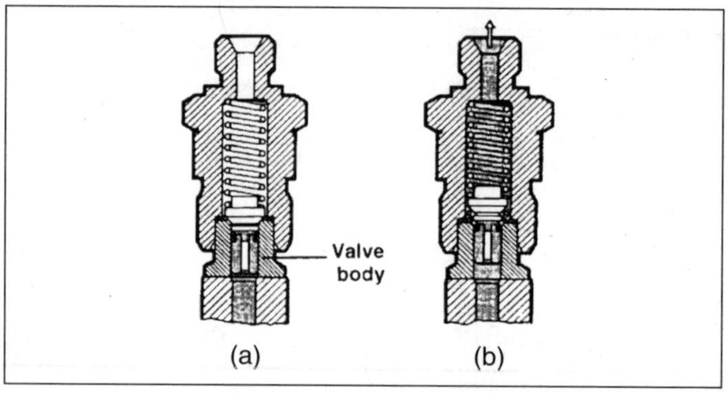

Figure 4.26 *Complete assembly of a constant-volume delivery valve: (a) closed and (b) during delivery. (Reprinted from Ref. 1, Fig. 12, courtesy of Robert Bosch GmbH.)*

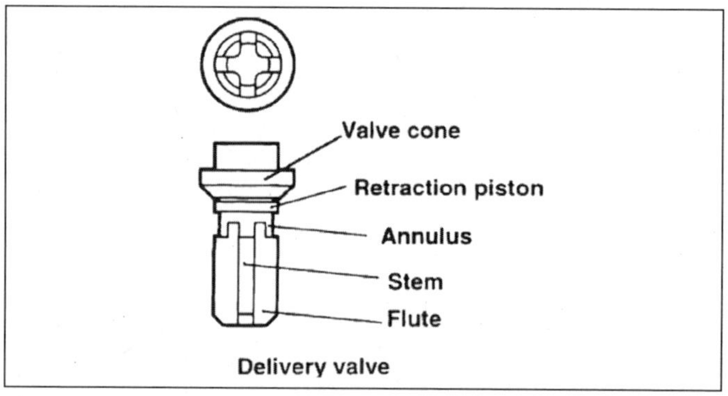

Figure 4.27 *Details of constant-volume delivery valve assembly. (Reprinted from Ref. 1, Fig. 12, courtesy of Robert Bosch GmbH.)*

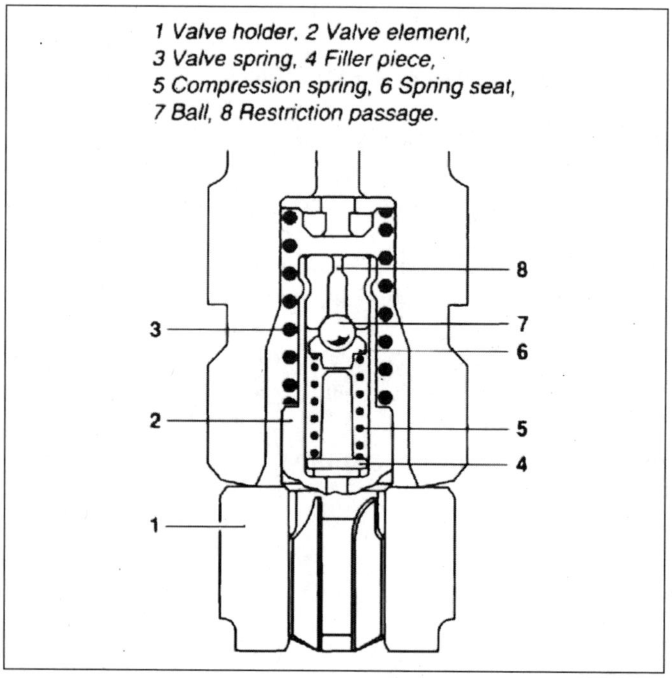

Figure 4.28 *Constant-pressure delivery valve. (Reprinted from Ref. 5, page 53, Fig. 13, courtesy of Robert Bosch GmbH.)*

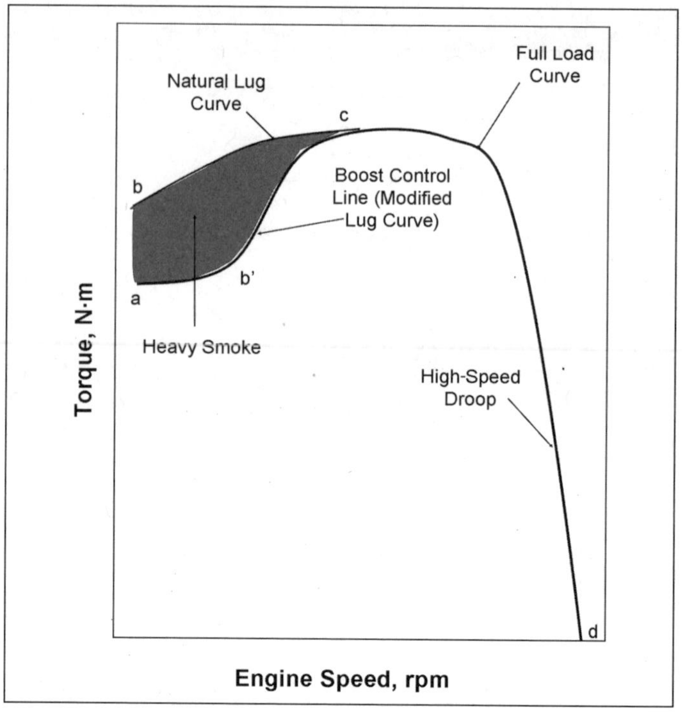

Figure 4.29 *Natural and modified torque curves for a diesel engine.*

as the in-line fuel injection pump, without trimming its fuel delivery at low engine speed and wide-open throttle conditions, heavy smoke would be emitted from its exhaust. The main reason for this phenomenon is the lack of adequate air charge at low speeds to prepare the fuel/air mixture for good combustion. For this reason, fuel system and combustion engineers have opted to reduce the pump fueling to better match the available air and avoid heavy smoke emission. To achieve this modified fuel rate, *boost control devices* (BCD) are used. These devices are also referred to as puff-limiters, aneroids, or manifold pressure compensators. They generally consist of a spring-loaded diaphragm enclosed in a housing. An opening in the BCD housing allows intake manifold pressure to be applied to the diaphragm on the side opposite from the spring (see Figure 4.30[(23)]). Increases in engine speed lead to higher airflow and higher manifold pressure, which when applied to the diaphragm would overcome the spring force. The resulting motion of the diaphragm causes further travel of the control rod (rack) and allows more fuel as speed and load increase.

The maximum amount of fuel that can be efficiently used in a diesel engine is limited by the amount of available air with which the fuel can effectively mix and combust without emitting excessive smoke. Many in the diesel engine design community refer to the maximum power produced by an engine without exceeding a certain smoke opacity as the *smoke-limited output* of that engine. Naturally, we would want to obtain the maximum power output from any engine, in other words maximize its specific power output. In some cases, such as in a naturally aspirated engine, charge air inducted per stroke is inversely proportional to engine speed. This phenomenon results

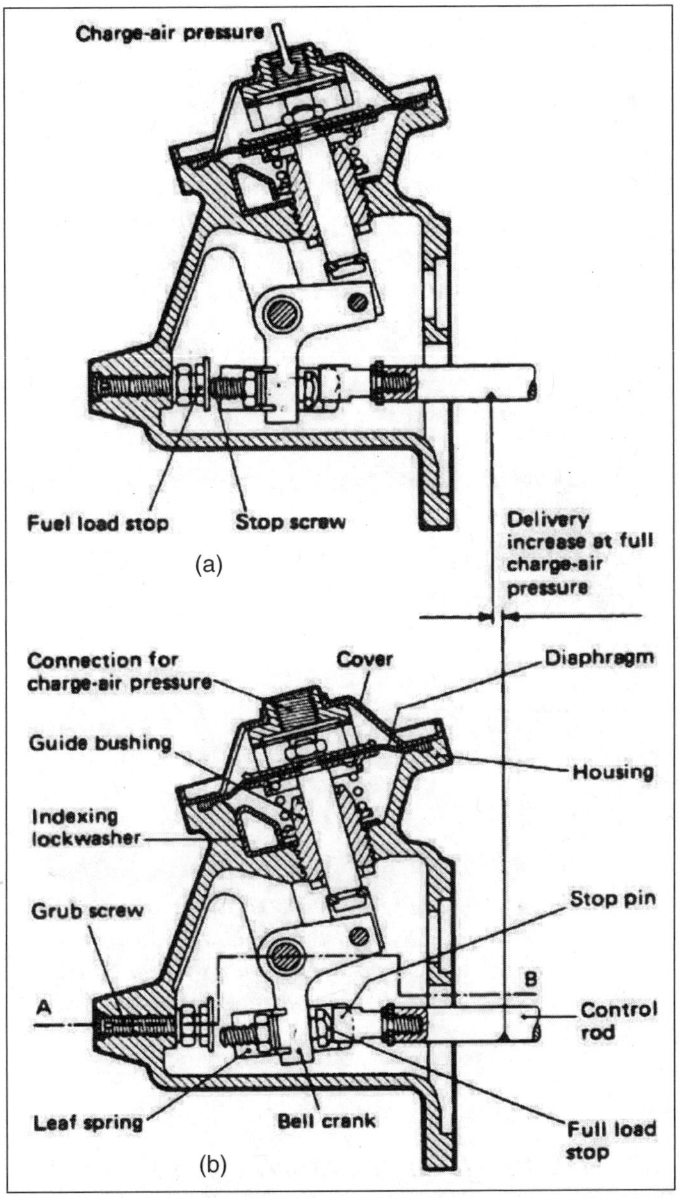

Figure 4.30 *Bosch boost control device: (a) full-load delivery without charge air pressure and (b) full-load delivery with full charge air pressure. (Reprinted from Ref. 1, Fig. 19, courtesy of Robert Bosch GmbH.)*

from decreasing volumetric efficiency with increasing engine speed. On the other hand, as engine speed increases, the fueling rate also increases due to a greater number of pump rotations, which leads to more pumping strokes per unit time. Therefore, it would be desirable to reduce maximum fuel as engine speed increases to avoid smoke formation. This function is performed by the *torque control capsule* that is part of the high-pressure pump governor assembly. Figure 4.31 describes the function of the torque control capsule. The process of lowering the pump fueling rate from line b to line c in Figure 4.31 is referred to as *positive torque control.*

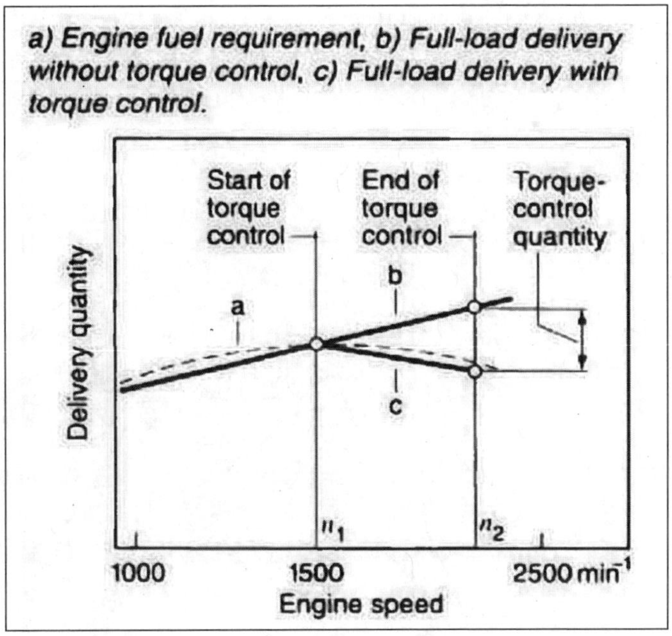

Figure 4.31 *Function of a torque control capsule. (Reprinted from Ref. 5, page 84, Fig. 11, courtesy of Robert Bosch GmbH.)*

Injection pump governors use a variety of torque control mechanisms and designs to achieve the desired torque shape. In turbocharged engines, charge air does not drastically reduce with engine speed. Therefore, one would think that reducing the fuel rate at high engine speeds may not be necessary to maintain a minimum A/F ratio for smoke-free operation. Yet, designers may be faced with other goals such as drivability issues. For instance, for on-highway applications it is desirable to provide truckers with a high torque ratio, that is, the ratio of peak to rated speed torque, also known as *torque backup.* This performance aspect (high-torque backup) allows the driver to climb long hills with fewer gear changes. To achieve higher torque backup, a cam profile is often used as an integral part of the governor to accurately meter fuel over the relevant engine speed range. An example of such design is the Bosch RQV...K variable-speed governor shown in Figure 4.32, where the full-load stop (item 1) is precision machined for accurate torque shaping.

Other control functions could be described regarding the performance of the in-line pump system. However, the treatment of the subject just given addresses the more common aspects of this type of injection system and displays the intricacies of the mechanical design required to perform those control functions. In-line pump systems, made by R. Bosch, Lucas (now Delphi), and others have served the diesel industry well even when equipped with mechanical controls as described earlier. Some variations in design were needed to meet the requirements of specific applications. Therefore, not all in-line pumps are created equal. Some may deliver fuel at low injection pressures that are required by prechamber (or IDI) diesel engine designs. Others, such as the R. Bosch P-type pump are capable of delivering up to 1200 bars at full load and rated speed conditions.

1 Full-load stop with rocker guide, 2 Control rack, 3 Adjusting nut, 4 Governor spring, 5 Flyweight, 6 Rocker, 7 Control lever, 8 Plate cam, 9 Guide block, 10 Fulcrum lever, 11 Slider, 12 Bell crank, 13 Sliding bolt, 14 Guide lever.

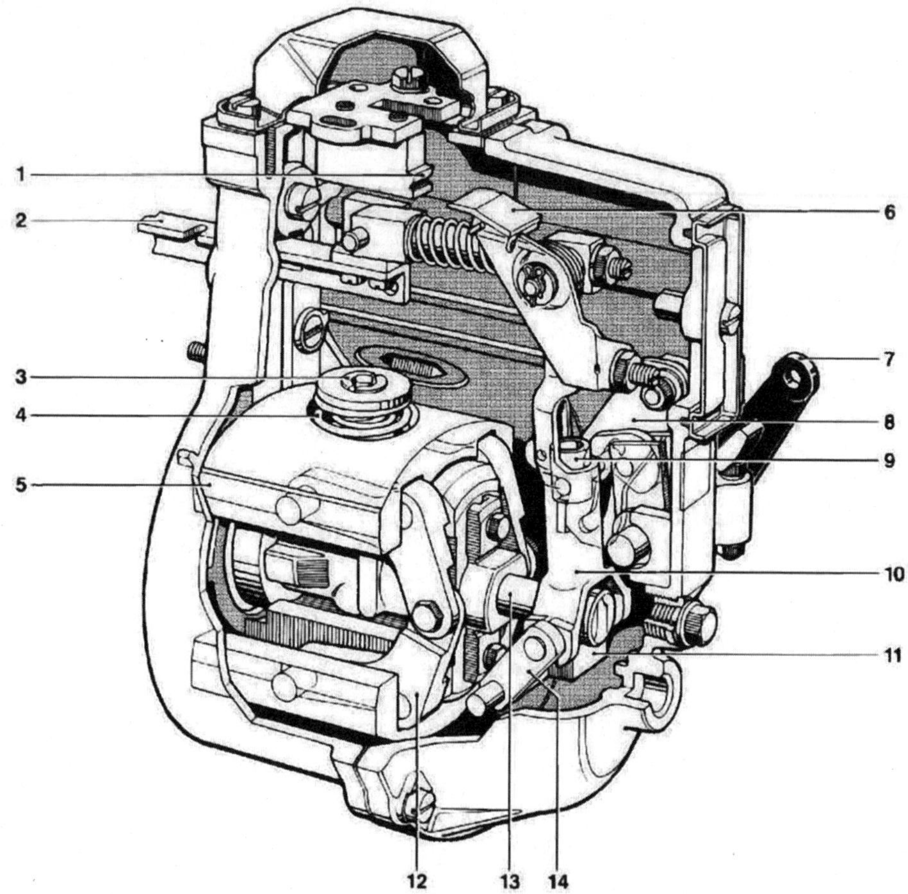

Figure 4.32 *Cutaway of a Bosch RQV...K variable-speed governor. (Reprinted from Ref. 5, page 103, Fig. 22, courtesy of Robert Bosch GmbH.)*

Another notable distinction in injection equipment is the pencil-type nozzle holder shown in Figure 4.33, a Roosa Master injector nozzle. The pencil-type injector was made famous by Roosa Master/Stanadyne and fits in very small holes (about 10 mm in diameter) through the cylinder head. Many designers preferred the pencil injector because it allowed them to install the nozzle tip closer to the center of the combustion bowl even in very difficult and tight passages through the cylinder head.

4.4.2 The P-L-N Distributor/Rotary Pump System

As in the case of the in-line pump injection system, the distributor pump performs the same functions. It allows a precisely metered amount of fuel to be injected in the combustion chamber at the right time for efficient combustion. It is also responsible for atomizing the fuel and assisting in the fuel/air

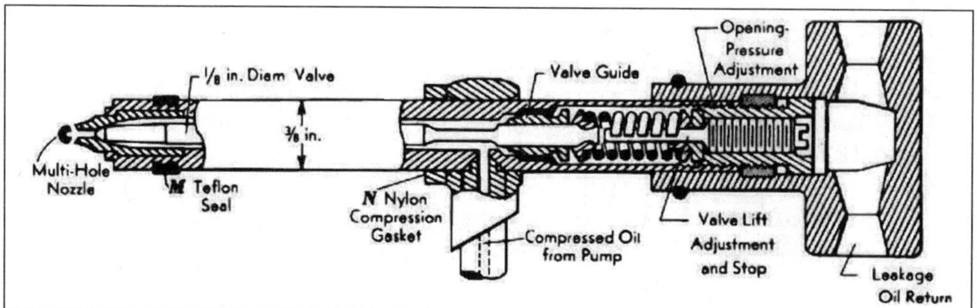

Figure 4.33 *Pencil-type injector by Roosa Master. (Reprinted from Ref. 7.)*

mixture preparation to achieve best performance, good fuel economy, and lowest exhaust emissions. Unlike the in-line pump system, distributor pumps have one pumping element (see Figure 4.8) serving multiple cylinders. This feature leads to a lower cost system than the in-line pump because the resulting pump has fewer parts, especially high-precision machined parts. Figure 4.34 shows the components of a distributor pump injection system, including a fuel tank (1), fuel line (2), fuel filter (3), low-pressure supply line to high-pressure distributor pump (4), high-pressure injection line (5), connected to injector nozzle holder (6) shown spraying fuel in a prechamber diesel engine. Overflow fuel from the high-pressure pump and leak-off fuel from the injectors flow back to the tank (1) through a return line (7). Of course, the prechamber or IDI engine sports a glow plug (8) to aid in starting the engine.

The distributor pump is either gear or belt driven. The drive is through a tapered shaft, item 1 in Figure 4.35. The driveshaft has a vane-type, low-pressure fuel supply pump whose pres-

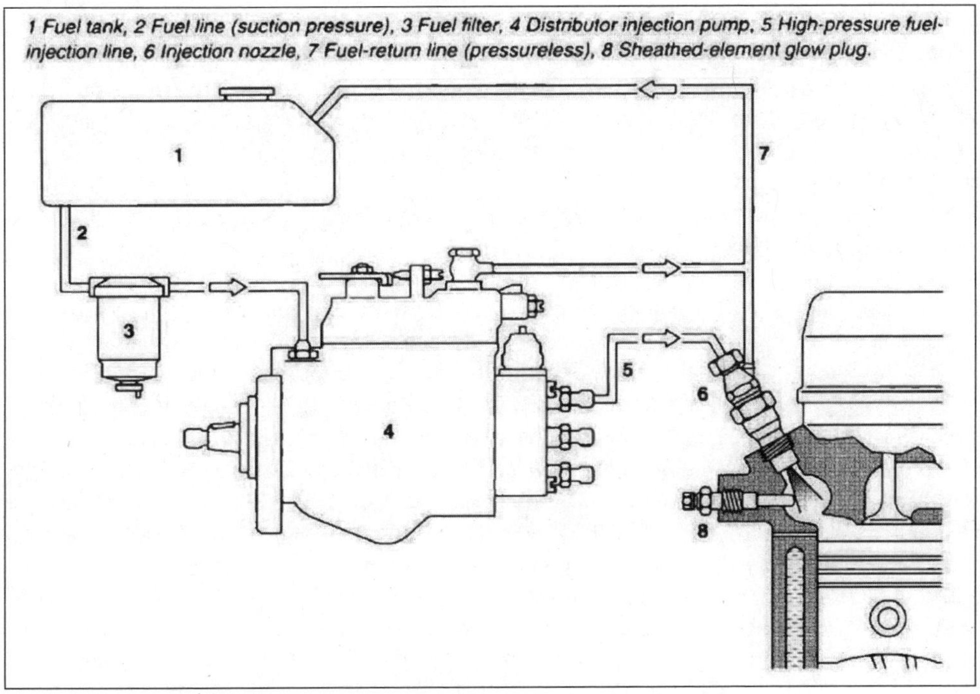

Figure 4.34 *Distributor pump injection system. (Reprinted from Ref. 5, page 152, Fig. 1, courtesy of Robert Bosch GmbH.)*

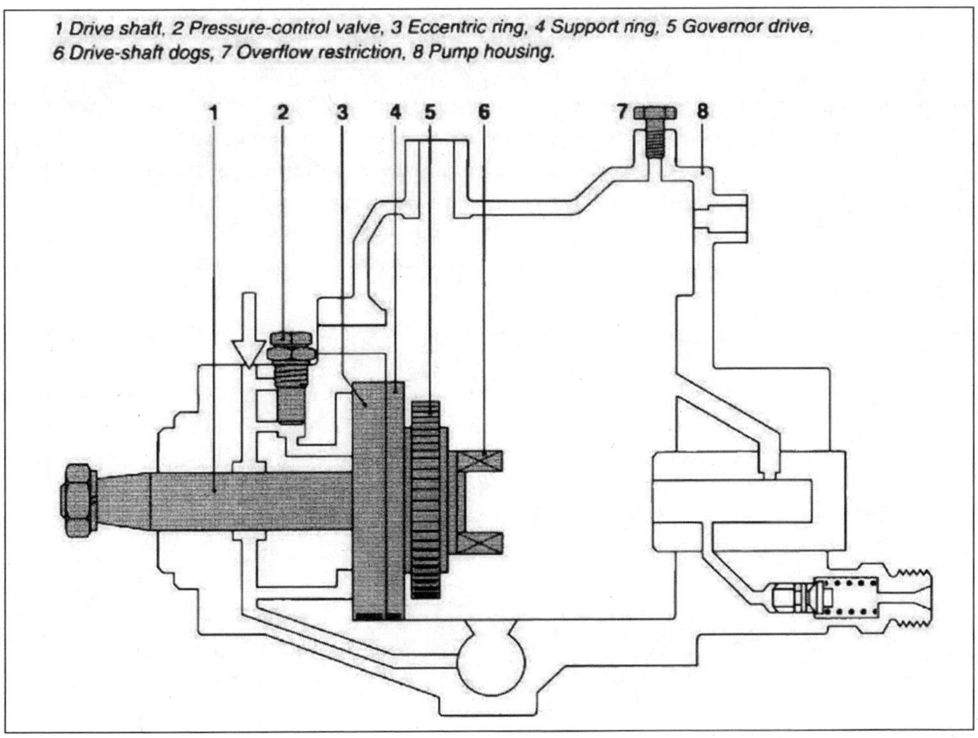

Figure 4.35 *Distributor pump with partial list of its components. (Reprinted from Ref. 5, page 153, Fig. 2, courtesy of Robert Bosch GmbH.)*

sure is regulated by a pressure control valve (2). The vane-type pump consists of an eccentric ring (3) and a support ring (4). Excessive pressure from the vane pump is discharged through a control valve (2) into the distributor pump housing. Next on the driveshaft is the governor drive gear (5) followed by the flexible drive dogs (6). These are only a fraction of the components that are housed inside the pump housing whose internal pressure is regulated by an overflow valve (7). The distributor pump housing (8) houses other components that will be described shortly.

Additional details regarding the other components that are driven by the pump driveshaft through a flexible coupling are given in Figure 4.36. A surface multilobe cam plate (1) is rotated through the driveshaft dogs. It is connected to the plunger on which slides a precision fit collar (2). The plunger (3) has extensive details including slots machined on its periphery near its end (4). These slots allow supply fuel from passage (5) to enter into the cavity (6) where fuel collects before the plunger travels to the right, thus reducing the volume of this cavity and leading the fuel to increase its pressure while moving through a central passage in the plunger. A crescent-shaped outlet port cut into the side of the plunger diverts the pressurized fuel to one of the outlet passages (7) and on to a delivery valve (8) to the high-injection pressure line and injector.

The distributor pump metering and delivery phases are explained in steps a through d in Figure 4.37. Step d describes the fuel entry into the high-pressure cavity. As the plunger travels to the right, the fuel entry is cut off, and high-pressure fuel exits through the outlet passage (shaped like a crescent) into another passage through the distributor head assembly and is ultimately directed to the injector. The end of injection comes as the plunger stroke positions the T-slot of the plunger past the sealing collar (see step c in Figure 4.37), thus dissipating the high-pressure fuel into the pump housing.

Steps a and b of Figure 4.37 describe the end of the fuel entry into the high-pressure cavity and the beginning and middle of the fuel delivery. Other control components such as governors, the BCD, and injection timing control work, in principle, in a fashion similar to that described in the case of the in-line pump. As combustion needs for high pressure grew quite important, the distributor pump met the challenge until it became obvious that another system was needed to help the engine meet stringent new emissions and performance targets.

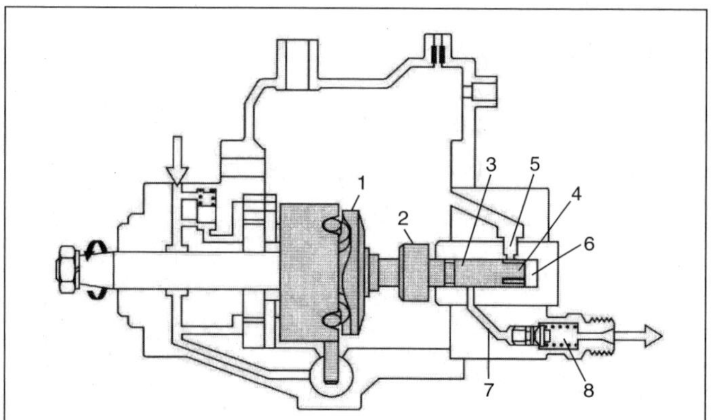

Figure 4.36 Additional details of the Bosch distributor pump. (Reprinted from Ref. 5, page 156, Fig. 7, courtesy of Robert Bosch GmbH.)

The R. Bosch distributor pump is not the only distributor/rotary design that caters to the needs of light-duty diesels. Other designs include the Stanadyne rotary pump, which is shown in Figure 4.38. A major difference between the Stanadyne rotary pump and the R. Bosch distributor pump is how high fuel pressure is generated. In the Stanadyne rotary pump, a radial cam driven by the pump driveshaft sends two floating pistons (slugs) toward each other. With fuel accumulated in the volume between the two pistons being squeezed into a smaller volume, its pressure increases as it flows toward the distributor section of the pump. Other distributor pumps also include the American Bosch pump, which was extensively used on International Harvester in-line, six-cylinder engines for some time.

4.4.3 The P-L-N Unit Pump System

The last of the P-L-N systems is the unit pump (UP), sometimes referred to as single pump. Unit pumps were primarily used in large engines such as locomotive, marine, and stationary engines. The reason for their popularity in those applications was the ability to shorten the high-pressure injection lines. Naturally, if an in-line pump were to be used for these applications, it would most likely be centrally installed relative to the engine. The distance from this central location to the farthest cylinder would be quite long, thus necessitating a long injection line to reach the injector of that cylinder. Longer high-pressure injection lines have the propensity for wave dynamics that could superpose and produce undesirable injection characteristics. This problem was solved by the UP system. One might think of UPs as unconnected or singular pumping elements from an in-line pump system, but separated from the multicylinder injection pump. Each pumping element is placed in proximity to its power cylinder. The short distance between the UP and the injector nozzle allows shortening of the length of the high injection pressure line and helps alleviate the problem of after-injection, secondary injection, or any other injection abnormalities in the injection system. Unit pumps operate on the same principle as the in-line pump.

In large engines, the governor is normally installed on the engine. Fuel quantity is controlled through mechanical linkages extending from the governor to the unit pumps. Injection timing can be adjusted for each cylinder through an intermediate member, as shown in Figure 4.39.

In Figure 4.39, a rocker mechanism (2, 3) is sandwiched between the injection cam (10) and tappet roller (7). By adjusting the eccentricity (2), the relative position of the tappet roller (7) to the injection cam (10) can be varied. This device helps engineers optimize fuel consumption and emissions, and also gives them the ability to modify injection timing according to fuel quality.

As mentioned earlier, unit pump systems are just for large diesel engines, but their use found favor in medium- and light-duty applications. Of course, modern unit pumps, like other injection systems, are now more sophisticated and have

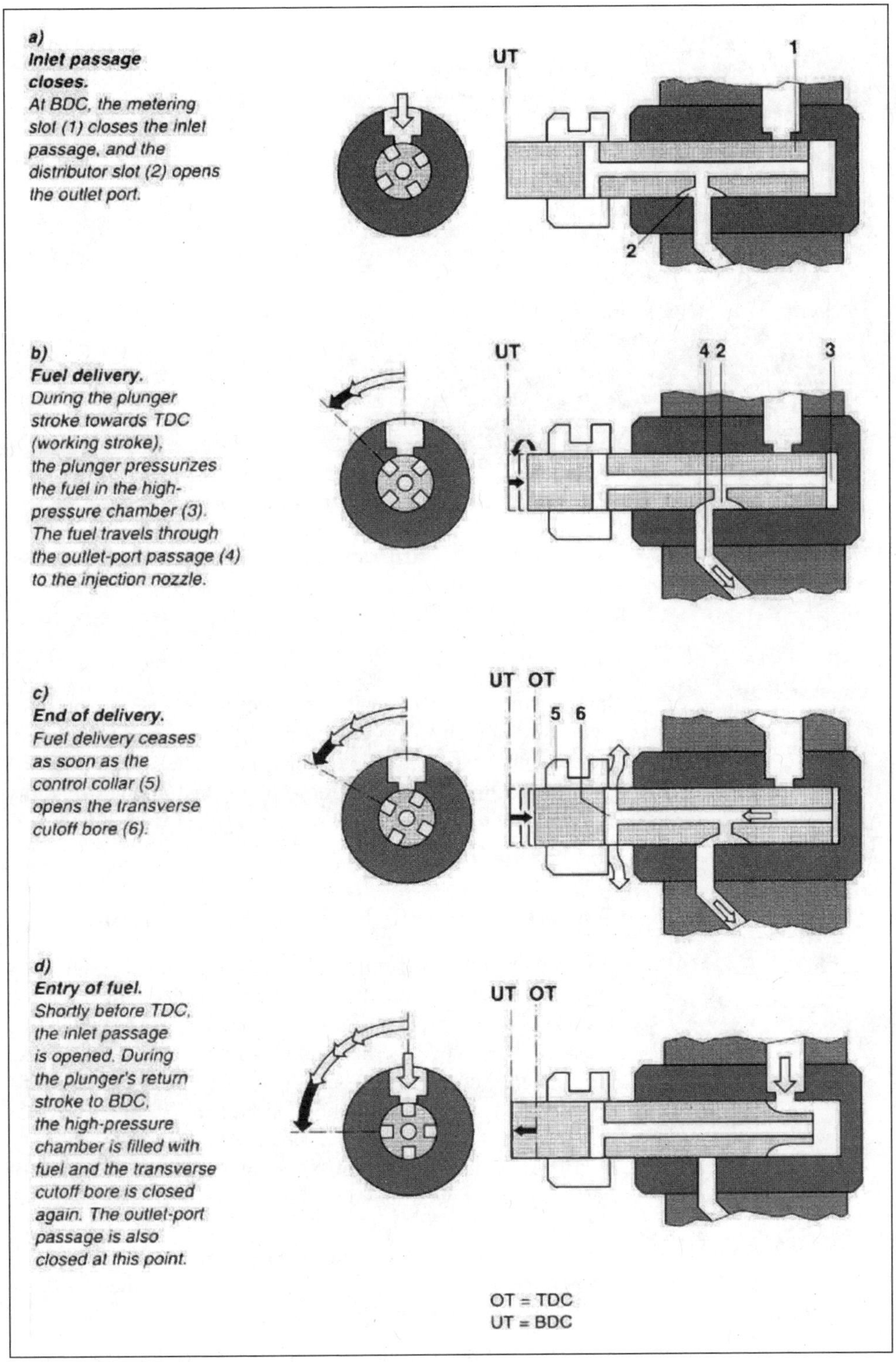

Figure 4.37 *Strokes and delivery phases of the Bosch distributor pump. (Reprinted from Ref. 5, page 159, Fig. 10, courtesy of Robert Bosch GmbH.)*

much more control flexibility because they are electronically controlled. An example of a unit pump application in the on-highway truck application is reported in Ref. [9] and has a Euro-American flavor. The engine is manufactured by Detroit Diesel in the United States, and is marketed by Mercedes Benz in Europe as the OM457LA. The same engine is marketed in the United States as the Series 55.

4.5 The Unit Injector System

The high-pressure injection line connecting pump and injector can enhance the possibility of wave superposition, lead to after-injections, and contribute to injection delays. Eliminating the high-pressure line would necessitate grouping both the pumping element and the nozzle in one embodiment called the unit injector and shown earlier in this chapter as Figure 4.5.

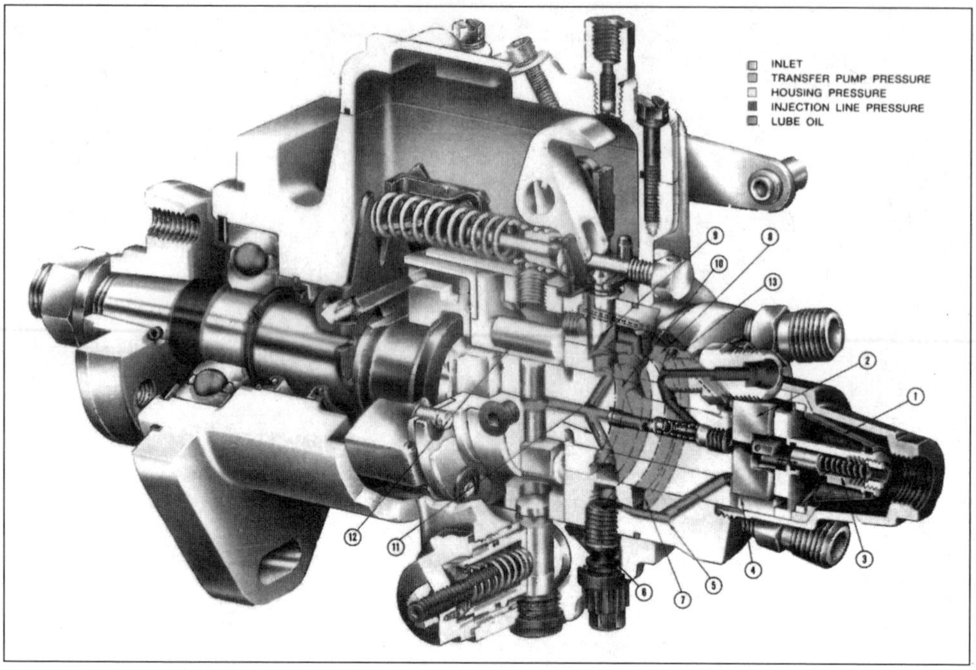

Figure 4.38 Cutaway of the Stanadyne rotary/distributor pump. (Reprinted from Ref. 8, courtesy of Stanadyne Corporation.)

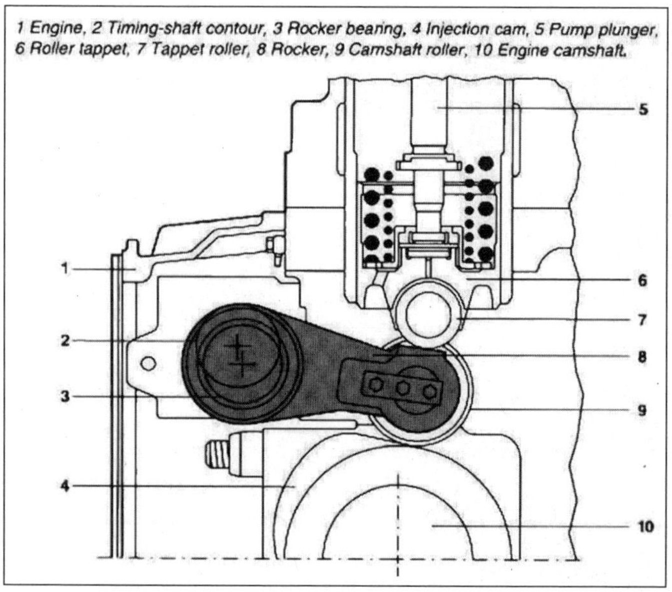

Figure 4.39 Timing mechanism in unit pump systems. (Reprinted from Ref. 5, page 143, Fig. 2, courtesy of Robert Bosch GmbH.)

One of the more famous applications of the unit injector is Detroit Diesel Corporation's two-stroke engine, which is still used in many of the transit buses in the United States. Unit injectors are driven by the engine camshaft and share the same fuel metering and timing control principles of the in-line pump system. Even though unit injectors do not have the line dynamics problems associated with P-L-N systems, they may still generate these dynamics within their own narrow passages. However, line dynamics difficulties in unit injectors are less troublesome than those of their P-L-N counterparts. Traditionally, unit injector systems have been able to deliver extremely high injection pressures. One of the main reasons for this unique capability is the absence of connections made possible by eliminating the high-pressure line.

4.6 The Common-Rail Injection System

A feature that is characteristic of the P-L-N and unit injector systems is that their fuel pressure depends on engine speed. It follows that when the engine is running at lower speeds, fuel pressure would be rather low, and the higher the engine speed, the higher the injection pressure. Such dependency on the engine speed for a very important injection system function is seen as restrictive to many engine and fuel system designers. Besides, tough new restrictions on emissions and customer demands for performance as well as good fuel economy require as much control flexibility as possible.

In the course of developing the modern diesel engine, engineers have discovered the needs and benefits of the following fuel injection equipment features:

- High injection pressure that can be controlled without dependence on engine speed
- High injection pressure independent of fuel quantity
- Precision in fuel metering or injected quantity
- Flexibility in injection timing or injection timing that could be varied according to parameters chosen by the designers
- Ability to perform multiple injections in one combustion cycle
- Ability to meter extremely small fuel quantities

- Ability to maintain minimum variation in injected quantities from one cylinder to another in multicylinder engines
- Ability to minimize cycle-to-cycle variability in injected quantity
- Ability to minimize cylinder-to-cylinder injection timing variations
- Ability to minimize cycle-to-cycle injection timing variability
- Ability to maintain accuracy of fuel metering and injection timing over the expected life of the engine
- Ability to interact with other subsystems to maximize performance and fuel economy and minimize exhaust emissions.

The last of these functions involves the combustible mixture preparation. For instance, the injector nozzle design must atomize the fuel to enhance its evaporation and interaction with the available air in the combustion chamber. It must also produce the smallest of fuel particles with the capability of penetrating deep into the combustion chamber volume and seeking more air/oxygen to enhance combustion. Granted, this is a long list of desired features that comes with a corresponding list of engineering challenges.

The common-rail injection system seems to meet many of the features listed and desired by design engineers. It is simple in concept in that it involves a high-pressure pump that can be driven at crank speed (engine or twice the camshaft speed). This high-pressure pump delivers fuel to a common rail. High-pressure injection lines connect the common rail to the engine injectors that are actuated mechanically via overhead cams or rocker arm mechanisms (see Figure 4.40).

Of course, Figure 4.40 is for a mechanically controlled, common-rail system that does not have the sophistication or capabilities afforded by today's electronically controlled systems, which are addressed later in this chapter. A complete common-rail system includes a low-pressure supply pump that draws fuel from the tank and feeds it to the high-pressure pump. High-pressure fuel delivered into the rail may bring along pressure pulsations; therefore, the volume of the common rail should be designed to dampen those pulsations. These pulsations result from the delivery characteristics of the multiple plungers on the radial fuel pump. Driving the pump at engine speed increases fuel pumping capacity and improves the potential to achieve high injection pressures. Control of fuel metering and injection timing is similar to that described earlier for the in-line pump and unit injector systems. In spite of the availability of these desirable features in common-rail systems, it is worth noting that such systems could not attain their potential without help from electronic controls. In fact, electronics were introduced in all types of fuel injection systems to expand their capabilities and improve their performance.

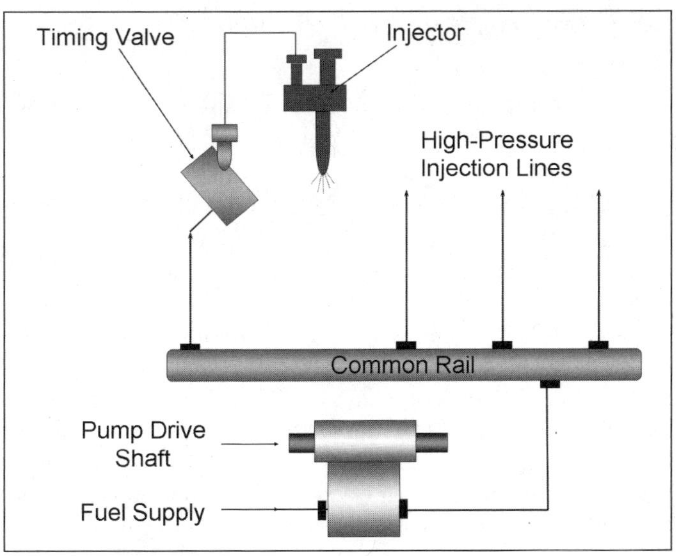

Figure 4.40 *Common-rail injection system.*

4.7 More on P-L-N Systems

Before moving on to the modern fuel injection systems, it is important to mention a few more words about P-L-N systems. The various types of fuel injection systems as presented in this chapter would have us think that there are no other variations on either in-line, distributor/rotary, or unit pump systems. Yet, in reality we have examples of other P-L-N systems that were designed, manufactured, and successfully used in commercial engines for many years. One example is the Cummins P-T injection system. This system was built by the Cummins Engine Company and used on their N-Series engines for many years. Even though the P-T system gives the general appearance of a typical P-L-N design, it has a slightly different philosophy for metering. In most of the injection systems addressed so far in this chapter, the fuel metering method was described. In general, a volume is filled with fuel, this fuel volume is pushed forward by a plunger that follows a certain cam profile, and the pulse of fuel creates pressure that lifts the needle off of its seat and allows injection to proceed. The end of injection takes place when the spill hole is suddenly uncovered and fuel pressure drops below the nozzle opening pressure. This process is often referred to as *real-time fuel metering*.

Another method of metering uses the side of the cam that is away from the injection segment to perform the metering process. In this method, fuel metering may start immediately following the injection and continue until the right amount of fuel has accumulated in a specific volume within the injector itself. At the designated time, a large injection cam/follower mechanism would apply extreme force on the metered and accumulated fuel to force it to be sprayed into the combustion chamber. Since the fuel was metered prior to its injection into the cylinder, the term *premetered method* was selected to describe this process. The Cummins P-T system is a P-L-N system that is premetered, and could be also considered a unit injector system since in the end the fuel pumping and injector nozzle are contained in the same body as the injector nozzle holder assembly. It may even be considered a form of common rail, since the pump delivers fuel to a rail that feeds each injector through a separate line.

4.8 Modernizing the Diesel Fuel Injection System

Although there were numerous motivators for modernizing the diesel fuel injection system, the need to meet emissions regulations was certainly one of the most important, if not the most important, motivator. Customer demand for better engine response and drivability also played a major role in advancing the fuel system's state of the art. Diesel engine designers and builders were not just satisfied with its superior fuel economy, but sought to continue to improve it to maintain its lead over its competition. Not only was the goal to improve its fuel economy standing, but to achieve this objective while also meeting the tough emission regulations that were imposed in the 1980s and 1990s, as well as early in the 21st century.

Some of the early gains in the evolution of the modern diesel injection system were realized through one extremely important feature, increasing injection pressures to unprecedented levels. However, engine designers felt their hands were tied and were unable to make other gains until they were able to introduce electronic control into the injection system. Introducing this technology made other advances easier to achieve. It was easier to customize torque outputs from the same base engine to meet the demands of several applications in the heavy-duty market. It was also easy to include many value-added features such as limiting maximum road speed for fleets where fuel consumption is directly related to driving habits. For the first time injected fuel quantity was trimmed according to fuel temperature for consistent power output regardless of fuel temperature, within certain limits. Electronics brought along the promise of reduced cost through simplifying many of the functions otherwise provided by cumbersome mechanical devices, and allowed the implementation of full-authority parameter control strategies. For instance, a drive-by-wire system allowed the implementation of low- or no-smoke strategies versus systems where the injection was solidly connected to the accelerator pedal. In addition to smoke-free strategies, electronic controls facilitated flexibility in injection timing and metering control, and reduced cycle-to-cycle and cylinder-to-cylinder variability.

Historically, passenger car diesels were first to adopt electronic controls around 1980. By 1985 most, if not all, of the diesel-powered passenger cars had adopted electronic controls in various degrees. The heavy-duty engine industry was a bit more reluctant to join the passenger car movement toward electronics. This cautious attitude is characteristic of the very conservative and durability/reliability-minded heavy-duty diesel industry. The only electric component to be found on the heavy-duty engine in the late 1970s was the fuel solenoid shutoff valve, and it accounted for a major share of the warranty cost. Despite this experience, the advantages of electronics were too promising to ignore even by the heavy-duty engine makers, who could not resist this wave of technology. Today, heavy-duty trucks sport many advanced electronic features such as global positioning and travel data logging and reporting. "Electronification" of the heavy-duty engine began in the late 1970s and continued through the 1980s. By the mid-1990s the entire inventory of new heavy-duty diesel engines built in the Western world for on-highway applications was electronically controlled. The heavy-duty off-highway applications did not adopt the trend toward electronic controls until the late 1980s/early 1990s, and became fully converted to electronic controls by the turn of the century.

4.9 Modernizing the Pump-Line-Nozzle System

Modernization of the P-L-N system went through an evolutionary process in which the initial steps were simply using electrical components to replicate those functions previously performed by mechanical components. The introduction of electronics to the diesel engine industry has been slow, largely due to the negative cost implications, as well as doubts about the reliability of electronics in the rugged applications of the diesel engine. Uncertainty about whether electronics would really be required to meet emission regulations while helping maintain good engine performance further delayed progress toward adopting electronics in heavy-duty diesel fuel systems. Emission regulations continued on an increasingly stringent path, forcing more demands on the fuel injection system. Further, early demonstrations of what electronics can do helped focus attention on these developments and direct more capital and resources into research efforts.

This section addresses the specific developments of the P-L-N fuel injection system. A somewhat detailed description is given for the "electronification" of the in-line pump with special treatment of some of its basic as well as new functions. Parallels are drawn to project the changes performed in the distributor/rotary pump system as well as the unit pump system. Introduction of electronic controls to both the unit injector and the common-rail systems is addressed. More importantly, several innovations in fuel injection equipment are reviewed with emphasis on their special features and contributions to the combustion process.

4.9.1 Electronic Control of the In-Line Pump System

For the in-line pump system, heavy-duty applications were the primary target for introducing the use of electronic controls. The 1988 emissions regulations in the United States and contemporaneous performance objectives were met without using electronic controls. By 1991, however, some North American manufacturers decided that the time for introducing electronics in commercial on-highway trucks and buses was ripe. Projects were established for the evaluation of engine electronic controls at most, if not all, engine companies in North America, as well as in Japan and Europe. For instance, R. Bosch, one of the largest manufacturers of in-line pump systems, introduced the P-3000 and P-7000 Series pumps, which were initially equipped with just fuel metering control and later with electronic injection timing control. These pumps were equipped with an electronically controlled rack (commonly known as an electronic governor) and code named RE-24. Some of the rack's control functions are described in this section as well as results of an engine evaluation of the electronic governor.

With electronic control of fuel quantity as implemented in the R. Bosch P-3000 pump equipped with an RE-24 electronic governor, some functions were substantially improved and new functions were added. The following is a summary of these various functions.

4.9.1.1 Improved Functions

Electronic controls allow for improvements in many of a diesel engine's functions, as described in the following list:

- *Key start/stop* was accomplished by using a solenoid to actuate the pump shutoff lever in the traditional in-line pump system. With electronic rack control, this function is performed by de-energizing the rack actuator (actuator for fueling) shown in Figure 4.41. This eliminates the need for a separate shutoff solenoid. In addition, an emergency shutoff solenoid is provided at the fuel entry point into the high-injection pressure pump. When the emergency solenoid is activated, the path of the supply fuel to the pump fill gallery is interrupted.
- *Torque curve shaping* is more easily achieved through the use of electronics. In some cases it may be difficult if not impossible to obtain the desired torque/speed relationship with a mechanical governor. Add-on devices such as two-stage boost controllers, torque capsules, puff limiters, and K-cams (as in the R. Bosch RQV...K governor) are used to approximate the desired torque shape. In the case of electronic rack control, this function is accomplished by regulating the current to the rack actuator according to a preset schedule stored in the memory of the microprocessor. This preset schedule is a function of the throttle position, engine speed, intake air pressure and temperature, and fuel temperature.
- *Flexible drivability characteristics* can be accomplished with relative ease. This means that more than one type of governor characteristic can be stored in memory and then retrieved as desired according to each application. Under certain driving conditions, an all-speed governor may be preferred, whereas other conditions may require a minimum/maximum governor scheme. Both control strategies can be accommodated if required for a certain application.
- *Achieving a zero droop (change of fuel rate/crankangle degree) at idle* can easily be achieved with electronic rack control compared to the mechanical governor, which relies on the dynamics of a mass/spring system.
- *Flexibility and accuracy in controlling smoke* are provided by electronic rack control through the use of anticipatory strategies that account for the turbocharger lag.
- Electronic rack control *provides tighter droop* (close to zero) through the intermediate speed range (idle to rated) for power takeoff (PTO) applications. This system offers added flexibility for the end speed (or high idle speed) regulation.
- The electronic system makes it more *difficult to tamper with* the fuel system calibration.

4.9.1.2 New Functions and Advantages

Electronic control also makes possible new functions and advantages for diesel engines:

- *Fuel temperature compensation* is nonexistent with the mechanical system. With the electronic rack controller, the rack position can be modified as a function of the fuel temperature.
- *Air temperature compensation* is provided with an electronic rack controller.
- *Coolant temperature* as an indicator of a cold or a warm engine condition can be used to control the starting (excess) fuel quantity.

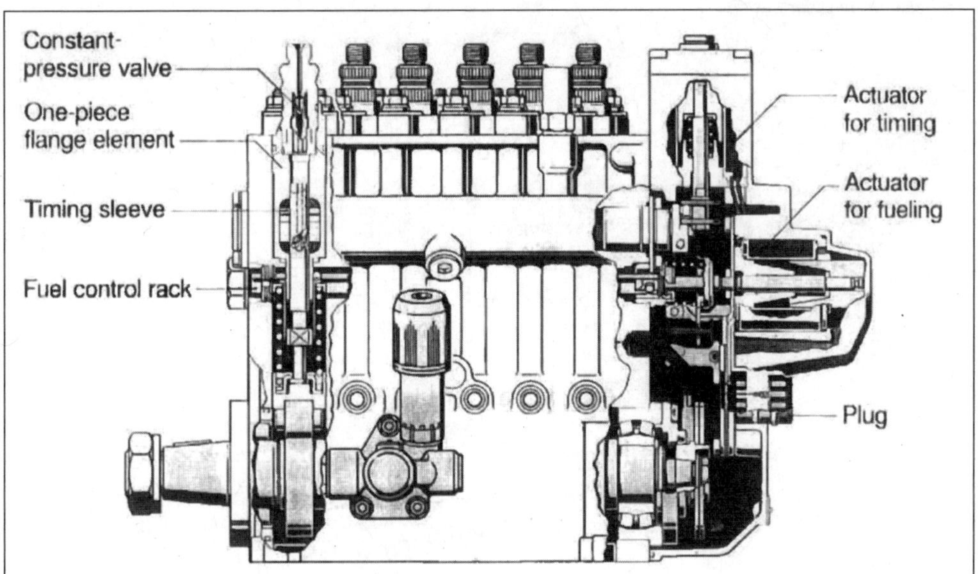

Figure 4.41 Electronically controlled in-line pump. (Reprinted from Ref. 5, page 65, Fig. 1, courtesy of Robert Bosch GmbH.)

- *Cruise control* can easily be implemented.
- *Road speed limiting* is another function that is easily provided. This feature has shown significant savings in fuel consumption.
- *Linking the fuel system control and other engine control devices*, such as variable-geometry turbochargers, can be achieved more effectively without the use of mechanical linkages.
- *On-board monitors and diagnostics* are effectively implemented.
- *Pump inventory reductions* are accomplished by means of a common hydromechanical pump (where applicable) and a blank programmable read-only memory (PROM) for end-of-line programming.

4.9.1.3 Description of the RE-24 Rack Controller

The R. Bosch RE-24 rack controller is more than just an electronic governor. It has the responsibility of metering fuel to the engine, as a function of a number of variables, throughout its speed and load ranges. Figure 4.41 shows the construction of the electrical components of the system. A fuel rack actuator (or linear solenoid coil) is in contact with the end of the control rack. Varying the current to the coil moves the control rod to increase or decrease fueling to the engine. A return spring always returns the rack to the no-fuel position when the coil is de-energized. An eddy current sensor with an induction ring on the control rack serves as a rack position sensor, which is also temperature compensated. An engine speed sensor is installed in the "governor" housing. The engine speed sensor is an electromagnetic sensor that works in conjunction with a toothed gear to indicate speed. The electrical components that are built into the "governor" housing are wired to a multipin, heavy-duty connector that is accessible from the back of the pump, as shown in Figure 4.41. The oil fill hole on the side of the "governor" housing provides visual means to verify the static "base" injection timing.

Figure 4.42 is a functional schematic representation of the RE-24 system. It shows other sensors that can be used for the acquisition of information required as input to the microprocessor. These sensors include an accelerator pedal position sensor, a boost pressure sensor, an air temperature sensor, a coolant temperature sensor, and a fuel temperature sensor. Other sensors may be added depending on the needs of the specific application. Inputs from these sensors are received by the microprocessor, which in turn computes a rack position. This rack position is relayed to the rack actuator, which in turn moves the rack to that position. The rack position sensor closes the loop by sending a signal to the microprocessor confirming the actual position of the rack, thus causing the microprocessor to take corrective action in case the rack is not at the desired position.

Figure 4.42 also shows elements of vehicle controls such as an exhaust brake and how diagnostics might be implemented. Several other sensors are used to feed the control unit information required for various other functions such as cruise control and positive traction control.

1 Fuel tank, 2 Supply pump, 3 Fuel filter, 4 In-line fuel-injection pump, 5 Electrical shutoff device (ELAB), 6 Fuel-temperature sensor, 7 Rack-travel sensor, 8 Actuator with linear-motion solenoid, 9 Pump-speed sensor, 10 Injector 11 Coolant-temperature sensor, 12 Accelerator-pedal sensor, 13 Switches for brakes, exhaust brake, clutch, 14 Operator panel, 15 Warning lamp and diagnosis connection, 16 Tachograph or vehicle-speed sensor, 17 ECU, 18 Air-temperature sensor, 19 Charge-air-pressure sensor, 20 Exhaust-gas turbocharger, 21 Battery, 22 Glow-plug and starter switch.

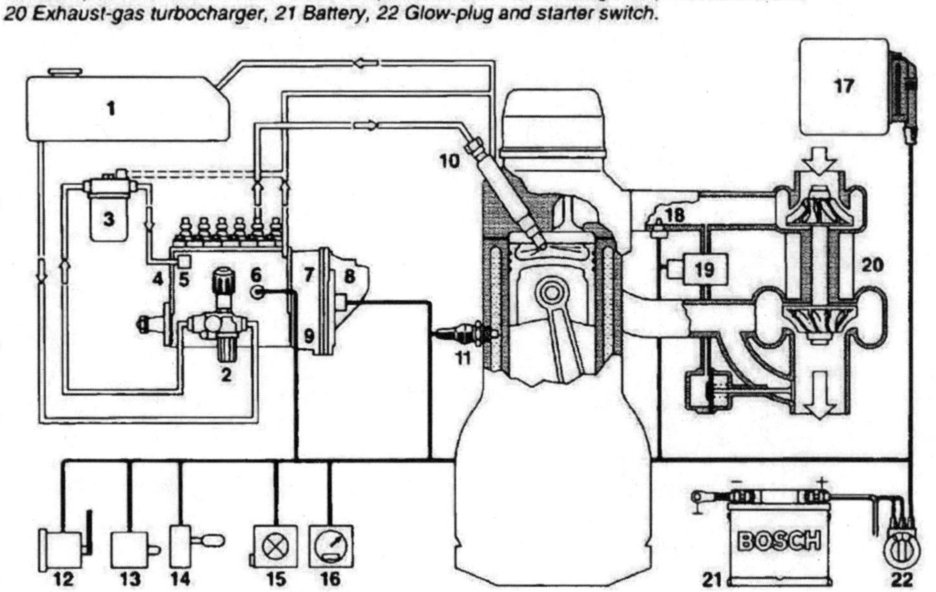

Figure 4.42 *Functional schematic of and electronic in-line pump system. (Reprinted from Ref. 5, page 135, Fig. 1, courtesy of Robert Bosch GmbH.)*

4.9.1.4 Function of the RE-24 Rack Controller

Control Overview

Figure 4.43 shows an overview of the rack control functional variables, which are usually stored table form in the microprocessor memory. Shown is a map that includes the following:

1. *Start fuel* is given in terms of a "rack position" command (RC), which corresponds to a "fuel quantity" command (FC). Two lines (1—solid and dotted) shown in Figure 4.43 indicate either a fuel command or a rack command, for cold engine start. Two additional lines (2—again, solid and dotted), show the same two variables for a warm engine start. A vertical dashed line (N1) shows the "engagement" speed below which no fuel is allowed into the engine.
2. *Full-load torque* is shown as line 3, which is determined for each engine according to the desired level of performance, that is, the torque output, at each speed. Dotted lines 4 and 5 indicate the limits that could be set on the full-load torque curve (3) to account for the cold fuel temperature (4) or the hot fuel temperature (5).
3. The *smoke limit*, for a given pressure and temperature condition, is designated by line 6. Lines 7 and 8 show the smoke limit under cold and warm intake manifold temperatures, respectively.
4. *Intermediate speed regulation* is designated by diagonal parallel lines (9).
5. *End speed regulation* is shown as line 10.
6. *Low-idle, rated, and high-idle speeds* are designated N2, N4, and N5, respectively.
7. The *reengagement speed* (N2′) is the speed at which a cold engine might stall, and the start fuel quantity (modified by fuel temperature) would be required to avoid stalling.
8. *Cutout speed* (line N3) is the speed indicating the end of the start process.

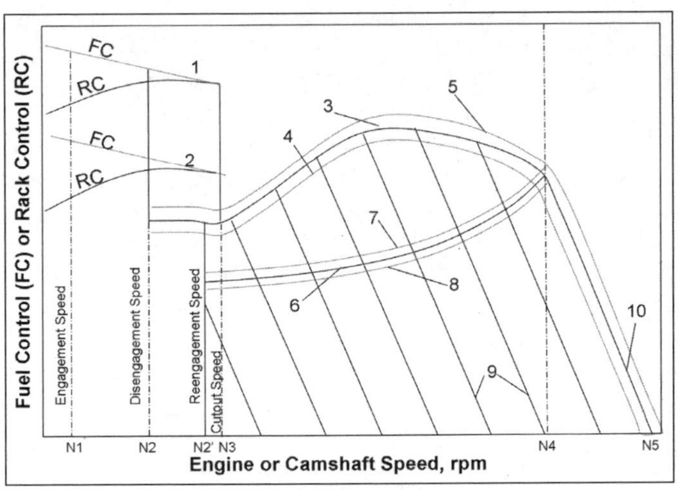

Figure 4.43 *Functional overview of electronic governor function and control.*

Fuel Quantity Regulation

Figure 4.44 is a block diagram of a sample logic used in the regulation of fuel quantity. Starting from the upper right-hand corner of Figure 4.44, a *throttle position,* or command A, and an *engine speed* signal B are inputs to the *governor limit* block C. For that engine speed and throttle command, the microprocessor will look up a rack command (or fuel command) value that is stored in its memory. This value is an output signal N from block C to the *minimum* block D. *Torque limit* F receives an engine speed signal B for which the microprocessor looks up a maximum allowable torque value (limit) for that speed. This value is an input G into the *minimum* block H.

Smoke limit E receives two inputs. The first is an engine speed input B, and the second is an *air density* signal I resulting from a computation in the *air density* block J based on the intake manifold pressure and temperature values. The output of block E (smoke limit) is input to the comparator block H, where a minimum value K is selected (wins). The latter is used as input into the *fuel temperature compensation* block L, which computes a *fuel quantity command* signal M based on engine speed, smoke limit, and a fuel temperature correction factor. *Comparator* D receives two signals M and N, and governed by a "minimum-win" logic, a fuel flow command (FC) is selected and transmitted to the rack actuator.

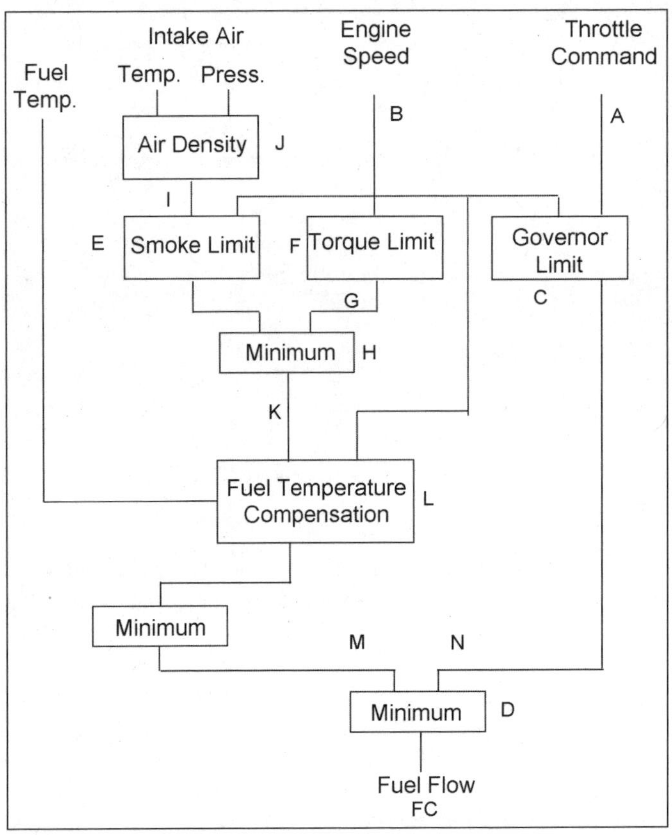

Figure 4.44 *Block diagram of the logic used for fuel quantity regulation.*

4.9.1.5 Electronic In-Line Pump Timing Control

Figure 4.45 shows the relationship between the plunger, barrel, and control sleeve in two positions. Figure 4.45(a) shows the spill port in the closed position that leads to the start of delivery, and Figure 4.45b shows the spill port in the opening position, which leads to the end of delivery (and ignition). The motion and position of the control sleeve is based on inputs from several sensors, as in the case of the fuel quantity regulations. In addition, a needle lift sensor serves as a monitor and sends a feedback signal for accurate timing control.

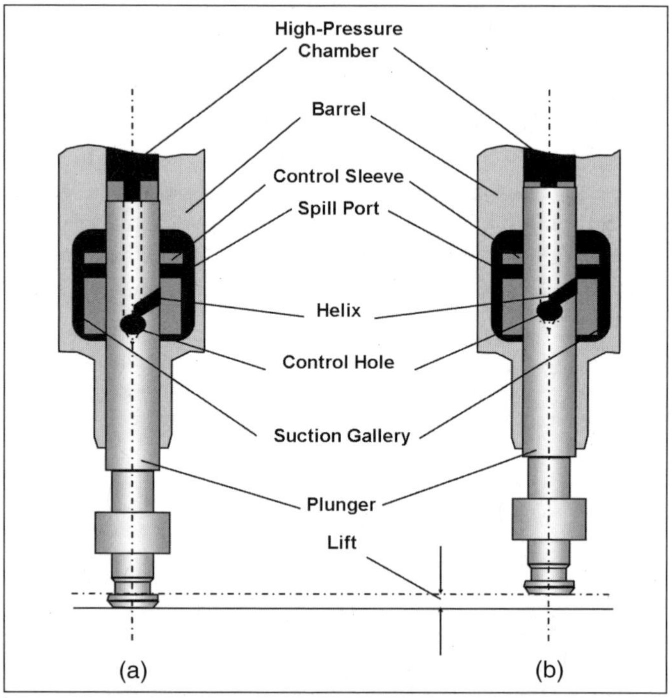

Figure 4.45 Control of injection timing/metering via control sleeve.

4.9.2 Electronic Control of the Distributor Pump System

Figure 4.46 shows the details of an electronically controlled distributor pump. The actuator (2) turns a control shaft at the end of which is an eccentric pin (2a). The eccentric pin engages the control collar (6) through a slot and causes it to move horizontally, thus sliding over the plunger and able to cover or uncover the spill ports (7). Control of fuel quantity is achieved in a fashion very similar to that of a mechanically controlled distributor pump. Injected fuel quantity can be controlled at any value from no fuel to the pump's maximum delivery capacity. Maximum delivery is often required in cold start conditions to overcome friction and assist the engine in achieving stable

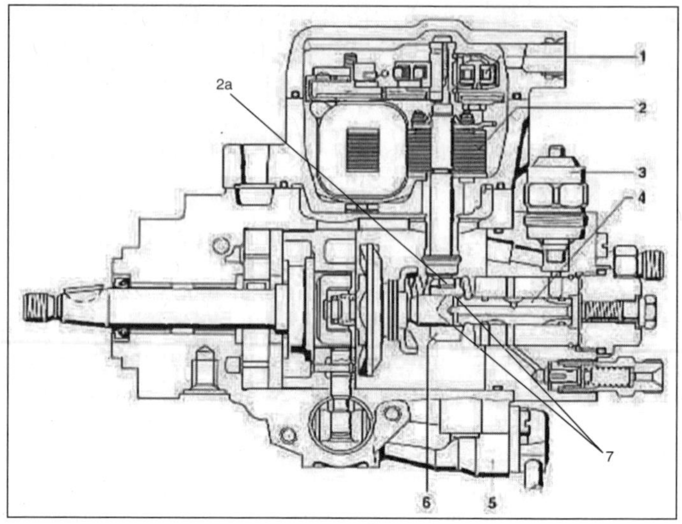

Figure 4.46 Electronically controlled distributor pump system. (Reprinted from Ref. 5, page 188, Fig. 4, courtesy of Robert Bosch GmbH.)

idle speed. The control shaft angular position at a given pump or engine speed is an input to the electronic control unit to determine injected fuel quantity at that speed. If no voltage is applied on the actuator, a return spring rotates the control shaft and uncovers the spill ports to prevent any fuel delivery.

The electronically controlled distributor pump system is equipped with an automatic advance and retard mechanism. This system uses internal pump pressure applied to a small piston in performing this operation. Because the internal pump pressure is proportional to engine or pump speed, a clocked solenoid valve modulates the timing device pressure to trim the basic start of injection. When the solenoid valve is open, the pressure on the timing-device piston is reduced to effectively retard injection timing. The opposite action causes injection timing to advance.

Figure 4.47 illustrates the effect of the start of injection on a number of dependent variables, including hydrocarbon, smoke, and nitrogen oxide emissions as well as fuel consumption for various brake mean effective pressures (p_e). This work was performed on a 2.4-L direct-injected, turbocharged engine.

The trend toward higher injection pressures, especially for direct-injected engines, continued through the last century. It accelerated, however, in the 1970s and later for the heavy-duty applications to meet the demand for cleaner and more fuel efficient engines. The high-speed, passenger car market adopted the same philosophy, and even though it started a decade after the heavy-duty applications, both light-duty and heavy-duty applications have similar injection pressures. This fact is illustrated in Figure 4.48, which illustrates the evolution in peak injection pressure by model year.

To meet this demand for high injection pressure while maintaining cost at a reasonable level, R. Bosch developed the VP44 pump in the early 1990s. This pump is a radial-piston rotary distributor design that allows peak injection pressures of

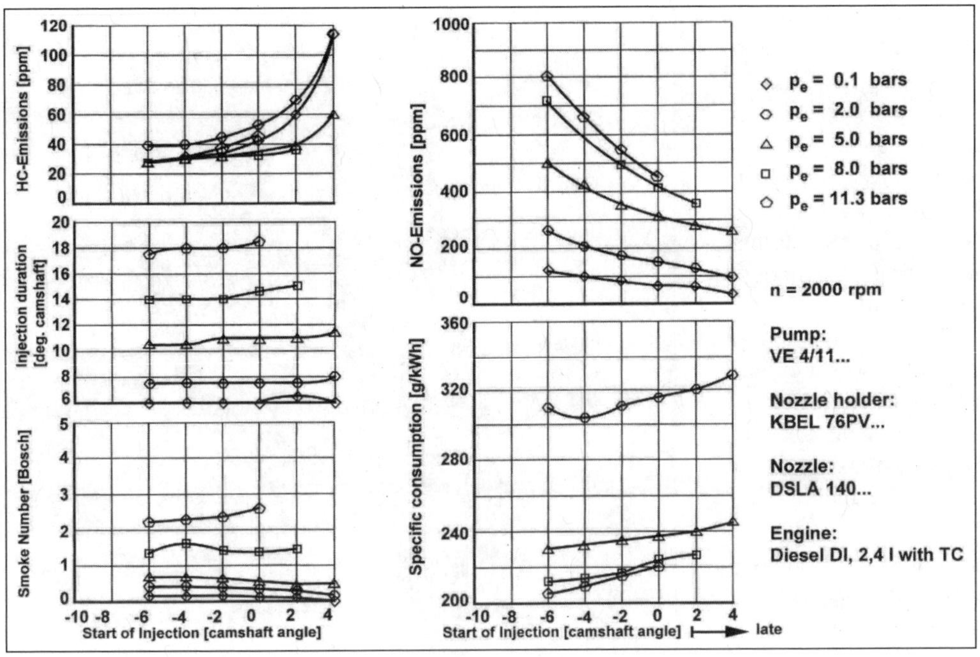

Figure 4.47 *Effect of injection timing on emissions and fuel consumption. (Reprinted from Ref. 3, Fig. 25, courtesy of Robert Bosch GmbH.)*

about 1400 bar. The pump operates according to the principle of the Roosa Master rotary distributor pump from the mechanical standpoint. Fuel injection timing and metering is achieved by means of a solenoid valve, and the electronic control unit is pump mounted as shown in Figure 4.49. The primary difference is that instead of relying on a control sleeve in the typical distributor pump system, a high-pressure solenoid valve closes off the high-pressure area from the fuel supply area. Closing and opening the solenoid valve, in conjunction with the proper pump cam phasing, determines injection timing and fueling rate. This system is also capable of pilot injection (preinjection), which reduces the noise level, especially at engine idling conditions.

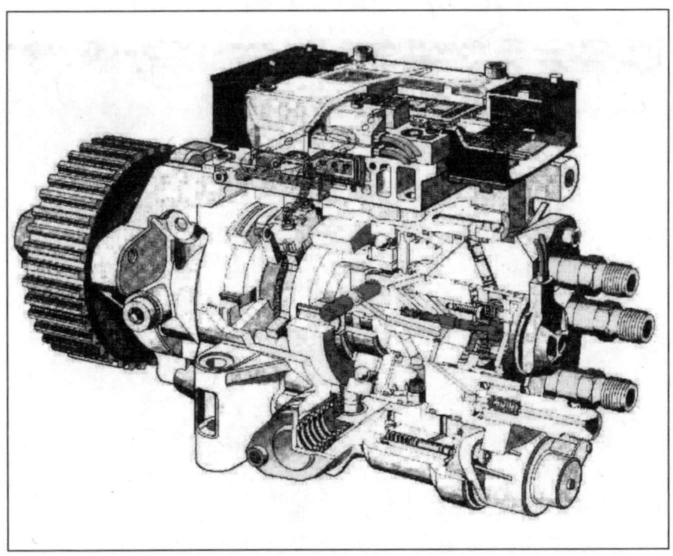

Figure 4.49 *Cutaway of the Bosch VP44 rotary-piston distributor pump. (Reprinted from Ref. 3, courtesy of Robert Bosch GmbH.)*

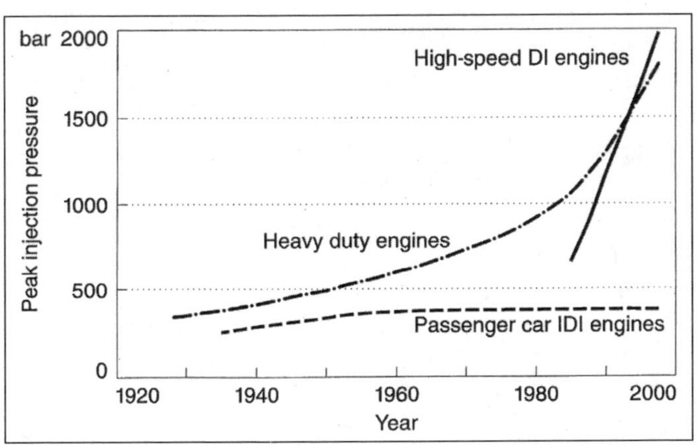

Figure 4.48 *Peak injection pressure versus model year. (Reprinted from Ref. 3, courtesy of Robert Bosch GmbH.)*

4.9.3 Electronic Control of the Unit Pump System

As illustrated earlier, Figure 4.9 shows a mechanically controlled unit pump. As also described earlier, there are many similarities between the function of an in-line pump and a unit pump. What separates these two systems is that the in-line pump has a central housing that encompasses multiple pumping elements (one for each cylinder), but the unit pump is an assembly that caters to one of many cylinders in an engine (Section 4.4.3).

A high-speed, electronically controlled solenoid valve is installed on the unit pump (see Figure 4.50). The opening of the solenoid valve allows the injection to commence. Its closing terminates the injection event. The time between its opening and closing determines the total fuel metered into a cylinder. As illustrated in Figure 4.50, unit pump systems consist of the pump housing with the attached solenoid valve, the high-pressure injection line, and the nozzle holder. Sensors monitor important parameters and transfer their signals to the electronic control unit, which in turn sends its command to the high-speed solenoid valve to either open or close. This process is done in an optimal way to ensure low exhaust emissions while maintaining good fuel economy.

The unit pump system is suitable for a number of different applications. Its electronics allow fault-detection capability, self-diagnosis, and the possibility of communicating with other compatible systems through various interfaces.

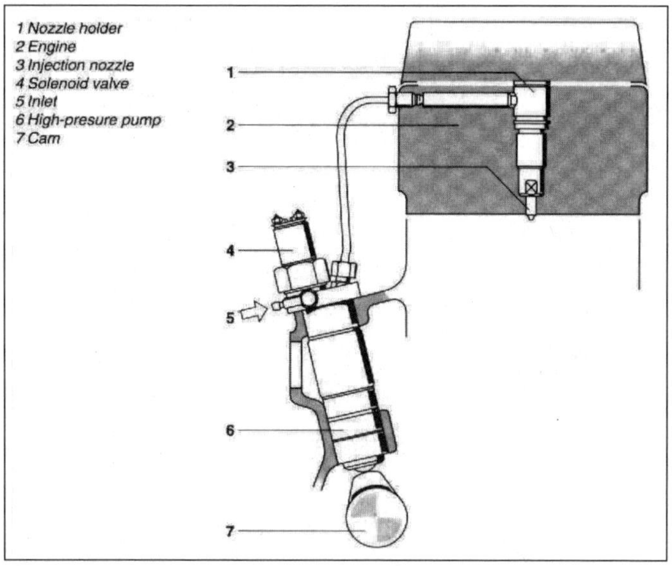

Figure 4.50 *Electronically controlled unit pump system. (Reprinted from Ref. 5, page 147, Fig. 2, courtesy of Robert Bosch GmbH.)*

4.10 Modernizing the Unit Injector System

As in the case of the unit pump, unit injectors also feature a high-speed solenoid valve installed alongside the body of the injector as shown in Figure 4.51. Unit injector assemblies include a pumping element and an injector nozzle. Missing in this system is the high-pressure pipe that connects the pumping element with the injector nozzle.

The electronic unit injector shown in Figure 4.51 has the capability of delivering up to 2000-bar injection pressure. It has full authority control of injection timing and fuel metering. It also provides two-stage lift, which allows for rate shaping and pilot injection. Modern unit injector controllers are also designed to communicate and control other engine subsystems such as exhaust gas recirculation, turbocharger waste gates, and positive traction. Cycle-to-cycle variability can be monitored and corrected for, and idle speed cylinder cutout is an added feature to provide even lower fuel consumption. The control unit also offers full diagnostics capability.

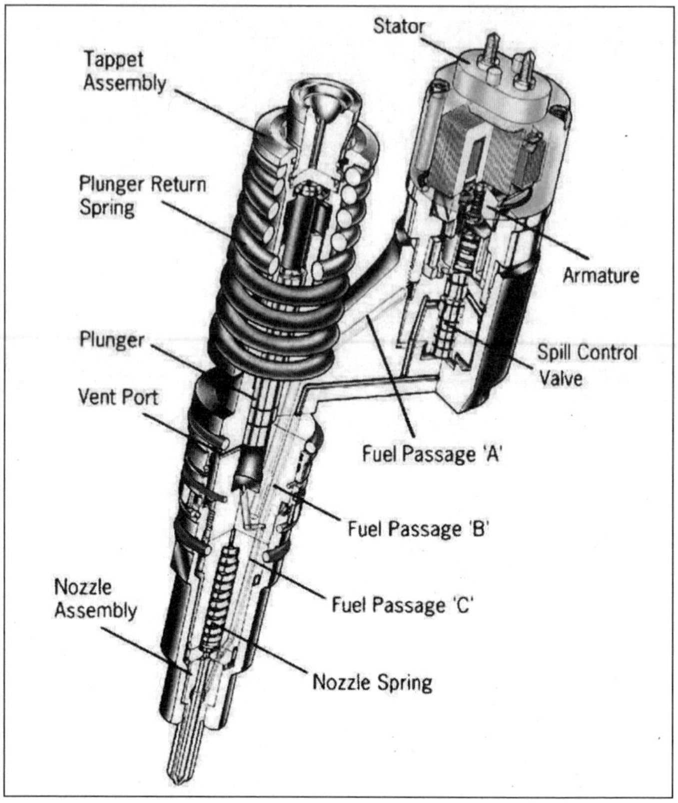

Figure 4.51 *Section of an electronically controlled unit injector. (Reprinted from Ref. 10, courtesy of Delphi Automotive Systems, formerly Lucas.)*

The unit injectors are normally installed in the cylinder head above the combustion chamber, with the pumping plungers driven by the engine camshaft via rocker arms. Fuel feed and spill are through integral passages in the cylinder head. Sensors provide information to the electronic control unit (ECU) on the relevant functions of engine operation, including the following:

- Accelerator pedal position
- Engine speed
- Camshaft position
- Intake manifold temperature
- Intake manifold pressure
- Coolant temperature.

This information is continuously compared with the optimum values stored in the ECU memory. The result of that comparison is translated into signals instructing the unit injector solenoid-actuated spill valve system to deliver the fuel at the timing required by the engine.

Each plunger moves through a fixed stroke under the action of the engine camshaft via a rocker arm and a plunger return spring. On the upward (filling) stroke, fuel at feed pressure passes from the cylinder head, through passage A, the open spill valve, and passage B into the chamber below the plunger. At a point determined by the ECU, an electrical pulse energizes the solenoid stator, moving the armature and thus closing the spill control valve. The plunger continues its downward stroke, causing pressure to build rapidly in the high-pressure passages. At a preset pressure, the nozzle opens and injection begins. When the stator is de-energized, the spill control valve is opened, causing the pressure to collapse, allowing the nozzle to close, and thus rapidly terminating the injection process. Injection timing and fuel quantity are determined by the starting point and duration of the electrical signal supplied to the stator by the electronic control unit.

4.11 Electronically Controlled Common-Rail System

High injection pressures that can be chosen independently of fuel quantity and engine speed, highly accurate injected fuel quantity, flexible injection timing control and pilot injection capability, small injection quantity capability, full electronic control, and high precision over the system's lifetime are all desirable qualities in a fuel injection system. The common-rail system fulfills these requirements, decoupling pressure generation and injection and allowing selection of the injection pressure independently of engine speed and injected fuel quantity, within certain limits. Combustion and its corresponding noise are improved by increasing fuel pressure up to 1500 bar and by introducing pilot or multiple injections. Common-rail systems can replace conventional injection systems for engines producing 30 to 50 kW/cylinder without requiring major engine modifications.

The modular common-rail system (Figure 4.52) consists of a fuel-quantity-controlled, high-pressure pump, fuel rail and high-pressure pipes, injectors, electronic control unit, and various sensors and actuators. The supply pump draws fuel from the tank and delivers it to the high-pressure fuel rail. The fuel volume in the rail damps the oscillations caused by the high-pressure pump and the injection process. A sensor measures pressure in the rail and provides the data to the ECU. The pump, sensor, and ECU comprise a closed loop for rail pressure control. The ECU compares the actual pressure with a set value, which depends on engine speed and injection quantity, and changes delivery of the pump to achieve minimal rail pressure deviation. The control of injection pressure is a completely new feature provided by the common-rail system.

Key components of the system are the electronically controlled injectors. They are supplied with fuel from the rail by short pipes. Their shape is quite similar to existing units, but injection is started by activating a solenoid valve inside each injector. A pulse from the ECU defines the start and end of the injection process. Injection quantity is determined by rail pressure, hydraulic flow of the nozzle, and the pulse length from the ECU.

The high-pressure pump for heavy commercial vehicles is a radial type with three pistons. To minimize its power consumption, its delivery can be adapted to the demand by a variable metering orifice. A solenoid valve positions a slide valve with a control edge to change the orifice and the fuel delivery to the pumping plungers. The required delivery is comprised of the injection quantity, the amount of leakage (especially in the injectors), and the amount of fuel in the solenoid valve for control of injection (control valve).

In a common-rail system, the fuel volume between the high-pressure pump and the injector acts as an accumulator. Therefore, the pressure in the rail is almost constant during the injection process. Only a small oscillation (less than 5 MPa) in pressure occurs. This oscillation is due to the pulsating delivery from the pump and the injection process at the injectors. To

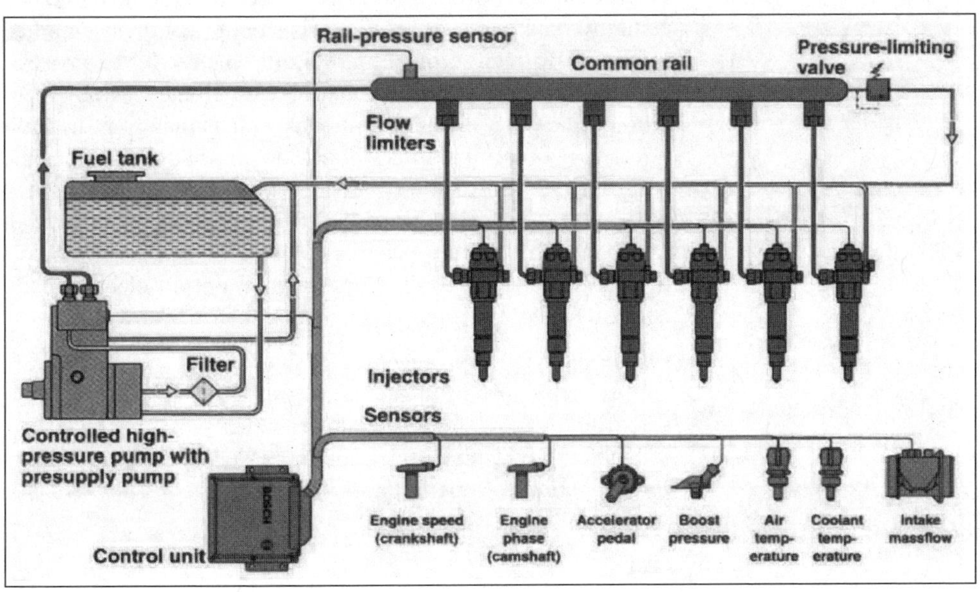

Figure 4.52 *Common-rail injection system. (Reprinted from Ref. 11, courtesy of Robert Bosch GmbH.)*

prevent an effect on the fuel injection value, the rail pressure is measured and the fuel injection quantity setpoint is calculated just before injection. The rail, a pipe containing a 60-cm^3 fuel volume, is a compromise between the prevention of rail pressure oscillations and assurance of sufficient pressure dynamics in the system.

The resistance of the rail pressure sensor element changes as a function of pressure. The injector (Figure 4.53) is controlled by a special solenoid-operated valve. It consists of an injector body similar to a conventional nozzle holder, a multihole nozzle, a valve body with two orifices controlling the pressure in the control chamber and a corresponding control piston, an armature device with a small ball to open and close the throttle (A in Figure 4.53), and a solenoid-coil assembly and the electrical connection for the ECU. The ECU oversees the functional control of the injectors (such as fuel pressure), engine (such as low idle), and vehicle (such as cruise control). Fuel-quantity calculation is similar to that in conventional electronic diesel injection systems. In normal conditions, the driver determines the fuel quantity with the accelerator pedal. From the fuel-quantity setpoint and the measured rail pressure, the control unit calculates the pulse length for the injectors. In addition, the ECU controls rail pressure by varying the duty cycle of the solenoid valve in the pump.

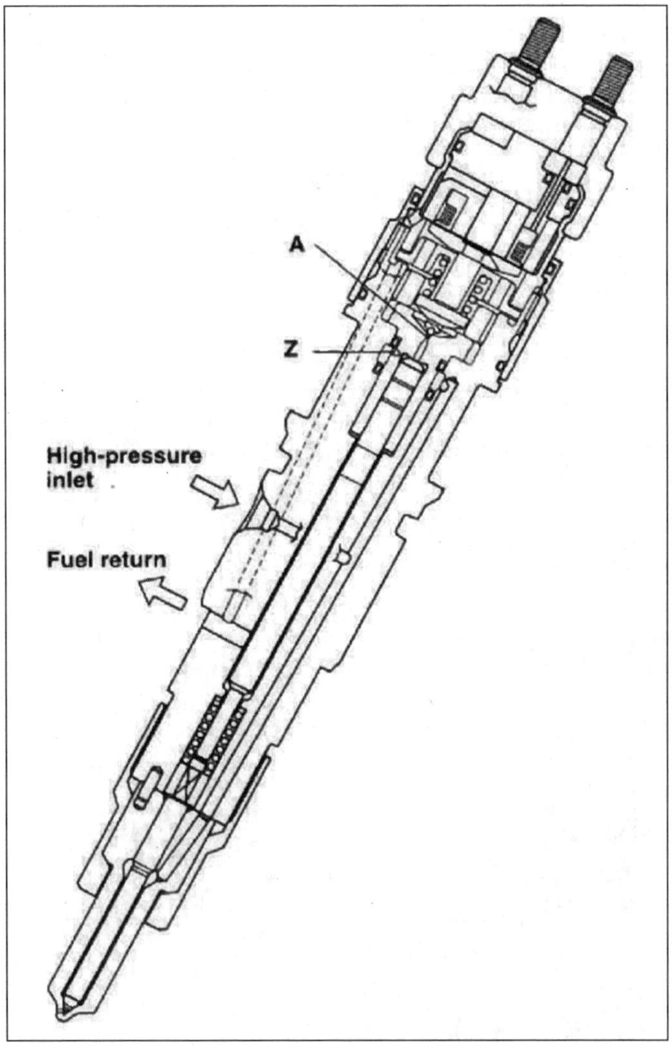

Figure 4.53 Common-rail injector. (Reprinted from Ref. 11, courtesy of Robert Bosch GmbH.)

The system's sensors are constantly monitored by the ECU (see Figure 4.54). A distinguishing feature of a common-rail system is the closed loop for rail pressure control. The duty cycle for the control of the high-pressure pump is changed by the ECU if the measured rail pressure deviates from the proper value. In the case of continual deviation, the fuel injection value for full load as well as the speed for the maximum speed governor are reduced. A valve mounted on the rail limits system pressure to avoid damage to the circuit.

In contrast to conventional systems, the rail pressure is always available at the nozzle seat. To avoid engine and injection system damage, additional measures and components are necessary. Flow limiters in the rail, one for each cylinder, stop fuel flow to the injectors in case of leakage or continuous injection.

The common-rail system maintains injection pressure regardless of other engine parameters. Figure 4.55 shows the effect of pressure variation on the injection quantity for different solenoid energizing times. Specific fuel quantities can be delivered with different energizing times at different pressures. The curves are smooth and follow a typical Bernoulli-type behavior. A minimum rail pressure of about 15 MPa is necessary for injection to occur in the case of Figure 4.55. This minimum pressure depends on the flow rates of the injector's two throttles. These throttles have to be manufactured carefully, because they also determine pressure dynamics in the control chamber and, therefore, the dynamics of the nozzle–needle motion.

4.12 Components of an Electronically Controlled Injection System

As mentioned earlier in this chapter, emission regulation of diesel exhaust is largely credited with modernizing the diesel engine. Other motivators are customer demand for improved fuel economy and better drivability. The fuel injection system had to be transformed to meet these requirements. In the process, engineers adopted electronic controls to facilitate the implementation of flexible injection timing as well as metering. In addition, many other functions were made possible with electronic controls. Several of these have already been mentioned and others are addressed in this section. There are, however, common features to these controls and a quick review of these features may be beneficial at this time. Most, if not all electronically controlled fuel injection systems include the following:

- Flexible control of injection timing
- Flexible control of injected quantity
- Ability to process additional parameters
- Tighter control tolerances and increased accuracy over very long periods of operation.

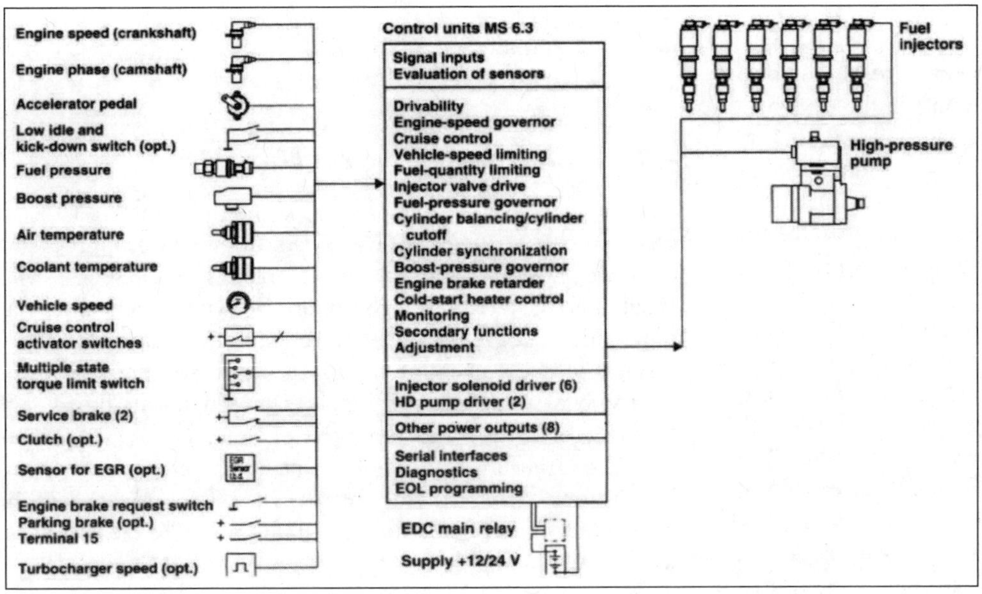

Figure 4.54 Electronic control layout of common-rail injection system. (Reprinted from Ref. 11, courtesy of Robert Bosch GmbH.)

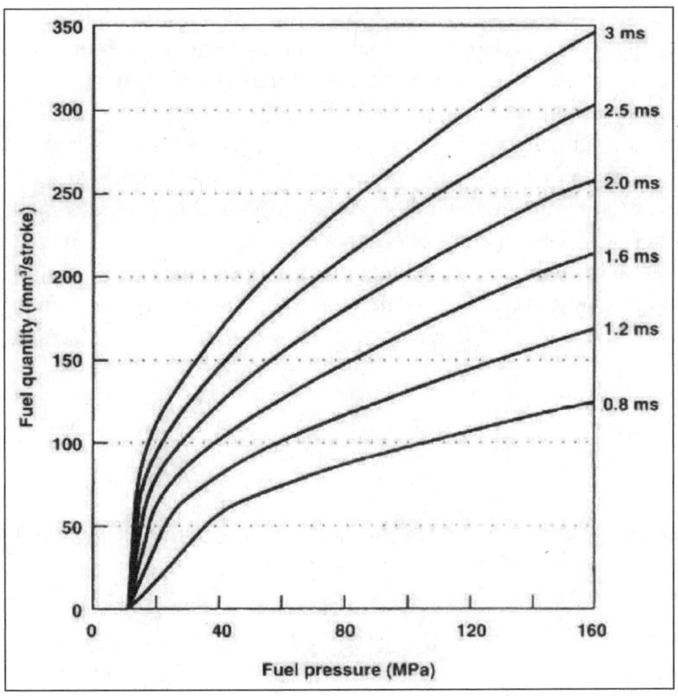

Figure 4.55 Injection quantity versus rail pressure for various control pulse widths. (Reprinted from Ref. 11, courtesy of Robert Bosch GmbH.)

The electronic system allows for electronic measurement, flexible and fast data processing, and closed-loop control with electrical actuators.

Diesel combustion is highly influenced by the following:

- Injected fuel quantity
- Start of injection
- Exhaust-gas recirculation (EGR) *when applied*
- Charge air temperature and pressure.

These controlled variables must be optimally adjusted for every working mode in order to ensure efficient diesel engine operation. This is primarily what the electronic system does for each engine operating condition and at various ambient conditions as well.

The electronic system consists mainly of three major blocks as shown in Figure 4.56:

1. Sensors that monitor operating conditions and record physical parameters in the form of electrical signals
2. Electronic control unit (ECU) with microprocessors that processes the information according to specific control algorithms and outputs corresponding electrical signals
3. Actuators that convert the ECU's electrical output signals into mechanical quantities.

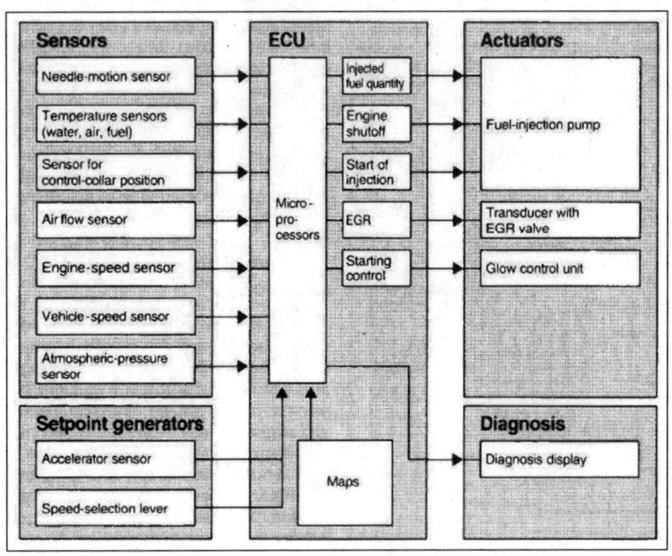

Figure 4.56 Electronic control system building blocks. (Reprinted from Ref. 5, page 186, Fig. 1, courtesy of Robert Bosch GmbH.)

4.12.1 Sensors

The positions of the accelerator pedal (driver demand) and the control collar in the injection pump (in the case of a distributor pump) are registered by the angle sensors. Engine speed and TDC are monitored and recorded by inductive sensors. Sensors with high measuring accuracy and long-term stability are used for pressure and temperature measurements. The start of injection is recorded by a sensor that is integrated into the nozzle holder.

Other sensors used to monitor conditions that may require trimming the primary injection timing or fuel metering functions are also incorporated. For instance, temperature sensors for cooling water provide information regarding cold-start conditions that may require additional fuel to overcome higher engine friction. Cold starting may also call for more advanced timing to improve startability, or in some cases increased exhaust back pressure to eliminate white smoke emissions associated with cold starting. Airflow sensors may be used in control algorithms that would limit fuel quantity to avoid smoke formation resulting from poor combustion. Sensing of the fuel temperature allows for the implementation of fuel quantity trimming to maintain power output regardless of ambient temperature. Sensing both airflow and air temperature allows the ECU to calculate air mass and determine the optimum fuel mass for good combustion efficiency.

Additional control features may include vehicle speed control. Whether controlling vehicle speed means limiting the allowable road speed for fuel economy purposes or for cruise control purposes, this function can easily be achieved by sensing vehicle speed and using this signal to control fueling accordingly. Sensing atmospheric pressure is extremely important for limiting fuel (derating engine output) to avoid increased smoke emissions at high altitudes in places such as Denver, Colorado, or Mexico City, Mexico.

4.12.2 Electronic Control Unit

The *electronic control unit*, sometimes referred to as electronic control module, utilizes digital microprocessor technology. The microprocessor interfaces with the sensors via input interface circuitry. It also interfaces with the actuators via output interface circuitry. A signal conditioning compartment is responsible for translating signals received from the sensors to microprocessor-compatible inputs. These inputs are processed through control logic and mapped parameters such as engine speed, load, airflow, and coolant temperature, and a desired fuel quantity and injection timing are calculated. The resultant control parameter is converted into a signal that is sent to the actuators to perform the final injection timing as fuel quantity control. A simplified example of this process is given in Figure 4.57 for a smoke limiting control strategy.

A speed signal is used in a control logic block that contains the maximum fuel required for full-load torque output. This output is converted to a signal sent to a comparator block that is designed to select the minimum mass of fuel. Besides the full-load torque signal, a maximum allowable fuel signal based on a manifold absolute pressure value sensed and calculated is also received by the minimum wins block. Two additional signals are also received by the minimum wins block: a mass fuel signal from the governor function and a maximum fuel signal from the engine protection circuitry/logic. The minimum wins comparator selects the lowest value of all the signals received after they have been processed through their various

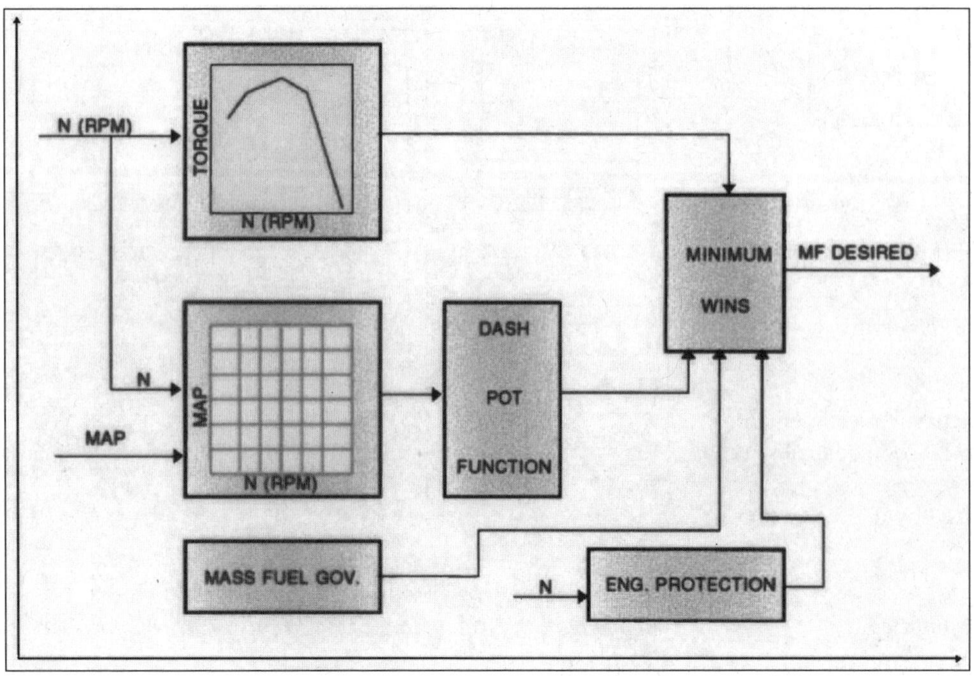

Figure 4.57 Example of smoke control logic.

logic circuits and sends this signal to the actuators to indicate the mass of fuel desired. This signal would guarantee that the maximum allowable safe power output (that is, the smoke-free output) is produced.

Obviously, ECUs have the ability to provide a tremendous amount of additional control. As seen from the diagram in Figure 4.56, engine shutoff, engine start, and EGR control are just a few of the other functions that can be handled by the ECU. Microprocessor technology is continually improving and becoming more capable of handling additional computations in reduced time and, therefore, more capable of handling additional control functions.

Precautions have to be taken to protect against electromagnetic interference. Inputs and outputs are short-circuit-proof and protected against spurious pulses from the vehicle's electrical system. Protective circuitry and mechanical shielding provide a high level of EMC (electromagnetic compatibility) against outside interference.

4.12.3 Actuators

Actuator technology is also developing at a rapid pace to match microprocessor developments. So far, electromagnetic solenoid valves have been the most common form of electrical actuation in electronically controlled fuel injection systems. Initially designers were faced with the problem of high power consumption since the early designs had to develop enough force to withstand high injection pressures. Newer designs, however, have featured clever mechanisms that use high pressure within the system to assist in improving valve sealing and actuation, while reserving the control function *only* to the electrical component. R. Bosch employs linear solenoid actuators to provide control rod (rack) actuation, and a rotary solenoid actuator to move the spill collar in the distributor pump. The response time of these electromagnetic solenoids is about 300 ms, which allows its performance to meet the requirements of electronic controls.

In the quest for continued improvement and to meet the desire for even faster actuator response, several companies have turned to piezoelectric actuators. Piezoelectric crystals provide motion when given a pulse of electrical energy. Piezoelectric crystals are often arranged in wafer form, and when several wafers are stacked and energized, they are capable of producing displacements that can be used to open or close tiny passages as found in injection systems. One of the early examples of using piezoelectric stacks was described in U.S. Patent Nos. 4,175,587 and 4,180,022, both issued in 1979. These patents were assigned to the Chrysler Corporation and described the invention and development of a piezoelectric stack–operated valve used to intercept the hydraulic connection between a distributor pump and its commutator assembly. By striking the stack with electrical energy, the inventors were able to obtain two injections, a pilot and a main injection, for each combustion cycle. This was made possible because of the fast response time of the piezoelectric valve. Reference [12] provides some details about piezoelectric-actuated injectors (see Figure 4.58).

This injector was originally intended to serve as a development tool to explore the potential offered by such a system. The benefits of the system are not restricted to high injection pressure, but include extremely fast and accurate valve actuation. The new injector offers a variety of features for future vehicle applications such as multiple injection, pilot injection, and split injection, which can result in reduced engine noise, smoke emissions, and fuel consumption.

Utilizing a pressure-balanced, three-way servo valve in combination with a small piezoactuator (0.5 cc in volume), this injector is capable of extremely low injection quantities of less than 1 mm^3/stroke, while maintaining an excellent injection quantity repeatability up to rail pressures of 23,000 psi. This injector is quite attractive for small DI diesel engines, where exact and fast fuel metering of very small quantities for pilot injection may be required.

Because of its low moving mass inside the injector, the dynamics are significantly improved compared to conventional high-pressure fuel injectors, resulting in a nozzle–needle transient time of about 0.25 ms, a marked improvement over electromagnetic solenoid valves. Based on its unique design, the injector requires only 0.04 W/stroke at low operating current and a maximum voltage of 200 V, resulting in lower energy requirements for operation than is exhibited by current gasoline MPI injectors. The German automotive supplier Siemens has already developed a diesel injection system using piezoelectric technology.

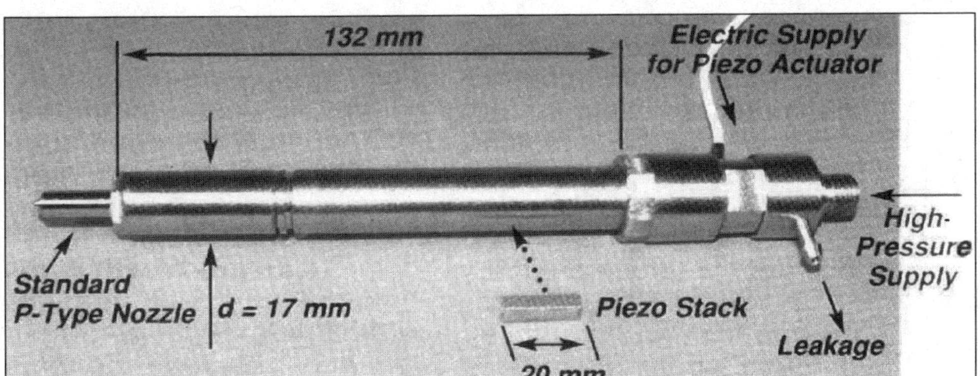

Figure 4.58 *Piezoelectric-actuated injector. (Reprinted from Ref. 12, courtesy of FEV Engine Technology.)*

4.13 Innovative Injection Systems

Pressure applied primarily by the need to design engines that met mandated emission regulations led major engine manufacturers to develop their own and often unique fuel injection systems. This statement is especially true in the heavy-duty engine field where examples of internally developed injection systems abound. For instance, Detroit Diesel Corporation developed its own line of electronically controlled unit injector systems in the 1980s in cooperation with GM's Rochester Products Division. The system was controlled by the DDEC electronic control unit, which went through several iterations. Another example is Caterpillar Inc.'s hydraulically actuated, electronically controlled unit injector system (HEUI). Although Caterpillar holds the original patent for this system, Navistar is credited with its commercial success as well as many patented improvements that were assigned back to Caterpillar.

In some cases, major heavy-duty engine manufacturers were able to license patented technology and further develop the concepts for their own engine line. An example is the Bendix Diesel Engine Controls unit injector system that was licensed by Cummins Engine Company. Several features of this patented design appear in the Cummins Quantum CELECT injection system. The Bendix invention, covered by U.S. Patent No. 4,281,792, was perhaps an idea that was ahead of its time. It called for high injection pressure at a time when this was strange to even veteran diesel engineers. It also called for full authority, drive-by-wire, totally flexible injection control at a time when the only electrical component on a heavy-duty diesel engine was the fuel shutoff solenoid.

Figure 4.59 shows a schematic diagram of a diesel fuel injection system that utilizes electronic control of an engine-driven unit injector. Elements of the system are described in the following sections.

4.13.1 The Bendix Diesel Engine Control Unit Injector System

Figure 4.60 shows the components of the Bendix electronic diesel fuel injector (BEDFI), which was designed for use in high-speed engines of the automotive type. Figure 4.60 also describes the operating principle of the BEDFI. Shown schematically are the operating plunger, item A (diagram 1, Figure 4.60); the floating piston, item B; the solenoid control valve, item C; a spring-loaded nozzle valve, item D; a spring-loaded fill check valve, item E; a bypass port, item F; and a fuel supply passage, item G. In diagram 1 the operating plunger, actuated by an injection cam/rocker system from the engine camshaft, is moving downward on a pumping stroke. Solenoid valve C is open, allowing fuel to be displaced from the injector cavity to the supply passage at low pressure. In diagram 2 the solenoid control valve is closed and fuel displaced by plunger A opens nozzle valve D and is injected into the combustion chamber; fill check valve E is closed. Injection continues until the upper edge of floating piston B passes bypass port F. Fuel displaced by the continued downward movement of the plunger flows to the supply passage through the bypass port as shown in diagram 3. Diagram 4 shows the injector return spring forcing the plunger upward. Check valve E opens and the floating piston is forced upward by fuel at supply pressure. When the floating piston covers the bypass passage, metering of fuel for the next stroke begins. Fuel will flow into the metering volume under the metering piston until the metering period is terminated when the solenoid valve is opened. Diagram 5 shows the metering piston held in a stationary position as fuel at supply pressure flows into the timing chamber region between the metering piston and the operating plunger; the plunger continues to rise to its upper position. A metered quantity of fuel is now trapped below the floating piston to be injected during the next cycle of operation.

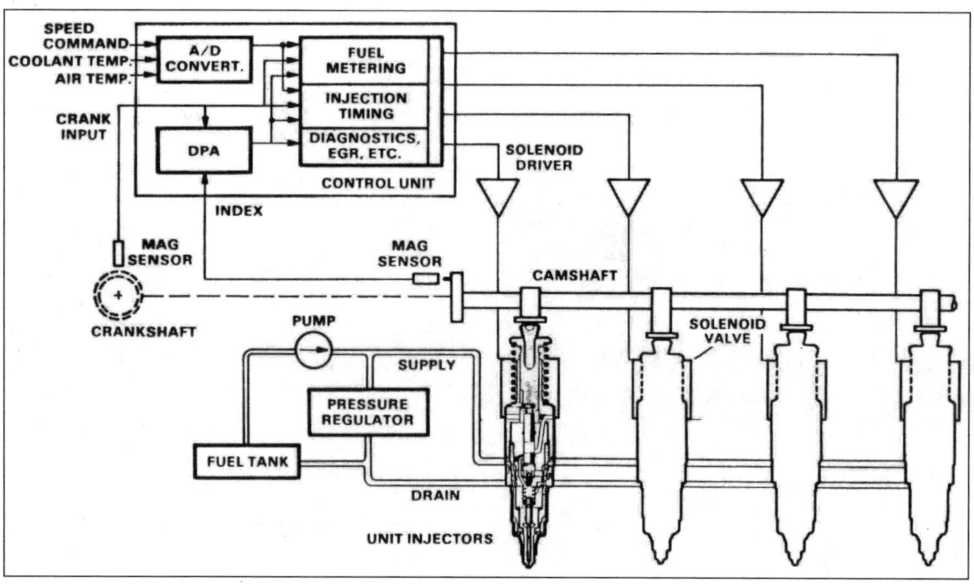

Figure 4.59 The Bendix injection system. (Reprinted from Ref. 13.)

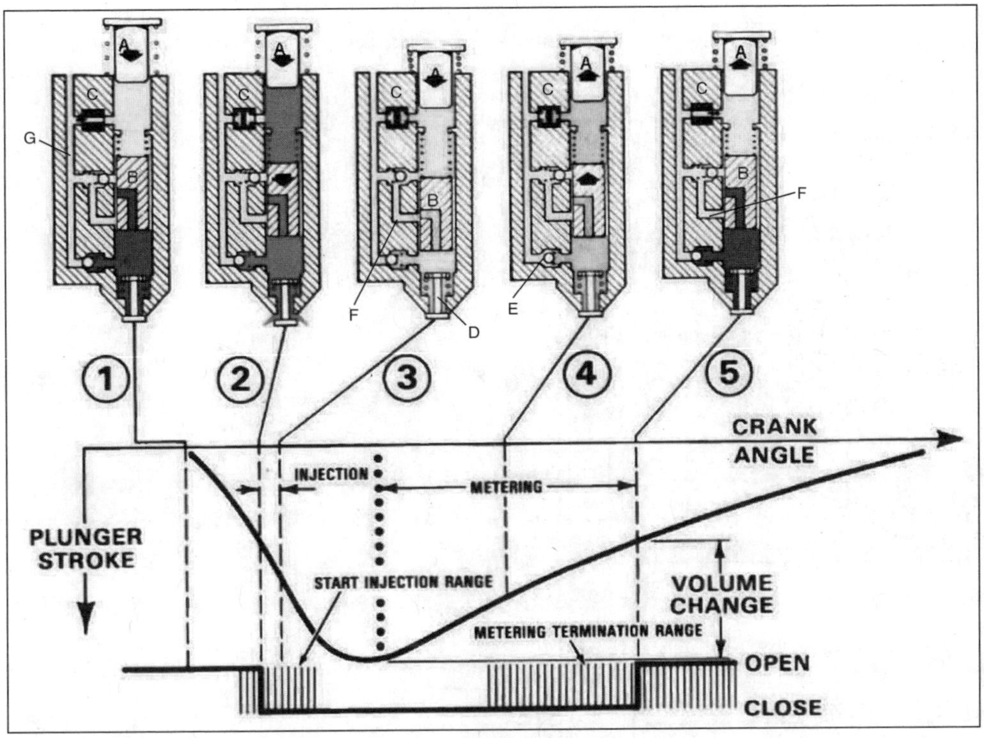

Figure 4.60 Operation of the Bendix injection system. (Reprinted from Ref. 13.)

The timing diagram of Figure 4.60 shows that the time at which injection begins is variable and controlled by the time in the cycle at which the solenoid control valve is closed. The metered fuel quantity is variable and controlled by the time in the cycle at which the solenoid valve is opened. These times are under the control of the ECU and can be varied independently.

The injector is driven by a rocker arm operated from the injection cam. The fuel supply to the injector is through a passage in the cylinder head; excess fuel is returned from the injector to the fuel supply system through another passage also in the cylinder head. The injectors are connected to the electronic control unit through a wiring harness. Fuel is supplied to the injector supply passage by a low-pressure transfer pump (generally) that is belt driven from the engine. Proper filters and strainers are required to protect the injectors and the transfer pump. A pressure-regulating valve is used to control the supply pressure to the injectors.

The electronic control unit provides the electrical signals to the injectors to control the solenoid valves at the proper times to obtain the desired injection timing and the desired quantity of fuel for a particular engine operating point. The ECU can be programmed to provide optimum timing and deliveries for such performance factors as fuel economy, emissions, and noise control over the speed and load range, and for best starting. In addition, it provides speed governing, and can be programmed for variable-speed governing or for idle and all-speed governing. The ECU is designed to receive a number of signals from engine-mounted sensors as well as the basic operator input signals, which establish an operating speed/load condition.

The Quantum CELECT Cummins system (see Figure 4.61) has a number of similarities to the Bendix patent. In particular, both systems feature a floating piston, a cavity on top of the floating piston (timing chamber), and a cavity below the floating piston (metering chamber), and both systems use the premetering versus the real-time metering concept. Functions made possible by the electronic control of the engine include optimized starting, low- and high-idle governing, automotive min/max-type governor function, all-speed governing, air/fuel control, injection timing control, engine protection through torque limiting, and a host of other engine and vehicle control features that are unavailable in mechanical systems.

4.13.2 The Hydraulically Actuated, Electronically Controlled Unit Injector

Caterpillar Inc. developed a diesel engine fuel system that is powered by hydraulics and controlled electronically. This hydraulic electronic unit injector (HEUI) requires no mechanical actuating or mechanical control devices and offers many advantages over conventional fuel injection systems. Inherent features of the HEUI fuel system include injection pressure control that is *independent* of engine load or speed, totally flexible injection timing, and full electronic control of injection parameters. The hydraulic actuating circuit is straightforward, using lubricating oil from the engine sump. Hydraulic lines may be internal to the engine or external. Figure 4.62 is a schematic cutaway showing the HEUI details.

The middle segment of the injector consists of the hydraulic intensifier piston, the plunger and barrel, and the plunger return

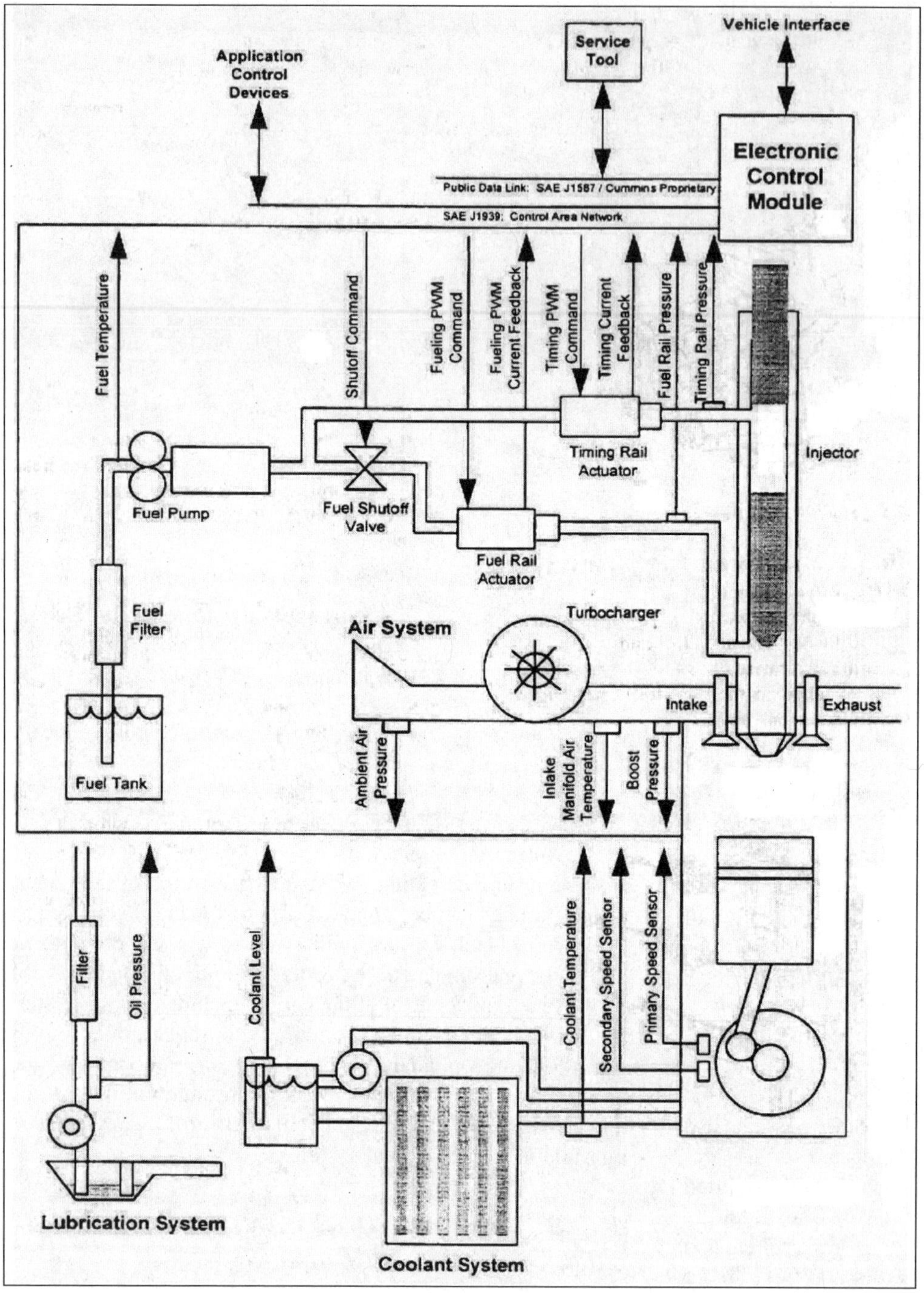

Figure 4.61 The Quantum CELECT injection system.(Reprinted from Ref. 14.)

spring. Intensification of the fuel pressure to desired injection pressure levels is accomplished by the ratio of areas between the intensifier piston and plunger. The intensification ratio can be tailored to achieve desired injection characteristics. Injection begins as high-pressure actuating oil is supplied to the top of the intensifier piston. As the piston and plunger move downward, the pressure of the fuel below the plunger rises. The piston continues to move downward until the solenoid is de-energized, causing the poppet to return to the lower seat, blocking oil flow. The plunger return spring returns the piston and plunger to their initial positions. As the plunger returns, it draws replenishing fuel into the plunger chamber across a ball check valve. The nozzle is typical of other diesel fuel system nozzles.

Fuel is supplied to the nozzle through internal passages. As fuel pressure increases, the nozzle needle lifts from the lower seat, allowing injection to occur. As pressure decreases at the end of injection, the spring returns the needle to its lower seat. Figure 4.63 is a diagram showing the major HEUI system components and how they are connected. The six main components are as follows:

- High-pressure oil pump
- Rail pressure control valve (RPCV)

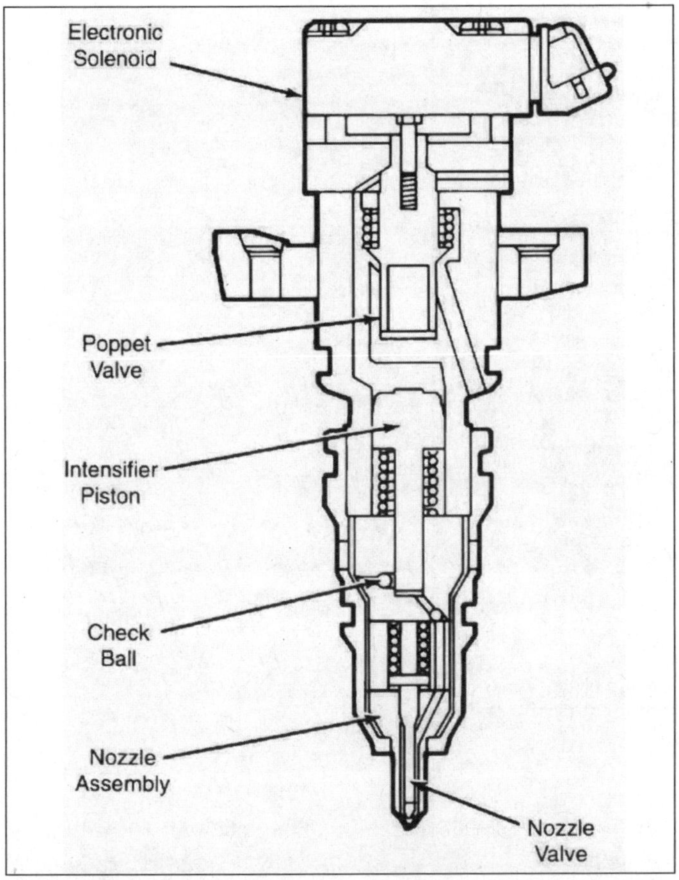

Figure 4.62 *Schematic cutaway of the HEUI injector. (Reprinted from Ref. 14. Artwork of HEUI fuel system courtesy of International Truck & Engine Corporation.)*

- Hydraulic unit injector
- Sensors
- Electronic control module (ECM)
- Fuel transfer pump.

Three stages of injection can be distinguished with the HEUI system: (1) the fill cycle, (2) injection, and (3) end of injection. Each stage is discussed in the following paragraphs.

Fill cycle: During preinjection all internal components are found in their spring-loaded (nonactuated) positions. The solenoid is not energized and the poppet valve is blocking high-pressure oil from entering the injector. The plunger and intensifier are at the top of their bore and the plunger cavity is full of fuel. Fuel pressure in the plunger cavity is the same as fuel rail pressure, approximately 40 psi. The fill cycle is illustrated in Figure 4.64.

Injection: When the ECM determines that the injector should be fired, the following sequence of events occurs:

- The ECM sends a fuel delivery control signal (FDCS) to the injector drive module (IDM).
- The IDM sends an electric current pulse to the injector solenoid.
- The solenoid is fully energized almost instantly, creating a strong magnetic pull on the armature.

Because the armature is connected to the poppet valve, it follows that the injection process begins when these events occur:

- The magnetic pull of the solenoid overcomes the spring tension holding the poppet closed.

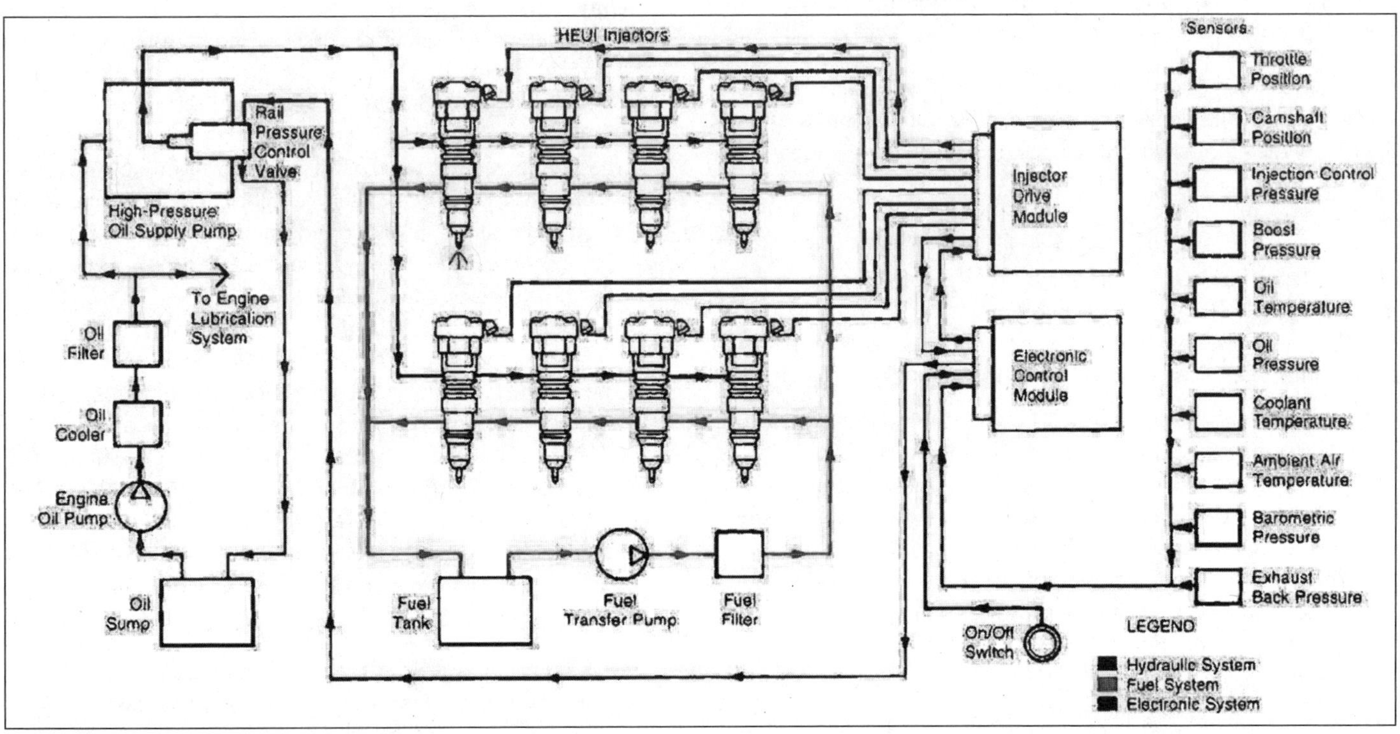

Figure 4.63 *Components of the HEUI injection system. (Reprinted from Ref. 15. Artwork of HEUI fuel system courtesy of International Truck & Engine Corporation.)*

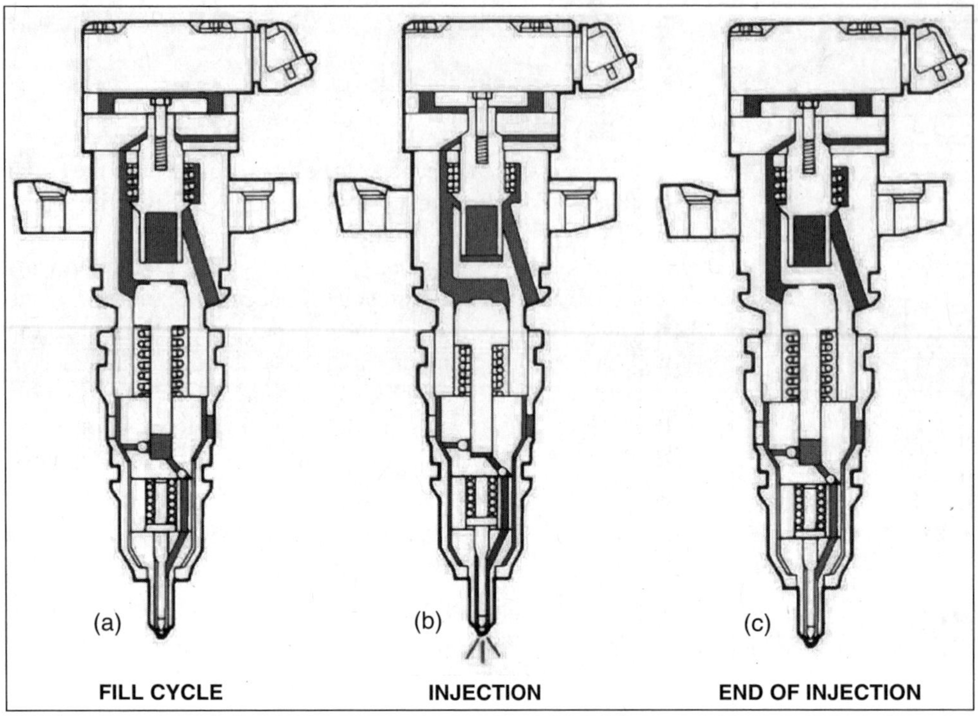

Figure 4.64 *HEUI system cycles: (a) fuel fill cycle, (b) injection, and (c) end of injection. (Reprinted from Ref. 15. Artwork of HEUI Fuel System courtesy of International Truck & Engine Corporation.)*

- The poppet is quickly raised off its seat.

When the poppet valve opens:

- The upper poppet land closes off the path to the drain.
- The lower land opens the poppet chamber to incoming high-pressure oil.
- High-pressure oil flows around poppet to the top of the intensifier piston.

Pressure on the top of the intensifier forces it down along with the plunger. The downward movement of the plunger pressurizes the fuel in the plunger cavity and nozzle. When the fuel pressure reaches a valve opening pressure (VOP) of about 2700 psi, the nozzle valve lifts off its seat and injection begins.

As long as the solenoid is energized, the poppet remains open and pressurized oil continues to flow in, pushing down the intensifier and plunger. Injection pressures range from 3000 to 21,000 psi depending on engine requirements. Injection continues until one of the following conditions occurs:

- The solenoid is de-energized, allowing the poppet spring to close the poppet and shut off the high-pressure oil.
- The intensifier hits the bottom of its bore. This action indicates maximum fuel delivery. The distance from the bottom of the intensifier to the bottom of its bore is a controlled distance to limit maximum fuel delivery. This feature is called *stroke limiting,* and it prevents electronic tampering from increasing the maximum fuel delivery beyond the maximum intended fuel rates.

End of injection: This process begins when the ECM terminates the fuel delivery control signal to the IDM. The IDM then terminates the electric pulse to the solenoid. The following events occur:

- The magnetic field of the solenoid collapses and is no longer able to overcome poppet spring tension to hold the poppet off its seat.
- The poppet closes, shutting off high-pressure oil from entering the injector.
- When the poppet is seated, the upper land of the poppet opens the poppet cavity to drain.
- Pressurized oil in the intensifier chamber and poppet chamber flows upward around the poppet seat, through the vent holes in the poppet sleeve, and out the adapter drain hole.
- The pressure of the fuel in the plunger cavity exerts an upward force on the plunger and intensifier. As the pressure of the pressurized oil above the intensifier drops, so does the downward force on the intensifier.
- The upward force from the pressurized fuel almost instantly becomes greater than the downward force on the intensifier so the downward motion of the intensifier and plunger stops.
- When the plunger stops, fuel flow also stops.
- With the check valve still open, the remaining pressure on the fuel pushes a tiny amount of additional fuel out of the orifice holes. This causes a large pressure drop, which lowers nozzle pressure below the valve closing pressure (VCP) of 1600 psi. Spring tension on the check valve now reseats it and injection stops.

The injection as well as the end of injection are illustrated in Figures 4.64b and c, respectively. To complete the total cycle we should revisit the fill cycle, which follows these steps:

When the check closes, injection stops and the fill cycle starts as follows:

- The poppet and intensifier cavities are open to atmospheric pressure through the poppet valve and adapter drain hole.
- Pressure drops very rapidly in the intensifier and plunger cavities to near zero.
- The plunger return spring pushes upward on the plunger and intensifier, forcing oil around the poppet, through the holes in the poppet sleeve, and out the adapter drain hole.
- As the plunger rises, pressure in the plunger cavity also drops to near zero. The 40-psi fuel rail pressure unseats the plunger fill check, allowing the plunger cavity to fill with fuel.
- The fill cycle ends when the intensifier is pushed to the top of its bore.
- The plunger cavity is full and the fill check returns to its seat.
- Pressure in the intensifier and poppet chambers is zero.
- The injection cycle is complete and is ready to begin again.

Through this description of the HEUI system and the definition of the various types of diesel fuel injection systems, it is obvious that HEUI is a unit injector system. The reason for this definition is that the high pressure is created in the same embodiment that houses the injector nozzle. However, the HEUI system also exhibits some of the characteristics of the common-rail system. One of the more important shared features between the HEUI and common-rail systems is their ability to develop pressure independently of the engine speed. Another shared feature is the potential of the HEUI system to provide multiple injections. To do so, the HEUI system had to be redesigned to allow for the solenoid valve to perform this function. This redesign produced the HEUI-B system, which is shown in a diagrammatic form in Figure 4.65. However, the most important feature of the new system may be its expanded rate-shaping capability. The HEUI-B allows the injection "shape" to be electronically selected, all with an eye toward optimizing performance and emissions in a variety of operating conditions. The HEUI-B can shape injection in one of three ways—ramp, square, or split—all of which are a function of the injection pressure vs. crank angle. A ramped rate shape has an increasing injection pressure from the start of injection until the end. Shaping injection in this way can lower NO_x and

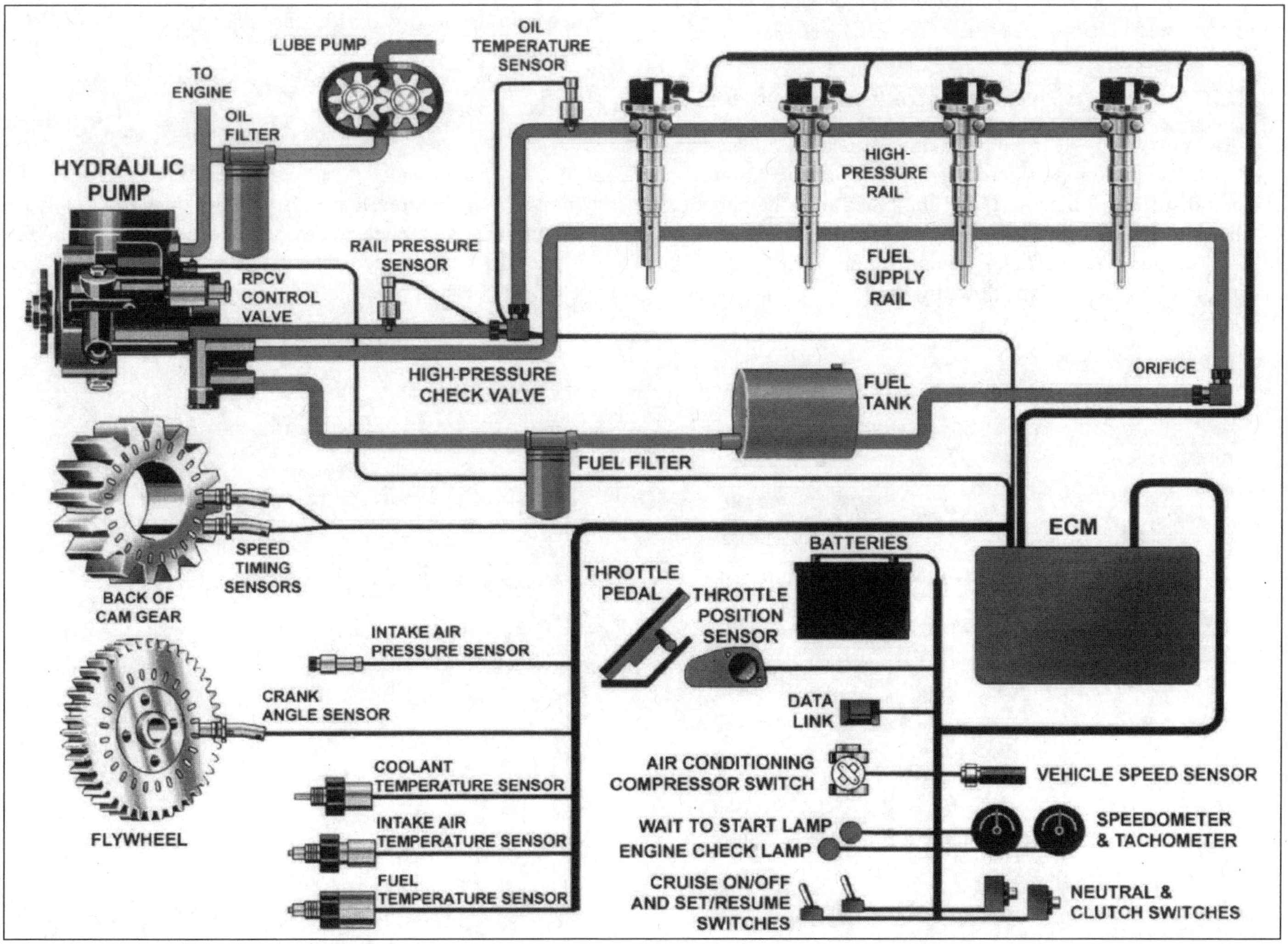

Figure 4.65 *The HEUI-B injection system. (Reprinted from Ref. 15. Artwork of HEUI fuel system courtesy of International Truck & Engine Corporation.)*

noise and provide improved BSFC at cylinder pressure limited timings, that is, with greater timing advance. The square rate shape has a flat pressure profile that produces benefits such as lower smoke levels, improved intermediate load BSFC (shorter duration, high MEP), and improved response because of the shorter duration and low smoke.

The primary difference between the original HEUI system and the HEUI-B is the use of a two-stage valve (see Figure 4.66). Split injection shapes the rate of both the pilot (quantity, timing and pressure) and the main fuel injection, which can produce lower combustion noise, reduced low load hydrocarbon (HC) levels, and lower levels of NO_X + HC. Both HEUI systems use a single solenoid with two wires, but the HEUI-B injector has a unique valving concept where the single solenoid controls both the oil pressure to the intensifier, as in the original system, and also the oil pressure to control the nozzle check. Rate shaping is accomplished by precise hydraulic and event control of the two valves, one of which acts as a pilot to the other. It should also be noted, that discrete valve events control the injection process. Neither valve depends on any sort of modulation during its travel to accomplish injection control. With this direct control of the nozzle check for both opening and closing, the various injection rate shapes are realized by independent control of opening and closing.

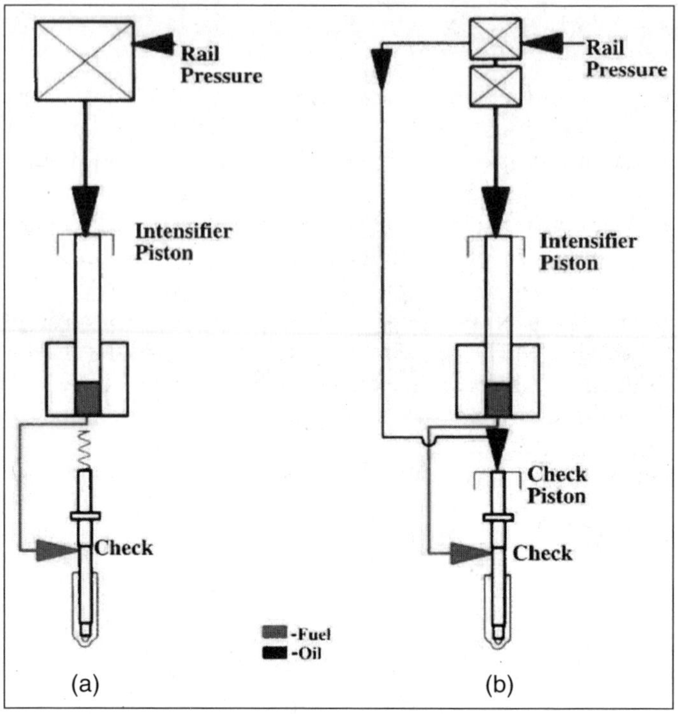

Figure 4.66 Operational differences between (a) the HEUI single two-wire solenoid poppet valve system and (b) the HEUI-B single two-wire solenoid, two-stage valve system. (Reprinted from Ref. 16.)

4.13.3 Cummins Accumulator Pump System

Engineers at the Cummins Engine Company conceived a fuel injection system that combines features of the common-rail and distributor pump systems, combined with the benefits of electronic controls. Figure 4.67 shows the flow schematic of the Cummins Accumulator Pump System (CAPS). In this system, the CELECT valve allows fuel to fill the volume above the plunger. As the plunger moves upward, the CELECT valve closes and acts as a check valve to seal the fill passage. The high-pressure fuel proceeds past a check valve to an accumulator where pressure pulses are dampened. High-pressure fuel

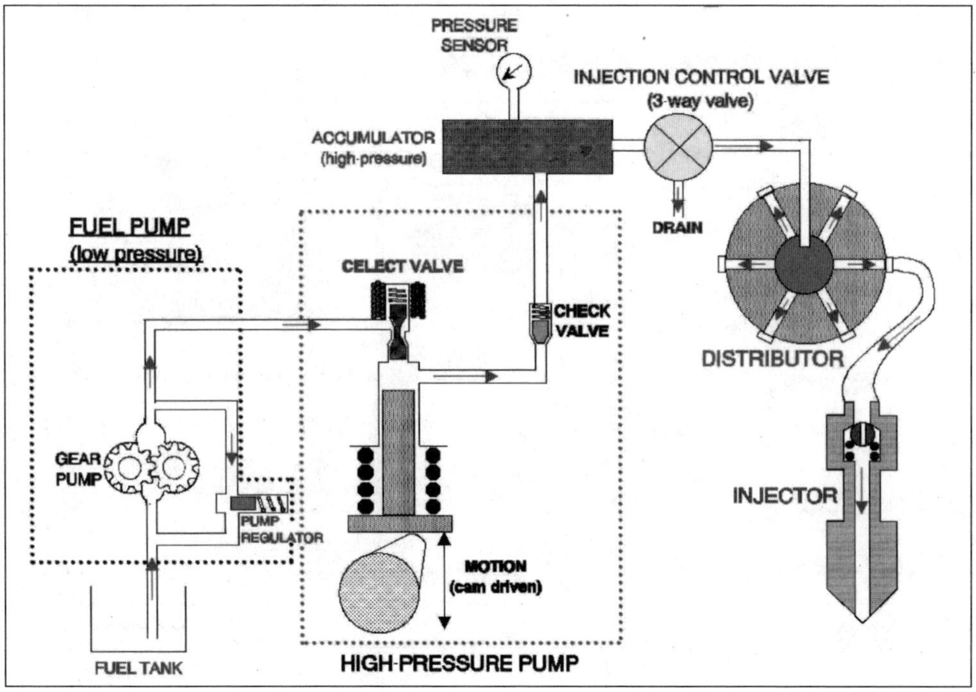

Figure 4.67 Cummins Accumulator Pump System. (Reprinted from Ref. 17, courtesy of Cummins Engine Company.)

proceeds from the accumulator to the commutator section (distributor) through a three-way, electronically controlled valve that is responsible for controlling the start and end of injection.

4.14 Conclusion

Although there is much more to be said and written about the diesel fuel injection system, this chapter attempted to provide a generalized overview. It is designed to acquaint the reader with the evolution of the system from the old mechanically controlled designs to the modern electronically controlled designs. Many heroic attempts were made along the way with the purpose of achieving tighter injection controls using mechanical systems. An example that was not covered is the dual-spring injector shown in Figure 4.68. This injector allowed for staged injection, which provided noise control at idle conditions. In presenting this material, both high injection pressure and injection control flexibility were credited as being major contributions toward the modern diesel engine. For some applications, however, injection pressure effects far outweighed the benefits of electronic control flexibility. So it is perhaps fitting to close this chapter with a graph showing injection pressure capability and projections of various injection systems (see Figure 4.69).

References

1. "Technical Instruction—Fuel Injection Equipment for Diesel Engines," Report VDT-UBP 001/15 B (2.72), R. Bosch, Germany, 1972.
2. Heisler, H., *Advanced Engine Technology,* Butterworth-Heinemann, New York, 1995.
3. Tschoeke, H., "The Fuel Injection System," Lecture Notes from SAE International Diesel Engine Design Academy, Ypsilanti, MI, July 1999. Photos courtesy of Robert Bosch GmbH.
4. "Fuel Injection Equipment for Diesel Engines—Fuel Injection Pumps PE and PF," 1st ed., Robert Bosch GmbH, Germany, 1971.
5. *Diesel Fuel Injection,* 1st ed., Robert Bosch GmbH, Germany, March 1994.
6. "DDC/MTU Series 2000 and 4000 Engines," Article on Industry Innovations, *Off-Highway Engineering,* SAE International, Warrendale, PA, December 1996.
7. Roosa, V., T. Hess, and J. Walker, "The Roosa Master Nozzle," SAE Paper 640183, SAE International, Warrendale, PA, September 1964.
8. "Operation and Instruction Manual Model DM Pump," Roosa Master Publication No. 99834, Rev. 1 (9-76), Stanadyne/Hartford Division, Box 1440, Hartford, CT 06102, USA.
9. Brady, R. N., *Modern Diesel Technology,* Prentice Hall, Upper Saddle River, NJ, 1996, p. 119.
10. EUI Systems brochure, Delphi Injection Systems (formerly Lucas), Troy, MI, 1988.

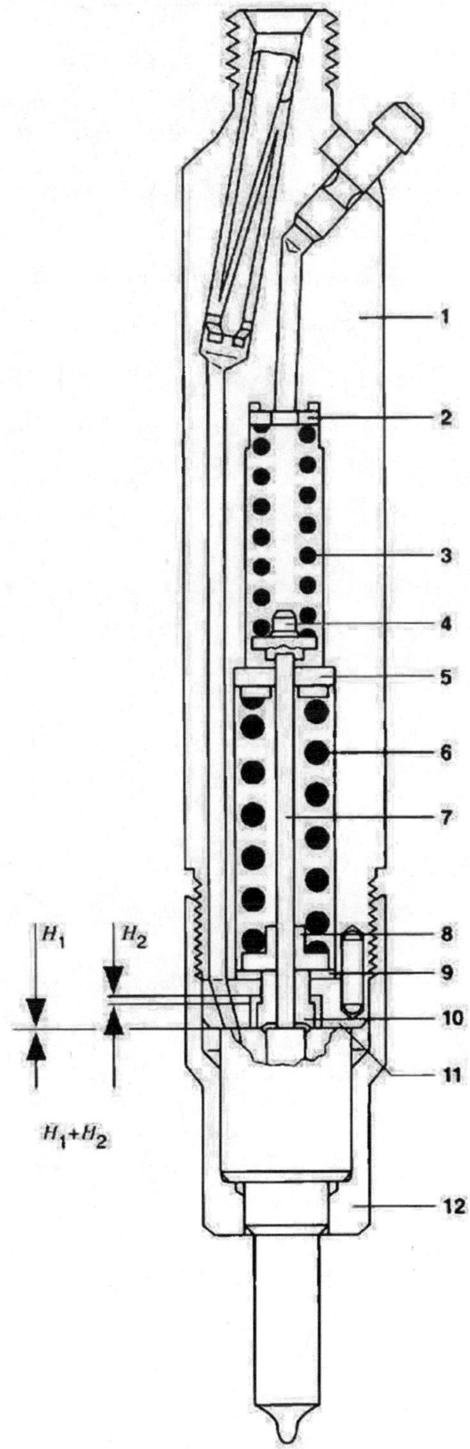

Figure 4.68 *Two-spring injector by Bosch. (Reprinted from Ref. 18, page 92, Fig. 10, courtesy of Robert Bosch GmbH.)*

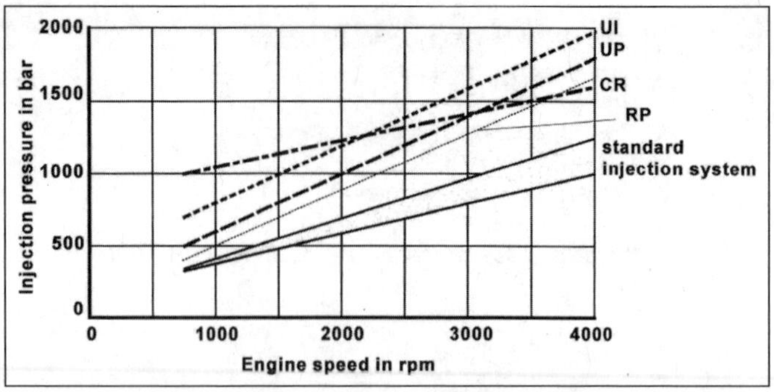

Figure 4.69 *Injection pressure capability of various injection systems. (Reprinted from Ref. 3, courtesy of Robert Bosch GmbH.)*

11. Boehmer, W., and K. Hummel, "Decoupling Pressure Generation and Injection," Article on Common-Rail Technology, *Off-Highway Engineering,* SAE International, Warrendale, PA, August 1997. Photos courtesy of Robert Bosch GmbH.
12. "High-Pressure, Piezo-Actuated Diesel Fuel Injector from FEV," *Diesel Progress,* North American ed., Volume LXVI, No. 10, Diesel & Gas Turbine Publications, Waukesha, WI, October 2000.
13. Khair, M., "The Diesel Explosion," MBA thesis, Michigan State University, 1982.
14. Amlung, R., J. P. Seger, T. Warlick, and D. Wilhelm, "Cummins Quantum CELECT Fuel System Description for the QSK 19 Diesel Engine," SAE Paper No. 961748, SAE International, Warrendale, PA, 1996
15. "HEUI—Fuel System Operation," CGE 472-1, Rev. 2/95, Navistar International Transportation Corporation, Chicago, IL, 1995.
16. Glassey, S. F., A. R. Stockner, and M. A. Flinn, "HEUI—A New Direction for Diesel Engine Fuel Systems," SAE Paper No. 930270, SAE International, Warrendale, PA, 1993.
17. Amlung, R., J. P. Seger, T. Warlick, and D. Wilhelm of Cummins Engine Company, Columbus, IN. "Electronics for Engine Control," *Off-Highway Engineering,* SAE International, Warrendale, PA, December 1996.
18. "Diesel Engine Management," Robert Bosch GmbH, Germany, 1999.

Chapter 5

Combustion in Diesel Engines

5.1 Introduction

Growing concern regarding energy resources and the environment has increased interest in the study of energy conversion machines and systems, as well as alternative sources of energy. In some cases, this concern has led to a serious debate about our way of life in the Western and industrialized world. Meanwhile, engineers and researchers have intensified their efforts toward radically improving traditional energy conversion machines such as the internal combustion engine with the goal of lowering its exhaust emissions and fuel consumption. This research has involved many, if not all, of the subsystems of the internal combustion engine. Because it is the most energy efficient of the conversion machines, the diesel engine has gained popularity in many applications. It has unseated the gasoline engine from the heavy-duty market and has begun its conquest of the medium-duty market with an ever-increasing share of the light-duty truck market in the United States. The U.S. Department of Energy is funding and supporting several programs aimed at "dieselizing" the sport utility vehicles (SUVs) so popular in the United States. Demand for small diesel engines has been phenomenal in the European countries, where consumers are more fuel economy-minded than their American counterparts.

Diesel engines have an excellent reputation for their low fuel consumption, reliability, and durability characteristics. They are also known for their extremely low hydrocarbon and carbon monoxide emissions. However, they have also been rejected by the driving public, especially in the United States, for their odorous and sooty exhaust, which is also characterized by high nitric oxide and particulate matter emissions. Because performance, fuel consumption, and emitted pollutants result from the combustion process, it is necessary first to study the mechanisms of combustion in diesel engines if we are to improve them.

Combustion in diesel engines is very complex and its detailed mechanisms are not well understood. Its complexity seems to defy researchers' attempts to unlock its many secrets despite the availability of modern tools such as the high-speed photography used in "transparent" engines, the computational power of contemporary computers, and the many mathematical models designed to mimic combustion in diesel engines. This chapter reviews the most established combustion models and principles to help readers gain an appreciation of diesel combustion and how it impacts performance and emission formation.

5.2 Components of the Combustion Process

The basic operating principle of diesel combustion is its unique way of releasing the chemical energy stored in the fuel. To perform this process, oxygen must be made available to the fuel in a specific manner to facilitate combustion. One of the most important aspects of this process is the mixing of fuel and air, a process often referred to as *mixture preparation*.

In diesel engines, fuel is injected into the engine cylinder near the end of the compression stroke [1], just a few crankangle degrees before top-dead-center (TDC). The liquid fuel is usually injected at high velocity as one or more jets of fuel through small orifices or nozzles in the injector tip. The fuel atomizes into small droplets and penetrates the combustion chamber. The atomized fuel absorbs heat from the surrounding heated compressed air, vaporizes, and mixes with that high-temperature, high-pressure air. As the piston continues to move closer to TDC, the mixture (mostly air) temperature reaches the fuel's ignition point. Instantaneous ignition of some premixed fuel and air occurs after the ignition delay period. This instantaneous ignition is considered the start of combustion (also the end of the ignition delay period) and is marked by a sharp cylinder pressure increase as combustion of the air/fuel (A/F) mixture takes place. Increased pressure resulting from the premixed combustion compresses the unburned portion of the charge and shortens the delay before its ignition. It also increases the evaporation rate of the remaining fuel. Atomization, vaporization, fuel vapor/air mixing, and combustion continue until all the injected fuel has combusted.

The lowest A/F ratio is often found at peak torque conditions in diesel engines. To avoid excessive smoke formation, the A/F ratio at peak torque is usually maintained above 25:1, well above the equivalence ratio (i.e., the chemically correct ratio). In turbocharged diesel engines, the A/F ratio at idle may exceed

160:1, which means that the overall A/F ratio in diesel engines is characteristically very lean. Therefore, excess air present in the cylinder after the fuel has combusted continues to mix with burning and already burned gases throughout the combustion and expansion processes. At the opening of the exhaust valve, excess air exits along with the combustion products, which explains the oxidizing nature of diesel exhaust. Although combustion occurs when vaporized fuel mixes with air to form a chemically correct mixture (stoichiometric ratio) locally, and the proper ignition temperature is reached, the overall A/F ratio is lean. In other words, the majority of the air inducted into the cylinder of a diesel engine is compressed and heated, but never engages in the combustion process. Oxygen in the excess air helps oxidize gaseous hydrocarbons and carbon monoxide, reducing them to extremely small concentrations in the exhaust gas.

It is clear from this discussion that the following factors play a primary role in the combustion process:

- The inducted charge air, its temperature, and its kinetic energy in several dimensions
- The injected fuel, its atomization, spray penetration, temperature, and its chemical characteristics.

Although these two factors are most important, other parameters may dramatically influence them and therefore play a secondary, but nevertheless important, role in the combustion process. Secondary parameters include the following:

- The intake port design has a strong influence on charge air motion (especially as it enters the cylinder) and ultimately the mixing rate in the combustion chamber. The intake port design may also influence charge air temperature. This may be accomplished by heat transfer from the water jacket to the charge air through the intake port surface area.
- The intake valve size controls the total mass of air inducted into the cylinder in a finite amount of time.
- The compression ratio influences fuel vaporization and consequently the mixing rate and combustion quality.
- The injection pressure controls the degree of fuel atomization.
- The geometry of the nozzle hole (i.e., length/diameter) controls the spray penetration.
- The spray geometry directly affects combustion quality through air utilization. For instance, a larger spray cone angle may place the fuel on top of the piston and outside the combustion bowl in open-chamber DI diesel engines. This condition would lead to excessive smoke (incomplete combustion) because it would deprive the fuel of access to the air available in the combustion bowl (chamber). Wide cone angles may also cause the fuel to be sprayed on the cylinder walls, rather than inside the combustion bowl where it is required. Fuel sprayed on the cylinder wall will eventually be scraped downward to the oil sump where it will shorten the lube oil life.
- The valve configuration controls the injector position. Two-valve systems force an inclined injector position, which implies an uneven spray arrangement that leads to compromised fuel/air mixing. On the other hand, four-valve designs allow for vertical injector installation, a symmetric fuel spray arrangement, and equal access to the available air by each of the fuel sprays.
- The top piston ring position controls the dead space between the piston top land (the area between the top piston ring groove and the top of the piston crown) and the cylinder liner. This dead space/volume traps air that is compressed during the compression stroke and expands without ever engaging in the combustion process.

Therefore, it is important to realize that the combustion system of the diesel engine is not limited to the combustion bowl, injector sprays, and their immediate surroundings. Rather, it includes any part, component, or system that may affect the final outcome of the combustion process.

5.3 Fuel and Air Distribution in the Fuel Spray of a DI Diesel

The combustion process in DI diesel engines is an unsteady heterogeneous, three-dimensional process [2]. The industry has developed a good conceptual understanding of diesel engine combustion, but difficulties remain in quantitatively describing many of its critical individual processes. Yet, some models that exist today are able to approximate the combustion process and arrive at resultant emissions levels that resemble those measured by experiment.

A study of high-speed photographic films of combustion in a DI diesel engine [3] shows that at the beginning of combustion, the spray has a shape similar to that schematically shown in Figure 5.1. The average distance between the droplets is expected to change with their location in the spray, and it is greatest near the edge downstream from the centerline of the spray, where the smaller droplets are concentrated. The

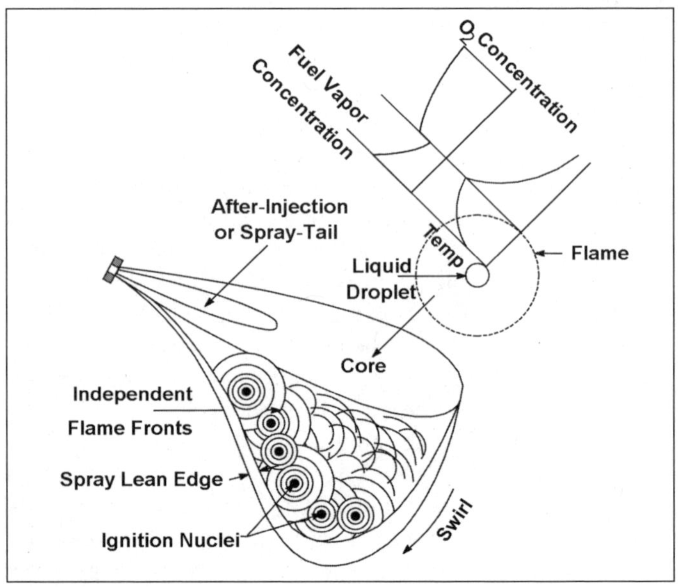

Figure 5.1 Schematic of a fuel spray mixing with air. (Redrawn from Ref. 2.)

average local A/F ratios and, consequently, the combustion mechanisms are therefore expected to vary from one location to another [2].

As one might expect, the local A/F ratio is highest along the centerline of the spray and diminishes as we move to the outer extremities of the spray cone. Of course, this A/F ratio varies along the spray centerline. This distribution varies with the radial distance from the nozzle hole and takes into account all the fuel, both in the liquid and vapor phase. However, at the downstream edge of the spray and at distances farther away from the spray core, the A/F ratio always approaches zero, and it increases as we move toward the core of the spray. We can assume that most of the droplets that are transported away from the core will be mostly or completely evaporated before the start of ignition [4], and that the mixture near the downstream edge of the spray consists of premixed fuel vapor and air. Further understanding of the model given in Figure 5.1 can be achieved by dividing it into several regions, as discussed next.

5.3.1 Lean Flame Region

Vapor concentration between the core and the downstream edge of the spray is not homogeneous, and the local A/F ratio may vary from zero to infinity [2]. Photographic studies on spray combustion in diesel engines [3, 5, 6] show that ignition starts in the spray envelope near the downstream edge of the spray. Ignition nuclei are usually formed at several locations where the mixture will most likely autoignite. This is illustrated in Figure 5.1. Once ignition starts, small independent nonluminous flame fronts propagate from the ignition nuclei and ignite the combustible mixture around them. This mixture, on a mass average basis, is lean [5]. The region in which these independent flames propagate is referred to as the *lean flame region* (LFR). In this region, nitric oxide is formed at high local concentrations as combustion proceeds to its completion. Under very light loads, little fuel is consumed and combustion temperatures may not be high enough to produce high nitric oxide concentrations.

5.3.2 Lean Flame-Out Region

Near the outer edge of the spray, the mixture is often too lean to ignite or to support stable combustion. This region is referred to as the *lean flame-out region* (LFOR). Within the LFOR, some fuel decomposition and partial oxidation products can be found [2]. The decomposition products are mainly lighter hydrocarbon molecules. The partial oxidation products include aldehydes and other oxygenates. It is believed that the LFOR is one of the major sources of unburned hydrocarbons as well as odorous constituents. The size of the LFOR depends on many factors, including the temperature and pressure in the chamber during the course of combustion, the air swirl, and the type of fuel. In general, higher temperatures and pressures extend the flames to leaner mixtures [7, 8] and thus reduce the LFOR size. Increases in temperature and pressure result from burning the remainder of the fuel spray and may have a secondary effect on the LFOR. We know that factors such as the overall A/F ratio, turbocharging, and coolant temperature affect cylinder gas pressure and temperature and, in turn, influence the size of the LFOR.

In the zone where the LFOR and LFR meet, primary combustion takes place where hydrocarbons react to form CO, H_2, and H_2O as well as various radical species such as H, O, and OH. Other unburned hydrocarbons, having a lower carbon number atoms than the initial fuel molecules, are formed. Reference [9] suggests that a paraffin molecule in a lean fuel flame undergoes a reaction according to the following relationship:

$$OH + C_nH_{2n+2} \rightarrow H_2O + C_nH_{2n+1} \rightarrow C_{n-1}H_{2n-2} + CH_3 \quad (5.1)$$

Reference [9] also indicates that because hydrocarbon radicals higher than ethyl are thermally unstable, the initial radical C_nH_{2n+1} usually splits off CH_3, thus forming the next lower ethyl-type compound as in Eq. (5. 1). The unsaturated hydrocarbons formed react rapidly with oxygen atoms in this region to form oxygenated hydrocarbons. If the reaction is completed, as in the LFR, recombination reactions occur and produce CO_2 and H_2O. However, if the flame is quenched, as in the LFOR, then unburned hydrocarbons, oxygenated hydrocarbons, CO, and other intermediate compounds are left without being completely burned [2].

5.3.3 Spray Core

Following the ignition and combustion in the LFR, the flame propagates toward the core of the spray. In this region, which is between the LFR and the core of the spray, the fuel droplets are larger. They gain heat by radiation from the already established flames and evaporate at a higher rate. The increase in temperature increases the rate of vapor diffusion, due to the increase in molecular diffusivity [2]. These droplets may be completely or partially evaporated. If they are completely evaporated, the flame will burn all the mixture within the rich ignition limit. The droplets that are not completely evaporated may be surrounded by a diffusion-type flame, as illustrated in Figure 5.1, and burn as individual droplets or evaporate to form a fuel-rich mixture. The combustion of this mixture depends on the aerodynamic factors that control the rate of mixing of the fuel and air rather than the droplet parameters [10, 11].

The combustion in the spray core generally depends on the local A/F ratio, which is affected mainly by the interaction between the spray core and the swirling air. Under part-load operation, adequate oxygen is available and combustion is usually completed, resulting in NO_x formation. The temperature of the flame depends on both the local mixture temperature before the start of combustion and the heat of combustion, which is partly a function of the concentration of the heavy compounds in the fuel used. The flame zone temperature is the major factor affecting NO_x formation.

Near full-load conditions, incomplete combustion occurs in many locations in the fuel-rich core. The authors of Ref. [9] suggested that in fuel-rich, saturated-hydrocarbon flames, the initial reaction is simply the H-atom stripping reaction, that is,

$$H + C_nH_{2n+2} \rightarrow H_2 + C_nH_{2n+1} \quad (5.2)$$

Recombination between the hydrocarbon radicals may occur, and hydrocarbons heavier or lighter than the original fuel molecules may be formed. This may explain why in diesel engines with a single-compound hydrocarbon fuel, the exhaust emissions contain molecules with both more and fewer carbon atoms than the original fuel [12, 13].

In addition to unburned hydrocarbons, carbon monoxide, oxygenated compounds, and carbon may be formed near full-load conditions. Nitrogen oxides are formed at low concentrations under these conditions.

5.3.4 Spray Tail

The last part of the fuel injected consists of large droplets [5] due to the relatively small pressure differential acting on the fuel near the end of the injection process. This is caused by a combination of decreased fuel injection pressure and increased cylinder gas pressure. The penetration of this part of the fuel, referred to as the *spray tail*, is usually poor. Under high-load conditions, the spray tail has little chance of entering regions with adequate oxygen concentration. The temperature of the surrounding gases, however, is fairly high (near the maximum cycle temperature), and the rate of heat transfer to these droplets is very high. These droplets, therefore, tend to evaporate quickly and decompose. The decomposed products contain unburned hydrocarbons and a high percentage of carbon molecules. Partial oxidation products include carbon monoxide and aldehydes.

5.3.5 After-Injection or Secondary Injection

Under medium and high loads, many injection systems produce after-injection (see Chapter 4). When this occurs the injector needle valve bounces off of its seat and opens for a short time after the end of the main injection. In general, the amount of fuel, delivered during after-injection is very small; however, it is injected late in the expansion stroke, under a relatively small pressure differential, and with very little atomization and penetration. This fuel is quickly evaporated and decomposed, resulting in the formation of CO, carbon particles (smoke), and unburned hydrocarbons. Henein and Bolt [14] conducted an extensive study of the effect of after-injection on exhaust smoke. Ways and means to deal effectively with after-injection as well as secondary injection are discussed in Chapter 4.

5.3.6 Fuel Deposited on the Walls

Some fuel sprays impinge on the walls. This is especially the case in small, high-speed DI engines because of the shorter spray path and the limited number of sprays. The rate of evaporation of the liquid film depends on many factors, including gas and wall temperatures, gas velocity, gas pressure, and properties of the fuel. The vapor concentration is maximum on the liquid surface and decreases with increased distance from the surface. Assuming that the surrounding gas has a high relative velocity and contains enough oxygen, the flame propagates to within a small distance from the wall. Combustion of the rest of the fuel on the walls depends on the rate of evaporation and mixing of fuel and oxygen. If the surrounding gas has a low oxygen concentration or the mixing is poor, evaporation occurs without complete combustion. Under this condition, the fuel vapor decomposes and forms unburned hydrocarbons, partial oxidation products, and carbon particles.

As the piston moves through the expansion stroke, the gases flow outward radially in an inverse squish motion to fill the expanding space between the piston top and the cylinder head. In shallow-bowl combustion chambers, most of the combustion process takes place in the bowl. In deep-bowl types, the reversed squish flow is more significant than in shallow-bowl chambers. In both cases the swirl motion continues, but due to frictional losses, it continues at a lower rotational velocity than during the compression stroke. The combined inverse squish and swirl components draw the fuel vapor and the partial oxidation products from the bowl toward the cylinder walls. Photographic high-speed films showed eddies and heavy smoke clouds outside the bowl from the areas where the sprays impinge on the wall. These eddies are heterogeneous, and on combustion they burn with a luminous flame due to the presence of carbon particles [5, 15].

Based on this model, the progression of combustion in the spray and the cylinder pressure and mass average temperature at the start of combustion in each region may be represented

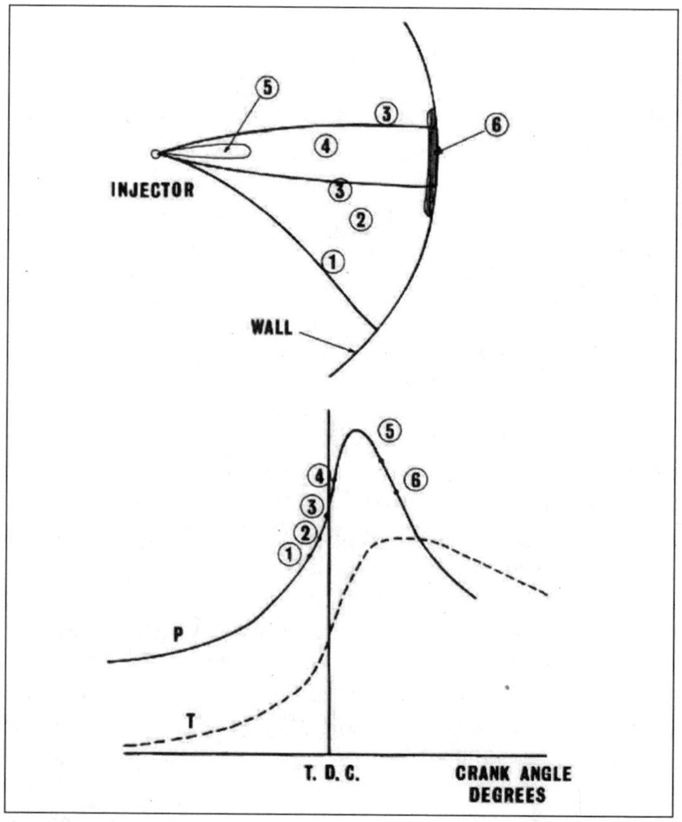

Figure 5.2 Progression of combustion, cylinder pressure, and temperature. (Reprinted with permission from Ref. 2.)

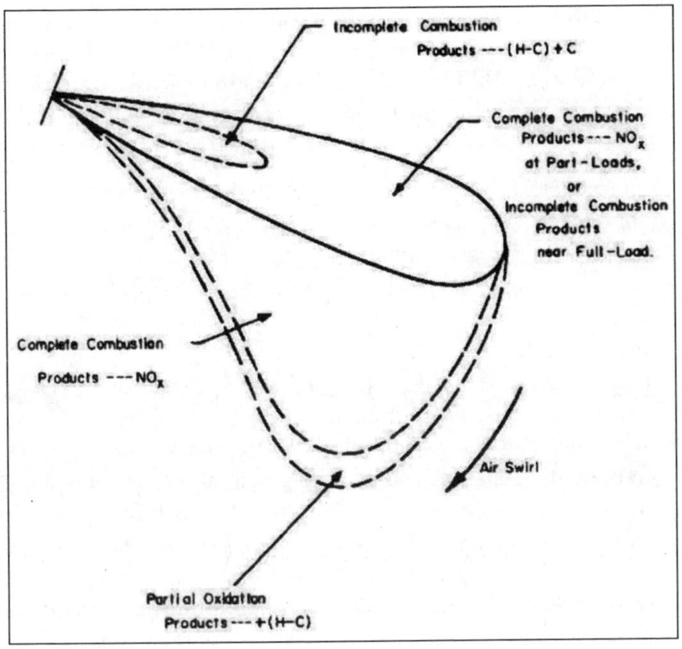

Figure 5.3 Emission species formation in fuel spray in air swirl. (Reprinted with permission from Ref. 2.)

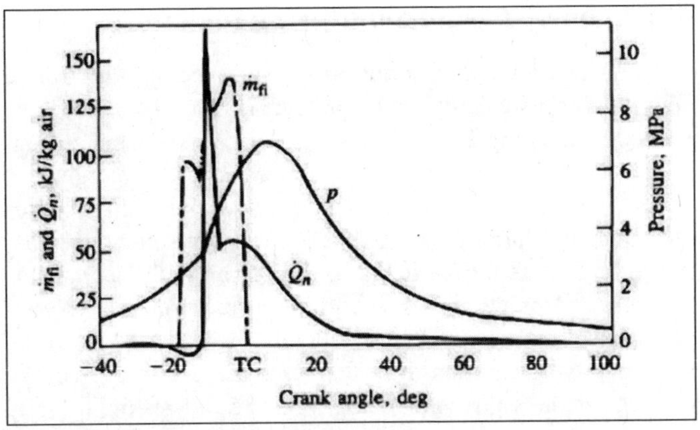

Figure 5.4 Heat release rate, cylinder pressure, and rate of fuel injection. (From Internal Combustion Engine Fundamentals *by John B. Heywood. ©1988. Reproduced with permission of The McGraw-Hill Companies.)*

as shown in Figure 5.2. The pollutant formation expected in the different regions of the spray without fuel impingement on the walls may be summarized as shown schematically in Figure 5.3.

5.4 Heat Release Rates in DI Engines

Many researchers have studied cylinder pressure traces [15–20] to determine heat release rates. Pressure traces obtained in open-chamber diesel engines are usually analyzed for this purpose. Figure 5.4 shows an example of a rate of net heat release diagram ($\dot{Q}_n$) together with cylinder pressure (p) and fuel injection rate ($\dot{m}_{fi}$) [1]. Details of the work done to obtain these traces are found in Ref. [15]. The initial sharp rise in the heat release rate results from burning the premixed portion of the fuel. During the ignition delay (ID), the spray is formed and the droplets at the leading edge of the spray are the first to evaporate, forming a nonhomogeneous, fuel/vapor/air mixture. As mentioned earlier, the complete evaporation of these droplets takes place in much less time than the ID. The negative heat release after the start of injection, at about 20° BTDC, is due mainly to the heat transfer from the hot air to the liquid fuel.

From Figure 5.4 it appears that ignition occurred at about 15° BTDC and the rate of burning reached its peak at about 10° BTDC. This example is indicative of how fast the premixed fuel portion burns in the cylinder. It took just about three crankangle degrees to reach the peak heat release rate. The end of ID, at 15° BTDC, coincides with the start of combustion in the LFR. Note that the flame observed, up to 10° BTDC, was of very low luminosity, indicating that the burning was essentially confined to the premixed portion of the jet.

At the point of maximum heat release rate, the cumulative heat release is about 5% of the total computed heat release. This percentage reflects the approximate portion of fuel burned soon after ignition of the total amount of fuel in this particular case. However, it is worth noting that this portion may vary according to the fuel's properties.

Usually, the first appearance of the orange, luminous flame is observed at about 10° BTDC. This flame does not spread to surround the tip of the jet until about 7° BTDC. During the remainder of the combustion process, the flame has been observed to have high luminosity due to the presence of carbon particles. High luminosity is a characteristic of diffusion-type flames.

5.5 Additional Details of Spray Formation

The interface between the fuel injection system and the combustion process is in the form of a fuel spray. In fact, it could be said that the complication and sophistication of the control of the fuel injection system is mainly for the purpose of delivering a proper spray into the cylinder. The component responsible for delivering the fuel spray is the injector nozzle. The spray is formed with the help of a differential pressure across its spray holes. The fuel is broken into droplets of different sizes and concentrations in the spray.

The combustion process depends a great deal on the development of the spray from the start of injection, even before the spray is fully developed. The behavior of that spray is very important to the combustible mixture formation and start of ignition. In addition, the behavior of the spray after the end of injection and fuel cutoff is very critical in terms of emission formation. The following subsections provide additional insight into spray formation during injection and its behavior after fuel cutoff.

5.5.1 Spray Formation during Injection

Figure 5.5 shows a spray formed by injecting fuel from a single hole in stagnant air. Upon leaving the nozzle hole, the jet becomes completely turbulent a very short distance from the point of discharge. Due to jet turbulence, the emerging jet becomes partly mixed with the surrounding air. Air becomes entrained and carried away by the jet, which results in increasing mass flow in the x-direction. Concurrently the jet spreads out in the y-direction and, according to the principle of conservation of momentum, the jet velocity decreases. Meanwhile, the velocity of the jet will further decrease as it moves in the x-direction due to frictional drag. Figure 5.5 gives the velocity distribution at two cross sections. The fuel velocity is highest at the centerline and decreases to zero at the interface between the zone of disintegration (or the conical envelope of the spray) and ambient air.

5.5.2 Spray Atomization

Spray atomization can generally be described as the breakup of the fuel jet as it exits the nozzle hole. The size of the droplets formed by this breakup is smaller than the nozzle hole's diameter. The degree of atomization increases due to the breakup of large droplets as the jet moves further along the x-axis (see Figure 5.5). The breakup and, hence, atomization continues as long as the Weber number exceeds a threshold value. The Weber number is defined as the ratio of the inertia forces to the surface tension forces and is described by the following equation:

$$W_e = \rho d^2 V^2/\sigma_d \tag{5.3}$$

where:

ρ = mass density
d = droplet diameter
V = upstream velocity
σ = surface tension.

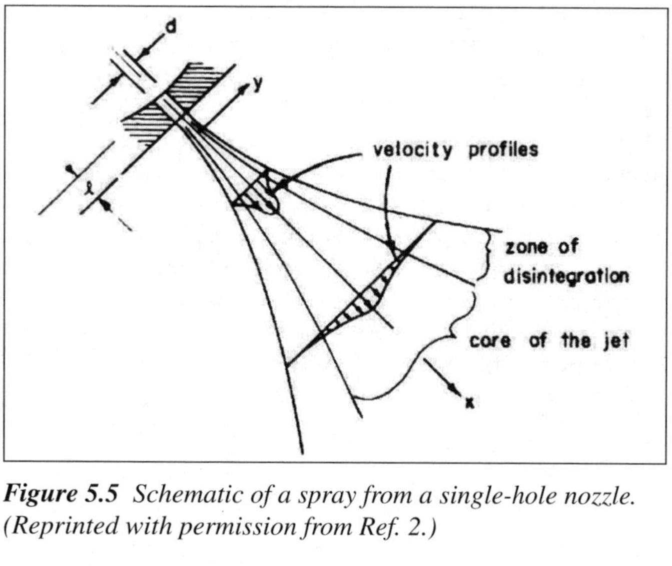

Figure 5.5 *Schematic of a spray from a single-hole nozzle. (Reprinted with permission from Ref. 2.)*

In Ref. [21], it was found that the threshold value was reached at the critical Weber number of 10.

Reference [1] divides the atomization of a jet into four distinct regimes: low, medium, high, and very high jet velocity regimes:

- At *low jet velocity*, breakup is due to the unstable growth of surface waves caused by surface tension and results in drops larger than the jet's diameter.
- As jet velocity is increased (*medium jet velocity*), forces resulting from the relative motion of the jet and the surrounding air augment the surface tension force and lead to drop sizes of the order of the jet's diameter. This is called the *first wind-induced breakup regime.*
- A further increase in jet velocity (*high jet velocity*) results in breakup characterized by divergence of the jet spray after an intact or undisturbed length downstream of the nozzle. In this *second wind-induced breakup regime,* the unstable growth of short-wavelength waves induced by the relative motion between the liquid and surrounding air produces droplets whose average size is much less than those of the jet's diameter.
- Further increases in jet velocity (*very high jet velocity*) lead to breakup in the atomization regime, where the breakup of the outer surface of the jet occurs at, or before, the nozzle exit plane and results in droplets whose average diameter is much smaller than those of the nozzle's diameter. Aerodynamic interactions at the liquid/gas interface appear to be one major component of the atomization mechanism in this regime.

Because atomization is responsible for creating a fog-like medium inside the combustion chamber, a subject of relevance is the size as well as distribution of these droplets. An indicator of the atomization process is how small a droplet can be and how high the number of small droplets is relative to the larger ones. A common indicator used in reference to atomization is the *Sauter mean diameter.* It is defined as follows:

$$D_{SM} = (\int D_d^3 \, dn)/(\int D_n^2 \, dn) \tag{5. 4}$$

where dn is the number of drops having diameter D_d. In other words, the Sauter mean diameter is the diameter of a droplet that has the same surface-to-volume ratio as that of the total spray [1, 2]. In DI diesel engines equipped with modern fuel injection systems that provide high injection pressures, the average droplet size ranges from 6 to 15 μm.

In essence, at the beginning of injection, the fuel is kept from advancing into the combustion chamber by the drag of air that starts to move along the spray axis. Fuel droplets following those in the spray front encounter less resistance and overtake the droplets encountering the highest resistance, forcing them toward the periphery of the spray. The droplets at the axis become less atomized since they do not encounter drag from the surrounding air. At the spray front, the droplets meet the highest aerodynamic resistance, but the spray penetrates the air because the droplets that are delayed at the front are continuously replaced by new, high-momentum ones. Accordingly, the droplets somewhere in the periphery of the spray are the earliest ones in the injection sequence. These evaporate very quickly and get mixed with the air. The first flame appears in the envelope at a distance from the nozzle dependent mainly on the injection pressure [2]. Because of the very thick spray in the intact zone, the flame never proceeds into this zone while the fuel is being injected (see Figure 5.5).

5.5.3 Spray Penetration

While good atomization helps the fuel droplet evaporate and mix with the air available in its immediate vicinity, combustion air remains substantially underutilized. For more air utilization, the droplets would have to travel farther into the combustion volume to reach air that is present across the combustion chamber. The faster the spray penetrates into the combustion volume, the greater the mixing rates as well as the air utilization [1, 2]. On the other hand, it is not desirable to have the spray penetrate so far that it would impinge on the combustion chamber walls. These surfaces are relatively cold and often contribute to lower mixing rates. Due to its importance to the combustion process, spray penetration has been investigated by many researchers [22–24]. Dent [25] developed a predictive relationship for spray penetration as described here:

$$S = 3.07(\Delta P/\rho_g)^{1/4}(td_n)^{1/2}(294/T_g)^{1/4} \qquad (5.5)$$

where ΔP is the pressure drop across the spray hole in pascals, ρ_g is the gas density in kilograms per cubic meter, t is the time after the start of injection in seconds, d_n is the nozzle hole diameter in meters, and T_g is the gas temperature in kelvins. From Eq. (5.5), it appears that the spray penetration is a function of the pressure across the nozzle, density of the air ρ_g, diameter of nozzle hole d_n, time t, and gas temperature T_g. Similar relationships have been derived by others [26]. Figure 5.6 shows the penetration versus time from start of injection for several values of injection pressure (P_o) and ambient pressure (P_a) [1].

5.5.4 Droplet Size Distribution

Having reviewed the importance of spray atomization and penetration for the proper operation of DI diesel engines, it is also necessary to review the importance of their influence on heat transfer characteristics. A greater number of small size droplets leads to a greater surface area that facilitates heat transfer from the hot compressed air to the small fuel volume contained in those droplets. It is also important to note that the injection process is not a steady, stabilized process during which pressure, effective spray hole geometry, and injection rate are fixed. In fact, they do vary from the start to the end of injection. Therefore, one might expect the droplet size distribution to vary during the injection process. To further complicate the combustion process, heat inside the combustion chamber is not uniform and this nonuniformity varies throughout combustion. Therefore, mixing rates vary according to many parameters such as droplet size distribution, heat distribution within the combustion chamber, spray penetration and atomization, fuel quality and its evaporation rates, as well as many other parameters. Our ability to model or to simply understand diesel combustion depends to a great extent on our ability to define each of the parameters involved in this process including droplet size distribution. Figure 5.7 is an example of the effect of injection pressure on droplet size as influenced by nozzle hole geometry (Figure 5.7a) and nozzle hole diameter (Figure 5.7b).

The droplet size distribution given in Figure 5.8 is for a fuel spray produced from a nozzle hole at different times from

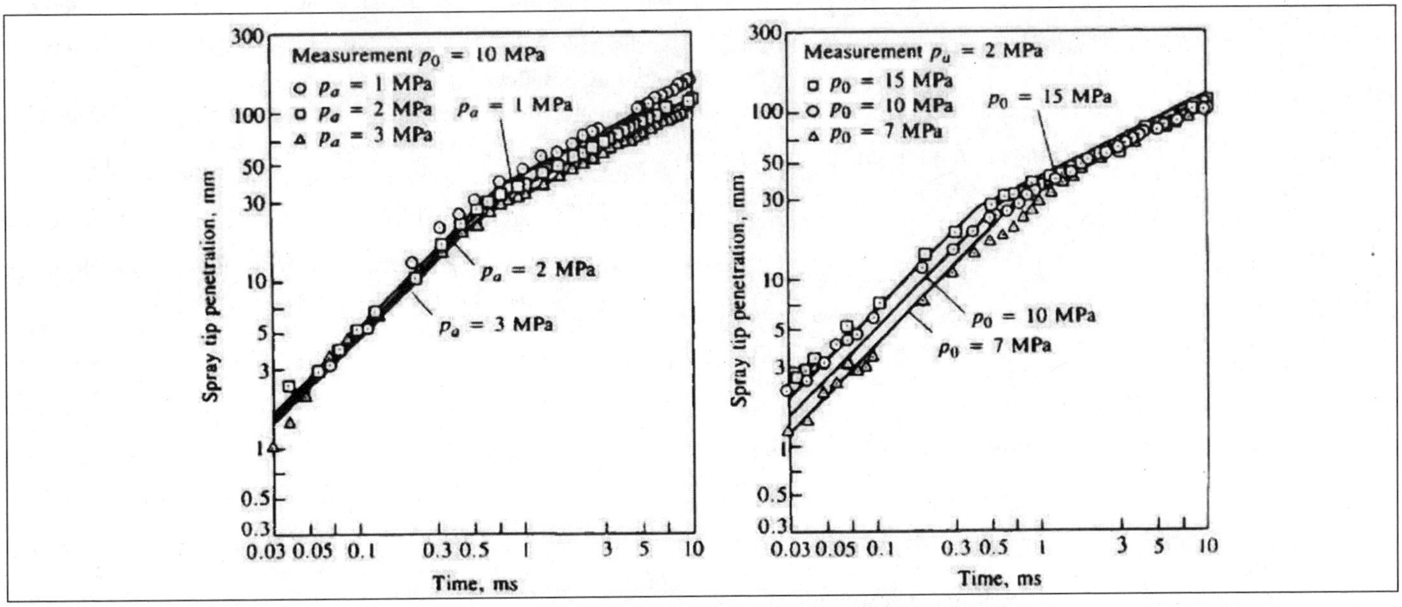

Figure 5.6 Spray tip penetration as a function of time—in ambient pressure. (From Internal Combustion Engine Fundamentals *by John B. Heywood. ©1988. Reproduced with permission of The McGraw-Hill Companies.)*

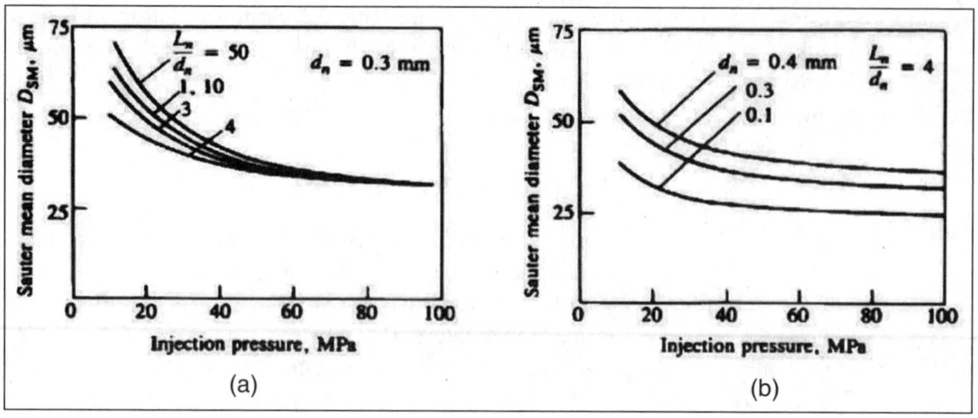

Figure 5.7 Effect of injection pressure on droplet size, as influenced by (a) nozzle hole geometry and (b) nozzle hole diameter. (From Internal Combustion Engine Fundamentals *by John B. Heywood. ©1988. Reproduced with permission of The McGraw-Hill Companies.)*

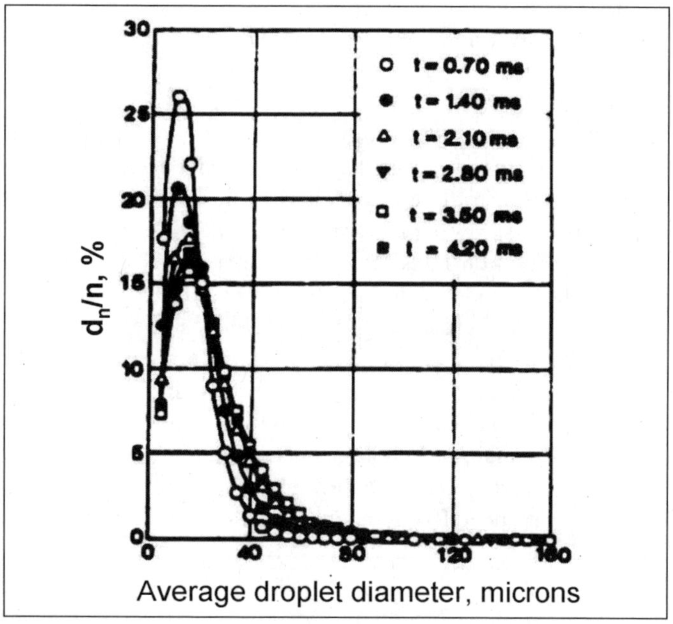

Figure 5.8 Transient change in droplet size distribution. (Reprinted from Ref. 27.)

the start of injection. At 0.70 ms injection duration, Figure 5.8 indicates that small droplets had a high frequency. At later times, larger droplet diameters had greater frequency than small droplets. In other words, as the injection continues, the smaller droplet population decreases as the larger droplet population increases, as a percent of the total number of droplets. Injection systems that sustain high injection pressure over most of the injection duration will most likely support a larger number of small droplets. This in turn improves the overall surface-to-volume ratio of the droplets, thus increasing their evaporation and mixing rates.

5.6 The Three Phases of Diesel Combustion

Diesel combustion is complex and includes both physical and chemical steps that have been described by many [1, 2]. The combustion process is usually described as consisting of three distinct phases as follows:

1. Ignition delay
2. Premixed combustion
3. Rate-controlled combustion.

5.6.1 Ignition Delay

Ignition delay in diesel engine combustion is the time required after the start of injection for the preignition processes to produce the ignition nuclei and detectable combustion. The duration of the ignition delay is one of the most important criteria, having a great effect on the combustion process, mechanical stresses, engine noise, and exhaust emissions. In contrast with gasoline engines, where combustion starts by an electric spark at one location, combustion in diesel engines starts by autoignition nuclei at numerous locations in the combustion chamber. Figure 5.9 summarizes the physical and chemical steps experienced prior, and leading to, autoignition.

The following *physical* processes are involved in the ignition delay period (ID):

- Spray breakup and droplet formation
- Fuel and air mixing
- Heating of the liquid fuel and evaporation
- Diffusion of the vapor into the air to form a combustible mixture.

These step are often referred to as *mixture preparation*, although this term may include more than just the steps listed here. Included in the mixture preparation is the air motion appropriated through the special intake port design, heat transfer to the air flowing through the intake port, air temperature history as well as temperature immediately prior to injection, and any other parameters that may affect the quality of the fuel/air mixture. Most, if not all, of these steps are physical in nature and are generally completed in an extremely short period of time.

The following *chemical* processes take place during the ignition delay period:

- Preflame oxidation of the hydrocarbon fuel that has been premixed

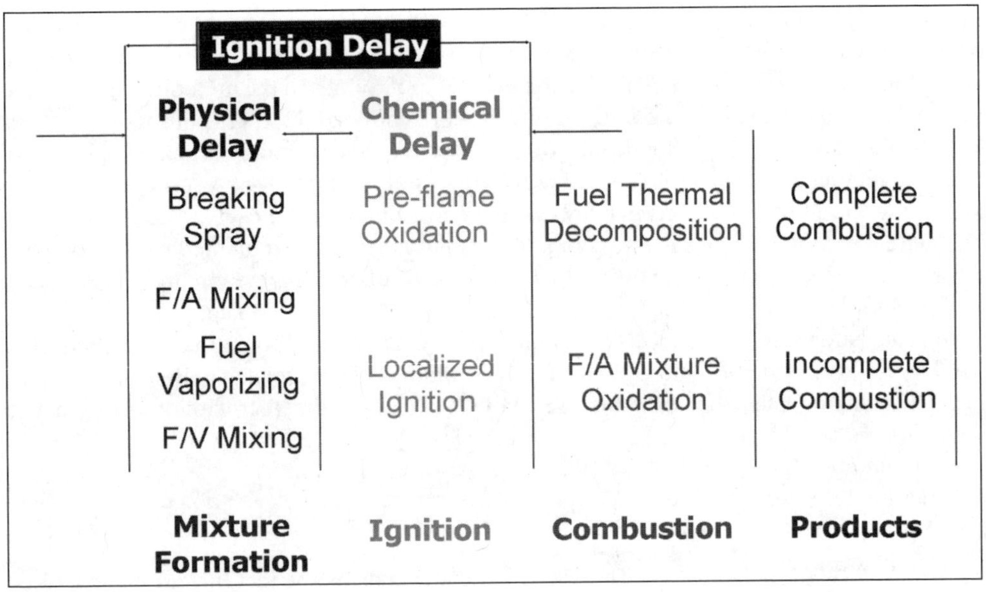

Figure 5.9 *Summary of steps before and after autoignition.*

- Localized ignition that takes place in several areas within the combustion chamber.

Following the localized ignition, enough heat is generated to lead to:

- The decomposition of the heavy hydrocarbons into lighter components and the formation of combustion radicals
- The preignition chemical reactions between the decomposed components and oxygen.

It is difficult to draw a distinct line separating the physical and chemical processes, because they overlap. The chemical processes start after the fuel vapor makes contact with the air. At the very early stages of injection, however, the mass of fuel vapor that undergoes chemical reaction is too small to cause any detectable combustion phenomena. Therefore, the early stages of preignition can be considered to be dominated by the physical processes that result in the formation of a combustible mixture. The later stages of preignition can be considered to be dominated by the chemical changes that lead to autoignition [2].

In actual diesel engines, the fuel temperature in the injector is higher than the room temperature because of the heat transfer from the injector body to the fuel and the work done on the fuel in the injection pump. To maintain the temperature of the injector body at a reasonable level, an oversupply of fuel is circulated through the injector and returned to the fuel tank. Therefore, the fuel spray is often formed with heated fuel and this has an effect on spray formation. Figure 5.10 illustrates the relationship between fuel temperature and the spray cone angle based on work reported in Ref. [28].

Fuel evaporation could be due to flash evaporation or heat transfer to the liquid fuel from the gases or wall surfaces. Flash evaporation results from the sudden drop in the fuel pressure as it is injected at very high pressures. These pressures usually exceed the critical pressures of the majority of the fuel constituents [2]. For example, the critical pressures of *n*-octane, 1-decane, and *n*-dodecane are 361, 311, and 272 psia, respectively. Actual diesel fuels contain heavier hydrocarbons than these and their critical pressures decrease with the increase in the carbon number (number of carbon atoms in the molecule). Upon injection, the fuel pressure drops to the cylinder air pressure, which is in the range of 350 to 850 psia [7]. Higher pressures can be reached in supercharged engines.

In addition to the fuel type (chemical structure), temperature, and pressure conditions, ignition delay is also affected by local A/F ratios. The final rate of combustion is dependent on the rate of fuel evaporation, but ultimately on the likelihood that the evaporated fuel will mix with air and form a stoichiometric air and fuel mixture that is ready to burn. An important point to remember is that although ignition occurs in the vapor phase, oxidation reactions can and do proceed in the liquid phase between the fuel molecules and oxygen dissolved in the fuel droplets [1].

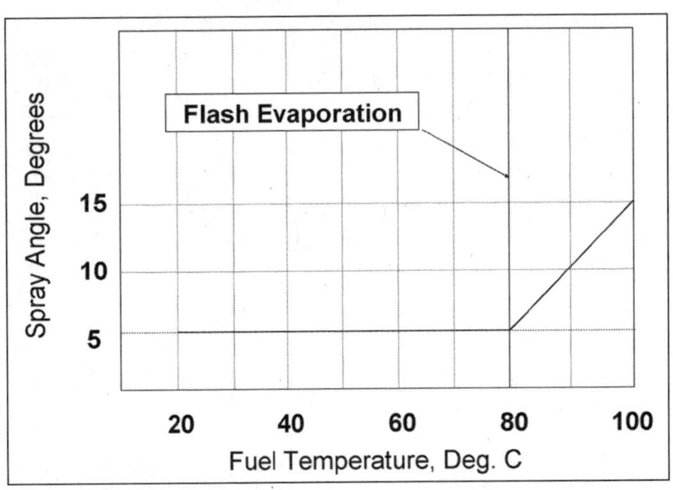

Figure 5.10 *Relationship between fuel temperature and spray angle. (Reprinted from Ref. 28.)*

Because the ignition characteristics affect the *ignition delay*, a fuel's ability to ignite is very important in determining diesel engine operating characteristics. Fuel conversion efficiency, smoothness of operation, misfire, smoke emissions, noise, and ease of starting are all dependent on ignition quality. The cetane number of a fuel is used to define its ignition quality. The cetane number is determined by comparing the ignition delay of the fuel with that of primary reference fuel mixtures in a standardized engine test. For low cetane fuels with too long an ignition delay, most of the fuel is injected before ignition occurs, which results in very rapid burning rates once combustion starts, with high rates of pressure rise and high peak pressures. Under extreme conditions, when autoignition of most of the injected fuel occurs, this produces an audible knocking sound, often referred to as *diesel knock*. For higher cetane number fuels, with shorter ignition delays, ignition occurs before most of the fuel is injected. The rates of heat release and pressure rise are then controlled primarily by the rate of injection and fuel/air mixing, and smoother engine operation results.

The cetane number scale is defined by blends of two pure hydrocarbon reference fuels. Cetane (*n*-hexadecane, $C_{16}H_{34}$), a hydrocarbon with high ignition quality, represents the top of the scale with a cetane number of 100. An isocetane, heptamethylnonane (HMN), which has a very low ignition quality, represents the bottom of the scale with a cetane number of 15. Thus, cetane number (CN) is given by

$$\text{CN} = \text{percent } n\text{-cetane} + 0.15 \times \text{percent HMN} \qquad (5.6)$$

The engine used in cetane number determination is a standardized single-cylinder, variable compression ratio engine with special loading and accessory equipment and instrumentation. The engine, the operating conditions, and the test procedure are specified by ASTM Method D613. Its operating requirements include the following:

- Engine speed of 900 rpm
- Coolant temperature of 100°C
- Intake air temperature of 65.6°C (150°F)
- Injection timing of 13° BTC
- Injection pressure of 10.3 MPa (1500 lb/in.^{2}).

The engine is operated at these conditions on the candidate fuel whose cetane number is to be determined, and the compression ratio is varied until combustion starts at TDC; that is, an ignition delay period of 13° (2.4 ms at 900 rpm) is produced. The procedure is then repeated using reference fuel blends. Each time a reference fuel is tried, the compression ratio is adjusted to give the same 13° ignition delay. When the compression ratio required by the actual fuel is bracketed by the values required by two reference blends differing by less than five cetane numbers, the cetane number of the fuel is determined by interpolation between the compression ratios required by the two reference blends.

To avoid the expense of determining the cetane number experimentally, many correlations predicting ignition quality based on the physical properties of diesel fuels have been developed [29, 30]. A calculated cetane index (CCI) is often used to estimate the ignition quality of diesel fuels (ASTM D976). It is based on API gravity and the midboiling point (i.e., the temperature at which 50% of the fuel is evaporated). It is applicable to straight-run fuels, catalytically cracked stocks, and blends of the two. Its use is suitable for most diesel fuels and gives numbers that correspond quite closely to cetane number. A diesel index is also used. It is based on the fact that ignition quality is linked to hydrocarbon composition: *n*-paraffins have high ignition quality, and aromatic and napthenic compounds have low ignition quality. The aniline point (the lowest temperature at which equal volumes of the fuel and aniline become just miscible; ASTM D611) is used, together with the API gravity, to give the diesel index:

$$\text{Diesel index} = \text{Aniline point (°F)} \times (\text{API gravity}/100) \qquad (5.7)$$

The diesel index depends on the fact that aromatic hydrocarbons mix completely with aniline at comparatively low temperatures, whereas paraffins require considerably higher temperatures before they are completely miscible. Similarly, a high API gravity denotes low specific gravity and high paraffinic quality, as well as good ignition quality. The diesel index usually gives values slightly above the cetane number.

5.6.2 *Premixed Combustion*

The term *premixed combustion* refers to the combustion of a portion of the fuel injected during the ignition delay period (indicated as *a* to *b* in Figure 5. 11). This portion of the fuel would have undergone atomization, evaporation, and the pre-ignition chemical reactions. This portion would have mixed with air in the chemically correct proportion and is ready to ignite once the proper temperature (autoignition temperature) is reached. When autoignition occurs, the premixed portion of the fuel burns at a very high rate (*b* through *c* in Figure 5. 11), producing high temperature as well as pressure in the combustion chamber [31]. The remaining fuel that is not involved in the premixed combustion has not been injected, evaporated, or mixed with air. It may be too lean or too rich to burn [31]. The characteristic noise of the diesel engine is also associated with this phase. It is generally accepted that the rate of pressure rise resulting from the premixed combustion is proportional to the noise intensity in diesel engines.

5.6.3 *Rate-Controlled Combustion*

The balance of the fuel that did not participate in premixed combustion represents the bulk of the fuel consumed during a complete cycle. In the rate-controlled combustion phase, this fuel controls the burning rate, which depends on the rate it mixes with air and acquires a condition that makes it ready to burn. This phase is characterized by a lower heat release peak than that reached in the premixed phase. The rate-controlled combustion phase is often referred to as *diffusion combustion* and is represented by the curve between *c* and *d* in Figure 5. 11.

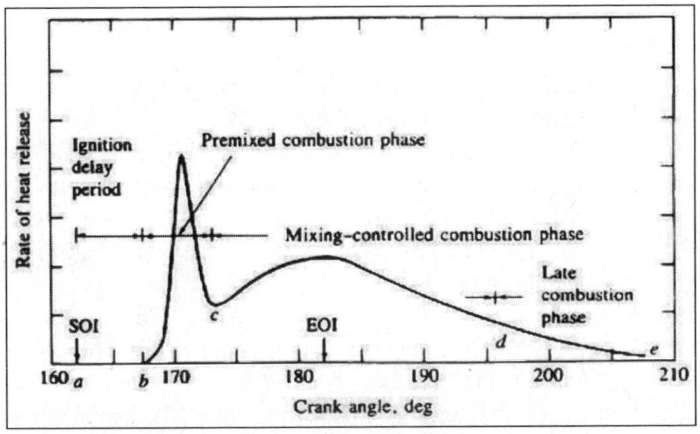

Figure 5.11 Phases of combustion in diesel engines. (From Internal Combustion Engine Fundamentals *by John B. Heywood. © 1988. Reproduced with permission of The McGraw-Hill Companies.)*

Figure 5.12 presents the combustion paths of three types of mixtures according to Professor J. V. Gerpen of Idaho University; rich, stoichiometric, and lean. For the stoichiometric ratio, combustion is complete and its products are generally water and carbon dioxide. For rich mixtures, there are two possibilities. If the mixture remains rich, combustion will be incomplete, and it will manifest this fact by producing soot. The second option is for the rich mixture to find a leaner mixture or additional air with which to mix, forming an overall stoichiometric mixture to produce a complete burn.

The third type of mixture treated in Figure 5.12 starts lean. This mixture will also have two potential paths. If it mixes with a leaner mixture or just air, it will not burn effectively, and will produce unburned hydrocarbons most likely in the gaseous phase. However, if the lean mixture mixes with a rich mixture

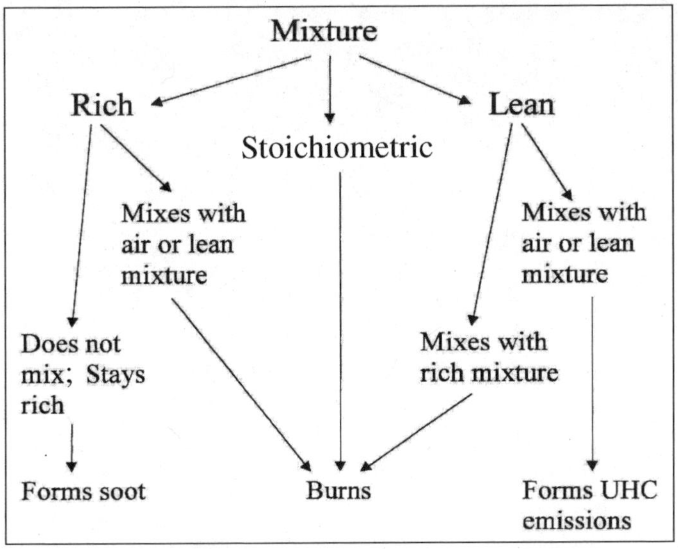

Figure 5.12 Three paths from mixture to combustion. (Reprinted with permission from Ref. 32.)

or more fuel and reaches a stoichiometric condition, it will then burn completely.

While the combustion process has been treated as consisting of three distinct phases, some would like to consider a fourth phase that describes the activity in the final stages just prior to opening the exhaust valve. In this final phase any remaining fuel that had not burned will continue to burn, perhaps at a much lower rate as shown in Figure 5.11. In addition, some of the fuel that may have burned in the rate-controlled phase could have formed carbon, having some energy yet to release if oxidized. Conditions in this phase are still not uniform from the temperature distribution as well as the chemical makeup points of view. Therefore, as long as motion still exists inside the cylinder, mixing will continue to occur and provide opportunities for the fuel as well as partially burned products to completely burn. This late combustion phase in represented by the segment *d* through *e* in Figure 5.11.

References

1. Heywood, J. B., *Internal Combustion Engine Fundamentals,* McGraw-Hill, New York, 1988.
2. Henein, N.A., and Patterson, D.J., *Emissions from Combustion Engines and Their Control,* Ann Arbor Science Publishers, Ann Arbor, MI, 1972.
3. Scott, W.M., "Looking in on Diesel Combustion," SP-345, SAE International, Warrendale, PA, 1968.
4. Henein, N.A., "Combustion and Emission Formation in Fuel Sprays Injected in Swirling Air," Paper No. 710220, presented at Society of Automotive Engineering Congress, Detroit, MI, 1971.
5. Rife, J., and J.B. Heywood, "Photographic and Performance Studies of Diesel Combustion with a Rapid Compression Machine," SAE Paper No. 740948, SAE International, Warrendale, PA, 1974.
6. Watts, R., and W.M. Scott, "Air Motion and Fuel Distribution Requirements in High-Speed Direct-Injection Diesel Engines," in *Diesel Engine Combustion Symposium,* Institute of Mechanical Engineers, London, 1970, pp. 167–177.
7. Schmidt, F.A.F., *The Internal Combustion Engine,* translated by R.W.S. Mitchell and J. Borne, Chapman and Hall, London, 1965.
8. Friedman, R., and W. Johnson, *J. Appl. Phys.* **21,** 8, 1950, pp. 791–795.
9. Fristrom, M., and A.A. Westenberg, *Flame Structure,* McGraw-Hill, New York, 1965.
10. McCreath, C.G., and N.A. Chigier, "Liquid-Spray Burning in the Wake of a Stabilizer Disc," presented at 14th Symposium on Combustion, The Combustion Institute, 1972.
11. Vanderpoel, R.E., Rife, J.M., and Rogowski, A.R., "More Efficient Combustion in Small Open Chamber Diesel Engines," SAE Paper No. 720775, 1972.
12. Barnes, G.J., "Relation of Lean Combustion Limits in Diesel Engines to Exhaust Odor Intensity," SAE Paper No. 680445, SAE International, Warrendale, PA, 1968.

13. Milks, D., C.W. Savery, J.L. Steinberg, and R.A. Matula, "Studies and Analysis of Diesel Engine Odor Production," presented at Clean Air Congress of the International Union of Air Pollution, Washington, DC, 1970.
14. Henein, N.A., and J.A. Bolt, "The Effect of Some Fuel and Engine Factors on Diesel Smoke," SAE Paper No. 690557, SAE International, Warrendale, PA, 1969.
15. Lyn, W.T., "Study of Burning Rate and Nature of Combustion in Diesel Engines," *Proceedings of the 9th Symposium on Combustion,* Institute of Mechanical Engineers, London, 1963, pp. 1069–1082.
16. Lyn, W.T., *Proc. Inst. Mech. Engineers (A.D.)* **1,** 1960–1961, pp. 34–46.
17. Austen, A.E.W., and W.T. Lyn, *Proc. Inst. Mech. Engineers (A.D.)* **1,** 1960–1961, pp. 47–62.
18. Shipinski, J., O.A. Uyehara, and P.S. Myers, "Experimental Correlation between Rate-of-Injection and Rate-of-Heat-Release in a Diesel Engine," ASME Paper 68-DGP-11, American Society of Mechanical Engineers, New York, 1968.
19. Grigg, H.C., and M.H. Syed, "The Problem of Predicting Rate of Heat Release in Diesel Engines," Paper No. 18, in *Symposium on Diesel Engine Combustion,* Institute of Mechanical Engineers, London, 1970.
20. Henein, N.A., and J.A. Bolt, "Kinetic Considerations in the Autoignition and Combustion of Fuel Sprays in Swirling Air," Paper No. A-i, in *Trans. CIMAC 9th Int. Congress on Combustion Engines,* Stockholm Sweden, 1971, pp. 1–36; also ASME Publication 72-DGP-8, 1972.
21. Hinze, J.O., *Appl. Sci. Res.* **Al,** 1949.
22. Sass, R., *Compressorless Diesel Engines,* Springer Verlag, Berlin, 1929.
23. Schweitzer, P.H., "Penetration of Oil Sprays," Bulletin No. 46, Pennsylvania State College, 1937.
24. Lyshevskiy, A.S., "Determination of the Length of an Atomized Fuel Spray," Tudy Novocherkassogo, Potitekhnicheskogo Instituta, **26,** 1955, pp. 391–401.
25. Dent, J.C., SAE Paper No. 710571, SAE International, Warrendale, PA, 1971.
26. Parks, M.V., C. Polonski, and R. Toye, "Penetration of Diesel Fuel Sprays in Gases," SAE Paper No. 660747, SAE International, Warrendale, PA, 1966.
27. Takeuchi, K., J. Senda, and S. Masayuki, "Transient Characteristics of Fuel Atomization and Droplet Size Distribution in Diesel Fuel Spray," SAE Paper No. 830449, SAE International, Warrendale, PA, 1983.
28. Oza, R.D., and J.F. Sinnamon, "An Experimental and Analytical Study of Flash-Boiling Fuel Injection," SAE Paper No. 830590, SAE International, Warrendale, PA, 1983.
29. Henein, N.A., and A.N. Fragoulis, "Correlation between Physical Properties and Autoignition Parameters of Alternate Fuels," SAE Paper 850266, SAE International, Warrendale, PA, 1985.
30. Gulder, O. L., B. Glavincevski, and G. F. Burton, "Ignition Quality Rating Methods for Diesel Fuels—A Critical Appraisal," SAE Paper 852080, SAE International, Warrendale, PA, 1985.
31. Gerpen, J. V., "Combustion in Diesel Engines—Modeling and Simulation," SAE Engineering Academies Lecture Notes, SAE International, Warrendale, PA, 2001.

Chapter 6

Emissions Formation in Diesel Engines

6.1 Emission Formation in Diesel Engines

Unlike spark-ignited engines, in which the combustible mixture is predominantly homogeneous, diesel combustion is heterogeneous in nature. Diesel fuel is injected in a cylinder full of high-temperature, compressed air. Emissions formed as a result of burning this uneven air/fuel (A/F) mixture depend on the prevailing conditions not only during combustion, but also during the expansion and especially prior to the exhaust valve opening. Mixture preparation during the ignition delay, fuel ignition quality, residence time at different combustion temperatures, expansion duration, and general engine design features play a very important role in emission formation. In essence, the concentration of the different emission species in the exhaust is the result of their formation and of their reduction in the exhaust system. Incomplete combustion products formed in the early stages of combustion may be oxidized later during the expansion stroke. Mixing of unburned hydrocarbons with oxidizing gases, a high combustion chamber temperature, and adequate residence time for the oxidation process permit more complete combustion. In most cases, once nitric oxide (NO) is formed it is not decomposed, particularly in the lean flame region (LFR) [1], but may increase in concentration during the rest of the combustion process if the temperature remains high or even increases. However, some NO decomposition may be experienced in the fuel-rich regions unless further oxidation takes place.

Because combustion is a complex series of chemical reactions, reaction rates, and heat transfer processes, the resultant emissions formation requires further analysis and understanding. A simplified model for heterogeneous combustion was presented in Figure 5.1 in Chapter 5. A more developed version of this model is presented in Figure 6.1, which indicates the sources of unburned hydrocarbons (HC) and NO in direct-injected diesel engines. The figure also details the species formed in both the premixed and mixing-controlled combustion phases [2, 3].

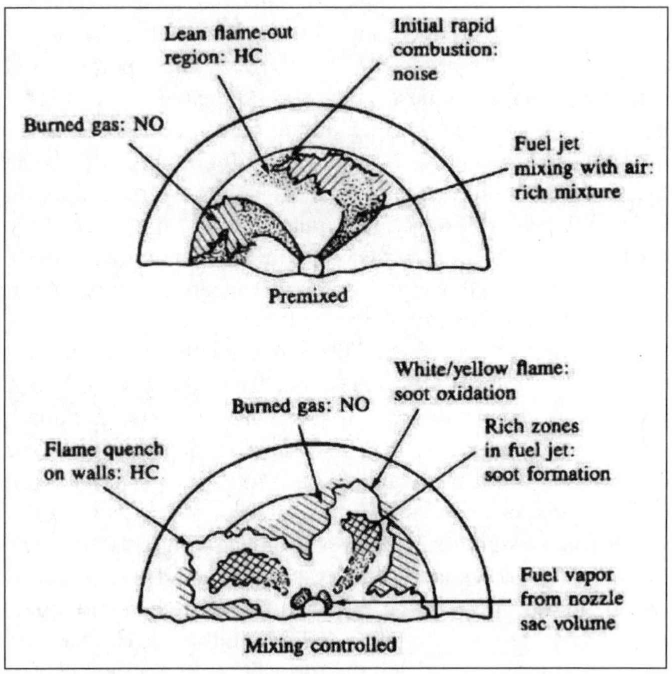

Figure 6.1 Pollutant formation mechanisms in a DI combustion system. (From Internal Combustion Engine Fundamentals *by John B. Heywood. ©1988. Reproduced with permission of The McGraw-Hill Companies.)*

6.2 Formation of Hydrocarbons in Diesel Engines

Hydrocarbons in diesel exhaust consist of either decomposed fuel molecules or recombined intermediate compounds [4]. Another source of HC in diesel exhaust is lubricating oil. The sources of HCs in DI diesel engines can be traced mainly to overmixing or undermixing of fuel and air and to large fuel droplet sizes at the end of injection. In general, unburned hydrocarbons are related to the lean flame-out region (LFOR), spray core, portion of fuel injected on the walls, spray tail, and after-injection. Obviously, engine design and operating variables play a major role in the formation and oxidation of the hydrocarbon molecules. The effects of some of these variables are reviewed in the following sections [5].

6.2.1 Air-to-Fuel Ratio

In naturally aspirated diesel engines, if the change in volumetric efficiency is neglected, the mass of air per cycle is almost constant. Changing power at fixed speed is accomplished by controlling the amount of fuel injected. Injector-to-injector differences may lead to fuel distribution variations, amount of fuel deposited on the walls, cylinder gas pressure and temperature, and injection duration. In addition, fuel injected during the ignition delay period (premixed portion) will mix with air and produce a wide range of equivalence ratios. Some of these ratios may be too lean to support stabilized combustion, others may be close to the stoichiometric ratio and ready to combust, and yet others may be too rich and too slow to evaporate, mix, and engage in the combustion process. At the onset of autoignition, mixtures that are most likely to burn are those closer to the stoichiometric ratio. Combustion will continue unless the localized A/F ratio decreases rapidly due to overmixing with surrounding air or if the flame is quenched by a thermal boundary layer [3, 6]. As for the very lean mixture, it may eventually engage in the combustion process if it meets and mixes with additional evaporated fuel.

The opposite is true of the large droplet sizes injected toward the end of injection duration. These droplets will need both additional air or lean mixtures and high temperatures to help them evaporate, mix, and subsequently engage in the combustion process. Evaporation, mixing, and combustion rates must be fast enough to consume most of the sprayed fuel in advance of significant expansion and cooling off during the expansion stroke; otherwise, excessive unburned hydrocarbons will be found in the exhaust. Figure 6.2 is a schematic representation of the premise presented for the three A/F ratios and describes their influence on the formation of unburned hydrocarbon emissions [2]. The explanation and the schematic shown in Figure 6.2 are specifically applicable to the portion of the fuel injected during the ignition delay period.

For fuel injected after the start of combustion, fuel pyrolysis or rapid oxidation by mixing with air leads to complete combustion. Conditions that may slow the rate of pyrolysis and oxidation will lead to localized rich mixtures and delays in completing combustion. If such conditions prevail, then exhaust gas will contain greater concentrations of unburned hydrocarbon. A schematic representation of the formation of unburned hydrocarbons from post-ignition delay fuel spray is given in Figure 6.3.

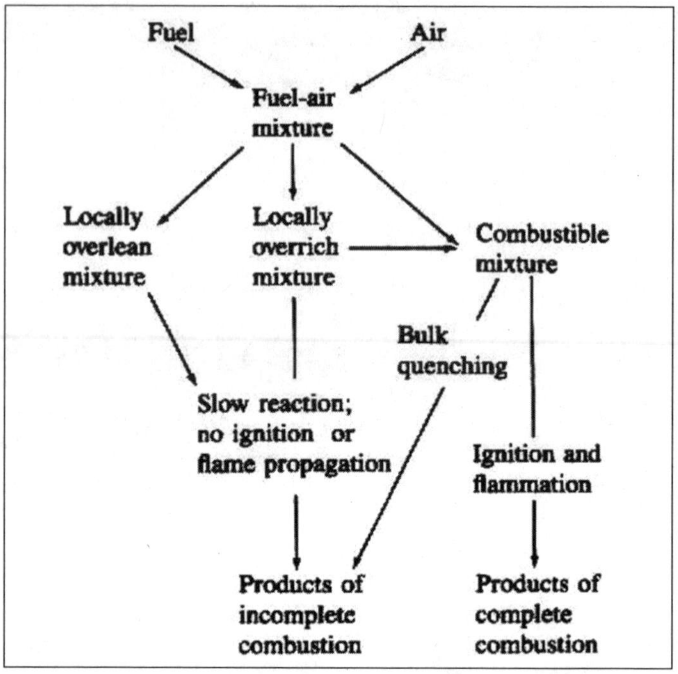

Figure 6.2 Schematic representation of hydrocarbon formation from fuel injected during ignition delay. (From Internal Combustion Engine Fundamentals *by John B. Heywood. ©1988. Reproduced with permission of The McGraw-Hill Companies.)*

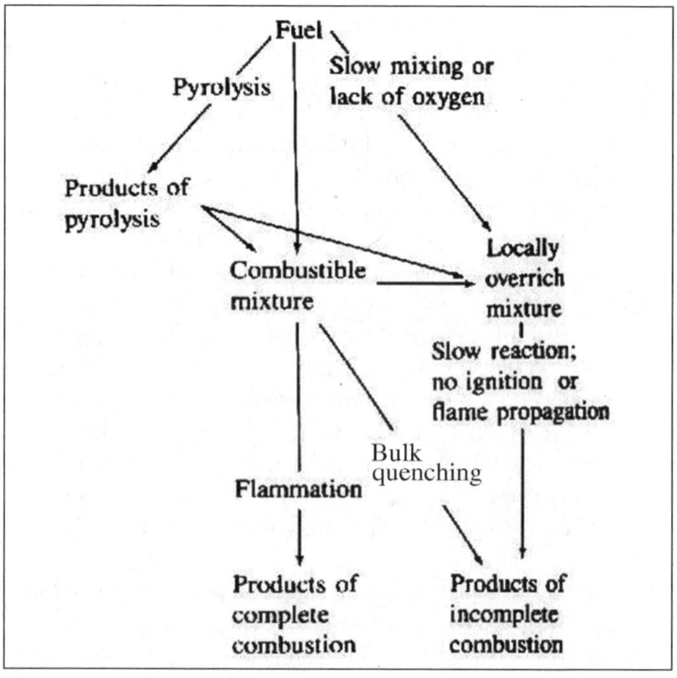

Figure 6.3 Schematic representation of hydrocarbon formation from fuel injected after the ignition delay. (From Internal Combustion Engine Fundamentals *by John B. Heywood. ©1988. Reproduced with permission of The McGraw-Hill Companies.)*

6.2.2 *Effect of Engine Load*

Having reviewed the typical progression of combustion in heterogeneous mixtures, it may useful to investigate the effects of operating conditions on hydrocarbon emissions formation. At idle and light load conditions, spatial combustion is more likely, meaning that the fuel does not reach the walls of the combustion bowl, but instead burns in the space between the nozzle tip and the combustion bowl perimeter. In addition, the fuel concentration in the spray core is rather small. In this case, unburned hydrocarbon emissions originate mainly from the LFOR. As combustion continues, increases in local temperatures in this region are very small and are associated with slower oxidation reaction rates. These reaction rates are further reduced due to the very low concentration of fuel molecules as they diffuse in the air around this region. The ratio of unburned hydrocarbons formed in this region to total fuel injected is highest at idling [1].

As engine load increases, a decreasing A/F ratio causes more fuel to be deposited on the walls and produces higher concentrations in the spray core. The unburned hydrocarbons formed in this case increase; however, there is sufficient oxygen in the mixture so that, with increased temperature, oxidation reaction rates are enhanced and hydrocarbon emissions reduced [3, 6, 7]. The effect of load on unburned hydrocarbons is shown graphically in Figure 6.4 [1, 7].

At full load, as well as in overload conditions, further decreases in the A/F ratio result in more unburned hydrocarbon molecules in the spray core and near the combustion bowl walls. In this case, the contribution of the LFOR to the total unburned hydrocarbon emissions is very small. The oxidation reactions are also limited due to the lack of oxygen in the locally rich fuel/air mixture, in spite of the very high temperatures reached and the overall lean mixture [7].

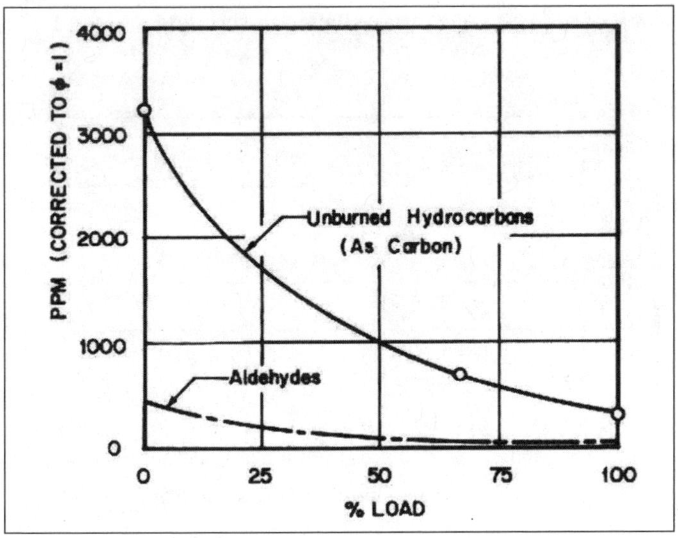

Figure 6.4 Effect of load on unburned hydrocarbon emissions. (Reprinted with permission from Ref. 1.)

It is also worth noting that the molecular structure of the hydrocarbon emissions tends to vary with the A/F ratio. At idle and light loads, hydrocarbons emitted from DI engines are related to the LFOR and consist mainly of the original fuel molecules. At high loads most of the hydrocarbons emitted originate from the fuel molecules in the spray core and on the combustion bowl walls, where temperatures are fairly high, and cause decomposition of the original fuel molecules. Because the A/F ratio in the spray core and near the combustion bowl walls is generally low (rich condition), there is a great possibility that some recombination reactions may occur between the hydrocarbon radicals and the intermediate compounds [8]. This results in higher concentrations of the heavier hydrocarbons. The process of recombination of the hydrocarbon compounds and radicals also result in compounds having a different structure than the original fuel [9, 10].

6.2.3 *Effect of After-Injection and Secondary Injection*

Injection system design problems such as after-injection and secondary injections (see Figure 4.25) tend to increase unburned hydrocarbon emissions [9, 10]. The mechanism of unburned hydrocarbon formation is a combination of slow evaporation rate (because these uncontrolled injections take place in the expansion stroke while the gas temperature is cooling off) and a mixture that is too rich to burn. Modern diesel engines equipped with exhaust emission control devices that require external heating for their function may use a controlled post-injection pulse of fuel. Unless extra care is taken in designing the timing and the quantity of this post-injected pulse of fuel, problems in the form of excessive unburned hydrocarbons, carbon monoxide, and particulate matter can arise. Injection system line dynamics as well as the mass-spring system (injector needle and return spring) must be carefully engineered over a wide range of engine speeds and loads to avoid these problems [1].

6.2.4 *Effects of Cold Starts and Misfires*

In cold environments, white smoke may form due to a misfire. This is a symptom of partially evaporated fuel that is struggling to evaporate and mix with air. Lack of heat from the cylinder walls, piston crown, or any of the adjacent engine components makes it difficult for the fuel to evaporate. Unless the compression ratio is raised by design or by control in variable-compression engines, misfires will be experienced. The resulting white smoke emission is an ultimate form of unburned hydrocarbons that disappears once the engine warms up [11]. These problems are generally aggravated by air-to-air charge air cooling, low compression ratio, injection timing retard, and coolant temperature below 77 to 85°F [2, 12]. This condition is the only operating condition in which a diesel engine experiences a misfire. Once the engine warms up, a misfire should not occur in diesel engines. Some engine designs include an intake charge heater that is energized a few seconds before cranking to assist in white smoke avoidance or elimination. Other engines, such as the Navistar (International Truck and Engine) 7.3-L DI engine, are equipped with a back-pressure valve that exerts a light load on the engine during warmup. This measure is usually combined with an increase in the idle speed to increase the back pressure and thus the load applied on the engine. It also increases the friction heat. Both measures combine to provide heat to aid in fuel spray evaporation and white smoke reduction.

6.2.5 *Effect of Turbocharging*

Turbocharging increases the mass of air inducted into the engine and allows a proportional increase in the injected fuel quantity, which leads to higher engine output. Air velocity through the intake port is higher in turbocharged engines relative to naturally aspirated engines. It follows that air motion through the intake port of either a swirl-supported or quiescent combustion system is more likely to enhance mixing and mixture preparation. In general, better mixing leads to higher reaction and oxidation rates and also results in higher combustion temperatures, which in turn causes a reduction in the concentration of hydrocarbon emissions. The reduction of hydrocarbon emissions with turbocharging has been observed by Perez and Landen [6] and reported by Henein and Patterson [1]. The effect of the A/F ratio on HC emissions is shown in Figure 6.5 for both DI and IDI engines.

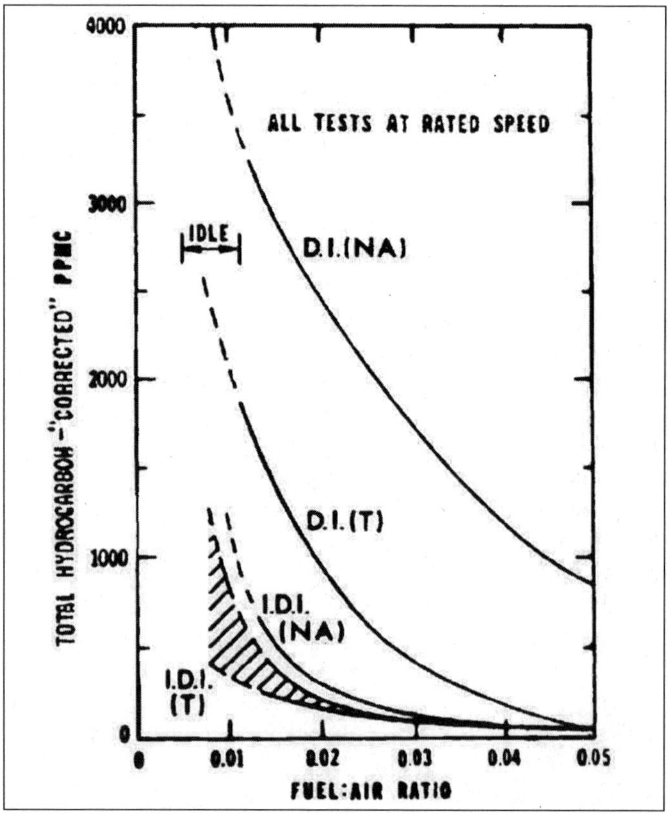

Figure 6.5 *Effect of the fuel-to-air ratio on HC emissions in DI and IDI turbocharged and naturally aspirated diesel engines. (Reprinted with permission from Ref. 1.)*

6.2.6 *Effect of Nozzle Opening Pressure*

The effect of nozzle opening pressure on performance and emissions was evaluated in a single cylinder research diesel engine [1, 13–16]. The engine was equipped with a toroidal combustion system and the results for unburned hydrocarbon emissions are given in Figure 6.6. Data in Figure 6.6 are given for operation at both rated (2500-rpm) and peak torque (1600-rpm) speeds at full-load conditions. It is apparent in both cases that by raising the nozzle opening pressure, HC emissions are increased. Most of the HC increase, however, is seen at nozzle opening pressures below 240 bar. It is believed that raising the nozzle opening pressure improves atomization and widens the lean flame-out region, which, as discussed earlier, is one of the major sources of unburned hydrocarbon emissions.

6.2.7 *Effect of Injection Timing*

The same single cylinder research engine used earlier in the nozzle opening study was used to investigate the effect of injection timing on performance and emissions. The results are given in Figure 6.7 for the same two engine conditions described in Section 6.2.6. Injection timing advance appears to have increased unburned hydrocarbon emissions in both cases. This finding agreed with similar results reported by Khan and Grigg [12]. The mechanism that could explain this phenomenon appears to be the longer ignition delay, which allows more fuel vapor and small droplets to be carried away with the swirling air, producing a wider LFOR. Another factor suggested by Khan and Grigg [12] is the increase in fuel impingement on the combustion bowl walls.

6.2.8 *Effect of Intake Swirl*

A review of swirl in DI diesel engines is given in Chapter 13. For purposes of this section, however, we can define swirl by the term *swirl ratio,* which is the ratio of the circumferential speed of air as it enters the cylinder and the engine speed. We

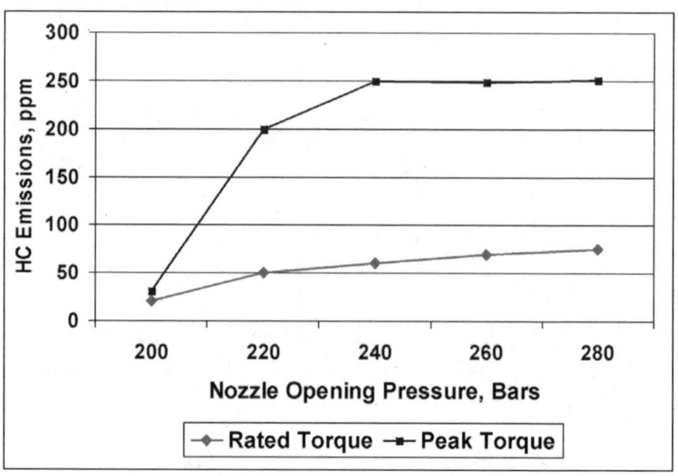

Figure 6.6 *Effect of nozzle opening pressure on unburned HC emissions.*

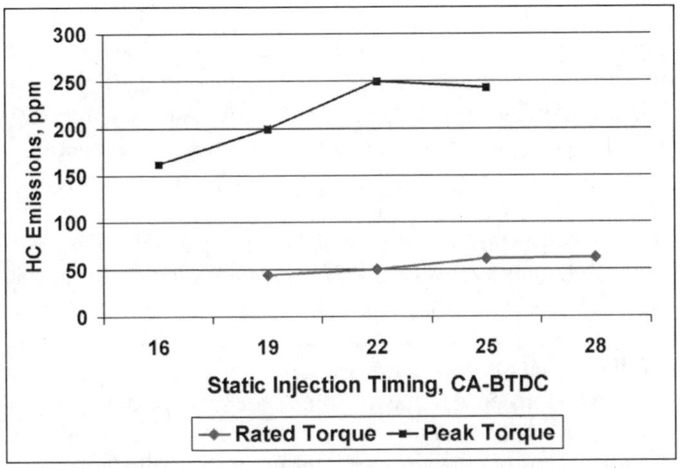

Figure 6.7 *Effect of injection timing on HC emissions.*

will also consider that increased swirl in DI engines generally improves the mixing and the hydrocarbon oxidation processes. Excessive swirl may, however, produce a wider LFOR as discussed earlier, or an overlap of the sprays and an increase in unburned hydrocarbon emissions. As explained earlier, swirl in DI engines may be changed by varying the ratio of the bowl diameter to its depth, known as the *aspect ratio.* The variation in the aspect ratio has been found by Watts and Scott [17] to have little effect on the fuel economy; it may, however, affect the pollutant formation. Because combustion bowl sizes and shapes vary widely, it is difficult to make a general statement about how the combustion bowl and its influence on swirl may affect hydrocarbon emissions.

6.2.9 Final Comments on Sensitivity of HC to the Fuel System

The fuel system plays a major role, not only in engine performance, but also in the formation of emissions from diesel engines. The point at which the fuel system comes closest to the combustion process is the injector nozzle hole. It follows then that the nozzle hole design is very important, especially in terms of its effect on the formation and emission of HCs. Nozzle hole diameter, length, and surface quality, as well as its machining process (mechanically drilled or by electronic discharge machining) and the shape and surface finish of its upstream inlet and downstream discharge side are all factors affecting HC emissions. These design parameters may directly affect the rate of injection, atomization of the fuel, and the spray penetration into the combustion bowl. Results reported in Ref. [12] highlight the influence of sac volume on hydrocarbons emissions. Sac volume (see Figure 6.8) is defined as the volume between the needle seat and the nozzle holes.

Watts and Scott [17] reported that reducing sac volume greatly reduced unburned hydrocarbon emissions. A conical sac (Figure 6.8b) has a smaller volume than a cylindrical sac (Figure 6.8a). Fuel accumulating in the sac leaks out (termed *nozzle dribble*) through the injector holes during the expansion stroke. This fuel does not atomize into the combustion chamber nor mix with the oxidants; therefore, it finds its way out to the exhaust system through the exhaust valves.

Secondary injections and after-injections (see Chapter 5) lead to unburned hydrocarbon emissions, especially at part-load conditions. Careful design of the mass (needle) and spring system dynamics can usually solve this problem. In addition, delivery valves are often used in conventional pump-line-nozzle systems to maintain the injection line residual pressure at a level that would prevent after-injections and secondary injections from occurring.

6.3 Formation of Carbon Monoxide in Diesel Engines

Carbon monoxide (CO) is a toxic gas that is capable of killing humans by displacing oxygen in their bloodstream. It is an intermediate product in the combustion of hydrocarbons [20]. It is formed mainly due to incomplete combustion, which is exacerbated by a lack of oxidants, temperature, and residence time. As combustion proceeds to completion, oxidation of CO to CO_2 occurs through recombination reactions between CO and various oxidants. If these recombination reactions are incomplete due to lack of oxidants or due to low gas temperatures, CO will be left without oxidation [1].

Combustion of fuel-rich mixtures usually produces high CO emissions, but because diesel combustion is lean and has an

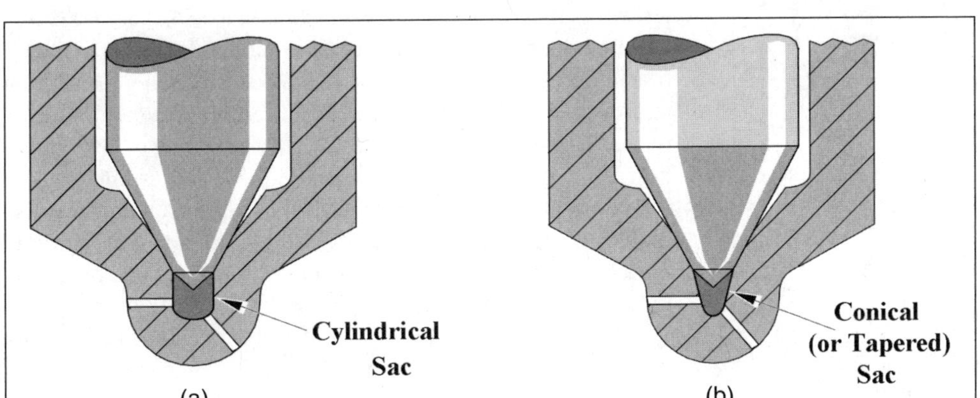

Figure 6.8 *Schematic representation of (a) cylindrical versus (b) conical sac volume.*

abundant amount of air, CO from diesel combustion is extremely low. In fact, it is so low that Professor John Heywood in his book titled *Internal Combustion Engine Fundamentals* [2] concluded that "CO emissions from diesels are low enough to be unimportant, therefore, they will not be discussed further."

During the various stages of spray combustion in DI diesel engines, conditions that promote high combustion temperature, high oxygen content, and high mixing rates are responsible for the oxidation of CO to CO_2. However, the main parameter affecting its formation remains a low A/F ratio. Figure 6.9 shows the relationship between the inverse of the A/F ratio—the fuel-to-air (F/A) ratio—and CO emissions for a turbocharged DI and a naturally aspirated IDI diesel engine.

6.4 Formation of Nitrogen Oxides in Diesel Engines

The term *nitrogen oxides* (NO_x) refers to nitric oxide (NO) and nitrogen dioxide (NO_2) in the exhaust of internal combustion engines. In DI diesels, NO constitutes 70% to 90% of the total NO_x. Nitric oxide is a by-product of combustion rather than a compound in transition like CO. Nitrogen makes up more than 75% of the intake charge air by volume. With the balance of the inducted air volume being oxygen and with help from heat resulting from burning HC fuel, NO_x is formed and remains until it is exhausted through the exhaust valve. The amount of fuel-borne nitrogen is negligibly small, so it is not considered a major contributor to the chemical reaction process. The formation of NO is well understood. It is accepted that NO is formed by the extended (or modified) Zeldovich mechanism:

$$O_2 + N_2 \rightleftharpoons NO + N \tag{6.1}$$

$$N + O_2 \rightleftharpoons NO + O \tag{6.2}$$

$$N + OH^{\cdot} \rightleftharpoons NO + H \tag{6.3}$$

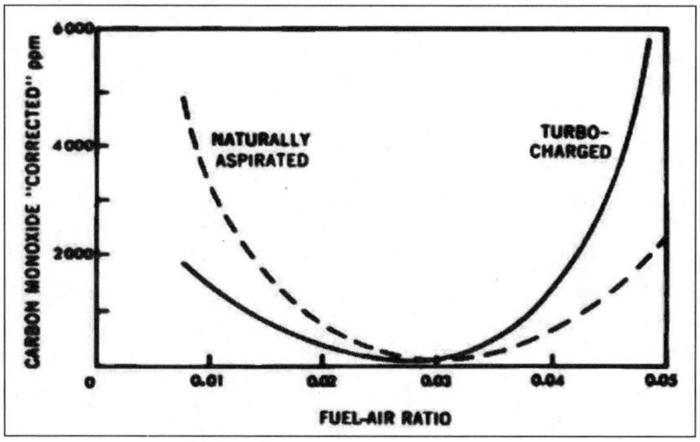

Figure 6.9 Effect on F/A ratio on CO emissions in Diesel Engines. (Reprinted with permission from Ref. 1.)

Essentially, when a molecule of O_2 is exposed to high temperature in an N_2 environment, NO is formed and an atom of nitrogen is left in an energetic and unstable state [Eq. (6.1)]. The nitrogen atom tries to reach a stable state by combining with oxygen present in the cylinder, while enough energy is available in the form of heat from the combustion process. This reaction, once again, forms more nitric oxide and leaves an atom of oxygen in an unstable state as shown in Eq. (6.2). In the modified Zeldovich mechanism, scientists suggested that radicals, such as $OH^{\cdot}$, are formed and react with the nitrogen atoms to form more NO and an atom of hydrogen as given by Eq. (6.3). It follows that all three equations show mechanisms by which NO can be formed and that the longer the residence time in this high NO formation regime, the greater the NO_x concentration. According to this mechanism, atomic nitrogen does not start the chain reaction because its equilibrium concentration during the combustion process is relatively low compared to the equilibrium concentration of atomic oxygen [1, 2]. Therefore, in diesel combustion the local NO formation in the spray is related to the local oxygen atom concentration and the local temperature. NO_2 forms from NO by reaction with excess oxygen and quenching by excess air in the cylinder can freeze NO levels at well above equilibrium concentrations. As the temperature decreases during the expansion stroke, the NO concentration does not decrease to the equilibrium concentration, because it is stable under these conditions. References [19] and [20] reported that in reciprocating engines the NO removal processes during the expansion stroke are very slow, and thus NO concentration remains nearly constant during expansion. This is particularly true in fuel-lean combustible mixtures [21].

The premixed portion of the fuel (mixed during the ignition delay) burns before the time of peak cylinder pressure and is particularly important for NO formation. After it has burned, it is compressed to a higher pressure and temperature and so reaches the highest temperature of any portion of the cylinder charge. For this reason, techniques to control NO_x focus on this early phase of combustion. Unfortunately, most of these techniques resort to reducing combustion temperatures. In so doing, they lead to penalties in hydrocarbon emissions, particulate matter (PM) emissions, and fuel consumption.

It is common to refer to the NO_x/PM trade-off or the NO_x/fuel consumption trade-off in diesel engines. These expressions point in part to the admission that the means to reduce NO_x emissions would invariably lead to increases in particulate emissions as well as fuel consumption, and vice versa. Most factors affecting the initial rate of heat release will affect the NO_x formation rate in the same direction [21]. Many control schemes attempt to reduce the amount of fuel burned in the premixed burning phase. These schemes include rate shaping using a lower injection rate early in the injection period followed by a considerable injection rate increase after the start of combustion. They also include pilot injection, which has a separate small injection to initiate combustion with a minimum of fuel followed by a main injection after combustion initiation.

Having reviewed NO_x formation, it is perhaps time to study the effects of some engine variables and other parameters on NO_x emissions.

6.4.1 Fuel-Air Ratio

Increasing the load in naturally aspirated diesel engines implies an increase in fuel for the same mass of air inducted at a constant speed. As more fuel is burned, more heat is generated and an increase in combustion temperature is experienced. The availability of a high combustion temperature provides adequate energy to help the chemical reactions described by the Zeldovich and modified Zeldovich mechanisms, thus causing more NO_x formation as long as N_2 and O_2 are present. It was suggested by Henein and Patterson [1] that a reduction in the F/A ratio to below 0.02 would actually reduce NO_x formation due to the drop in combustion temperature by air dilution. Figure 6.10 shows the increase in maximum combustion pressure (P_{max}), exhaust temperature (T_{exh}), brake mean effective pressure (BMEP), and observed NO as a function of the F/A ratio. It is worth noting that smoke also increases as the F/A ratio increases, but NO, corrected to an equivalence ratio of 1, decreased as a function of the F/A ratio within the test range [1]. The increase in observed NO emission did not appear to be proportional to the F/A ratio, indicating that other factors are affecting NO formation.

6.4.2 Effect of Nozzle Opening Pressure

Increasing the nozzle opening pressure tends to improve atomization. With smaller fuel particles due to improved atomization, combustion efficiency improves, releasing higher heat and raising the combustion temperature. In general, higher combustion temperatures lead to higher NO formation [1].

6.4.3 Effect of Injection Timing

Injection timing advance appears to extend the ignition delay. The reason for this observation is that fuel is injected into a lower pressure medium having lower temperature. The longer the ignition delay, the greater the portion of fuel injected during that period, and the better the chance for it to mix with air. In other words, the earlier the fuel is injected, the longer the ignition delay (see Figure 6.11) and the greater the premixed portion of the fuel prior to ignition. Higher NO formation is usually related to the premixed portion of the fuel.

As injection timing is retarded, the opposite effect is experienced. Ignition delay tends to be shorter and ultimately, the premixed portion of the fuel is reduced. Reducing the premixed fuel portion leads to lower NO formation and is illustrated in Figure 6.11. Since retarding injection timing has a desirable effect on NO formation, it has become an effective means to control NO_x emissions. An important side effect to this action is the increase in fuel consumption, as shown in Figure 6.11, unless other measures are taken to avoid this loss in fuel economy.

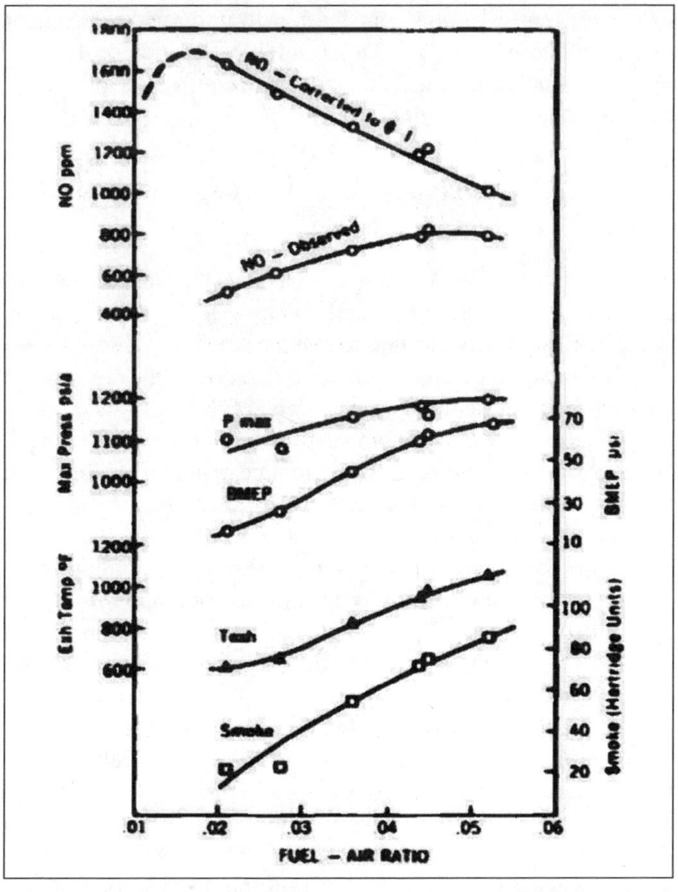

Figure 6.10 *Effect of F/A ratio on NO formation. (Reprinted with permission from Ref. 1.)*

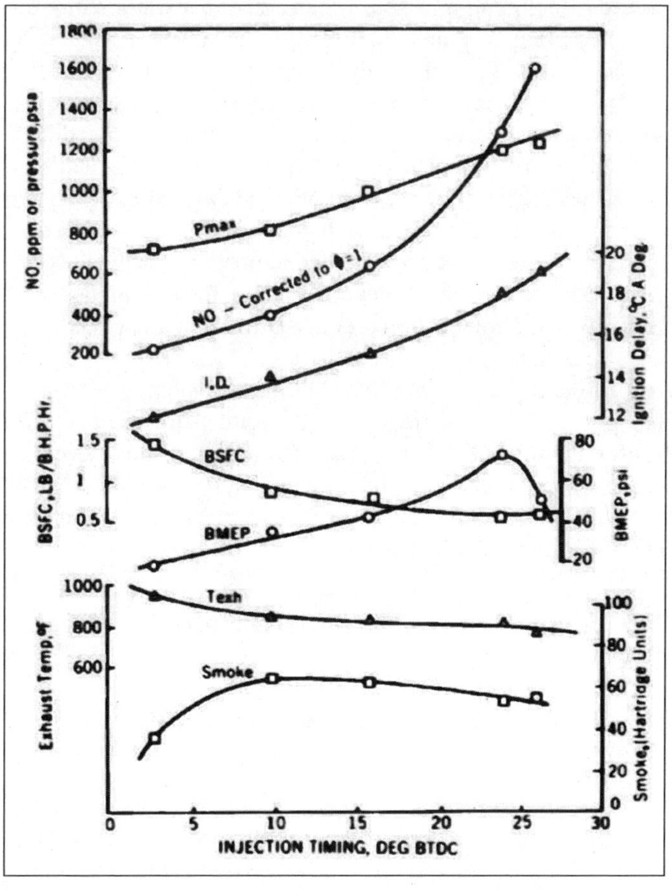

Figure 6.11 *Effect of injection timing on NO and other parameters. (Reprinted with permission from Ref. 1.)*

6.4.4 Effect of Cetane Number

Fuels with a high cetane number easily evaporate and are quick to ignite. Therefore, they are characterized by shorter ignition delay periods. Consequently, they have smaller premixed fuel portions and lower NO emissions for the same BMEP compared to fuels having lower cetane number [20]. With low cetane fuel, ignition delay is usually long and more fuel is present in the LFR when combustion starts. This larger quantity of premixed fuel produces a higher gas temperature upon combustion early in the cycle, and more NO is formed in the LFR.

6.4.5 Effect of Swirl

Air motion in the cylinder affects fuel and air mixing rates. Although high swirl, or a high degree of air motion, is generally desirable, overswirling can be experienced. A good indication of overswirling is a deteriorating fuel economy and an increase in HC, CO, and particulate matter emissions. Generally, an increase in air swirl improves mixing and promotes efficient combustion and more NO formation. The mechanism responsible for the increased NO formation is the higher initial heat release rate [20].

6.4.6 Effect of Intake Charge Dilution

Intake air charge dilution is accomplished through several methods including nitrogen, water, and exhaust gas recirculation. The effect of charge air diluents is usually seen as reduced NO formation. Two mechanisms are referred to when explaining the reason for reduced NO formation. The first is simply oxygen displacement, meaning that the normal oxygen inducted through the fresh air charge is reduced or displaced by a diluent that is inert or has less oxygen than fresh air. Reducing oxygen in the cylinder leads to sluggish combustion (lower rate) and, therefore, lower peak pressure and temperature as well as NO emission. The second explanation is that the diluent acts as a heatsink in the combustion process. Diluents with high specific heat, such as water, N_2, or CO_2, are most effective in reducing NO formation. Their high specific heat combined with their much lower temperature (relative to combustion temperature) causes them to absorb heat from combustion, thus lowering combustion temperature and reducing NO emissions. Chapter 16 is dedicated to this subject and specifically deals with the application of EGR in diesel engines.

6.5 Formation of Particulate Matter in Diesel Engines

6.5.1 Constituents and Sources of Particulate Matter

Perhaps the most controversial of all regulated emissions, particulate matter is defined as any matter in the exhaust of an internal combustion engine that can be trapped on a sampling filter medium at 125°F (52°C) or less [22]. As such, diesel PM is not a well-defined physical species. Rather, it is a complex emission, including a number of components, which are discussed in more detail in Chapter 7.

The preceding definition of PM uses the all-encompassing phrase "any matter" to describe what can be considered part of the total PM emissions. In general, particulate matter originates from the organic and inorganic substances inducted into the engine along with the fuel and air. One of the major constituents of PM is the carbonaceous matter resulting from the heterogeneous combustion process in diesel engines. Small particles or combustion nuclei form as a result of both diffusion and premixed flames, and in the case of premixed flames by both rich and lean conditions. Research on these particles, including when they are formed in the flame and their chemical composition, is ongoing. Dust (in air) or inorganic material (in the fuel or fuel additives) may appear as particulates in the exhaust (ash, oxides, etc.). Trace metals from engine component wear may be carried by the lube oil and eventually migrate to the exhaust where they can be trapped on the particulate sampling filter.

The presence of sulfur in the fuel and lube oil contributes to the formation of sulfate particulates. An added problem associated with sulfates is their retention of humidity, which may more than double their mass.

Finally, high boiling hydrocarbons and their derivatives are also included in the PM. This material, referred to as the soluble organic fraction (SOF), is composed mainly of lube oil–derived hydrocarbons. Loss of oil control resulting from improper cylinder bore honing or piston ring wear can be a major source of oil consumption and contribution to unburned oil content in particulate matter [23]. Other soluble organic constituents include unburned fuel, in spite of the high combustion efficiency displayed by diesel engines.

6.5.2 Carbonaceous Particulate Release and Oxidation

Figure 6.12 gives a summary of the mechanisms involved in carbonaceous particulate release and its oxidation. The top line of the chart shows that an increase in the amount of incompletely mixed fuel and its equivalence ratio is one of the causes of net carbonaceous particulate release. A good example of incomplete combustion is partially combusted carbon in the form of soot. In other words, carbonaceous particulate results from diffusion burning rather than combustion of a well-mixed fuel and air mixture whose equivalence ratio is close or equal to stoichiometric. Diffusion burning would obviously be enhanced when more fuel is injected after the start of ignition. Another possibility for more diffusion burning is through a greater unmixed portion of the fuel injected during the ignition delay period. Factors that may affect diffusion burning are, therefore, the ignition delay period and the physical and chemical events that take place in it as well as the rate and duration of injection. Higher initial injection rates tend to increase the premixed portion of the fuel and reduce carbonaceous particulates. Longer injection durations tend to increase diffusion combustion by increasing the fuel injected after the start of ignition. Designers of diesel combustion systems may have a few tools to avoid excessive diffusion combustion and carbon release. Increasing the cetane number improves the evaporation and mixing processes and in turn reduces diffusion burning and particulate formation. Another design parameter that could help is the compression ratio, since increasing it would provide more heat that could be used by the injected fuel to evaporate, mix with air, and burn as premixed rather than diffusion combustion [24].

Another approach to the particulate control issue is its oxidation following its formation. The lower half of the chart in Figure 6.12 presents a few tools to enhance the oxidation of carbon after its release during combustion. Increasing the time available for the combustion process may give additional opportunity for incomplete combustion products to find extremely lean pockets with which to mix. In the presence of high temperature in the cylinder, oxidation is then enhanced and exhaust carbonaceous particulate emissions are reduced. Providing more time for the combustion and oxidation process can be achieved through reduced engine speed.

6.5.3 Summary of Potential Sources of Particulate Matter

Figure 6.13 summarizes the sources of particulate matter in diesel engines. They are listed randomly and not necessarily in order of contribution to the total PM problem. Loss of oil control, combustion quality, cold starting emission of white smoke, unburned fuel, and general fuel quality problems are among those sources of particulate matter.

Loss of oil control can result from many areas within the internal combustion engine. Improper cylinder liner honing, lack of piston ring pack sealing, lube oil seeping through excessive valve guide and stem clearances, turbocharger lubrication and bearing malfunction, extended diesel engine idling, and excessive piston-to-liner clearances are all contributors to high oil consumption. Lube oil jets are usually an integral part of heavy-duty diesel engine applications. These jets are aimed at the underside of the combustion bowl to cool the entire piston crown area. Excessive flow through these jets has been shown to cause high oil consumption. In addition, sudden accelerations or changes from low to high loads are prone to generate more unburned lube oil, since ring pack sealing may be less than desirable. At low load, combustion pressure is low and causes a decrease in the sealing force applied to the back of the piston ring. The lower the sealing force, the lower the sealing and the higher the oil consumption. This condition is shown schematically in Figure 6.14.

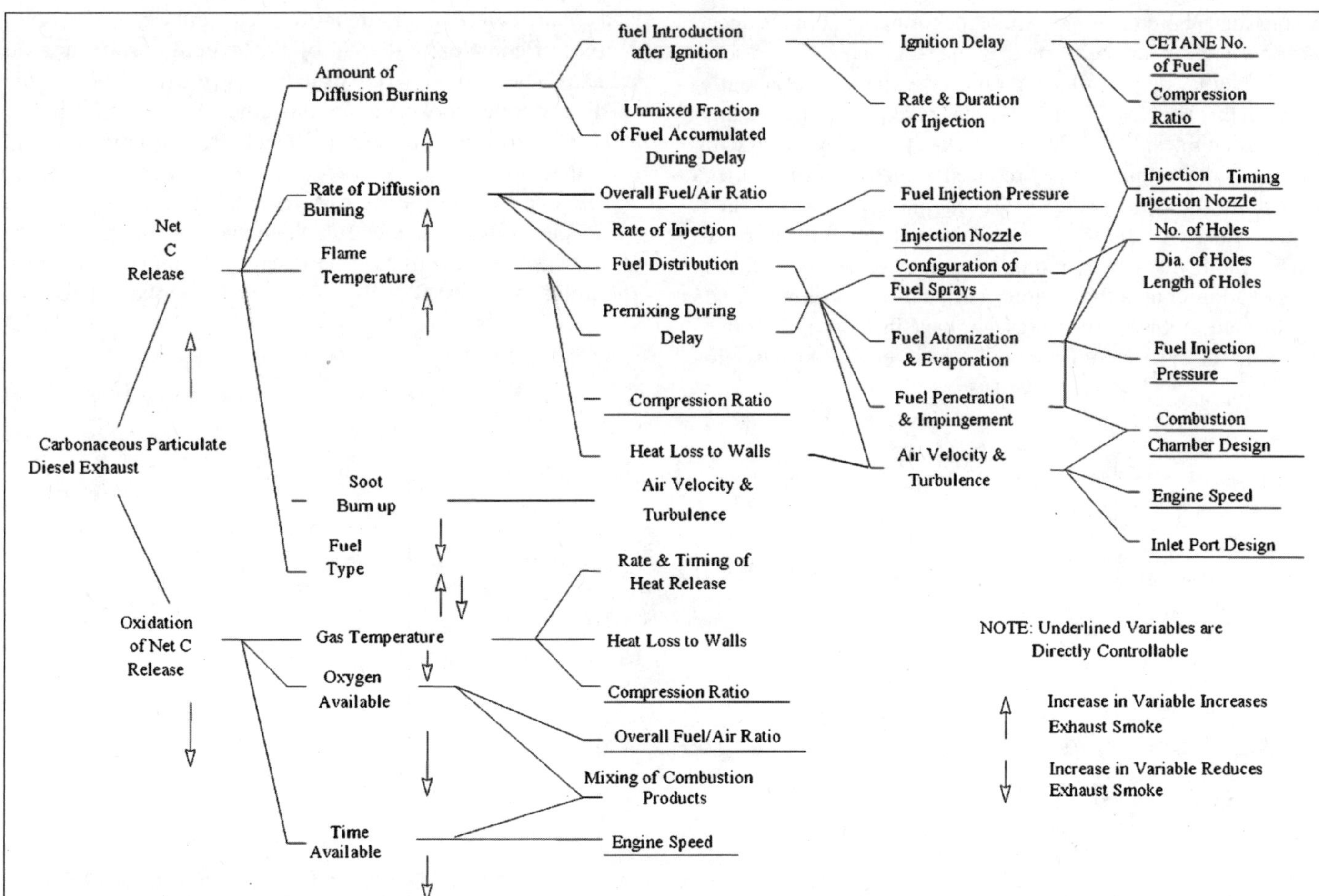

Figure 6.12 *Summary of diesel engine variables that affect soot particle formation and oxidation. (Based on Ref. 24.)*

1. LOSS OF OIL CONTROL
2. BLACK SMOKE FROM EXCESSIVE RICHNESS
- **OVERFUELING**
- **POOR COMBUSTION**
- **TRANSIENTS**

3. WHITE SMOKE
- **COLD START**
- **MISFIRE**

4. OTHER HYDROCARBON ORIGINS
- **NOZZLE DRIBBLE**
- **SECONDARY INJECTIONS**
- **AFTER-INJECTIONS**

5. FUEL QUALITY
- **SULFUR CONTENT**
- **AROMATIC CONTENT**
- **CETANE NUMBER**

Figure 6.13 *Summary of sources of particulate matter. (Reprinted from Ref. 23.)*

Although combustion-related sources of particulate were considered in the previous section, other engine operational reasons may also encourage particulate formation. For instance, hard accelerations are performed by an instantaneous increase in engine fueling. In general, the air that is available to efficiently mix with this increased fuel is less than is required. In naturally aspirated engines, the ability to quickly fill the cylinder with charge air may be hindered by reduced volumetric efficiencies, whereas the turbocharger lag in turbocharged engines may cause an instantaneous drop in charge air. In both engine configurations, a momentary lack of charge air is experienced and this leads to additional carbonaceous particulate emission.

In cold ambients, compression heat in most DI engines with a 17:1 or 18:1 compression ratio is unable to raise the compressed air temperature enough to help the fuel evaporate, mix with air, and burn. Therefore, fuel is partially evaporated and exits the exhaust port and system in the form of white smoke. This problem is especially aggravated by additionally retarding injection timing and overcooling charge air, and is primarily caused by misfires.

The contribution of unburned hydrocarbons to the development of particulate matter was discussed earlier. Nozzle dribble, after-injections, and secondary injections are all major sources of hydrocarbon-based particulates [25]. Yet, besides lube oil, fuel itself is the main source of particulates if it does not have the right ingredients to support good combustion or it is not being used properly. A high sulfur and aromatic content combined with a low cetane number is the best list of ingredients for high particulate formation and tailpipe emissions [26–29].

6.6 Smoke in Diesel Engines

Smoke in diesel engine exhaust is an indication of poor combustion, resulting from an overrich A/F ratio or partially evaporated fuel during cold start conditions [30–33]. Smoke emission, especially if it is black, is the diesel's worst enemy because it is emitted in plain view of the public, giving the diesel its reputation of being a *dirty* engine. Social and environmental pressures are leading diesel engine manufacturers to design and produce engines that are nearly smoke free. Most industrialized countries have therefore introduced regulations of varying degrees of complexity to control smoke emissions from road vehicles. The regulations are in addition to the relatively simple existing visible smoke regulations covering industrial plants and have involved much development both of test methods and instrumentation.

Smoke may be in the form of particles, either solid or liquid (aerosols), suspended in the exhaust gases [34–39]. It obstructs, reflects, or refracts light; causes soiling of structures; and limits visibility. Diesel engine exhaust smoke can be classified as follows:

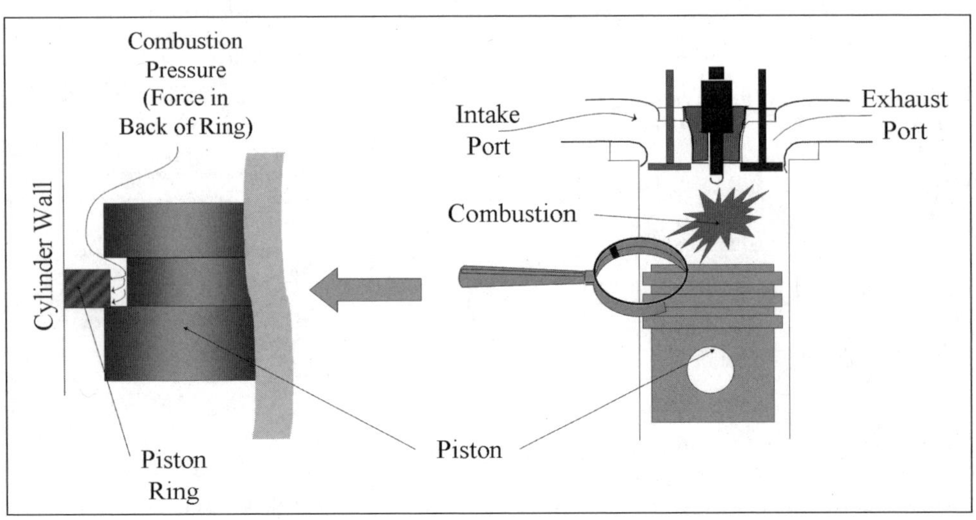

Figure 6.14 *Schematic illustration describing lack of piston ring sealing. (Reprinted from Ref. 23.)*

1. White in appearance under direct illumination, and consisting of a mixture of fuel and lubricating oil particles in an unburned or partly burned state. This form of smoke is sometimes referred to as *liquid* smoke or *fog* (see Figure 6.15).
2. Black in appearance and consisting of solid particles of carbon from otherwise complete combustion of fuel. This form of smoke is often referred to as *hot* or *solid* smoke.

In cases where excessive lube oil consumption is experienced, the smoke emitted takes on a shade of blue or gray. Blue or gray smoke is therefore an indication that an engine has reached the point where major maintenance is required. However, unburned fuel can also appear as blue smoke if the droplet size is about 0.5 μm. The white component, on the other hand, is mainly a result of too low a temperature in the combustion chamber during the fuel injection period. It has a droplet size of about 1.3 μm. This phenomenon is mainly experienced during cold starting, in low ambient temperatures, or at high altitude. White smoke disappears as the engine warms up. Figure 6.16 shows a photograph of white smoke emitted from an experimental truck in cold start mode [40–44].

White smoke can also result from fuel injected too late in the combustion cycle or can even be an indication of a design fault, in the sense that the compression ratio is too low, or has been optimized for an inappropriate combination of operating conditions [23]. A few manufacturers opted to use electric inlet or intake manifold heaters. These are typically turned on perhaps 30 s before cranking the engine in cold start conditions. They also cycle between being on and off until the water jacket temperature indicates 77 to 80°F. Recent limits imposed on NO_x emission have dictated that injection timing be retarded and that air-to-air intercooling be used. Both measures lead to increased likelihood of white smoke formation. There are no regulations in place limiting white smoke even though it is quite visible and widespread at truck stops especially in colder climates.

The main causes of excessive black smoke emission are either poor maintenance of air filters and/or fuel injectors, or incorrect setting of the fuel injection pump/system. Such smoke consists mainly of carbon particles or coagulates of a wide range of sizes, from 0.02 to more than 0.12 μm mean diameter [42]. This size distribution depends to some extent on the type of combustion system, which also affects the onset of smoke emission as fuel input quantity is increased. Figure 6.17 gives an example of black smoke emitted under accelerating conditions.

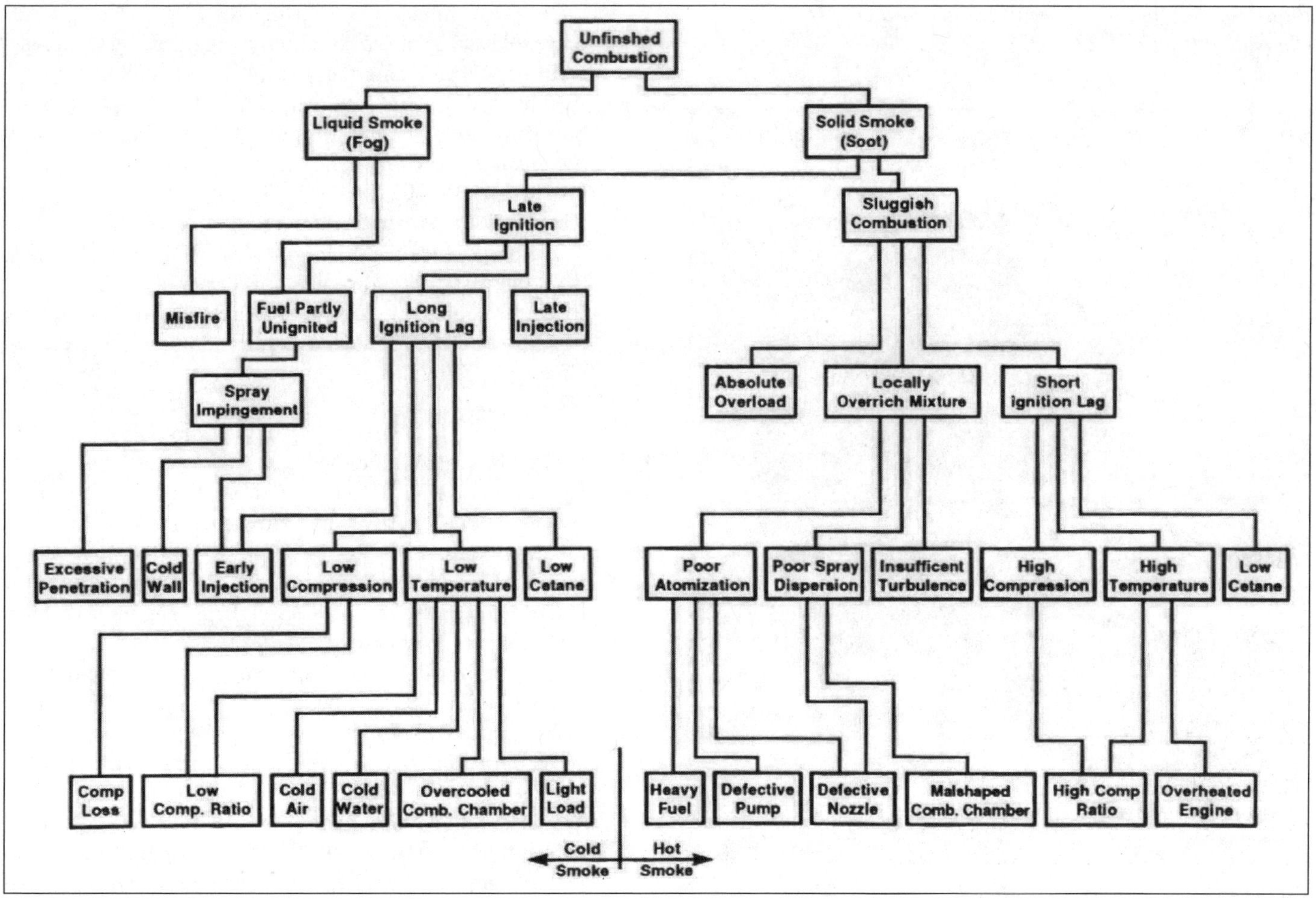

Figure 6.15 *Classification of smoke in diesel engines.*

Figure 6.16 White smoke emitted during a cold start. (Reprinted from Ref. 23.)

Figure 6.17 Black smoke emitted under accelerating conditions. (Reprinted from Ref. 23.)

References

1. Henein, N.A., and Patterson, D.J., *Emissions from Combustion Engines and Their Control,* Ann Arbor Science Publishers, Ann Arbor, MI, 1972.
2. Heywood, J. B., *Internal Combustion Engine Fundamentals,* McGraw-Hill, New York, 1988.
3. Marshall, W.F., and R.W. Burn, "Factors Influencing Diesel Emissions," SAE Paper No. 680528, SAE International, Warrendale, PA, 1968.
4. Henein, N., "Combustion, Engine Economy, Performance, and Emissions," Lecture Notes at Ford Tractor Operations, October 16–November 20, 1985.
5. Haägen-Smit, A. J., "The Control of Air Pollution," *Sci. Am.* **210,** 1, 1965, p. 25.
6. Perez, J.M., and E.W. Landen, "Exhaust Emission Characteristics of Precombustion Chamber Engine," SAE Paper 680421, SAE International, Warrendale, PA, 1968.
7. Burn, R.W., "Combustion in Heterogeneous Mixtures," in *12th International Symposium on Combustion,* The Combustion Institute, Pittsburgh, PA, 1969, pp. 677–687.
8. Bascom, R.C., L.C. Broering, and D.E Wulfhorst, "Design Factors that Affect Diesel Emissions," 1971 SAE Lecture Series: Engineering Know-How in Engine Design, SAE International, Warrendale, PA, March 1971.
9. Henein, N.A., "Combustion and Emission Formation in Fuel Sprays Injected in Swirling Air," Paper No. 710220 presented at Society of Automotive Engineering Congress, Detroit, MI, SAE International, Warrendale, PA, 1971.
10. Aaronson, A.E., and R.A. Matula, "A Study in Hydrocarbon Emissions in Lean Mixtures," in *13th International Symposium on Combustion,* The Combustion Institute, Pittsburgh, PA, 1971, pp. 471–481.
11. Khair, M.K., "Air-to-Air Intercooling of the Ford 7.8 L Mid-Range Truck Diesel Engine," SAE Paper No. 870534, SAE International, Warrendale, PA, 1987.
12. Khan, I.M., and H.C. Grigg, "Progress of Diesel Combustion Research," CIMAC, 9th International Congress on Combustion Engines, Stockholm, Sweden, Paper No. A-18, May 1971.
13. Schweitzer, P.H., "Penetration of Oil Sprays," Bulletin No. 46, Pennsylvania State College, 1937.
14. Lyshevskiy, A.S., "Determination of the Length of an Atomized Fuel Spray," Tudy Novocherkassogo, Potitekhnicheskogo Instituta, **26,** 1955, pp. 391–401.
15. Parks, M.V., C. Polonski, and R. Toye, "Penetration of Diesel Fuel Sprays in Gases," SAE Paper No. 660747, SAE International, Warrendale, PA, 1966.
16. Hiroyasu, H., T. Kadota, and M. Arai, "Supplementary Comments: Fuel Spray Characterization in Diesel Engines," in *Proceedings of the Symposium on Combustion Modeling in Reciprocating Engines,* General Motors Research Center, MI, 1978, pp. 369–408.
17. Watts, R., and W.M. Scott, "Air Motion and Fuel Distribution Requirements in High-Speed Direct-Injection Diesel Engines," in *Diesel Engine Combustion Sympo-*

sium, Institute of Mechanical Engineers, London, 1970, pp. 167–177.

18. Merrion, D.F., SAE Paper No. 680422, SAE International, Warrendale, PA, 1968.
19. Lavoie, G.A., J.B. Heywood, and J.C. Keck, *Combust. Sci. Technol.,* 1970, pp. 313–326.
20. Starkman, E.S., and H.K. Newhall, "Direct Spectroscopic Determination of Nitric Oxide in Reciprocating Engine Cylinders," SAE Paper 670122, SAE International, Warrendale, PA, 1967.
21. Komiyama, K., and J.B. Heywood, "Predicting NO Emissions and Effect of Exhaust Gas Recirculation in Spark-Ignition Engines," SAE Paper 730475, SAE International, Warrendale, PA, 1973.
22. *Federal Register* 21(61), Part II, 1966.
23. Khair, M. K., "Emission Formation in Diesel Engines," SAE Engineering Academies Lecture Notes, SAE International, Warrendale, PA, 2001.
24. Broome, D., and I. Khan, *The Mechanism of Soot Release from Combustion of Hydrocarbon Fuels,* Institution of Mechanical Engineers, London, 1971.
25. Milks, D., C.W. Savery, J.L. Steinberg, and R.A. Matula, "Studies and Analysis of Diesel Engine Odor Production," presented at Clean Air Congress of the International Union of Air Pollution, Washington, DC, 1970.
26. Henein, N.A., and J.A. Bolt, "The Effect of Some Fuel and Engine Factors on Diesel Smoke," SAE Paper No. 690557, SAE International, Warrendale, PA, 1969.
27. Lyn, W.T., "Study of Burning Rate and Nature of Combustion in Diesel Engines," *Proceedings of the 9th Symposium on Combustion,* Institute of Mechanical Engineers, London, 1962, pp. 1069–1082.
28. Lyn, W.T., *Proc. Inst. Mech. Engineers (A.D.)* **1,** 1960–1961, pp. 34–46.
29. Austen, A.E.W., and W.T. Lyn, *Proc. Inst. Mech. Engineers (A.D.)* **1,** 1960–1961, pp. 47–62.
30. Shipinski, J., O.A. Uyehara, and P.S. Myers, "Experimental Correlation between Rate-of-Injection and Rate-of-Heat-Release in a Diesel Engine," ASME Paper 68-DGP-11, American Society of Mechanical Engineers, New York, 1968.
31. Grigg, H.C., and M.H. Syed, "The Problem of Predicting Rate of Heat Release in Diesel Engines," Paper No. 18, in *Symposium on Diesel Engine Combustion,* Institute of Mechanical Engineers, London, 1970.
32. Henein, N.A., and J.A. Bolt, "Kinetic Considerations in the Autoignition and Combustion of Fuel Sprays in Swirling Air," Paper No. A-i, in *Trans. CIMAC 9th Int. Congress on Combustion Engines,* Stockholm Sweden, 1971, pp. 1–36; also ASME Publication 72-DGP-8, 1972.
33. Hinze, J.O., *Appl. Sci. Res.* **Al,** 1949.
34. Gerpen, J. V., "Combustion in Diesel Engines—Modeling and Simulation," SAE Engineering Academies Lecture Notes, SAE International, Warrendale, PA, 2001.
35. Lipkea, W.H., J.H. Johnson, and C.T. Vuk, "The Physical and Chemical Character of Diesel Particulate Emissions—Measurement Techniques and Fundamental Considerations," SP-430, SAE Paper No. 780108, SAE International, Warrendale, PA, 1978.
36. Broome, D., and I. Khan, *The Mechanism of Soot Release from Combustion of Hydrocarbon Fuels,* Institution of Mechanical Engineers, London, 1971.
37. Khan, I.M., "Formation and Combustion of Carbon in a Diesel Engine," Paper No. 3, *Diesel Engine Combustion Symposium,* Institute of Mechanical Engineers, London, April 1970.
38. Palmer, H.B., and C.F. Cullis, "The Formation of Carbon from Gases," Vol. 1 of *Chemistry and Physics of Carbon,* P.L. Walker and P.A. Thrower (Eds.), Dekker, New York, 1965, p. 265.
39. "Diesel Particulates," Unrestricted Report DP78/1452, Ricardo Consulting Engineers, Shoreham-By-Sea, UK, 1978.
40. Kittelson, D.B., D.F. Dolan, R.B. Diver, and E. Aufderheide, "Diesel Exhaust Particle Size Distributions—Fuel and Additive Effects," SAE Paper 780787, SAE International, Warrendale, PA, 1978.
41. Khatri, N.J., and J.H. Johnson, "Physical Size Distribution Characterisation of Diesel Particulate Matter and the Study of the Coagulation Process," SAE Paper 780788, SAE International, Warrendale, PA, 1978.
42. Kittleson, D.B., "Engines and Nanoparticles," Paper Draft Submitted to *AAAR Journal,* October 14, 1997.
43. Johnson, H.J., "Automotive Emissions," Michigan Technological University, 1988.
44. Gurevich, M., and Mauderly, J., "Health and Environmental Effects of Engine Emissions," Lecture presented at The DEER 2000 Conference, San Diego, CA, 2000.
45. Challen, B., and R. Baranescu, *Diesel Engine Reference Book,* 2nd ed., SAE International, Warrendale, PA, 1999.

Part II

Diesel Emissions

Chapter 7:
Characterization of Emissions

Chapter 8:
Health and Environmental Effects

Chapter 9:
Measurement of Emissions

Chapter 10:
Emission Regulations

Chapter 7

Characterization of Emissions

7.1 What Are Diesel Emissions?

7.1.1 Diesel Combustion Products

A diesel engine, like other internal combustion engines, converts chemical energy contained in the fuel into mechanical power. Diesel fuel is a mixture of hydrocarbons that theoretically produce only carbon dioxide (CO_2) and water vapor (H_2O) during combustion. Indeed, diesel exhaust gases are primarily composed of CO_2, H_2O, and the unused portion of engine charge air. Concentrations of these gases in diesel exhaust vary depending on the engine and its load and speed conditions, typically in the following ranges:

- CO_2: 2% to 12%
- H_2O: 2% to 12%
- O_2: 3% to 17%
- N_2: balance.

None of the principal diesel combustion products, with the exception of CO_2 for its greenhouse gas properties, have adverse health or environmental effects.

Diesel emissions also include pollutants that can either be toxic to humans or cause other negative environmental effects. The diesel pollutants are by-products of diesel combustion that originate in several nonideal processes that occur during real combustion. These processes include incomplete combustion of fuel, reactions between mixture components under high temperature and pressure, combustion of engine lubricating oil and oil additives, and combustion of nonhydrocarbon components of diesel fuel, such as sulfur compounds or various fuel additives. The total concentration of pollutants in exhaust gases from today's new diesel engines amounts to a fraction of a percent.

Some diesel emissions are regulated in the United States, Europe, and several other countries. Regulated emissions include the following compounds:

- *Diesel particulate matter* (DPM) also referred to as PM (particulate matter) or TPM (total particulate matter), regulated by the mass of emitted particles
- *Nitrogen oxides* (NO_x)
- *Hydrocarbons* (HCs) including either the total hydrocarbons (THCs) or only the nonmethane hydrocarbons (NMHCs)
- *Carbon monoxide* (CO).

Because diesel engines operate with "internal" mixture formation and with compression ignition, where combustion takes place during and after fuel injection, diesel emissions are somewhat different from those observed in spark-ignited engines. An example comparison of regulated emissions from a light-duty diesel engine with those from a gasoline engine is shown in Figure 7.1. Both engines were tested on the European test cycle (4-cylinder MY1992 engines, 1.7-L displacement).

Emissions of carbon monoxide and hydrocarbons from diesel engines are significantly lower than those from gasoline engines. Diesel NO_x emissions are usually lower than those from gasoline engines but higher than those from gasoline engines equipped with a three-way catalyst. A real drawback of the diesel engine is its high PM mass emissions, which are practically absent from gasoline exhaust gases. Emissions of

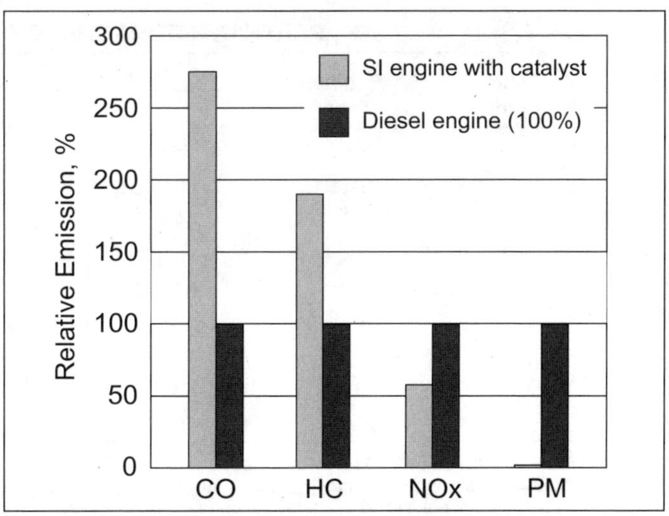

Figure 7.1 *Comparison of regulated emissions from spark-ignited (SI) and diesel engines.* (Based on data from Ref. 1.)

PM and NO_X are the focus of today's diesel emission control technologies.

Several nonregulated pollutants or suspected pollutants can be found in diesel exhaust, usually at concentration levels much lower than those of the regulated emissions. Some of them are part of the complex diesel particulate matter emission; others are totally separate species in the gas phase. The list of unregulated diesel emissions includes the following compounds:

- *PAH,* polynuclear aromatic hydrocarbons, which are heavy organic compounds found mostly in the DPM but some PAHs are also present in the gas phase.
- *SOF,* soluble organic fraction, which constitutes part of the DPM
- *Aldehydes,* R-CHO, which are derived from hydrocarbons (formaldehyde, HCHO, is regulated in some applications)
- *Sulfur dioxide*, SO_2, from sulfur present in the fuel
- *Nitrous oxide*, N_2O (nitrous oxide is not included in NO_X)
- *Dioxins:* although there is no conclusive proof at this moment, diesel emissions are suspected to be a potential source of dioxin emissions
- *Metal oxides:* Several engine lubricating oil additives include metalorganic compounds resulting in some metal oxide emissions including phosphorus, zinc, and calcium. Fuel additives researched as a means of diesel emission control may result in emissions of iron, copper, cerium, or other metals.

TABLE 7.1
GASEOUS EMISSIONS FROM DIESEL ENGINES

Time Period	Engine Type	NO_X (g/kWh)	HC (g/kWh)	CO (g/kWh)
1970s	4-NA	6.7–12.7	2.1–4.5	3.5–5.5
	4-TC	6.9–16.7	0.2–2.9	1.5–2.4
	2-DI	21.1–27.1	0.4–2.7	3.0–11.3
1980s	4-NA	5.8–10.1	1.5–3.3	2.7–4.8
	4-TC	11.2–12.5	1.6–1.75	1.5–2.1
	2-DI	6.0–7.7	0.6–1.7	1.3–3.8
1990s	4-NA	3.21–0.22	0.15–0.22	3.21–3.28
	4-TC	3.21–3.73	0.15–0.75	0.89–1.72
	2-DI	3.58–3.66	0.08–0.38	1.12–1.19

Source: Data from Ref. 2.
Key: 4-NA, four-stroke, naturally aspirated engines
4-TC, four-stroke, turbocharged engines
2-DI, two-stroke, direct-injected engines

7.1.2 Diesel Emission Levels

Tables 7.1 and 7.2 list emission levels in grams per kilowatt-hour from different time periods for different types of diesel engines. The numbers were compiled from various publications and grouped into engine-type categories [2]. The compilation includes data obtained by different test procedures. Because results from different test cycles are not always comparable, the given ranges are approximate. Data generated after 1974 include U.S. Environmental Protection Agency (EPA) certification data for gaseous emissions. Data generated after 1988 include EPA certification data from the FTP transient cycle for particulate matter and for gases. Three engine types are listed in the tables:

- 4-NA: four-stroke, naturally aspirated engines
- 4-TC: four-stroke, turbocharged engines
- 2-DI: two-stroke, direct-injected engines.

An impressive reduction of exhaust emissions has been achieved during the last 20 years. DPM emissions are approximately 10 times lower (by mass) today than they were in the 1970s. This reduction has been made possible mainly due to the progress in diesel engine design and control. Low sulfate

TABLE 7.2
PARTICULATE MATTER EMISSIONS FROM DIESEL ENGINES

Time Period	Engine Type	TPM (g/kWh)	SOF (g/kWh)	Carbon (g/kWh)	Sulfates,[a] mg/kWh
1970s	4-NA	2.1–3.0	0.22–0.31	1.8–2.6	20–120
	4-TC	0.6–2.5	0.08–0.28	0.5–2.2	40–120
	2-DI	0.8–1.9	0.08–0.2	0.7–1.7	20–50
1980s	4-NA	0.25–0.6	0.05–0.36	0.2–0.25	15–70
	4-TC	0.28–0.58	0.07–0.26	0.2–0.31	27–55
	2-DI	0.4–0.6	0.03–0.06	0.35–0.55	20–50
1990s	4-NA	0.18–0.19	0.03–0.09	0.02–0.04	<0.5
	4-TC	0.11–0.39	0.03–0.09	0.02–0.04	<0.5
	2-DI	0.1–0.25	0.03–0.09	0.02–0.04	<0.5

Source: Data from Ref. 2.
Key: 4-NA, four-stroke, naturally aspirated engines
4-TC, four-stroke, turbocharged engines
2-DI, two-stroke, direct-injected engines
a. In the 1990s, the data fuel sulfur level was 0.05%, while the 1970s and 1980s data were for typical sulfur levels of 0.3–0.35%. The unit of measure, mg/kWh, was misprinted in the original source.

emissions for engines built during the 1990s are due to the low sulfur certification fuel that is used today.

7.2 Gaseous Emissions

7.2.1 Nitrogen Oxides

Nitrogen oxides as defined by emission regulations include nitric oxide (NO) and nitrogen dioxide (NO_2). The physical properties of both gases are listed in Table 7.3 [3]. Concentrations of NO_x in diesel exhaust are typically between 50 and 1000 ppm. If concentrations are given in mass units, NO_x is usually expressed as NO_2 equivalent.

Nitric oxide (NO) is a colorless and odorless gas that can be synthesized directly from nitrogen and oxygen under high temperature and pressure:

$$N_2 + O_2 = 2NO - 182.4 \text{ kJ/mol} \qquad (7.1)$$

The negative heat effect represents an endothermic reaction. NO is produced accordingly to reaction (7.1) in the engine cylinder, where temperature and pressure are high. At low temperature and pressure, the chemical equilibrium moves to the left side of Eq. (7.1). Thermodynamically, nitric oxide has a tendency to decompose to nitrogen and oxygen under the conditions experienced in diesel exhaust. The rate of decomposition, however, equals practically zero, and NO_x control from diesel engines remains an unsolved problem.

In older technology engines, approximately 95% of nitrogen oxides were composed of NO and only 5% of NO_2. The proportion of NO_2 in total NO_x in newer, turbocharged diesel engines can be as high as 15% or more. NO can be easily oxidized by oxygen into nitrogen dioxide at ambient conditions:

$$2NO + O_2 = 2NO_2 + 113.8 \text{ kJ/mol} \qquad (7.2)$$

This reaction occurs spontaneously (but not instantaneously) in the NO/air mixture after exhaust gases are discharged into the atmosphere. NO_2 is a very toxic red-brown gas with an unpleasant irritating odor. NO_2 is extremely reactive and exhibits strong oxidation properties. NO_2 reactions that occur in various types of emission control catalysts include oxidation of hydrocarbons, carbon monoxide, and diesel particulates.

TABLE 7.3
PHYSICAL PROPERTIES OF NITRIC OXIDE AND NITROGEN DIOXIDE

Name	Nitric oxide	Nitrogen dioxide
Formula	NO	NO_2
Formula weight	30.01	46.01
Appearance	Colorless gas	Red-brown gas
Density	1.0367 (rel. to air)	—
Melting point	–161°C	–9.3°C
Boiling point	–151°C	21.3°C

Nitrogen oxides are highly active ozone precursors; hence, they play an important role in smog chemistry. Besides diesel particulates, NO_x is one of the most critical pollutants found in diesel exhaust.

7.2.2 Hydrocarbons

Hydrocarbons (HCs) found in the gas phase of diesel exhaust are a mixture of many hydrocarbon species derived from diesel fuel and from lubricating oil. Fuel hydrocarbons are characterized by shorter carbon chains, whereas lube oil hydrocarbons contain the heavier material. The generic chemical formula for hydrocarbons, C_nH_m, represents a molecule with n atoms of carbon and m atoms of hydrogen.

In engine emission standards, HCs are commonly regulated as either total hydrocarbons (THC) or as nonmethane hydrocarbons (NMHCs), the latter category excluding the simplest hydrocarbon methane (CH_4) due to its different atmospheric reactivity from longer chain hydrocarbons. Even though the concentration of methane in diesel exhaust is very low, certain HC emission standards from diesel engines are expressed as NMHC for the sake of compatibility with other emission regulations. In the United States, there is no regulatory analytical method for measuring methane in the exhaust gas from diesel engines. The EPA recommends that 2% can be subtracted from the measured THC value to obtain an NMHC value in heavy-duty diesel engines [4].

Hydrocarbons, especially those with longer carbon chains, may have a characteristic or irritating odor. Some of them, such as benzene, are toxic and/or carcinogenic. Various hydrocarbon derivatives, for example, aldehydes, are also present in diesel exhaust gases. Many aldehydes have irritating odor and/or are toxic. Some of them, such as formaldehyde, are classified as carcinogens.

Concentrations of gaseous hydrocarbons in diesel exhaust range from approximately 20 to 300 ppm. Emission rates in grams per mile for several hydrocarbon and aldehyde species are shown in Table 7.4 [5]. The data have been compiled from several literature sources dated between 1986 and 1991. Light- and heavy-duty diesel emission rates are compared with those from gasoline engines.

Diesel exhaust hydrocarbons are divided between the gas phase and the particulate (liquid or adsorbed) phase. There is no clear distinction between volatile and nonvolatile hydrocarbon species. As a guideline, compounds with a vapor pressure of above 0.1 mmHg in standard conditions (20°C, 760 mmHg) can be considered volatile. Volatile diesel hydrocarbons contain aliphatic and aromatic species with up to approximately 24 carbon atoms in their molecule. Hydrocarbon emission regulations refer to the volatile gas-phase HCs. The particle-phase hydrocarbons are referred to as SOF and are described in Section 7.3.

Hydrocarbons can be oxidized by oxygen to produce carbon dioxide and water [Eq. (7.3)]—this is one of the fundamental reactions that occurs in emission control catalysts. Under mild oxidizing conditions, hydrocarbons form aldehydes or ketones [Eq. (7.4)]:

TABLE 7.4
HYDROCARBON AND ALDEHYDE EMISSION RATES

Compound	Diesel, g/mile (g/km)		Gasoline, g/mile (g/km)	
	HD	LD	Catalyst	Noncatalyst
Total HC	3.65 (2.28)	0.23 (0.14)	1.8 (1.2)	5.4 (3.4)
Methane	NA	0.01 (0.008)	0.26 (0.16)	0.27 (0.17)
Ethylene	NA	0.04 (0.03)	0.14 (0.09)	0.3 (0.2)
Acetylene	NA	NA	0.04 (0.02)	0.26 (0.16)
Propylene	NA	0.01 (0.008)	0.04 (0.03)	0.15 (0.09)
n-Pentane	NA	NA	0.03 (0.02)	0.09 (0.06)
iso-Pentane	NA	NA	0.07 (0.04)	0.27 (0.17)
n-Decane	0.01 (0.007)	NA	0.003 (0.0016)	
n-Dodecane	0.027 (0.017)	NA	0.003 (0.002)	
Benzene	0.024 (0.015)	0.02 (0.015)	0.06 (0.04)	0.31 (0.19)
Toluene	0.01 (0.007)	0.006 (0.004)	0.1 (0.07)	0.7 (0.45)
Xylenes	0.006 (0.004)	0.002 (0.001)	0.08 (0.05)	0.96 (0.6)
Ethyl benzene	0.005 (0.003)	0.001 (0.0006)	0.02 (0.01)	0.21 (0.13)
Naphthalene	0.01 (0.007)	0.003 (0.002)	NA	NA
Total aldehyde	NA	0.03(0.02)	0.04 (0.03)	NA
Formaldehyde	NA	0.02 (0.01)	0.025 (0.015)	0.06 (0.04)
Acetaldehyde	NA	0.007 (0.004)	0.01 (0.007)	NA
Acrolein	0.053 (0.033)	0.01 (0.006)	0.002 (0.001)	NA
Benzaldehyde	NA	NA	0.003 (0.002)	NA

Source: Data from Ref. 5.
Key: HD, heavy-duty diesel; LD, light-duty diesel; NA, data not available.

$$C_nH_m + (2n + m/4)O_2 = nCO_2 + (m/2)H_2O \quad (7.3)$$

$$R\text{-}CH_3 + O_2 = R\text{-}CHO + H_2O \quad (7.4)$$

In the atmosphere, hydrocarbons undergo photochemical reactions with NO_x leading to formation of smog and ground-level ozone. Different hydrocarbons exhibit different activity as ozone precursors. Methane (CH_4), the simplest hydrocarbon, does not exhibit any activity and is not considered an ozone precursor. This is the reason why many regulations exclude methane from regulated hydrocarbon emission by setting standards for nonmethane hydrocarbons (NMHCs).

7.2.3 Carbon Monoxide

Carbon monoxide (CO) is an odorless, colorless, and very toxic gas of about the same density as that of air. At high concentrations CO is very flammable, burning in air with a bright blue flame. Table 7.5 lists the physical properties of CO. CO emissions from today's new diesel engines are relatively low. Carbon monoxide concentrations in diesel exhaust range from approximately 10 to 500 ppm.

At elevated temperatures or over an oxidation catalyst, carbon monoxide can be oxidized by oxygen to form carbon dioxide, as follows:

$$2CO + O_2 = 2CO_2 + 565.6 \text{ kJ/mol} \quad (7.5)$$

The reaction produces high heat effect, which, in the case of CO-rich exhaust gases, can cause a significant increase in the gas temperature in catalytic reactors designed to oxidize CO. Adiabatic oxidation of 1% of CO in the exhaust stream raises the gas temperature by approximately 100°C.

7.2.4 Sulfur Dioxide

Sulfur dioxide (SO_2) is an unregulated diesel emission that originates from the sulfur in fuel and in engine lubricating oil. SO_2 is a colorless gas with a characteristic, irritating odor; other physical properties are listed in Table 7.6. Sulfur dioxide can be oxidized to sulfur trioxide (SO_3), which is the precursor of the sulfuric acid responsible for sulfate particulate emissions. The majority of sulfur in raw diesel exhaust exists as SO_2. Only approximately 2% to 4% of fuel sulfur is emitted as SO_3 from the engine.

TABLE 7.5
PHYSICAL PROPERTIES OF CARBON MONOXIDE

Name	Carbon monoxide
Formula	CO
Formula weight	28.01
Appearance	Colorless, odorless, tasteless, noncorrosive gas
Density	1.250 kg/m^3 (0°C, 1 atm) 0.968 (rel. to air)
Melting point	–205.0°C
Boiling point	–191.5°C
Flammability limits	12.5-74.2 vol. %

TABLE 7.6
PHYSICAL PROPERTIES OF SULFUR DIOXIDE

Name	Sulfur dioxide
Formula	SO_2
Formula weight	64.06
Appearance	Colorless gas
Density	2.264 (rel. to air)
Melting point	–75.5°C
Boiling point	–10.0°C

The exhaust concentration of sulfur dioxide is in direct proportion to the fuel sulfur level. In fact, SO_2 concentrations can be calculated from the fuel consumption and its sulfur content with good accuracy. Such calculation, based on an air-to-fuel (A/F) ratio of 20, which is typical for diesel engines operating at full-load conditions, is shown in Figure 7.2. As seen in the graph, diesel fuels of 500 ppm S produce exhaust SO_2 levels of about 20 ppm.

As the levels of sulfur in fuels decrease, engine lubricating oils become an important source of SO_2 in diesel exhaust. Diesel lube oils typically contain 4000 to 10,0000 ppm sulfur, primarily as part of their additive package. Antiwear additives typically contain zinc, sulfur, and phosphorus in the form of zinc dithiophosphates. Many detergent additives also contain alkyl sulfonates [6].

The lowest oil consumptions published in the literature amount to about 0.075% of the fuel. Real-life data, as well as oil consumption standards, typically oscillate between 0.1% and 0.2% of fuel [7–9]. Two cases of lube oil contribution to the exhaust SO_2 levels are shown as horizontal lines in Figure 7.2. The lower line corresponds to a lubricating oil consumed at a rate of 0.1% of fuel and containing 4000 ppm S. Produced SO_2 levels are comparable to those associated with a 5 ppm S diesel fuel. The upper oil line represents the "worst case" scenario with a 10,000 ppm S oil consumed at a rate of 0.2% of fuel. In this case, the oil-derived sulfur level is higher than that produced by diesel fuel of 15 ppm S content.

The levels of sulfate particulates also depend on the fuel sulfur content. When high-sulfur (0.25% S) fuel is used, approximately 0.050 g/bhp-h of sulfate particulates is generated. Sulfate particulate emissions decrease to about 0.01 g/bhp-h when 500 ppm sulfur fuel is used. Catalytic devices, such as oxidation catalysts or catalyzed particulate filters, may convert a significant portion of SO_2 into SO_3 and substantially increase sulfate particulate emissions, as discussed in Chapters 19 and 22.

7.2.5 Nitrous Oxide

Nitrous oxide (N_2O) has been known as laughing gas. It is a colorless gas with a sweet smell that exhibits narcotic activity on humans. N_2O has a potential environmental impact due to its strong greenhouse gas properties, as well as its attack on stratospheric ozone. Table 7.7 lists the physical properties of N_2O.

Nitrous oxide is an unregulated diesel emission. Even though chemically it is an oxide of nitrogen, nitrous oxide has been excluded from the regulated nitrogen oxide (NO_x) emissions. The engine-out N_2O emissions from heavy-duty diesel engines are typically very low levels of about 3 ppm [10]. This number corresponds to about 0.03 g/kWh when measured on a transient engine test cycle (FIGE). For comparison, N_2O emissions from

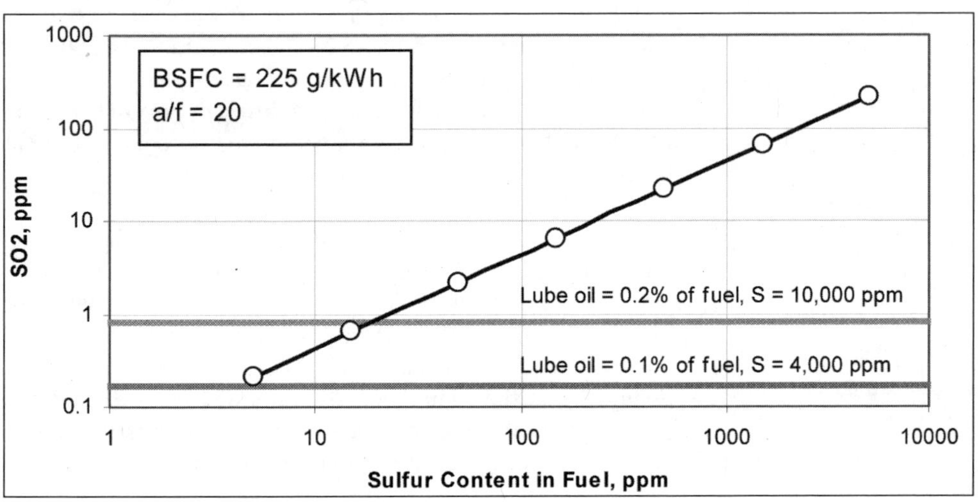

Figure 7.2 *Exhaust SO_2 as a function of fuel sulfur content.*

TABLE 7.7
PHYSICAL PROPERTIES OF NITROUS OXIDE

Name	Nitrous oxide
Formula	N_2O
Formula weight	44.02
Appearance	Colorless gas of a sweet smell
Density	1.530 (rel. to air)
Melting point	–102.3°C
Boiling point	–90.7°C

gasoline engines equipped with a three-way catalyst range from 12 to 35 ppm.

Certain emission controls, such as Pt-based or systems in some configurations, may increase N_2O emissions from diesel engines. There is a consensus among emission control manufacturers that, even though it is not a regulated emission, emission control technologies should not increase N_2O levels. Therefore, the tendency to generate nitrous oxide is an important factor in the development of catalyst systems.

7.2.6 Hydrogen

Hydrogen (H_2) is a colorless gas that has no direct environmental or health impact but can play a role in a number of reactions in emission control catalysts, possibly including some diesel catalysts. Its physical properties are shown in Table 7.8. Normally, under lean conditions, hydrogen is not present in diesel exhaust. Gasoline exhaust typically contains approximately 0.3 mol of H_2 per mole of CO [11].

At ambient temperatures, hydrogen exhibits a relatively low chemical activity. At higher temperatures and/or in the presence of catalysts, hydrogen shows very strong reducing properties. It reacts with oxygen to form water [Eq. (7.6)]. This reaction, extremely slow at room temperatures, reaches a noticeable rate after the gases are heated to 180°C. A further heating of the mixture to 450°C causes an explosion.

$$2H_2 + O_2 = 2H_2O + 571.9 \text{ kJ/mol} \qquad (7.6)$$

TABLE 7.8
PHYSICAL PROPERTIES OF HYDROGEN

Name	Hydrogen
Formula	H_2
Formula weight	2.016
Appearance	Colorless, odorless gas
Density	0.06948 (rel. to air, 20°C)
Melting point	–259.1°C
Boiling point	–252.7°C

Hydrogen may be generated in emission control catalysts from water vapor and other exhaust components in the water shift reaction and the steam reforming reaction. These processes are well known and have been used commercially in coal gasification technologies.

In the water shift reaction of Eq. (7.7), carbon monoxide reacts with water vapor producing carbon dioxide and hydrogen:

$$CO + H_2O = CO_2 + H_2 \qquad (7.7)$$

In the steam reforming reaction, water vapor reacts with hydrocarbons to produce hydrogen and carbon dioxide:

$$C_nH_m + 2nH_2O = (2n + m/2)\, H_2 + nCO_2 \qquad (7.8)$$

7.3 Diesel Particulate Matter

7.3.1 What Are Diesel Particulates?

Particulate matter—perhaps the most characteristic of diesel emissions—is responsible for the black smoke traditionally associated with diesel-powered vehicles. The diesel particulate matter emission is usually abbreviated as PM or DPM, the latter acronym being more common in occupational health applications. Diesel particulates form a very complex aerosol system. Despite a considerable amount of basic research, neither the formation of PM in the engine cylinder, nor its physical and chemical properties or human health effects are fully understood. Nevertheless, on the basis of what is already known, PM is perceived as one of the major harmful emissions produced by diesel engines. Diesel particulates are subject to diesel emission regulations worldwide and, along with NO_x, have become the focus in diesel emission control technology.

Contrary to gaseous diesel emissions, PM is not a well-defined chemical species. The definition of particulate matter is in fact determined by its sampling method, the detailed specification of which is an important part of all diesel emission regulations. PM sampling involves drawing an exhaust gas sample from the vehicle's exhaust system, diluting it with air, and filtering it through sampling filters. The mass of particulate emissions is determined based on the weight of PM collected on the sampling filter. It is quite obvious that any changes in the procedure, for example, using a different type of sampling filter or different dilution parameters, may produce different results.

Standardization of sampling methods is of utmost importance if results from different laboratories are to be comparable. Such standards have been developed for the of PM mass in the area of public health regulations (i.e., emission standards for diesel engines and vehicles) worldwide. Whenever reference to "particulate matter" is made throughout this book, it is quietly assumed that the sampling was performed from diluted and cooled exhaust, in accordance with these standards (unless explicitly stated otherwise). However, the reader should realize that other sampling procedures and corresponding definitions of PM are perfectly possible. Ongoing research in Europe is aimed at developing standardized measuring methods based on particle number emissions, rather than mass, for the inclusion in future emission standards in addition to the PM mass metric [12]. So far no common standard has been reached in the area of diesel occupational health regulations, where a number of different measuring methods—and corresponding DPM definitions—exist in parallel.

Diesel particulate matter, as specified by the EPA's procedures and most other standards and regulations worldwide, is sampled by filtering diluted diesel exhaust at temperatures not higher than 125°F (52°C). This cooling effect is typically achieved with laboratory dilution ratios in the range of 3:1 to 20:1. Devices used in the laboratory to produce the mixture of air with diesel exhaust gas are known as *dilution tunnels.* The intention of this procedure is to simulate conditions at which diesel particulates are released from vehicles into the atmosphere. The substance that is sampled and regulated is supposed to correspond to the diesel soot that is suspended in the ambient air. The fiberglass filters that are used for laboratory PM sampling capture solid particles, as well as liquid droplets, or mist, which condense from exhaust gases during the dilution process. In effect, the definition of PM extends to "any matter"—all solid and liquid material (condensate)—present in the diluted and cooled diesel exhaust. We should emphasize that the preceding definition of diesel particulates is to a large degree arbitrary. Because the atmospheric dilution ratios of PM (about 500 to 1000) are much higher than those used in laboratory dilution tunnels, the simulation of the atmospheric dilution is far from perfect [13].

Diesel particulates are composed of elemental carbon particles that agglomerate and adsorb other species to form structures having complex physical and chemical properties. Diesel particulates have a bimodal size distribution. They are a mixture of nuclei mode and accumulation mode particles, as shown schematically in Figure 7.3. Nuclei mode particles are very small—according to most authors, their diameters are between approximately 0.007 and 0.04 μm. More recent studies redefine the nuclei mode particle size range to be even smaller, from 0.003 to 0.03 μm [14], thus making them comparable to certain large molecules. Nuclei mode particles are often referred to as *nanoparticles,* although these two terms are not the same. Nanoparticles are usually defined as particles below 50 nm in diameter (0.05 μm). This is an arbitrary definition, not related to the physical properties of diesel exhaust; nanoparticles include practically all nuclei mode particles, but may also contain a certain fraction of the accumulation mode.

The nature of nuclei mode particles is still being studied in research laboratories. It is believed that nuclei mode particles are primarily volatile and consist mainly of hydrocarbon and hydrated sulfuric acid condensates that are formed from gaseous precursors as the temperature decreases in the exhaust system and after mixing with cold air, be it in the laboratory dilution tunnel or in the ambient air. These volatile particles are very unstable; their concentrations depend strongly on dilution conditions such as dilution ratio and residence time. A small amount of nuclei mode particles may consist of solid material, such as carbon or metallic ash from lube oil additives [13, 15]. Nuclei mode particles constitute the majority of the particle number—on the order of 90%—but only a few percent of the PM mass.

Accumulation mode particulates are formed by agglomeration of primary carbon particles and other solid materials, accompanied by adsorption of gases and condensation of vapors. They are composed mainly of solid carbon mixed with condensed heavy hydrocarbons (Figure 7.3), but may also include sulfur compounds, metallic ash, and cylinder wear metals. The diameters of the accumulation mode particles are between approximately 0.04 and 1 μm with a maximum concentration between 0.1 and 0.2 μm [13, 16]. Most PM mass emission (but only a small proportion of the total particle number) is composed of agglomerated particles.

7.3.2 Composition of Diesel Particulates

Based on analysis performed by a combination of physical and chemical methods, particulate matter is traditionally divided into three main fractions, which can be further subcategorized, as follows:

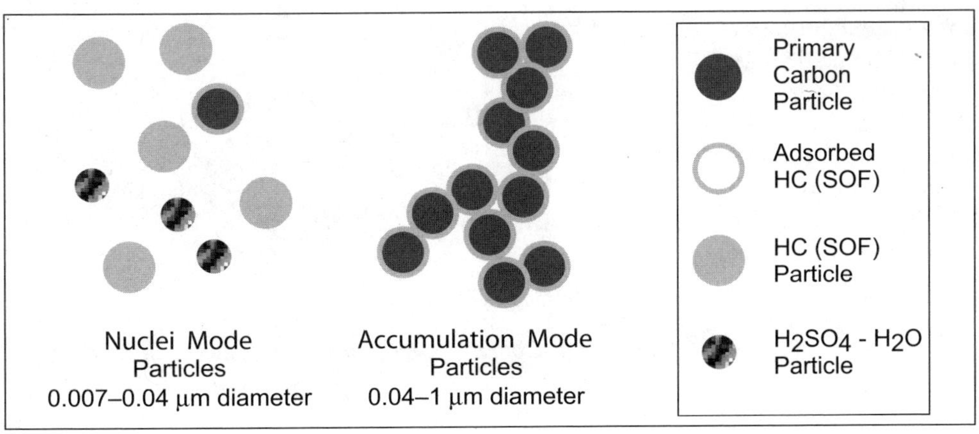

Figure 7.3 *Schematic of diesel particulate matter.*

1. Solid fraction (SOL)
   a. elemental carbon
   b. ash
2. Soluble organic fraction (SOF)
   a. organic material derived from engine lubricating oil
   b. organic material derived from fuel
3. Sulfate particulates (SO_4)
   a. sulfuric acid
   b. water

According to that classification, the total particulate matter (TPM) can be defined as follows:

$$TPM = SOL + SOF + SO_4 \tag{7.9}$$

Particles leaving the engine are composed primarily of solid-phase carbon material (SOL). Both individual (nuclei mode) and agglomerated carbon particles are formed in the combustion chamber. In the exhaust system, depending on the temperature, the particles undergo limited oxidation and further agglomeration. Some particles are deposited on the exhaust pipe walls due to thermophoretic forces (i.e., mass transfer driven by temperature gradient). Other PM precursors including hydrocarbons, sulfur oxides, and water are present in the hot diesel exhaust as gases or vapors.

Another source of solid material in diesel exhaust is the metal ash compounds that are derived from lubricating oil additives and from engine wear. Nucleation of volatile ash constituents is believed to take place during the expansion stroke in the engine cylinder. The ash nuclei can then agglomerate to form accumulation mode particles. The relative proportion of ash generally increases in new engines, due to the presence of fewer carbon particulates and a lower TPM mass.

The physical and chemical properties of PM change once the exhaust gases enter the dilution tunnel, are mixed with air, and cooled to below 52°C. Heavy hydrocarbons, which are derived from lubricating oil and unburned fuel, condense or adsorb onto the surface of carbon particles, forming the organic portion of PM (SOF). If the amount of carbon particles that can act as a "sponge" is insufficient, hydrocarbons will nucleate, forming increased numbers of volatile (liquid) nuclei mode particles. In the dilution tunnel, the total hydrocarbon material from the combustion chamber finally becomes proportioned between particulate (SOF) and gas-phase hydrocarbons (at least in theory; in practice, a portion of diesel hydrocarbon material may be measured and accounted for twice: in the particulate phase and in the gaseous phase).

Sulfuric acid in diesel exhaust is derived from the fuel sulfur. Sulfur leaves the combustion chamber primarily in the form of sulfur dioxide (>95%) and as a small proportion of sulfur trioxide (typically 2% to 5%; this proportion may be significantly increased by). In the presence of water, SO_3 reacts to produce sulfuric acid:

$$SO_3 + H_2O = H_2SO_4 \tag{7.10}$$

Sulfate particulates are formed in the dilution tunnel through a heteronucleation process from the molecules of H_2SO_4 and water. During PM measurements, sulfate particulates are deposited on the filters together with the carbonaceous material. It was once believed that sulfuric acid is attached to or associated with carbon particles. Later research found that sulfate particles may be also separate from carbon particles [17]. It is now envisioned that sulfate particulates, while existing in the accumulation mode mixed with carbon and organic SOF material, are also an important source of volatile ($H_2SO_4 - H_2O$) nuclei mode particles [14]. Depending on the availability of metal-based compounds, sulfuric acid may also form solid (nonvolatile) sulfate salts.

The composition of PM varies greatly depending on the engine technology, test conditions, and, in the case of sulfate particulates, the sulfur content in the fuel. An example PM composition from a post-1994 U.S. heavy-duty diesel engine is illustrated in Figure 7.4.

Diesel particulates also include (usually as a part of the SOF) certain other classes of heavy hydrocarbon or hydrocarbon-derived material of special interest, for example, the polynuclear aromatic hydrocarbons (PAH) or dioxins. Concentrations of these substances are orders of magnitude lower in comparison to the primary PM components, making them "invisible" in Figure 7.4. Due to extremely high biological activity in their pure state, the presence of some of these compounds in diesel exhaust attracted a lot of attention. PAHs and dioxins are discussed in more detail later in this chapter.

7.3.3 Solid Fraction

7.3.3.1 Carbon

The solid fraction of diesel particulates is composed primarily of elemental carbon, sometimes also referred to as *inorganic carbon.* This carbon, not chemically bound with other elements, is the finely dispersed "carbon black" or "soot" substance responsible for black smoke emissions. The carbonaceous PM fraction results from the heterogeneous combustion process in diesel engines, where solid particle precursors are formed in both diffusion and premixed flame. Hexagonal arrays of carbon atoms in soot particles form platelet-like structures. The platelets are arranged in layers, typically two to five, to form

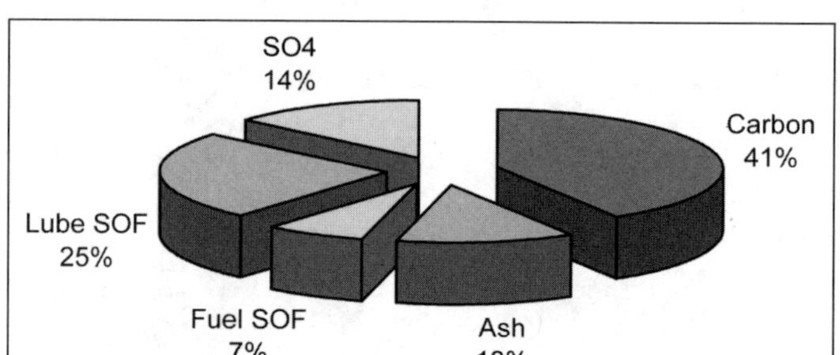

Figure 7.4 *Composition of diesel particulate matter for an HD diesel engine, U.S. FTP transient cycle. (Based on data from Ref. 13.)*

leaf-shaped graphite crystallites. The crystallites are randomly packed with their planes generally parallel to the surface of the primary (i.e., nuclei mode) particle. These three structures are shown schematically in Figure 7.5.

The primary particles agglomerate in the cylinder, while traveling through the exhaust system, and after discharge into the atmosphere. The structure of an agglomerated diesel particle is shown in the transmission electron microscope (TEM) picture of Figure 7.6. A primary particle in an agglomerate can be seen in Figure 7.6a. The particle is formed of numerous graphite crystallites, a few nanometers in size, and ordered to yield an onion-like structure. Both the size and the arrangement of the crystallites are in reasonably good agreement with the representation in Figure 7.5. The agglomerated particle in Figure 7.6b exhibits a grape cluster–like structure formed of hundreds of nuclei mode particles.

As is apparent from Figure 7.6, while nuclei mode particulates are nearly spherical, a sphere is a poor approximation for the shape of the accumulation mode particle. Agglomerated particulates often form chain-like elongated structures having no well-defined "diameter." This presents a certain difficulty in , since particle sizes have to be expressed using various types of equivalent particle diameters, which can differ depending on the measurement principle.

7.3.3.2 Ash

Another important component of the solid fraction of PM is metallic ash. In new engines producing less carbon particulates, the relative importance of noncarbon, solid particulate emissions increases. Particulate matter from U.S. post-1994 HD diesel engines contains 10% and more of ash [19]. This proportion may increase even further in future engines.

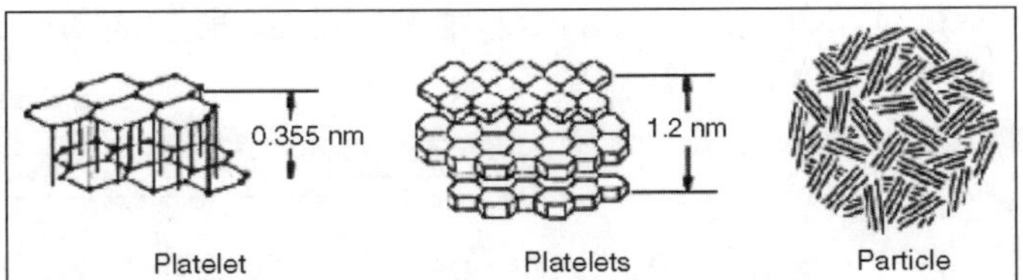

Figure 7.5 *Structure of primary carbon particle. (Modified from Ref. 18.)*

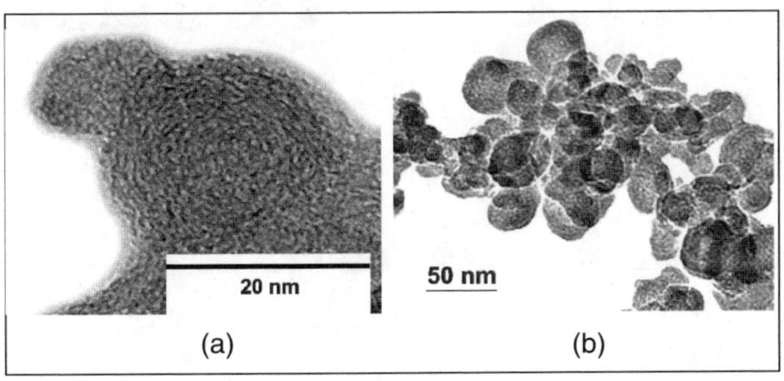

Figure 7.6 *TEM pictures of an agglomerated carbon particle. (Courtesy of NMI.)*

Ash emissions receive a lot of attention from the developers of diesel particulate filter materials, which must be formulated to resist corrosion by ash compounds. In general, diesel exhaust ash consists of a mixture of the following components [20]:

- Sulfates, phosphates, or oxides of calcium (Ca), zinc (Zn), magnesium (Mg), and other metals that are formed in the engine's combustion chamber from burning of additives in the engine lubricating oil. These chemicals are present in lube oil as detergents, dispersants, acid neutralizers, antioxidants, corrosion inhibitors, antiwear and extreme pressure additives, and so on. Lube oil ash emissions can be modeled based on the known oil ash content (typically around 1.5%) and oil consumption rate (typically 0.1% to 0.2% of the diesel fuel consumption).
- Metal oxide impurities resulting from the engine wear, which are carried into the combustion chamber by the lube oil. These include oxides of iron (Fe), copper (Cu), chromium (Cr), and aluminum (Al).
- Iron oxides resulting from corrosion of the engine exhaust manifold and other exhaust system components. Depending on the metallurgy of the exhaust system materials, these particles may also include chromium, nickel, and aluminum.

Furthermore, if metallic are used to facilitate the regeneration of particulate filters, exhaust gases will contain the corresponding metal oxide ash. Metals used for that purpose include cerium, iron, and strontium. Metallic additives could become a source of high numbers of solid nuclei mode particle emissions [14].

An example composition of ash from a heavy-duty diesel vehicle is listed as "Ash A" in Table 7.9. The ash was collected in a particulate filter over a 100,000-km-long experiment. Ash analysis revealed the presence of $(Zn,Mg)_3(PO_4)_2$ and $CaSO_4$, as agglomerates of primary particles that were 100 to 500 nm in size. The BET surface area (see page 563) of the ash was 10 m^2/g. Ashes B and C were prepared by doping Ash A with iron and cerium, respectively, to simulate the use of additives for filter regeneration. The amount of additive was modeled based on the assumed concentrations of 10 ppm Fe and 40 ppm Ce in the fuel.

TABLE 7.9
COMPOSITION OF DIESEL PARTICULATE ASH

Oxide	Ash A	Ash B	Ash C
	Weight Percent of Oxide in Ash		
CaO	29.6	20.7	10.4
ZnO	9.9	6.9	3.5
MgO	5.5	3.9	1.9
SO_3	38.8	27.2	13.6
P_2O_5	15.8	11.1	5.5
Fe_2O_3	0.41	30.3	0.14
CeO_2	0.0	0.0	65.0

Source: Data from Ref. 20.

7.3.4 Soluble Organic Fraction

7.3.4.1 Composition

Hydrocarbons adsorbed on the surface of carbon particles and/or present in the form of fine droplets form the soluble organic fraction of diesel particulates. The adjective *soluble* takes its origin from the analytical technique that utilizes extraction with solvents to isolate the organic fraction of particulates. Sometimes this fraction is also referred to as *volatile organic fraction* (VOF). VOF is usually measured by vacuum evaporation and its interpretation is very close to that of SOF. In some laboratories, however, the evaporated fraction may also include water-soluble material. The VOC should be then adjusted for sulfates, which are determined by chemical analysis. If no such adjustment is made, the interpretation of VOC should instead be the sum of SOF and sulfate particulates, SO_4.

We should emphasize again that the SOF fraction becomes liquid only after cooling to below 52°C in the laboratory dilution tunnel. At the temperatures in diesel exhaust, most of the SOF compounds exist as vapors, especially at higher engine loads when temperatures are high. One needs to keep this realization in mind to understand the changes that occur to diesel particulates in such aftertreatment devices as oxidation catalysts or particulate filters.

The proportion of SOF in the total PM may vary significantly between engines. Particulates with low SOF content are called *dry particulates.* PM having a high SOF content is called *wet particulate.* In wet particulates, the organic fraction may constitute more than 50% of the total PM, indicating a multilayer hydrocarbon adsorption on the surface of the particles. In dry particulates, the SOF content may be as low as 10% and less. In a given engine, the SOF is strongly dependent on the operating conditions (i.e., test modes and duty cycles). Typically, SOF content is highest at light engine loads when exhaust temperatures are low. This is illustrated in Figure 7.7, which presents total PM emissions (represented by the size of the circles) and their SOF content, as measured at a number of steady-state conditions from a 2.8-L DI diesel engine [21]. All measurements at exhaust temperatures up to 200°C (measured at the inlet to a catalyst) showed SOF content in excess of 50% of total PM. At high engine loads of above 70% and temperatures of 400°C and more, the SOF dropped to less than 5%.

This temperature trend was also confirmed by cold start emission tests, which were reported to produce approximately 25% higher SOF levels than hot start emissions [22]. Two-stroke engines also tend to have higher SOF emissions than their four-stroke counterparts. The fluctuations in SOF with engine technology and with engine test cycle may be extremely important in developing PM control strategies, since solids behave differently from the SOF in both diesel and in .

The SOF is typically composed of lube oil–derived hydrocarbons, with a small contribution from the higher boiling end diesel fuel hydrocarbons. Diesel fuel is composed of hydrocarbons having a boiling range equal to that of aliphatic hydrocarbons with 12 to 20 atoms of carbon in their molecules. Diesel lube oil is a continuum of many compounds with boiling

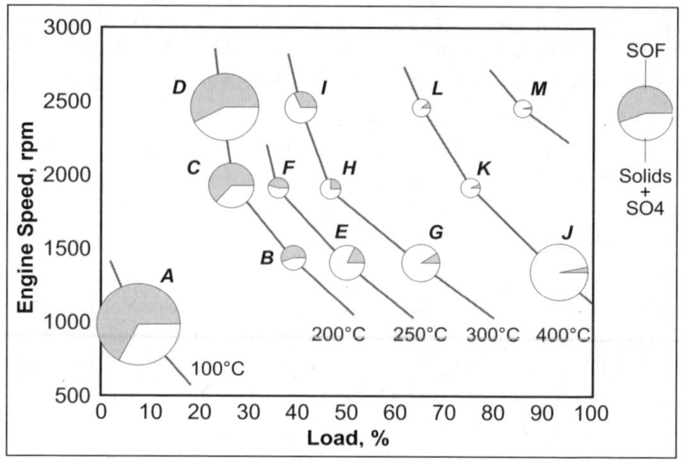

Figure 7.7 PM composition at different engine operating conditions for a 2.8-L, DI, turbocharged diesel engine with 0.38 wt. % S in fuel; PM emission/SOF content (g/bhp-h/%): A, 2.48/69%; B, 0.66/55%; C, 1.12/65%; D, 1.60/58%; E, 0.85/21%; F, 0.52/46%; G, 0.90/10%; H, 0.54/25%; I, 0.71/33%; J, 1.41/2.7%; K, 0.51/8.7%; L, 0.45/11%;and M, 0.45/4.0%. (Based on data from Ref. 21.)

points comparable to normal aliphatic hydrocarbons in the range between C18 and C36. Figure 7.8 shows an example chromatogram of diesel particulate SOF compared with chromatograms of diesel lube oil and diesel fuel. The particulates were collected from a diesel engine running at low speed and high load [23]. The chromatogram of SOF sample closely resembles that of diesel lube oil hydrocarbons.

Lube oil hydrocarbons are also believed to be the major contributor to nuclei mode SOF. Experiments were reported in which the volatility of nuclei mode particles was found to be similar to that of C24 through C32 normal alkanes, implying the lube oil origin [14].

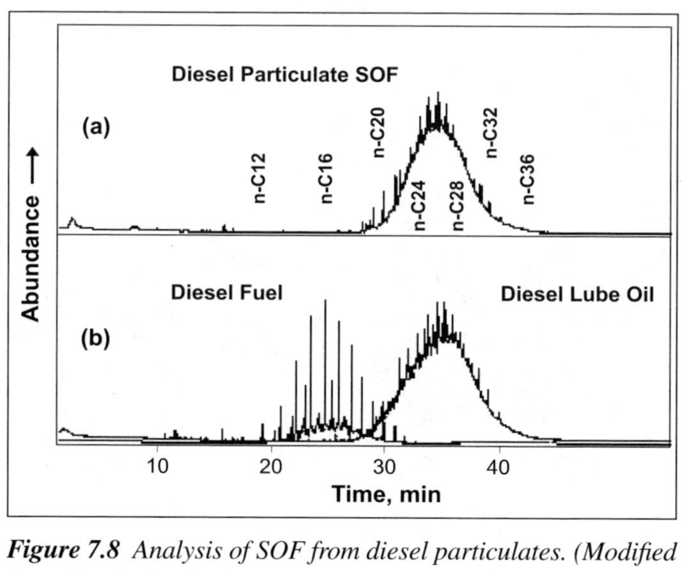

Figure 7.8 Analysis of SOF from diesel particulates. (Modified from Fig. 5 in Ref. 23.)

The proportion of lube oil–derived SOF also shows considerable variability. A study with pre-Euro 1 light-duty vehicles found that the lube oil–derived SOF material contributed anywhere from between 19% and 88% of the total PM emission [24].

7.3.4.2 Polynuclear Aromatic Hydrocarbons

The SOF fraction contains most of the polynuclear (or polycyclic) aromatic hydrocarbons (PAHs) and nitro-PAHs emitted with diesel exhaust gases. PAHs are aromatic hydrocarbons with two or more (up to five or six) benzene rings joined in various, more or less clustered forms (Figure 7.9). They attracted special attention because of their mutagenic and, in some cases, carcinogenic character. There was a proliferation of scientific publications on PAHs in the late 1980s. Judging by the number of publications, there is less interest in the PAHs today. However, these compounds are still studied by air quality agencies, notably by the EPA in the United States under its "air toxics" regulations.

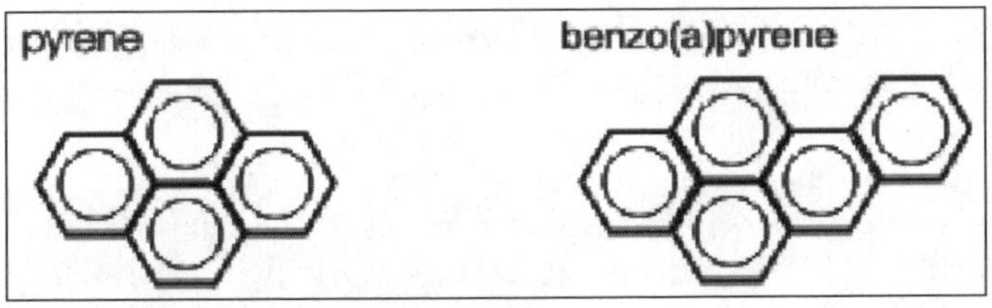

Figure 7.9 *Example PAH compounds.*

PAHs include tens of compounds, some of them having very complex structures (Table 7.10). They may also contain cyclopentane rings and heterogeneous rings with atoms of nitrogen (PANH) or sulfur (PASH).

The EPA has introduced the term *polycyclic organic matter* (POM), defined as a class of air toxic compounds with more than one benzene ring and a boiling point of 100°C or higher. The POM class is almost identical to the PAH compounds. A group of seven polynuclear aromatic hydrocarbons [benz(a)anthracene, benzo(b)fluoranthene, benzo(k)fluoranthene, benzo(a)pyrene, chrysene, 7,12-dimethy lbenz(a)anthracene, and indeno(1,2,3-cd)pyrene], all identified as probable human carcinogens, are often used by the EPA as surrogates for the entire group of POM compounds.

PAHs in the exhaust gas are split between gas and particulate phase. A fraction of the light, two-ring compounds is present in the gas. The most harmful compounds of four or more rings can be found almost exclusively in the organic fraction (SOF) of diesel particulate matter.

TABLE 7.10
PAHS IDENTIFIED IN DIESEL PARTICULATE EXTRACTS

Compound	Molecular Mass	Compound	Molecular Mass
Acenaphthylene	152	Pyrene	202
Dibenzofuran	168	Ethylmethylphenanthrene	220
Fluorene	166	Methylfluoranthene	216
Methylfluorene	180	Ethylmethylphenanthrene	220
Methyldibenzofuran	182	Benzo(a)fluorene	216
Dibenzothiophene	184	Methylpyrene	216
Phenanthrene	178	Naphtobenzothiophene	234
Anthracene	178	Ethylpyrene	230
Benzo(h)quinoline	179	Benzo(g,h,i)fluoranthene	228
Acridine	179	Benz(a)anthracene	228
Dimethylfluorene	194	Chrysene	228
Methyldibenzothiophene	198	Methylnaphtobenzothiophene	248
Dimethylfluorene	190	Methylchrysene	242
Methyldibenzothiophene	198	Binaphtyl	254
Methylphenanthrene	192	Benzofluoranthene	252
4h-Cyclopenta(d,e,f)phenanthrene	190	Benzo(e)pyrene	252
Ethyldibenzothiofene	212	Benzo(a)pyrene	252
Ethylphenanthrene	206	Dibenz(a,h)anthracene	278
Fluoranthene	202	Benzo(b)chrysene	278
Benzacenaphthylene	202	Benzo(g,h,i)perylene	276
Benz(d,e,f)dibenzothiofene	208	Coronene	380

Source: Data from Ref. 25.

PAHs are present in diesel fuel. Their concentrations vary between 1.5% and 2.5% in different fuels. Some of them (0.2% to 1% of the total fuel PAH) survive combustion and can be found in the exhaust gas [26]. It is also believed that some of the heaviest PAH compounds are generated by pyrosynthesis in the engine cylinder. Another phenomenon occurring in the engine is the formation of nitroderivatives of PAHs due to high NO_x concentrations.

Emissions of PAHs typically constitute a fraction of a percent of the total PM emission, with many studies reporting at about 0.5% of total PM [27]. Concentrations of particular PAH compounds vary from 0 to about 250 μg/mile, with total PAH emission reaching anywhere from 1 to more than 10 mg/mile, depending on engine technology. In volumetric units, most single PAH species are measured between 0 and 100 μg/m^3, with some species reaching above 150 μg/m^3 [26].

7.3.4.3 Dioxins

Dioxins is the generic term for a special group of chlorinated polynuclear hydrocarbon compounds characterized by extremely high toxicity, suspected carcinogenicity, and resistance to biological breakdown. Even though the term *dioxin* refers chemically to a simple heterocyclic ring with two atoms of oxygen and no chlorine atoms, its meaning in environmental sciences extends over three families of chlorinated compounds: the chlorinated dibenzo-*p*-dioxins (CDDs), chlorinated dibenzofurans (CDFs), and certain polychlorinated biphenyls (PCBs). Sometimes the term *dioxin* is also used to refer to the well-studied and one of the most toxic dioxins, the 2,3,7,8-tetrachlorodibenzo-*p*-dioxin (TCDD; Figure 7.10). Dioxins are created primarily during incineration of chlorine-containing wastes, manufacture of some chlorine-containing chemicals (PVC, herbicides, etc.), chlorine bleaching of wood pulp and paper, and in some natural processes.

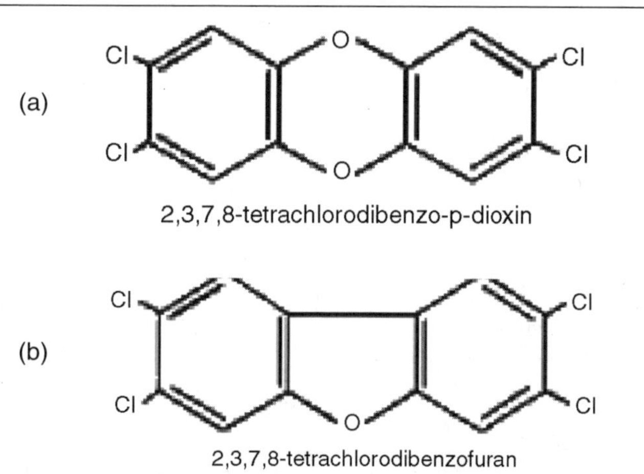

Figure 7.10 Example dioxin compounds: (a) TCDD and (b) TCDF.

Studies detecting CDD/CDFs in used motor oils [28], followed by discoveries of elevated levels of CDD/CDFs in European traffic tunnels, raised concerns that motor vehicles, including diesel engines, may be contributing to nationwide ambient dioxin levels [29]. In its draft dioxin health assessment document [30], the EPA reviewed the existing literature and provided estimates for dioxin emission factors from internal combustion engines, as listed in Table 7.11 (in picograms per kilometer and picograms per liter of fuel).

Various dioxin compounds may have greatly different toxicities. Emissions of dioxin mixtures, such as those listed in Table 7.11, are typically expressed using toxicity equivalency factors (TEQ), in which the concentration of each compound is corrected based on its toxic effect, relative to the toxicity of TCDD and TCDF. In other words, the TEQ data represent a calculated concentration (or mass) of TCDD/TCDF that would have a toxic effect equivalent to that of the real mixture.

Available literature data show a considerable spread of emission factors. Data reviewed by the EPA varied anywhere from "nondetectable" to more than 600 pg TEQ/km, resulting in a low confidence rating assigned to the factor of 172 pg TEQ/km shown in Table 7.11. A more recent study by EPA authors [31] investigated dioxin emissions from six diesel trucks and measured emission factors from 8 to 164 pg TEQ/km, with an average of 47 pg TEQ/km.

According to the EPA inventories, diesel engines are only a minor source of dioxin emission in the United States, being attributed 1.1% of the total atmospheric release (33.5 g TEQ per year in 1995, compared to a total from all sources of 2888 g TEQ) [30].

TABLE 7.11
DIOXIN EMISSION FACTORS

Fuel/Vehicle	Toxicity Equivalency Factor, pg/km	Toxicity Equivalency Factor, pg/L	Fuel Consumption, km/L
Diesel trucks[a]	172	946[b]	5.5
Unleaded gasoline (w/catalyst)	1.5	14.9	10
Leaded gasoline	45	450	10

Source: Data from Ref. 30.

a. Highway trucks only; no data existed for nonroad engines.

b. Based on assumed fuel consumption of 5.5 km/L (fuel consumption not reported in the analyzed studies).

We must emphasize that certain catalytic combustion additives may increase emissions of dioxins by orders of magnitude, as was reported for copper compounds [32]. Therefore, fuel additives/fuel-borne catalysts must always be evaluated for their dioxin formation activity.

7.3.5 Sulfate Particulates

Sulfate particulates are composed primarily of hydrated sulfuric acid and, as such, are mostly liquid. Formation of sulfate particulates requires an interaction between molecules of H_2SO_4 and H_2O. The process is theoretically modeled as heteromolecular nucleation [33]. According to this theory, both sulfuric acid and water vapor can be undersaturated and still produce particles. The nucleation starts with small molecular clusters, which then grow into more stable nuclei particles. Most stable sulfate particulates are formed when the molar ratio of H_2O to H_2SO_4 equals 8:3 (i.e., 2.67:1 or 0.49:1 by weight). Nuclei sulfate particles are composed of approximately 8000 molecules of H_2O and 3000 molecules of H_2SO_4. It is believed that sulfate particulates are separate from carbon particles and are present in the exhaust gas primarily as nuclei mode particles.

Formation of particles in a dilution tunnel depends on the H_2SO_4 and H_2O vapor pressures, which are functions of the following:

- Fuel sulfur level
- Conversion of fuel sulfur to SO_3
- Air/fuel ratio
- Dilution tunnel temperature
- Dilution tunnel relative humidity.

Dilution tunnel sampling variability can thus be minimized by maintaining a constant relative humidity and using a constant sulfur level fuel. Note that the PM sampling procedures currently in place do not have any special requirements regarding the relative humidity in the dilution tunnel, while, according to the heteronucleation theory, the rate of sulfate particulates nucleation strongly depends on that parameter.

In addition to sulfuric acid, sulfate particulates may also include sulfate salts. The most common salt is calcium sulfate ($CaSO_4$), which can be formed in reactions between H_2SO_4 and calcium compounds originating from lube oil additives. Various sulfates may be also produced in reactions between sulfuric acid and exhaust system components. Sulfate particulates having a high salt content would bind less water, because most sulfate salts are less hydrophilic than sulfuric acid.

As discussed earlier, the TPM emission is determined by weighing the total mass of material collected on the sampling filter. As a consequence, the sulfuric acid, sulfate salts, and combined water are all counted as part of the TPM. The exact amount of water that is combined with sulfate particulates changes as sulfuric acid on the filter reaches an equilibrium with moisture in the surrounding atmosphere. Sampling filters are typically preconditioned at controlled conditions until their mass stabilizes. Assuming that sulfate particulates are pure sulfuric acid, the amount of combined water may be determined from Figure 7.11 depending on the relative humidity during preconditioning of the sampling filter [34]. If filters are preconditioned at typical conditions of 50% relative humidity and 25°C, each gram of sulfuric acid is associated with 1.32 g of water. This corresponds to an average hydration ratio of 7.19, that is, the chemical composition of sulfate particulates would be $H_2SO_4 \times 7.19\ H_2O$ (the molecular mass of H_2SO_4 and H_2O is 98.08 and 18.016, respectively).

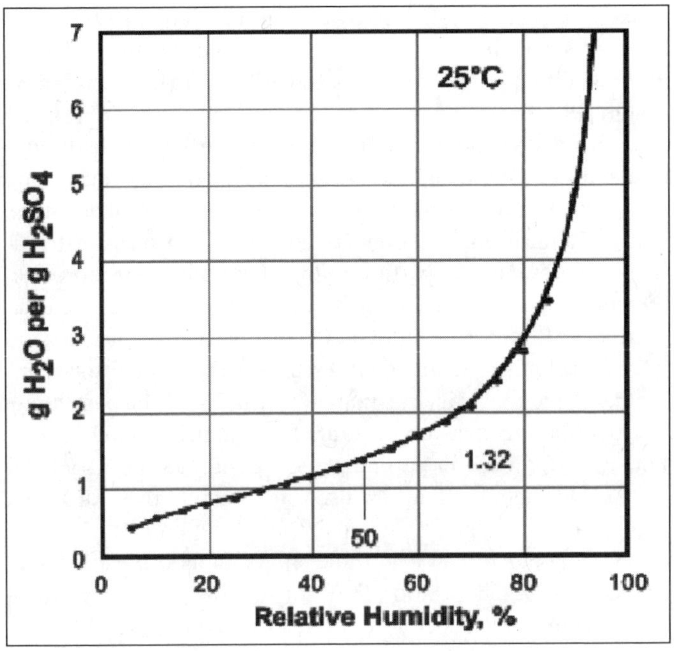

Figure 7.11 *Mass of bound H_2O per gram of H_2SO_4. (Based on a figure in Ref. 34.)*

Samples collected for the determination of PM emissions can be chemically analyzed for sulfates. Care should be taken while interpreting the results. Different methods may be used for the analysis and reporting of results for sulfates—an unregulated diesel emission. Many laboratories would report the result as mass of the SO_4 group (molecular mass of 96.06 g/mol). Certain corrections are necessary to estimate the corresponding overall contribution of sulfates to the TPM emission. To be exact, the SO_4 result should be first multiplied by the molecular mass ratio of 98.08/96.06 to convert to H_2SO_4. If calcium sulfate or other salts were present in the sample, the contribution of the cations remains unknown, unless a separate chemical analysis is performed. However, one can easily adjust the H_2SO_4 figure for combined H_2O by adding the water portion, which, as determined from Figure 7.11, often contributes more to the total PM mass than sulfuric acid alone.

7.3.6 Particle Size Distribution

An increased interest in diesel particle size distribution has been sparked by medical research which indicates that adverse health effects from exposure to particulates may be increasing with decreasing particle size, even if the particles are composed of toxicologically inert materials. In the United States, these concerns have been reflected by the introduction in 1997 of new ambient air quality standards for particulates smaller than 2.5 μm in addition to the existing standards for particulates below 10 μm (PM10).

Size distributions of diesel exhaust particulates exhibit a bimodal character with two concentration peaks that correspond to the nuclei and accumulation mode particles. The nuclei mode is usually fairly insignificant (a few percent) if mass distributions are presented. In particle number representation, however, the nuclei mode often accounts for more than 90% of the total particulate count. Since the late 1990s, a number of emission laboratories have undertaken PM size distribution investigations. Unlike the total particulate mass quantification, the particulate size distribution measurements are far from being standardized. Due to differences in particulate sampling and the variety of measuring methods that are used, test results from different laboratories have not been consistent. An important objective of the ongoing research is to develop standard measurement procedures, which would allow for reliable quantification of particle number emissions and size distributions [12, 14].

An increased interest in the size of diesel particulates was triggered by a study funded by the Health Effects Institute (HEI), which investigated two heavy-duty diesel engines to find increased numbers of nanoparticles in the newer generation engines [35]. That observation, although based on a single engine measurement, implied that PM emissions from new technology engines might be of more health concern than earlier engines, despite their lower PM mass. Later studies have generally not confirmed these findings, suggesting that they were characteristic to the particular engine tested by the HEI. A very comprehensive study sponsored by the Coordinating Research Council (CRC E-43) concluded that newer, low PM mass engines generally produce fewer particle numbers than their older predecessors [14]. More light has been also shed on the nature of nuclei mode particulates, indicating that health concerns in relation to nanoparticle emissions from diesel engines, although real, may have been exaggerated. For instance, it is now believed that nuclei mode particles from diesel engines are composed primarily of liquid condensates, as opposed to solid material, making them less likely to cause harm to human lungs. Furthermore, the smallest size particles are very short lived in the atmosphere. A 90% reduction (and nearly 100% in the range below 0.1 μm) of total particle number concentrations occurs within a few minutes or a 100- to 1000-m distance [36]. Thus, diesel and other mobile particle sources influence particle number concentrations in the ambient air within a limited area, mainly near roadways.

Discussion of particulate number emissions and size distributions is not possible without more background on particle formation, characterization, and measurement, all of which involve many unresolved issues. These topics are discussed in more detail in the next section.

7.4 Diesel Exhaust Particle Size

7.4.1 Diesel Particle Size Distribution

Since the mid-1990s, particle size distributions from internal combustion engines have been receiving increased attention due to the possible adverse health effects of fine and ultrafine particulates. Diesel emission control strategies, based on both engine design and aftertreatment, are being examined and reevaluated for their effectiveness in the control of the finest fractions of diesel particulates and particle number emissions. However, a fair performance assessment of various control technologies can be possible only if the research community reaches a consensus on the definition of and the measurement techniques for the smallest fractions of diesel particulates. The determination of particle sizes and numbers is much more sensitive to the measuring techniques and parameters than the quantification of particulate mass emissions. Dilution and sampling methods are key variables that must be taken into account to ensure accurate and repeatable results. On the other hand, particle sizing instruments exist that have significantly better sensitivities than the gravimetric measurement, thus presenting an attractive alternative for the PM emission measurement in future engines, provided standardized measuring methods are developed.

Ambient particulate matter is divided by most authors into the following categories based on their aerodynamic diameter (the aerodynamic diameter is defined as the diameter of a 1 g/cm^3 density sphere of the same settling velocity in air as the measured particle):

- PM_{10} (particulates having an aerodynamic diameter of less than 10 μm)
- Fine particles having diameters less than 2.5 μm
- Ultrafine particles having diameters less than 0.1 μm or 100 nm
- Nanoparticles, which are characterized by diameters of less than 50 nm.

A typical size distribution of diesel exhaust particulates is shown in Figure 7.12. (Note that a logarithmic scale is used for particle aerodynamic diameter.) Nearly all diesel particulates have sizes of significantly less than 1 μm. As such, they represent a mixture of fine, ultrafine, and nanoparticles. Due to the current PM sampling techniques (diluted exhaust, temperature <52°C), diesel particulate matter includes both solids, such as elemental carbon and ash, and liquids, such as condensed hydrocarbons, water, and sulfuric acid. Formation of particulates starts with nucleation, which is followed by subsequent agglomeration of the nuclei particles. The nucleation occurs both in the engine cylinder (carbon, ash) and in the dilution tunnel (hydrocarbons, sulfuric acid, water), through homogeneous and heterogeneous nucleation mechanisms.

Size distributions of diesel particulates have a well-established bimodal character that corresponds to the particle nucleation and agglomeration mechanisms, with the corresponding particle types referred to as the nuclei mode and the accumula-

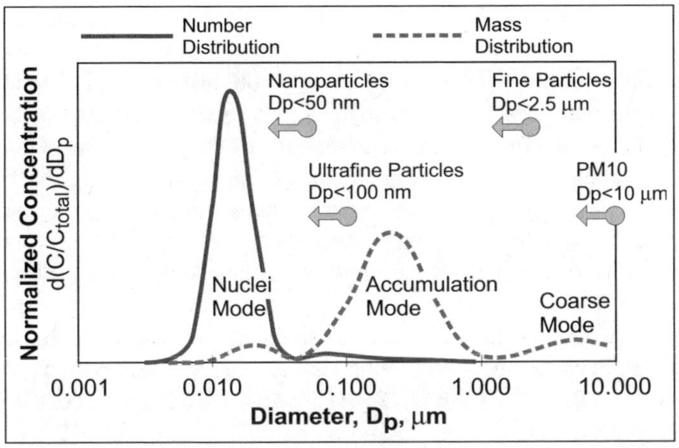

Figure 7.12 *Diesel particulate size distribution.*

tion mode. Size distributions are usually presented using either particle mass or particle number weighting. In each representation normal-logarithmic distribution curves are produced, as shown in Figure 7.12. Both the maximum particle concentration and the position of the nuclei and accumulation mode peaks, however, depend on which representation is chosen. In mass distributions, the majority of the particulates (i.e., the particulate mass) is found in the accumulation mode. In number distributions, on the other hand, most particles are found in the nuclei mode. In other words, diesel particulate matter is composed of numerous small particles holding very little mass, mixed with relatively few larger particles that contain most of the total mass. A small fraction of diesel particulates reside in a third, coarse mode (Figure 7.12). These three particle modes can be characterized as follows:

- *Nuclei mode:* The diameter of the original nucleus, such as that formed during sulfuric acid nucleation, is about 1 nm [37]. Today's measuring techniques are capable of detecting a minimum particle size of approximately 3 nm. According to various definitions, the diameters of nuclei mode particles were generally less than 40 to 50 nm (0.04 to 0.05 μm). Based on particle size research in the 1990s technology heavy-duty diesel engines, it has been postulated that the nuclei mode extends through sizes from 3 to 30 nm (0.003 to 0.03 μm) [14, 38]. All of the size ranges listed place nuclei mode particles entirely within the nanoparticle range. The maximum concentration of nuclei mode particles occurs at 10 to 20 nm. The nuclei mode, depending on the engine technology and particle sampling technique, typically contains only 0.1% to 10% of the total PM mass, but it often includes more than 90% of the total particle count. Sometimes the nuclei mode particles present as much as 99% of the total particulate number. Nuclei mode particles are composed mostly of volatile condensates (hydrocarbons, sulfuric acid) and contain little solid material.
- *Accumulation mode:* The accumulation mode is made of submicron particles of diameters ranging most often from 30 to 500 nm (0.03 to 0.5 μm nm) [14], with a maximum concentration between some 100 and 200 nm (0.1 and 0.2 μm). As shown in Figure 7.12, the accumulation mode extends through the fine, ultrafine, and the upper end of the nanoparticle range. Accumulation mode particles are made of solids (carbon, metallic ash) intermixed with condensates and adsorbed material (heavy hydrocarbons, sulfur species).
- *Coarse mode:* These particles with aerodynamic diameters above 1 μm (1000 nm) contain 5% to 20% of the total PM mass and practically no contribution to particle numbers [14]. The coarse particles are not generated in the diesel combustion process. Rather, they are formed through deposition and subsequent re-entrainment of particulate material from the walls of the engine cylinder, exhaust system, or the particulate sampling system.

Other particle weightings that may be used include particle surface (which would produce a curve located between the mass and number weightings in Figure 7.12) and particle volume weighting, which is proportional to the mass weighting.

7.4.2 Particle Distributions from New Diesel Engines

Interest in particulate size distributions and nanoparticle emissions from diesel engines was sparked by reports that newer technology engines—designed for low PM mass emissions—may still generate high particle numbers. The most significant study indicating such possibility, involving measurement of particle size distributions from an older and a newer generation heavy-duty diesel engine, was published by the Health Effects Institute [35]. As indicated by later research based on more comprehensive data from several engine models, old and new technology engines produce in general nuclei modes of similar magnitude, and the PM mass reductions in new engines are due to smaller accumulation modes [14]. It is believed that emissions from the new engine in the HEI study—an experimental 1991 design—have not reflected general emission trends in new technology engines. That study, however, must be credited with prompting quality research by the diesel industry, academia, and governments, leading to a greatly increased understanding of particulate emissions from internal combustion engines.

The HEI study compared two 10-L Cummins engines: a 1988 model year LTA 10-300 engine with a 1991 LTA10-310 [33]. The 1991 engine had a higher injection pressure, a higher turbo boost pressure, and a larger intercooler than its 1988 counterpart. The engines were operated at several steady-state modes. Modes 9 and 11, which are referenced here, were at rated speed and at 75% and 25%, respectively, of the maximum torque. Figure 7.13 illustrates the particle size distribution measured from both engines at Mode 9, presented in the volumetric units of $\mu m^3/cm^3$.

Particle size distributions from both engines showed the bimodal character typical for diesel particulates. The peak concentrations for nuclei and accumulation mode particles were at about 0.02 μm and about 0.2 μm, respectively. In the 1988 engine, most of the generated particulates were in the

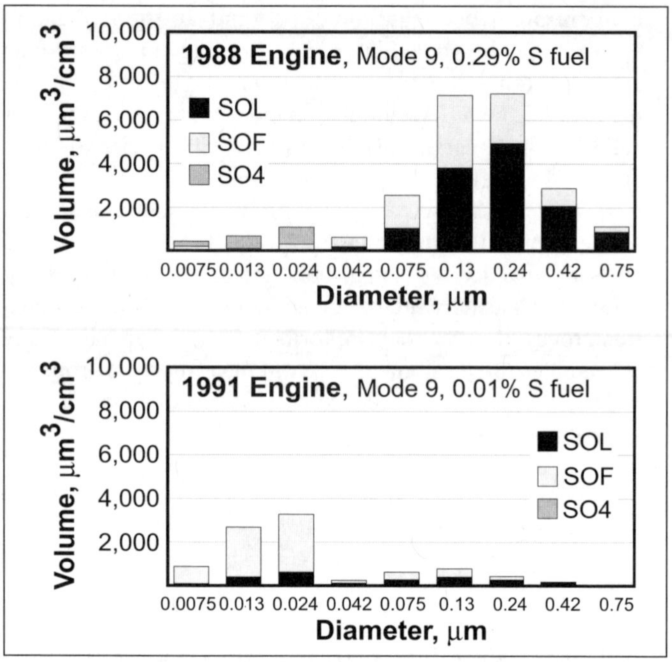

Figure 7.13 *Particle size distribution from two diesel engines (HEI study). (Based on data from Ref. 33.)*

accumulation mode. The 1991 engine showed a more than 90% PM mass emission reduction in the accumulation mode relative to the 1988 engine, however, its nuclei mode was greatly increased, up to as much as 40% of the total particle volume at certain conditions.

Particle numbers emitted from each engine in Modes 9 and 11 are compared in Table 7.12 (all data with 0.01% S fuel). Even though the total PM volume (as well as mass) was reduced in the new engine by about 80%, a significant increase in the total number of particles was measured at both test modes. In Mode 11 the particle numbers from the new engine were more than 2 orders of magnitude higher than that from the old engine. One important reason for the high particle numbers was the high SOF content in the 1991 engine, which resulted in the formation of nuclei mode hydrocarbon particles.

Increased particle numbers were also reported by some other studies. However, in most cases the observed increases in number emissions were not as dramatic as those shown in Table 7.12. For instance, two Liebherr diesel engines were compared in the VERT study: an old 914 TI (6.11-L, 2000-rpm, 105-kW, DI, turbocharged engine) and a new 924 TI E (6.64-L, 2000-rpm, 140-kW, DI, turbocharged and intercooled engine) [39]. In most operating modes increases in particle numbers between 15% and 50% were reported, with a sixfold increase under one operating mode. The maximum particle concentrations were of the order of 107 particles/cm3.

Investigation of particle numbers from heavy duty engines was a part of the comprehensive diesel aerosol sampling study E 43, funded by the U.S. Coordinating Research Council (CRC) [14]. A particle number comparison from various engine technologies compiled by the study is shown in Figure 7.14. The three E 43 sets of data (Cummins ISM and CAT 3406) were collected in highway chase experiments, where a mobile emission laboratory was following a diesel truck on a road, sampling air behind its exhaust plume. The chart also includes a curve representing the composite from the HEI data for the 1991 engine discussed earlier. The comparison shows that the model year 1999–2000 engines in the E 43 study produced fewer nanoparticles than the 1991 HEI engine.

TABLE 7.12
NUMBER OF PARTICLES FROM TWO DIESEL ENGINES (HEI STUDY)

	1988 Engine	1991 Engine
Mode 9		
Total particles (particles/cm^3)	$3.75 \cdot 10^7$	$1.26 \cdot 10^9$
Total volume (µm^3/cm^3)	$2.42 \cdot 10^4$	$3.58 \cdot 10^3$
Number fraction in nuclei mode	0.58	0.99
Volume fraction in nuclei mode	0.018	0.41
NMD* in nuclei mode	0.011	0.011
NMD* in accumulation mode	0.30	0.17
Mode 11		
Total particles (particles/cm^3)	$4.58 \cdot 10^7$	$6.69 \cdot 10^9$
Total volume (µm^3/cm^3)	$2.56 \cdot 10^4$	$5.76 \cdot 10^3$
Number fraction in nuclei mode	0.44	0.99
Volume fraction in nuclei mode	0.022	0.44
Number mean diameter in nuclei mode, µm	0.015	0.017
Number mean diameter in accumulation mode, µm	0.16	0.12

Source: Data from Ref. 33.

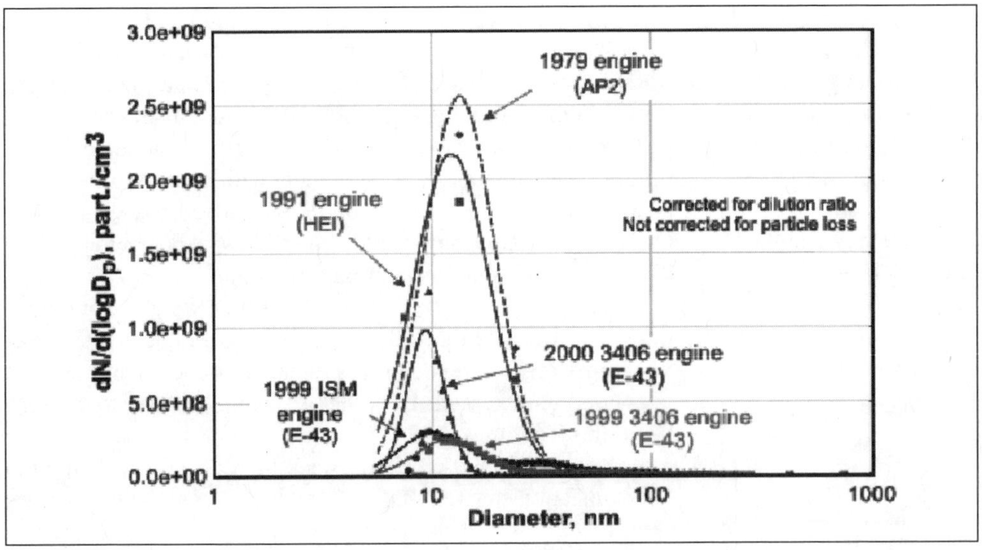

Figure 7.14 Comparison of particle number emissions from different engines. AP2: 1979 Cat NTC350 engine, 3000 ppm S fuel; HEI: 1991 Cummins LTA10 310, 100 ppm S; E 43: 1999 Cummins ISM (California calibration), 1999 Cat 3406E (CA), 2000 Cat 3406E (EPA), 326 406 ppm S. (Based on Fig. 70 in Ref. 14.)

To put things into better historical perspective, the chart also shows a set of data collected in another CRC study (AP2), involving a chase experiment with a model year 1979 engine [40]. Particle numbers produced by the 1979 engine were somewhat higher than those in the 1991 HEI engine—a clear indication that nanoparticle emissions have been on the roads for a long time and are not a new phenomenon caused by new technology engines.

Generally decreasing particle number trends with advanced engine technology were reported in Euro I, Euro II, and Euro III engines when tested over regulatory emission test cycles [41, 42]. Comparisons of particle number emissions were also conducted in older and newer technology diesel passenger cars in studies funded by ACEA [43]. It was found, based on steady state and ECE+EUDC cycle tests, that the size distribution of diesel particulates was similar in the older and more advanced technology vehicles. It was also shown that particulate mass could be correlated with particle numbers.

Diesel particle counts reported in different studies span a wide range of approximately 1x106 to 1x109 particles/cm3. These results would suggest that there is large variability in particle counts between different engine technologies. However, due to the lack of standard measuring methods, comparisons of data from different laboratories have only limited validity. Until sampling and measuring procedures mature, valid particle number comparisons should be based on data generated within a single research program, using the same set of measuring techniques and equipment. Comparing data from different laboratories may require a number of corrections, such as for different dilution ratios, and is not always possible.

7.4.3 Spark-Ignited Engines

At first, particulate emission research focused on the characterization and control of diesel PM emissions. This was due to the fact that PM mass emissions from spark-ignited (SI) engines were small when compared to those from diesel engines. Prior to the introduction of diesel PM legislation in the 1990s, the SI engines used to have PM emissions of the order of 1% of diesel particulate emissions. However, recent trends to shift focus from particle mass to particle numbers resulted in renewed interest in particulate emissions from SI engines.

Research on the characterization of the SI particulate emissions [44] and on comparison of diesel and SI particles [45, 46] seems to be reaching a consensus on the following issues:

- SI engine particulates, such as those produced by port fuel injected (PFI) gasoline engines, are composed of much smaller particles than those from diesel engines. In some reports, the SI engine particles were found to be almost entirely in the nanoparticle range (<50 nm).
- Particle numbers in the SI engines are about 2 orders of magnitude lower than those in diesel engines when measured at low engine speeds or over standard engine test cycles. The particle number emissions increase significantly at rich air to fuel ratios, for example, during high-speed highway cruising. At such conditions, comparable particle numbers may be emitted from SI and diesel engines.
- SI particulate numbers, when plotted over time, show repeated patterns of high concentration "spikes," whereas the diesel engines emit steady numbers of particles. This would indicate a different mechanism of particle formation in both types of engines.
- SI engine particles appear to be composed of ash (calcium, magnesium, copper, iron, sodium, etc.), lube oil, and other volatile components [47]. Their elemental carbon content is low in comparison to that of diesel engines.
- Three way catalytic converters decrease the concentrations of the smallest particles. Different types of fuel additives may either increase or decrease the particulate number concentrations.

A comparison of data from different literature reports, expressed in number of emissions per kilometer, shows that automotive diesel engines emitted between about 2.4×1014 and 8×1013 particles/km, while emissions between 1.2×1014 and 3×1010 particles/km were observed from gasoline engines [46, 48]. Expressed per unit of gas volume, new PFI SI engines emit number concentrations from about 1×105 particles/cm3 (low load) to 1×107 particles/cm3 (high load), while older SI engines can emit concentrations in excess of 108 particles/cm3 [49].

A large study to compare diesel and gasoline particulate number emissions was conducted by CONCAWE [48, 50]. In that study four diesel cars (DI and IDI) and three gasoline PFI cars were tested using a matrix of three diesel and two gasoline fuels. The vehicles were tested on several constant-speed conditions and over the ECE+EUDC test cycle. Example results are shown in the following figures. Figure 7.15 illustrates the total particle counts averaged over the tested fuels at selected test modes. These results are in agreement with the previously listed characteristics of SI engine particles. Indeed, the SI particle numbers increase significantly at high-speed driving, while diesels tend to emit constant particle numbers.

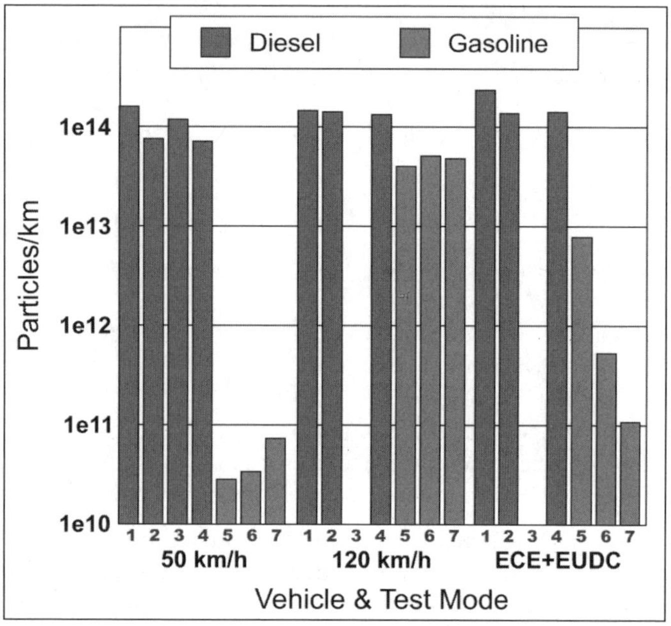

Figure 7.15 *Particulate number emissions from diesel and gasoline cars. (Based on data from Refs. 48 and 50.)*

Figure 7.16 presents selected number weighted distributions for diesel and gasoline cars. The particular lines reflect variability for different cars and fuels. Gasoline engine particulates are characterized by generally smaller sizes than those from diesel engines.

Gasoline-fueled cars have been increasingly adopting the direct injection engine (GDI) technology. Cars with GDI engines are already commercially available in Japan and Europe. The fuel injection strategy adopted by those engines depends on the engine operating condition. Typically, under low-load conditions, fuel is injected during the compression stroke, resulting in a stratified charge with an overall A/F ratio of 30:1 to 40:1. Under moderate and high-load conditions, fuel is injected during the intake stroke, resulting in a homogeneous charge. At moderate loads the A/F ratios range from 20:1 to 24:1, whereas at high loads the engine enters into closed loop stoichiometric feedback mode [51].

The GDI injection strategy at low loads bears some similarity to the diesel injection process, whereas at high loads it is identical to the traditional PFI gasoline engine. GDI emission studies indicate that the complex injection strategy produces equally complex particle emission behavior. When tested at high engine loads, under stoichiometric control operation, GDI engine emissions are similar to those of PFI gasoline engines. At lower loads on the other hand, when the engine enters the

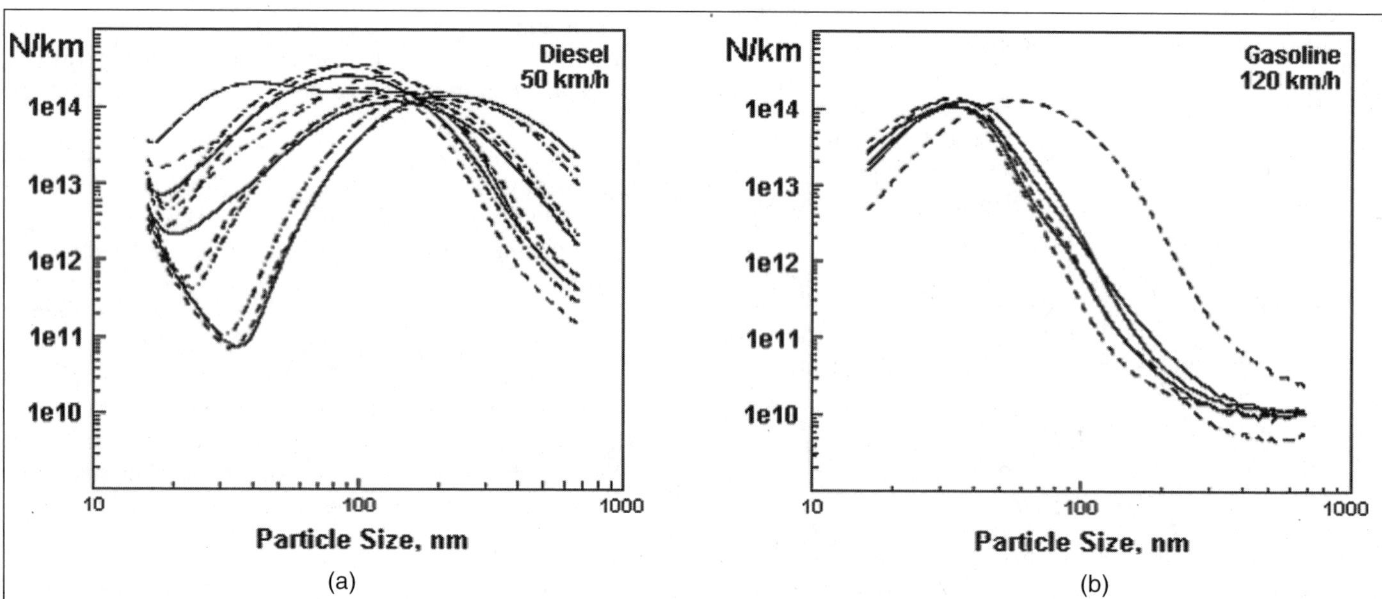

Figure 7.16 *Particle size distributions from (a) diesel and (b) gasoline cars. (Reprinted from Ref. 50.)*

stratified combustion regime, its PM emissions may resemble those from the diesel engine. In one GDI vehicle study, when tested at high engine load (stoichiometric operation), particle numbers reached 9.3×107 particles/cm3 with the characteristic spiking time resolved pattern [49]. At lower engine loads—in the stratified, lean burn mode—particle counts were much higher, reaching up to 2.8×108 particles/cm3. They also exhibited steady state, diesel like time distributions. Furthermore, the GDI engine produced agglomeration mode particles of a maximum concentration at about 100 nm at all but one operating mode. This is again similar to the diesel engine and unlike the PFI engine, which generates particulates primarily in the nanoparticle size range (<50 nm). High particle numbers for GDI engines have been reported by numerous other studies [52, 53]. The latter investigation also compared the FTP 75 cycle PM mass emissions from an equivalent PFI gasoline engine, a GDI gasoline engine, and a 1995 diesel IDI vehicle, reporting about 2, 10, and 100 mg/mile, respectively. By taking a position between the PFI and diesel, the GDI engine becomes a potentially significant source for not only particle numbers, but also PM mass emissions.

The chemical composition of the GDI particles still remains an open question. Considering the similarity of the stratified combustion to the diesel cycle, it is likely that GDI particulates contain a significant proportion of elemental carbon.

High emissions of small size particles have been also reported from natural gas engines. In a comparison of a diesel and CNG bus engine emissions, the diesel bus was shown to have 1 to 2 orders of magnitude higher particle counts in the size range above 0.09 μm. However, at 0.04 μm (nanoparticle size range) the CNG and diesel emitted comparable particle numbers [54].

7.4.4 Influence of Particulate Sampling

Contrary to the gravimetric measurements of total diesel particulate emissions, which are a well-established routine measurement of good repeatability, size distributions measured at different laboratories show considerable variability. Reported distributions, especially those weighted by particle numbers, frequently disagree by several orders of magnitude. That spread of experimental results is related to the following issues:

- Different measuring equipment and measurement errors
- The influence of exhaust gas sampling and dilution parameters on the properties and size distributions of diesel particulates.

The measurement of diesel particle size distributions is a relatively new analytical discipline. As increasingly more experience is being accumulated, the analytical methods and instruments are becoming more reliable and accurate. The precision, repeatability, measuring ranges, response to transient conditions, and, generally, the quality of measurement has been improving and will continue to improve.

The exhaust gas sampling issues and their influence on the particles are more serious considerations. There is wide agreement that sampling parameters, such as dilution ratio, dilution rate, and residence time, have a critical impact on the properties and sizes of the particulates [38, 55, 56], to the point that different setups of the same measurement system can produce conflicting data. That influence is especially strong with regard to the number concentrations and distributions, as opposed to the mass/volume particle concentrations. Another consideration is that exhaust gas dilution which takes place in the dilution tunnel(s) is not a good simulation of atmospheric dilution. In effect, conditions in the dilution tunnel may easily produce artifacts, or artificial particulates, which are not normally encountered when dilution takes place in the atmosphere.

Exhaust particulates in the tailpipe are very different from those sampled in dilution tunnels or those present in the ambient atmosphere. At high exhaust temperatures, the particles are composed primarily of solids, including elemental carbon and ash. When the gases are discharged to the atmosphere and mixed with ambient air, the particulate phase increases through nucleation, adsorption, and condensation, which are driven by decreasing temperatures. Nucleation is a formation of new particles, solid or liquid, from gas-phase components. The rate of nucleation rapidly increases when gas components approach their saturation. Although the nucleation is usually not very significant for mass concentrations, it can increase the number concentrations of particulates by an order of magnitude. The nuclei particles may grow due to coagulation, which occurs both in the hot exhaust gas and during the dilution. Other mechanisms of particle size growth include adsorption of gas molecules and condensation of vapors on the surface of particles.

Once finally released into the atmosphere, the particles are by no means stable. The processes that occur include atmospheric evaporation and desorption of the volatile particulate fractions, as well as changes in the particle sizes. Several independent studies found that the nanoparticles disappear with aging and their ambient concentrations quickly decrease with time. Particle lifetimes, τ_{90} and τ_{99}, are defined as the times necessary for the reduction of the total number concentrations from their initial plume values by 90% and 99%, respectively. An investigation of the lifetime of diesel aerosol found that for typical urban conditions τ_{90} is of the order of a few minutes and τ_{99} is of the order of 20 to 30 min (during these time periods, particulates can travel with the wind over a distance of 100 to 1000 m and 0.5 to 10 km, respectively) [36].

Important factors that influence gas-to-particle conversions, including nucleation, adsorption/condensation, and coagulation, include the following:

- Dilution ratio
- Cooling
- Residence time
- Temperature
- Humidity
- Ambient aerosols.

As a consequence, the gas-to-particle conversion processes are very sensitive to correct simulation of atmospheric dilution. The critical parameters include saturation ratios of volatile components, which influence the nucleation of new particles, and residence times, which influence particle coagulation. Typical atmospheric dilution ratios of vehicle exhaust gases are between 500 and 1000. Because dilution tunnels used for PM sampling operate at dilution ratios of about 10, the simulation is far from being perfect. The dilution tunnel saturation ratios are much higher than those in the atmospheric dilution, while the dilution rates are slower and the residence times are longer. As a result, particulate size distributions measured in a dilution tunnel may be very different from what would be seen in the atmosphere. Furthermore, different number concentrations and distributions may be measured whenever dilution ratios or residence times are changed.

Low dilution ratios lead to an increase in the total particulate mass due to increased SOF fraction. That effect has been known for quite a long time [55]. The influence of dilution ratio and dilution tunnel residence time on number-weighted size distributions can be illustrated by the results of a study with a Perkins diesel engine [37]. Diesel particulates were sampled from a Perkins T4.40, four-cylinder, DI, turbocharged and aftercooled engine operated with low sulfur fuel (<0.05% S). The particulate sampling system included a primary and secondary dilution tunnel of adjustable dilution ratio. The influence of dilution ratio on size distributions was measured by changing the primary dilution rate between 4 and 84 at a fixed secondary dilution rate. For dilution ratios greater than 60, both the number and volume concentrations were found to be essentially independent of the dilution ratio. For smaller dilution ratios, number and volume concentrations increased with decreasing dilution ratio. Particles smaller than about 30 nm were more strongly influenced than the larger ones. For example, the number concentration of 8-nm particles increased 2.8 times when the dilution ratio was decreased from 59 to 4. The increased number concentrations of small particles can be explained by nucleation of hydrocarbons, sulfur compounds, and water in the dilution tunnel. Particle nucleation is favored by higher saturation ratios, which occur at lower dilution ratios.

The effect of residence time on particle number and size found in the same study is shown in Figure 7.17. The residence times were measured as the time between primary and secondary dilution. The lowest particle numbers were recorded with the shortest residence time of 0.04 s. Because the residence time was probably too short to allow for condensation and particle nucleation, these particles may have been composed primarily of solids. The nuclei mode peak below a 10-nm diameter might include oil derived ash particles. As the dilution ratio was increased to 0.4 s, the number of small particles increased by 1 order of magnitude (note the logarithmic scale in Figure 7.17). This was the effect of formation of new particles through nucleation in the dilution tunnel. Further increases in the residence time lead to coagulation and size growth of small particles, as indicated by the decreasing number concentration and increasing particle diameter.

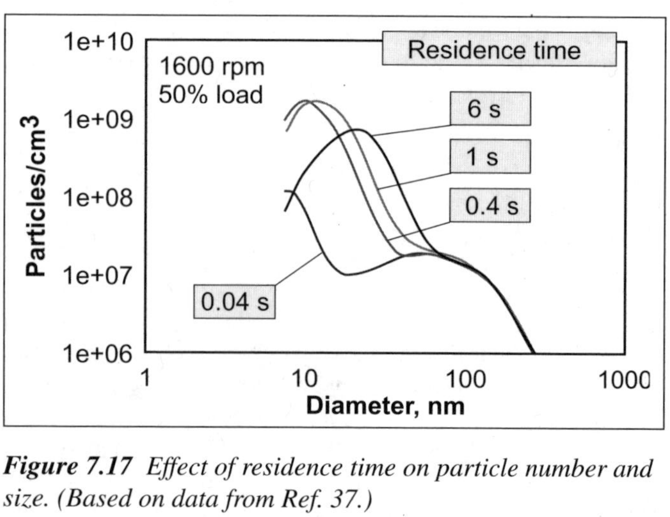

Figure 7.17 Effect of residence time on particle number and size. (Based on data from Ref. 37.)

Clearly, in the sampling setup used in the preceding study, one could select the sampling parameters to obtain a wide range of number concentrations of small particles. For instance, for 10-nm-diameter particles, number concentrations spanning more than 1 order of magnitude could be realized by simply adjusting the dilution tunnel residence time.

Particle distributions measured from dilution tunnels are further influenced by sampling problems related to wall interactions. These problems include both (1) sample losses and (2) sample additions. Sample losses are due to deposition of particles on the walls of the sampling system by mechanisms such as these: diffusion, inertial deposition, and thermophoresis.

Due to high diffusion coefficients, particles smaller than about 15 nm are subject to large losses in sampling systems and in instruments. Sample losses of more than 50% are not uncommon in diesel particle measurement systems, but are often not reported in the literature [14].

Sample additions are caused by re-entrainment and/or outgassing of material deposited on the walls. Outgassing may happen whenever the engine load is increased and hot gas enters the sampling system at a lower temperature. At such conditions condensation may occur at the walls. After some time the wall temperature will increase, causing the condensates to evaporate and add to the sample material. In addition, certain materials or components used in sampling systems, such as silicone rubber hoses, were reported to produce ultrafine particle emissions under certain conditions and should be avoided [56].

Measurement problems may be also caused by storage and release of particulates in exhaust systems. It was shown that diesel particulates can be stored in practically all components of the exhaust system, including pipe walls, mufflers, and catalytic converters [57]. Storage and release from the exhaust system did influence measured particle size distributions [7].

7.4.5 Nanoparticles: Composition and Formation

Generally, diesel nanoparticles include the same constituents found in the total particulate matter emissions, including elemental carbon, ash, hydrocarbons, sulfuric acid, and water. Contrary to the total PM emission, diesel nanoparticles cannot be chemically analyzed to precisely determine their composition. No sampling/analytical procedure is available that would allow us to separate a sufficient mass of the nanoparticle material for such an analysis. As a result, the exact formulation of diesel nanoparticles has to be studied through indirect experiments. A consensus has been forming that diesel nanoparticles are mostly volatile. In most cases, they are composed primarily of hydrocarbon and sulfuric acid condensates, with a small contribution of solid material, such as ash and carbon.

The nanoparticle composition can be dramatically changed by exhaust emission control measures, such as diesel particulate filters or fuel additives. Very high numbers of nanoparticles were measured downstream of particulate filters in the VERT study [58]. A special sampling technique was developed to distinguish between solid particles and volatile condensates. The diluted sample passed through a heated vapor trap (thermodenuder), filled with active carbon, before entering the particle size analyzer. Depending on the vapor trap temperature, different fractions of particle condensates were driven off the sample. An example of such condensate analysis is presented in Figure 7.18. The data were generated on a 100-kW, DI diesel engine equipped with a particulate filter at full load with low-sulfur fuel (<0.04% S). The exhaust gas temperature in the particulate trap was 460°C. The temperatures in the active carbon vapor trap were changed between 120 and 250°C, as shown in the graph.

The high peak of 30 to 40-nm particles disappears from the graph when the sample is heated to 120°C. This is an indication that these particles included volatile materials, presumably hydrocarbons and hydrated sulfuric acid. As the temperature was increased to 250°C, the high nanoparticle emissions disappeared almost completely (note the logarithmic scale in the graph). It can be concluded that the high nanoparticle emissions observed downstream of particulate filters were composed primarily of condensates (strictly speaking, this conclusion was valid only within the measuring system detection limit of 15 nm, below which solid particles still could be found).

In the same study, particle distributions with an iron based fuel additive were analyzed from an engine without a particulate filter. Distributions were found with two particle concentration peaks: (1) the additive ash particulate peak at about 20 nm and (2) the carbon particles peak at about 90 to 100 nm. As shown in Figure 7.19, the additive caused a substantial increase in the nanoparticle emissions. Analysis with the heated vapor trap did not significantly change these results, confirming the mostly solid character of nanoparticles in that case. These results indicate that metallic fuel additives, if used, may make a significant contribution to solid ash nanoparticle emissions.

Nanoparticle volatility experiments were also conducted in the CRC E 34 study [14]. The setup involved a Cummins ISM engine and a nano tandem differential mobility analyzer (nano-TDMA). The instrument consisted of two differential mobility analyzers (DMAs) with a heating section between the two DMAs. It allowed a single particle size to be selected with the first DMA, which was then heated, and the resulting size change determined by the second DMA.

Several particle sizes representing the bottom size end of the accumulation mode (50 nm) and the nuclei mode (down to 7 nm) were subjected to the experiment. It was expected that in this size range, representing the overlap between accumulation and nuclei modes, a mix of solid and liquid particles would be found. Upon heating, solid particles consisting primarily of carbon and volatile particles consisting primarily of hydrocarbons would be differentiated. Volatile particles would shrink until they totally disappear. Solid particles, on the other hand, would shrink only a little, because any volatile material on their surface would be removed, and then their sizes would stabilize. The results of the experiment, for initial sizes of 50, 30, 12, and 7 nm, are shown as a function of temperature in Figure 7.20. The 50-nm particles split into a mainly solid mode that shrinks to a diameter of 43 nm, and a volatile mode that rapidly shrinks to 15 nm. The 30- and 12-nm particles show a similar behavior, being differentiated into solid and volatile fractions. However, the 7-nm particles continue shrinking upon heating, not showing evidence of containing any solid material. (Volatile particles do not disappear due to a possible instrument artifact around its detection limit.)

The volume fraction that remained in the 12 to 30-nm particles after heating was also determined in the study. It was

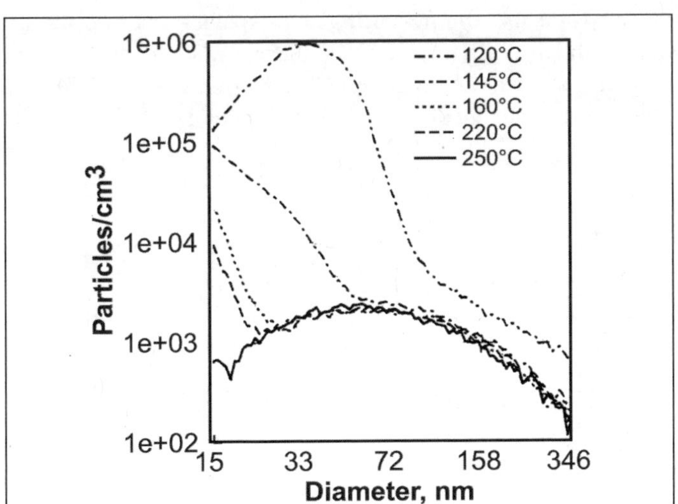

Figure 7.18 Solid particles and condensates in diluted diesel exhaust. (Based on a chart received from Andreas Mayer, TTM, Switzerland.)

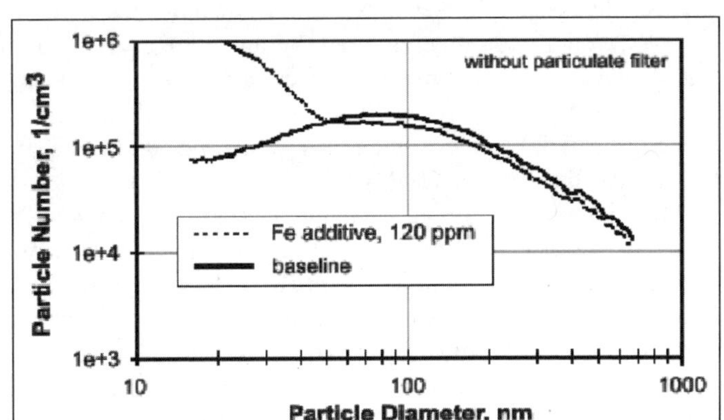

Figure 7.19 Nanoparticle emissions with a fuel additive. (Based on a chart received from Andreas Mayer, TTM, Switzerland.)

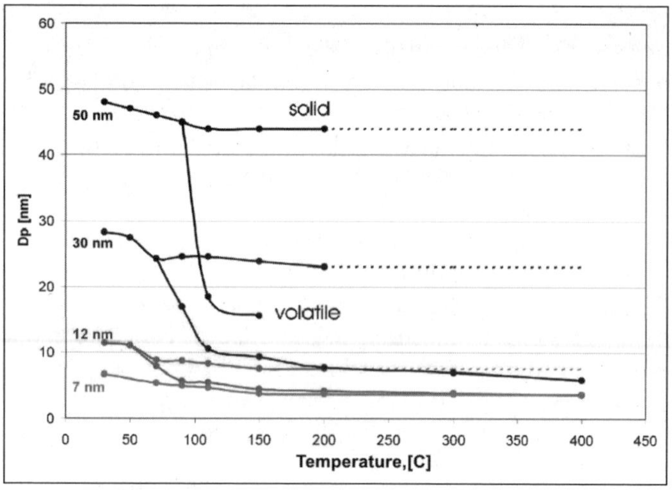

Figure 7.20 Volatility of diesel nanoparticles. (Based on a chart received from D. B. Kittelson.)

found that the volatile fraction amounted to 97% to 98% of the particle volume, leaving only 2% to 3% for possible nonvolatile, solid cores in the tested particles. Similar heating experiments, as well as evaporation calculations, were conducted for particles made of pure alkanes of different carbon chain length. It was determined that the shrinkage of particles from diesel engines was in the same range as for C28 through C36 normal alkanes, thus corresponding to the hydrocarbons in the engine lube oil. These results suggest that hydrocarbons that make up diesel nanoparticles are derived mostly of the engine lube oil. Note, however, that some fuel effect studies found increased nanoparticle numbers with more volatile fuels, indicating that fuel-derived HCs may also present a noticeable contribution [41].

Various hypothesis have been formulated to explain the increased particle numbers that were seen in certain studies with new diesel engines. It was once suspected that the high particulate numbers were related to high injection pressures, which are used by engine manufacturers as a strategy for meeting emission targets. That theory, however, has not been confirmed. On the contrary, a continuous decrease in mass and number emissions was shown when injection pressure increased from 400 to 1000 bar [59]. In another study, the injection pressure had only small influence, and only under some engine operating conditions, on the particle size distribution [60].

It is believed that increased number emissions are a simple consequence of (1) a high concentration of nanoparticle precursors (HC, SO4), combined with (2) a decreased mass of accumulation mode particles. Normally, the accumulation particles act as a "sponge" for the condensation and/or adsorption of volatile materials. In the absence of that sponge, gas species that are to become liquid (or solid) will nucleate to form large numbers of small particles. The driving force for the gas-to-particle conversion is the saturation ratio, defined as the ratio of the partial pressure of a species to its saturated vapor pressure. For the constituents of the SOF or sulfuric acid, the maximum saturation ratios occur during dilution and cooling of the exhaust and are typically achieved at dilution ratios between 5 and 30 [19]. The relative rates of condensation/adsorption and nucleation are a function of the saturation ratio [61], as shown in Figure 7.21.

The rate of adsorption/condensation is a linear function of the saturation ratio. The rate of nucleation, very low at low saturation ratios, becomes very high in the proximity of the saturation. Therefore, low saturation ratios favor adsorption/condensation, whereas high saturation ratios favor nucleation. The rate of adsorption/condensation is proportional to the surface area of available particulate matter. Because the large quantities of soot agglomerates present in the exhaust of old technology engines provide ample surface area, the vapors may be adsorbed and condensed. If such adsorption/condensation indeed occurs, the saturation ratio never rises high enough to produce nucleation. An opposite picture can be seen in systems with few accumulation mode particulates. Due to the low available soot surface area, little condensation occurs and the nucleation becomes more likely. This is especially true if elemental carbon emissions are reduced more than the SOF and SO4 emissions.

The nucleation theory discussed explains the high particle number emissions seen downstream of particulate traps (which remove solid accumulation mode particulates, but not necessarily nanoparticle precursors), as well as high number emissions from engines of high SOF fraction, such as the 1991 Cummins engine in the HEI study (Figure 7.13). It is also consistent with the observation of high particle numbers from gasoline engines, which are believed to be composed primarily of liquid condensates [46]. In all of these cases, the low quantities of agglomerated particulates present in the gas cannot provide enough surface area for the condensation/adsorption of volatile material. As the species approach saturation, high numbers of small particulates are produced through nucleation.

A similar nucleation mechanism may also apply to the formation of ash particulate. In that case, ash nucleation takes place inside the engine during the expansion stroke, rather than in the dilution tunnel. Once these ash nuclei are formed, they may serve as heterogeneous nucleation sites for SOF and other species during dilution and cooling in the exhaust. It was

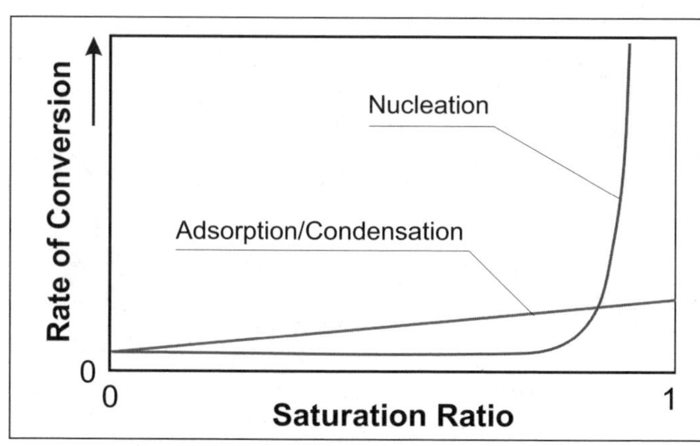

Figure 7.21 Relative rates of gas-to-particle conversion.

suggested that some particles in the high-SOF nuclei mode may include ash cores [19].

7.4.6 Engine Nanoparticle Issues

Findings of medical research indicate that small size particulates are more harmful to humans than the larger ones. Ambient concentrations of particulate matter below 10 μm (PM10) have been regulated by law in the United States for many years. In 1997, the EPA introduced new additional ambient air quality standard for particulates of aerodynamic diameter below 2.5 μm (PM2.5). That standard was in part based on health studies that showed a strong correlation between mortality and the concentration of fine particles below 2.5 μm.

It is clear that all diesel particulates fall under the PM2.5 category. The increased interest in diesel nanoparticles has been sparked primarily by (1) their suspected deep penetration into human lungs and (2) a belief that nanoparticles may be more harmful than larger particles—an extrapolation of the findings in the 10 to 2.5-μm size range. Much more medical research and diesel emission studies are needed to fully characterize the diesel nanoparticle emissions and their impact on human health. Some of the issues related to the health effects, regulations, and measuring techniques that will have to be addressed in the future include these:

1. If nanoparticles indeed present a higher health hazard than larger, agglomerated diesel particles, diesel PM emission regulations may have to include size distributions or number concentrations rather than be expressed only by the total particulate mass.
2. In the available medical research, the term particles refers to solids. There is probably no health study on the effects of volatile particulates. It is now known that diesel nanoparticles are normally composed mainly (>90%) of liquid condensates. The issue is this: Are volatile particulates a health hazard? Their short life span (τ_{90} of a few minutes) makes 90% of the ultrafine particle count disappear within 100 to 1000 m from their source [36]. This estimate indicates that the effect of diesel engines and other mobile ultrafine particle sources on ambient particle number concentrations will be limited to areas near roadways.
3. Under certain operating conditions, spark-ignited engines emit an amount of nanoparticles comparable to that emitted by diesel engines. The SI engine particulates are believed to be mainly volatile. If volatile particulates are a health hazard, the way in which spark-ignited engine emissions are regulated will have to be revisited.
4. If the health hazards are related to solids rather than to condensates, it would make sense to change the definition of particulate matter to include only solid materials. Such a change would seriously affect regulations, sampling and measuring techniques, and emission control technologies.
5. A similar approach has been suggested for pragmatic, rather than health, reasons. Sampling techniques that include volatiles cause measurements to be less repeatable. Considering that current laboratory dilution methods poorly simulate atmospheric conditions, artificial particulates may be produced that otherwise would not exist in the atmosphere. Therefore, diesel emission sampling and regulations should focus on the material found in the tailpipe, that is, on solid particulates, rather than trying to simulate atmospheric dispersions of emissions. This philosophy is being adopted in the European Union, where future particle number/size measuring methods will likely be based on solid material only [12]. In the United States, on the other hand, research is conducted to develop laboratory dilution methods that would accurately reproduce atmospheric dilutions, involving both solid and volatile material [14].
6. Whatever PM definition is adopted—including or excluding volatile fractions—standards must be developed that precisely specify sampling and analytical procedures for the measurement of particle numbers and sizes.

References

1. "Diesel Fuel Injection," Robert Bosch GmbH, 1994.
2. Sawyer, R.F., and J.H. Johnson, "Diesel Emissions and Control Technology," in Diesel Exhaust: A Critical Analysis," Health Effects Institute, Cambridge, MA, 1995.
3. Perry, R.H., Perry's Chemical Engineers' Handbook, 6th ed., McGraw Hill, New York, 1984.
4. "Control of Emissions of Air Pollution from Highway Heavy Duty Engines; Final Rule," U.S. Federal Register 62, 203, October 21, 1997, pp. 54693–54730, retrieved from www.epa.gov/otaq/regs/hd hwy/1997frm/frhwy.pdf.
5. "Health Assessment Document for Diesel Emissions," EPA/600/8 90/057C, U.S. Environmental Protection Agency, Washington, DC, February 1998, retrieved from www.epa.gov/ncea/diesel.htm.
6. "Diesel Emission Control Sulfur Effects Program, Phase I Interim Report No. 1," U.S. Department of Energy, Washington, DC, August 1999, retrieved from www.ott.doe.gov/decse/pdfs/interim.pdf.
7. Andrews, G.E., et al., "Diesel Particle Size Distribution Changes Along a Practical Exhaust System during Cold Start in a Passenger Car IDI Diesel," SAE Paper No. 2000 01 0514, SAE International, Warrendale, PA, 2000.
8. Andrews, G.E., et al., "The Influence of an Oil Recycler on Lubricating Oil Quality with Oil Age for a Bus Using In-Service Testing," SAE Paper No. 2000 01 0234, SAE International, Warrendale, PA, 2000.
9. "ACEA European Oil Sequences," European Automobile Manufacturers Association, Brussels, Belgium, September 1999, retrieved from www.acea.be/ACEA/20020618PublicationsOilSequences.pdf.
10. Havenith, C., and R.P. Verbeek, "Transient Performance of a Urea DeNOx Catalyst for Low Emissions Heavy Duty Diesel Engines," SAE Paper No. 970185, SAE International, Warrendale, PA, 1997.
11. Heck, R.M., and R.J. Farrauto, Catalytic Air Pollution Control: Commercial Technology, Van Nostrand Reinhold, New York, 1995.

12. Andersson, J., "UK Particle Measurement Programme: Heavy Duty Methodology Development," Report DP02/2493, Ricardo Consulting Engineers, Shoreham-By-Sea, UK, July 31, 2002, retrieved from www.ricardo.com/chemistry/UK_PMP_HD_Programme.pdf.
13. Kittelson, D.B., "Engines and Nanoparticles: A Review," J. Aerosol Sci. 5/6, 29, 1998, pp. 575–588.
14. Kittelson, D.B., W.F. Watts, and J. Johnson, "Diesel Aerosol Sampling Methodology—CRC E 43: Final Report," University of Minnesota, Report for the Coordinating Research Council, August 19, 2002, retrieved from www.crcao.com/reports/recentstudies00 02/UMN%20Final%20E 43%20Report.pdf.
15. Tobias, H.J., D.E. Beving, P.J. Ziemann, H. Sakurai, M. Zuk, P. McMurry, D. Zarling, R. Waytulonis, and D. B. Kittelson, "Chemical Analysis of Diesel Engine Nanoparticles Using a Nano DMA/Thermal Desorption Particle Beam Mass Spectrometer," Environ. Sci. Technol. 35, 2001, pp. 2233–2243.
16. Brown, J.E., M.J. Clayton, D.B. Harris, and F.G. King, Jr., "Comparison of the Particle Size Distribution of Heavy Duty Diesel Exhaust Using a Dilution Tailpipe Sampler and an In Plume Sampler During On Road Operation," J. Air Waste Management Assoc. 50, 2000, pp. 1407–1416.
17. Walters, R.B., et al., 1988. "A Generator for the Production of Sulfuric Acid Coated Diesel Soot Aerosols," Atmospheric Environ. 22, 1, pp. 17–23.
18. Broome, D., and I. Khan, The Mechanism of Soot Release from Combustion of Hydrocarbon Fuels, Institution of Mechanical Engineers, London, 1971
19. Abdul Khalek, I.S., et al., "Diesel Exhaust Particle Size: Measurement Issues and Trends," SAE Paper No. 980525, SAE International, Warrendale, PA, 1998.
20. Merkel, G.A., W.A Cutler, and C.J. Warren, "Thermal Durability of Wall Flow Ceramic Diesel Particulate Filters," SAE Paper No. 2001 01 0190, SAE International, Warrendale, PA, 2001.
21. Horiuchi, M., K. Saito, and S. Ichihara, "The Effects of Flow Through Type Oxidation Catalysts on the Particulate Reduction of 1990's Diesel Engines," SAE Paper No. 900600, SAE International, Warrendale, PA, 1990.
22. Wachter, W.F., "Analysis of Transient Emission Data of a Model Year 1991 Heavy Duty Diesel Engine," SAE 900443, SAE International, Warrendale, PA, 1990.
23. Voss, K.E., et al., "Catalytic Oxidation of Diesel Particulates with Base Metal Oxides," in Catalysis and Automotive Pollution Control III, A. Frennet A. and J.M. Bastin (Eds.), Elsevier, Amsterdam, 1995, pp. 499–515.
24. Cartellieri, W. P., and P. Tritthart, "Particulate Analysis of Light Duty Diesel Engines (IDI and DI) with Particular Reference to the Lube Oil Particulate Fraction," SAE Paper No. 840418, Society of Automotive Engineers, Warrendale, PA, 1984.
25. Mills, G.A., Ph.D. Thesis, University of Southampton, UK, 1983.
26. Williams, P.T., et al., "The Relation between Polycyclic Aromatic Compounds in Diesel Fuels and Exhaust Particulates," Fuel 65, August 1986, pp. 1150–1157.
27. Rogge, W.F., L.M. Hildemann, and M.A. Mazurek, "Sources of Fine Organic Aerosol: Noncatalyst and Catalyst Equipped Automobiles and HD Diesel Trucks," Environ. Sci. Technol. 27, 1993, pp. 636–651.
28. Ballschmiter, K., H. Buchert, R. Niemczyk, A. Munder, and M. Swerev, "Automobile Exhausts versus Municipal Waste Incineration as Sources of the Polychloro Dibenzodioxins (PCDD) and Furans (PCDF) Found in the Environment," Chemosphere 15, 7, 1986, pp. 901–915.
29. Jones, K.H., "Diesel Truck Emissions, an Unrecognized Source of PCDD/PCDF Exposure in United States," Risk Anal. 13, 3, 1993, pp. 245–252.
30. "Exposure and Human Health Reassessment of 2,3,7,8 Tetrachlorodibenzo p Dioxin (TCDD) and Related Compounds," Draft Document, U.S. Environmental Protection Agency, Washington, DC, September 2000, retrieved from www.epa.gov/ncea/pdfs/dioxin/part1and2.htm.
31. Gullett, B.K., and J.V. Ryan, "On Road Emissions of PCDDs and PCDFs from Heavy Duty Diesel Vehicles," Environ. Sci. Technol., 36, 2002, pp. 3036–3040.
32. Heeb, N.V., "Influence of Particulate Trap Systems on the Composition of Diesel Engine Exhaust Gas Emissions," EMPA Report No. 167985, Duebendorf, Switzerland, January 11, 1998.
33. Baumgard, K.J., and J.H. Johnson, "The Effect of Fuel and Engine Design on Diesel Exhaust Particle Size Distribution," SAE Paper No. 960131, SAE International, Warrendale, PA, 1996.
34. "Chemical Methods for the Measurement of Nonregulated Diesel Emissions," SAE Recommended Practice J1936, SAE International, Warrendale, PA, 1995.
35. Bagley, S.T., K.J. Baumgard, L.D. Gratz, J.J. Johnson, and D.G. Leddy, "Characterization of Fuel and After Treatment Device Effects on Diesel Emissions," Report No. 76, Health Effects Institute, Cambridge, MA, 1996, retrieved from http://healtheffects.org/Pubs/st76.htm.
36. Capaldo, K., and S. Pandis, "Lifetimes of Ultrafine Diesel Aerosol," Final Report E 43, Department of Chemical Engineering, Carnegie Mellon University, Pittsburgh, PA, December 1, 2001.
37. Abdul Khalek, I.S., and D.B. Kittelson, "The Influence of Dilution Conditions on Diesel Exhaust Particle Size Distribution Measurements," SAE Paper No. 1999 01 1142, SAE International, Warrendale, PA, 1999.
38. Hall, D.E., et al., "Measurement of the Number and Mass Weighted Size Distributions of Exhaust Particles Emitted from European Heavy Duty Engines," Report 01/51, CONCAWE, Brussels, Belgium, January 2001, retrieved from www.concawe.be/Download/Reports/Rpt_01 51.pdf.
39. Mayer, A., et al., "VERT: Diesel Nano Particulate Emissions: Properties and Reduction Strategies," SAE Paper No. 980539, SAE International, Warrendale, PA, 1998.

40. Kittelson, D.B., et al., "Characterization of Diesel Particles in the Atmosphere," AP 2 Project Group Final Report, Coordinating Research Council, Alpharetta, GA, 1998.
41. Wedekind, B., J. Andersson, D. Hall, R. Stradling, C. Barnes, and G. Wilson, "DETR/SMMT/CONCAWE Particle Research Program: Heavy Duty Results," SAE 2000 01 2851, SAE International, Warrendale, PA, 2000.
42. Andersson, J., and B. Wedekind, "DETR/SMMT/CONCAWE Particulate Research Programme," Summary Report DP01/0515, Ricardo Consulting Engineers, Shoreham-By-Sea, UK, May 2001, retrieved from www.ricardo.com/downloads/SummaryReport.pdf.
43. "ACEA Programme on Emissions of Fine Particles from Passenger Cars [2]," European Automobile Manufacturers Association, Brussels, Belgium, July 2002, retrieved from www.acea.be/ACEA/20020702PublicationPMReport.pdf.
44. Graskow, B.R., et al., "Exhaust Particulate Emissions from Two Port Fuel Injected Spark Ignition Engines," SAE Paper No. 1999 01 1144, SAE International, Warrendale, PA, 1999.
45. Greenwood, S., "Particulate Size Analysis," MIRA Automobile Abstracts, MIRA Automotive Information Centre, Warwickshire, UK, February 1996.
46. Graskow, B.R., et al., "Characterization of Exhaust Particulate Emissions from a Spark Ignition Engine," SAE Paper No. 980528, SAE International, Warrendale, PA, 1998.
47. Andrews, G.E., and F.M. Ahamed F.M., "The Composition of Spark Ignition Engine Steady State Particulate Emissions," SAE 1999 01 1143, SAE International, Warrendale, PA, 1999.
48. "A Study of the Number, Size and Mass of Exhaust Particles Emitted from European Diesel and Gasoline Vehicles under Steady State and European Driving Conditions," Report No. 98/51, CONCAWE, Brussels, Belgium, February 1998.
49. Graskow, B.R., et al., "Exhaust Particulate Emissions from a Direct Injection Spark Ignition Engine," SAE Paper No. 1999 01 1145, SAE International, Warrendale, PA, 1999.
50. Hall, D.E., et al., "A Study of the Size, Number and Mass Distribution of the Automotive Particulate Emissions from European Light Duty Vehicles," SAE Paper No. 982600, SAE International, Warrendale, PA, 1998.
51. Kazunari, K., et al., "Mixing Control Strategy for Engine Performance Improvement in a Gasoline Direct Injection Engine," SAE Paper No. 980158, SAE International, Warrendale, PA, 1998.
52. Hall, D.E., and C.J. Dickens, "Measurement of the Number and Size Distribution of Particles Emitted From a Gasoline Direct Injection Vehicle," SAE Paper No. 1999 01 3530, SAE International, Warrendale, PA, 1999.
53 Maricq, M.M., et al., "Particulate Emissions from a Direct Injection Spark Ignition (DISI) Engine," SAE Paper No. 1999 01 1530, SAE International, Warrendale, PA, 1999.
54. Eastlake, A., "The Latest Developments in Heavy Duty Vehicle Aftertreatment Testing for Real World Emissions and Fuel Economy," SAE Paper No. 1999 01 0470, SAE International, Warrendale, PA, 1999.
55. Ahlvik, P., L. Ntziachristos, J. Keskinen, and A. Virtanen, "Real Time Measurements of Diesel Particle Size Distribution with an Electrical Low Pressure Impactor," SAE Paper No. 980410, SAE International, Warrendale, PA, 1998.
56. Maricq, M.M., et al., "Vehicle Exhaust Particle Size Distributions: A Comparison of Tailpipe and Dilution Tunnel Measurements," SAE Paper No. 1999 01 1461, SAE International, Warrendale, PA, 1999.
57. Andrews, G.E., et al., "The Transient Storage and Blow Out of Diesel Particulate in Practical Exhaust Systems," SAE Paper No. 2001 01 0204, SAE International, Warrendale, PA, 2001.
58. Mayer, A., "VERT—Curtailing Emissions of Diesel Engines in Tunnel Sites," TTM Report W11/12/97, TTM, Niederrohrdorf, Switzerland, November 12, 1997.
59. Jing, L., et al., "Einfluss des Einspritzdrueckes auf die Partikelemission des DI Dieselfahrzeuges," Zeitshrift fuer das Messwesen, 3, 1996, pp. 7–10.
60. Pagan, J., "Study of Particle Size Distributions Emitted by a Diesel Engine," SAE Paper No. 1999 01 1141, SAE International, Warrendale, PA, 1999.
61. Kittelson, D.B., "Nanoparticles and Engines," presented at Technical Meeting on Mobile Source Aerosol Sampling and Measurement Techniques, EPA, Ann Arbor, MI, February 1998.

Chapter 8

Health and Environmental Effects

8.1 Overview

8.1.1 Toxic Compounds in Diesel Exhaust

The principal toxic gas compounds found in diesel exhaust include carbon monoxide (CO), nitric oxide (NO), nitrogen dioxide (NO_2), and sulfur dioxide (SO_2). The biological activity and toxic characteristics of these compounds have been studied for years and are relatively well understood.

In recent years, emission of diesel particulate matter (DPM) has become one of the major health concerns among all diesel emissions. Medical research on the adverse health effects of DPM is still in its initial phase. Many opinions are controversial and many questions have not been answered, awaiting the results of ongoing and future studies. These uncertainties as to the effects of DPM and its components are also reflected in the lack of a precise, universal definition of diesel particulates. Although practically all of the public health/engine emission regulations define DPM as a mix of solids, organics, and sulfates, such definitions as total carbon (i.e., excluding sulfates) or elemental carbon (i.e., excluding sulfates and organics) have been proposed and/or implemented by various occupational health regulations.

Diesel emissions contain numerous other compounds that are present in diesel emissions in smaller quantities than the above discussed emissions (i.e., DPM, CO, NO, NO_2 SO_2), but still may pose a health threat to humans. The most important substances in this group include polynuclear aromatic hydrocarbons (PAHs), nitro-PAHs, aldehydes, and selected other hydrocarbons and their derivatives. In their pure state, several of these species have been classified as human carcinogens (e.g., benzo[a]anthracene, benzo[a]pyrene, benzo[b]fluoranthene, benzo[k]fluoranthene, chrysene, dibenzo[a,h]anthracene, and indeno[1,2,3-c,d]pyrene). Even though their concentrations in diesel exhaust are orders of magnitude lower in comparison to the principal diesel pollutants, they are still seen as a potential serious health hazard. In the United States, the U.S. Environmental Protection Agency (EPA) included "polycyclic organic matter" (POM) in the list of urban hazardous air pollutants (HAPs) [1]. POM, defined as compounds with more than one benzene ring and a boiling point of 100°C and higher, includes practically all of the diesel PAH material.

Most of the heavy organic compounds, such as PAHs, are found in the particulate phase of diesel emissions. This association, in combination with their very low concentrations, makes it rather difficult to differentiate between the health effects of the solid DPM fraction and the particular organic species. A common approach is to study the effects of DPM as a whole. Diesel particulates, including both the solid and organic phase, have been identified as a toxic air contaminant in California [2].

Some health studies take an even more simplistic approach, investigating the effects of "whole diesel exhaust," which includes both gaseous pollutants and particulates. Diesel particulates are frequently used in these studies as an indicator of the diesel exhaust exposure, but no effort is made to analyze which components of the exhaust gases are responsible for particular health effects. From the perspective of diesel emission control, this approach is not practical. Blaming the entire diesel exhaust for adverse health effects is not useful in setting emission control targets or selecting suitable control technologies. After all, diesel exhaust gas is composed of more than 99% of nontoxic materials, including nitrogen, oxygen, water vapor, and carbon dioxide. A consensus in newer publications is that the particulate phase in diesel exhaust, including solid inorganic carbon and the associated organic material, has the greatest effect on health [3].

Major diesel emission components that have adverse health or environmental effects are listed in Table 8.1 [4]. A summary of the possible biological impacts and atmospheric reaction products (secondary pollutants) is also provided. This summary refers to biological activity of the chemical compounds in their pure state. The biological effects of particular compounds may or may not occur at the concentration levels that are found in diesel exhaust.

TABLE 8.1
BIOLOGICAL IMPACT OF DIESEL EMISSION COMPONENTS

Emission Component	Atmospheric Reaction Products	Biological Impact
Gas Phase		
Carbon monoxide	—	Highly toxic to humans; blocks oxygen uptake.
Nitrogen oxides	Nitric acid, ozone	Nitrogen dioxide is a respiratory tract irritant and major ozone precursor. Nitric acid contributes to acid rain.
Sulfur dioxide	Sulfuric acid	Respiratory tract irritation. Contributor to acid rain.
Carbon dioxide	—	Major contributor to global warming.
Saturated hydrocarbons (alkanes, $<C_{19}$)	Aldehydes, alkyl nitrates, ketones	Respiratory tract irritation. Reaction products are ozone precursors (in the presence of NO_x).
Unsaturated hydrocarbons (alkenes, $<C_5$)	Aldehydes, ketones	Respiratory tract irritation. Some alkenes are mutagenic and carcinogenic. Reaction products are ozone precursors (in the presence of NO_x).
Formaldehyde	Carbon monoxide, hydroperoxyl radicals	Formaldehyde is a probable human carcinogen and an ozone precursor (in the presence of NO_x).
Higher aldehydes (e.g., acrolein)	Peroxyacyl nitrates	Respiratory tract and eye irritation; causes plant damage.
Monocyclic aromatic compounds (e.g., benzene, toluene)	Hydroxylated and hydroxylated-nitro derivatives	Benzene is toxic and carcinogenic in humans. Some reaction products are mutagenic in bacteria (Ames assay).
PAHs (<5 rings) (e.g., phenanthrene, fluoroanthene)	Nitro-PAHs (<5 rings)	Some of these PAHs and nitro-PAHs are known mutagens and carcinogens.
Nitro-PAHs (2 and 3 rings) (e.g., nitronaphtalenes) [[AU: Should there be an "h" after the "t" in preceding word?]]	Quinones and hydroxylated-nitro derivatives	Some reaction products are mutagenic in bacteria (Ames assay).
Particulate Phase		
Elemental carbon	—	Nuclei adsorb organic compounds; size permits transport deep into the lungs (alveoli).
Inorganic sulfates	—	Respiratory tract irritation.
Aliphatic hydrocarbons (C_{14}–C_{35})	Little information; possibly aldehydes, ketones, and alkyl nitrates	Unknown.
PAHs (4 rings and more) [e.g., pyrene, benzo(a)pyrene]	Nitro-PAHs (4 rings and more), nitro-PAH lactones	Larger PAHs are major contributors of carcinogens in combustion emissions. Many nitro-PAHs are potent mutagens and carcinogens.
Nitro-PAHs (3 rings and more) (e.g., nitropyrenes)	Hydroxylated-nitro derivatives	Many nitro-PAHs are potent mutagens and carcinogens. Some reaction products are mutagenic in bacteria (Ames assay).

Source: Based on Ref. 4.

8.1.2 Inventories and Exposures

The overall air pollution effect in any given area is always caused by emission contributions from a number of sources. Besides internal combustion engines, important pollution contributors are the power generation industry, municipal waste incinerators, wood/coal stoves and open fires, as well as noncombustion industrial sources. Emissions from these sources, which are attributed to human activities, are referred to as *anthropogenic*. Air pollutants are also generated from a number of nonanthropogenic, natural sources. Examples include biological processes in soil and oceans, vegetation, animals, volcano eruptions, fires, or wind erosion.

To control pollution, one must know which of these sources of pollutants are significant contributors in a given area. Only if such information is available, can pollution be effectively prevented by applying adequate control measures to the emission sources that matter. Therefore, emission modeling and determination of emission inventories from various sources is routinely performed by government environmental regulatory agencies. Emission inventory studies are also undertaken by research organizations or academia.

Estimates of worldwide natural and anthropogenic emissions for selected air pollutants, including relative share of traffic pollution within the anthropogenic sources, are listed in Table 8.2.

For some pollutants, such as NO_x, human-made emissions present the biggest contribution to their total inventory. For other compounds, their anthropogenic sources are small in comparison to the natural release. Available data in the literature are not sufficient to provide similar balances for global release of all engine pollutants [5]. We do know, however, that based on inventories for anthropogenic sources internal combustion engines in mobile on- and off-road applications are significant contributors to the total human-made emission of NO_x, CO, HC, and PM. In today's industrialized countries, mobile sources contribute as much as 50% to the total inventory of NO_x, more than 70% of CO, as well as sizable portions of hydrocarbon (HC) and particular matter (PM) emissions. Within the mobile source category, diesels are responsible for about 70% of PM10 emissions. Because diesel particulates are characterized by their small particle size (almost entirely below 1 μm), contributions of diesels to PM2.5 inventories tend to be even higher. Mobile source NO_x emissions are split at about 50% between diesels and spark-ignited engines. The spark-ignited engines are responsible for most of the HC and CO emissions, with diesels emitting only about 10% and 5% of these pollutants, respectively [15].

Emission models that are used to generate inventory estimates are built on several assumptions as to vehicle populations, duty cycles, or dependence of emission factors on vehicle age, maintenance, or weather conditions. Because these assumptions are constantly verified and changed, emission inventories are never final. It is common to see quite significant changes in different editions of the same inventory or, even more so, differences between inventories generated by different organizations. Different local conditions (traffic density and structure, weather, etc.) will have a strong influence on local emission inventories. For example, even though most U.S. inventories attribute a higher proportion of PM pollution to diesels than to gasoline vehicles, a study conducted in the Denver, Colorado, area found that the contribution of gasoline engines to the PM2.5 pollution was almost three times higher than that of diesels [7].

The relative contribution to the global emission inventory from a given class of vehicles has an influence on the structure of respective emission standards. An exemplary illustration of such influence is the U.S. emission standard for railway locomotives, adopted in 1997 [8]. The EPA estimated that emissions from railway locomotives contributed 5.5% to nationwide, all-sources NO_x emissions. At the same time, the locomotive contribution to nationwide PM emissions was estimated at 0.1%. As a result, the primary target of the final emission standard was NO_x. The standard established very relaxed PM limits. In fact, the first two tiers of the PM standard were more relaxed than the existing locomotive emission levels at the time the standard was adopted.

The adverse health effects of a given substance depend on the toxicity of the particular compound and the local exposure conditions: concentration and time. Human exposure to diesel exhaust is the fundamental piece of information needed for such tasks as health hazard assessment, dose-response analysis, or during epidemiological diesel exhaust studies.

Diesel particulates are frequently used as a "tracer" that represents the exposure to diesel exhaust. DPM exposures in ambient air vary between about 0.2 and 4 μg/m^3, with the U.S. national average of 1.8 μg/m^3 [6]. Exposures at the workplace may be higher, but vary significantly between different occupational settings. For example, truck drivers experience exposures of 4 to 6 μg/m^3, while underground miners are exposed to DPM concentrations of 100 to 1700 μg/m^3 [9].

8.1.3 *Effects on Health and Environment*

Human health effects from exposure to toxic air pollutants can be divided into acute and chronic. Acute effects, caused by exposure to high concentration of pollutants, are more likely to occur in occupational environments. Chronic effects, which are associated with prolonged exposure to low concentrations of pollutants, are more typical for urban air pollution.

TABLE 8.2
WORLDWIDE EMISSIONS FROM NATURAL AND ANTHROPOGENIC SOURCES

Emission	Natural	Anthropogenic			
	Mt/yr	Total Mt/yr	Passenger Cars, %[b]	Trucks, %[b]	Other Traffic,[a] %[b]
Carbon dioxide, CO_2	770,000	28,000	5.5	6	6.5
Methane, CH_4	225	380	0.3	0.2	—
Nitrous oxide, N_2O	35	16	1.6	1.6	1.8
Nitrogen oxides, NO_x	80	110	9	11	10

Source: Data from Ref. 5.

a. Includes air, open sea, and other traffic.

b. Relative to total anthropogenic emission.

Acute health effects related to exposures to high concentrations of toxic gases are fairly well researched and understood. However, acute gas poisoning is rarely the case with diesel exhaust. It can happen only if the diesel engine is operated in an enclosed area with insufficient ventilation. Even then, the toxic effect of diesel exhaust is by far less than that of spark-ignited engines, be it a gasoline or a natural gas–powered unit, operated in similar conditions. This lower toxicity of diesel emissions is a result of the inherently low carbon monoxide levels in diesel exhaust. The low CO emission and safe handling of fuel are advantages of the diesel engine that make it the powerplant of choice for all confined space applications.

Diesel emissions receive more attention in the context of health effects of air pollution in large cities, both in North America and in Europe. There is evidence that many air pollutants, including ambient PM, cause various acute and chronic health effects, including increased morbidity, increased hospital admissions, or increased risk of heart and respiratory diseases.

Diesel particulate emissions have also been suspected to show carcinogenic properties, leading to an increased risk of lung cancer. Even though there seems to be a consensus that DPM exposure causes a small relative increase in the cancer risk, there is no agreement about the absolute risk factors. Animal studies have shown different reactions to diesel particulates by different animal species. There is no agreement on whether the animal studies can be extrapolated to humans or whether the limited existing human data are appropriate for estimating cancer risk factors [10]. Certainly, much more research is needed in this area.

In addition to their effects on human health, air pollutants can adversely affect the environment and the property values of people. Diesels, due to their high relative contribution to the total NO_x emission inventories, play an important role in the photochemical smog phenomenon that is plaguing all large metropolitan areas worldwide.

A positive environmental effect of the diesel engine is its good fuel efficiency and low "greenhouse gas" emission, including both CO_2 and N_2O. Global warming attributed to the increased combustion of fossil fuels has become a serious environmental concern. In December 1997, the Kyoto Protocol to the 1992 Climate Change Treaty was approved by a number of countries. According to the agreement, greenhouse gas emissions shall be reduced by 38 countries to the 1990 levels between 2008 and 2012. The average reduction amounts to about 5% (United States: 7%; EU: 8%; Japan: 6%). Increasing the proportion of diesel engines is a simple and effective means of reducing CO_2 emissions. The greenhouse gas benefit of diesels, however, may be offset by the climate effects of soot (black carbon) emission, which is a suspected nongreenhouse gas global warming agent [11].

Other environmental benefits of the diesel engine include low HC emissions, including tailpipe, cold start, and evaporative emissions. Hydrocarbons (or volatile organic compounds, VOCs) are another precursor of photochemical smog and ground-level ozone.

8.1.4 Summary

Diesel engines have a number of environmental benefits and drawbacks relative to the spark-ignited (SI) engine. An environmental "balance sheet" of the diesel engine is shown in Table 8.3.

The way in which diesel engines are regulated in different countries reflects the relative importance of these benefits and drawbacks, as weighted by local regulatory authorities. However, in view of the increasing evidence on the adverse health effects of particulate matter and the worldwide air quality problems, future diesel engines will have to adopt measures for much more effective control of the DPM and NO_x emissions.

8.2 Diesel Emission Inventory

8.2.1 Diesel Engines and Air Pollution

Today's emissions from mobile sources, including cars, trucks, and off-road vehicles, are estimated to be responsible for about 50% of nationwide NO_x emissions, about 40% to 50% of HC (or VOC) emissions, and as much as 80% to 90% of CO emissions in urban areas. Similar contributions are seen in the United States and in other developed countries, including Europe and Japan. A large portion of these emissions comes from gasoline-fueled cars and light trucks. However, the relative importance of NO_x and PM emissions from diesel engines has been increasing. This increase has been caused by (1) the introduction of three-way catalyst technology for SI engines in the 1980s, which lowered gasoline NO_x emissions, and (2) increasing trends in diesel vehicle population and diesel fuel usage.

The passenger car market in Western Europe has experience an increased penetration of diesel-fueled vehicles. With the exception of Sweden, where diesels almost disappeared, sales of diesel cars in most countries show an increasing trend. As illustrated in Table 8.4, the proportion of diesel cars sold in 2002 in Austria, Belgium, and France exceeded 60% of all new automobiles in these countries [12, 13].

TABLE 8.3
ENVIRONMENTAL/HEALTH IMPACT OF DIESEL ENGINES

Benefits	Drawbacks
Good fuel economy/high efficiency Low "greenhouse gas" emissions, both CO_2 and N_2O Low hydrocarbon emissions Low carbon monoxide emissions Low cold start and evaporative emissions Safe handling of fuel (relative to gasoline, natural gas, LPG)	High PM mass emissions (relative to SI) High nitrogen oxides (NO_x) emissions (relative to three-way catalyst-equipped SI engines)

TABLE 8.4
MARKET PENETRATION OF NEW DIESEL CARS IN EUROPE, PERCENT OF NEW CARS SOLD

	1980	1985	1990	1995	2002[a]
Austria	3.3	13.7	25.7	42.6	69.0
Belgium	13.8	26.4	32.9	46.8	63.5
France	9.9	15.0	33.0	46.5	62.0
Germany	8.0	22.1	10.8	14.6	37.0
Italy	8.1	25.1	7.3	10.3	42.0
Sweden	6.0	2.2	0.6	2.8	6.7
United Kingdom	0.4	3.6	6.4	20.2	22.5
Europe (weighted average)	7.2	15.6	13.9	22.1	39.3

a. New registrations for January–June 2002.

Even though diesel passenger cars are not used very much in the United States (market penetration 0.03% in 1994), the consumption of diesel fuel relative to the total fuel consumption by all highway vehicles has been steadily increasing. The same increasing trend has been noted in Japan (Figure 8.1 [12]).

In the past, when diesel fuels contained high levels of sulfur, diesel engines were also important contributors to the sulfur dioxide (SO_2) emission inventory. Lowering the fuel sulfur level from about 0.3–0.5% to 0.05%, which occurred in the mid-1990s in the United States and in Europe, was an effective means of controlling SO_2 emissions from diesel-fueled engines. In 1999, the contribution of highway diesel engines to the nationwide SO_2 inventory in the United States was only 0.6%. However, the contribution of nonroad engines, which still used high sulfur diesel fuel, amounted to 2.7% [14]. The relative importance of diesels in SO_2 pollution may be higher in countries that still use diesel fuels that have a high sulfur content.

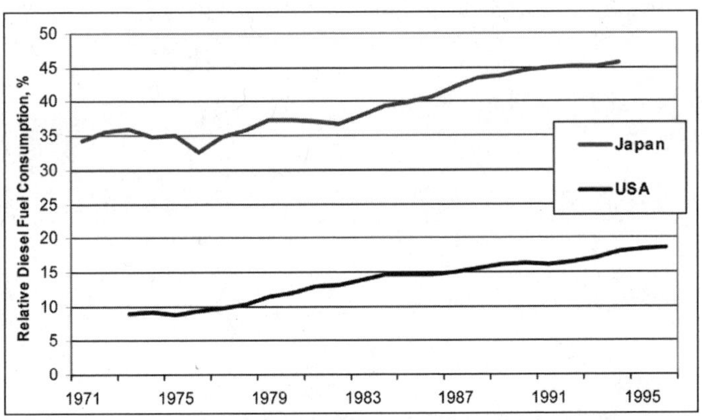

Figure 8.1 *Diesel fuel consumption in the U.S. and Japan as a fraction of total highway fuel consumption. (Data from Ref. 12.)*

8.2.2 United States

8.2.2.1 Regulated Diesel Emissions

National emission inventories are computed in the United States by the U.S. Environmental Protection Agency. They are published in the periodically updated "Emission Trends Document" [14, 15]. These inventories cover a number of criteria pollutants, which are not always identical with the regulated diesel emissions. Among regulated diesel emissions, CO and NO_x are also designated as ambient air pollutants. Emissions of diesel PM are part of the inventory of either PM10 (ambient particulates of diameter less than 10 μm in diameter) or PM2.5 (diameter less than 2.5 μm). Because diesel particulate sizes are typically less than 1 μm, the entire diesel PM emission should be a part of both the PM10 and PM2.5 inventories, with the relative proportion of diesels somewhat higher in the PM2.5 metric. Hydrocarbons, the other regulated diesel emission, can be considered a part of the VOC inventory. The relative contribution of diesel engines to the nationwide emission inventory is shown in Figures 8.2 through 8.5. According to the data in these figures, as well as to several other inventories, diesel engines are significant contributors to the nationwide NO_x and PM emissions, whereas their emissions of CO and HCs (VOCs) are low in comparison to those from spark-ignited engines. The NO_x and PM emission inventories, grouped by engine classes, are shown in Tables 8.5 and 8.6. Emissions are expressed in short tons, where 1 short ton = 2000 lb = 907.18 kg. The term *gas* refers to spark-ignited (gasoline-fueled) engines.

Diesel engines, including those used in the railroad industry, contribute more than 50% of NO_x emitted from all mobile sources. A high proportion of these emissions comes from nonroad equipment. Among highway vehicles, diesels contribute about 17% of total mobile source NO_x emissions, while 44% is contributed by gasoline-fueled highway engines.

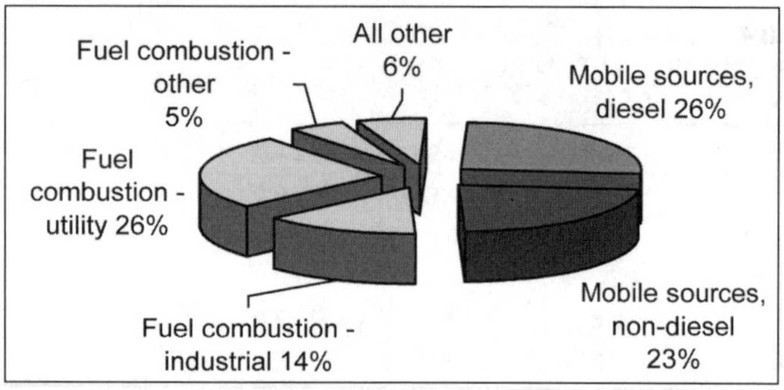

Figure 8.2 *U.S. NO_x emission inventory in 1997. "Mobile sources, diesel" includes railway locomotives and marine vessels.*

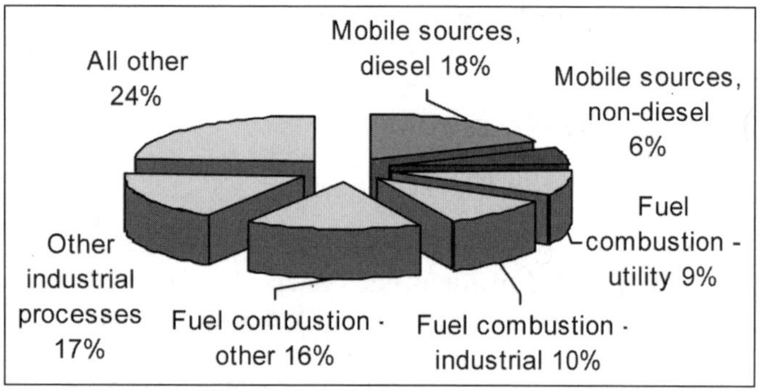

Figure 8.3 *U.S. PM10 emission inventory in 1997. PM10 data include exhaust, brake, and tire wear emissions. "Mobile sources, diesel" includes railway locomotives and marine vessels.*

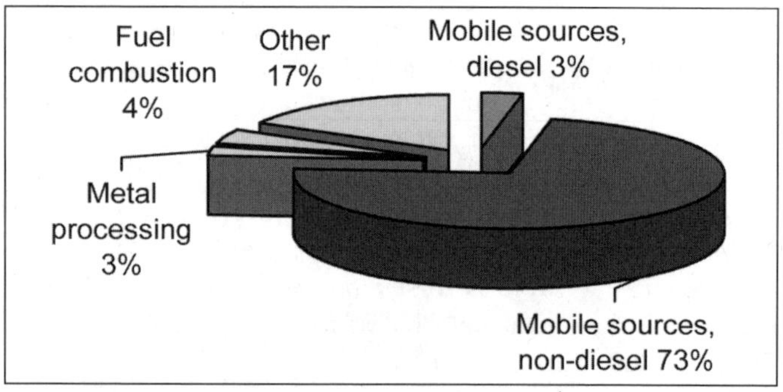

Figure 8.4 *U.S. CO emission inventory in 1997. "Mobile sources, diesel" includes railway locomotives and marine vessels.*

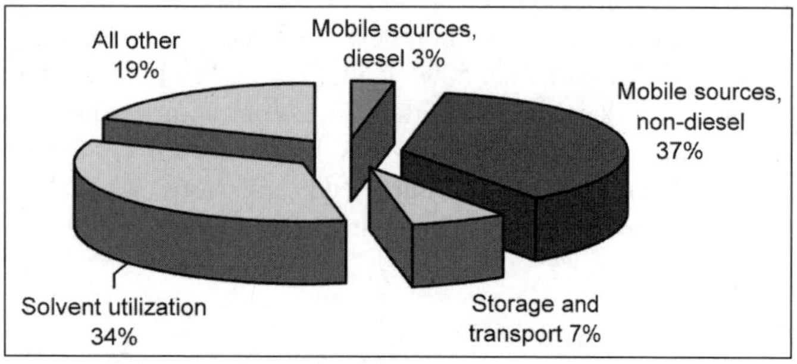

Figure 8.5 *U.S. VOC emission inventory in 1997. "Mobile sources, diesel" includes railway locomotives and marine vessels.*

TABLE 8.5
U.S. NITROGEN OXIDES EMISSION INVENTORY, 1997 (THOUSAND SHORT TONS PER YEAR)

Category	NO_x Emissions	% of Total	% of Mobile
Total, all sources	23,582	100	—
Total, mobile sources[a]	11,595	49	100
• Gas[b]	5,313	22.5	45.8
• Diesel[b]	4,919	20.9	42.4
• Railroad[c]	949	4.0	8.2
• Marine vessels[c]	235	1.0	2.0
On-road vehicles	7,035	30	61
• Gas	5102	21.6	44.0
• Diesel	1932	8.2	16.7
Nonroad vehicles[d]	4,560	19	39
• Gas[b]	211	0.9	1.8
• Diesel[b]	2,987	12.7	25.8

a. Also includes aircraft.
b. Excluding marine vessels, railroad, and aircraft.
c. All fuels.
d. Also includes marine vessels, railroad, and aircraft.

TABLE 8.6
U.S. PM10 EMISSION INVENTORY, 1997 (THOUSAND SHORT TONS PER YEAR)

Category	PM10 Emissions	% of Total	% of Mobile
Total, all sources[a]	3,112	100	—
Total, mobile sources[b]	734	24	100
• Gas[c]	156	5.0	21.3
• Diesel[c]	479	15.4	65.3
• Railroad[d]	27	0.9	3.7
• Marine vessels[d]	31	1.0	4.2
On-road vehicles	268	9	37
• Gas	105	3.4	14.3
• Diesel	163	5.2	22.2
Nonroad vehicles[e]	466	15	64
• Gas[c]	51	1.6	6.9
• Diesel[c]	316	10.2	43.1

a. "Controllable" PM10 emissions, not including natural sources, fugitive dusts, etc. Total PM10 emissions including natural sources amounted to 33,581 thousand tons.
b. Also includes aircraft. Mobile source inventories include exhaust, brake, and tire wear emissions.
c. Excluding marine vessels, railroad, and aircraft.
d. All fuels.
e. Also includes marine vessels, railroad, and aircraft.

Diesels are responsible for an even higher proportion of PM emissions (Table 8.6). In total, including railroad and marine vessels, they contribute more than 70% of the total mobile sources PM10. Nonroad diesel engines have again a much higher contribution than their highway counterparts. PM emission from highway diesel engines is responsible for about 22% of the total mobile source PM10, compared with 14%, which can be attributed to spark-ignited highway engines.

In contrast, diesel emissions are relatively insignificant in the overall inventory of VOCs, including hydrocarbons and derivatives, and carbon monoxide (see Figures 8.4 and 8.5). These data are shown in more detail in Tables 8.7 and 8.8. Both the VOC and CO emissions from mobile sources can be attributed almost entirely to spark-ignited engines. Diesel engines are responsible for about 10% of mobile source VOCs and for less than 5% of total CO from all mobile sources.

All particulate matter inventories just discussed refer to directly emitted diesel PM. In addition to the directly emitted PM, secondary particulate pollution is generated in the atmosphere through the formation of nitrates, such as ammonium nitrate, from NO_X. The EPA estimates that 1 kg of secondary nitrate PM is generated from approximately 25 kg of original NO_X emissions. For instance, based on the estimated 4,560,000 tons of NO_X emitted by nonroad vehicles (mostly by diesels) in 1997, these engines contributed about 180,000 tons of secondary nitrate PM emission in addition to their 466,000 tons of original PM10 (Tables 8.5 and 8.6).

8.2.2.2 Nonregulated Emissions

Sulfur dioxide is a nonregulated diesel emission, but an important regulated ambient air pollutant. The U.S. inventory for SO_2 is shown in Figure 8.6 and given in absolute numbers in Table 8.9 [14].

At the time of the inventory (1999), the maximum sulfur level was specified at 500 ppm wt. for highway diesel fuels and 0.5% (5000 ppm) for nonroad fuels. The high sulfur level in nonroad fuels is reflected in the higher contribution of nonroad diesels to the SO_2 inventory, relative to highway engines. However, with the total contribution of diesel vehicles at only 3.3%, and the total contribution of all vehicle categories at 6.9%, mobile sources in general are relatively insignificant contributors to the nationwide SO_2 inventory. Most certainly, lowering sulfur levels in diesel fuels below 500 ppm cannot be justified by the global, environmental SO_2 reduction benefit. Indeed, further reductions in diesel sulfur are being introduced primarily to enable sulfur-intolerant emission control technologies, which, in turn, may bring further reductions of NO_X and PM emissions.

TABLE 8.7
U.S. VOC EMISSION INVENTORY, 1997 (THOUSAND SHORT TONS PER YEAR)

Category	VOC Emissions	% of Total	% of Mobile
Total, all sources	19,214	100	—
Total, mobile sources[a]	7,660	40	100
• Gas[b]	6,702	34.9	87.5
• Diesel[b]	672	3.5	8.8
• Railroad[c]	50	0.3	0.7
• Marine vessels[d,c]	50	0.3	0.7
On-road vehicles	5,230	27	68
• Gas	4,991	26.0	65.2
• Diesel	239	1.2	3.1
Nonroad vehicles[d]	2,430	13	32
• Gas[b]	1,711	8.9	22.3
• Diesel[b]	433	2.3	5.7

a. Also includes aircraft.
b. Excluding marine vessels, railroad, and aircraft.
c. All fuels.
d. Also includes marine vessels, railroad, and aircraft.

TABLE 8.8
U.S. CARBON MONOXIDE EMISSION INVENTORY, 1997 (THOUSAND SHORT TONS PER YEAR)

Category	CO Emissions	% of Total	% of Mobile
Total, all sources	87,451	100	—
Total, mobile sources[a]	67,012	77	100
• Gas[b]	62,991	72.0	94.0
• Diesel[b]	2,809	3.2	4.2
• Railroad[c]	115	0.1	0.2
• Marine vessels[d,c]	85	0.1	0.1
On-road vehicles	50,257	57	75
• Gas	48,749	55.7	72.7
• Diesel	1,508	1.7	2.3
Nonroad vehicles[d]	16,755	19	25
• Gas[b]	14,242	16.3	21.3
• Diesel[b]	1,301	1.5	1.9

a. Also includes aircraft.
b. Excluding marine vessels, railroad, and aircraft.
c. All fuels.
d. Also includes marine vessels, railroad, and aircraft.

8.2.2.3 Emission Trends

The contribution of diesel engines to the nationwide emission inventory increased quickly in the 1970s and 1980s, before the introduction of the first diesel emission standards in the United States. After such standards were introduced, the contribution of diesels declined and stabilized, both as absolute numbers and as a percentage of the total mobile sources of emissions. These emission trends are illustrated in Figures 8.7 and 8.8 (1997 data, railway and marine vessels not included in diesel emissions [15]).

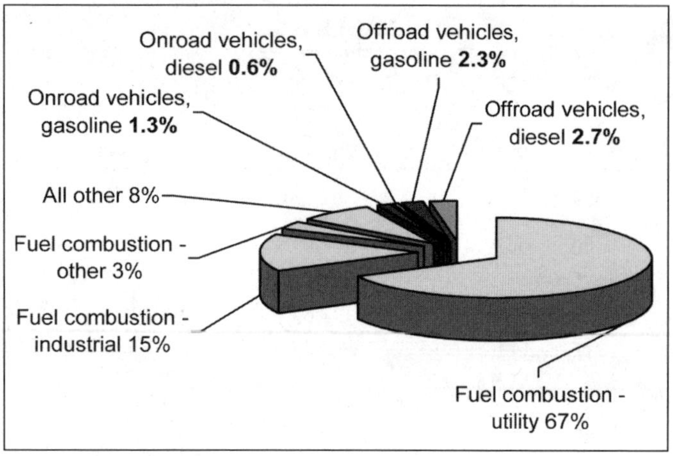

Figure 8.6 U.S. SO_2 emission inventory in 1999.

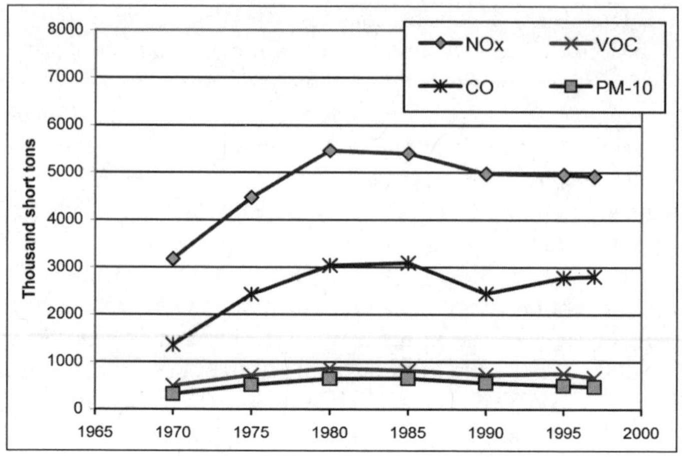

Figure 8.7 Diesel emission trends in the United States.

TABLE 8.9
U.S. SULFUR DIOXIDE EMISSION INVENTORY, 1999
(THOUSAND SHORT TONS PER YEAR[a])

Category	SO_2 Emissions	% of Total	% of Mobile
Total, all sources	18,867	100	—
• Fuel combustion, utility	12,698	67.3	—
• Fuel combustion, industrial	2,805	14.9	—
• Fuel combustion, other	588	3.1	—
• All other	1,477	7.8	—
Total, mobile sources	1,299	6.9	100
• Gasoline	674	3.6	51.8
• Diesel	625	3.3	48.2
On-road vehicles	363	1.9	27.9
• Gasoline	245	1.3	18.8
• Diesel	118	0.6	9.1
Nonroad vehicles	936	5.0	72.1
• Gasoline	429	2.3	33.0
• Diesel	507	2.7	39.1

a. 1 short ton = 907.18 kg.

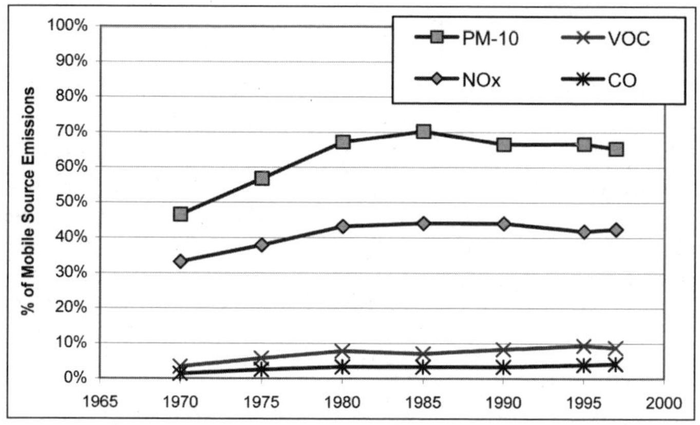

Figure 8.8 Diesel emissions as a percentage of the total mobile source emissions.

8.2.2.4 Local Inventories

In California, highway diesels are believed to be the most polluting engine category. Highway diesel vehicles were estimated to generate 59% of the total statewide diesel PM emissions, while mobile nonroad diesel engines contributed 36%, and stationary diesel engines 5%. A detailed inventory of California diesel PM emissions is provided in Table 8.10 [2]. The data in the table do not include PM emissions generated by gasoline engines or other sources. The total diesel exhaust PM emission amount was estimated to be 56% of the total PM emission generated by combustion of fuels in California.

Contrary to the California estimates given in Table 8.10, not all sources confirm the high proportion of diesel PM in the overall PM inventories. In 1998, a large air quality study, known as the Northern Front Range Air Quality Study, was completed by researchers at Colorado State University [7]. Emission inventories for locations in the Denver, Colorado, area showed that PM2.5 emissions from diesel engines produced only 10% of the total (human-made + natural) PM2.5, whereas gasoline vehicles contributed 29%. It was found that the majority of gasoline PM emissions was caused by "smoky," poorly maintained vehicles (about 2.5% of the fleet) and during cold starts. PM emissions from well-maintained gasoline engines operating in hot, stabilized mode amounted only to 3% of the total PM2.5. The low proportion of diesel PM emissions was also caused by the small population of diesels, estimated at 4% to 5% of vehicles on the road.

8.2.3 European Union

Emission inventories in European countries are reported in source categories that do not always allow us to separate diesel from gasoline emissions. Tables 8.11, 8.12, and 8.13,

TABLE 8.10
INVENTORY OF DIESEL PM EMISSIONS IN CALIFORNIA, TONS/YEAR

	1990	1995
Onroad Diesel Vehicles		
Heavy-duty trucks and nonurban buses	36,340	19,610
Light-duty passenger cars	1,590	720
Urban buses	1,270	860
Light/medium-duty trucks	430	230
Nonroad Mobile Diesel Engines		
Nonfarm equipment	7,690	7,790
Heavy-duty farm equipment	1,970	1,990
Ships	1,030	1,080
Trains	1,250	1,300
Recreational/commercial boats	620	660
Mobile refrigeration units	280	270
Stationary Area Sources		
Oil and gas production	110	100
Industrial	610	680
Stationary Point Sources		
Industrial	440	470
Electric utilities	230	260
Commercial/institutional	230	250
TOTAL	54,090	36,270

Source: Based on Ref. 2.

TABLE 8.11
U.K. NO_x AND PM10 EMISSION INVENTORY, 1997

Category	NO_x		PM10	
	kton/yr	%	kton/yr	%
Total[a]	1835	100	184	100
Road transport	883	48	43[b]	23
• Gasoline	485	26	10	5
• Diesel	398	22	33	18
Nonroad[c]	74	4	2	1

Note: NO_x reported as NO_2.
a. All anthropogenic sources.
b. Includes tire and brake wear.
c. Excluding marine vessels, railroad, aircraft, and military.

TABLE 8.12
U.K. CO AND VOC EMISSION INVENTORY, 1997

Category	CO		VOC	
	kton/yr	%	kton/yr	%
Total[a]	5090	100	2130	100
Road transport	3797	74	496	23
• Gasoline	3635	71	411	19
• Diesel	162	3	85	4
Nonroad[b]	315	6	40	2

Note: VOC category excludes methane.
a. All anthropogenic sources.
b. Excluding marine vessels, railroad, aircraft, and military.

TABLE 8.13
U.K. GREENHOUSE GAS INVENTORY (CO_2, N_2O), 1997

Category	CO_2		N_2O	
	kton/yr	%	kton/yr	%
Total[a]	147,700	100	192.1	100
Road transport	31,730	21	10.9	6
• Gasoline	18,900	13	8.9	5
• Diesel	12,830	9	2.0	1
Nonroad[b]	1,500	1	1.2	1

a. All anthropogenic sources.
b. Excluding marine vessels, railroad, aircraft, and military.

which show emission inventories in the United Kingdom, list respective contributions of diesels and gasoline engines within the road transport category. However, the share of diesels is not separated from that of gasoline in the nonroad category. All data are been taken from the official U.K. emission inventory database for the year 1997 [16].

The relative contribution of mobile sources to the global emission inventory in the United Kingdom is very similar to that in the United States. In each country, mobile sources represent about 50% of total NO_x and more than 75% of CO. The split between diesel and gasoline engines within mobile sources is also similar. The biggest difference can be found in the relative contribution of nonroad diesel engines. In the United States, nonroad diesels are estimated to produce more NO_x and PM emissions than highway diesel engines. In the United Kingdom, the contributions from nonroad engines are small. It is not clear whether these numbers reflect differences in the structure and usage of engines or result from different methodologies used to compute the inventories.

8.3 Exposure to Diesel Exhaust

8.3.1 Introduction

Diesel exposure information is of vital importance for diesel epidemiological studies or for diesel risk analyses. However, because diesel emissions constitute only a small portion of the total air pollutant mix, accurate assessment of exposure to diesel exhaust or its components is very difficult. Combustion aerosols from diesel and spark-ignited engines are remarkable similar in many ways; once they are mixed, it is very difficult to determine the source of the exposure. Combustion of other materials, such as liquid and gas fuels, coal, or tobacco, as well as industrial emissions, produces many of the same chemical components that are present in diesel exhaust. In addition, there are several natural sources, both combustion and noncombustion, of respirable particles that are found in the ambient air. Further complications are caused by atmospheric transport and transformation processes that may change the original chemical composition and physical properties of diesel emissions, creating new compounds of different toxic properties.

People spend most of their time indoors. Therefore, accurate estimates of average exposures require a consideration of the amount of time spent in different environments and the respective concentrations of pollutants. Because exposure levels may change over time, they are often averaged and presented as *time-weighted exposures*. Time-weighted averaging provides a satisfactory representation for human exposure, but it may not account for short-duration exposure spikes or "burst modes" of exposure. The burst mode of exposure, which may happen, for example, in the case of failure of a ventilation hood, spill container, or enclosure, is very important because in such a failure the concentrations of pollutants can increase by orders of magnitude.

From the historical perspective, changes in diesel engine and fuel technologies also influence human exposures and the toxic properties of diesel emissions. New engines not only produce fewer pollutants, but also may produce pollutants of different chemical and/or physical properties. For example, the particulate emissions from different engines may have a different size distribution or different proportion of metallic ash from lube oil additives. In engines equipped with particulate filters, PM emissions are almost exclusively composed of volatile material, mostly SOF. If oxidation catalysts are used, the NO_x emissions may have increased the proportion of nitrogen dioxide. It is important that models which estimate future exposure levels properly address the changing character of emissions from new diesel engines.

The following sections discuss typical ranges of exposure to diesel emissions for both ambient (environmental) and occupational settings. We should emphasize that the human health response is not necessarily directly related to exposures. Rather, it is dependent on the dose to a particular target organ, for example, to the lungs. A discussion of exposure and dose issues that puts exposures in perspective can be found in the literature [17].

8.3.2 Exposure Estimation

The following three groups of methods have been developed to estimate diesel emission exposures [2]:

- *Tracers:* In comparison with other combustion emissions, diesel emissions contain a higher proportion of elemental carbon and fine particles. These constituents are frequently used as a marker of exposure (tracer) or as a surrogate index for the diesel particulate matter. These methods are commonly used in occupational health air quality surveys. Other DPM tracers that have been used include particle-associated diesel fuel additives, PAHs and PAHs ratios, and lubricating oil combustion products. Known quantities of artificial tracers that are not normally found in the atmosphere can be also added to the fuel. An example of such a tracer is the rare earth element dysprosium, which was used in a study to quantify diesel exposure in Vienna, Austria [18].
- *Modeled exhaust emissions:* Emission exposures can be estimated by mathematical modeling based on assumed traffic intensity, vehicle types, and their emission factors. Researchers at Volkswagen conducted such analyses and estimated PM concentrations in urban street canyons and near U.S. expressways [19]. Several advanced emission models are being developed by the EPA. Exhaust emission modeling can be used to create estimates of past, current, and future exposure levels.
- *Source apportionment:* Source apportionment is a technique for identifying the sources and source emissions that contribute to the total ambient concentrations of a given pollutant. Such an apportionment is usually based on PM10 or elemental carbon concentrations measured in a given area during ambient air quality surveys. Then, a chemical balance model is applied that matches the chemical profiles of ambient particulate samples with known chemical profiles of particulates from specific types of sources. This method requires that adequate emissions inventory and ambient particulate data be available for the specific area under investigation. Source apportionment can be used for future DPM exposure estimates based on future emission inventories. The source apportionment method was used by the California ARB for exposure assessment in the identification of DPM as a toxic air contaminant [2].

8.3.3 Diesel Particulate Matter

8.3.3.1 Ambient Exposure

Ambient (nonoccupational) exposure to diesel exhaust occurs worldwide, primarily in urban areas. The level of exposure depends on geographic areas, on the number and type of diesel engines in the area, and on atmospheric patterns of emissions dispersal. The highest exposure levels recorded in heavily populated urban areas that have a high usage of diesel vehicles usually reach about 10 $\mu g/m^3$, but exposures in excess of 20 $\mu g/m^3$ have been reported. Ambient exposures in areas of low pollution range from less than 1 to 2 $\mu g/m^3$. The following example data provide an overview of the ranges of ambient DPM exposure.

A study conducted in Vienna, Austria, a city with a large number of diesel-fueled passenger cars, found that 12% to 33% of the total suspended particulates was DPM [18]. The DPM was relatively well mixed in the atmosphere, with few differences between commercial and residential areas. DPM concentrations during the study varied from 5 to 23 μg/m^3, with a typical concentration about 11 μg/m^3.

In a mathematical simulation by Volkswagen, particulate matter exposure from diesel and gasoline light-duty vehicles was estimated. In a street canyon environment, the estimated exposure was 3.9 to 8.8 μg/m^3. For a U.S. expressway, the concentrations ranged from 7.14 μg/m^3 at 4 m from the side of the road to 2.57 μg/m^3 at 100 m from the road [19].

The EPA estimated the DPM exposures using two methods [20]. According to one of the methods, a model based on fleet average emission factors, the total concentration of particulates from diesel exhaust in ambient air in urban areas for 1995 was 2 μg/m^3, the concentration in rural areas was 0.6 μg/m^3, and the nationwide average concentration was 1.1 μg/m^3. In the second method, using ambient monitoring data, total suspended particulate matter for 1990 was determined to be 48 μg/m^3. The assumption was made that 5% of that number is associated with diesel exhaust. Multiplication of the total by 5%, and adjusting for time spent indoors, resulted in an integrated estimate of 1.5 μg/m^3. Estimated exposure levels until the year 2010 (method 1) are listed in Table 8.14.

A California ARB analysis shows the average ambient outdoor DPM to range from 0.2 to 3.6 μg/m^3 across 14 California air basins with a population weighted average of 3.2 μg/m^3 [2]. The same source gives winter period DPM estimates in three California locations of 4 to 22 μg/m^3.

8.3.3.2 Occupational Exposure

Occupational exposures to diesel exhaust and diesel particulates tend to be higher than ambient exposures. They range from the ambient levels (4 μg/m^3) to as high as 1700 μg/m^3 among different occupational groups. Miners in underground dieselized mines experience by far the highest exposures. Other occupational groups exposed to diesel emissions include forklift truck operators, truck drivers, railroad workers, dock workers, bus garage mechanics, fire station workers, and more. According to the National Institute for Occupational Safety and Health (NIOSH), 1.35 million people have been exposed to diesel emissions in the workplace in the United States [21]. Table 8.15 shows approximate ranges of DPM exposure (8-h average) among different occupational groups, as reported by several surveys [2, 9].

Example DPM exposures in underground mining are illustrated in Figure 8.9, which compares samples collected in a number of Canadian mines [22]. The exposures vary from less than 0.05 mg/m^3 to more than 1.5 mg/m^3, with an average exposure of 0.29 mg/m^3. This variation reflects differences in exposure levels at different work places and, to a degree, differences among mines. It is believed that Canadian mines, relative to the worldwide average, exhibit a high level of mining technology, equipment maintenance, and occupational health

TABLE 8.14
ESTIMATED ANNUAL AMBIENT DPM CONCENTRATIONS IN THE UNITED STATES, μg/m^3

Year	Rural	Urban	U.S. National	California
1990	1.1	2.0	1.8	3.2
1995	0.6	1.2	1.1	2.1
2000	0.4	0.7	0.6	2.2
2010	0.2	0.4	0.4	2.3

Source: After Ref. 2; California data are used.

TABLE 8.15
OCCUPATIONAL EXPOSURE TO DIESEL PARTICULATES, μg/m^3

Group	Exposure Level
Underground miners	100–1700
Urban fire stations	100–480
Forklift operators, dock workers, railroad workers	20–100
Truck drivers	4–6

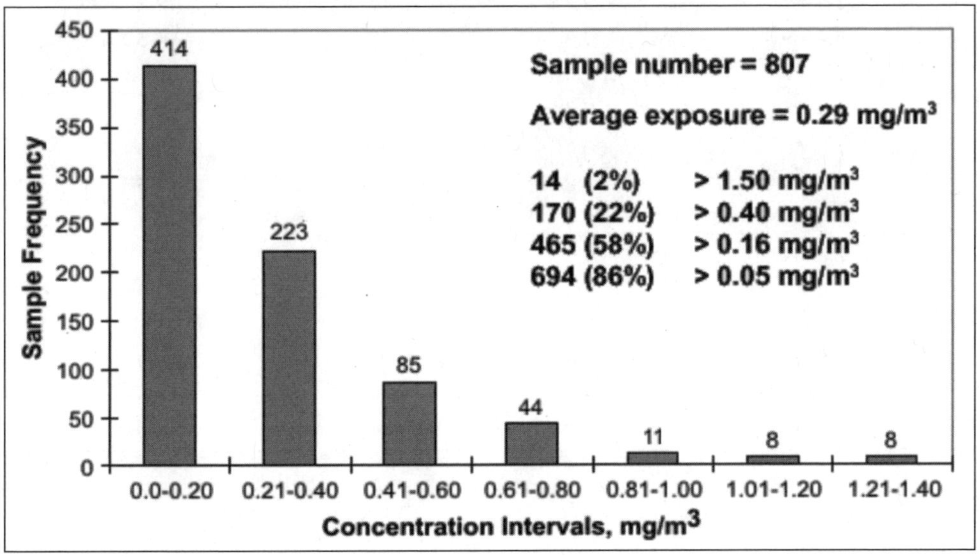

Figure 8.9 Diesel particulate matter exposure in Canadian mines. RCD method, CANMET Sudbury Laboratory, January–December 1998. (Modified from Ref. 22.)

measures; higher exposures are likely to be encountered in less advanced mining operations.

8.3.4 Polynuclear Aromatic Hydrocarbons

In a study conducted in the Baltimore Harbor Tunnel, particulate- and vapor-phase samples were collected from the passing traffic and analyzed for PAHs and related compounds [23]. No breakdown of traffic into numbers of diesel- and gasoline-powered vehicles was given. Table 8.16 lists the range of particle-phase PAH concentrations and the mean particle- and vapor-phase PAH concentrations for 48 samples collected in the tunnel. The last column in the table specifies the ratios of mean particle-phase PAH concentrations to that of benzo[e]pyrene (B[e]P), which is considered to be a nonreactive PAH.

Alkyl-substituted phenanthrenes in the tunnel samples had relatively high concentrations compared with those of not substituted phenanthrene. Because DPM extracts were previously shown to have high relative concentrations of methyl and dimethylphenanthrenes [24], it is believed that these compounds were emitted by diesel trucks in the tunnel. Several of the higher molecular weight PAHs may be associated with the contribution of gasoline-fueled emissions.

Table 8.17 presents some of the results of ambient air sampling for PAHs conducted during a summertime photochemical air pollution episode in Glendora, California [25]. The sampling was conducted at a site affected by motor vehicle emissions (less than 1 km from the heavily traveled freeway I-210 and downwind of Los Angeles). Data in the table were averaged from three daytime and three nighttime samples.

TABLE 8.16
PAH CONCENTRATIONS IN BALTIMORE HARBOR TUNNEL

Compound	Concentration, ng/m^3			
	Range, Particles	Mean, Particles	Mean, Vapor	Ratio to B[e]P[a]
Phenanthrene	4.3–56	18.0	132	4.3
Anthracene	0.6–12	2.9	18	0.6
3-Methylphenanthrene	3.9–58	13.9	70	3.3
2-Methylphenanthrene	5.3–74	19.0	—	4.6
2-Methylanthracene	0.6–12	3.0	5.3	0.7
9- and 4-Methyl-phenanthrene and 4H-cyclopenta[def]-phenanthrene	4.7–50	12.9	71	3.0
1-Methylphenanthrene	2.6–43	9.8	43	2.3
2,6-Dimethylphenanthrene	4.7–62	14.0	30	3.4
2,7-Dimethylphenanthrene	3.4–38	9.2	16	2.2
1,3-, 2,10-, 3,9-, and 3,10-Dimethyl- and phenanthrene	9.5–119	26.0	61	6.3
1,6- and 2,9-Dimethylphenanthrene	4.5–63	14.0	27	3.3
1,7-Dimethylphenanthrene	3.9–41	10.2	20	2.4
2,3-Dimethylphenanthrene	3.5–41	9.3	16	2.2
Fluoranthene	6.4–69	20.0	16	4.5
Pyrene	9.7–76	27.0	26	6.3
Benzo[ghi]fluoranthene	3.2–26	9.6	ND[b]	2.1
Cyclopenta[cd]pyrene	7.6–65	20.0	ND	4.6
Benz[a]anthracene	1.9–29	7.6	ND	1.5
Chrysene/triphenylene	2.9–47	12.0	ND	2.4
Benzofluoranthenes[b,j,+k]	2.2–44	10.6	ND	2.1
Benzo[e]pyrene	1.5–19	5.0	ND	1.0
Benzo[a]pyrene	1.3–26	5.8	ND	1.1
Indeno[1,2,3-cd]pyrene	0.3–15	4.6	ND	0.9
Benzo[ghi]perylene	1.8–18	8.0	ND	1.6
Coronene	1.0–10	4.7	ND	0.9

Source: Data after Ref. 23.

a. Mean ratios (particulate) to particulate phase B[e]P.

b. None detected.

TABLE 8.17
AVERAGE AMBIENT CONCENTRATIONS OF PAHs IN GLENDORA, CALIFORNIA

PAH	Total Concentration, ng/m^3	Ratio to B[e]P[a]
Phenanthrene	20.0	—[b]
Anthracene	1.0	—[b]
Fluoranthene[c]	5.6 (0.26)	0.27
Pyrene[c]	4.1 (0.35)	0.37
Benzo[ghi]-fluoranthene	0.26	0.28
Cyclopenta[cd]-pyrene	0.09	0.1
Benz[a]anthracene	0.2	0.22
Chrysene/triphenylene	1.0	1.1
Benzofluoranthenes[b,j+k]	1.6	1.7
Benzo[e]pyrene	0.94	1.0
Benzo[a]pyrene	0.33	0.35
Indeno[1,2,3-cd]-pyrene	1.6	1.7
Benzo[ghi]perylene	3.8	4.0
Coronene	2.8	3.0

Source: Data after Ref. 25.
a. Ratios of particle-phase PAHs to particle-phase B[e]P.
b. No particle-phase PAHs found.
c. PAHs are distributed between gas and particulate phases; numbers in parentheses represent particle concentrations.

Concentrations of PAHs measured in Glendora were lower than those measured in the Baltimore tunnel, due to the higher distance from the emission source and higher atmospheric dilutions. The different concentration ratios of particle-bound PAH to those of B[e]P, which were seen in these two studies, may indicate either contributions from sources other than motor vehicles in Glendora or photochemical transformations of PAHs in the atmosphere.

PAH exposure assessments were also conducted in occupational settings. Table 8.18 shows PAH exposures measured in two underground mines, compared with ambient levels in the city of Sudbury, Ontario [9]. The mine concentrations, much higher than ambient, are generally comparable to those measured in the Baltimore tunnel.

TABLE 8.18
PAH CONCENTRATIONS IN CANADIAN MINES

PAH	Mean Concentration, ng/m^3		
	Nickel Mine	Salt Mine	City
Pyrene	80.2	256.0	14.6
Flouranthene	5.6	19.7	1.1
Benz[a]anthracene	8.9	18.5	1.6
Benzo[e]pyrene	18.2	62.2	5.3
Perylene	0.5	0.3	0.1
Benzo[k]fluoranthene	2.6	5.4	0.6
Benzo[a]pyrene	1.9	1.2	0.3
Benzo[ghi]perylene	6.9	9.3	0.9
Dibenzo[def,mno]chrysene	0.5	0.6	0.1
Dibenzo[ah]anthracene	1.9	3.5	0.3
Naphto[1,2,3,4-def]-chrysene	4.6	8.4	1.1
Benzo[rst]pentaphene	0.9	1.9	0.1
Total PAHs	132.7	387.0	26.1

Source: Data after Ref. 9.

8.3.5 Hydrocarbons and Aldehydes

Ambient levels of volatile hydrocarbon and aldehyde species were measured along a section of U.S. Highway 70 near Raleigh, North Carolina [26]. Traffic volume during sampling was determined by visual counting (1050 ± 10% vehicles per hour in each direction) and was classified into four groups: (1) light-duty gasoline and diesel, (2) heavy-duty gasoline, (3) heavy-duty diesel, and (4) motorcycles. Typical distributions were 91.5%, 3.2%, 5.1%, and 0.2%, respectively.

Table 8.19 lists the mean concentrations from four roadsides for selected hydrocarbons and aldehydes. The data are expressed in ppb C and as a percentage contribution of individual hydrocarbons and aldehydes to total nonmethane hydrocarbons (NMHC) and total aldehydes, respectively.

8.4 Health Effects of Gas Phase Components

8.4.1 Introduction

The "classic" toxic gas compounds found in diesel exhaust include carbon monoxide (CO), nitric oxide (NO), nitrogen dioxide (NO_2), and sulfur dioxide (SO_2). CO and NO_x (i.e., the sum of NO and NO_2) are regulated diesel emissions. SO_2 emissions from diesel engines, even though not directly regulated, are controlled by diesel fuel standards, which legislate increasingly lower sulfur content. In most countries, these gases also have maximum exposure limits established by occupational health authorities. These gases are present in diesel exhaust in relatively high quantities, from a few parts per million for SO_2 to as high as 1000 ppm and higher for NO_x. Their health effects vary from acute poisoning, resulting from exposure to high concentrations of gases, to long-term urban air pollution. Although acute poisoning by CO, NO_2, or SO_2 is fairly well understood, their air pollution effects have not been quantified and are still subject of studies by medical sciences.

The task of quantifying the health effects of air pollution and correlating them with the exposure to particular pollutants is very difficult. Air pollution is seldom caused by a single compound. Insofar as pollutants have common sources, fluctuations in their ambient concentrations tend to be correlated with each other. It can, therefore, be difficult to ascertain which pollutant is chiefly responsible for the adverse effects of pollution, or whether they are attributable to the mixture as such. The difficulty is further increased by imprecisions in

TABLE 8.19
CONCENTRATIONS OF INDIVIDUAL HYDROCARBONS AND ALDEHYDES MEASURED AT A ROADSIDE

Compound	Concentration, ppb C	Contribution, %[a]
Hydrocarbons		
Ethane	16.30	1.81
Ethylene	64.30	7.15
Acetylene	50.90	5.65
Propane	7.90	0.88
Propylene	22.60	2.51
n-Butane	15.80	1.75
1-Butene	5.70	0.64
n-Pentane	25.40	2.82
iso-Pentane	53.00	5.89
Methylcyclopentane	10.40	1.15
Methylcyclohexane	4.70	0.53
n-Decane	3.00	0.33
Benzene	29.00	3.23
Toluene	59.30	6.60
m- and *p*-Xylenes	53.10	5.90
o-Xylene	12.70	1.41
Ethylbenzene	12.00	1.33
Total NMHCs	900.00	100.00
Total paraffins	369.20	41.00
Total olefins	164.50	18.20
Total aromatics	252.00	28.00
Total unidentified NMHC	63.60	7.10
Aldehydes		
Formaldehyde	6.74	1.05
Acetaldehyde	3.00	18.40
Acrolein	1.20	7.30
Benzaldehyde	2.31	3.88
Total aldehydes	16.38	100.00

Source: Data after Ref. 26.

a. Percent contribution of individual hydrocarbons to total NMHCs and of individual aldehydes to total aldehydes, based on ppb C.

estimating exposures, influence of other variables (e.g., cigarette smoking), and so forth. For these reasons, the results of various epidemiological studies have not been entirely consistent. Much more work is still needed before the health effects of air pollution and the importance of particular pollutants are better understood.

8.4.2 Carbon Monoxide

8.4.2.1 Legal Exposure Limits

In the United States, the legal limits of CO at the workplace are established by the Occupational Safety and Health Administration (OSHA). Recommendations by the American Conference of Governmental and Industrial Hygienists (ACGIH) provide an important reference for all occupational safety and health regulatory authorities.

Maximum CO concentrations in the ambient air have been recommended by the World Health Organization (WHO). Legal ambient CO limits for urban areas are set in the United States by the EPA as part of the National Ambient Air Quality Standards (NAAQS), based on a comprehensive review of the pertinent scientific information [27]. Example exposure limits for carbon monoxide are compared in Table 8.20.

8.4.2.2 Acute Poisoning

Carbon monoxide is an extremely toxic gas. Exposures to CO concentrations in excess of 1% can be fatal within minutes. The mechanism and the effects of acute carbon monoxide poisoning have been studied since the 19th century and are fairly well understood [28].

The mechanism of CO poisoning is related to the affinity of carbon monoxide to the blood hemoglobin (Hb). Hemoglobin is the blood component that binds oxygen molecules in the lungs, in the form of oxyhemoglobin, HbO_2, and carries them to the various organs, tissues, and cells in the human body. The affinity of hemoglobin for CO is more than 200 times stronger than that for oxygen. As a result, if CO is present in the lungs, it preferentially binds to the hemoglobin molecules to form carboxyhemoglobin, HbCO, which depletes the blood of its oxygen-carrying capacity. Thus, the distribution of oxygen by hemoglobin may be impaired or entirely blocked in the presence of CO.

The blood levels of CO can be related to the partial pressure of CO and O_2 in the air by the following equilibrium equation, which was first published at the turn of the century by John Scott Haldane:

$$[HbCO]/[HbO_2] = C \cdot p_{CO}/p_{O2} \tag{8.1}$$

where:

p_{CO} = partial pressure of CO
p_{O2} = partial pressure of oxygen
C = haldane constant, $C = 210$.

The level of CO in the blood, expressed as a percentage of the total hemoglobin that is bound to CO (i.e., carboxyhemoglobin – HbCO), increases with time until it reaches a plateau. That plateau concentration of HbCO corresponds to its equilibrium level given by Eq. (8.1) for a given CO concentration.

The symptoms of CO poisoning can vary from dizziness and headaches to coma and death. A correlation between the severity of symptoms and the blood CO levels is shown in Table 8.21. The reader should remember that the toxic reaction of living organisms varies greatly among individuals and, therefore, conclusions about the ambient CO concentrations and the blood CO levels from symptoms, or vice versa, should be drawn with caution. Note that the intent of the present ACGIH threshold limit value (TLV) of 25 ppm is to maintain the HbCO concentration at a level of less than 3.5%.

TABLE 8.20
CARBON MONOXIDE EXPOSURE LIMITS

Authority	Limit	Value
Occupational Safety and Health Limits		
U.S. OSHA	Permissible exposure limit (PEL), 8-h average	50 ppm
U.S. OSHA	Ceiling value, 15-min average	200 ppm
ACGIH	Threshold limit value (TLV), 8-h average	25 ppm
Ambient Air Quality Limits		
EPA	8-h average	9.4 ppm
EPA	1-h average	35 ppm
WHO	8-h average	10,000 μg/m^3

TABLE 8.21
SYMPTOMS OF CO POISONING

Blood HbCO Level	Typical Symptoms
Less than 5%	No obvious symptoms
5%–10%	Decreased tolerance for exercise in persons with existing pulmonary disease Decreased angina threshold Decreased threshold for visual stimuli
10%–20%	Headaches Dizziness Confusion Decreased visual acuity
20%–30%	Severe headache Nausea Dizziness Increased respiration Irritability Impaired judgment Visual disturbance
30%–40%	Vomiting Decreased awareness Cardiac irregularities Muscle weakness
40%-50%	Fainting
50%-60%	Convulsions Paralysis
60%-70%	Coma Usually death in a few minutes
More than 70%	Immediately fatal

Source: Data after Ref. 28.

8.4.2.3 Air Pollution Effects

Acute Effects

For cigarette smokers the most important source of CO is cigarette smoke. For nonsmokers, ambient CO may also have health effects, particularly in patients with preexisting cardiorespiratory disease. Studies from several large cities, both in the United States and Europe, found an association between ambient CO levels and such acute effects as (1) mortality, (2) case fatality, and (3) hospital admissions. Day-to-day fluctuations in CO concentrations were positively correlated with mortality by several studies. An increase of 10 ppm in the 8-h average CO concentration resulted in an increase of all-cause mortality between 4% and 11.5 %.

An earlier study in the Los Angeles metropolitan area [29] investigated the association between the case fatality of myocardial infarction (i.e., the proportion of people with cases of the disease who died) and CO levels. The case fatality tended to be higher in hospitals that were in the more polluted areas, and the difference in the case fatality between polluted and unpolluted areas became evident only when CO concentrations were relatively high.

Several studies conducted in American cities, in Canada, and in the United Kingdom found a positive correlation between high CO levels and hospital admissions for congestive heart failure, myocardial infarction, and other cardiovascular diseases. The increases in hospital admissions ranged from 10% to 37% in response to a 10-ppm rise in the ambient CO concentration. The pathological mechanisms responsible for these effects are not known. It can be speculated that hearts that are already failing are particularly sensitive to any interference with their oxygen supply. The results of some of the mortality and hospital admissions studies are summarized in Table 8.22 [30].

Chronic Effects

Some reports indicate that chronic exposure to CO increases the risk of ischemic heart disease. However, the evidence is not clear and there is disagreement as to the contribution of CO to the cause of heart disease [30].

8.4.3 Sulfur Dioxide

8.4.3.1 Exposure Limits

The U.S. occupational health exposure limits for SO_2 are as follows:

TABLE 8.22
CHANGES ASSOCIATED WITH A RISE IN CO CONCENTRATION OF 10 PPM

Outcome	Area	Increase
All-cause mortality	Los Angeles	4%
	Los Angeles	5%
	Los Angeles	7%
	Athens	11.5%
Cardiovascular admissions	London	23%[a,b]
	Tucson	17%[b]
	7 American cities (over 65)	10%–37%[c]
	10 Canadian cities (over 65)	22%[c]

Source: Data after Ref. 30.
a. Acute myocardial infarction.
b. All cardiovascular diseases.
c. Congestive heart failure.

- ACGIH TLV: 2 ppm (8 h); short-term exposure limit (STEL): 5 ppm (15 min)
- OSHA PEL: 2 ppm (8 h); STEL: 5 ppm (15 min).

8.4.3.2 Poisoning Symptoms

Sulfur dioxide causes irritation to eyes, nose, and throat. Typical symptoms of SO_2 exposure include choking, coughing, rhinorrhea, reflex bronchoconstriction, and eye and skin burns.

8.4.3.3 Air Pollution Effects

It has long been recognized that SO_2 is a potent respiratory irritant when inhaled acutely in the laboratory at levels achieved during exceptional air pollution conditions, especially in patients with asthma. During the last years, SO_2 has come to be regarded as less important as a pollutant from the health point of view than previously, though recent studies have clearly identified this gas as a continuing cause of adverse effects on health.

SO_2 both causes and aggravates symptoms, particularly in patients with preexisting asthma. In association with particles, it appears to increase mortality. The current estimates of the acute effects are that each 50 μg/m^3 (17 ppb) rise in the 24-h average concentration raises the death rate by 3% for all causes, 4% for cardiovascular diseases, and 5% for respiratory diseases [30].

8.4.4 Nitric Oxide

8.4.4.1 Exposure Limits

The occupational health and safety exposure limits of NO are as follows:

- ACGIH TLV: 25 ppm (8 h)
- OSHA PEL: 25 ppm (8 h).

8.4.4.2 Poisoning Symptoms

NO poisoning symptoms include irritation of the eyes, nose, and throat; drowsiness; and unconsciousness. Inhalation may result in delayed pulmonary edema.

8.4.4.3 Air Pollution Effects

NO is an important precursor for other substances of much higher toxicity, including NO_2 and ground-level ozone. Nitric oxide undergoes a series of chemical reactions when released into the atmosphere. Ozone is produced by photochemical reactions between NO and hydrocarbons. Another reaction involves the oxidation of NO to NO_2. Because the toxicity of NO_2 is much higher than that of NO, most air pollution studies focus on the effects of nitrogen dioxide.

8.4.5 Nitrogen Dioxide

8.4.5.1 Exposure Limits

Most occupational health and safety exposure limits for NO_2 are at or below 5 ppm. Various ambient air quality standards stipulate NO_2 limits between 0.050 ppm (50 ppb) and 0.500 ppm (500 ppb), depending on the exposure period. Some example limits are listed in Table 8.23.

8.4.5.2 Poisoning Symptoms

Nitrogen dioxide is an extremely toxic gas. Exposures to NO_2 concentrations of 100 ppm are dangerous, and exposures to 200 ppm may be fatal.

Symptoms of acute NO_2 poisoning may include coughing, mucoid frothy sputum and dyspnea, chest pain, pulmonary edema, cyanosis, tachypnea and tachycardia, or eye irritation. Inflammation of lungs may cause slight pain, but resulting edema several days later may cause death. Literature reports

TABLE 8.23
NITROGEN DIOXIDE EXPOSURE LIMITS

Authority	Limit	Value, ppm
Occupational Safety and Health Limits		
U.S. OSHA	PEL, ceiling value	5
ACGIH	TLV, 8-h average	3
ACGIH	STEL, 15-min average	5
Ambient Air Auality Limits		
EPA (NAAQS)	Annual average	0.052
Environment Canada	Annual average (maximum acceptable)	0.050
Environment Canada	24-h average (maximum acceptable)	0.110
Environment Canada	1-h average (maximum acceptable)	0.210

indicate a range of possible adverse health effects resulting from NO_2 exposures as low as 100 ppb (0.1 ppm). The following symptoms have been reported [31]:

- Respiratory irritation and breathing problems, particularly in people with asthma and bronchitis, can develop at nitrogen dioxide exposures as low as 100 ppb.
- Children appeared to be susceptible to bronchitis and breathing difficulties associated with chronic exposures to nitrogen dioxide concentrations around 100 ppb.
- Odor perception and the ability to adapt to dim illumination after looking at a bright light have also been impaired in human subjects at levels between 100 and 200 ppb.
- In animals, exposure to nitrogen dioxide levels of about 500 ppb for 4 h reduced resistance to bacterial and viral infection.

8.4.5.3 Air Pollution Effects

Nitrogen dioxide is probably the most confusing of the common air pollutants. High levels of NO_2 are periodically recorded in urban areas. Its effect on health, including those resulting from low-concentration exposures, has been demonstrated. Despite this, epidemiological studies that have looked into the effects of the mixture of air pollutants commonly found in ambient air have tended not to show that nitrogen dioxide contributes much to the overall effects. Other pollutants, including sulfur dioxide and particles, tend to figure more strongly in the analyses and may be masking the effects of nitrogen dioxide.

The U.K. Department of Health summarized the effects of NO_2as follows [30]:

- The effects of NO_2 are among the most difficult to assess at the moment because they appear to be dependent on a wide range of modifying influences.
- All-cause mortality is increased by approximately 3.5% per 100 μg/m^3 (53 ppb) on days with elevated NO_2 levels. The principal difficulty is in estimating whether this represents a substantial increase in the number of deaths over a year or not.
- There is little evidence that admissions to hospitals are increased by increased levels of NO_2 when all respiratory causes are looked at together. There are, however, apparent increases in admissions for chronic obstructive pulmonary disease and asthma.
- Estimates of chronic effects, many of which use unquantified and indirect measures of exposure, suggest that the effects of NO_2 exposure could be substantial. They are, however, not yet based on a very firm footing.

8.5 Health Effects of Diesel Particulates

8.5.1 *Introduction*

8.5.1.1 Definition of Diesel Particulates

Diesel particulate matter is a complex mixture of solid and liquid material (compare Section 7.3). The particles in diesel exhaust are of special concern because, due to their respirable size, they can penetrate deep into human lungs. The composition of DPM includes many species that are known for their adverse health effects, including several carcinogens.

Primary materials found in the solid phase of DPM include inorganic carbon and metal ashes, while the liquids include high boiling hydrocarbons, water, and sulfuric acid. According to most legal definitions of diesel particulates and the corresponding sampling and measurement methods, diesel particulates are sampled from exhaust gases that are diluted with air and cooled, typically to 52°C. These sampling procedures have been introduced in an attempt to simulate atmospheric dilution effects on diesel emissions.

Although it is known today that the products of real-life atmospheric dilutions may differ from what is sampled in a laboratory dilution tunnel, a consensus has been reached with regard to the preceding definition among worldwide public health authorities. Accordingly, the total particulate matter is traditionally divided into the following fractions:

1. Solids (inorganic carbon, ash)
2. Organics (soluble organic fraction, SOF)
3. Sulfates (hydrated sulfuric acid, metal sulfates).

There is no such consensus, however, in regulating diesel particulate exposure limits in various occupational health environments. The proposed DPM exposure limits for U.S. underground metal mines are expressed as total carbon, excluding the sulfate fraction from DPM. Occupational health regulations in Germany define DPM as elemental carbon, excluding both sulfates and the organic fraction.

These examples are helpful in realizing that what constitutes DPM, contrary to such diesel emissions as nitrogen oxides or carbon monoxide, is not well defined. Not surprisingly, medical research on the health effects of DPM is much more difficult and has provided fewer answers than that on simple chemicals. In particular, medical sciences have not yet established a firm opinion as to what components of DPM are responsible for any particular health effects: inorganic carbon, organics such as PAHs, or a combination of both.

8.5.1.2 DPM Exposure

For all chemicals the severity of health effects, whether they are acute poisoning or air pollution effects, depend on the concentration of a given species in the ambient air. Therefore, the real objective of epidemiological studies on diesel exhaust is to learn whether DPM or its components produce health effects at the levels of exposure that are encountered in practice.

Humans are exposed to diesel particulates in both ambient and occupational settings. Relatively high DPM exposures occur in many urban areas having large populations and high traffic densities. However, DPM concentrations found in various occupational health environments, such as in bus repair garages or warehouses that use diesel forklifts, are usually much higher than those in ambient air. By far the highest DPM exposures are found in dieselized underground mines. Approximate ranges of exposure to diesel particulates in different environments are listed in Table 8.24 [4].

TABLE 8.24
HUMAN EXPOSURE TO DIESEL PARTICULATES

Environment	Range of Exposure, μg/m^3
Urban Areas	
Less polluted (monthly average)	1–4
Heavily polluted (monthly average)	10
Occupational Groups	
Truck drivers	4–6
Bus garage workers, forklift operators, etc.	20–100
Underground miners	100–1700

Source: Data from Ref. 4.

8.5.2 Noncancer Health Effects

8.5.2.1 PM10

Diesel particulates are a contributor to the ambient PM10 (particles less than 10 μm in diameter) levels. PM10 is a known air pollutant and its health effects have been extensively studied. A number of PM10 epidemiologic studies has been reviewed by the California Air Resources Board [2]. The studies confirm an association between ambient particulate matter and such adverse health outcomes as mortality, respiratory hospital admissions, emergency room visits, restricted activity days and respiratory symptoms for adults, lower respiratory tract illnesses for children, asthma attacks, and chronic diseases. Statistically significant relationships have been found among these studies using various alternative measures of particulate matter including PM10, fine particles (<2.5 μm), total suspended particles (TSP), British smoke, coefficient of haze (COH), and sulfates.

The studies have covered different chemical compositions of the particulate matter, various populations, and various climates. Associations have been found on a daily basis between concentrations of particulates and mortality in several metropolitan areas including London; Steubenville, Ohio; Philadelphia; Santa Clara County; Los Angeles; Detroit; Birmingham, Alabama; Utah Valley; and St. Louis. Studies also indicate that country-wide mortality rates across the United States are related to annual averages of particulate matter.

8.5.2.2 Noncarcinogenic DPM Studies

Numerous experimental studies have been conducted to examine the noncarcinogenic effect of diesel exhaust and diesel particulates, as opposed to the total ambient particulates from all sources (PM10) [2, 4]. These studies tried to examine the association of occupational exposure to diesel exhaust with respiratory morbidity and mortality of nonmalignant character. They included short-term and long-term exposure investigations among several occupational groups: miners, bus garage workers, dock workers, railroad workers, heavy equipment operators, as well as general population. Some of the studies reported on the following findings:

- Short-term exposures caused symptoms of irritation and compromised pulmonary function.
- Removal of long-term DPM exposure from workplace air improved the pulmonary function of workers.
- Exposure to diesel particulates increased human allergic responses.

However, most epidemiologic studies were inconclusive and have not found consistent evidence of an effect of diesel exhaust exposure on nonmalignant respiratory diseases or related mortality. The California ARB concluded in its report [2] that available data from studies on humans is not sufficient for deriving either a short- or a long-term health risk guidance value.

Numerous species of animals have been also subjected to noncancer diesel exhaust studies. About 50 such studies have been summarized and discussed by Watson and Green [32]. The conclusion was that chronic inflammation and cell hyperplasia were the two most sensitive effects of biological significance. These and other effects were found at high DPM dose levels, much higher than the human exposures. Below those dose levels, an effect of diesel exhaust exposure was not detectable.

Exposures to very high concentrations of diesel exhaust, which may happen when diesel engines are operated in confined spaces of insufficient or no ventilation, may cause dizziness, drowsiness, headaches, nausea, decrement of visual acuity, and decrement in forced expiratory volume. These symptoms can be attributed to the toxic properties of gaseous pollutants (carbon monoxide, hydrocarbons and their derivatives), rather than to DPM.

8.5.3 Carcinogenicity of Diesel Exhaust

8.5.3.1 Animal Studies

Carcinogenic effects of diesel particulates have been investigated in inhalation studies with laboratory animals, primarily rodents, including rats, mice, and hamsters, but also with higher animals, such as monkeys. A review of long-term animal studies was conducted by NIOSH [21]. The results of most studies appear to be in agreement on the following responses:

- Prolonged exposure of rats to high concentrations of diesel particulates (2000 to 10,000 μg/m^3) initiates a dose-dependent progression of cellular changes that eventually leads toward the development of benign and malicious lung tumors. The underlying pathophysiological process is termed *lung overload.* It occurs when the dose rate for particulates is so high that it overwhelms the long-term pulmonary clearance process. Rats exposed to lower DPM concentrations do not develop tumors.
- Experiments conducted with other rodents, including hamsters and mice, as well as with monkeys, seem not to produce tumors. Under prolonged exposure to high doses of diesel particulates, these animals show such symptoms of lung overload as impaired particle clearance, inflammation, cell proliferation, and fibrosis, but not cancer.
- Tumors develop in rats when exposed not only to diesel particulates, but also to high doses of any fine insoluble particles, without the associated organic material. Exposure to ultrafine carbon black, titanium dioxide, or talc particles caused cancer in rats (but not in mice or hamsters).

Medical research tries to explain cancer development in terms of mechanisms. Two possible pathways of carcinogenesis are (1) genotoxic and (2) nongenotoxic mechanisms. Genotoxic carcinogens act by altering the information encoded in DNA. They cause genetic changes, including mutations or changes in chromosome structure, leading ultimately to tumor development. Most known human carcinogens can be classified as genotoxic. Nongenotoxic carcinogens lead to tumors either by killing cells and inducing compensatory cell proliferation, or by increasing rates of cell proliferation. In either case, the increase in cell division associated with internal or external stimulation is believed to increase the risk of genetic errors, leading to cancer.

Diesel particle extracts and PAHs are known to exhibit mutagenic activity on bacteria and mammalian cells, as well as chromosomal damage and cell transformation. In view of the animal tests, it is not clear whether this *in vitro* mutagenicity is relevant to the carcinogenic risk in humans from *in vivo* exposures to diesel particulates. Experiments with rats showed very convincingly that tumors were developed under exposure to pure carbon black and other substances that are free of any organic mutagenic materials. The observation that mutagenic compounds of diesel exhaust do not appear to play a role in tumor development in rats suggests that the carcinogenicity of diesel exhaust may be in fact nongenotoxic. These findings have shifted the focus in the medical research from the organic fraction and PAHs to the inorganic carbon portion of DPM.

An important issue under discussion is the validity of extrapolating animal responses to humans. It is not clear which, if any, rodent species may have tumor responses that would be useful for quantitative predictions of human lung cancer. Rats appear to be very sensitive models for (diesel) particulate-induced carcinogenicity. It is not understood why the rat response is so different from that of other animals.

8.5.3.2 Human Studies

More than 30 occupational epidemiological studies have been conducted on workers exposed to diesel emissions after 1950, when the railroad, trucking, and public mass transit industries widely converted to diesel engines. Railroad, dock, trucking industry, and bus garage workers were the most extensively studied occupational groups. Underground miners, even though they experience the highest exposure to diesel exhaust, have not generally been targeted for diesel epidemiological studies because they are usually exposed to other pollutants such as dusts, metals, and radon progeny.

A thorough review of human epidemiological studies was conducted by the U.S. Health Effects Institute (HEI) [33]. The epidemiological data show that long-term exposure to diesel exhaust in a range of occupational settings is associated with small increases (in the 1.2 to 1.5 range) in the relative risk of lung cancer occurrence, mortality, or both. These statistics have been obtained primarily in studies carried out among railroad workers and truck drivers. A general difficulty in these studies is presented by confounding cigarette smoking, which is associated with a high risk of lung cancer. Some of the studies (but not all) tried to introduce an adjustment for cigarette smoking.

Some epidemiological studies suggest that, in addition to lung cancer, an elevated risk of bladder cancer may be linked to diesel exhaust exposure. The evidence for bladder cancer, however, is not as consistent as that for lung cancer.

The preceding conclusions and risk factors are based on studies of male workers. None of the conducted epidemiological studies examined the risk of diesel particulate exposure for women or potentially susceptible populations such as infants, children, or people with health disorders.

8.5.3.3 Risk from Exposure to Diesel Particulates

Risk assessments for diesel particulates are carried out by various organizations on an ongoing basis. The assessments depend on scientific information from the available human and animal studies, as well as on the judgment of the group making the classification. An overview of the carcinogen classifications of diesel particulates is shown in Table 8.25. Risk assessments marked as drafts are preliminary and can be changed in final versions of the respective documents.

The World Health Organization published a monograph that evaluates the risk for human health and the environment from exposure to diesel emissions [3]. With respect to the protection of human health, it was stated that the reviewed data support the conclusion that inhalation of diesel exhaust could increase the risk of both cancer and noncancer diseases. Lifetime cancer risk estimates that were considered reasonable in the monograph range from 1.6×10^{-5} to 7.1×10^{-5} per 1 $\mu g/m^3$ of DPM exposure, with a geometric mean of 3.4×10^{-5}. WHO recommended that diesel emission be controlled, particularly in urban environments. In the occupational environment, adequate ventilation must be provided to prevent excessive exposure (what constitutes "excessive exposure" was not quantified).

In 1998 the state of California identified diesel exhaust particulates as a toxic air contaminant under the California air toxics law. The health hazard assessment report concluded that "at current ambient concentrations, diesel exhaust may cause an increase in the likelihood of cancer" [2]. The report also listed the following unit risk estimates for diesel exhaust:

- Cancer lifetime unit risk estimates based on human studies ranged from 1.3×10^{-4} to 2.4×10^{-3} $(\mu g/m^3)^{-1}$ with a geometric mean of 6×10^{-4} $(\mu g/m^3)^{-1}$.
- Risk estimates based on animal studies varied from 1×10^{-5} to 3×10^{-4} $(\mu g/m^3)^{-1}$ with a geometric mean of 6×10^{-5} $(\mu g/m^3)^{-1}$.

The Scientific Review Panel, which reviewed the California report, concluded that 3×10^{-4} $(\mu g/m^3)^{-1}$ is a reasonable estimate of unit risk expressed in terms of diesel particulate. Based on available information, a level of diesel exhaust exposure below which no carcinogenic effects are anticipated has not been identified.

In 1999, the Health Effects Institute examined a number of published epidemiological studies of diesel emissions and lung cancer for possible use in support of quantitative risk assessment related to diesel exposure (i.e., whether the data from human epidemiological studies can be used to develop reliable estimates of the magnitude of any risk for lung cancer) [10]. Epidemiological studies reviewed by the HEI included four railroad worker studies (including two studies by Garshick) and three teamster studies. In conclusion, the HEI recommended against using the current railroad worker data as the basis for quantitative risk assessment in ambient settings. The HEI also recommended that further scrutiny is needed on the teamster data. The epidemiological studies examined by HEI were used by the state of California in support of the identification of diesel particulates as a toxic air contaminants. According to the

TABLE 8.25
RISK ASSESSMENT OF DIESEL EXHAUST

Organization	Year	Human Evidence	Animal Evidence	Classification
National Research Council	1981	Not convincingly demonstrated	Negative	—
National Institute for Occupational Safety and Health [21]	1988	Limited	Confirmatory	Potential occupational carcinogen
International Agency for Research on Cancer	1989	Sufficient	Limited	Probable human carcinogen
U.S. Environmental Protection Agency (draft)	1990	Limited	Sufficient	Probable human carcinogen
U.S. Environmental Protection Agency (draft)	1994	Limited	Sufficient	Probable human carcinogen
California Environmental Protection Agency (draft)	1994	Sufficient	Sufficient	Human carcinogen
U.S. Environmental Protection Agency (draft) [6]	1998	Limited	Sufficient	Probable human carcinogen ("highly likely to be carcinogenic")
California Environmental Protection Agency [2]	1998	Sufficient	Sufficient	Human carcinogen
U.S. National Institute of Environmental Health Sciences (9th Report on Carcinogens) [34]	1998	—	—	Reasonably anticipated to be a human carcinogen
U.S. Environmental Protection Agency [20]	2002	Limited	Sufficient	Probable human carcinogen ("likely to be carcinogenic")

Source: Data from Refs. 2, 4, 6, 20, 21, and 34.

work by HEI, the California risk analysis was based on human epidemiological data that are not appropriate for that purpose.

Perhaps the most comprehensive, scientifically accurate, and objective summary of current human knowledge on the health effects of diesel emissions was published by the EPA in 2002. In the final version of its "Health Assessment Document for Diesel Engine Exhaust" the EPA classified diesel exhaust as "likely to be carcinogenic" [20]. It was concluded, based on review of epidemiological studies, that occupational exposures to diesel exhaust particulates cause a small increase in the risk of lung cancer, by the factor of about 1.2 to 1.5. The EPA believed the cancer hazard was also applicable to ambient (i.e., environmental) exposures, even though none of the studies examined by the EPA showed that a lung cancer hazard was indeed present at environmental levels of exposure. No specific cancer unit risk estimate for diesel exhaust was adopted or recommended in the EPA assessment (in accordance with the HEI recommendation, and in conflict with the position of the California EPA). Animal (rat) cancer studies were deemed not clear for human hazard prediction and unsuitable for environmental exposure risk estimates. Due to a number of uncertainties, the available human epidemiological studies data were found to be unsuitable for quantitative risk assessment.

In June 1993, the EPA determined an inhalation reference concentration (RfC) of 5 $\mu g/m^3$ for noncancer effects of diesel exhaust. The same RfC has been adopted by California [2] as its chronic inhalation reference exposure level (REL). The RfC of a chemical is defined as a daily exposure to the human population, including sensitive subgroups, that is likely to be without appreciable risk of deleterious effects during a lifetime of exposure.

In its 1995–1996 Notice of Intended Changes, the ACGIH proposed a threshold limit value for diesel exhaust particulates in the work environment of 0.15 mg/m^3. This proposal was later lowered to 0.05 mg/m^3 and then to 0.02 mg/m^3, the last TLV measured as elemental carbon (EC). The proposed carcinogenicity classification was A2, "suspected human carcinogen." None of these proposed TLVs have been finalized.

8.5.4 Future Issues

Scientific evidence has shown that ambient particulates of smaller diameters (i.e., less than 2.5 μm) are more harmful to humans than coarse particles having diameters of less than 10 μm. This finding was reflected by the introduction of a new ambient air quality standard for particles below 2.5 μm (PM2.5) by the EPA in 1997.

Diesel particulates, with practically all particles being smaller than 1 μm, are entirely within the PM2.5 category. They are composed of accumulation mode particles that have diameters between approximately 50 nm and 1 μm and nuclei mode particles that have diameters smaller than 50 nm. These less-than-50-nm particles are referred to as *diesel nanoparticles* (Section 7.4). There is a notion, and a concern, that nanoparticle emissions present a higher health risk than that posed by diesel particles of larger sizes. At this time, the notion of the harmful character of nanoparticle emissions is to a large degree a simple extrapolation of the findings from the 1- to 10-μm size range into the submicron sizes. It will be a big challenge for the medical research community in the coming future to explain the health effects of nanoparticles from diesel and other sources. If the nanoparticle concerns indeed prove legitimate, emission standards would need to account for particle numbers in addition to particle mass emissions.

8.6 Environmental Effects of Emissions

8.6.1 Pollutants and Air Quality

Air pollutants are substances that adversely affect the environment by interfering with climate, the physiology of plants, animal species, and entire ecosystems, as well as with human property in the form of agricultural crops or human-made structures. Air pollutants can originate from natural or anthropogenic (human-made) sources, or both. Examples of natural sources of pollution include volcanic eruptions or wind erosion. Emissions from internal combustion engines are an exemplary source of anthropogenic pollution. Some sources of pollution, such as forest fires, can be related to both natural phenomena and human activities.

Atmospheric reactions can transform primary pollutants into different chemical species. These reactions can produce both harmless compounds and secondary air pollutants that may be more harmful than their precursors. The world's most important pollutants, their sources, and known or suspected environmental effects are listed in Table 8.26.

Governments and international organizations have been taking actions to protect the quality of air. Ambient air quality standards and guidelines, issued by environmental protection authorities, are instrumental in achieving this objective. An example of such legislation is set by the U.S. National Ambient Air Quality Standards (NAAQS) adopted by the EPA. The NAAQS apply to both human health (primary standard) and public welfare (secondary standard). Primary standards protect sensitive members of the human population from the adverse health effects of criteria air pollutants. Secondary standards protect the public welfare from any known or anticipated adverse effects associated with the presence of a pollutant in the ambient air. Welfare effects include effects on soils, water, crops, vegetation, human-made materials, animals, wildlife, weather, visibility, climate, damage to and deterioration of property, and hazards to transportation, as well as effects on economic values and personal comfort and well-being.

Under the U.S. Clean Air Act of 1990, the NAAQS set maximum concentrations for six criteria pollutants: ozone, carbon monoxide, nitrogen dioxide, lead, particulate matter, and oxides of sulfur. Urban areas are required to achieve attainment with regard to ambient concentrations of these criteria pollutants. NAAQS limits for selected "classical" pollutants, as well as the limits established by California and air quality guidelines from the WHO, are given in Table 8.27. The particular limits are average over different periods of time, as indicated in the table.

TABLE 8.26
AIR POLLUTANTS, THEIR SOURCES, AND THEIR EFFECTS

Pollutant	Natural Source	Anthropogenic Source	Environmental Effects
Nitrogen oxides ($NO + NO_2$)	Lightning, soil bacteria	High-temperature fuel combustion from motor vehicles, industry, and utilities	Primary pollutants that produce photochemical smog, acid rain, and nitrate particulates; destruction of stratospheric ozone; human health impact
Particulates	Forest fires, wind erosion, volcanic eruption	Coal, waste, and fossil fuel burning	Reduced atmospheric visibility; human health impact; black carbon particulates contribute to global warming
Sulfur dioxide	Volcanic eruptions and decay	Coal combustion, ore smelters, petroleum refineries, diesel engines burning high-sulfur fuels	Acid rain; human health impact
Ozone	Lightning, photochemical reactions in the troposphere	Secondary pollutant produced in photochemical smog	Damage to plants, crops, and human-made products; human health impact
Carbon monoxide	Unnoticeable	Rich and stoichiometric combustion, mainly from motor vehicles	Human health impact
Carbon dioxide	Animal respiration, decay, release from oceans	Fossil fuel and wood combustion	Most common greenhouse gas
Nonmethane hydrocarbons (VOCs)	Biological processes	Incomplete combustion, solvent utilization	Primary pollutants that produce photochemical smog
Methane	Anaerobic decay, cud-chewing animals, oil wells	Natural gas leak and combustion	Greenhouse gas
Chlorofluorocarbons (CFC)	None	Solvents, aerosol propellants, refrigerants	Destruction of stratospheric ozone

Source: Data from Ref. 35.

TABLE 8.27
SELECTED AMBIENT AIR QUALITY STANDARDS

Pollutant	U.S. NAAQS	California	WHO
Ozone (1 h)	0.12 ppm	0.09 ppm	—
Ozone (8 h)	0.08 ppm	—	120 μg/m^3
Carbon monoxide (1 h)	35 ppm	20 ppm	30,000 μg/m^3
Carbon monoxide (8 h)	9.4 ppm	9 ppm	10,000 μg/m^3
PM10 (24 h)	150 μg/m^3	50 μg/m^3	—
PM10 (1 yr)	50 μg/m^3	30 μg/m^3	—
Nitrogen dioxide (1 h)	—	0.250 ppm	200 μg/m^3
Nitrogen dioxide (24 h)	—	—	—
Nitrogen dioxide (1 yr)	0.052 ppm	—	40 μg/m^3
Sulfur dioxide (24 hr)	—	—	125 μg/m^3
Sulfur dioxide (1 yr)	—	—	50 μg/m^3

Air quality indexes have been devised to characterize the severity of air pollution in one composite measure. These indexes are based on concentrations of several criteria pollutants and their respective concentration limits.

The Air Quality Index (AQI, formerly the Pollutant Standards Index, PSI) used by the EPA is shown in Table 8.28. An AQI value for each of the pollutants is calculated by means of a linear approximation between breakpoints listed in the table. The final AQI is the highest value calculated for each pollutant. In other words, the pollutant with the highest relative concentration determines the AQI.

To achieve the air quality objectives, one must identify and control the sources of particular air pollutants. Environmental protection agencies compile emission inventories, which provide information on the relative importance of various sources of pollution. Motor vehicles are usually shown to be responsible for a high proportion of the total NO_x, HC, CO, and PM pollution. Within the motor vehicle category, NO_x is usually equally divided between SI and diesel engines; HCs and CO are produced primarily by the SI engines; and most of the PM emission is attributed to diesels.

8.6.2 Photochemical Smog

Smog is a mixture of air pollutants that plagues most of the large metropolitan areas in the world. The primary components of smog are nitrogen oxides, sulfur oxides, hydrocarbons, carbon monoxide, and fine particulates (including secondary particulates such as nitrates formed in the atmosphere from NO_x). Ground-level ozone is generated in the photochemical smog as a secondary air pollutant. The most obvious manifestation of smog is a brownish-gray haze that causes irritation to the human respiratory system. Common symptoms include coughing, choking, and stinging eyes.

The sources of smog pollution include road traffic; nonroad vehicles such as planes, ships, and trains; and industrial emissions, as well as small businesses and households where polluting products are used. Diesel trucks, buses, and off-highway vehicles contribute significantly to smog, emitting NO_x and particulates, as well as, in smaller quantities, SO_2 and hydrocarbons, which are precursors of ozone.

Ozone (O_3) which is a "benchmark" pollutant for measuring the severity of smog, forms in photochemical reactions from nitrogen oxides and hydrocarbons. The development of smog occurs in daily cycles. It starts with a buildup of the primary pollutants in the atmosphere: NO_x and hydrocarbons. Gradually, ozone is generated through a complex set of chemical reactions involving NO_x, hydrocarbons, and sunlight. As a result, concentrations of primary pollutants decrease and the concentration of ozone increases. The highest ozone levels, as recorded during warm summer afternoons, are favored by strong sun, high temperatures, and an inversion layer in the atmosphere.

Even though there is wide agreement on the general smog development pattern just discussed, we should emphasize that detailed mechanisms are far from being understood. For instance, heavy-duty diesel traffic, due to its significant contribution to the NO_x emission inventory, has been suspected of being one of the most important sources of smog in Southern California. Air quality measurements, however, consistently show that the highest levels of ozone are measured during weekends, during the time when little diesel traffic occurs [37]. Decreased ozone levels are measured during weekdays, when most diesel trucks are in operation. It is currently believed that ozone levels are not determined by the levels of NO_x or other pollutants alone, but by the concentration ratios between pollutants. At the California atmospheric conditions, decreased NO_x/VOC (HC) ratios—as experienced during periods of lower diesel traffic—actually cause an increase in ozone concentration [38].

TABLE 8.28
U.S. AIR QUALITY INDEX

AQI	Category	O_3 (8 h), ppm	O_3 (1 h), ppm	PM2.5 (24 h), mg/m^3	PM10 (24 h), μg/m^3	CO (8 h), ppm	SO_2 (24 h), ppm	NO_2, ppm
0–50	Good	0.000–0.064	—	0.0–15.4	0–54	0.0–4.4	0.000–0.034	—
51–100	Moderate	0.065–0.084	—	15.5–40.4	55–154	4.5–9.4	0.035–0.144	—
101–150	Unhealthy for sensitive groups	0.085–0.104	0.125–0.164	40.5–65.4	155–254	9.5–12.4	0.145–0.224	—
151–200	Unhealthy	0.105–0.124	0.165–0.204	65.5–150.4	255–354	12.5–15.4	0.225–0.304	—
201–300	Very unhealthy	0.125–0.374	0.205–0.404	150.5–250.4	355–424	15.5–30.4	0.305–0.604	0.65–1.24
301–400	Hazardous	—	0.405–0.504	250.5–350.4	425–504	30.5–40.4	0.605–0.804	1.25–1.64
401–500	Hazardous	—	0.505–0.604	350.5–500.4	505–604	40.5–50.4	0.805–1.004	1.65–2.04

Source: Data after Ref. 36.

Ozone is an oxidizing compound with extreme chemical activity. As its concentration rises, ozone reacts with other constituents of smog, cleaning the air and slowly disappearing from the atmosphere. Ozone is playing here a role as nature's "cleansing agent," which is released to remove the impurities from air. Unfortunately, its chemical attack is not limited to pollutants. Ozone also attacks and damages human lung tissue, aggravates respiratory diseases, and makes people susceptible to respiratory infections. Furthermore, ozone inhibits the growth of plants and can cause widespread damage to crops and forests.

The only effective strategy for reducing ozone levels in urban areas is the control of primary smog pollutants: hydrocarbons (VOCs) and NO_x. In view of the relationship between ozone levels and the NO_x/VOC ratio, pollutant reductions need to be coordinated. As is apparent from the Southern California experience, NO_x reductions not accompanied by corresponding control of VOCs may be counterproductive from the ozone control point of view.

8.6.3 *Acid Rain*

Acid rains originate from acidic gases such as SO_2, NO_x, and hydrochloric acid, which are emitted by sources of airborne pollution. In the atmosphere, the acidic gases are converted to secondary pollutants by dissolution in water and/or by oxidation. The reaction products include sulfurous, sulfuric, and nitric acids. These acids precipitate with rainfall, snow, fog, and dew, and are deposited on vegetation, soils, and the materials of human-built structures.

"Pure" rain is not a pH-neutral solution. Due to the atmospheric CO_2, its pH typically amounts to about 5.6. In the "acid rains," the pH may decrease to 4 to 5, sometimes even below 4. Acid rains are believed to play an important role in (1) acidification of lakes and waterways, (2) the death of forests, and (3) damage to building stone, concrete, and other human-made structures.

8.6.3.1 Acidification of Waters

The consequences of acidification of waters are as follows [39]:

- Reduction or disappearance of fish species (trout, salmon) and amphibians
- Changes in the character of plankton
- Substitution of bacteria by fungus in decomposition processes
- Increased growth of peat moss in lakes.

Acidification of water is also responsible for dissolution of heavy metals, which remain insoluble at higher pH levels. This may cause such metals as mercury, cadmium, lead, or copper to be introduced into drinking water.

8.6.3.2 Death of Forests

The damage caused to forests by industrial pollution sparked the awareness of the media and general public about the problem of air pollution. Death of forests is not a new phenomenon; extensive damage to forests in different parts of Europe was already being reported in the 1970s and 1980s. More than 500,000 hectares (2.2 million acres) of forest disappeared in Czechoslovakia, where high sulfur lignites were burned in industrial power plants [39].

Death of forests is caused by a combination of interrelated pollution and nonpollution factors, including the following:

- *Chemical factors:* Airborne pollution, such as SO_2 and NO_x emissions and also ozone, causes chemical attacks on foliage.
- *Climatic factors:* Droughts and late frosts can contribute to the damage, especially if the trees are already weakened by other factors.
- *Silvicultural factors:* Planting of new trees is often aimed at maximizing lumber yields. The selection of tree species for lumber may not be best suited for their survival in a given environment, soil, and climate conditions. High stand densities increase competition for water between neighboring trees.
- *Pathological factors:* Ecological changes, such as acidification of soil or water depletion, can favor the development of pathogenic species.

8.6.4 *Global Warming*

Global warming is an increase in the earth's surface temperature, which has been observed over several decades. In the 20th century, the global surface temperature increased by more than 0.5°C. Even though disagreement exists among the scientific community as to whether this warming trend is caused by human activities, as opposed to being a part of earth's natural climate change cycle, increasingly more evidence suggests that global warming is, at least in part, a result of anthropogenic climate forcing agents [11]. These anthropogenic climate forcing agents are substances that are accumulating in the atmosphere due to increased combustion of fossil fuels and other human activities. Carbon dioxide (CO_2) is considered the most important substance in this class.

Different opinions exist as to the importance and severity of the global warming phenomenon. Furthermore, predictions of the future repercussions of global warming depend on a number of assumptions, including the future trends in the emissions of CO_2 and other climate forcing substances. There is a notion that global warming is leading to the most dramatic climate change in the history of mankind. The warming of the climate would cause melting of glacial ice sheets, rising sea levels, changes in agricultural patterns, and intensified weather phenomena, such as tornadoes and floods. Alternative scenarios, however, predict only moderate climate changes in the next 50 years [40].

The earth's surface temperature we experience is the result of an energy balance between radiation received from the sun and radiation re-emitted into space, with some 30% of the received energy being reflected back. The most important class of climate forcing agents responsible for global warming are *greenhouse gases*—substances present in the atmosphere in trace concentrations that can capture the reflected energy. These gases are transparent for the high-frequency solar radiation, but "opaque" for the low-frequency infrared radiation re-emitted by the earth. This is a one-way filter effect, similar to that observed with the glass roof of a greenhouse.

In addition to carbon dioxide, the most important greenhouse gases include water vapor, freons [chlorofluorocarbons (CFCs)], methane, nitrous oxide, and tropospheric ozone. The properties of these gases, excluding water vapor, are listed in Table 8.29.

The greenhouse gas properties of particular compounds vary significantly. Freons, for example, are up to almost 20,000 times more powerful than carbon dioxide. Even though their concentration is negligible compared with the other gases, freons are responsible for about 15% of the total greenhouse effect.

Whereas freon emissions originate exclusively from human activity, the other greenhouse gases have several natural sources, as shown earlier in Table 8.26. Carbon dioxide, the most important greenhouse gas, is produced primarily by natural sources. The anthropogenic emission of CO_2 is estimated at 3.5% to 4% of its total release [5, 35].

Climate forcings can be quantified and expressed in watts per square meter (W/m^2). For example, all of the human-made greenhouse gases now cause a global forcing of more than 2 W/m^2 [40]. It is an equivalent of placing one miniature 2-W light bulb over every square meter of the earth's surface, or of increasing the brightness of the sun by about 1%. For comparison, the climate forcing during the Ice Age 20,000 years ago is estimated at -6.6 ± 1.5 W/m^2 relative to the current interglacial period. This forcing and the 5°C cooling of the Ice Age imply a climate sensitivity of about 0.75°C per 1 W/m^2. Figure 8.10 shows estimates of climate forcings since 1850 [41]. Positive forcings (in dark gray) contribute to global warming, negative forcings (in light gray), such as reflective aerosols, counteract the global warming.

TABLE 8.29
GREENHOUSE GASES

Gas	Formula	Current Inventory	Rate of Increase per Year	Warming Effectiveness per Mole	Relative Greenhouse Effect Contribution
Carbon dioxide	CO_2	356 ppm	+0.4%	1	50%
Methane	CH_4	1.74 ppm	+0.6%	23	20%
Nitrous oxide	N_2O	0.31 ppm	+0.25%	270	5%
CFC-11	$CFCl_3$	0.26 ppb	+2.1%	14,000	5%
CFC-12	CF_2Cl_2	0.47 ppb	+2.3%	19,500	10%
Tropospheric ozone	O_3	0.03 ppm	-0.51%	—	10%

Source: Data after Ref. 35.

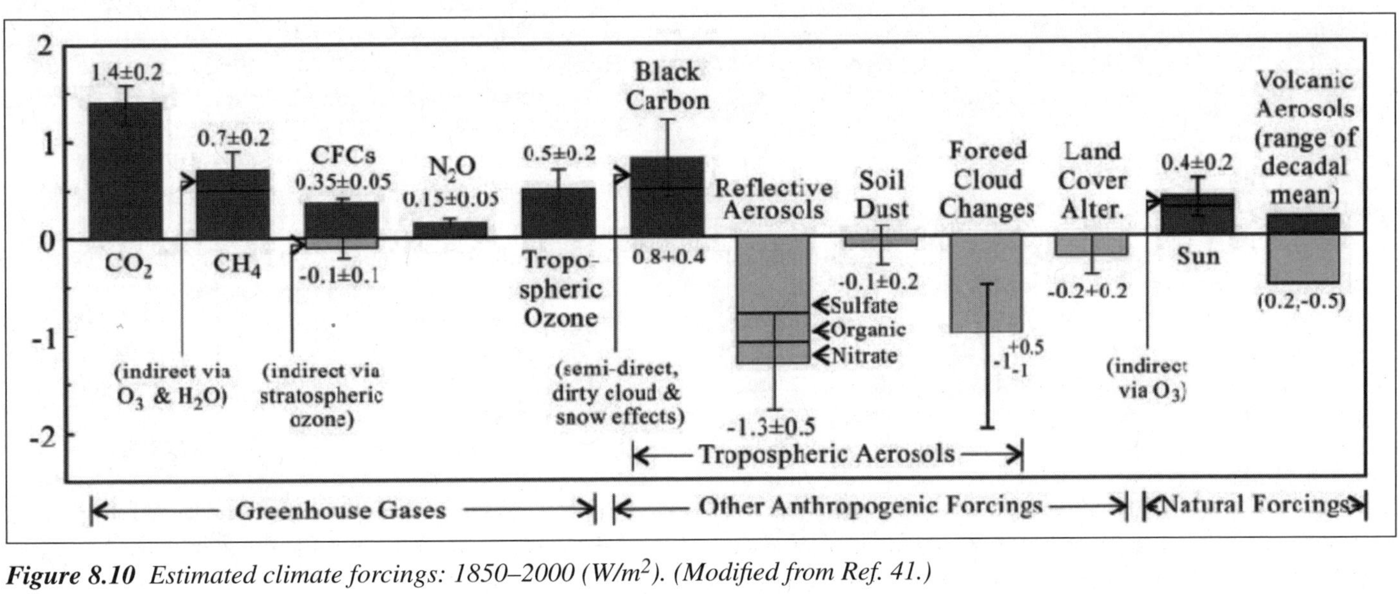

Figure 8.10 *Estimated climate forcings: 1850–2000 (W/m^2). (Modified from Ref. 41.)*

The sum of all positive forcings is 4.3 W/m^2; the sum of all negative forcings is –2.7 W/m^2. The resulting net forcing is 1.6 W/m^2. Most climate simulations provide similar net forcing figures, but typically assume fewer negative forcings than are shown in Figure 8.10, and also less positive forcing from black carbon, O_3, and methane [11]. The global warming contribution from black carbon, which is obviously related to diesel particulate emission, should be discussed in more detail.

Since the late 1990s, black carbon (soot) aerosols formed by incomplete combustion, including those from diesel engines, have been suspected of causing a positive climate forcing. Black carbon contributes to global warming through absorption of sunlight and through heating of the lower atmosphere. Black carbon may be also causing indirect forcings through its effect on snow and sea ice (reduction of reflectiveness), as well as by reduction of the negative forcing by sulfates, nitrates, and organic carbon aerosols. The total climate forcing of black carbon is estimated to be from +0.25 W/m^2 [11] to some 0.5 to 1.0 W/m^2 [41].

Diesel engines offer better thermal efficiency and fuel economy than their spark ignited counterparts, which leads to lower CO_2 emissions. According to estimates from the late 1990s, life cycle emissions of CO_2 from diesel engines were 14% lower than those from gasoline vehicles [42]. In another study, the well-to-wheel greenhouse gas emissions from diesel were estimated at about 13% less than those from gasoline [43]. This represents a significant greenhouse gas benefit of the diesel engine. However, in view of the climate forcing potential of carbon black, the overall global warming benefit of diesels appears to be uncertain. Indeed, only diesel engines with PM emissions controlled to low levels would offer full global warming advantage.

Despite the disagreement and uncertainty among scientists, global warming has become a serious environmental concern and a political issue. Policies to prevent further increases in greenhouse gas emissions are being considered. Some have already been implemented by most countries in the world. In 1997, a number of countries signed the Kyoto Protocol, which calls for an average reduction of greenhouse gas emissions (including CO_2, CH_4, CFCs, and N_2O) by 5.2% below 1990 levels by 2008–2012 [44]. By December 2001, the Kyoto Protocol has been signed by 84 countries and ratified by 46 countries.

8.6.5 Reduced Atmospheric Visibility

Reduced air visibility can be caused by natural sources, such as fog, volcanic eruptions, forest fires, or sand storms, as well as by air pollution. The air pollution causes special concern because, contrary to the natural source effects, its effect is permanent.

Because light is diffracted on suspended particles, reduced visibility is strongly related to the particulate matter pollution. Carbon particles generated by diesel engines are an important factor in darkening of skies in large urban areas. According to a study in the Netherlands, sulfate and nitrate particles have an even stronger effect than that of carbon particles [45]. Nitrate particulates can be formed in the atmosphere as a secondary pollutant from NO_x. Thus, a part of the nitrate pollution can be also traced back to diesel engines.

References

1. "National Air Toxics Program: The Integrated Urban Strategy," Notice 64 FR 38706, U.S. Environmental Protection Agency, Washington, DC, July 19, 1999, retrieved from www.epa.gov/ttn/uatw/urban/fr19jy99.pdf.
2. "Proposed Identification of Diesel Exhaust as a Toxic Air Contaminant," Report by the Staff of the California Air Resources Board and the Office of Environmental Health Hazard Assessment, April 22, 1998.
3. "Diesel Fuel and Exhaust Emissions: Environmental Health Criteria 171," World Health Organization, Geneva, 1996.
4. "Diesel Exhaust: A Critical Analysis of Emissions, Exposure, and Health Effects," Health Effects Institute, Cambridge, MA, April 1995.
5. Lenz H.P., and C. Cozzarini, *Emissions and Air Quality,* SAE International, Warrendale, PA, 1999.
6. "Health Assessment Document for Diesel Emissions (Draft)," EPA/600/8-90/057C, U.S. Environmental Protection Agency, Washington, DC, February 1998, retrieved from http://cfpub.epa.gov/ncea/cfm/recordisplay.cfm?deid=51822.
7. "Northern Front Range Air Quality Study," Report, Colorado State University, Denver, CO, 1998.
8. "Emission Standards for Locomotives and Locomotive Engines," Final Rule 63 FR 18978, U.S. Environmental Protection Agency, Washington, DC, April 16, 1998, retrieved from www.epa.gov/oms/regs/nonroad/locomotv/frm/loco1.pdf.
9. Watts, W.F., "Assessment of Occupational Exposure to Diesel Emissions," in *Diesel Exhaust: A Critical Analysis*, Health Effects Institute, Cambridge, MA, 1995, pp. 107–123.
10. "Diesel Emissions and Lung Cancer: Epidemiology and Quantitative Risk Assessment," Health Effects Institute, Cambridge, MA, June 1999, retrieved from www.healtheffects.org/Pubs/DieselEpi-C.pdf.
11. Houghton, J.T., *et al.* (Eds.), *Climate Change 2001,* Intergovernmental Panel on Climate Change, Cambridge University Press, Cambridge, UK, 2001.
12. Walsh, M.P., "Global Trends in Diesel Emissions Control—A 1999 Update," SAE Paper No. 1999-01-0107, SAE International, Warrendale, PA, 1999.
13. Schindler, K.-P., "Advances in Diesel Engine Technologies for European Passenger Vehicles," presented at 8th Diesel Emissions Reduction Conference (DEER), San Diego, CA, August 2002, U.S. Department of Energy, Washington, DC, 2002, retrieved from www.orau.gov/deer2002/Session8/Schindler.pdf.
14. "National Air Quality and Emissions Trends Report, 1999," Report EPA 454/R-01-004, Office of Air Quality Planning and Standards, U.S. Environmental Protection Agency, Washington, DC, March 2001, retrieved from www.epa.gov/oar/aqtrnd99.

15. "National Air Pollutant Emission Trends Update, 1970-1997," Report EPA-E-98-007, Office of Air Quality Planning and Standards, U.S. Environmental Protection Agency, Washington, DC, December 1998.
16. "National Atmospheric Emissions Inventory in the UK," Database maintained for the Department of the Environment by AEA Technology Environment, 1997, retrieved from www.aeat.co.uk/netcen/airqual/naei.
17. McCellan, R.O., "Health Effects of Diesel Exhaust: A Case Study in Risk Assessment (1985 Stokinger Lecture)," *AIHA Journal* **47,** 1986, p. 1.
18. Horvath, H., *et al.,* "Diesel Emissions in Vienna," *Atmos. Environ.* **22,** 7, 1988, pp. 1255–1269.
19. Lies, K.H., "Unregulated Motor Vehicle Exhaust Gas Components," Research and Development, Volkswagen AG, Wolfsburg, Germany, 1989.
20. "Health Assessment Document for Diesel Engine Exhaust," Report EPA/600/8-90/057F, U.S. Environmental Protection Agency, Washington, DC, May 1, 2002, retrieved from http://cfpub.epa.gov/ncea/cfm/recordisplay.cfm?deid=29060.
21. "Carcinogenic Effects of Exposure to Diesel Exhaust," Current Intelligence Bulletin No. 50, DHHS (NIOSH) Pub. No. 88-116, NTIS Pub. No. PB-88-252-994, National Institute for Occupational Safety and Health, National Technical Information Service, Springfield, VA, 1988.
22. Grenier, M., and D. D'Aguiar, "DPM Exposure in Canadian Mines," in *Proceedings of the Mining Diesel Emissions Conference (MDEC),* CANMET, Toronto, November 1999.
23. Benner, B.A., Jr., G.E. Gordon, and S.A. Wise, "Mobile Sources of Atmospheric Polycyclic Aromatic Hydrocarbons: A Roadway Tunnel Study," *Environ. Sci. Technol.* **23,** 1989, pp. 1269–1278.
24. Yu, M-L., and R.A. Hites, R.A., "Identification of Organic Compounds on Diesel Engine Soot," *Anal. Chem.* **53,** 1981, pp. 951–954.
25. Atkinson, R., et al., "A Survey of Ambient Concentrations of Selected Polycyclic Aromatic Hydrocarbons (PAH) at Various Locations in California [Final Report]," Contract no. A5-185-32, California Air Resources Board, Sacramento, 1988.
26. Zweidinger, R.B., J.E. Sigsby, S.B. Tejada, F.D. Stump, D.L. Dropkin, W.D. Ray, and J.W. Duncan, "Detailed Hydrocarbon and Aldehyde Mobile Source Emissions from Roadway Studies," *Environ. Sci. Technol.* **22,** 1988, pp. 956–962.
27. "Air Quality Criteria for Carbon Monoxide," Report EPA 600/P–99/001F, U.S. Environmental Protection Agency, Washington, DC, June 2000, retrieved from www.epa.gov/ncea/coabstract.htm.
28. Penney, D.G., *Carbon Monoxide,* CRC Press, Boca Raton, FL, 1996.
29. Cohen, S.J., *et al.,* "Carbon Monoxide and Survival from Myocardial Infarction," *Arch. Environ. Health* **19,** 1969, pp. 510–517.
30. Holgate, S.T. (Chairman), U.K. Department of Health, "Quantification of the Effects of Air Pollution on Health in the United Kingdom," The Stationery Office, London, 1998.
31. Hilborn, J., and M. Still, "A State of the Environment Report: Canadian Perspectives on Air Pollution," SOE Report No. 90-1, Environment Canada, Ottawa, September 1990.
32. Watson, A.Y., and G.M. Green, "Noncancer Effects of Diesel Emissions: Animal Studies," in *Diesel Exhaust: A Critical Analysis,* Health Effects Institute, Cambridge, MA, 1995, pp. 139–164.
33. Cohen, A.J., and M.W.P. Higgins, "Health Effects of Diesel Exhaust: Epidemiology," in *Diesel Exhaust: A Critical Analysis,* Health Effects Institute, Cambridge, MA, 1995, pp. 251–292.
34. "10th Report on Carcinogens: Diesel Exhaust Particulates," National Toxicology Program, U.S. National Institute of Environmental Health Sciences, Research Triangle Park, NC, retrieved from http://ehp.niehs.nih.gov/roc/tenth/profiles/s069dies.pdf.
35. Sher, E., "Environmental Aspects of Air Pollution," in *Handbook of Air Pollution,* E. Sher (Ed.), Academic Press, Boston, 1998, pp. 27–41.
36. "Air Quality Index Reporting," Final Rule 64 FR 42530, U.S. Environmental Protection Agency, Washington, DC, August 4, 1999, retrieved from www.epa.gov/ttn/oarpg/t1/fr_notices/airqual.pdf.
37. Fujita, E.M., "OHVT Weekday/Weekend Ozone Study in the South Coast Basin," presented at 7th Diesel Emissions Reduction Workshop (DEER), Portsmouth, VA, U.S. Department of Energy, Washington, DC, August 2001, retrieved from www.osti.gov/hvt/deer2001/fujita.pdf.
38. Fujita, E.M., W. Stockwell, D. Campbell, R. Keislar, J. Sagebiel, W. Goliff, B. Zielinska, and D. Lawson, "Weekend/Weekday Ozone Study in the South Coast Air Basin," presented at 8th Diesel Emissions Reduction Conference (DEER), San Diego, CA, August 2002, U.S. Department of Energy, Washington, DC, retrieved from www.orau.gov/deer2002/Session5/fujita.pdf.
39. Degobert, P., "Automobiles and Pollution," SAE/Editions Technip, SAE International, Warrendale, PA, 1995.
40. Hansen, J.E., "The Forcing Agents Underlying Climate Change," Statement before the Committee on Commerce, Science and Transportation United States Senate, 2001, retrieved from www.giss.nasa.gov/research/forcings/senate/Senate-complete.pdf.
41. Hansen, J.E., and M. Sato, "Trends of Measured Climate Forcing Agents," *Proc. Natl. Acad. Sci.* **98,** pp. 14778–14783, 2001, retrieved from www.giss.nasa.gov/gpol/papers/2001/2001_HansenSato.pdf.
42. Walsh, M.P., "Global Diesel Emission Trends," *Automotive Engineering,* June 1998, pp. 114–118.

43. "Well-to-Wheel Energy Use and Greenhouse Gas Emissions of Advanced Fuel/Vehicle Systems—North American Analysis," General Motors Corporation/Argonne National Laboratory Report, Argonne, IL, June 2001, retrieved from www.transportation.anl.gov/ttrdc/pdfs/TA/163.pdf.
44. "Kyoto Protocol to the United Nations Framework Convention on Climate Change," United Nations Framework Convention on Climate Change, 1997, retrieved from http://unfccc.int/resource/convkp.html.
45. Diederen H., *et al.,* "Visibility Reduction by Air Pollution in the Netherlands," *Atmos. Environ.* **19,** 1985, pp. 377–383.

Chapter 9

Measurement of Emissions

9.1 Overview

9.1.1 Introduction

Emissions can be characterized, regulated, or controlled only if they can be accurately measured. The increased health and environmental concerns about diesel emissions resulted in the development of a wide range of measurement techniques of different levels of sophistication, equipment cost, and accuracy that suited a variety of applications. Many techniques, especially those used for regulatory purposes, are highly standardized to produce comparable results even if performed at different testing laboratories. The main types of diesel emission measurements can be grouped as follows:

- *Laboratory testing:* regulatory testing, emissions research, and engine and emission control system development
- *Field testing:* mobile emission laboratories, on-vehicle measurements, inspection and maintenance programs, remote emission measurement, emission-assisted equipment maintenance, and occupational health measurements.

Laboratory emission testing methods generally use very complex and sophisticated equipment to deliver the highest possible accuracy and repeatability. Test methods used for regulatory purposes, such as engine/vehicle emission certification or compliance testing, are highly standardized. Detailed descriptions of the measurement setup, type of equipment, and test procedures are an important part of every emission regulation. These standard methods cover the measurement of regulated pollutants, which traditionally include CO, HC, NO_x, and PM, all of which are measured in units of mass. Thanks to the existence of standard methods, measurements of regulated pollutants taken at different laboratories usually correlate very well. Emission research laboratories also perform measurements of a number of unregulated emissions, using techniques that may be subject to various nonregulatory standards and recommended procedures, or else using nonstandard methods. The focus in recent years has been on the characterization of diesel particulate matter through the determination of such parameters as particle number concentrations and size distributions. Several laboratories have developed highly accurate methods for these and other nonregulated emissions. However, if nonstandard methods are used, results from different laboratories may not be comparable. This is especially true in measurements of particulate matter, which, being a mix of many chemical compounds, is defined by its measurement method.

Measurement equipment and methods used for measuring engine emissions in the field vary widely, from mobile laboratories, with capabilities comparable to those of stationary emission laboratories, to simple, low-budget tools that can offer only very approximate results. Examples of high-end mobile laboratories include the diesel aerosol sampling rig built by the University of Minnesota [1] or the transportable heavy-duty vehicle chassis dynamometer and emission test facility developed and operated by the University of West Virginia [2].

A new class of vehicle emission test equipment, first developed in the late 1990s, is the highly accurate, onboard emission analyzers [3]. These analyzers can be installed on a vehicle to measure real-life emissions with a laboratory level of accuracy (Figure 9.1). They include laboratory-class analyzers, packaged in a portable unit, and a simplified sampling system (wet-based analyzers are typically used). Onboard units can provide results as both raw gas concentrations and mass-based emission factors, such as grams per kilometer or grams per kilowatt-hour. The exhaust gas flow rate and other parameters needed to convert concentrations into mass-based emissions can be obtained from the engine control module. On-vehicle testing opens several new possibilities for both engine developers and emission regulatory agencies. It can be used for compliance testing of new and aging vehicles (including NTE limits and "emission defeat" strategies), development of mobile source emission models, development of engine control strategies, aftertreatment systems, and onboard diagnostic (OBD) systems.

Another area of emission measurement is vehicle maintenance. Several U.S. states and a number of countries worldwide operate mandatory vehicle inspection and maintenance (I&M) programs, where vehicles must pass a periodic emission check. The methods vary greatly, from relatively sophisticated chassis dynamometer tests (IM240 in the United States) to a simple

Figure 9.1 *On-vehicle emission test system on a passenger car. (Courtesy of Sensors Inc.)*

emission measurement at engine idle condition. In the case of heavy-duty diesel engines, the most common procedure is to measure smoke opacity during engine acceleration. Technologies for remote exhaust emission sensing are being developed [4] that are also likely to be used in future I&M programs.

All of the above-mentioned areas of emission testing were concerned with the measurement of the tailpipe emissions from an engine or vehicle. For several applications, including occupational health practice, the focus is on the ambient air exposure to pollutants, including diesel exhaust pollutants. While certain principles of emission sampling and analysis have general validity, the measurement of ambient diesel aerosol has developed into a separate area of expertise, as discussed later in Section 9.4.

9.1.2 Emission Test Cycles

Engine emissions depend critically on engine operating conditions, such as speed, load, or transients. Before emissions can be measured, the engine has to be subjected to the desired set of conditions, which must be accurately controllable and reproducible. In all cases where highly accurate emission measurements are desired, such as in laboratory measurements, this task is achieved by running an engine or vehicle over a standardized test cycle on an engine or vehicle dynamometer.

The dynamometer is a device that can apply controlled speed and load conditions on an engine or vehicle. Dynamometers can be divided into two classes [5]: (1) chassis (rolling road) dynamometers, for testing of vehicles, and (2) engine dynamometers, for engine testing. In the chassis dynamometer, the complete vehicle is placed on rollers and operated by a driver following a prescribed test cycle as closely as possible. The task of the dynamometer is to reproduce the load and inertia of the vehicle. The inertia of the vehicle during speed changes is simulated by a set of flywheels that are coupled to the rollers through a clutchable chain drive. The mechanical energy transmitted by the wheels to the rollers is absorbed using a brake, which is integrated with the rollers through a shaft. The three common types of brakes are as follows:

- *Water brake,* in which the brake rotor passes its energy to a stream of pressurized water
- *Eddy current brake,* which uses electromagnetic braking with eddy currents dissipated by the rotor into a stream of cooling water
- *Electric brake,* which uses a DC or asynchronous electric motor/generator, which can return the absorbed energy to the grid. Contrary to the other two types, the electric brake can not only absorb, but also supply power to the vehicle, for example, to drive the wheels during deceleration.

Chassis dynamometers are most commonly used for testing of passenger cars and other light-duty vehicles.

In the engine dynamometer, the engine is coupled directly to the brake. In this configuration, the dynamometer has to simulate the behavior of the complete vehicle and transmission system, including the extremely rapid torque and speed response during transients. The same three general types of brakes just listed can be used in engine dynamometers. Electric dynamometers (DC or AC) are required to run test cycles that include motoring segments, where the dynamometer is driving the engine, such as the U.S. Federal Test Procedure (FTP) transient or the European Transient Cycle (ETC). Engine dynamometers are most commonly used for testing of heavy-duty engines, as illustrated in Figure 9.2.

Engine (emission) test cycles are sequences of speed and load conditions performed on a dynamometer. Emissions measured on chassis dynamometers are usually expressed in grams of pollutant per unit of traveled distance, for example, grams per kilometer or grams per mile. Emissions measured over an engine dynamometer test cycle are typically expressed in grams of pollutant per unit of mechanical energy delivered by the engine, such as g/kWh or g/bhp-hr. Depending on the character of speed and load changes, cycles can be divided into

Figure 9.2 *Heavy-duty engine on engine dynamometer testing stand.*

steady-state cycles and transient cycles. Steady-state cycles are sequences of constant engine speed and load modes. Emissions are analyzed for each test mode. Then the overall emission result is calculated as a (weighted) average from all test modes. In a transient cycle the vehicle (engine) follows a prescribed driving pattern that includes accelerations, decelerations, changes of speed and load, and so forth. An example of a transient test cycle—the FTP transient test used for emission certification of heavy-duty engines in the United States—is illustrated in Figure 9.3. The final test results can be obtained either by analysis of exhaust gas samples collected in plastic bags over the duration of the cycle or by electronic integration of a fast-response, continuous emission measurement.

In theory, emission test cycles reflect certain real-life driving patterns and associated emission levels as closely as possible. For instance, the FTP transient test shown in Figure 9.3 includes segments that were developed based on the analysis of freeway and nonfreeway driving by heavy-duty vehicles in Los Angeles and New York City. In practice, it is not possible to account for the variety of real-life driving conditions in a limited set of test cycles. A large body of literature exists to indicate that emissions depend on a large number of factors, many of them related to the driver's behavior, as follows [6]:

- Acceleration is one of the most influential factors for CO, HC, and PM emissions, especially at high speed. PM emissions in diesel engines measured over transient cycles can be 30% to 50% or more higher compared to steady-state measurements.
- Speed and load have been long recognized as critical factors influencing emissions. Generally, all emission levels increase with increased speed. High loads produce high NO_x levels, as well as increases in other emissions.
- Auxiliary loads, such as air conditioning, ventilation equipment, or cabin heating can cause additional emissions not accounted for in most test cycles.
- Driver-related factors that affect emissions include smoothness and consistency of vehicle speed and sharp accelerations. In passenger cars, studies found that aggressive driving, which accounts for only 2% of the driving time, may produce 40% of total emissions [6].

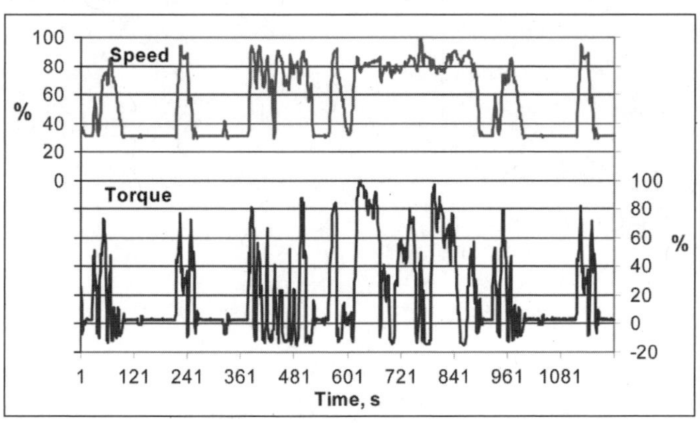

Figure 9.3 *U.S. FTP transient test cycle.*

- Road-related factors are not accounted for by legislative test cycles, which all assume straight roads with zero gradient. In reality, road ascents require increased engine loads, producing consistently high NO_x emissions, while descents result in low NO_x levels. Preliminary light-duty vehicle data indicate that horizontal road curvatures require increased engine torque and thus have an influence on emissions [6].

New test cycles are continuously being developed to account for an increasingly wider variety of driving patterns. One of the consequences is lack of uniformity among existing regulatory test methods. Regulatory authorities in different countries have not been unanimous in adopting emission test procedures and many types of cycles are in use. In general, specific engine emissions that were measured on different test cycles may not be comparable even if they are expressed or recalculated into the same units of measure. This should be kept in mind whenever comparing emission standards from different countries.

In many cases of field emission measurements, a dynamometer is not available. Simplified methods that are often used to ensure repeatable engine operating conditions include steady-state measurements at idle conditions and at low and/or high engine speeds. With certain types of vehicles, it is also possible to apply a relatively repeatable level of engine load through the engagement of certain powertrain components or auxiliaries. For instance, in emission-assisted maintenance procedures developed for underground mining vehicles in the United States, engines are to be loaded using a torque converter stall method [7]. A portable gas analyzer is then used to measure CO to determine if engine maintenance is needed. The analyzer is also used to measure CO_2 or O_2 to verify proper engine loading.

9.1.3 Sampling and Analysis

9.1.3.1 Overview

The measurement of gases and particulate matter in engine exhaust is divided into two steps: (1) sampling and (2) analysis. The task of the exhaust gas sampling system is to provide a sample of gas from the vehicle tailpipe or from the engine exhaust manifold to the analyzers. The sample must be uniform, must be representative of the product analyzed, and must correspond to the actual emissions from the engine as tested on the driving cycle. Both gas and PM samples can be either collected for later analysis or analyzed in real (or near-real) time by *in situ* instruments. In the case of gases, samples are often collected in plastic bags. If continuous analysis is used, the gas sample is preconditioned in the sampling system as needed and fed to the analyzers. In the traditional method of measurement of diesel particulates, which still dominates in regulatory measurements, a PM sample is collected on a sampling filter, which is then analyzed gravimetrically. More recently, a number of *in situ* instruments for the measurement of various characteristics of diesel PM have been also developed.

An important function of the sampling system is to condition the exhaust gas sample to meet the inlet sample specifications of the particular analytical instruments. Gas analyzers typically require particulate matter to be removed from the sample. Thus, particulate filters are necessary components of the sampling system. Some of the analytical instruments (so-called "dry" instruments, such as most IR analyzers) require that the sample be cooled and dried. Water vapor should be condensed and removed from the sample in a way that minimizes losses of other gases due to their solubility in the condensate. Several instruments, for example, the flame ionization detector (FID) hydrocarbon analyzer and chemiluminescence NO_x analyzer, are available in heated (wet) versions to prevent losses due to condensation in the sampling system. All sampling system elements that are in contact with the sample (sampling lines, pumps, filters, valves) are also heated and maintained at temperatures of about 180 to 200°C.

Diesel particulate matter sampling, according to most engine and vehicle emission regulations, is done from diluted exhaust gas. This is achieved by turbulent mixing of exhaust gases with air in a dilution tunnel. The dilution air is drawn through a polished and heated stainless steel tube, typically 250 to 300 mm in diameter, at a velocity of about 10 m/s. The exhaust gas from the engine enters the upstream end of the tube (the dilution tunnel) and turbulent mixing takes place. The full-flow dilution tunnel is a bulky and expensive device. Partial-flow mini-dilution tunnels have been developed to address this problem [8]. Commercially available mini-dilution tunnels have been certified to be fully equivalent with the U.S. Environmental Protection Agency (EPA) full-flow dilution systems under steady-state particulate sampling conditions. A substantial amount of research is being conducted to develop partial dilution systems that would be equivalent to full-flow dilution under transient test conditions [9, 10].

Exhaust gas analysis performed on a steady-state test is usually done with the gas sample continuously flowing through instruments. The sampling system has to redistribute the sample to all analyzers at the required sample flow rates. For analysis on transient cycles, exhaust gases are either collected in plastic Teflon bags for later steady-state analysis, or continuous measurements are taken by fast-response analyzers.

9.1.3.2 Constant Volume Sampling

The constant volume sampling (CVS) method has been adopted by most countries for regulatory exhaust emission testing. In the CVS method the exhaust gases are diluted with air to maintain a constant total flow rate (air + exhaust) under all running conditions. The constant flow rate enables an easy and accurate calculation of mass-based emission factors, as required by emission regulations, based on the concentration values measured by analyzers. On the negative side, very sensitive analyzers must be used that are capable of measuring low levels of emissions from the diluted exhaust.

The heart of this type of sampling system is the CVS unit, which draws filtered air into the dilution tunnel. The CVS unit can be either a positive displacement pump (PDP) or, more commonly, a vacuum pump that makes the gas flow through a critical flow venturi nozzle (CFV). Critical flow orifices are simple and inexpensive devices widely used for flow rate control. Their working principle is based on the phenomenon that, once a certain level of pressure drop is reached and the speed of sound is achieved in the venturi throat, the velocity will not increase with a further reduction in downstream pressure [11]. A layout of a generic CVS sampling system is shown in Figure 9.4. This sampling setup can be used to perform the U.S. FTP transient cycle. Exhaust gases collected under transient conditions are mixed with air in the full-flow primary dilution tunnel. A sample for particulate measurement is drawn from that tunnel into a small secondary dilution tunnel, further mixed with air, and collected on particulate filters. Samples for continuous gas-phase measurements are drawn from the primary dilution tunnel. The sampling lines must be heated to about 190°C to avoid condensation and sample losses. Gas samples for bags are drawn from downstream of the heat exchanger. In the case of gases that may suffer losses in the heat exchanger (HC, NO_x),

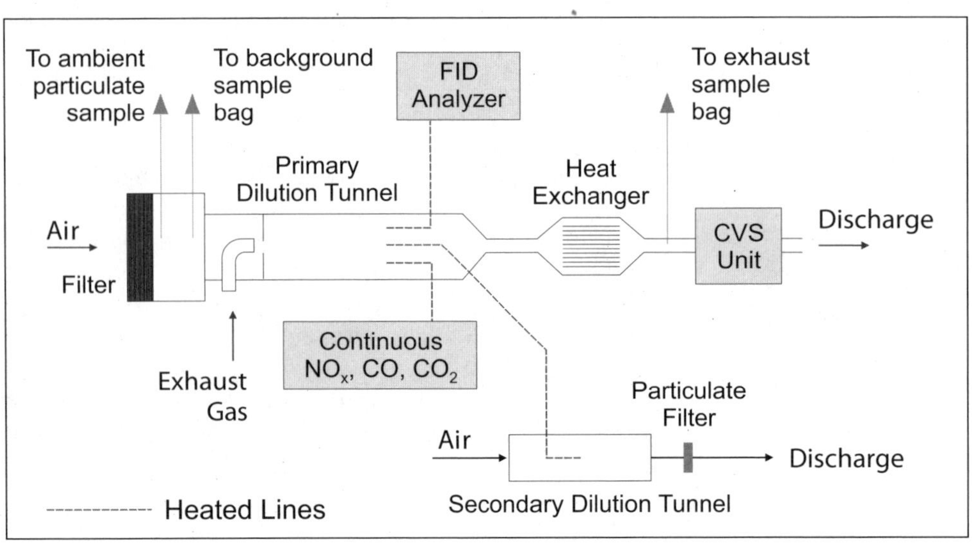

Figure 9.4 Constant volume sampling system.

heated bag sampling from upstream of the heat exchanger is the preferred laboratory practice.

The highest risk of sample loss during sampling and analysis exists with hydrocarbons that have a high boiling point. Diesel HC analysis should be performed by a continuous, heated analyzer even if other gases are measured from bags by steady-state instruments. An integrator is used with the continuous analyzer to calculate the cumulative emissions over the cycle time.

Raw (undiluted) exhaust gas measurements are sometimes performed for noncompliance testing. Undiluted sampling can also be used for all regulated pollutants other than particulates in nonroad engines. Generally, raw exhaust sampling requires more sophisticated sample conditioning systems. Because of high concentrations of moisture and particulates and high temperatures, it is not an easy task to prevent losses of material through condensation or by dissolving acid gases in the condensate.

Specialized sampling methods have been developed in the measurement of ambient aerosols, particularly in the area of measurement of workers' exposure to diesel particulates in occupational health environments. In sampling particulates from the ambient air, one collects a mixture of particles that may originate from a number of sources, both diesel and nondiesel. Occupational sampling methods commonly distinguish between diesel and other-source particulate matter based on the small, submicron size of diesel particles. This is achieved by particle size classifiers, such as cyclones and impactors, which are a part of the particulate sampling train in occupational health applications.

9.1.3.3 Methods of Analysis

Analytical methods for regulated emissions are specified in the pertinent emission regulations. For regulatory measurement of gas compounds, the commonly used methods are:

- *CO*, CO_2: nondispersive IR analysis (NDIR)
- *HC:* FID
- NO_x: chemiluminescence analysis
- O_2: paramagnetic analyzer.

The mass of diesel particulate matter is measured gravimetrically, by weighing the mass increase of the sampling filter, after preconditioning at controlled temperature and humidity conditions. A host of other, nonregulated parameters of diesel particulates, such as particle numbers, sizes, and size distributions, are measured using various nonstandardized methods and instruments.

Many other nonregulated emissions, some related to the particulate phase (soluble organic fraction, sulfates, polynuclear aromatic hydrocarbons) and others to the gas phase (carbonyl compounds, aldehydes), are measured using specialized laboratory methods described in the literature [12, 13].

9.2 Gas-Phase Measurements

9.2.1 Techniques of Gas Analysis

Methods that are commonly used for the analysis of exhaust gas constituents are listed in Table 9.1 [5]. All of these techniques are based on a selected physical property of the measured gas. In some cases, the measured species undergoes a chemical reaction in the analyzer but even then only an accompanying physical property is directly measured. Concentrations of gases are calculated by comparing the instrument signal with that generated using a calibration gas of precisely known concentration.

The analytical techniques listed in Table 9.1 have been adopted or developed for measurements of exhaust gases from internal combustion engines. They offer reasonably high accuracy and insensitivity to mutual interference between gas species. The use of many of these methods has been stipulated by various exhaust emission regulations for certification and compliance testing.

In addition to the methods listed in Table 9.1, electrochemical cells for measuring of many gases have been developed. Many portable analyzers utilize such electrochemical sensors. Instruments of this type are commonly used in occupational health and safety practice for ambient air quality evaluation. Portable electrochemical cell exhaust gas analyzers are also available for emission measurements in field conditions. Electrochemical sensor instruments are usually less accurate and, as

TABLE 9.1
TECHNIQUES OF EMISSION ANALYSIS

Technique	Gas	Typical Range	90% Response Time
Nondispersive infrared	CO	0–3000 ppm	2–5 s
	Co_2	0–20%	2–5 s
Chemiluminescence	NO_x	0–10,000 ppm	1.5–2 s
Flame ionization detector	Total HC	0–10,000 ppm	1–2 s
Fast FID	Total HC	0–2000 ppm	1–2 ms
Fourier transform infrared (FTIR)	NO_x, some HC, etc.	Various	5–15 s
Paramagnetic analyzer	O_2	0–25%	1–5 s

Source: Data from Ref. 5.

a rule, are not allowed by regulated emission test procedures. The basic sources of error when using electrochemical instruments are:

- *Cross-sensitivities:* Electrochemical cells may be sensitive to other compounds present in the measured gas. For example, the NO cross-sensitivity of a CO cell may be as high as 30%, depending on the quality of the cell.
- *Deterioration of the cell:* The life span of electrochemical cells is usually 1 to 2 years and a gradual decline in performance is frequently observed.
- *Simple sampling systems:* Simple sampling in portable instruments may cause several errors related to the temperature or humidity of the measured gas. If condensation occurs in the sampling line, acid gases will dissolve in the condensate and disappear from the gas phase.

Because of their slow response, today's methods of emission analysis are not very suitable for true transient engine testing. Analyzers designed for steady-state work must be accurate, sensitive, and stable, and thus they tend to have slow response times and to be well dampened. Analyzers for transient work must respond very quickly, preferably within a few milliseconds. With the exception of the fast FID hydrocarbon analyzer, there are at present no very fast response sensors for emission testing. A further problem in transient testing arises from the time and distance lags associated with the positioning of the exhaust gas sampling points, located downstream of the engine. Mathematical techniques are used for reconstructing the true signal from the instrument signal, taking into account sampling delays and instrument response characteristics [14].

9.2.1.1 Nondispersive Infrared

The nondispersive infrared analyzer (NDIR) makes use of the property of certain gases of selectively absorbing infrared radiation over a narrow range of wavelengths. Molecules that are formed of atoms of at least two different elements (CO, NO, NO_2, SO_2, etc.) absorb infrared radiation by converting the light energy into vibration/rotation energy of the molecules, which can be detected in the form of heat. NDIR instruments use the total absorption over a given wavelength range, rather than using a monochromator (prism or grating) to separate the light. The spectral range used is the near and middle infrared (a 0.75- to 30-μm wavelength).

One of several possible NDIR instrument layouts is illustrated in Figure 9.5. The infrared emitters (E) are heated filaments that act like blackbody radiators over a wide range of wavelengths. The detection receiver (D) is a volume of gas identical to that to be measured in the analyzed gas mixture. Thus, the receiver remains insensitive to absorption by other chemicals as long as their absorption ranges do not overlap. Most of the instruments use twin beams generated by two identical infrared sources. The incident infrared beams are modulated by a rotating chopper (M).

One beam passes through an absorption cell (A) containing the analyzed gas. The other beam passes through a reference cell (R) filled with an inert gas, such as nitrogen. After passing

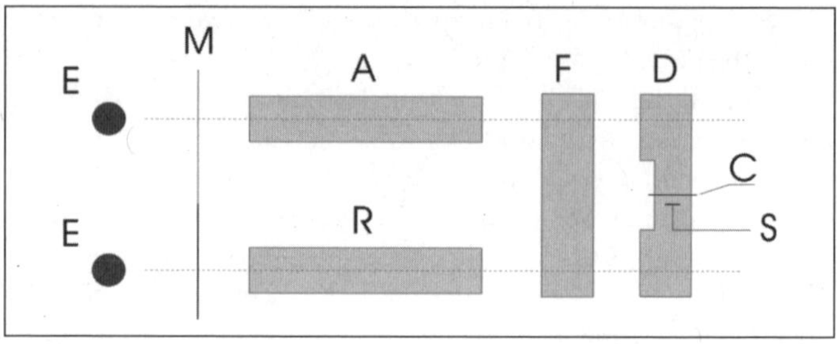

Figure 9.5 *NDIR analyzer.*

through a compensation cell (F) the two beams pass through two chambers filled with the gas to be measured in its pure state that comprise a differential sensor (D). It is a Golay pressure sensor consisting of two identical chambers separated by a diaphragm (C) that moves between plates of a capacitor. The difference in heat received in each chamber due to the presence of the analyzed gas on the path of one of the half-beams causes increased pressure on one side of the diaphragm. The pressure is modulated by the rotating chopper (M) turning at a frequency of about 10 Hz. The output signal (S) from the cell represents the variations of capacitance of the system. The two detector chambers are connected by a capillary in some instruments to compensate for ambient temperature gradients.

NDIR is sensitive to pressure, so this parameter must remain constant during the calibration and analysis. It is also sensitive to water vapor. Materials used in some IR analyzers (alkaline earth halogenide windows) are also sensitive to water. Gas samples for IR analysis are thus prepared cold and dry.

NDIR analyzers can be used to measure many gases, including CO, CO_2, CH_4, NO, N_2O, SO_2, HCl, and hydrocarbons. They are mainly used for CO and CO_2 in diesel and automotive exhaust analysis.

9.2.1.2 Chemiluminescence Analyzer

The chemiluminescence detector (CLD) utilizes a chemical reaction between nitric oxide and ozone:

$$NO + O_3 = NO_2^* + O_2 = NO_2 + O_2 + \text{photon} \qquad (9.1)$$

Ozone is produced in the instrument by a high-voltage discharge in oxygen. The excited nitrogen dioxide molecules spontaneously return to the normal state. This transition is accompanied by emission of red light with photon emissions in the 0.6- to 3-μm band. The light emission is filtered to eliminate interference from other gases, such as CO, SO_2, and unsaturated hydrocarbons, and measured by a photomultiplier. The signal is proportional to the NO concentration in the sample.

Total NO_x can be also measured by the same instrument. The gas sample passes through a stainless steel or molybdenum furnace where, at elevated temperature, NO_2 is decomposed to NO and oxygen. NO_2 concentration may be calculated as a difference between the total NO_x and NO results.

CLD instruments offer a wide measuring range with a fast response time. Their drawbacks are that interference is caused by water vapor and carbon dioxide and they require complex and expensive equipment, including such additional components as vacuum pumps and a deozonizer [15].

9.2.1.3 Flame Ionization Detector

The flame ionization detector (FID) depends on the production of free electrons and positive ions during the combustion of hydrocarbons in a nonionized hydrogen flame. A current, in the range of 10 pA, is generated between the burner and an electrode raised to 100 to 300 V. The current is a function of the number of carbon atoms that pass through the flame.

"Hydrocarbons" as measured by the FID analyzer include any compounds containing carbon. The results of measurements are reported as parts per million by volume of carbon (ppm C or ppm C1) measured in comparison with an aliphatic hydrocarbon reference, usually propane or hexane. The reader should note that "ppm C1" is not a real volumetric unit, neither does it have any relation to the true volumetric concentration of hydrocarbons in the sample. Rather, it is a customary unit corresponding to the quantity of carbon atoms in the analyzed gas. The FID response for organic compounds may also deviate from real concentrations, depending on the particular molecule. This is especially true for oxygenated compounds, such as alcohols, aldehydes, ketones, nitro-compounds, or ethers.

The flame in an FID instrument is usually generated by combusting a 40% H_2/60% He mixture in air. The FID analyzer is influenced by any oxygen present in the sample. To minimize this effect, zero and reference gases are used that have an O_2 content close to that of the gas to be measured. FID instruments are unaffected by the water vapor in the analyzed gas.

9.2.1.4 Fast FID

The fast FID is a miniaturized version of the regular FID that is capable of a response time measured in milliseconds and can thus be used for true transient testing as well as for in-cylinder and exhaust manifold measurements.

9.2.1.5 Fourier Transform Infrared

The Fourier transform infrared (FTIR) analyzer generates interferograms of two IR beams of changing optical path lengths. The beams originate from the same source. One of them is reflected on a fixed mirror and the second on a vibrating mirror. Differences in optical paths are induced by the amplitude of the vibrating mirror. When the beams recombine, interference fringes (interferogram) are created that contain all of the information on the spectral distribution of the source radiation. When that radiation is sent to the gas mixture, spectral components corresponding to the absorption spectra of the different gases are absorbed. The light intensity distribution is attenuated by the presence of the chemicals. Results are generated by a computer that runs the mathematical Fourier transform to obtain the intensity distribution as a function of wavelength from the intensity distribution as a function of the optical path. Particular gas concentrations are obtained from the intensity distribution on the basis of their known IR absorption spectra. Concentration distributions of different chemicals may be obtained simultaneously and selectively in the FTIR instrument.

Although the response time of FTIR instruments is rather slow (more than 5 s) the actual measurement cycle is less than 1 s, thus allowing real-time measurements. The method is particularly useful for analysis of NO, NO_2, CO, CO_2, short-chain alcohols, hydrocarbons, and aldehydes (e.g., methanol, ethanol, formaldehyde), as well as several other compounds. FTIR is not suitable for measuring the longer chain HCs that are typical of diesel exhaust.

9.2.1.6 Magnetic Methods

Some gases including oxygen and nitrogen oxides exhibit paramagnetic properties. When a paramagnetic gas is subjected to a heterogeneous magnetic field, it undergoes a force directed toward increasing fields. This magnetic force can be measured in a number of ways, which is the principle of a paramagnetic gas analyzer. In practice, paramagnetic analyzers are used to measure oxygen concentration in diesel or automotive exhaust gases.

9.2.2 Analysis of Exhaust Pollutants

9.2.2.1 Carbon Monoxide and Carbon Dioxide

CO and CO_2 are commonly measured by the NDIR technique [16]. The accuracy of measurement is essentially governed by the calibration. Samples for analysis are cooled and dried.

Heated NDIR (h-NDIR) analyzers are being developed, mainly for onboard measurement applications, which allow for measurement under wet gas conditions [17]. Wet-based NDIR analyzers must correct for the interference from water vapors by calculation.

Good results in measuring CO and CO_2 are also obtained by the FTIR analyzer.

9.2.2.2 Hydrocarbons

Total hydrocarbons (THCs) are measured by the FID technique. A heated FID should be used, especially for diesel applications, to avoid condensation and loss of heavier hydrocarbons. Nonmethane hydrocarbons (NMHCs) can be measured by chromatographic separation and subtraction of methane from the THCs.

NDIR analyzers were formerly also used to measure exhaust hydrocarbons. Today this method has been abandoned because it delivered poor results. NDIR instruments yielded different responses for different hydrocarbon species. Table 9.2 compares the average responses of an NDIR and FID for several hydrocarbons [18]. Both instruments were calibrated with *n*-hexane. It is apparent that the NDIR response was too low for all classes of unsaturated hydrocarbons.

TABLE 9.2
RELATIVE RESPONSE OF NDIR AND FID ANALYZERS

Hydrocarbon	NDIR	FID	Hydrocarbon	NDIR	FID
Paraffins			**Olefins**		
Methane	30	110	Ethylene	9	100
Ethane	100	102	Propylene	31	100
Propane	103	100	1-Butene	53	—
i-Butane	106	100	1-Pentene	57	—
i-Pentane	99	100	1-Hexene	61	—
n-Pentane	104	102	**Acetylenics**		
n-Hexane	100	100	Acetylene	1	99
n-Heptane	97	101	Methylacetylene	16	99
Aromatics			**Ethylacetylene**	**32**	**99**
Benzene	2	98			
Toluene	13	98			

Source: Data from Ref. 18.

9.2.2.3 Nitrogen Oxides

Nitrogen oxides (NO_x) are measured using the chemiluminescence method. A standard chemiluminescence instrument requires a cold, dry sample to avoid condensation and dissolving of NO_2 in the condensate. Heated versions of chemiluminescence analyzers operate with hot and wet samples. The measured gas in this analysis is NO. A furnace to reduce NO_2, which allows us to measure total NO_x, is a standard feature on most analyzers. NO_2 concentrations can be calculated from the difference of NO_x and NO.

FTIR spectroscopy also yields very good accuracy for measuring NO_x. It measures NO and NO_2 directly and simultaneously. Total NO_x can be calculated as a sum of NO and NO_2.

Ultraviolet (UV) spectroscopy is still another method used to measure NO_x in vehicle exhaust [15]. In comparison to CLDs, UV analyzers may offer less noise, no water and CO_2 interference, and simpler and less expensive equipment.

9.3 Particulate Matter Measurements

9.3.1 Introduction

9.3.1.1 Particle Measurement Parameters

Diesel particulate matter (DPM) is not a well-defined substance; rather, it is a complex mixture characterized by widely changing chemical composition and physical properties. The properties of DPM change not only depending on the fuel, engine technology, operating conditions, and exhaust aftertreatment, but also with time, because the particulates undergo transformation once released in the atmosphere or while in the sampling train or in the measurement apparatus. Examples of such transformation include particle coagulation, evaporation and/or condensation of volatile compounds, and many more. PM transformation occurring in the sampling/measurement equipment may result in formation of artifacts, that is, compounds that are not present in real life, either in the undiluted engine-out exhaust gas or in the atmosphere around diesel vehicles. Due to this complexity, contrary to the gaseous emissions, a single absolute measure of DPM emission does not exist. Because the very definition of diesel particulate matter is in fact determined by the PM measuring technique that is used, all measurements of DPM have a somewhat arbitrary character.

A large number of measurement techniques have been developed or adapted from other fields to measure various parameters of diesel particulates. The ultimate choice of a method depends on its quality (accuracy, repeatability, cost, etc.) and suitability for its purpose (emission legislation and enforcement, emission research, etc.), but also on the importance of the particular PM parameter from the health effect perspective. The critical diesel particulate measurement parameters are listed in Table 9.3 [19].

Particle mass is the most common particle parameter used for the purpose of regulating PM emissions from diesel engines. Practically all emission standards from new diesel engines worldwide are expressed in terms of *total particulate matter* (TPM) emission, as determined by gravimetric methods. Regulating diesel particulate emissions as TPM is also compatible with ambient air quality standards, which in most areas (including the United States and Europe) are expressed using the total mass of particulates below a certain diameter. Examples of such ambient standards in the United States are the National Ambient Air Quality Standards (NAAQS) for particulates smaller than 10 μm (PM10) and smaller than 2.5 μm (PM2.5). For the purpose of air quality modeling, it is very important that ambient pollutant levels and engine emissions be expressed in a compatible manner.

TABLE 9.3
CRITICAL PARTICLE MEASUREMENT PARAMETERS

Parameter	Comments
Particle mass	Total particle mass, including a number of chemical compounds in both solid and liquid fractions; may not be the best indicator for possible health impact
Particle size	Instruments measure different properties that are translated into different particle diameters (aerodynamic diameter, electrical mobility diameter, etc.)
Particle number concentration	Majority of diesel particles resides in ultrafine size range that penetrates deep into the lungs
Particle surface area	Possibly a more accurate indication of degree of adverse impact on health
Particle composition	Compounds adsorbed to particles may have adverse health effects

Source: Data from Ref. 19.

More recent medical research has indicated that particles of smaller sizes may have more adverse health effects than those of large diameters. As a result, we have been witnessing increased interest in the measurement of particulate numbers and particle size-related characteristics. At this time, particle size and number parameters are measured primarily for research purposes. However, clean air authorities in the European Union have initiated programs to develop measurement methods for certain size-related characteristics of DPM [20], which could be incorporated, in addition to the TPM limit, in diesel engine emission standards.

9.3.1.2 Types of Measuring Techniques

Particulate matter measuring techniques can be divided into two main classes: (1) collecting techniques and (2) *in situ* techniques. In the collecting techniques particles are first deposited on a sampling filter, followed by an analysis of the collected material. Most conventional PM measuring methods, including gravimetric analyses, are based on this principle. Collected PM samples can be analyzed for different parameters using a number of analytical procedures. However, during sample collection, the particles may be changed and many of their properties (including sizes and numbers) may be quite different than in the aerosol phase. Another disadvantage of collecting methods is a long delay from the time samples are collected to the time the results of measurement are available.

In the *in situ* techniques, the analysis is performed directly in the aerosol phase flowing through the measuring apparatus. In these techniques changes to the particle properties are minimized, allowing for measurement of such parameters as particle numbers and sizes. The capability to perform continuous or quasi-continuous measurement is another advantage of *in situ* methods. On the negative side, *in situ* techniques typically provide an indirect measurement of a given property, which is then obtained through a calculation. For example, particle sizes can be calculated based on a measurement of the aerodynamic, electrical mobility, or various optical properties of the aerosol, often resulting in a poor correlation between different methods and instruments. *In situ* methods may be also used to measure particle mass, for example, based on smoke opacity measurement, but the accuracy of such measurements would probably not be as good as that of gravimetric methods.

Instruments exist that have certain aspects of both the collecting and *in situ* techniques and thus cannot be assigned to either category. For example, cascade impactors classify particles *in situ*, in the aerosol phase, but collect size-fractioned samples for later analysis.

In situ techniques are becoming increasingly popular for the measurement of diesel particulate properties, partly because of the interest in particle sizing. Many of the techniques newly introduced into the engine emission area have been known of for a very long time in other fields. For example, commercial mobility analyzers became available in 1967 and the principle of the condensation particle counter dates back to the 19th century [20].

9.3.2 *Gravimetric Method*

9.3.2.1 PM Mass Emission

The actual mass of total particulate matter present in a given volume of exhaust gas is determined by gravimetric methods. The majority of regulatory and compliance testing requires gravimetric diesel particulate matter determination in conjunction with the constant volume sampling (CVS) technique. The exhaust gases from the engine are mixed with an amount of air in a device known as a *dilution tunnel.* The minimum amount of mixing air must be sufficient to cool the air/gas mixture to a specified temperature. A sample of gas from the dilution tunnel is drawn through Teflon-coated fiberglass filters. The filters are than weighed on a very high precision balance placed in a controlled-temperature enclosure with controlled relative humidity. Because the volumes of dilution air and of the sample passing through the filters are carefully controlled and measured, the total mass of particulates emitted in the course of the test can be calculated. The final results are usually reported in grams of PM per kilometer of distance traveled by the vehicle (g/km) or in grams of particulates per kilowatt-hour of energy produced by the engine (g/kWh), depending on the test procedure. The equivalents in the U.S. system of units were gram per mile (g/mi) and gram per brake-horsepower-hour (g/bhp-hr).

By definition, *particulates* are measured as any material deposited on the Teflon-coated fiberglass filter from the dilute exhaust gases sampled at a temperature lower than 52°C. Because of this definition, diesel particulate matter includes not only solids but also liquid material that would condense in the form of mist or droplets at 52°C, such as sulfuric acid or high-boiling hydrocarbons (see Section 7.3).

9.3.2.2 PM Analysis

Chemical Composition

As an extension of the gravimetric method, PM samples collected on the fiberglass filters can be subjected to analysis using physical and chemical methods to determine several unregulated emissions, such as the soluble organic fraction (SOF), polynuclear aromatic hydrocarbons (PAHs), and sulfates.

The SOF of diesel particulates is usually measured after extraction of the PM sample in a Soxhlet extractor [21]. The most commonly employed solvent is dichloromethane (DCM), the chemical recommended by the EPA. Other solvent systems that are sometimes used include toluene, benzene/ethanol, and benzene/methanol mixtures. The results obtained may differ according to the choice of solvent, resulting in different relative shares of the extracted compounds. Using toluene may favor the extraction of heavy PAH when compared with extraction using DCM. The addition of alcohols to the solvent mixture leads to some extraction of sulfates [22].

The SOF extract can be further analyzed for PAHs. The solvent is first removed by low-temperature distillation with inert gas. The solvent-free sample is then further fractioned by distillation and tested for the PAHs using such methods as gas chromatography (GC), high performance liquid chromatography (HPLC), and mass spectrometry.

The sulfate content of PM is typically determined by ion chromatography. The sample is prepared by water extraction of the filters containing particulate matter. The extract is then filtered and analyzed on an ion chromatograph calibrated with a standard potassium sulfate solution [12].

An alternative method sometimes used to determine the organic portion of PM is vacuum evaporation. The fraction obtained by evaporation is called the volatile organic fraction (VOF). A drawback of this method is that water and sulfate material evaporate as well and are usually included in the VOF result.

Biological Activity

The biological activity of PM is usually evaluated by the Ames bacteria test of the SOF fraction [23]. The Ames test was developed by Bruce Ames at the Biochemistry Department of the University of California, Berkeley. Selected strains of *Salmonella* bacteria are exposed to the tested material. Some of the bacteria mutate as a result of the exposure. The relative number of revertants is the result provided by the Ames test, which gives a measure of the mutagenicity of the test sample.

The Ames test is a mutagenicity assay performed on bacteria, that is, prokaryotic cells. Although not precisely explained or understood, there is a correlation between the Ames mutagenicity and the longer term carcinogenicity in higher animals and humans, that is, in eukaryotic cells. A summary of Ames results for many chemicals and the correlation between Ames mutagenicity and carcinogenicity can be found in a work by Hollstein and McCann [24].

9.3.3 Particle Number and Size Analysis

9.3.3.1 Particle Number Concentration

The condensation particle counter (CPC), also called the condensation nucleus counter (CNC), is the most common instrument used to determine number concentrations of diesel particulates [20]. As shown in Figure 9.6, upon entering the CPC the aerosol stream is saturated with alcohol (typically butanol) vapor. As the mixture is cooled in the condenser tube, the vapor becomes supersaturated and condenses on particles. As a result, the particles grow to a diameter of about 10 μm, allowing for optical detection [25, 26]. The particle size detection limit in the CPC is related to the increasing saturation ratio that is required with decreasing particle diameters. Modern CPCs have detection limits of around 10 nm [20].

CPCs can be operated in two modes: (1) the counting mode and (2) the opacity mode. In the counting mode, pulses of scattered light from individual particles are counted. This mode provides the most accurate measurements, but can be used only at low particle concentrations. In the opacity mode, used for concentrations of typically above 104/cm^3, number concentrations are determined from the total scattering intensity. This mode, generally subject to a bigger error, requires that all particles grow to the same diameter and that the optical system be frequently calibrated [27]. The CPC instrument is very sensitive to the inside temperature, which affects the degree of supersaturation, as well as to the positioning and vibration. For these reasons, it is suitable primarily for laboratory measurements; its use in field equipment is more challenging.

The CPC is used for particle detection in many aerosol size distribution measurement instruments.

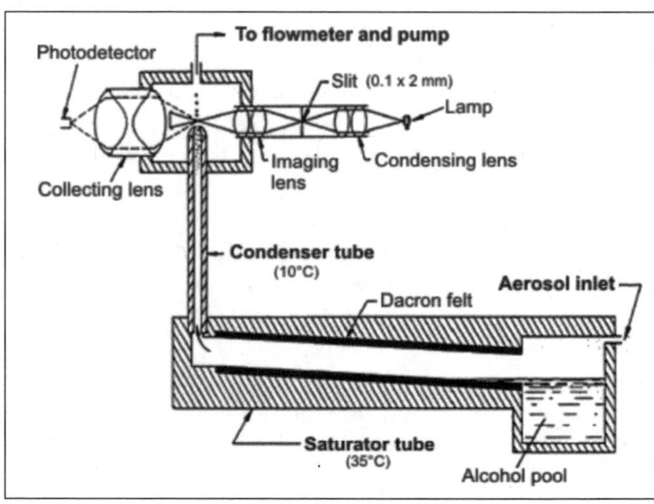

Figure 9.6 Condensation particle counter. (Based on Ref. 20, Fig. 11.)

9.3.3.2 Particle Size and Size Distributions

Classification of Methods

Particle size analysis determines the dimensional distributions of the particles making up the diesel aerosol. Size distributions are typically presented as particle number or particle mass distributions. Sometimes distributions of particle volume- or surface area-weighted concentrations are also determined. Various techniques used for this purpose can be classified as follows [28]:

- *Aerodynamic methods,* utilizing the inertia of accelerated particles, are used in various types of cascade impactors. One of the common instruments—the electric low-pressure impactor (ELPI)—is classified in this group.
- *Electrical mobility analyzers* determine particles' mobility diameters based on the measurement of charged particle mobility in an electrical field. Another widespread analyzer—the scanning mobility particle sizer (SMPS)—belongs to this class.
- *Diffusion batteries* determine particles' mobility diameters based on the measurement of their diffusion properties.

Example instruments based on these three methods are listed in Table 9.4 [19, 20].

TABLE 9.4
DIESEL PARTICLE SIZING INSTRUMENTS

Instrument	Size range	Comments
Cascade impactors (aerodynamic method)		
Micro-orifice uniform deposit impactor (MOUDI)	0.056–10 μm	• Not real time; gravimetric and chemical analysis performed after collection
Nano-MOUDI	0.010–10 μm	• Same as MOUDI. • Low pressure may cause volatile particles to evaporate
Electrical low-pressure impactor (ELPI)	0.032[a]–10 μm	• Near real time with time resolution from 2–15 s. • Can measure ambient concentrations, but low pressure may cause volatile particles to evaporate
Electrical mobility analyzers		
Electrical aerosol analyzer (EAA)	0.003–1 μm	• Time resolution between 2 and 3 min; 11 size intervals • Strongly distorted results if concentration and size change during scan • More sensitive to these changes than SMPS • Requires high dilution (>200:1) to prevent electrometer saturation • Results for diameters <10 nm are questionable unless number concentrations are high, due to diffusional losses • Preclassification stage must be used to remove particles >1.0 μm for diesel aerosol measurements
Differential mobility particle sizer (DMPS)	0.01–0.5 μm	• Discrete changes in electric field of DMPS make sampling for entire range slower than for SMPS. • Biased results if concentration and size change during scan. • Preclassification stage must be used to remove particles >1.0 μm for diesel aerosol measurements.
Scanning mobility particle sizer (SMPS)	0.010–0.7 μm	• 2 to 4 min scan time for 100 size intervals; can ramp DMA electric field to reduce scan time • Time resolution of a few seconds for a single channel • Scan times as short as 30 s can be used with lost resolution • Biased results if concentration and size change during scan • Preclassification stage must be used to remove particles >1.0 μm for diesel aerosol measurements
Nano-differential mobility analyzer (nano-DMA)	3–50 nm	• Time resolution of about 10 s required to scan all intervals. • Not fully characterized and developed.
Diffusion batteries		
Diffusion battery (DB)	0.01–0.5 μm	• Slow sampling time requires steady-state conditions • Strongly distorted results if concentration and size change during scan
Electrical diffusion battery (EDB)	5–300 nm	• Fast and simple instrument, but size resolution is not very good • Not fully characterized and developed

[a] Lower detection limit of 0.007 μm with added electrical filter stage.

Source: Data from Refs. 19 and 20.

The measurement ranges of selected particle sizing instruments are illustrated graphically in Figure 9.7 [19]. Note that most instruments are not capable of covering the entire diesel particulate size range. More than one analyzer is usually required to determine size distributions over the entire diesel range.

Measurement of "Wet" and "Dry" Particles

A large proportion of diesel particles in the smallest size ranges (nanoparticles) are made up from liquid condensates of volatile materials such as hydrocarbons, sulfuric acid, and water. Liquids can be bound to solid particles or form pure liquid droplets, thus potentially affecting particle size and number. Most particle sizing instruments are sensitive to both solid and liquid particles. Based on the result of a measurement, the contribution of liquid and solid material to size distributions it is not known. However, one can choose to eliminate vapors—the precursors of liquid particles—from the gas sample before they enter the sizing instrument.

The volatile material can be removed from the sample in a commercially available preconditioning instrument known as a *thermodesorber* or *thermodenuder.* In the thermodesorber (Figure 9.8) the sample is first heated to a well-defined temperature, typically on the order of 250°C, in order to desorb volatiles from the particulates and evaporate condensates that may have been present in the sample. Then the gas is passed through a cooled section filled with activated carbon, which collects the vaporized volatiles. Thermodesorbers also allow us to study the composition and nature of the volatile material [29]. "Dry" particle size measurements taken with thermodesorbers, although not a good representation of atmospheric aerosols, have the advantage of yielding much more reproducible results than measurements of "wet" particulates [20].

On the negative side, thermodenuders affect not only volatiles, but also solid particles, resulting in a significant additional source of sample loss. Thermodenuder sample losses of as much as 20% to 30% have been reported in the literature [30]. The particle losses have been attributed to thermoforetic forces, where particles from the hot sample gas migrate toward the low-temperature adsorption part of the instrument. This effect

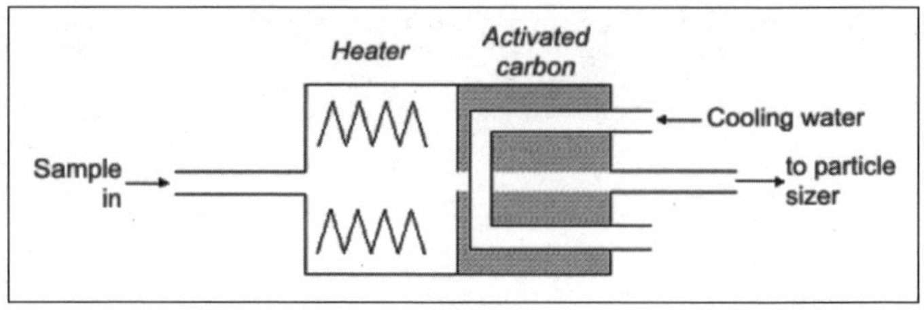

Figure 9.8 Schematic of thermodenuder. (Based on Ref. 20, Fig. 5.)

is especially prominent in the smallest particle size range [31]. Thus, thermodenuders should be designed to minimize the sample exposure to temperature gradients. To aid in minimizing errors, thermodenuders must be frequently cleaned.

Catalytic strippers, in which the hot sample is passed over an oxidation catalyst, provide an alternative method of removing volatile particle precursors, possibly resulting in less measurement error. However, only organic material (hydrocarbons)—as opposed to sulfuric acid and water—can be removed in the catalytic stripper.

Aerodynamic Methods

Aerodynamic techniques rely on the measurement of the relaxation time constant (τ) of an accelerated particle, which depends on the particle mass (m) and mobility (b), as follows [26]:

$$\tau = m \cdot b \tag{9.2}$$

The equivalent particle diameter, which is determined based on the result of the measurement, is known as the *aerodynamic diameter* and is defined as the diameter of a unit density (1 g/cm^3) spherical particle that has the same settling velocity as the measured particle.

The two main classes of aerodynamic instruments are (1) aerodynamic particle sizers (APS) [25] and (2) impactors. The aerodynamic particle sizer uses a process of linear acceleration of particles in a system of nozzles, followed by

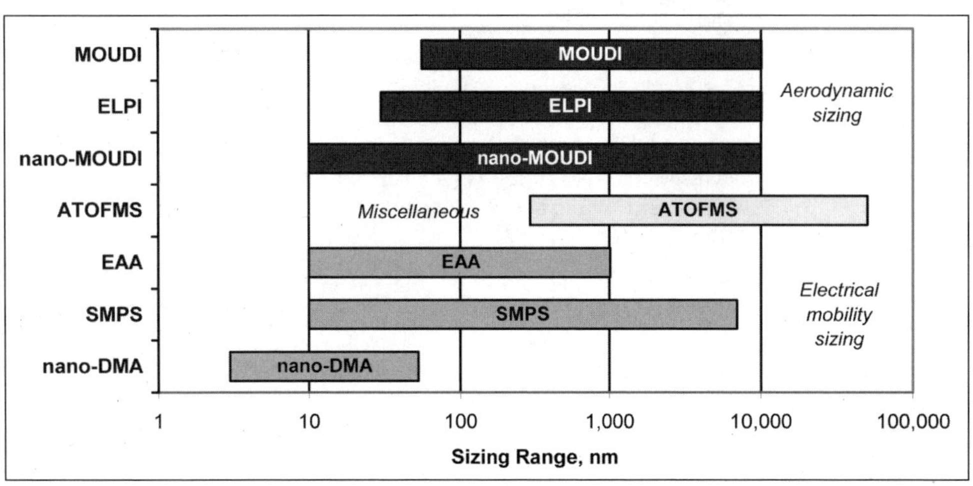

Figure 9.7 Dynamic range of particle sizing instruments. (Based on data from Ref. 19.)

optical detection using a laser beam. Particles entering the APS are accelerated to their terminal velocity. Smaller particles of greater aerodynamic mobility reach higher terminal velocities. Because of its optical detection method, the APS has a relatively high particle size detection limit of around 0.5 μm, making the instrument unsuitable for diesel aerosol measurements [20].

In impactors, the aerosol stream is sent through an orifice on an impaction plate, which forces the stream to change direction by 90 degrees. Bigger particles with more inertia cannot follow the current lines and impact against the plate. Multistage impactors, known as *cascade impactors* (Figure 9.9), are used to classify the particulates by decreasing size. By utilizing more stages, cascade impactors can be sensitive down to particle diameters of about 10 nm. However, as the gas sample is drawn through the instrument by a vacuum pump, particle loss through evaporation of volatile material may occur in the low-pressure stages.

Cascade impactors typically incorporate 6 to 12 stages. Particulates collected in the impactor stages are weighted to determine the size distribution. The samples can be also analyzed chemically or using other methods. Impactors may require long sampling times to collect sufficient mass of particulate.

The following are examples of frequently used cascade impactors:

- *Micro-orifice uniform deposit impactor (MOUDI):* The MOUDI is a 10-stage cascade impactor (the number of stages can be altered) that provides a mass weighted size distribution of an aerosol [32]. The cutoff of the lowest stage is 56 nm. Particles are classified by aerodynamic diameter on aluminum foil stages or substrates for subsequent gravimetric analysis. Depending on the amount and nature of the collected particles, the substrates may be used with or without grease.
- *Nano-micro-orifice uniform deposit impactor (nano-MOUDI):* This is a modified MOUDI that is enhanced for operation in the small particle region. It has 13 stages that can work with particles down to the 10-nm size [33].
- *Andersen Mark III impactor:* This is an 8-stage impactor, with a range from 0.4 to 10 μm, commonly used but of high lower detection limit [34].
- *Low-pressure impactor (LPI):* The Berner 10-stage LPI measures down to 16 nm [35].

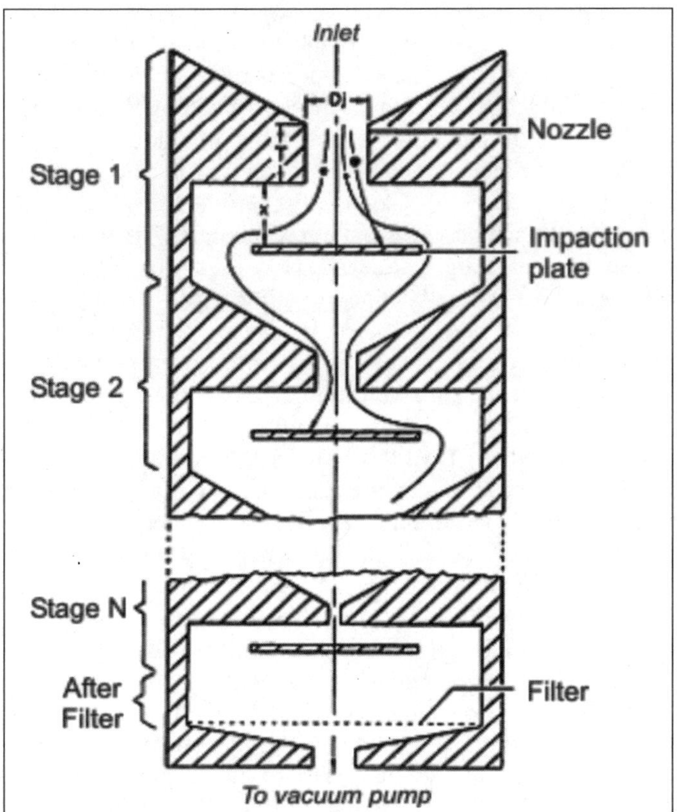

Figure 9.9 *Cascade impactor. (Based on Ref. 20, Fig. 13.)*

Conventional impactors exhibit sensitivity problems, are susceptible to measurement artifacts, and are not suitable for transient tests. Some comparisons between MOUDI and gravimetric measurements show differences up to a factor of 3 [36]. Several modified impactor concepts have been proposed to overcome these problems [20].

The electrical low-pressure impactor (ELPI) is perhaps the most widespread instrument, especially in Europe, among such modified impactor designs [37]. In the ELPI, prior to entering the impactor, particles are charged by a corona discharge (Figure 9.10). Number-weighted size distributions are determined based on the measurement of the current produced by particles precipitated in each stage and on the particle charging efficiency in the corona charger. The ELPI allows for near-real-time measurement with a scan time as low as 2 s, which allows for its use in transient testing. Collected particulates may be subjected to subsequent gravimetric analysis, providing mass-weighted distributions. The measuring range in the original 12-stage impactor configuration is 32 nm to 10 μm.

The lower detection limit on the order of 30 nm is not really sufficient for diesel particles. Improved lower size detection could be achieved in the ELPI by adding additional impactor stages, but a more powerful vacuum pump and even lower pressures would be required. To avoid lowering the pressure, a filter stage enclosed in an isolated Faraday's cage with electrical current measurement has been developed to measure particles below 30 nm. A particle measurement limit of 7 nm was possible in an ELPI configuration with 11 impactor stages followed by an electrical filter stage [38].

Particles are classified in the ELPI based on their aerodynamic diameter, but the charging efficiency is determined by the mobility diameter. This may lead to calibration problems with the ELPI [20].

Electrical Mobility Analyzers

In mobility analyzers particles are first charged, then the aerosol is classified in a high electric field according to the electrical mobility of the particles. The particle size distribution is obtained on the basis of the relationship between mobility and sizes. The measured parameter is the electric mobility z,

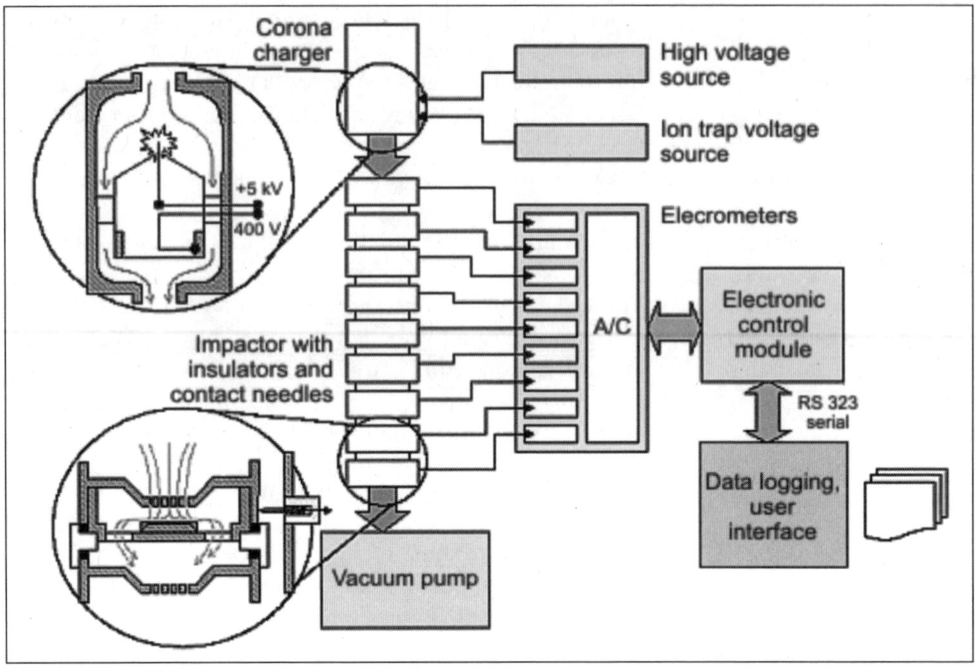

Figure 9.10 *Electrical low-pressure impactor. (Reprinted with permission of the author from Ref. 20, Fig. 14.)*

$$z = b \cdot q \tag{9.3}$$

where b is the particle mechanical mobility and q is the particle charge. The ultimate size parameter determined from electrical mobility measurements is the equivalent mobility diameter. The electrical mobility diameter is a close approximation of the *Stokes diameter*, which is defined as the diameter of a spherical particle having the same density and settling velocity as the measured particle. In general, the mobility diameter is different from the aerodynamic diameter measured in aerodynamic particle sizers. The aerodynamic diameter provides a good representation of the particle behavior in such processes as impaction and settling, that is, processes characteristic for larger particles with high inertial forces (high Stokes numbers). The mobility or Stokes diameter, on the other hand, is more representative for describing diffusional processes, which dominate in small particles. The correlation between the aerodynamic and mobility diameters depends, among other factors, on particle density. Because densities of diesel particles are typically less than 1 g/cm^3, their aerodynamic diameter is less than the mobility diameter. Some researchers believe the mobility diameter is more representative than the aerodynamic diameter to describe the movement of small diesel particles in the human lungs [20].

Historically, the electrical aerosol analyzer (EAA) was the first very common instrument used both in the United States and Europe for diesel particulate measurement. Particles in the EAA are electrically charged and then routed through an electric mobility analyzer that only allows particles of a specified electric mobility to pass through and be measured by a downstream electrometer [23, 39]. The EAA provides number-weighted particle size distributions. With time, the EAA has been gradually replaced by more advanced differential mobility analyzers.

A schematic of the differential mobility analyzer (DMA) is shown in Figure 9.11. The DMA analysis uses bipolar diffusion charging. Particles are charged to their well-defined equilibrium charge (Boltzmann equilibrium) and passed through an electric mobility analyzer that only allows particles of a specified mobility to pass through and be measured downstream. Due to its increased sensitivity over an electrometer, the CPC is usually used for particle counting. As an alternative, a Faraday cup can be used if the concentration is high enough for the measurement of the electric charge [40]. For correct consideration of multiple charging, the sample must contain no particles larger than the measured range. Large particles must be thus removed from the sample, which is usually done by an impactor.

Differential mobility analysis may be performed in two modes of operation, resulting in two corresponding types of instruments: (1) the differential mobility particle sizer (DMPS) and (2) the scanning mobility particle sizer (SMPS). When operated as a DMPS, voltage in the mobility analyzer is increased stepwise to measure a size distribution [41]. At each discrete voltage level, a different interval of electric mobility particles can pass through the instrument and be measured. It takes about 20 min for a complete size distribution run. In the alternate configuration, as an SMPS, the analyzer voltage is scanned in a continuous exponential ramp instead of steps [42]. This reduces the time required to about 30 to 60 s, depending on the concentration and required precision. The reduction in time is accompanied by a slightly reduced accuracy. The SMPS is used more commonly than the DMPS because of its faster scanning ability. When using an SMPS, note that the correct

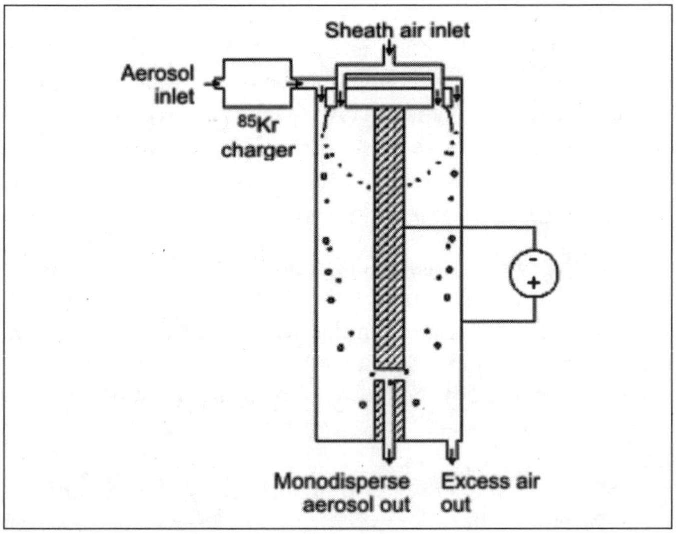

Figure 9.11 *Differential mobility analyzer. (Based on Ref. 20, Fig. 15.)*

scanning time must be chosen and the size distribution must remain constant during the scan.

The SMPS time resolution of up to 1 min is still insufficient for true transient testing. A DMA can also be operated at a constant voltage, measuring only one particle size range. The response time with fixed voltage is only a few seconds, thus allowing transient tests. To analyze very small particles, a

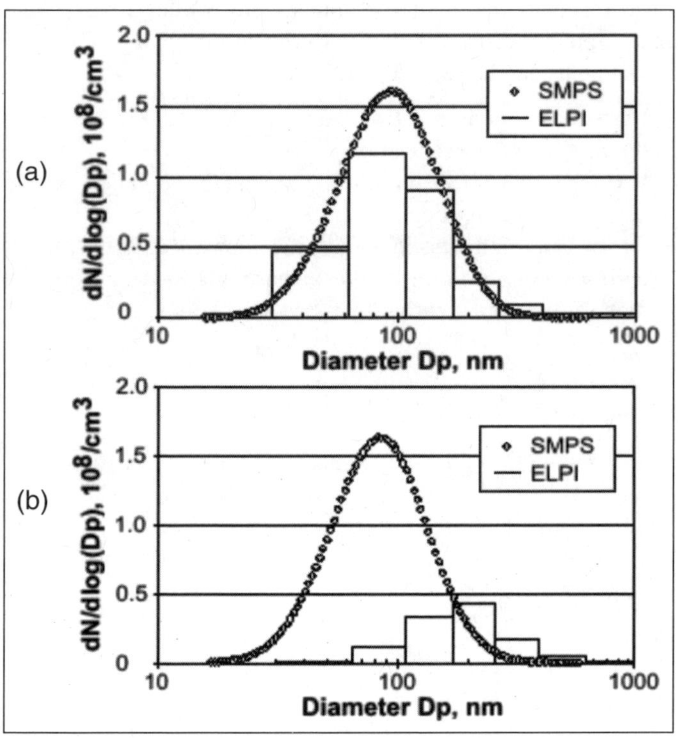

Figure 9.12 *Comparison of size distribution measurements with SMPS and ELPI: (a) fuel sulfur < 100 ppm, low engine load, and (b) fuel sulfur 500 ppm, high engine load. (Based on Ref. 20, Fig. 24.)*

nano-DMA instrument has been developed (TSI). It extends the small particle sizing range from some 10 nm in the conventional SMPS down to 3 nm.

Several studies compared particle size resolutions obtained with different SMPS instruments, as well as results obtained with SMPS and other types of instruments. A comparison of 11 SMPS systems using diesel particles showed differences of up to 10% in the particle size and up to 20% in particle numbers between different systems [43]. Figure 9.12 shows the result of a parallel size distribution measurement taken with an SMPS and an ELPI, both operated with a CVS dilution tunnel [20]. At certain conditions, the two instruments show fairly good agreement, as illustrated in Figure 9.12a. At other conditions, however, the correlation is poor, as shown in Figure 9.12b. The differences are attributed to different particle composition, density, and other properties. The test conditions in Figure 9.12b favored formation of high numbers of liquid particles made of sulfuric acid.

Note that, strictly speaking, comparisons of "raw" data from the two instruments are not valid, since one compares the aerodynamic diameter (ELPI) with the mobility diameter (SMPS), which are different physical properties. It was shown in the literature that the correlation between instruments may improve if the ELPI aerodynamic diameters are converted into mobility diameters [44]. Particle density, however, which is also a function of particle size, is a significant source of uncertainty in such calculations. The correlation between the SMPS and ELPI remains an important issue in diesel particle sizing research. It is a fairly complex topic, the detailed discussion of which is outside the scope of this chapter.

Generally, an electrical mobility analysis is considered a high-performance particle sizing technique that produces good particle size resolution. Despite their complexity and high cost, SMPS instruments have been widely used in emission laboratories. To account for the less than perfect correlation with other methods while facing the lack of standardized particle sizing procedures, some laboratories use more than one instrument employing, for example, both an SMPS and an ELPI in the same study.

Diffusion Batteries

Particle diffusion coefficient D and its mobility b are related to each other by the following Einstein formula:

$$D = kT \cdot b \qquad (9.4)$$

where:

- D = diffusion coefficient, m^2/s
- k = Boltzmann constant, $1.3807 \cdot 10^{-23}$ J/K
- T = absolute temperature, K
- b = mechanical mobility, s/kg.

Because of this correlation, a measurement of particle sizes based on the diffusion coefficient, as performed in diffusion batteries, ultimately yields the same information as mobility analysis, that is, the mobility diameter [20].

Particles in a diffusion battery (DB) are classified according to their diffusion coefficients. Because such phenomena as diffusion and Brownian motion are well pronounced in very small particles, this method is especially useful for analyzing particles below 300 nm [20]. Diffusion batteries were combined with impactors to cover larger particle sizes of up to 10 μm [45]. The method relies on the particles being fixed to the walls when carried by the gas stream in a small-section duct. In the analyzer, gas is passed through channels (e.g., tubes) of different diffusion lengths (e.g., cross sections). The fraction of the aerosol that has not been stopped by a given tube diameter is measured downstream of the tube. Often a CPC is used to determine the input and output number concentrations.

Different types of DBs, such as the parallel tube, plate, or screen type, may be used [46], as illustrated in Figure 9.13. Typically, valves are applied to change the diffusion path length (e.g., to pass the gas over a different number of screens or screen sections) for the determination of different particle size ranges. In this method the measurement is stepwise and thus relatively slow. It is also required that particle concentrations and size remain constant during the scan. Another approach is to use a parallel flow setup, which allows for a fast measurement and real-time response, but this approach requires several CPCs [47].

In the electrical diffusion battery (EDB), shown in Figure 9.14, the particles are charged prior to entering a four-stage, screen-type diffusion battery [48]. Charged particles deposited in each stage are detected by measuring the resulting electrical current. A backup filter following the fourth stage collects the remaining particles. This device allows for a fast (seconds) measurement of the size distribution in the size range below about 300 nm. Even though an electrical measurement is used, the sizing occurs by diffusion and therefore is not affected by multiple charging of particles. The pressure drop across the diffusion battery is very small, not allowing for evaporation of volatile material.

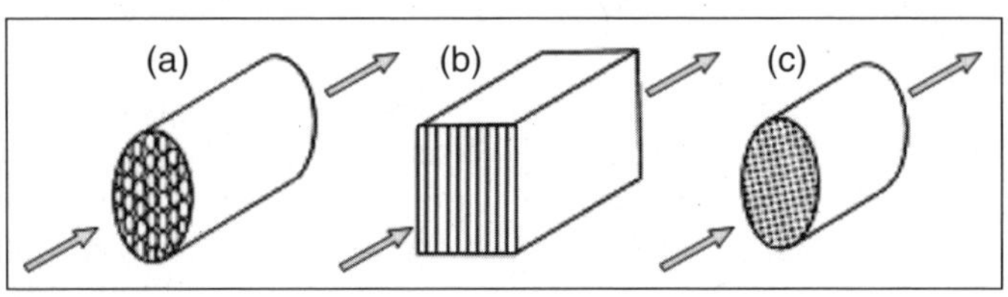

Figure 9.13 *Types of diffusion batteries: (a) parallel tube package, (b) parallel plate, and (c) screen type. (Based on Ref. 20, Fig. 16.)*

Other Particle Sizing Instruments

Photon correlation spectroscopy utilizes fluctuations in the light intensity diffused by the particles in an aerosol [49]. A laser beam is sent through the diesel aerosol present in the dilution tunnel. Particle diameters are obtained through a statistical analysis of the fluctuations of the light diffused by the aerosol. A short measurement time (1 to 10 s) allows for transient testing.

Miscellaneous other instruments, designed primarily to measure other parameters such as particle mass or chemical composition, may also provide some particle sizing capacity. Two such instruments, the aerosol time-of-flight mass spectrometer (ATOFMS) and the time-resolved, laser-induced incandescence (TIRE-LII), are described in the following sections.

9.3.4 Instrumental Analysis of Other PM Parameters

9.3.4.1 Types of Instruments

Numerous other techniques and instrumentation have been developed that allow for *in situ* measurement of various parameters of aerosols. This section provides a brief summary of instruments that could be considered for diesel particulate measurements. Based on the measured particle parameter, these instruments can be categorized as follows [20]:

- Mass measurement instruments, for *in situ* measurement of the total mass concentration
- Particle surface instruments, measuring the so-called "active surface" of the particulates [50]
- Optical instruments, in most cases sensitive to the black, carbonaceous portion of diesel particulates, returning the mass of elemental carbon (EC) concentration

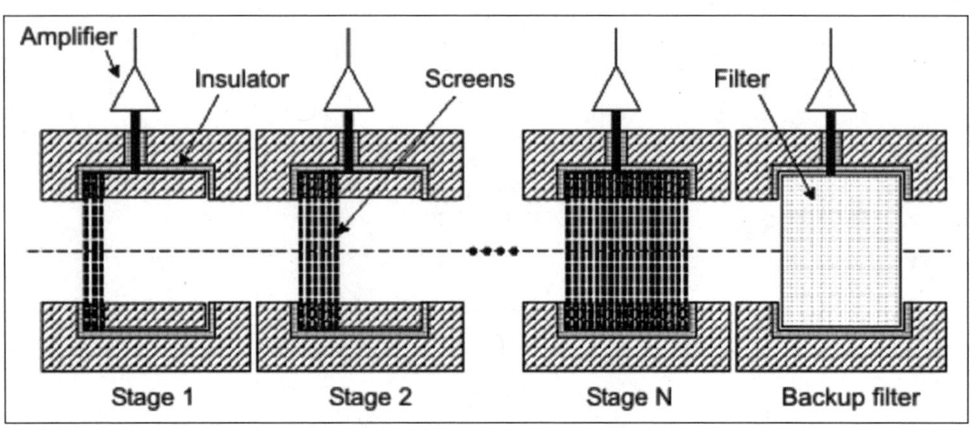

Figure 9.14 *Electrical diffusion battery. (Based on Ref. 20, Fig. 17.)*

- Other instruments, typically measuring more than one parameter by a combination of detectors and operating principles (ATOFMS, NanoMet)

9.3.4.2 Overview of Instruments

Particle Mass Measurement

- *Tapered element oscillation microbalance (TEOM):* Measures directly the total mass concentration of particulates. The mass of particles is determined based on the change in the resonance frequency of a tapered element [51], which carries the sampling filter on its top (Figure 9.15). The TEOM can be used for ambient air measurements with a detection limit of 2 μg/m^3 for a 24-hour integrated sample.
- *Quartz crystal microbalance (QCM):* This instrument uses the same working principle as the TEOM, but particles are deposited by electrostatic precipitation on an oscillating quartz crystal. The QCM, very popular in the 1970s, disappeared later due to various technical problems. Development of a new, faster QCM has been reported [52].

Particle Surface Measurement

- *Epiphaniometer:* Measures surface concentration of an aerosol by passing the particles through a chamber with a radioactive lead isotope that attaches to the particles. The particle surface area is directly proportional to the measured level of radioactivity of the particles [53]. The epiphaniometer is a very sensitive, but slow instrument; for an integration time of 30 min, the detection limit is 0.003 μm^2/cm^3.

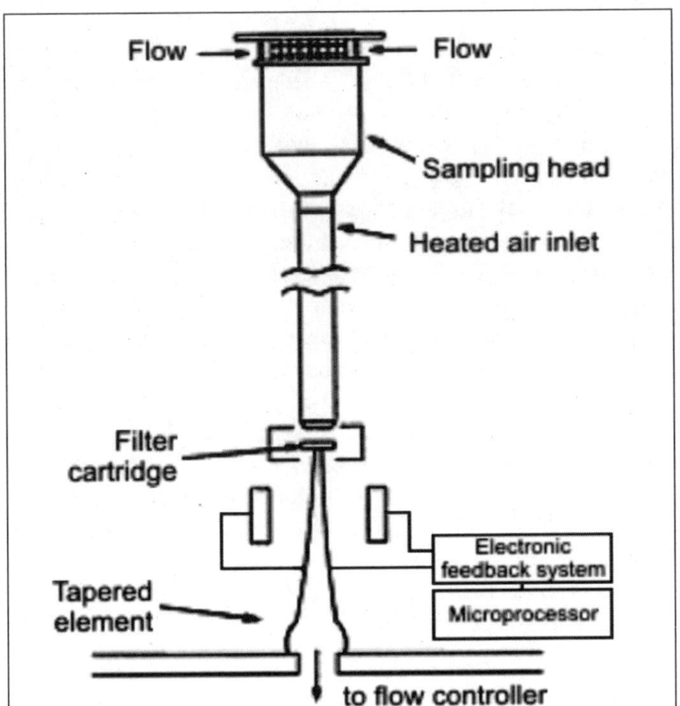

Figure 9.15 TEOM, ambient air sampling version. (Based on Ref. 20, Fig. 20.)

- *Diffusion charging sensor (DC):* Measures surface concentration by charging particles with positive ions from a corona discharge. The charged gas then flows through a measuring filter. The active surface of the particles is proportional to the current measured on the filter. The DC yields the same information as an epiphaniometer, but is much faster at the expense of being significantly less sensitive. The response time is sufficiently short to allow transient tests. The detection limit of about 1 μm^2/cm^3 allows for both emission and ambient air measurements in urban areas.

Optical Instruments (Elemental Carbon)

- *Time-resolved laser-induced incandescence (TIRE-LII):* Measures the radiation of particles heated by a short (nanosecond) laser pulse. The elemental carbon mass can be determined from the intensity of radiation. In addition, the time dependence of the cooling of the particles yields information on the primary particle size, based on a thermal model of the particle radiation [54]. The TIRE-LII is a very fast and sensitive method, capable of transient and low-emission engine testing. As shown in Figure 9.16, the LII can be applied in the undiluted gas, including direct exhaust pipe measurement. The system is relatively compact, but fairly expensive.

It must be emphasized that particle sizes obtained from the LII refer to the primary, carbon nuclei particles, and not to the actual diesel particulates that form aggregates. The LII-determined size is a kind of volume-based diameter of the primary particles, which is also different from the mobility diameter determined in mobility analyzers. The LII can also be calibrated for particle mass measurement. Then, based on the mass, the primary particle size, and an assumed particle density, a number concentration can be calculated. Again, the calculated value would be the concentration of primary particles,

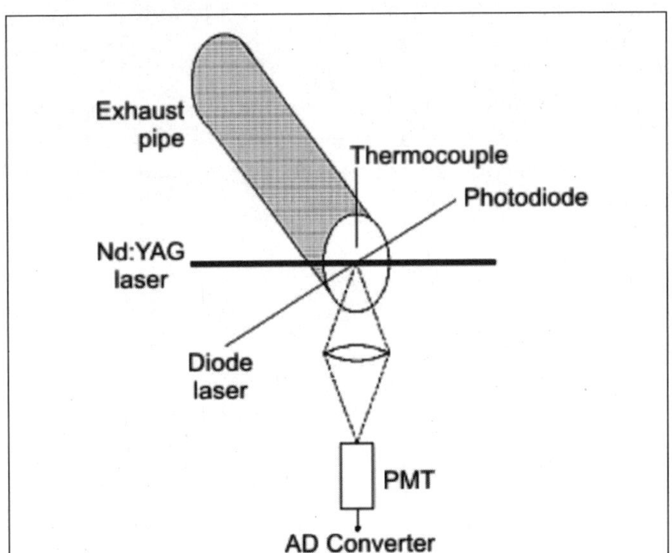

Figure 9.16 TIRE-LII measurement at the end of the tailpipe. The diode laser extinction measurement is used for calibration. (Based on Ref. 20, Fig. 18.)

not the concentration of emitted particles at various states of agglomeration.

- *Aethalometer:* Measures the attenuation of a white-light beam through a transparent filter that is holding the collected particles [55]. Results are expressed as the mass of elemental carbon. This instrument's accuracy suffers from variation of light absorption coefficients due to light scattering and from cross sensitivity to non-EC material. The aethalometer performs continuous measurements, but it is too slow for transient cycles.
- *Photoelectric aerosol sensor (PAS):* Operates through irradiation of particles with UV light of enough intensity to cause the particles to emit electrons due to the photoelectric effect [56]. The electrons are quickly removed from the gas by an electric field (see the ion trap in Figure 9.17), whereas the positively charged particles are collected and the charge is measured by an electrometer. Because photoemission involves absorption of a photon by the particle bulk material and emission of an electron through the particle surface, the resulting charge on the particles (and the PAS signal) depends on surface properties and a material coefficient.

While the PAS is highly sensitive to diesel aerosol, it is not entirely clear what kind of material or material property triggers its response. In ambient air studies, the PAS signal was correlated with the total concentration of particle-bound PAHs and was, in turn, the basis for the determination of surface area-weighted particle concentrations. Newer studies indicate that in the case of diesel engine measurements the PAS signal may be closely related to the elemental carbon concentration [57]. The PAS needs to be calibrated for the measurement of diesel EC. The PAS provides a continuous measurement with a time resolution of about 1 s. Its detection limit of 100 ng/m^3 is sufficient for ambient air measurements in urban areas.

- *Photoacoustic spectroscopy:* Involves particle heating by absorption of light from a chopped laser beam, following by the measurement of the resulting acoustic wave [58]. Results are given as EC mass. This method is under development; no commercial instruments are yet available. Problems are caused by absorption of gaseous species (water vapor, NO_x). If cross-sensitivities with gaseous species are overcome, photoacoustics could be a relatively simple, direct, and fast method.

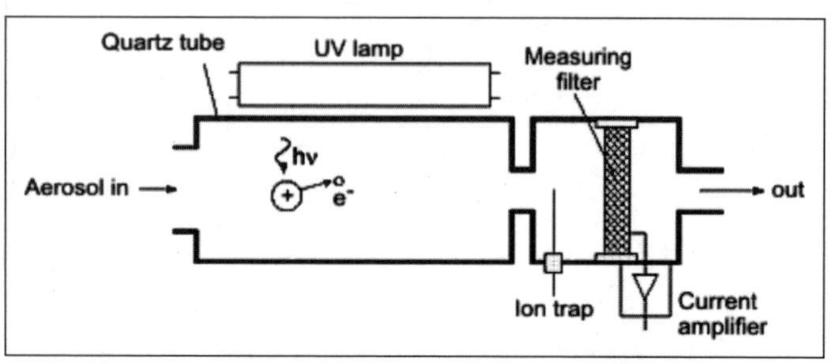

Figure 9.17 Photoelectric aerosol sensor. (Based on Ref. 20, Fig. 19.)

Other Instruments

- *Aerosol time-of-flight mass spectrometer (ATOFMS):* This instrument is a mass spectrometer with the additional capability of measuring particle sizes. Mass spectrometry can be used for particle composition analysis. A strong laser pulse evaporates the particle, completely or in part, following by a mass spectrometry analysis of the vapors. The ATOFMS, manufactured by TSI, measures the aerodynamic diameter of entering particles using an APS . The accelerated particles are sized based on the time of flight between two lasers [59]. Each individual particle is then pulsed by a laser. The chemical composition of the resultant desorbed and ionized material is determined in a dual-polarity mass spectrometer. The ATOFMS range is 0.3 to 10 μm.
- *NanoMet:* The NanoMet is a very compact instrument that uses a rotating disk dilution system combined with two sensors: a diffusion charging sensor (DCS) and a photoelectric aerosol sensor (PAS) [60]. The simultaneous operation of the two sensors yields both the active surface (DCS) and the active surface × material coefficient (PAS). By dividing these two signals, NanoMet calculates the material coefficient, which is characteristic of the particle source. In the case of diesel particles, this allows us to, for example, determine if nucleation of volatiles took place or not. Thanks to the use of a diluter, the measurable concentration range stretches from engine raw emissions to ambient air and/or occupational exposure measurements.

9.3.5 Smoke and Smoke Opacity

9.3.5.1 Overview

Smoke and smoke opacity meters measure the optical properties of diesel exhaust. These instruments have been designed to quantify the visible black smoke emission by utilizing such physical phenomena as the extinction of a light beam by scattering and absorption. In general, smoke and opacity meters are much simpler (some of them very simple) and less costly in comparison to most of the instruments described in the preceding sections. They are used to evaluate smoke emissions in locations outside the laboratory, such as in maintenance shops or in the field. In fact, the smoke opacity measurement is the only relatively low-cost and widely available method to measure a PM-related emission parameter in the field. For this reason, opacity limits are used in most inspection and maintenance programs for diesel engines. Smoke opacity limits may be also included as auxiliary limits in new engine emission standards.

In view of the demands of advanced, low-emission diesel engines, the following areas of concern can be identified in the smoke opacity meter technology:

- *Insufficient resolution:* Smoke levels in low-emission, smokeless diesel engines (e.g., Euro 4) are near the resolution of a conventional opacity meter. Resolution, stability, and noise have to be improved to allow opacity measurements in advanced diesel engines.

- *Cross-sensitivity to nitrogen dioxide:* Catalytic exhaust aftertreatment devices, such as oxidation catalysts or particulate filters, can increase the percentage of NO_2 in the total NO_x from less than 10% to as much as 40% and more. Nitrogen dioxide absorbs green light, which is used in conventional opacity meters. Opacity meters read between 0.00016 and 0.00024 m^{-1} per ppm NO_2, depending on the sensor bandwidth [61]. A 300-ppm NO_2 concentration in a modern engine—not an unrealistic assumption—would cause a 0.06 m^{-1} opacity reading, which is equivalent to 40% of the Euro 3 EEV pass level, intended to be for soot emissions. For engines fitted with catalytic particulate filters, nearly all opacity signal will be caused by nitrogen dioxide.
- *Insensitivity to small particles:* A substantial portion of diesel particles have diameters below 200 nm. Particles of 200 nm in diameter or greater block green light in proportion to their cross-sectional surface area. Particles of 50 nm in diameter, however, block only about 15% of their surface area [61]. This means that opacity readings depend on particle sizes and will be underestimated if smaller particles are measured.

Various attempts have been made to improve the sensitivity of diesel opacity meters, for instance by using multiple light path systems with mirrors [62]. The NO_2 cross-sensitivity could be eliminated by switching to a different light wavelength, for example, to red light. Red, however, is less sensitive to small particles than green. A switch from green to ultraviolet, in turn, could improve the "visibility" of small particles [61]. In view of these conflicting solutions, a simple switch to a different wavelength would not be sufficient. Rather, future opacity meters will likely be multiple-wavelength devices, utilizing a range of sensors to separate the various effects.

Smoke opacity readings generally do not correlate well with other PM measurement parameters. Numerous correlations between opacity or smoke readings and PM mass that have been developed can provide only approximate results. Because opacity readings may be affected by sulfates, HCs, and water vapor, as well as by PM composition (SOF) or physical conditions (e.g., coagulation), no accurate correlation is possible.

9.3.5.2 Measuring Principle

The following terms are used in optical smoke measurement techniques:

- *Opacity* is the fraction of incident light that is adsorbed or scattered by the smoke. It depends on path length. The opacity is 100% for an infinite path length.
- *Transmittance* is the fraction of incident light that passes through the smoke. The sum of transmittance and opacity is equal to one (100%).
- *Reflectance* is the fraction of incident light that is reflected from a surface.
- *Smoke number* is the fractional reduction in reflectance by a smoke filter due to the blackening of its surface by soot.
- *Extinction coefficient* is the inverse of the path length over which light intensity decreases by the factor of e (2.718). The extinction coefficient is proportional to the smoke density.

There are two groups of smoke meters, which utilize different measuring principles:

- Opacity meters that measure smoke in the exhaust gas
- Smoke number meters that optically evaluate soot collected on paper filters.

The principle of an opacity meter is shown in Figure 9.18. The measurement involves a light source and a light detector. The opacity of the measured column of smoke is calculated from the light intensities E_0 and E:

$$\text{Opacity} = 1 - E/E_0 = 1 - \exp[-KL] \qquad (9.5)$$

where:

E_0 = light intensity before the smoke column, W/cm^2
E = light intensity after the smoke column, W/cm^2
K = extinction coefficient, 1/cm
L = light path length, cm.

Some opacity meters, such as the Public Health Service (PHS) opacity meter or Celesco, are designed for *in situ* measurements. They simply clamp on the vehicle tailpipe and evaluate the light attenuation across the discharged exhaust gas stream. These instruments may be disturbed by ambient conditions, such as bright sunlight. Other instruments, such as the Hartridge or Bosch RTT100 opacity meters, include an integral sampling system. A special pump draws the exhaust gas sample through a measuring cell, inside the instrument, where the actual optical measurement is performed. The actual light path length (L) depends on the type of instrument. In sampling opacity meters, the light path is typically about 400 mm.

Opacities are typically expressed as percentage values. A value of 0% opacity represents clear air; 100% opacity is the infinite smoke. As a general reference, opacities of below 2% are not visible to the bare eyes, whereas opacities higher than 5% present clearly visible plumes of smoke. The result of the opacity measurement can be also given as the extinction coefficient K. Opacity meters typically have a resolution of K = 0.01 m^{-1}. Opacity meter terms and principles are defined by the ISO 11614 standard.

Figure 9.19 illustrates the measurement principle employed in the paper filter smoke meters. First, a volume of exhaust gas is drawn through a paper filter, which is then optically evaluated.

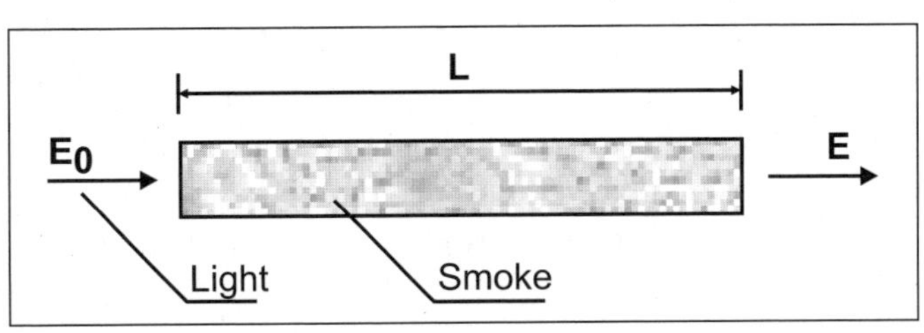

Figure 9.18 *Opacity measurement.*

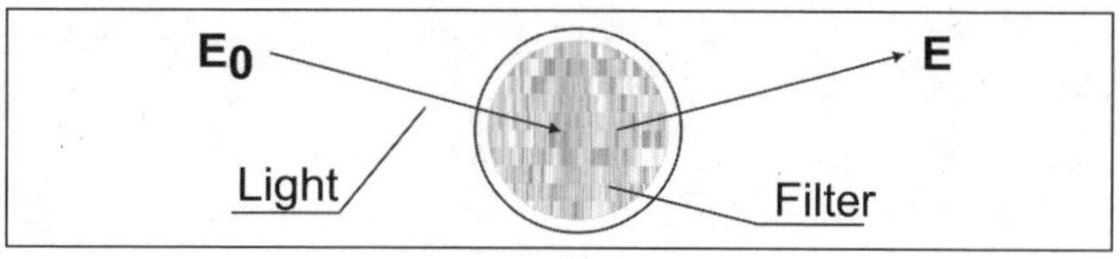

Figure 9.19 *Smoke number measurement.*

The reflectance of the filter is reduced by the soot. The fractional reduction of filter reflectance on a linear scale of 0 to 10 or 0 to 100 is the smoke number [see Equation (9.6)]. For example, if a clean filter reflects 50% of incident light and the smoke-laden filter reflects 40%, the smoke number of the filter is 20 on a scale of 0 to 100 and 2 on a scale of 0 to 10.

$$\text{Smoke number} \sim 1 - (E/E_0)/(E_{clean}/E_0) \qquad (9.6)$$

where:

E_0 = light intensity incident on the filter, W/cm^2
E = light intensity reflected from soot laden filter, W/cm^2
E_{clean} = light intensity reflected from a clean filter, W/cm^2.

Standard smoke number methods include the Bosch smoke number, the ASTM D 2156 (Bacharach) with a smoke number scale from 0 to 10, and the SAE (ARP 1179) with a scale from 0 to 100.

9.3.5.3 Opacity and PM Mass Fraction

Several methods for converting opacity or smoke number measurements into PM mass units are discussed by Homan [63]. We should emphasize that the optical methods are less accurate than the gravimetric analysis. The conversion equations may not apply if any of the following occurs:

- Exhaust gases contain the white or blue diesel smoke of transient or cold start conditions.
- The smoke is formed from pyrolyzed droplets of heavy liquid fuel.
- The smoke contains sulfuric acid (sulfate particulates).
- Noncarbonaceous solids, for example, from fuel additives, are present in the smoke.

The following formulas allow us to calculate the PM density in the exhaust gas (g/cm^3) and its mass fraction from the smoke opacity and temperature. The equations utilize data published by Green [64].

$$T_E = 1 - O_E \qquad (9.7)$$

$$K = -\ln(T_E)/L \qquad (9.8)$$

$$\rho_{exh} = 0.001\ PM/RT \qquad (9.9)$$

$$S_p = 86400 \text{ for } T < 553\ K \qquad (9.10a)$$

$$S_p = 348.13T - 106117 \text{ for } 553\ K < T < 623\ K \qquad (9.10b)$$

$$S_p = 110769 \text{ for } T > 623\ K \qquad (9.10c)$$

$$\rho_s = K/S_p \qquad (9.11)$$

$$Y = \rho_s/\rho_{exh} \qquad (9.12)$$

where:

OE = opacity, dimensionless
TE = transmittance, dimensionles
ρ_{exh} = density of exhaust gas, g/cm^3
P = pressure, Pa (barometric pressure = 101,325 Pa)
M = molecular weight of exhaust gas, kg/kmol (approximately 29 kg/kmol)
R = gas constant, 8314.34 J/(kmol K)
T = temperature, K
S_p = optical cross-section of soot, cm2/g
ρ_s = soot density in exhaust gas, g/cm^3
Y = mass fraction of soot in sampled gas, dimensionless.

9.4 Measurement of Ambient Diesel Aerosol

9.4.1 Aerosol Sampling and Analysis

9.4.1.1 Overview

A number of measurement methods have been developed for the determination of ambient levels of suspended particulate matter, as well as gaseous pollutants, for the purpose of air quality surveys in ambient (i.e., environmental) and occupational settings. In both cases the purpose of the measurement may be either (1) determining compliance with ambient air quality or occupational health standards or (2) air quality research. Air quality research takes many directions; for instance, it might be used to evaluate the health effects of pollution, to study the chemical composition and physical properties of air pollution, or to apportion pollution constituents to their emitting sources. A substantial amount of research is also performed to evaluate the measurement methods themselves, to correlate results obtained from different methods, and to improve and develop new air sampling and measurement techniques.

As was the case with PM measurement in engine emissions, methods for ambient particulate measurement can be classified as (1) collecting or (2) *in situ* techniques. Traditionally, ambient air measurements have been dominated by collecting methods, which involve the two following steps:

1. Sampling (collecting) of the particulate matter on a filter
2. Analysis of the collected material.

Methods for sampling and analysis, often developed independently, are combined in order to determine concentrations of measured compounds. In general, this sampling/analysis approach is similar to that used in measurement of engine emissions. However, while the methods used for engine emission measurement are very similar worldwide, a wide range of different methods have been developed for different ambient air sampling applications. Keep in mind that in the case of complex air pollutants that are not well-defined chemical species, such as diesel particulates or ambient particulate matter, the emission is defined by the measurement method. A consensus has been reached for particulate sampling and analysis in engine tailpipe emission testing, where the PM definition covers all material collected on the filter at a standard temperature and similar dilution ratios, including solids (inorganic carbon and ash), SOF, and sulfates, as determined gravimetrically. That substance is commonly referred to as total particulate matter (TPM). There is no single sampling method to measure ambient exposures to diesel particulates. As a consequence ambient particulates measured using various methods may be different, both in terms of quality (i.e., chemical composition and physical properties) and quantity (e.g., mass). In fact, different occupational health regulatory authorities have formulated exposure limits for diesel particulates using different definitions, such as "elemental carbon" or "total carbon." Making comparisons between data obtained using different methods or formulating correlations between exposure figures and engine emissions requires that a way to correlate methods be established.

Another important difference between engine emission and ambient sampling is that engine sampling produces a sample composed of "pure" diesel particulate matter, whereas ambient air samples contain a mixture of particulates from a number of human-made and natural sources, which may include varying proportions of diesel and/or other combustion particulates. Determination of DPM exposures based on the measurement of total suspended particulates requires apportioning a certain fraction of the measured particulate exposure to the diesel source. This is not an easy task; in essence, accurate discrimination between diesel and nondiesel particulates based exclusively on measurement is not possible. Many methods of diesel exposure estimation , especially those used in environmental settings, utilize various forms of complex mathematical modeling based on a number of assumptions in addition to or in lieu of measurement.

In most occupational settings, the total suspended particulates can be traced back to fewer sources than in the case of ambient air studies, and their composition is known or can be deducted from the processes in use. Diesel particulates are also subject to an increasing number of occupational air quality standards (in contrast to ambient air standards, which regulate the total suspended particulates such as PM10 or PM2.5). For these reasons, a number of "diesel-specific" sampling and analysis methods have been developed for use in occupational health air surveys. Most of these methods attempt to differentiate between diesel and other-source particulates based on two characteristics of diesel particles: They have (1) a high proportion of elemental carbon and (2) a high proportion of fine particles. The EC content of total particulates can be determined through analysis, whereas fine particle fractions can be obtained through size-selective sampling techniques. The accuracy of measurement depends on the particular application. For instance, EC may be a good surrogate for diesel particulates in noncoal mines, where no other significant sources contribute to inorganic carbon emissions. The same method is not likely to yield precise results in coal mines, due to interferences from coal dust, or in any environment with a significant concentration of nondiesel combustion aerosols.

With few exceptions, the current measurement methods for ambient aerosols are dominated by the use of collection on sampling filters followed by mass (gravimetric) analysis. In addition to gravimetric analyses, samples may be subjected to chemical analyses for specific elements or compounds. There is a great deal of commonality between methods used in general environments and in the workplace. However, the mass concentrations involved are typically an order of magnitude greater in occupational settings compared to the general environment [65]. Sampling and analysis of ambient aerosols have become a complex area of knowledge. Sophisticated sampling methods have been developed that utilize inertial, gravitational, centrifugal, and thermal mechanisms of particle collection. There is an increasing trend toward using direct-reading, *in situ* techniques utilizing such principles as optical particle detection. Detailed discussion of measurement methods and instruments for ambient aerosols, as well as their underlying science, can be found in the literature [11, 25, 66].

Most sampling methods developed specifically for diesel aerosols, which are derived from general aerosol sampling techniques, are also based on filter collection and gravimetric determination of the particulate mass. In the past, gravimetric measurement of diesel particulates was the basis for nearly all health studies. A growing body of literature suggests that a toxic response for some insoluble materials may be associated with surface area and/or particle number, rather than with particle mass [67–69]. These alternative metrics are believed to be particularly appropriate for the characterization of combustion aerosols, which are composed of much finer particles than general ambient aerosols. It can be expected that an increasing number of particle size and number related methods and instruments will be used for ambient particulate measurement.

9.4.1.2 Biologically Relevant Sampling

Ambient air aerosol sampling is ultimately concerned with measuring those aspects of aerosols that lead to specific health effects. The biological effects resulting from deposition of an aerosol in the respiratory tract depend on the dose received and on the response of the human organism. Sampling methods are thus designed to measure a "potential dose" by sampling particle fractions in relevance to their deposition in various areas of the respiratory system. The human respiratory system, a size-selective aerosol sampler on its own, can be divided into three zones, as follows [65]:

1. Nose and mouth (nasopharyngeal region); particles entering this zone are termed *inhalable particles*.

2. Trachea and the upper lungs (tracheobronchial region); entered by thoracic aerosol.
3. Alveolar region; particles entering the alveoli region are termed the *respirable aerosol.*

The respirable aerosol is a subfraction of the thoracic aerosol, which is, in turn, a subfraction of the inhalable aerosol.

Sampling standards have been developed that describe the penetration characteristics of aerosol particles through the respiratory system as a function of the particle aerodynamic diameter. International workplace sampling conventions [70] and the PM10 and PM2.5 ambient conventions are illustrated in Figure 9.20. These sampling conventions provide a basis for estimating the aerosol concentration potentially available to cause harm within the respective areas of the respiratory system. As seen in the figure, the probability of a particle entering the mouth and nose (inhalable particles) is about 50% for particles having a 100-μm aerodynamic diameter, and increases to nearly 100% for particles below a few microns. Note that nearly all diesel particulates have diameters below 1 μm and, thus, the diesel aerosol is almost totally respirable. These sampling conventions underlie many of the industrial hygiene aerosol sampling methods and instruments.

9.4.2 Measurement in Ambient Air

Most ambient air quality standards regulate concentrations of total suspended particulate matter having diameters of less than 10 μm (PM10) and less than 2.5 μm (PM2.5). The following are the current U.S. National Ambient Air Quality Standards (NAAQS) for PM [71]:

- *PM10:* annual average, 50 μg/m^3; 24-h average (99th percentile), 150 μg/m^3
- *PM2.5:* annual average, 15 μg/m^3; 24-h average (98th percentile), 65 μg/m^3.

Ambient levels of diesel particulates in general environments are not regulated. Due to their submicron size, suspended diesel particles are a subfraction of both the PM10 and PM2.5 particulate classes. Typical ambient exposures to diesel particulates range from 1 to 2 μg/m^3 in areas of low pollution to more than 10 μg/m^3 in areas of high pollution, as discussed in Chapter 8.

Figure 9.21 illustrates the components of a generic PM sampling system for the measurement of the total suspended particulates [72]. Practical implementations require integration of the components specific to the sampling objectives; some of the components (e.g., denuders) are often omitted. Equipment used for regulatory compliance sampling must meet certain requirements, which in the United States are outlined in the federal reference methods (FRM) adopted by the EPA.

Air enters the size-selective inlet where particles larger than desired are removed. The size-selective inlet is a very important component of the system, which is responsible for compliance with the pertinent sampling standards such as PM10 or PM2.5 (compare Figure 9.20). Inlets are characterized by sampling effectiveness curves showing the fraction of particles that passes through as a function of diameter. An important parameter related to the sampling effectiveness is the 50% cut point (d_{50}), defined as the diameter at which half of the particles pass through the inlet. Inertial classifiers typically employed as size-selective inlets include various types of impactors or cyclones. Some systems also utilize polycarbonate or polyurethane foam filters. Depending on their flow rates, inlets (and sampling systems) are classified into high volume

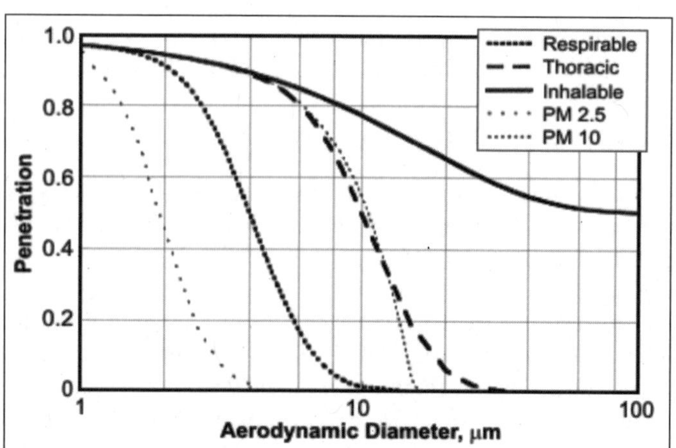

Figure 9.20 PM sampling conventions. (Reprinted with permission of John Wiley & Sons, Inc., from Ref. 65, Fig. 25-1. © 2001.)

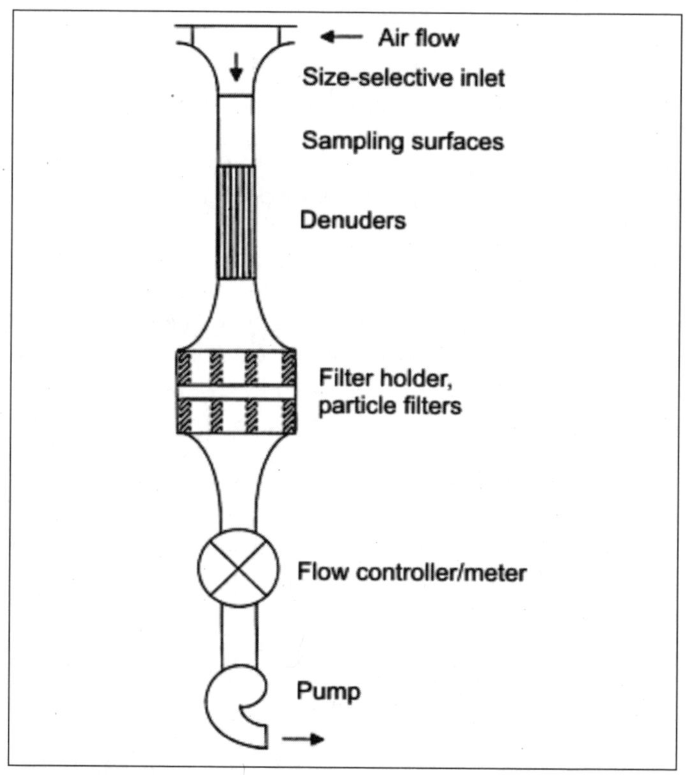

Figure 9.21 Aerosol sampling system. (Reprinted with permission of John Wiley & Sons, Inc., from Ref. 72, Fig. 27-1. © 2001.)

(1000 L/min), medium volume (100 L/min), low volume (10-20 L/min), and mini-volume (5 L/min). The flow rate must be selected to provide an adequate mass of deposit for analysis. A comprehensive listing of commercially available inlets has been compiled by Watson [72].

Sampling surfaces (inlets and transfer tubes) are most commonly made of anodized aluminum. Many materials absorb or react with certain gases, especially with nitric acid vapors, thereby changing the gas-particle equilibrium of particulate ammonium nitrate. Some volatile organic species may also interact with wall materials. Teflon-coated sampling surfaces are offered to minimize these interactions. Denuders may be placed in the air stream to remove selected gases while transmitting particles. Gas molecules diffuse to the denuder walls, while particles of lower mobility pass through. Denuder walls may be covered with chemicals that retain specific gases. Denuders can later be extracted in a solvent for laboratory determination of the average gas concentration. The most common use of denuders is to obtain an accurate measurement of nitrate particulates [72].

Sampling filter media are selected for specific applications based on their mechanical and chemical stability, particle sampling efficiency, flow resistance, loading capacity, artifact formation, compatibility with analytical methods, and cost. No universal filter medium fits all applications. Filter media most commonly used for atmospheric particle and gas sampling include a Teflon membrane, quartz fiber, nylon membrane, cellulose fiber, glass fiber, Teflon-coated glass fiber, and polycarbonate membrane. The various types of filters have been discussed in detail by Chow [73]. Filters are held by holders, which consist of a frame that seals the edge of the filter so that air is forced to pass through its porous center. A support grid downstream of the filter prevents it from being sucked through the holder by the pump vacuum. The filter holder can be configured as in line or open faced. In in-line filters, particles are concentrated in the center, causing bias when analyses are performed on a portion of the filter. Holders can be often adapted to fit many sampling systems and can accommodate a number of typical filter diameters, such as 25, 37, and 47 mm [72].

None of the general ambient air (i.e., nonoccupational) sampling methods would be capable of direct measurement of suspended DPM. Research studies that have investigated ambient diesel exposures used various mathematical methods for the determination (apportioning) of the diesel fraction in the total suspended particulates [74, 75]. Ambient air studies often try to estimate concentrations of diesel particulates by identifying substances that can be used as markers or "tracers" for diesel exposure. Once such a tracer is chosen, its content in the total particulate sample is determined by analysis. Diesel tracers that have been used include diesel fuel additives, PAHs and their ratios, and lube oil combustion products. Adding artificial tracers (dysprosium) not normally found in the atmosphere to the diesel fuel, difficult as it may sound, has been also attempted [76]. Because diesel particulates are characterized by higher carbon content than most ambient dusts, elemental carbon has become the most common tracer for diesel exhaust particulates. Standard EC analysis methods, developed both in the United States and in Europe, are described in the following sections. The EC method has been successfully used in certain occupational applications, where practically all inorganic carbon particulates can be attributed to diesel engine emissions. Unfortunately, this is not necessarily true in general environments, where combustion aerosols exist from such sources as industrial or spark-ignited engine combustion. Splitting the ambient PM inventories between diesel and gasoline sources has been notorious for its difficulties and controversies, with newer studies indicating that the gasoline engine contributions may have been underestimated [77]. Therefore, the EC method must be used with extreme caution in environmental settings.

Gravimetric methods also lack the sensitivity that would be required to measure ambient diesel exposures. More sensitive measurement can be offered by *in situ* instruments capable of measuring ambient levels of particulate mass, elemental carbon, or other properties of particulate matter [41]. The instruments that should be mentioned include these:

- Aethalometer (EC mass)
- Real-time EC analyzer (R&P)
- Quartz crystal microbalance (by Booker, total PM mass).

A number of *in situ* PM instruments were described earlier in this chapter, when discussing PM measurements in engine emissions. Several of those analyzers can be used for ambient air measurements.

9.4.3 Measurement in Occupational Health Settings

9.4.3.1 Diesel Pollutant Levels

The list of pollutants that are subject to occupational health exposure studies includes compounds that have established exposure limits—permissible exposure limits (PEL) or threshold limit values (TLV)—by regulatory authorities, as well as compounds that have no formal PELs but are suspected to have adverse health effects. A number of the gases present in diesel exhaust (including carbon monoxide, carbon dioxide, nitric oxide, nitrogen dioxide, sulfur dioxide, and formaldehyde) have long been recognized as workplace air pollutants and have exposure limits widely adopted by authorities. In view of its potential adverse health effects, diesel particulate matter became one of the most common targets of more recent occupational health exposure surveys. Some countries have legislated general occupational exposure limits for DPM (Germany, 0.1 mg/m^3 EC [78]), in others DPM exposure limits apply to underground mining (Canada, United States), an occupational setting that has the highest diesel exposures. The American Conference of Governmental Industrial Hygienists (ACGIH) has proposed a TLV recommendation of 0.02 mg/m^3 (EC) for diesel exhaust particulates. Approximate ranges of occupational exposure limits legislated or proposed for diesel compounds by different regulatory authorities, along with ACGIH TLV recommendations [79], are listed in Table 9.5.

Methods that are developed and used for measurements of air quality at the workplace have to be sensitive enough to allow

TABLE 9.5
RANGES OF EXPOSURE LIMITS FOR DIESEL POLLUTANTS

Pollutant	Typical Range	ACGIH TLV[a]	Unit
Diesel particulate matter[b]	0.02–1.5[c]	0.02[d]	mg/m^3
Carbon monoxide, CO	25–50	25	ppm vol.
Carbon dioxide, CO_2	5000–10,000	5000	ppm vol.
Nitric oxide, NO	25	25	ppm vol.
Nitrogen dioxide, NO_2	1.0–5.0	3	ppm vol.
Sulfur dioxide, SO_2	2.0–5.0	2	ppm vol.
Formaldehyde, HCHO	0.016–2.0	0.3	ppm vol.

a. 8-h time-weighted average (TWA).
b. Standards based on different definitions, such as total carbon, elemental carbon, or respirable combustible dust (RCD).
c. RCD, underground mines, Canada.
d. Elemental carbon, proposed (Notice of Intended Changes for 2001).

for accurate and repeatable measurements of the respective components at concentrations below their legislated exposure limits. In the case of DPM, as seen in Table 9.5, exposure limits can vary significantly. Introduction of more stringent exposure limits may require the adoption of new, more sensitive measurement methods.

9.4.3.2 Aerosol Sampling in the Workplace

Traditionally, aerosols in the workplace have been categorized as fumes (i.e., fine particles and agglomerates generated through combustion and vapor condensation), smokes (solid and liquid particles from incomplete combustion), sprays (liquid aerosols of large particle size, produced through mechanical means), and mists (liquid aerosols of finer particles, usually produced through condensation and atomization) [65]. These definitions are somewhat academic; most practical aerosol samplers would not differentiate between fumes, smoke, or mist, collecting all types of particles, both solid and liquid.

Interestingly, from the early 1900s through the 1950s, particle number was the dominant metric for the measurement of workplace aerosols, as determined by such samplers as the konimeter, impinger, and thermal precipitator coupled with microscope analysis [65]. Later, similar to general environment methods, collection on filters or impaction plates followed by gravimetric analysis became the dominant technique for occupational aerosol measurement. As mentioned earlier, this may revisited in the future with possible return of the particle number metric.

Two general types of sampling are used in the workplace: (1) fixed location sampling (also called static or area sampling) and (2) personal sampling (where the sampler is placed on the worker). Industrial hygiene aerosol samples are generally collected on a filter or on an impenetrable impaction substrate. Most current samplers comply with the size-selective sampling conventions (see Figure 9.20). For example, a respirable dust sampler would typically include a preclassifier (e.g., cyclone) with a 50% cut point of 4 μm to remove coarse particles. The air sample is drawn through the sampler by a sampling pump. The pumps are typically either diaphragm- or piston-type pumps powered by a rechargeable battery pack. Sampling filters can be made of such materials as cellulose, glass fiber, PVC membrane, Teflon membrane, silver membrane, or polycarbonate. Filters may be held in cartridges inside the sampler or may be mounted directly into the sampling head. Impaction substrates are typically made of Mylar or aluminum foil (usually coated with grease or oil to prevent particle bounce). The selection of a collection substrate depends on the sampling equipment used and the following analysis. Gravimetric analysis, which requires good weight stability, may be affected by a number of factors, for example, by the adsorption of water onto substrates and holders.

Cascade impactors have been widely used in occupational settings for the determination of mass-weighted particle size distributions. One of the most commonly used instruments has been the micro-orifice uniform deposit impactor. A number of direct-reading, *in situ* instruments have a place in occupational exposure assessment of combustion aerosol, such as the condensation particle counter, diffusion charging sensor, and other instruments discussed in Section 9.3.

9.4.3.3 Elemental Carbon Analysis

In occupational health environments with no nondiesel sources of inorganic carbon emissions, elemental carbon can be considered a very selective marker of diesel exposure. For the estimation of diesel particulate exposure, the EC analysis is performed on the respirable fraction of ambient aerosol (50% cut point of 4 μm; Figure 9.20). Elemental carbon accounts for about 50% of the mass of the total DPM, but this percentage may vary depending on the engine, its duty cycle, aftertreatment devices, and other factors. EC is easily detected by means of thermal-optical analysis. Elemental carbon analyses provide much better accuracy than do gravimetric methods.

The EC analysis has to differentiate between elemental carbon and organic carbon (OC). The organic carbon basically corresponds to the SOF fraction of diesel particulates. In occupational settings, however, the OC may also include oil mist, hydraulic fluids, and other materials not necessarily associated with diesel exhaust. The sum of EC and OC, known as total carbon, has been also used to formulate diesel exhaust exposure limits by some regulatory authorities (e.g., diesel exposure limits in U.S. mines [80]).

The analytical method used for EC in the United States, which was originally developed for atmospheric aerosols [81], was adopted by the National Institute for Occupational Safety and Health (NIOSH) for occupational health environments [82]. The two different types of carbon are detected and quantified by a two-stage thermal analysis with optical correction for the filter charring effect (the so-called thermal-optical method). In the first stage of the analysis, the sample is placed in an oven in an oxygen-free inert atmosphere. As the oven temperature is increased, the organic carbon material (hydrocarbons and derivatives) is evaporated and subsequently oxidized to CO_2, reduced to CH_4, and quantified by an FID analyzer. The second stage of analysis begins by reducing the combustion chamber temperature, introducing oxygen, and reheating to 750°C. During this stage, elemental carbon and char are oxidized to CO_2, reduced to CH_4, and detected.

The charring effect is related to certain organic compounds that pyrolytically decompose (rather than evaporate) during the first stage of analysis, and blacken the filter. Oxidation of the char causes a false-positive report on the EC during the second stage. The NIOSH method employs an optical correction device (He-Ne laser and a photodiode receptor) to compensate for the charring effect. The EC analysis method used in Europe, although similar in principle, is performed under a fixed set of analytical parameters and does not correct for the char error [83].

The estimated limit of detection of the elemental carbon method is 2 μg of either EC or OC per filter, which translates to an ambient concentration of about 1 μg/m^3, assuming 2 m^3 of sample air volume. This sensitivity is 100-fold better than that of gravimetric analyses.

In some German regulations, DPM is also defined as total carbon, as determined by a coulometric analysis of a fine dust particulate sample [84, 85].

9.4.4 Measurement in Mines

9.4.4.1 Mine Aerosols

Underground mining is a specific occupational health application characterized by uniquely high exposures to diesel particulates. Mining deserves special attention due to the substantial amount of aerosol research that has been conducted in mines, resulting in the development of certain diesel particle measurement methods specific to the mine application.

Mine aerosols typically originate from three sources: (1) comminution, (2) re-entrainment, and (3) combustion. Comminution aerosols are generated by mechanical crushing of the mined material by drill bits, crushers, and mining machine picks. The aerosol composition is generally similar to that of the host ore. In the case of coal mines, the aerosol includes a significant portion of coal dust. Aerosol that results from comminution may settle on roadways or in return air entries, becoming subject to re-entrainment through mobile equipment activities or by ventilation.

Combustion is another major source of aerosols in mines. Combustion aerosols are smaller in size and have higher number concentration than the coarser comminution aerosol. Major sources of combustion aerosols in mines include diesel engine particles and condensates, blasting agent fumes, and welding fumes. The diesel aerosol is of particular concern because it is almost entirely respirable in size, with more than 90% of the particles, by mass, having an aerodynamic diameter of less than 1 μm [86]. In the United States, mine ventilation rates were dictated in the past by concentrations of gaseous pollutants (CO, CO_2, NO, NO_2, SO_2). With the adoption of exposure limits for DPM by the Mine Safety and Health Administration [80], diesel aerosol concentrations may become the factor that controls the required ventilation rates.

9.4.4.2 Size-Selective Sampling

Diesel exhaust particulates exhibit trimodal size distributions, which include nuclei, accumulation, and coarse modes, as discussed in Chapter 7. Research conducted in mines confirmed that diesel aerosol is distinguished from nondiesel particles based on its smaller particle size. Figure 9.22 shows aerosol size distributions measured in (a) diesel-equipped and (b) all-electric coal mines [87]. In the dieselized mine the modal character of the size distribution is clearly discernible (even though in the mass-weighted representation the nuclei mode is suppressed compared with the accumulation mode). In the electric mine, on the other hand, practically all aerosol is found in the coarse mode (diameters of greater than 1 μm). When compared, the Figure 9.22 indicates that diesel particulates make a strong contribution to accumulation mode aerosols in the diesel-equipped mine.

The size-selective (SS) sampling method, developed by researchers of the former U.S. Bureau of Mines, is based on a notion that submicron particles found in mines are composed primarily of diesel particulates [88]. Other dusts, including the coal dust in coal mines, are composed of particles with larger aerodynamic diameters. In the SS method, diesel aerosols are separated from mechanically generated noncombustion aerosols based on their different aerodynamic particle sizes.

The first step in the SS method involves typical respirable dust sampling, as defined by the use of a cyclone with a 50% preclassification cut point of 4 μm. Diesel particles have a mass median aerodynamic diameter of 0.2 μm and 90% of these particles have diameters of less than 1 μm. Hence, the respirable sampling collects all diesel particles, as well as nondiesel particles that fall into the respirable range. The diesel and nondiesel fractions are then separated using inertial impaction on a greased aluminum foil substrate with a separation cut size of 0.8 μm. The inertial impaction removes nearly all (>90%) nondiesel particles and a small percentage (<15%) of large diesel particles. The submicron particles are collected on a PVC filter downstream of the impaction substrate. Mass concentrations in each of the size ranges are determined by gravimetric analysis of the collection media (i.e., the impaction substrate and the filter). The diesel aerosol fraction is assumed to be equal to the mass of the submicron sample. The total respirable dust can be calculated as the sum of the two size fractions.

A personal diesel exhaust aerosol sampler based on size-selective sampling developed for use in underground coal mines is shown in Figure 9.23 [86]. A 10-mm Dorr-Oliver cyclone is used as the respirable dust preclassifier. Its second stage is a

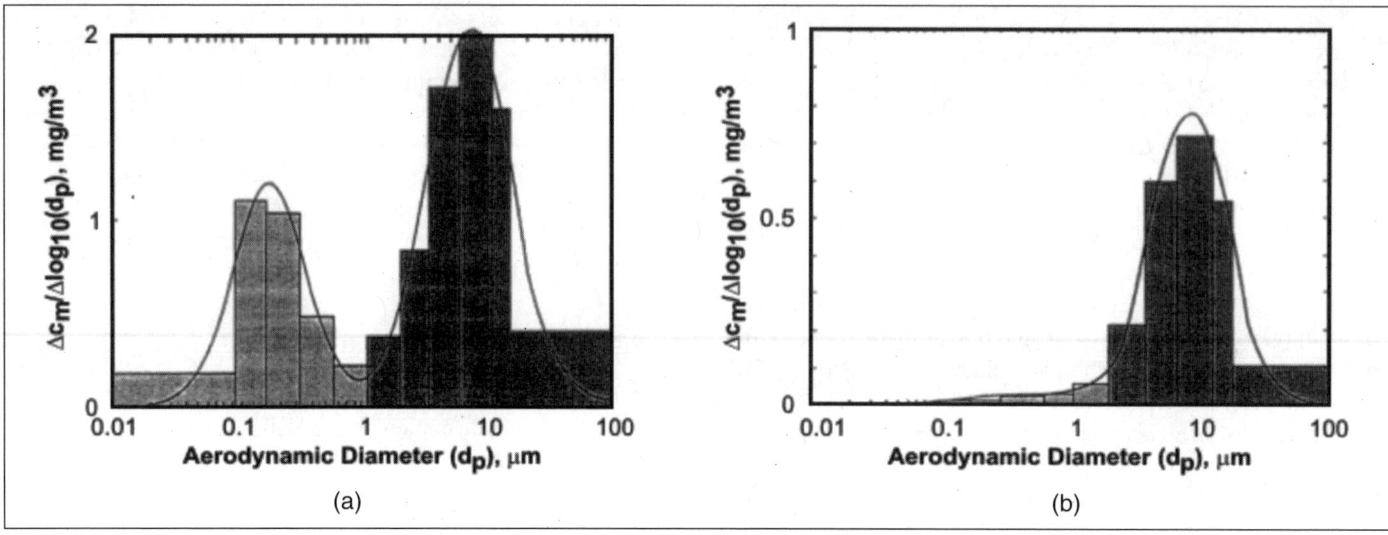

Figure 9.22 *Size distribution of mine aerosols: (a) diesel-equipped mine and (b) all-electric equipped mine. (Based on Ref. 87.)*

four-nozzle impactor with a 50% cut point of 0.8 µm. The third stage is a filter that collects the diesel fraction of the aerosol. The sampler operates at a 2 L/min sampling rate.

9.4.4.3 Diesel Aerosol Analysis

Three analytical methods have been used for measurement of diesel particulate exposures in underground mines, as follows:

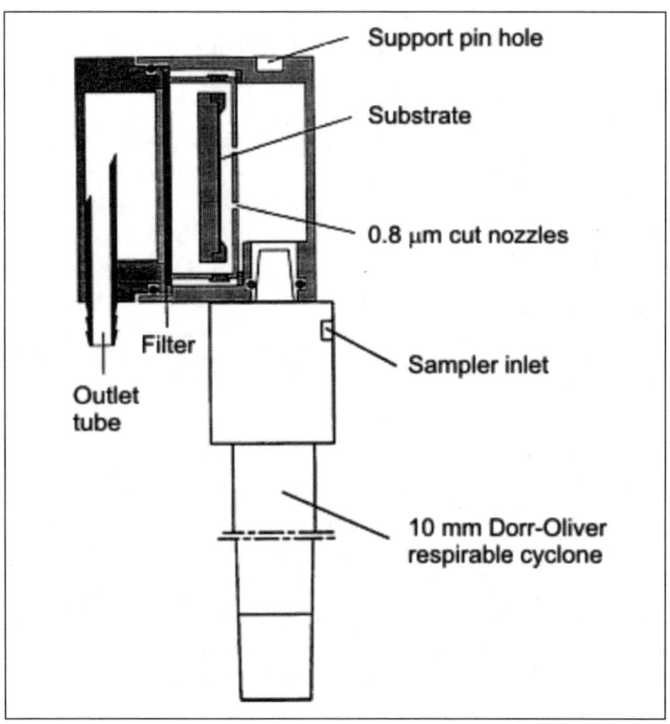

Figure 9.23 *Personal diesel exhaust sampler. (Reprinted with permission of John Wiley & Sons, Inc., from Ref. 86, Fig. 26-8. © 2001.)*

1. Gravimetric analysis
2. Carbon content analysis (EC/OC)
3. Combustible fraction analysis (RCD).

Gravimetric Analysis

The size-selective method coupled with gravimetric analysis is effective in environments where submicron aerosol sources are well defined and have relatively high concentrations. At a sample flow of 2 dm^3/min and an accuracy of the gravimetric analysis of 0.1 mg, the theoretical detection limit of the aerosol below 0.8 µm is 0.3 mg/m^3 ± 50% [88]. This detection limit is a worst-case estimate based on the experience from U.S. mines. In situations where the concentrations of diesel aerosol are lower, or when submicrometer particles are also produced by nondiesel sources, analytical methods of greater specificity and sensitivity are required to quantify the diesel aerosol fraction.

Carbon Content (EC/OC)

Due to its detection limits, the gravimetric analysis cannot be used to support exposure standards of less than about 0.3 mg/m^3. Because control technologies exist and exposure standards were adopted that reduce exposures to below that level [80], a more sensitive analytical technique was needed. NIOSH researchers proposed the use of the elemental carbon analysis [82]. The EC method is typically a more selective marker of DPM than the total carbon (TC = EC + OC), because the OC can originate from several nondiesel sources [86].

A simple sampling train to obtain samples for EC analysis consists of a 10-mm Dorr-Oliver cyclone, followed by a 37-mm precombusted ultrapure tissue-quartz fiber filter mounted in a plastic cassette. Alternatively, the cyclone can be followed by the size-selective impactor to remove larger, mechanically generated aerosols of diameters above 0.8 µm [86].

Respirable Combustible Dust

The respirable combustible dust (RCD) method is used to estimate DPM concentrations in Canadian noncoal mines [89]. In the RCD method, total respirable dust (TRD) is collected from the air sample on a 25- or 37-mm, 0.8-μm silver membrane filter (or a glass fiber filter) after passing through a 10-mm Dorr-Oliver cyclone respirable dust preclassifier. The sample flow rate is controlled by a personal sampling pump at 1.7 dm^3/min. The TRD is determined gravimetrically by weighing the silver membrane filter before and after the sample is collected. The RCD portion of TRD is determined gravimetrically from the amount of material removed from the silver membrane filter during controlled combustion at 400°C (500°C for the glass fiber filters) for 1 to 2 h. A correction is made for loss of the filter media during the combustion.

An estimate of the DPM concentration can be made by multiplying the RCD number by an empirical correction factor of 0.67 for the contributions of nondiesel carbonaceous materials, such as the drill oil mist. Because coal dust would be also included in the RCD, that method is inappropriate for estimating DPM levels in coal mines.

Comparison of Methods

The methods of DPM measurement just discussed have been compared based on studies in underground metal mines [90, 91]. Linear regression equations were developed to correlate the results. Table 9.6 provides a comparison of methods that was generated using the experimental correlations. The variability of diesel aerosols somewhat limits the universal validity of this comparison. Nevertheless, the data are very useful as an estimate of the relationship between methods, as well as for placing different DPM exposure standards in perspective.

9.4.4.4 Other Methods

Source Apportionment

The use of source apportionment analyses has been proposed for use in underground mines based on the size-selective measurements [92]. This technique relates elemental, mineral, or chemical components found in the aerosol sample to the same components in the sources of the aerosol (in the case of coal mines: coal, rock dust used as fire retardant, and diesel aerosol). The contribution of each source to the sampled aerosol is determined using mathematical techniques based on chemical mass balance analysis.

TABLE 9.6
RELATIONSHIP OF DPM MEASURING METHODS, mg/m³

Total Carbon	Elemental Carbon[a]	RCD[b]	SS[c]
0.25	0.125	0.27	0.26
0.50	0.25	0.62	0.59
0.60	0.30	0.75	0.75
0.75	0.275	0.96	0.93
1.00	0.50	1.30	1.27

Source: Data from Refs. 90 and 91.
a. Calculated as 50% of total carbon.
b. 0.8-μm silver membrane filter.
c. <0.8-μm fraction.

Real-Time Instruments

Real-time measurement of mine dusts remains a future objective for aerosol research. Personal exposure monitoring is preferable to area monitoring in mining. Personal real-time monitoring is believed to be the preferred strategy to meaningfully approximate workers' exposure [86]. Some of the approaches to development of mine dust monitors have been based on the following instruments:

- Light-scattering instruments, called *photometers* or *nephelometers,* use a light source to illuminate the sample aerosol and a light sensor to measure the scattered light, which is related to the dust concentration.
- A *tapered element oscillating microbalance (TEOM)* was being adapted for use as a personal dust monitor [51].
- A *dust dosimeter* is a newer instrument based on a dust detector tube [93]. The instrument, consisting of a pump with an integral pressure transducer and a disposable detector tube, similar to a gas detector tube, correlates pressure drop across the tube with mass accumulation.

9.4.5 Gas-Phase Measurements

9.4.5.1 Instruments

Concentrations of gaseous pollutants in the workplace are typically measured using electrochemical cell-based instruments. These instruments are quickly replacing the chemical detector tubes that have been used in the past for ambient air quality checks. The detector tubes were characterized by very poor accuracy (with error, in some cases, as high as 35%) and high costs due to the need to use a new tube for each measurement.

Many of the electrochemical instruments offer an accuracy of about 5% or better. However, three important sources of additional error occurs during measurements of diesel pollutants:

- Cross-sensitivity of gas sensors is one source of additional error. All electrochemical cells interfere with other gases present in the measured sample. The cross-sensitivities could be as high as 25% or more. Because diesel exhaust is a mixture of gases, sensor cross-sensitivities may cause significant errors. Sensor cross-sensitivity information is available from the cell manufacturer (many instruments use electrochemical cells from City Technologies, United Kingdom). Only high-quality cells having low cross-sensitivities are suitable for diesel emission surveys.
- Temperature changes were reported to cause a ±15% error in the reading.
- Pressure changes were reported to cause a ±20% error in the reading [94].

9.4.5.2 Air Quality Index

The Air Quality Index (AQI) is a single expression designed to characterize exposure to all major pollutants, both gases and particulates. It is built from the actual concentrations of pollutants divided by their respective exposure limits. The following expression is the AQI definition after the Canada Centre for Mineral and Energy Technology [95]:

$$AQI = CO/50 + NO/25 + RCD/2 + 1.5(SO_2/3 + RCD/2) + 1.2(NO_2/3 + RCD/2) \quad (9.13)$$

where:

CO, NO, SO2, NO2 = concentrations of the respective gases, ppm
RCD = respirable combustible dust, mg/m^3.

An equivalent expression for raw exhaust gas is known as the Exhaust Quality Index (EQI). The AQI/EQI approach has never gained wide acceptance in diesel emission applications, but it is still used in underground mine engine approvals in Canada.

References

1. Kittelson, D.B., J. Johnson, W. Watts, Q. Wei, and N. Bukowiecki, 2000. "Diesel Aerosol Sampling in the Atmosphere," SAE Paper No. 2000-01-2122, SAE International, Warrendale, PA, 2000.
2. Clark, N.N., *et al.,* "Design and Operation of a New Transportable Laboratory for Emissions Testing of Heavy-Duty Trucks and Buses," *Int. J. Vehicle Design: Heavy Vehicle Systems* **2,** 3/4, 1995, pp. 285–299
3. Jetter, J., *et al.,* "Development of an Onboard Analyzer for Use on Advanced Low Emission Vehicles," SAE Paper No. 2000-01-1140, SAE International, Warrendale, PA, 2000.
4. Baum, M.M., E.S. Kiyomiya, S. Kumar, A.M. Lappas, and H.C. Lord, "Multi-Component Remote Sensing of Vehicle Exhaust by Dispersive Absorption Spectroscopy," SAE Paper No. 2000-01-3103, SAE International, Warrendale, PA, 2000.
5. Plint, M., and A. Martyr, *Engine Testing: Theory and Practice,* Butterworth-Heinemann, Oxford, UK, 1995.
6. Samuel, S., L. Austin, and D. Morrey, "Automotive Test Drive Cycles for Emission Measurement and Real-World Emission Levels—A Review," *Proc. Inst. Mech. Engineers: Part D, J. Automobile Eng.* **216,** 2002, pp. 555–564.
7. Spears, M.W., "An Emissions-Assisted Maintenance Procedure for Diesel-Powered Equipment," Report #USDI/1432 CO369005, National Institute for Occupational Safety and Health, Pittsburgh, PA, 1997, retrieved from www.cdc.gov/niosh/mining/eamp/eamp.html.
8. Silvis, W., N. Kreft, G. Marek, and W. Schindler, "Diesel Particulate Measurement with Partial Flow Sampling Systems: A New Probe and Tunnel Design That Correlates with Full Flow Tunnels," SAE Paper No. 2002-01-0054, SAE International, Warrendale, PA, 2002.
9. Schweizer, T., and H.J. Stein, "A New Approach to Particulate Measurement on Transient Test Cycles: Partial Flow Dilution as Alternative to CVC Full Flow Systems," SAE Paper No. 2000-01-1134, SAE International, Warrendale, PA, 2000.
10. Khalek, I.A., *et al.,* "Performance of Partial Flow Sampling Systems Relative to Full Flow CVC for Determination of Particulate Emissions under Steady-State and Transient Diesel Engine Operation," SAE Paper No. 2002-01-1718, SAE International, Warrendale, PA, 2002.
11. Cohen, B.S., and S.V. Hering (Eds.), *Air Sampling Instruments for Evaluation of Atmospheric Contaminants,* 9th ed., American Conference of Governmental Industrial Hygienists, Cincinnati, OH, 2001.
12. Perez, J.M., R.E. Jass, and D.G. Leddy, "Chemical Methods for the Measurement of Unregulated Diesel Emissions," CRC Report No. 552, Coordinating Research Council, Inc., Atlanta, GA, August 1987
13. "Chemical Methods for the Measurement of Nonregulated Diesel Emissions," SAE Recommended Practice J1936, SAE International, Warrendale, PA, 1995.
14. Beaumont, A.J., *et al.,* "Signal Reconstruction Techniques for Improved Measurement of Transient Emissions," SAE Paper No. 900233, SAE International, Warrendale, PA, 1990.
15. Baronick, J.D., *et al.,* "Evaluation of an UV analyzer for NOx Vehicle Emission Measurement," SAE Paper No. 2001-01-0213, SAE International, Warrendale, PA, 2001.
16. "Measurement of Carbon Dioxide, Carbon Monoxide, and Oxides of Nitrogen in Diesel Exhaust," SAE Standard, SAE International, Warrendale, PA, June 1995
17. Nakamura, H., N. Kihara, M. Adachi, and K. Ishida, "Development of a Wet-Based NDIR and Its Application to Onboard Emission Measurement System," SAE Paper No. 2002-01-0612, SAE International, Warrendale, PA, 2002.
18. Dobson, J.G., *et al.,* "Anwendung von FID Detektoren fuer die Analyse kontinuirlisher Prozesse," *Archiv. Technisches Messen* **361,** 1966, pp. R25–R44.
19. Kittelson, D.B., M. Arnold, and W.F. Watts, "Review of Diesel Particulate Matter Sampling Methods," Final Report, Center for Diesel Research, University of Minnesota, Minneapolis, January 14, 1999, retrieved from www.me.umn.edu/centers/cdr/reports/EPAreport3.pdf.
20. Burtscher, H., "Literature Study on Tailpipe Particulate Emission Measurement for Diesel Engines," Report for Particle Measurement Programme, BUWAL/GRPE, March 2001, retrieved from www.akpf.org/pub/burtscher_bericht.pdf.
21. Perez, J.M., "Cooperative Development of Analytical Methods for Diesel Emissions and Particulates, Solvent Extractables, Aldehydes, and Sulfate Methods," SAE Paper No. 840413, SAE International, Warrendale, PA, 1984.
22. Perez, J.M., *et al.,* "Informational Report on the Measurement and Characterization of Diesel Exhaust Emissions,"

Coordinating Research Council Report No. 517, Alpharetta, GA, 1980.

23. Bagley, S.T., K.J. Baumgard, L.D. Gratz, J.H. Johnson, and D.G. Leddy, "Characterization of Fuel and Aftertreatment Device Effects on Diesel Emissions," Research Report #76, Health Effects Institute, Cambridge, MA, September 1996, retrieved from http://healtheffects.org/Pubs/st76.htm.
24. Hollstein, M., J. McCann, "Short-Term Tests for Carcinogens and Mutagens," *Mutation Res.* **65,** 1979, pp. 133–226.
25. Baron, P.A., and K. Willeke (Eds.), *Aerosol Measurement: Principles, Techniques, and Applications,* 2nd ed., John Wiley, New York, 2001.
26. Hinds, W.C., *Aerosol Technology,* J. Wiley & Sons, New York, 1999.
27. Wiedensohler, A., *et al.,* "Intecomparison Study of the Size Dependent Counting Efficiency of 26 Condensation Particle Counters," *Aerosol Sci. Technol.* **27,** 1997, pp. 224–242.
28. Lipkea, W.H., *et al.,* "The Physical and Chemical Character of Diesel Particulate Emissions, Measurement Techniques and Fundamental Considerations," SAE Paper No. 780108, SAE International, Warrendale, PA, 1978.
29. Burtscher, H., *et al.,* "Separation of Volatile and Non-Volatile Aerosol Fractions by Thermodesorption: Instrumental Development and Applications," *J. Aerosol Sci.* **32,** 2001, pp. 427–442.
30. Andersson, J., 2002. "UK Particle Measurement Programme: Heavy-Duty Methodology Development," Report DP02/2493, Ricardo Consulting Engineers, Shoreham-By-Sea, UK, July 31, 2002, retrieved from www.ricardo.com/chemistry/UK_PMP_HD_Programme.pdf.
31. Kawai, T., "Influence of Thermodenuder Dimensions on Nanoparticle Measurement," in *Proc. 6th ETH Conf. on Nanoparticle Measurement,* Zurich, August 2002.
32. Marple, V.A., K.L. Rubow, and S.M. Behm, "A Microorifice Uniform Deposit Impactor (MOUDI): Description, Calibration, and Use," *J. Aerosol Sci. Technol.* **14,** 1991, pp. 434–446.
33. Marple, V.A., K.L. Rubow, and B.A. Olson, "Low Pressure Stages for the Micro-Orifice Uniform Deposit Impactor (MOUDI)," in *4th Int. Aerosol Conference Abstracts,* R.C. Flagan (Ed.), Vol. 2, Los Angeles, CA, 1994.
34. Bell, S.P., C.J. Dickens, E.A. Feest, A.P. Smith, and A.M. Tope, "Experimental Assessment of Particulate Measurement Instrumentation," AEA Report AEAT-4978, AEA Technology, Abingdon, UK, 1999.
35. Hillamo, R.E., and E.I. Kauppinen, "On the Performance of the Berner Low Pressure Impactor," *J. Aerosol Sci.* **14,** 1991, pp. 33–47.
36. Andersson, J., and B. Wedekind, "DETR/SMMT/CONCAWE Particulate Research Programme," Summary Report DP01/0515, Ricardo Consulting Engineers, Shoreham-By-Sea, UK, May 2001, retrieved from www.ricardo.com/downloads/SummaryReport.pdf.
37. Keskinen, J., K. Pietarinen, and M. Lehtimaki, "Electrical Low Pressure Impactor," *J. Aerosol Sci.* 23, 1992, pp. 353–360.
38. Marjamaki, M., *et al.,* "Electrical Filter Stage for the ELPI," SAE Paper No. 2002-01-0055, SAE International, Warrendale, PA, 2002.
39. Liu, B.Y.H., and D.Y.H. Pui, "On the Performance of the Electrical Aerosol Analyser," *J. Aerosol Sci. 6,* 1974, pp. 249–264.
40. Winklmayr, W., G.P. Reischel, A.O. Lindner, and A. Berner A., "A New Electromobility Spectrometer for the Measurement of Aerosol Size Distributions in the Size Range from 1 to 100 nm," *J. Aerosol Sci.* **22,** 3, 1991, pp. 289–296.
41. Pui, D.Y.H., and D.L. Swift, "Direct-Reading Instruments for Airborne Particles," Chap. 16 in *Air Sampling Instruments for Evaluation of Atmospheric Contaminants,* 8th ed., B.S. Cohen and S.V. Hering (Eds.), American Conference of Governmental Industrial Hygienists, Cincinnati, OH, 1995, pp. 337–368.
42. Wang, S.H., and R.C. Flagan, "Scanning Electrical Mobility Spectrometer," *J. Aerosol Sci.* **13,** 1990, pp. 230–240.
43. Schlatter, J., "Internationaler Ringversuch mit Partikelgrössenmessgeräten," Report No. 2000-230-436, Swiss Federal Office of Metrology, Bern, 2000.
44. Ahlvik, P., L. Ntziachristos, J. Keskinen, and A. Virtanen, "Real Time Measurements of Diesel Particle Size Distribution with an Electrical Low Pressure Impactor," SAE Paper No. 980410, SAE International, Warrendale, PA, 1998.
45. Boulard, D., and M. Diouri, "A New Inertial and Diffusional Device (SDI 2000)," *J. Aerosol Sci.* **19,** 1988, pp. 927–930.
46. Scheibel, H.G., and J. Porstendörfer, J., "Penetration measurements for tube and screen-type diffusion batteries in the ultrafine particle size range," *J. Aerosol Sci.* **15,** 1984, pp. 673–682.
47. Cheng, Y.S., and C.H. Yeh, "Analysis of Screen Diffusion Battery Data," *Am. Ind. Hyg. Assoc. J.* **45,** 8, 1984, pp. 556–561.
48. Burtscher, H., H.L. Scherrer, and H.C. Siegmann, "The Electrical Diffusion Battery for Dynamic Classification of Nanoparticles," in *Proc. 4th ETH Conference on Nanoparticle Measurement,* A. Mayer (Ed.), ETH, Zurich, 2001.
49. Vanhobel, F., *et al.,* in *Proc. SIA Int. Conf.,* Lyon, June 13–14, 1990 (in French), Paper No. #90089, 1990, pp. 109–114.
50. Siegmann, K., and H.C. Siegmann, "Fast and Reliable *in situ* Evaluation of Particles and Their Surfaces with Special Reference to Diesel Exhaust," SAE Paper No. 2000-01-1995, SAE International, Warrendale, PA, 2000.
51. Patashnick, H., and G. Rupprecht, "Personal Dust Exposure Monitor Based on the Tapered Element Oscillating Microbalance," *Bull. Mines* OFR 56-84, NTIS PB 84-173749, 1983.
52. Booker D., "Development of a Real-Time Transient Cycle Mass Monitor," in *Proc. 3rd ETH Conference on Nanoparticle Measurement,* Paper #20, A. Mayer (Ed.), ETH, Zurich, August 2000.
53. Gaeggeler, H.W., *et al.,* "The Epiphaniometer, A New Device for Continuous Aerosol Monitoring," *J. Aerosol Sci.* **20,** 5, 1989, pp. 557–564.

54. Schraml, S., S. Will, A. Leipertz, T. Zens, and N. DAlsonso, "Performance Characteristics of TIRE-LII Soot Diagnostics in Exhaust Gases of Diesel Engines," SAE Paper No. 2000-01-2002, SAE International, Warrendale, PA, 2000.
55. Liousse, C., H. Cachier, and S.G. Jennings, "Optical and Thermal Measurements of Black Carbon Aerosol Content in Different Environments: Variation of the Specific Attenuation Cross Section, Sigma," *Atmos. Environ.* **27A,** 1993, pp. 1203–1211.
56. Burtscher, H., and H.C. Siegmann, "Monitoring PAH-Emissions from Combustion Processes by Photoelectric Charging," *Combust. Sci. Technol.* **101,** 1994, pp. 327–332.
57. Dahmann, D., T. Mosimann, and U. Matter, "Validation of On-Line Sensors for Monitoring Occupational Exposures from Diesel Engines," *J. Aerosol Sci.* **31,** Suppl. 1, 2000, pp. S21–S22.
58. Krämer, L., Z. Bozoki, and R. Niessner, R., "Setup, Calibration and Characterization of a Mobile Photoacoustic Soot Sensor," *J. Aerosol Sci.* **31,** Suppl. 1, 2000, pp. S72–S73.
59. Silva, P.J., and K.A. Prather, "On-Line Characterization of Individual Particles from Automobile Emissions," *Environ. Sci. Technol.* **31,** 1997, pp. 3074–3080.
60. Kasper, M., U. Matter, and H. Burtscher, "NanoMet: On-Line Characterization of Nanoparticle Size and Composition," SAE Paper No. 2000-01-1998, SAE International, Warrendale, PA, 2000.
61. Jones, M., "Advanced Opacity Meters: Their Potential Role in Future Emission Testing Legislation for Diesel Vehicles," in *Proc. 6th ETH Conference on Nanoparticle Measurement,* ETH, Zurich, August 2002.
62. Zahoransky, R.A., and Z. Wizard Z., "On-Line/In-Line Measurements of Particle Emissions by Optical Multi-Wavelength Technique," in *Proc. 3rd ETH Conference on Nanoparticle Measurement,* ETH, Zurich, August 2000.
63. Homan, H.S, "Conversion Factors among Smoke Measurements," SAE Paper No. 850267, SAE International, Warrendale, PA, 1985.
64. Green, G.L., and D. Wallace, "Correlation Studies of an In-Line, Full-Flow Opacity meter," SAE Paper No. 801373, SAE International, Warrendale, PA, 1980.
65. Maynard, A.D., and P.A. Jensen, "Aerosol Measurement in the Workplace," in *Aerosol Measurement, Principles, Techniques, and Applications,* 2nd ed., P.A. Baron and K. Willeke (Eds.), John Wiley & Sons, New York, 2001, pp. 779–799.
66. Vincent, J.H., *Aerosol Science for Industrial Hygienists,* Pergamon, Bath, UK, 1995.
67. Vincent, J.H., and C. F. Clement, "Ultrafine Particles in Workplace Atmospheres," *Phil. Trans. R. Soc. Lond. A.* **358,** 2000, pp. 2673–2682.
68. Donaldson K., X.Y. Lee, and W. MacNee, "Ultrafine (Nanometre) Particle Mediated Lung Injury," *J. Aerosol Sci.* **30,** 1998, pp. 553–560.
69. Oberdorster, G., J. Ferin, and B.E. Lehnert, "Correlation between Particle Size, *in vivo* Particle Persistence, and Lung Injury," *Environ. Health Perspect.* **102,** S5, 1994, pp. 173–179.
70. *Air Quality—Particle Size Fraction Definitions for Health-Related Sampling,* ISO Standard 7708, International Standards Organization, Geneva, 1995.
71. Environmental Protection Agency, "National Ambient Air Quality Standard for Particulate Matter: Final Rule," *Federal Register* **63,** 1998, pp. 6032–6037.
72. Watson, J.G., and J.C. Chow, "Ambient Air Sampling," in *Aerosol Measurement, Principles, Techniques, and Applications,* 2nd ed., P.A. Baron and K. Willeke (Eds.), John Wiley & Sons, New York, 2001, pp. 821–844.
73. Chow, J.C., "Critical Review: Measurement Methods to Determine Compliance with Ambient Air Quality Standards for Suspended Particulates," *JAWMA* **45,** 1995, pp. 320–382.
74. "Proposed Identification of Diesel Exhaust as a Toxic Air Contaminant," Report by the Staff of the California Air Resources Board and the Office of Environmental Health Hazard Assessment, April 22, 1998.
75. "Multiple Air Toxics Exposure Study II (MATES II)," MATES II Report, California South Coast Air Quality Management District, 2000, retrieved from www.aqmd.gov/matesiidf/matestoc.htm.
76. Horvath, H., *et al.,* "Diesel Emissions in Vienna," *Atmos. Environ.* **22,** 7, 1988, pp. 1255–1269.
77. Lawson, D.R., "OHVT Gasoline/Diesel PM Split Study," 7th Diesel Emissions Reduction Workshop (DEER), Portsmouth, VA, August 2001, U.S. Department of Energy, Washington, DC, retrieved from www.osti.gov/hvt/deer2001/lawson1.pdf.
78. "TRGS 554—Dieselmotoremissionen (DME)," BArbBl. (German Ministry of Labor), March 2001, pp. 112.
79. "Threshold Limit Values for Chemical Substances and Physical Agents and Biological Exposure Indices," American Conference of Governmental Industrial Hygienists, Cincinnati, OH, 2001.
80. Mine Safety and Health Administration "Diesel Particulate Matter Exposure of Underground Metal and Nonmetal Miners, Final Rule (30 CFR Part 57)," *Federal Register* **66,** 13, January 19, 2001, pp. 5706, retrieved from www.msha.gov/REGS/FEDREG/FINAL/2001finl/01-996.pdf.
81. Hering, S.V., *et al.,* "Comparison of sampling methods for carbonaceous aerosols in ambient air," *J. Aerosol Sci. Technol.* **12,** 1990, pp. 200–213.
82. "Elemental Carbon (Diesel Exhaust) 5040," in *Manual of Analytical Methods* (NMAM), 4th ed., National Institute for Occupational Safety and Health, Atlanta, GA, March 15, 1996, retrieved from www.cdc.gov/niosh/nmam/pdfs/5040f3.pdf.
83. Dahmann, D., H.H. Fricke, and H.D. Bauer, "Diesel Engine Emissions in Workplace Atmospheres in Germany," *Occ. Hygiene* **3,** 196, pp. 225–262.
84. Heymann, C., "Verfahren zur Bestimmung von Kohlenstoff im Feinstaub—anwendbar fur Partikelfoermige Dieselmotor-Emissionen in Arbeitsbereichen," Carl Heymanns Verlag KG, Cologne, 1995.

85. Bauer, H., D. Dahmann, and H. Fricke, "Coulometric Determination of Diesel Engine Emissions in Non-Coal Mines," *Staub-Reinhaltung der Luft* **51,** 1991, pp. 319–325.
86. Cantrell, B.K., and J.C. Volkwein, "Mine Aerosol Measurement," in *Aerosol Measurement, Principles, Techniques, and Applications,* 2nd ed., P.A. Baron and K. Willeke (Eds.), John Wiley & Sons, New York, 2001, pp. 801–820.
87. Cantrell, B.K., and K.L. Rubow, "Measurement of Diesel Exhaust Aerosol in Underground Coal Mines," IC 9324, Bureau of Mines, 1992, pp. 11–17.
88. Cantrell, B.K., and W.F. Watts, "Diesel Exhaust Aerosol: Review of Measurement Technology," in *Proc. Canadian Adhoc Diesel Committee Plenary Conf.,* CANMET, Ottawa, Canada, 1996, pp. 7.1.
89. Gangal, M., J. Ebersole, J. Vallieres, and D. Dainty, "Laboratory Study of Current (1990/91) SOOT/RCD Sampling Methodology for the Mine Environment," Mining Research Laboratory, Canada Centre for Mineral and Energy Technology (CANMET), Ottawa, 1990.
90. Watts, W.F., and G. Ramachandran, "Diesel Particulate Matter Sampling Methods: Statistical Comparison," Final Report submitted to the Diesel Emission Evaluation Program, November 6, 2000, retrieved from www.deep.org/reports/watts_final.pdf.
91. Watts, W.F., B.K. Cantrell and J.S. Stachulak, "Diesel Aerosol Sampler Comparison in a Metal Mine," in *Proc. Canadian Adhoc Diesel Committee Third Plenary Conf.,* CANMET, Ottawa, Canada, 1997, pp. II.1.1.
92. Cantrell, B.K., "Source Apportionment Analysis Applied to Mine Dust Aerosols: Coal Dust and Diesel Emissions Aerosol Measurement," in *Proc. 3rd Mine Ventilation Symp.,* Society for Mining, Metallurgy, and Exploration, Inc., Littleton, CO, 1987, pp. 495–501.
93. Volkwein, J.C., *et al.,* "Laboratory Evaluation of a Pressure Differential-Based Respirable Dust Detector Tube," *Appl. Occup. Environ. Hyg.* **15,** 2000, pp. 158–164.
94. Gangal, M., "Diesel Exhaust Measurement Technology," in *Proc. Canadian Adhoc Diesel Committee Plenary Conf.,* CANMET, Ottawa, Canada, 1996, pp. 8.1.
95. Gangal, M., *et al.,* "Monitoring and Assessment of Air Quality in Dieselized Mines," Report MRL 86-81(OP), Canada Centre for Mineral and Energy Technology (CANMET), Ottawa, June 1986.

Chapter 10

Emission Regulations

10.1 Historical Background

Emissions result from a wide range of natural occurrences and human activities. When certain pollutants become highly concentrated in specific areas, they can cause harm to the human life and/or the environment. Realizing this fact—and faced with a significant increase in concentrations of pollutants, especially in the industrialized world—several concerned groups have advanced the cause of clean environment. A significant increase in automotive emissions in the United States, Europe, and Japan led to special awareness of the potential harm of automobile-sourced pollutants. Regulatory agencies have been empowered by local and state government and legislative bodies to control and limit known harmful emissions from several sources.

From the historical perspective, visitors to the city of London, England, have long known the characteristic smell and heavy air associated with automobile exhaust and foggy weather. The combination of smoke emanating from stationary sources, buses, and taxi cabs, with fog, helped coin the word *smog*. Many suspected that smog had adverse health effects on the general population, but specifically on London Transport workers [1].

It was first thought that the pollution problem in the Los Angeles, California, area was similar to the phenomenon observed in London. Subsequent studies, however, showed that this was not the case. Symptoms of the problem in California included vegetation damage, eye and throat irritation, and reduced visibility. The problem became serious enough that some civic groups, aided by the media, pressured the California legislature into passing into law regulations limiting emissions in Los Angeles County in 1947. These initial regulatory efforts did not directly associate the environmental degradation with automobile exhaust. Therefore, regulatory enforcement was applied to stationary emissions sources such as steel mills and refineries. Still, the benefits were felt in the form of improved visibility and dustfall reductions compared to the levels that had prevailed in 1940. Eye and throat irritation still persisted, leading to more research into the health effects of the Los Angeles pollution problem [2].

Research by Professor A. J. Haägen-Smit of the California Institute of Technology demonstrated that, in the presence of sunlight and nitrogen dioxide, hydrocarbon compounds reacted to form oxidation products. These products included ozone, to which adverse health effects could be attributed. Against this background of research findings, public awareness, and legislative action, the Automobile Manufacturers Association formed the Vehicle Combustion Product Committee in 1953 to further define this pollution problem and propose solutions. At the same time California increased its effort to control hydrocarbon emissions from stationary sources. This activity culminated in the first ever air quality standards shown in Table 10.1.

The first emissions policing agency, the California Motor Vehicle Pollution Control Board (CMVPCB), was created in 1960. It was given authority to set emissions limits on future vehicles sold in the state and to ensure compliance with those limits. Crankcase blowby was being regulated by the CMVPCB by 1963, and the automobile industry was using positive crankcase ventilation systems in California to control blowby in 1961. This solution was applied nationwide in 1963 models [1].

The federal government has traditionally lagged behind the state of California in its environmental efforts. In 1955 the Department of Health, Education, and Welfare was empowered to provide assistance in problems related to air pollution control activities. This resulted in the Clean Air Act of 1963,

TABLE 10.1
AIR QUALITY STANDARDS

Pollutant	Parts per Million for 1 Hour		
	"Adverse" Level	"Serious" Level	"Emergency" Level
Carbon monoxide		120	240
Ethylene	0.5		
Hydrogen sulfide	0.1	5	
Sulfur dioxide	1	5	10
Hydrocarbons Nitrogen dioxide Oxidant ozone aerosols	0.15 on "oxidant index"	Not established	Not established

which stimulated regulatory action by state and local agencies. By 1965, the Clean Air Act authorized the writing of national standards for motor vehicle emissions control. This action led to requiring emission control devices on vehicles sold nationwide by 1968. Compliance with these standards was determined by placing vehicles on chassis dynamometers and operating them through a prescribed cycle that had initially been adopted by the state of California. That same cycle was subsequently adopted by the federal government and commonly called the Federal Test Procedure and shown in Figure 10.1 [3].

The first air quality standards adopted by the State of California established three limits for the pollutants listed in Table 10.1. The first level or limit was called *adverse,* and indicated the level at which vegetation damage and human sensory irritation occurred. The second level was called *serious* and indicated the level at which human bodily functions would be altered as well as chronic diseases spread. The third was the *emergency* level at which acute sickness or death might occur in certain population groups who could be especially vulnerable [4].

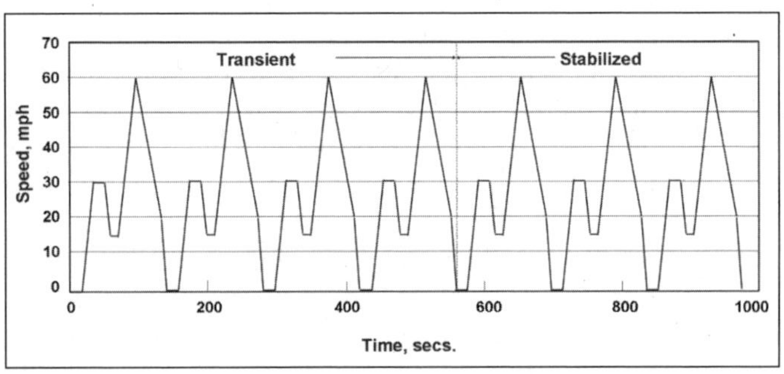

Figure 10.1 *First automotive emission test cycle in the United States. (Based on data from Ref. 3.)*

Mass measurement of exhaust emissions is not an easy or convenient task. Therefore, exhaust gas volumetric measurement and concentrations of pollutants were utilized to assess a vehicle's compliance with the standards for 1970 and 1971. This meant that exhaust gas pollutant concentrations were measured then multiplied by a calculated volume according to a preset formula to arrive at an equivalent mass of pollutants emitted into the atmosphere. For new vehicles in 1972–1974, a new cycle, test procedure, and emissions measurement system were developed. A constant volume sampling (CVS) system was utilized from which an exhaust sample was introduced into a plastic bag. The latter was then discharged into instruments that measured the pollutant concentrations of the bag contents.

As shown in Table 10.2, emission limits were imposed on hydrocarbons (HCs) and carbon monoxide (CO) from 1968 through 1970. Control of nitrogen oxides did not start until 1971 when photochemical smog reactions were recognized as serious health hazards. Evaporative emissions controls were introduced in California as well as the rest of the nation in 1971 and 1972 [1].

For diesel engines, particulate matter was the last emission species to be regulated. The first limit of 0.6 g/bhp-hr was introduced for heavy-duty engines only in 1988, in spite of proposed limits 13 years earlier [1]. The reasons given for introducing particulate emission limits were health and aesthetics. In those early days of particulate regulations, many speculated about the

TABLE 10.2
EARLY FEDERAL REGULATORY LIMITS

Year	HC	CO	NOx	PM	*Federal Register* Reference
FTP Cycle					
1968	275 ppml (3.2 g/mi)[a]	1.5% (33 g/mi)[a]	—	—	Vol. 21, No. 61, Pt. II, March 30, 1966
1970	2.2 g/mi (180 ppm)[a]	23 g/mi (1%)[a]	—	—	Vol. 33, No. 108, Pt. II, June 4, 1968
1971	2.2 g/mi	23 g/mi	4.0 g/mi[b]	—	
CVS-1 Cycle					
1971	4.6 g/mi	47 g/mi			Vol. 35, No. 219, Pt. II,I November 10, 1970
1972	3.4 g/mi	39 g/mi	4.0 g/mi		
1973–1974	3.4 g/mi	39 g/mi	3.0 g/mi		
CVS-3 Cycle					
1975	0.41 g/mi	3.4 g/mi	3.0 g/mi	0.1 g/mi[c]	Vol. 36, No. 128, Pt. II, July 2, 1971
1976	0.41 g/mi	3.4 g/mi	0.4 g/mi	0.1 g/mi[c]	

Note: Evaporative losses: 6.0 g per test in 1970 in California and 1971 nationwide, and 2.0 g per test beginning 1972 nationwide.

a. Equivalent for a 4000-lb vehicle.

b. California requirements only.

c. Proposed.

need for catalysts to control and reduce emission levels from diesel engines [1]. Yet, engine designers were able to develop combustion systems that were capable of reducing particulate emission to extremely low levels without the need for additional exhaust control devices.

Evaporative emissions limits were introduced in the early 1970s, and although they did not seem to present a serious problem for diesel engine-powered vehicles, they did represent another degree of regulation for automotive emissions. Since those modest beginnings, regulatory agencies in the United States, Europe, and Japan have introduced several layers of increasingly tough standards, not the least of which are the heavy-duty diesel standards for on-highway model year 2007 applications.

Since the middle to late 1950s, the regulatory process has developed into a significant air quality policy in the industrialized world. Many new engine and vehicle categories have been included under the umbrella of emission regulations. Significant advances and research have been fueled by the need to meet clean air requirements while maintaining performance and fuel economy. The modern diesel engine not only meets very strict emission limits and standards, but also outperforms older models in responsiveness and fuel economy.

10.2 Types of Emission Regulations

Over a decades-long period of evolution, engine emission-related standards and regulations have become a complex system of legislation established by different government branches and government levels. In addition to the emission standards for new engines or vehicles, programs exist and are being developed that try to control emissions from in-use diesel engines. Diesel engines operated in confined spaces, such as in buildings or in mines, may be also regulated, directly or indirectly, by occupational health regulations. In general, diesel-related emission regulations and programs can be divided into the following categories:

- Emission standards for new engines and/or vehicles ("tailpipe" emission standards)
- Occupational health regulations
- Emission control retrofit programs for existing vehicles
- Inspection and maintenance programs.

Tailpipe emission standards for new engines/vehicles specify the maximum amount of pollutants allowed in exhaust gases discharged from a diesel engine. Tailpipe emission standards are usually adopted and enforced by government ministries responsible for the protection of environment, such as the Environmental Protection Agency (EPA) in the United States. The duty to comply with these standards is on the equipment (engine) manufacturer. Typically all equipment has to be emission certified before it is released to the market. The standards also specify the required emission durability and/or emission warranty, which may be expressed as the minimum mileage or time period over which the engine must meet the standard.

In most countries regulated diesel emissions include the following:

- *Diesel particulate matter* (PM) mass emission, as measured by gravimetric methods. Sometimes diesel smoke opacity is also regulated as an auxiliary standard.
- *Nitrogen oxides* (NO_x), as the sum of nitric oxide (NO) and nitrogen dioxide (NO_2). Other oxides of nitrogen that may be present in exhaust gases—for example, nitrous oxide (N_2O)—are not regulated.
- *Hydrocarbons* (HCs), regulated either as total hydrocarbon (THC) emissions or as nonmethane hydrocarbons (NMHCs). One combined limit for HC + NO_x is sometimes used instead of two separate limits.
- *Carbon monoxide* (CO).
- Formaldehyde (HCHO) is regulated by some diesel emission standards.

Emissions are measured over an engine or vehicle test cycle, which is an important part of every emission standard. Regulatory test procedures are necessary to verify and ensure compliance with the various standards. These test cycles are supposed to create repeatable emission measurement conditions and, at the same time, simulate real driving conditions for a given application. Analytical methods that are used to measure emissions are also regulated by the standard.

Occupational health regulations specify the required air quality in the workplace. Most commonly, occupational ambient air quality standards specify maximum allowable concentrations of air contaminants called permissible exposure limits (PELs) . If engines are operated indoors, their emissions may accumulate in the ambient air to levels exceeding their respective PELs, thus forcing various emission control measures. Sometimes, in addition to the PELs, occupational health regulations may introduce tailpipe emission standards, engine approval programs, and other engine-related control measures.

Occupational health regulations are set and enforced by occupational health and safety authorities such as the Occupational Safety and Health Administration (OSHA) and Mine Safety and Health Administration (MSHA) in the United States. The duty to comply is on the engine user (warehouse operator, mine operator, etc.) who has to make sure that the emission control measures that have been employed are adequate for the type and number of pieces of polluting equipment. In most cases, engine or equipment manufacturers do not have any direct obligations with regard to the occupational health and safety air quality standards.

Gases found in diesel emissions including carbon monoxide, nitric oxide, nitrogen dioxide, sulfur dioxide, and many other compounds have their PELs set by occupational health and safety authorities. Since the 1990s, exposure limits for diesel particulates have been developed and adopted by occupational health authorities worldwide. Occupational exposure limits for diesel particular matter (DPM) had been adopted in the United States for underground mining (noncoal), but not for general occupational environments.

Emission control retrofit programs are becoming increasingly popular as a measure to reduce emissions from existing diesel engine fleets, especially in large urban areas that are experiencing severe air quality problems. Such retrofit programs involve installation of emission control devices, for example, catalytic converters or diesel particulate filters, on in-use heavy-duty diesel engines. Some retrofit programs are obligatory, while others have a voluntary character. In voluntary programs, a system of incentives, for example tax incentives and grants, is established to promote retrofits. As a rule, emission control equipment must be formally approved for use in retrofit programs, to ensure that real emission benefit is obtained. The approval typically involves determination of the emission performance based on testing in an engine emission laboratory. Some retrofit programs also require demonstration of a certain emission durability and/or emission warranties, while others issue approvals based exclusively on the performance of new equipment.

Examples of heavy-duty diesel retrofits include the following programs:

- The EPA's *Urban Bus Rebuild/Retrofit* (UBRR) program, started in 1995, is intended to reduce the ambient levels of particulate matter (PM) in urban areas and is limited to 1993 and earlier model year urban buses operating in metropolitan areas with 1980 populations of 750,000 or more, whose engines are rebuilt or replaced after January 1, 1995 (40 CFR Part 85 Subpart O). Every time the affected bus engines are rebuilt, their PM emissions have to be reduced to different levels, subject to the availability of certified emission control equipment. Most certified devices are diesel oxidation catalysts and engine kits. No diesel particulate filters were certified under the UBRR program, presumably due to the difficulties in meeting the 150,000-mile emission durability requirement and the $7,940 equipment cost cap.
- The Swedish *Environmental Zones* program, started in 1996, establishes a maximum allowable age (generally 8 years) for heavy-duty diesel vehicles traveling in the biggest cities in Sweden: Stockholm, Gothenburg, Malmö, and Lund. Older vehicles must be retrofitted with emission controls to enter specified "Zones." Currently allowed emission control devices include particulate filters/oxidation catalyst systems, which must reduce PM emissions by at least 80%, and 35% efficient NO_x control devices. An estimated 3000 vehicles were retrofitted with emission control systems during the first 3 years of the program. In 1999, the program was extended to off-road vehicles, with a required 80% PM emission reduction (particulate filters) or an 80% HC emission reduction (oxidation catalysts).
- *California diesel retrofit programs* are based on a comprehensive risk reduction plan for diesel particulate emissions adopted by the California Air Resources Board in 2000 [5]. Among other measures, the plan calls for retrofitting of 90% of all in-use diesel engines in California, both mobile and stationary, with diesel particulate filters. The retrofits, to be completed by 2008, would be regulated by a number of rules pertaining to different vehicle categories.
- The EPA's *Voluntary Diesel Retrofit Program* (VDRP) is being developed to promote retrofits of heavy-duty diesel engines with EPA-verified emission control technologies in exchange for the State Implementation Plan (SIP) emission credits to states [6]. The program adopted a very wide definition of a retrofit, which includes retrofitting engines with a catalytic converter or a diesel particulate filter, engine upgrade, engine replacement, and use of clean fuels and/or fuel additives.

Inspection and maintenance (I&M) programs involve a periodic emission check from in-use vehicles. I&M programs are often maintained by lower level governments, such as by states in the United States or by provinces in Canada. These programs target poorly maintained, high-emission vehicles. Most I&M programs affect cars, but some of them also extend to heavy-duty diesel vehicles.

I&M programs rely on their own emission test methods and limits, which are not related to the new engine emission standards. The emission test is often very simple, such as a steady-state measurement of selected pollutant concentration. In the case of diesel vehicles, the test is typically limited to a smoke opacity measurement.

In the case of a failure of an I&M emission check, the vehicle operator is usually required to do the necessary service and/or repairs to bring it into compliance. I&M programs are enforced through vehicle registration systems, whereas passing the emission test is required to obtain the renewal of registration.

10.3 Emission Test Cycles

Engine and vehicle test cycles are standardized sequences of engine speed and load conditions, designed to simulate certain real-life operating conditions for the determination of certain engine parameters, such as emissions or fuel economy. Test cycles can be divided into (1) steady-state cycles and (2) transient cycles. In steady-state cycles the engine is operated through a number of stabilized speed and load modes. Emissions are measured over the cycle at each mode and averaged, typically using different weighting factors for different modes. In transient cycles the vehicle (engine) follows a prescribed driving pattern, which includes accelerations, decelerations, and changes of speed and load. Emissions are calculated either through integration of continuous, real-time measurements, or through analysis of gas samples collected in sampling bags.

Engine test cycles are performed on an engine dynamometer. Their emission results are most often expressed relative to the mechanical work performed by the engine and given in units of grams per kilowatt-hour or grams per brake-horsepower-hour (g/bhp-hr). Vehicle cycles are conducted on a chassis dynamometer, with emissions expressed in grams of pollutant per unit of traveled distance and given in grams per kilometer or grams per mile.

A large number of cycles have been developed for both regulatory and emission research purposes. The following sections provide brief descriptions of the most important cycles used for regulating new engine and vehicle emissions in the United States and the European Union (EU). More comprehensive test cycle information is available in the literature and on the Internet [7].

10.3.1 U.S. Heavy-Duty FTP Transient Cycle

The Federal Test Procedure (FTP) heavy-duty transient cycle is currently used for emission testing of heavy-duty on-road engines in the United States (CFR Title 40, Part 86.1333). The transient test was developed to take into account the variety of heavy-duty truck and bus traffic in American cities, including traffic in and around the cities on roads and expressways. The cycle includes "motoring" segments and, therefore, requires a DC or AC electric dynamometer capable of both absorbing and supplying power.

The transient cycle consists of four phases: The first is a NYNF (New York nonfreeway) phase typical of light urban traffic with frequent stops and starts, the second is the LANF (Los Angeles nonfreeway) phase typical of crowded urban traffic with few stops, the third is a LAFY (Los Angeles freeway) phase simulating crowded expressway traffic in Los Angeles, and the fourth phase repeats the first NYNF phase. It comprises a cold start after a vehicle has been parked overnight, followed by idling, acceleration, and deceleration phases, and a wide variety of different speeds and loads sequenced to simulate the running of the vehicle that corresponds to the engine being tested. There are few stabilized running conditions, and the average load factor is about 20% to 25% of the maximum horsepower available at a given speed.

The cycle is carried out twice and the second repetition is made with a warm start after a stop of 1200 s (20 min) on completion of the first cycle. The equivalent average speed is about 30 km/h and the equivalent distance traveled is 10.3 km for a running time of 1200 s. The variation of normalized speed and torque with time is shown in Figure 10.2.

Heavy-duty diesel engines tested on the FTP cycle produce medium to high exhaust gas temperatures. Generally, the temperature is at a medium level between 250 and 350°C, but there are some hot sections in which temperatures reach as high as 450°C.

10.3.2 U.S. FTP-75 Cycle

The FTP-75 cycle is a chassis dynamometer, transient cycle that has been used for emission certification of light-duty vehicles in the United States (Figure 10.3). The FTP-75 cycle consists of the following segments:

- Cold start phase
- Transient phase
- Hot start phase.

The following are basic parameters of the cycle:

- Distance traveled: 11.04 miles (17.77 km)
- Duration: 1874 s
- Average speed: 21.2 mph (34.1 km/h).

The emissions from each phase are collected in a separate Teflon bag, analyzed, and expressed in grams per mile (grams per kilometer). The weighting factors are 0.43 for the cold start, 1.0 for the transient phase and 0.57 for the hot start phase.

10.3.3 European Stationary Cycle

The European Stationary Cycle (ESC) test cycle (also known as OICA/ACEA cycle) was introduced, together with the European Transient Cycle and the European Load Response tests, for emission certification of heavy-duty diesel engines in the European Union starting in the year 2000 [8]. The ESC is a 13-mode, steady-state procedure that replaces the earlier R-49 test. Since October 1998 this cycle has also been used in the United States for testing of certain heavy-duty engines in addition to the FTP transient test.

The engine is tested on an engine dynamometer over a sequence of steady-state modes (see Table 10.3 and Figure 10.4). The engine must be operated for the prescribed time in each mode, completing engine speed and load changes in the first

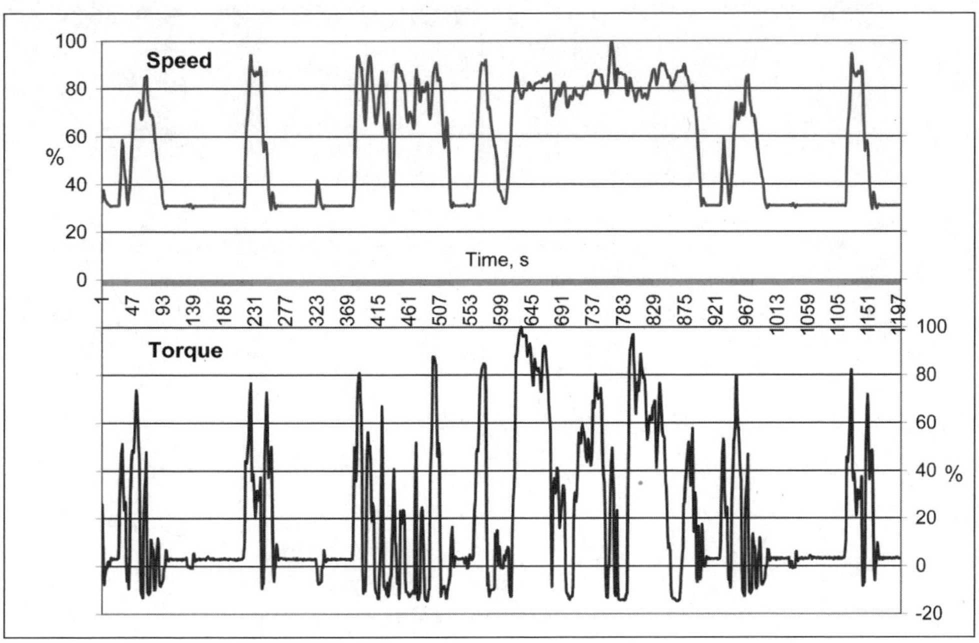

Figure 10.2 FTP transient cycle. (Based on data from the Code of Federal Regulations.*)*

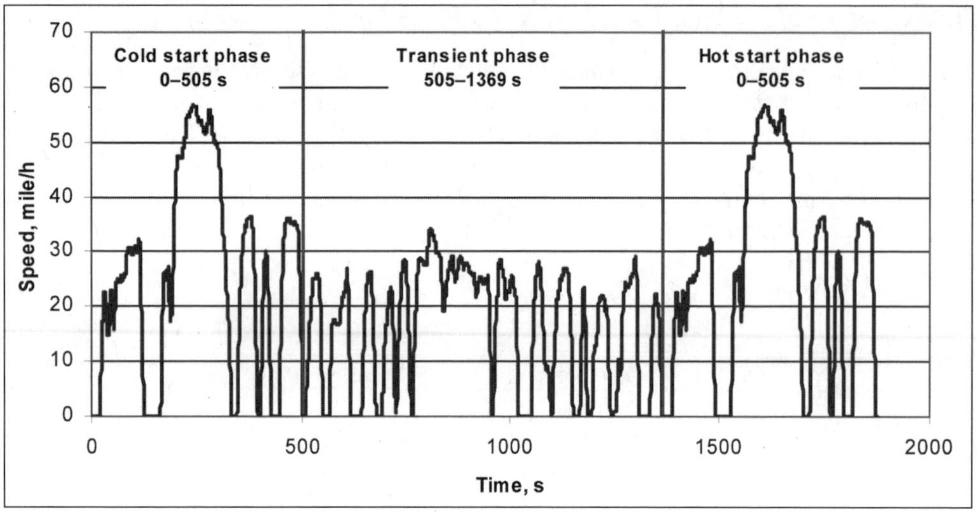

Figure 10.3 *FTP-75 cycle. (Based on data from the* Code of Federal Regulations.*)*

20 s. The specified speed shall be held to within ±50 rpm and the specified torque shall be held to within ±2% of the maximum torque at the test speed. Emissions are measured during each mode and averaged over the cycle using a set of weighting factors. Particulate matter emissions are sampled on one filter over the 13 modes. The final emission results are expressed in grams per kilowatt-hour.

During emission certification testing, the certification personnel may request additional random testing modes within the cycle control area (Figure 10.4). Maximum emissions at these extra modes are determined by interpolation between results from the neighboring regular test modes.

The engine speeds are defined as follows:

1. The high-speed *nhi* is determined by calculating 70% of the declared maximum net power. The highest engine speed where this power value occurs (i.e., above the rated speed) on the power curve is defined as nhi.

TABLE 10.3
ESC TEST MODES

Mode	Engine Speed	% Load	Weight Factor, %	Duration, min
1	Low idle	0	15	4
2	A	100	8	2
3	B	50	10	2
4	B	75	10	2
5	A	50	5	2
6	A	75	5	2
7	A	25	5	2
8	B	100	9	2
9	B	25	10	2
10	C	100	8	2
11	C	25	5	2
12	C	75	5	2
13	C	50	5	2

TABLE 10.4
SUMMARY OF ECE AND EUDC CYCLE PARAMETERS

Characteristics	Unit	ECE 15	EUDC
Distance	km	4×1.013 = 4.052	6.955
Duration	s	4×195 = 780	400
Average speed	km/h	18.7 (with idling)	62.6
Maximum speed	km/h	50	120

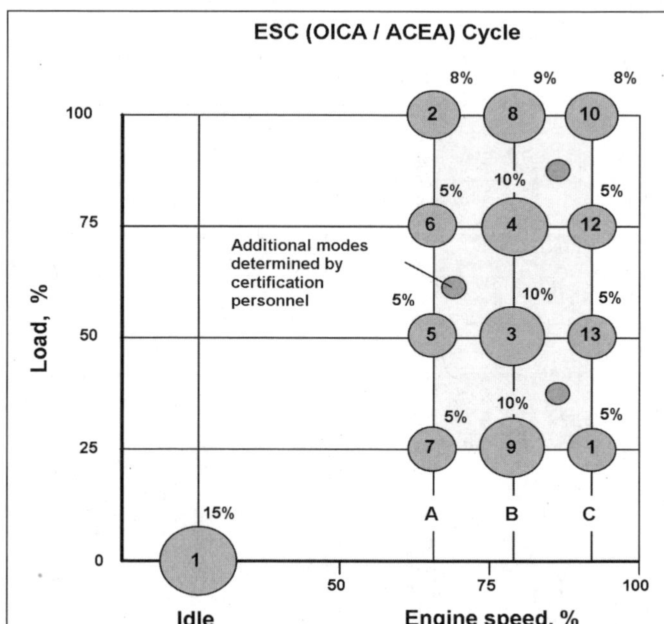

Figure 10.4 *European Stationary Cycle (ESC). (Based on data from the* Euro III Directive.*)*

2. The low-speed *nlo* is determined by calculating 50% of the declared maximum net power. The lowest engine speed where this power value occurs (i.e., below the rated speed) on the power curve is defined as nlo.
3. The engine speeds A, B, and C to be used during the test are then calculated from the following formulas:

$$A = nlo + 0.25(nhi - nlo)$$
$$B = nlo + 0.50(nhi - nlo)$$
$$C = nlo + 0.75(nhi - nlo)$$

The ESC test is characterized by high average load factors and very high exhaust gas temperatures.

10.3.4 European Transient Cycle

The European Transient Cycle (ETC) test (also known as the FIGE transient cycle) was introduced, together with the ESC, for emission certification of heavy-duty diesel engines in Europe starting in the year 2000 [8]. The ETC cycle has been developed by the FIGE Institute, Aachen, Germany, based on real road cycle measurements of heavy-duty vehicles [9]. The final ETC cycle is a shortened and slightly modified version of the original FIGE proposal.

Different driving conditions are represented by three parts of the ETC cycle, including urban, rural, and motorway driving. The duration of the entire cycle is 1800 s. The duration of each part is 600 s. Part 1 represents city driving with a maximum speed of 50 km/h, frequent starts and stops, and idling. Part 2 is rural driving starting with a steep acceleration segment. The average speed is about 72 km/h. Part 3 is motorway driving with average speed of about 88 km/h.

The FIGE Institute developed the cycle in two variants: as a chassis and an engine dynamometer test. For the purpose of engine certification, the ETC cycle is performed on an engine dynamometer. The engine speed and torque curves are shown in Figure 10.5.

10.3.5 ECE 15 + EUDC

The ECE + EUDC test cycle is performed on a chassis dynamometer. The cycle is used for emission certification (type approval) of light-duty vehicles in Europe. It is also known as the MVEG-A cycle. The cycle definition can be found in Ref. [10]. The entire cycle includes four ECE segments (Figure 10.6) repeated without interruption, followed by one EUDC segment (Figure 10.7). Before the test, the vehicle was allowed to soak for at least 6 h at a test temperature of 20 to 30°C. It was then started and allowed to idle for 40 s. Effective in the year 2000, that idling period was eliminated; that is, engine starts at 0 s and the emission sampling begin at the same time. This modified cold start procedure is sometimes referred to as the "new European driving cycle." Emissions are sampled during the cycle according to the CVS technique and then analyzed and expressed in grams per kilometer for each of the pollutants.

The ECE cycle is an urban driving cycle, also known as UDC. It was devised to represent city driving conditions, for example, in Paris or Rome. It is characterized by low vehicle speed, low engine load, and low exhaust gas temperature.

The Extra Urban Driving Cycle (EUDC) segment has been added after the fourth ECE cycle to account for more aggressive, high-speed driving modes. The maximum speed of the EUDC cycle is 120 km/h. An alternative EUDC cycle for low-powered vehicles has been also defined with a maximum speed limited to 90 km/h.

Table 10.4 includes a summary of the parameters for both the ECE and EUDC cycles.

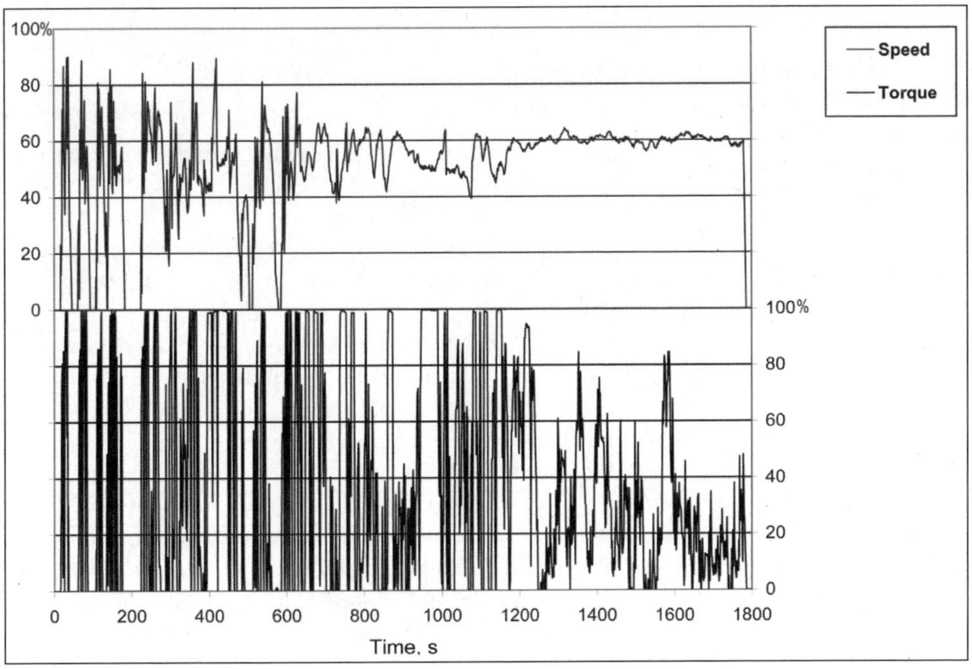

Figure 10.5 ETC test cycle. (Reprinted with permission from DieselNet.)

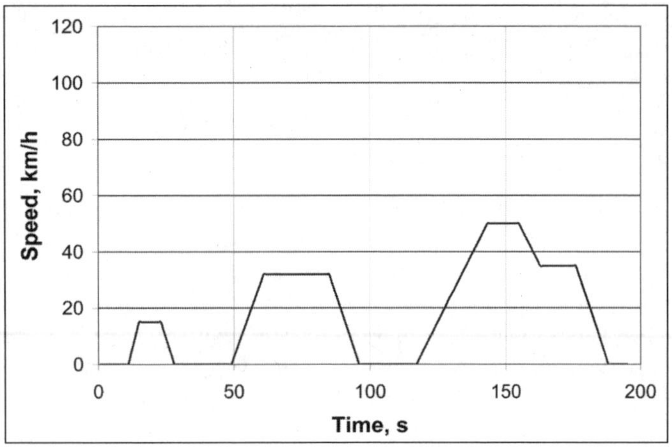

Figure 10.6 ECE 15 cycle.

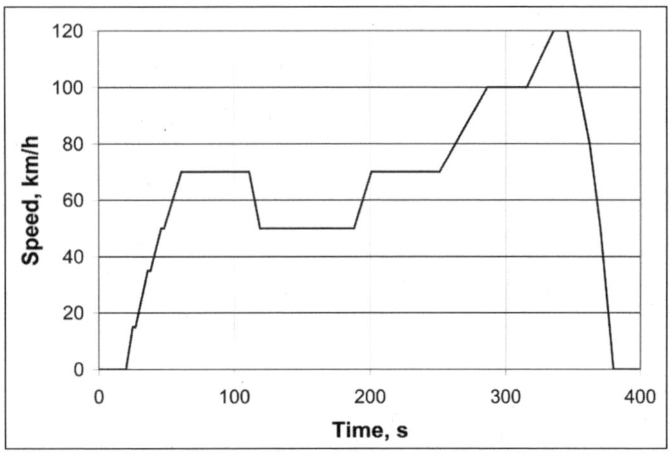

Figure 10.7 EUDC cycle.

10.3.6 ISO 8178

The ISO 8178 standard has been used worldwide, including the United States and the EU, for emission testing and engine certification in various nonroad engine applications. ISO 8178 is actually a collection of many steady-state test cycles (type C1, C2, D1, etc.). Each of these cycles represents a sequence of several steady-state modes with different weighting factors. The particular engine modes and their weighting factors for B-type (11 mode) test cycles are listed in Table 10.5.

10.4 Selected Diesel Emission Standards

10.4.1 Overview

Since the early 1990s, due to increased health and environmental concerns, diesel emissions have been subject to increasingly more stringent emission standards worldwide. Today, emissions are the driving factor and a significant technological challenge in the development of new diesel engines. The most stringent standards are being adopted in the United States and the European Union.

The evolution of the U.S. and European emission standards for PM and NO_X from heavy-duty diesel engines is shown in Figures 10.8 and 10.9. The U.S. limits, expressed in the regulations in grams per brake-horsepower-hour, have been recalculated to grams per kilowatt-hour to enable comparison with the EU standards. Regulatory authorities in different countries have not been unanimous in adopting emission test procedures, and many test cycles are in use. Because exhaust emissions depend on the engine speed and load conditions, specific engine emissions that were measured on different test cycles may not be comparable even if they are expressed or recalculated into the same units of measure. This should be kept in mind whenever comparing emission standards from different countries.

The most important of the coming emission standards include the U.S. 2007 (NO_X = 0.2 g/bhp-hr, PM = 0.01 g/bhp-hr) and the Euro IV/Euro V limits (NO_X = 3.5/2.0 g/kWh, PM = 0.02 g/kWh). All earlier emission standards for heavy-duty diesel engines, including Euro III and U.S. 2004, are met using advanced engine technologies, mostly without the use of exhaust gas aftertreatment. The Euro IV/V (2005/2008) and the U.S. 2007 standards have been adopted by regulators to force widespread use of advanced exhaust gas aftertreatments, such as diesel particulate filters and NO_X catalysts, on new diesel engines.

A brief overview of the most important U.S. and European emission standards is provided in the following sections. This summary is a snapshot of the emission regulation status as of mid-2002. Because emission standards evolve and change continuously, this summary will be soon outdated. More comprehensive and up-to-date information on diesel emission standards can be obtained from the respective regulatory authorities. Summary information on diesel emission regulations is also available from sources on the internet [11].

10.4.2 United States

10.4.2.1 Heavy-Duty Highway Engines

The following emission standards apply to new diesel engines used in heavy-duty highway vehicles. The current federal definition of a compression-ignition (diesel) engine is based on the engine cycle, rather than the ignition mechanism, with the presence of a throttle being used as an indicator to distinguish between Diesel cycle and Otto cycle operation. Regulating power by controlling the fuel supply in lieu of a throttle corresponds with lean combustion and the diesel-cycle operation. (This allows for the possibility that a natural gas–fueled engine equipped with a spark plug could be considered a compression-ignition engine.)

Heavy-duty vehicles are defined as vehicles having a gross vehicle weight rating (GVWR) of more than 8500 lb in the federal jurisdiction and more than 14,000 lb in California (model year 1995 and later). Under the federal light-duty Tier 2 regulation (phased in beginning in 2004), vehicles having a GVWR up to 10,000 lb used for personal transportation have

TABLE 10.5
WEIGHTING FACTORS OF B-TYPE ISO 8178 TEST CYCLES

Mode Number	1	2	3	4	5	6	7	8	9	10	11
Torque, %	100	75	50	25	10	100	75	50	25	10	0
Speed	Rated speed					Intermediate speed					Low idle
Off-Road Vehicles											
Type C1	0.15	0.15	0.15	—	0.10	0.10	0.10	0.10	—	—	0.15
Type C2	—	—	—	0.06	—	0.02	0.05	0.32	0.30	0.10	0.15
Constant Speed											
Type D1	0.30	0.50	0.20	—	—	—	—	—	—	—	—
Type D2	0.05	0.25	0.30	0.30	0.10	—	—	—	—	—	—
Locomotives											
Type F	0.25	—	—	—	—	—	—	0.15	—	—	0.60
Utility, Lawn and Garden											
Type G1	—	—	—	—	—	0.09	0.20	0.29	0.30	0.07	0.05
Type G2	0.09	0.20	0.29	0.30	0.07	—	—	—	—	—	0.05
Type G3	0.90	—	—	—	—	—	—	—	—	—	0.10
Marine Application											
Type E1	0.08	0.11	—	—	—	—	0.19	0.32	—	—	0.30
Type E2	0.20	0.50	0.15	0.15	—	—	—	—	—	—	—

Marine Application Propeller Law

Type E3, Mode #	1	2	3	4	
Power, %	100	75	50	25	
Speed, %	100	91	80	63	
Weighting factor	0.2	0.5	0.15	0.15	
Type E4, Mode #	1	2	3	4	5
Power, %	100	80	60	40	0
Speed, %	100	71.6	46.5	25.3	Idle
Weighting factor	0.06	0.14	0.15	0.25	0.4
Type E5, Mode #	1	2	3	4	5
Power, %	100	75	50	25	0
Speed, %	100	91	80	63	Idle
Weighting factor	0.08	0.13	0.17	0.32	0.3

Notes:
Engine torque is expressed in percent of maximum available torque at a given engine speed.
Rated speed is the speed at which the manufacturer specifies the rated engine power.
Intermediate speed is the speed corresponding to the peak engine torque.

been reclassified as "medium-duty passenger vehicles" (MDPV; primarily larger SUVs and passenger vans) and are subject to the light-duty vehicle legislation. Therefore, the same diesel engine model used for the 8500- to 10,000-lb vehicle category may be classified as either light or heavy duty and certified to different standards, depending on the application.

Current federal regulations require certification of heavy-duty engines, as opposed to vehicle certification. Consequently,

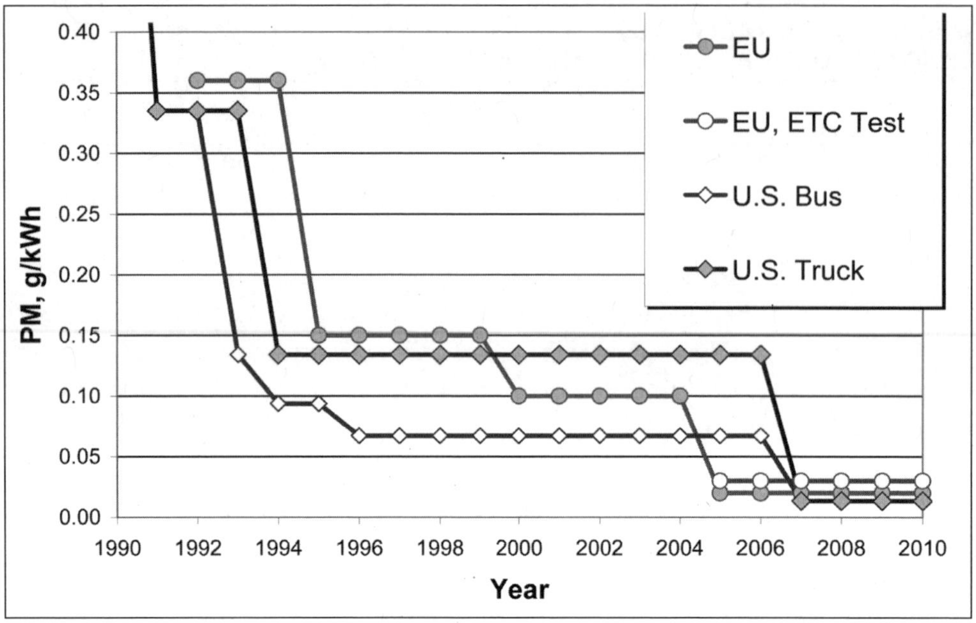

Figure 10.8 *Trends in PM emission standards for heavy-duty diesel engines.*

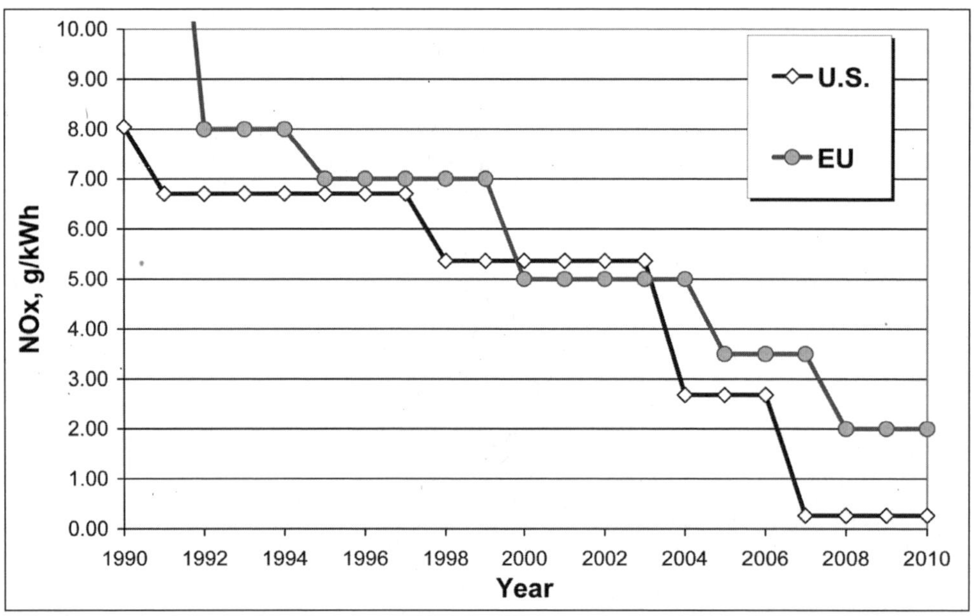

Figure 10.9 *Trends in* NO_x *emission standards for heavy-duty diesel engines.*

the basic standards are expressed in grams per brake-horsepower-hour and require emission testing over the FTP transient engine dynamometer cycle. Additional emission testing requirements, first introduced in 1998, include the (1) Supplemental Emission Test (SET) and (2) the Not-to-Exceed (NTE) limits. These tests were introduced for most signers of the 1998 Consent Decrees between the EPA and engine manufacturers for the 1998–2004 period. Federal regulations require that the supplemental testing be done by all engine manufacturers effective 2007. In California, the tests are required for all engines effective model year 2005.

The SET is a steady-state test that was introduced to help ensure that heavy-duty engine emissions are controlled during steady-state type driving, such as a line-haul truck operating on a freeway. The test is identical to the EU 13-mode ESC schedule, which in the United States is commonly referred to as the "Euro III" cycle).

The NTE limits have been introduced as an additional instrument to make sure that heavy-duty engine emissions are controlled over the full range of speed and load combinations commonly experienced in use. The NTE approach establishes an area (the "NTE zone") under the torque curve of an engine where emissions must not exceed a specified value for any of the regulated pollutants. The NTE test procedure does not involve a specific driving cycle of any specific length (mileage or time). Rather it involves driving of any type that could occur within the bounds of the NTE control area, including operation under steady-state or transient conditions and under varying ambient conditions. Emissions are averaged over a minimum time of 30 s and then compared to the applicable NTE emission limits.

Model Year 1987–2003

Model year 1987–2003 U.S. federal (EPA) and California (ARB) emission standards for heavy-duty diesel truck and bus engines are summarized in Tables 10.6 and 10.7. Applicable to the 1994 and following year standards, sulfur content in the certification fuel has been reduced to 500 ppm wt.

Model Year 2004 and Later

In October 1997, the EPA adopted new emission standards for model year 2004 and later heavy-duty diesel truck and bus engines. These standards reflect the provisions of the Statement of Principles (SOPs) signed in 1995 by the EPA, California ARB, and the manufacturers of heavy-duty diesel engines. The goal was to reduce NO_X emissions from highway heavy-duty engines to levels approximately 2.0 g/bhp-hr beginning in 2004. Manufacturers have the flexibility to certify their engines to one of the two options shown in Table 10.8. All emission standards other than NMHC and NO_X applying to 1998 and later model year heavy-duty engines (Table 10.6) will continue at their 1998 levels.

The federal 2004 standards for highway trucks are harmonized with California standards, with the intent that manufacturers can use a single engine or machine design for both markets. However, California certifications for model years 2005–2007 additionally require the Supplemental Emission Test and NTE limits of 1.25 times the FTP standards. California also adopted a different standard for urban bus engines.

TABLE 10.6
EPA EMISSION STANDARDS FOR HEAVY-DUTY DIESEL ENGINES, g/bhp-hr

Year	HC	CO	NO_X	PM
Heavy-Duty Diesel Truck Engines				
1990	1.3	15.5	6.0	0.60
1991	1.3	15.5	5.0	0.25
1994	1.3	15.5	5.0	0.10
1998	1.3	15.5	4.0	0.10
Urban Bus Engines				
1991	1.3	15.5	5.0	0.25
1993	1.3	15.5	5.0	0.10
1994	1.3	15.5	5.0	0.07
1996	1.3	15.5	5.0	0.05[a]
1998	1.3	15.5	4.0	0.05[a]

a. In-use PM standard 0.07.

TABLE 10.7
CALIFORNIA EMISSION STANDARDS FOR HEAVY-DUTY DIESEL ENGINES, g/bhp-hr

Year	NMHC	THC	CO	NO_X	PM
Heavy-Duty Diesel Truck Engines					
1987	—	1.3	15.5	6.0	0.60
1991	1.2	1.3	15.5	5.0	0.25
1994	1.2	1.3	15.5	5.0	0.10
Urban Bus Engines					
1991	1.2	1.3	15.5	5.0	0.10
1994	1.2	1.3	15.5	5.0	0.07
1996	1.2	1.3	15.5	4.0	0.05

1998 Consent Decree

In October 1998, a court settlement was reached between the EPA, Department of Justice, California ARB, and engine manufacturers (Caterpillar, Cummins, Detroit Diesel, Volvo, Mack Trucks/Renault and Navistar) over the issue of high NO_X emissions from heavy-duty diesel engines during certain driving modes. Since the early 1990s, the manufacturers used engine control software that caused engines to switch to a more fuel-efficient (but higher NO_X) driving mode during steady highway cruising. The EPA considered this engine control strategy an illegal "emission defeat device."

Provisions of the Consent Decree included the following:

- Civil penalties for engine manufacturers and requirements to allocate funds for pollution research
- Upgrading of existing engines to lower NO_X emissions
- Steady-state SET with a limit equal to the FTP standard and NTE limits of 1.25 × FTP (with the exception of Navistar)
- Meeting the 2004 emission standards by October 2002, 15 months ahead of time (with some exceptions for Navistar).

Model Year 2007 and Later

On December 21, 2000, the EPA signed emission standards for model year 2007 and later heavy-duty highway engines (the California ARB adopted virtually identical 2007 heavy-duty engine standards in October 2001). The rule includes two components: (1) emission standards and (2) diesel fuel regulation. The first component of the regulation introduces new, very stringent emission standards, as follows:

- *PM:* 0.01 g/bhp-hr
- *NO_X:* 0.20 g/bhp-hr
- *NMHC:* 0.14 g/bhp-hr.

TABLE 10.8
EPA EMISSION STANDARDS FOR MODEL YEAR 2004 AND LATER HEAVY-DUTY DIESEL ENGINES, g/bhp-hr

Option	NMHC + NO_X	NMHC
1	2.4	n/a
2	2.5	0.5

The PM emission standard takes full effect in the 2007 heavy-duty engine model year. The NO_X and NMHC standards are phased in for diesel engines between 2007 and 2010. The phase-in is on a percent-of-sales basis: 50% from 2007 to 2009 and 100% in 2010. Emission certification requirements also include the SET test, with limits equal to the FTP standards, and NTE limits of 1.5 × FTP standards.

Effective for model year 2007, the regulation also eliminates the earlier crankcase emission control exception for turbocharged heavy-duty diesel engines. Crankcase emissions from these engines are treated the same as (i.e., added to) other exhaust emissions. Manufacturers are expected to control crankcase emissions by routing them back to the engine intake or to the exhaust stream upstream of the exhaust emission control devices.

The diesel fuel regulation limits the sulfur content in on-highway diesel fuel to 15 ppm (wt.), down from the previous 500 ppm. Refiners were required to start producing the 15 ppm S fuel beginning June 1, 2006. Refiners can take advantage of temporary compliance options allowing them to continue producing 500 ppm fuel in 20% of the volume of diesel fuel they produce until December 31, 2009.

Ultra-low-sulfur diesel fuel has been introduced as a "technology enabler" to pave the way for advanced, sulfur-intolerant exhaust emission control technologies, such as catalytic diesel particulate filters and NO_X catalysts, which will be necessary to meet the 2007 emission standards.

The EPA estimates the cost of reducing the sulfur content of diesel fuel will result in a fuel price increase of approximately 4.5 to 5 cents per gallon. The EPA also estimates that the new emission standards will cause an increase in vehicle costs between $1,200 to $1,900 (for comparison, new heavy-duty trucks typically cost up to $150,000 and buses up to $250,000).

10.4.2.2 Mobile Off-Road Diesel Engines

These standards cover mobile off-road (nonroad) diesel engines of all sizes used in a wide range of construction, agricultural, and industrial equipment and in some marine applications. Examples of regulated applications include farm tractors, excavators, diesel lawn tractors, bulldozers, logging equipment, portable generators, road graders, forklifts, and sailboat auxiliary propulsion units. Excepted are engines used in locomotives and underground mining equipment, and also engines over 37 kW (50 hp) used in marine vessels.

The EPA's off-road engine regulations are structured as a three-tiered progression to low emission standards. Each tier involves a phase-in (by horsepower rating) over several years. Tier 1 standards were phased in from 1996 to 2000. More stringent Tier 2 standards for all engine sizes took effect from 2001 to 2006, and yet more stringent Tier 3 standards for engines rated over 37 kW (50 hp) will be phased in from 2006 to 2008. The Tier 3 standards are expected to lead to implementation of emission control technologies similar to those that will be used by manufacturers of highway heavy-duty engines (i.e., trucks and buses) to comply with the 2004 highway engine standards.

The emission standards are listed in Table 10.9. The regulation is in the metric system of units; that is, the standards are expressed in grams of pollutant per kilowatt-hour. Emissions are measured on a test cycle that is nominally the same as the ISO 8178 C1, eight-mode steady-state test cycle. Other ISO 8178 test cycles are allowed for selected applications, such as constant-speed engines (D2 cycle), variable-speed engines rated under 19 kW (G2 cycle), and marine engines (E3 cycle).

The EPA is considering the adoption of new Tier 4 standards, which would force the use of exhaust gas aftertreatment on nonroad engines, as well as the introduction of a new transient test cycle. The Tier 4 standards are likely to become effective around 2009–2010.

10.4.3 European Union

10.4.3.1 Heavy-Duty Highway Engines

The European regulations for new heavy-duty diesel engines are commonly referred to as Euro I through V. The Euro I standards for medium and heavy-duty engines were introduced in 1992. The Euro II regulations came to the forefront in 1996. These standards applied to both heavy-duty highway diesel engines and urban buses. The urban bus standards, however, were voluntary.

In 1999, the European Parliament and the Council of Environment Ministers adopted the final Euro III standard (Directive 1999/96/EC of December 13, 1999, amending the heavy-duty diesel emissions directive, Directive 88/77/EEC) and also adopted Euro IV and V standards for the years 2005 and 2008. The standards also set specific, stricter values for extra-low-emission vehicles (also known as "enhanced environmentally friendly vehicles" or EEVs) in view of their contribution to reducing atmospheric pollution in cities.

It is expected that the emission limit values set for 2005 and 2008 will require all new diesel-powered heavy duty vehicles to be fitted with exhaust gas aftertreatment devices, such as particulate traps and $DeNO_X$ catalysts. Table 10.10 contains a summary of the emission standards and their implementation dates.

Changes in the engine test cycles were introduced in the Euro III standard (year 2000). The old steady-state engine test cycle ECE R-49 will be replaced by two cycles: the ESC cycle and the European Transient Cycle. Smoke opacity is measured on the European Load Response (ELR) test. For the type approval of new vehicles with diesel engines according to the Euro III standard (year 2000), manufacturers have the choice between either of these tests. For type approval according to the Euro IV (year 2005) limit values and for EEVs, the emissions have to be determined on both the ETC and the ESC/ELR tests.

Emission standards for diesel engines that are tested on the ETC test cycle, as well as for heavy-duty gas engines, are summarized in Table 10.11.

TABLE 10.9
U.S. EPA NONROAD ENGINE EMISSION STANDARDS, g/kWh (g/bhp-hr)

Engine Power	Tier	Model Year	NO_x	HC	NMHC + NO_x	CO	PM
kW < 8	Tier 1	2000	–	–	10.5 (7.8)	8.0 (6.0)	1.0 (0.75)
(hp < 11)	Tier 2	2005	–	–	7.5 (5.6)	8.0 (6.0)	0.80 (0.60)
8 <= kW < 19	Tier 1	2000	–	–	9.5 (7.1)	6.6 (4.9)	0.80 (0.60)
(11 <= hp < 25)	Tier 2	2005	–	–	7.5 (5.6)	6.6 (4.9)	0.80 (0.60)
19<= kW < 37	Tier 1	1999	–	–	9.5 (7.1)	5.5 (4.1)	0.80 (0.60)
(25 <= hp < 50)	Tier 2	2004	–	–	7.5 (5.6)	5.5 (4.1)	0.60 (0.45)
37 <= kW < 75	Tier 1	1998	9.2 (6.9)	–	–	–	–
(50 <= hp < 100)	Tier 2	2004	–	–	7.5 (5.6)	5.0 (3.7)	0.40 (0.30)
	Tier 3	2008	–	–	4.7 (3.5)	5.0 (3.7)	–
75 <= kW < 130	Tier 1	1997	9.2 (6.9)	–	–	–	–
(100 <= hp < 175)	Tier 2	2003	–	–	6.6 (4.9)	5.0 (3.7)	0.30 (0.22)
	Tier 3	2007	–	–	4.0 (3.0)	5.0 (3.7)	–
130 <= kW < 225	Tier 1	1996	9.2 (6.9)	1.3 (1.0)	–	11.4 (8.5)	0.54 (0.40)
(175 <= hp < 300)	Tier 2	2003	–	–	6.6 (4.9)	3.5 (2.6)	0.20 (0.15)
	Tier 3	2006	–	–	4.0 (3.0)	3.5 (2.6)	–
225 <= kW < 450	Tier 1	1996	9.2 (6.9)	1.3 (1.0)	–	11.4 (8.5)	0.54 (0.40)
(300 <= hp < 600)	Tier 2	2001	–	–	6.4 (4.8)	3.5 (2.6)	0.20 (0.15)
	Tier 3	2006	–	–	4.0 (3.0)	3.5 (2.6)	–
450 <= kW < 560	Tier 1	1996	9.2 (6.9)	1.3 (1.0)	–	11.4 (8.5)	0.54 (0.40)
(600 <= hp < 750)	Tier 2	2002	–	–	6.4 (4.8)	3.5 (2.6)	0.20 (0.15)
	Tier 3	2006	–	–	4.0 (3.0)	3.5 (2.6)	–
kW >= 560	Tier 1	2000	9.2 (6.9)	1.3 (1.0)	–	11.4 (8.5)	0.54 (0.40)
(hp >= 750)	Tier 2	2006	–	–	6.4 (4.8)	3.5 (2.6)	0.20 (0.15)

TABLE 10.10
EU EMISSION STANDARDS FOR HEAVY-DUTY DIESEL ENGINES, g/kWh (SMOKE IN M^{-1})

Tier	Date and Category	Test Cycle	CO	HC	NO_x	PM	Smoke
Euro I	1992, <85 kW	ECE R-49	4.5	1.1	8.0	0.612	
	1992, >85 kW		4.5	1.1	8.0	0.36	
Euro II	1996.10		4.0	1.1	7.0	0.25	
	1998.10		4.0	1.1	7.0	0.15	
Euro III	1999.10, EEVs only	ESC and ELR	1.5	0.25	2.0	0.02	0.15
	2000.10	ESC and ELR	2.1	0.66	5.0	0.10 0.13[a]	0.8
Euro IV	2005.10		1.5	0.46	3.5	0.02	0.5
Euro V	2008.10		1.5	0.46	2.0	0.02	0.5

a. For engines of less than 0.75 dm^3 swept volume per cylinder and a rated power speed of more than 3000 min^{-1}.

TABLE 10.11
EMISSION STANDARDS FOR DIESEL AND GAS ENGINES, ETC TEST, g/kWh

Tier	Date and Category	Test Cycle	CO	NMHC	CH_4[a]	NO_x	PM[b]
Euro III	1999.10, EEVs only	ETC	3.0	0.40	0.65	2.0	0.02
2000.10	ETC	5.45	0.78	1.6	5.0	0.16 0.21[c]	
Euro IV	2005.10	4.0	0.55	1.1	3.5	0.03	
Euro V	2008.10	4.0	0.55	1.1	2.0	0.03	

a. For natural gas engines only.
b. Not applicable for gas-fueled engines at the year 2000 and 2005 stages.
c. For engines of less than 0.75 dm^3 swept volume per cylinder and a rated power speed of more than 3000 min^{-1}.

10.4.3.2 Off-Road Engines

The first European legislation to regulate emissions from off-road mobile equipment was promulgated on February 27, 1998, in Directive 97/68/EC. The regulations for off-road diesels were introduced in two stages: Stage I was implemented in 1999 and Stage II was implemented from 2001 to 2004, depending on the engine power output. The equipment covered by the standard includes industrial drilling rigs, compressors, construction wheel loaders, bulldozers, off-road trucks, highway excavators, forklift trucks, road maintenance equipment, snowplows, ground support equipment in airports, aerial lifts, and mobile cranes. Agricultural and forestry tractors have the same emission standards but different implementation dates, as specified in Directive 2000/25/EC of May 22, 2000 [12]. Engines used in ships, railway locomotives, aircraft, and generating sets are not covered by the standard and are not currently regulated in the EU.

Stage I and Stage II emissions are not allowed to exceed the amounts shown in Table 10.12. The Stage I emissions are engine-out limits and shall be achieved before any exhaust aftertreatment device. Emissions are measured on the ISO 8178 C1 8-mode cycle and expressed in grams per kilowatt-hour.

TABLE 10.12
EU EMISSION REGULATIONS FOR OFF-ROAD DIESEL ENGINES, g/kWh

Net Power	Date	CO	HC	NOx	PM
Stage I					
130–560 kW	1999.01	5.0	1.3	9.2	0.54
75–130 kW	1999.01	5.0	1.3	9.2	0.70
37–75 kW	1999.04	6.5	1.3	9.2	0.85
Stage II					
130–560 kW	2002.01	3.5	1.0	6.0	0.2
75–130 kW	2003.01	5.0	1.0	6.0	0.3
37–75 kW	2004.01	5.0	1.3	7.0	0.4
18–37 kW	2001.01	5.5	1.5	8.0	0.8

10.5 Selected Diesel-Related Occupational Health Regulations

10.5.1 United States

Occupational health and safety regulations are set in the United States at the federal level by two agencies within the Department of Labor: OSHA, which is responsible for general occupational environments, and the MSHA, which is responsible for mining

A table of threshold limit values (TLVs) for chemical substances and physical agents found in the workplace is published annually by the American Conference of Governmental Industrial Hygienists (ACGIH). The TLVs set by ACGIH are developed as recommendations or guidelines to be used in the practice of industrial hygiene, which establish the maximum ambient concentrations of toxic chemical substances. Formally, they are not legal standards. However, regulatory authorities follow the ACGIH recommendations in their law-making activities. Diesel emission exposure or the use of diesel engines may be additionally regulated by occupational health or mining authorities in particular states.

10.5.1.1 Exposure Limits for Gases

The exposure limits for selected gaseous pollutants found in diesel exhaust are listed in Table 10.13. Limits set by OSHA are known as permissible exposure limits (PELs). Both OSHA PELs and MSHA TLVs are legally enforceable limits. Limits shown in the "OSHA 88" column were adopted as a final rule in 1988, but were later remanded by the courts and have no legal significance. The TLVs by ACGIH are industrial hygiene recommendations. All limits are 8-h time-weighted averages (TWAs), unless marked as ceiling values.

10.5.1.2 Exposure Limits for Particulates

Legal exposure limits for DPM have been adopted by the MSHA for underground metal/nonmetal mines. Currently, there

TABLE 10.13
EXPOSURE LIMITS FOR GASEOUS POLLUTANTS (ppmv, 8-h TWAs)

Substance	CAS #	OSHA PEL	OSHA 88[a]	MSHA TLV	ACGIH TLV
CO	630-08-0	50	35	50	25
CO_2	124-38-9	5000	5000	5000	5000
NO	10102-43-9	25	25	25	25
NO_2	10102-44-0	5 (ceiling value)	1[b]	5	3
HCHO	50-00-0	0.75			0.3 (ceiling value) A2[c]
SO_2	7446-09-5	5	2	5[d]/2[e]	2

a. Not legal limits (PELs adopted in 1988 were later remanded by court).
b. 15-min, short-term exposure limit (STEL).
c. A2, a suspected human carcinogen.
d. For metal/nonmetal mines.
e. For coal mines.

are no legal exposure limits for DPM in general occupational settings in the United States. DPM exposure limits have been proposed, but not yet adopted, by ACGIH.

10.5.1.3 Nonmining Environments

In its 2001 Notice of Intended Changes, the ACGIH proposed a TLV of 0.02 mg/m^3 for diesel exhaust particulates measured as elemental carbon (EC), with proposed carcinogenicity classification A2, "Suspected Human Carcinogen." This EC-based TLV is practically equivalent to the previously proposed TLV of 0.05 mg/m^3, presumably as total diesel particulate matter (the EC fraction typically constitutes about 40% of the total diesel particulate mass). The first DPM exposure limit of 0.15 mg/m^3 was originally introduced by ACGIH in its 1995–1996 Notice of Intended Changes. Since that time DPM has remained on the list of intended changes but the TLV has never been finalized.

10.5.1.4 Underground Mining

On January 19, 2001, the MSHA published final diesel regulations for underground metal/nonmetal (i.e., noncoal) and for underground coal mines. The metal/nonmetal mine rule adopts exposure limits for diesel particulates. The MSHA has not introduced exposure limits for coal mines, due to the lack of a suitable DPM measuring method in the presence of coal dust.

The underground metal/nonmetal mine rule establishes a concentration limit for diesel particulate matter and requires mine operators to use engineering or work practice controls to reduce DPM to that limit. The final limit is phased in over a 5-year period. Starting July 19, 2002, mines had to observe an "interim" DPM concentration limit of 400 μg/m^3. Effective January 19, 2006, that level had to be reduced to 160 μg/m^3. For the purpose of ambient sampling (according to NIOSH method 5040), DPM is defined as total carbon (TC). This definition includes both elemental and organic (i.e., hydrocarbon derived) carbon, and excludes inorganic ash and sulfates from the TLV. (Note that this occupational health definition of diesel particulate matter is different from the definition used in engine tailpipe emission regulations.)

10.5.1.5 Mining Diesel Engine Regulations

In addition to setting exposure limits for air pollutants, as just summarized, the MSHA has established other health protection measures with regard to the use of diesel engines underground.

Diesel engines to be used in underground mining require formal MSHA approval. The approval process involves emission testing on an engine dynamometer using the ISO 8178 C1 test cycle (mining engines are exempted from the EPA off-road emission standard and certification). Based on the test results, ventilation rates are calculated for each approved engine model. The ventilation rates are computed as the quantity of additional ventilation air that has to be provided to the mine in order to dilute raw exhaust emissions to the respective TLV levels (see the MSHA data in Table 10.13). This mechanism ensures that mine operators increase the amount of ventilation as more engines are introduced underground. A particulate index (PI) is also calculated for DPM dilution to an arbitrary ambient level of 1 mg/m^3. Because the PI has only an advisory role, the PI-based ventilation rates are not enforceable.

10.5.1.6 Coal Mining Regulations

MSHA coal mine diesel regulations published on January 19, 2001, set a specific engine emission limit of 2.5 g/h of DPM for both permissible and nonpermissible equipment. This limit was to be phased in over a 48-month period. Coal mine operators may use a combination of controls to comply with the emission limit.

It is expected that, in most cases, meeting the limit will require the use of particulate filters. Only the smallest engines will be able to meet the 2.5 g/h limit with no emission aftertreatment. Permissible vehicles, which are equipped with water scrubbers or heat exchangers, are expected to use disposable, paper-based particulate filter cartridges, while nonpermissible equipment would use catalytic particulate filters.

10.5.1.7 Diesel Fuel Requirements

Coal mining regulations adopted in 1997 introduce the following diesel fuel requirements:

- Maximum sulfur content of 500 ppm wt.
- Flashpoint greater than 100°F
- Fuel additives have to be EPA registered.

The January 2001 regulations introduced the same fuel sulfur (500 ppm) and additive (EPA registration) requirements for metal/nonmetal mines. The EPA "low-sulfur" fuel (500 ppm S) is also used for engine approval testing.

10.5.2 Germany

Historically, Germany was the first country to introduce stringent occupational exposure limits for diesel particulates. By "stringent" we mean exposure limits that required implementation of serious emission control measures, including the use of diesel particulate filters, by equipment operators participating in construction projects inside buildings or tunnels. Credit should be given here to Canadian underground mine regulators, who many years ago introduced the first in the world DPM exposure limit. However, that limit of 1.5 mg/m^3 must be considered very relaxed and, from today's industrial hygiene perspective, not adequate.

Germany has two types of occupational exposure limits:

- MAK (Maximale Arbeitsplatzkonzentration) is the maximum concentration of a substance in the ambient air in the workplace that has no adverse effect on the workers' health.
- TRK (Technische Richtkonzentration) is the concentration of a substance in the ambient air in the workplace that can be achieved using technically available measures.

Once established, both types of limits are legally binding and have to be observed in the workplace. Table 10.14 summarizes the exposure limits for the major diesel exhaust pollutants.

The diesel particulate emissions (DME - Dieselmotorenemissionen) are defined as elemental carbon and are limited in size to about 5μm (EN 481). Specific rules pertaining to the use of diesel engines in occupational health environments were first introduced on April 30, 1993 [13]. These regulations introduced a general TRK exposure limit for diesel particulates of 0.2 mg/m^3 and a limit of 0.6 mg/m^3 for underground noncoal mines and tunneling. The particulates were defined as total carbon (EC + OC) and determined by a coulometric analysis.

The regulation was modified in 1996. The TRK limits for diesel particulates in the workplace and in mining/tunneling were lowered to 0.1 and 0.3 mg/m^3, respectively (compare Table 10.14). The definition of diesel particulates was changed to elemental carbon. Coal mines are exempted from these limits due to the difficulties in differentiating between diesel particulates and coal dust.

TABLE 10.14
MAK EXPOSURE LIMITS FOR DIESEL EXHAUST POLLUTANTS

Substance	ppmv	mg/m^3	Category[a]
CO	30	33	4
CO_2	5000	9000	4
NO	25	30	—
NO_2	5	9	1
HCHO	0.5	0.6	1
SO_2	2	5	1
Diesel particulates (EC), tunneling and noncoal mining	—	0.3[a]	4
Diesel particulates (EC), other applications	—	0.1[a]	—

Source: Data from Technische Regeln für Gefahrstoffe TRGS 900.

a. All values in the table have to be met as workshift, time-weighted averages. The following short-term exposure categories apply: 1, ceiling value, not to be exceeded at any point in time; 4, ceiling value = 4× the limit value.

b. TRK limit.

References

1. Henein, N.A., and Patterson, D.J., *Emissions from Combustion Engines and Their Control,* Ann Arbor Science Publishers, Ann Arbor, MI, 1972.
2. Haägen-Smit, A. J., "Chemistry and Physiology of Los Angeles Smog." *Ind. Eng. Chem.* **44,** 6, 1952, p. 1342.
3. *Federal Register* 21(61), Part II (1966).
4. Haägen-Smit, A. J., "The Control of Air Pollution," *Sci. Am.* **210,** 1, 1964, p. 25.
5. "Risk Reduction Plan to Reduce Particulate Matter Emissions from Diesel-Fueled Engines and Vehicles," California Air Resources Board, October 2000, retrieved from www.arb.ca.gov/diesel/documents/rrpfinal.pdf.
6. "Voluntary Diesel Retrofit Program," U.S. Environmental Protection Agency, Washington, DC, retrieved from www.epa.gov/otaq/retrofit, April 2002.
7. "Emission Test Cycles," DieselNet.com, retrieved from www.dieselnet.com/standards/cycles, April 2002.
8. *Official Journal of the European Communities*, L44, 16.2.2000, Euro III, Directive 1999/96/EC of December 13, 1999. p 1-155.

9. FIGE Report 104 05 316, FIGE Institute, Aachen, Germany, January 1994.
10. EEC Directive 90/C81/01.
11. "Emission Standards," DieselNet.com, retrieved from www.dieselnet.com/standards.html, April 2002.
12. *Official Journal of the European Communities,* L173, July 12, 2000.
13. Diesel Motor Emissions (DME), TRGS 554, German Ministry of Labor (BMA), 12 March 2001.

Part III

Diesel Fuels

Chapter 11:
Petroleum Diesel Fuel

Chapter 12:
Alternative Diesel Fuels

Chapter 11

Petroleum Diesel Fuel

11.1 What Is Diesel Fuel?

11.1.1 Components of Diesel Fuel

Diesel fuels are mixtures of hydrocarbons with boiling points in the range of 150 to 380°C. They are obtained from the distillation of crude oil. Refineries are increasingly blending conversion products (crack components) with the primary distillation streams in order to meet the appropriate product specifications and to meet the market demands for different proportions of gasoline, diesel fuel, and other petroleum products.

In the primary refining unit, operating under atmospheric pressure, the process parameters are set to give maximum distillation without cracking. The quantity and quality of the streams drawn off depend on the chemical composition of the crude oil being used. Crude oils of different hydrocarbon type, including paraffinic (aliphatic), naphthenic (cycloaliphatic), and aromatic hydrocarbons, produce diesel fuels of different cetane number, energy content, cloud point, sulfur content, and so forth. They also yield proportions of gas, gasoline, middle distillate, and residual fuel oil that are usually different from the product demand patterns in particular markets. The only way to balance the refinery production pattern with market demands is through downstream conversion processes. In these conversion processes large molecules are broken into smaller ones by application of heat, pressure, or catalysts. Most refineries have vacuum distillation and other conversion units, such as thermal cracking (visbreaking and coking), catalytic cracking, and hydrocracking units, to increase the yield of desired products by cracking unwanted heavy fractions.

The high content of paraffinic hydrocarbons in the crude oil yields fuel with good ignition qualities, although it may have trouble meeting low-temperature specifications, especially with high-wax-content crude oils. Cracking processes create blend components that have a lower wax content, but oils from catalytic and thermal cracking also have a lower ignition quality. The final blend has to be a careful choice of particular components if it is to meet the fuel specifications.

11.1.2 Diesel Fuel Properties

The requirements placed on diesel fuels are specified in standards, such as the ASTM D975 in the United States and EN 590 in the European Union (EU). The most important parameters specified by the standards include:

- Cetane number (and index)
- Viscosity
- Cold behavior
- Flash point
- Volatility
- Lubricity
- Sulfur
- Additives.

11.1.2.1 Cetane Number

The readiness of a fuel to ignite when injected into the engine cylinder is indicated by its cetane number; the higher the number, the easier it is to ignite. Cetane (*n*-hexadecane, $n\text{-}C_{16}H_{34}$), which has very good ignition qualities, is assigned the cetane number of 100. Another hydrocarbon used for reference fuel blends is heptamethyl nonane with an assigned cetane number of 15. The cetane number of a fuel is determined by comparing its ignition quality with that of a reference fuel on a standard engine test. The CFR Cetane Engine, ASTM D613, is the most widely accepted method. When a fuel has the same ignition quality as a mixture of the two reference fuels, cetane and heptamethyl nonane, its cetane number is derived from the following equation:

$$\text{Cetane number} = \%\ n\text{-cetane} + 0.15\ (\%\ \text{heptamethyl nonane}) \quad (11.1)$$

The minimum cetane number of diesel fuels is 40 in the United States and 45 in Japan and Germany. Even higher cetane numbers of around 50 are desirable for optimum operation of modern engines (quiet operation, low particulate emissions). A high proportion of paraffins increases the fuel cetane number. A high proportion of aromatics of the types found in cracked fuel components impairs the ignition quality and lowers the cetane number.

Cetane number can be also improved by using fuel additives. 2-Ethylhexyl nitrate (2-EHN)—at dose levels typically varying from 0.05% to 0.20% and more—has become the universal cetane number improver due to its good performance and low cost. Correlations that allow us to calculate the cetane number increase as a function of 2-EHN concentration can be found in the literature [1].

11.1.2.2 Cetane Index

The CFR engine method of determining the cetane number has poor precision and, obviously, requires that an appropriate test engine stand be available. The cetane index—a calculated value derived from fuel density and volatility properties—has been introduced as an estimate of the cetane number. It is now widely used to monitor and control diesel ignition quality. The cetane index is now being incorporated into diesel specifications as an additional test to control base fuel quality and to limit the extent of cetane number improvement by means of cetane improver additives. For example, the pan-European specification for automotive diesel fuel, EN 590:1993, specifies a minimum cetane number of 49 and a minimum cetane index of 46.

The formula used to calculate the cetane index (CI) is modified from time to time, as fuel composition evolves. The 1980 revision of the equation (ASTM D976) is given by the following formula:

$$CI = 454.74 - 1.641416\cdot\rho + 0.00077474\cdot\rho^2 - 0.554\cdot t_{50} + 97.803\log^2(t_{50}) \tag{11.2}$$

where ρ is the density at 15°C in kg/m^3, and t_{50} is the midboiling temperature in °C, corrected to standard barometric pressure.

A new equation for the CI was developed in 1988 and is now included in the ASTM Standard Book as ASTM D4737. The new equation is also based on fuel density and volatility but uses four variables: the density and three distillation temperatures (10%, 50%, and 90%), as shown in Eq. (11.3):

$$\begin{aligned}CI = {} & 45.2 + 0.0892(t_{10} - 215) + 0.131(t_{50} - 260) + \\ & 0.0523(t_{90} - 310) + 0.901\cdot b(t_{50} - 260) - \\ & 0.420\cdot b(t_{90} - 310) + 0.00049(t_{10} - 215)^2 - \\ & 0.00049(t_{90} - 310)^2 + 107.0\cdot b + 60.0\cdot b^2\end{aligned} \tag{11.3}$$

where t_{10}, t_{50}, and t_{90} represent 10%, 50%, and 90% recovery temperatures, respectively, in °C, corrected to standard barometric pressure, and b is the factor given by Eq. (11.4).

$$b = \exp[-3500(\rho - 850)] - 1 \tag{11.4}$$

where again ρ is the density at 15°C in kg/m^3.

Cetane index formulas have been developed for "natural" cetane numbers. If cetane number improver additives are used, the calculated cetane index will underestimate the cetane number. The cetane number improvement obtained using additives can only be measured by engine testing.

11.1.2.3 Viscosity

The viscosity of a fluid indicates its resistance to flow; the higher the viscosity, the greater the resistance to flow. A dynamic (or absolute) viscosity of a fluid (in Pa·s) is defined as the force (in newtons) required to move an area of 1 m^2 at a speed of 1 m/s past a parallel surface 1 m away and separated from it by the fluid. A kinematic viscosity is calculated by dividing the dynamic viscosity by the fluid density.

The viscosity of diesel fuel is important in terms of the operation of fuel injection equipment, which has to accurately measure the small quantities of fuel to be injected. As viscosity decreases with increasing temperature, the tolerance band between maximum and minimum viscosity values should be kept as small as possible.

A high viscosity at low temperature could reduce the flow of fuel and result in incomplete filling of the metering chamber. If the fuel is very viscous, the possibility of pump distortion arises due to the heat generated by the shearing action in the small clearances. A low-viscosity fuel can significantly increase leakage from the pumping elements, particularly at low speeds. Such a situation can arise when attempting a hot restart after a brief engine shutdown following a period of operation at high load. The temperature of the already hot fuel injection equipment is further increased by heat soak-back from the engine, reducing the fuel viscosity to the point that leakage may make restarting impossible until the fuel system cools down.

11.1.2.4 Cold Behavior

Most diesel fuels include a significant proportion of paraffinic components, which include waxes. At low temperatures, the precipitation of paraffin crystals (wax) can cause clogging of the fuel filter and an interruption in fuel supply. Depending on the fuel properties, the start of paraffin precipitation can be as early as 0°C. Consequently, winter diesel fuels must be specially selected, or treated, in order to ensure problem-free operation in cold weather.

The following is a list of some of the properties that have been defined to describe the cold behavior of diesel fuels:

- The *cloud point* (CP) is the temperature at which wax first becomes visible when the fuel is cooled (ASTM D2500).
- The *pour point* is the temperature at which the amount of wax out of solution is sufficient to gel the fuel (ASTM D97).
- The *cold filter plugging point* (CFPP) is the lowest temperature at which fuel will pass through a fine wire mesh screen. The CFPP is widely used in Europe (European standard EN116:1981). It does not correlate well with the

low-temperature performance of North American fuels in North American equipment and is not in the ASTM Standards Book.

Flow improvers are usually added to fuels at the refinery. Although the flow improvers do not prevent precipitation of paraffins, they limit their crystal growth. The crystals are so small that they can still pass through the filter pores. As a result, filterability can be extended to lower temperatures.

The resistance to cold can be improved even further by additives that prevent the precipitation of paraffin crystals. The winter diesel fuels that are available today guarantee cold resistance to at least –22°C. Additional measures are filter heating or addition of petroleum products to the diesel fuel. An addition of regular gasoline can also delay precipitation. However, due to the very low cetane number of gasoline, the ignition quality is impaired and the flash point considerably reduced.

The cloud point of a fuel is also affected by its distillation characteristics. The maximum 90% distillation point is, therefore, limited to 315°C in Canada, a country with extremely cold winters; a 90% distillation point as high as 379°C is possible in tropical or subtropical areas.

11.1.2.5 Flash Point

The flash point is the temperature at which a combustible liquid gives off just enough vapor to produce a vapor/air mixture that will ignite when a flame is applied. The flash point is measured in a standardized apparatus using standard test methods, such as ASTM D93 or ISO 2719.

The importance of the flash point is primarily related to safe handling of the product. If the flash point is too low, there could be a fire hazard. For this reason mandatory minimum limits on the flash point have been set by government agencies and insurance companies. Typical minimum values for automotive diesel fuels range from 38°C in the United States to 56°C in some European countries.

The flash point of a diesel fuel is not significant for engine performance. Variations in flash point do not influence autoignition temperature or other combustion characteristics.

11.1.2.6 Volatility

The volatility characteristics of diesel fuel are expressed in terms of the temperature at which successive portions are distilled from a sample of the fuel under controlled heating in a standardized apparatus. One of the most widely used methods is the ASTM D86. The distillation or boiling range of the fuel depends on the fuel chemical composition and, therefore, influences other properties such as viscosity, flash point, autoignition temperature, cetane number, and density.

The fuel sample is placed in the distillation apparatus and heated. The vapors, which form as the temperature increases, are condensed and collected in a cylinder graduated in percentage of the initial volume of the liquid. The information recorded during distillation includes:

- Initial boiling point (IBP)
- End point (EP) or final boiling point (FBP)
- Percent of condensate recovered
- Percent of nonvolatile residue.

The temperatures corresponding to the amounts distilled as the test proceeds are plotted in a distillation graph, as illustrated in Figure 11.1.

11.1.2.7 Lubricity

The lubricating properties of diesel fuel are very important for the fuel injection equipment. Many injection pumps and injectors rely on the lubricity of the fuel to protect their components from excessive wear. However, throughout the 1980s, little progress was made in the development of lubricity standards, specifications, and testing methods. Although lubricity issues were addressed in some special applications, such as aviation fuels or arctic conditions, most fuel injection equipment was simply designed based on the lubricity level of commercial diesel fuel [2].

The importance of lubricity specifications was first recognized when, effective in 1993, California regulated low-aromatics diesel fuel, and 2 years later, when Sweden introduced fuels with almost no aromatics and sulfur (10 ppm S). The Swedish fuel led to catastrophic lubricity-related injection equipment failures, which occurred after using less than one tank of fuel. No major reported lubricity problems have surfaced in California, but the effect of slightly lower lubricity fuels on the life of equipment is difficult to assess.

As fuel suppliers became more knowledgeable and treated their low-sulfur fuels with lubricity additives, other regions in the world avoided the Swedish equipment failures. Fuel injection manufacturers have also incorporated design changes that compensate for the use of lower lubricity fuels. However, as fuels of lowered sulfur and aromatics content are introduced, fuel lubricity will become more important. Reduced clearances between components that may be needed to provide higher injection pressures are also likely to require higher lubricity. As a result, future diesel fuels will increasingly rely on lubricity additives to protect fuel injection equipment. Standard lubricity test

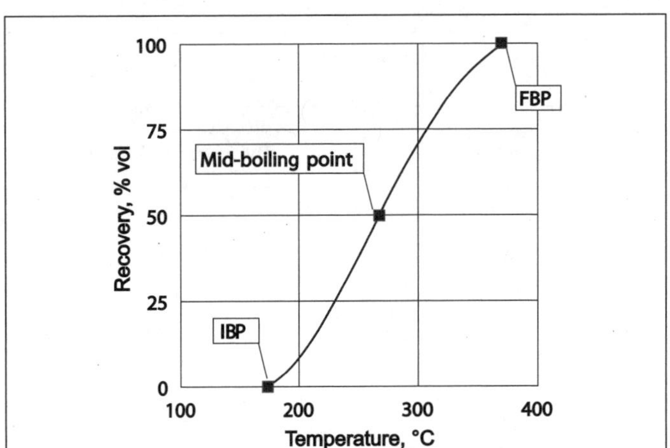

Figure 11.1 *Typical distillation curve of diesel fuel.*

methods are vital to guide equipment manufacturers, because excessive levels of lubricity additives may cause sediment and gum formation and other harmful effects.

Despite the growing importance of lubricity, there is still no consensus on the test method and no single lubricity test has been adopted by all technical groups or regions. Some of the lubricity tests are listed in Table 11.1, with the BOTD and HFRR methods attracting the most attention as candidates for future standards. Technical problems with lubricity tests include a poor correlation between methods and a different response for natural and additive-induced lubricity. A comprehensive overview of the ongoing development work in this area, which has been conducted by the SAE, ISO, ASTM, and others, has been published by Nikanjam [2].

11.1.2.8 Sulfur

Diesel fuels contain chemically bound sulfur. The amount of sulfur depends on the crude oil quality and the components used in blending the fuel. In particular, some crack components have high sulfur content. Refineries can reduce the sulfur content of diesel fuel by treatment with hydrogen.

Sulfur increases the lubricating properties of diesel fuels. Therefore, fuels having low sulfur levels typically require lubricity additives to avoid potential damage to fuel injection equipment [4].

Sulfur has a number of negative effects on diesel fuel, which can be categorized as follows [5]:

Emissions

- *Sulfur dioxide emissions:* Most of the sulfur is converted in the engine into sulfur dioxide, a substance that has a negative environmental impact.
- *Sulfate particulate emissions:* A fraction of the sulfur is oxidized to sulfur trioxide. The SO_3 binds with water, forms sulfuric acid, and contributes to the total particulate matter (PM) emissions.

Corrosion and Wear

- *Corrosion of exhaust system components by sulfur condensates:* This effect is especially troublesome in exhaust gas recirculation coolers [6, 7].
- *Increased wear:* Increased wear of engine parts may happen as a result of corrosion by internally formed acid.

Exhaust Aftertreatment

- The SO_2/SO_3 shift is increased very significantly if are used in the emission control system, resulting in increased PM emissions.
- Sulfate particulates are also generated in catalytic particulate filters, both in the and traps. At certain conditions, the benefit of reducing the carbon fraction of diesel PM can be more than offset by the generated sulfates.
- Deactivation of by sulfur is one of the most important obstacles in the implementation of this technology.
- Many catalysts are, reversibly or not, poisoned by sulfur. Sulfur may cause problems with future emission control systems that are currently not known.

To minimize these adverse effects, many are advocating a reduction in the level of sulfur in diesel fuels worldwide. Until early 1990s, the sulfur level in diesel fuel was not subject to environmental regulations. The maximum sulfur level in good-quality fuels, as listed by fuel specifications, was at about 0.5% (5000 ppm = 5000 mg/kg). In the 1990s, environmental regulations limited the maximum sulfur level to about 500 ppm; this type of diesel fuel was typically referred to as *low-sulfur diesel.* Further pressures from the increasingly stringent diesel emission standards, such as the Euro 4 and US2007, will require maximum sulfur levels to be limited to 50 to 10 ppm. Diesel fuels of typically 15 ppm S (but not more than 50 ppm) are commonly referred to as *ultra-low-sulfur diesel* (ULSD). In Europe, diesel (and gasoline) fuels of maximum sulfur content of 10 ppm are termed *sulfur-free fuels.*

TABLE 11.1
SELECTED FUEL LUBRICITY TEST METHODS

Method		Developed by	Reference
BOCLE	Ball-on-cylinder lubricity evaluator	Lubrizol	ASTM D5001
SLBOCLE	Scuffing load ball-on-cylinder lubricity evaluator		ASTM D6078
HFRR	High-frequency reciprocating rig		ASTM D6079
BOTD	Ball on three disks	Falex	Ref. [3]
BOTS	Ball on three seats		

11.1.2.9 Additives

Quality improvement of diesel fuels by the addition of additives, which has been a common practice with gasoline fuels for many years, has recently become popular. Multiple-effect additive packages can be used, incorporating some of the following additives:

- Ignition improvers raise the cetane number and facilitate quieter combustion.
- Antioxidants and stabilizers prevent high-temperature gum-forming reactions and sediment formation.
- Detergents prevent injection nozzle clogging.
- Corrosion inhibitors prevent corrosion of metal parts in the event water is entrained into the fuel system.
- Antistatic additives are occasionally added to diesel fuel to avoid the risk of an explosion due to a charge of static electricity building up during fast rates of pumping.
- Antifoaming agents are used to facilitate problem-free tank filling.

- Dehazers and demulsifiers may occasionally be needed if the fuel becomes hazy due to the presence of finely dispersed droplets of water.
- Lubricity additives become more common, especially with low-sulfur fuels that have low levels of natural fuel lubricants.
- Biocides are sometimes used to prevent the growth of bacteria and fungi in fuel tanks.

The overall concentration of additives is generally below 0.1%, so that the physical properties of the fuel, such as density, viscosity, and volatility, are not changed.

11.1.3 Typical Specifications

11.1.3.1 Overview

Significant differences exist in the average quality of diesel fuel in different parts of the world. These differences arise from several reasons, such as crude oil type, the balance between gasoline, jet fuel, and diesel fuel demand, and local climatic conditions. One of the most striking differences is the cetane quality of U.S. diesel fuel, which is much lower than that in other parts of the world. The main reason for the low cetane number is the high demand for gasoline relative to diesel fuel in the United States. U.S. refineries have been configured for maximum gasoline production. As a result, diesel fuel in the United States contains a higher fraction of cracked components of inherently low cetane quality, which are by-products of gasoline manufacture.

More recently, fuel quality has been further influenced by national and local regulations aimed at reducing emissions. Many countries introduce tax incentives for cleaner burning, better quality fuels (primarily lower sulfur, but also lower density, aromatics, higher cetane) to offset their higher production costs. Examples of fuel specifications and tax incentives [8] in the 1990s are listed in Table 11.2.

Table 11.3 shows typical diesel fuel characteristics found in the United States, Japan, and Germany during a fuel quality survey carried out in winter 1994 [9]. The survey values (see the "min," "mean," and "max" columns) were generated from 20 (U.S. and Germany) to 26 (Japan) fuel samples. The survey data are compared with diesel fuel standards in the respective countries.

11.1.3.2 Sulfur Content

Sulfur has a special position among other diesel fuel properties, due to its negative impact on a number of emission control technologies. Environmental legislation limiting the maximum content of sulfur in diesel fuel has been driven first by the impact of sulfur on PM emissions (sulfate particulates) and, later, by its interference with various emission catalysts. At the time when the typical sulfur specification amounted to 0.5%, sulfur dioxide emissions from the diesel engines still had a noticeable contribution to the global SO_2 emission inventory, thus creating some additional pressure on lowering the sulfur level in diesel. Ultra-low-sulfur fuel, however, having a maximum sulfur content between 50 and 10 ppm has been introduced as a "technology enabler" for catalytic diesel particulate filters and NO_x control catalysts, as summarized in Table 11.4. The associated SO_2 emission benefit is small and may be important only in some specific environments, such as in "street canyons" where there is a high traffic density but slow movement of air.

TABLE 11.2
DIESEL FUEL SPECIFICATIONS

	Sulfur (max. ppm)	Cetane No. (min.)	Total Aromatics (max. % vol.)	Density g/cm^3	T90/95	Tax Incentive
US No. 2 (ASTM D975)	500	40	—	—	338	—
CARB[a]	500	40	10	—	338	—
EU 1996 (EN 590)	500	49	—	0.82–0.86	370	—
Japan No. 2	500	45[b]	—	—	350	—
Japan No. 3	500	45[b]	—	—	330	—
Sweden Class I	10	50[b]	5	0.80–0.82	285	97[c]
Sweden Class II	50	47[b]	20	0.80–0.82	295	54[c]
Finland Sw II	50	47[b]	20	0.80–0.82	295	34
Denmark Sw I	10	50/47[b]	5	0.80–0.82	285	85
Denmark City Bus	500	50	—	0.82–0.855	325	50
UK City Diesel	10	49[b]	—	0.80–0.83	—	37.5

Source: Data from Ref. 8.

a. Or fuel must show emissions equivalent or better compared to CARB reference fuel of 500 ppm S, 48 cetane number, 10% aromatics, 1.4% polyaromatics, 0.83–0.86 g/cm^3 density, max T90 of 321°C.

b. Cetane index.

c. Initial tax incentive at introduction (1991); current tax incentive is lower (e.g., $76/ton Swedish Class I in 1998).

TABLE 11.3
DIESEL FUEL STANDARDS AND SURVEY DATA

Property	U.S. East Coast				Japan Grade 2				Germany			
	ASTM D975	Min.	Mean	Max.	JIS K2204	Min.	Mean	Max.	DIN 51601	Min.	Mean	Max.
Density, kg/m^3 @15°C		833.2	846.4	851.9		823.1	835.0	844.5	820-860	824.9	834.3	844.6
Viscosity, mm^2/s @20°C									2.0-8.0	3.278	3.953	5.120
@30°C					2.5 min.	3.092	3.859	4.872				
@40°C	1.9–4.1	2.066	2.348	2.711								
Sulfur, % m/m	0.05 max.	0.013	0.027	0.036	0.2 max.	0.068	0.154	0.194	0.2 max.	0.051	0.124	0.196
Cetane number	40 min	41.7	44.9	49.8	45 min	47.0	54.9	59.0	45 min	48.0	51.6	53.6
Cetane index												
1980 equation		43.5	45.1	50.2		52.8	55.7	58.0		47.8	50.2	53.9
1988 equation		43.3	45.2	50.1		52.7	56.5	59.9		48.2	50.2	53.8
Cloud point, °C		−19	−14	−7		−11	−5	−1		−13	−9	−6
Pour point, °C		−45	−29	−24	−7.5 max.	−33	−16	−8		−41	−32	−26
CFPP, °C		−36	−22	−16	−5 max.	−12	−9	−6	−15 max.	−32	−28	−21
Wax,[a] % m/m		1.0	1.6	2.1		2.8	4.0	6.1		0.7	1.4	2.8
Distillation, °C									*b*			
IBP		148	174	206		151	172	209		150	173	186
20%		209	221	229		215	242	260		198	216	235
50%		242	253	264		274	285	298		241	257	280
90%	282–338	305	312	322	330–350	318	336	347		309	328	363
FBP		340	344	352		339	360	375		329	355	388

Source: Data from Ref. 9.

a. 10°C below cloud.

b. 65% max. @250°C, 85% min. @350°C

TABLE 11.4
REDUCTION OF SULFUR CONTENT IN DIESEL FUEL

Fuel Type	Sulfur Level	Reason for Reducing Sulfur	Targeted Emission	Emission Standards
Standard diesel	≥0.5% (5000 ppm)	n/a	n/a	United States before 1993 European Union before 1996
Low-sulfur diesel	500–350 ppm	Sulfate PM control SO_2 control	PM, SO_2	US1994: 0.1 g/bhp-hr PM EURO 2/3: 0.25/0.1 g/kWh PM
Ultra-low-sulfur diesel	50–10 ppm	To enable NO_x adsorbers To enable catalytic DPFs Sulfate PM control	NO_x, PM	US1997: 0.2 g/bhp-hr NO_x; 0.01 g/bhp-hr PM EURO 4/5: 0.02 g/kWh PM
Sulfur-free diesel	≤10 ppm[a]	Post-EURO 5		

a. EU: proposed phase-in from 2005 *to* 2009. Simultaneous introduction of sulfur-free gasoline would enable lean NOx control from spark-ignited engines to reduce CO_2 emissions through increased fuel economy.

All major steps in reducing the sulfur level in fuel have been synchronized in (and in fact are an integral part of) diesel emission standards, both in the United States and in the European Union. In the past, reductions of sulfur were necessary to accomplish the regulated reductions in PM emissions. In the future, it is believed that ultra-low-sulfur fuels will also enable new NO_x control technologies.

The first significant reduction of sulfur in diesel fuel took place in the United States in October 1993, when a 500 ppm S cap was implemented. That move was coordinated with the 1994 0.1 g/bhp-hr standard for diesel PM. Sulfate particulate emission levels in pre-1994 engines operated with fuels of about 0.3% S could be as high as 0.07 g/bhp-hr or more. Obviously, a total PM standard of 0.1 g was not feasible without lowering the sulfur content. The next reduction in sulfur level in the United States was scheduled for mid-2006, just before the 2007 standards of 0.01 g/bhp-hr PM and 0.2 g/bhp-hr NO_x are to take place. The sulfur was lowered to 15 ppm to enable advanced PM and NO_x control technologies on new diesel engines. A similar evolution of fuel quality changes along with emission standards can be seen in the European Union, as illustrated in Figure 11.2.

Although ultra-low-sulfur fuel of 50 ppm S is legislated in the European Union effective 2005, a number of European countries introduced fuels with sulfur levels between 10 and 50 ppm in the late 1990s and early 2000s by means of tax incentive programs. Ultra-low-sulfur fuels are available in such countries as Sweden, the United Kingdom, Ireland, Denmark, the Netherlands, Germany, and Belgium.

Motor vehicle manufacturers have developed a World-Wide Fuels Charter, a document that lists specifications of diesel and gasoline automotive fuels that are desired to harmonize fuel quality worldwide in accordance with the needs of vehicle and emission control technologies [10]. The charter calls for the introduction of "sulfur-free" fuels, both diesel and gasoline. Governments have been following the charter recommendations; the EU's proposed legislation called for the availability of sulfur-free fuels in 2005 and a complete switch to 10 ppm S diesel and gasoline in 2009.

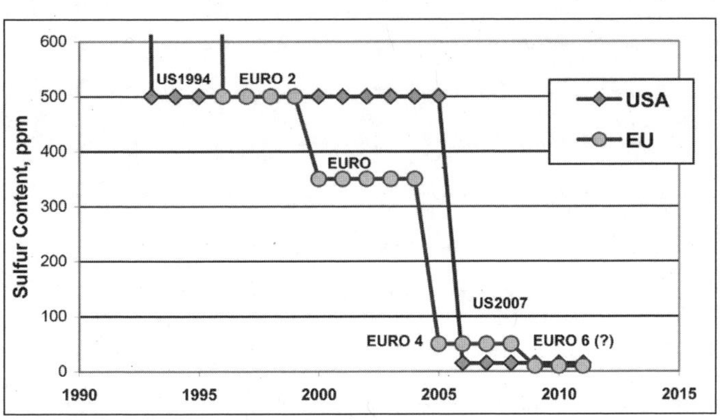

Figure 11.2 Diesel fuel sulfur limits driven by emission legislation. Corresponding emission standards marked in the chart area.

All numbers in Table 11.4 and in Figure 11.2 refer to highway diesel fuels. Sulfur levels in nonroad diesel fuels still remain at higher levels (0.5% maximum in the United States).

11.2 Fuel Properties and Emissions

11.2.1 Introduction

Historically, fuel properties have been continuously changing for various reasons, including crude oil prices, crude oil quality, refinery technologies, relative demand for diesel and gasoline fuel, and changing engine technologies. In recent years environmental considerations and emission legislation have become increasingly more important in the formulation and properties of fuels. The interaction mechanisms between fuel quality, engine technologies, and emissions need to be understood to find the most effective approach toward low-emission diesel engines. A number of research studies have been carried out to investigate the influence of fuel properties on emissions. The most comprehensive programs include the European Programme on Emissions, Fuels and Engine Technologies (EPEFE) [11] and the American Auto/Oil Air Quality Improvement Research Program (AQIRP) [12]. Numerous other studies have been conducted by the oil and engine industries, research institutes, and universities. A bibliography of publications selected for modeling of fuel emission effects in heavy-duty engines has been published by the U.S. Environmental Protection Agency (EPA) [13].

Despite the wealth of experimental data, the influence of some of the fuel properties on emissions is still not clear. The following is a list of considerations that make the interpretation of results and the comparison of data from different studies difficult:

- Intercorrelation of fuel properties: The properties of diesel fuel that influence emissions are usually intercorrelated. An example of this is density, aromatics content, and cetane number. Diesel fuels blending streams that contain high levels of aromatics are high in density and also have low cetane numbers. To study the effect of a specific fuel property on diesel emissions, care must be taken to decouple the change in a particular fuel property from changes in other properties of the test fuel. Some studies have not decoupled the fuel properties adequately. If a number of fuel properties are changed simultaneously, it is not possible to ascribe any emission changes to a change in one property.
- Engine technologies: Diesel engine technology has evolved in different directions around the world. In the 1990s, at the time when most of the previously mentioned studies were conducted, heavy-duty engines in the United States had large displacements and already featured a high degree of electronic control. In Europe mechanical engine control still dominated. The engines were more highly rated and had smaller displacements. In Japan large displacement, naturally aspirated engines dominated the market. All of these different engine technologies tend to show somewhat different emission sensitivities to the fuel quality. It is also almost certain that the emission response of future engine technologies will

be different from those currently in production. The biggest difference in the fuel quality impact on emissions has been found between heavy-duty and light-duty engines [14]. Apparently, the results from heavy-duty engine studies cannot be extrapolated to the light-duty engines, or vice versa, and the two engine classes should be discussed separately.

- Emission test cycles: Engines for different geographical markets are emission certified using different engine test cycles. Most research on the influence of fuel quality on emissions has focused either on U.S. technology engines tested on the U.S. transient FTP cycle or on EU engines tested on the ECE R-49 cycle. The EPEFE study attempted to make a comparison between these two test cycles [15]. Considering the magnitude of effects found in the study and the spread of effects across the EU fleet that was tested, the effects of fuel quality on emissions from the U.S. and EU sets of data are generally similar. Despite the different test cycles and different rates of pollutant formation, general extrapolation of fuel effects from one data set to another appears to be possible.
- *Aftertreatment technologies:* Meeting future emission standards may require more extensive use of exhaust gas aftertreatment technologies, such as diesel oxidation catalysts, lean NO_x catalysts, diesel particulate filters, or other techniques. The influence of fuel quality on these technologies is generally unknown. One exception is the fuel sulfur, which has been thoroughly tested for its influence on the performance of diesel catalysts. If an efficient aftertreatment device is used, it will become the primary driver on tailpipe emissions. From the emissions point of view, the properties of fuel would have only secondary importance. Thus, the primary fuel issue would be its compatibility with particular aftertreatment technologies.

11.2.2 Heavy-Duty Engines

The most important properties that influence emissions from heavy-duty diesel engines include the following:

- Sulfur content
- Cetane number
- Aromatics (total and polyaromatics)
- Density
- Volatility (T90/T95)
- Oxygenates.

A comprehensive review and correlation of literature data on the fuel impact on heavy-duty engine emissions, including more than 40 fuel emission studies, has been published by Lee *et al.* [8].

11.2.2.1 Sulfur Content

During combustion, the organic sulfur compounds in diesel fuel are first degraded to such molecules or radicals as H_2S, HS, S, or S_2 and then oxidized via the SO radical to SO_2. The reactions are very fast, resulting in practically complete conversion of sulfur to SO_2. Volumetric concentrations of SO_2 in the exhaust gas can be easily calculated based on the mass balance of sulfur entering and exiting the combustion chamber. Typical levels for a 0.35% S fuel would range from below 50 ppm (or less in very lean engines) at idle to some 150 ppm near full load (equivalence ratio λ of about 1.4), as illustrated in Figure 11.3.

Other than the changes in the levels of exhaust SO_2, changing fuel sulfur content does not cause any observable effects on gaseous engine-out emissions. The emissions of hydrocarbons (HCs), CO, and NO_x are practically independent from the fuel sulfur level. This can be illustrated by the data shown in Figure 11.4, where emissions of CO were constant at all sulfur levels, HC increased slightly (3% to 8%) with increased sulfur level, and NO_x slightly decreased [17]. Average fuel consumption (BSFC) increased with increased sulfur, but the data showed considerable scatter.

Direct impact of fuel sulfur on emissions is limited to emissions of particulate matter (PM). A fraction of the fuel sulfur is converted to sulfur trioxide and sulfuric acid that, due to the current definition of diesel particulate matter by regulatory

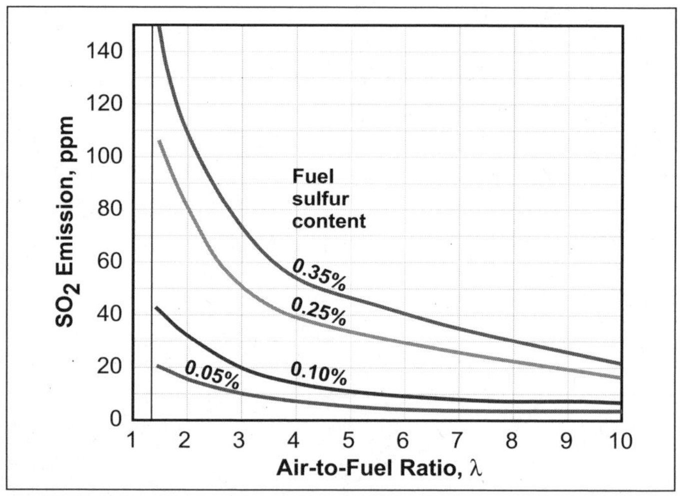

Figure 11.3 *SO_2 emissions at different fuel sulfur levels. (Based on data from Ref. 16.)*

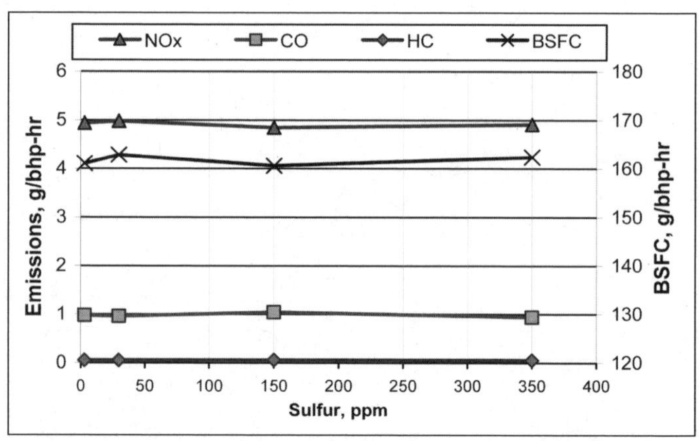

Figure 11.4 *Fuel sulfur effects on gaseous emissions and BSFC. Engine: CAT 3126, six-cylinder, 7.2-L, 20- kW (275-hp) engine at 2200 rpm. Cycle: ESC (OICA). (Based on data from Ref. 17.)*

authorities, is accounted for as PM emission. That emission is known as *sulfate particulates.* Studies have shown that the amount of sulfur converted to sulfate is usually about 2% of the fuel sulfur content [18], and rarely exceeds 4%. This conversion rate depends only slightly on the engine technology. We should emphasize that the impact of sulfur is limited to sulfur particulates. There is no evidence that reduced sulfur levels have any influence on the carbonaceous portion of PM and on the black smoke.

Because sulfates are just one of several components of the PM emissions, lowering fuel sulfur levels has only limited the potential as a means of PM control. The reduction of diesel fuel sulfur levels from 0.30% to 0.05%, as legislated in the United States in 1994, yielded relatively large benefits of about 0.04 to 0.08 g/bhp-hr PM reductions. However, further reductions of fuel sulfur from the current 0.05% to lower levels has only a small incremental PM reduction benefit of about 0.008 to 0.016 g/bhp-hr.

An important aspect of the fuel sulfur levels is the compatibility of fuels with various emission control strategies. It has been shown that several aftertreatment technologies, including diesel oxidation catalyst, CRT and catalyzed particulate filter, or NO_X adsorber/catalyst systems, are intolerant to sulfur. Their performance is either impaired or totally blocked by sulfur compounds. Compatibility with exhaust aftertreatment, rather than a direct emission impact, is the prime driver in lowering the sulfur content in diesel to levels below 500 ppm.

11.2.2.2 Cetane Number

Cetane number influences the ignition delay in diesel engines. Increasing the cetane number reduces the ignition delay, as well as the size of premixed combustion. This results in lower NO_X formation rates due to a slower combustion pressure rise, which gives more time for cooling through heat transfer and dilution and leads to lower local gas temperatures. CO and HC emissions have been also reported to decrease with increasing cetane numbers. Both natural and additive improved cetane numbers appear to have the same emission benefit.

Example test data on the influence of cetane number on gaseous emissions [9] is shown in Figure 11.5. Emissions are expressed in parts per million, averaged over the Japanese six-mode test cycle.

With the increasing requirements for lower NO_X, combustion events in new engines are being designed with little premixed burn. This is likely to challenge the traditional understanding of cetane impact on NO_X emissions. Indeed, for engines with little premixed combustion, the impact of cetane number on NO_X is small or nonexistent. Also the CO and HC emission benefits appear to be lower with new, low-emission engines.

The effect of increasing cetane number on PM emissions is engine specific. In the majority of engines, cetane number has no influence on PM emissions. In some engines a small but beneficial effect has been observed. In other engines increasing cetane numbers lead to increased PM emissions [8]. Other effects of cetane number on engine performance involve an increase in engine noise with lower cetane number. Some increase in fuel

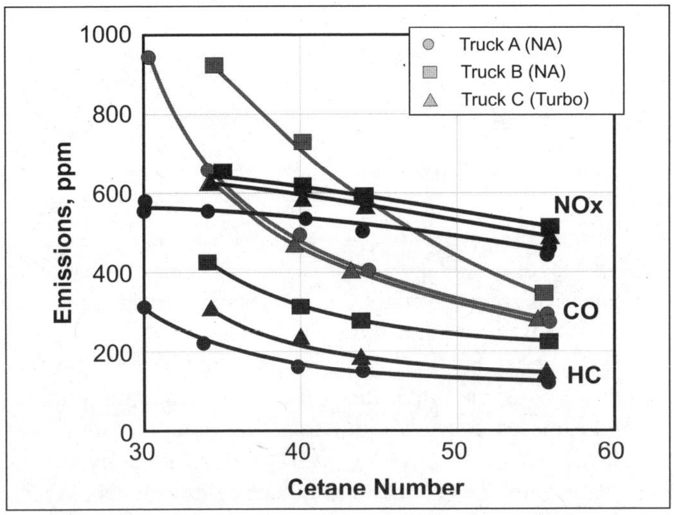

Figure 11.5 Effect of cetane number on heavy-duty diesel emissions. (Based on data from Ref. 9.)

consumption with higher cetane number may also occur due to the lower heating value of the higher cetane blends.

11.2.2.3 Aromatics

The effect of *total aromatics* on emissions has been much debated. Many studies in the past have failed to decouple the changes in aromatics from changes in density, T90, cetane, and so forth. They reported decreases in the HC and PM emissions with decreasing aromatics content, which were apparently caused by changes in other fuel properties. More recent work that has broken the intercorrelations of aromatic content with other fuel properties shows mixed results regarding the effects on HC, CO, and PM emissions. Most studies indicate that the aromatics content has no influence on HC, CO, or PM.

Decreasing the total aromatics from 30% to 10% does, however, produce a small benefit (0% to 5%) in NO_X emissions. This effect has been attributed to the high flame temperature associated with aromatic compounds. By reducing aromatics the flame temperature would drop, leading to a lower NO_X production rate. Another possible mechanism is related to the high C/H ratios in aromatic compounds. A high C/H ratio leads to a higher amount of CO_2 relative to H_2O during combustion. Because CO_2 has a higher tendency to dissociate at high temperatures, high aromatics fuels will produce higher concentrations of oxygen radicals, O·, which, in turn, will collide with nitrogen molecules, yielding higher production rates of NO.

More recent studies have attempted to isolate the influence of *polyaromatics* or polycyclic (di- and tri-) aromatics from that of the total aromatics. A reduction in the polyaromatics appears to reduce HC and NO_X emissions but has no effect on CO emissions. It also shows a significant benefit for PM emissions in older high-emissions engines (>0.1 g/kWh). In modern low-emission engines, reductions in the polyaromatics has little effect on PM emissions. These effects of polyaromatics are attributed, like aromatics, to their high flame temperatures and high C/H ratios. The C/H ratios in polyaromatics are even higher than those in monoaromatics.

11.2.2.4 Density

Density is an important fuel property with respect to volumetric fuel economy and maximum power. In general, the energy per unit volume increases with increasing density. Studies that decouple the effects of density from those of other fuel properties (e.g., aromatics) prove that it is also an important factor influencing regulated diesel emissions.

It is critical that the engine power output be kept constant while assessing the influence of fuel density on emissions. If comparisons are made without adjusting to a constant power output, then lower density fuels will produce lower maximum power and, thus, lower emissions. In such cases lower NO_x emissions are observed due to lower peak pressure and temperature, as well as lower PM emissions due to the overall leaning of the air/fuel mixture.

Reducing the fuel density from about 0.86 to 0.81 g/cm^3 has a beneficial effect on PM emissions. That effect can be as high as 20% in old polluting engines (PM > 0.25 g/kWh). It becomes gradually lower as the engine-out particulate emissions decrease. Modern engines (PM < 0.1 g/kWh) show very little or no change in PM emissions in response to changes in fuel density.

Reducing the density also produces a small reduction in NO_x but a large increase in HC emissions and a small increase in CO. The increase in HC is particularly well pronounced in engines that are already high emitters (PM > 0.15 g/kWh). The effect of density on emissions is related to complex physical interactions with the fuel injection system that can lead to changes in the dynamic timing and mass injection flow rate [15].

11.2.2.5 Volatility

The back-end volatility of diesel fuel, expressed as the 90% or 95% distillation recovery temperature (T90/T95), has some effect on emissions. When the volatility is reduced, a slight increase in HC and CO emissions and a small decrease in NO_x emissions are observed. There is no effect on PM emissions when the volatility is reduced. Given the small nature of these effects, diesel fuel volatility should be considered a minor factor in determining emission performance.

It should be expected that the volatility effect will depend on the composition of the back-end. For example, heavy alkane waxes that have high cetane numbers would give some emission benefits, compared to low-cetane polyaromatics.

11.2.2.6 Oxygenates

There is an increasing interest in adding various oxygenates, such as biodiesel or alcohols, to diesel fuel. That interest stems from the availability of oxygenates from biomass feedstocks. Emission benefits are sometimes claimed for such fuel blends.

The majority of emission studies fail to decouple the addition of oxygenate from changes in other fuel parameters (e.g., density) that occur as the fuel is diluted by the oxygenate. In addition, blending oxygenate into the fuel reduces the fuel energy density and, thus, changes the emission levels. The engine must be recalibrated to its original power output before valid comparisons can be made. Therefore, current oxygenate emission data should be considered tentative.

Studies on the emission influence of oxygenates seem to indicate some decrease in PM emissions, a slight decrease in CO, and an increase in HCs when the oxygenate content is increased from 0% to 4%.

11.2.2.7 Summary

The effects of fuel properties on heavy-duty engine emissions are summarized in Table 11.5. Trends shown in the table are based on multiple engine tests over different engine test cycles. Note that large variations may be seen in individual engines. The effects listed in Table 11.5 have been established by decoupling the influence of particular fuel properties. In real life, changing one fuel property will usually lead to several changes in other parameters. Note that, in such cases, the emission changes attributed to changes in the particular fuel properties are not necessarily additive.

With the exception of sulfur, fuel quality effects have much less significance in modern, clean diesel engines than they used to have in older, more polluting engines. This is a well-pronounced tendency for all of the discussed fuel properties. Apparently, very little or no emission improvement can be achieved in modern, clean diesel engines by changing the fuel quality. However, as new technologies are developed and incorporated into heavy duty diesel engines, the fuel sensitivities may change and the trends discussed may need to be reappraised.

Both natural and fuel additive–induced changes in fuel properties appear to have similar emission benefits. In light of the decreasing fuel quality sensitivity of modern engines, it is

TABLE 11.5
INFLUENCE OF FUEL PROPERTIES ON HEAVY-DUTY DIESEL EMISSIONS

	HC	CO	NO_x	PM
Reduced sulfur	**0**	**0**	**0**	⇓⇓...↓[a]
Increased cetane number	⇓⇓...**0**[b]	⇓⇓...**0**[b]	⇓	**0**
Reduced total aromatics	**0**	**0**	⇓[c]	**0**
Reduced polyaromatics	⇓	**0**	⇓[c]	⇓⇓...**0**[b]
Reduced density	⇑⇑	⇑	⇓	⇓⇓... **0**[b]
Reduced T95	↑	↑	↓	**0**
Increased oxygenate[d]	⇑	↓	**0**	⇓

Source: Data from Ref. 8.

Key: ⇓⇓, large effect; ⇓ , small effect; ↓, very small effect; **0** , no effect.

a. Effect disappears on low-emission engines.

b. Smaller effects are observed with low sulfur levels.

c. Polyaromatics are expected to give a bigger reduction than monoaromatics.

d. Tentative results, require confirmation by future work.

believed that the effects of fuel additives would show similar regularities. Various additives, such as cetane improvers or combustion enhancers, are likely to be an effective means for reducing emissions from old, polluting engines that use poor-quality diesel fuels. The effect of combustion-enhancing fuel additives on emissions from modern engines is likely to be minimal.

A fuel's sulfur content has a special position among other fuel properties. Although only minor engine out emission improvements can be achieved by further reducing current sulfur levels (~0.05%), several candidate diesel emission control technologies are intolerant to sulfur and would require that sulfur levels be lowered significantly.

11.2.3 Light Duty Engines

The EPEFE program is the most comprehensive study available on the influence of fuel properties on the emissions from light duty diesel engines [19]. All test vehicles and engines were complying with the 1996 European emission standards (PM = 0.08, HC + NO_x = 0.7, CO = 1.0 g/km). A matrix of 11 diesel fuels was covering the following changes in fuel properties:

- Cetane number: 50–58
- Density: 0.855–0.828 g/cm^3
- Polyaromatics: 8–1 wt. %
- T95: 370–325°C.

The fuel quality effects on light and heavy duty diesel engines were quite different. This is illustrated in Figure 11.6, which compares the emission changes corresponding to a simultaneous change in fuel properties across the entire tested fuel matrix (i.e., as specified in the preceding list) [14]. Only PM emissions changed in the same direction in response to the fuel change. Emissions of NO_x, HC, and CO changed in different directions. According to that data, the fuel sensitivity of light duty engines appears to be larger than that of the heavy duty engines (with the exception of NO_x emissions).

A summary of light duty engine emission responses to changes in particular fuel properties as found in the EPEFE program [11] is given in Table 11.6. Note that these trends have been determined from a much smaller base of experimental data than the heavy duty responses listed in Table 11.5.

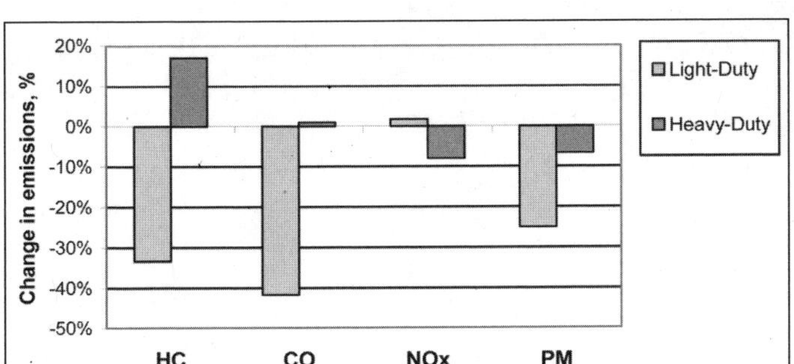

Figure 11.6 Fuel effects on light- and heavy-duty engine emissions. (Based on data from Ref. 14.)

TABLE 11.6
INFLUENCE OF FUEL PROPERTIES ON LIGHT-DUTY DIESEL EMISSIONS

	HC	CO	NO_x	PM
Reduced sulfur	0	0	0	⇓
Increased cetane number	⇓⇓	⇓⇓	0	⇓
Reduced density and polyaromatics[a]	⇓	⇓	0	⇓⇓
Reduced T95	↓	↓	↓	↓

Source: Data from Ref. 11.

Key: ⇓⇓, large effect (10–20%); ⇓ , small effect (2–10%); ↓ , very small effect (<2%); **0**, no effect.

a. Influence of these parameters was not decoupled.

These trends are representative of European diesel car engine technology in 1996. They may or may not extrapolate to other light duty diesel engines, such as other generations of automotive engines or light-duty industrial diesel engines in different geographical areas.

References

1. Thompson, A., *et al.,* "Prediction and Precision of Cetane Number Improver Response Equations," SAE Paper No. 972901, SAE International, Warrendale, PA, 1997.
2. Nikanjam, M., "Diesel Fuel Lubricity: On the Path to Specifications," SAE Paper No. 1999-01-1479, SAE International, Warrendale, PA, 1999.
3. Voitik, R.M., and R. Ning, "Diesel Fuel Lubricity by Standard Four Ball Apparatus Utilizing Ball on Three Disks, BOTD," SAE Paper No. 950247, SAE International, Warrendale, PA, 1995.
4. Batt, R., J.A. McMillan, and I.P. Bradbury, "Lubricity Additives—Performance and No-Harm Effects in Low Sulfur Fuels," SAE Paper No. 961943, SAE International, Warrendale, PA, 1996.
5. Ruzicka, N., and T. Liebscher, "Possible Exhaust Gas Aftertreatment Concepts for Passenger Car Diesel Engines with Sulphur-free Fuel," SAE Paper No. 1999-01-1328, SAE International, Warrendale, PA, 1999.
6. McKinley, T., "Modeling Sulfuric Acid Condensation in Diesel Engine EGR Coolers," SAE Paper No. 970636, SAE International, Warrendale, PA, 1997.
7. Kreso, A.M., J.H. Johnson, L.D. Gratz, S.T. Bagley, D.G. Leddy, "A Study of the Vapor- and Particle-Phase Sulfur Species in the Heavy-Duty Diesel Engine EGR Cooler," SAE Paper No. 981423, SAE International, Warrendale, PA, 1998.
8. Lee, R., J. Pedley, and C. Hobbs, "Fuel Quality Impact on Heavy Duty Diesel Emissions: A Literature Review," SAE Paper No. 982649, SAE International, Warrendale, PA, 1998.

9. Owen, K., and T. Coley, *Automotive Fuels Reference Book,* 2nd ed., SAE International, Warrendale, PA, 1995.
10. "World-Wide Fuels Charter," European Automobile Manufacturers Association, April 2000, retrieved from www.acea.be/acea/WWFCharter042000.pdf.
11. MacKinven, R., and M. Hublin, "European Programme on Emissions, Fuels and Engine Technologies—Objectives and Design," SAE Paper No. 961065, SAE International, Warrendale, PA, 1996.
12. Burns, V.R., *et al.,* "Description of Auto/Oil Air Quality Improvement Research Program," SAE Paper No. 912320, SAE International, Warrendale, PA, 1991.
13. "Bibliography of Data Sources for Use in Developing a Model of Diesel Fuel Emission Effects," Office of Transportation and Air Quality, U.S. Environmental Protection Agency, Washington, DC, January 25, 2001, retrieved from www.epa.gov/otaq/models/analysis/bibliogr.pdf.
14. Rickeard, D.J., R. Bonetto, and M. Singer, "European Programme on Emissions, Fuels and Engine Technologies (EPEFE)—Comparison of Light and Heavy Duty Diesel Engine Studies," SAE Paper No. 961075, SAE International, Warrendale, PA, 1996.
15. Signer, M., *et al.,* "European Programme on Emissions, Fuels and Engine Technologies (EPEFE)—Heavy Duty Diesel Study," SAE Paper No. 961074, SAE International, Warrendale, PA, 1996.
16. Leyer, J., *et al*, "Heavy-Duty Emission Control Using Oxidation Catalysts," Institute of Mechanical Engineers, C435, February 1991, pp. 163–191.
17. DECSE, "Phase I Interim Data Report No. 4: Diesel Particulate Filters," U.S. Department of Energy, Washington, DC, January 2000, retrieved from www.ott.doe.gov/decse/pdfs/interim4.pdf.
18. Cowley, L.T., R.J. Stradling, and J. Doyon, "The Influence of Composition and Properties of Diesel Fuel on Particulates Emissions from Heavy-Duty Engines," SAE Paper No. 932732, SAE International, Warrendale, PA, 1993.
19. Hublin, M., P.G. Gadd, D.E. Hall, and K.P. Schindler, "European Programmes on Emissions, Fuels and Engine Technologies (EPEFE)—Light Duty Diesel Study," SAE Paper No. 961073, SAE International, Warrendale, PA, 1996.

Chapter 12

Alternative Diesel Fuels

by Peter Ahlvik and W. Addy Majewski

12.1 Overview of Alternative Diesel Fuels

12.1.1 Environmental Driving Forces

Air pollution is generally considered today to be one of the main driving forces for the use of alternative fuels. Presumably, this will continue to be of importance for a long time. However, it is likely that the focus could shift significantly in the future. For alternative fuels, future driving forces are of greater importance than the current focus.

Shindler *et al.* at Volkswagen have predicted that the emissions of greenhouse gases (GHGs) and energy security will be of greater importance than exhaust emissions in the future [1]. This is shown schematically in Figure 12.1. Note that the relative importance of exhaust emissions compared to the sum of GHGs and energy would be significantly less in 2025 than it is today.

The rationale for predicting a shift in focus, as indicated in Figure 12.1, is that the production capacity of the conventional oil resources will be a limiting factor much earlier than the point at which all oil resources are depleted. There are indications that this scenario could take place as soon as the 2010–2020 time frame, as shown in Figure 12.2. Use of unconventional oil resources, alternative fuels, and increasing energy efficiency could extend the resources but, inevitably, this scenario will appear sooner or later.

The use of conventional fossil fuels is likely to cause climate change, which might lead to an ecological disaster in certain areas. Therefore, a reduction of fossil carbon dioxide emissions well above 50%, maybe approaching 80%, is needed to stabilize the climate. In such a scenario, the need to reduce carbon dioxide emissions from the transport sector cannot be avoided. This task cannot be accomplished by increasing the

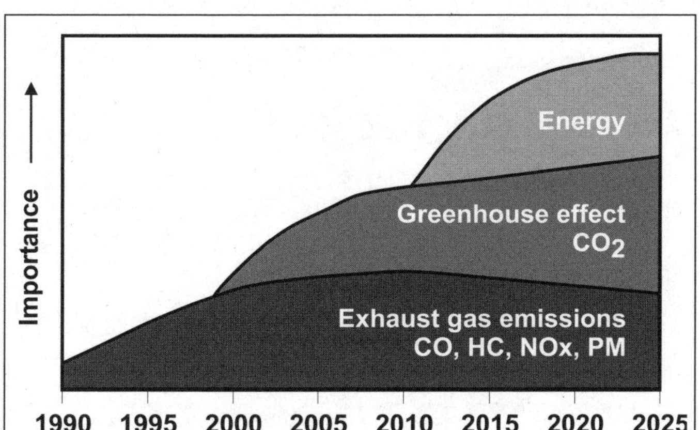

Figure 12.1 Environmental driving forces. (Based on Ref. 1.)

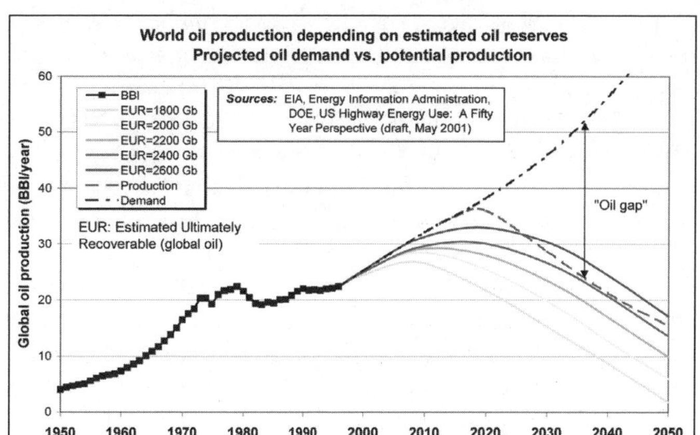

Figure 12.2 Global oil production and demand depending on the size of oil reserves. (Adapted from Ref. 2.)

energy efficiency alone; instead, the substitution of renewable fuels for fossil fuels is necessary. Hydrogen and fuel cells might be a long-term option for light-duty vehicles, but they are not easily utilized in heavy-duty vehicles due to the bulkiness of the fuel tank. Furthermore, fuel production and especially fuel distribution are issues yet to be solved for hydrogen. Therefore, fuel options other than hydrogen must also be considered. Liquid fuels are easy to distribute, store, and handle and are, therefore, prime candidates for heavy-duty vehicles.

12.1.2 Alternative Fuel Options

Diesel engines can use many different alternative fuel options. Some of the most important options are as follows:

- Synthetic diesel fuel
- Biodiesel, i.e., esters of fatty oils (often vegetable oils)
- Dimethyl ether (DME) and other heavier ethers
- Alcohols, i.e., ethanol and methanol
- Methane, i.e., natural gas and biogas
- Hydrogen.

Several other diesel fuel options and fuel blends are possible but are not discussed here. Some examples of blended fuels are emulsions of diesel fuel and alcohols (e.g., E-Diesel, Diesohol) and diesel–water emulsions (see Chapter 15). Levulinic acid, which can be converted to ethyl or methyl esters, has recently been proposed as a new diesel fuel candidate.

Synthetic diesel fuels are often referred to as Fischer-Tropsch diesel fuel (FT diesel fuel or FTD) or gas-to-liquid (GTL) fuel. FT diesel and biodiesel can often be used directly as blending components in diesel fuel or as neat fuels (with some additives) in modern diesel engines (provided that the fuel is compatible with the fuel system elastomers). The fuel properties of FT diesel and biodiesel result in a fuel flexibility for diesel engines that is not offered by many other fuel options. Fuel flexibility is a very important feature during the introductory phase for a new fuel.

Dimethyl ether (DME) is a "natural" diesel fuel with high cetane number and it has an interesting emission potential. However, DME has to be used in dedicated engines, so it does not offer fuel flexibility.

Alcohol fuels have a very low octane number and need a positive ignition source or additives (ignition improvers) in the fuel to raise the cetane number to be used in diesel engines. Alcohol fuels have some emission advantages regarding NO_x and particulate emissions, but some emission components can also be higher than for diesel fuel. As for DME, no fuel flexibility can be expected for alcohols. The use of methane is discussed separately later in this chapter.

Hydrogen could potentially be used in internal combustion engines. However, hydrogen is an ideal fuel for fuel cells and offers a higher efficiency in such energy converters. Therefore, the use of hydrogen in internal combustion engines could be viable during a bridge-over period.

12.1.3 Resource Use and Energy Efficiency

To be able to replace as much conventional diesel fuel as possible, the largest possible feedstock or energy resources have to be used. Many unconventional sources, such as nuclear power (fission or fusion), photovoltaics, and methane hydrates, are not of much concern for the near and mid-term future. Natural gas and biomass, however, are two potential feedstocks for the near and mid-term future.

Natural gas will most certainly be used as a feedstock for fuel production in the future. The most efficient utilization of natural gas is an important issue. Natural gas could be used directly or converted to liquid (methanol, FT diesel) or gaseous (DME, hydrogen) fuels. Conversion of natural gas to liquid fuels is of particular interest for remote natural gas resources.

In the biofuels case, the availability of biomass resources is crucial. In terms of availability, great variation is seen from country to country. McCormick at the National Renewable Energy Laboratory (NREL) refers to a study regarding the potential of renewable diesel fuel options in the United States [3]. The results are shown in Figure 12.3. As can be seen in this figure, FT diesel has the greatest potential, i.e., almost covering the total use of diesel fuel in the on-road sector. Oxygenates also have significant potential but new sources for biodiesel must be developed.

Fermentation of sugar cane, corn, and grain to ethanol is a well-established technology. Lignocellulosic biomass would have a greater feedstock potential, provided that breakthrough in the production technology can be made.

The production of fuels from synthesis gas (syngas) is a particularly interesting route. Natural gas, any kind of biomass, and even coal could be used as feedstocks for syngas production. DME, methanol, FT diesel, hydrogen, and methane could be produced in this process. The syngas step would be common for all fuels. The production of biofuels via the syngas route is shown in Figure 12.4. If hydrogen could be produced in a viable way, it could be added in the process. Surplus hydrogen could enable additional CO_2 to be used. This would be an alternative to direct use of hydrogen.

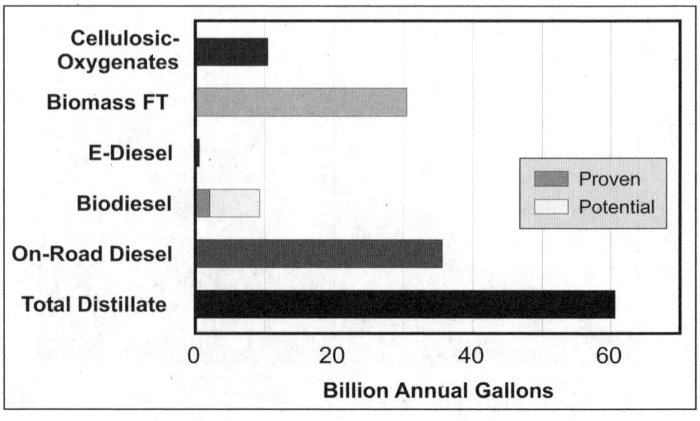

Figure 12.3 *Energy security implications of renewable diesel fuel options. (Based on data from Ref. 3.)*

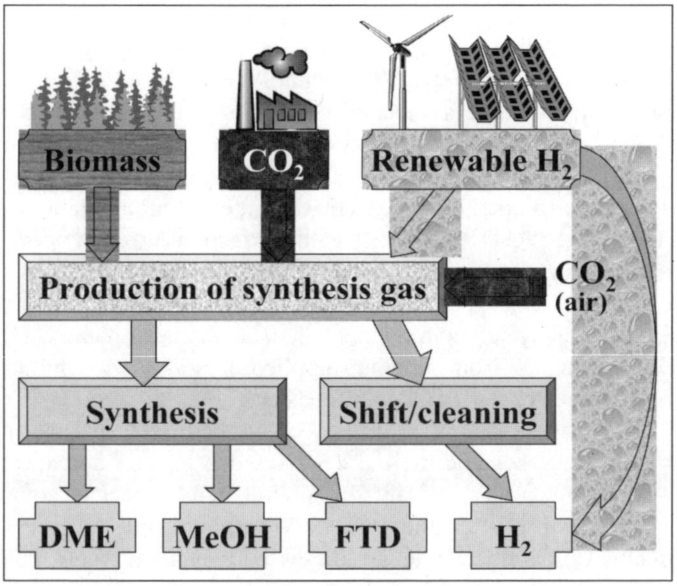

Figure 12.4 *Production of biofuels from synthesis gas.*

Several life cycle assessments for various alternative fuels have been carried out during the last couple of years. As mentioned earlier, efficiency will be a very important factor in the future. An example of the well-to-wheel efficiency in passenger cars for the 2010–2015 time frame for fuels from biomass feedstock [4] is shown in Figure 12.5. In this figure, only the best combinations of fuels and drivetrains have been included.

DME, hydrogen, and methanol appear to be the fuels with the highest efficiency (Figure 12.5) using the assumptions made in the study. Ethanol and FT diesel have a somewhat lower efficiency. Because great losses are experienced in several steps of the chain, the total efficiency in absolute numbers is low; around 11% for best cases. It is hardly surprising that all of the best options comprise only hybrid drive systems. It also is not surprising that no Otto engine is among the best combinations; fuel cells and/or the diesel engine provide the most efficient powertrains for all fuels.

Fuel distribution is a very important factor for an alternative fuel that is intended for high market penetration. Widespread use is necessary to significantly influence GHG emissions and energy security. A finely branched, low-cost distribution network to make a fuel available everywhere will require liquid fuels that are easily handled (such as petrol and diesel fuel today). Duplication of such a network for handling liquefied gases under pressure or cryogenic liquids will hardly be acceptable, due to cost reasons. Niche applications of such fuels will be the remaining possibility. The problems in hydrogen distribution are so severe that a recent study proposed the focus be shifted from "hydrogen economy" toward "methanol economy" [5].

12.2 Synthetic Diesel Fuel

12.2.1 Introduction

Diesel fuels and other petroleum products are traditionally manufactured by refining of crude oil. However, they can be also produced synthetically from various carbon-bearing feedstocks. The feedstock that attracts the most attention today is natural gas. Worldwide reserves of natural gas are estimated at 140,000 billion cubic meters (bcm), while annual global consumption is only 2100 bcm [6]. There are many remote ("stranded") natural gas reserves around the world that for various reasons are not readily accessible by pipelines. In several other locations natural gas that accompanies crude oil ("associated gas"), but is not utilized, is flared in huge quantities. Last, but not least, producing high-quality liquid fuels from natural gas is an alternative to all other methods of gas utilization. Because liquid fuels are several times less expensive to transport than gas, conversion

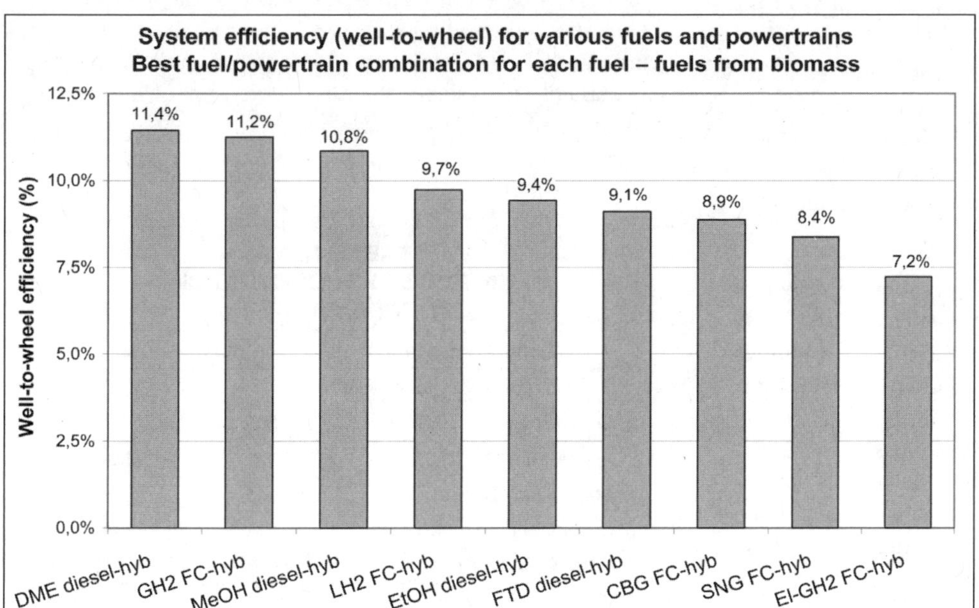

Figure 12.5 *Well-to-wheel efficiency for the best combinations of biofuel/drivetrains. DME diesel-hyb: DME, diesel engine hybrid; GH2 FC-hyb: compressed hydrogen, fuel cell hybrid; MeOH diesel-hyb: methanol, diesel engine hybrid; LH2 FC-hyb: cryogenic (liquid) hydrogen, fuel cell hybrid; EtOH diesel-hyb: ethanol, diesel engine hybrid; FTD diesel-hyb: Fischer-Tropsch diesel fuel (synthetic fuel), diesel engine hybrid; CBG FC-hyb: compressed biogas, fuel cell hybrid; SNG FC-hyb: compressed synthetic natural gas; El-GH2 FC-hyb: compressed hydrogen, produced from electric (local), fuel cell hybrid.*

to liquids provides an opportunity to expand the use of natural gas while lowering the transportation costs.

The first and best known synthetic fuel technology is the Fischer-Tropsch (FT) process, which was developed in the 1920s in Germany. Commercial use of FT fuels, besides the two historical incidents of use by World War II Germany and South Africa during economic embargo periods, has been extremely limited. Nevertheless, FT research work has been continued by several companies, leading to the development of a mature technological process. Today, the major FT technology players include large oil companies, such as ExxonMobil, Shell, and Sasol. Research is also sponsored by governments, which perceive synthetic fuels as an important option for future alternative fuels [7]. Furthermore, small development-stage companies exist (e.g., Syntroleum or Rentech) that develop and license FT processes to others.

Because of the natural gas focus, synthetic fuel processes are frequently referred to as gas-to-liquid, or GTL, technologies. That term, however, does not cover all synthetic fuel technologies, because liquid fuels may be produced, and have been produced, from almost any carbonaceous feedstock that is either gaseous, liquid, or solid. Coal is a good example of a solid feedstock that was used for the manufacture of FT diesel fuel in the past. On the other hand, the term GTL is sometimes also used in relation to non-FT fuels, for example dimethyl ether, which can be produced from natural gas feedstock as well. The discussion in this section is limited to fuels produced by means of Fischer-Tropsch synthesis, with a focus on natural gas as the most important feedstock.

Synthetic diesel fuels are important and attractive for several reasons:

- Synthetic fuels are compatible with existing engines, so there is no need to make engine modifications.
- Synthetic fuels are compatible with conventional diesel (that is, they have a comparable energy density, can be mixed with petroleum diesel, can be transported as liquid in existing petroleum infrastructure).
- The fuels can be designed to have very good properties for both engine performance and emissions.
- Synthetic fuels can be used neat or as a valuable blending stock, to improve the properties of petroleum fuels.
- The sulfur content is practically zero, making synthetic fuels compatible with a range of sulfur-sensitive exhaust gas aftertreatment technologies, such as NO_x adsorbers or the CRT filter.

On the other hand, environmental concerns may present an obstacle in the commercialization of synthetic fuels. FT fuels manufactured from natural gas bring no discernible greenhouse gas benefit relative to petroleum diesel (unless the feedstock gas was flared before the production started). Only FT fuels made from biomass can provide a life cycle CO_2 emission benefit.

Potential locations for commercialization of GTL plants are in regions with ample low-cost gas resources, such as the Middle East, West Africa, and the North Slope of Alaska. Fields like those on Alaska's North Slope contain plenty of natural gas but are far from market. The Trans-Alaska Pipeline System offers the opportunity to transport GTL products through the existing pipeline and provide high-quality synthetic hydrocarbons to world markets. GTL technology could be important in locations where associated gas is reinjected or flared for lack of nearby markets. In these locations GTL plants could produce hydrocarbons that could be conveniently refined or, if upgraded, shipped directly to market in conventional tankers. Integration of GTL technology with production and other operations offers additional incentives. Use of the by-products of the GTL process, such as steam, power, and nitrogen, can further enhance its overall commercial value. On the other hand, GTL fuels produced from pipeline-supplied natural gas would not be competitive due the higher feedstock cost.

Process economy, as influenced by the high capital costs of FT processes and the market risks due to the fluctuation of crude oil prices, presents the main barrier for wider commercialization of synthetic fuels. Whenever crude oil prices decline, economic challenges increase for synthetic fuels. The sensitivity of FT fuel costs to crude oil prices is illustrated in Table 12.1. It is generally agreed that FT production is feasible only at very low gas prices on the order of \$0.50/MMBtu (1 MMBtu = 1 million Btu = 1055 MJ), unless a premium is paid for environmentally friendly fuel [8]. It is also estimated that crude oil prices must be above \$20 to \$25 per barrel for FT plants to be profitable.

An important economic benchmark for comparing FT technology is the capital cost of building a manufacturing plant. For a refinery, that cost is in a range of \$12,000 to \$14,000 per daily barrel, while the cost of the various existing FT technologies is in the \$20,000 to \$30,000 per daily barrel range [9]. All of the figures from Table 12.1 reflect the cost situation in the 1990s. Advances in technology continue to drive down the cost of manufacturing high-quality liquid products from natural gas. As shown in Figure 12.6, costs have been reduced by more than 50% since 1990 as a result of advancements in slurry hydrocarbon synthesis (HCS), improved syngas generation options, and new technologies for upgrading HCS products [10].

Other, non-FT technologies that aim at producing less expensive synthetic fuels are also being researched. One such venture, involving Catalytica and Syntroleum, is developing a new class of catalysts for the direct oxidation of methane into methanol and liquid hydrocarbons.

TABLE 12.1
UNIT COST PRODUCTION OF GTL FUEL, \$/BARREL

Cost Component	Refinery	GTL
Natural gas (@\$0.50/MMBtu)		\$4.00
+ Crude oil (@\$17/Bbl)	\$17.00	
+ Operating costs	2.50	3.00
Total Cash Costs	19.5	7
+ Capital recovery, taxes	6.50	12.00
Total Cost to Produce	\$26.00	\$19.00

Source: Data from Ref. 7.

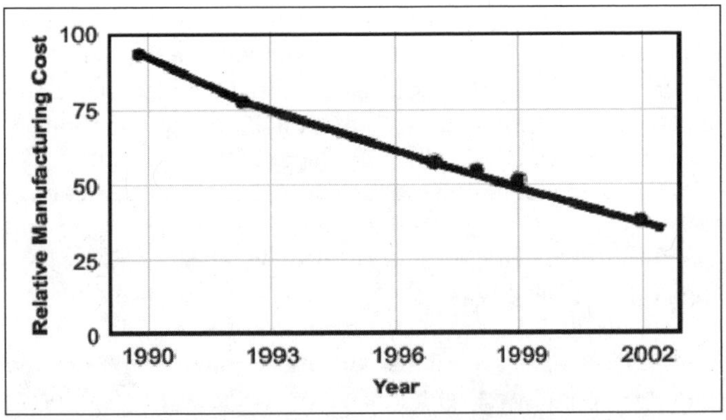

Figure 12.6 *GTL Process Relative Manufacturing Cost*

Direct liquefaction of coal is another method that can be used to produce petroleum products (the FT synthesis using coal is also referred to as the "indirect liquefaction of coal"). In this process, coal is converted to liquid hydrocarbons in a single-step operation. Hydrogen is added to the coal during the conversion process to upgrade the liquid products, giving them characteristics comparable to those of petroleum. Research continues to improve the process economy.

12.2.2 Fisher-Tropsch Process

12.2.2.1 History

The FT process was named after two German scientists who developed it in the 1920s: Franz Fischer and Hans Tropsch [11]. Shortly thereafter the process was employed by German companies in full-scale industrial plants to manufacture synthetic fuels, mainly diesel, from coal for use by the Germans during WWII. With Germany's very limited crude oil resources it became critical to find a means to obtain fuel for tanks, planes, and motor vehicles. The Fischer-Tropsch synthesis, based on the abundant coal resources as the feedstock, was the solution. It is estimated that some 4,500,000 barrels of synthetic fuels was made during each of the war years in Germany [9].

In 1944 the United States passed a Synthetic Fuels Act that charged the U.S. Bureau of Mines (BOM) with the responsibility of developing synthetic fuels. When the Allies won WWII, the U.S. BOM actually dismantled two German plants and brought them back to the United States. One of the two plants was reassembled in Louisiana and produced around 100 barrels per day of FT products for several years.

Another important thread in the development of FT technology started in South Africa in the late 1940s when the South African government began FT research. That work was conducted through a company that became later the South African Coal, Oil, and Gas Corporation, or Sasol. The first Sasol plant was build in 1955 and additional plants were started in the 1980s. All plants used coal as a feedstock. Sasol's main objective was to make gasoline from local coal resources, because, at the time these plants were built, South Africa was subject of an international embargo and was unable to import gasoline. The South African plants are still in operation, producing around 160,000 barrels per day of FT products [9].

The most recent chapter in the history of FT synthesis started in the 1970s, after the Arab oil crisis. At that time, with oil prices rising again, another synthetic fuels act was passed in the United States. Several massive FT research programs, as well as numerous FT pilot plants, were started by various oil companies. The interest in the technology, and the amount of research dollars spent, has followed the fluctuations in crude oil prices. Initially, the most technologically advanced players were Exxon and the Royal Dutch/Shell Group. Since the late 1990s, nearly all major oil companies, including BP, ExxonMobil, and ChevronTexaco, have announced plans to commercialize the GTL process.

12.2.2.2 Process Description

FT fuels are manufactured in a three-step procedure:

1. *Syngas generation:* A carbonaceous feedstock (e.g., natural gas) is converted into synthesis gas (syngas) composed of carbon monoxide and hydrogen.
2. *Hydrocarbon synthesis:* The syngas is catalytically converted into a mixture of liquid hydrocarbons and wax, producing a "synthetic crude." This step is the actual Fischer-Tropsch synthesis.
3. *Upgrading:* The mixture of FT hydrocarbons is upgraded through hydrocracking and isomerization and fractionated into middle distillate (diesel) fuels, which can travel in an oil pipeline or oil tanker.

The feedstock used for the generation of the syngas can be almost any material containing carbon. As already mentioned, it can be natural gas or coal, but it can also be biomass or a range of petrochemical by-products (refinery bottoms, bitumens, tars, coke). Feedstocks are converted into the syngas in the presence of oxygen and steam through such processes as partial oxidation (POX) or catalytic reforming. Typical composition of synthesis gases is H_2, 35% to 75%; CO, 15% to 45%; and CO_2, 2% to 10%. The exact composition depends on the particular syngas manufacturing process.

The syngas is introduced into a catalytic reactor, where the Fischer-Tropsch synthesis occurs. Two types of FT catalysts have been used: iron-based and cobalt-based catalysts. The choice of catalyst is very important in determining what feedstocks can be converted to synthetic fuels. Iron has been used mostly with coal feedstock (WWII Germany, South Africa). Cobalt is considered to be more suitable for natural gas processes and its use is being pursued by both ExxonMobil and Shell [12].

The FT products are a mix of liquid and wax hydrocarbons in a ratio of around 1:1. If needed, the waxes can be converted to liquid hydrocarbons by cracking with hydrogen. The synthetic FT oil can be distilled into two fractions: naphtha and middle distillates. The naphtha is a good feedstock for ethylene production, while the distillates produce jet fuel, kerosene, and diesel fuel.

Fischer-Tropsch products are composed almost exclusively of paraffins and olefins. They contain very few aromatic compounds. Furthermore, they are practically free from sulfur, as well as from other compounds that are found in crude oils, such as nickel, vanadium, or nitrogen.

12.2.3 FT Fuel Properties

Because a number of FT catalyst and reactor technologies exist, the properties of FT diesel fuels are likely to vary. The FT fuel properties can be also designed to suit a particular application by selecting the production process parameters. In general, in comparison to petrodiesel, synthetic fuels have the following qualities:

- Very good ignition characteristics (e.g., the cetane number may reach 75 or higher)
- No (or very low) sulfur
- Very low aromatics and polyaromatics content
- Colorless and almost odorless appearance
- Energy density similar to that of petrodiesel.

On the negative side, FT fuels are usually distinguished by poor lubricity and poor cold flow properties.

A very important feature of synthetic fuels is their compatibility with existing diesel engines. The only adjustment that may be required is increasing the lubricity of fuel in order to prevent excessive wear of the fuel injection system. That can be achieved using commercial lubricity additives. Doping the FT fuel with biodiesel, which has excellent lubricity properties, is also possible.

Table 12.2 lists an example analysis of 100% Shell FT diesel fuel without and with a lubricity additive. The analysis is compared to the fuel quality requirements from an engine manufacturer; in this case, Caterpillar specifications were used [13].

12.2.4 Emissions with Synthetic Fuel

12.2.4.1 Regulated Air Pollutants

A number of experimental studies showed emission benefits when comparing FT diesel fuels (neat or blends) with petrodiesel [13]. These findings are in agreement with known fuel effects on emissions. High cetane number, low aromatics, and low sulfur—the basic properties of FT fuels—favor reductions in the emissions of several diesel exhaust pollutants in both heavy- and light-duty engines.

Various studies evaluate different FT fuels, which are manufactured by different processes and characterized by different properties. Therefore, exact comparison of the results is usually not possible. In some studies differences in the engine power output are observed, due to differences in the energy density of the tested FT fuel and petrodiesel. If the engine is not recalibrated to its original power rating, the validity of such emission comparisons may be limited. The emission effects of FT fuels, as is the case with other diesel fuels, tend to be engine specific. A general trend appears to be that older technology engines can achieve more emission benefits by using better fuels than do new engines, which have optimized combustion systems. Note that fuels, including FT diesel, have only a secondary impact on emissions; the primary diesel emission drivers are the combustion system and, if present, exhaust gas aftertreatments.

The potential emission impact of FT fuel and its blends in heavy-duty engines can be illustrated by the results of a Sasol fuel study, conducted on a 1991 DDC Series 60, 12.7-L diesel engine [14]. The engine was tested on the FTP Transient test cycle. The FT fuel was blended with conventional U.S. No. 2 grade diesel fuel. Emissions with the blends were compared with the emissions using No. 2 diesel fuel and with California diesel fuel. Quite substantial emission reductions were achieved with the FT fuel, as presented in Figures 12.7 and 12.8. It was found that blends of FT and No. 2 petroleum fuels in a 40:60 proportion produced emissions equivalent to those with California fuel.

A second stage of the Sasol study evaluated the emission impact of FT fuel in a newer, model year (MY) 1999 engine (DDC, 12.7-L, turbocharged, intercooled, 321 kW @ 2100 rpm) [15]. This time, the emission reductions were 37% in HC, 37% in CO, 17% in NO_X, and 37% in particulate matter (PM), relative to the U.S. No. 2 fuel. Compared with the California fuel, FT emissions were lower by 18% in HC, 33% in CO, 11% in NO_X, and 32% in PM. California diesel emissions were matched or improved with a 50% FT blend in No. 2 diesel. A comparison with the older, MY 1991 engine results (Figures 12.7 and 12.8) shows that the HC and NO_X reduction by FT fuel was smaller in the model year 1999 engine, but the PM reduction was larger. PM was reduced in the 1991 engine through reductions in both soluble organic fraction (SOF) and solid carbon, while in the 1999 engine, PM was reduced primarily by carbon emission reductions alone.

Several FT emission studies on HD engines have been conducted under the auspices of the U.S. Department of Energy. The following are summaries of two such projects [16]:

- White-GMC WG64T Class 8 tractors (80,000 lb GVW) powered with MY 1996 to 1997 Caterpillar 3176B, 10.3-L diesel engines were tested on a chassis dynamometer over the WVU 5-mile cycle [13]. The use of Shell FT fuel in place of California diesel in these trucks led to lower levels of all four regulated emissions. NO_X emissions were reduced by an average of 12%, PM by 24%, CO by 18%, and THC by 40%.
- Six 40-foot urban buses powered by DDC 6V92 two-stroke diesel engines were operated on Mossgas FT diesel and No. 2 diesel and tested on a chassis dynamometer over the CBD test cycle [17]. Three of the buses were equipped with diesel oxidation catalysts. The catalyst-equipped buses emitted an average 8% less NO_X and 31% less PM when operated on FT diesel. NO_X emissions were reduced by 5% and PM by 20% in buses without catalysts.

According to data published by Shell, FT fuels can provide higher (40% to 60%) emission benefits for CO, HC, and PM in light-duty (LD) vehicles, compared to only 5% to 30% in heavy-duty (HD) engines [6]. Only small NO_X benefits were

TABLE 12.2
ANALYSIS OF SHELL FT DIESEL FUEL

Property	ASTM Method	100% FT Diesel	FT Diesel with Lubricity Additive	Fuel Requirements from Engine Manufacturer
Flash point, °C	D93	72		Legal limit
Cloud point, °C	D2500	3		Max.not above the lowest ambient temperature
Water and sediment, wt. %	D1796	<0.02		0.1 max.
Carbon residue, wt. %	D524	0.02		1.05 max.
Ash, wt. %	D482	<0.001		0.02 max.
Distillation, °C	D86			
T10		260		282 max.
T90		331		360 max.
Kinematic viscosity, cSt @ 40°C	D445	3.57	3.57	1.4 min., 20 max.
Sulfur	D5453	<5 ppm[a]		3 wt% max.
Corrosion, 50°C/3 h	D130	1A		No. 3 max.
Cetane number	D613	>74[b]		40 min.
Density @ 15°C	D4052	0.7845		
API gravity @ 15°C, °API	D287	54		30 min., 45 max.
Pour point, °C	D97	0		6° min. below ambient temperature
SFC aromatics, wt. %	D5186			
Mono-		0.1		
Di-		0.1		
Poly-		0.1		
FIA, vol. %	D1319			
Aromatics		0.1		35 max.
Olefins		0.1		
Saturate		99.8		
Gum content, mg/100 mL	D381	0.2		10 max.
Lubricity SLBOCLE, g	D6078	1700	4050	3100 min.
Lubricity HFRR, μm	D6079	420/540/570	210	380 max.
Carbon/hydrogen, wt. %	D5291			
Carbon		84.91		
Hydrogen		14.97		
Nitrogen		0.67		
Residual		−1.09		
Oxygen (by diff.)		Negligible		
Heat of combustion, Btu/gal	D240			
Gross		132,600		
Net		123,600		

Source: Data from Ref. 13.
a. Below minimum measurement capability.
b. Above maximum measurement capability.

seen with FT fuels. The Shell data for different diesel engine technologies, generated over the ESC test for HD and the ECE+EUDC test for LD vehicles using CEN96 reference fuel, are listed in Table 12.3.

12.2.4.2 Life Cycle Analysis

From the life cycle greenhouse gas (global warming) emission perspective, FT fuels produced from natural gas are approximately equivalent to petroleum fuels [6] (Figure 12.9). The carbon efficiency of the Shell SMDS process is currently lower than that of a state-of-the-art refinery. This can be offset by benefits upstream and in vehicle usage. Some FT fuels may provide up to about a 5% CO_2 benefit relative to petroleum diesel due to their higher heat value and higher H/C ratio. As shown in Figure 12.9, FT fuels show favorable life cycle emissions of other gases, including NO_x and SO_2.

While presenting a potentially very attractive means of gas utilization, in view of global warming concerns, natural

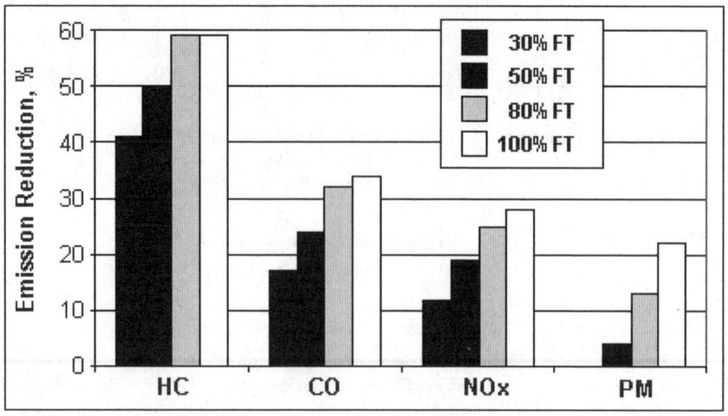

Figure 12.7 Emission reductions using FT diesel compared to No. 2 diesel. (Based on data from Ref. 14.)

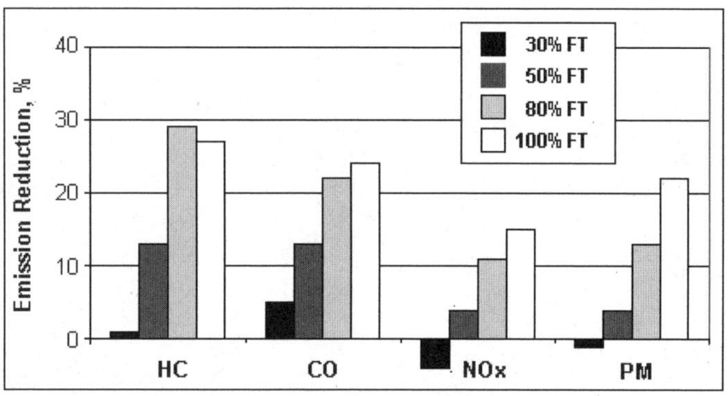

Figure 12.8 Emission reductions using FT diesel compared to California diesel. (Based on data from Ref. 14.)

gas–derived FT fuels are not the most attractive fuel alternative for the future. However, as mentioned before, FT fuels can be produced from synthesis gases obtained from renewable biomass. Biomass-derived FT fuels, characterized by a clear CO_2 advantage, can play an important role in the future fuel market. Biomass-FT resources in the United States were estimated at some 25 billion gallons per annum, quite a significant figure when compared with the total U.S. distillate fuels market of 60 billion gallons [3].

TABLE 12.3
EMISSION BENEFITS FOR NEAT SHELL SMDS FUEL (%)

	Light-Duty Vehicles			Heavy-Duty Vehicles		
Emission	Euro 1	Euro 2	Euro 3	Euro 1	Euro 2	Euro 3
CO	40	53	75	22	5	16
HC	45	63	62	13	23	<9
PM	42	39	41	18	18	34...10
NO_x	10	5	5	16	15	5...19

Source: Data from Ref. 6.

12.3 Biodiesel

12.3.1 Introduction

Many vegetable oils and animal fats, mostly fatty acid triglycerides, have been suggested and investigated as diesel fuel substitutes. The common sources of oil include soybean, rape, sunflower, coconut, palm, and used frying oil, but methods have been also developed to make biodiesel from such exotic materials as oils naturally produced by certain species of algae [18]. Vegetable oils are renewable energy sources and, as such, have been supported by several pieces of legislation and government-sponsored initiatives in the United States and in Europe. These oils are converted into methyl esters before they are used as diesel fuel. In the United States, the ASTM (American Society for Testing and Materials) Biodiesel Task Force has adopted the following description of biodiesel [19]: *Biodiesel is defined as the mono alkyl esters of long-chain fatty acids derived from*

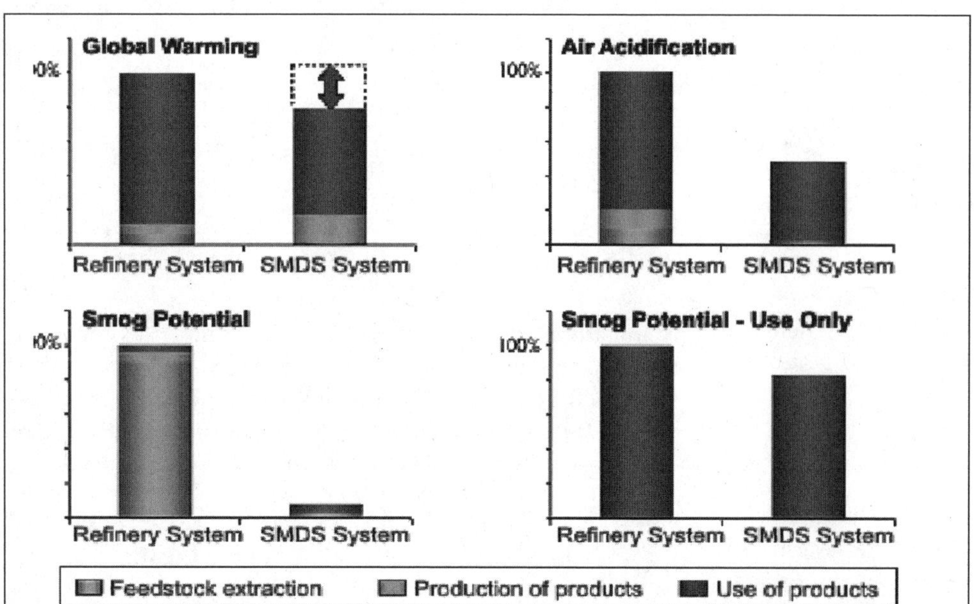

Figure 12.9 Life cycle analysis for Shell SMDS fuel. (Reprinted from Ref. 6.)

renewable lipid feedstocks, such as vegetable oils and animal fats, for use in compression ignition (diesel) engines.

The mono alkyl ester definition eliminates pure vegetable oils as well as monoglycerides and diglycerides from consideration as biodiesel. During the 1970s and 1980s a fair amount of research was conducted with pure vegetable oils and partially estrified oils in their neat (or pure) form and with blends of petrodiesel. Eventually, biodiesel research focused on esters due to their superior properties (virgin oils tend to have a significantly lower cetane number and higher viscosity than ester biodiesel). The use of pure or partially estrified oils also caused a variety of engine and injector problems. We must emphasize that unprocessed or partially estrified vegetable oils are not considered suitable as diesel fuels and do not meet the requirements set by existing biodiesel standards and specifications.

The most common source of biodiesel is soybeans in the United States and rapeseed in Europe. Depending on the feedstock, biodiesel may be referred to by a number of more specific terms and corresponding acronyms, the most common ones as follows:

- Soybean methyl ester (SME or SOME)
- Rape methyl ester (RME)
- Fatty acid methyl ester (FAME), a collective term including both of the preceding entries
- Vegetable oil methyl ester (VOME).

Biodiesel, due to its renewable character, greenhouse gas emission reduction potential, and a generally favorable life cycle analysis (as discussed later), is an attractive alternative diesel fuel. High prices often present a barrier for more widespread biodiesel use, especially at the onset of commercialization. In markets with no government subsidies, the initial retail price of biodiesel is typically about two to three times higher than that of petrodiesel. As the volumes increase and the technology for producing fuel-grade materials matures, the price may be reduced. This is illustrated in Figure 12.10, which shows the evolution of biodiesel prices in the United States [3]. The initial commercialization of biodiesel or its blends requires some form of government subsidy, tax incentive, or legislation. Examples of such government support include a preferential biodiesel taxation introduced in Germany or the classification of biodiesel as an alternative fuel under the Energy Policy Act in the United States.

A common approach used to reduce the cost of biodiesel is to blend it with petroleum diesel fuels. Such blends are already commercialized as diesel fuel in many Western European countries and in the United States. The most common blend in the United States has been 20% biodiesel and 80% conventional petrodiesel, sometimes referred to as B20 (under the same convention, neat biodiesel is termed B100). Blending with petrodiesel allows the manufacturer to control cost increases, as well as to minimize the degree of fuel system modification that may be required in engines that were not designed for biodiesel fueling. Blends of 5% biodiesel in petrodiesel are broadly accepted for use in existing diesel engines by engine and fuel injection equipment manufacturers [20].

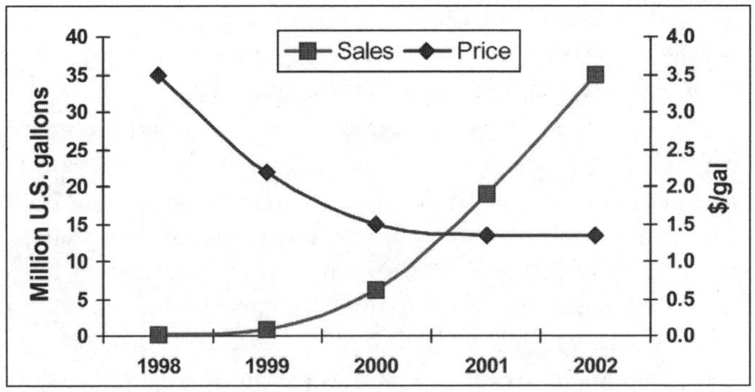

Figure 12.10 *Sales and price of biodiesel in the United States. 1 U.S. gallon = 3.785 liter. (Reprinted from Ref. 3.)*

The biggest long-term barrier to the widespread application of biodiesel is its limited production potential. In the United States, biodiesel resources are estimated at about 2 billion gallons per annum (7.6 billion liters), a fairly insignificant figure when compared to the global U.S. distillate fuel market of some 60 billion gallons [3]. Unless ongoing R&D identifies new sources of triglycerides for its production, biodiesel can be considered only a niche market alternative fuel option.

12.3.2 Manufacturing Process

In general, biodiesel can be manufactured using one of the following processes:

- Base-catalyzed transestrification of the oil with alcohol
- Direct acid-catalyzed estrification of the oil with alcohol
- Conversion of the oil to fatty acids, and then to alkyl esters by means of acid catalysis.

The most commonly used process, due to is superior economy, is the base-catalyzed reaction. In this process, biodiesel is produced through a transestrification reaction of a natural oil triglyceride (vegetable oil or animal fat) with a short-chain alcohol (typically methanol) in the presence of a catalyst (usually sodium or potassium hydroxide), as shown in the following equation:

$$(CH_2OOCR)(CHOOCR')(CH_2OOCR'') + 3CH_3OH \rightarrow (CH_2OH)(CHOH)(CH_2OH) + 3RCOOCH_3 \quad (12.1)$$

The reaction occurs stepwise, with one fatty acid chain being removed first (forming one monoalkyl ester and a diglyceride), the second fatty acid removed next (forming two monoalkyl esters and a monoglyceride), and, lastly, reaction of the third fatty acid. The resulting products are glycerin and three monoalkyl esters (biodiesel) with carbon chains of typically 12 to 22 atoms [21]. Many oil and fat feedstocks used for manufacturing of biodiesel produce fairly uniform straight carbon chains of 16 to 18 atoms. The similarity of the chemical species that form biodiesel results in a very narrow boiling temperature range, often resembling the boiling point of a pure compound, as opposed to the wide distillation temperature range of petrodiesel. The biodiesel "boiling point" is typically anywhere between 330 and 357°C, depending on the feedstock [22].

Many oils used for the manufacture of biodiesel contain varying amounts of free fatty acids. In some feedstocks, for example, those in used oils, the fatty acid content may be as high as 30%. The fatty acids may be partially estrified by methanol, but in the base-catalyzed process they form quantities of fatty acid soap by-product. A formation of potassium soap is shown in Eq. (12.2). Various methods have been developed to convert these soaps back to fatty acids, according to Eq. (12.3), for recycling to the reactor. Alternatively, in some processes the feedstock is preprocessed with alkali to remove free fatty acids prior to the transestrification with methanol.

$$RCOOH + KOH \rightarrow RCOOK + H_2O \qquad (12.2)$$

$$RCOOK + HCl \rightarrow RCOOH + KCl \qquad (12.3)$$

A general biodiesel production schematic is shown in Figure 12.11. The primary components and the catalyst are fed into the reactor, where the triglycerides are reacted with methanol to produce the ester, glycerol, and fatty acid soaps. The soaps, which are usually washed out of the ester with water, can be either disposed of or converted back to fatty acids and recycled to the reactor. The glycerol fraction can be further treated to recover high-purity glycerin and a small amount of potassium fertilizer material.

The critical items in the determination of biodiesel quality and production process monitoring are as follows:

- Complete reaction to the monoalkyl esters (monitored via the total and free glycerin tests)
- Removal of free glycerin (free glycerin test)
- Removal of residual catalyst (sulfated ash test)
- Removal of reactant alcohol (assessed by flash point measurement)
- Absence of free fatty acids (acid value test).

12.3.3 Properties and Specifications

The final ASTM specification D6751 for biodiesel was adopted in the United States in 2002 [22] (superseding the earlier, provisional specification PS121 [23]). The D6751 standard covers biodiesel (B100) for use as a blend component with petroleum diesel fuels. The specification is based on the existing ASTM standard for petrodiesel, D975, which was modified by elimination of items not applicable to biodiesel and by addition of items specific to biodiesel. Development work on new analytical methods for a number of biodiesel properties, such as oxidative stability, was initiated in relation to the ASTM biodiesel standard [24]. Requirements for biodiesel blend stock, as defined by ASTM, are listed in Table 12.4.

Draft biodiesel standards also exist in a number of European countries. In this group of standards, one of the most complete biodiesel specifications is provided by the German draft standard DIN EN 14214, *Automotive Fuels—Fatty Acid Methyl Esters (FAME) for Diesel Engines—Requirements and Test Methods.*

The average properties of neat biodiesel fuels compiled from various literature data are compared with those of petroleum diesel in Table 12.5. The biodiesel data included soybean and rapeseed oil-based fuels, as well as animal-based biodiesels. The petrodiesel properties were based on fuels sold in the United States outside California.

Properties of different biodiesel fuels reported in the literature tend to be similar, with the exception of the cetane number. Figure 12.12 illustrates the variation in cetane number in various neat biodiesel fuels, compared with petroleum diesel [25]. One point, indicated by a square in the top left part of the chart, represents the properties of a virgin oil fuel, which are considerably different from those of ester-based biodiesels.

In comparison with petrodiesel, biodiesel is characterized by:

- No sulfur or ultralow sulfur content
- No aromatics contents (and no polycyclic aromatic hydrocarbons [PAHs])
- About 11% oxygen content (petrodiesel contains no oxygen)

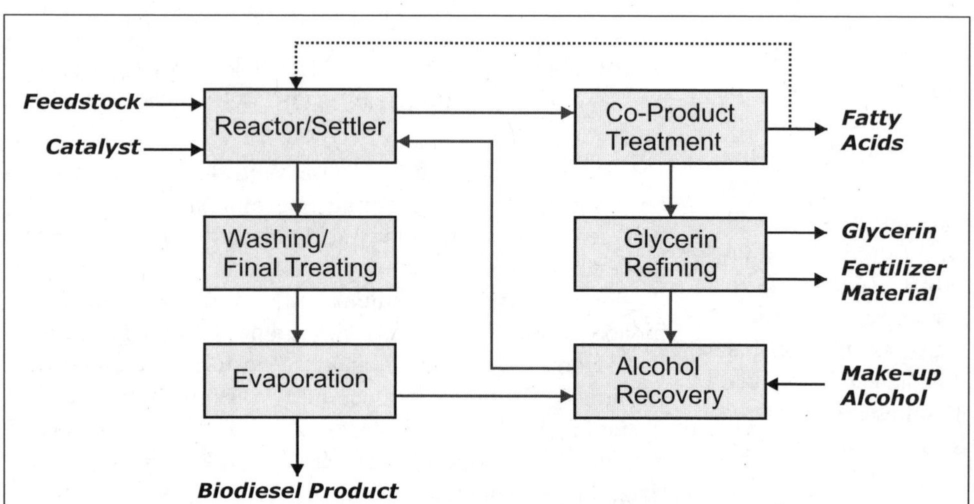

***Figure 12.11** Simplified diagram of biodiesel production process. (Based on data from Ref. 21.)*

TABLE 12.4
U.S. BIODIESEL BLEND STOCK SPECIFICATIONS (ASTM D6751)

Property	ASTM Method	Value
Flash point, °C	D93	130.0 min.
Water and sediment, vol. %	D2709	0.050 max.
Kinematic viscosity @ 40°C, mm²/s	D445	1.9–6.0
Sulfated ash, wt. %	D874	0.020 max.
Sulfur, wt. %	D5453	0.05 max.
Copper strip corrosion	D130	No. 3 max.
Cetane number	D613	47 min.
Cloud point, °C	D2500	Report to customer
Carbon residue (on 100% sample), wt. %	D4530	0.050 max.
Acid number, mg KOH/g	D664	0.80 max.
Phosphorus content, wt. %	D4951	0.001 max.
Free glycerin, wt. %	D6584	0.020 max.
Total glycerin, wt. %	D6584	0.240 max.
Distillation temperature, 90% recovered, °C	D1160	360 max.

Source: Data from Ref. 22.

TABLE 12.5
COMPARISON BETWEEN BIODIESEL AND PETRODIESEL PROPERTIES

Property	Biodiesel	Petrodiesel
Natural cetane number	55	44
Sulfur, ppm	54	333
Nitrogen, ppm	18	114
Aromatics, vol. %	0	34
T10, °C	331	217
T50, °C	343	263
T90, °C	352	317
Specific gravity	0.88	0.85
Kinematic viscosity @ 4.4°C, mm²/s	6.0	2.6

Source: Data from Ref. 25.

- Higher cetane value (typically 45 to 60)
- Lower heating value
- Better lubricity
- Higher viscosity
- Higher freezing temperature
- Higher flash point
- No toxicity or low toxicity
- Biodegradability
- Different corrosive properties.

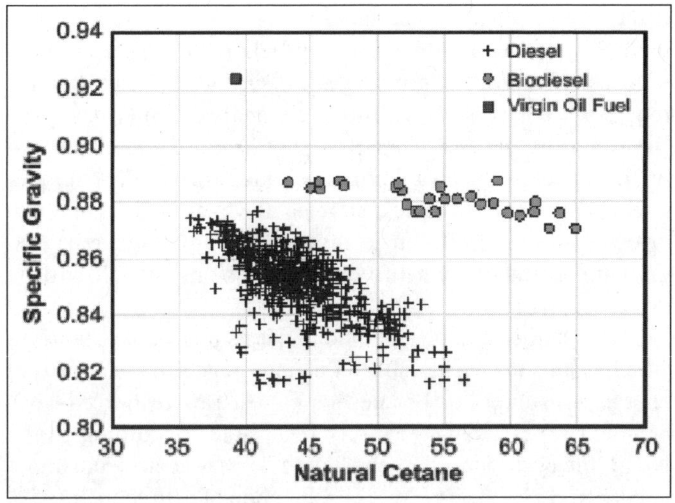

Figure 12.12 *Cetane variability in biodiesel and petrodiesel fuels. (Reprinted from Ref. 25.)*

Some of these characteristics, such as the high cetane value or good lubricity, are obvious advantages of biodiesel, while others, including the lower heating value, high freezing temperature, corrosion properties, or flow properties (at low temperature), are drawbacks.

The typical heating value (LHV) for biodiesel is 32.6 MJ/dm³ (117,000 Btu/gal), as compared to 36.2 MJ/dm³ (130,000 Btu/gal) for the U.S. No. 2 diesel [3]. As a result, a loss of engine power of about 8% is measured with neat biodiesel, and of about 2% with the B20 blend [26]. A fuel consumption penalty as high as 13% and more was reported with heavy-duty engines over the U.S. FTP Transient test cycle [26].

Sulfur content in biodiesel depends on the feedstock used for its manufacturing. Practically sulfur-free biodiesel is obtained from vegetable oils. Fuels manufactured from waste oils or animal fats may contain small amounts of sulfur, typically below 50 ppm. For example, the typical sulfur content of soybean-based biofuel is below 1 ppm, whereas biodiesel from waste restaurant grease (yellow grease) has about 30 ppm sulfur [3].

The biodegradability of biodiesel, an obvious advantage from the environmental point of view (fuel spills, and so on), is also a drawback for everyday engine use. The higher the concentration of biodiesel in a fuel blend, the more susceptible the fuel is to degradation and water absorption.

Last but not least, biodiesel changes the character and can dramatically increase the intensity of the odor of diesel exhaust. Authors who developed methods to quantify exhaust odors measured an elevation of odor levels of up to 350% when using RME fuel [27].

12.3.4 Biodiesel Emissions

12.3.4.1 Life Cycle Analysis

Life cycle analyses look at the entire picture of how a fuel is made, from "cradle to grave." The life cycles begin with the extraction of all raw materials to make petroleum diesel and biodiesel, and end with using the fuels in a vehicle application. Examination of global issues, such as CO_2 emissions, requires a comprehensive life cycle analysis. Such analysis, sponsored by the U.S. Department of Energy, was conducted by the National Renewable Energy Laboratory [28]. A number of emission issues of local character, however, such as in urban or occupational health environments, require an examination of tailpipe, rather than life cycle, emissions, which will be discussed later.

The major advantage of biodiesel is its renewable character, which results in considerable reductions in petroleum and fossil energy consumption. Biodiesel's largest raw resource—soybeans or rapeseeds—is entirely renewable. According to the NREL life cycle analysis, substituting B100 for petroleum diesel in buses reduces the life cycle consumption of petroleum by 95% [28]. In terms of energy, biodiesel yields about 3.2 units of fuel product energy for every unit of fossil energy consumed in its life cycle. By contrast, 1.2 units of fossil resources are needed to produce 1 unit of petroleum diesel.

Due to the primarily renewable character of carbon in the fuel, biodiesel provides substantial reductions in life cycle carbon dioxide emissions. Carbon is biologically cycled when plants such as soybean crops convert atmospheric CO_2 to carbon-based compounds through photosynthesis. Biomass-derived fuels participate in the relatively rapid biological cycling of carbon to the atmosphere (via engine tailpipe emissions) and from the atmosphere (via photosynthesis). Fossil fuel combustion, on the other hand, releases carbon that took millions of years to be removed from the atmosphere. For this reason, the net effect of shifting from fossil fuels to biomass-derived fuels is to reduce the amount of CO_2 in the atmosphere.

An estimate of CO_2 emissions for petroleum and biodiesel fuels based on a heavy-duty bus application is shown in Table 12.6 [28]. Figures listed as "life cycle CO_2" include a number of nontailpipe contributions, such as industrial emissions that occurred during the production and distribution of fuels. For example, the difference between the total life cycle and the total tailpipe emission for petrodiesel represents primarily emissions from the refinery, which contribute 9.6% of the total CO_2 emissions. The small tailpipe fossil CO_2 listed for biodiesel is attributed to the methanol contribution.

The net reduction of the life cycle CO_2 due to biodiesel use amounts to 78% (compare figures in the "Total life cycle fossil CO_2" column). This effect is entirely due to the biological recycling of CO_2; the tailpipe emissions are actually higher for biodiesel than for petrodiesel.

According to the same source, other air pollutant emissions from petro- and biodiesel can be compared as follows (urban bus application). Note again that these numbers represent life cycle emissions, which include emissions in the production of fuels and are different from tailpipe emissions.

- Carbon monoxide reduced with B100 by 35%
- Total particulate matter reduced by 32%
- Sulfur oxides reduced by 8%
- Nitrogen oxides increased with B100 by 13% (mostly due to a tailpipe increase)
- Hydrocarbons increased by 35% (but reduced at the tailpipe)

12.3.4.2 Tailpipe Emissions: General Trends

Numerous studies have been conducted throughout the world on the tailpipe exhaust emissions from biodiesel fuels and their blends. The results of these studies are often inconclusive or contradictory, making it difficult to quantify emission trends. The observed scatter of data may be attributed to the following reasons:

- Variability exists between the properties of various biodiesel fuels, especially in their cetane number, as well as between properties of the petroleum blending stock in the case of blends.
- Studies have been conducted on different types of engines (heavy-duty on-road, nonroad, light-duty) which, in general, may show different emission trends with biodiesel.
- Studies use different test cycles, including an array of steady-state and transient testing conditions which result in obvious emission differences.
- For a valid emission comparison, engines should be recalibrated to their original power output to account for the lower heating value of biodiesel—a requirement that is too often neglected.

TABLE 12.6
CO_2 EMISSIONS FOR PETROLEUM DIESEL AND BIODIESEL (g CO_2/bhp-hr)

Fuel	Total Life Cycle Fossil CO_2	Total Life Cycle Biomass CO_2	Total Life Cycle CO_2	Tailpipe Fossil CO_2	Tailpipe Biomass CO_2	Total Tailpipe CO_2	Total CO_2 from Tailpipe
Petroleum diesel	633.28	0.00	633.28	548.02	0.00	548.02	86.54%
Biodiesel B100	136.45	543.34	679.78	30.62	543.34	573.96	84.43%

Source: Data from Ref. 28.

- Last but not least, optimal utilization of biodiesel would require that the engine combustion process be specifically tuned for that fuel. In practice, studies test biodiesel in engines designed and calibrated for petroleum diesel fuels. If engines were specifically calibrated for biodiesel, higher emission benefits would likely be possible, as suggested by both commonsense and by the high degree of engine specificity among the existing emission data.

Degradation effects present another uncertainty related to biodiesel emissions. It can be assumed that all of the published studies were carried out using fresh biodiesel fuels. However, biodiesel is susceptible to degradation, the emission effect of which remains unknown.

The apparent influence of biodiesel on emissions relative to petrodiesel is summarized in Table 12.7.

There is wide agreement in the literature that biodiesel and its blends decrease the emissions of both CO and HC. This effect is attributed to the oxygen content in biodiesel, which enables more complete oxidation in the engine cylinder. The magnitudes of these reductions vary, with some reports showing a 40% CO emission reduction and total elimination of HCs with neat biodiesel [26]. It was also suggested that higher CO and HC emission reductions are seen at higher engine loads. A study investigating the effect of biodiesel on DI engine emissions at different engine operating conditions found an increase of both CO and HC at low loads [29].

NO_x emissions are increased with biodiesel, typically by more than 10% [26, 30, 31]. This effect is associated by most authors with the high oxygen content of biodiesel, but no convincing combustion mechanism has been adopted. NO_x emissions in B20 biodiesel blends are typically increased by 2% to 4% [3].

TABLE 12.7
EFFECT OF BIODIESEL FUEL ON TAILPIPE EMISSIONS

Emission	Effect of Biodiesel[a]
Regulated Emissions	
Carbon monoxide	Decrease
Hydrocarbons	decrease
Nitrogen oxides	Increase
Total particulate matter (TPM)	Decrease
Unregulated Emissions	
Carbon particulates	Decrease
Organic particulates (SOF)	Increase
Sulfate particulates	Decrease
Visible smoke	Decrease
PAH, nitro-PAH	Decrease
Aldehydes	Inconclusive

a. Increase: biodiesel increases emissions relative to petrodiesel; decrease: biodiesel decreases emissions relative to petrodiesel.

Biodiesel decreases the carbon particulate emissions and increases the SOF. As a result, the visible smoke and opacity are decreased. The effect of biodiesel on TPM depends on the composition of diesel particulates from the particular engine. Most studies reported a decrease in TPM emission with biodiesel, in some cases by as much as 25% to 50% [26]. Some authors, however, saw an increase in TPM [32]. If increased TPM emissions occur, they are caused by an increased SOF fraction, which was reported to be as high as 60% to 90% at low engine loads and 10% to 40% at high loads [29]. Because biodiesel does not contain sulfur, the sulfate emissions are very low. The shift of PM emission toward higher SOF content, as well as the absence of sulfur, makes biodiesel compatible with diesel oxidation catalysts, which can maximize its PM benefit.

Emissions of PAH and nitro-PAH were found by most authors to be significantly lower than those observed with petrodiesel [30, 33]. Also the toxicological impact of diesel exhaust on animal lung tissue was reported to be reduced with increasing supplementation of the fuel with biodiesel [34].

Biodiesel emissions of aldehydes appear to depend on the type of engine and test cycle. Both aldehyde decreases [30] and increases [32], relative to petrodiesel, have been reported.

An increase in nanoparticle number emissions was reported with RME biodiesel [32]. This increase might have been caused by the relatively high SOF emissions with biodiesel, resulting in nucleation of hydrocarbon condensates and formation of nanoparticles.

A comprehensive summary of biodiesel emission effects was compiled by the U.S. Environmental Protection Agency (EPA) as a guidance for states in claiming emission credits for the use of biodiesel and its blends [25]. The EPA selected 39 literature studies, out of 80 studies included in the analysis, which were conducted on the FTP Transient test cycle using commercial, heavy-duty, on-road engines and commercial fuels. Figure 12.13 illustrates the correlation between the percentage

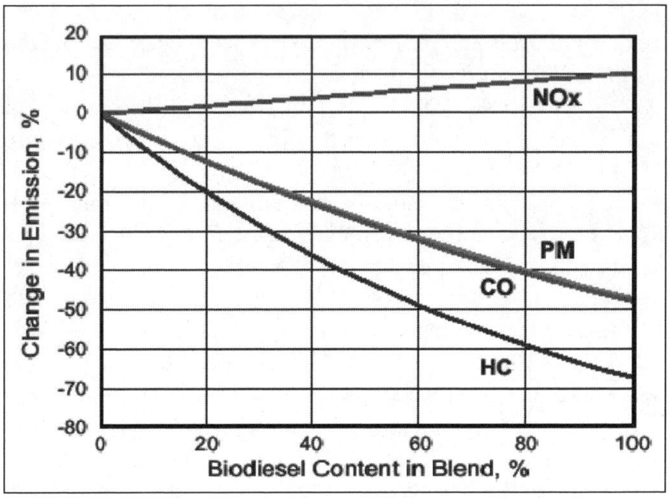

Figure 12.13 *Average impact of biodiesel on emissions from heavy-duty engines subjected to the U.S. FTP Transient test. (Reprinted from Ref. 25.)*

emission impact and the biodiesel content in the blend. These correlations are statistically valid, although the source data showed significant scatter.

Light-duty and nonroad engine emissions were predicted poorly by the above correlations, but the data set was too limited to allow for statistically valid generalizations. The steady-state test cycle used for nonroad engine testing (not properly representing the real-life operating conditions for most nonroad machinery) was likely an important source for the different emission response.

The EPA analysis also looked into the biodiesel effect on several substances classified as toxic air pollutants, including formaldehyde and acetaldehyde, benzene, 1,3-butadiene, *n*-hexane, toluene, and others. In some cases (aldehydes, naphthalene, ethylbenzene, xylene), statistically significant reductions were determined with increasing biodiesel content in blends. The magnitude of the emission reduction, however, was considerably smaller than the HC effect. In other cases, no clear emission trends could be found.

In summary, biodiesel and its blends generally reduce most of diesel emissions. The only regulated emission that has been shown to consistently increase with biodiesel is nitrogen oxides. Even though the NOx increases are small, they did cause concerns about biodiesel use, especially in environmentally sensitive areas. Various strategies have been suggested to mitigate the NO_x increase, such as using a lower emitting base fuel for biodiesel blending, adding a cetane improver additive to the biodiesel blend, or determining what properties of biodiesel can be modified to lower NO_x emissions [35].

12.3.4.3 Example Studies

An emission comparison between low-sulfur petrodiesel (S < 0.05%) and RME was performed on a Volvo THD 103KF bus engine [36]. It was a Euro II, six-cylinder, DI, turbocharged, intercooled diesel engine rated for 170 kW @ 2000 rpm. Measurements were taken at several steady-state engine modes. Regulated emissions with both fuels are shown in Table 12.8. The results were calculated as weighted averages from 5 test modes with weighting factors selected so as to approximate the results of the full R-49 13-mode test.

TABLE 12.8
REGULATED EMISSIONS WITH RME FUEL AND PETROLEUM DIESEL, g/kWh

Emission	RME Biodiesel	Petroleum Diesel
NO_x	6.81	5.54
HC	0.32	0.55
CO	0.69	0.79
TPM	0.27	0.24
Fuel consumption	262	229

Source: Data from Ref. 36.

Increased NO_x and TPM emissions and decreased CO and HC were observed using the RME fuel. The increase in TPM was due to high SOF emissions with the RME fuel. It was concluded in the study that particulate emissions from RME fuel were composed mainly of SOF. Opacity measurements using RME were about 50% lower than those with diesel fuel, confirming low carbonaceous particulate emissions with biodiesel. Measurements of other unregulated emissions indicated about 10 times lower PAH levels with RME, comparable benzene-toluene-xylene emissions with both fuels, and higher aldehyde emissions using RME. Concentrations of formaldehyde and acetaldehyde, as measured at 50% engine load and intermediate speed, were about two times higher using biodiesel compared to petrodiesel.

In another experimental study, emission performance of soy methyl ester fuel was compared with that of the conventional diesel fuel [37]. Evaluations were done on a Caterpillar 3304 PCNA, 7-L, four-cylinder, IDI diesel engine. Emission tests were performed on two transient cycles: a "light-duty" and a "heavy-duty" cycle with maximum torques of 50% and 90%, respectively, of the engine rated torque. The fuels tested included a commercially available low-sulfur, No. 2 petroleum diesel fuel (D2), and distilled (>99% pure) soy methyl esters. Properties of the fuels are listed in Table 12.9.

Particulate matter emissions measured with petrodiesel (D2) and with soy methyl ester (SME) are shown in Figure 12.14. The graph compares emissions from light-duty and heavy-duty test cycles, without and with a diesel oxidation catalyst. The SME fuel greatly decreased (>50% reduction) the nonvolatile (NVOL) constituents of TPM, which are primarily the carbonaceous fraction of diesel particulates. The volatile organic (VORG) constituents of TPM (approximately equivalent to SOF) were increased by the soy fuel, but the decreases in NVOL were high enough to cause an overall decrease in TPM emissions. The diesel oxidation catalyst was effective at reducing the VORG for both diesel and soy fuels over the light-duty cycle. On the heavy-duty cycle, due to high exhaust temperatures, the

TABLE 12.9
PROPERTIES OF SOY METHYL ESTER FUEL

Property	Soy Methyl Esters	Diesel Fuel D2
Cetane number (D287)	54.7	43.2
Lower heating value, MJ/kg	37.0	42.7
C:H:O molar ratio	19:34:2	16:30:0
Kinematic viscosity @ 40°C, mm^2/s	3.0	2.4
Sulfur, wt. %	0.00	0.01
Oxygen, wt. %	11.0	0
Aromatics, wt. %	0	39
Cloud point (D2500), °C	-2	-19
Flash point (D93), °C	160	68

Source: Data from Ref. 37.

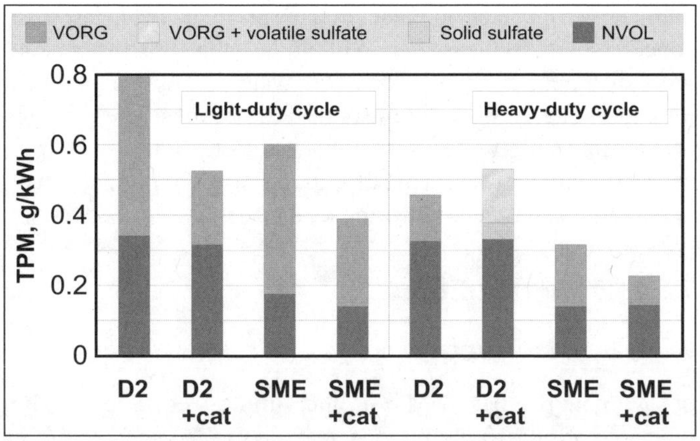

Figure 12.14 DPM emissions with soy methyl ester fuel. (Based on data from Ref. 37.)

DOC produced sulfate particulates and increased TPM emissions when petroleum diesel fuel was used. For the soy fuel that contained no sulfur, the catalyst efficiently reduced TPM emissions even at high exhaust temperatures.

Perhaps the most comprehensive laboratory evaluation of regulated and unregulated diesel emissions was conducted under the auspices of the U.S. National Biodiesel Board [26, 33]. Tests were carried out using three heavy-duty engines: a 1997 Cummins N14 (276 kW), a 1997 DDC Series 50 (205 kW), and a 1995 Cummins B5.9 (119 kW). Emissions were measured over the FTP Transient test cycle. The major results can be summarized as follows:

- Neat biodiesel was associated with lower engine-out CO (by 40%), HC (reduced below detection level), and PM (by 25% to 50%).
- NO_x emissions increased by 12% on two of the tested engines. However, no NO_x increase was recorded on the Cummins B5.9 engine, which was the only engine with a low-pressure, pump-line-nozzle fuel injection system. The authors commented that the NO increase mechanisms with biodiesel are not entirely understood and require more research.
- Biodiesel reduced emissions of speciated vapor phase HCs in the C_1 to C_{12} range and emissions of aldehydes and ketones, PAHs, and nitro-PAHs.
- Biodiesel blends resulted in lower emission effects, roughly proportionally to the biodiesel content.

Biodiesel blends were also investigated in underground mines as a potential strategy to control PM emissions [38]. The test vehicle, a Wagner ST-8a scooptram powered by a Deutz F12L 413FW diesel engine rated at 204 kW at 2400 rpm, was operated in a nonproducing section of an underground mine. During the first week of the evaluation, the vehicle used low-sulfur, No. 2 diesel fuel (D2). During the second week, the scoop was operated on a 58% (by mass) blend of soy-based biodiesel fuel and a low-sulfur D2. During both weeks the scoop exhaust system was equipped with a pair of identical diesel oxidation catalysts (DOCs) designed to reduce the SOF fraction of diesel particulates, as well as gaseous HC and CO.

Air samples collected in the test section demonstrated that the combination of the blended biodiesel fuel and DOC decreased total carbon emissions by about 21.4%, compared to the combination of petrodiesel and DOC. Elemental carbon was reduced by 28.6% and organic carbon was reduced 6.0%. There was a slight increase in NO_2 concentrations measured at the downwind location and a corresponding decrease in NO emissions.

12.3.5 Compatibility with Petroleum Diesel Engines

Whenever the fuel composition is changed in a fuel system, material compatibility is a major concern. Fuel system designers incorporate materials based on laboratory testing and the available historical data. Thus, changes in fuel composition and the introduction of alternative fuels often create unforeseen problems in seals, gaskets, O-rings, and metallic components in the fuel system. The consequences of changing the fuel composition without prior compatibility studies are best illustrated by the reduction of aromatics in low-sulfur diesel fuel in California. This "minor" change resulted in numerous problems due to O-ring shrinkage, including leakage in pumps and other fuel system components [39].

Material compatibility of biodiesel and its blends with elastomers and metals was investigated in an experimental laboratory study [40]. Samples of tested elastomers were stored in biofuel blends at 51.7°C for up to 694 h. The tensile strength, elongation, hardness, and swelling were determined for each sample after the storage period. It was concluded that the physical properties of nitryle rubber, nylon 6/6, and high-density polypropylene were affected by biodiesel and its blends while Teflon, Viton 401-C, and Viton GFLT appeared to have good resistance to biodiesel.

Metals evaluated in the study included copper, steel, brass, aluminum, and bronze. Samples of metals were stored in the biodiesel and its blends for 6 months at 51.6°C. The samples were rated visually and the total acid number (TAN) of the fuels was determined. Copper-containing metals exhibited severe corrosion when exposed to biodiesel and its blends. They also formed large amounts of gum when stored at elevated temperatures. The steel and aluminum samples did not exhibit gum formation, but did exhibit exceptionally high TANs, which could corrode various parts of the fuel system.

Faced with the increasing usage of biodiesel and with incidents of related engine problems, the major manufacturers of diesel fuel injection equipment (FIE) have reached a common position on the use of biodiesel [20]. The FIE manufacturers stated that blends of up to 5% of biodiesel in mineral diesel fuel should not create any serious engine problems, provided the biodiesel fuel meets existing national biodiesel standards at the point of sale. Injection equipment for blends of greater than 5% biodiesel, including 100% biodiesel fuel, can be supplied by the manufacturers. However, using high-percentage biodiesel blends in engines that were not designed for such fuel or using biodiesel fuels that are not compliant with applicable national standards will render the FIE manufacturer's guarantee null and void. The key biodiesel issue for the FIE manufacturers is resistance to oxidation. Aged or poor-quality biodiesel fuel may contain organic acids, free water, peroxides, and products of polymerization that can attack many components, reducing the service life of fuel injection equipment. The list of potential issues identified by FIE manufacturers is given in Table 12.10.

The 5% biodiesel limit in fuel blends is becoming a widely adopted industry standard. For instance, fuel blends containing up to 5% biodiesel are accepted by engine manufacturers who allow for use of biodiesel in their products [41]. The main concern, again, is the oxidation of biodiesel leading to a gradual increase in contamination and free water content. To prevent moisture, engine operators using biodiesel must keep storage and vehicle tanks as full as possible, protect storage tanks from extreme temperatures, avoid extended storage of biodiesel fuel, and conduct routine monitoring of the fuel's water content.

12.4 Dimethyl Ether

12.4.1 Introduction

Dimethyl ether (DME) is the simplest ether, consisting of two methyl groups bonded to a central oxygen atom, as

TABLE 12.10
FUEL INJECTION EQUIPMENT: POTENTIAL PROBLEMS WITH BIODIESEL

Fuel Characteristic	Effect	Failure Mode
Fatty acid methyl esters (general)	Causes some elastomers including nitrile rubbers to soften, swell, or harden and crack	Fuel leakage
Free methanol in biodiesel	Corrodes aluminum and zinc Low flash point	Corrosion of FIE
Biodiesel process chemicals	Potassium and sodium compounds Solid particles	Blocked nozzles
Dissolved water in biodiesel	Reversion of biodiesel to fatty acid	Filter plugging
Free water in mixtures	Corrosion Sustains bacteria Increases the electrical conductivity of fuel	Corrosion of FIE Sludging
Free glycerine, monoglyceride, and diglyceride	Corrodes nonferrous metals Soaks cellulose filters Sediments on moving parts and lacquering	Filter clogging Injector coking
Free fatty acid	Provides an electrolyte and hastens the corrosion of zinc Salts of organic acids Organic compounds formed	Corrosion of FIE Filter plugging Sediments on parts
Higher modulus of elasticity	Increases injection pressure	Potential for reduced service life
High viscosity at low temperature	Generates excessive heat locally in rotary distributor pumps Higher stressed components	Pump seizures Early life failures Poor nozzle spray atomization
Solid impurities	Potential lubricity problems	Reduced service life
Aging Products		
Corrosive acids (formic and acetic)	Corrodes all metallic parts	Corrosion of FIE
Higher molecular organic acids	Similar to fatty acid	
Polymerization products	Deposits especially from fuel mixes	Filter plugging Lacquering formation in hot areas

Source: Data from Ref. 20.

expressed by its chemical formula CH_3OCH_3. DME, most commonly produced from natural gas, provides a possible avenue for the utilization of various natural gas reserves that are inaccessible by pipelines, in competition to such technologies as the manufacturing of synthetic fuels through Fischer-Tropsch synthesis. DME has replaced CFC gases (Freons) as an environmentally friendly and safe aerosol propellant, which is one of its major current applications. Other potential uses of DME include a substitute for other fuels in power generation and in the household and, in the future, a source of hydrogen for fuel cells [42]. In the 1990s, DME was produced worldwide in a quantity of 100,000 to 150,000 tons per annum [43].

With a chemical structure somewhat similar to that of methanol, DME contains oxygen and no carbon–carbon bonds, thus seriously limiting the possibility of forming carbonaceous particulate emissions during combustion. However, unlike methanol, DME has a high enough cetane number to perform well as a compression-ignition fuel. Also unlike methanol, DME is a gas at ambient temperature and pressure, so it must be stored under pressure as a liquid similar to liquefied petroleum gas (LPG). When used as a diesel fuel, DME provides reduced NO_x and PM emissions, but increased CO and HC, but these can easily be controlled by an oxidation catalyst. Preliminary studies in the 1990s concluded that it should be possible to achieve ULEV emissions using a properly designed DME-based fuel injection system with an HSDI engine and an oxidation catalytic converter in a passenger car [44].

The physical properties of DME (density, viscosity, lubricity, etc.) are so different from diesel fuel that the entire fuel system must be redesigned [45]. Although it seems clear that the use of DME, perhaps like some other alternative fuels, would result in much larger emissions reductions than is possible with diesel fuel, the emission benefit comes at the price of a specific level of complexity of the fuel storage and injection system, including the need to carry a pressurized fuel tank on board the vehicle. Furthermore, it is not clear that the apparently inherent emission advantage of DME can offset the fuel's lack of established supply and fueling infrastructure. From today's perspective, DME fuel is more likely to be used in certain niche applications, rather than provide a widescale alternative to liquid diesel fuels.

12.4.2 Manufacture of DME

DME is most commonly produced from synthesis gas, composed of H_2, CO, and CO_2, in a catalytic process that is very similar to the production of methanol. The most cost-effective feedstock for both DME and methanol is natural gas at remote locations (as opposed to pipeline natural gas), but many other carbon-containing feedstocks are possible. The first stage is the production of synthesis gas via steam reforming (from natural gas) or via gasification (from coal, oil residues, biomass, wood). The second stage is methanol synthesis, in which, using a special catalyst, methanol can be simultaneously converted to DME.

DME production technologies have been developed by a number of manufacturers, such as Haldor Topsoe (HT) [46, 47] and NKK [42, 48]. The HT process, illustrated in Figure 12.15, consists of the following main steps:

- Desulfurization
- Autothermal reforming (ATR)
- CO_2 adjustment
- Combined methanol and DME synthesis
- Final DME purification.

The feedstock natural gas is first desulfurized to protect the catalysts in the downstream units. Recycled hydrogen is added to the natural gas to convert heavy, refractory sulfur compounds into hydrogen sulfide, which is subsequently absorbed by activated ZnO. The autothermal reformer is a key element in the production technology. It consists of a refractory lined pressure vessel with a burner, a combustion chamber, and a catalyst bed. A more detailed description can be found in Ref. [46].

The desulfurized natural gas is reacted in the reformer with oxygen and steam to produce synthesis gas, according to the following reactions:

$$CH_4 + 1.5\,O_2 \rightarrow CO + 2\,H_2O \qquad (12.4)$$

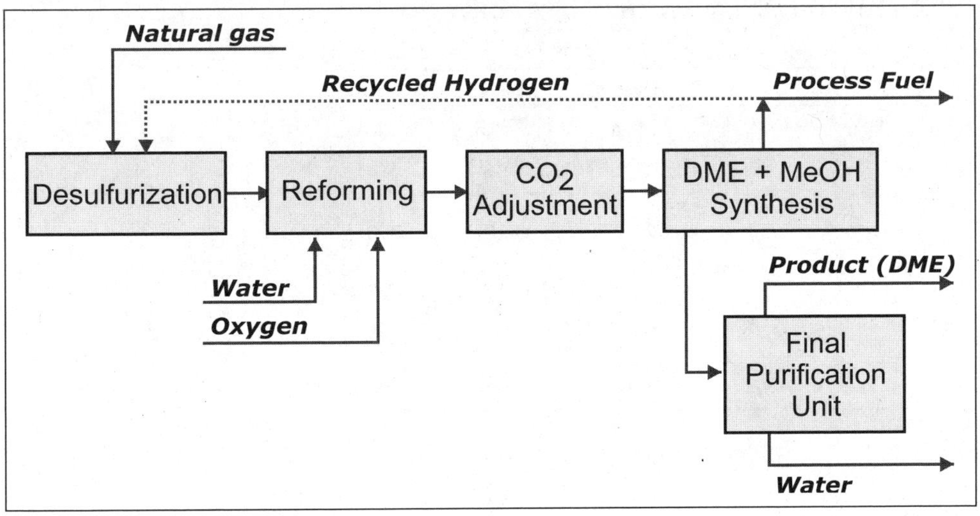

Figure 12.15 *Simplified diagram of the DME production process.*

$$CH_4 + H_2O \rightarrow 3\,H_2 + CO \quad (12.5)$$

$$CO + H_2O \rightarrow CO_2 + H_2 \quad (12.6)$$

Subsequently, some CO_2 is removed in an absorption tower in order to adjust the synthesis gas composition to an optimum value, as defined by the module $(H_2 - CO_2)/(CO + CO_2)$ of slightly above 2. In the following synthesis unit, composed of three adiabatic reactors in series, the following reactions take place over a multifunction catalyst:

$$3\,H_2 + CO_2 \rightarrow CH_3OH + H_2O \quad (12.7)$$

$$H_2O + CO \rightarrow H_2 + CO_2 \quad (12.8)$$

$$2\,CH_3OH \rightarrow CH_3OCH_3 + H_2O \quad (12.9)$$

Most of the produced DME/methanol/water mixture is then condensed and separated. The unconverted synthesis gas is split into two streams, one of which is used as the recycled hydrogen, the other as process fuel. The DME product is purified by distillation. A number of advancements have been introduced to the HT manufacturing process, including changes in the ATR design and elimination of the CO_2 removal step before the synthesis reactor [47]. The process can produce neat DME or a raw fuel-grade DME, which can be composed of specific blends of DME, methanol, and water.

The DME production technology developed by NKK is similar in principle to the HT process. The main process units are (1) a reformer for the generation of synthesis gas, (2) a CO_2 absorber, and (3) a slurry bed reactor for the DME synthesis followed by (4) a distillation unit for separation of methanol and DME purification. The slurry bed reactor contains a high-boiling-point solvent as the reaction medium, in which fine catalyst particles are mixed. More information on the NKK process can be found in Ref. [48].

12.4.3 DME as Automotive Fuel

12.4.3.1 Properties

Dimethyl ether (CAS #115-10-6) is a colorless gas at room temperature with an ethereal odor, has a vapor pressure of 0.593 MPa (4450 mmHg) at 25°C, and is highly flammable in air (3.4% to 18%) [49]. It can be easily handled and stored as liquefied gas, in the form of a colorless clear liquid, similar to LPG. DME has no corrosive effect on metals, but it is an excellent solvent that can dissolve a number of elastomers, including those used in conventional diesel fuel systems.

Due to its low autoignition temperature, dimethyl ether can be used as a fuel for diesel (compression ignition) engines. DME is characterized by a relatively high cetane number of 55 to 60, but an inferior heating value compared to diesel fuel. The physical and chemical properties of DME are compared with those of other fuels in Table 12.11 [50]. The autoignition temperature, 350°C in Table 12.11, depends on test conditions; the value of 235°C is also reported for DME [42].

The properties listed in Table 12.11 refer to chemically pure DME. An internationally recognized specification for fuel-grade DME developed by Haldor Topsoe is listed in Table 12.12 [47]. Fuels meeting this standard should not cause additional compatibility problems with materials used in fuel systems compared to the use of pure DME.

12.4.3.2 Energy Efficiency

The overall energy efficiency of engine fuels is best expressed as the "well-to-wheel" efficiency. The well-to-wheel energy efficiency is a product of (1) the well-to-filling station efficiency (including fuel recovery, production, and distribution), (2) the

TABLE 12.11
COMPARISON OF DME WITH OTHER FUELS

Property	Fuel			
	DME	Diesel	LNG	Methanol
Formula	CH_3OCH_3	N/A	CH_4	CH_3OH
Boiling temperature, °C	-25	~150–380	–162	65
Liquid density @ 20°C, kg/m^3	660	800–840	420	790
Lower heating value, MJ/kg	28.4	42.5	49.4	19.5
Cetane number	55–60	40–55	N/A	~5
Autoignition temperature,[a] °C	350	206[b]	732	470

Source: Data from Ref. 50.
a. With air at 0.1 MPa.
b. *n*-Cetane.

TABLE 12.12
SPECIFICATION OF FUEL-GRADE DME

Component	Maximum Content, Wt. %
Water	0.01
Methanol	0.05
Methyl ethyl ketone	0.20
Higher alcohols	0.05
Higher ethers	0.05
Ketones	0.05

Source: Data from Ref. 47.

engine efficiency, (3) the transmission efficiency (including auxiliary power consumption), and (4) the vehicle weight correction factor (to account for weight differences due to differences in fuels, fuel tank weight, and engine weight).

The well-to-wheel efficiency of DME in a heavy-duty urban bus application is lower than that of diesel, comparable to LPG and CNG, and higher than that of gasoline or methanol [43]. The respective efficiency numbers are 30% for diesel; 21% to 24% for CNG, LPG, and DME; and 20% for gasoline. For light-duty applications, the direct injection diesel has the highest energy efficiency of about 26%. DME (19%) falls in between the DI gasoline engine (21%) and the LPG, CNG, and gasoline engines (18%, 17%, and 17%, respectively).

12.4.3.3 Safety

From the safety point of view, DME is comparable to LPG and diesel. European regulations put DME in the same safety class as LPG. The toxicity and environmental aspects of DME were extensively evaluated by a number of companies (DuPont, Akzo-Nobel) and government agencies because of the application of DME as an alternative to CFCs for spray cans. The conclusion was that the toxicity for both humans and laboratory animals is low. DME concentrations of 8% to 12% can cause a narcotic effect [43]. For propane and butane, concentrations of 5% to 10% cause a similar effect. Human exposures to DME concentrations greater than 14.4% resulted in unconsciousness after 26 min [49].

Current DME exposure standards, for both the American Industrial Hygiene Association (AIHA) and the German MAK, are set at 1000 ppm (8-h TWA). DME is also considered environmentally benign. Comprehensive information on the toxicity and other environmental properties of DME is available in Ref. [49].

12.4.3.4 Cost

The cost of DME fuel on a large production scale should be comparable to the cost of methanol. The production setup of DME in terms of both the technical arrangement of the production process and in the infrastructure aspects is very similar to that of methanol. It is estimated that DME could be up to 10% less expensive than methanol due to the higher efficiency of the production process and possible larger scale production [46]. Compared to diesel fuel, the cost of DME produced from natural gas equals 90% to 135% of the cost of diesel fuel [43]. The cost of renewable DME, such as that manufactured from biomass, would be about three times higher than that of diesel fuel. These numbers have been corrected for energy efficiency.

The competitiveness of DME relative to petrodiesel is determined by the crude oil prices. The cost of diesel is normally nearly proportional to the cost of crude. A case study on very large DME plants (7000 MTPD DME), in which the DME is made from inexpensive natural gas, concluded that DME can be competitive when the crude oil price is above $20 per barrel (USD) [47]. The analysis was based on 1999 conditions, including the assumption that remote natural gas is available at a price below $1 per gigajoule.

12.4.4 DME Engines

DME engine developments are currently taking place in Europe and in North America. These developments are focused on fuel injection hardware. Although the injection pressure for DME is much lower than that for DI diesel (about 20 and >120 MPa, respectively), this is still a challenge because of the very low viscosity of DME in combination with its high vapor pressure. Because DME injection has a strong influence on the combustion process, the DME conversion process is much more difficult than the well-known LPG and NG conversion processes. Because of the totally different injection parameters and fuel properties, it is most likely that dedicated DME engines will be developed rather than flexible, dual-fuel diesel-DME units. As is most often the case with alternative fuels, special materials have to be used for fuel system components. This is especially important in the case of DME fuel systems, because most elastomers used in diesel engine fuel systems swell or dissolve when they come in contact with DME. Figure 12.16 shows an example DME storage and injection system, developed for a prototype DME-fueled bus [47]. The fuel system consists of the following main components:

- Fuel tanks with level indicators and integrated submerged fuel pumps for low-pressure supply of liquid DME
- Valves, connectors, and fuel lines, including safety-enhancing devices (pressure relief valves)
- Purge tank for storage of gaseous DME
- Pneumatic control unit for the purge system
- Air-driven purge compressor
- Electrical control unit and driver interface
- Wiring harness.

The fueling system is similar to that for a typical LPG system, with the addition of the fuel purge system. DME fueling takes place through a standard LPG receptacle on one side of the bus. The filling of the two vehicle's fuel tanks (only one fuel tank is shown in Figure 12.16) is monitored by the tank system engine control unit (ECU) and the electrical filling valve in the tanks. Distribution of DME to the engine is also

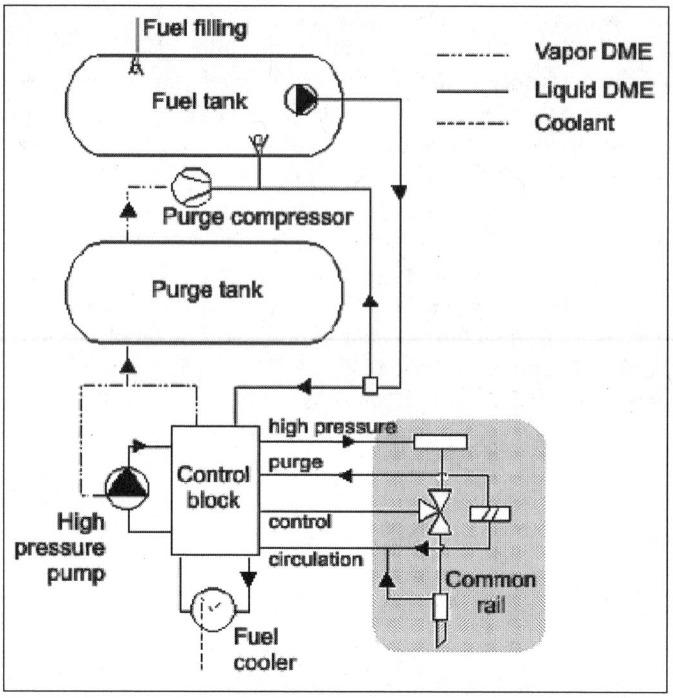

Figure 12.16 *DME fuel system layout. (Reprinted with permission of the Danish Road Safety Agency from Ref. 47, Fig. 20.)*

monitored by the ECU. A special ECU function is provided that, after the ignition key is turned on, ensures even distribution of the fuel into the two tanks. During engine operation, the fuel is pumped from the tank at a working pressure of 1.05 MPa above vapor pressure and a flow rate of approximately 93 L/h. It is circulated back to the tank, where it passes a takeoff point at which the fuel injection system is supplied with guaranteed liquid DME. The purge system must be provided for safety reasons. It is independent of electricity and signals from the other systems and uses compressed air from the vehicle brake system for its operation.

Because DME provides insufficient lubrication in the injection system, a lubrication additive must be added. The Lubrizol LZ539N additive was used in the study because it had the best wear-reducing ability of the available additives, despite its tendency to separate from DME and settle over time [47]. For safety reasons, ethyl mercaptan was added to the fuel as an odor additive. Initially 1000 ppm Lubrizol LZ539N and 20 ppm ethyl mercaptan were added to DME. Work continued to minimize the use of both additives and, hence, minimize the potential negative impact on exhaust emissions. The performance of engines converted to DME fuel was reported to be roughly equivalent to that of the diesel engine before conversion. This is illustrated in Figure 12.17, which shows that the thermal efficiency (brake specific fuel consumption) of diesel and DME was similar [48]. An improvement in engine noise was also recorded with DME.

12.4.5 DME Fuel and Emissions

12.4.5.1 Greenhouse Gas Emissions

Among all greenhouse gases, carbon dioxide emissions have been subject to the most comprehensive analyses. The well-to-wheel CO_2 emission is determined by the CO_2 emission during production and distribution of fuel and the CO_2 produced during the combustion of the fuel. An important factor is also the vehicle efficiency, which determines the fuel consumption.

The CO_2 efficiency of diesel and DME is very similar. In heavy-duty engine applications, if the well-to-wheel CO_2 emissions of the diesel engine equal 1.00, the emissions of DME are 1.02 [43]. For comparison, the relative emissions would be also 1.00 for lean-burn CNG, 1.09 for stoichiometric CNG, and 1.46 for gasoline. According to other sources, CO_2 emissions with DME, relative to diesel, would range between 0.96 and 1.14 [50].

Well-to-wheel CO_2 emissions for renewable DME would be only 17% relative to diesel. However, as mentioned earlier, the current cost of renewable DME would be prohibitive [43].

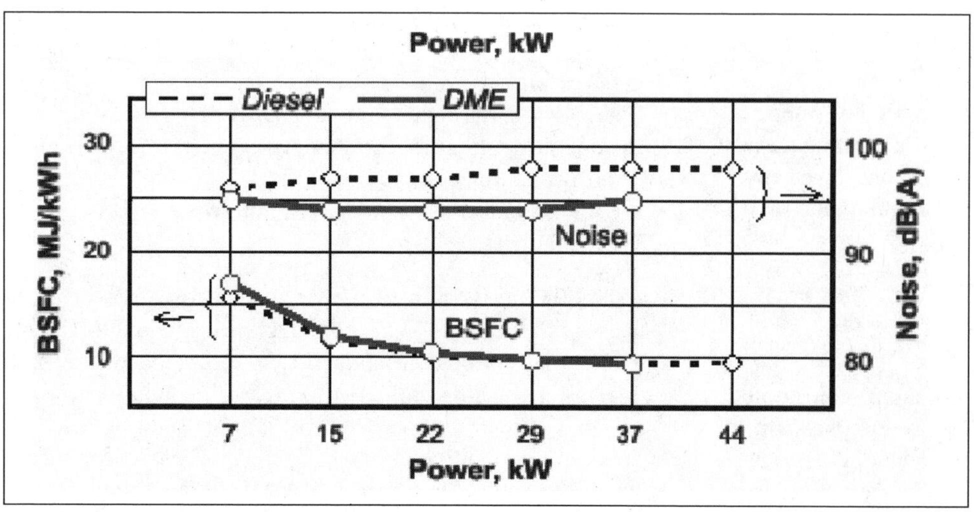

Figure 12.17 *DME and diesel engine performance comparison at a constant speed of 2000 rpm. (Based on data from Ref. 48.)*

12.4.5.2 Emissions of Air Pollutants

DME is characterized by a low autoignition temperature and by instantaneous evaporation when injected into the cylinder. These characteristics make the DME engine very suitable for optimization through the "rate shaping" of the injection event. It is possible to obtain a very gradual pressure buildup by reducing the amount of fuel injected during the ignition delay period. This results in low peak pressures and temperatures, which, in turn, produce low NO_x emissions and low noise. Because the fuel evaporation is very fast, particulate emissions are also lower than those from diesel engines. This makes the DME engine very suitable for further NO_x reductions through exhaust gas recirculation (EGR). Emissions of CO and gas-phase HC, which may increase with the use of DME, are easy to control by means of a diesel oxidation catalyst.

Because DME fuel contains no or few carbon–carbon bonds, there are no or low PAH or benzene-toluene-xylene emissions during combustion. Aldehyde emissions from DME were reported to be lower than those from diesel engines [43].

Emissions from a heavy-duty engine powered with DME were tested in a study by Haldor Topsoe, AVL, Navistar, and Amoco [51, 52]. Two engines were investigated: a 7.6-L, turbocharged V8 Navistar engine and a single-cylinder, 2-L, AVL test engine. Emissions from both engines are compared with a production Euro II diesel engine in Figure 12.18 [43].

NO_x emissions from the DME engines were 40% to 70% lower than from the Euro II engine. Similarly, the PM emissions from the DME Navistar engine were about 50% lower than those from the diesel engine. A significant increase in CO emissions was observed from the DME engine. NO_x and smoke emission benefits from the use of DME measured in a different study [48] are illustrated in Figure 12.19. These data were measured from a light-duty, 2-ton diesel truck converted to DME by NKK Corporation in 1998 (claimed to be the first DME vehicle in the world).

12.5 Natural Gas

12.5.1 Utilization of Natural Gas

Natural gas, comprised mostly of methane, is not a "natural" diesel fuel due to its high octane and low cetane number. Thus, conversion of an engine from the Diesel cycle (compression ignition) to the Otto cycle (spark ignition) is one option. Because the focus in this book is on diesel engines, the latter alternative is not covered in detail. Natural gas, biogas, and pure methane could be treated as relatively similar fuels from combustion

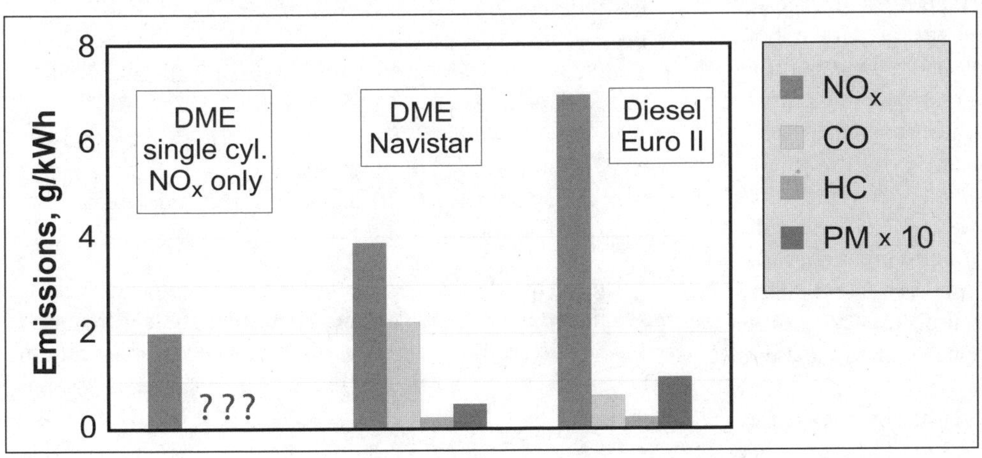

Figure 12.18 Exhaust emissions with DME for the ECE R-49 13 mode test. (Based on data from Ref. 43.)

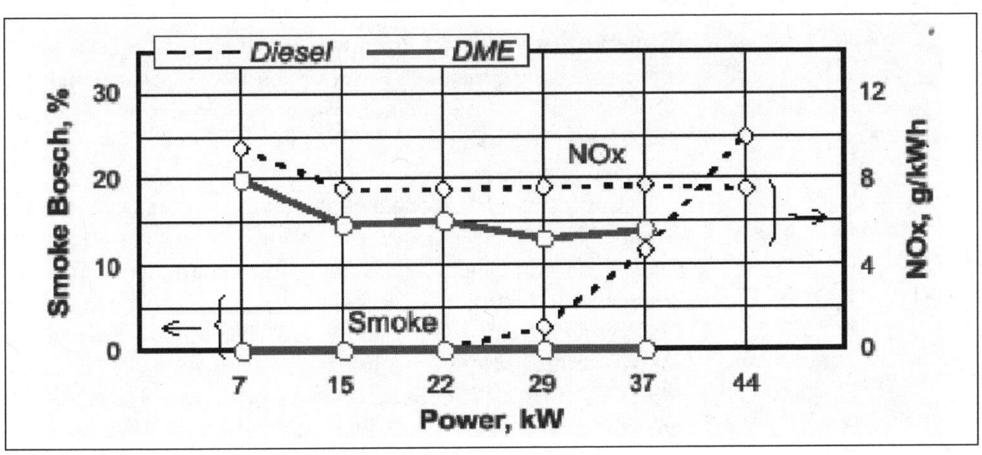

Figure 12.19 Smoke and NO_x emissions with DME at a constant speed of 2000 rpm. (Based on data from Ref. 48.)

viewpoint, although it should be noted that the composition is different.

The various engine technology options for using natural gas can be grouped into the following three groups:

- Spark ignition natural gas (SING), Otto cycle
- Direct injection natural gas (DING), Diesel cycle
- Dual fuel natural gas (DFNG), Diesel cycle.

SING engines could be of the lean-burn or stoichiometric type. The latter can utilize the three-way catalyst (TWC) emission technology and, therefore, provide the greatest emission potential. Problems with thermal stresses and low power density have favored the use of the lean-burn combustion system in heavy-duty engines. However, most light-duty SING engines are stoichiometric. A specific advantage of SING engines is low particulate emissions in comparison with diesel engines without particulate filters. The NO emissions are generally lower (lean-burn) or significantly lower (TWC) than those created by conventional diesel engines.

DING and DFNG engines use late-cycle direct injection and, therefore, they are characterized as engines utilizing the Diesel cycle. Both engine types need some kind of ignition source, because methane has a very low cetane number. The DING engine uses a small quantity of diesel fuel or a glow plug as ignition sources. Because the injection system for diesel fuel does not have the capability of greater injection quantities, this option has no dual-fuel properties. On the other hand, optimization of the pilot injection can be made to achieve lower emissions. An injection system combining the injection of both diesel fuel and natural gas has been developed by Westport Inc. [53, 54]. A DFNG engine mixes natural gas before induction to the cylinder and uses diesel fuel as the ignition source. Because conventional diesel injection is used in this case, this engine has dual-fuel properties. A kind of "hybrid" version of both the mentioned systems (often referred to as PING) uses pilot injection of diesel fuel and premixing of natural gas. Because the combustion in DING and DFNG engines is not completely premixed combustion, soot formation can occur. Thus, future very stringent particulate emissions may necessitate the use of a particulate filter.

Utilization of natural gas could be either in compressed form as CNG or in liquefied form as LNG. The difference regarding engine performance and emissions between these options is relatively small. In the LNG case, injection of liquid fuel is possible.

12.5.2 Emission Comparison

12.5.2.1 Criteria for Comparisons

Comparing exhaust emissions and the impact of emissions on health and the environment is not an easy task. First, the conditions for the comparisons must be established. Some of the options for such comparisons are these:

- Comparison among "average" fleets
- Technology-neutral comparisons
- Cost-effectiveness comparisons.

The first alternative necessitates numerous measurements on fleets representative of each fuel and, in many cases, economical constraints limit the number of possible measurements. This is even more pronounced if measurements of unregulated emissions and tests of biological activity in the exhaust are of interest.

A technology-neutral comparison is attractive, but such comparisons are not often possible to accomplish. For example, in the past, it was often common to compare alternative fuels with exhaust gas aftertreatment with diesel fuel without any aftertreatment. It is difficult to distinguish the impact of the fuel from the technology influence in such cases.

Ideally, cost effectiveness should be considered in some way in the comparisons. A certain incremental cost—compared to the baseline—could be kept constant for all engine/fuel combinations investigated. The problem is that the incremental cost for some options (e.g., gaseous fuels) is often so high that it is impossible to find an equal level for the comparison. Another way would be to compare the cost effectiveness of decreasing the emissions of a specific emission component.

12.5.2.2 Selected Studies

The discussion here is limited to some projects that studied the emissions from heavy-duty vehicles fueled by CNG and diesel fuel:

- Emission tests in Sweden during the 1990s on CNG/biogas, ethanol, and diesel fuel with various emission control devices
- The fleet test of low-sulfur diesel fuel and aftertreatment on heavy-duty vehicles in California organized by BP
- An ongoing study by the California Air Resources Board (ARB) on CNG and diesel buses
- A study by SwRI on CNG and a diesel bus with and without a particulate filter.

During the 1990ds, many emission measurements on transit buses fueled with diesel fuel and alternative fuels were carried out in Sweden at the research facilities of MTC near Stockholm. A summary and assessment of results from these tests was published in an SAE paper in 2000 [55] and at the DEER Workshop in 2001 [56]. A technology-neutral approach was utilized in the comparisons of the engine/fuel options.

The purpose of the BP project (initiated by Arco, now part of BP) was to validate ultralow sulfur diesel fuel and exhaust aftertreatment devices [57–63]. Tests have been carried out with and without aftertreatment devices as well as with various diesel fuel specifications. A CNG bus without aftertreatment has also been tested.

California ARB is conducting a study on exhaust emissions from CNG and diesel buses with various exhaust gas aftertreatments [64–66]. The study is continuing and more results are expected in the future.

SwRI has carried out a study on a CNG bus and a diesel bus in two configurations [67]. Forty-one compounds listed by the ARB as "toxic air contaminants" were analyzed. The CNG bus was compared with a diesel bus with particulate filter, ultralow sulfur fuel, and two electronic control strategies (low emitting and conventional). This study was funded by International Truck and Engine Corporation.

12.5.2.3 Regulated Emissions, Fuel Consumption, and CO_2

NO_x emissions were generally lower for CNG in most of the studies cited. A great variability was seen for CNG in the tests in Sweden resulting in higher NO_x emissions for the average bus population. Therefore, a classification of the CNG results was made to highlight the potential with a properly tuned control system. Tests by ARB also showed a considerable variation, even though newer technology engines were tested. The diesel bus tested at SwRI, in both conventional and low-emitting configurations, emitted fewer NO_x emissions than CNG. The relative NO_x advantage of either type of engine will be influenced by the increasing use of EGR on diesel engines. EGR systems for heavy-duty diesel engines were first introduced as retrofit systems in Sweden in 1998, and later as an OEM solution. During recent years extensive tests of EGR systems have been conducted in the United States. Series production of EGR-equipped, heavy-duty engines started in the fall of 2002. Using EGR and other advanced engine technologies, NO_x emissions can be cut by some 50%, thus reducing the (apparent) advantage of lower NO_x emissions for CNG. However, improvements on CNG engines can also be expected in the future. Note that test cycles could also have a significant impact on NO_x emissions. The considerable formation of NO_2 that takes place with some diesel particulate filters is also an undesirable effect that should be minimized.

HC and NMHC emissions are generally lower for diesel than for CNG. This is usually also the case for CO emissions but it should be noted that very low CO emissions could be achieved with an oxidation catalyst on CNG as well.

Particulate mass emissions from uncontrolled diesel engines are higher than from CNG. If particulate filters are used, diesel PM emission can be reduced to the same level as for CNG and, in some cases, lower. Ultralow or sulfur-free diesel fuel is essential for the use of some types of particulate filters.

Fuel consumption is generally lower for diesel fuel than for CNG, because the Diesel cycle is more efficient than the Otto cycle. CNG contains more hydrogen than diesel fuel. Therefore, the CO_2 emissions are sometimes lower for CNG but, because the test cycle and engine technology also have a great influence, the opposite is sometimes the case.

12.5.2.4 Unregulated Emissions

The discussion about unregulated emissions has been limited to a few selected emission components. Several PAHs and polycyclic aromatic compounds (PACs) are mutagenic, carcinogenic, or suspected carcinogens. Because diesel fuel contains PAHs, it is not surprising that both PAHs and PACs can be found in diesel exhaust. In general, PAHs and PACs can be also formed in the engine. However, studies on the origin of diesel PAH emissions have found that tailpipe emissions are totally dominated by PAHs in the fuel [68, 69]. Interestingly, PAHs and PACs have also been found in the exhaust from CNG engines. In this case, the origin of these compounds has not been properly explained.

In general, diesel fuel with a high PAH content and an engine without any aftertreatment have the highest PAH/PAC level. Tests in Sweden showed lower levels of heavier (tri+) PAC for diesel fuel with catalytic aftertreatment devices (catalysts or particulate filters) than for CNG/biogas with oxidation catalyst. Most of the other studies cited have shown a lower or approximately similar level for diesel fuel in comparison to CNG. Note, however, that many of the U.S. tests on CNG have been performed without an oxidation catalyst and this has to be considered in the assessment of the results.

Heavy PAH emissions are generally considered to have higher cancer potency than lighter PAH emissions. Because the level of diaromatic PAHs can be relatively high in diesel fuel, the exhaust emissions of these emissions components are higher than for tri+ PAHs. This can be a confounder in the interpretation of the results. Ideally, weighting of the individual PAHs/PACs should be made according to their cancer risk factor. However, this is not yet possible due to lack of data. Selective comparison of groups of PAHs is another, simpler option. Some results for 12 tri+ PAH species from the study at SwRI and some selected results from MTC are compared in Figure 12.20. In the results from MTC, the impact of a cleaner diesel fuel (EC1 vs. EPEFE) and aftertreatment devices (catalyst and particulate filter) can be compared with the results on CNG and biogas (CBG). The tests at SwRI compare conventional diesel without aftertreatment (CD) with low-emission diesel (LED) and CNG.

The results from MTC show the severe impact of a clean diesel fuel and aftertreatment. With aftertreatment, the level for diesel fuel is lower than that for CNG and biogas. The comparison between conventional diesel and low-emission diesel is even more dramatic in the SwRI case. In the latter case, the level is also considerably lower than for CNG, although it should be noted that the CNG bus did not have an oxidation catalyst. In general, the tests at SwRI show a lower level than at MTC. One reason is probably because all engines are from a newer generation, but the test cycles could also have an influence.

Considerably higher carbonyl emissions have been detected with CNG without aftertreatment in comparison to diesel fuel with and without aftertreatment. The carbonyl emissions in CNG exhaust are comprised mostly of formaldehyde. An oxidation catalyst has a considerable impact (more than 95% reduction) on the carbonyl emissions from CNG buses.

The results on 1,3-butadiene emissions are somewhat inconclusive, due to problems with measurement accuracy. In most cases, the highest level has been found for CNG without aftertreatment. The level of 1,3-butadiene has generally been low or below the detection level for diesel.

Diesel fuel without a particulate filter has a significantly higher level of particulate number emissions and emissions of nanoparticles than CNG and diesel fuel with a particulate filter.

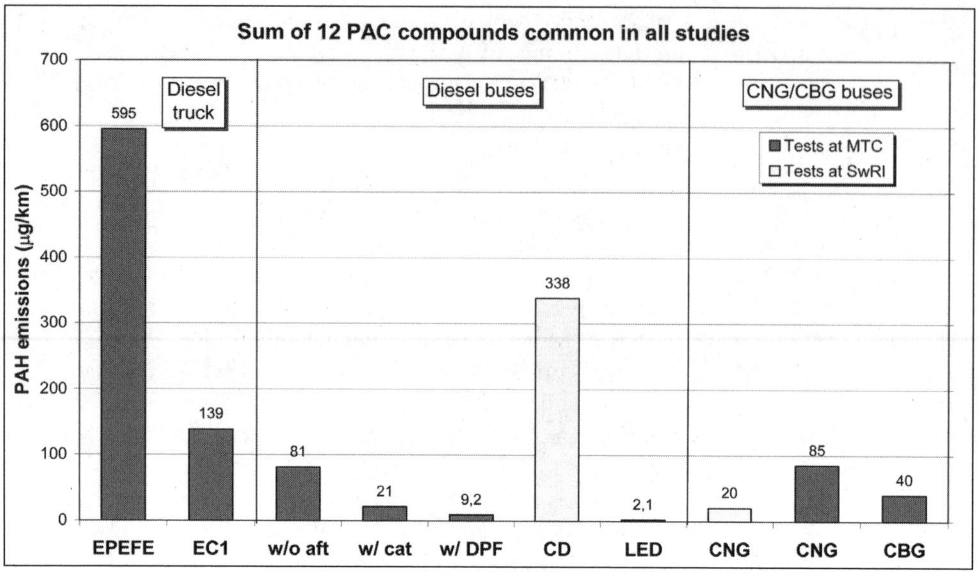

Figure 12.20 *Comparison of some PAH emission results from MTC and SwRI.*

In general, CNG and diesel with DPF are at a similar level. In some cases, for example, at idle for CNG, the high emissions of nanoparticles are of some concern. However, as discussed in Section 7.4, measurement of particle number emissions is a very difficult task and more development is needed in this area before any firm conclusions can be drawn. Recent and ongoing work should provide more insight into these phenomena [70].

Analyses show that the particles from CNG and diesel with a particulate filter are comprised mostly of organic carbon (i.e., of volatile material). Emissions of metal compounds in the particles can be significantly reduced with a particulate filter. Maybe the most important finding regarding the emissions of ultrafine and nanoparticles is that these emissions can be reduced dramatically with a particulate filter. Note that some combustion concepts for gaseous fuels—based on relatively few data so far—are likely to generate significantly higher particulate (mass and number) emissions than the concepts tested until now. Consequently, the question of whether a particulate filter may have to be used on CNG engines in the future remains.

12.5.2.5 Biological Activity

In the relatively low number of tests performed in Sweden on the cleanest configurations (i.e., gaseous fuels and diesel with aftertreatment), very low levels of biological activity—as determined by the Ames and TCDD tests—were measured. Therefore, it has been difficult to distinguish between the various fuel options. A general conclusion is that the biological activity can be greatly reduced compared to an old diesel baseline.

Preliminary results on the California ARB website show a similar or lower level of biological activity for diesel with particulate filter in comparison with CNG with an oxidation catalyst [66]. CNG without an oxidation catalyst had the highest level. Also in this area more results are necessary before any firm conclusions can be drawn.

12.5.2.6 Discussion

The cited tests, as well as other results, show that the emission level for many of the toxic emission components can be reduced to very low levels. One route is to use gaseous fuels; another option is clean diesel fuel in combination with aftertreatment devices. Currently, the measurement accuracy is not satisfactory for many of these emission components. Methods to control the background levels in the dilution tunnel have to be developed. Contamination of the measurement equipment from previous tests with "dirtier" fuel/vehicle combinations has to be considered. The engine itself can also become contaminated with pollutants from tests with dirtier fuels. For instance, it has been shown that the transport of PAHs in the engine and previous contamination with dirtier fuel could become a significant factor for clean diesel fuels [68, 69].

As will be discussed in later chapters, fuel reformulation (i.e., reduction of the sulfur level in diesel fuel) can enable the use of new aftertreatment technology. This is probably the most important impact of cleaner fuels. It has also been shown that a reduction in the PAH level can significantly reduce the PAC emissions in the exhaust as well as the biological activity. However, when aftertreatment devices are used, the level is so low anyway that it is difficult to determine this impact. To establish the "ultimate" emission level, the cleanest fuels possible exemplified by FT diesel and pure methane (e.g., biogas) could be used. Whether the reductions possible by these "radical" fuel reformulations justify a significantly higher cost is another issue.

Note that most of the recent emission tests have been carried out on relatively new vehicles. The impact of aging on the emission aftertreatment has yet to be determined.

An overall conclusion that could be drawn from the available results so far is that it is possible to reduce the environmental and health impact from virtually all engine/fuel options in the future. The emission level for a "clean" fuel option appears to be more influenced by the emission control technology than the fuel itself. Although significant reductions of emissions have been made, some issues remain. However, if the emission levels of the future U.S. emission standards are fulfilled, the difference between fuel options will be diminished. Therefore, it is likely that fuel efficiency and energy security will be of increasing importance in the future.

References

1. Shindler, K.-P., "Advances in Diesel Engine Technologies for European Passenger Vehicles," Paper presented at 8th Diesel Emissions Reduction (DEER) Conference, San Diego, CA, August 2002, U.S. Department of Energy, Washington, DC.
2. Birky, A., *et al.,* "Future U.S. Highway Energy Use: A Fifty Year Perspective," Draft Report, U.S. Department of Energy, Washington, DC, May 3, 2001.
3. McCormick, R., "Renewable Diesel Fuels: Status of Technology and R&D Needs," Paper presented at 8th Diesel Emissions Reduction (DEER) Cnoference, San Diego, CA, August 2002, U.S. Department of Energy, Washington, DC, retrieved from www.orau.gov/deer2002/Session4/McCormick.pdf.
4. Ahlvik, P., and Å. Brandberg, "Sustainable Motor Fuels—Introduction of Biofuels," Publication 2002:144, The Swedish National Road Administration, 2002, retrieved from www.vv.se/publ_blank/bokhylla/miljo/2002_144/2002_144.pdf.
5. Eliasson, B., and U. Bossel, "The Future of the Hydrogen Economy: Bright or Bleak?" *Fuel Cell World,* Lucerne, July 5, 2002.
6. Clark, R.H., "An Evaluation of Shell Gas-to-Liquids Diesels—The Environmental Benefits," Paper presented at 8th Diesel Emissions Reduction (DEER) Conference, San Diego, CA, August 2002, U.S. Department of Energy, Washington, DC, retrieved from www.orau.gov/deer2002/Session4/clark.pdf.
7. Shen, J., *et al.,* "Opportunities for the Early Production of Fischer-Tropsch (F-T) Fuels in the U.S.—An Overview," Paper presented at 8th Diesel Emissions Reduction (DEER) Conference, San Diego, CA, August 2002, U.S. Department of Energy, Washington, DC, retrieved from www.orau.gov/deer2002/Session4/shen.pdf.
8. Morita, Y., "Marketability of GTL from Natural Gas," The Institute of Electrical Engineers of Japan, November 2001, retrieved from www.aboutdme.org/gtl_marketability_from_n_gas1.pdf.
9. Howard, "Fischer-Tropsch Technology," Howard, Weil, Labouisse, Friedrichs Inc., New Orleans, LA, December 1998.
10. Kaufmann, T.G., R.A. Fiato, G.C. Lahn, and R.F. Bauman, "Gas-to-Liquids Technology Provides New Hope For Remote Fields," *Lubricants World,* October 2000.
11. Fischer, F., and H. Tropsch, "Verfahren zur Gewinnung mehrgliedriger Paraffinkohlenwasserstoffe aus KohlenOxyden und Wasserstoff auf katalytischem Wege," German Patent DRP 484337, 1926.
12. Eilers, J., S.A. Posthuma, and S.T. Sie, "The Shell Middle Distillate Synthesis Process (SMDS)," *Catalysis Letters* **7,** 1990.
13. Norton, P., *et al.,* "Emissions from Trucks using Fischer-Tropsch Diesel Fuel," SAE Paper No. 982526, SAE International, Warrendale, PA, 1998, retrieved from www.osti.gov/hvt/982526.pdf,
14. Schaberg, P.W., *et al.,* "Diesel Exhaust Emissions Using Sasol Slurry Phase Distillate Fuels," SAE Paper No. 972898, SAE International, Warrendale, PA, 1997.
15. Schaberg, P.W., *et al.,* "Comparative Emissions Performance of Sasol Fischer-Tropsch Diesel Fuel in Current and Older Technology Heavy-Duty Engines," SAE Paper No. 2000-01-1912, SAE International, Warrendale, PA, 2000.
16. Clark, N.N., *et al.,* "On-Road Use of Fischer-Tropsch Diesel Blends," SAE Paper No. 1999-01-2251, SAE International, Warrendale, PA, 1999, retrieved from www.osti.gov/hvt/1999-01-2251.pdf.
17. Norton, P., *et al.,* "Emissions from Buses with DDC 6V92 Engines Using Synthetic Diesel Fuel," SAE Paper No. 1999-01-1512, SAE International, Warrendale, PA, 1999, retrieved from www.osti.gov/hvt/1999-01-1512.pdf.
18. Sheehan, J., *et al.,* "A Look Back at the U.S. Department of Energy Aquatic Species Program—Biodiesel from Algae," NREL/TP-580-24190, National Renewable Energy Laboratory, Golden, CO, 1998, retrieved from www.afdc.doe.gov/pdfs/3814.pdf.
19. Howell, S., "U.S. Biodiesel Standards—An Update of Current Activities," SAE Paper No. 971687, SAE International, Warrendale, PA, 1997.
20. "Fatty Acid Methyl Ester Fuels as a Replacement or Extender for Diesel Fuels," FIE (Fuel Injection Equipment) Manufacturers Common Position Statement: Delphi, Stanadyne, Denso, Bosch, June 2000, retrieved from www.dieselnet.com/tech/fuels/fie_fame_position_2000_06.pdf.
21. Berry, W.W., "Developing Biodiesel Fuels for Market—NOPEC Used Oil to Fuel Process," Paper presented at SAE Clean Diesel Engine Technology TOPTEC, St. Pete Beach, FL, September 1999.
22. *"Standard Specification for Biodiesel Fuel (B100) Blend Stock for Distillate Fuels,"* D6751-02, American Society for Testing and Materials, 2002.
23. *"Provisional Specification for Biodiesel Fuel (B100) Blend Stock for Distillate Fuels,"* PS121-99, American Society for Testing and Materials, 1999.
24. Stavinoha, L.L., and S. Howell, "Potential Analytical Methods for Oxidative Stability Testing of Biodiesel and Biodiesel Blends," SAE Paper No. 1999-01-3520, SAE International, Warrendale, PA, 1999.
25. "A Comprehensive Analysis of Biodiesel Impacts on Exhaust Emissions," Draft Technical Report EPA420-P-02-001, U.S. Environmental Protection Agency, Washington, DC, October 2002, retrieved from www.epa.gov/otaq/models/analysis/biodsl/p02001.pdf.

26. Sharp, C.A., S.A. Howell, and J. Jobe, "The Effect of Biodiesel Fuels on Transient Emissions from Modern Diesel Engines—Part I: Regulated Emissions and Performance," SAE Paper No. 2000-01-1967, SAE International, Warrendale, PA, 2000.
27. Hamm, E., G. Hohenberg, U. Standt, and P. Zelenka, "The Effect of Fuel Specifications and Different Aftertreatment Systems on Exhaust Gas Odour and Non-Regulated Emissions at Steady-State and Dynamic Operation of DI-Diesel Engines," SAE Paper No. 1999-01-3559, SAE International, Warrendale, PA, 1999.
28. Sheehan, J., *et al.,* "An Overview of Biodiesel and Petroleum Diesel Life Cycles," NREL/TP-580-24772, National Renewable Energy Laboratory, Golden, CO, 1998, retrieved from www.afdc.doe.gov/pdfs/3812.pdf.
29. Choi, C.Y., *et al.,* "Effects of Biodiesel Blended Fuels and Multiple Injections on DI Diesel Engines," SAE Paper No. 970218, SAE International, Warrendale, PA, 1997.
30. Sharp, C., "Summary of Results from NBB/US EPA Tier I Health and Environmental Testing for Biodiesel," Paper presented at Canadian Mining Diesel Conference (CANMET), Toronto, Ontario, October 1998.
31. Grägg, K., "Effects of Environmentally Classified Diesel Fuels, RME and Blends of Diesel Fuels and RME on the Exhaust Emissions," Report 9209B, MTC, Sweden, 1994.
32. Schroeder O., J. Krahl, A. Munack, and J. Buenger, 1999. "Environmental and Health Effects Caused by the Use of Biodiesel," Draft Technical Report EPA420-P-02-001, SAE 1999-01-3561, SAE International, Warrendale, PA, 1999.
33. Sharp, C.A., S.A. Howell, and J. Jobe, J., "The Effect of Biodiesel Fuels on Transient Emissions from Modern Diesel Engines—Part II: Unregulated Emissions and Chemical Characterization," SAE Paper No. 2000-01-1968, SAE International, Warrendale, PA, 2000.
34. LePrieur, E., *et al.,* "Toxicological Impact of Diesel Fuel Supplementation with Rapeseed Methylester (RME) on the Lung Toxic Potential of Diesel Engine Exhausts," SAE Paper No. 2000-01-2060, SAE International, Warrendale, PA, 2000.
35. McCormick, R., J.R. Alvarez, M.S. Graboski, K.S. Tyson, and K. Vertin, "Fuel Additive and Blending Approaches to Reducing NOx Emissions from Biodiesel," SAE Paper No. 2002-01-1658, SAE International, Warrendale, PA, 2002.
36. Hansen, K.F., and M.G. Jensen, "Chemical and Biological Characteristics of Exhaust Emissions from a DI Diesel Engine Fueled with Rapeseed Oil Methyl Ester (RME)," SAE Paper No. 971689, SAE International, Warrendale, PA, 1997.
37. McDonald, J., *et al.,* "Emission Characteristic of Soy Methyl Ester Fuels in an IDI Compression Ignition Engine," SAE Paper No. 950400, SAE International, Warrendale, PA, 1995.
38. Watts, W.F., *et al.,* "Evaluation of Biodiesel Fuel and Oxidation Catalyst in an Underground Metal Mine," Report for the DEEP Technical Committee, University of Minnesota, September 24, 1998, retrieved from www.deep.org/reports/inco_bio.pdf.
39. Cusano, C.M., *et al.,* "Changes in Elastomer Swell with Diesel Fuel Composition," SAE Paper No. 942017, SAE International, Warrendale, PA, 1994.
40. Besee, G.B., and J.P. Fay, "Compatibility of Elastomers and Metals in Biodiesel Fuel Blends," SAE Paper No. 971690, SAE International, Warrendale, PA, 1997.
41. "John Deere Approves Eco-Friendly Biodiesel Fuel for Its Products," Press release, Deere & Company, Lenexa, KS, February 21, 2002, retrieved from www.deere.com/en_US/newsroom/2002/releases/corporate/020221_biodiesel.html.
42. "DME Project," NKK Corporation, retrieved August 1, 2002, from www.nkk.co.jp/en/environment/dme.
43. Verbeek, R., and J.V. Van der Weide, "Global Assessment of Dimethyl-Ether: Comparison with Other Fuels," SAE Paper No. 971607, SAE International, Warrendale, PA, 1997.
44. Kapus, P., and W. Cartellieri, "ULEV Potential of a DI/TCI Diesel Passenger Car Engine Operated on Dimethyl Ether," SAE Paper No. 952754, SAE International, Warrendale, PA, 1995.
45. McCandless, J., and S. Li, "Development of a Novel Fuel Injection System (NFIS) for Dimethyl Ether and Other Clean Alternative Fuels," SAE Paper No. 970220, SAE International, Warrendale, PA, 1997.
46. Hansen, J.B., *et al.,* "Large Scale Manufacture of Dimethyl Ether—A New Alternative Diesel Fuel From Natural Gas," SAE Paper No. 950063, SAE International, Warrendale, PA, 1995.
47. Hansen, J.B., and S.E. Mikkelsen, "DME as a Transportation Fuel," Danish Road Safety & Transport Agency, July 2001, retrieved from www.fstyr.dk/publikation/Publikationer/dme_eng.pdf.
48. Ohno, Y., N. Inoue, T. Ogawa, M. Ono, T. Shikada, and H. Hayashi, "Slurry Phase Synthesis and Utilization of Dimethyl Ether," NKK Technical Review No. 85, 2001, pp. 23–28, retrieved from www.nkk.co.jp/en/gijutu/rev/85/pdf/85_04.pdf.
49. "Robust Summary for Dimethyl Ether," DuPont Submission to the U.S. Environmental Protection Agency, October 11, 2000, retrieved from www.epa.gov/opptintr/chemrtk/dimethr/c12794b.pdf.
50. Basu, A., M.J. Gradissi, J.G. Massin, and T.H. Fleisch, "Dimethyl Ether: Amoco's New Clean Diesel Option," Paper presented at the 1st Diesel Engine Emissions Reduction (DEER) Workshop, San Diego, CA, July 1995.
51. Fleisch, T.H., and P.C. Meurer, "DME—The Diesel Fuel for the 21st Century?" Paper presented at the AVL Conference on Engines and the Environment, Graz, Austria, 1995.
52. Fleisch, T.H., *et al.,* "A New Clean Diesel Technology: Demonstration of ULEV Emissions on Navistar Diesel Engine Fueled with Dimethyl Ether," SAE Paper No. 950061, SAE International, Warrendale, PA, 1995.
53. Oullette, P., *et al.,* "Performance and Emissions of a Two-Stroke Engine Fueled Using High Pressure Direct Injection

of Natural Gas," SAE Paper No. 981160, SAE Internation, Warrendale, PA, 1998.

54. Bartuuch, B., Hilger, U., "Direct Injection Natural Gas (DING): A Diesel-Derived Combustion System for Low Emissions and High Fuel Economy," SAE Paper No. 1000-01-282, SAE International, Warrendale, PA, 2001.
55. Ahlvik, P., and Å. Brandberg, "Relative Impact on Environment and Health from the Introduction of Low Emission City Buses in Sweden." SAE Paper No. 2000-01-1882, SAE International, Warrendale, PA, 2000.
56. Ahlvik, P, "Swedish Experiences from Low Emission City Buses: Impact on Health and Environment," Paper presented at the 7th Diesel Emissions Reduction (DEER) Workshop, Portsmouth, NH, 2001, retrieved from www.ecotraffic.se.
57. LeTavec, C., J. Uihlein, J. Segal, and K. Vertin, "EC-Diesel Technology Validation Program Interim Report," SA Paper No. E 2000-01-1854, SAE International, Warrendale, PA, 2000.
58. Clark, N., J.E. Boyce, W. Xie, M. Gautam, D.W. Lyons, *et al.,* "Class 8 Trucks Operating On Ultra-Low Sulfur Diesel with Particulate Filter Systems: Regulated Emissions." SAE Paper No. 2000-01-2815, SAE International, Warrendale, PA, 2000.
59. Lev-Onm M., C. LeTavec, J. Uihlein, K. Kimura, T.L. Alleman, *et al.,* "Class 8 Trucks Operating on Ultra-Low Sulfur Diesel with Particulate Filter Systems: A Fleet Start-Up Experience," SAE Paper No. 2000-01-2821, SAE International, Warrendale, PA, 2000.
60. Chatterjee, S., C. McDonald, R. Conway, H. Windawi, K. Vertin, *et al.,* "Emission Reductions and Operational Experiences with Heavy Duty Diesel Fleet Vehicles Retrofitted with Continuously Regenerated Diesel Particulate Filters in Southern California," SAE Paper No. 2001-01-0512, SAE International, Warrendale, PA, 2001.
61. Lev-On, M., C. LeTavec, J. Uihlein, T.L. Alleman, D.R. Lawson, *et al.,* "Chemical Speciation of Exhaust Emissions from Trucks and Buses Fueled on Ultra-Low Sulfur Diesel and CNG," SAE Paper No. 2002-01-0432, SAE International, Warrendale, PA, 2002.
62. LeTavec, C., J. Uihlein, K. Hallstrom, K. Vertin, K. Chandler, *et al.,* "Year-Long Evaluation of Trucks and Buses Equipped with Passive Diesel Particulate Filters," SAE Paper No. 2002-01-0433, SAE International, Warrendale, PA, 2002.
63. Lev-On, M., C. LeTavec, J. Uihlein, K. Kimura, T.L. Alleman, *et al.,* "Speciation of Organic Compounds from the Exhaust of Trucks and Buses: Effect of Fuel and After-Treatment on Vehicle Emission Profiles." SAE Paper No. 2002-01-2873, SAE International, Warrendale, PA, 2002.
64. Ayala, A., N.Y. Kado, R.A. Okamoto, B.A. Holmén, P.A. Kuzmicky, R. Kobayashi, and K.E. Stiglitz, "Diesel and CNG Heavy-Duty Transit Bus Emissions over Multiple Driving Schedules: Regulated Pollutants and Project Overview," SAE Paper No. 2002-01-1722, SAE International, Warrendale, PA, 2002.
65. Ayala, A., and B.A. Holmén, "Ultrafine PM Emissions from Natural Gas, Oxidation-Catalyst Diesel, and Particle-Trap Diesel Heavy-Duty Transit Buses," *Environ. Sci. Technol.* **36,** 23, 2002, pp. 5040–5050.
66. California ARB CNG and Diesel Transit Bus Emission Research, retrieved January 2003 from www.arb.ca.gov/research/cng-diesel/cng-diesel.htm.
67. Ullman, T.L., L.R. Smith, and J.W. Anthony, "Exhaust Emissions from School Buses in Compressed Natural Gas, Low Emitting Diesel, and Conventional Diesel Engine Configurations," SwRI Report 08.05303, 2002.
68. Froelund, K., and J. Schramm, "PAH-Transport in Diesel Engines," SAE Paper No. 972960, SAE International, Warrendale, PA, 1997.
69. Schramm, J., S. Hori, and T. Abe, "The Emission of PAH from a DI Diesel Engine Operating on Fuels and Lubricants with Known PAH Content," SAE Paper No. 940342, SAE International, Warrendale, PA, 1994.
70. Kittelson, D., W. Watts, and J. Johnson, "Diesel Aerosol Sampling Methodology—CRC E-43—Final Report," August 19, 2002, retrieved from www.crcao.com/reports/recentstudies00-02/UMN%20Final%20E-43%20Report.pdf.

Part IV

Emission Control through Engine Design

Chapter 13:
Steps Toward the Modern Diesel Engine

Chapter 14:
Emerging Technologies and the Modern Diesel Engine

Chapter 15:
Water Addition to Diesel Combustion

Chapter 16:
Exhaust Gas Recirculation in Diesel Engines

Chapter 17:
Controls for Modern Diesel Engines

Chapter 13

Steps Toward the Modern Diesel Engine

13. 1 Introduction

Increasing public awareness about the depletion of natural resources became the focus of political debate in the United States in the mid-1970s. This issue gained importance when oil from the Middle East was used as a political weapon against the United States and other industrialized nations [1]. Alternative sources of energy became one of the most talked about subjects, besides energy conservation, passenger car downsizing, car pooling, and so forth. Motivated by the oil embargo crisis, many engine manufacturers decided to *dieselize* or at least offer a diesel option to their customers not only conserve energy, but also to allow their customers to capitalize on reduced fuel consumption and operating costs.

The most common view of the public about diesel engines is rather negative. Smoke emissions, as discussed in Chapter 6, are emitted in plain view of the public and are a major obstacle to their wider acceptance. Public pressure led the legislative bodies in the industrialized world to mandate increasingly stringent emission limits on many sources including the transportation sector especially for on-highway vehicles. Diesel engines enjoyed modest regulation until the 1990s when these regulations became increasingly difficult to meet. Intense research and development, however, led to advances in engine technology that have made it possible for the diesel engine to meet these very low emissions limits. Future standards have generated interest in other technologies such as exhaust aftertreatment and fuel composition and quality, as well as additional engine improvements.

To accommodate energy conservation as well as clean air requirements, engine manufacturers made several design choices while being concerned with the increased production costs and competition. These design choices involved many of the engine components and gradually assisted the engine to transition from its old reputation as a slow, smoky, dirty, noisy, heavy, and smelly engine to a responsive and even exciting engine to drive in some European light-duty vehicles [2, 3]. In general, the most impressive gains were achieved through changes in the combustion chamber design, improved fuel system characteristics, implementation of low-temperature charge air cooling, and special attention to lube oil consumption. Several companies felt that electronic engine controls would be needed to meet the 1991 emissions standards for heavy-duty diesel engines. As time progressed, this notion was abandoned as several engine manufacturers managed to meet those regulations without the use of electronics. A few major diesel engine manufacturers used electronics for optimizing performance while maintaining good fuel economy.

This chapter reviews the technologies adopted and implemented by the diesel engine industry during the transitional decades of the 1980s and 1990s. Starting with measures designed for NO_x reduction, this chapter discusses the efforts to improve mixture preparation, improve combustion efficiency, and recover fuel economy after it had suffered from NO_x reduction measures that led to increased fuel consumption.

13.2 Reduction of Nitrogen Oxides

While modern technologies have added to the sophistication of NO_x emission controls, the beginnings were quite modest and perhaps not as efficient as we know them today. However, reviewing these modest beginnings may help us understand and appreciate today's emerging technologies.

13.2.1 Injection Timing Retard

To better understand NO_x and its relationship to injection timing, tests were conducted on various engines to investigate the effect of engine control parameters on performance and emissions. Figure 13.1 is an example from Gill [4] and shows the effect of injection timing on NO_x emissions. A remarkably similar relationship is presented by Khair [2] and shown in Figure 13.2. Note that injection timing in Figure 13.1 is given as dynamic timing versus Figure 13.2, where injection timing is shown as static timing. Dynamic timing for these engines was about 10° crankangle. Both references involved engines that were being developed to meet the 1991 heavy-duty emission standards and describe what was then state-of-the-art technology.

The reason for decreasing NO_x emissions is shown in Figure 13.3 where a steady decrease in cylinder pressure is experienced when retarding injection timing [4]. A more detailed explanation is given in Figure 13.4 using the pressure crankangle diagram to show the directional change in cylinder pressure when retarding injection timing. It must be stressed that Figure 13.4 does not represent real data.

13.2.2 Intake Charge Cooling

Cooling charge air has a similar effect on cylinder pressure as injection timing retard. However, its impact is much smaller than that of timing retard [4]. Figure 13.5 is actual engine data showing the effect of intake charge cooling, also measured in terms of intake manifold temperature, on NO_x emissions [4]. Figure 13.6 is a schematic representation of the effect of intake charge cooling on cylinder pressure [2].

13.2.3 Combined Effect of Injection Timing and Intake Charge Cooling

By way of summary, both injection timing retard and charge air cooling are beneficial to achieve low NO_x emissions. The mechanism at work in both cases appears to be lower combustion pressure, which leads to lower peak combustion temperature and, thus, lower NO_x emissions. Figure 13.7 is derived from actual engine data for which the intake manifold temperature was controlled by varying the coolant flow in a simulated air-to-air intercooler.

Retarding injection timing from 18° to 10° BTDC reduced transient cycle NO_x emissions from about 8.5 to about 5.0 g/bhp-hr. Applying additional cooling to charge air and dropping the intake manifold temperature from 200 to 120°F provided even more reductions in NO_x emissions (from about 5.0 to 4.0 g/bhp-hr).

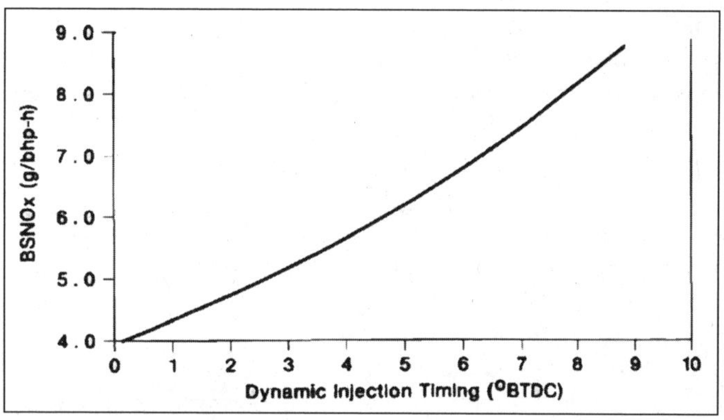

Figure 13.1 Effect of injection timing on NO_x emissions. (Reprinted from Ref. 4.)

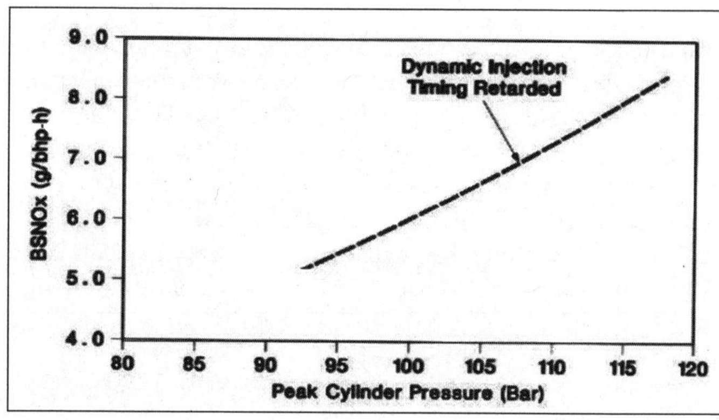

Figure 13.3 Effect of cylinder pressure on NO_x emissions. (Reprinted from Ref. 4.)

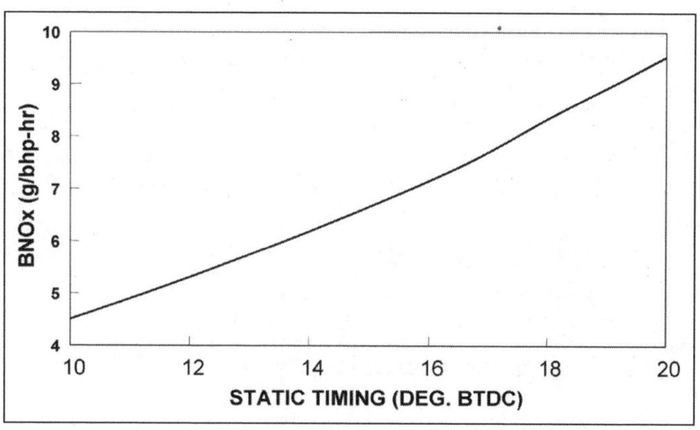

Figure 13.2 Effect of injection timing on NO_x emissions. (Reprinted from Ref. 2.)

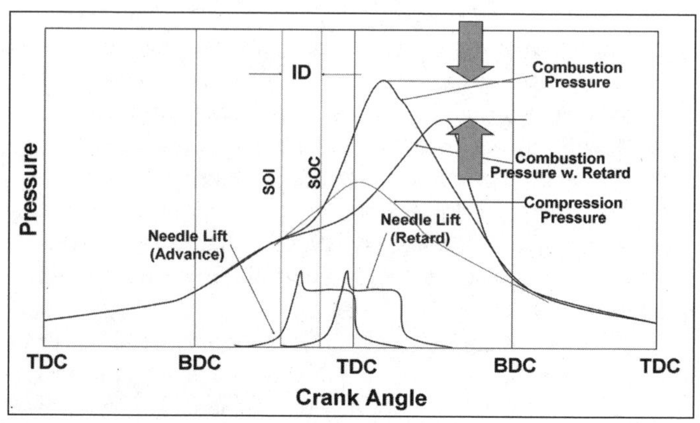

Figure 13.4 Schematic representation showing the relationship between injection timing and cylinder pressure. (Reprinted from Ref. 2.)

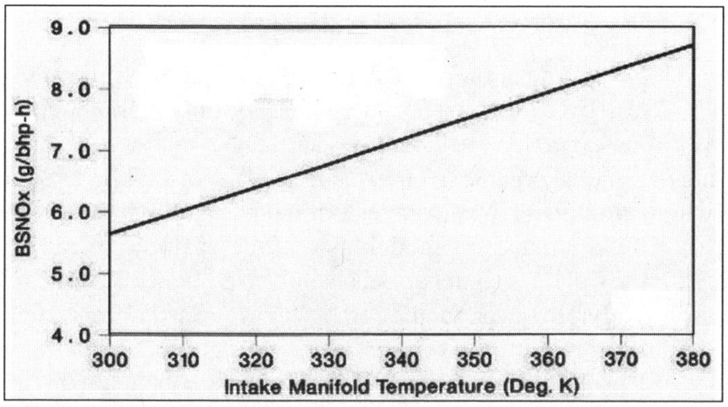

Figure 13.5 *Effect of intake manifold temperature on NO_x emissions. (Reprinted from Ref. 4.)*

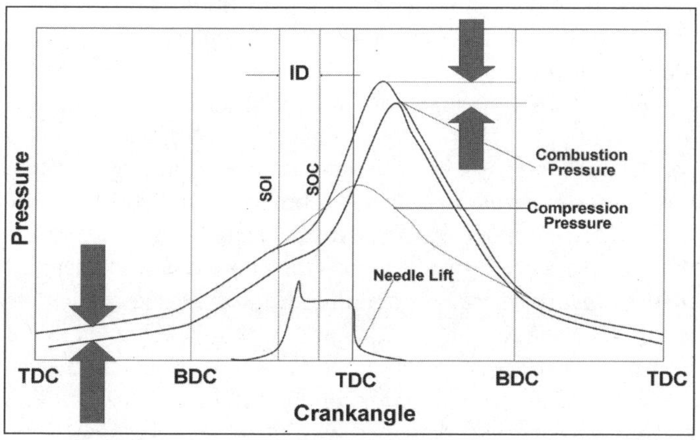

Figure 13.6 *Schematic representation showing how intake manifold temperature affects cylinder pressure. (Reprinted from Ref. 2.)*

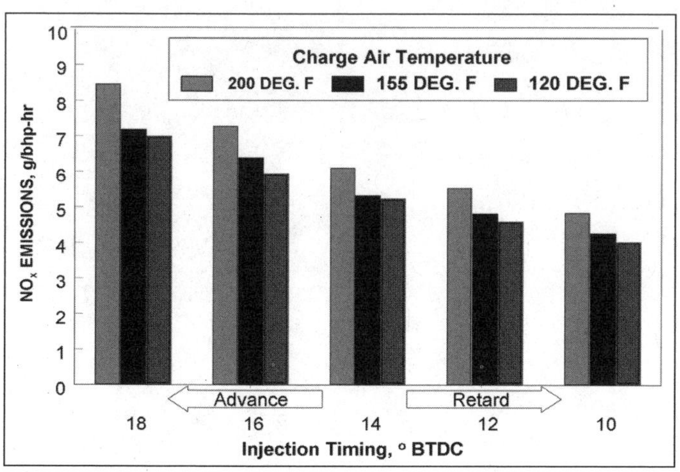

Figure 13.7 *Combined effect of injection timing and charge air cooling on NO_x emissions. (Reprinted with permission from Ref. 5, Fig. 3.)*

13.2.4 Combined Effect of Injection Timing Retard and Charge Air Cooling on Fuel Consumption

It is true that the combined effect of injection timing retard and charge air intercooling helped diesel engines meet the NO_X standards that were in effect in the early 1990s. However, if there are any reasons for the diesel engine to exist, fuel economy and durability would be at the top of the list. Conscious of this fact, engine design engineers always strive to not only preserve, but also improve, the engine's fuel economy advantage. If no other measures were taken other than retarding injection timing for NO_X control, a substantial loss in fuel economy would be experienced. Therefore, just simply retarding injection timing is not quite acceptable for the diesel engine. Figure 13.8 shows the trade-off between NO_X emissions controlled via injection timing retard and brake-specific fuel consumption for a medium-duty DI diesel engine [4].

Reducing NO_X emissions from about 10.7 to about 4.5 g/bhp-hr caused a 6% loss in fuel economy in engine designs from the late 1980s and early 1990s [5]. Reasons for this loss in fuel economy are attributed to the loss in peak combustion pressure that leads to reduced cycle work (the area between the combustion diagram and the compression diagram in Figures 13.4 and 13.6). To compensate for the loss of work (power), more fuel is provided to produce the desired work and this, in turn, leads to more fuel consumption.

A 6% loss in fuel economy for medium- or heavy-duty engine applications is totally unacceptable to the trucking industry, which sometimes survives by virtue of its fuel savings. It is necessary then for additional measures to be taken not only to recover, but also to improve, fuel economy.

13.2.5 Effect of Injection Pressure on Fuel Consumption

Basic research conducted on a single-cylinder engine showed that increasing peak injection pressure from 700 to 1000 bar had a significant impact on fuel consumption. The experiment was performed at different injection timings and

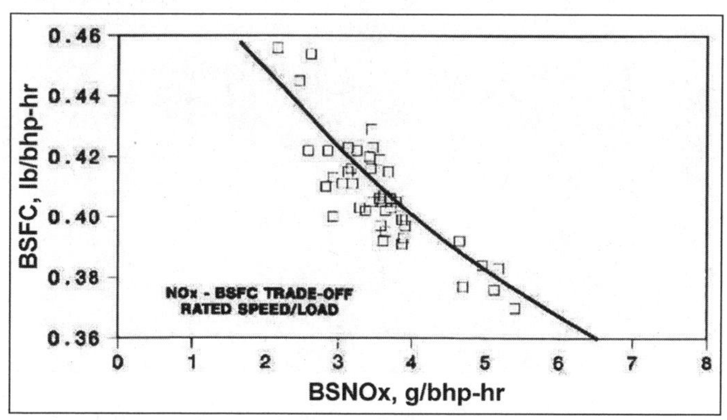

Figure 13.8 *Trade-off between NO_x and BSFC due to injection timing retard. (Reprinted from Ref. 4.)*

resulted in fuel consumption reduction at all injection timings. Because the experiment was conducted on a single-cylinder engine, the indicated specific fuel consumption was used to take into account the increased frictional and parasitic loads of the engine accessories. Figure 13.9 illustrates the effect of injection pressure on fuel consumption at various NO_x concentrations [2].

Applying high injection pressure in multicylinder engines confirmed the benefits of fuel economy that were demonstrated in single-cylinder experiments. Figure 13.10 illustrates the mechanism of recovering work through high injection pressure even when retarding injection timing to control NO_x emissions.

High injection pressure was obtained through faster injection cam profiles in the pump-line-nozzle systems that were prevalent in the early 1990s. In addition, small hole nozzles combined with a faster cam profile helped achieve peak injection pressures in excess of 1000 bar. Of course, adopting unit injector systems gives the highest injection pressure potential, especially when coupled with small injection hole size as discussed in Chapter 4.

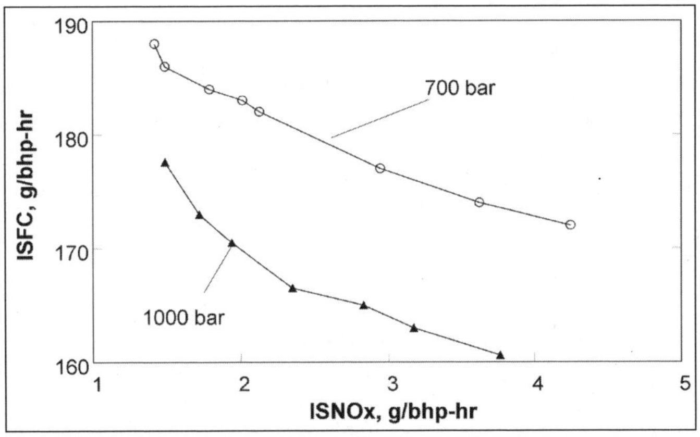

Figure 13.9 Effect of injection pressure on fuel consumption. (Reprinted from Ref. 2.)

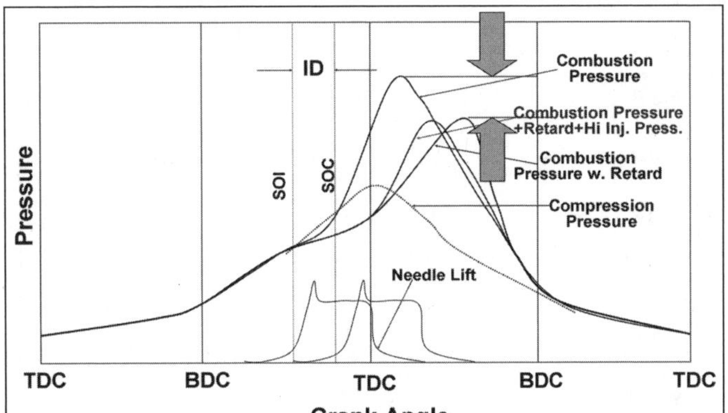

Figure 13.10 Schematic Illustration of the Benefit of High Injection Pressure in Increasing Cycle Work. (Reprinted from Ref. 2.)

13.3 Reduction of Particulate Matter

Figure 13. 11 shows a breakdown of particulate fractions for a typical 1988 California production heavy-duty diesel engine. As discussed in Section 7.3, diesel particulate matter is divided into the soluble organic fraction (SOF) and the insoluble fraction (which are obtained by Soxhlet extraction in dichloromethane [6] or by alternative methodology). In the example of Figure 13.11, the split into solubles and insolubles is about 39 and 61%, respectively. However, for the combustion system developer, this first breakdown into two fractions is not entirely satisfactory because each fraction contains subfractions associated with the fuel and lube oil. To further analyze the SOF for fuel and lube oil contributions, a gas chromatographic technique, described in Cartellieri and Tritthart [7], is used routinely for research and development activities concerned with particulate reduction. In the example of Figure 13.11, lube oil contributes 73% to the organic soluble fraction and 29% to the total particulates.

Turning to the insoluble fraction, the major portion is dry soot, a product of "too-rich-to-burn" fuel combustion. However, there is evidence that lube oil also contributes to this fraction [7]. In Figure 13.11, the insoluble lube oil fraction was estimated at 0.02 g/bhp-hr (5% of total particulate). A precise method for determining the carbonaceous matter originating from the lube oil is described by Mayer, Lechman, and Hilden [8]. They developed a carbon-14 radioactive tracer technique to determine the contribution from the engine's oil sump to the exhaust particulate matter collected on a filter. They analyzed particulate matter from a light-duty swirl chamber diesel engine operating at speeds and loads typical of light-duty driving conditions. The most noteworthy outcome of their work was the fact that all radioactivity was recovered with the SOF; no radioactivity remained on the extracted filters. Mayer *et al.* concluded that lube oil had significantly contributed to the particulate matter

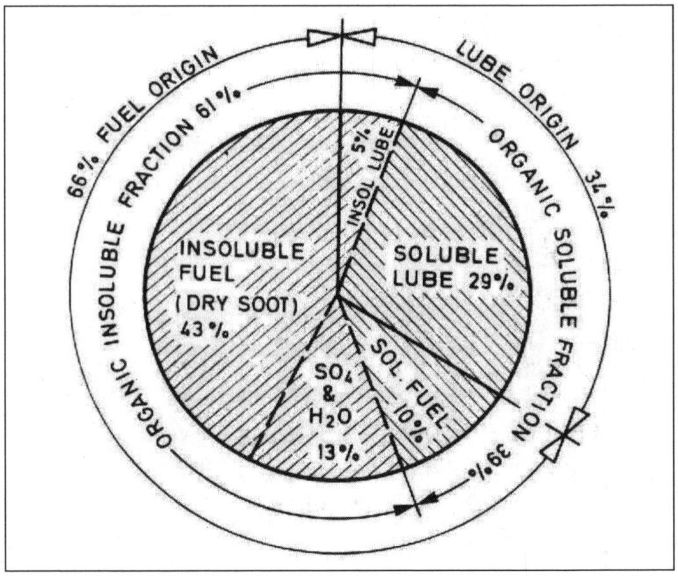

Figure 13.11 Constituents of diesel exhaust particulate matter. (Reprinted from Ref. 6.)

after most of their chemical analysis had been completed. While this conclusion may apply to light-duty driving conditions, it may not fully apply to heavy-duty operating conditions.

The insoluble (dry soot) carbon fraction in Figure 13.11 may be determined by subjecting the dichloromethane nonextractable portion to a second extraction in isopropanol/water [6]. The majority of the nonextractable portion of this second extraction process may be taken as carbonaceous matter, but wear metals and ash from the fuel and lube oil additives may also be included.

The preceding analysis of particulate matter is in general agreement with the findings of Khair [5], where the organic insoluble fraction was found to be 47% of total particulate matter. The total SOF in this case was about 41% versus 39%, as reported by Cartellieri *et al.* [6]. The findings from Ref. [5] are illustrated in Figure 13.12 and will be used as reference for simplicity.

Having determined the composition of particulate, one can clearly see that the carbonaceous and unburned fractions are by far the most significant components of particulate. Therefore, working to reduce these two components might bring about the biggest reward. Reducing carbonaceous particulate is a task that involves more than one of the engine subsystems. For instance, without proper mixing of fuel and air, combustion efficiency is expected to suffer and carbonaceous particulate will be emitted. However, mixing of fuel and air involves both the fuel system as well as the induction system. Therefore, attention must be given to at least these two systems if we are to reduce carbon in particulate emissions. Additionally, unburned lube oil may have many sources such as cylinder bore honing, piston ring pack specifications, valve stem seals, turbocharger seals, and other miscellaneous engine components. Properly engineering these components is imperative if particulate matter is to be reduced through engine design.

In the next sections, particulate matter reductions through the following systems will be reviewed:

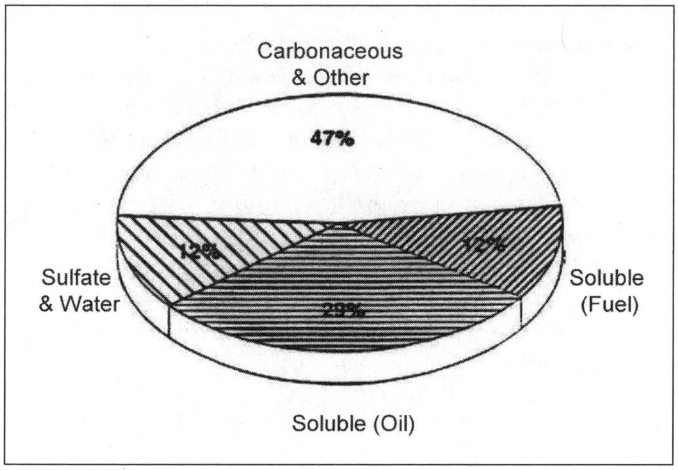

Figure 13.12 Graphical representation of the constituents of typical particulate matter . (Reprinted with permission from Ref. 5,Fig. 4.)

- Air management
- Combustion
- Oil consumption control
- Fuel injection.

13.3.1 Reduction of Particulate Matter through Air Management

Engine air management begins at the intake air filter and ends with the delivery of air into the cylinder. How charge air is managed through the entire system will have a significant impact on the products of combustion. Airflow, its temperature and pressure, as well as its kinetic energy, are all important parameters for controlling the mixing process. The influence of turbocharger performance, charge air cooling, intake manifold design, and intake port design are addressed in the following subsections.

13.3.1.1 Turbocharging

Thermodynamically, turbochargers are attractive because they recover energy that would have otherwise been exhausted into the atmosphere. In turbomachinery, low pressure ratios, reduced airflows, and low mechanical efficiencies are associated with low engine speeds [9] (see Figure 13.13). Engine air supply is critical at low speeds if smoke and soot are to be controlled.

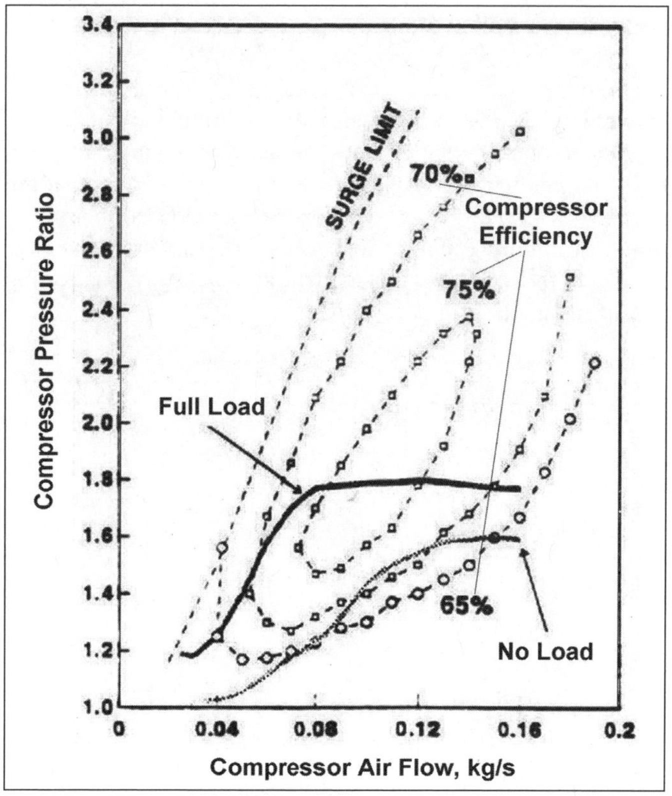

Figure 13.13 Performance characteristics of turbocharger compressor. (Reprinted with permission from Ref. 5,Fig. 5.)

Using an unoptimized turbocharger at low engine speeds may not provide adequate air-to-fuel ratios to control smoke. Using an unoptimized turbocharger at high engine speeds may cause excessive swirling in the combustion chamber. This distorts the injection plumes and hurts fuel economy. To meet 1991 emissions standards, it was extremely important to maximize the air-to-fuel ratios at low engine speeds to control black smoke with increasing loads. Higher turbocharger speeds had to be limited to avoid the negative effect of overswirling on the mixing process and to extend the turbocharger's life.

To optimize the match of turbochargers to engine needs, some manufacturers used variable geometry turbochargers in their development efforts. This "flexible" device can provide a range of pressure ratios for a given engine speed. Variable-geometry turbochargers can, therefore, significantly reduce smoke, especially at low speed conditions. They can also improve engine output at low speeds within a specified smoke limit. Figure 13.14 shows the impact of increasing the air-to-fuel (A/F) ratio on full-load smoke for an engine at a fixed speed of 1000 rpm. Figure 13.15 shows the fuel economy and smoke improvements at speeds below 2000 rpm. These improvement were recorded at higher torque outputs made possible by the variable-geometry turbocharger.

Reference is made to Figure 3.23 of Chapter 3 for additional results showing smoke improvements at speeds below 2000 rpm. Figure 3.23 gives an expanded view of the smoke characteristics between the fixed-geometry (i.e., conventional) turbocharger and its variable-geometry counterpart. Meaningful smoke reductions were a good incentive for carefully matching the conventional turbocharger to the specific needs of the engine.

Having recorded the advantages of improved engine "breathing" at low speeds, especially at high load conditions, engineers sought to achieve these results through upgrading the turbocharger to match engine needs. The incentive for avoiding the use of variable-geometry turbochargers (VGTs) in production was mainly economical because VGTs were expensive devices. Other concerns included the complexity of controlling VGTs and the lack of durability data.

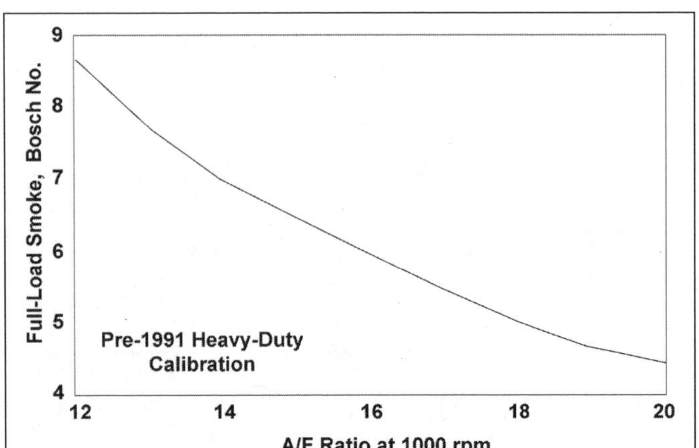

Figure 13.14 Effect of A/F ratio on full-load smoke. (Reprinted with permission from Ref. 5, Fig. 6.)

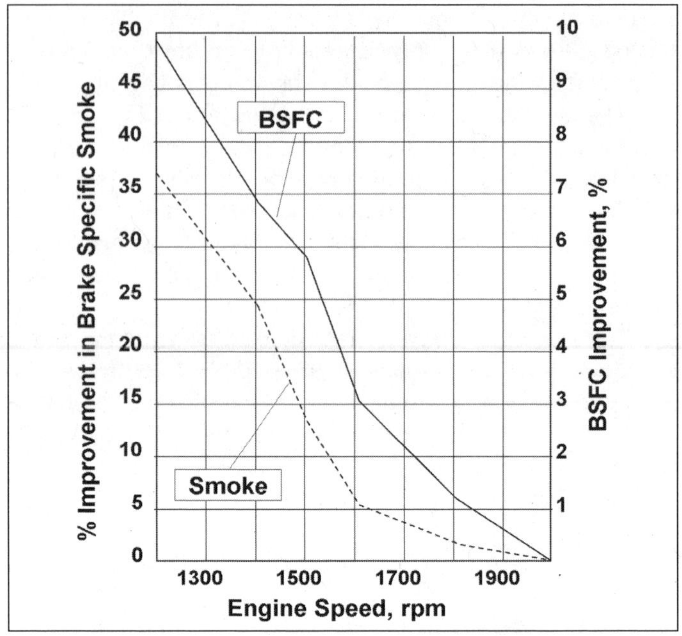

Figure 13.15 Brake-specific smoke and fuel consumption improvements with a variable-geometry turbocharger. (Reprinted with permission from Ref. 5, Fig. 7.)

The compressor selection process is complicated by the fact that it is not possible to simultaneously maximize turbocharger efficiency at all engine operating conditions. The need to maintain adequate compressor surge margin may further restrict compressor selection. More importantly, however, is that a fundamental trade-off exists between the competing requirements of having adequate boost and an adequate A/F ratio at the engine peak torque speed while avoiding excessive boost, which causes peak cylinder pressures to exceed engine structural limits at high speed. Besides these steady-state design trade-offs, the engine designer must also carefully consider the turbocharger transient operation. A major concern is the "turbocharger lag." This phenomenon refers to the inability of the turbocharger to respond instantaneously to changes in engine speed and load because of its rotating inertia (see Chapter 3).

Turbocharger lag can be a problem under transient operation because the compressor flow may not be sufficient to meet engine requirements during the lag period. This temporary shortage of air may cause excessive exhaust smoke and can limit engine responsiveness. To minimize the effects of turbocharger lag, it is desirable to reduce the turbine A/R. (See Section 3.7, Chapter 3). This increases airflow and keeps the rotational speed of the turbocharger high, but may result in excessive boost and cylinder pressure under high engine speed and load conditions.

A turbocharger technology that has received interest from heavy-duty engine designers is the use of ceramic turbine wheels. These wheels have little effect on steady-state turbocharger performance, but because they have much less inertia than conventional metallic rotors, they significantly reduce turbocharger lag. Engine response and transient smoke characteristics can be improved without resorting to an A/R that overboosts the engine at high speed and without applying waste gates or variable-geometry turbocharging. Ceramic turbine wheels are fairly commonplace on the smaller turbochargers used on passenger car engines, however, their usage on larger heavy-duty diesel turbochargers has been limited because of costs and the difficulties of producing defect-free ceramic castings in the larger size. As ceramic casting technology improves, wider utilization of ceramic turbine wheels on heavy-duty engines can be anticipated.

Several turbocharger technologies are available to address these design trade-offs. One of these technologies is wastegating. A waste gate is a passage in the turbine housing that allows exhaust gas to bypass the turbine and pass directly into the vehicle exhaust system as described in Chapter 3. The waste gate is closed at low intake manifold pressure and is opened at manifold pressures above some preset trigger level. In essence, the waste gate is a boost limiting device and, as such, the valve opening pressure and the flow passage must be designed to ensure that engine boost and cylinder pressure limits are not exceeded.

Use of a waste gate allows the designer to optimize the engine/turbocharger match for low-speed and transient operation without being constrained by high-speed boost limits. In most cases this results in the selection of a lower turbine A/R than would otherwise be possible. The lower A/R ensures sufficient boost and airflow to meet low-speed engine demands, and because it increases turbine speed, it reduces turbocharger lag and improves turbocharger and engine response.

Although wastegating is an effective way to allow the designer to maximize low-speed and transient engine performance, the addition of the waste gate adds to the cost and complexity of the turbocharger and necessitates additional development to prove the reliability and durability of the waste-gate valve and valve actuator. In addition, "wasting" of exhaust energy results in reduced overall engine efficiency at high speeds. If this trade-off is not acceptable, the designer may choose to go to the next level of turbocharger design sophistication, which is the variable-geometry turbocharger discussed earlier. With a VGT, movable components are incorporated in the turbine that allow the turbine flow area or the flow angle of the gas entering the turbine to be continuously varied. The variable-geometry feature allows turbine performance to be optimized over a wide range of operating conditions. Figures 13.16 and 13.17 illustrate two types of VGTs.

Figure 13.18 illustrates an example of what was achieved through the optimization of the turbine housing. Use of a turbine housing with a smaller A/R ratio (X2) as opposed to a baseline A/R ratio (X1) reduced smoke levels, especially at lower engine speeds. Similar improvements were achieved by careful optimization of the compressor aspect ratio.

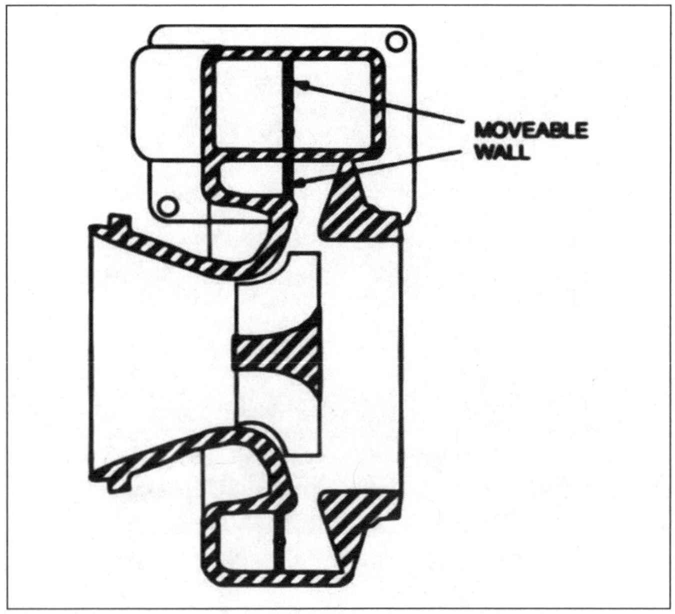

Figure 13.16 Movable wall variable-geometry turbocharger. (Reprinted from Ref. 10.)

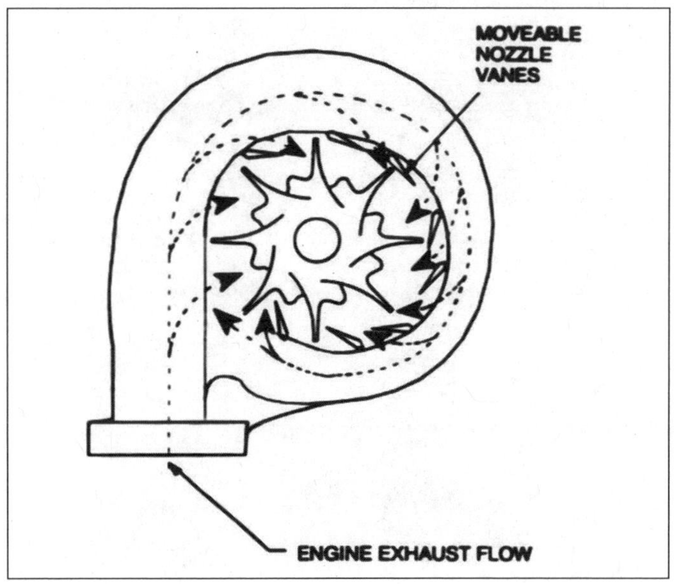

Figure 13.17 Movable nozzle vanes turbocharger. (Reprinted from Ref. 10.)

13.3.1.2 Charge Air Cooling

Charge air cooling improves the specific power output of an engine by increasing charge air density. Benefits derived from this improved specific power output are better fuel economy and reduced exhaust emissions. Prior to the 1990s, most aftercooled (for purposes of this chapter aftercooling and intercooling are used interchangeably) engines utilized jacket-water cooling in which charge air temperatures were thermostatically controlled to about 195°F. Figure 13.19 shows a water heat exchanger installed in the intake manifold of a 7.8-L, 210-hp, medium-duty DI diesel engine [2]. However, for the 1991 engines, the use of air-to-air aftercooling (see Figure 13.20) became quite popular for medium-duty and heavy-duty diesel engines, and some light-duty engines included air-to-air intercooling

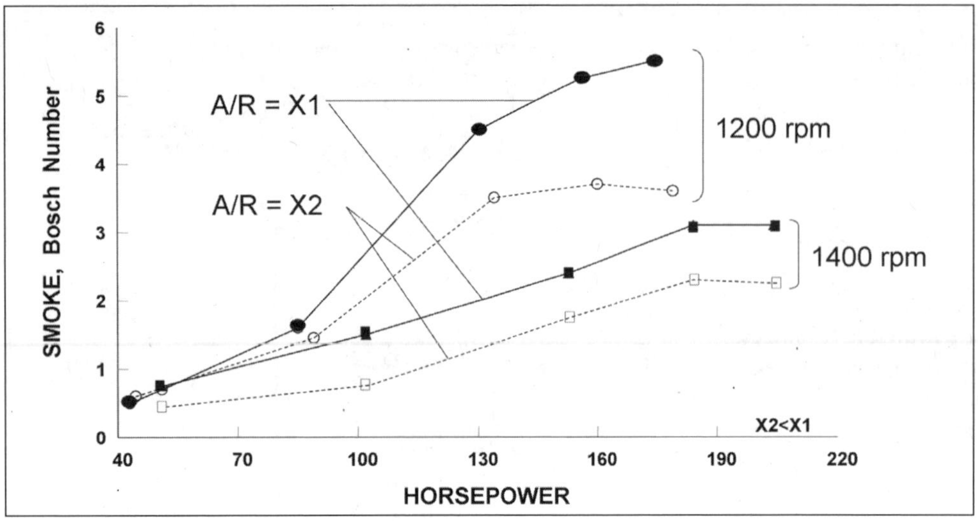

Figure 13.18 *Effect of turbine A/r Ratio on smoke from a medium-duty DI diesel engine. (Reprinted with permission from Ref. 5, Fig. 9.)*

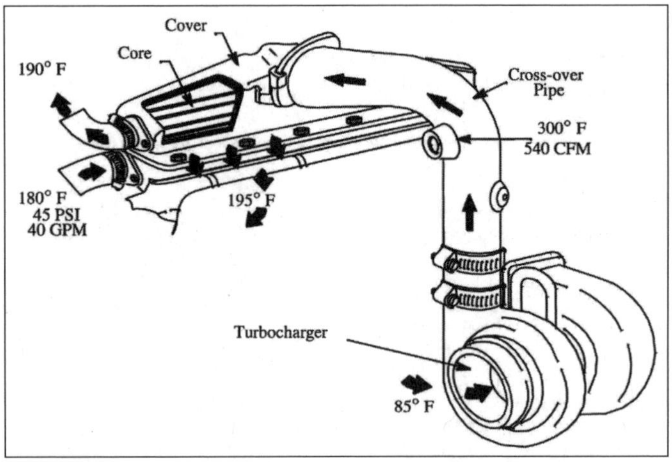

Figure 13.19 *Jacket-water cooling of charge air. (Reprinted from Ref. 2.)*

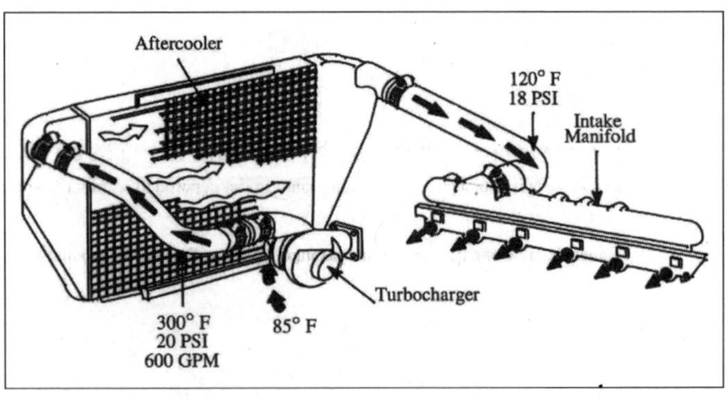

Figure 13.20 *Schematic representation of air-to-air charge air cooling. (Reprinted from Ref. 2.)*

in production. Figure 13.21 shows the improvement in fuel economy associated with three charge air temperatures (200, 155, and 120°F). These temperatures were selected to represent a water jacket aftercooled engine, an engine with intermediate charge air temperature, and an air-to-air aftercooled engine, respectively. By adopting air-to-air intercooling, intake manifold temperatures of 25 to 35°F above ambient were achieved, which led to higher air densities. Air-to-air aftercooling also reduces the thermal loading of the engine by limiting combustion temperature extremes. This leads to extended engine life and improved reliability.

The net effect of low-temperature charge air cooling was increased air density that led to increased mass of air available to the combustion process and increased particulate oxidation. In addition, lower intake manifold temperature led to lower peak cylinder pressure and temperature. Lower fuel consumption

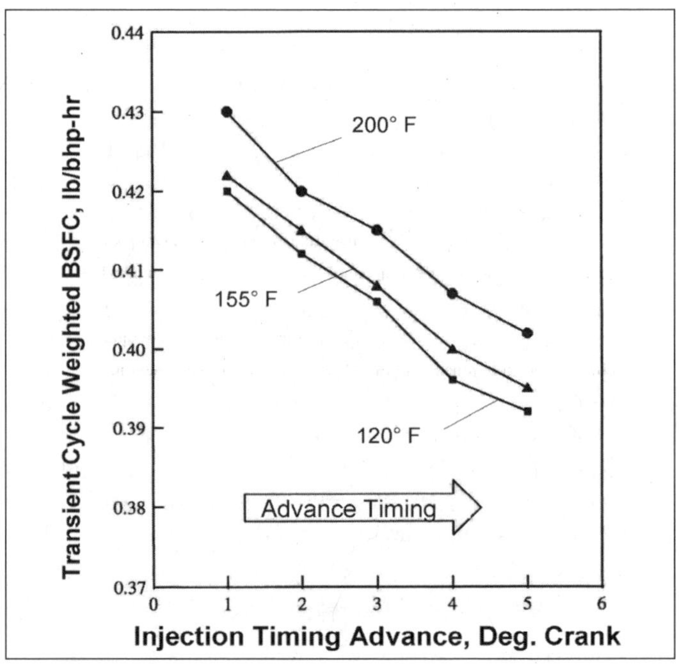

Figure 13.21 *Effect of charge air temperature and injection timing on transient cycle fuel consumption. (Reprinted from Ref. 2.)*

associated with low-temperature charge air cooling indicates that more of the fuel is converted into brake power, that is, the unburned fuel fraction is reduced as well as its corresponding particulate matter [10].

Air-to-air charge cooling design for many diesel engine applications offers the best compromise between charge cooling potential, system cost, and complexity. The heart of the system is the air-to-air heat exchanger. The heat exchanger consists of a core of cross-flowing finned passages for the charge air, and the ambient cooling air. Inlet and outlet header sections for the charge air cap each side of the heat exchanger, as shown in Figure 13.22.

To achieve maximum charge cooling, the air-to-air heat exchanger is normally placed just ahead of the engine radiator (Figure 13.23). If the vehicle does not have air conditioning, the heat exchanger and the radiator usually have the same frontal area. In parallel flow designs, the radiator and air-to-air heat exchanger are placed side-by-side or stacked vertically as shown in Figure 13.23. It is always desirable to minimize the cooling airflow restriction since this maximizes the opportunity to achieve engine cooling with ram air and minimizes the parasitic engine power losses associated with cooling fan operation. Series designs place a premium on minimizing the cooling air restriction. Because of the interrelationships with the vehicle configuration, the heat exchanger design may need to be customized for each engine/vehicle combination.

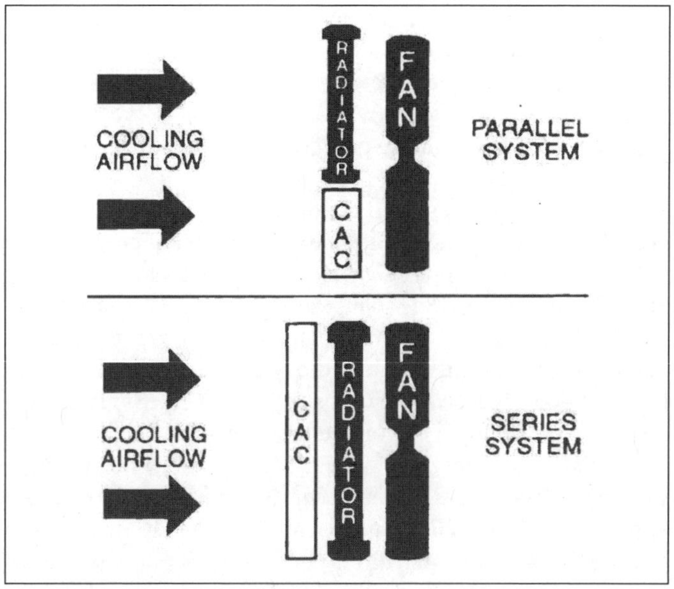

Figure 13.23 *Air-to-air and radiator mounting options. (Reprinted from Ref. 10.)*

13.3.1.3 Intake Manifold Design

While it is difficult to relate intake manifold designs to particulate matter, they do contribute to good mixture preparation, without which combustion may be incomplete, leading to considerable particulate emissions. In fact, in a multicylinder engine, charge air maldistribution in just one cylinder can create a poor mixture in that cylinder and produce heavy smoke and particulate emissions that would reflect on the entire engine [2].

Although intake manifolds appear to be simple components, they play a major role in engine performance and deserve careful attention from engine designers. It is important for the intake manifold to distribute the air evenly to the cylinders. Considerable bench testing and development led to intake manifold designs that have even flow distribution to all cylinders with minimum pressure losses. The intake port shape and manifold design are often tested as a total system rather than as two separate components. In some cases, effort was expended in reducing the impact of consecutive intake valve openings on the pressure waves in the intake manifold. Segmenting the intake manifold, then grouping certain ports to the proper manifold segment can prevent the interference of unsteady airflow into the combustion chamber [5].

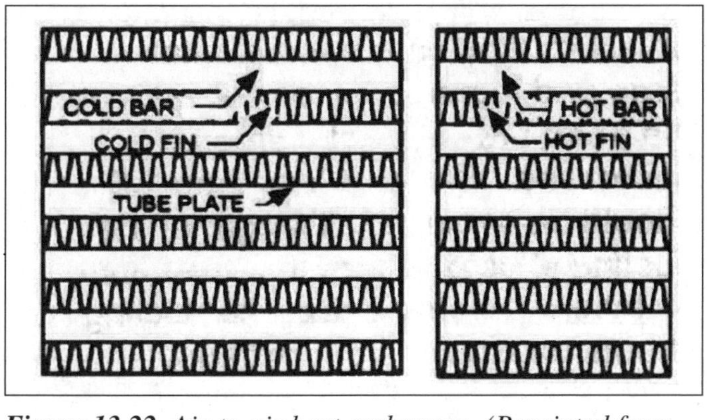

Figure 13.22 *Air-to-air heat exchanger. (Reprinted from Ref. 10.)*

To assist with engine breathing, the designer may wish to consider the use of tuned manifolds. Tuned manifolds are specially designed to make use of pressure waves initiated by the engine valve events. In the case of the exhaust, cylinder blowdown occurs when the exhaust valve opens, creating a compression pressure pulse. In a properly tuned manifold, the compression pulse is reflected and returned to the exhaust valve as a rarefaction wave prior to the valve closing. Low pressure from the rarefaction wave improves cylinder scavenging and reduces exhaust stroke pumping work [10].

On the intake side, rarefaction is generated at that valve opening and is returned to the intake valve as a compression pulse just before valve closing. The pulse improves cylinder charging and reduces intake stroke pumping work. Because tuning relies on the timed arrival of reflected pressure disturbances, benefits are obtained only at certain engine speeds. Typically, designs are optimized at the engine's peak torque speed because the greatest need for increased airflow occurs at this speed [10].

13.3.1.4 Intake Port Design

The mixing of fuel and air in a diesel engine has a very strong influence on its performance and emission characteristics.

Diesel emission control engineers are always challenged by the fuel and air management to achieve optimum performance. Flow bench fixtures are used to select intake port designs that impart good swirl or air rotation while maintaining good breathing quality. Two types of intake port designs are common. The first is known as the tangential port and the second as the helical port. The former gives good breathing qualities, but very little circumferential air motion to help in the mixing process. The helical port provides more air motion, thus assisting in the air and fuel mixing. Mixing using helical port designs comes at the cost of more pressure loss than that experienced with tangential ports. An illustration of the two types of ports is found in Heywood [11] and shown in Figure 13.24. Some engine manufacturers adopted a quiescent combustion system using a tangential port as opposed to the helical design associated with swirl ports. The quiescent combustion system is mostly matched with higher fuel injection pressures to optimize combustion efficiency. Other manufacturers chose swirl-supported designs that provide air motion, especially at low engine speeds where it is most needed. At high engine speeds, a variety of methods are used to avoid overswirling, which can lead to increased soot and particulate emissions as well as fuel consumption.

It is very difficult to qualify and measure swirling flow in an actual operating engine. Therefore, steady flow tests are often used to characterize swirl generated in intake ports. Engine designers use flow benches to design the right port type and shape for their engines. An example of these flow benches is shown schematically in Figure 13.25. The core of this test fixture is its impulse meter, which consists of a honeycomb matrix (A) (see inset of Figure 13. 25), placed in a steady flow stream through cylinder (B) and normal to its axis. The impulse meter arrests all swirl so that reaction torque (C) equals the total angular momentum flux irrespective of axial or tangential velocity distributions. Cylinder head (D) is placed atop a cylinder (B) to allow one intake port to be tested at a time. A bell-mouth insert (E) is placed upstream of the cylinder head to provide smooth, nonturbulent entry of inducted air (m˙).

Air inducted into the fixture has a flow rate of m˙, and usually at controlled pressure and temperature conditions, P_o and T_o, respectively. A water manometer (F) is used to regulate the pressure drop across the intake port to 10 inches of water

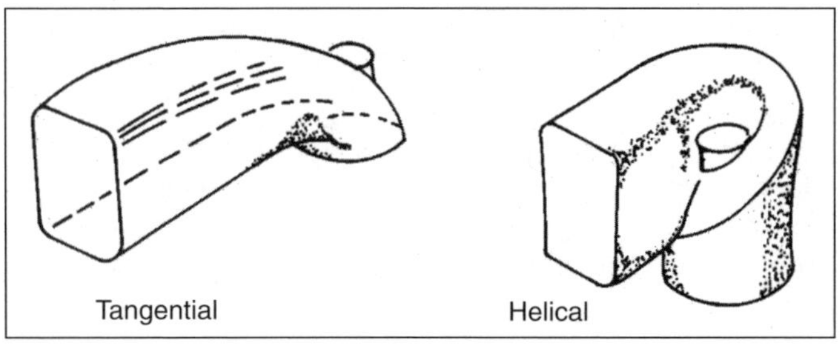

Figure 13.24 Schematic representations of tangential and helical intake ports. (From Internal Combustion Engine Fundamentals *by John B. Heywood. ©1988 Reproduced with permission of The McGraw-Hill Companies.)*

by modulating the delta P control valve (G). The flow bench shown in Figure 13.25 uses the "suction" method, meaning that conditioned air (at P_o and T_o) is sucked through the intake port into cylinder (B), then rectified through the honeycomb matrix (A) by means of a motor-driven blower (H) before it is exhausted back to atmosphere. A hot-wire anemometer (I) is provided to monitor airflow. Swirl torque from the flow rectifier and airflow signals are indicated on the torque readout (J) and the flow readout (K), respectively. An arbor-mounted valve adjustment mechanism (L) is equipped with a cam and return spring assembly (M) and with a micrometer (N) to depress the valve and measure its lift.

The total torque exerted by the swirling flow equals the flux of angular momentum through the plane coinciding with the flow-straightener upstream face [11]. A swirl coefficient is defined that essentially compares the flow's angular momentum with its axial momentum. When swirl measurements are made in an operating engine, a swirl ratio is normally used to define the swirl. It is defined as the angular velocity of a solid-body rotating flow ω_s, which has equal angular momentum to the actual flow, divided by the crankshaft angular rotational speed ($2\pi N$), expressed as follows:

$$R_s = \omega_s / 2\pi N \tag{13.1}$$

During the induction stroke in an engine, the airflow, valve open area, and consequently the angular momentum flux into the cylinder vary with crankangle. In rig tests, however, airflow and valve open area are fixed and the angular momentum passes through the cylinder continuously. Steady-state impulse torque meter flow bench data can be used to estimate engine swirl. Typically, flow and swirl measurements are conducted at fixed valve lift increments then integrated to obtain an average value of flow and torque for an intake port design. A mathematical derivation is found in Monaghan and Pettifer [12] and described in Eqs. (13.2) and (13.3) as modifiers of Eq. (13.1):

$$\omega_s = \frac{8}{B^2}\left(\int_{\theta_1}^{\theta_2} T d\theta\right) \Big/ \left(\int_{\theta_1}^{\theta_2} m d\theta\right) \tag{13. 2}$$

$$R_s = \frac{\omega_s}{2\pi N} = \pi\eta_v BL\left[\int_{\theta_1}^{\theta_2}(A_v C_D d\theta)\right] \Big/ \left[\int_{\theta_1}^{\theta_2}(A_v C_D) d\theta\right]^2 \tag{13. 3}$$

where B is the cylinder bore, T is the swirl torque, θ_1 and θ_2 are crankangles at the start and end of the intake process, N is the engine speed, η_v is the volumetric efficiency, L is the valve lift, $m^\bullet$ is the air mass flow rate, and $A_v\ C_D$ is the effective valve flow area at each crankangle.

Flow and swirl characterization performed on rigs such as that shown in Figure 13.25 are typically conducted at various valve lifts. An intake port test performed for a 7.8-L Ford Tractor Operations diesel engine produced the results shown in Figure 13.26. The port was helical and designed for a swirl-supported combustion system, and the test was performed at

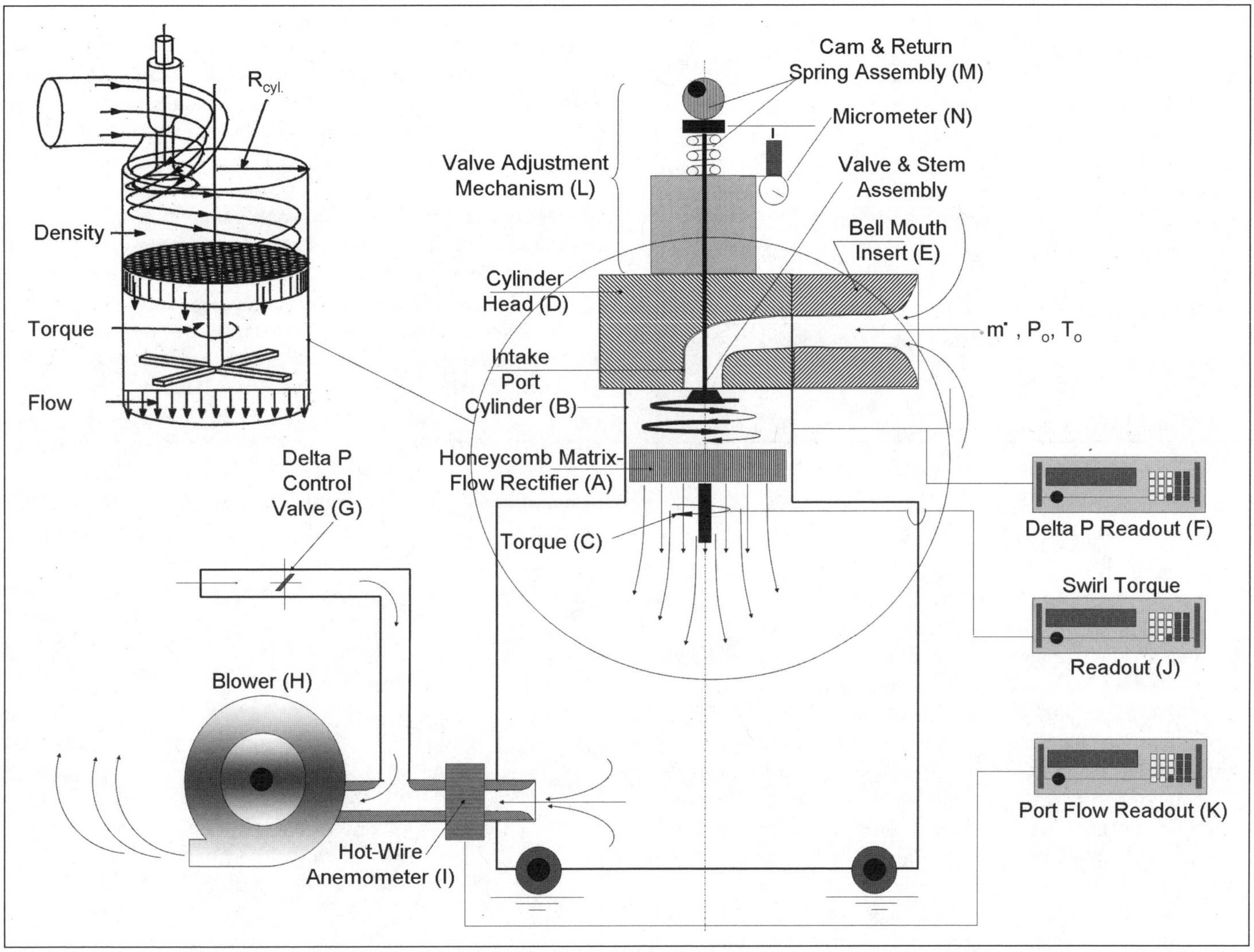

Figure 13.25 *Schematic representation of intake port flow and swirl bench.*

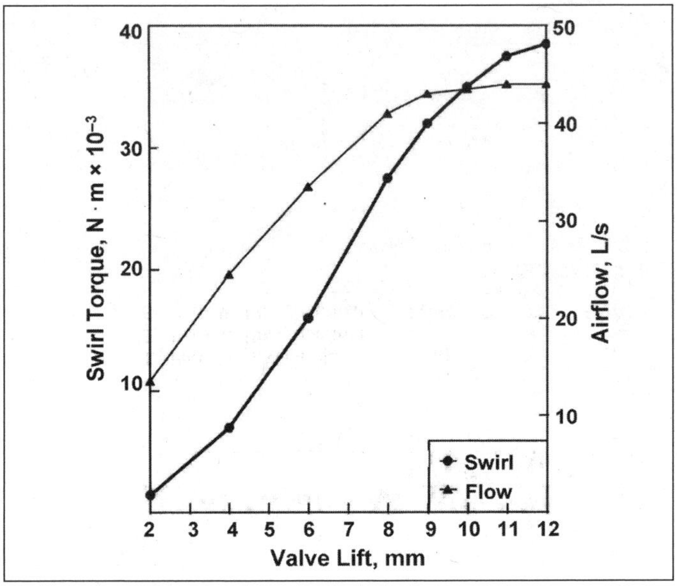

Figure 13.26 *Flow and swirl characteristics of a helical intake port. (Reprinted from Ref. 15.)*

valve lift increments of 2 mm where both flow and swirl torque measurements were taken. These measurements were taken at a constant 10-inch H_2O pressure drop across the port. It is often common to narrow down the cross-sectional area of the port as flow approaches the valve stem area. This action is required to increase the kinetic energy of charge air as it enters through the intake valve opening. Optimization of the swirl ratio and charge airflow is usually required prior to finalizing intake port designs.

While swirl and flow research and development is tedious and quite involved, the result should be an intake port design that enhances complete combustion by developing the proper air and fuel mixture. Cartellieri and Herzog [6] describe work done at AVL on an 11-L, DI, turbocharged and intercooled, six-cylinder heavy-duty diesel engine. This engine was available in two-valve and four-valve configurations. The two-valve engine was used with a swirl-supported combustion system, whereas the four-valve engine served as the quiescent combustion engine. A description of the combustion system for these two configurations is given in Table 13.1; steady-state as well as transient

TABLE 13.1
COMBUSTION SYSTEMS DESCRIPTION

Parameter	Swirl-Supported Configuration	Quiescent Configuration
Number of valves/cylinder	2	4
Intake port design	Helical	Tangential
Swirl ratio	2.25	1
Bowl shape	Reentrant	Open
Fuel injection system	Pump-line-nozzle system	Pump-line-nozzle system
Injector location	Inclined/offset	Vertical/concentric
Number of spray holes	5	8
Compression ratio	19.5:1	17.5:1

Source: Data from Ref. 6.

particulate emissions results and their breakdown from these two engine configurations are given in Table 13.2.

Note that both systems were calibrated to produce about the same transient cycle NO_x emission of about 4.5 g/bhp-hr. Examination of the steady-state particulate matter results disclosed that the dry soot fraction was about 32% and 35% of total particulate for the swirl-supported versus the quiescent configuration, respectively. This result indicated that, generally, combustion efficiency for both systems was quite similar, which demonstrates their ability to mix fuel and air equally in a fashion that supports good combustion efficiency. Other particulate matter composition differences exist between the two configurations, but these differences are not involved with combustion quality and mixture preparation. Cartellieri and Herzog [6] concluded that, in practice, well-tuned combustion systems of either type provide similar dry soot emissions at the same NO_x level. Hence, neither of the two systems holds a definite advantage over the other. They do mention, however, that engines having smaller displacements are best served by swirl-supported combustion because it offers an additional degree of freedom in optimizing air utilization over a wide speed range. They also recommend quiescent combustion for engines with larger displacements and operating over a smaller speed range since the mixing energy is provided almost exclusively by the spray characteristics made possible by an extremely high fuel injection pressure.

In general, an increase in swirl improves the mixing process; reduces hydrocarbons, particulate matter, and smoke emissions; and enhances fuel economy. However, excessive swirl may produce spray overlap as illustrated schematically in Figure 13.27. Spray overlap is counter to good mixing, leading to pockets of rich zones where air is unable to penetrate and to fuel that has difficulty evaporating. The result is higher emissions and most likely poor fuel economy. It is important to note, therefore, that in swirl port development, the objective is not necessarily designing a high swirl ratio, but a swirl port whose design details match those of the fuel injection system in order to maximize mixing and combustion efficiency. Because this objective is difficult to achieve for engines with a wide speed range, some companies adopted intake systems having two intake ports, such as Komatsu's 15.2-L engine design, as seen

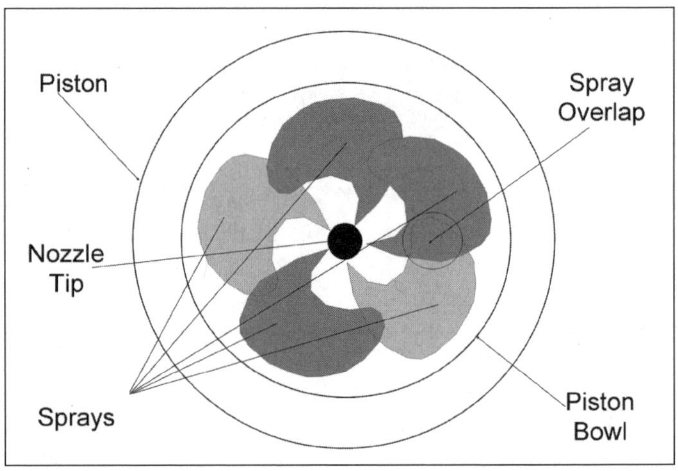

Figure 13.27 *Schematic representation of spray overlap.*

TABLE 13.2
STEADY-STATE AND TRANSIENT PARTICULATE EMISSIONS RESULTS AND THEIR BREAKDOWN FOR SWIRL-SUPPORTED AND QUIESCENT COMBUSTION SYSTEMS

	Swirl-Supported Configuration		Quiescent Configuration	
Parameter	Steady-State	Transient	Steady-State	Transient
Total particulate	0.210	0.252	0.240	0.279
Organic insoluble fraction	0.080	0.100	0.135	0.154
Soluble organic fraction	0.130	0.152	0.105	0.125
Lube oil organic fraction	0.092	0.110	0.093	0.097
Fuel organic fraction	0.038	0.042	0.012	0.028
Dry soot[a]	0.067	—	0.086	—

Source: Data from Ref. 6.
a. Estimated from steady-state Bosch smoke measurements.

in Figure 13.28 [14]. Other companies, such as Isuzu, adopted a novel design with a swirl destroyer for their 9.8-L engine, as shown schematically in Figure 13.29 [15].

The objective of these intake port designs and systems is mostly to provide high swirl ratios at low engine speeds and reduced swirl at high engine speeds to avoid spray overlap. In dual-port systems, the intake air stream with its momentum favors the helical port, which enhances mixing at low engine speeds. At high engine speeds, intake air flowing through the intake manifold favors the tangential port, which reduces swirl while maintaining good spray/air mixing characteristics. In the case of Isuzu's subport design, a solenoid valve-controlled charge air bleed is introduced at a predetermined speed in the opposite direction of the swirl. This action destroys the swirl and reduces it to preserve good mixing.

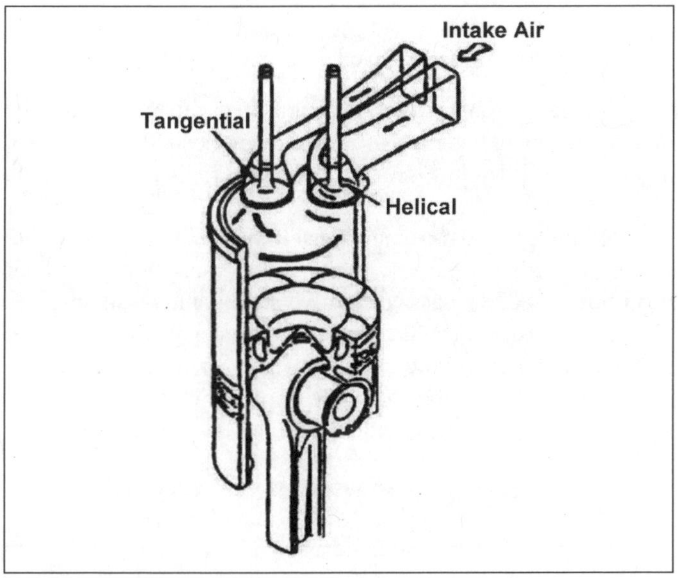

Figure 13.28 *Dual-intake port design for the Komatsu 15.2-L engine. (Reprinted from Ref. 14.)*

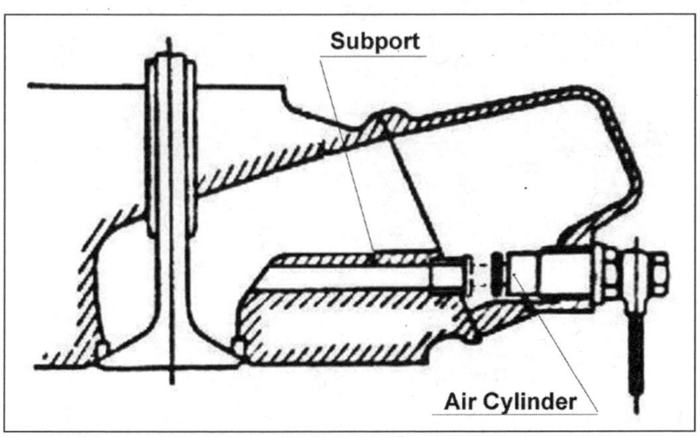

Figure 13.29 *Intake port featuring a subport design for the Isuzu 9.8-L engine. (Reprinted from Ref. 15.)*

13.3.1.5 Intake Manifold and Port Combination

So far intake manifold and intake port designs have been treated independently for their influence on mixing of fuel and air, combustion efficiency, and particulate emissions in a diesel engine. Yet, work done at Ford New Holland [14] and other engine and consulting companies produced results that indicated a strong effect on charge air swirl and flow characteristics from combining both the intake manifold and intake port. No matter how well optimized the intake port design might be, it has to produce the proper swirl and flow characteristics when the intake manifold is installed upstream from it. In fact, components of the intake system consisting of inlet piping, intake manifold, and intake ports ought to work in concert to deliver charge air to each cylinder in a way that would ensure proper mixing.

Figure 13.30 shows the results of bench testing a new six-cylinder head shortly after its machining to verify its port flow and swirl characteristics. When the six ports were tested without the manifold they produced maximum swirl torque ranging from about 26 to 28 × 10^{-3} N • m. The variance was normally acceptable for production in the early 1990s. Yet, when these ports were tested with manifold A installed, the swirl torque of port 1 showed a drastic drop.

One of the probable causes for this drop was the position of the intake manifold inlet. Charge air stream entering the intake manifold had to make a sharp turn, as indicated by the arrow in Figure 13.30, manifold A, to proceed into port 1. Manifold B was designed as an alternative to manifold A. The designers made a couple of changes in an attempt to improve swirl in port 1. The first change is obvious from the inset of Figure 13.30, where the inlet position was turned sideways to include cylinder 1 in the charge airflow path as suggested by the arrows. The second change was a slight increase in the manifold volume to allow easy access for the air into cylinder 1. The result was a small improvement, but still not enough to restore swirl in cylinder 1.

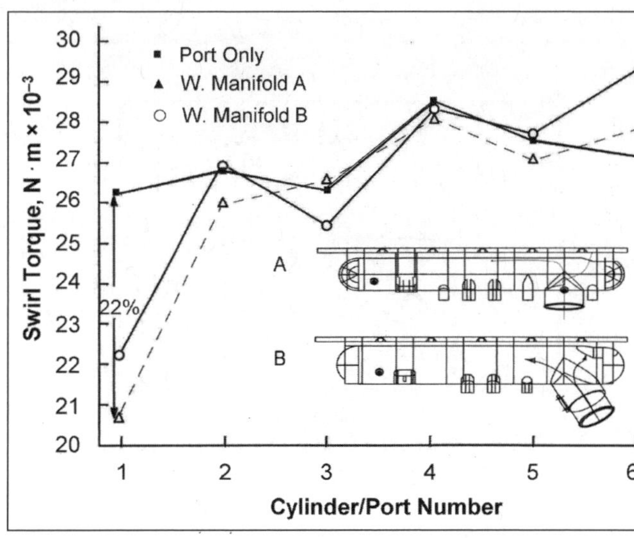

Figure 13.30 *Swirl characteristics of intake port with two different intake manifold designs. (Based on Ref. 13.)*

Although the problem could have been completely corrected, packaging constraints under-the-hood dictated that manifold B was chosen as a compromise for the final production design.

13.3.2 Reduction of Particulate Matter through Combustion System Design

13.3.2.1 Combustion Bowl Shape

A key goal in the design of the combustion bowl is for good mixing of fuel and air to be achieved. Turbulence in the air motion within the combustion bowl is found to be beneficial to the mixing process [6,16]. Swirl induced by the intake port can be enhanced to create more turbulence during the compression stroke through proper design of the crater, or bowl, in the piston crown. Designs of combustion bowls in the early 1990s are generally described as having a "reentrant" bowl shape. Figure 13.31 is a representative design of a reentrant bowl (solid line) with the traditional "straight-sided" combustion bowl design shown by a dashed line. Elements for the optimization of this design include consideration of different ratios of radii of the combustion bowl to the cylinder bore, different chamfer angles of the reentrant lip, and different aspect (bowl diameter/bowl depth) ratios [5].

The reentrant bowl features sides that slant inward toward the central axis of the piston. This feature improves mixing and turbulence of the air and fuel in the combustion space, which leads to generally reduced exhaust emissions. Tests have shown, however, that there is a progressive and significant deterioration in piston life as the bowl shape becomes more reentrant (the outer angle of the bowl becomes more severe). This flank angle should not exceed 15° to the vertical unless the piston is cooled by an oil jet or reinforced [17]. The sharper the edge of the reentrant bowl, the more likely it is to sustain cracks that may spread and eventually lead to total piston failure.

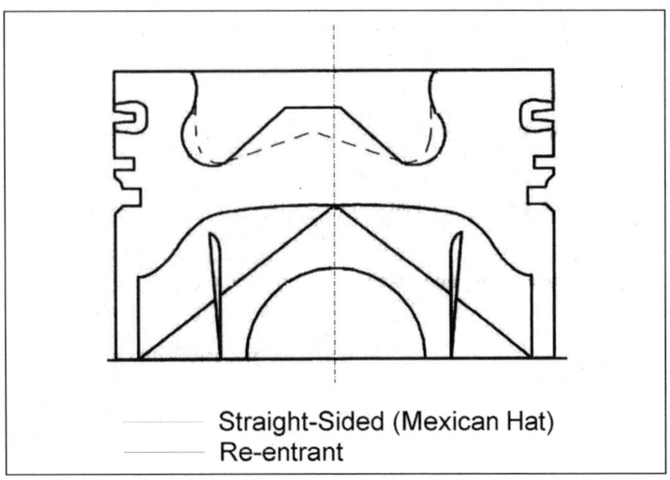

Figure 13.31 Comparison between a straight-sided and a reentrant bowl. (Reprinted with permission from Ref. 5, Fig. 13.)

Selecting the correct angle for the reentrant bowl is always a challenge and normally turns into a compromise. As stated earlier, the sharper the edge of the reentrant bowl, the worse it is for piston durability. Yet, for emission reduction purposes, the sharper the edge of the reentrant bowl, the lower the smoke characteristic as shown in Figure 13.32 [18]. It is thought that as the air enters the combustion bowl, it is sheered by the reentrant bowl's sharp edge, thus creating turbulence and eddies. This turbulent motion assists in mixing air with vaporized fuel that results from spraying finely atomized fuel on the hot surface of the reentrant bowl. Swirling air entering the cylinder from the intake port increases its swirl intensity as it enters the combustion bowl. The combined air motion resulting from the intake port swirl and the air entering the combustion bowl provides the mixing intensity necessary to support good combustion efficiency. As the piston moves away from top-dead-center (TDC), the swirling motion takes on an axial component and becomes a tumbling motion.

13.3.2.2 Combustion Bowl Location

In combustion systems featuring two-valve designs (one intake and one exhaust valve), injectors are usually installed at an angle as shown in Figure 13.33 [19]. The inclination angle of the injector is dictated by the many cavities in the cylinder head. Cylinder head designers must find room to accommodate intake and exhaust valves, intake and exhaust ports, cylinder head bolts, cooling passages, valve stems, and in some cases glowplugs. Positioning the injector at an angle results in nonsymmetrical nozzle hole drilling, which ensures that the fuel is sprayed in the combustion bowl. In addition to the nonsymmetrical drilling of the nozzle holes, the combustion bowl has

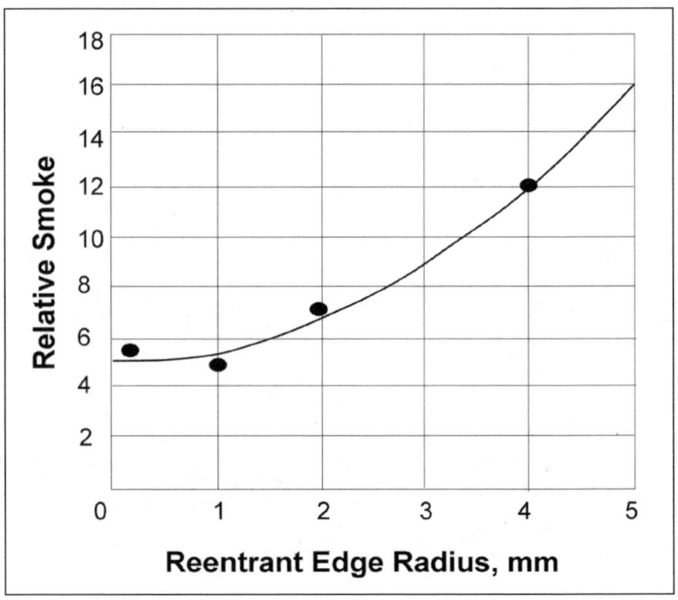

Figure 13.32 Effect of reentrant edge radius on smoke. (Based on Ref. 18.)

to be offset from the center of the cylinder bore to allow for concentricity between the nozzle tip and the combustion bowl. The lack of symmetry in drilling the nozzle holes leads to uneven spray lengths as shown in Figure 13.33, which in turn leads to uneven utilization of the compressed air in the combustion bowl. Moving toward the modern diesel engine design, two-valve combustion systems attempted to reduce the offset between the center of the combustion bowl and the cylinder bore for better air utilization [2, 5].

Figure 13.34 shows the effect on smoke and hydrocarbons (HCs) of matching of the fuel spray length and combustion chamber diameter as tested on various small size engines [20]. In these engines, smoke at high load and HCs at low load are the critical conflicting parameters. The fuel spray length is the measured length (L) of the liquid phase of fuel spray. Spray length is normalized using the combustion bowl radius (R) and is used in Figure 13.34 as a design parameter. Although smoke and HC emissions are influenced by various other factors, Figure 13.34 suggests that 0.7 to 0.8 of L/R value would be a good compromise for light-duty diesel engines.

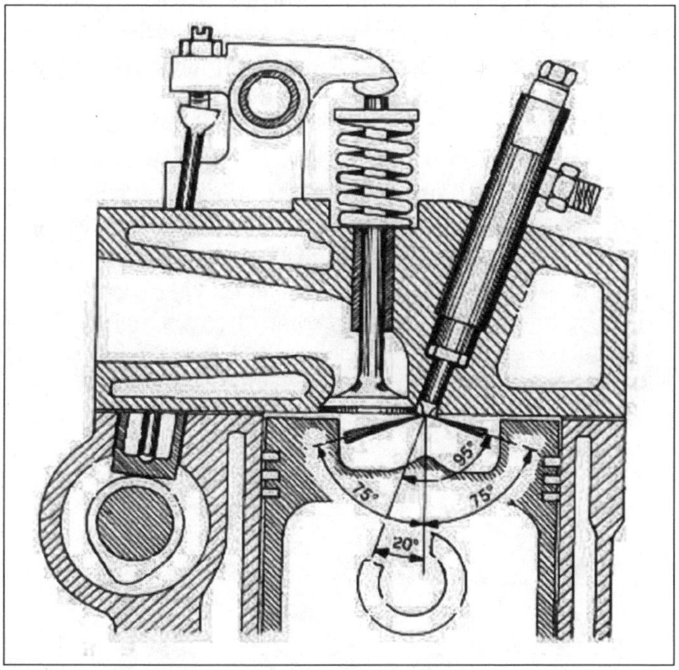

Figure 13.33 Injector installation and combustion bowl offset in a two-valve combustion system. (From Advanced Engine Technology *by Heinz Heisler. Reprinted by permission of Butterworth-Heinemann.)*

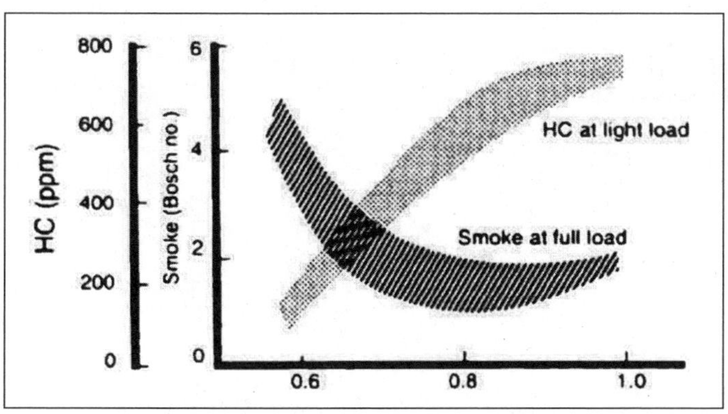

Figure 13.34 Matching of fuel spray to combustion bowl design. (Reprinted from Ref. 20.)

These results indicate the importance of the air motion in the combustion chamber and optimum matching of the spray and combustion chamber dimensions. However, we should also emphasize that combustion chamber shape design has typically used a trial-and-error method that relied mainly on the engineer's experience and intuition. Analytical methods have now replaced the traditional methods with greater degrees of success and lower development costs. For instance, Figures 13.35a and b show two basic combustion bowl designs. Design (a) is a deep-dish, straight-sided, Mexican hat, swirl-supported design. Design (b) is a shallow-bowl, quiescent-type combustion system. A grid layout represents points at which calculations are made to predict a number of variables that give comparative results between various designs. Computer-aided analytical designs investigate the movement of air within the cylinder and combustion bowl, the formation of fuel sprays, and subsequent combustion. Additionally, they analyze the thermal cycle and renewal of the cylinder's charge, as well as the interaction between fuel injection and air movement, which determines the development of efficient combustion [21]. Further development of analytical technologies and their practical use are necessary in the future.

The *compression ratio* is another engine design parameter that affects emission control. In general, higher compression ratios cause a reduction in PM emissions and improved fuel economy, but also cause an increase in NO_X emissions. However, higher compression ratios require a stronger engine

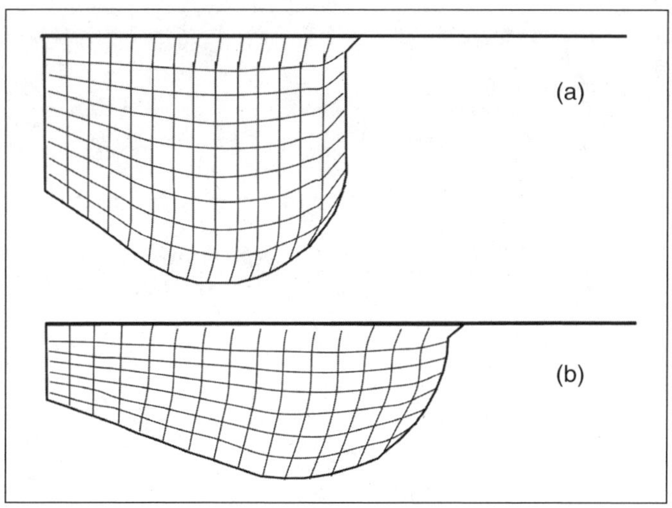

Figure 13.35 Combustion bowl designs with grid for computer-aided design analyses: (a) deep-dish, straight-sided, Mexican hat, swirl-supported design and (b) shallow-bowl, quiescent-type design. (Based on Ref. 21.)

structure, which may increase weight and cost. The increased engine weight and frictional losses somewhat offset the fuel economy benefit of higher compression ratios, especially at very high compression ratios. Conversely, lower compression ratios generally cause a reduction in NO_x emissions, while causing an increase in PM emissions and decreased fuel economy. In addition, lower compression ratios lead to cold-starting difficulties [22].

By raising the compression ratio, ignition delay is reduced, which in turn reduces the premixed portion of the fuel and allows more injection timing retardation to control NO_x while maintaining a good NO_x/BSFC trade-off. White smoke problems are often aggravated by a combination of retarded injection timing and colder charge air. Raising the compression ratio from the 16.5:1 to 18.5:1 leads to significant reductions in white smoke especially during cold start and warm-up operation [5].

13.3.3 Reduction of Particulate Matter through Oil Consumption Control

Various investigations [23, 24] have demonstrated that lube oil is a significant contributor to diesel particulate emissions. In fact, for the diesel engine designs of the early 1990s, the lube oil particulate fraction eventually determined whether the particulate standard was met or not. To achieve low oil consumption, oil must be controlled in three major areas:

- Cylinder wall
- Intake and exhaust systems
- Lube oil formulation.

The first of these three areas has always been considered by engine manufacturers during their design and development phases. Lube oil formulation as it relates to oil consumption control continues to be a cooperative task between lube oil and engine companies. Discovery of the lube oil impact on particulate emissions added more emphasis on its formulation, and as more stringent emissions are mandated this trend is expected to continue.

Highest on the priority list of oil consumption control is the attention given to the *cylinder wall* area. This area is by far the most involved and entails several technologies and component design interactions. Zelenka *et al.* [25] assembled the summary chart shown in Figure 13.36. The parameters included in that chart give us an idea of the immensely difficult task of designing pistons, piston rings, cylinder liners, combustion systems, and manufacturing processes that work in concert to reduce oil consumption while at the same time achieve performance and emissions objectives. The same authors [25] singled out cylinder *bore distortion* for its predominant effect on oil consumption and emphasized the importance of using advanced finite element techniques to minimize bore distortion, thereby minimizing its influence on lube oil consumption. The influence of bore distortion is illustrated in their paper by a comparison between two engines of nearly identical bore/stroke design (see Figure 13.37). The engines were of the one liter/cylinder heavy-duty class that is quite common for truck applications. Both engines had very similar combustion system designs. In Figure 13.37, Engine A shows very little distortion in two different planes, whereas Engine B is highly distorted at the 10-mm plane below the fire

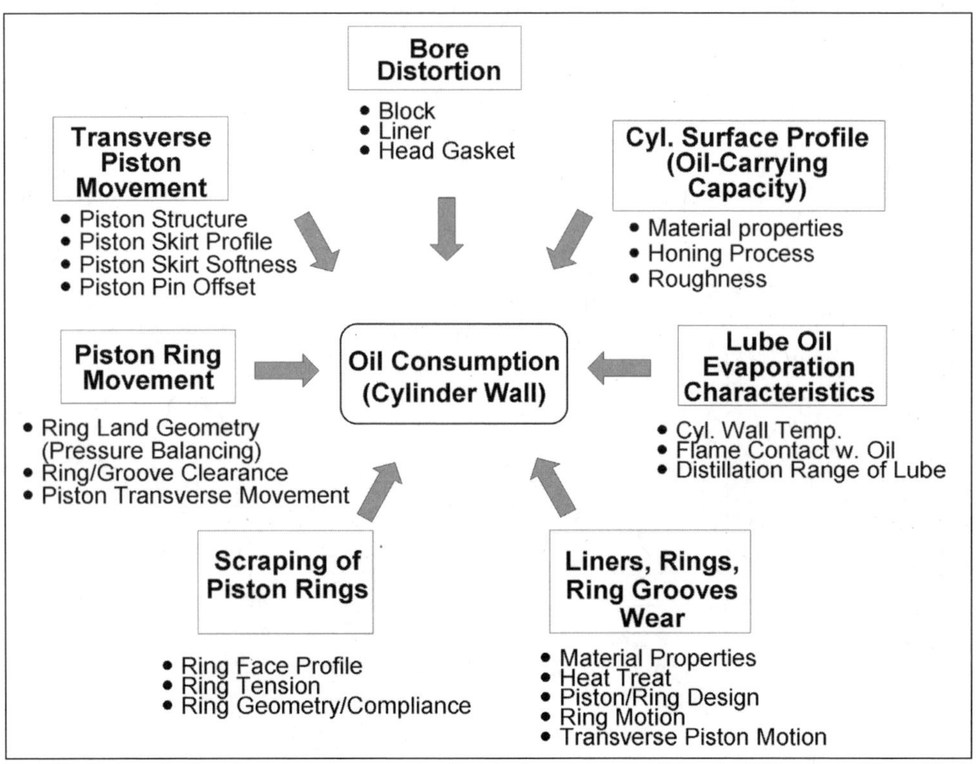

Figure 13.36 *Criteria and design parameters affecting lube oil consumption from the cylinder wall area. (Reprinted from Ref. 25.)*

deck. This position is quite critical since the top piston ring reaches its TDC and reversal point at that plane. In transient testing, Engine A produced lube oil–derived particulate emissions of 0.048 g/bhp-hr versus 0.16 g/bhp-hr for Engine B.

Piston rings are used to seal the combustion gases within the combustion chamber, assist in heat transfer from the piston to the cylinder wall, and control lubrication in the cylinder wall area. Typical piston ring pack designs incorporate three or more rings:

- A top compression or fire ring
- One or more secondary compression rings
- One or more oil control rings.

The need for a second compression ring is dictated by the ring gap that is necessary for ring installation. Normally, ring gaps from consecutive rings are staggered to avoid creating a leak path from the compression volume to the crankcase [10].

In general, it is preferable to minimize the total number of rings because they are a major source of friction. Major influences on piston ring design include ring bearing face and side wear, fatigue fractures, thermal resistance, and dynamic characteristics. Piston rings are designed to closely fit in piston ring grooves to minimize gas leakage, or blowby, and to properly meter lubrication. To minimize wear and to prevent excessive oil consumption or undesired lubricant contribution to combustion emissions, piston rings and their grooves are designed to minimize ring sticking and optimize ring dynamic motion. Piston ring design is very complex and a vast body of knowledge has been generated over the years on all aspects of their design.

Piston rings seal the combustion chamber by contact between the piston ring face and the cylinder wall and the ring sides and the top or bottom of its groove. This contact is the result of forces from the spring action of the ring and the combustion gas pressure behind and on top of the ring, as illustrated in Figure 13.38. The spring forces, called the contact pressure, depend on the ring material and dimensions, the camber (or circumferential profile), and the size of the free, or uncompressed, gap in the ring. The contact pressure can either be constant or variable depending on the camber, and should be minimal to prevent excess frictional losses. Typically, uniform contact pressure is desired for all engine operating conditions. However, due to thermal distortion of the rings and cylinder liner, or special considerations such as liner ports in two-stroke engines, variable contact pressures are specified at room temperature. In all cases, the ring should not have an opening between the ring face and the cylinder wall when installed in the cylinder, a specification known as "light-tight." Ring stress calculations are normally conducted to ensure that the opening and closure stresses are below material limits when the ring is opened to slip over the piston and compressed to enter the cylinder. There are five basic types of compression ring cross sections

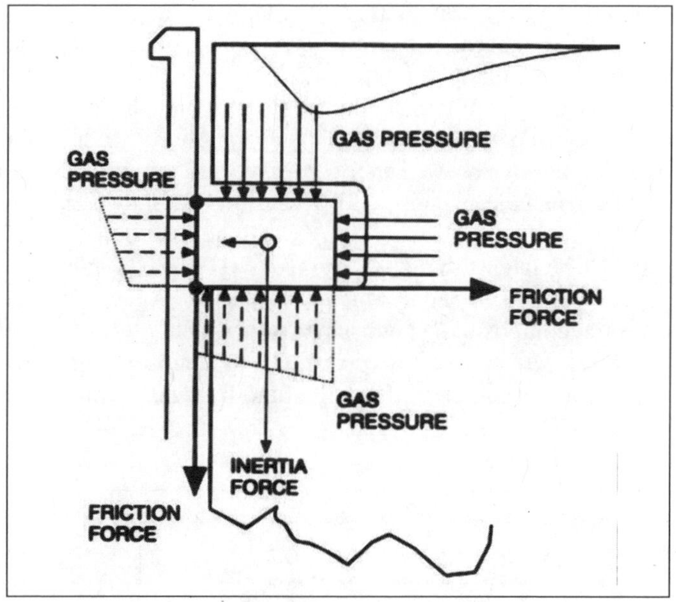

Figure 13.38 *Effect of gas pressure application on a compression ring. (Reprinted from Ref. 10.)*

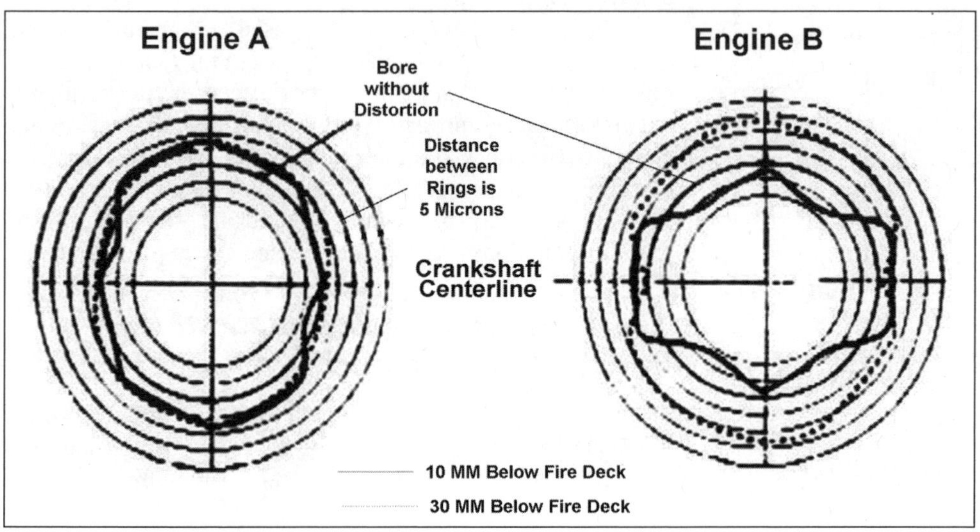

Figure 13.37 *Bore distortion comparison between two generally similar heavy-duty engines. (Reprinted from Ref. 25.)*

and two basic oil control ring types. The compression ring cross sections are rectangular, taper faced, keystone, internally beveled, and L shaped. The rectangular ring is geometrically simple, thus less expensive, and provides a sufficient seal for most applications.

Piston ring design is complex and requires substantial experience, and it is further complicated by bore distortion. Attempts to make piston rings light-tight at room temperature do not guarantee the same fit at the high temperatures to which they are exposed at operating engine conditions. Once again, experience is supplemented with advance finite element analysis to predict the shape of piston rings within distorted bores. Work performed at Ricardo Engineers [26] produced such predictions for a top piston ring within a distorted bore at room temperature. Even at room temperature, three major gaps existed in the work presented by Ricardo. These gaps can be a major source of lube oil in particulate matter. To overcome this potential problem, piston ring manufacturers use more compliant ring materials to ensure light-tight rings in an attempt to reduce oil consumption without jeopardizing durability.

An experiment was conducted to determine the influence of ring pack compliance on particulate emissions. Rings from two suppliers were used in a controlled test using the same mid-range DI diesel engine [5]. The results are shown graphically in Figure 13.39 where ring C was less compliant than ring G. The latter showed an average of 0.04 to 0.07 g/bhp-hr advantage in U.S. Environmental Protection Agency (EPA) transient composite PM emissions over ring C.

Liner surface finish is very critical to engine durability as well as oil consumption. Honing of the liner ID is conducted

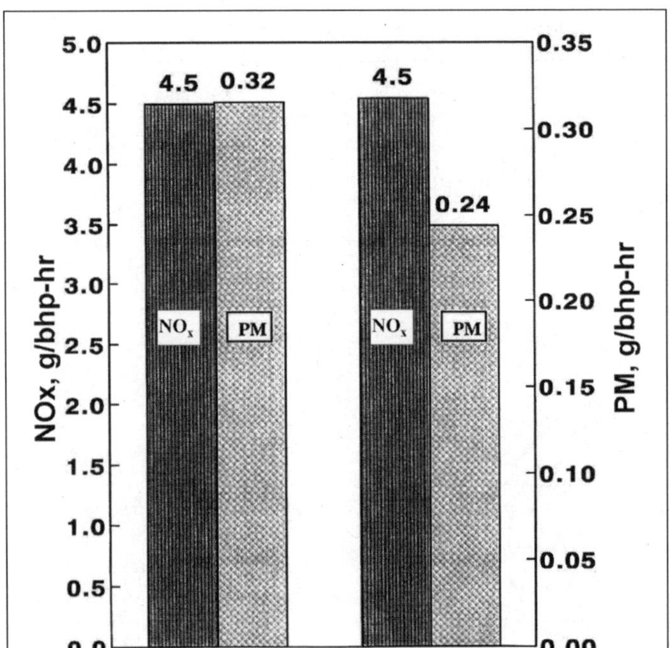

Figure 13.39 *Effect of ring compliance on PM emissions. (Reprinted with permission from Ref. 5.)*

to achieve the proper surface finish. Correct honing produces a cylinder liner surface finish that exhibits a crosshatch pattern. Each engine manufacturer specifies in its service manual what angle of crosshatch pattern and what surface finish are desired. Surface finish is usually stated as being in the region of 20 to 35 rms (root mean square), which is simply a mathematical term indicating the average irregularity in millionths of an inch (0.000001 in.). The actual microinch surface finish is controlled by proper selection of the honing stone (grit) used. For example, both roughing and finishing stones are available with grits ranging from 70 to 600; the 70-grit stones are regarded as roughing types, 150-grit as finishing, 220-grit as a medium finishing set, 280-grit as a polishing set, 400-grit as a fine set, and 600-grit as an extra-fine finishing set. It is worth noting that the rougher the stone grit used, the larger the microinch surface finish.

Figure 13.40 shows the particulate emissions for two identical pre-1991 engines. Engine A met the 1988 federal particulate emission standard of 0.6 g/bhp-hr with a reserve margin of more than 30%, whereas Engine B exceeded the particulate matter limit. Analysis revealed that both engines had similar insoluble particulate emissions as indicated in Figure 13.40. However, the soluble organic fraction of Engine B was considerably larger than that of Engine A. Note that the fuel-based particulate of the SOF for both engines was practically the same. The oil-based particulate of Engine B was twice as high as that of Engine A. Investigation of the cause for the high oil particulate disclosed that the bore honing process was out of specification, which led to excessive oil consumption. Therefore, special manufacturing control measures are needed to ensure strict compliance with the bore honing specifications in terms of pattern angularity, depth, and consistency [5].

The second major source of oil contribution to particulate emission is through the *intake and exhaust systems*. At first glance this categorization may seem rather unusual; however, significant leakage paths contribute lube oil through the intake and exhaust systems. For instance, valve stem seals, both on the intake as well as the exhaust sides, can cause lube oil intrusion in the intake or exhaust ports. Turbocharger seals are often the source of lube oil in the intake manifold. Blowby returned into the inlet system is another potential source of oil-based contribution to the particulate problem.

The *formulation of lube oil* to assist in meeting the very strict emissions requirements is the last of the factors affecting oil consumption. This topic not only affects engine durability, but also impacts future use of postcombustion exhaust emission control devices such as diesel oxidation catalysts. Therefore, the development of new lube oil formulation has to be carried out in close cooperation between engine and oil companies. Testing these formulations is a very tedious and costly process [25].

13.3.4 Reduction of Particulate Matter through the Fuel Injection System

Considerable effort has gone into the development of the fuel injection system. Components had to be selected and carefully matched to the combustion system. Although many of the

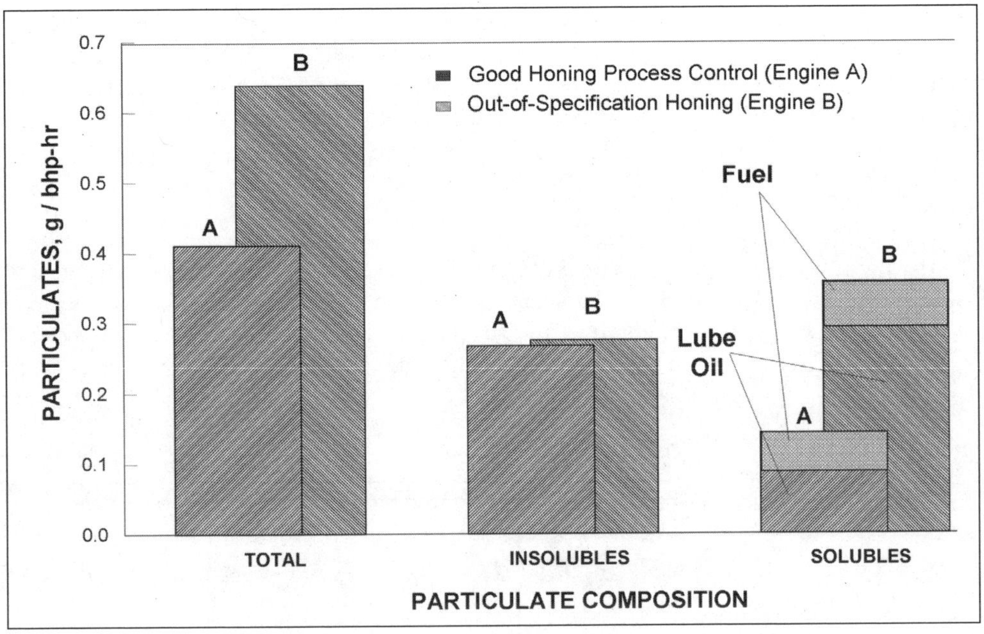

Figure 13.40 *Effect of bore honing on particulate emissions. (Reprinted with permission from Ref. 5, Fig. 14.)*

parameters discussed in this section were addressed earlier, the following fuel system variables are reviewed:

- Injection timing
- Injection pressure
- Injection duration
- Nozzle hole configuration
- Nozzle opening pressure

Injection timing was the most effective tool by which NO_X emissions were reduced [27]. Figures 13.1 and 13.2 show the effect of injection timing on NO_X emissions for an engine having an ignition delay of about 10 crankangle degrees. Unfortunately, retarding of the injection timing conflicts with good fuel economy and low particulate emissions. To reduce the fuel economy penalty associated with retarding of the injection timing, engineers adopted measures to reduce ignition delay such as using a high compression ratio and higher injection pressures [9].

The effect of *injection pressure* on engine performance and emissions has been recognized since the early 1970s. However, some manufacturers felt that injection rate shaping was more effective in reducing emissions than injection pressure. A large body of work in this area has been published [4]. Changes in injection pressure affect the rate of fuel injection, injection duration, atomization, and spray penetration. Therefore, it is very difficult to separate the effect of injection pressure alone from that of rate shaping. Studies conducted to investigate this subject show that increasing injection pressure leads to improved fuel economy (Figure 13.9) and smoke characteristics [28] (Figure 13.41). For diesel engines designed for the early 1990s, full-load rated speed injection pressures were between 1000 and 1200 bar versus pre-1990s levels of 650 to 700 bar [5].

Injection durations were traditionally extremely short and injections were designed to be completed close to TDC where maximum combustion efficiency could be realized. With the introduction of new combustion developments, more emphasis was placed on the proper matching of injection duration to the combustion system. Engines equipped with swirl-supported combustion systems, an injection duration at rated speed, and load conditions of approximately 25° to 30° crankangle were found to be optimum. Of course, this by no means suggests that

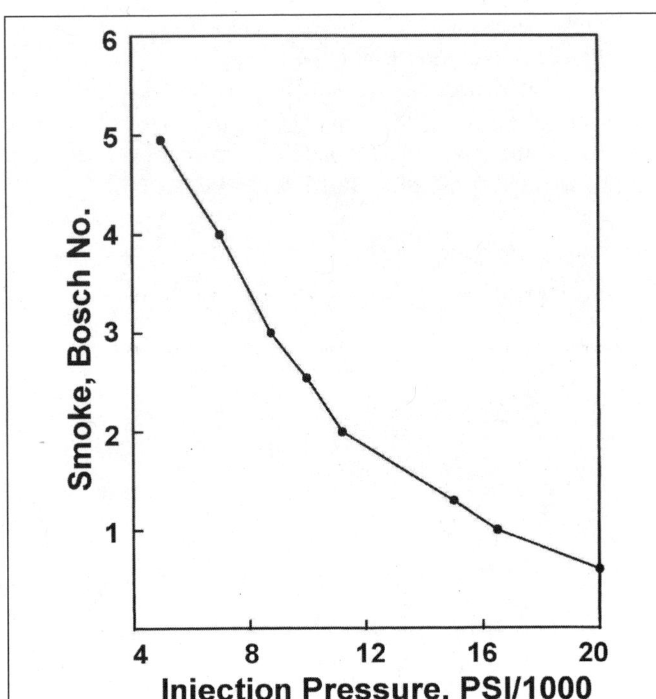

Figure 13.41 *Effect of injection pressure on smoke emissions. (Reprinted with permission from Ref. 5, Fig. 18.)*

injection duration is an independently controlled parameter. It is, however, a product of several choices related to the injection system and engine power output. These choices include nozzle hole size, number of nozzle holes, injection pressure, engine speed, and injected fuel quantity [5].

The *injector nozzle* is the point of fuel delivery to the combustion chamber. Its design and placement in the cylinder head (angle of inclination) are of paramount importance to engine performance and emissions characteristics. Several parametric studies have been conducted and lead to the following generalized statements describing optimized components of the nozzle configurations for the 1990s DI swirl-supported engines:

- Nozzle sac volumes must be reduced to a minimum. In most cases the so-called cylindrical sac (shown in Figure 13.42) was adequate, but in some cases a conical design was necessary. Conical sac volume is much less than its cylindrical sac counterpart, resulting in reduced nozzle dribble and its associated particulate emissions [5].
- The number of injector holes must be optimized for the specific combustion system as well as for the injection pressure characteristics to maximize air utilization.
- The nozzle hole diameter must be selected to provide the proper fuel atomization. Figure 13.43 illustrates CO and particulate emission results for two different injector nozzle hole sizes. Injector nozzle B, with a hole size of 0.26 mm, exhibited higher CO and particulate emissions than injector nozzle A, which had a hole size of 0.25 mm diameter.
- The length of the nozzle hole must be chosen for its contribution to fuel spray penetration, which together with proper atomization helps in maximizing air utilization. In general, the optimum hole length was found to be 2.7 times the hole diameter [5].
- The spray cone (included) angle must be selected with regards to the specific combustion bowl design. The reentrant type bowl design requires a smaller cone angle than the traditional type of bowl design [28]. Typical reduction in the cone angle for the reentrant bowl was about 10° from that of conventional straight-sided combustion bowl designs.
- A significant amount of interaction exists between the various parameters already mentioned. The effect of some of these variables can readily be seen. However, in certain cases appreciable testing must be performed and results statistically analyzed to make the proper selection. By applying this development process, nozzle opening pressures were raised from the pre-1990s level of 240 bar to 270 to 290 bar [6].

The preceding review of the fuel system components and their characteristics for the 1990s engines is by no means a comprehensive account of all the changes. Some very important choices were made in the areas of injection in-line pump delivery

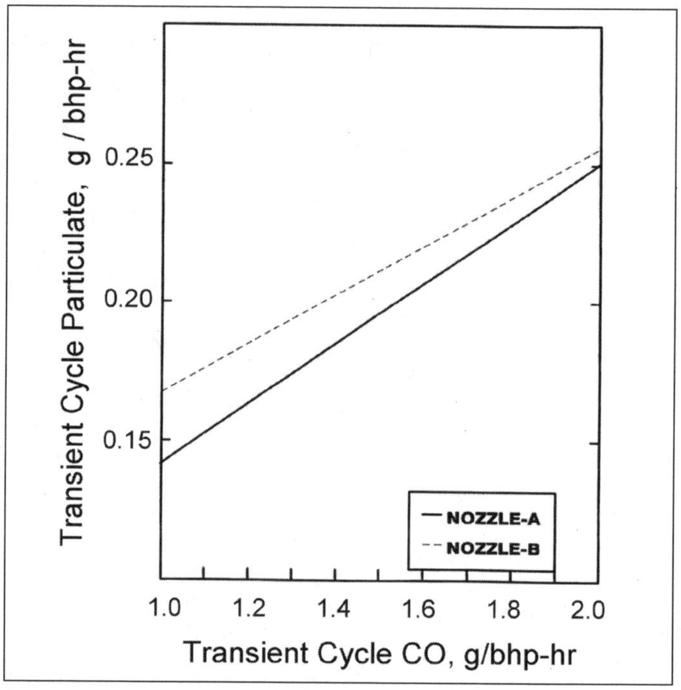

Figure 13.43 *Effect of nozzle hole size on emissions. (Reprinted with permission from Ref. 5, Fig. 21.)*

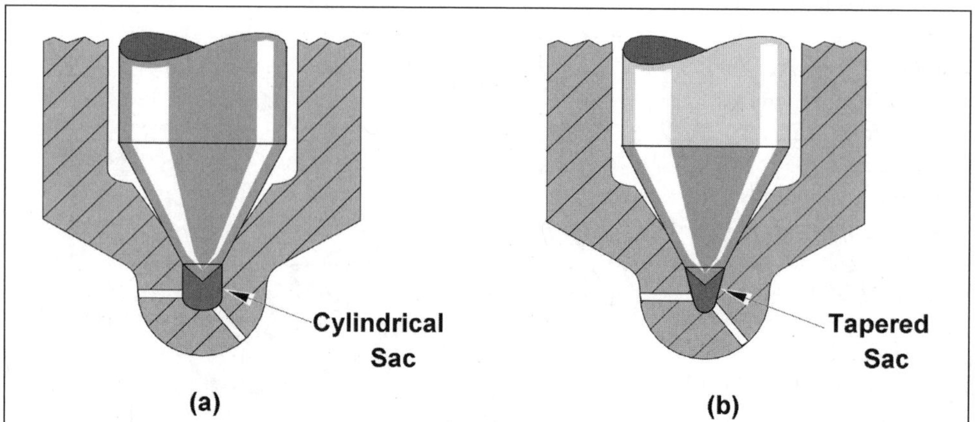

Figure 13.42 *Comparison between (a) cylindrical and (b) tapered (conical) sac volumes. (Reprinted with permission from Ref. 5, Fig. 20.)*

valves, high-pressure injection lines, and the number and size of spill ports. In addition, shaping the torque curve between peak torque and rated speed, and the selection of a smoke limiting device were especially important in reducing emissions. Manufacturers that opted to include electronic controls had the additional task of optimizing the injection timing schedule. The optimization was relative to several variables such as speed, load, engine coolant temperature, and ambient temperatures. In spite of the vast improvement achieved, much remained to be done before the diesel engine could meet the emissions limit challenge of the future.

References

1. Owen, K., and T. Coley, *Automotive Fuels Reference Book,* 2nd ed., Society of Automotive Engineers, Warrendale, PA, 1995.
2. Khair, M.K., "Diesel Engine Technology," SAE Seminar, April 1993.
3. Broge, J.L., "The Diesel Is Coming, The Diesel Is Coming," *Automotive Engineering International*, SAE International, Warrendale, PA, January 2004.
4. Gill, A., "Design Choices for 1990s Low Emission Diesel Engines," SAE Paper No. 880350, SAE International, Warrendale, PA, February 1988.
5. Khair, M.K., "Progress in Diesel Engine Emissions Control," ASME Paper No. 92-ICE-14 , January 1992.
6. Cartellieri, W.P., and P.L. Herzog, , "Swirl Supported or Quiescent Combustion for 1990s Heavy-Duty DI Diesel Engines—An Analysis," SAE Paper No. 880342, SAE International, Warrendale, PA, March 1988.
7. Cartellieri, W.P., and P. Tritthart, "Particulate Analysis of Light-Duty Diesel Engines (IDI & DI) with Particular Reference to the Lube Oil Fraction," SAE Paper No. 840418, SAE International, Warrendale, PA, 1984.
8. Mayer, W. J., D.C. Lechman, and D.L. Hilden, "The Contribution of Engine Oil to Diesel Exhaust Particulate Emissions," SAE Paper No. 800256, SAE International, Warrendale, PA, SAE International, Warrendale, PA, 1980.
9. Henein, N., and D. Patterson, "Combustion Engine Economy Performance and Emissions," Lecture Notes to Ford Tractor Operations, October 16, 1985.
10. Merrion, D.F., "Diesel Engine Design for the 1990s," The Fortieth L. Ray Buckendale Lecture, SAE SP-1011, SAE International, Warrendale, PA, 1994.
11. Heywood, J.B., *Internal Combustion Engine Fundamentals,* McGraw-Hill, New York, 1988.
12. Monaghan, M.L., and H.F. Pettifer, "Air Motion and Its Effect on Diesel Performance and Emissions," SAE Paper No. 810255, SAE International, Warrendale, PA, 1981.
13. Khair, M.K., "A Study of Ford New Holland's Truck Engine Intake Manifold," Internal Report No. B87013, July 1987.
14. Ueno, N., and F. Kawashima, "New Komatsu SA 6041 Engine," SAE Paper No. 860461, SAE International, Warrendale, PA, 1986.
15. Okada, I., and O. Hashimoto, "Isuzu New 9.3L Diesel Engine with Variable Geometry Turbocharger," SAE Paper No. 860460, SAE International, Warrendale, PA, 1986.
16. Dani, A.D., U.P. Nagpurkar, and P.A. Lakshminarayanan, "Universal Mixing Correlations for the Performance and Emission of Open Chamber Diesel Combustion Supported by Air Swirl,"
   Swirl," SAE Paper No. 900446, SAE International, Warrendale, PA, 1990.
17. Brady, R.N., *Modern Diesel Technology,* Prentice Hall, Englewood Cliffs, NJ, 1996.
18. Osenga, M., "EPA Proposes Stationary Diesel Emissions Regulations," *Industry News*, Diesel Progress—North American Edition, August 2005.
19. Heisler, H., *Advanced Engine Technology,* Butterworth-Heinemann, Boston, 1995.
20. Hikosaka, N., "A View of the Future of Automotive Diesel Engines," SAE Paper No. 972682, SAE International, Warrendale, PA, 1997.
21. Bersia, C., G. Bodritti, and C. Sparti, "Computer Applications in Diesel Engine Development," Iveco-Fiat SpA, Truck Technology International, Sterling Publications Limited, London, 1988.
22. "Emission Control Technology for Diesel Trucks," Report to Congress, U.S. Environmental Protection Agency, Washington, DC, October 1993.
23. Cartellieri, W.P., and W.F. Wachter, "Status Report on Preliminary Survey of Strategies to Meet US-1991 HD Diesel Emission Standards without Exhaust Gas Aftertreatment," SAE Paper No. 870342, SAE International, Warrendale, PA, 1987.
24. Wachter, W.F., and W.P. Cartellieri, "Wege zur Erfüllung zukünftiger Emissionsgrezwerte bei Lkw-Dieselmotoren" (Ways to Meet Future Heavy-Duty Diesel Emissions Standards), Paper presented at 8th International Vienna Engine Symposium, May 1987.
25. Zelenka, P., W. Kriegler, P.L. Herzog, and W.P. Cartellieri, "Ways Toward the Clean Heavy-Duty Diesel," SAE Paper No. 900602, SAE International, Warrendale, PA, 1990.
26. Needham, J., "What Is the Emissions Potential of the Heavy-Duty Diesel Engine?" SAE TOPTEC Presentation, May 14–15, 1992, SAE International, Warrendale, PA.
27. Khair, M.K., "Air-to-Air Intercooling of the Ford 7.8 L Mid-Range Truck Diesel Engine," SAE Paper 870534, SAE International, Warrendale, PA, 1987.
28. Wachter, W.F., "Analysis of Transient Emission Data of a Model Year 1991 Heavy-Duty Diesel Engine," SAE Paper No. 900443, SAE International, Warrendale, PA, 1990.

Chapter 14

Emerging Technologies and the Modern Diesel Engine

14.1 Introduction

Diesel-powered trucks, trains, and boats are at the core of the U.S. supply and distribution network, moving 94% of all freight in the United States. Each day, diesel-powered equipment moves 16 million ton of freight. In addition, throughout the world, and especially in Europe, diesel engines power an increasing number of passenger cars and light-duty trucks [1]. The inherent benefits of diesel technology, efficiency, and durability have led to the diesel's emergence in meeting future energy and environmental goals. However, there are challenges for the automotive and diesel industries.

In December 2000, the U.S. Environmental Protection Agency (EPA) announced future diesel engine emission regulations for heavy-duty trucks and buses and plans for cleaner diesel fuel requirements that are unprecedented. The requirements call for maximum particulate emissions from new heavy-duty diesel engines of 0.01 g/bhp-hr by the 2007 model year. The limit for NO_x is 0.20 g/bhp-hr and for nonmethane hydrocarbons (HCs), 0.14 g/bhp-hr, to be phased in from 2007 to 2010. The new diesel engine emissions limits are in addition to an existing set of limits that were met in 2004 [2].

Reduced sulfur content in diesel fuel is crucial to emissions reductions. That is why the EPA, starting in 2006, required refiners to reduce the sulfur content from the current limit of 500 ppm to 15 ppm, which will help reduce particulate matter (PM) and NO_x emissions. There is by no means a universal view that the new 2007 standards, including all of the durability and testing requirements, can be met in the time required. There are also concerns about the availability of the "cleaner" fuel in substantial volumes in time for the 2007 standards. The new U.S. standards, in addition to more stringent requirements around the world, will require unprecedented technological innovation by the heavy-duty diesel engine industry. An example of advanced technology has already been set by the light-duty applications in Europe and Japan. Still, additional new approaches are required to meet emission standards that are not just limited to the United States, but also in other countries around the world [3]. Advances in combustion systems, catalysts, and exhaust filtration systems will have to be applied to powertrains to meet the new standards. However, the new fuel rules create opportunities for emissions improvements in all vehicles, whether in existing commercial truck and bus fleets or passenger cars and trucks.

It is becoming increasingly evident to those engaged in diesel engine technologies that the clean diesel engine is clearly a viable technology for the future with the addition of "cleaner" fuels and exhaust treatment systems. However, getting there involves overcoming significant challenges for all parties, including fuel suppliers, engine manufacturers, and the makers of emissions treatment and filtration technologies [1].

This chapter reviews the technologies being developed and considered by researchers for future diesel engines. Progress made toward meeting emission standards is highlighted together with examples featuring modern diesel engines from around the world.

14.2 A Platform for Future Heavy-Duty Diesels

A rare alliance between the U.S. government and industry was formed in the mid-1990s for the benefit of clean air and the environment. This alliance was referred to as the Partnership for a New Generation of Vehicles (PNGV) and had as its primary goal the development of vehicles that are three times as fuel efficient as the then-current vehicles without sacrificing affordability, comfort, safety, or emissions control. If these goals were achieved, it would have meant significant progress toward addressing climate change and other highly publicized political and technical issues. The choice of powertrain was not limited to conventional internal combustion engines, but the scope of possible technologies was broad enough to include various advanced internal combustion engines, fuel cells, gas turbines, or Stirling engines [4].

Scientists from many disciplines were invited to analyze the benefits and trade-offs of several powertrains as part of the PNGV program. They concluded that the ambitious goal for fuel economy set by the PNGV program can be best met by a hybrid diesel/electric vehicle. Naturally, this decision was anticlimactic to those who insisted that diesels were environmentally hazardous engines and therefore must go. A closer look at the heavy-duty diesel engine can shed some light on its performance relative to the PNGV targets. Table 14.1 provides an interesting fuel economy and payload-based emissions comparison between heavy-duty Class 8 applications and a PNGV-type vehicle [5].

When viewed in light of the comparison in Table 14.1, heavy-duty diesels do not appear to be the villain most people think they are. Although a heavy-duty vehicle handles six times the load of a PNGV, it emits only 40% of the PNGV NO_X target on the basis of its payload ratio. For particulate matter (PM) emissions, the ratio is even more favorable for the heavy-duty diesel engine. Yet, scientists and engineers working on heavy-duty emissions reduction programs are not satisfied with the performance reported in Table 14.1 and are feverishly trying to develop new technologies that will not only achieve extremely low emission targets, but also improve performance and energy use.

In 1998, heavy-duty engines in the United States met a composite (one-seventh of a cold start plus six-sevenths of a hot start) NO_X/PM emission standard of 4.0/0.1 g/bhp-hr in the EPA transient cycle. The emphasis here is on NO_X and PM emissions, because HC and CO emissions present little or no challenge to emissions control engineers. The next level of control was in 2002 when engine manufacturers faced a NO_X + HC limit of 2.5 g/bhp-hr while PM was held at 0.1 g/bhp-hr. For 2007, however, EPA and the California Air Resources Board (CARB) have established a record low NO_X limit of 0.20 g/bhp-hr combined with a PM limit of 0.01 g/bhp-hr that defies the current capabilities in measurement accuracy and repeatability. Even with these incredibly low limits, early demonstrations have successfully shown that technology exists to allow heavy-duty diesels to meet these stringent limits.

TABLE 14.1
HEAVY-DUTY DIESEL PERFORMANCE COMPARED TO PNGV TARGETS

Parameter	HDD	PNGV Target	HDD/PNGV Ratio
Payload weight, lb	50,000	2000	25
Fuel economy, mpg	8	80	0.1
Payload-based fuel economy, ton-mile/gal	200	33	6
NO_X, g/kg fuel (2004)	12	5	2.4
Payload-based NO_X, g/ton-mile	0.19	0.48	0.40
Particulate, g/kg fuel (2004)	0.63	1.0	0.63
Payload-based particulate, g/ton-mile	0.01	0.10	0.10

Source: Data from Ref. 5.

14.3 Platform for Future Light-Duty Passenger Car Diesels

The light-duty market for diesel engines is too small to present a significant environmental role in the United States. One of the main reasons for this is the price of gasoline. Despite heavy taxation by local and federal agencies, gasoline prices in the United States have remained relatively low. By contrast, gasoline prices in Europe and Japan have risen dramatically and are two to three times their corresponding prices in the United States. This fact led to a rise in demand for diesel-powered cars in Europe and the engine/car companies have responded to this customer demand by offering an excellent array of engines featuring new technologies and achieving the target of 100 km/3 L of fuel. What makes this a more significant feat is that this new breed of engines has low emissions and demonstrates excellent responsiveness and performance.

Notable among these new vehicles and by virtue of being the first is the Volkswagen Lupo 3L TDI. It is described by Volkswagen as "officially the most fuel-efficient and least polluting of all the world's gasoline- and diesel-engine cars" [6], a claim it can make without argument after finishing a tour around the world in 80 days. What makes this latest Lupo trip special was its team's aim to complete the officially required distance on just 1000 L of fuel. In fact, the record-breaking Lupo used just 792 L of diesel for its journey of 33,333 km which ended in August 2000 [6]. In this trip, the frugal Lupo used 174 gallons to travel 20,700 miles, maintaining an overall average fuel consumption of 118 mpg.

Newer diesels, especially in Germany, are viewed as environmentally friendly and having benefits in many areas such as in the area of carbon dioxide, the greenhouse gas, which has certainly become the subject of ever more anguished political debate. Modern diesel engines outperform modern gasoline engines in many aspects, except for idle noise and particulate emissions. Yet with all of the innovation and technical accomplishments achieved so far, these new engines/cars are still some distance from achieving emission targets. For instance, for production cars sold in the United States the NO_X/PM trade-off is at 0.50/0.05 g/km, which is a Euro III standard, when tested through the European light-duty cycle that includes the high-speed extraurban driving cycle. By 2005, these limits changed to 0.025/0.025 g/km for NO_X/PM respectively [7].

14.4 Platform for Future Light-Duty Truck Diesels

Data issued by the National Automotive Dealers Association (NADA, McLean, Virginia) confirm that Americans are buying trucks (and other truck-like vehicles) in ever-increasing numbers. In 1998, the organization reports, the popularity of vans, SUVs, and pickups helped increase light-duty truck sales for the seventh consecutive year, to a record 7.4 million units. For this survey, NADA considered a light truck to be one that weighed 14,000 pounds or less. This meant that the light-duty trucks' share of the new-vehicle market was a record 47.8% [8].

According to the 1999 Freedonia Group study "Diesel Engines," U.S. demand for motor-vehicle-type diesel engines was forecast to rise 4.8% per year, to 7.6 billion dollars in 2003. Unit demand, projected to increase 2.5% per year, will be restrained by decelerating growth in truck and bus production, the report noted; however, continued growth in U.S. motor vehicle production will ensure growth for the diesel engine industry through 2000. Lower emissions limits, as was the case for heavy-duty as well as light-duty passenger cars, will prompt the use of newer, more technologically advanced engine designs, as will continued pressure for improved fuel economy and lower maintenance cost [8].

14.5 Technology Options for Modernizing Diesels

A quick review of technologies employed in the past may help in determining those remaining areas that are yet to be explored and perhaps implemented in future engines. An excellent summary and road map for past improvements made during the last three decades is shown in Figure 14.1. Since 1974 heavy-duty diesel engine NO_X emissions have been reduced by about 85% while PM was reduced by more than 90%, according to Flynn [9].

As shown in Figure 14. 1, the most effective technology for NO_X reduction in the late 1980s and early 1990s was injection timing retard. Next to injection timing retard was reducing the intake manifold temperature (IMT). These two technologies were discussed in detail in the previous chapter. Shorter heat release rates (HRRs) were achieved partly through higher injection pressures. This technology was partially implemented in the late 1980s, but became the norm in the early 1990s. Lower friction has always been a concern of the engine designer as part of continuous engine design improvement. Retarding injection timing and reducing IMT combined to increase cold start emissions of HCs and especially white smoke. A higher compression ratio was adopted to avoid cold-temperature white smoke and HC emissions during the warm-up phase. Closer attention was given to the engine's breathing quality, leading to higher boost (higher intake manifold pressure). Heavy-duty engines for 1991 featured generally retarded injection timing, low IMT, and high intake manifold pressure. Some of these engines adopted simple injection timing control algorithms that provided moderate injection timing flexibility, and thus allowed fuel consumption optimization while still meeting emissions limits.

A point of departure from the road map described in Figure 14.1 is the technology implemented in heavy-duty engines in the early to mid-1990s. Combustion system research led to more reliance on swirl-supported intake ports and combustion bowl designs. This technology helped medium heavy-duty engines (1.0 to 1.5 L/cylinder) achieve the 1991/1994 emission standards without additional exhaust emission control devices (aftertreatment). However, some of the higher speed medium-duty engines, such as the Navistar 7.3 L, the Caterpillar 3116, and the Cummins B series engines, were equipped with an oxidation catalyst for particulate control to provide a margin that safeguarded against production variability and engine aging. Continued improvement in in-cylinder designs coupled with advances in electronic controls allowed some of the same engines to shed their oxidation catalysts and still meet the 1998 heavy-duty emission standards.

As suggested in Figure 14.1, the next level of emission standards will require medium- and heavy-duty diesel engines to use exhaust gas recirculation. This technology has been used in

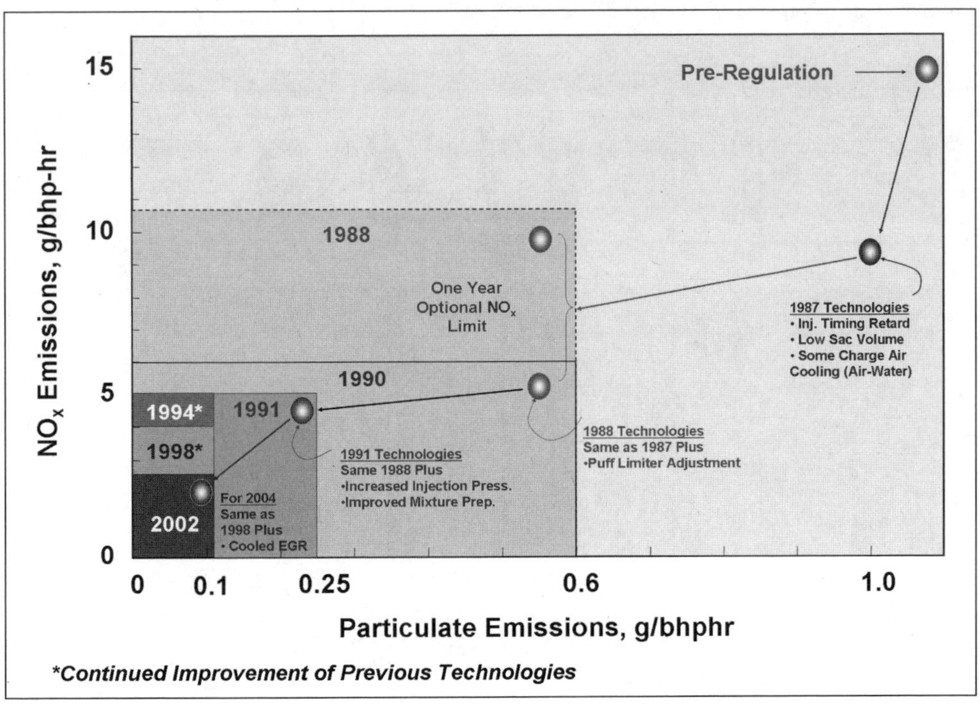

Figure 14.1 Evolution of engine emissions controls.

light-duty diesels for sometime since these engines rarely operate at full load and can beneficially use large amounts of EGR at part load where air-to-fuel (A/F) ratios are rather high (greater than 50:1). This direction is in agreement with Wall [5], who predicted that to obtain NO_x emissions below 4.0 g/bhp-hr and maintain good fuel economy, EGR and improved fuel injection systems would need to be employed (see Figure 14.2).

It is worth noting that the preceding summary of past technologies is rather simplified, and engines produced in the late 1980s and during the 1990s were excellent performers. The characteristic black smoke of pre-1991 had all but been eliminated, drivability and engine response improved, emissions were reduced, and at the same time fuel consumption had not increased; in fact, in many cases it had decreased.

New standards require additional and substantial reductions in NO_x and PM emissions. These new standards are equally stringent for both light- and heavy-duty applications. New technologies as well as improvements of the old technologies are required if the diesel engine is to survive. Exploring the technology landscape reveals that the diesel engine industry is investigating several fronts:

- In-cylinder and engine-based design changes
- Fuels and lubes formulations designed to reduce engine-out emissions and to be friendly to postcombustion emission control devices
- Postcombustion exhaust emission control devices.

In-cylinder and engine-based design changes cover a number of the engine subsystems, including the fuel injection, induction, combustion, valve train, engine cooling, and even the electrical system as well as accessories.

Prominent among the fuel formulations are those involving low sulfur, low aromatics, and high cetane number. In the area of lube oil, low sulfur content formulations are being investigated and tested as well as formulations especially designed for engines equipped with EGR systems. Postcombustion exhaust emission control devices fall into two categories. The first includes oxidation catalysts and particulate filters and is designed to reduce particulate emissions. The second category is mainly responsible for NO_x emission reduction and includes lean NO_x catalysts ($DeNO_x$), lean NO_x traps (NO_x adsorbers), and selective catalytic reduction (SCR). Both NO_x and PM exhaust treatment will be addressed in detail in the second part of this book.

14.6 Fuel Injection System Technology Options

The characteristics of the injection system are among the most important factors influencing emissions and performance of DI diesels. Many investigations have been conducted to assess the effect of injection characteristics such as start of injection, the shape of the rate of injection versus time curve, and injection spray characteristics' effects on mixture formation and consequently on DI diesel combustion. Some of these characteristics were discussed in detail in Chapter 5.

As a result of these investigations, a so-called "ideal" shape for the rate of injection versus cam angle in the engine speed and load map was defined (see Figure 14.3). The "ideal" shape of the rate of injection should be realized by pressure modulation, without throttling fuel flow near the entrance to the spray holes to reduce the injection rate during the initial ramp (boot) period [10]. ("Boot" indicates a low rate of injection at the start, followed by the main portion of injection at a higher rate. It differs from pilot injection in that it is continuous, rather than interrupted; see Figure 14. 4.) This requirement was set up to avoid any influence of the throttling devices (e.g., as with two spring nozzle holders) on fuel atomization and spray formation, and to ensure the best possible conformity of all individual fuel sprays. Thus the needle should stay fully opened during the entire injection period while the injected quantity is modulated by the injection pressure.

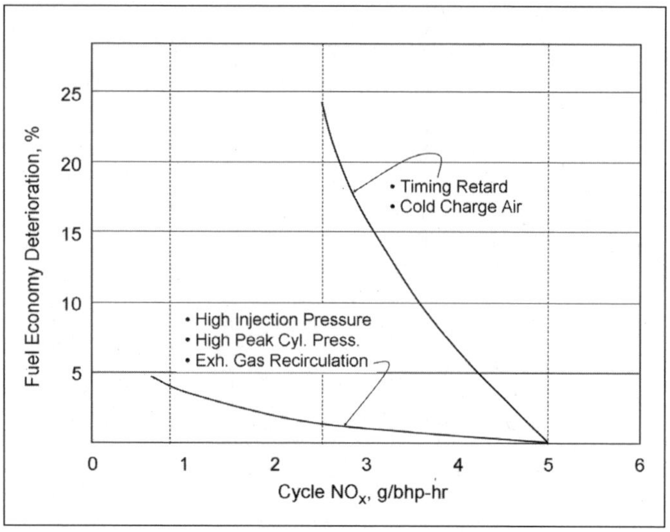

Figure 14.2 Technologies for low NO_x emissions and good fuel economy.

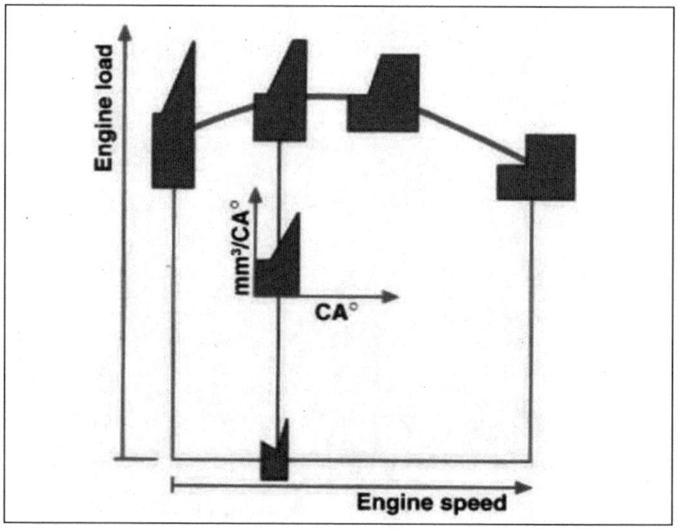

Figure 14.3 Ideal injection rate shape for various engine conditions. (Based on Ref. 10.)

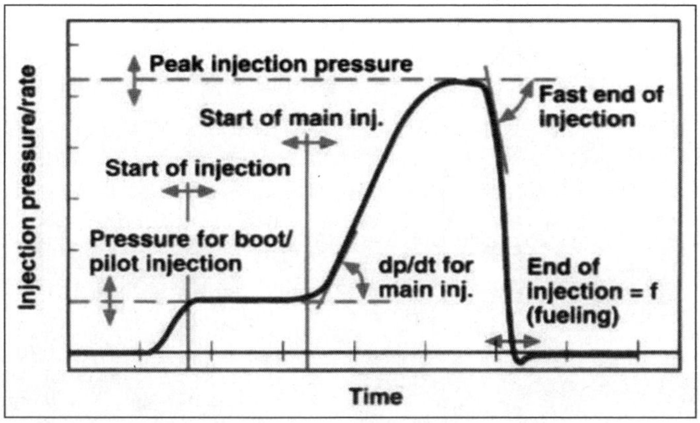

Figure 14.4 *Staged injection in the shape of a boot. (Reprinted from Ref. 10.)*

To verify the "ideal" rate of injection quickly and simply, tests were conducted with a high-speed, direct-injection (HSDI), single-cylinder research diesel engine. To realize an injection rate shape like the one shown in Figure 14.4, the injection system consisted of two separate injection pumps: one with a small pumping rate for the boot part of the injection and one with a high rate for the main part of the injection. The pumps were operating via check valves on one nozzle holder.

In Figure 14.5, the NO_x/particulate trade-off with this injection system for one representative operating point (2000 rpm, 11 mm^3/stroke, equivalent to 200 kP_a BMEP of a multicylinder engine) is shown in comparison to a standard and a high-pressure unit-injector type. With this type of injection rate control (IRC), the NO_x/particulate trade-off could be shifted to lower NO_x levels while keeping nearly the same particulate level as with the high-pressure fuel injection equipment (FIE). In addition, the BMEP level was increased by the application of IRC. The increased BMEP level is an indicator for improved combustion efficiency, because fueling was kept constant for all cases shown in Figure 14.5 [10].

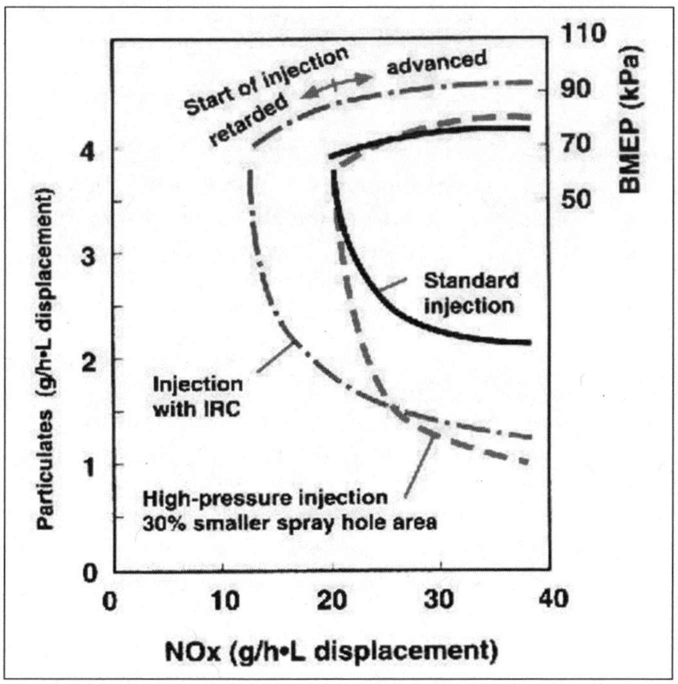

Figure 14.5 *NO_x/PM trade-off comparison for various injection schemes. (Reprinted from Ref. 10.)*

The following requirements are defined as the objective for a modern fuel injection system:

- High pressure potential (>200 Mp_a)
- Realization of the "ideal" rate of injection via pressure modulation alone
- Capability of realizing staged (boot) as well as pilot injection where the pilot is separated from the main injection
- Flexibility in selecting the start of boot, or pilot injection, and of main injection, as well as the end of injection (fueling)
- Flexibility in selecting a pressure level during the boot or pilot part of the injection
- Moderate initial pressure rise of main injection
- Sharp cutoff at the end of injection
- Flexibility in selecting the peak injection pressure
- Options to automatically change the rate shape according to engine speed and load.

The requirement for high-pressure potential dictates that the fuel injection system be compact, with small trapped volumes in the high-pressure fuel section. The required flexibility called for a fully electronically controlled system. Such a system capability is shown in Figure 14.6 where separate pilot and main injections, as well as various ramp rates including boot, pilot and boot, and pilot alone, were demonstrated.

Further, when the traces in Figure 14. 6 are examined, the pilot injected quantity shown was measured and found to be in the range of 3 to 4 mm^3, and the minimum stable separation between the pilot and main injection was in the range of 2° to 3° cam at rated speed. Figures 14.6 and 14.7 show some side benefits of rate shaping realized with this injection system. Among these is a combination of pilot and boot injection (Figure 14.6, middle trace) with fast or with slow end injection (Figure 14.6, middle vs. upper trace). Additionally, a post-injection was realized (Figure 14.7, lower trace) again in combination with boot injection (Figure 14.7, middle trace) and with fast or with slow end of injection (Figure 14.7, middle versus upper trace).

This fuel injection system was tested on an AVL single cylinder engine having the following specifications:

- Bore: 124 mm
- Stroke: 165 mm
- Displacement: 2.0 L
- Combustion bowl: Reentrant
- Compression ratio: 17.7:1
- Swirl ratio: 1.8
- Cylinder head with four valves and central injector position
- Injection cam rate: 0.356 mm/° constant
- Injection nozzle: 8 holes × 0.18-mm diameter
- Nozzle opening pressure: 20 MP_a for all tests.

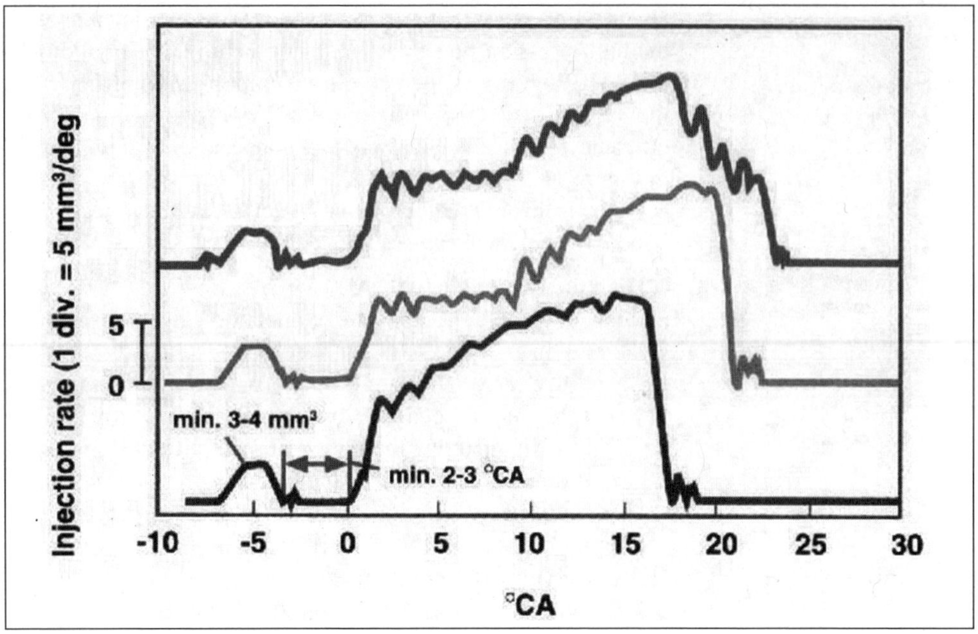

Figure 14.6 *Demonstration of the flexibility of a modern injection system. (Reprinted from Ref. 10.)*

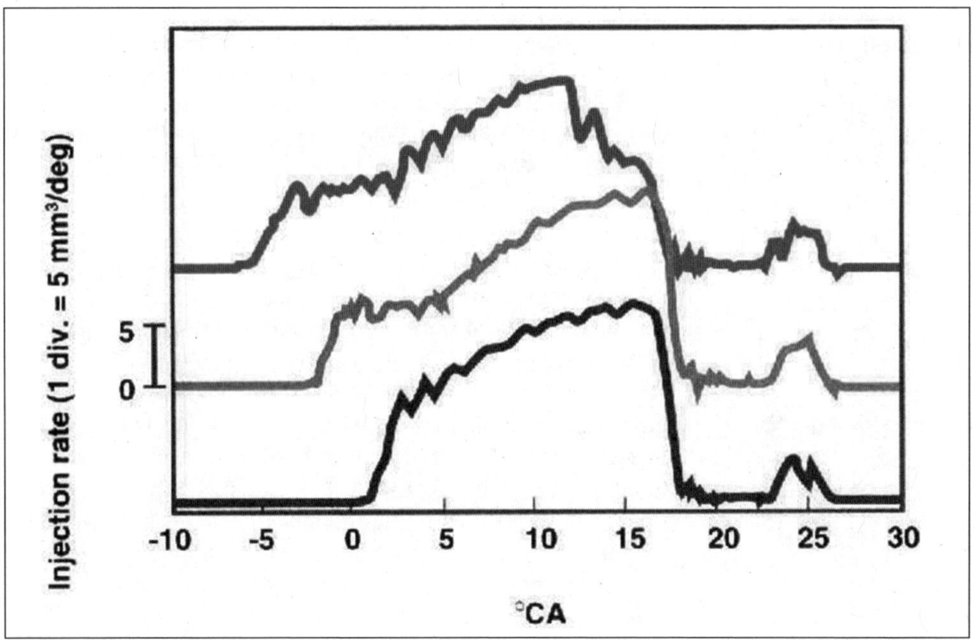

Figure 14.7 *Various combinations of injection shapes with post-injection capability demonstrated. (Reprinted from Ref. 10.)*

The following test conditions were maintained:

- Engine speed: 1000 rpm
- Injection quantity: 200 mm^3/stroke
- Boost pressure: 95 kPa gauge
- Intake manifold temperature: 35°C
- Air excess ratio: 1.62:1.

Figure 14.8 shows some of the test results in terms of injection timing and rate shape, and their influence on the combustion diagram. It is worth noting that the needle lift trace is almost identical among the three cases tested with the exception of injection timing. Yet, injection pressure for the three cases was different and the injected rate was directly controlled by the injection pressure rate shape. The influence on the combustion pressure rate shape was distinctly different and led to different NO_x formation rates and smoke opacity as shown in Figure 14. 9.

The foregoing example gives us a glimpse into the tedious parametric work needed to develop the right fuel injection system for future diesel engines. It is not enough to provide a

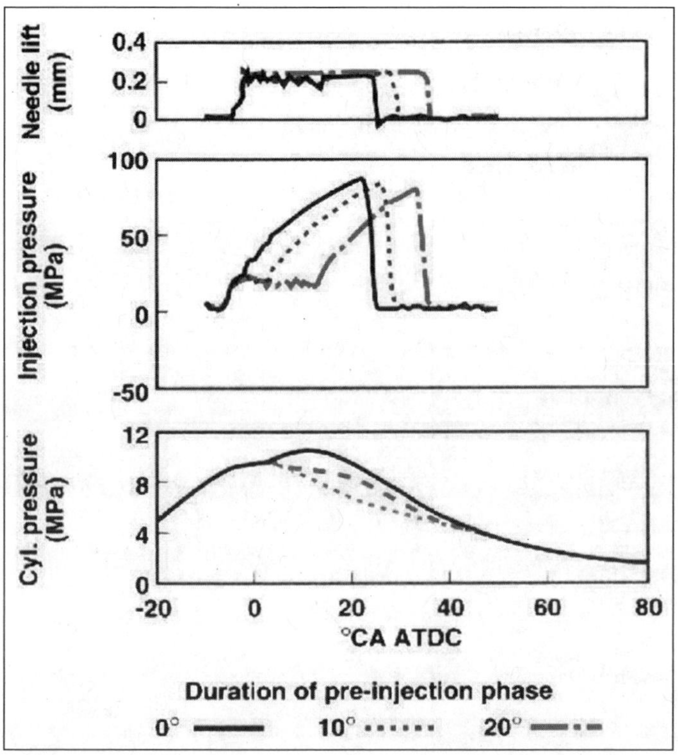

Figure 14.8 Single-cylinder engine test results. (Reprinted from Ref. 10.)

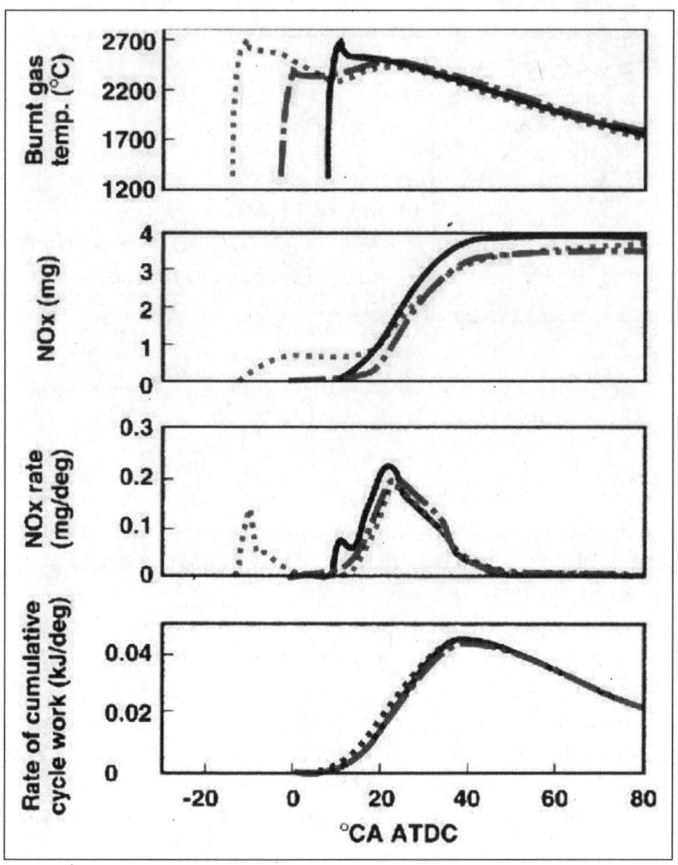

Figure 14.9 Effect of injection rate shape on emissions. (Reprinted from Ref. 10.)

very flexible injection system, but each of its features has to be tailored to produce the end result in terms of engine performance and emissions.

Diesel fuel injection systems are very complex, but quite versatile systems as described in Chapter 5 as well as the preceding sections. Each component of the system plays a very important part in overall system performance. From the low-pressure supply to the nozzle hole size, each part has to be engineered so that its performance complements the entire system. For instance, the fuel injection pressure in common-rail and other current systems is five times higher than in prior systems. These new high-pressure systems are in the 20,000 to 30,000 psi (1400 to 2000 bar) range [11]. An absolute requirement of the high-pressure fuel injection systems is a finer level of fuel filtration. The higher pressure causes the closely fitted components of the fuel injection pump and injectors to be more susceptible to wear damage from dirt particles. If water passes through the fuel injection system, the reduction in lubricity could cause the precision pumping and metering components to score, scar, or seize. The extreme high pressure is present at all times. If an injector tip is damaged as a result of dirt particles or water and the injector valve does not seat and seal completely, unwanted fuel under high pressure can be delivered to the cylinder, resulting in poor engine performance. If a malfunction causes an injector valve to stick open, fuel could dump continuously into the cylinder with catastrophic results. Replacing and repairing fuel injection components is very costly. Leading manufacturers of high-pressure fuel injection components and systems, along with major engine and equipment OEMs, have mandated that their systems be protected by fuel filters that have very high water separation and dirt removal capabilities and ratings.

On the other end of the fuel injection system, the nozzle and its engineering are no less demanding than the rest of the fuel system components. Hole size, its machining, and its length in relation to its diameter are a few of the elements that require close attention. Emerging technologies are considering sophisticated engineering for the nozzle. The surface finish of the hole itself was found to affect injection quality and emissions [12]. Nozzle holders are designed to provide the boot injection shape through needle throttling or prelift, assisted by two springs as shown in Figure 14.10 [12].

Because the nozzle hole is the last component of the fuel injection system that communicates with the combustion process, its details play a crucial role in the spray formation and characteristics as well as the mixing process. The nozzle hole diameter has been decreasing throughout the 1990s. Modern engines have nozzle hole diameters ranging between 0.15 and 0.20 mm that produce sprays of atomized fuel whose particle size is as small as 5 μm [12]. A design objective that has long eluded fuel systems designers is the variable area nozzle. Once developed, this nozzle will have the capability of varying the fueling rate in real time. With reduced flow area at the beginning of injection, the fueling rate would be low; and if it is desired to increase the rate after the start of combustion, then the full nozzle area would be available for the increased rate. Three different designs considered by R. Bosch were reported by Tschoeke [12] and shown in Figure 14.11.

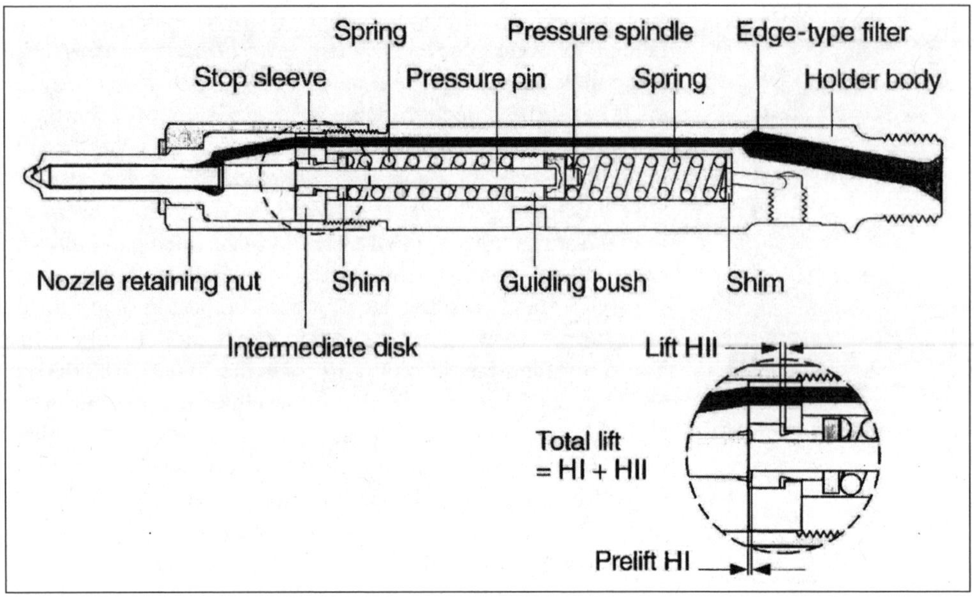

Figure 14.10 *Nozzle holder assembly with two springs and prelift capability. (Reprinted from Ref. 12, courtesy of Robert Bosch GmbH.)*

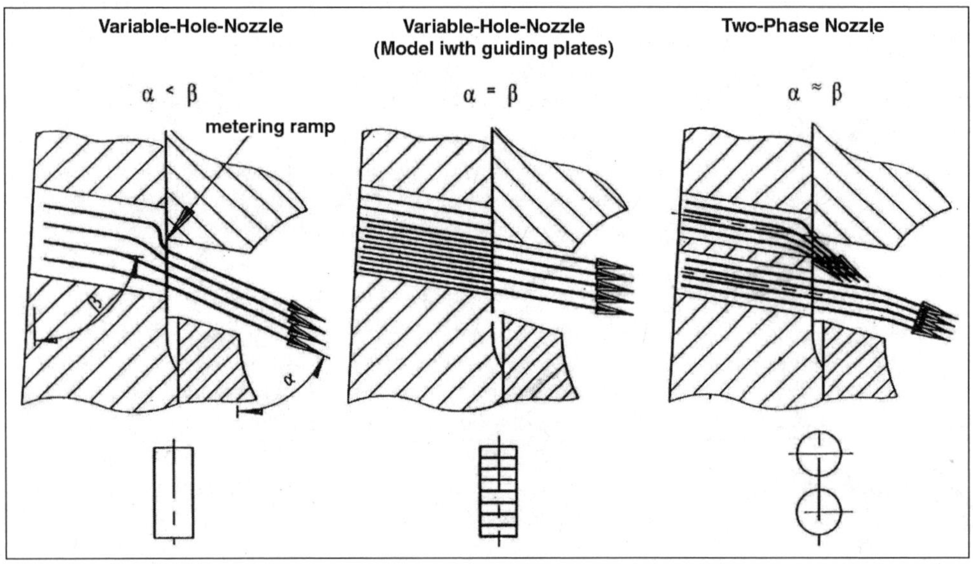

Figure 14.11 *Schematics of three variable hole area nozzles. (Reprinted from Ref. 12, courtesy of Robert Bosch GmbH.)*

The function of unit injector, common-rail, and unit pump systems (see Chapter 5) is primarily determined by electromagnetically controlled valves, which close and open the path for the high pressure to proceed to the injector tip. Electromagnetic solenoid valves are normally used in these systems. Solenoid valve designs with the lowest moving mass in the smallest space, and the largest initial electromagnetic force, are used [13]. The typical switch times of the valve for a medium-duty/heavy-duty truck application are [13] as follows:

- Closing period at rated power: 650 μs
- Opening period at rated power: 300 μs

However, the modern high-speed diesels that are powering the majority of new passenger cars in Europe and Japan have rated speeds of between 4500 and 5500 rpm. The need for rate shaping and fast control over each injection event is leading designers to piezoelectric technology. Piezoelectric stacks are appearing as the actuator of choice because their response time (150 to 250 μs) is a fraction of that of the electromagnetic solenoid valves [14].

Having highlighted the importance and capabilities of a modern diesel fuel injection system and its impact on emissions and performance, it is worth noting that such a system needs to be supported by a highly sophisticated control mechanism unit capable of performing literally thousands of calculations in a matter of milliseconds. The injection systems of the past were very complicated as seen from Chapter 4. Even so, they controlled two basic functions: injection timing and fuel quantity. Early attempts at modernizing the fuel injection system saw a very gradual change to electrical add-on devices merely mimicking the function of the mechanical actuators and devices they replaced. Today's electronic controllers are extremely sophisticated, having interlocking circuits and algorithms that model the performance of engine subsystems and can predict engine functions in advance. They operate self-check operations and alert the operator about malfunctions, maintenance schedules, and so forth, while keeping track of the many engine and exhaust system treatment functions.

14.7 Combustion System Technology Options

Combustion in a DI diesel engine, whether a heavy-duty or high-speed, light-duty engine, is the result of careful design and control of many components and engine parameters. As discussed in Chapter 5, it is difficult to limit the combustion sytem to just a piston bowl shape or an intake port swirl ratio. Spray details—starting at the fuel supply system and moving on toward the injector tip and nozzle hole geometry—compression ratio, air utilization, number of valves and the timing of their events, heat exchange throughout the cylinder volume at various engine load and speed conditions, fuel specifications, and many other parameters contribute to the success or failure of the combustion system in performing as desired. As the combustion is considered many in the industry, turn to the injection spray and some may even consider the fuel particle as the tiniest component of that spray. Flynn [9] presented a simplified model of the temperatures and chemistry that take place when a spray of relatively cold fuel entrains air and combusts. Figure 14.12 is a graphic representation summarizing the combustion process.

When exposed to heat, fuel vaporizes a short distance from the spray nozzle and creates a reacting gas jet. Because temperatures are still relatively low at that point, the chemical reactions are slow. As temperatures rise to 825 K, low-temperature reactions take place. These reactions increase in intensity at temperatures above 1100 K and consume all available fuel and oxygen in the vicinity. According to Flynn [9], three different and critical regimes can be identified in this process:

- 800 to 850 K where activity is mainly in the weakly bonded hydrogen
- 1050 to 1100 K where hydrogen-peroxide is decomposed
- 1500 to 1600 K where chain branching is experienced, yielding 1900 to 2000 K flame temperatures.

Flynn also observed that fuel burning is a two-stage process, a rich premixed reaction occurs upon spray entry and then a peripheral diffusion flame completes the fuel burning. In the first stage, carbonaceous particulate is formed, while the higher temperatures that characterize the second stage consume particulate formed in the first stage, but produce NO_x because of flame temperatures exceeding 1900 K. He also noted that attempts to cool this flame will reduce NO_x, but cause incomplete combustion. Herzog [15] summarized the root cause of the NO_x/PM trade-off problem in the schematic shown in Figure 14.13, which indicates where today's combustion systems are (see circled area on Figure 14.13) and where the target is for future technologies.

Modern design methods and emerging technologies that hold the best promise of achieving the target area shown in Figure 14.13 include the following:

- Improved air charging
- Improved air utilization
- Proper selection of the bore/stroke ratio
- Improve cylinder breathing—by adopting four- versus two-valve designs
- Optimize mixture formation–by centralizing injector location, by providing flexible/variable swirl, or by matching injection spray and swirl
- Proper selection of compression ratio
- Careful design of intake and exhaust port surface to achieve efficient heat exchange

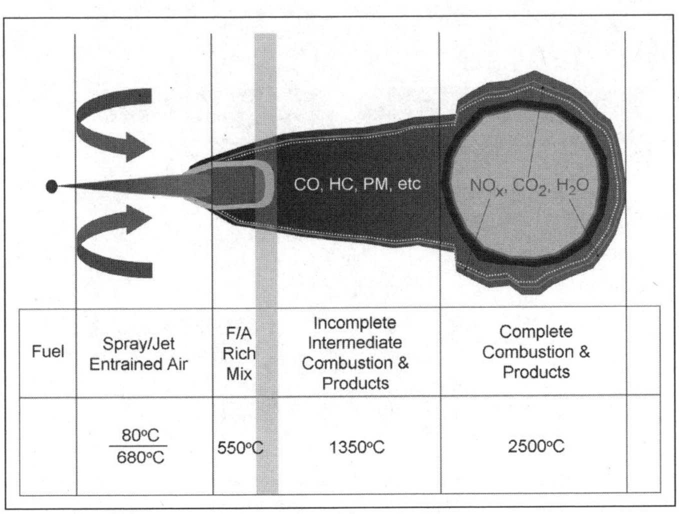

Figure 14.12 Temperature and chemistry evolution in diesel combustion.

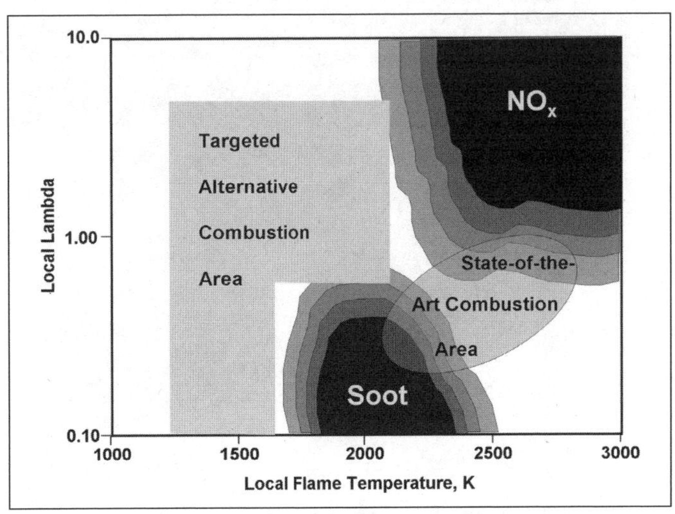

Figure 14.13 Target for future combustion system performance. (Reprinted from Ref. 15, courtesy of AVL List GmbH.)

- Careful design of combustion bowl to maintain/enhance the fuel and air mixing process
- Minimize combustion chamber dead space
- Injection rate shaping/multiple injections per combustion cycle.

It is worth noting that this list of topics is very interrelated. From the initial conceptual design to the final development details, continuous improvement and matching of these areas and parameters is required for the best possible combustion characteristics.

Since the task of designing future diesel engines is very demanding, designers are resorting to tools such as *finite element analysis* (FEA), *computational fluid dynamics* (CFD), and *computer-aided design* (CAD) for assistance. These analytical and design tools help speed up the cycle for engine design by minimizing the number of hardware iterations and determining design flaws not only early in the process, but sometimes even before the design is made final. Figure 14.14 is an example of a CAD mesh for a cylinder and its ports. Figures 14.15a and b show the results of an analysis of the influence of swirl on the oxygen field during the expansion stroke, and the influence of swirl on the gas temperature field for the same engine speed and load conditions [16]. In many instances, these mathematical models can even be used to test the designed components under dynamic conditions to predict stress risers, heat concentrations, and potential failure modes prior to the hardware stage, thus reducing the risk of early failure if the hardware were to be built and tested.

The use of mathematical models is not confined to hardware design only, but also to fluid mechanics. Charge air motion is simulated as well as injection spray mixing with surrounding charge air. By injecting air or oxygen-enriched air directly into the combustion chamber and late in the combustion cycle, particulate emissions can be reduced [17]. Additionally, by combining this technique with optimized fuel injection char-

Figure 14.14 *CAD mesh of ports and cylinder. (Reprinted with permission from Ref. 16.)*

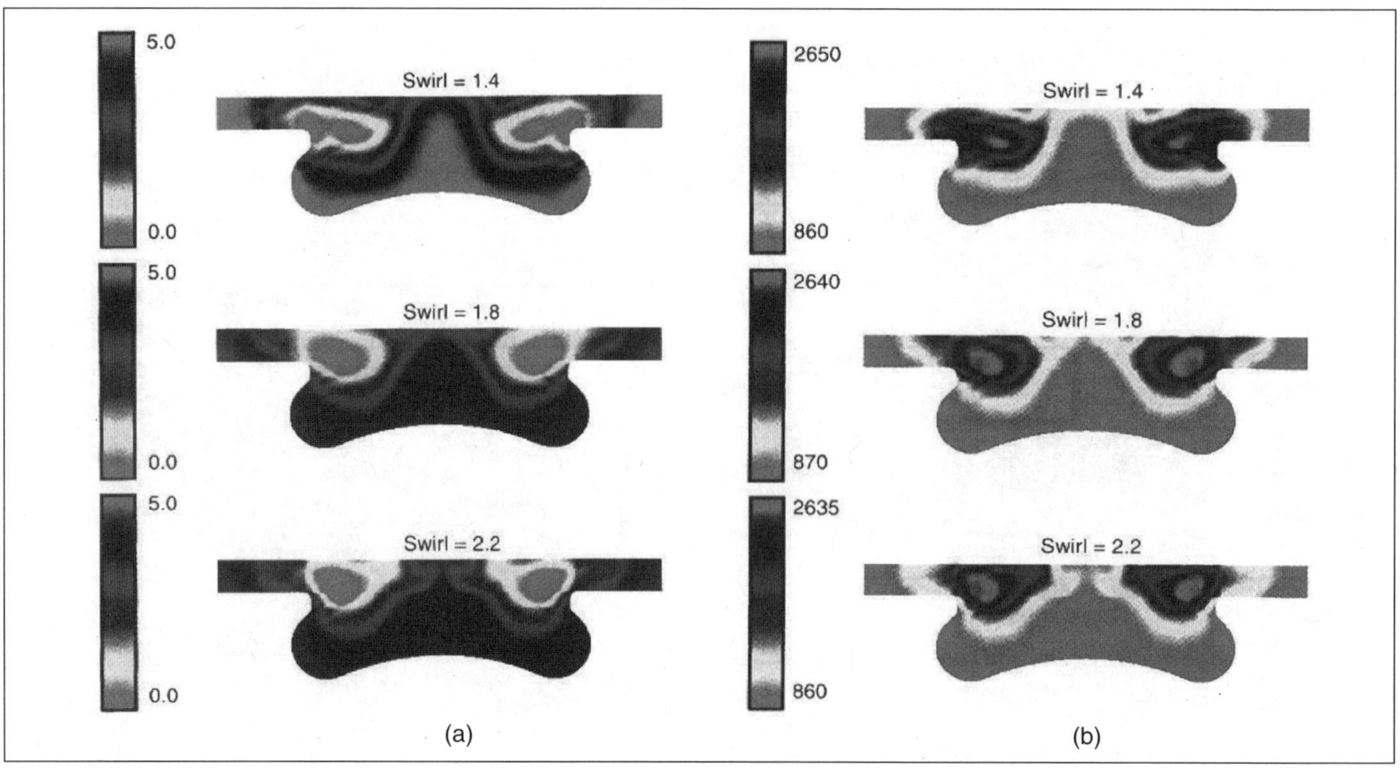

Figure 14.15 *Influence of swirl on (a) oxygen field and (b) gas temperature field during expansion stroke. (Reprinted with permission from Ref. 16.)*

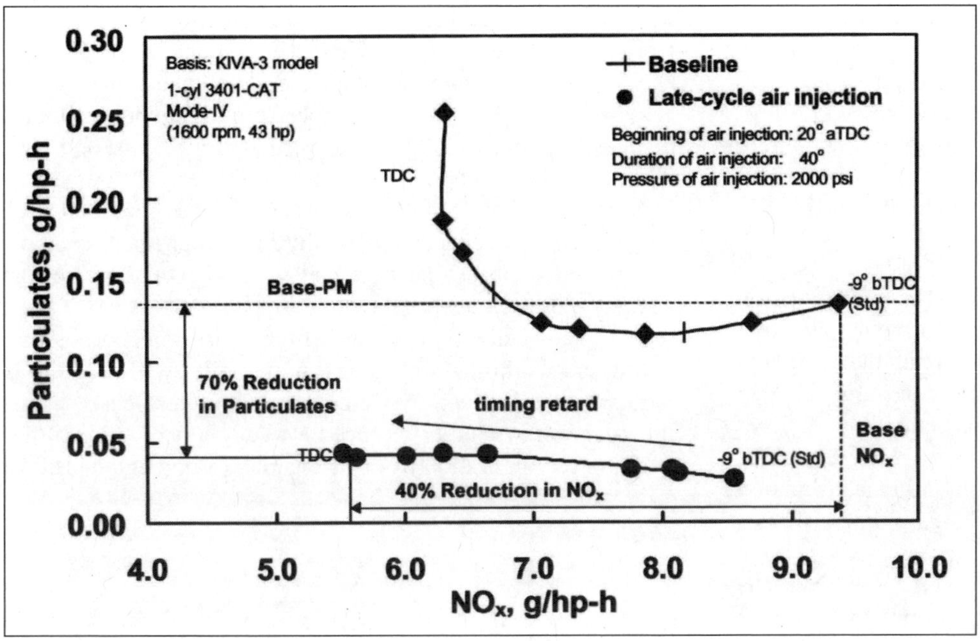

Figure 14.16 Simultaneous reduction of NO_x and PM with optimized late-cycle air injection and fuel injection timing. (Reprinted with permission from Ref. 17.)

acteristics, both NO_x and particulate emissions can be reduced simultaneously. Extensive analytical studies using the KIVA-3 CFD model developed at the University of Wisconsin [17] have been conducted. These studies have shown the importance of the gas injection characteristics on the technique's effectiveness in reducing emissions. They have additionally shown a simultaneous reduction of NO_x and particulate emissions (see Figure 14.16) when the fuel injection characteristics are optimized in conjunction with late-cycle air injection.

Increasing charge air pressure is an effective measure to increase power density in almost all modem high-performance and heavy-duty diesel engines. Improving specific power output in DI engines by using exhaust turbochargers has been common practice for some time. Recent reduced emissions and high-performance requirements led many OEMs to adopt advanced turbochargers as well as mechanically driven superchargers to *improve air charging*. Variable geometry turbochargers and turbochargers with waste gates are more common on production engines today than in the past. Generally, higher charging pressure without increasing exhaust back pressure, good turbocharger efficiency over the operating range, and rapid charge pressure development are good qualities to aim for in designing intake charge systems that support good combustion [18].

Diesel combustion is known to be very lean with air-to-fuel (A/F) ratios of 25:1 at peak torque, 30:1 at rated speed/maximum power conditions, and over 150:1 at idle for turbocharged engines. Yet, this extra air does not enter into the combustion process. Rather, it is heated during combustion and escapes through the exhaust valve, causing the diesel exhaust to be extremely lean. A goal for the engine designer is to make the best use possible of charge air. In general, this objective is described by *improving charge air utilization*, which dictates the design of several combustion chamber components. In fact, a major issue to consider in optimizing a DI diesel combustion system is the proportion of available air participating in the combustion process [19]. The K-factor, calculated as the ratio of the piston bowl volume to the total combustion chamber volume, is a widely used measure of the proportion of air available for combustion. Reducing engine displacement leads to a decrease in the relative K-factor and hence a tendency for deterioration in the combustion behavior. For a given displacement and at constant compression ratio, the K-factor can be improved by selecting a longer stroke.

Selection of the *bore-to-stroke* for the engine is usually based on the K-factor consideration and several others, including engine packaging, porting and valving, and so forth. A particularly key issue in setting the maximum bore-to-stroke ratio is the very challenging cylinder head packaging required to accommodate four-valve per cylinder design and a common-rail fuel injection system with a centrally located injector. Cylinder heads are complicated to design due to their many passages including water cooling, cylinder head holddown bolts, intake and exhaust ports, injectors, glowplugs, valves and valve stems, valve recesses and seats, as well as other passages that are needed for exhaust gas recirculation in some designs [20].

Compared with a two-valve engine, four-valve engines offer the following key advantages:

- Optimized mixture formation resulting from the *central, vertical injector*. Computations also predict that the mean in-bowl swirl intensity for a *centrally located bowl* is increased by about 10% compared with that in a typical eccentric bowl for a two-valve engine for the same inlet swirl level. This results in reduced demand for intake swirl or the opportunity to increase the bowl diameter, which is advantageous for small-displacement engines [21].
- Symmetric and more uniform fuel distribution in the chamber resulting from nearly equal flow and spray penetration from

each nozzle hole. Nonvertical injectors often show up to 20% variation in flow from hole to hole [20].

- Potential for increasing the number of nozzle holes since symmetry exists between injector and combustion bowl [20].
- More efficient gas exchange through better port flow characteristics.
- The potential for port deactivation, which offers load- and speed-dependent swirl adaptation.

The beneficial characteristics of four-valve combustion systems result not only in improved performance and fuel economy but also provide the basis for realizing the extremely high exhaust gas recirculation (EGR) ratios required to calibrate modern and future engines for lowest NO_x emissions [22].

As mentioned earlier, combustion in a DI diesel engine is heavily dependent on the quality of the mixture formation process. For a given *combustion bowl design,* mixture formation is determined primarily by the *intake swirl intensity* and the fuel injection process. In small-displacement engines, the two are equally important, with mixture formation occurring partly through air entrainment into the fuel spray and partly through swirl-assisted evaporation of fuel that has impinged on the wall of the combustion chamber. The utilization of a common-rail injection system, which has the capability to provide high-pressure fuel over the entire operating range, may change the requirements for swirl to achieve good mixture formation. The increased fuel pressure may offset the drop in swirl at low engine speeds [20]. Yet, as engine displacement increases, optimum swirl intensity has a tendency to decrease [20] (see Figure 14.17).

Figure 14.17 shows a broad range of experimental data that clearly indicate the trend toward increasing swirl requirement in high-speed DI diesel engines. This aspect of the design reduces the amount of wall wetting and also assists in fuel evaporation from the chamber walls as cylinder displacement is reduced.

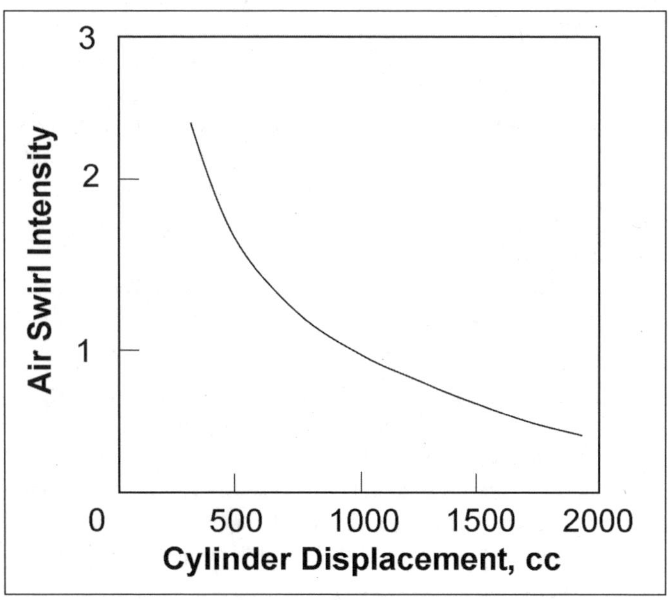

Figure 14.17 *Intake swirl requirement as a function of displacement. (Reprinted from Ref. 20.)*

Figure 14.18 shows a comparison of combustion chamber cross-sections for a 1.9-L, high-speed engine having a cylinder displacement of 475 cc [23]. It is worth noting that the optimum design was 2a, which was matched with two swirl ports as shown in Figure 14.19. *Flexible/variable swirl* is one more feature of the modern diesel engines, especially small, high-speed engines. This added flexibility can be achieved through dual-intake port designs [23] such as that shown in Figure 14.19.

In dual-intake port designs, high swirl is made possible at low speeds through the helical port, while swirl intensity is reduced at high speed by using a tangential port design. Figure 14.20 shows the difference between the two ports of the dual-port configuration from a Renault 375-hp engine [16]. In

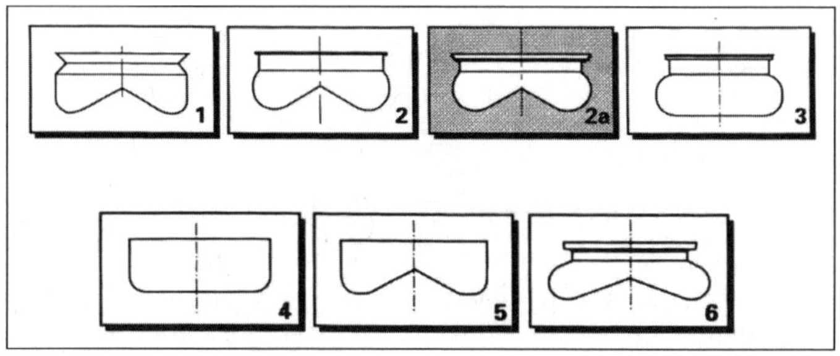

Figure 14.18 *Experimental combustion bowls for the Volkswagen 1.9-L engine. (Reprinted from Ref. 23.)*

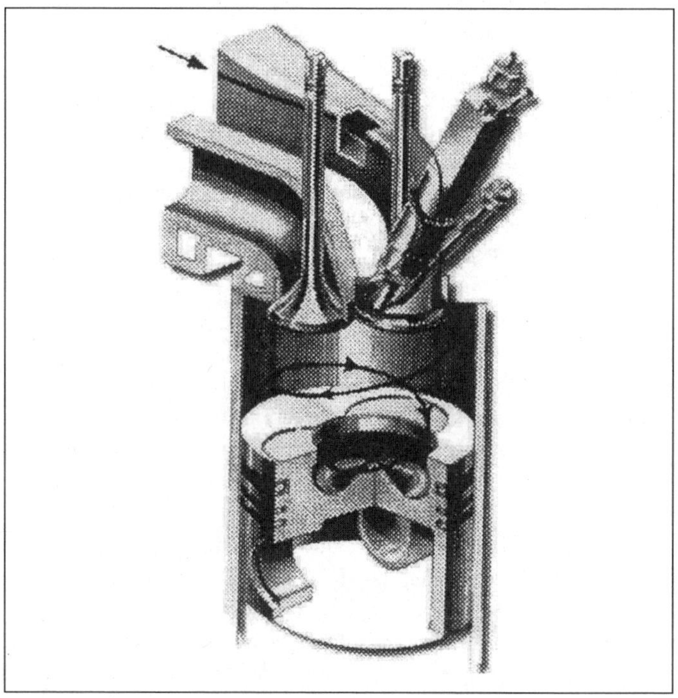

Figure 14.19 *Dual-intake ports for swirl intensity control. (Reprinted from Ref. 23.)*

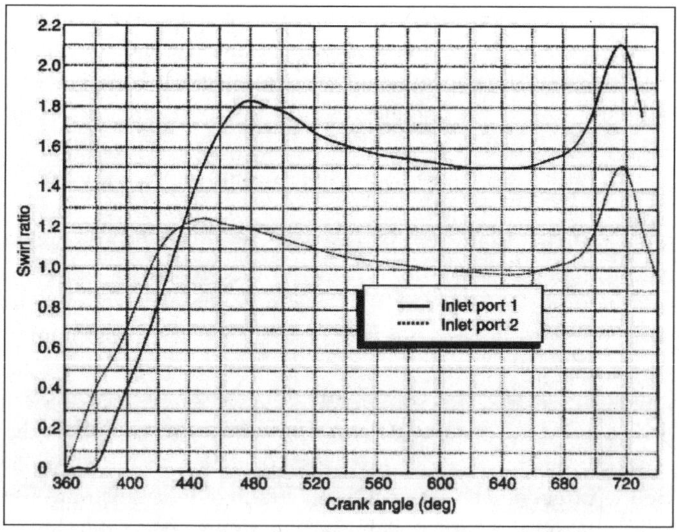

Figure 14.20 Swirl ratio difference during intake and compression strokes for dual-port design. (Reprinted with permission from Ref. 16.)

general, *reentrant bowl designs* are used with swirl-supported systems for small, high-speed and medium-duty engines. Large heavy-duty diesel engines normally use *shallow bowl designs* and low swirl intensity ports (tangential). The emphasis, however, is always on the mixture formation in support of high combustion efficiency.

The efforts expended in designing enough flexibility in both intake port and injection spray characteristics give control engineers as many tools as possible to *match spray and swirl* to achieve optimum mixture formation at any speed and load condition. Figure 14.17 illustrates the higher potential influence of spray-wall interaction on the combustion process in small, high-speed engines.

Increasing the length and *surface of intake ports* exposes charge air to heat from cooling water. Heating the charge is contrary to measures taken to reduce NO_x emissions. Therefore, the length of intake ports and their surface area exposed to heating should be minimized. *Dead space* in the combustion chamber retains air that is compressed during the compression stroke at the expense of work by the piston. This trapped air expands in the expansion stroke well after the combustion process is complete. This air is subject to work by the piston and yet it never enters into the combustion process.

A high *compression ratio* is required for cold start purposes, especially in cold ambient conditions. The higher the compression ratio, the higher the thermal efficiency and the higher the frictional losses. Therefore, the compression ratio is always a compromise between ease of starting in cold weather and good fuel economy through reduced frictional losses. Another benefit for high compression ratios is providing more heat during compression for the injected fuel to evaporate and subsequently mix with air. Better evaporation tends to reduce the ignition delay period and allows further injection timing retard without sacrificing thermal efficiency.

14.8 Air Charging System Technology Options

This topic was addressed earlier and the need for a new line of supercharging equipment was outlined. One of the technical solutions that is emerging in the area of supercharging is a technology that combines turbochargers and electrically driven superchargers. An add-on electronically controlled, high-speed electric motor–powered supercharger was one of the early developments in this new combination of turbochargers and electrically driven superchargers. It was used experimentally and in retrofit kits to supplement engine boost during acceleration conditions to bridge the turbocharger lag period and to improve low-end performance characteristics [24].

The product, as tested in several diesel applications, is described as an electronic demand charger by Turbodyne. The electronic demand charger (EDC) offers the opportunity to significantly reduce the insoluble fraction of exhaust particulate matter formation. The EDC addresses overall emissions formation by increasing the A/F ratio at low-speed operating conditions in order to primarily reduce smoke and particulate emissions. The system (see Figure 14.21) introduces additional air to the engine manifold via a simple valving mechanism. This additional air not only makes the A/F ratio leaner, but also provides additional boost to the engine, which improves engine performance by eliminating the traditional turbocharger lag.

The application of a motor-assisted device, such as the electronically actuated EDC, provides a means of supplying air to the engine cylinder at low idle or low loads in preparation for the next acceleration of the engine, before an increased amount of fuel is injected. When used in series with a standard turbocharger, the EDC supplies boost pressure to the engine until there is enough energy in the engine exhaust to accelerate the turbocharger. When the turbocharger is capable of supplying sufficient air, the unit is de-energized.

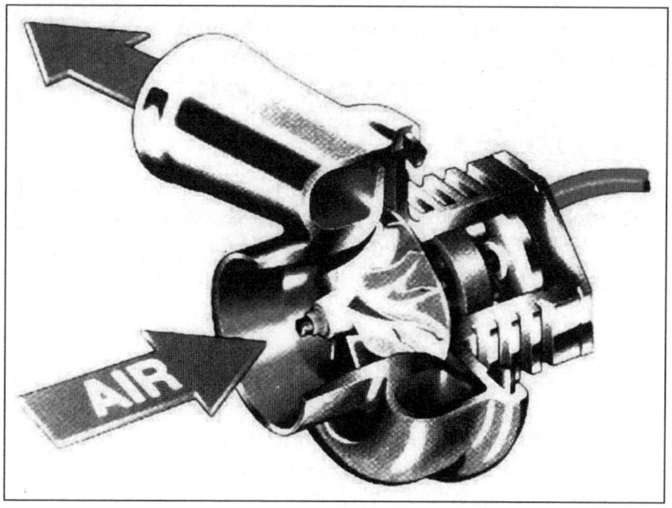

Figure 14.21 Cutaway view of the Turbopac electronic demand charger. (Reprinted from Ref. 24.)

In diesel applications, the EDC is an inlet air compressor installed independently of the turbocharger between the air intake and the engine manifold. The system is comprised of a compressor housing, compressor wheel, brushless electric motor, motor controller, engine interface controller, and a simple valving mechanism. On turbocharged diesels it is installed in series with the turbocharger. In this instance, the assembly consists of a compressor, a motor controller in a separate housing, a 5-in. (12.7-cm)-diameter check valve, associated intake tubing and connecting sleeves, a manifold pressure shutoff switch, a throttle switch and bracket, a throttle switch actuation arm, and a wiring harness. Traditionally, efforts to reduce engine-out emissions have concentrated on fuel management. The EDC focuses instead on air management with the move toward optimization of the A/F ratio at low-end operating conditions.

In turbocharged engines, when the throttle is opened to accelerate the engine, there is a time lag before the exhaust energy level rises sufficiently to activate the turbocharger turbine. Over the years turbochargers have been made smaller and lighter in order to minimize this turbocharger lag period. Some manufacturers have resorted to ceramic turbine wheels because of their low inertia in the hope of reducing turbocharger lag. The EDC can be installed on both turbocharged and nonturbocharged vehicles. In nonturbocharged vehicles, it is installed between the air intake system and the engine manifold.

On diesel turbocharged vehicles, the compressor and check valve are configured in parallel, and together they feed in series with the existing turbocharger inlet. The check valve is oriented so that intake air will be drawn through valve and the EDC during periods of high intake air demand. During periods of low intake airflow, the check valve will close and the EDC will pressurize the turbo inlet.

In configurations that utilize a preexisting electronic engine controller, the unit can be integrated into the existing controller system. The interface controller then activates the EDC during predefined engine operating modes indicated by the sensors. The electronic control module can also interface with engine management systems in similarly equipped vehicles.

The EDC compressor must accelerate rapidly to effectively reduce turbocharger lag and PM formation. This acceleration is accomplished by means of the brushless permanent magnet motor, which produces high torque, accelerating the compressor shaft at a rate of more than 100,000 rpm per second. To further reduce the time lag, the compressor "idles" at 10,000 rpm, waiting for the command to accelerate to 40,000 rpm. The rapid pressure rise was tested with the compressor discharge pressure monitored on a digital oscilloscope and the results of this test indicated that the compressor was able to produce 80% of its final pressure in 250 ms, corresponding to 2.5 engine revolutions at a 600-rpm engine idle speed.

A second-generation design called the Dynacharger, which integrates the high-speed motor into the center section of the turbocharger for OEM applications, is in its advance development phases. An EDC was installed on a 277-hp (206-kW) output, two-stroke Detroit Diesel 6V-92TA engine with mechanical unit injectors. The engine was installed on an engine dynamometer at the Southwest Research Institute [24] and tested for emissions using the EPA heavy-duty engine transient test procedure. No adjustments to the fuel control governor or injector timing were made. A significant reduction in particulate emissions was a direct result of the enhanced air supply to the engine during low-speed accelerations. Additional research is currently under way to further enhance this possible simultaneous reduction in PM and NO_x while utilizing the EDC in a "dual-use" format (air pump and EGR pump). The EDC provided 20-kPa intake manifold pressure within 0.25 s after the throttle moved from idle position. Without the EDC, the turbocharger required 2 to 3 s to provide the same boost level. The EDC was programmed to run at high speed for 8 s after the throttle was moved from the idle position. This was sufficiently long for the engine exhaust-driven turbocharger to spool up to provide the remainder of the air requirement. Another test was conducted to measure the EDC's effect on smoke emissions. Figure 14.22 illustrates the EDC smoke reduction capability. The figure shows the smoke results of the first two accelerations during the EPA transient cycle. It is evident that the EDC was able to almost completely eliminate smoke emissions.

The battery voltage and current were monitored during EPA transient testing to determine the power and energy requirements of the EDC. It was determined that the EDC consumed 13.8 amp-h of battery energy at 24 V during the 20-min test. The energy consumption was 330 watt-h. The total energy requirement from the engine during the 20-min test was 14.8 kWh, thus the energy required to drive the EDC was 2.2% of the total engine power. If the vehicle alternator efficiency is assumed to be 60%, this would be the equivalent of 1.6-kW average mechanical power. Reducing the high-speed time period from 8 s to 5 s should result in a corresponding reduction in the EDC power consumption [24].

Enhancing the air supply to diesel engines has proven to result in significant improvements in engine low-speed torque, efficiency, and exhaust emissions. An EDC has demonstrated that this improvement can be made in a device retrofitted to existing engine configurations and is an effective alternative to traditional approaches that are associated with decreased engine performance, turbocharger lag, fuel penalties, and PM/NO_x trade-offs.

In addition to electrically assisted turbochargers, variable-geometry turbochargers and hydraulically assisted turbochargers are being considered in some new engine designs. Described as the *sequential hydro-supercharging system* (SHS), the system was developed to provide increased airflow to the engine in a controllable fashion over the engine's entire speed and load ranges [25]. An "ideal" charge air system would produce constant BMEP over the entire engine speed range. Based on thermodynamic analysis, a turbodiesel typically needs a supercharging pressure ratio of about 1.4 in the low rpm range and about 1.2 in the high rpm range [25]. For the same turbodiesel, the charge airflow increases by about a factor of 4, while the engine rpm increases only by a factor of 3. For a typical 7-L engine, charge airflow is about 7.0 kg/min at 800 rpm and about 28.0 kg/min at 2400 rpm. A variable-speed SHS can accomplish these requirements with a high degree of flexibility. In addition, a SHS can adjust for optimum airflow at a variety of engine speed

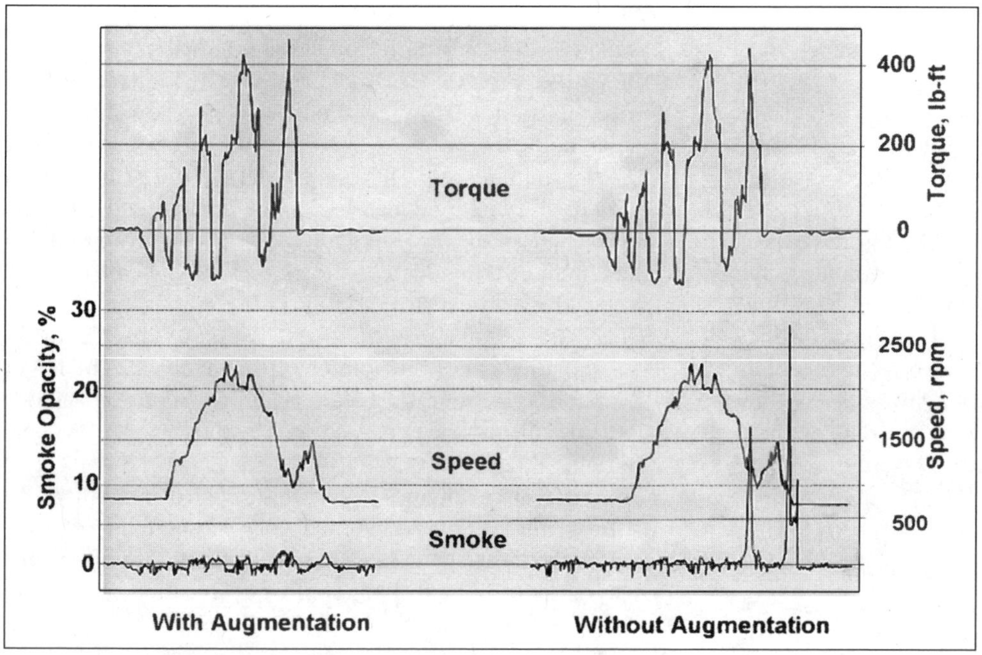

Figure 14.22 Effect of the EDC on smoke emission. (Reprinted from Ref. 24.)

and load conditions. Based on engine thermodynamic analysis, each individual SHS compressor remains at near optimum thermodynamic efficiency throughout the full range of engine speeds and loads. The hydraulic turbine efficiency is nearly 80% and the hydraulic pump efficiency is nearly 90% [25].

Figure 14.23 shows the calculated engine pressure differential for three cases, an SHS, a waste-gated turbocharger, and a positive-displacement supercharger. The SHS provides for substantial positive differential throughout the full speed range. At high rpm, the waste-gated case shows a negative differential mainly due to exhaust waste gating. A 15% exhaust bypass flow was used for the waste-gated case. For the positive-displacement supercharger, the pressure differential is lower than with the SHS system in the high rpm range. This is due to the lack of supercharging in the high rpm range [25].

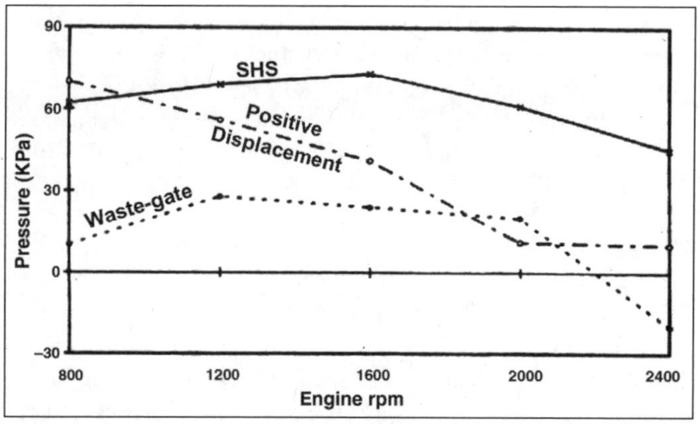

Figure 14.23 Pressure differential for three charging types over various engine speeds. (Reprinted from Ref. 25.)

Innovations in engine air charging also include a new turbocharger that will increase efficiency. The design is called the *twin scroll,* signifying that the exhaust inlet is split into two parts. One carries exhaust from cylinders 1 and 4, while the other serves cylinders 2 and 3, in a four-cylinder diesel engine. This design prevents exhaust from being drawn back into the cylinders [26]. The new system achieves greater efficiency especially at partial-load conditions. It also provides more air near and at full load to assist in achieving a 15% to 20% reduction in fuel consumption.

14.9 Air Induction System Technology Options

Air charging is one of the main components of the overall air induction system. Intake manifolds, intake ports, intake valves, and their actuation are some of the other components making up the remainder of the induction system. In the greater scheme of charge air preparation, one must also include the charge air cooler as well as intake manifold heaters, if they are used. A consent decree, which announced a government/industry agreement specifically between the U.S. Department of Justice and several heavy-duty engine manufacturers, further tightened the limits on emission controls. The new NO_x limits led most, if not all, heavy-duty engine manufacturers to consider and develop EGR systems for use in heavy-duty applications. Light-duty passenger car applications had already been using EGR for the same purpose. The most common EGR designs introduce a portion of the exhaust through the inlet system (inlet to the turbocharger compressor, intake manifold, or intake port). Therefore, EGR systems ought to be included in the discussion of future technologies, particularly for heavy-duty applications. Due to its importance, all of Chapter 16 is devoted to this topic.

Supercharging in all its varieties is used mainly to improve the engine's power output. Yet, supercharging increases charge air pressure as well as its temperature, therefore decreasing its density. *Charge air cooling* (intercooling/aftercooling) is used to reduce charge air temperature and return density to a high value. More dense charge air occupies less volume in the combustion chamber and increases the mass of air, as well as its oxygen content, thus allowing greater combustion efficiency [27]. In developing charge air cooling, air-to-air intercoolers are currently universally accepted in on-highway heavy-duty applications as well as many medium- and light-duty applications. Further studies have been conducted to optimize these devices in the following areas:

- Sizing of the heat exchanger to provide the proper cooling of the charge air at full load and rated speed conditions, even in high ambient temperatures and low vehicle speeds. The generally accepted practice is to size these intercoolers to control charge air temperature at rated speed and load conditions to about 30 to 35°F above ambient [27].
- Minimizing the pressure drop across the intercooler ducting through carefully designing the connections to and from the turbocharger assembly. Generous duct sizing and minimum bends and joints are required to achieve this design goal.
- Minimizing the allowed amount of system leakage.
- Coordinating the function of the air-to-air intercooler and the radiator for each to accomplish its required cooling effectiveness without interfering with the function of the other.

The last task gained special importance in heavy-duty engines when EGR was introduced to this class of engines. EGR coolers, using jacket water as the cooling medium, rejected more heat, which resulted in more cooling load to the radiator. Another system is discussed in Section 14.13 and is designed to minimize pressure drop through the heat exchanger, but requires additional components to operate.

As previously discussed, extreme care is given to developing the proper air motion through the induction system to achieve the desired mixture in support of efficient combustion. Designing the ideal swirl intensity through shaping of the intake port or ports may be a futile exercise if, when installing the *intake manifold* upstream of the cylinder head, that swirl intensity would be radically altered (see Figure 13.34). Therefore, attention must be given to designing and developing intake manifolds that would be transparent to the intake port and have sufficient volume to maintain a good supply of air. Lack of intake manifold volume may lead to instantaneous smoke increases when hard accelerations are performed. In low-speed, high-output engines plenum-type manifolds appear to be more popular than log-type manifolds. In smaller high-speed, low-output engines log-type manifolds are preferred. Several design parameters are useful to intake manifold designers. The following is a description of three parameters mentioned in Ref. [27]:

- *The angle of turn*: This is the angle through which the induction air stream must turn from a longitudinal to a transverse runner (see Figure 14.24). The designer must strive to minimize this angle and improve uniformity of flow among the intake ports.
- *The flow path length*: That is the length of the path from the point of entry into the manifold to the point of entry into the intake port as shown in Figure 14.24. Modest variations in the length of the induction stream may produce discernible flow and swirl intensity variations [28].
- *The cross-sectional area along the flow path*: This parameter is also indicative of the volume of the manifold. The larger the manifold volume, the better its ability to damp pulsations and flow disturbances caused by the intake valve cyclical motion.

Modern diesel engines feature more control flexibility than their progenitors. This added flexibility was certainly demonstrated in the fuel management area with common-rail systems displaying not only excellent control of injection timing and metering, but also having the ability to control the injection pressure. In addition to these features, we have also seen the system's capability in providing more than one injection, in fact, multiple injections per combustion cycle [29]. Moreover, common-rail systems display an added feature of control that is the shape of the main injection. Likewise, the design of air management systems is moving in the same direction of providing more flexibility in controlling swirl intensity. This feature was made possible in part through the four-valve configuration in which each of the two intake valves acts as a gatekeeper of air supplied through an intake port with different characteristics than the other. The tangential port (the low-swirl-intensity port)

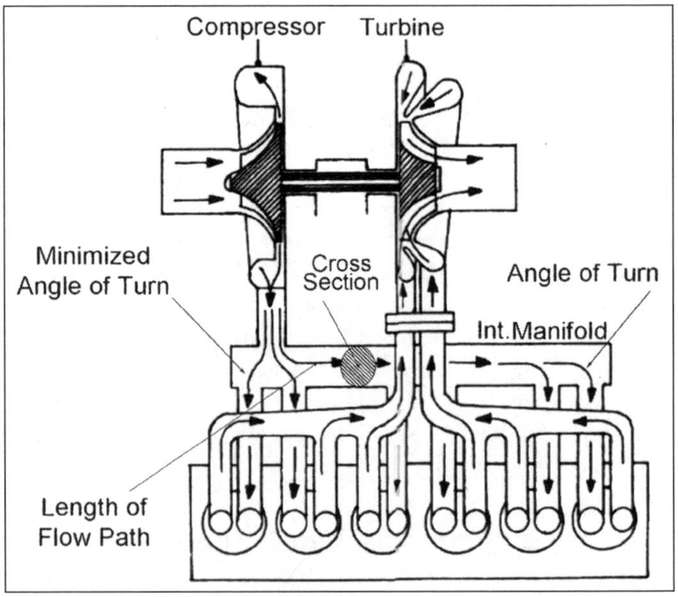

Figure 14.24 *Intake manifold design parameters. (From* Advanced Engine Technology *by Heinz Heisler. Reprinted by permission of Butterworth-Heinemann.)*

is now equipped with a throttle valve which, when opened, lowers the swirl intensity and vice versa (see Figure 14.25).

14.10 Variable Valve Actuation Technology Options

Also known as *variable valve timing* (VVT), *variable valve actuation* (VVA) is presented here as the last technology in the air induction system. In keeping with the same trend of developing as much flexibility in the engine subsystems as possible, controlling intake valve timing and motion is a technology that is receiving increased interest from engine designers. In fact, German car manufacturers such as VW [30] and BMW [31] have already introduced VVA in production.

As controller area networks (CANs) for vehicles are becoming commonplace, engine designers are pursuing advanced, next-generation VVA technologies to achieve the goal of direct, cycle-to-cycle, digitally controlled air management. A partial list of engine benefits resulting from cycle-to-cycle internal combustion engine air management includes these:

- Reduced fuel consumption
- Lower emissions
- Improved starting and idle
- Faster warm-up
- Increased low-speed torque
- Reduced transient response time (or reduced turbo lag)
- Engine braking (compression release)
- Exhaust gas temperature control for aftertreatment systems
- Advanced combustion processes using multiple fuels.

Significant promise is held in the application of increasingly flexible air systems, but there is still debate as to which valve train architecture will be the basis of future engines. A systematic grouping quickly divides technologies into two groups: camshaft systems and camless systems. Further technologies may be classified as primarily mechanical, electrical, or hydraulic (or various combinations). Companies in the United States and Europe are wrestling with the question of which system would have the highest benefits and lowest cost as well as liability. A system of choice at Jacobs Vehicle Systems is the digitally controlled, lost motion, electrohydraulic VVA system [32]. With experience in precision valve actuation for compression release engine braking in heavy-duty diesel engines, VVA is a natural progression in developing technology for precision air management in diesel engines. The low-lift secondary exhaust valve that opens shortly before TDC compression dissipates the power and acts as a vehicle speed retarder. This form of hydromechanically created compression release valve event is an early form of VVA technology. In pursuit of increased engine retarding power, variable timing of the compression release event was needed for ultimate performance across the engine speed range. In addition, application of the same hydraulic principles used for engine braking could improve engine performance and emissions control, and also enable further increases in retarding power at all engine speeds.

Advanced development work in this area is showing some benefits for high-speed diesel engines [32]:

- Reduced fuel consumption (4% to 5% average in various engine speed/load conditions)
- Increased low-speed torque (up to 4 bar)
- Improved startability
- Improved transient response (20% quicker versus the standard engine at low rpm)
- Improved braking power (80% of positive power).

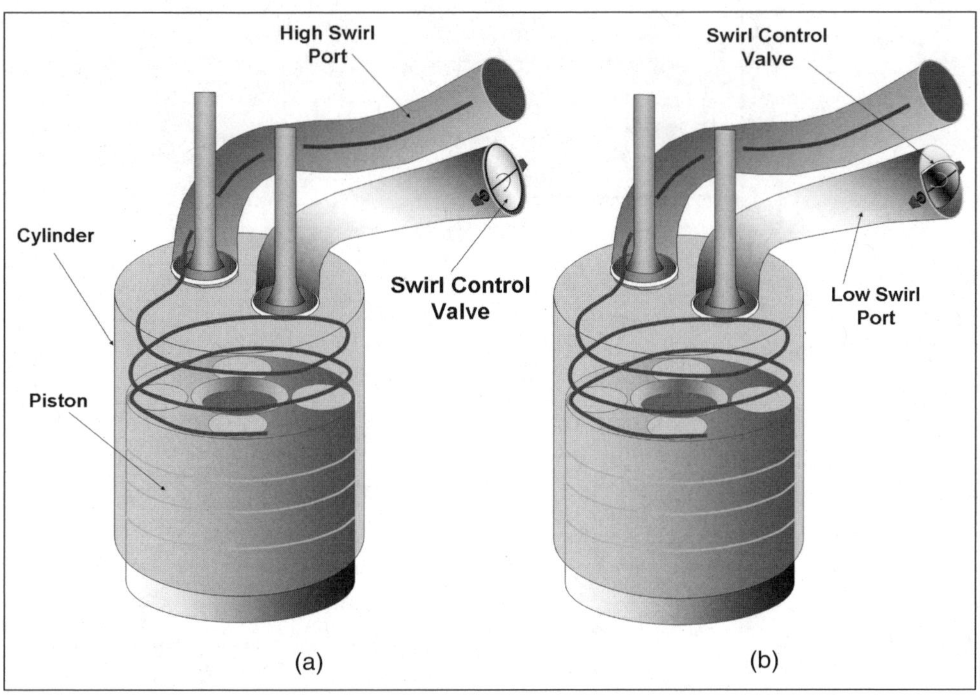

Figure 14.25 Throttling mechanism for variable swirl control: (a) valve closed, high swirl, and (b) valve open, low swirl.

While development and optimization of these systems and their controls continues, added care must be directed toward avoiding catastrophic failure (the cam profile can prevent valve-piston contact); improving reliability (valve seating velocity and valve-piston clearance requirements met without sophisticated controls); providing repeatability (less subject to trigger-valve variability); improving controllability (can provide secondary events without separate triggering); providing limp home (limited lost motion); adding capability; and reducing parasitic losses. Additional capabilities of VVA systems include controlling of valve lift within cam-profile limits to enable early valve closing (Miller cycle benefits) and secondary valve events (e.g., compression release braking, residual gas control also known as *internal EGR*). Additional benefits include valve-lift cancelation (cylinder cutout variable displacement) and lift variation for reduced valve train power consumption, fuel economy, and engine breathing [32].

Alternatives to the hydromechanical VVA system is FEV's electromechanical system [33]. This system employs four electromechanical actuators to control the cylinder's four-valve system. In theory, this arrangement gives total flexibility in controlling intake and exhaust valves. Of course, limitations in valve timing and valve lift are imposed to prevent catastrophic failures and obviously avoid contact with the piston. A schematic representation is given in Figure 14.26 of the valve control system, valve duration capability, and valve timing control range.

Figure 14.27 shows how variable valve actuation can improve cycle work. Figure 14.27a is a graphical representation of a conventional pressure–volume diagram, showing the negative work (work exerted by the piston on the gas). Naturally, net cycle work is the resultant work after subtracting the negative work from that imparted by the gas (the working medium) on the piston. Variable valve timing allows flexibility in valve timing, as shown in Figure 14.27b, where early closing of the intake valve leads to a major reduction in the negative work. This valve timing control is used to increase BMEP and specific power output and to reduce fuel consumption.

While early closing of the intake valve may be beneficial for increased specific power output and fuel economy, late opening of the intake valve can also be beneficial from an emission control point of view [33]. Figure 14.28 shows how late opening of the intake valve might cause increased residual gas in the cylinder, thus reducing NO_x emissions. Obviously, when combining late intake valve opening with early intake valve closing, one might derive improved engine output, lower fuel consumption, and reduced NO_x emission.

Of course, VVA is not limited to the systems described here and it seems that more designs are forthcoming. For instance, BMW has also adopted a VVA system referred to as *VALVETRONIC* in its new gasoline eight-cylinder engine [34]. A significant feature of the VALVETRONIC system is the additional variability of the inlet valve lift, which allows for load control without the need for a throttle [34].

14.11 Combined Charge Air and Variable Valve Actuation Technology Options

In the 1980s, the fuel injection system went through intense development and experienced unprecedented innovation. Advanced fuel injection systems were directly responsible for the tremendous success in performance improvements and emissions reductions. During this period, little attention was given to the air management technology other than designing swirl ports and refining them to achieve the desired mixing of fuel and air

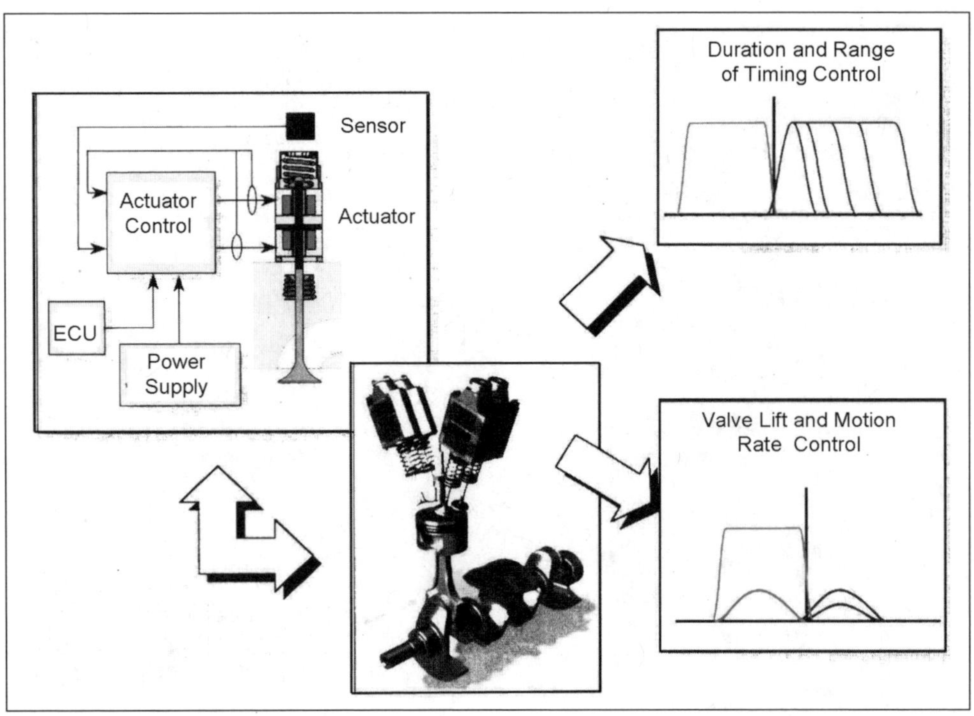

Figure 14.26 VVA as a tool for combustion optimization. (Reprinted with permission from Ref. 33.)

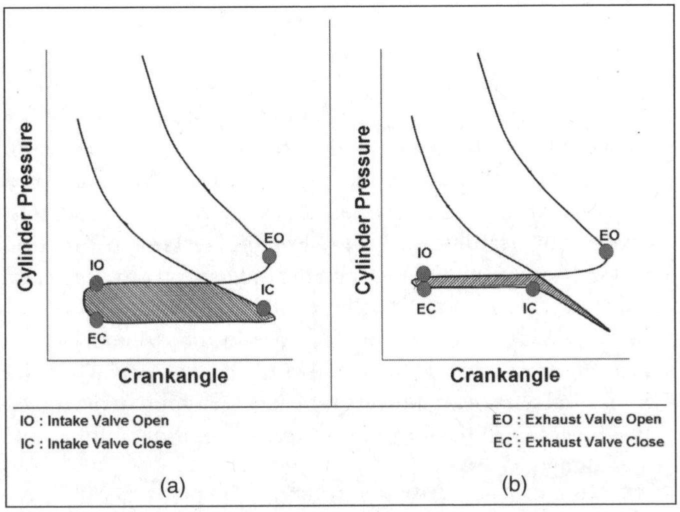

Figure 14.27 *VVA as a tool for reducing negative work. (Reprinted with permission from Ref. 33.)*

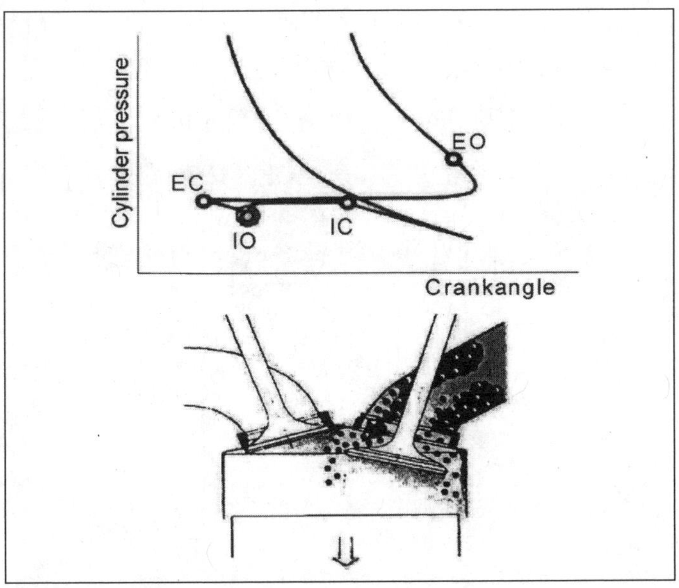

Figure 14.28 *Method for combined improved output, reduced fuel consumption, and reduced NO_x emission through late opening and early closing of the intake valve. (Reprinted with permission from Ref. 33.)*

in support of good combustion efficiency. Continued customer demand for better performance and lower fuel consumption combined with the legal requirement for low emissions challenged diesel engine design engineers to come up with more innovative solutions in air management to match that achieved in fuel management.

A new wave of supercharging machines was further developed to improve air charging of diesel engines. Fixed-geometry turbochargers equipped with waste-gates, variable-geometry (or nozzle) turbochargers, and hydraulically and electrically assisted turbochargers were the next crop of charging devices to replace the aging fixed-geometry turbochargers that had already been improved to better match combustion requirements. With advances in intake and exhaust valve movement control and charge air systems, a new series of ideas combining these two technologies began to emerge.

Pulse charging using an air pulse valve (APV) is an innovative concept for optimizing the gas exchange in diesel engines. By means of a very fast valve flap in the intake manifold, a substantial increase in engine torque can be realized across the speed range, especially at low engine speeds. The basic idea behind the APV process is to influence the charge cycle by means of a rapidly switching valve located upstream of the intake valve. With this additional valve, the air mass flow is controlled either in response to demand, or the air mass is increased compared to the naturally aspirated engine. The increase in air mass flow, so-called "dynamic supercharging," is controlled by the opening and closing event of the APV valve independent of the intake valve lift and the piston motion. The intake valve opens and the piston starts the intake stroke, while the APV stays closed. The air in the intervening space between the intake valve and APV flap is expanded into the combustion chamber. At a sufficient expansion level, the APV flap is opened and the air flows into the cylinder at a high velocity. Coinciding with this process, a suction wave moves from the flap to the manifold. It then reflects a positive pressure wave that moves toward the combustion chamber. At the piston head, the inflowing air is slowed down and again reflected. Dynamic supercharging uses the effects of the conversion of kinetic energy into potential energy in the reflection process, as well as wave oscillations in the intake due to the valve timing. With the air pulse valve, reverse flow can be prevented and the high pressure is kept in the combustion chamber. The process also allows for increases in the mass flow of the incoming air and for prevention of the emergence of returning airflows at the end of the intake process at other points in the engine speed range.

The APV process was patented in 1987, outlining two basic possibilities for its implementation [35]. The simpler method is the use of a controlled check valve that is opened by a defined pressure level between the intake system and the combustion chamber caused by the piston as it moves downward. As the valve opens, a spring is compressed, thus storing the energy for the subsequent closing event. The system can then be operated by the opening pressure difference and the closing spring force as the flow decreases or as reverse flow begins. Additional advantages are provided by electromechanical control of the flap movement. This allows the charging effects to be used in a more purposeful manner, for example by late APV opening, in order to obtain sufficient decompression and therefore higher inflow velocities. Results obtained using such a system have been presented by Schatz [36] and show the considerable potential for increasing the cylinder charge.

The performance of the APV can be extended by using a flap in combination with an external drive, as shown schematically in Figure 14.29. The flap actuation is independent of the pressure differential between the manifold and the combustion chamber as well as the air velocity and the flow direction at the air pulse valve. This makes it possible to use more efficient flap timing in order to increase dynamic supercharging, while unthrottled load control with the APV can also be carried out. As described in Ref. [36], the application of a flap with an external actuator offers a further process step toward obtaining a higher cylinder charge by double switching the APV flap in the intake stroke. By the double activation of the flow processes of the air in the intake system during the intake valve lift, a further increase in dynamic supercharging efficiency is possible compared to single activation of the APV. Moreover, such an active flap system makes numerous process engineering options, such as these, available:

- *Options for charge exchange control*: The flap in the intake manifold equipped with an independently controllable fast actuator offers a variety of opportunities for significantly influencing the charge exchange of the engine. The functions include the control of the cylinder charge as well as thermodynamic effects to produce an increase or decrease in the temperature of the air in the cylinder [36]. Figure 14.29 is a schematic representation of the system, explaining its operation and showing the action of both the APV and intake valve during the cycle.
- *Dynamic supercharging*: Switching of the APV during the intake stroke increases the quantity of air in the cylinder. Due to the processes outlined here, the air in the intake pipe is accelerated and decelerated at defined points in time, thus producing oscillating waves. For the system to function properly, it is crucial to contain the induced positive pressure waves within the combustion chamber. The increase in air density leads to the desired increase in the cylinder charge (see Figure 14.29). Double activation of the APV while the intake valve is open, however, makes very high demands on the dynamics of the flap and also on the reproducibility of its switching processes.
- The effect of the APV is felt without delay when the load requirement changes, for example during acceleration. Another advantage of the APV is that the dynamic charging process is available when starting the engine. Furthermore, charge preparation in the combustion chamber is supported

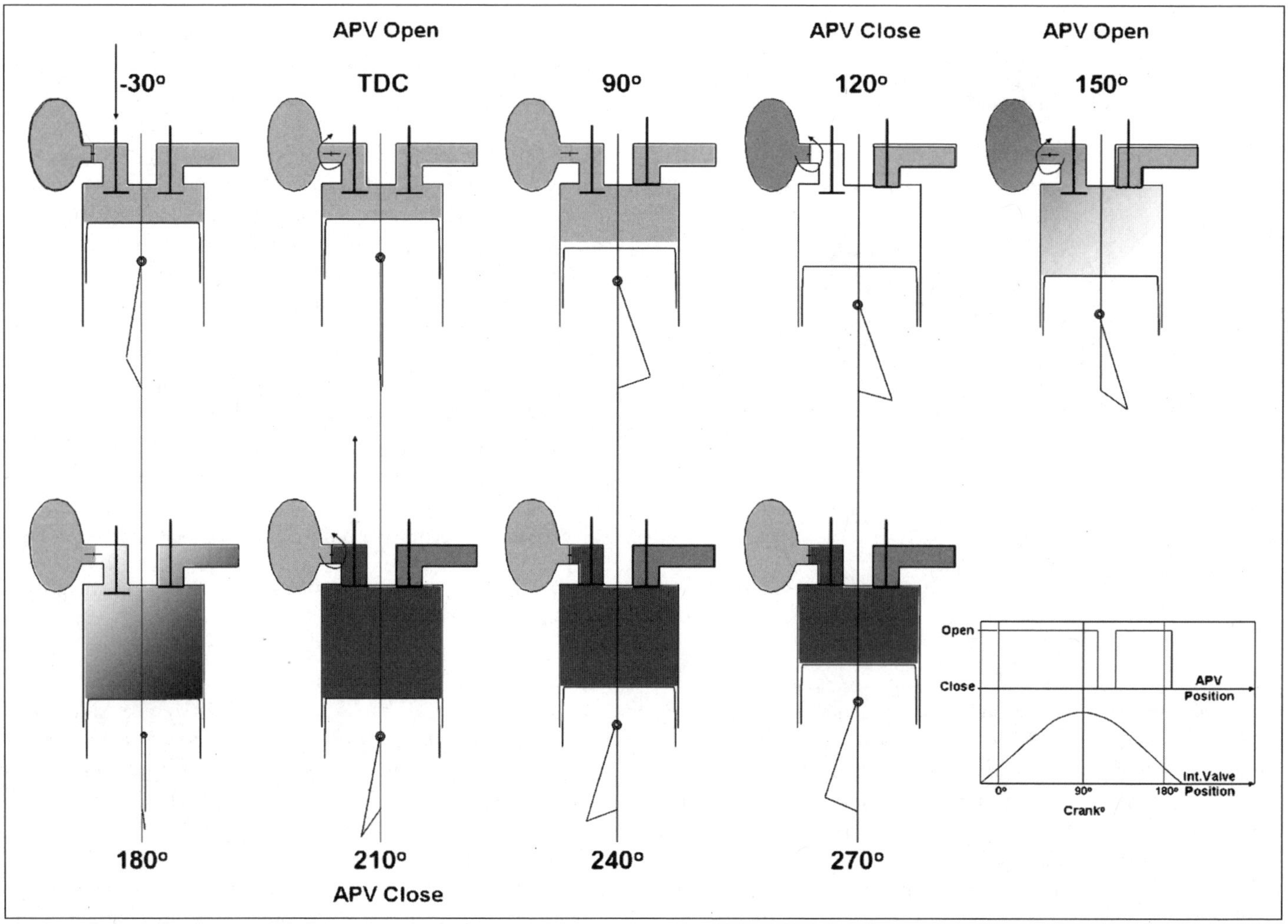

Figure 14.29 APV operation.

by the high air velocity in the inlet phases. The resulting homogenization of the fuel/air mixture and the acceleration of the combustion process offers a capacity for reducing the emission of unburned hydrocarbons during engine starting and warm-up as well as under regular operating conditions, an important emission control action, especially for gasoline engines.

- *Ram effect of supercharged engines*: The APV offers additional ways of improving the characteristics of supercharged engines equipped with superchargers or turbochargers [37]. Due to the high air mass flow at low engine speeds, low-end torque is increased, and turbocharged engines in particular have an additional advantage from the faster turbine acceleration caused by the intensified flow rate. In the engine starting process, the higher mass flow causes a higher compression end temperature and, therefore, improved cold start behavior. Due to the increase in airflow through the APV even at low engine speeds, the charge pressure can be reduced, leading to a more moderate load in the compressor. This allows the use of smaller compressors, and turbochargers with variable geometry may no longer be needed for handling the engine's part load requirements.
- *EGR control*: Analogous to the ram effect in supercharged engines, the EGR supply can also be controlled on a cylinder-by-cylinder basis using the APV. Exhaust gas is drawn into the combustion chamber while the APV is closed. In a second step, the APV opens and the fresh air required for combustion is delivered. This makes it possible to achieve a defined stratification in the combustion chamber, thus offering further flexibility with regard to inflow velocity and charge dilution.
- *Heat charging*: The aim of the so-called heat charging process is an increase in air temperature in the combustion chamber. This has a positive effect on the mixture preparation in diesel engines in cold start conditions. This process can be initiated from the moment the starter turns the engine. With sufficient rise in temperature, it would be possible, in principle, to omit the glowplug in a diesel engine. The increase in temperature is brought about by a change in the state of the aspirated air. First, the pressure in the cylinder decreases rapidly when the APV is closed during the intake stroke, while the mixture of air and exhaust gas in the cylinder is expanded. Because sufficient heat can be transferred via the cylinder walls, this process may be regarded as being isothermal. The cylinder is filled quickly when the APV is opened. Subsequently, the air is accelerated to high velocity due to the high negative pressure in the cylinder, and results in an increase in air temperature. The level of the generated temperature difference depends on the initial timing and on the duration of the opening of the APV, as well as on the leakage at the closed flap. The resulting increases in mass flow, temperature, and flow velocity of the air improve the combustion quality and allow higher fuel injection rates. This yields higher exhaust temperatures at cold start, reduced combustion noise, and faster warm-up of the engine. To achieve this, a fast-acting air pulse valve is required to allow high flow velocity and to prevent backflow of the air.
- *Cold charging of supercharged engines*: The APV can also be used to lower the air temperature in the combustion chamber of supercharged engines. The compressed air is precooled by the heat exchanger, locked in the cylinder by early closing of the APV, and then expanded by the piston movement. This action causes a further reduction of the air temperature in the combustion chamber and in turn decreases the temperature and pressure at the end of the compression stroke. This reduces NO_X emission levels as well as auto-ignition temperatures. Consequently, the charge air pressure may be increased for an additional rise in engine torque and power.
- *Cylinder deactivation*: Alternating cylinder cutoff during partial-load operation is another control method offered by the APV. It is achieved simply by closing the APV flap during the intake stroke. Deactivated cylinders can be periodically reactivated to prevent the combustion chamber from cooling down, as is possible with electromagnetic valve systems [33].

14.12 New Combustion Technology Option

Research efforts have been under way in Japan, Europe, and the United States to investigate a new compression ignition combustion concept that drastically reduces NO_X and smoke emission. One example of such research is *premixed lean diesel combustion* (PREDIC) pursued by the New ACE Institute in Japan. Another term for this technology is *homogeneous charge compression ignition* (HCCI). This concept uses diesel fuel and a high compression ratio like that of a diesel engine [38]. As shown in Figure 14.30, a conventional diesel engine emits 400 ppm of NO_X when the fuel is injected immediately after the top-dead-center (TDC) of the compression stroke. If the fuel is injected much earlier than it is in a conventional engine, it forms a homogeneous-like air/fuel mixture in the cylinder and leads to a drastic drop in NO_X emissions. As shown in the figure, when the fuel is injected at 64° BTDC, NO_X emissions are reduced to 250 ppm, and when 78° BTDC is selected, NO_X emissions dropped to 20 ppm. Observation of the combustion process showed no luminous flame, which indicated that lean combustion was taking place. This combustion, however, can be realized in only very limited operating conditions, mostly part-load operation and within a limited range of intake temperatures. Therefore, more research is required to determine its future practical application.

Even so, this is quite an interesting approach that may achieve very low NO_X combustion and low smoke emissions without sacrificing fuel economy. In addition, in the case of 64° BTDC injection, a very high rate of heat release right after TDC is observed without knocking and this should provide significant improvements in thermal efficiency [38].

14.13 Engine Accessories and Subsystem Component Technology Options

So far we have dealt with the major systems that directly affect engine performance and emissions. Yet as one might think, it is difficult to ignore the potential influence any of the subsystems might have on engine performance. When properly designed

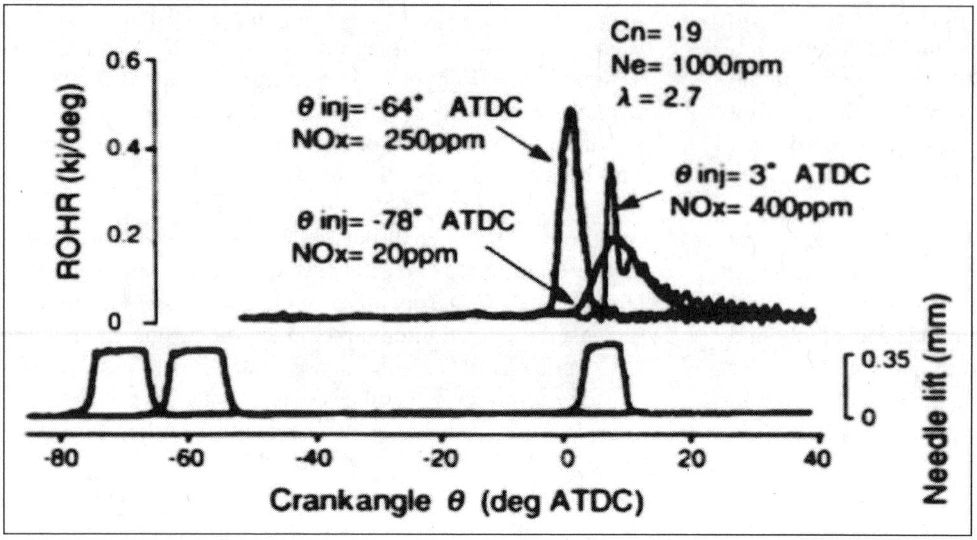

Figure 14.30 *Rate of heat release of PREDIC system. (Reprinted from Ref. 38.)*

all of the engine systems should work together to accomplish the ultimate in performance within the limits imposed by emission regulations. To this end, engine designers, manufacturers, and component suppliers have attempted to streamline their efforts and contribute to the ultimate goal of superior engine performance coupled with low emissions. In this section a review of some technological advances made or being considered is presented. Some of these ideas may not necessarily be novel, but they warrant a second look and additional consideration in designing future diesel engines.

It is generally recognized that lower fuel consumption and emissions coupled with good performance and minimum change to the current vehicle structure are desirable objectives. In keeping with these objectives, especially in high-speed, light-duty engine applications, a lightweight, downsized engine is preferable. Such a selection may lack adequate torque at low engine speeds and that could lead to bad drivability and responsiveness. To overcome this potential difficulty, *electric air boost assist* can be used to allow additional fuel to be injected and to produce more power at the operating regime where the deficiency is anticipated.

Another innovation of the late 1990s is a *flywheel-mounted electrical device* often called the *electric flywheel*, which can perform several functions. The electrical flywheel is generally used as a generator while the engine is producing excess power at high-speed/high-load conditions. Electricity generated at these operating conditions can be used in the reverse direction as torque-assist at low engine speeds. This function is referred to as *mild hybridization* [39] and is considered a much less expensive alternative to total hybridization.

The electric flywheel also doubles as a starter, thus reducing component clutter and eliminating items such as starter mounting brackets and flywheel gear, and it is placed directly on the crankshaft between the engine and the transmission [40]. Other functions that can be performed by this electrical flywheel include stop/start operation, regenerative braking, engine balancing, multivoltage electrical system coordination, power provision for diesel particulate filter regeneration, and exhaust temperature increases during idle and partial-load engine operating conditions to facilitate efficient exhaust emission device functioning. Figure 14.31 summarizes the example given and shows the logic of justifying the use of such device in graphical form [39]. Engineers at Ricardo have claimed that by downsizing the engine and operating it at high load they have realized a 20% fuel economy improvement in a 2.0-L diesel engine and about a 30% weight savings. Another 3% improvement in fuel economy was experienced through the traffic light stop/start function in addition to 5% to 7% attributed to regenerative braking [39].

Although the advantages of a flywheel-mounted electrical devices are attractive, onboard non-engine-mounted generators are also gaining in popularity because of their benefits. A belt-driven induction generator is another system that can be mounted in the engine compartment. Although this type of generator can produce up to 8.5 kW of continuous, highly regulated clean 120/240-V ac power, an optional battery charger can provide 200 amp of 14-V dc power at the same time [41]. One future use for this device would be assisting diesel exhaust emission control equipment. The obvious application would be powering an electric heater for regenerating a particulate filter or a NO_x adsorber catalyst. One of the problems with some regenerative catalysts is that there is not enough power (or heat) to effectively "light off" the catalyst. Having an additional source of power that does not significantly affect vehicle performance might provide part of the answer.

Efforts to improve diesel combustion continue to intensify in the quest for cleaner energy conversion through compression ignition. As discussed earlier with regard to emerging induction technology options, inducting more air into the engine is regarded as beneficial. Novel ideas in air management systems were described earlier in this chapter. When more air is provided in the inlet ducts, extra care needs to be exercised to avoid pressure losses in the system until the charge air is delivered to the combustion chamber. *Indirect charge air cooling* is a process in

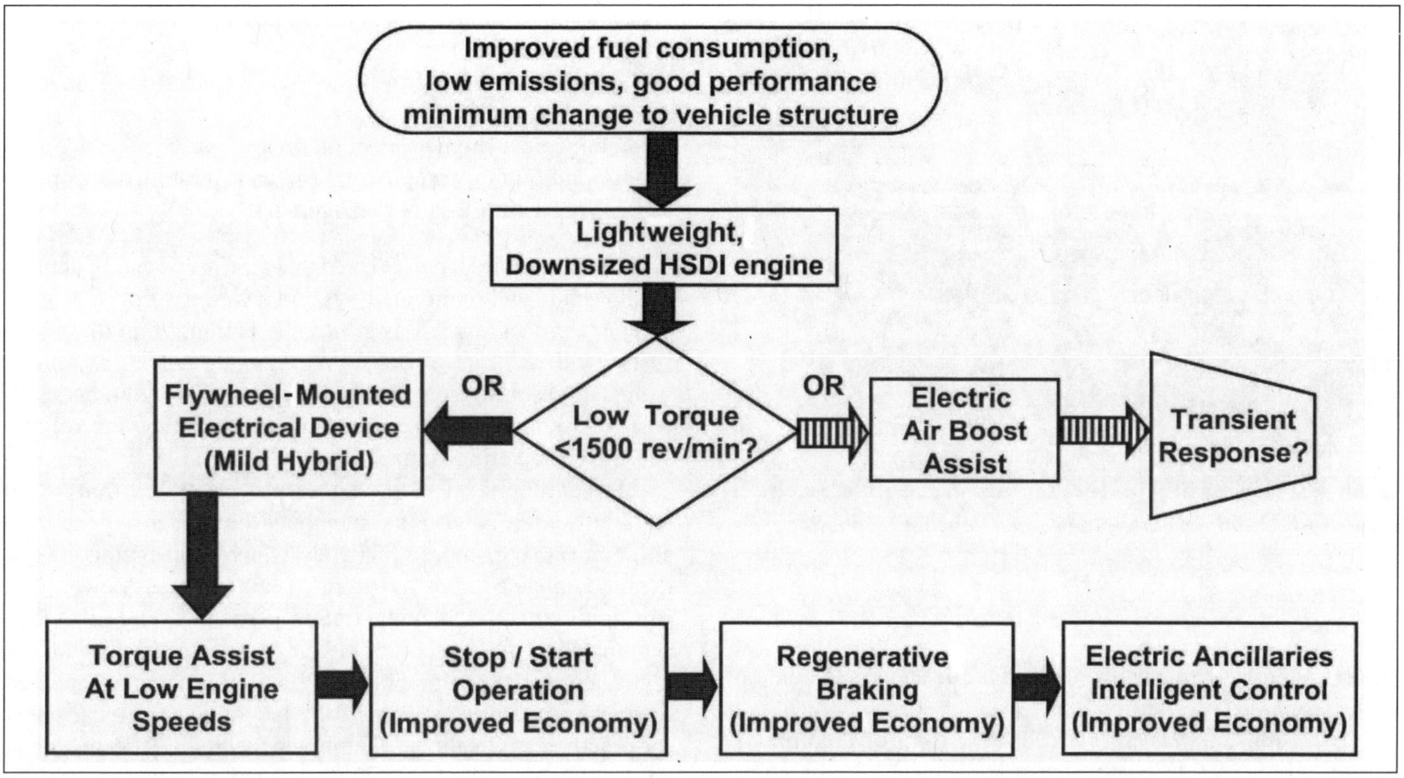

Figure 14.31 *Flow diagram and graphical representation for future powertrain design. (Reprinted with permission from Ref. 39.)*

which charge air is not cooled with air, as is usually the case, but by using normal engine coolant in a low-temperature circuit. The charge air can thus be cooled to the same or lower temperature than with current air-cooled systems, but with a significantly smaller drop in pressure of the charge air [42]. This means that a larger mass of air is available for combustion in the engine. This primarily performance-enhancing effect can then be used to reduce fuel consumption and emissions, because standard performance can be obtained using a smaller, more economical engine operated more efficiently at high BMEP. The benefit of indirect charge air cooling is 40% to 65% less pressure loss [42]. The lower pressure loss encountered with indirect charge-air cooling results from:

- A charge-air cooler with better flow characteristics, which is possible because of more efficient heat transfer by the coolant
- Significantly shorter charge air ducts, because the charge air cooler can be located near the engine.

The charge air cooler operates either with a separate cooling circuit or with one that is part of the main cooling circuit. With an integrated circuit, part of the coolant is diverted from the engine cooling circuit, and, after cooling the charge air, returns to it. In both cases, a small amount of coolant (about 5% to 10% of the throughput of the main circuit) is cooled to a temperature close to the ambient air temperature, in order to cool the charge air down to its low target temperature. If electrically driven coolant pumps are used in the system, it is even possible to warm up the charge air in cold engines. This measure can be used to reduce the characteristic white smoke that is emitted from diesel engines during cold starts.

Although this system may not be new to the field of charge air cooling, it provides an additional degree of freedom in combustion control with the ability to use an electrically driven, electronically controlled coolant pump. The water-to-water cooler is designed with generous dimensions to avoid excessive pressure drop and can be located in proximity to the intake manifold, thus saving additional pressure drop losses in the intake system. Using the charge air cooler as a heater for cold start and warm-up conditions may have limited success, especially in cold ambient conditions. *Electric intake manifold heaters* [43] that combine short warm-up times and high operating temperatures are finding more acceptance in modern designs (see Figure 14.32).

Dissipation of energy in the form of heat through the engine radiator represents a major loss in internal combustion engines. Many sources contribute to the radiator cooling load, with friction being one of the major contributors. *Friction-reducing coatings* are often applied to moving parts that slide, roll, or rotate, such as engine components. However, these coatings may be difficult, time consuming, or even impossible to deposit on materials such as plastics, ceramics, and certain metals. In

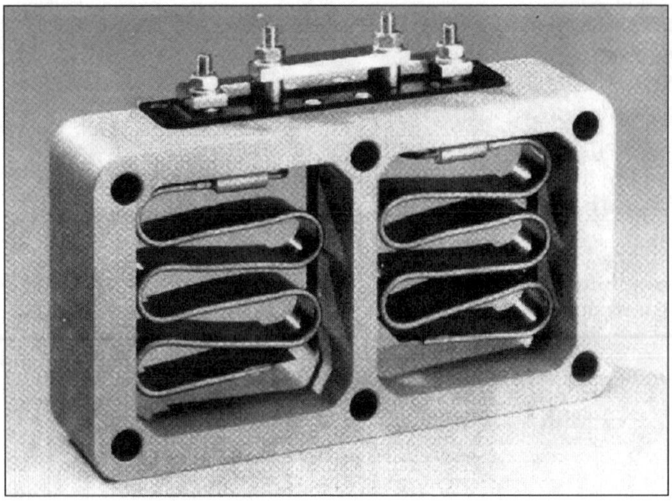

Figure 14.32 Intake manifold heater. (Reprinted from Ref. 43.)

addition, these coatings may not be durable enough to hold up under strenuous use.

Ultra-low-friction carbon (UFC), a form of amorphous carbon, has a friction coefficient of less than 0.001; by comparison, Teflon's friction coefficient is about 0.04. Thin (1-μm) UFC films can be deposited on virtually any substrate by ion beam deposition, sputtering, or chemical vapor deposition at low temperatures (room temperature to 200°C), without the risk of damaging heat-sensitive materials [44].

UFC coatings combine unusual mechanical and chemical properties that make them ideal for a wide variety of engineering applications:

- Very low friction
- Extreme hardness
- High wear resistance
- Corrosion resistance
- Virtual transparency
- Good electrical conductor or insulator.

The combination of extreme hardness with ultra-low friction in UFC films provides long-term wear resistance and lubrication to reduce material and energy losses. Specific benefits include the following:

- Near-perfect finished products that do not require secondary machining or grinding
- Increased lifetime and improved performance in rolling, sliding, and rotating components
- Ultra-hard and extremely slick coatings for sliding parts and machining of high-precision surfaces
- Higher productivity, lower costs, and improved environmental compliance (reduction or elimination of flammable and hazardous metalworking fluids).

UFC coatings could be used in many automotive applications:

- Piston rings (reduced wear at higher cylinder pressures)
- Gears and bearing components (improved durability; smaller, light components)
- Cams and cam-roller followers (increased resistance to rolling-contact fatigue)
- Fuel injectors (higher injection pressures)
- Air-conditioning compressors (greater durability of components used with R- 134a refrigerant).

Recent studies have revealed that UFC films can be applied to aluminum and titanium alloys and a variety of plastics [44]. One new process uses ions to literally hammer thin films onto surfaces. This process yields coatings that adhere extremely well. The resulting coated parts provide low friction and high wear resistance in mechanical applications involving rolling, sliding, and rotating contacts.

Researchers have known since the 1970s that when oxygen-rich air is fed into an engine, the higher oxygen ratio allows fuel to burn more completely, which reduces emissions of the incomplete combustion products and increases power [45]. But, until recently, the reduction in particulate emissions was accompanied by an increase in NO_x emissions because of the higher combustion temperatures. Lack of a compact onboard oxygen source also prevented the use of *oxygen enrichment* for practical applications. Membrane technology developed at Argonne National Laboratory (ANL) seems to have overcome both challenges. The breakthrough came when ANL researchers modestly increased the oxygen in the engine air supply (23% to 25% by volume), retarded the fuel injection timing, and increased the rate of fuel flow to the engine. Using this combination of conditions, researchers were able to reduce particulate emissions by 60%, reduce NO_x emissions by 15%, and increase gross engine power by 18% [45].

Once the concept was proven, ANL and Compact Membrane Systems (CMS) developed a low-cost permeable *membrane* to separate ambient air into oxygen-rich and nitrogen-rich air before entering the engine. The membrane works by a process called *solution diffusion,* in which pressurized ambient air is passed across tubes coated with a special polymer that separates the air into two streams: one rich in oxygen and the other in nitrogen. The oxygen-rich air is directed to the engine to optimize combustion. It is necessary, however, to minimize the size and power requirements for the membrane. The current version of the membrane is 10 times shorter and uses 60% less power than its predecessors. The goal is to develop a membrane about the size of a car's air filter [45].

14.14 Concluding Remarks

As long as research continues, the list of emerging technologies cannot be concluded and must be left open ended. Although the foregoing review of emerging technologies may be substantial, it falls short of doing justice to the body of knowledge and experience that has been accumulated over the years, especially in the last three decades. As mentioned early in this chapter, experience with exhaust gas recirculation and the body of work to be covered is too large to contain within this chapter. Therefore, Chapter 16 is devoted solely to this subject and, although EGR itself is not novel, many novel ideas are involved with its implementation. Interest in EGR has certainly taken center stage in the last few years since the 1998 consent decrees between the U.S. Department of Justice and several diesel engine manufacturers.

Another large body of work involves the catalyst industry and a long list of products that will be integrated into the overall control of the engine. In fact, a parallel can be drawn between the three-way catalyst in the control of emissions from the gasoline engine and what is eventually anticipated to be accomplished in the diesel engine. This area is certainly an emerging technology that is currently being developed and is expected to play a crucial role in the future of the diesel engine. This body of science and technology is certainly worth devoting several chapters to their review and discussion. The second part of this book is set aside for the *postcombustion emission control devices*, their principles of operation, their advantages and disadvantages, and their performance especially when combined in integrated systems.

With the advent of electronics and their massive use in automotive technology, more of the emerging technologies are based on the capabilities afforded by electronic controls. From *intelligent sensors* to sophisticated control algorithms, *controls technologies* are experiencing a revolutionary change and opening new opportunities for automotive systems. Chapter 17 addresses this specialty and gives a broad overview of state-of-the-art controls technology.

References

1. Jost, K., "A Diesel in Your Future?," *Automotive Engineering International,* Volume 109, No. 2, February 2001, pg. 6.
2. "Heavy-Duty Engine Emission Standards for Highway Trucks and Buses," EPA420-F-96-006, U.S. Environmental Protection Agency, Washington, DC, June 1996.
3. Walsh, M.P., "International Regulatory Developments," Walsh Report, November 2001.
4. Gore, A., "VP's Announcement of Joint Industry–Government Research Plan to Produce the World's Cleanest Diesels," Remarks by Vice President Al Gore at the 6th Vice Presidential Symposium—The Partnership for a New Generation of Vehicles, Washington, DC, July 22–23, 1997.
5. Wall, J.C., "Heavy-Duty Diesel Manufacturer's Perspective on the Viability of Diesel as the Powerplant for PNGV," Paper presented at the 6th Vice Presidential Symposium—The Partnership for a New Generation of Vehicles, Washington, DC, July 22–23, 1997.
6. Harvey, M., "Closing in on Diesel's Last Taboo," *Financial Times,* August 12, 2000, pg. xiv.
7. Schindler, K.P., "The Future of Diesel," *Auto Emissions,* No. 14, Corning Incorporated, Fall 1999.
8. Carnes, K., "The Diesels Are Coming," *Lubricants World,* July 2000, pg. 28.
9. Flynn, P., "Diesels—Promises and Issues," Paper presented at the Diesel Engine Emissions Reduction (DEER) Workshop, 8th DEER Conference, San Diego, CA, August 2000.
10. Erlach, H., F. Chmela, W. Cartellieri, and P. Herzog, "Pressure-Modulated Injection of an HD Diesel Engine," SAE Paper No. 952059, April 1996.
11. "Fuel Filtration for High-Pressure Injection Systems," *Diesel Progress,* North American Edition, March 2001, pgs. 32–36.
12. Tschoeke, H., "The Fuel Injection System," Diesel Engine Design Academy Lecture Notes, September 2001.
13. Lauvin, P., A. Löffler, A. Schmitt, W. Zimmermann, and W. Fuchs, "Electronically-Controlled High Pressure Injector System for Diesel Engines," SAE Paper No. 911819, SAE International, Warrendale, PA, 1991.
14. Buchholz, K., "Why Diesels, Why Now?", *Automotive Engineering International*, August 2003, pgs. 65–66.
15. Herzog, P., "Technologies for Euro Standards," Paper presented at ISATA Conference, Firenze, 1997.
16. Gaillard, P., and M. Hedna, "Renault's Euro 3 Engine," *Automotive Engineering International,* Volume 109, No. 2, February 2001, pg. 140.
17. Mather, D., "Late-Cycle Injection of Air/Oxygen-Enriched Air to Reduce Diesel Exhaust Emissions," Paper presented at the Diesel Emissions Reduction (DEER) Conference, 8th DEER Conference, San Diego, CA, August 2000.
18. Nefischer, P., F. Steinparzer, and W. Stütz, "Gas Exchange System of the New BMW V8 Diesel Engine," Lecture Notes, BMW Motoren GmbH, Austria, 2000.
19. Herrmann, H.O., M. Dümholz, and T. Lichtenberg, "The Influence of Stroke-to-Bore Ratio on the Combustion Process of a Passenger Car 4-Valve Dl Diesel Engine," Paper presented at the 17th International Engine Symposium, Vienna, Austria, 1996.
20. Boggs, D.L., R.C. Belaire, B. Bartunek, M. Dürnholz, and H.-J. Ecker, "A Small Displacement Dl Diesel Engine Concept for High Fuel Economy Vehicles," SAE Paper No. 972680, SAE International, Warrendale, PA, 1997.
21. Scheid, E., and M. Dümholz, "High Speed DI Diesel Engines for Commercial Vehicle and Passenger Car Applications," Paper presented at SAE Congress, Beijing, China, August 25–28, 1996.
22. Dümholz, M., H.O. Hermann, and H. Endres, " Development Potential of 4-Valve High Speed DI Diesel Engines," Paper presented at 4th Aachen Colloquium for Vehicle and Engine Technology, Aachen, Germany, 1993.
23. Georgi, B., S. Hunkert, J. Liang, and M. Willmann, "Realizing Future Trends in Diesel Engine Development," SAE Paper No. 972686, SAE International, Warrendale, PA, 1997.
24. Christ, A., and D. Page, "Advances in Air Charging Technologies," SAE Paper No. 972675, SAE International, Warrendale, PA, 1997.
25. Kapich, D.D., "Sequential Hydro-Supercharging Systems for Turbodiesels," SAE Paper No. 961744, SAE International, Warrendale, PA, 1996.

26. Grosse, J., "New Engines for Renault in Quest for Green Machine," *FT Automotive World,* January 2000, pg 23.
27. Khair, M.K., "Engine Technology Projections for 1991," Technical Planning Document, Report No. B86104, Ford Tractor Operations, March 1986.
28. Heisler, H., *Advanced Engine Technology,* Butterworth-Heinemann, Boston, 1995.
29. Pischinger, F., "European Perspective on IC Engines for Passenger Cars," Paper presented at the 6th Vice Presidential Symposium—The Partnership for a New Generation of Vehicles, Washington, DC, July 22–23, 1997.
30. Hadler, J., Westphal, C., Schmidt-Loose, k., Köhn, S., Scher, U., "Der V-10 TDI Motor," *Motortechnische Zeitschrift*, EXTRA – 25 Jahre Dieselmotoren von Volkswagen, May 2001, pgs. 38–39.
31. Steinparzer, F., Mattes, W., Hall, W., Bock, C., "Die Dieselantriebe der neuer BMW 7er-Reihe," *Motortechnische Zeitschrift*, October 2002, pgs. 790–803.
32. Mauriello, B., "Daydream," *Engine Technology International,* January 2002, pgs. 32–34.
33. Salber, W., H. Kemper, F. Van Der Staay, and T. Esch, "Der Elektromrchanische Ventiltrieb-Systembaustein für zukünftige Antriebskonzepte-Teil II," *Motortechnische Zeitschrift,* January 2001, pgs. 44–55.
34. Hirschfelder, K., Hofmann, R., Jägerbauer, E., Schausberger, C., and Schopp, J., "The New BMW Eight-Cylinder Petrol Engine-Part 1," Article in Motortechnische Zeitschrift, September 2001.
35. Elsäßer, A., W. Schilling, J. Schmidt, K. Brodesser, and O. Schatz, "Pulse Charging and Load Control," *Motortechnische Zeitschrift,* December 2001.
36. Schatz, O., Patent DE 43 08 931 C2, Germany, 1993.
37. Schatz, O., and T. Steidele, "Pulse Charging—A New Approach for Dynamic Charging," Paper C382/116, 2nd International Conference on New Developments in Powertrain and Chassis Engineering, Strasbourg, June 14–16,1989.
38. Hikosaka, N., "A View of the Future of Automotive Diesel Engines," SAE Paper No. 972682, SAE International, Warrendale, PA, 1997.
39. Gordon, R., "Demonstrating Future C-Class Vehicle Performance and Emissions through Advanced Powertrain Technologies," Paper presented at Engine Expo 2001, Stuttgart, Germany, June 2001.
40. "Siemens Starter/Generator for 2002," *Automotive Engineering International,* January 2001, p. 56.
41. Osenga, M., "Turning Vehicles into Gen-Sets," *Diesel Progress,* North American Edition, August 2001, pgs. 15–17.
42. Banzhaf, M., D. Hendrix, M. Schmidt, and E. Pantow, "Ecological Advances in Engine Cooling," *Automobil Technische Zeitschrift,* September 2001, pgs. 28–29.
43. "What's New at SAE '98," *Automotive Engineering International,* January 1998, pgs. 37–38.
44. Fenske, G., "Ultra-Low-Friction Coatings Could Be an Engine's Best Friend," *Trans Forum* (Research Review), Argonne National Laboratory, Argonne, IL, Winter 1998.
45. Poola, R., "New Technology Promises Cleaner Diesels," *Trans Forum* (Research Review), Argonne National Laboratory, Argonne, IL, Winter 1999.

Chapter 15

Water Addition to Diesel Combustion

15.1 Addition of Water to Diesel Process

15.1.1 Methods of Water Addition

Addition of water into the diesel combustion process is a known method for reducing NO_x and, in some implementations, simultaneously reducing NO_x and particulate matter (PM) emissions. The very notion of introducing water into the cylinder of a diesel engine, however, may sound controversial. After all, engineers have always taken great care to accomplish the exact opposite and protect the combustion chamber from water contamination, be it from the fuel or from water condensation in intake air coolers. The controversy around water addition is founded on the observation that water droplets impinging on the cylinder walls can immediately destroy the film created by the lubrication oil. This danger, however, although very real, is posed exclusively by liquid water. Once water is evaporated, it can no longer affect the lube oil film [1]. Thus, water addition methods which ensure that water droplets cannot contact the cylinder liner surface may be considered harmless. Further concerns have been raised that increased concentrations of water vapor in the engine cylinder may result in condensation of water and/or sulfuric acid, leading to corrosion problems. Apparently, these suspicions are not justified either, because the dew point of sulfuric acid at a very high water-to-fuel ratio of 1:1 is increased by only up to 15°C [2]. Considering the temperatures in diesel combustion, condensation in the combustion chamber is not possible at any time.

In general, water can be introduced into the diesel combustion process using one of the following methods:

- Emulsified fuel
- In-cylinder water injection
- Water injection into the intake air.

These methods are shown schematically in Figure 15.1.

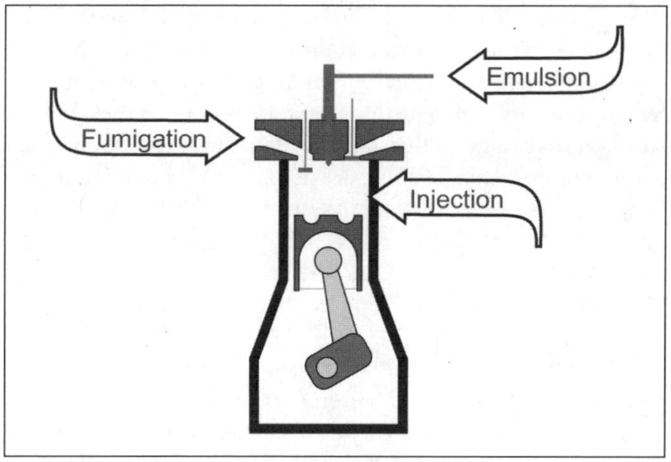

Figure 15.1 *Water addition methods.*

Emulsion is a system consisting of two immiscible liquids, one of which is finely dispersed in the other. In all water/diesel fuel emulsions of practical importance, water is dispersed in the form of fine droplets in the continuous diesel fuel phase. This type of emulsion, schematically shown in Figure 15.2, is often referred to as *water-in-fuel* emulsion. In the opposite configuration, with fuel dispersed in the continuous water phase, water would be much more likely to contact the cylinder liner surface and other metal parts, leading to corrosion and engine problems.

In practice, running an engine on a water-in-fuel emulsion makes it possible to reduce NO_x by up to about 50% with the required water quantity being about 1% for each percentage point of NO_x reduction [1]. The limiting factor for water emulsions is the delivery capacity of the injection system. If emulsions are to be used without engine modifications (e.g.,

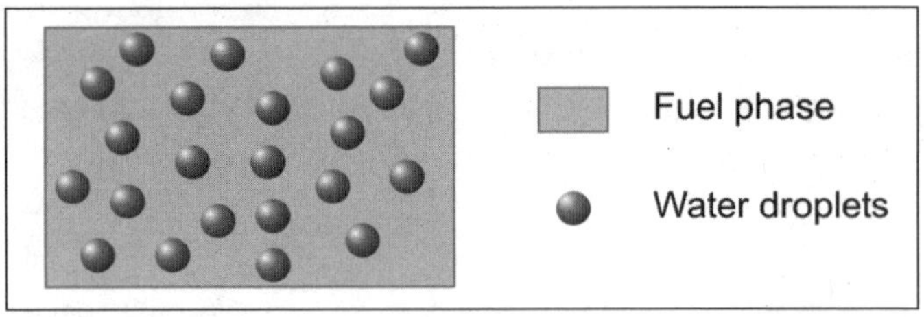

Figure 15.2 *Water-in-fuel emulsion.*

to substitute regular fuel in existing engines), the maximum quantity of water and the degree of NO_x reduction are both limited to some 10% to 20%. Even then, the engine may not be able to reach its rated power, running in effect at a slightly derated condition.

Emulsions are distinguished among other methods of water addition by the fact that water, being incorporated into the fuel spray droplets themselves, is introduced directly into the combustion flame area where emissions are formed. In addition to the NO_x benefit, which in all methods is attributed primarily to the lowering of combustion temperature by water, emulsions result in enhanced fuel spray atomization and mixing. Enhanced mixing that extends throughout the diffusion flame can result in quite impressive reductions in PM emissions. As a result, water-in-fuel emulsions are one of the rare diesel emission control strategies that can simultaneously reduce NO_x and particulate emissions without or with only a small fuel economy penalty. Reduction of PM emissions by emulsions has not yet been as thoroughly researched as NO_x reduction. Nevertheless, as will be discussed later, the achievable effectiveness of PM reduction appears to be more than twice the level of NO_x reduction.

In-cylinder injection of water requires a separate, fully independent injection system, preferably under electronic control. This method offers the ability to inject very large quantities of water without the need to derate the engine. This system also allows water injection to be switched on and off, as may be needed, without affecting engine reliability. Direct water injection needs to be carefully optimized with respect to injection timing, water consumption, emissions, and other parameters. This flexibility in optimizing parameters allows NO_x reductions to be achieved that are similar to those seen in emulsion systems, despite the fact that water is not introduced directly into the diesel flame area as an integral part of the spray. However, PM emission reductions, if any, do not match those with emulsified fuels. The complex development work required for water injection systems in different engine types makes this approach suited for OEM rather than for retrofit applications.

Fumigation, meaning the introduction of water into the intake air, is the simplest method of water addition. This method offers very little control over the injection parameters such as timing or spatial coordinates. For this reason, observed NO_x reductions tend to be lower than those with emulsions or direct injection. Fumigation typically reduces NO_x emissions by 10% for each 20% water addition to the fuel [1].

If the fumigated water does not completely evaporate in the intake air, it will impinge on the cylinder walls, causing disintegration of the lube oil film and engine damage. A safer approach is to fumigate water vapor rather than liquid. Water vapor may be generated using waste heat from the engine, such as from the exhaust gas and/or from the compressed charge air. Another possibility is to use steam, which may be available in certain stationary engine applications.

Regardless of the method of water addition, consideration must be given to the logistics of providing the water supply. The use of emulsifying agents allows emulsions to be prepared that can remain stable for a number of days or even weeks. In this case, vehicles may be simply fueled with emulsions in place of regular fuel. Such application of emulsions is obviously limited to vehicle fleets that are centrally fueled from one facility where the emulsion would be prepared. Other water addition methods would require that water tanks and handling systems be installed on the vehicle. The obvious drawback of such a system is the large quantity of water that is needed for NO_x reduction, which would require large tanks and frequent replenishment. This is likely the main reason why water addition technologies attract more attention in stationary and marine applications, where supplying large quantities of water is less problematic. However, most systems for oceangoing ships would work on freshwater only, thus requiring additional freshwater generation equipment.

15.1.2 Impact on Combustion Process

Addition of water, often in large quantities, has a profound effect on the diesel combustion process. The phenomena that occur are not always clearly understandable, and contradicting reports are easily found in the literature. Most likely, many of the water effects are engine specific, that is, different magnitude or even direction in the change of a given parameter is observed in different engines. In general, the effects of water on combustion can be summarized as follows:

- Addition of water increases the mass of air/fuel mixture and the associated air quantity per same supplied energy and per unit time, resulting in improved atomization and mixing. The improved atomization effect is frequently attributed to droplet microexplosions, especially in the case of fuel emulsion sprays [3], but there is no universal agreement as to the effectiveness of this mechanism in the diesel engine [4]. The improved mixing is due to the increased momentum of the vaporized fuel jet, which can entrain more air. Accelerated mixture formation and increased ignition delay are observed, leading to increased premixed combustion. This effect is consistently seen in all types of diesel engines [5, 6].
- The increased premixed flame causes a higher maximum rate of heat release, higher rate of pressure rise, and higher maximum combustion pressure. Typical magnitudes for these phenomena are illustrated in Figure 15.3 for regular fuel and for a water-in-fuel emulsion.

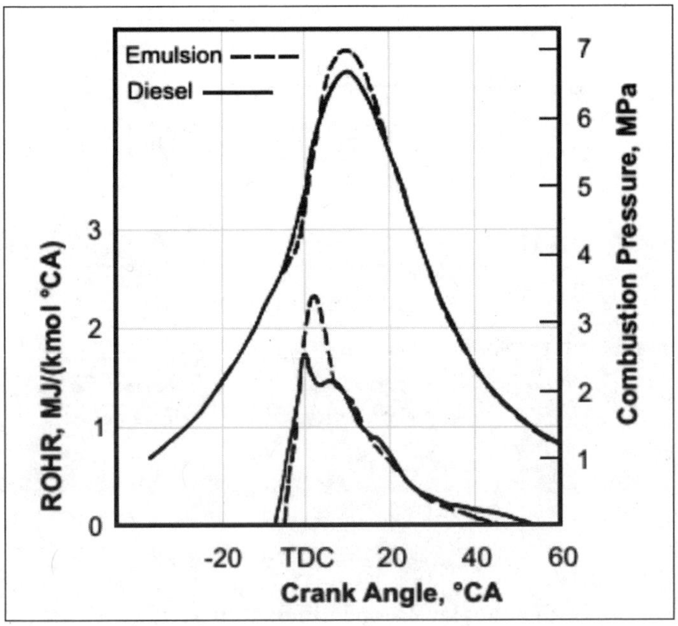

Figure 15.3 *Influence of water on rate of heat release (ROHR) and combustion pressure in a single-cylinder DI engine; injection timing, 18.5° BTDC; BMEP, 0.52 MPa; nozzle opening pressure, 19.6 MPa; water content, 33.8%. (Reprinted from Ref. 7.)*

- From the NO_x control perspective, the increased heat release rate and pressure are counterproductive effects. Water addition, however, also produces reduced peak combustion temperatures, the prime effect responsible for NO_x reduction in all methods of water addition [4]. Combustion temperatures are lowered thanks to the heat absorption effect due to an increase in both the gas mass and its specific heat [8]. (Remarkably, this effect bears a close similarity to the NO_x reduction mechanisms through exhaust gas recirculation.)
- Improved mixing and microexplosion phenomena are responsible for accelerated rate-controlled (diffusion) combustion, resulting in a decrease in the total combustion duration. This assists in reducing NO_x formation from the diffusive flame.
- Chemical effects are also suspected, such as an accelerated water gas reaction due to the increased partial pressure of the water [8].

Some water addition systems, in particular those based on emulsions, as opposed to direct water injection or fumigation, have shown high reductions in PM emissions and smoke. This is usually explained by the increased mixing and increased local air excess ratio in the rate-controlled flame, leading to better oxidation of carbon, and by chemical effects that reduce formation of soot precursors in the premixed flame [9]. However, PM reduction mechanisms by water are generally less well understood than those in NO_x reduction. A number of studies on water-in-fuel emulsion systems have measured increased PM emissions with increased water content [10].

Not many studies are available that focus on hydrocarbon (HC) and carbon monoxide (CO) emissions in water addition systems. The existing evidence indicates that HC emissions, both in the gas phase and as particulate soluble organic fraction (SOF), are increased by water addition, sometimes by as much as a factor of 2 or even 3 [4]. Less consistency is seen with CO emissions, with some studies measuring an increase, others a decrease in carbon monoxide. Evidently, many emissions depend on the particular engine and engine parameters. For example, a study of a DI engine showed that CO was reduced by emulsions at retarded injection timing, but unchanged or slightly increased at advanced injection [7]. Interestingly, the same tendency was found for smoke, which was most effectively reduced by emulsions at retarded injection timing. Hydrocarbons, on the other hand, were increased by emulsions, especially at advanced injection timing.

Water addition is one of the few engine emission control strategies that allows for reducing NO_x (and, in some cases, PM) with little or no fuel economy penalty. Existing studies seem to be divided on the issue of engine thermal efficiency with water addition. Some authors measured a fuel economy penalty [1], others a BSFC improvement [7], still others no significant change [6]. In the case of emulsions, it was shown that the BSFC can be improved by proper selection of the injection pump [11]. Theoretically, an improvement of engine efficiency with water addition seems to be inconsistent with the decreased combustion temperature effect. One study suggests the explanation, based on experiments with water injection into the prechamber on an IDI engine, that only localized peak temperature areas are quenched by water, while the mean combustion chamber temperature and, thus, the thermal efficiency and the BSFC remain practically unchanged [6].

15.2 Fumigation of Water into Intake Air

Spraying of liquid water into the engine intake air is certainly the most easy to implement method of water addition. It has been shown, however, that the associated NO_x reduction benefit is smaller than that achieved with the other methods [1, 12]. In addition, water fumigation carries the most risk, among all methods, of destroying the engine lube oil film by unevaporated water droplets that may reach the cylinder wall surface. For these reasons, injection of liquid water into the engine intake air stream attracts little attention today.

Corrosion problems may be minimized and possibly higher NO_x reductions may be achieved through fumigation of water vapor, in place of liquid water, into the charge air. It is a known fact that NO_x is reduced by simply increasing the humidity of the intake air. Increasing the absolute humidity of the engine intake air by 0.01 kg/kg was reported to result in about a 20% reduction in NO_x, as illustrated in Figure 15.4 [8]. This effect is roughly equivalent to using water-in-fuel emulsion of 20% water content. Effects of ambient air humidity are also responsible for higher NO_x levels from diesel engines in summer months in comparison to winter.

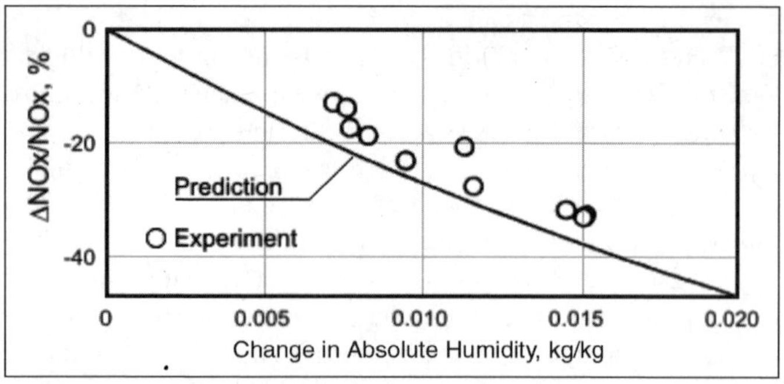

Figure 15.4 NO_x *reduction as a result of an increase in absolute humidity. (Reprinted from Ref. 8.)*

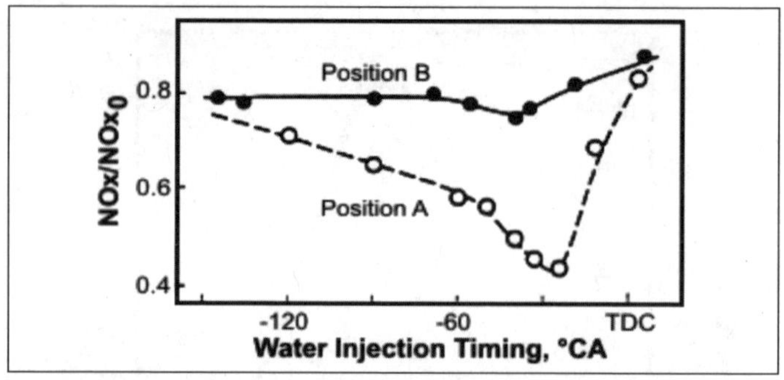

Figure 15.5 Effect of water injection timing and location on NO_x *reduction. Water:fuel = 0.6; n=1500 rpm; BMEP = 0.6 MPa; injection into the prechamber of IDI engine. (Reprinted from Ref. 6.)*

The use of steam is most commonly researched in laboratories as a method of introducing water vapor into the intake air [13]. A potential drawback of steam systems may be an increase in PM emissions, which were reported to almost double at high engine loads [14]. Engine intake air may be also humidified by contact with hot water in a special humidification column. A system that utilizes this principle, which has been developed for ships, is described later in this paper.

15.3 Direct Injection of Water

In this method, water is supplied directly into the combustion chamber through a dedicated nozzle. The water injection nozzle can be provided as a separate injector or it may be part of a tandem, water-in-fuel injector. In either case, the injection of water must be controlled independently from fuel injection with regard to timing, rate, and quantity. Very careful process optimization is required to ensure that water is supplied into the right area at the right time. To maximize NO_x reductions, water must be supplied into the flame area at the time when emissions are formed. From the spray formation perspective, the best congruence of water and fuel sprays is achieved when the two liquids are injected in parallel. Such parallel injection, however, does not optimally correspond to the period when most of the NO_x is formed, believed to be during the first quarter of combustion. For this reason, part of the water should be injected at the beginning of combustion. Late injection of water is not only useless from the NO_x control perspective, but was also reported to increase fuel consumption [1].

The sensitivity of NO_x reduction to the injection geometry and timing is illustrated in Figure 15.5 [6], where positions A and B denote different configurations of the fuel and water injection nozzles. The optimization must also include other emissions, such as CO, HC, and smoke/PM, which likewise depend on the water injection timing, but may not follow the NO_x reduction pattern. A slight reduction of PM emission was also measured in this study.

Optimized direct water injection systems for NO_x control show water consumption levels that are comparable to those seen in emulsions. Figure 15.6 shows NO_x reductions at different water consumption levels (and different fuel and water nozzle configurations) as measured on a large, two-stroke marine diesel engine [1]. The fuel consumption was not referenced in the paper, but one can assume BSFC of 165 to 175 g/kWh as typical for this engine category. If so, the system shown in Figure 15.6 (arrangement 1) requires a water-to-fuel ratio of just below 1, or 50% water in the total water-in-fuel mix, to achieve an approximately 50% reduction in NO_x.

Direct water injection appears not to be very effective in controlling PM. Most studies report only very small or no reduction of PM emissions. More sophisticated direct injection techniques, such as stratified water-in-fuel injection, have been proposed that may facilitate simultaneous NO_x and PM control in direct water injection systems [15]. In stratified systems, a special injector produces spray that is composed of a sequence of fuel and water phases. This concept may be considered a borderline method between the direct water injection and emulsion methods.

Stratified diesel fuel–water–diesel fuel (DWD) injection may be illustrated by a system developed by Daimler in cooperation with Robert Bosch [16]. The DWD system consists of a

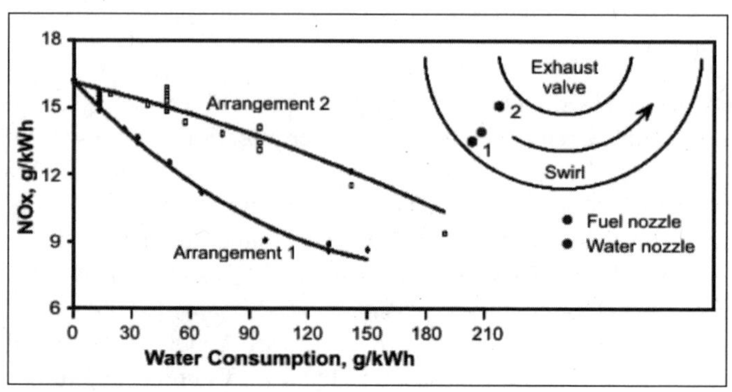

Figure 15.6 NO_x *reductions at different water consumption and injection nozzle arrangements. (Reprinted with permission from Ref. 1. Courtesy of Wartsila.)*

water supply system and a specially designed injection nozzle, as shown in Figure 15.7.

With the injector entirely filled with diesel fuel at low pressure from the preceding injection, water is forced through a one-way valve, a coaxial hole, and connected radial holes in the injector needle into the space surrounding it, whereby the diesel fuel above the water exits and is returned to the injection pump. The amount of water entering the nozzle can be adjusted to the requirements of varying engine loads and speeds. At the start of injection, that is, with increasing pressure in the nozzle, the one-way valve is closed and a sequence of diesel fuel, water, and diesel fuel is injected into the combustion chamber. Distance (d) between the water exits and the needle seat determines the amount of diesel fuel injected ahead of water. Increasing the distance (d) implies that with a given total injection volume, the injection of water is retarded and the subsequently injected fuel volume is diminished. The application of DWD injection combined with EGR resulted in 60% to 75% lower NO_X emissions, accompanied by an additional reduction in PM emissions.

15.4 Fuel Emulsions

15.4.1 Types of Emulsions

In emulsion systems, water is mixed with diesel fuel and the mixture is supplied through one injection nozzle. Depending on the method of mixing, either phase can be dispersed in the other. As mentioned before, in order to minimize water contact with internal engine surfaces, diesel fuel should form the continuous phase in which water should be dispersed in the form of small droplets. Water-in-fuel emulsions will also minimize the potential corrosion problems in the fuel system. Due to the differences in density and other physical properties, water-in-fuel emulsions are not stable. With time, the water droplets will coalesce, increasing their size, and settle down in the tank to form a pure water phase underneath the diesel fuel. In a NO_X control system, the emulsion must have sufficient stability to pass through the injection process, that is, the spray should be composed of fuel drops incorporating tiny droplets

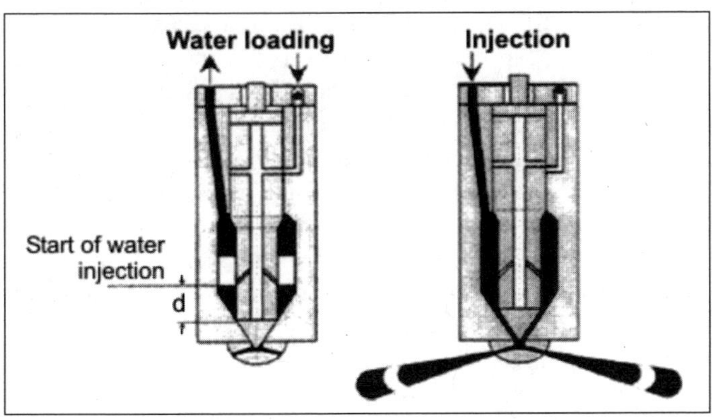

Figure 15.7 *Stratified DWD injection nozzle. (Reprinted from Ref. 16.)*

of water. This objective can be achieved using one of the following two approaches:

- *Unstabilized emulsion*—prepared on-vehicle, upstream of the injector
- Stabilized emulsion—prepared off-board and then used for vehicle fueling.

Unstabilized emulsions separate quickly and must be prepared online in the engine fuel injection system immediately before the injection. This type of system involves two tanks, one for fuel, another for water, followed by two pumps and a mixing device. In one laboratory study, mixing was carried out as a two-stage process [17]. First, the liquids were mixed in a premixer, where the internal phase (water) was injected through nozzles into the continuous phase (fuel) swirling through a cylindrical chamber. The mixture was then fed, through a gear pump, to an emulsifier. The emulsifier was a static mixing device featuring a system of nozzles and channels, where the blend's drop dimensions were decreased due to jet diffraction, impact, friction, and high pressure rates. The homogeneous emulsion was then fed to the engine fuel injector.

The higher the density difference between the two liquids in emulsion, the faster the separation. Therefore, more stable emulsions are possible with heavy fuels, such as those used in marine applications. Preparing unstabilized emulsions from lighter fuels, such as those used for highway vehicles, is more problematic. The use of unstabilized emulsions requires complex modifications to the engine and the injection system, which is an obvious drawback of this technique.

Stabilized emulsions involve mixing of fuel and water with the addition of small quantities (typically 1% to 3% of the fuel) of chemicals that facilitate the formation of emulsion (emulsifiers) and prevent or delay separation once the emulsion is prepared (stabilizers). The emulsion, prepared in an off-board mixing device, is used for fueling the vehicle, which in this case has only one tank and a single train fuel injection system. Emulsions must remain stable for a sufficient period of time to prevent separation in the fuel tank. In practical terms the required stability period must be at least several days, preferably several weeks.

If an engine designed for diesel fuel is to be fueled with emulsions, the injection system has to deliver more fluid. This can be achieved through increased injection duration, but the capacity of the injection system is limited. The engine has to be either derated or the water content and the maximum achievable NO_X reduction must be limited to typically less than 20%. To obtain the maximum NO_X reduction also at full load, the injector nozzle may need to be changed to adjust for the longer injection duration, provided the fuel pump has sufficient capacity and the camshaft has sufficient strength [11]. With emulsions of high water content, redesign of the pump, the camshaft, camshaft drive, and so forth, may be also necessary. Once the injector design has been adapted to handle the increased quantity of injected liquid, the fuel consumption, as well as component temperatures, may be penalized when running without water [1].

An important limitation on the use of emulsions is imposed by the maximum temperature in the fuel injection system. If the temperature exceeds the boiling point of water, the emulsion may be destroyed through the evaporation of water droplets. The effectiveness of the emulsifiers may be also compromised at high temperatures. In practice, emulsions can be used in engines equipped with pump-line-nozzle systems, but may not be suitable for common-rail engines, where fuel temperature in the return lines may significantly exceed 100°C.

The use of emulsions may be also limited by low ambient temperatures in cold climate areas, where freezing of the water portion may occur. The low-temperature limit of emulsions may be extended by adding antifreeze agents (e.g., methanol) to the formulation.

15.4.2 Impact on Combustion and Engine Performance

Properties of emulsified fuels differ significantly from the properties of pure diesel fuel, resulting in numerous engine effects. Literature reports indicate that water droplets in the emulsion are relatively uniform and scattered within a range of 2 to 5 μm, with an average diameter of around 1 μm. There is a tendency for the number of larger droplets to increase with increased water content in the emulsion, as shown in Figure 15.8 [5].

Fuel viscosity is an important variable affecting spray characteristics in diesel combustion. Water-in-fuel emulsions show increased viscosity with increasing water content. This, in turn, changes the size distribution of fuel droplets in the spray toward larger sizes, as shown in Figure 15.9 [5].

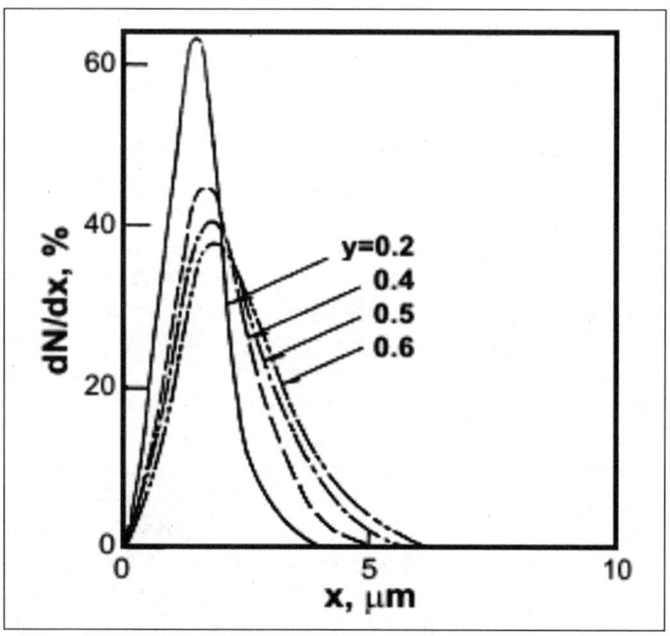

Figure 15.8 Effect of water fraction on water droplet size distribution, where y is mass ratio of water to the total emulsion. (Reprinted from Ref. 5.)

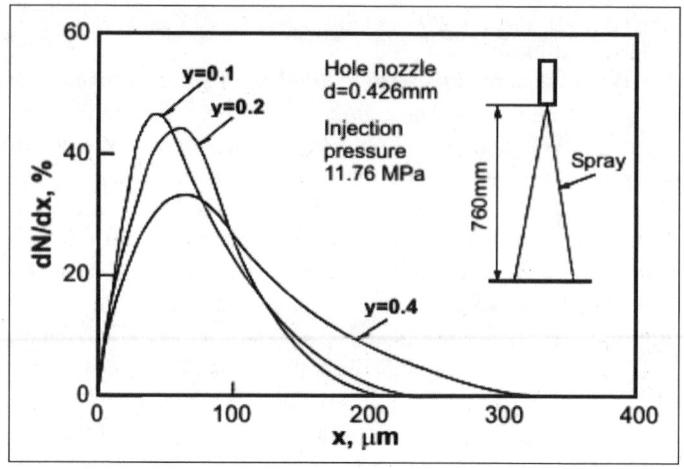

Figure 15.9 Effect of emulsified fuel on fuel spray droplet distribution, where y is the mass ratio of water to the total emulsion. (Reprinted from Ref. 5.)

A correlation between the relative fuel viscosity and dispersion parameters of the emulsified fuel is given by Equation (15.1) [5]:

$$\mu_r = 0.192 \times 10^{0.439 S_p/S_e} \tag{15.1}$$

where:

μ_r = the ratio of the emulsified fuel viscosity (as measured by a rotary-type viscometer) to the viscosity of diesel fuel

S_p = the sum of the surface areas of distributed water droplets (assumed to be spherical)

S_e = the surface area of emulsified fuel droplet (spherical).

The density of emulsified fuel should in theory increase in proportion to the water content. Some published data indicate that emulsions may contain air bubbles, making their bulk density a little less than the sum of its fuel and water components. The air bubble content was estimated to be between 1×10^{-4} and 2×10^{-4} of the water content [5].

Compared with pure diesel, emulsified fuels tend to increase the mean excess air ratio in the spray and its final penetration. Other spray characteristics such as spray angle and volume were reported to deteriorate [7]. The Sauter mean diameter of spray drops also deteriorates (increased; compare Figure 15.9). Due to the higher amount of injected liquid and the changes in its properties, the injection rates are lower and the injection period is longer. Depending on the particular engine and injection system, the overall duration of combustion may be either shorter (due to the shortened rate-controlled flame) or longer. Some authors reported dramatic increases in the duration of combustion, as high as 10° crankangle, as measured on large stationary engines with water addition of 30% [11].

Rough, unstable engine operation during transients has been reported with emulsions [5]. The effect was explained by fluctuations in injection timing. The increased viscosity of the emulsified fuel was believed to decrease the leakage quantity at the injection pump, which advanced the start of fuel injection. On the other hand, the emulsion contained a certain quantity of air bubbles, which decreased its bulk density and the bulk modulus of elasticity, causing a retardation in the opening of the injection nozzle. The relative magnitude of these effects changed following the random scatter of air bubbles and water droplets in the fuel, causing fluctuations in the start of injection.

Water-in-fuel emulsions have been reported to increase engine noise levels, especially at low speeds in IDI engines [5] and at advanced injection timings in DI engines [7].

15.4.3 Impact on Emissions

Before discussing the emission performance with water-in-fuel emulsions, a general comment should be made about comparing emissions from different fuels or fuel additives. As is commonly known, emissions change with engine operating conditions, in particular, with engine load and speed. To compare fuel effects, a test is done with a baseline fuel, followed by tests with fuels under study. However, the change of fuel properties is likely to have an impact on the engine. Quite often after switching to a different fuel, the engine delivers a different amount of power. Care must be taken that after the fuel change the engine is operated at the same conditions and, importantly, delivers the same mechanical work. Otherwise, the measured emissions could not be the basis for comparisons of emission effects of the tested fuels or additives; rather, they would reflect a combined effect of the fuel change and the change in engine operating point. In the case of engine dynamometer test cycles, the engine must be recalibrated with each fuel to deliver the same rated power in every test.

The same holds true for engines operated on emulsions; any valid emission comparison must be based on a common reference point. This can be illustrated by Figure 15.10, which shows measured levels of NO_x concentration reduction using water-in-fuel emulsions [17]. The lower line (higher NO_x reductions) was based on measurements where the total mass flow of emulsion was equal to the flow of diesel fuel in the baseline test. In the upper line measurements, the fuel portion of emulsion matched the fuel flow in the baseline test. Power delivered by the engine (assuming constant thermal efficiency) is proportional to the amount of diesel fuel which was consumed. Therefore, the NO_x reductions shown in the lower line are in fact a combined effect of engine derating and emulsions. NO_x reduction that can be attributed to the effect of emulsion alone is represented by the upper, equal fuel energy line.

Considering the large percentage of water used in emulsions and the limited capability of fuel injection systems, it is often not possible to recalibrate the engine to its original power rating when using emulsified fuels. As a result, measurements with emulsions are taken with derated engines. To make valid comparisons, engines should be also derated to the same power for baseline tests, or some other method must be used to distinguish between the effects of emulsion and engine derating. Regrettably, in many publications found in the literature, comparisons made between data collected at different engine conditions are interpreted as the effect of emulsions. If the expected magnitude of error is not estimated, such comparisons must be considered invalid.

The typical effect of water emulsions on NO_x is shown in Figure 15.11, which also illustrates the combined effect of retarded injection timing [1]. The measurements, both baseline and for emulsified fuel, were taken at 75% of the nominal maximum load of the engine. Solid lines in the figure refer to constant injection timing, while the dashed lines represent constant water content in the emulsion. As seen in the graph, increasing the water content in the fuel has an inversely proportional effect on NO_x emissions. Running the engine with 40% water allows for reductions in NO_x of about 45%. Some additional reduction is possible through the use of retarded injection. There is a small fuel economy deterioration with increased water content; at 40% water, the BSFC penalty amounts to about 1.2%. It is

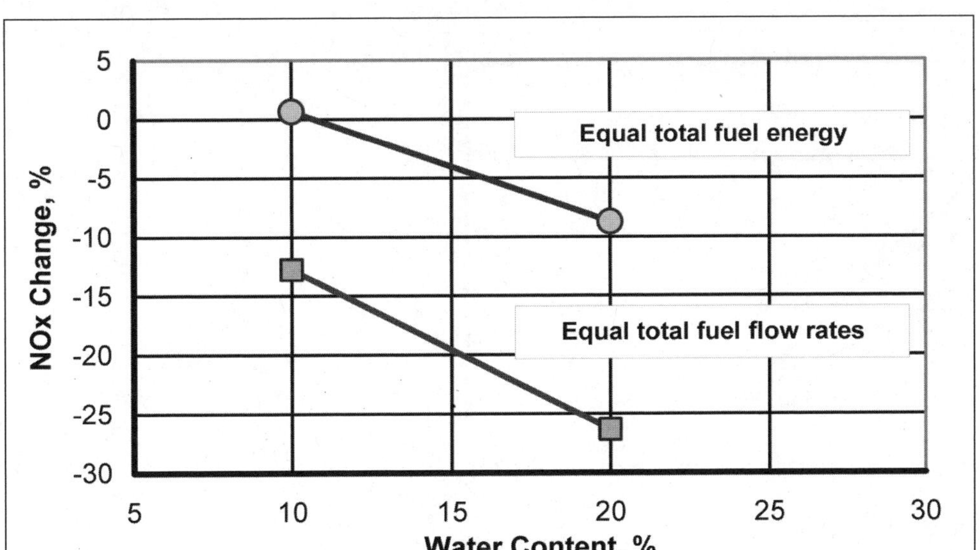

Figure 15.10 *NO_x reduction for different fuel feeding options. (Reprinted from Ref. 17.)*

evident from the figure that, for the same NO_x benefit, the use of retarded injection timing results in a much higher BSFC penalty than the use of emulsion.

Injection timing appears to be an important variable, which can be used to optimize the effectiveness of emulsions. In tests on a Caterpillar 3406B marine diesel engine, a 20% water emulsion had practically no NO_x effect at injection timing of 30° BTDC, but was very effective at 17° BTDC [9]. The effect of emulsion on reducing PM emissions was also higher at the retarded injection timing.

Water-in-fuel emulsions may be also an effective means of controlling diesel particulates. Many reports, both from the older literature [4] and from newer experimental studies [9], show very high PM reductions. The apparent percentage of PM reductions in emulsion systems is much higher than the simultaneous NO_x reductions, perhaps by a factor of 2. The published data, however, show such a degree of scatter that we are in no position to make generalizations. Reports vary from some 20% PM emission reductions at 40% water to 70% PM reductions at 10% water. Some authors compare emulsion test results from derated engines with baseline data at full ratings; such reports have limited or no reference value. The PM performance of emulsions is also obscured by the differences in sulfur levels between studies and the unknown contribution of sulfates.

To make the PM effect even more confusing, some studies measured increased PM levels with emulsified fuels. Total particulates increased with added water in a study on a MAN 10L67GBE-S, 16,255 kW @ 125 rpm, two-stroke diesel engine [11]. Lower PM increases were measured with a larger injection pump, which is an indication that PM reduction may be sensitive to the type of engine and injection system. Large increases in the total combustion period, which were measured in the study of Ref. [11], may also be related to the increased PM, because the PM reduction effect is related to improved mixing and a shorter diffusion flame period with emulsions. An increase of PM with emulsions appears to also be more likely in large, two-stroke, stationary engines, whereas more reports on decreased PM come from smaller, four-stroke engines.

Emulsions were reported to be compatible with diesel oxidation catalysts, which can bring an additional PM reduction benefit through the oxidation of the SOF [9]. Catalysts may be especially beneficial in emulsion systems that increase the SOF of diesel particulates or gaseous emissions such as HCs and/or CO.

15.5 Practical Embodiments

15.5.1 Commercial Status

The high quantities of water required to achieve any significant NO_x reduction are a major drawback of all water addition methods. It is especially true in highway applications, where refilling of high-volume water tanks or the use of high-water-content emulsions is not practical. More commercial interest in water addition methods has been observed in the stationary and marine engine markets. The accumulated experience indicates that water addition can be a viable NO_x emission control strategy. In particular, water addition systems can be designed to operate with little or no fuel economy penalty and without causing adverse engine effects, as demonstrated in a number of large engine installations. In one such application, a 20-MW diesel powerplant had been operated for more than 10 years on fuel of 30% water content with no technical problems [11]. A number of water addition systems for large marine engines were under development in the late 1990s, in anticipation of the international IMO MARPOL NO_x emission limits for oceangoing ships.

More recently, commercial water-in-fuel emulsions have been developed for mobile applications. Without engine modifications, the maximum water content is limited by the capability of the injection system, setting a practical limit of about 20% on the maximum NO_x reduction. However, a simultaneous,

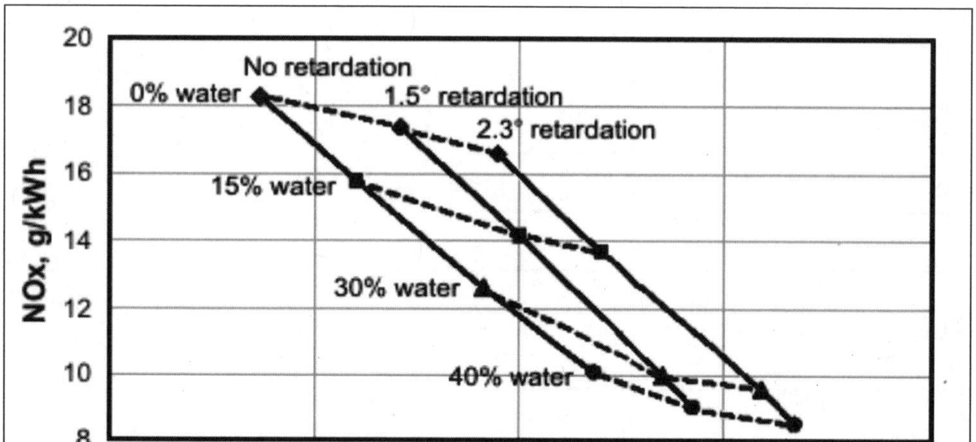

Figure 15.11 Influence of water content on water-in-fuel emulsion and injection timing on NO_x emissions for a Sulzer 7RTA84T two-stroke engine at 75% load. (Reprinted with permission from Ref. 1. Courtesy of Wartsila.)

relatively high PM emission reduction may enable commercialization of emulsions in certain mobile engine applications.

Descriptions of three different commercial water addition systems follow. These example products were selected to illustrate commercial implementations of all of the three different water addition methods: emulsion, direct injection, and fumigation.

15.5.2 PuriNOx Emulsion

The Lubrizol Corporation, a manufacturer of lubrication oil additives and related chemicals, has developed a water emulsion system that is being commercialized under the trade name *PuriNOx*. The PuriNOx fuel involves mixing of diesel fuel and water with emulsifying and stabilizing additives made by Lubrizol. It is designed for DI heavy-duty diesel engines in centrally fueled fleets.

A special PuriNOx blending unit is approximately the size of a shipboard container: 6 m wide by 2.4 m deep and tall [18]. It becomes functional with electrical supply and diesel fuel and water line connections. The capacity of the blending unit is 20 million liters of emulsified fuel per year. The blending unit can be installed by the operator of a centrally fueled fleet of diesel vehicles such as an urban bus garage. The PuriNOx blend can be used to fuel existing vehicles without the need for engine modifications.

PuriNOx fuel includes 20% water blended with 77% diesel fuel and a 3% additive package (additive 1121A). To enable the use of water blends at low ambient temperatures, winter PuriNOx fuel incorporates methanol. A winter formulation was developed consisting of 16.8% water and 5.7% methanol, blended with 74% diesel fuel and a 3.5% winter additive package [19]. The water content may be limited at levels of less than 20% by regulatory fuel specifications. For instance, PuriNOx formulations sold in the United Kingdom and Italy contain only 12% water, resulting in lower emission reductions than the 20% blend. PuriNOx blends have an opaque, white appearance that resembles milk. Water droplet size in PuriNOx is below 1 μm. The emulsion has good stability and can be stored in a tank without agitation for up to 3 months without water separation. Engines fueled with the emulsion experience approximately 15% power loss.

According to the manufacturer, PuriNOx fuel can offer NO_x reductions of up to 30% and PM emission reductions of up to 50% [18]. Emission reductions appear to be highly dependent on the engine type. PuriNOx has been verified by the California Air Resources Board to provide emission reductions of 14% for NO_x and 62.9% for PM [20]. The U.S. Environmental Protection Agency (EPA), based on existing literature data, established emission reduction factors for PuriNOx emulsions of 20% water content depending on the engine type and category, as listed in Table 15.1 (EPA draft, subject to change) [21]. PuriNOx provided consistent reductions of NO_x between 9% and 20%, with higher reductions in nonroad engines. PM emissions were reduced between 16% and 58%, with higher reductions in highway engines. HCs were increased by PuriNOx in all engine classes.

TABLE 15.1
EMISSION IMPACTS OF PURINOX
(% REDUCTION)

	NO_x	PM	HC	CO
Highway Engines				
Light-heavy	9.0	55.6	–120.2	13.4
Medium-heavy	10.2	51.1	–119.1	–25.2
Heavy-heavy	12.9	58.2	–87.8	33.3
EGR	10.2	55.6	–103.2	13.4
Nonroad Engines				
0–100 hp	19.3	23.3	–99.4	–34.7
100–175 hp	17.0	16.8	–80.1	13.4
175–300 hp	18.8	16.8	–72.8	13.4
300+ hp	20.2	16.8	–30.0	13.4

The PuriNOx emulsion is compatible with catalysts, which can control the increased HC emissions and maximize the PM benefit. A combination of the emulsion with a diesel oxidation catalyst was reported to reduce PM emissions by 70% [22]. The manufacturer has also been investigating optimization of emission reductions by combining PuriNOx with retarded injection and catalysts, concluding that both NO_x and PM were controlled more effectively at retarded injection timing [9].

15.5.3 Wartsila Water Injection System

Wartsila, a Finland-based manufacturer of marine diesel engines, has developed a direct water injection system for large, oceangoing vessels [23]. The first systems were installed on ships around 1999. The system is commercially available for new Wartsila engines or as retrofit equipment for existing engines.

The Wartsila system reduces NO_x emissions by 50% to 60% (down to typically 4 to 6 g/kWh for marine diesel oil and 5 to 7 g/kWh for heavy diesel oil) operating at an injected water-to-fuel ratio of 0.4 to 0.7 (which corresponds to 28% to 41% water in the water-in-fuel mix). NO_x reductions are most efficient at engine loads of 40% and above. The engine can be operated in a "no-water" mode if desired. It can be switched to the "no-water" mode at any engine load. In alarm situations, the switch to "no-water" mode is automatic.

The main components of the Wartsila system are shown in Figure 15.12. Filtered, clean water is fed to the engine by two water pressure units, each 1 m × 1 m × 1.7 m in size. The first low-pressure unit supplies 0.35 MPa water pressure to the second high-pressure unit, which in turn supplies 20 to 40 MPa water pressure to the combined fuel water injectors in each cylinder. A "water fuse," a safety device that shuts off the flow of water into the cylinder if the water needle becomes stuck, is installed in the water line at each cylinder. The entire system is electronically controlled, with water injection timing advanced relative to the fuel timing. A small fuel economy penalty is

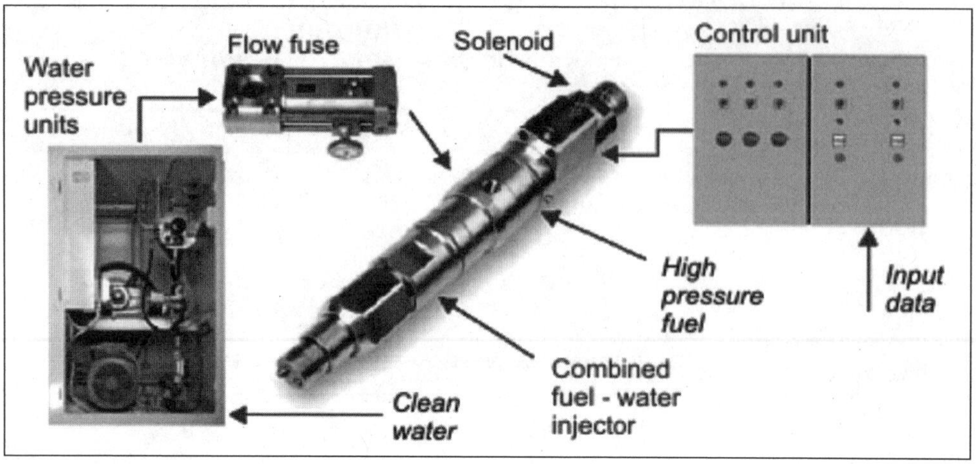

Figure 15.12 Components of the Wartsila direct water injection system. (Reprinted with permission from Ref. 23. Courtesy of Wartsila.)

associated with the water process and the parasitic load from the water pumps.

An important component in the system is the dual-fluid injector, which allows for simultaneous delivery of water and fuel into the cylinder. The injection of each fluid is controlled separately through dedicated solenoid valves. Fuel and water enter the cylinder through a dual nozzle with two separate needles (Figure 15.13).

15.5.4 Pielstick Humid Air Motor

The Humid Air Motor (HAM) is a water vapor fumigation system developed for ocean ships by the French engine manufacturer Pielstick [24]. It is now offered as an option for a number of Pielstick four-stroke marine engines.

The HAM system can achieve up to 70% NO_x reduction with no increase in smoke or HC emissions. In this system engine charge air is humidified using hot water. The HAM humidification tower can replace the engine intercooler, or other waste heat sources can be used, such as jacket cooling or exhaust gas heat. In the configuration shown in Figure 15.14, both the exhaust gas and the charge air heat are used in the HAM humidifier.

Water in the HAM process is completely evaporated, preventing entraining of water droplets into the cylinder. The system has been designed to tolerate low quality water, including seawater. Thanks to the evaporation process, there is no induction of impurities from water into the engine cylinder. The HAM system is started simultaneously with the engine, but stopped 15 min before the engine shuts down, to allow for drying of the system.

References

1. Holtbecker, R., and M. Geist, "Exhaust Emissions Reduction Technology for Sulzer Marine Diesel Engines: General Aspects," Wartsila NSD Switzerland Ltd., Winterthur, July 1998, retrieved from www.wartsila.com/english/pdf/emissions_tech_review.pdf.

Figure 15.13 Combined nozzle for injecting fuel and water into the cylinder of a large two-stroke diesel engine. (Reprinted with permission from Ref. 23. Courtesy of Wartsila.)

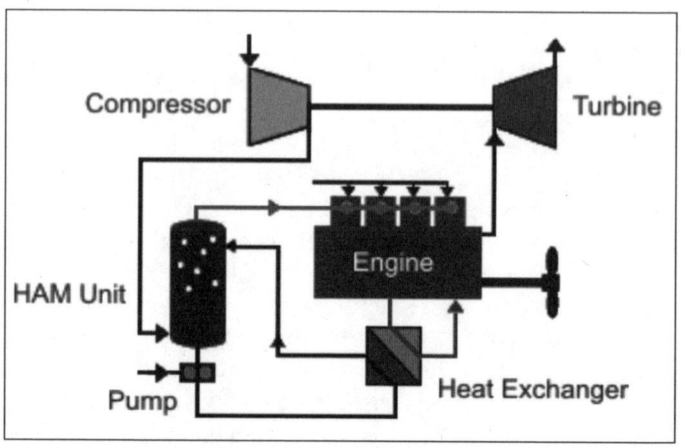

Figure 15.14 Schematic of Pielstick HAM system.

2. Vollenweider, J., M. Geist, and M. Schaub, "Residual Fuels in Emission-Controlled Diesel Engines: Background, Developments and Operational Results," in *Proceedings of the 1995 CIMAC Congress,* International Council on Combustion Engine (CIMAC), Frankfurt, Germany.
3. Sheng, H., L. Chen, and C. Wu, "The Droplet Group Micro-Explosions in W/O Diesel Fuel Emulsion Sprays," SAE Paper No. 950855, SAE International, Warrendale, PA, 1995.
4. Andrews, G.E., *et al.,* "The Reduction in Diesel Particulate Emissions Using Emulsified Fuels," SAE Paper No. 880348, SAE International, Warrendale, PA, 1988.
5. Sawa, N., and S. Kajitani, "Physical Properties of Emulsion Fuel (Water/Oil-Type) and Its Effect on Engine Performance under Transient Operation," SAE Paper No. 920198, SAE International, Warrendale, PA, 1992.
6. Miyamoto, N., *et al.,* "Significant NOx Reductions with Direct Water Injection into the Sub-Chamber of and IDI Diesel Engines," SAE Paper No. 950609, SAE International, Warrendale, PA, 1995.
7. Tsukahara, M., Y. Yoshimoto, and T. Murayama, "W/O Emulsion Realizes Low Smoke and Efficient Operation of DI Engines without High Pressure Injection," SAE Paper No. 890449, SAE International, Warrendale, PA, 1989.
8. Ishida, M., and Z. Chen, "An Analysis of the Added Water Effect on NO Formation in DI Diesel Engines," SAE Paper No. 941691, SAE International, Warrendale, PA, 1994.
9. Langer, D.A., *et al.,* "Maximizing the Effectiveness of Water Blended Fuel in Reducing Emissions by Varying Injection Timing or Using After-Treatment Devices," SAE Paper No. 2001-01-0513, SAE International, Warrendale, PA, 2001.
10. Mollenhauer, K., and P. Zelenka, "Combustion of Water-in-Fuel Emulsions in Stationary Operating Diesel Engines," *Motortechnische Zeitschrift,* No. 47, Heft 1, 1986, pp. 3–7.
11. Henningsen, S., "Influence of the Fuel Injection Equipment on NOx Emissions and Particulates on a Large Heavy-Duty Two-Stroke Diesel Engine Operating on Water-in-Fuel Emulsion," SAE Paper No. 941783, SAE International, Warrendale, PA, 1994.
12. Odaka, M., *et al.,* "Effects of EGR with a Supplemental Water Injection to Control Exhaust Emissions from Heavy-Duty Diesel Powered Vehicles," SAE Paper No. 910739, SAE International, Warrendale, PA, 1991.
13. Mellor, A.M., and J.P. Mello, "NOx Emissions from Direct Injection Diesel Engines with Water/Steam Dilution," SAE Paper No. 1999-01-0836, SAE International, Warrendale, PA, 1999.
14. Andrews, G.E., "NOx Control Using Water, SCR and SNCR," Diesel Particulate and NOx Emissions Course (University of Leeds), Ann Arbor, MI, October 2001.
15. Miyano, H., *et al.,* "Development of Stratified Fuel-Water Injection System for Low-NOx Diesel Combustion," Paper D24, in *Proceedings of the 20th CIMAC Congress,* International Council on Combustion Engine (CIMAC), Frankfurt, Germany, 1993, pp. 1–20.
16. Wirbeleit, F., C. Enderle, W. Lehner, A. Raab, and K. Binder, "Stratified Diesel Fuel–Water–Diesel Fuel Injection Combined with EGR—The Most Efficient In-Cylinder NOx and PM Reduction Technology," SAE Paper No. 972962, SAE International, Warrendale, PA, 1997.
17. De Vita, A., "Multi-Cylinder DI Diesel Engine Tests with Unstabilized Emulsion of Water and Ethanol in Diesel Fuel," SAE Paper No. 890450, SAE International, Warrendale, PA, 1989.
18. "The Lubrizol Corporation—Systems—PuriNOx," Lubrizol Corporation, retrieved February 25, 2002 from www.lubrizol.com/purinox.
19. U.S. Environmental Protection Agency, "Proposed Alternative Tier 2 Requirements for PuriNOx Diesel Fuel," *Federal Register* **67,** 98, May 21, 2002, pp. 35808–35809, retrieved from www.epa.gov/fedrgstr/EPA-AIR/2002/May/Day-21/a12707.htm.
20. Simeroth, D.C., "Verification Letter, PuriNOx Fuel," California Air Resources Board, January 31, 2001, retrieved from www.arb.ca.gov/fuels/diesel/altdsl/PuriNOxveri.pdf.
21. "Impacts of Lubrizol's PuriNOx Water/Diesel Emulsion on Exhaust Emissions from Heavy-Duty Engines," Draft Technical Report EPA420-P-02-007, U.S. Environmental Protection Agency, Washington, DC, December 2002, retrieved from www.epa.gov/otaq/models/p02007.pdf.
22. Brown, K.F., *et al.,* "Opportunity for Diesel Emission Reductions Using Advanced Catalysts and Water Blend Fuel," SAE Paper No. 2000-01-0182, SAE International, Warrendale, PA, 2000.
23. "Emission Control: Direct Water Injection," Product brochure, Wärtsilä Finland Oy, retrieved from www.wartsila.com/english/pdf/en_direct_water_inj.pdf.
24. Olsson, L.-O., "The H.A.M. System: A Simple Way to Reduce NOx in Diesel Engines of All Sizes," Paper No. P519, Institution of Diesel and Gas Turbine Engineers, Bedford, UK.

Chapter 16

Exhaust Gas Recirculation in Diesel Engines

16.1 Introduction

It has been established throughout the previous chapters that the diesel engine is more fuel efficient than other powerplants. However, admittedly, this powerplant has difficulty simultaneously meeting the desired NO_x and particulate matter (PM) standards. Traditionally, design changes aimed at reducing one of these two exhaust gas species have led to an increase in the other. This physical characteristic, which is known as the *NO_x/PM trade-off* remains the subject of intense research efforts.

The Consent Decrees Settlement of October 1998 between the U.S. Department of Justice and several diesel engine manufacturers clarified prior U.S. Environmental Protection Agency (EPA) interpretations of the Clean Air Act. Terms of that settlement covered fines imposed on each engine manufacturer, but also specified advancing the proposed emission standards for 2004 to 2002. In addition, a new steady-state emission test and limits were defined and made part of the new standards.

Engine manufacturers have developed systems for exhaust gas recirculation (EGR) to achieve these new NO_x standards. Using EGR in diesel engines has real advantages at low engine load conditions where high air-to-fuel (A/F) ratios (50:1 to 100+:1) dominate. At high engine load conditions, A/F ratios are greatly reduced and may reach values as low as 20:1. These low A/F ratios usually lead to excessive smoke formation. Recirculating exhaust with a high soot content, whether at low or high load conditions, has adverse consequences on engine durability and lube oil quality.

This chapter addresses the subject of EGR, its basic definition, its various configurations, and its effect on emissions and engine performance. Components of a typical EGR system are also described together with analyses of their important design details. Features of the control strategy for successful implementation of an EGR system are also given. The adverse consequences of using excessive EGR are discussed and general information about certain EGR system applications is supplied.

16.2 Definition

Simply put, exhaust gas recirculation is the method by which a portion of engine's exhaust is returned to its combustion chambers via its inlet system (see Figure 16.1). This method is designed to extract heat from the combustion process, thus lowering its temperature and reducing NO_x. This method also involves displacing some of the oxygen inducted into the engine with its fresh charge air, thus reducing the rate of NO_x formation.

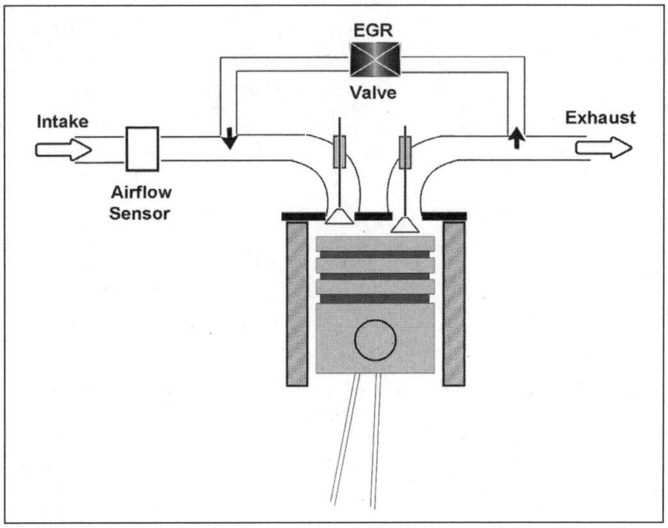

Figure 16.1 *Schematic representation of an EGR system for heavy-duty diesel engines.*

Exhaust gas recirculation is certainly not a new technology for gasoline or light-duty diesel engines. However, its application in heavy-duty diesels was not required until the new emissions limits were conceived. With the prospect of tighter NO_x limits (2.0 g/bhp-hr or 2.7g/kWh), engine manufacturers have tested and developed sophisticated EGR systems for NO_x control in direct-injected, heavy-duty diesel engines. It has been shown that EGR is a very effective method for NO_x reduction. As alluded to earlier, two principles are believed to control NO_x reduction when using EGR. The first principle is that of reducing peak combustion temperatures where EGR acts as a heat sink. In this case, the heat absorbed by EGR is thought to be directly proportional to the product of EGR flow rate (m), the specific heat at constant pressure (C_p), and the temperature differential between combustion temperature and that of EGR (ΔT). Therefore, the absorbed heat (Q) can be expressed as follows:

$$Q = m \times C_p \times (\Delta T) \qquad (16.1)$$

Exhaust gas consists mostly of H_2O and CO_2 with specific heat at constant pressure of 4.18 and 0.82 kJ/kg K, respectively. By contrast, the specific heat at constant pressure of N_2 is 1.04 kJ/kg K. It is customary to meter EGR volumetrically; therefore, when densities of these diluents are taken into consideration, their impact on NO reduction can be seen in Figure 16.2. In addition, it can be concluded that EGR cooling would increase the temperature differential term in Eq. (16.1), thus increasing the EGR heat absorbing capacity and further reducing NO_x.

The second and more important principle by which EGR reduces NO_x emissions is by displacing some of the oxygen induced with the fresh air charge. Looking at this in a simplified way, thermal NO_x is formed as a function of N_2, O_2, combustion temperature, and residence time in an environment of high NO_x formation, modified by the dissociation of NO and NO_2; that is:

$$\frac{d(NO)}{dt} = K_1(N_2, O_2) - K_2(NO, NO_2) \qquad (16.2)$$

where K_1 and K_2 are reaction rate constants that are strong functions of combustion temperature. Controlling any of the basic variables (N_2, O_2, T_{comb}, and t_r) would control the rate of NO_x formation. Therefore, reducing the fresh charge air oxygen constant by means of EGR reduces NO_x formation through reducing one of the four factors contributing to NO_x formation.

One of the negative consequences of using EGR is its adverse influence on particulate emissions. A test conducted on a Series 60, 11-L, turbocharged, intercooled Detroit Diesel engine where EGR was systematically increased from level A through level D (given in Figure 16.3) shows the corresponding total particulate matter (TPM) emission increase during a series of EPA heavy-duty transient tests. Analysis of the particulate samples indicated that the soluble organic fraction (SOF) remained constant and concluded that the insoluble fraction (mostly carbonaceous) of the particulate increased as EGR rates increased from level A to D. Oxidation-type catalytic converters were originally designed to reduce particulate SOF and, unfortunately, do not curb the insoluble fraction, which is mostly carbonaceous matter. The only available aftertreatment alternative for the insoluble fraction became the diesel particulate filter (DPF) commonly known as diesel particulate trap.

16.3 EGR Configurations

Figure 16.1 implies that a pressure differential exists between the exhaust and intake manifold without which EGR could not flow from the former to the latter. In turbocharged heavy-duty diesel engines, it is sometimes difficult to introduce EGR into the intake manifold because the intake manifold pressures are usually greater than exhaust system pressures. To circumvent this dilemma, exhaust is intercepted at a point upstream of the turbocharger (see Figure 16.4) where pressure is higher than that of the intake manifold. A portion of the exhaust flow is returned to the engine cylinders through an electronically controlled EGR valve after being cooled, as shown in Figure 16.4. This approach is commonly referred to as high-pressure loop (HPL) EGR.

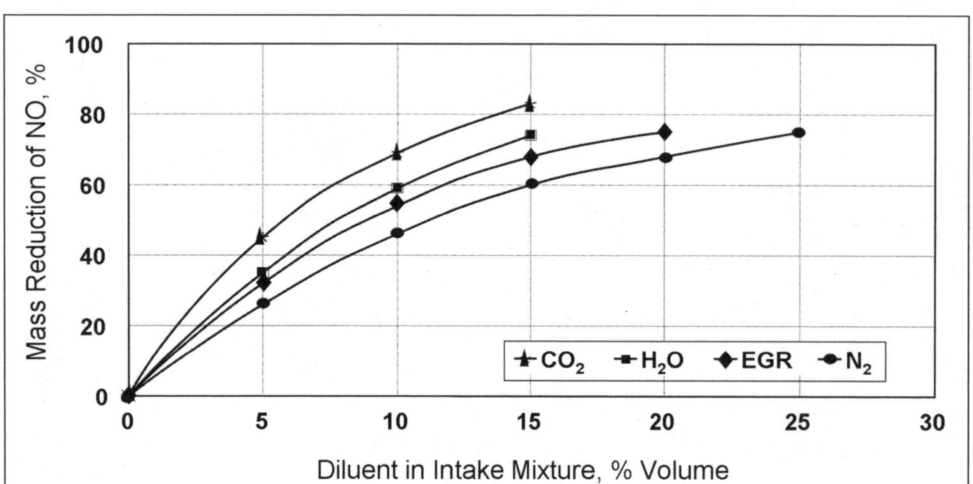

Figure 16.2 Effect of various diluents on NO reduction.

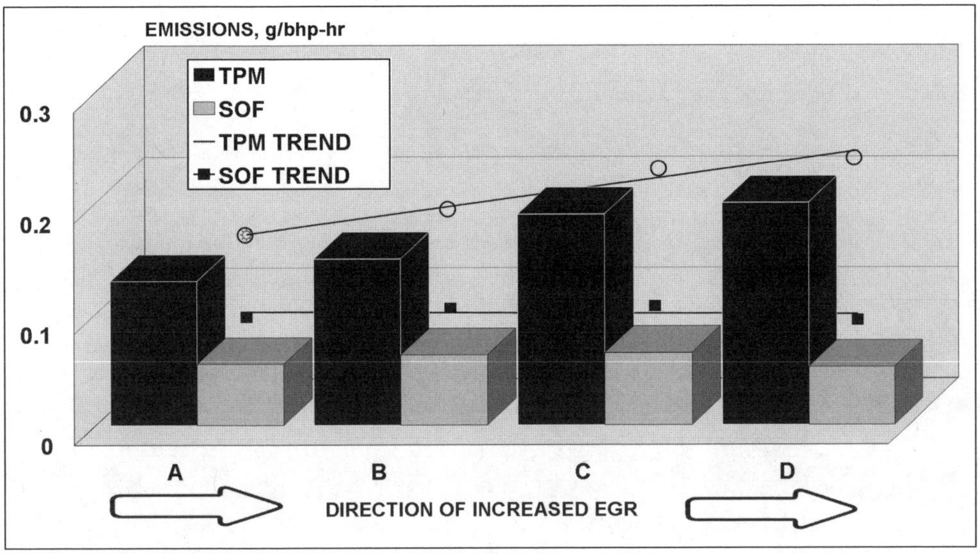

Figure 16.3 *Effect of EGR on particulate emission.*

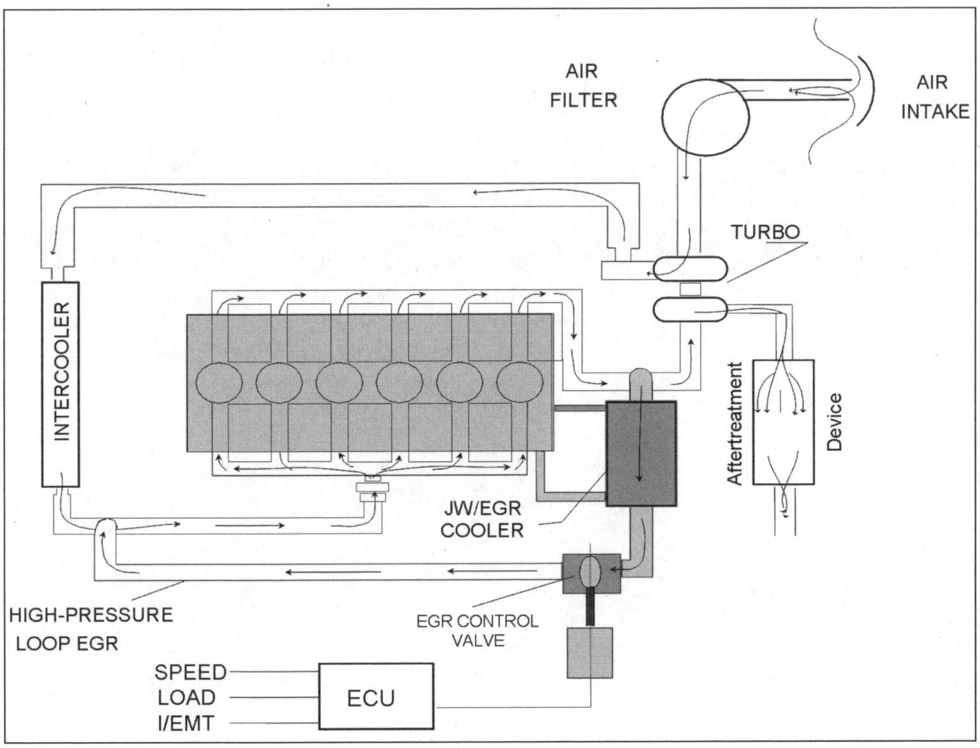

Figure 16.4 *HPL EGR configuration schematic.*

HPL EGR applied to several heavy-duty diesel engines has proven effective in reducing NO_x to the EPA FTP level of 2.0 g/bhp-hr (for heavy-duty engines). The penalty associated with such a system, however, was a recorded increase in fuel consumption and an associated increase in PM emissions. In HPL EGR implementations, turbocharger matching is usually revised to compensate for turbocharging effectiveness when a portion of the exhaust is intercepted and its energy rerouted away from the turbine wheel. This action often results in a reduced fuel economy penalty as well as the anticipated adverse impact on particulate emission.

EGR extracted from the exhaust manifold is usually routed through an EGR cooler, an EGR control valve, and then introduced in the intake manifold as illustrated in Figure 16.4.

To increase the pressure drop between the exhaust and intake manifolds, thus facilitating EGR flow into the engine inlet duct, some researchers have utilized a venturi (see Figure 16.5). The use of such designs normally increases the kinetic energy of EGR, allowing more of it to flow with lower pumping losses.

Another scheme for EGR implementation in heavy-duty diesel engines is the low-pressure loop (LPL) EGR system. This system is often employed in conjunction with trap-based

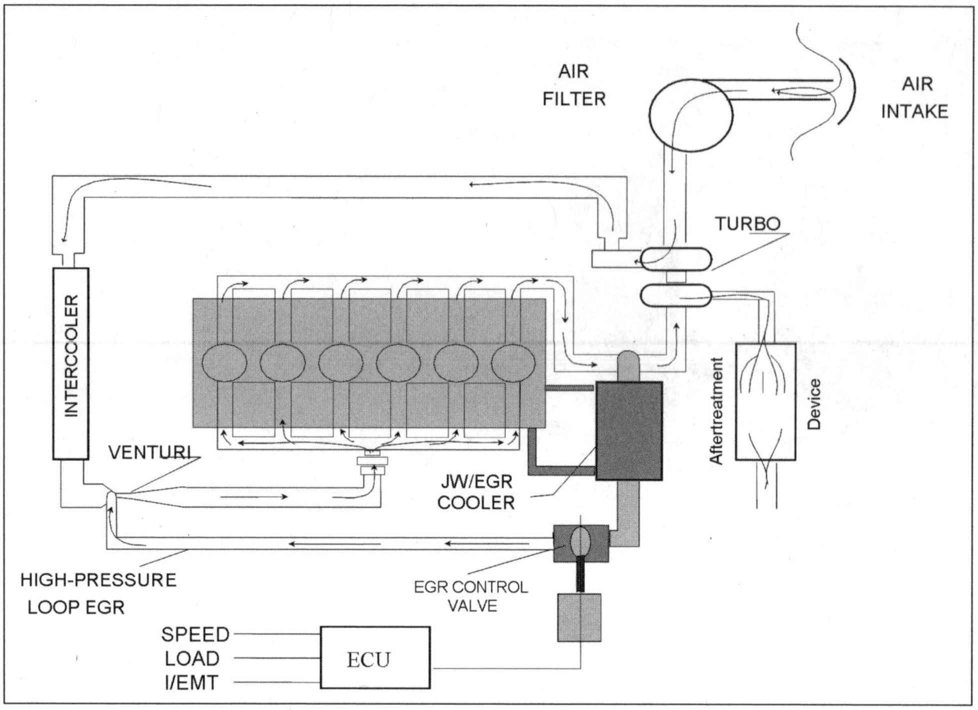

Figure 16.5 Schematic for HPL EGR configuration with venturi.

aftertreatment systems, where several benefits may accrue. Rather than sourcing EGR from a preturbine location (as in the HPL EGR case), LPL EGR systems use exhaust that has been filtered through diesel particulate filters (DPFs). Figure 16.6 shows a schematic representation of a LPL EGR system in which EGR was recirculated from a point downstream from the DPF. This alternate configuration sought to preserve the turbocharger performance by supplying exhaust gas from a point downstream of the trap (see Figure 16.6), thus allowing all of the exhaust to be utilized in the turbine. At this location, exhaust gas pressure is at a lower level than that of the intake manifold. To promote flow, EGR is introduced back in the engine just upstream of the turbocharger compressor. The pressure difference between a point downstream of the trap and upstream of the turbocharger is generally adequate for the EGR flow rates needed to reduce EPA FTP NO_x to the 2.0 g/bhp-hr level for heavy-duty diesel engines.

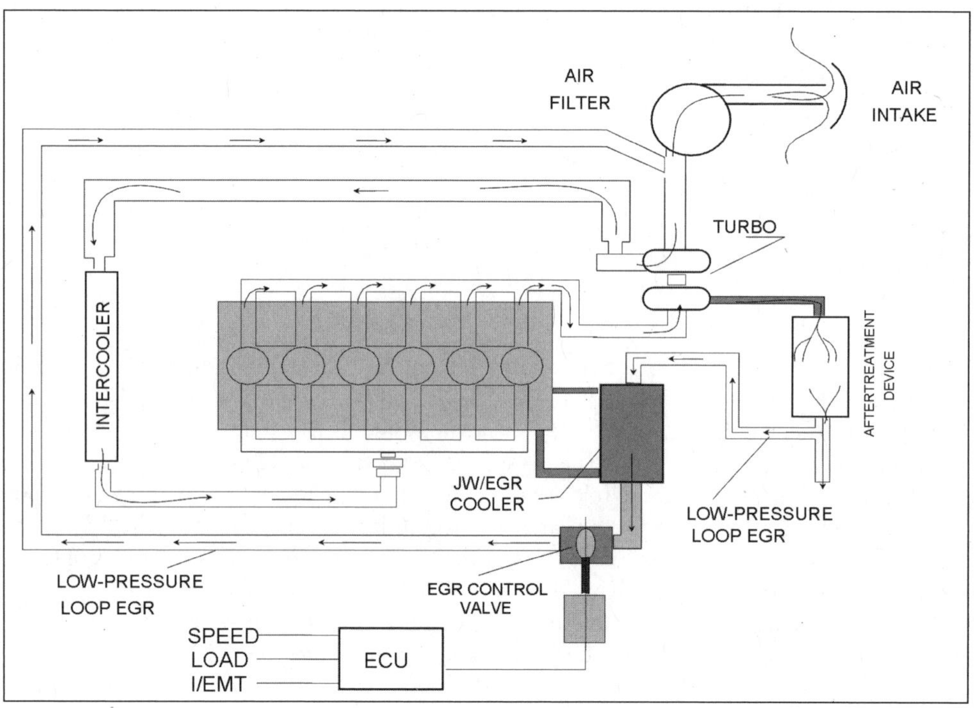

Figure 16.6 Schematic representation of a LPL EGR system for heavy-duty diesel engines.

Advantages of the LPL EGR system include the following:

1. Lower fuel consumption from that of the HPL configuration as a result of better turbocharger performance than that of the HPL configuration. However, if the HPL EGR system is properly executed, the turbocharger would have to be revised to account for the loss of exhaust energy resulting from rerouting some exhaust to the EGR circuit.
2. With the presence of a trap, the LPL EGR supplies predominantly filtered exhaust to the inlet of the engine through the turbocharger compressor. Therefore, engine durability can be better preserved.
3. Exhaust gas downstream of the trap is cooler than that provided from upstream of the turbocharger (as in the HPL case). Therefore, a LPL EGR would have a higher heat absorbing capacity for flow rates similar to those of the HPL EGR.
4. Due to the higher heat absorbing capacity of a LPL, an opportunity exists to reduce EGR cooling requirements (reduced size EGR cooler) and provide a more compact unit. In addition, the EGR cooler would have less heat rejected in the engine water jacket and, therefore, less cooling load for the radiator to handle.
5. Reducing the amount of EGR cooling may help prevent exhaust system condensation and potential erosion of the EGR cooler, EGR valve, EGR piping, and turbocharger compressor wheel.
6. Better EGR and fresh charge air mixing result from introducing the mixture upstream of the turbocharger compressor.

Despite the many advantages listed here, the LPL EGR system was not favored over the HPL system for a number of reasons:

1. Even though the LPL EGR is sourced downstream from the trap it is not entirely free from carbonaceous material since the filter trapping efficiency is less than 100%.
2. With carbonaceous material still remaining in the recirculated exhaust stream, its impact on the compressor wheel as it turns at a high rate of speed may potentially erode it.
3. In cases where air-to-air intercoolers are used, carbonaceous and other unfiltered matter flowing through the compressor would likely be trapped in the narrow cooler passages. If left to accumulate over time, then airflow to the engine would be reduced, leading to performance, emission, and fuel economy deterioration.
4. Unburned oil vapors as well as any unburned fuel adsorbed on the surface of carbonaceous particles accumulating in the inlet system may emit carbon monoxide (CO) gas when exposed to higher temperatures. When this happens, fresh and cooled air to the engine would be displaced by CO, which is detrimental to combustion efficiency.
5. Plumbing for the LPL EGR arrangement is often awkward and cumbersome.

Oftentimes, engines may be equipped with waste-gated (bypass) turbochargers. These devices are designed to accommodate engine airflow requirements throughout the engine speed range and particularly at those engine speeds below peak torque. Above peak torque speeds, some of the exhaust flow to the turbocharger is usually bypassed to preserve fuel spray and air mixing quality in the combustion chamber. Bypassing exhaust around turbochargers, as shown in Figure 16.7, generally reduces exhaust manifold pressure as well as the pressure differential between exhaust and intake manifolds. In such cases, a hybrid system combining features of a HPL and LPL is often used (see Figure 16.8). In the schematic given in Figure 16. 8, EGR is

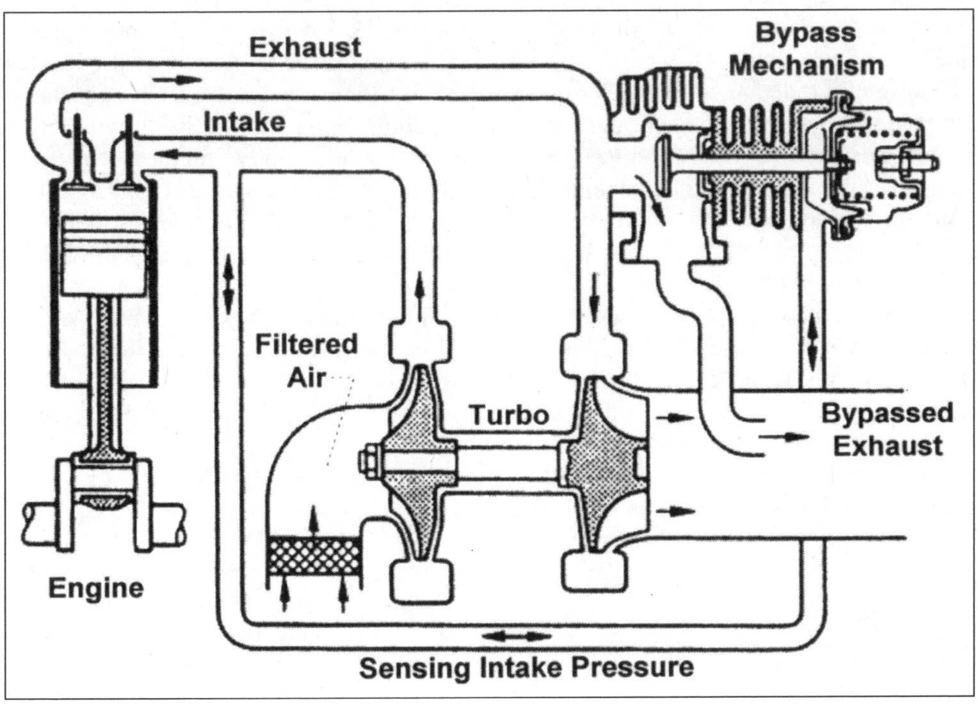

Figure 16.7 *Schematic diagram of a waste-gated (bypass) turbocharger. (Reprinted from Ref. 1 with permission from Robert Bosch GmbH.)*

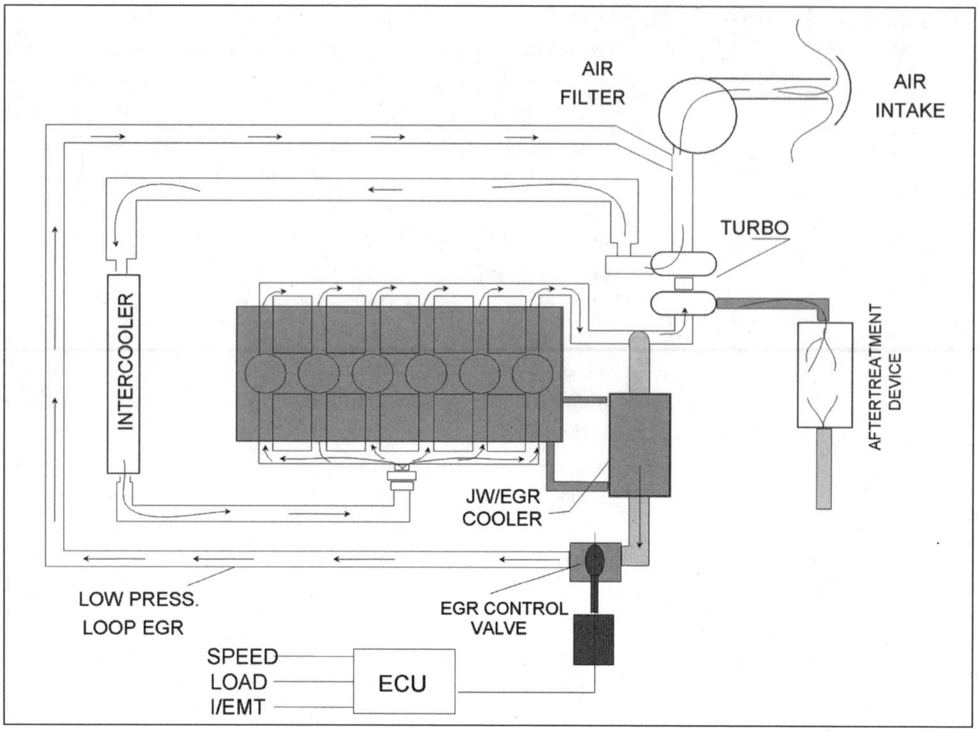

Figure 16.8 *Schematic of a hybrid HPL/LPL EGR configuration.*

sourced from a point upstream from the turbine (preturbine) as in a HPL configuration, and delivered precompressor as in a LPL configuration. Although this system has the same adverse features of the LPL EGR system, it provides an adequate pressure differential between the exhaust and intake manifolds. This allows EGR rates that result in a substantial NO_x reduction without needing a pump or application of excessive exhaust back pressure to drive EGR into the engine.

As mentioned earlier, the pressure differential between the exhaust and intake manifolds plays a major role in controlling EGR rates. To provide adequate EGR rates while avoiding the need for a pump, a venturi is often used at the point of EGR entry as shown in Figure 16.5. The presence of the venturi reduces pressure at the point of EGR entry by converting the potential energy of EGR to kinetic energy, thus assisting its flow from the exhaust to the intake manifold. Pressure expansion reduces the EGR temperature as it flows through the venturi, adding a small side benefit to the use of this device.

One major drawback common to all EGR configurations discussed so far is the large residual volume between the EGR valve and the combustion chamber. The presence of this volume has a negative impact on smoke control, especially when rapid accelerations are required. As the engine responds to a step increase in fuel, the EGR remaining in the piping between the EGR valve and the combustion chamber is usually inducted with the fresh air into the engine. This action may lead to sluggish acceleration accompanied by a puff of smoke. The problem is further compounded by the *turbo-lag* phenomenon, defined as the time between injecting fuel to accelerate and when the turbocharger speed spools up and begins delivering more air to the working cylinders. During this turbo-lag, the combustion process can use all of the fresh air it can get to efficiently burn the increasing acceleration fuel rate. Yet, it is during this incremental time that the engine faces an air shortage due to turbo-lag and also inducts the residual EGR remaining in the piping leading from the EGR valve to the cylinders. For this reason, researchers have devised new schemes to minimize this EGR residual volume. These schemes are sometimes referred to as fast-acting EGR systems.

An example of a fast-acting EGR system conceived by Southwest Research Institute (SwRI) is shown schematically in Figure 16.9. Note that EGR is delivered to the intake port at a point close to the intake valve. The EGR valve is also located very close to the intake port, thus reducing the volume of exhaust an engine has to consume, especially during acceleration. The fast-acting EGR system has greatly minimized this EGR volume, making it easier for the engine to accelerate without interference from residual EGR. A reservoir with a shell-and-tube EGR cooler is also provided for storing cooled EGR. In some EGR designs, such as small high-speed passenger cars, the piping between the cooler and EGR control valve is buried inside the cylinder head nestled between the cooling water passages within the cylinder head. This design eliminates the need for a separate EGR cooler, even though EGR cooling effectiveness may suffer somewhat.

The other part of the fast-acting EGR system involves its source point. If EGR is sourced from a preturbine point, then it fits the description of HPL EGR. On the other hand, if it is sourced from downstream of the trap, it would be a LPL EGR system, which might require an EGR pump to elevate its pressure above that of the intake manifold as shown in Figure 16.10.

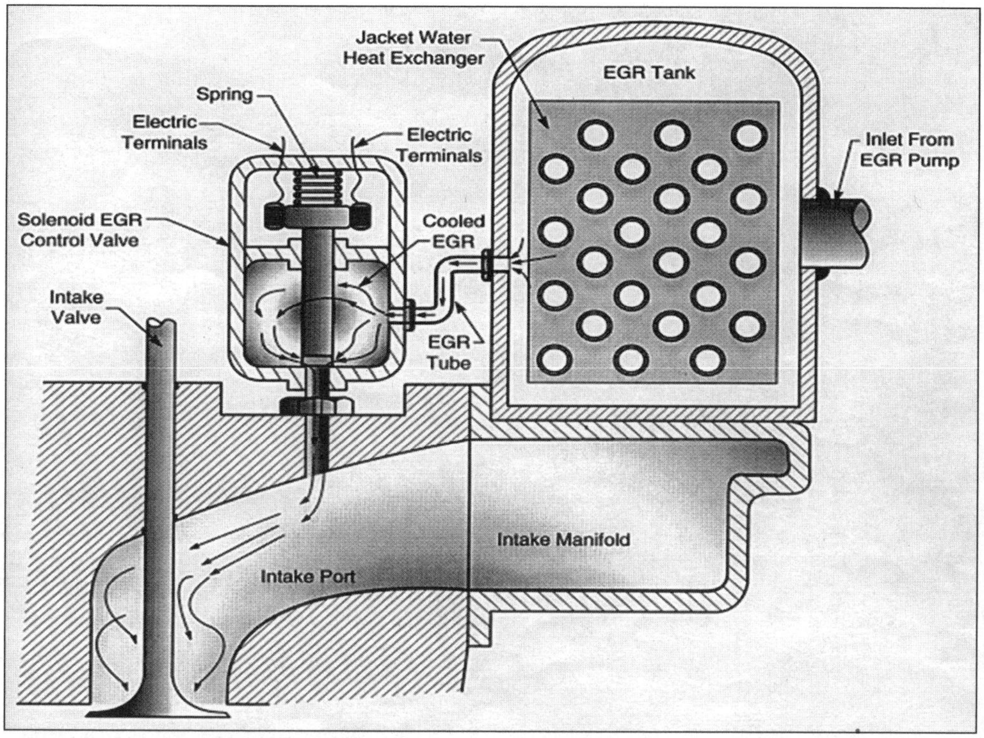

Figure 16.9 *Schematic representation of a fast-acting EGR system.*

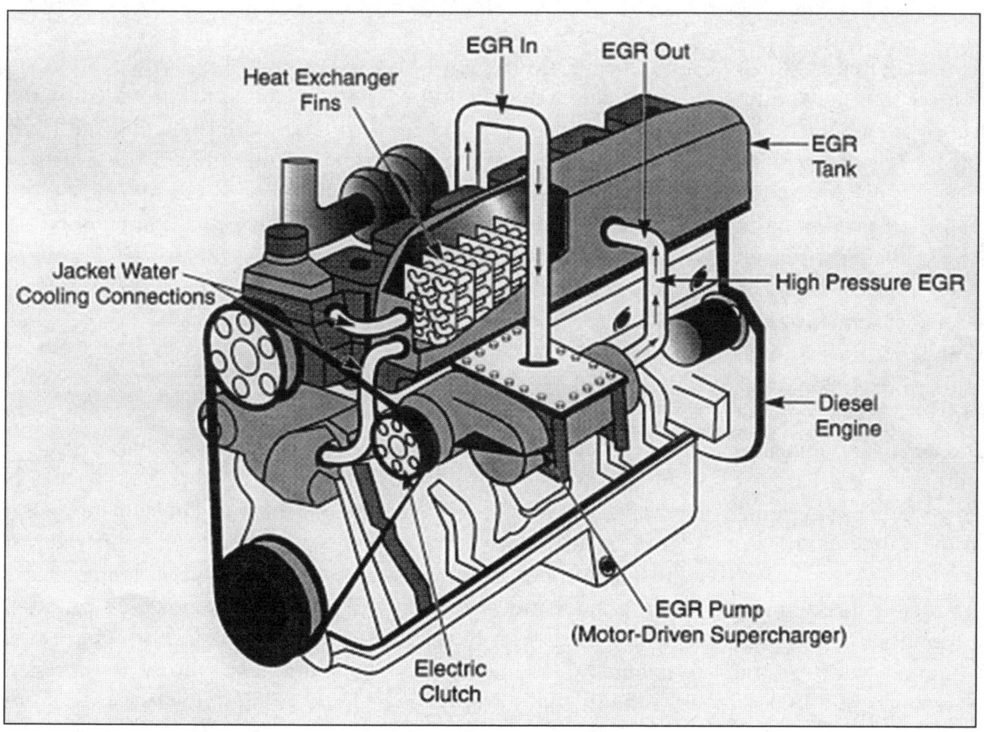

Figure 16.10 *EGR pump (HPL OR LPL) to overcome intake manifold pressure.*

16.4 System Components

A photograph of a hybrid HPL/LPL EGR system fitted on a DDC Series 60 heavy-duty engine is shown in Figure 16.11. The EGR system was designed to use preturbocharger exhaust supplied from the middle segment of the exhaust manifold (inset of Figure 16.11). Preturbine (HPL) exhaust was routed via an EGR shutoff valve, an EGR rate control orifice, and an EGR cooler to the engine inlet system. The EGR shutoff valve was designed to stop EGR during engine acceleration conditions. Input from the throttle position sensor was used to detect engine acceleration. Upon detecting acceleration, a signal was sent to

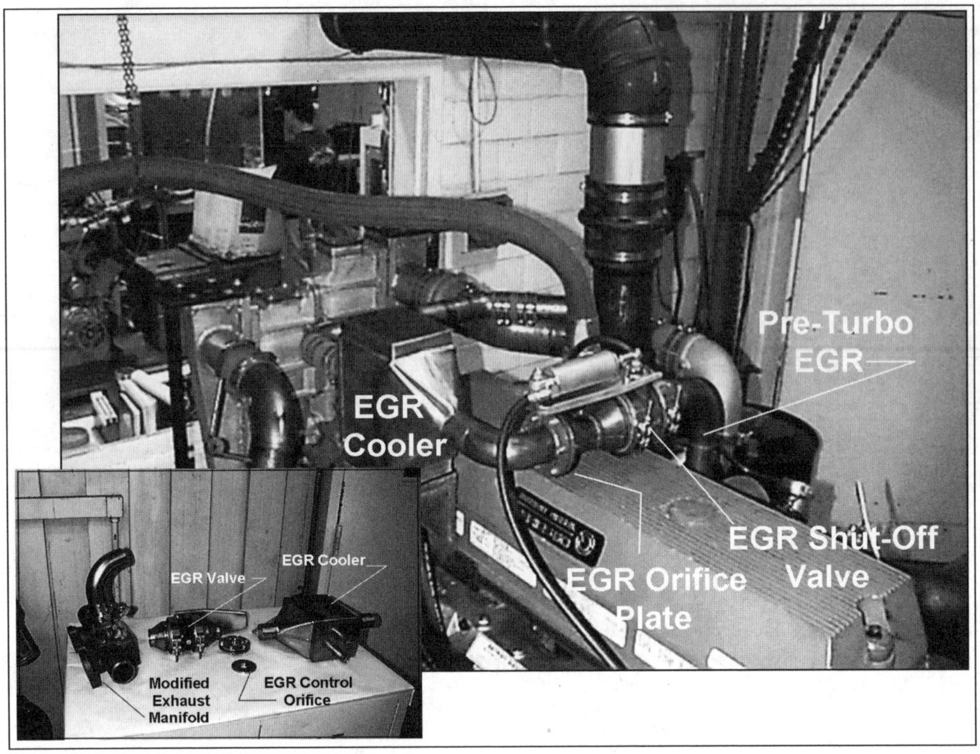

Figure 16.11 Top view of an EGR system on a DDC Series 60 heavy-duty diesel engine.

a three-way solenoid valve that activated an air cylinder, which in turn set the EGR valve to its shutoff position, thus preventing EGR flow. In engine operation other than acceleration, the EGR rate was controlled by a properly sized orifice interacting with the pressure differential between the exhaust manifold and the inlet system. From the rate control orifice, EGR proceeded to a shell-and-tube heat exchanger. Engine jacket water was used as cooling medium in the EGR cooler.

Components of the hybrid HPL/LPL EGR system installed on the Series 60 engine were described to illustrate the essential elements in an EGR system. Most, if not all, of the heavy-duty diesel engine manufacturers introduced their engines equipped with cooled EGR systems to meet mandated NO_x emission standards for model year 2002. Production-style EGR valves have been developed. Other designs are considering integrating the EGR valve and intake manifold and the valve and turbocharger housing.

Controlling EGR is easily accommodated through the engine control module (ECM). Many functions are shared between engine and EGR valve control. Engine speed, torque (often substituted for by intake manifold pressure), throttle position, and intake manifold temperature are just a few of the shared signals between both engine and EGR control. Even though the EGR components outlined earlier were for an experimental setup, most of these components are almost identical to those intended for use in serial production. The exception is EGR valve modulations, which are performed using linear solenoids or stepper motors in production. Most of the EGR coolers are of the shell-and-tube design and use engine coolant for the cooling medium. In some applications, such as the 7.3-L Navistar engine, an airflow sensor is used to control EGR rates. At given speed and load conditions, the fresh air signal provided by the airflow sensor for NO-EGR is reduced by opening the EGR valve. The signal corresponding to the reduced fresh airflow is used as feedback, indicating the proper EGR rate or leading to further adjustment in the EGR valve position to obtain a predetermined EGR rate. In general, a production EGR system may not need any fully dedicated sensors, but may share control signals with existing sensors.

16.5 EGR Cooling and EGR Coolers

The need for cooling EGR was expressed earlier in Eq. (16.1), where heat absorbed from the combustion process is proportional to EGR rate, specific heat at constant pressure, and the difference between combustion and EGR temperatures. It follows then that the greater the difference in temperature, the greater the heat absorbed from the combustion process, which leads to a lower rate of NO_x formation. In addition, a cooler EGR temperature means that less volume is occupied in the inlet system. A lower EGR volume displaces a smaller fraction of fresh filtered intake air, thus displacing less O_2, which helps maintain combustion efficiency. It is desirable to cool EGR prior to its introduction in the cylinder. A schematic representation of an EGR cooler is shown in Figure 16.12. Exhaust gas flows through the tubes of a shell-and-tube heat exchanger while jacket water cooling water flows around the outside of these tubes. Cooler effectiveness is measured by the ratio of the temperature drop of EGR across the cooler to the available cooling capacity.

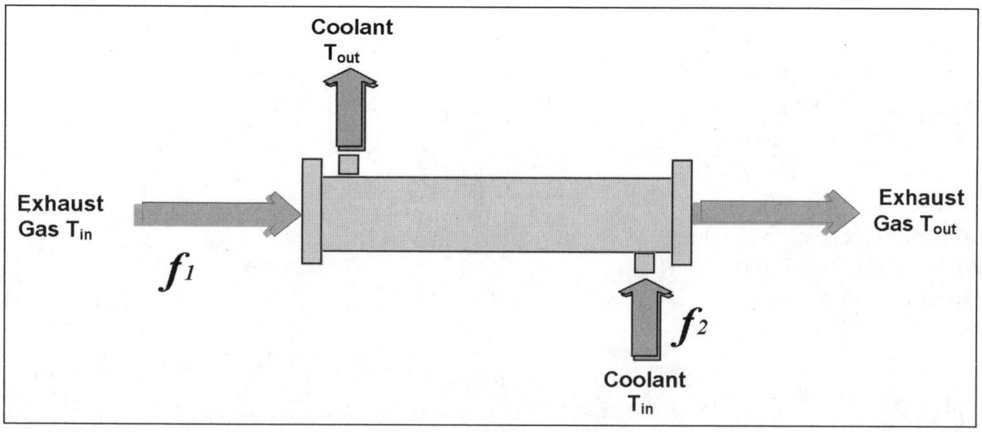

Figure 16.12 *Schematic diagram of an EGR cooler.*

Segmented baffles are part of the cooler construction, designed to minimize coolant pressure drop. Tubes are made of materials selected for high heat conductivity, strength, and corrosion resistance. Stainless steel with chromium, nickel, and molybdenum is a popular material. However, care should be exercised regarding the content of nickel and its susceptibility to the presence of corrosive sulfuric acid in the exhaust.

EGR coolers may feature one-, two-, or three-pass cooling flow paths. Selecting a greater number of flow paths is a trade-off between pressure drop and the desire for large temperature reductions. A major disadvantage of the greater number of flow paths is the associated pressure drop, which is critical when designing EGR systems. Having small pressure differences between the exhaust and intake manifolds does not permit additional pressure losses in the EGR system, such as those experienced in a cooler. Figure 16.13 gives an example of the pressure drop associated with various gas flow paths through an EGR cooler.

Fouling of the EGR cooler is a major concern because it contributes to lower cooler effectiveness and pressure loss. Experiments conducted on prototype coolers have shown that effectiveness degradation is usually limited to the first 20 to 30 h of EGR cooler use. Figure 16.13 indicates that the cooler effectiveness drop between 20 and 30 h of service is less than that experienced between the first and second 10-h segments.

Cooling of EGR may not always be desirable. For instance, in cold weather conditions, during which EGR is cooled below the dew point temperature, condensate may form and mix with exhaust containing sulfur compounds. The mixture of condensate and exhaust may lead to the formation of sulfuric acid, which could erode components of the EGR as well as engine systems. Some prefer a design that would bypass the EGR cooler during cold temperatures and during idle/no-load conditions. This practice has often improved idle/no-load and part-load fuel economy and also reduced hydrocarbon and carbon monoxide emissions.

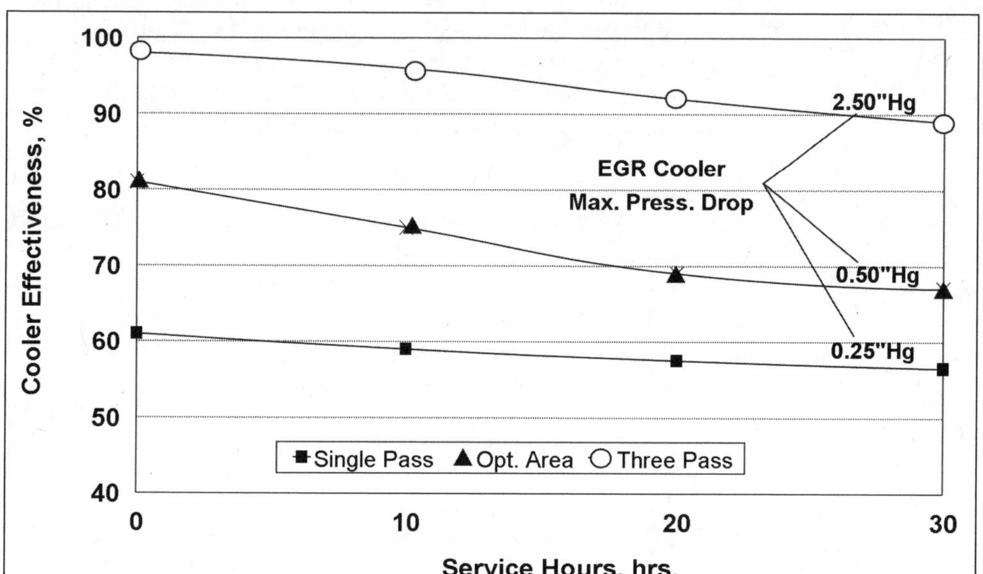

Figure 16.13 *Effectiveness and pressure drop for one-, two-, and three-pass EGR cooler versus time.*

16.6 EGR Control Strategy

Engine manufacturers often describe their EGR systems as being quite sophisticated and claim that accurate EGR rates are required to carefully maximize NO_X reduction without disproportionately increasing particulate emission. Although this statement is generally true, we have found that some engines are less sensitive to small variations in EGR rates than others. However, certain elements were found to be essential to achieve good NO_X and PM control. These control elements include shutting off EGR during acceleration, greatly reducing or eliminating EGR at or near-full-load conditions, and shutting off EGR when A/F ratios are lower than 25:1. Observing these principles generally leads to good engine performance, fuel economy, and reasonably controlled particulate emissions. In addition to the NO_X/PM control strategy, attention must be given to durability-related issues. Although there are many aspects to the durability of an engine using EGR, a limit on the EGR rate may be required to keep exhaust temperatures within reasonable limits, mainly to protect the turbocharger components.

In modern control schemes, two concepts are usually considered: open-loop and closed-loop control. Open-loop control systems are often based on the results of extensive engine mapping. For instance, mapping an uncontrolled engine would produce a three-dimensional relationship between engine speed, torque, and NO_X. The same exercise would also produce information related to fuel consumption at each of the engine speed and load conditions. By developing these relationships, one can then determine, by modeling or through experiment, the required EGR rate for a certain NO_X reduction within a threshold limit of fuel consumption penalty. An open-loop control system (see Figure 16.14) can be manually adjusted to set the EGR rates, thus allowing complete flexibility when mapping an engine. From this information, a look-up table can be generated that would be used by the open-loop control system to essentially provide the EGR rate based on the EGR actuator calibration (see Figure 16.15). It is then assumed that the desired NO_X reduction will be achieved. Such an assumption can be verified through experiment to ensure the adequacy of the open-loop control system. Another variation of open-loop control is shown in Figure 16.16 where a look-up table is added to allow the controller to respond to changes in the controlled system in a mechanical manner without manual input to the system.

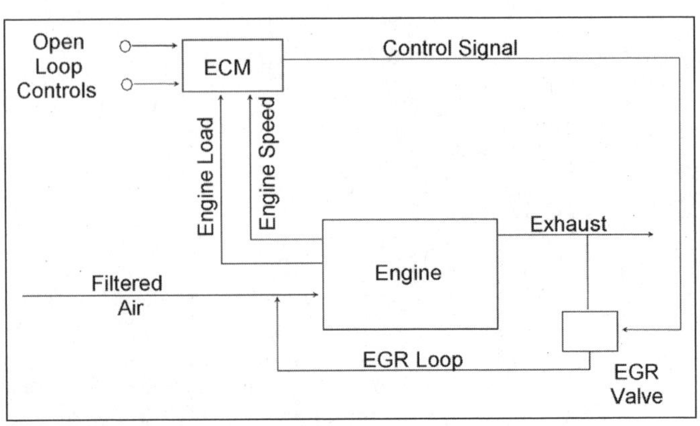

Figure 16.14 EGR open-loop control diagram.

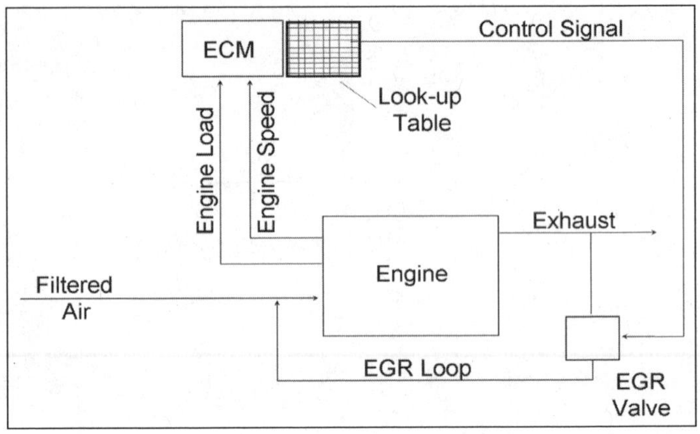

Figure 16.15 Open-loop control scheme with look-up table.

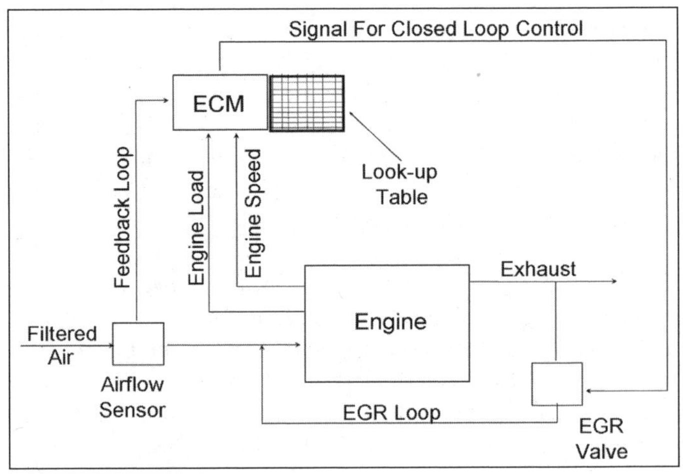

Figure 16.16 EGR closed-loop control diagram.

In closed-loop systems (see Figure 16.16), a scheme similar to that of the open-loop system is used. In addition, the value of the quantity being controlled, in this case it may be NO_X or another indicator of its reduction, is compared to the desired value and an adjustment is made in response to the difference between the desired and the actual values of the controlled quantity.

Mathematical modeling and computational fluid dynamics can play a major role in reducing the experimental phase of EGR development. Mathematical tools have become more popular as computers and their computational capabilities have improved. Predictive models for NO_X as a function of engine speed and load, intake and exhaust manifold pressure and temperature, fuel consumed, injection timing, and A/F ratios have been used to predict NO_X reduction with different EGR rates. For each speed and load condition, a baseline NO_X concentration and fuel consumption are calculated. The calculation is then repeated for the same point using an A/F ratio based on a given EGR rate and a new NO_X concentration is generated. If the new calculated NO_X matches the desired concentration, then it is used in an engine performance predictive model that would estimate fuel consumption for that EGR rate. This process (represented schematically in Figure 16.17) is repeated for a set of engine conditions simulating a standardized test cycle consisting of various engine speeds and loads in a transient behavior. The modeling procedure is iterative and can be repeated several times for the same engine speed and load condition until the desired NO_X concentration is reached. Evidently, this iteration can be automated for a specific engine.

Schematics showing the EGR system and its control suggest that a number of inputs are normally required for its operation. These inputs are provided to the electronic control module (ECM) by way of sensors. Sensors for EGR system control include:

- Temperature sensors for exhaust, EGR upstream of the EGR cooler, and intake manifold
- Pressure sensors for the intake manifold and exhaust back-pressure

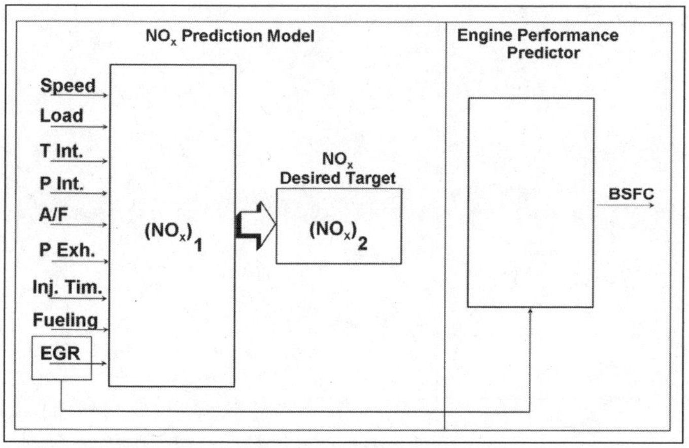

Figure 16.17 *Schematic diagram for EGR, NO_X, and BSFC predictions.*

- EGR valve position feedback
- Mass airflow
- Wide-ratio, lean O_2 sensors (in some cases).

Signals from mass airflow as well as EGR valve position sensors are very commonly used as control feedback parameters. Pneumatic and electric servo mechanisms are also popular for EGR valve actuation. Intake and exhaust restrictions have also been used to force more EGR into engines. Both schemes have to be carefully implemented so as not to drop below the threshold of the minimum A/F ratio to avoid excessive smoke formation.

Cylinder-to-cylinder EGR distribution is one of the most important factors that should be taken into account when designing EGR systems. This aspect is especially challenging when working with limited space, a vee-engine layout, and high-speed engines. Individual CO_2 probes for each intake port are often used in the early development stages of a new EGR system to ensure even distribution. As these details are observed, EGR systems development should lead to good NO_X reduction without excessive smoke and carbonaceous particulate formation.

16.7 Effects of EGR on Combustion Parameters and Emissions

The effect of EGR on fuel spray angle, fuel atomization, fuel penetration, O_2 concentration, and exhaust emissions has been studied by Arcoumanis *et al.* [2] at the Imperial College of Science. Their research was conducted on a small, 1.9-L, direct-injected (DI) diesel engine equipped with an experimental EGR system. Results of their investigation confirmed that EGR displaced a portion of the fresh filtered air, thus displacing a portion of the O_2 entrained into the engine. The relationship between EGR rate and O_2 displaced from the fresh filtered air is shown in Figure 16.18 and indicates that increasing EGR from 0% to 50% reduced O_2 from 21% to 14%.

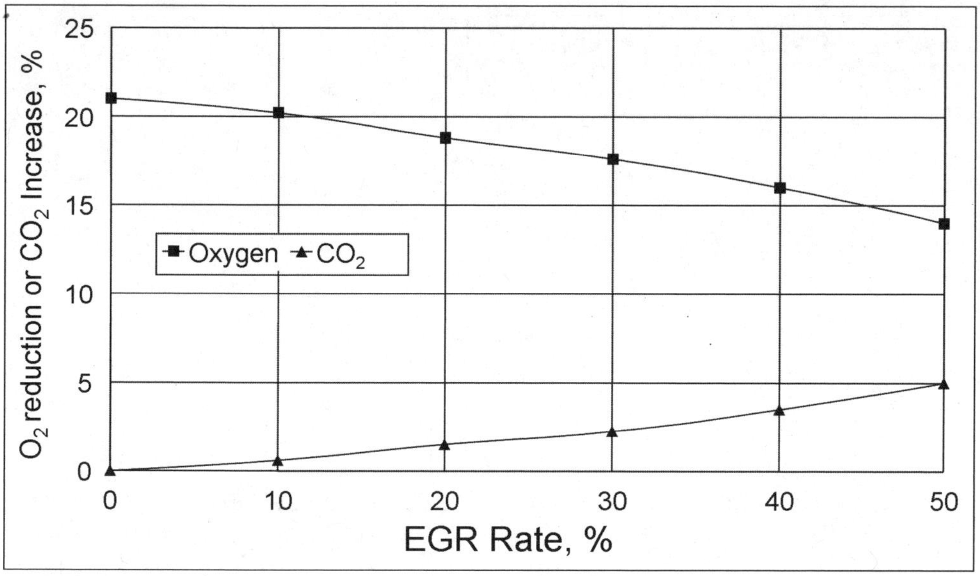

Figure 16.18 *Effect of EGR on fresh oxygen and carbon dioxide.*

Also shown in Figure 16.18 is the effect of EGR on CO_2 indicating that increasing EGR from 0% to 50% would increase CO_2 by about 5%. With O_2 being displaced (reduced), the soot oxidation rate drops and leads to higher concentrations of carbonaceous particulate matter. Arcoumanis *et al.* [2] have also found that the increase in CO_2 and soot was also accompanied by an increase in CO and HC. Interestingly, the influence of EGR on the fuel spray angle, fuel atomization, and fuel spray tip penetration was found to be minimal. As expected, the temperature in the intake manifold, where EGR and fresh air are mixed, increased and density decreased. Figure 16.19 shows the effect of EGR rate on the temperature of the mixture in the intake manifold.

Among the other observations, auto-ignition sites as well as their size increased at higher EGR rates. Conversely, flame luminosity and flame temperature were reduced. At 50% EGR, the flame temperature was reduced by about 100° K [3]. Because NO_x formation is strongly flame temperature dependent, it is suggested that a reduced combustion flame temperature is the major reason for the NO_x reduction.

In Figure 16.20, the effect of EGR rate on NO_x reduction at various engine load conditions is illustrated. Even though we pointed out earlier that cooling EGR may not always be desirable, it must be noted that if EGR is to be used at higher engine loads, then it should be cooled. Cooled EGR will displace less of the fresh air volume, thus maintaining an adequate overall A/F ratio to sustain good combustion efficiency.

Central to NO_x reduction via EGR is its impact on fuel consumption, especially if other NO_x reduction methods such as injection timing retard are considered. This issue was the subject

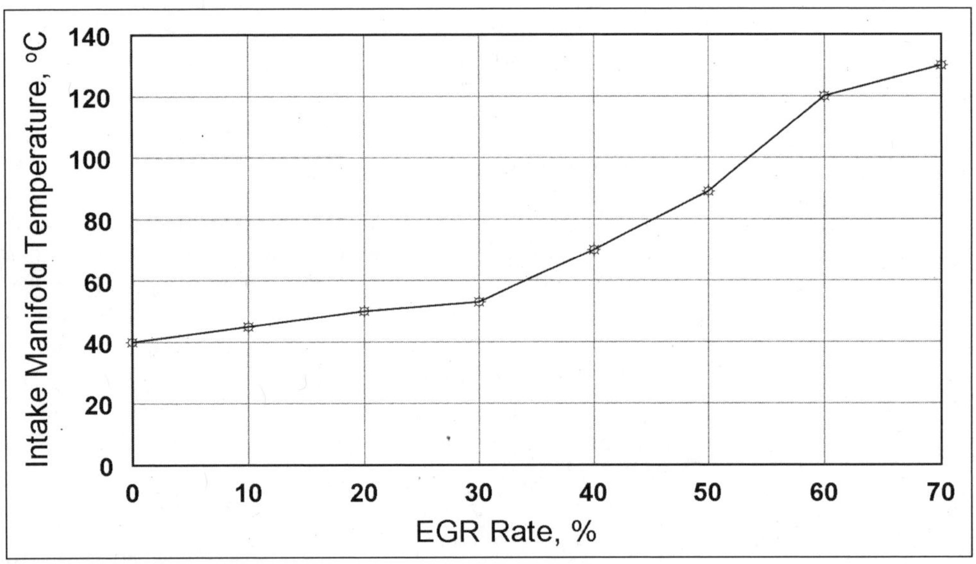

Figure 16.19 Effect of EGR rate on intake manifold temperature.

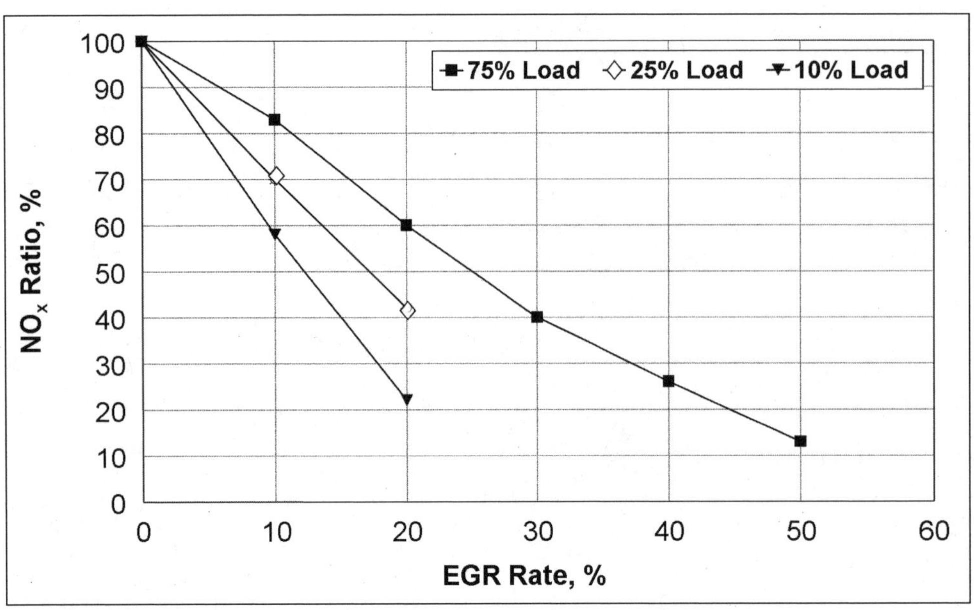

Figure 16.20 Effect of EGR rate on NO_X reduction at various engine loads.

of a study conducted on a heavy-duty diesel engine. It concluded that using EGR to reduce NO_X emissions from a baseline of about 4.0 to 2.8 g/bhp-hr was more efficient than achieving the same reduction through injection timing retard. Results of this study are illustrated in Figure 16.21, while Figure 16.22 describes the corresponding impact on PM emissions.

While the results of this study are specific to one heavy-duty engine with its sensitivity to either injection timing retard or a particular EGR configuration, they do illustrate a general preference for EGR as a system to reduce NO_X with a reasonable BSFC penalty and PM trade-off. Figure 16.23 underscores the importance of the A/F ratio in EGR system implementations and compares the NO_X reducing capability of injection timing retard versus two EGR configurations.

In Figure 16.23, the baseline NO_X/PM emissions of a medium-duty diesel engine were about 5.2/0.15 g/bhp-hr. Injection timing retard was adjusted to obtain the lowest possible NO_X and still maintain stable combustion. This effort resulted in 4.0 g/bhp-hr NO_X while PM emissions remained essentially unchanged. The same engine was fitted with a cooled HPL EGR system that enabled it to achieve NO_X as low as 2.0 g/bhp-hr (composite EPA FTP). Because this was done without revising the engine turbocharger, a serious A/F ratio drop was experienced, which in turn led to significant PM emission deterioration. The EGR configuration was then changed to a LPL EGR, which improved the A/F ratio and consequently the NO_X/PM trade-off. As indicated in Figure 16.23, the LPL EGR system was capable of reducing NO_X down to even 1.5 g/bhp-hr with

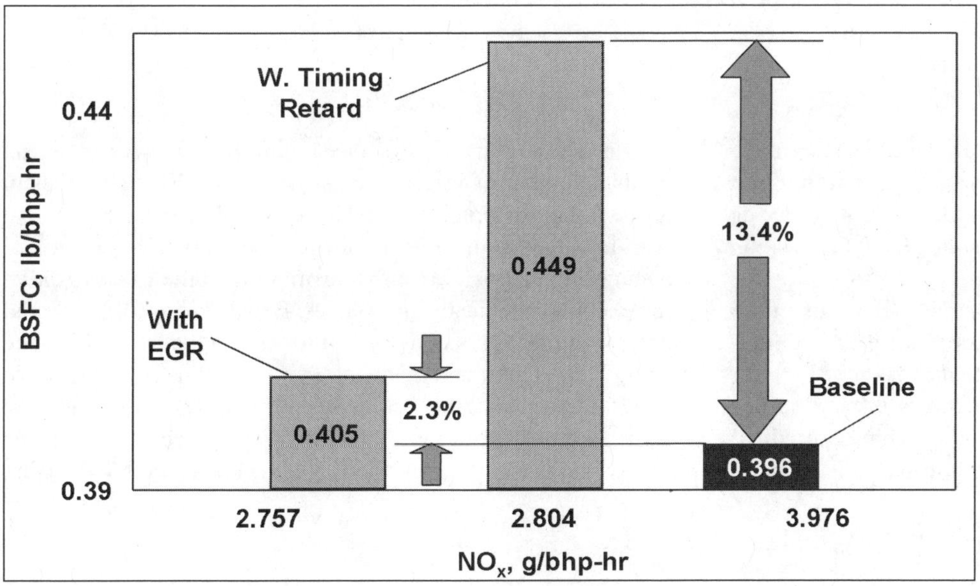

Figure 16.21 Effect of EGR versus injection timing retard on BSFC.

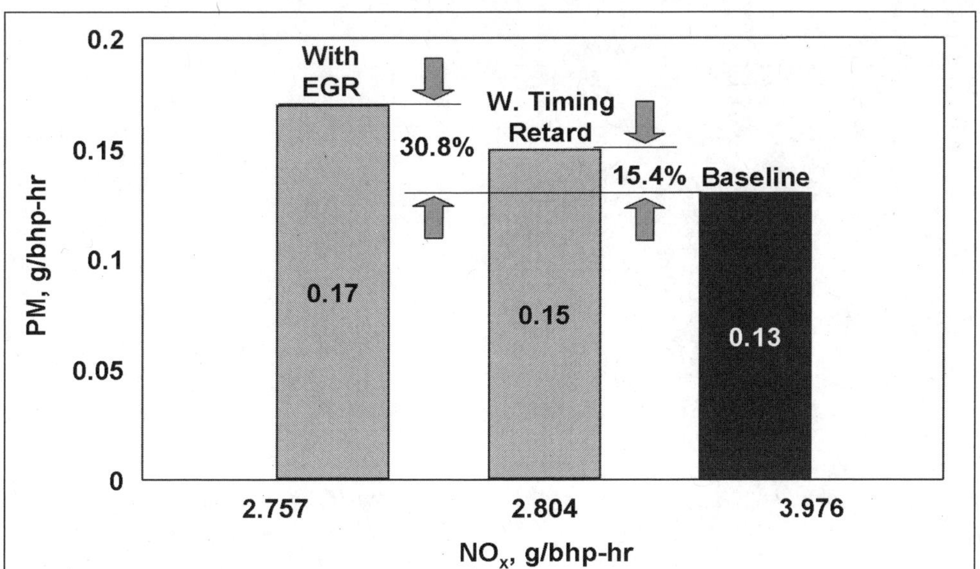

Figure 16.22 Effect of EGR versus injection timing retard on PM emissions.

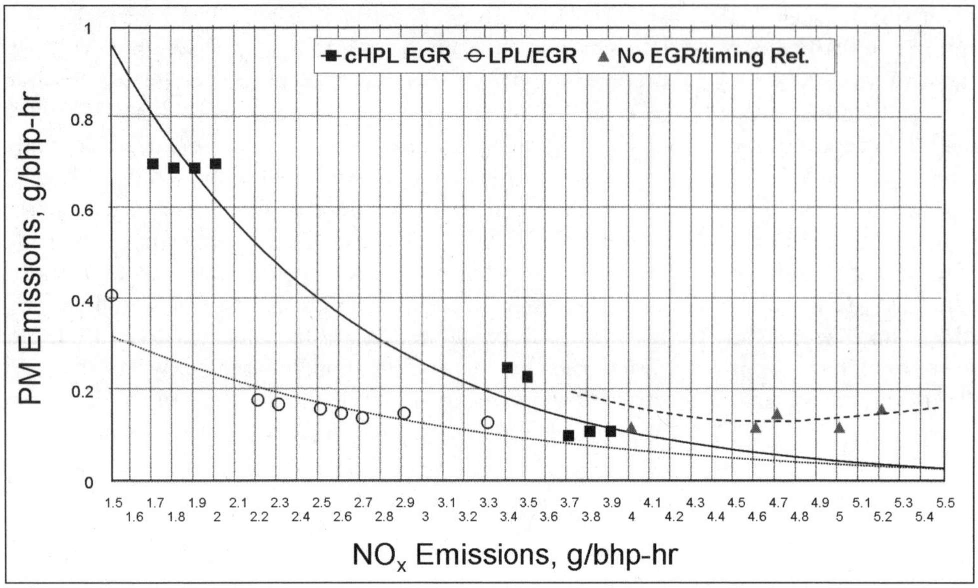

Figure 16.23 Effect of injection timing retard, cooled HPL EGR, and LPL EGR on NO_X/PM trade-off.

PM emissions in the 0.3 to 0.4 g/bhp-hr range versus more than 0.9 g/bhp-hr for the extrapolated PM concentration of the cooled HPL EGR configuration. It is worth noting that further A/F ratio optimization on this engine resulted in PM emissions of 0.105 g/bhp-hr.

Motivated by the need for A/F ratio control throughout the engine operating regime, several concepts were considered. The most common approach and one that is favored by both light- and heavy-duty engine manufacturers uses variable-geometry turbochargers. These devices are able to produce varying levels of boost pressure depending on their variable nozzle vane angular position (see Figure 16.24). An annular ring holds a number of variable vanes through individual pivots in an assembly that intercepts exhaust on its way from the exhaust manifold to the turbine wheel. An electronically controlled actuator is responsible for the positioning of the vanes to direct the exhaust onto the turbine blades. Exhaust manifold potential energy (pressure) is converted into varying degrees of kinetic energy (speed) depending on the angular position of the variable vanes. In response, the turbine speed increases as the vanes are turned toward their closed position and vice versa. As turbine speed increases, so does the resultant boost, thus increasing

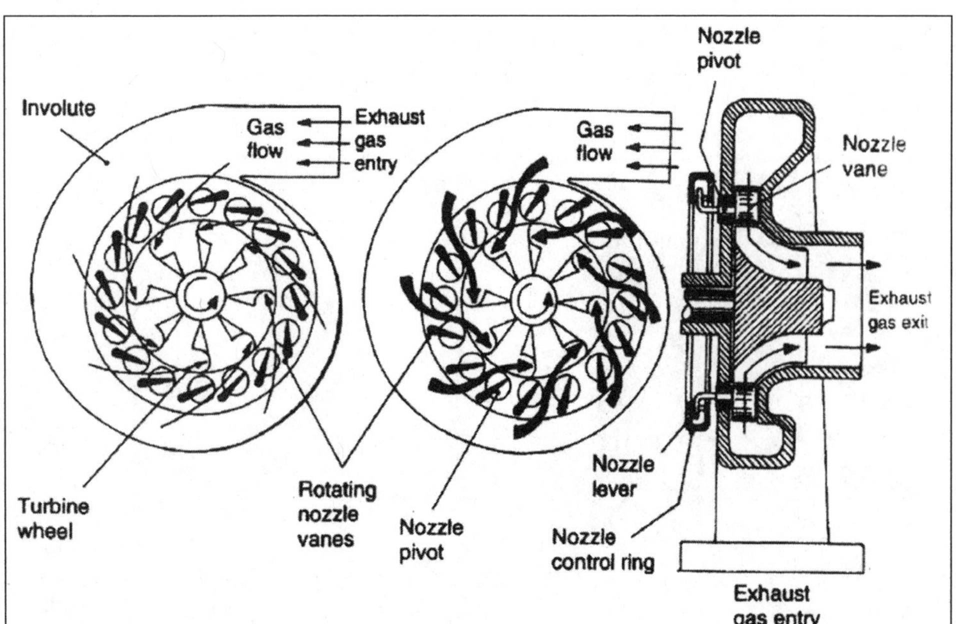

Figure 16.24 Schematic diagram of a variable-geometry turbocharger. (Reprinted from Ref. 4.)

airflow into the intake manifold. Hence, variable-geometry turbochargers (VGTs) can be used to modulate airflow at any engine speed and load condition to optimize the A/F ratio as a function of EGR rate.

In addition to controlling airflow, VGTs affect exhaust manifold pressure by virtue of the restriction caused by the position of the variable vanes. This feature can also be used as an added tool to control the pressure differential between the exhaust and intake manifolds, thus providing additional EGR rate control in conjunction with a fixed-orifice EGR valve. Of course, such a system provides more control flexibility, but requires more sophisticated control and calibration efforts. A well-designed EGR/VGT system can provide smoke-free operation at any speed and load condition by maintaining A/F ratios above a designated minimum threshold.

Recent developments in electrically and hydraulically driven superchargers have been rather helpful in providing additional control of A/F ratio. US Patents 5,771,868 (see Figure 16.25) and 5,927,075, describe A/F ratio control systems that operate in concert with EGR systems for the purpose of minimizing NO_X and PM emissions from diesel engines while improving engine response, acceleration smoke, and fuel economy. Both patents advocate the dual use of superchargers as air pumps as well as EGR pumps.

The system shown schematically in Figure 16. 25 is capable of controlling both EGR and power augmentation. Exhaust flows from the exhaust manifold (A) onto the turbine (B) of the turbocharger assembly and on through a diesel particulate filter (C) for particulate matter control. Exhaust gas is then split with one portion exiting the exhaust system (D) and the other recirculated back to the engine intake manifold. EGR passes through a heat exchanger (E) to reduce its temperature and thus maximize its NO_X reduction effectiveness. An electronically controlled diverter valve (F) serves as a junction point between post-filtered/cooled EGR (G) and fresh filtered charge air (H). Diverter valve F controls the charge mixture supplied to an electric motor-driven EDC (I). This mixture will depend on the engine working regime.

At a low idle speed and part-load conditions, diverter valve F would allow a mixture of fresh air and exhaust gas into the EDC. The proportions of that mixture would be determined by the NO_X reduction algorithm and the EGR rate needed for the desired NO_X control. For engine speeds below peak torque speed and as the load increases, the diverter valve would gradually reduce the proportion of EGR and increase that of the fresh air. With 100% fresh air allowed at full-load conditions, the EDC can be used in its power augmentation mode where excessive smoke can be averted and better fuel consumption can be observed. Thus, the combination of a diverter valve, such as valve F, and an EDC can control both EGR and smoke-limited power output at speeds below those of peak torque. Above peak torque speeds, the regular engine turbocharger is usually well matched to the engine charge air requirements for smoke-controlled operation. Therefore, the EDC (I) would be used for EGR rate control *only* for engine speeds above those of peak torque. Two check valves (K and L) are arranged to control the charge flow into the intake manifold and prevent backflow through the engine turbocharger or the EDC. Intake manifold charging would be through check valve K at speeds below peak torque speed, and through check valve L above peak torque speed when operating at full-load conditions. Implementation of this system on a medium-duty diesel engine led to smoke-free operation during hard acceleration (see Figure 16.26) as well as reduced NO_X by almost 50% in the EPA heavy-duty transient emission test.

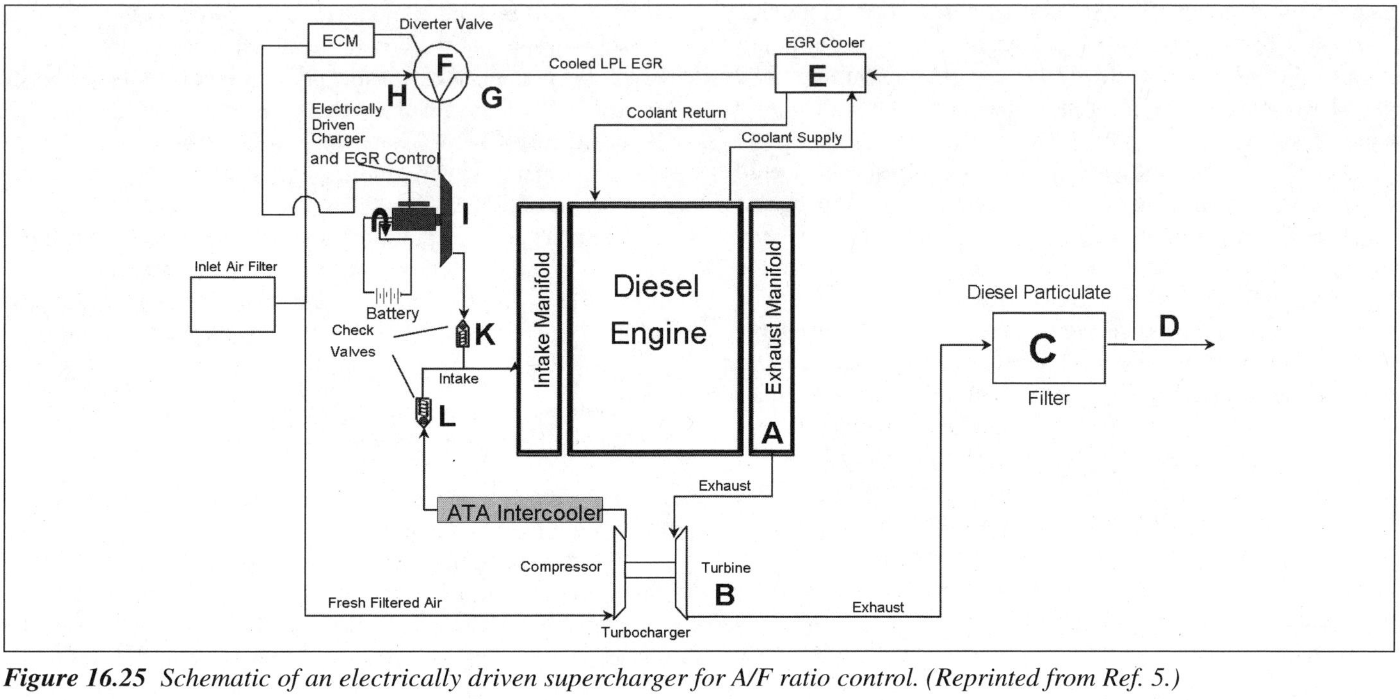

Figure 16.25 *Schematic of an electrically driven supercharger for A/F ratio control. (Reprinted from Ref. 5.)*

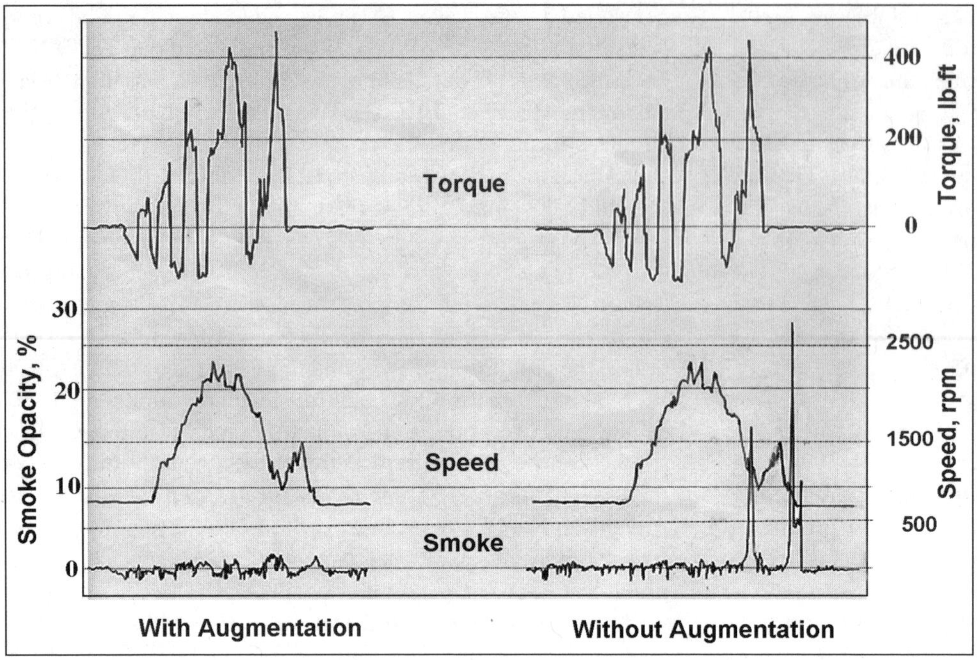

Figure 16.26 Smoke-free operation through A/F control.

16.8 Effect of EGR on Diesel Engine Wear and Durability

16.8.1 Introduction

Two of the diesel engine's strong attributes are its frugal use of fuel and its durability. These two features of the diesel engine have made it the premier powerplant for commercial transportation in industrialized as well as developing countries. In addition to this basic advantage, diesel engines emit extremely low levels of engine-out unburned hydrocarbon and carbon monoxide emissions. Among other good attributes that result from their superior fuel economy is their reduced net contribution to the global warming problem through low carbon dioxide (CO_2) emissions.

Having restated the case for diesel engines, one would hope that the public, regulators, and end users will recognize these good attributes and become determined to preserve them. In one sense, the move by the legislating bodies to tighten the limits on NO_x and PM emissions has contributed, more than anything else, to the new array of technologies now available to diesel engine manufacturers. A word of caution, however, is that any new technology adopted for the future of diesel engines, especially those used in heavy-duty markets, should not harm the engine's fuel economy superiority or its durability. Now that EGR has become a NO_x control feature on heavy-duty diesels, some are warning against its potential adverse effects on engine wear and durability. This concern has been the major motivator for many investigations researching the effect of EGR on engine wear and its inevitable consequence on engine durability. This segment of the chapter reports on the most recent work carried out in this area.

16.8.2 Background

When addressing the topic of engine wear, concern is immediately focused on the moving parts, such as pistons, ring pack, bearings, and shafts as well as their adjacent components such as cylinder liners, cams and camshafts, rocker arms, and the valve train. The concern is primarily associated with the mating surface condition when less-than-ideal lubrication is present due to the use of EGR. Having established the importance of lube oil quality to wear between moving parts, it is helpful to study the effect of EGR on lube oil quality. To be more specific, it would be useful to study how EGR would affect lube oil in view of its added soot content as a potential primary factor. Other points of interest are to determine how EGR would affect wear patterns and to identify whether some wear would lead to increases in lube trace metal content and the influence these trace metals may have on oil quality and its secondary effect on wear.

One of the more thorough studies conducted in this field was performed by Dennis *et al.* [6]. In their work, researchers from Perkins Engine Company and their colleagues at Loughborough University studied EGR effects on the following:

1. Oil condition
2. Cylinder parts, including piston and ring pack
3. General effect due to engine speed and load.

Their study also included an evaluation of two fuel sulfur levels and might be of special value.

Wear measurements and evaluations were conducted using irradiation techniques in which the engine components being observed were radiologically treated. The treatment targeted the surface of these components to a known depth, thus allowing standard operation of the engines to continue without major intrusions from instrumentation. The test engines were run in until their wear stabilized before starting the investigation.

16.8.3 Discussion

The effect of EGR on oil quality was based on periodic analyses of oil between changes. It highlighted iron content, soot loading, viscosity changes, oxidation, and changes in total base number (TBN). Figures 16.27 and 16.28 show the effect of EGR on iron and soot, respectively, in lube oil. It is clear that serious differences appeared after a little over 1,000 h into the experiment. It is worth pointing out that the EGR levels used in this test were about 28%, 16%, and 5% at 50%, 75%, and 100% load, respectively. In addition, the EGR system was a HPL configuration used initially without EGR cooling.

Examination of Figure 16.27 discloses that at each oil change the iron content dropped back to the same low levels. It was theorized that the reasons for the increased differences in iron and soot levels at extended service hours was because wear was sufficient enough to allow more soot to be transported into the sump. It is also thought that more soot had migrated between the piston rings and the liner, leading to excessive wear that might explain the higher lube oil iron content.

Higher EGR concentrations that might be required for lower NO_x emissions in the future are expected to further aggravate lube oil soot concentrations. This situation will also lead to additional iron in the lube oil since the mechanism for

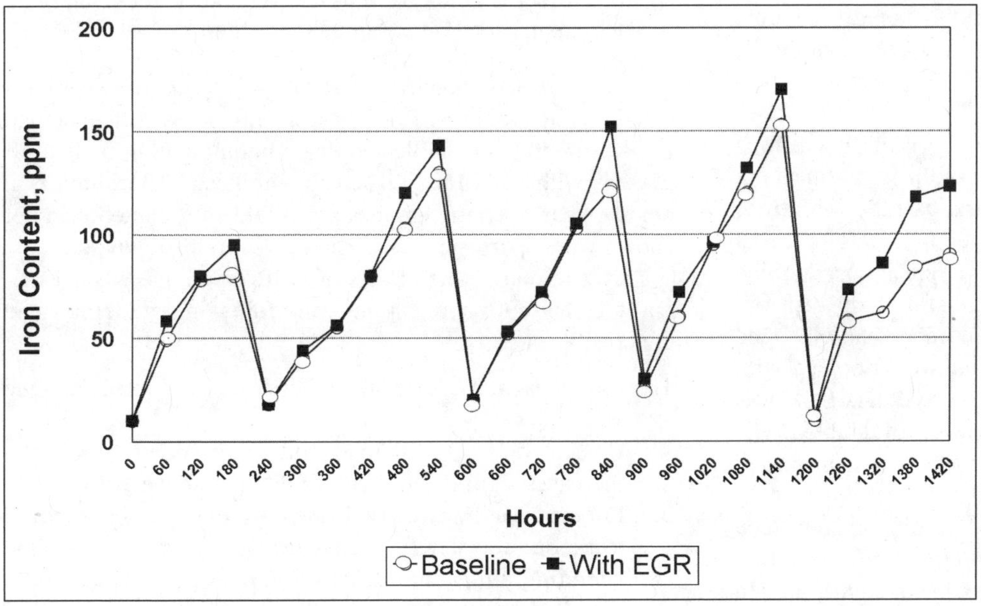

Figure 16.27 *Comparison of iron content in lube oil. (Adapted from Ref. 6.)*

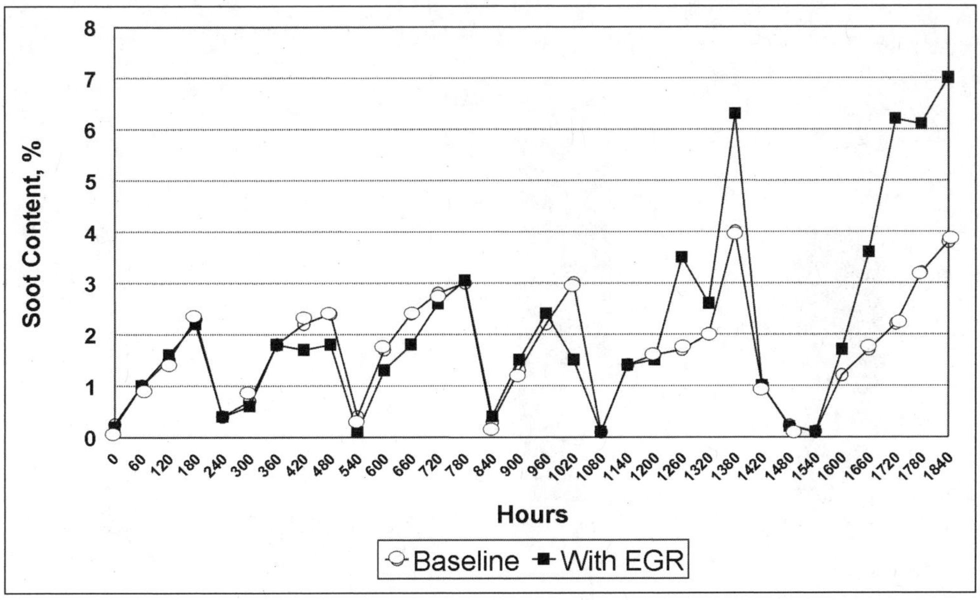

Figure 16.28 *Comparison of soot content in lube oil. (Adapted from Ref. 6.)*

liner and piston ring wear will increase in intensity and produce more wear. However, cleaner engines in terms of engine-out soot concentrations should alleviate this problem somewhat. With the modern trend toward higher injection pressure (above 2,000 bar), combustion efficiency should improve and, thus, the balance between PM formation and its oxidation should be favorable to engine-out soot control. It is interesting to note from the study that differences in lube oil viscosity, oxidation, and TBN between the engine operating with and that operating without EGR were not significant.

Tests were also conducted at steady-state, rated speed, and full-load conditions using 0%, 10%, and 15% EGR concentrations and two lube oil conditions: aged and fresh. These tests disclosed that with fresh oil, ring wear showed no increase when EGR was increased from 0% to 10%. However, an increase in ring wear of about 30% was registered when EGR was increased from 10% to 15%. With aged oil, ring wear showed significant increases. Even without EGR, aged oil led to more than double the ring wear of fresh oil, increasing by about 16% with EGR increasing to 10%, and by 48% when EGR was increased from 10% to 15% (see Figure 16.29).

Liner wear was also assessed for the same conditions described for the ring wear tests. Results of this assessment are shown graphically in Figure 16.30 and indicate that with fresh oil, the liner wear rate was unchanged when EGR was increased from 0% to 10%. However, the liner wear rate almost doubled when EGR was increased from 10% to 15% with fresh oil. Significant deterioration was observed in liner wear with aged oil. As in the case of ring wear, the liner wear doubled with aged oil from its rate with fresh oil at 0% EGR. As EGR was increased to 10%, the liner wear rate more than doubled. The same effect was again experienced when EGR was increased from 10% to 15%.

It is interesting to note from the study that differences in lube oil viscosity, oxidation, and TBN between the engine operating with and that operating without EGR were not significant. In other words, except for the higher iron and soot levels in the lube oil, no other difference was noted in lube oil characteristic parameters.

Further investigations showed a significant correlation between ring and liner wear and carbon concentration in the lube oil. Three mechanisms were thought of whereby carbon would cause increased wear:

1. Carbon can absorb lube oil antiwear additives that help maintain its ability to reduce friction and wear when exposed to heat for extended periods.
2. Carbon may have some abrasive quality that removes the antiwear layer from parts experiencing relative motion and friction.
3. Increased in-cylinder carbon may increase the acidity of local lube oil, thus increasing its inability to protect ring and liner surfaces from corrosive wear. It is suggested that this effect is neutralized as carbon comes in contact with the greater lube oil volume in the sump.

Additional work on the effect of EGR on engine wear indicated that uncooled EGR led to more engine wear than cooled EGR. Research also indicated that combining high fuel sulfur content with aged lube oil led to significant deterioration in engine wear, even though the same fuel sulfur content had not shown worse performance when used with fresh lube oil.

To summarize the effects of EGR on engine wear (especially heavy-duty applications), the following guidelines and conclusions are given:

1. Engine wear increases with increased EGR rates and lube oil age.
2. Engine wear at full load and rated speed conditions increases significantly with EGR rates above 10%.
3. Engine wear increases with uncooled EGR. This is believed to be due to increased soot formation, increased engine operating temperatures, and a thinner oil film thickness between the liner and ring.

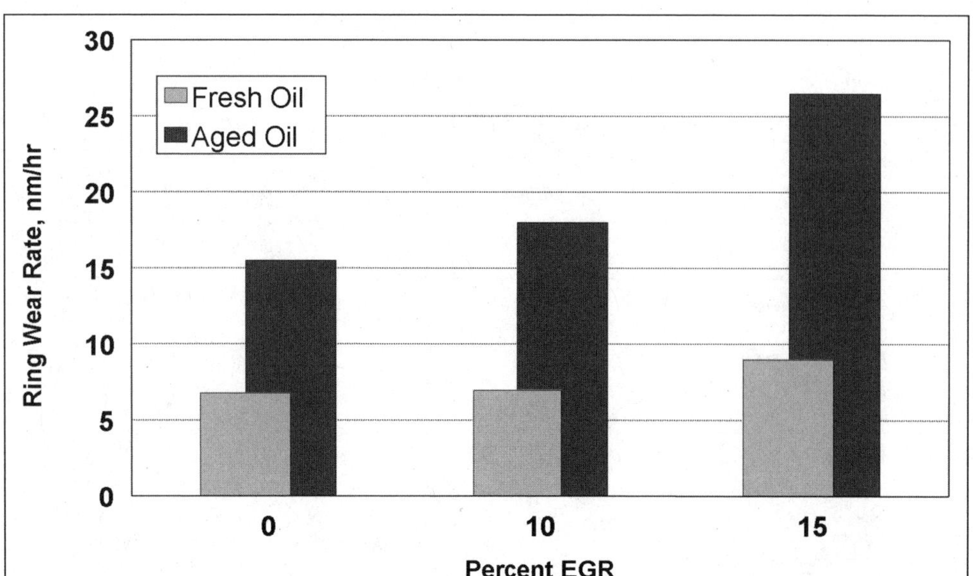

Figure 16.29 Ring wear rate for three EGR levels and two lube oil conditions.

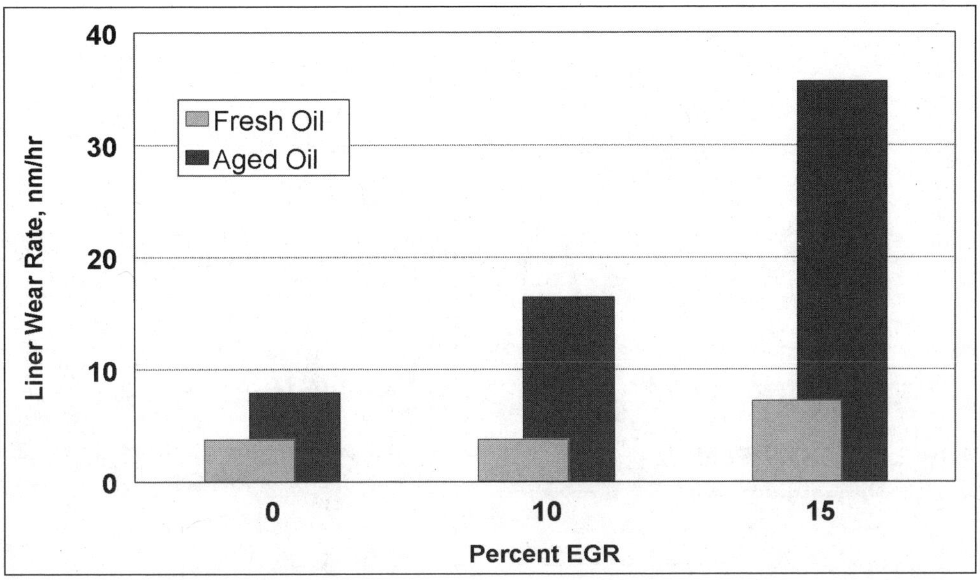

Figure 16.30 Liner wear rate for three EGR levels and two lube oil conditions.

4. Lube oil aging is greatly affected with exposure to carbon, which increases with higher EGR rates.
5. Use of EGR did not seem to adversely impact engine wear at part load, where A/F ratios are usually high.
6. Carbon carried with EGR was identified as the main reason for lube oil quality degradation and engine wear.
7. The combination of high fuel sulfur, aged lube oil, uncooled EGR, high EGR rate, high engine speed and load, and high engine coolant temperature is most severe and detrimental for engine wear and long-term durability.

16.9 Alternative In-Cylinder Technologies to EGR for Low NO_X in Diesel Engines: A Review

Many technologies for NO_x reduction have been considered during the last two decades, including :

- Pilot injection, and injection rate shaping [7–9]
- Intake charge air cooling [10–13]
- Injection timing retard [10–13]
- Water emulsion/injection [14, 15].

Some other technologies were adopted for their impact on PM reduction rather than NO_x reduction. However, note that reducing the PM emissions provided engineers with more freedom in controlling NO_x than would otherwise have been possible. For instance, increasing the injection pressure led to reducing PM emissions through better mixing and higher air utilization. In the meantime, the higher injection pressure reduced ignition delay and allowed engineers additional injection timing retard, leading to lower NO_x emissions without the anticipated fuel economy loss usually associated with injection timing retard.

16.9.1 Effect of Pilot Injection and Injection Rate Shaping

Researchers at Nagasaki University have shown that pilot injection reduced ignition delay by 50% from that of the conventional injection systems [7]. This ignition delay reduction allowed more injection timing retard with good combustion stability. Significant reductions in flame temperature, NO_x, and fuel consumption were also reported by the same researchers. These improvements were attributed to a slower rate of combustion pressure rise, resulting in better mechanical efficiency and lower cooling loss. Their work, however, documented an increase in smoke emissions and pointed to the need for flexible control of pilot injection timing and quantity as a function of engine speed and load. Similar results were reported earlier by researchers from the ACE Institute Co., Ltd., and JARI, Inc., in Japan [8]. These researchers achieved 35% NO_x reduction and 60% to 80% smoke reduction simultaneously without a fuel consumption penalty. They achieved these results by combining pilot injection with high injection pressures.

Pilot or staged injection research and development results indicated that an optimum injection strategy would be a slow injection rate (small pilot) for NO_x control. The latter portion of the injection would be characterized by a fast injection rate for particulate control, where most of the fuel would be injected at the highest combustion temperature near top-dead-center (TDC). Work by researchers at the University of Wisconsin, Madison [9], using multiple injections (more than two), generally supported the preceding conclusions. Figures 16.31 and 16.32 illustrate the effects of low injection rate on NO_x reduction and the adverse impact of higher rates after the bulk of the fuel had released its heat, respectively.

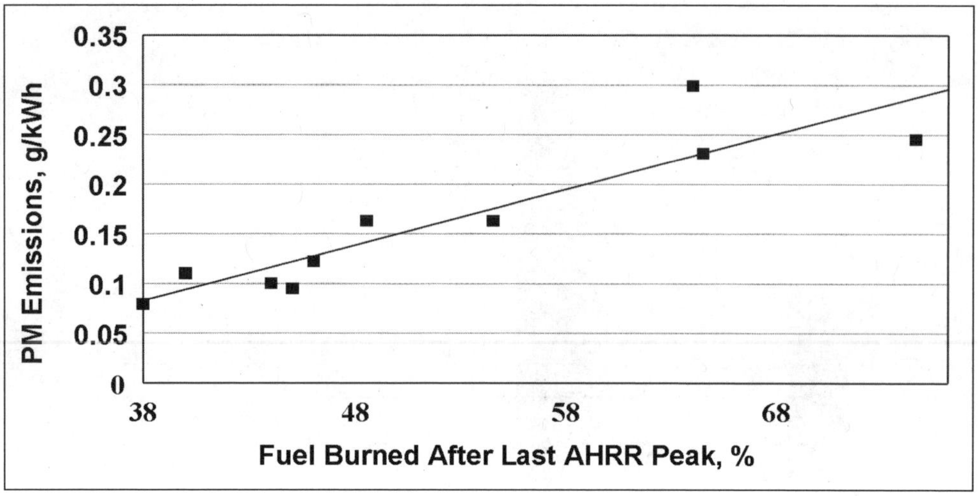

Figure 16.31 *Relationship between PM and the quantity of fuel injected after peak apparent heat release rate. (Adapted from Ref. 9.)*

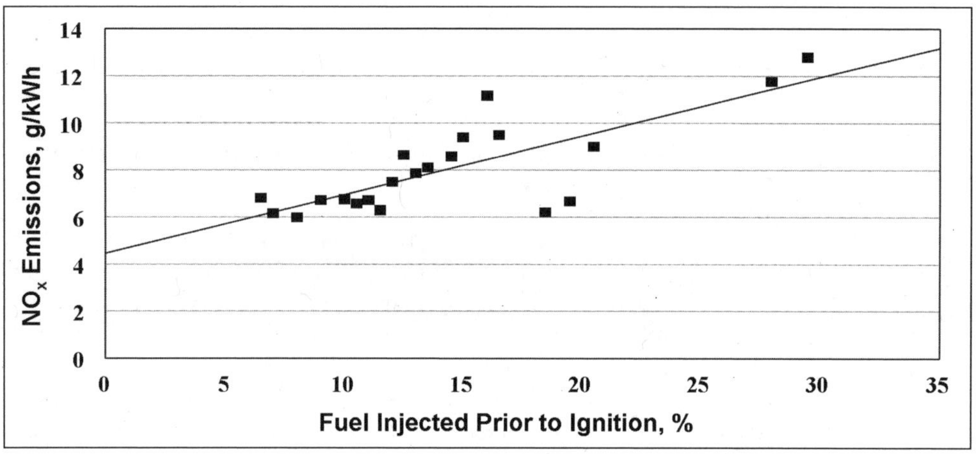

Figure 16.32 *Relationship between NO_X and the quantity of fuel injected prior to ignition at 75% engine load. (Adapted from Ref. 9.)*

16.10 Effect of Intake Charge Air Cooling

Most if not all published data agree that for high specific output diesel engines (high BMEP), there is a beneficial effect for lower temperature charge air [10–12]. Many engine manufacturers replaced their water-jacket systems with air-to-air intercooled systems [10, 13]. For a typical North American 7.5-L, turbocharged diesel engine, a fuel consumption improvement of about 2.5% was seen when the intake charge air temperature was reduced from 200°F to 120°F (93°C to 49°C) [13]. Meanwhile, a NO_x reduction of 17.5% was reported at the same injection timing [13]. Figure 16.33 gives the effect of rated speed intake manifold temperature on EPA heavy-duty transient test NO_x emissions.

16.11 Effect of Injection Timing Retard

The effect of retarding injection timing on NO_x emission in diesel engines has been well documented [11–13]. Retarding injection timing by an 8° to 10° crankangle was found to help most heavy-duty DI diesel engines to meet the 1991 NO_x standards of 5.0 g/bhp-hr (6.8 g/kWh) from their actual pre-1991 calibrations of 9 to 10.5 g/bhp-hr (12.2 to 14.2 g/kWh). Adverse NO_x emission behavior was noted, however, where injection timing was extended beyond this margin [12]. This phenomenon was attributed to an increase in the premixed fuel fraction due to a significant increase in ignition delay. Figure 16.34 ties together the influence of both injection timing retard and intake charge air cooling on NO_x emissions.

16.12 Issues Related to EGR Control Systems

EGR systems would invariably include one or more control valves and one or more EGR coolers. The remainder of the EGR control system consists of piping, flanges, and gaskets. With sulfur present in diesel fuel, it is normal to find sulfur dioxide and sulfuric acid in the exhaust. The same exhaust stream is also laden with soot, unburned oil, unburned fuel, and other trace metals. These exhaust constituents may cause erosion in the EGR system components; therefore, the challenge is to select, design, and develop reliable and trouble-free EGR systems. The following subsections outline some of the issues to be addressed when designing these systems.

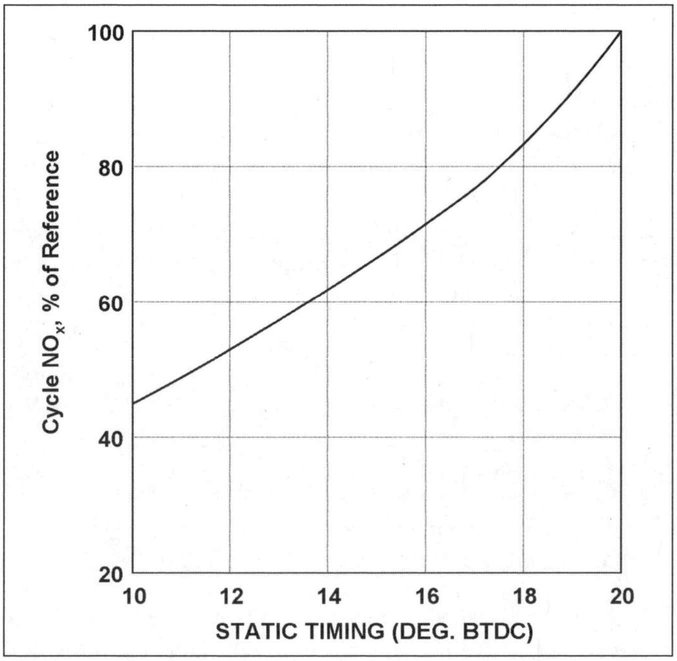

Figure 16.33 *Effect of charge air temperature on NO_x emissions. (Adapted from Ref. 13.)*

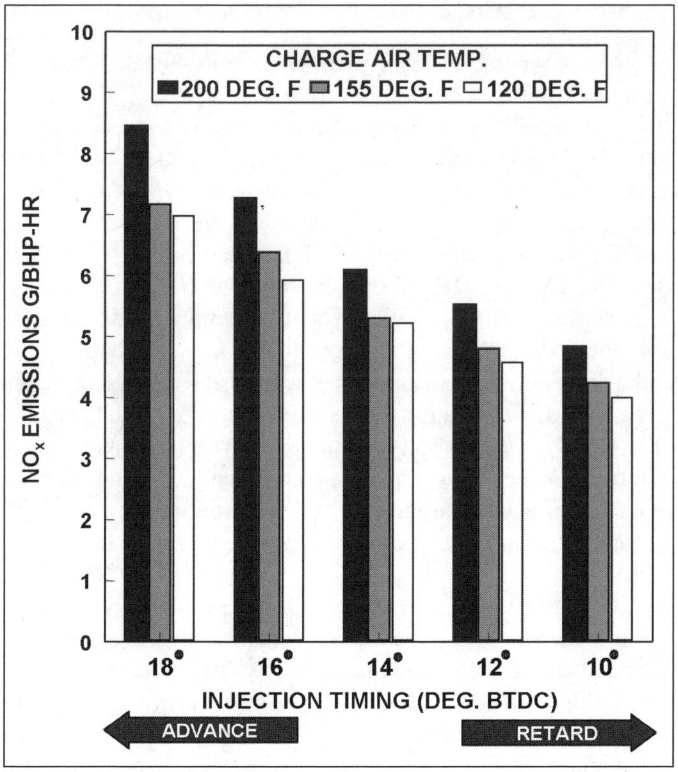

Figure 16.34 *Combined effect of charge air temperature and injection timing on NO_x emissions (Reprinted with permission from Ref. 13 .)*

16.12.1 Material Buildup within EGR Systems

Carbonaceous material deposited on the walls of the EGR system pipes, coolers, and valves reduces the capacity and efficiency of the system. Carbonaceous deposits restrict EGR flow, thus reducing its heat-absorbing capacity during combustion and its ability to reduce NO_X. In addition, these deposits act as insulation, preventing efficient heat transfer from the exhaust gas to the cooling medium. As heat transfer from the exhaust gas to the cooling medium is reduced, the temperature of the gas entering the combustion chamber will be higher than desired. This higher temperature will reduce the ability of the gas to absorb combustion heat and reduce NO_X emissions.

16.12.2 EGR Contaminants

The presence of sulfur in the EGR system in the form of sulfuric acid must be viewed as a major threat to the integrity of its components. A very important design consideration is the material selection of the EGR cooler. Erosion in any of the tubes would introduce water (cooling medium) to the engine and perhaps cause a catastrophic failure.

16.12.3 Engine Durability

Introducing unfiltered exhaust gas into the intake manifold, and eventually into the combustion chamber, may cause accelerated deterioration of the cylinder bore and piston rings. The LPL EGR system equipped with a trap has a clear advantage over the HPL EGR in this regard. However, attention must be exercised to avoid water condensation as EGR is introduced upstream of the turbocharger compressor. Extensive testing at various speeds and loads for a given engine should be conducted to determine those regimes that lead to adverse conditions. Excessive EGR cooling may be desirable for NO_X control, but could lead to water condensation that would reduce the turbocharger's efficiency and life.

16.12.4 Application/Test Cycle

Effort should be expended in simulating various applications through a carefully designed experiment in a controlled environment. High exhaust gas temperatures coupled with maximum EGR cooling may represent an extreme condition and establish a design target for EGR coolers. However, a large-capacity cooler may not be desirable at idle and low loads, especially in extreme cold weather applications. These two extremes present designers with a challenge to optimize the EGR cooler heat transfer capacity or design a cooler bypass as needed.

16.12.5 EGR Cooler Design

As mentioned earlier, two mechanisms come to mind when explaining the effectiveness of EGR in NO_X reduction. The first mechanism is the dilution of the amount of oxygen that is normally introduced with fresh air. The second mechanism is the use of EGR as a heat sink. The capacity of the latter aspect is generally proportional to the EGR rate, its specific heat, and the temperature differential between EGR and peak combustion temperature. It follows then that the colder the EGR, the more effective it would be for NO_x reduction. However, as mentioned earlier, there is a risk of forming water condensation and introducing it into the compressor of the turbocharger. On the other hand, a small EGR cooler capacity may not be as effective in NO_x reduction. The control strategy should be a major determinant of when and how EGR/cooling is to be used.

16.12.6 EGR System Packaging

For medium- and heavy-duty diesel engine applications, packaging for durability and long-term reliability must be extremely high on the priority list. Several EGR valve design configurations are being considered by diesel OEM suppliers. The first is a design in which the EGR valve is integral to the intake manifold. Another design has the EGR valve installed in the piping of the system. Several entities are considering integrating the EGR cooler with the engine oil cooler. Others are considering an independent EGR valve installation separate from any other system or engine components. Some EGR valves are of the poppet style, while others have adopted a rotary-type design. Each of these designs has its strong points as well as some shortcomings. Rigorous durability/reliability testing should be performed prior to introducing these packaging alternatives to the diesel engine market.

Another significant packaging feature is the location of the EGR cooler in the total package. In principle, the EGR cooler can be located anywhere between the source of the recirculated exhaust and the point of its introduction into the engine. However, mounting the EGR cooler closer to either of these two locations would have to be investigated to determine the best trade-off between efficiency in NO_x reduction, space availability, and the propensity to form or avoid water condensation.

16.12.7 EGR Valve Control Considerations

Automotive style EGR valves are mechanical components designed to control EGR flow mainly to achieve emissions control. In recent developments associated with medium- and heavy-duty diesel work, more emphasis has been placed on EGR control flexibility. For this reason, electronic controls and solenoid-actuated EGR valves are fulfilling this need. The proximity of electrical solenoids to hot exhaust has been identified as a problem and a limitation for valve performance and durability. The solution to this problem may involve relocating these valves to a cooler environment, such as closer to the engine intake manifold. Assessment of the advantages of such a solution should be carried out and design changes should be proposed and investigated.

16.12.8 Piping

The material, size, and route of the EGR system piping will play a significant role in total system heat and efficiency losses. This fact is especially important in the case of the LPL configuration in view of the length of pipe extending from the exhaust side to the inlet of the turbocharger. Although the preceding is not necessarily a comprehensive list of system design-related issues, it does represent a significant selection of the more important concerns that require special attention and effort. The effort is large enough to be supported by various industries and suppliers such as:

- EGR valve suppliers
- EGR cooler suppliers
- Engine OEMs
- Exhaust systems suppliers
- Turbocharger manufacturers
- Oil Companies

and others. It is felt that current and future production EGR systems will be well developed and will have durability and reliability features commensurate with heavy-duty, diesel engine applications.

16.13 Demonstration of Low NO_x and PM Using EGR and a Passive Trap System

A demonstration using two heavy-duty North American diesel engines was carried out at SwRI as part of an internal research project. This demonstration was performed using a 5.9-L, B-Series Cummins and an 11-L, Series 60 Detroit Diesel heavy-duty diesel engines. A Corning wallflow EX-80, 10.5-inch-diameter × 12-inch-long cordierite trap was used. The catalyst was a cerium-based fuel additive supplied by Rhône-Poulenc, Inc. [16–19]. Extensive development and applied research work led to a PC-based controller and an EGR strategy to reduce NO_x emissions to between 2.0 and 2.5 g/bhp-hr (2.7 and 3.4 g/kWh) on both engines. Analysis of the PM sample filters assisted in estimating the carbonaceous PM content typical of an EPA heavy-duty emission transient cycle. A corresponding additive concentration was then recommended by the supplier to maintain quasi-continuous trap regeneration.

Exhaust gas recirculation was implemented in a high-pressure loop configuration. A schematic diagram of the HPL EGR system is given in Figure 16.1 and Figure 16.4, where EGR is taken from a point upstream of the turbocharger turbine and introduced into the intake manifold. Results of hot-start, EPA heavy-duty transient emissions tests and cycle average specific fuel consumption for the Series 60 and the B-Series engines are tabulated in Tables 16.1 and 16.2, respectively. Starting from NO_x levels of 3.89 and 4.67 g/bhp-hr (5.3 and 6.35 g/kWh) for the Series 60 and B-Series engines, respectively, both engines were successful in achieving 2.02 and 2.29 g/bhp-hr (2.73 and

TABLE 16.1
SERIES 60 TRANSIENT EMISSIONS RESULTS

Test Description	Emissions, g/bhp-hr				BSFC, lb/bhp-hr
	HC	CO	NO_X	PM	
Baseline	0.08	1.19	3.89	0.085	0.400
Inactive HPL EGR	0.09	1.30	3.82	0.085	0.398
Active HPL EGR	0.25	2.91	2.36	0.204	0.403
Active HPL EGR + DPF	0.27	4.26	2.29	0.014	0.416

TABLE 16.2
B-SERIES TRANSIENT EMISSIONS RESULTS

Test Description	Emissions, g/bhp-hr				BSFC, lb/bhp-hr
	HC	CO	NO_X	PM	
Baseline	0.55	2.27	4.67	0.14	0.424
Active HPL EGR	0.83	4.20	2.58	0.35	0.421
Active HPL EGR + trap	1.36[a]	6.60	2.02	0.06	0.447

a. High hydrocarbons and CO attributed in part to elevated engine backpressure resulting from restricted trap.

3.10 g/kWh) NO_X emissions. In both cases, PM engine-out emissions had more than doubled (from 0.085 to 0.204 g/bhp-hr for the Series 60 and from 0.14 to 0.35 g/bhp-hr for the B-Series). However, using a trap and cerium-based fuel additive, these PM levels were reduced to 0.014 and 0.060 g/bhp-hr (0.02 and 0.08g/kWh), respectively. FTP cycle-weighted average fuel consumption increased by 4% for the Series 60 engine and by 5.5% for the B-Series engine. Both engines experienced significant increases in hydrocarbon and CO emissions. Although EGR contributed to this increase, the presence of the trap appeared to have added to the problem. This development was especially noticeable in the case of the B-Series engine with HC emissions where an older trap already laden with carbonaceous material was used without regenerating it prior to the test. For the same reason it appeared that NO_X emissions were exceptionally low as a result of the increased back pressure, which must have led to additional residual (internal) EGR.

Another demonstration program, using SwRI's internal research funds, was performed on a Navistar 7.3-L, heavy-duty diesel engine. This program involved HPL EGR with a Johnson Matthey CRT. The Navistar engine was a V-8, turbocharged, intercooled, direct-injected diesel engine equipped with a hydraulically actuated, electronically controlled unit injector system (HEUI). The CRT was tested originally on DDC, 6V92TA, two-stroke diesel engines to investigate the potential retrofit market. Permission to use the CRT was received from Johnson Matthey in November 1995 prior to its test on the Navistar engine.

The results of this demonstration were extremely encouraging and, as indicated in Figure 16.35, both HC and PM emissions were practically eliminated. Carbon monoxide emission was reduced to a trace, and NO_X + HC for the transient test was 2.29 g/bhp-hr. When tested in a LPL EGR configuration, the transient cycle average fuel consumption was within 2% from that of the baseline engine without EGR or trap.

These demonstrations were cited here to indicate the magnitude of emission reduction EGR can achieve in heavy-duty diesel engine applications. Realizing that one of the adverse consequences of using EGR is increasing particulate emissions, diesel particulate filter technologies were included for their synergistic contribution in reducing particulate emissions. From these sample EGR applications, it is apparent that NO_X reduction on the order of 40% to 50% can be achieved with modest penalties in particulate emission control that would require DPFs. The fuel consumption penalty for a well-developed EGR system is expected to disappear as development work continues to further improve EGR system design and control. Other alternatives to in-cylinder control of NO_X by means of EGR have also been extensively investigated. These alternatives involve the use of auxiliary emission control devices such as lean NO_X catalysts and traps, as well as selective catalytic reduction systems that use either ammonia or urea as reductants. Several references are cited for additional reading [18, 20–26]; however, this subject is extensively covered in Part V of this book.

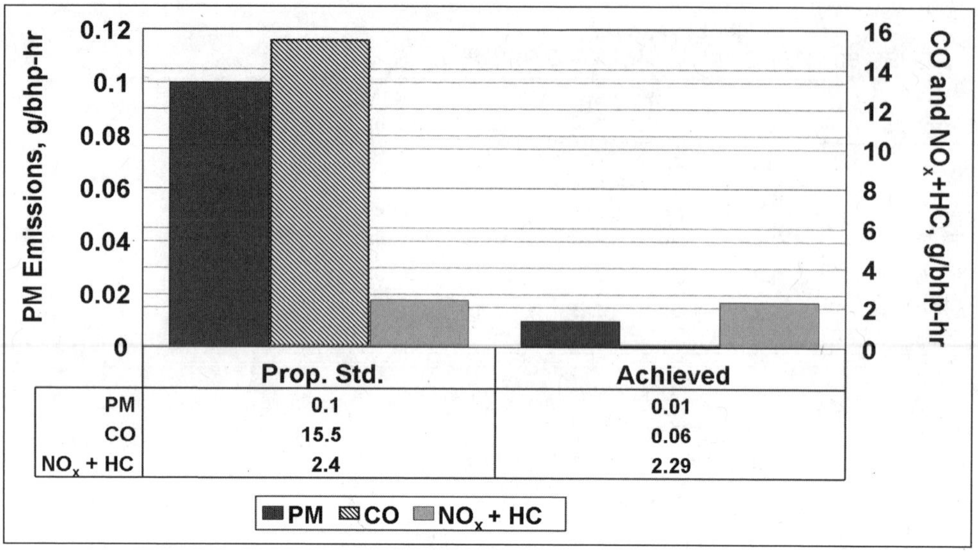

Figure 16.35 Demonstration results of Navistar engine with EGR-CRT technologies.

REFERENCES

1. Chapter on superchargers, in *Automotive Handbook,* 2nd ed., R. Bosch GmbH, Germany, 1986, pp. 355–359.
2. Arcoumanis, C., Schindler, J.-P., "Mixture Formation and Combustion in the DI Diesel Engine," SAE Paper No. 972681, SAE International, Warrendale, PA, 1997.
3. Arcoumanis, C., Nagwaney, A., Hentschel, W., Rarpke, S., "Effect of EGR on Spray Development, Combustion and Emissions in a 1.9L Direct-Injection Diesel Engine," SAE Paper No. 952356, SAE International, Warrendale, PA, 1995.
4. Hawley, J.G., Wallace, F.W., and Cox, A., Pease, A.C., Horrocks, R.W., Bird, G., "Reduction of Steady-State NO_x Levels from an Automotive Diesel Engine Using Optimized VGT/EGR Schedules," SAE Paper No. 1999-01-0835, SAE International, Warrendale, PA, 1999.
5. Khair, M., "Turbocharging Systems for Internal Combustion Engines," U.S. Patent No. 5,771,868, June 30, 1998.
6. Dennis, A.P., C.P. Garner, and D. Taylor, "The Effect of EGR on Diesel Engine Wear," SAE Paper No. 1999-01-0839, SAE International, Warrendale, PA, 1999.
7. Ishida, M., Z.-L. Chen, G.-F. Luo, and H. Ueki, "The Effect of Pilot Injection on Combustion in a Turbocharged D.I. Diesel Engine," SAE Paper No. 941692, SAE International, Warrendale, PA, 1994.
8. Shundoh, S., K. Masanori, K. Tsujimura, and S. Kobayashi, "NO_x Reduction from Diesel Combustion Using Pilot Injection with High Pressure Fuel Injection," SAE Paper No. 920461, SAE International, Warrendale, PA, 1992.
9. Tow, T., D. Pierpont, and R. Reitz, "Reducing Particulate and NO_x Emissions by Using Multiple Injections in a Heavy Duty D.I. Diesel Engine," SAE Paper No. 940897, SAE International, Warrendale, PA, 1994.
10. Khair, M., "Air-to-Air Intercooling of the Ford 7.8L Mid-Range Truck Diesel Engine," SAE Paper No. 870534, SAE International, Warrendale, PA, 1987.
11. Gill, A., "Design Choices for 1990's Low Emission Diesel Engines," SAE Paper No. 880350, SAE International, Warrendale, PA, 1988.
12. Herzog, P., L. Burgler, E. Winklhofer, P. Zelenka, and W. Cartellieri, "NO_x Reduction Strategies for DI Diesel Engines," SAE Paper No. 920470, SAE International, Warrendale, PA, 1992.
13. Khair, M., "Progress in Diesel Engine Emission Control," Paper 92-ICE-14, The American Society of Mechanical Engineers, New York, 1992.
14. Tsukahara, M., and Y. Yoshimoto, "Reduction of NO_x, Smoke, BSFC, and Maximum Combustion Pressure by Low Compression Ratios in a Diesel Engine Fueled by Emulsified Fuel," SAE Paper No. 920464, SAE International, Warrendale, PA, 1992.
15. Melton, R., Jr., S. Lestz, R. Quillian, Jr., and E. Rambie, "Direct Water Injection Cooling for Military Engines and Effects on the Diesel Cycle," SAE Paper No. 750129, SAE International, Warrendale, PA, 1975.
16. Lemaire, J., and M. Khair, "Effect of Cerium Fuel Additive on the Emissions Characteristics of a Heavy-Duty Diesel Engine," SAE Paper No. 942067, SAE International, Warrendale, PA, 1994.
17. Dementhon, J., B. Martin, P. Richards, M. Rush, D. Williams, L. Bergonzini, and P. Morelli, "Novel Additive for Particulate Trap Regeneration," SAE Paper No. 952355, SAE International, Warrendale, PA, 1995.
18. Hammerle, R., D. Ketcher, R. Horrocks, G. Lepperhoff, G. Huthwohl, and B. Luers, "Emissions from Diesel Vehicles with and without Lean NO_x and Oxidation Catalysts and Particulate Traps," SAE Paper No. 952391, SAE International, Warrendale, PA, 1995.

19. Lepperhoff, G., H. Luders, P. Barthe, and J. Lemaire, "Quasi-Continuous Particle Trap Regeneration by Cerium-Additives," SAE Paper No. 950369, SAE International, Warrendale, PA, 1995.
20. Sumiya, S., G. Muramatsu, N. Matsumura, K. Yoshida, and R. Schenck, "Catalytic Reduction of NO_x in Diesel Exhaust," SAE Paper No. 920853, SAE International, Warrendale, PA, 1992.
21. Feeley, J., M. Deeba, and R. Farrauto, "Abatement of NO_x from Diesel Engines: Status and Technical Challenges," SAE Paper No. 950747, SAE International, Warrendale, PA, 1995.
22. Nakahira, T., K. Tsujimura, T. Kakegawa, K. Aika, M. Takiguchi, M. Ikeda, and S. Ichihara, "Catalytic Engine NO_x Reduction of Diesel Engines with New Concept Onboard Ammonia Synthesis System," SAE Paper No. 920469, SAE International, Warrendale, PA, 1992.
23. Luders, H., R. Backes, G. Huthwohl, D. Ketcher, R. Horrocks, R. Hurley, and R. Hammerle, "An Urea Lean $_{NOx}$ Catalyst System for Light Duty Diesel Vehicles," SAE Paper No. 952493, SAE International, Warrendale, PA, 1995.
24. Kawanami, M., M. Horiuchi, E. Lox, J. Leyrer, and D. Psaras, "Advanced Catalyst Studies of Diesel NO_x Reduction for On-Highway Trucks," SAE Paper No. 950154, SAE International, Warrendale, PA, 1995.
25. Kharas, K., and J. Theis, "Performance Demonstration of a Precious Metal Lean NO_x Catalysts in Native Diesel Exhaust," SAE Paper No. 950751, SAE International, Warrendale, PA, 1995.
26. Rao, V., H. Cikanek, and R. Horrocks, "Diesel Particulate Control System for Ford 1.8L Sierra Turbo-Diesel to Meet 1997/2003 Particulate Standards," SAE Paper No. 940458, SAE International, Warrendale, PA, 1994.

Chapter 17

Controls for Modern Diesel Engines

17.1 Introduction

From the previous chapters it is obvious that modern diesel engines are much more sophisticated than ever before. Charge air is no longer just aspirated into the cylinders, it is boosted. Boosting diesels is no longer accomplished through just a simple supercharger; it is accomplished by means of a combination of exhaust-driven turbochargers and mechanically, electrically, or hydraulically driven superchargers. Its fuel injection system is not only providing much higher injection pressures, but also multiple injections, and each has an optimized rate to accomplish a specific purpose. In addition, the modern injection system is not only satisfied with fulfilling the traditional functions of injection timing and metering control, but it also provides flexibility in controlling injection pressure. We have also learned that charge air swirl is part of the overall mixing process and mixing preparation, which is extremely important to combustion efficiency. Modern diesels are not satisfied with a fixed swirl ratio and prefer to have variable swirl with varying characteristics throughout the speed and load range.

Besides the air and fuel management, modern diesels feature variable valve systems as well as new engine cooling techniques that allow faster engine warm-ups to reduce cold start emissions. Electrically driven oil pumps and flywheel-mounted electric starter/generator sets are finding new applications in modern diesels. Although these advances are excellent improvements for diesel engines, their full impact will not be felt unless intelligent controls are utilized.

17.2 Background

In general, the control system of a diesel engine is responsible for maintaining performance at its optimum while at the same time keeping the engine from exceeding certain emission limits. For instance, good engine performance may be had when injection timing is relatively advanced. Yet, this timing may not be appropriate for maintaining NO_x emissions below the mandated limit. The control action will then be to retard timing to a place where the engine can comply with NO_x emission limits without necessarily exceeding particulate emission limits.

To perform its function, the control system needs three components. *Sensors* to obtain a measurement of a physical variable through direct measurement or a combination of measurement and computation. For instance, electromagnetic sensors can produce an electrical signal any time their magnetic field is interrupted. Gear teeth on the perimeter of a flywheel interrupting the magnetic field of a sensor can be used to indicate a speed that is proportional to the frequency of the gear teeth interrupting the sensor's magnetic field. A *soft sensor* delivers a value through an intermediate computation [1]. These sensors should be able to measure a range of physical and chemical quantities in a short enough time to meet the control requirements of high-speed diesel engines. In addition, sensors have to survive the environment in which they are to perform their function. Yet, they need to be produced at a reasonable cost and provide automotive-type durability.

Electrical signals produced by sensors are relayed to the second major component of the control system, the *processor*. The processor is often described as the brain of the control system where a control action that will keep system performance at the required level is determined on the basis of calculations. The processor can be electronic, but many of these processors can simply be mass-spring devices that control basic functions such as speed in engines. However, processors based on purely mechanical or hydraulic devices tend to be cumbersome and bulky. For this reason, the majority of modern control systems are now fitted with electronic processors that are also programable for added flexibility.

The third of the three components of a control system is the *actuator*. An actuator is a device that receives an order from the processor (the brain) to perform a certain function or a required control action. In most cases this control function requires that the actuator either close or open a flow path or move a system control component a specific distance. Because of this function, actuators are generally likened to muscles in the human body. A very obvious and fundamental actuator in diesel engines is its fuel injection system, which controls fueling to each cylinder. In the past fueling was adjusted by setting the pump rack, an action that controlled fueling to all cylinders simultaneously. Modern systems allow full authority control over injection timing as well as fuel metering to each cylinder independently on a cycle-by-cycle basis.

17.3 Basic Operating Principles of the Control System

As we consider the basic operating principles of control systems, they can be categorized into two types: (1) open-loop and (2) closed-loop systems. In an *open-loop* control system, determination of the engine function is normally made by an observer (human). The observer has access to monitors that keep track of the vital functions of an engine. By human logic a determination may be made to change the engine conditions, such as increasing speed or power. The observer then proceeds by effecting the proper control function that would result in the desired change. By observation, once again, we determine whether or not the desired change had taken place. In *closed-loop* control, a measurement of the quantity being controlled is explicitly compared with the desired value. The difference, which is often referred to as *error*, is used as the basis of the control action.

It is apparent that for modern automatic and unsupervised control, the open-loop concept would not be selected. As mentioned earlier, in closed-loop controllers calculations are made on the basis of the error or the difference between a demand signal and the value measured from the system. An all-speed governor is a common example of such a feedback control system. The form of control computation decides the dynamic behavior of the controlled system and may be as simple as the multiplication of the error by a constant [1].

17.4 Variation in Closed-Loop Control Systems

As is often the case, there are variations to the basic principles of operation of control systems. The closed-loop system is no different and features the *feed-forward* technique designed to compensate for a measured disturbance. If a measurement is available, such as pedal movement, and its effect on the system to be controlled can be calculated (such as an increase in emissions), then feed-forward is an excellent way of ensuring a rapid response to avoid undesirable consequences. It is important to retain a feedback element in the control system to compensate for the inevitable inaccuracies that exist even in the best executed feed-forward scheme [1].

In Figure 17.1 a closed-loop control system is in effect when a measurement f is compared with a demand d. The error e is

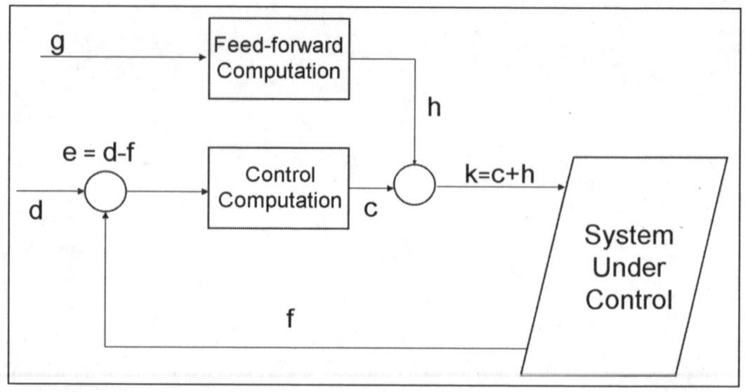

Figure 17.1 *Closed-loop control with a feed-forward system. (Based on Ref. 1.)*

calculated as d – f and then included in a simple calculation to form the control signal c. In a simple feedback system, c is sent direct to an actuator that responds by positioning itself accordingly. For feed-forward, a measurement g is given as advance warning of a coming change. A computation of the effect of this variable is done and delivers a value h, which is the amount by which the actuator must be moved to compensate for the change implied by the measurement g.

In the case presented in Figure 17.1, feedback provides a clear advantage by letting the basic closed-loop controller evaluate whether the feed-forward loop was (1) necessary and (2) correct, and also (3) whether the feed-forward action had accomplished the desired change. For instance, a successful performance for the feed-forward loop could have been the reduction of the effect of a disturbance to the output. If a sudden load is anticipated by the feed-forward loop, then the computation would estimate the fuel required to maintain engine speed. The feedback signal would track any change in engine speed and report it to the base control computation loop. If the error is maintained at zero, then the feed-forward computation and control function was successful.

Obviously, the more involved and sophisticated the feedback, the more likely it is that the disturbance will not be felt. By contrast, an open-loop controller would let the disturbance pass through unmodified until a regular inspection of the speed led to manual control correction. In some control schemes, the controller contains a calculation in which the error is integrated and the resulting integral scaled and applied as a control action; in such as case, the controller is said to include an *integral* component. This feature is capable of eliminating steady-state error in the control response.

17.5 Variation in Open-Loop Control Systems

In an engine control system, measured values of speed and torque are used to construct a control table of injection timing values. At any speed and load condition, the corresponding injection timing is calculated using an interpolation algorithm and translated into the output value to set the injection timing. Of course, there are minor changes in speed and load as injection timing is changed as well as minor changes in injection timing as speed and load are changed. This is where the feedback loop in closed-loop systems can come in handy—it can limit cycling where the injection timing hunts around a small area of the look-up table, which is a characteristic that calibration engineers must eliminate. When the control system relies on a series of look-up tables it is referred to as *scheduled* control (see Figure 17.2). In essence, it is basic open-loop control, but having a predetermined observer, a look-up table, instead of perhaps a human observer.

Modern engine control systems have a large number of tables and demand substantial effort to fill in the look-up tables (i.e., "calibrate" them). While full closed-loop control looks like an attractive alternative, its scope is severely limited in the absence of reliable sensors for closed-loop control. Scheduled control permits control, while the sensors can be used to deduce the engine's operating condition. The prognosis is for the availability of sensors to progressively shift control more to a closed-loop situation [1].

17.6 The New Principle in Control Systems—A Model-Based Approach

In look-up table controls, the desired control behavior is represented as values that correspond to any speed and load condition. For instance, in an EGR control system, the look-up table may have EGR valve positions that correspond to each speed and load condition under the full load curve of an engine.

Consider instead a different control approach. The new approach would be based on modeling the components of the EGR system (see Figure 17.3), which could include:

- Exhaust manifold
- EGR cooler
- EGR valve
- Air filter
- Turbocharger compressor
- Mixing chamber for fresh air and EGR
- Air-to-air intercooler
- Intake plenum/manifold
- Intake and exhaust valves

and the dynamics for each of these control elements. Of course, other effects will be felt, such as the pumping action the engine contributes to the overall dynamic behavior as well as intake and exhaust valve dynamics. Cook and Powell have considered the dynamics of the system components in their work titled "Modeling on an Internal Combustion Engine for Control Analysis" [2]. An example of developing a model for an EGR system is discussed in following sections.

17.6.1 EGR System Elements

Ambient air passing through the *air filter* suffers a drop in pressure, producing a flow dependent on that pressure drop. With its pressure reduced, air enters the *inlet duct* or *intake plenum* on its way to the turbocharger compressor. The intake

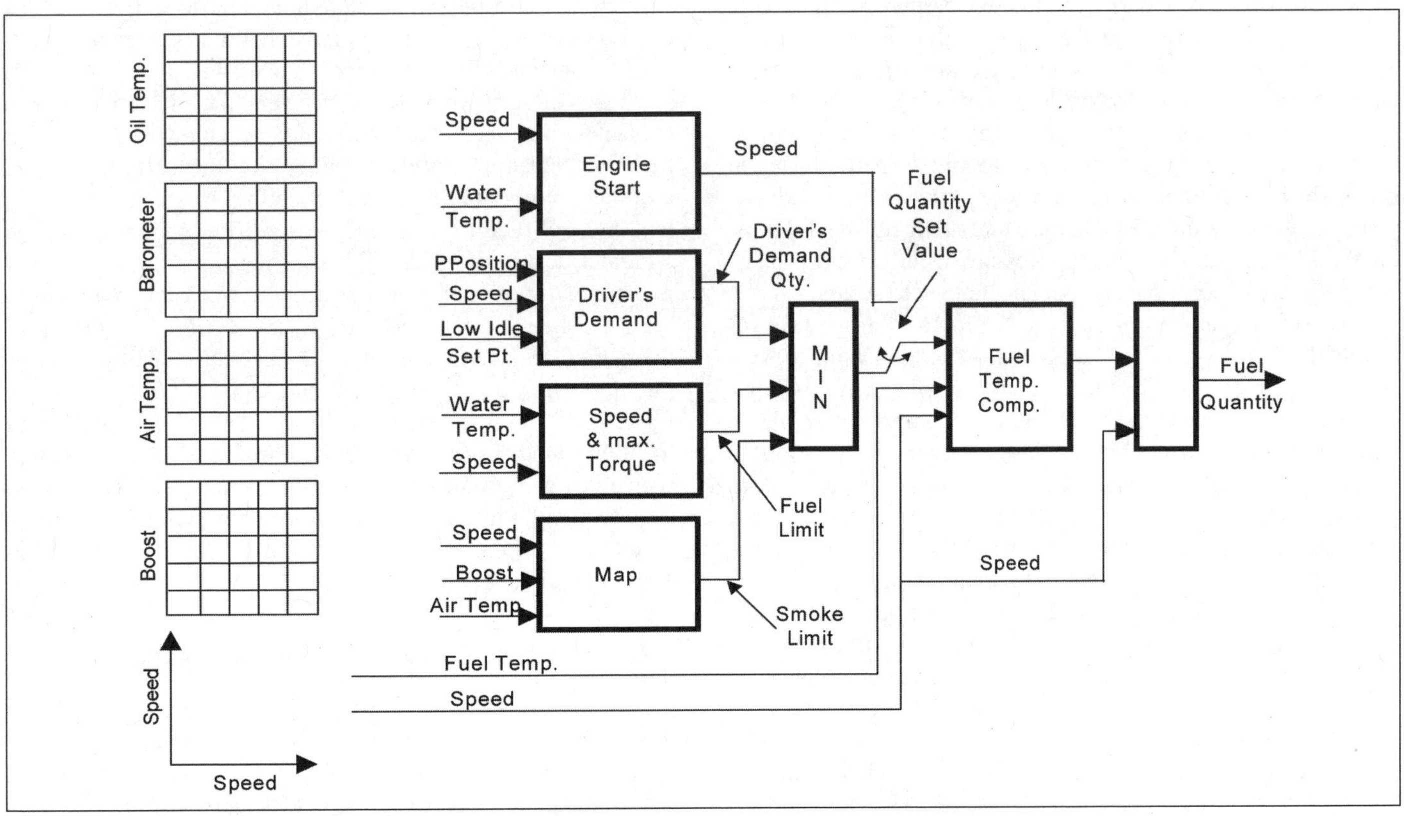

Figure 17.2 *Control strategy based on look-up tables. (Based on Ref. 1.)*

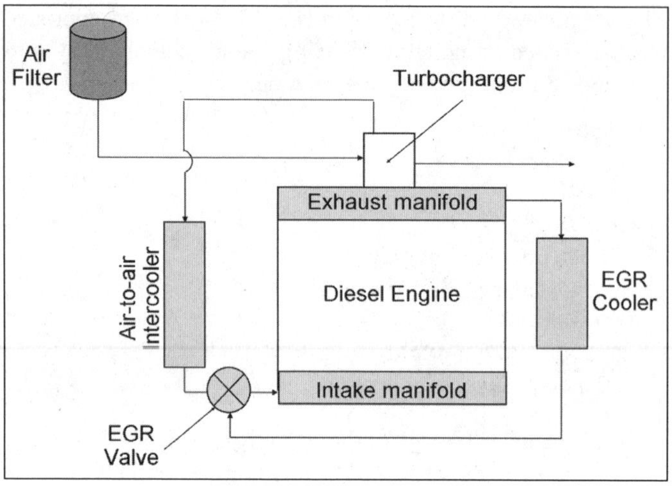

Figure 17.3 *Components of an EGR system.*

plenum acts as a buffer, resulting in an accumulator effect, but does not retain charge air. The *turbocharger compressor* is a centrifugal machine having a complex set of dynamic behaviors that relate to its inertia and its flow pattern. Air exiting the compressor has a different makeup of density, temperature, and flow characteristics, which it carries to the *air-to-air intercooler,* where it recovers density as it exchanges heat with the cooling medium.

Cooled charge air exiting the air-to-air intercooler passes through an *EGR valve* on its way to the intake manifold. The EGR valve is designed to have minimum pressure drop, allowing it to respond quickly to a control signal with minimum friction at various temperatures. It may also have some flow hysteresis, which must be captured for modeling accuracy.

Exhaust gas recirculated from the exhaust manifold enters the EGR valve where it is metered and mixed (partially) with the fresh air prior to entering the *intake manifold* as a mixture. Once in the intake manifold, the mixture of air and exhaust gas continues to mix and is subject to pressure disturbances emanating from the *intake valve* and combustion cycle events.

The *exhaust manifold* undergoes a totally different set of events. It can be considered a plenum where pulsating waves of exhaust arrive in accordance with the blowdown events from each of the *exhaust valves*. Some pressure damping takes place in the exhaust manifold, but the desire is to preserve as much of the pulsating energy as possible for the turbine, so that it can perform at high efficiency. The *turbine wheel* is also a rotating machine with flow characteristics of its own. The turbine discharges its exhaust gas into the exhaust system at or near atmospheric conditions. In a modern diesel engine, a variable-geometry turbocharger may be used, in which a variable-position vane intercepts flow exiting from the exhaust manifold and conditions it (changes its kinetic and potential energy) prior to impacting the turbine wheel.

Depending on the pressure differential between the exhaust and intake manifolds, EGR flows from the exhaust manifold through the *EGR cooler* where it loses some pressure and exchanges heat with the cooling medium. It then proceeds to the EGR valve as described earlier.

We have described in general terms what the components of an EGR system are, what impacts each of the components, and how each component impacts others. We can now write a mathematical model that defines all of the potential interactions based on the individual component's dynamic behaviors. Of course, such dynamic behavior is always modified by an ever-changing set of boundary conditions.

17.6.2 Some Principles of Modeling

In the discipline of controls, linear models are most popular because of their simplicity. They oftentimes include a separate, nonlinear element such as hysteresis. In practice, such linear models are relatively easy to represent and combine and can be accurate over a wide range of engine operations.

Control engineers try to maintain simplicity in their models, but not at the expense of precision and repeatability. The degree of complexity has to be balanced between having a simple but ineffective control or an extremely complex but cumbersome control. Normally, good engineering practices, experience, and assistance from parametric studies and test results are needed to successfully design a good control system. It is quite legitimate to lump various model elements together to produce an overall system model that is quite representative of system function.

By definition, a model is some mathematical representation of system behavior. The more accurate this representation is, the more effective the controller. Modeling has been done with extensive use of differential equations. Once boundary conditions are established, equations are written for various components of the system to derive their physical characteristics. For instance, if a workpiece is placed such that one of its sides is in contact with a hot surface and the adjacent side is in contact with a cold surface, differential equations can be written to derive the temperature distribution over that surface. Solving a limited number of differential equations may not represent a lengthy or difficult process. However, if the model is to have a greater degree of accuracy, a finer mesh requiring a greater number of differential equations needs to be written. To perform such a task, a computer would be very helpful, and the faster the computational capability of the computer, the more accurate the definition of the system behavior can be. An example for this method is referred to as *parametric* model, and one of the more popular of these models is the *identification* method, which is a variation of the least squares method. This method is quite popular in statistical analysis and is used extensively to fit a relationship between dependent and independent variables in experimental results [3].

17.6.3 Discrete and Continuous Models

Analysis of engine performance parameters and their dynamics is normally done with continuous time analysis. For instance, a typical transient test conducted according to U.S. Environmental Protection Agency (EPA) specifications involves variations of torque and speed over a 20-min cycle. A thorough analysis of system and controls can be done in continuous time and is thoroughly presented in introductory texts on control system design [4].

Today's controls are almost exclusively implemented in a digital form with a microprocessor as its basic element. Digital control systems utilize a different design approach that has been well documented in a number of references such as those by Franklin, Powell, and Emami-Naemi [5] and Astrom and Wittenmark [6]. One important aspect of the modeling of reciprocating engines is the impulsive nature of injection and combustion that leads to behavior that has a discontinuous character. Early control studies of engines showed erratic results at high speeds. The reason for this behavior was that at high engine speeds, the engine events were close to the Nyquist frequency [1] and there was consequent interference from sidebands. The solution is to synchronize the sampling with the engine speed and avoid such interference.

The inherently sampled nature of engine operation was developed in two important papers [7, 8] that make an important contribution to the controls analysis of reciprocating engines. The papers were both written specifically for compression ignition engines. The authors' analysis is extensive and contains several useful concepts:

- Torque development in an engine is inherently impulsive. The development of torque begins after the injection, and continues sometimes into the next injection/torque event.
- This continuing torque development can be represented by a partial first-order hold (Figure 17.4). The torque can be approximated as constant in the first sample interval, then as a diminished but constant value for the first part of the next interval.
- The impulses are distinct in a four-cylinder engine. There is some overlap in a six-cylinder engine, and there is substantial overlap in an eight-cylinder engine. The net effect is that the eight-cylinder engine is inherently less stable than the six-cylinder engine, a phenomenon that can be observed in the respective root locus diagrams of the two engine configurations [1].

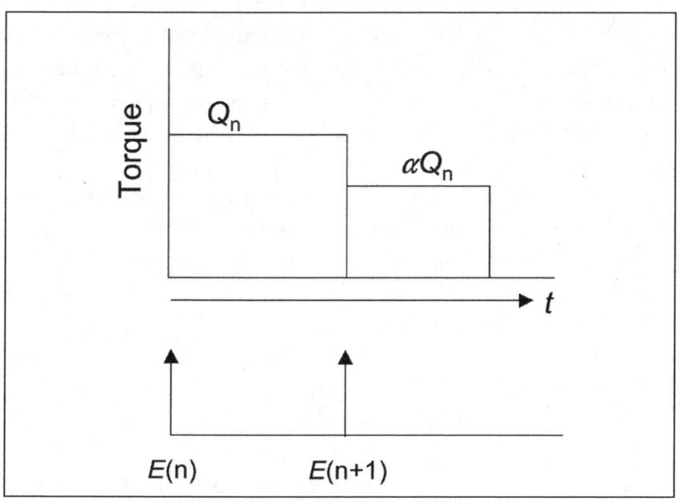

Figure 17.4 Torque development in a diesel engine. (Based on Refs. 7 and 8.)

Figure 17.4 shows the concept of a partial first-order hold that describes the torque development of each cylinder in an engine. In fact, some experiments were conducted at Bendix Diesel Engine Controls in which a signal was obtained and digitized to analyze the impulsive flywheel motion that results from the torque development. Figure 17.5 shows the results of this experiment, which was conducted on a four-cylinder Volkswagen diesel engine [9]. While the general observation is that in an engine the flywheel is rotating at a steady speed, it is in fact rotating in a pulsating pattern as shown in Figure 17.5. By referencing the trace in Figure 17.5, control engineers at Bendix were able to infer injection timing and fueling for each cylinder.

Analysis of such a trace can yield information regarding when the piston began its downward acceleration. From this determination, an injection timing is inferred by referencing the start of piston acceleration to a set top-dead-center (TDC) reference. Comparative analysis is then conducted by the electronic control unit to determine the injection timing for each individual cylinder. In injection systems where individual cylinder control of the fuel injection is available, adjustments can be made to equalize the effective injection timing in all cylinders. Likewise, the rate and amount of acceleration of each flywheel impulse can be used to infer the fueling in each cylinder. Once again, the electronic control unit is capable of adjusting the cylinder-to-cylinder fueling rate for smoother engine operation. Practical implementation of this system involved logging these pulses over 10 consecutive cycles and effecting the control change in the eleventh cycle. Among the benefits of such a system is the ability to lower the engine's idle/no-load speed and reduce the engine noise level. Limited investigations were conducted on the digital period analyzer's effect on emissions [9].

17.6.4 Data Conversion and Diagnostics Models

Such models can be also used as an analytical tool for diagnostics. In this section, discussion will be limited to diagnostic procedures that are related to controls. In part, this analytical redundancy tool refers to the property of a system that broadcasts details of its internal structure as its measurement is changing. Figure 17.6 shows the structure of a nonlinear observer whose function is to reconstruct certain unmeasurable or difficult-to-measure parameters. Such parameters, though not measured, can be derived or inferred from other measurements. The controlled device in the example of Figure 17.6 is the inlet subsystem of the engine, and the output P is the pressure. The real measurement is inherently slow and usually noisy, and to be useful it must be filtered. The result is that when the measurement becomes available it will be too late to be effectively processed and useful to the control function.

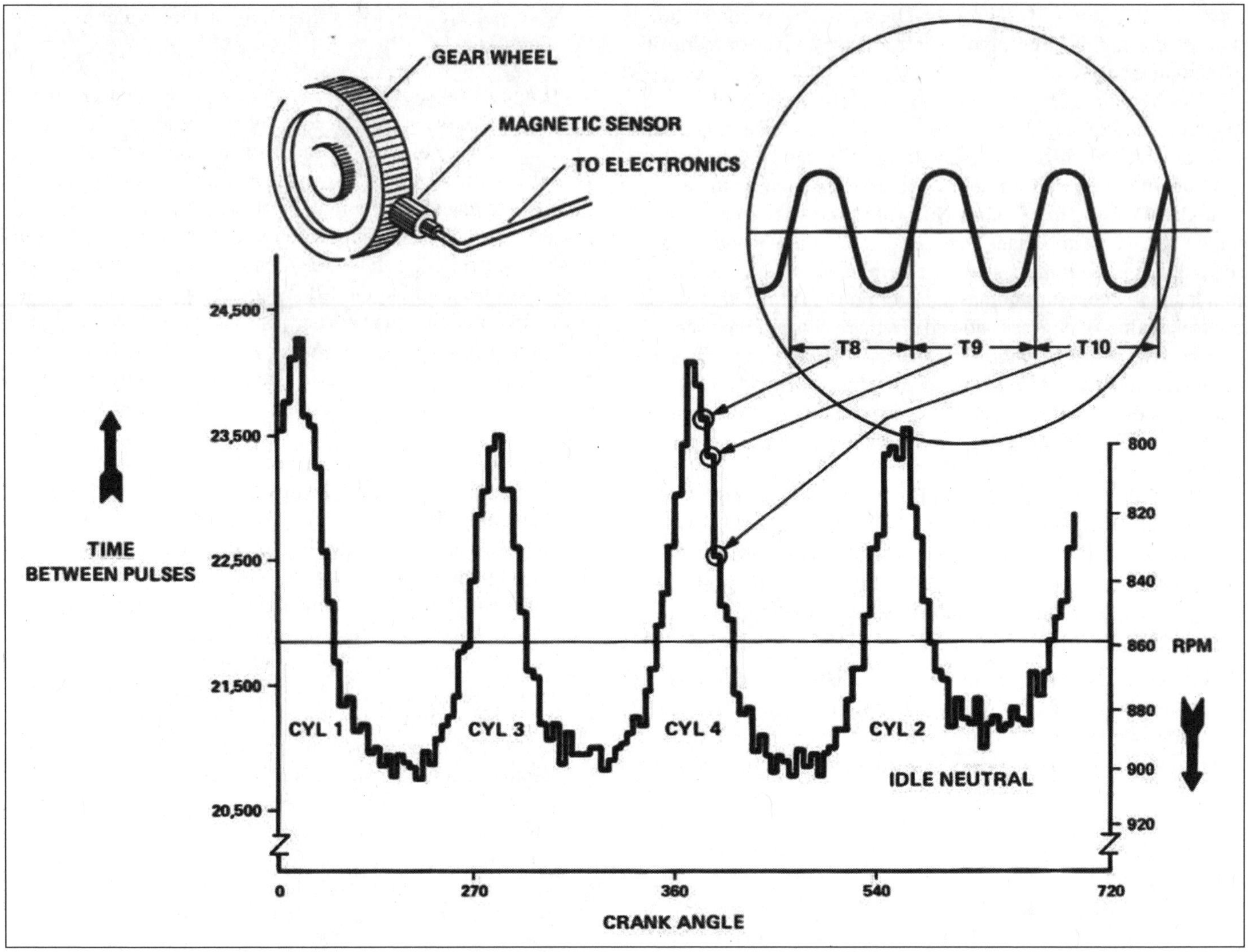

Figure 17.5 *Torque pulse development in a four-cylinder diesel engine. (Reprinted from Ref. 9.)*

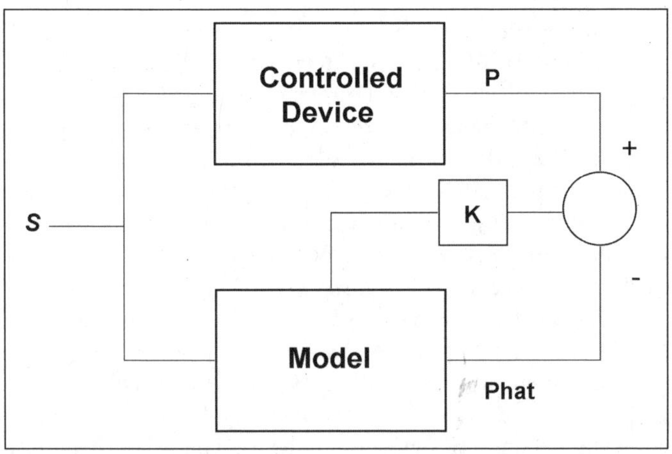

Figure 17.6 *Diagram of a nonlinear observer for engines. (Reprinted from Ref. 10.)*

The "model" is that of the manifold, and it is formulated such that when it receives an input from the throttle it computes the manifold pressure and makes the estimate available to the control system. In the short term, the accuracy can be very high but may drift in the longer term. The feedback loop is designed to compensate for this drift if and when it occurs.

While new approaches toward control are being developed, older and more traditional approaches have not been completely abandoned. For instance, the *proportional-integral-differential* (PID) control approach is still considered the classical way most controllers are designed. Yet *adaptive controllers*, which change as the system they are controlling changes, are also prevalent in many control mechanisms. An example for using adaptive controls in diesel engines is implemented in the electronically controlled fuel injection system where nozzle hole size may be changing over time. Sensing this change and processing it via a digital period analyzer (described earlier), the adaptive control would then change the injection parameter to limit power to the specified output even though the nozzle hole size may have increased.

An important feature of modern control systems is their *flexibility* and *robustness*. These features allow the controller to continue to operate correctly even in the presence of signal errors and noisy sensors. *Continuous* and *discrete* models were described earlier and stress the importance of sampling rate. The higher the sampling rate, the more accurate the controller may be, but the more cumbersome the computation and the greater the need for more computing ability. *System identification* models rely heavily on a mathematical model of the system they are controlling.

17.7 Challenges in Engine Control Systems

When confronted with a control problem, the designer needs to answer a series of fundamental questions about the system to be controlled. In practice, a control problem is likely to be solved with a standard control solution, but this does not absolve the designer from the need to understand the system to be controlled. Indeed the choice of a standard solution (from a library of possible solutions) is dictated by the underlying system.

Control *nonlinearities* are one of the most common problems presenting a real challenge to control engineers [11]. This problem is usually compounded by the *many sources for information* onboard an engine, with each source carrying the likelihood of nonlinearity. Another challenge is the presence of *dependent subsystems*. For instance, turbocharger speed depends on the exhaust gas condition at the inlet to the turbine as well as the compressor performance on the inlet gas. Changes in the state of the inlet and exhaust manifold conditions may result from the injection timing and fueling and, because the exhaust system is dependent on what happens in the cylinders, the turbocharger will respond in a dependent way to those changes. Of course, control engineers have to have a good understanding of how dependent the turbocharger is to changes in fuel injection in order to properly design the control system.

The choice of a *centralized* or *decentralized* control system has always been a challenge to control engineers. In a centralized system, the computational capability may become quite restricted and cumbersome. Decentralized systems (also called satellite or modular controllers) make more computational power available to the subsystems they control. They work in concert with the main ECM and share/borrow speed and load signals from the central electronic control unit through a *control-area-network* (CAN). A simplified structure of an engine controller is illustrated in Figure 17.7. Another challenge to control engineers is *variability* in build and type. Engines as well as electronic control modules are produced according to tight specifications and tolerances. Yet, out-of-control production irregularities will invariably lead to engines with bad performance or excessive emissions. Process control becomes extremely important in the production of modern engines and their controllers. The type of variability affects its solution. Variability derived from external sources such as engine vibration may require one solution, whereas variability derived from electromagnetic interference may require a very different solution.

17.8 Neural Networks

In-cylinder measures and engine design changes aimed at reducing emissions and maintaining good performance and fuel economy are reaching the point of diminishing returns. Research and development efforts are currently focusing on new areas from which more emission reduction can be had. New fuels or fuel formulations as well as exhaust emission controls (after-treatment) are receiving the majority of attention.

A theoretical modeling of engine emissions is very difficult because it involves extensive thermodynamic and chemical equations that describe the many interactions and processes [13]. In addition, local conditions that have a significant influence on emission formation are often unknown or not well defined. Modeling based on experimentally measured input–output data can be used alternatively to build an empirical model. However, that model may grow too complex as the main influential parameters grow in number. The problem becomes too involved and can hardly yield itself to visualization through multidimensional look-up tables.

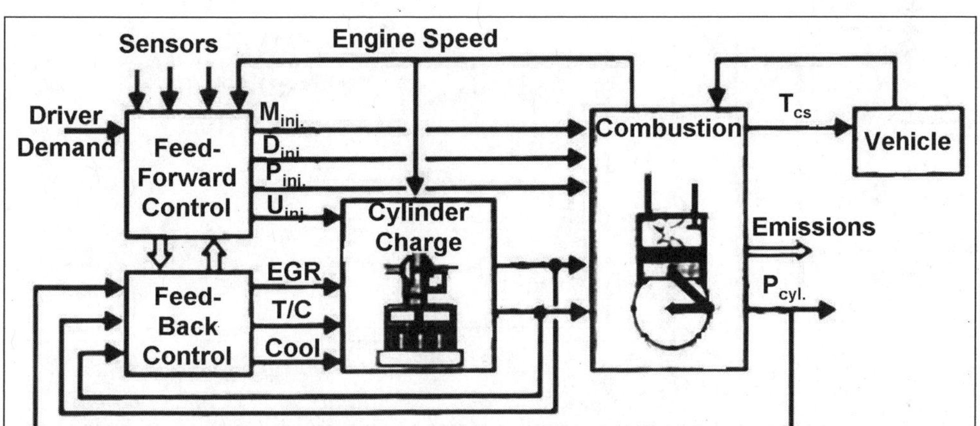

Figure 17.7 Simplified structure of an engine control system. (Reprinted from Ref. 12.)

Neural networks have the ability to approximate multidimensional nonlinear relations in a very compact way [14]. Among the most common types of neural networks are *multilayer perceptrons* (MLPs) and *radial basis function* (RBF) networks with specific advantages and disadvantages [15]. In practice, the identification itself is usually an iterative procedure: First, the experiment (i.e., the measurements) has to be designed carefully. Adequate model inputs have to be selected, the whole area of interest has to be considered to avoid extrapolation, a suitable sampling time has to be determined for dynamic measurements, and so forth. The measured data has then to be postprocessed (i.e., outliers or drift effects have to be corrected), the sensor dynamics should be compensated, and adequate filtering should be applied [12]. The training of the neural network is then performed by minimizing the error between the process and the model (see Figure 17.8). Finally, the model's performance is validated by means of generalization data (data that were not used for the training). If the validation results are not satisfying, the identification process has to be repeated with changes in the model structure or the data processing, or it may even have to be based on new measurements.

Let's look at an example for the special fast neural network *LOLIMOT* (LOcal Linear MOdel Tree). LOLIMOT is based on the idea of approximating a nonlinear function by a superposition of several linear functions, each of them being valid only locally, that is, points x and y in Figure 17.9 [13]. Line a–b represents a nonlinear function that can be described by a large number of linear functions (two are shown in Figure 17.9) that are valid only at points x and y. The transition between submodels is determined by Gaussian membership functions [12].

The training algorithm can therefore be generated in two steps: First, input spaces u_1 and u_2 in Figure 17.10 have to be partitioned into submodels, considering the nonlinearities of the process output [16]. Second, the parameters of each linear submodel have to be estimated. An incremental tree construction algorithm subdivides the input space into so-called (hyper)rectangles where local linear models are estimated. After starting with one global linear model for the whole input space, this global model of the first iteration is tentatively split in both dimensions, horizontally and vertically. In the second iteration, the parameters of the four resulting submodels are estimated in order to calculate the output error of the two possible overall models. Only the cut leading to the lower output error is actually performed leading to the subspaces 2–1 and 2–2 in the second iteration. Now, the local output errors of both submodels are calculated and only the one with the highest output error (here, 2–1) is chosen for subdivision in the next iteration. There, a horizontal cut of model 2–1 leads to the lowest overall error in the third iteration. The algorithm stops if either the error between measured data and the model output is below a prespecified limit or if a maximum number of submodels is reached [12].

The model output of a net with M submodels and n inputs is calculated according to the following equation:

$$y = \sum_{i=1}^{M} \left(w_{0i} + w_{1i} \cdot x_1 + \ldots + w_{ni} \cdot x_n\right) \cdot \Phi_i \left(x, c_i, \sigma_i\right) \quad (17.1)$$

where $w_{0i} \ldots w_{ni}$ represent the parameters of the i'th linear model, $x_1 \ldots x_n$ are the model inputs, and Φ_i is the normalized

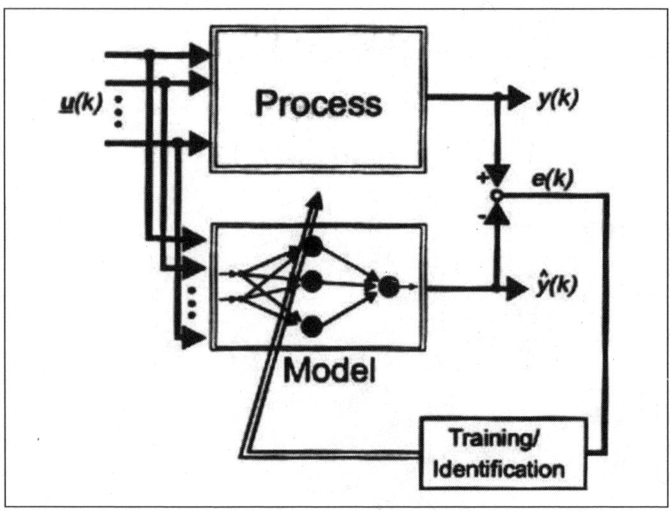

Figure 17.8 *Process identification by means of neural networks. (Reprinted from Ref. 12.)*

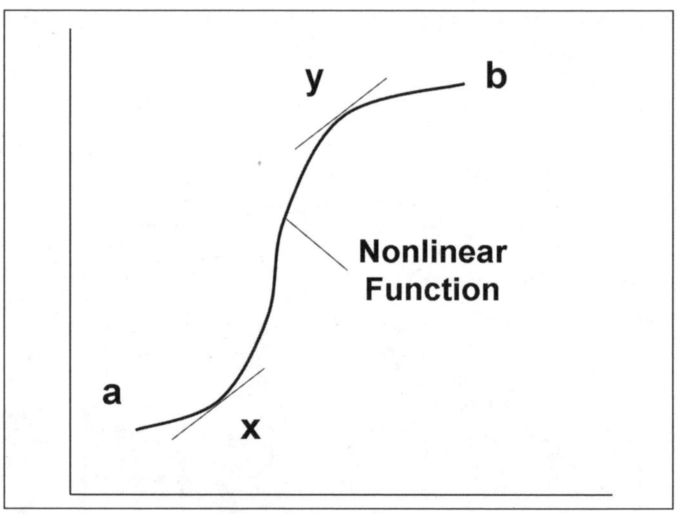

Figure 17.9 *Example of nonlinear function represented by linear functions whose solution is valid at one point only. (Based on Ref. 13.)*

Gaussian of the i'th submodel. The value of the Φ_i is determined by Eq. (17.2):

$$\Phi_1\left(x,c_i,\sigma_1\right)=\frac{z_i}{\sum_{j=1}^{M} z_j},z_j\ =$$

$$\exp\left(-\frac{1}{2}\left(\frac{\left(x_1-c_{1j}\right)^2}{\sigma_{1j}^2}+\frac{\left(x_2-c_{2j}\right)^2}{\sigma_{2j}^2}\right)+\ldots+\frac{\left(x_n\ -c_{nj}\right)^2}{\sigma_{nj}^2}\right) \tag{17.2}$$

with the c_j being the center coordinates and σ_j representing the standard deviations. The model is interpretable because for each point of the input space the sum of all validity functions is normalized to one, and thus the validity values of the submodels represents the percentage to which this point belongs to the corresponding submodels.

The identification of dynamic systems can be performed by applying time-delayed inputs and outputs as additional model inputs as shown in Figure 17.11. The dynamics are realized by external cascades of filters q^{-1}. Thus, the dynamic identification can be transformed into a static approximation problem.

17.9 Conclusion

As the internal combustion engine and, in particular, the modern DI diesel moves toward providing more flexibility, pressure will be applied on the control system to derive more performance, less fuel consumption, and lower emissions. These goals cannot be achieved unless sophisticated electronics are used. From simple open-loop to complex model-based and neural network controllers, the choice is wide open and so is the degree of sophistication. Each of these control concepts has its advantages and disadvantages. Controls engineers work closely with calibration engineers to make the right decision and obtain the optimum combination of performance, fuel economy, and emission control.

To facilitate the computational process, microprocessors are becoming faster and featuring greater computational capacity. Their durability is being tested with greater rigor in various extreme conditions. To make even better use of these capabilities, new sensors are being designed with more smarts than ever before. This fact allows system self-diagnosis and improves the ability for adaptive control.

References

1. Challen, B., and R. Baranescu (Eds.), *Diesel Engine Reference Book,* 2nd ed., SAE International, Warrendale, PA, 1999.
2. Cook, J.A., and B.K. Powell, "Modelling of an Internal Combustion Engine for Control Analysis," *IEEE Control Systems Magazine,* August 1988.
3. Norton, J.P., *An Introduction to Identification,* Academic Press, San Diego, 1986.
4. Richards, R. J., *An Introduction to Dynamics and Control,* Longman, 1979.

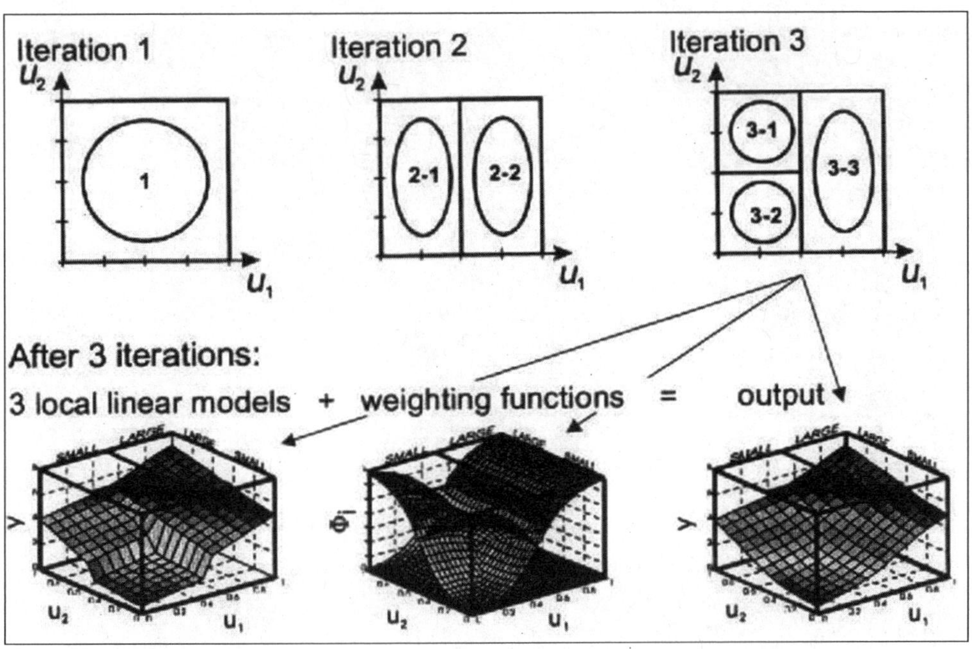

Figure 17.10 Diagrammatic representation of the use of local linear models and weighting factors to build model output. (Reprinted from Ref. 12.)

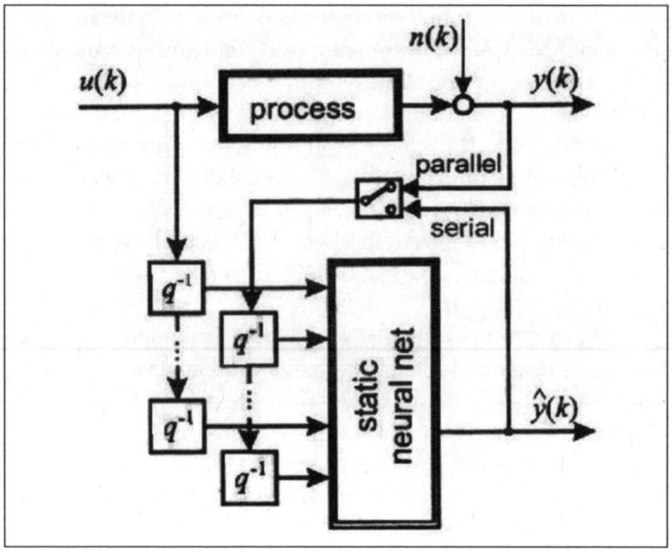

Figure 17.11 *Diagrammatic representation of dynamic neural net with external dynamics. (Reprinted from Ref. 12.)*

5. Franklin, G.F., J.D. Powell, and A. Emami-Naemi, *Feedback Control of Dynamic Systems,* Addison-Wesley, Reading, MA, 1986.
6. Astrom, K.J., and B. Wittenmark, *Computer Controlled Systems,* Prentice-Hall, Upper Saddle River, NJ, 1984.
7. Hazell, P.A., and J.O. Flower, "Sampled-Data Theory Applied to the Modelling and Control Analysis of Compression Ignition Engines—Part I," *Int. J. Control* **13,** 3, 1971, pp. 594–562.
8. Hazell, P.A., and J.O. Flower, "Sampled-Data Theory Applied to the Modelling and Control Analysis of Compression Ignition Engines—Part II," *Int. J. Control* **13,** 4, 1971, pp. 609–623.
9. Seitz, W., "Digital Period Analysis," Internal Report, Bendix Diesel Engine Controls, 1980.
10. Hendricks, E., T. Vesterholm, and S. Sorenson, "Non-Linear Closed-Loop, SI Observers," SAE Paper No. 92237, SAE International, Warrendale, PA, 1992.
11. Bernard, C., *Diesel Engine Design Academy,* Lecture Notes, SAE Engineering Academies, Warrendale, PA, 2001.
12. Hafner, M., "Steps Towards an Optimization of the Dynamic Emission Behavior of IC Engines," SAE Paper No. 2001-01-1793/4214, SAE International, Warrendale, PA, 2001.
13. Pitsch, H., H. Barths, and N. Peters, "Three-Dimensional Modeling of NO_x and Soot Formation in DI-Diesel Engine Using Detailed Chemistry Based on a Flamelet Approach," SAE Paper No. 962057, SAE International Congress, SAE International, Warrendale, PA, 1996.
14. Nelles, O., "Nonlinear System Identification with Local Linear Neuro-Fuzzy Models," Dissertation, Darmstadt Technische Universität, Germany, 1999.
15. Ayoubi, M., "Nonlinear System Identification Based on Neural Networks with Locally Distributed Dynamics and Application to Technical Processes," Dissertation, Darmstadt Technische Universität, Germany, 1996.
16. Tagaki, T., and M. Sugeno, "Fuzzy Identification of Systems and Its Application to Modeling and Control," *IEEE Trans. Systems, Man and Cybernetics* **15,** 1985, pp. 116–132.

Part V

Exhaust Gas Aftertreatment

Chapter 18:
Emission Control Catalysts

Chapter 19:
Diesel Catalysts

Chapter 20:
Diesel Particulate Filters

Chapter 21:
Diesel Filter Materials

Chapter 22:
Diesel Filter Systems

Chapter 23:
Plasma Exhaust Treatment

Chapter 18

Emission Control Catalysts

18.1 Overview

18.1.1 Background

Environmental catalyst technologies have their roots in air pollution problems from light-duty vehicles, mainly gasoline-fueled passenger cars. In many countries, car emissions were the primary target of early air pollution legislation, dating back to 1970s and 1980s. In the United States, the federal regulations are based on four acts: the Clean Air Act (CAA) of 1967, the CAA of 1970, and its two amendments adopted in 1977 and in 1990 [1].

Following the adoption of the CAA and the resulting vehicle emission standards, oxidation catalysts were commercialized in the United States in the mid-1970s to control emissions of carbon monoxide (CO) and hydrocarbons (HCs) from spark-ignited (SI) gasoline engine applications. Three-way catalyst technology (see Figure 18.1), introduced in the 1980s, made it also possible to control NO_x emissions from SI engines at levels that are significantly lower than NO_x emissions from diesel engines. Finally, in the 1990s, oxidation catalysts were introduced to diesel-fueled cars in Europe, as well as to heavy-duty diesel bus and (on a limited scale) truck engines in the United States. Since as early as the 1970s, diesel catalysts have been also used in some occupational environments, such as underground mining.

Following their success in mobile engine applications, catalyst technologies were introduced for stationary applications, for the control of volatile organic compounds (VOCs) and NO_x emissions. The list of catalyst applications covers such emission sources as chemical plants, painting and coating processes, ovens, printing, dry cleaning, power generation, and, last but not least, stationary engines. Examples of catalyst technologies

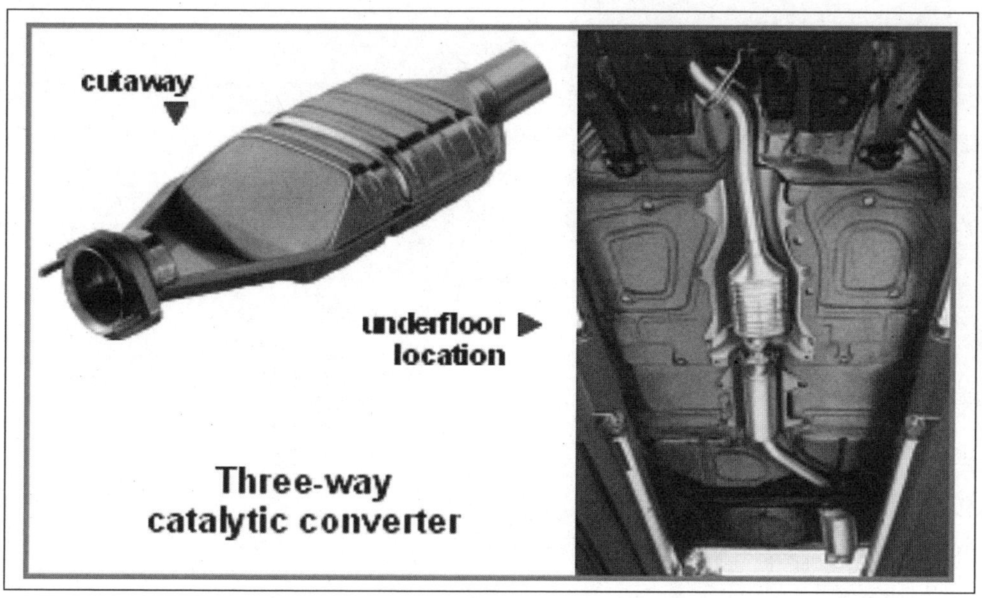

Figure 18.1 *Catalytic converter on a gasoline car.*

for stationary engines include nonselective catalytic reduction (NSCR) of NO_x from rich-burn natural gas engines and selective catalytic reduction (SCR) of NO_x by ammonia from diesel engines (Figure 18.2).

The wide acceptance and popularity of catalysts for emission control from internal combustion engines can be attributed to a number of advantages:

- Simplicity and passive character of most catalyst systems
- Negligible adverse effect on the engine or fuel economy (low pressure drop)
- Maintenance-free or low-maintenance operation
- Proven durability
- Relatively low cost.

Catalysts also have a number of disadvantages and potential issues, which need to be carefully observed when designing emission control systems, including the following:

- Dependence of performance on temperature. In the case of internal combustion engines, this translates to the dependence of performance on the engine operating conditions and duty cycle.
- Loss of activity through poisoning and thermal deactivation.
- Possibility of undesired reactions. In some cases, secondary emissions may be formed over the catalyst.
- Insufficient market availability of some catalytic materials, such as some noble metals, may cause unpredictable price fluctuations and supply shortages.

In this chapter we will discuss general topics pertaining to emission control catalysts, starting with catalyst fundamentals, through catalytic components, materials, and catalytic converters. Specific catalyst technologies as applied to diesel engines are discussed in Chapter 19.

18.1.2 Catalytic Reactors

Catalysts are substances that have the ability to accelerate certain chemical reactions between exhaust gas components. (A more precise catalyst definition is given in Section 18.2.) In emission control catalysis, solid catalysts are used to catalyze gas-phase reactions. The catalytic effect and the observed reaction rates are maximized by providing good contact between the gas phase and the solid catalyst. In catalytic reactors, this is usually realized by providing a highly catalytic surface area by finely dispersing the catalyst on a high specific surface area carrier (support).

Catalytic reactors for emission control from internal combustion engines are known as *catalytic converters.* The term *catalyst* is also frequently used as a synonym for the catalytic converter. Other terms for catalytic converters, which should be generally avoided, include such names as *exhaust purifier, catalytic purifier, exhaust scrubber,* or *PTX.* The term *purifier* is not very specific (isn't the diesel particulate filter an exhaust gas purifier too?). The term *scrubber* is derived from water scrubbers, which are sometimes used as safety devices on diesel engines operating in explosive atmospheres. As such, this term is clearly misleading. The term *PTX,* derived from a brand name of catalytic converters by Engelhard, is still used in many underground mines (in the same way the term *hoover* is sometimes used when referring to a vacuum cleaner).

The most common catalytic converter design is illustrated in Figure 18.3. It utilizes a ceramic honeycomb monolithic substrate (support), featuring many small parallel channels running in the axial direction. An alternative design utilizes a metallic substrate made of thin metal foil formed into a honeycomb structure, and it provides the same functionality. In either design, the ceramic or metallic substrate does not have any catalytic activity of its own. Rather, the monolith walls are coated with the active catalyst layer; the function of the monolith substrate is to provide a highly geometric surface area, such that the catalyst, which has been applied to the channel walls, can contact the gas phase, which flows through the channels.

Figure 18.2 SCR catalyst system on a stationary diesel engine. (Courtesy of Hug Engineering.)

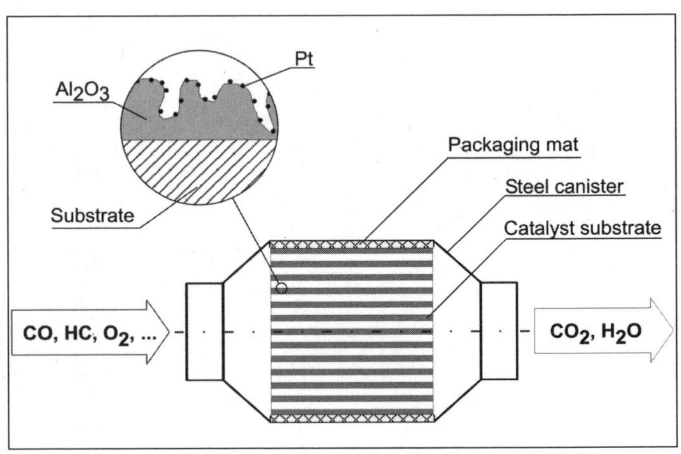

Figure 18.3 Schematic of catalytic converter.

In most cases, the catalytic coating includes noble metal catalysts impregnated on a porous, high surface area layer of inorganic carrier, called *washcoat* (the term *catalyst support* may also be used to denote the ceramic/metallic substrate, as well as the carrier/washcoat material, which can be confusing). Substances that are widely used for catalyst carriers include activated aluminum oxide (alumina, Al_2O_3 in Figure 18.3) and silicone oxide (silica). Common noble metals catalysts deposited on these carriers are platinum (Pt in Figure 18.3), palladium, and their mixtures.

Catalyzed substrates are packaged into steel shells to form catalytic converters, as shown in Figure 18.3. In the case of ceramic substrates, packaging mats made of ceramic fibers are used to protect the substrate and to evenly distribute the pressure from the shell. The mats often include vermiculite, which expands at high temperatures, thus compensating for the thermal expansion of the shell and providing adequate holding force under all operating conditions.

Catalyst can be also packaged inside mufflers, which are then referred to as *catalyst mufflers* or *catalytic mufflers*. In this case, one steel shell holds both the catalyst and the noise attenuation components, such as baffles and perforated tubing. Catalyst mufflers can offer a more space-saving design compared to the combination of a catalytic converter and a muffler.

Catalytic converters are installed as simple, passive devices in the vehicle's exhaust system, so the entire exhaust gas stream passes through the channels of the monolithic substrate contacting the catalyst before being discharged from the tailpipe. However, catalysts can also be part of a fairly complex system that involves various active strategies, such as injection of reactants in front of the catalyst or sophisticated engine control algorithms. Examples include a number of diesel catalyst systems being developed for the reduction of NO_x, which are discussed in Chapter 19. The attributes of simplicity and passive character that have been listed among the advantages of catalysts may no longer apply to those systems.

18.1.3 Conversion of Pollutants: Basic Terms

18.1.3.1 Conversion Efficiency

The effectiveness of catalytic converters in reducing a given emission is typically measured as the *conversion efficiency*. If an exhaust gas component (A) reacts over a catalyst with another component (B) according to the following equation:

$$A + B \rightarrow \text{Products}, \qquad (18.1)$$

then the conversion efficiency of A, expressed as percent, can be defined as follows:

$$\text{Conversion efficiency} = (C_i - C_o)/C_i \cdot 100\% \qquad (18.2)$$

where C_i is the concentration of A at the catalyst inlet, and C_o is the concentration of A at the catalyst outlet.

18.1.3.2 Light-Off Temperature

The rate of chemical reactions, including catalytic reactions, always increases with temperature. The strong dependency of conversion efficiency on temperature is a characteristic feature of all emission control catalysts. A typical relationship between the catalytic conversion rate of a pollutant and the temperature is shown as the solid line (A) in Figure 18.4. The conversion, near zero at low temperatures, increases slowly at first and then more rapidly, to reach a plateau at high gas temperatures. When discussing combustion reactions, the term *light-off temperature* is commonly used to characterize this behavior. The catalyst light-off is the minimum temperature necessary to initiate the catalytic reaction. Due to the gradual increase of the reaction rate, the preceding definition is not very precise. By a more precise definition, the light-off temperature is the temperature at which conversion reaches 50%. That temperature is frequently denoted T_{50}. When comparing activities of different catalysts, the most active catalyst will be characterized by the lowest light-off temperature for a given reaction.

18.1.3.3 Catalyst Window

In some catalyst systems, increasing the temperature may increase the conversion efficiency only up to a certain point, as illustrated by the dashed line (B) in Figure 18.4. Further temperature increases, despite increasing reaction rates, cause a decrease in the catalyst conversion efficiency. The declining efficiency is usually explained by other competing reactions that deplete the concentrations of reactants or by thermodynamic reaction equilibrium constraints.

The temperature range corresponding to the high conversion efficiency is frequently called the *catalyst temperature window*. This type of conversion curve is typical for selective catalytic processes. Good examples include selective reduction of NO_x by hydrocarbons or ammonia.

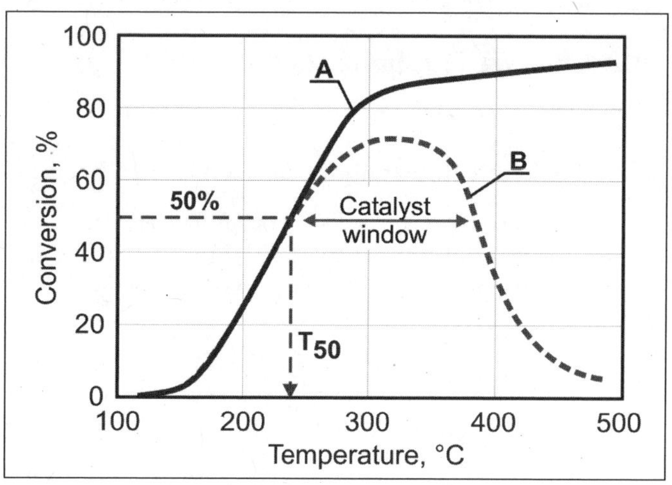

Figure 18.4 *Catalytic conversion as a function of temperature.*

18.1.3.4 Space Velocity

Another important variable influencing the conversion efficiency is the size of the reactor. The gas flow rate through a catalytic reactor is commonly expressed, relative to the size of the reactor, as *space velocity* (SV). The space velocity is defined as the volume of gas, measured at standard conditions (STP), per unit time per unit volume of the reactor, as follows:

$$SV = V / V_r \tag{18.3}$$

where V is the volumetric gas flow rate at STP (m^3/h), and V_r is the reactor volume (m^3).

SV has the dimension of reciprocal time, which is commonly expressed in 1/h. In various catalytic emission control applications, the space velocities range from 10,000 to 300,000 1/h. Space velocities for monolithic reactors are calculated on the basis of their outside dimensions, for example, on the diameter and length of a cylindrical ceramic catalyst substrate. Because this method does not take into account the geometric surface area of the substrate, cell density, wall thickness, or catalyst loading, it is not always appropriate for catalyst comparisons. Nevertheless, it is a commonly used and widely accepted industry standard.

18.1.3.5 Noble Metal Loading

Noble metals used in automotive catalysts include platinum (Pt), palladium (Pd), and rhodium (Rh). The amount of noble metal contained in a catalyzed substrate is a very important parameter, both for catalyst performance and for its cost. It is common to specify metal loadings by providing the total loading of the mixture in grams of metal per cubic foot of the catalyst substrate (based on its outside dimensions) and the ratio of the metals in the order Pt:Pd:Rh. This can be illustrated by the following examples:

- 50 g, 5:0:1 denotes a catalyst composed of platinum and rhodium in the mass ratio of 5:1 at a total loading of 50 g/ft^3.
- 10 g, 1:0:0 denotes a platinum catalyst at a 10 g/ft^3 loading.

Note that 1 g/ft^3 is equal to 0.03531 g/L in metric units.

18.2 Catalyst Fundamentals

18.2.1 Introduction

18.2.1.1 Chemical Reaction Kinetics and Equilibrium

A chemical reaction that results from the simultaneous combination of a molecules of A, b molecules of B, and c molecules of C, . . ., can be represented by the following equation:

$$aA + bB + cC + \ldots \leftrightarrows rR + sS + \ldots \tag{18.4}$$

The reaction rate for the forward reaction may be written as

$$r = k \cdot C_A{}^a \cdot C_B{}^b \cdot C_C{}^c \ldots \tag{18.5}$$

where:
- r = rate of reaction, kmol/(m^3·s)
- k = reaction rate constant, 1/s
- C_i = concentration of reactant i, kmol/m^3.

Equation (18.5) defines the reaction rate constant (k), which, under ideal conditions, depends only on temperature and the nature of the reaction. Another important term is the order of the reaction, defined as the sum of the exponents a, b, c, . . . , of Eq. (18.4). This sum equals to 1 for a first-order reaction, 2 for a second-order reaction, and so forth.

Some chemical reactions may not proceed completely to the end, that is, to the right side of Eq. (18.4). Rather, a thermodynamic equilibrium may be established in which certain concentrations of reactants A, B, C, . . . , will coexist in the system with certain concentrations of products R, S, ... From the chemical kinetics point of view, this is represented as a dynamic balance, where the original reactants A, B, C, . . . , are being re-created in a reverse reaction between R, S, ... This bidirectional character of the process is indicated by the double-headed arrow in the reaction equation. The rate of the reverse reaction can be written as

$$r' = k' \cdot C_R{}^r \cdot C_S{}^s \ldots \tag{18.6}$$

The net rate of the reaction is then equal to

$$r_0 = r - r' = k \cdot C_A{}^a \cdot C_B{}^b \cdot C_C{}^c \ldots - k' \cdot C_R{}^r \cdot C_S{}^s \ldots \tag{18.7}$$

The net reaction rate equals zero at equilibrium, $r_0 = 0$, thus

$$(C_R{}^r \cdot C_S{}^s \ldots)/(C_A{}^a \cdot C_B{}^b \cdot C_C{}^c \ldots) = k/k' = K \tag{18.8}$$

where K is the reaction equilibrium constant, which depends on temperature. It can be calculated from the thermodynamic functions of the system. Once known, the equilibrium constant allows for calculation of the concentrations of reaction products at equilibrium.

Reaction progress can be limited by reaction kinetics or by thermodynamic equilibrium. If the reaction rate constant (k) is low, the reaction, even if thermodynamically allowed, may take a long time to complete. Some reactions are so slow that their rates practically equal zero, and the processes could take years to complete. Fortunately, reaction rates may be increased by the use of catalysts.

On the other hand, if the reaction reaches its equilibrium, any further reaction progress is thermodynamically impossible. It is important to realize that catalysts have no effect on the reaction equilibrium. From the kinetics point of view, if a catalyst increases the forward reaction rate [Eq. (18.5)], it also increases, to the same degree, the reaction rate of the reverse reaction [Eq. (18.6)].

18.2.1.2 Catalyst Definition

A catalyst is usually defined as a substance that influences the rate of a chemical reaction but is not one of the original reactants or final products; that is, it is not consumed or altered in the reaction. In several known catalytic reaction mechanisms, the catalyst forms intermediate compounds with reactants. Many other catalytic processes are not explained or understood in their entirety. Neither are the principles governing the selection and preparation of catalysts for specific purposes. Many of the developments in this field are achieved through elaborate exploration programs involving trials of countless materials. Catalysts are widely used in chemical and petrochemical processing to facilitate reactions that otherwise are too slow or that require high temperatures to yield good efficiencies. Catalysts are also used to convert harmful components of engine exhaust gases, such as hydrocarbons, into harmless substances, such as carbon dioxide and water vapor.

During a chemical reaction the reactants must pass through an energy barrier, called the activation energy (E), before the final product is produced. The diagram shown in Figure 18.5 is an example of the energy path on the way from reactants to products. In this thermodynamic model, the catalyst provides a path with lower activation energy in the transformation of reactants to products, which increases the net rate of the reaction. The difference between the energy states of the reactants and products is the heat of reaction. In this example, the enthalpy of products is lower than that of reactants, indicating an exothermic reaction with positive heat effect (i.e., with generation of heat during the reaction). In general, catalysts also can be used for endothermic reactions with negative heat effects.

It is important to recognize that a catalyst does not change the net enthalpy and net free energy of reaction. A catalyst can affect only the rate of approach to the final product state, that is, the kinetics of reaching equilibrium. It does not change the energies of the initial and final states, nor the equilibrium itself. Thus, the reaction enthalpy (ΔH), reaction free energy (ΔG), and the equilibrium constant (K_e) are not altered by the presence of a catalyst.

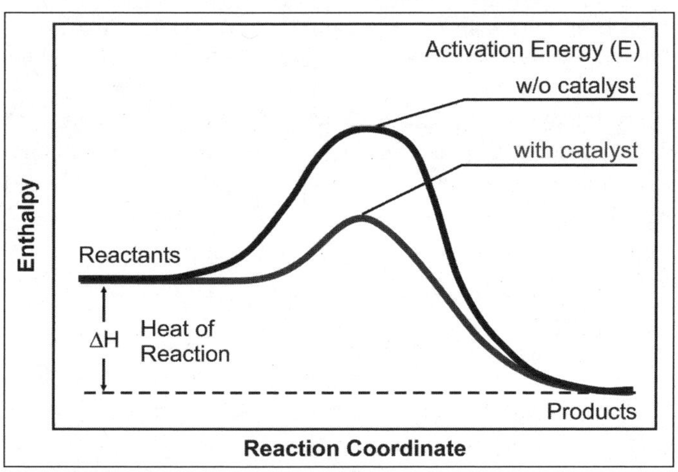

Figure 18.5 *Reaction energy paths.*

18.2.1.3 Activity and Selectivity

Catalysts can be characterized by two fundamental properties: activity and selectivity. *Activity* relates to the catalyst's ability to enhance reaction rates. If two catalysts influence the rate of the same chemical reaction, the more active catalyst will produce higher reaction rates.

Selectivity refers to the catalyst's ability to accelerate certain reactions in preference to others. This behavior is illustrated by the influence of different catalysts on the decomposition of ethyl alcohol [2]. In the presence of an activated alumina (Al_2O_3) catalyst, ethylene and water are formed, whereas, with a copper (Cu) catalyst, acetaldehyde and hydrogen predominate:

$$C_2H_5OH = C_2H_4 + H_2O \quad (Al_2O_3, 354°C) \tag{18.9a}$$

$$C_2H_5OH = CH_3CHO + H_2 \quad (Cu) \tag{18.10}$$

The selectivity of a catalyst often depends on temperature. A catalyst that is very selective at a given temperature may lose its selectivity when the temperature is changed. For example, the decomposition of ethyl alcohol in the presence of alumina follows Eq. (18.9a) at 354°C, but at a lower temperature of 269°C ether is the principal product according to reaction Eq. (18.9b):

$$2C_2H_5OH = (C_2H_5)_2 + H_2O \quad (Al_2O_3, 269°C) \tag{18.9b}$$

In many processes, the catalyst selectivity is its most important property. A selective catalyst can accelerate only the desired reactions and minimize the formation of undesired by-products that would result if all reactions were equally increased in rate. However, there is a price to pay for increasing the selectivity of a catalyst. An empirical rule formulated by Sabatier states that a catalyst selectivity improvement is always accompanied by a deterioration in its activity [3]. In other words, modifying a catalyst to be less active toward the formation of some undesired by-products (i.e., increasing its selectivity) will always result in some decline in its activity toward the desired reaction product. The Sabatier rule, although never proven or universally recognized by the science of catalysis, has been widely confirmed by experiences gained during the development of selective catalysts for diesel emission control.

18.2.1.4 Classification of Catalyst Systems

Catalyst systems are divided into homogeneous and heterogeneous catalysts. The heterogeneous catalysts are further divided into supported and unsupported catalysts. These terms are defined as follows:

- A homogeneous catalyst is a catalyst that is soluble in the reaction medium, for example, a water-soluble substance catalyzing water-phase reactions.
- In a heterogeneous catalyst system, the catalyst is in a different phase than the reactants, for example, a solid catalyst catalyzing gas-phase reactions.

- A supported catalyst consists of a carrier substance having a suitable form and a large surface on which the proper catalyst is deposited.
- In an unsupported catalyst, the entire catalyst structure (particle, bead, various extruded shapes, etc.) consists of the catalytic material.

In most catalytic emission control processes, the gaseous reactants pass through solid heterogeneous supported catalysts. In most systems, monolithic support structures are coated with a refractory oxide carrier (e.g., alumina), on to which a noble metal catalyst (e.g., platinum) is impregnated. The use of unsupported heterogeneous catalysts, usually in the form of extruded honeycomb monoliths, is limited to a few base metal catalyst applications (e.g., vanadia/titania SCR catalysts).

18.2.1.5 Heterogeneous Catalyst

It is assumed that reactions in heterogeneous catalysis occur on the surfaces of the solids at points of high chemical activity that are known as *active centers.* The activity of a catalyst surface is then proportional to the number of active centers per unit area. In many cases the concentration of active centers is relatively low, as indicated by the extremely small quantities of "poisons" that are sufficient to destroy the activity of a catalyst. Because the catalytic reaction rate depends on the number of active sites, much attention is given to increasing the catalyst surface area and maximizing the number of active sites through a fine dispersion of catalytic components into this surface. To accomplish this, catalytic components are usually dispersed on high-surface-area, porous oxide supports (carriers) including Al_2O_3, SiO_2, TiO_2, and SiO_2-Al_2O_3 combinations. The carriers themselves may or may not be catalytically active, but in either case they do play a major role in maintaining the overall stability and durability of the finished catalyst.

A model of a heterogeneous catalyst is shown in Figure 18.6 using alumina (Al_2O_3), which is the most common carrier used for emission control catalysts, as an example. Figure 18.6 presents a few select pores of a high-surface-area Al_2O_3. The diameters of the pores range from 20 to 100 Å. Platinum (Pt) has been deposited into the pores by solution impregnation. The Pt particles or crystallites are shown as dots.

As mentioned earlier, the alumina layer that is bonded to a monolithic honeycomb support is called a *washcoat.* Depending on the type of Al_2O_3 and its thermal history, the internal surface of the washcoat is rich in surface OH^- groups (not shown). These

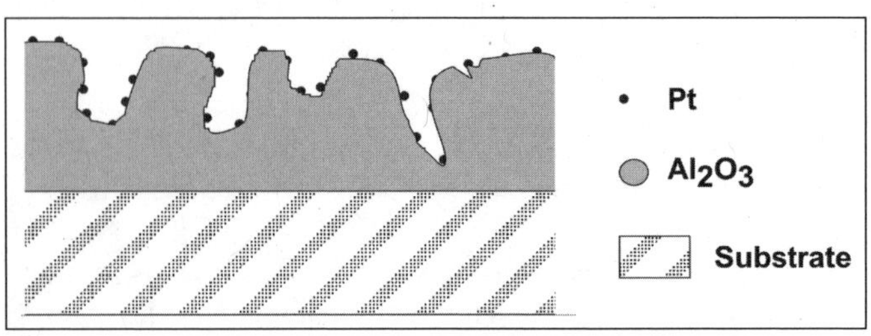

Figure 18.6 *Catalytic sites dispersed on a carrier.*

OH^- groups, which cover the entire surface including the walls of each pore, represent sites onto which one can chemically or physically bond a catalytic substance.

The physical surface area of the washcoat is the sum of all internal areas of the oxide including all the walls of each pore. The catalytic components are bound on these internal walls and at the OH^- sites. The catalytic surface area is the sum of all the areas of the active catalytic components—in this example, platinum. The smaller the crystallite size of the active catalytic material, the higher the catalytic surface area. It is also assumed that the higher the catalytic surface area, the higher the rate of reaction for a process controlled by kinetics rather than by mass transfer.

The washcoat and noble metal systems in today's catalysts are more complex than that discussed in the preceding example. As a rule, the washcoat includes several substances that act as stabilizers or promoters. In such systems complex interactions may occur among the washcoat components, as well as between the washcoat and the noble metal(s). These interactions should be controlled and used to increase the catalyst activity and durability rather than to cause a loss of activity.

The following interactions between washcoat and catalyst components can be used to enhance the activity and/or durability of the catalyst system:

- Stabilization
- Synergism
- Segregation.

Stabilizing substances increase the catalyst's durability by preventing its deterioration due to the loss of surface area at elevated temperatures.

Synergism between washcoat components occurs whenever the performance of the mixture is better than that of each of the components separately. Many synergisms are utilized in emission control catalysts. An example of a synergistic effect is the interaction between gamma-alumina and barium oxide, a common washcoat stabilizer. After calcination at 1000°C, the specific surface area of alumina is about 90 m^2/g. The corresponding surface area of barium oxide is less than 1 m^2/g. Due to the synergistic effect, a mixture of 3% BaO in alumina can have a surface area as high as 130 m^2/g [4].

Sometimes catalytic components have to be segregated to avoid undesired effects such as alloying between noble metals or reactions between a noble metal and a washcoat component. For example, an alloy can be formed between palladium and rhodium in a three-way automotive catalyst. In a multicomponent washcoat, such alloying metals can be physically segregated through impregnation on different washcoat species.

18.2.2 Conversion Rate in Heterogeneous Catalyst

The complete reaction process in a heterogeneous catalyst can be divided into several subprocesses or steps. These subprocesses are of a physical (mass transfer) and chemical (reaction kinetics) nature, each of them with its own rate. The apparent overall rate of the reaction is always lower than that of its lowest rate step. Depending on the particular system and physical conditions, any of the subprocesses may be limiting the overall reaction rate.

Lets imagine a gas stream that contains a reactant (A), flowing through a bed of heterogeneous catalyst. In the catalyst, the reactant (A) is converted to a product (B), as shown by Eq. (18.11):

$$A \rightarrow B \tag{18.11}$$

The complete conversion process of A into B involves the following physical and chemical steps:

1. Bulk diffusion of reactants. Reactant A must contact the catalytic sites at the outer surface of the carrier. It must diffuse through the boundary layer of gas in the proximity of the catalyzed carrier.
2. Pore diffusion of reactants. The majority of A molecules must diffuse through the porous carrier network toward the internally dispersed active sites.
3. Adsorption of reactants. After molecule A arrives at the catalytic surface, it must chemisorb onto a catalytic site.
4. Catalytic reaction. Catalytic reactions, which occur at the active sites, involve chemisorbed A, energetically located at the peak of the activation energy profile, which then converts to product B.
5. Desorption of products. After the reaction, product B desorbs from the catalytic sites.
6. Pore diffusion of products. Product B diffuses through the pore network to the outside surface.
7. Bulk diffusion of products. Having arrived at the outside surface, product B must diffuse through the boundary layer into the bulk gas.

These steps can be divided into the following three classes, which are characterized by different mechanisms and reaction rates:

- Chemical reaction mechanisms include the catalytic reaction itself (step 4), as well as the chemisorption of reactants and desorption of products (steps 3 and 5, respectively). The kinetics of all of these chemical steps follows an exponential dependence on temperature; that is, exp(–E/RT), where E is the activation energy. For chemical reaction, as well as for chemisorption, E is typically greater than 40 kJ/mol (10 kcal/mol) [5].
- Pore diffusion includes steps 2 (reactants) and 6 (products). The rate of these processes depends primarily on the size and shape of both the pore and the diffusing reactant and product. The rate of pore diffusion varies approximately with the square root of k, where k is the rate constant for the chemical conversion reaction. Its activation energy is approximately half that of a chemical reaction or about 25 to 40 kJ/mol (6 to 10 kcal/mol).
- Bulk mass transfer includes steps 1 (reactants) and 7 (products). The mass transfer rate in this regime is a function of the specific molecules, the dynamic of the flow conditions, and the geometric surface area (external) of the catalyst. Bulk molecular diffusion rates vary approximately with $T^{3/2}$ and usually have apparent activation energies of 8 to 16 kJ/mol (2 to 4 kcal/mol).

Any of the preceding seven steps can be rate limiting and can control the overall rate of reaction in a heterogeneous catalyst. The rate-limiting mechanism can be determined based on the differences in activation energies for the reaction, pore diffusion, and mass transfer. Figure 18.7 illustrates the conversion of a gas-phase reactant A to product B as the temperature is increased. At low temperatures, the rate of chemical reaction is slow relative to diffusion. That area is a reaction kinetics controlled regime. As the temperature increases, the reaction steps with high activation energies and exponential dependence increase the fastest, and control of the overall rate shifts from chemical to pore diffusion. Finally, at high temperatures, both the chemical and pore diffusion rates become sufficiently fast that bulk mass transfer, having a small relative dependence on temperature, becomes rate limiting.

The slope of the conversion curve in Figure 18.7 can give a qualitative picture of the rate-controlling mechanisms. The lower part of the curve, with fast increasing slope, is indicative of chemical reaction control (steps 3, 4, or 5). The relatively flat, temperature-insensitive portion reflects bulk mass transfer control (steps 1 or 7). The intermediate portion is characteristic of pore diffusion control (steps 2 or 7).

The chemical reaction steps can be modeled using the following chemical kinetics equations. Assuming a first-order reaction, the reaction rate expression, Eq. (18.5), can be simplified as follows:

$$r = k \cdot C \tag{18.12}$$

where C is the concentration of reactant A in kmol/m^3. The reaction rate constant (k) increases with temperature, reflecting the fact that reactions increase their rates at high temperatures.

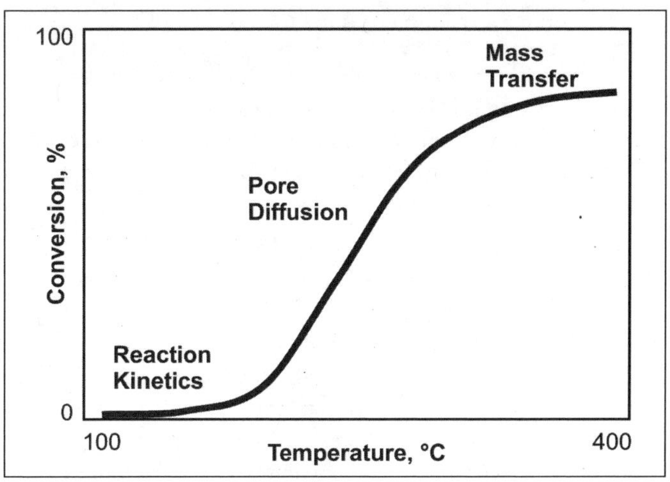

Figure 18.7 *Regions for kinetics, pore diffusion, and bulk mass transfer reaction rate control.*

The relationship between k and T is given by the Arrhenius equation:

$$k = k_0 \cdot \exp(-E/RT) \tag{18.13}$$

or

$$\ln(k) = \ln(k_0) - E/(RT) \tag{18.14}$$

where:
- R = universal gas constant,
- R = 8.31434 kJ/kmol·K
- T = absolute temperature, K.

Experimentally determined reaction rates can be plotted in coordinates of ln(k) vs. (1/T), a representation known as an Arrhenius plot. According to Eq. (18.14), the experimental data will form a straight line with a slope (–E/R) and intercept of $\ln(k_0)$. An Arrhenius plot is commonly used to determine activation energies of chemical reactions. If reaction rates corresponding to the data in Figure 18.7 were plotted in Arrhenius coordinates, the resulting graph would show differences in activation energies for mass transfer, pore diffusion, and reaction kinetics. The differences would be visible as different slopes for line sections corresponding to the respective regimes. Obviously, no straight lines would be observed in transitional areas, for example, for the transition between the reaction kinetics and pore diffusion limited regimes.

For reactions in a monolithic, flow-through reactor, one can solve a material balance equation to determine the relationship between the initial and final concentrations of the reactants. The basic material balance equation across any reactor, assuming one-dimensional plug flow, steady-state, and isothermal operation, is given by Eq. (18.15):

$$v\,(dC/dx) = -r \tag{18.15}$$

where:
- v = velocity, m/s
- x = length, m
- r = rate of reaction or mass transfer, kmol/(m^3·s).

If the process is reaction kinetics controlled and the reaction is of first order, the reaction rate (r) is given by Eq. (18.12). By combining and integrating the equations, one obtains the following relationship between the inlet (C_i) and outlet (C_o) concentrations of reactant A:

$$\ln(C_o/C_i) = -k\tau \tag{18.16}$$

where τ is the residence time (length of reactor/linear gas velocity) in seconds. Equation (18.16) can be used to calculate conversions in a catalytic reactor, which operates in the reaction kinetics regime, provided the rate constant (k) and the corresponding activation energy (E) are known. However, there is no universal method to calculate activation energy for a reaction with a given catalyst. Rather, activation energies of reactions are determined experimentally for each catalyst.

For a bulk mass transfer controlled process, the rate of mass transfer is given as:

$$r = K_g a C \tag{18.17}$$

where:
- K_g = mass transfer coefficient, m/s
- a = geometric surface area (of the monolith substrate) per unit volume, 1/m.

The relationship between the inlet and outlet concentrations of reactant A takes the following form:

$$\ln(C_o/C_i) = -K_g a\tau \tag{18.18}$$

The fractional conversion of A can be then calculated as:

$$(C_i - C_o)/C_i = 1 - \exp(-K_g a\tau) \tag{18.19}$$

The mass transfer coefficient (K_g) can be calculated from various correlations developed from experimental data. One such correlation for high-surface-area catalysts on monolithic substrates [6] can be expressed as follows:

$$N_{Sh} = 0.834(d_h/L)\, N_{Re}^{0.544}\, N_{Sc}^{0.56} \tag{18.20}$$

where:
- d_h = hydraulic diameter of the honeycomb monolith channel, m
- L = honeycomb monolith length, m
- N_{Sh} = Sherwood number, dimensionless
- N_{Re} = Reynolds number, dimensionless
- N_{Sc} = Schmidt number, dimensionless.

The dimensionless numbers are defined as follows:

$$N_{Sh} = K_g a/D \tag{18.21}$$

$$N_{Re} = (W/A_F)\, d_h/\mu \tag{18.22}$$

$$N_{Sc} = \mu/\rho D \tag{18.23}$$

where:
- D = diffusivity of the reactant in air, m^2/s
- W = total gas mass flow rate, kg/s
- μ = gas dynamic viscosity, kg/m·s
- A_F = open frontal area of the catalyst monolith, m^2
- ρ = gas density at operating conditions, kg/m^3.

From the preceding equations, the conversion of a given component can be estimated for a known monolith geometry and exhaust gas flow. First, the diffusivity of the component (D) has to be calculated using methods available in the literature. (The system can be considered a binary mixture of a given component and air; methods by Wilke and Lee or by Fuller give acceptably accurate results [7].) Next, N_{Re} and N_{Sc} are calculated from Eqs. (18.22) and (18.23). Honeycomb size parameters for several monolith configurations are listed in the sections describing ceramic and metallic monolith substrates. Next, N_{Sh} is calculated from Eq. (18.21) and K_g from an experimental correlation, for example, from Eq. (18.20). Finally, the fractional conversion can be calculated from Eq. (18.19).

The accuracy of the overall calculation depends on the accuracy of the correlation for K_g that was used. We should emphasize that the calculated conversions are valid for the mass transfer controlled regime. Therefore, they are valid only for sufficiently high exhaust temperatures. As the temperature decreases, pore diffusion and kinetics start controlling the process and observed conversions will be lower than those calculated for the mass transfer region.

18.2.3 Catalyst Deactivation

18.2.3.1 Modes of Catalyst Deactivation

Theoretically, catalysts are not used or consumed during catalytic reactions. In all practical applications, however, catalysts progressively deactivate with use. The following are major modes of deactivation of monolithic emission control catalysts:

- Poisoning and fouling
- Thermal deactivation (sintering)
- Reactions between catalytic species and the carrier
- Washcoat losses.

Catalyst poisoning is one of the most important sources of catalyst deactivation. It occurs when substances that are present in exhaust gases chemically deactivate the catalytic sites or cause fouling of the catalytic surface. Poisons in exhaust gases from internal combustion engines may be derived from lube oil components or from the fuels.

Thermal deactivation is caused by loss of active catalytic sites due to high-temperature sintering of the catalyst or of the carrier.

Interactions between different catalyst species or between catalyst species and carrier components are another temperature-induced mode of catalyst deactivation. An example of such interactions is the reaction between rhodium and CeO_2 in the automotive three-way catalyst. This type of problem can be prevented by using alternative carriers and special washcoat technologies that physically separate the reacting components.

Catalyst deactivation may also occur due to a physical washcoat loss through erosion and attrition. That mechanism may also be important for emission control catalysts because of the high gas velocities, temperature changes, and differences in thermal expansion between the washcoat and substrate materials.

In practical applications, deactivation of a catalyst is usually caused by a combination of several of the deactivation modes just discussed.

18.2.3.2 Poisoning

A catalyst poison is a substance that deposits on the surface of the catalyst, rendering it less active or inactive. A common cause of catalyst deactivation results from fuel- or lube oil-derived contaminants present in the exhaust gases. There are two basic mechanisms by which poisoning occurs: (1) selective poisoning and (2) nonselective poisoning.

In selective poisoning, a catalyst poison directly reacts with the active site or the carrier to render it less active or completely inactive. In the nonselective poisoning, masking of sites and pores occurs due to a deposition of fouling agents onto the catalyst carrier. Fouling results in a performance loss caused by decreased accessibility of reactants to active sites. Catalyst poisons can be further classified as either temporary or permanent.

Selective poisoning is illustrated in Figure 18.8. In this discriminating process, a poison reacts directly with an active site, decreasing its activity or selectivity for a given reaction. Some poisons (e.g., Pb, Hg, Cd) chemically react with the catalytic component (e.g., Pt) to form a catalytically inactive alloy. Such poisoning leads to permanent catalyst deactivation. Some poisons merely adsorb (chemisorb) onto sites (e.g., SO_2 on Pt) and block that site from further reaction. These mechanisms are, at least partially, reversible. The catalyst activity can be restored by desorbing the poison through heat treatment, washing, or simply by removing the poison from the exhaust gas.

Selective poisoning of the catalyst species results usually in a loss of active catalytic sites with no change in activation energy, because the remaining sites can function as before. When the carrier reacts with a constituent in the gas stream to form a new compound [e.g., $Al_2(SO_4)_3$], pores are usually partially blocked, resulting in increased diffusional resistance, causing a decrease in the apparent activation energy.

Nonselective poisoning (Figure 18.9) is also called *fouling* or *masking*. In this nondiscriminating process, high-molecular-weight materials deposit physically on the outer surface of the catalyst. It has a double effect of decreasing the available catalytic surface area and blocking access to the pores. In effect, fewer active sites are available for reactants and there is an increase in the diffusional resistance in pores. An example of

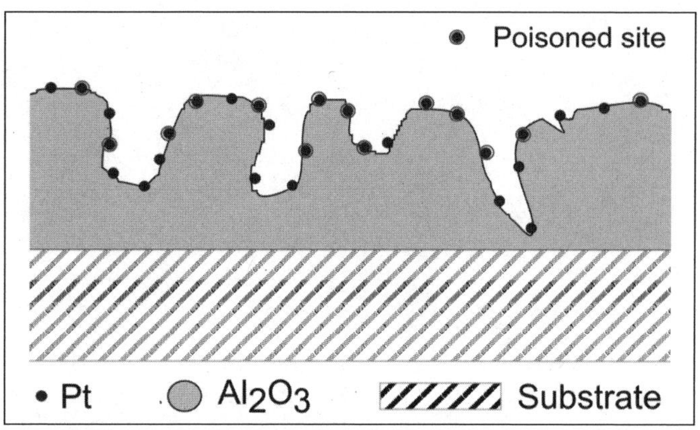

Figure 18.8 *Selective catalyst poisoning.*

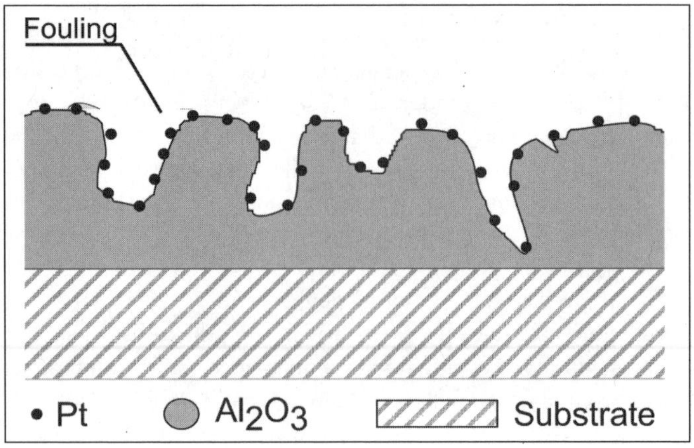

Figure 18.9 Catalyst fouling.

a masking agent is phosphorus from the engine lube oil. Phosphorous compounds produce a glaze over the outer washcoat surface leading to an irreversible loss of catalytic activity.

18.2.3.3 Thermal Deactivation

Thermal catalyst deactivation, known as sintering, can be attributed to changes in the structure of the catalytic component (catalyst sintering) or to changes in the structure of the carrier (washcoat sintering). These processes are influenced by high temperatures, but also by the nature of the catalytic species, the carrier, and the exhaust gas environment.

The catalytic component in an active catalyst is highly dispersed on the carrier. It is common that the fine catalyst crystallites are unstable and grow to better defined crystals. As this process proceeds, the crystals grow larger. The catalytic surface area decreases, leaving fewer catalytic sites on the surface of the crystal available to the reactants. Many active sites are also buried within the crystal with fewer active sites participating in the reaction. Sites may be altered, affecting catalyst selectivity as well. This mechanism is called catalyst sintering (Figure 18.10). Sintering rates of Pt can be reduced by certain rare earth stabilizers (CeO_2, La_2O_3) that are added to the washcoat.

Washcoat sintering is defined as a loss of the internal pore structure, leading to a decrease in the physical surface area of the carrier (Figure 18.11). In the case of Al_2O_3 washcoat, sintering is associated with loss of H_2O from the alumina surface. As washcoat sintering occurs, the pore openings get progressively smaller, introducing greater pore diffusion resistance. Thus, a chemically controlled reaction may gradually become limited by pore diffusion. The presence of this phenomenon is again manifested by a progressive decrease in the apparent reaction activation energy. In extreme cases, the pores close completely, making the catalytic species within the pore inaccessible to the reactants. The presence of stabilizers such as BaO, La_2O_3, SiO_2, and ZrO_2 can slow the rate of sintering in certain carriers. The stabilizers are believed to form solid solutions with the carrier surface decreasing their surface reactivity.

Another mechanism for carrier change involves conversion to a new crystal structure. For example, when gamma-Al_2O_3 converts to delta-Al_2O_3, there is a significant stepwise decrease in the internal surface area from about 150 to below 50 m^2/g.

Certain catalytic sites may occur at interfaces between metal (e.g., Pt) and surface support (e.g., alumina). As both phases sinter, the number and type of interactions change, affecting the catalyst performance.

18.2.3.4 Deactivation Modes and Conversion

Deactivation of catalyst changes the conversion versus temperature curve of a fresh catalyst. Various possible shifts in the character of the curve are illustrated in Figure 18.12 [5]. The shifts in the profile can provide insight into the prevailing mechanism of deactivation.

The way in which different deactivation modes affect the catalytic conversion profile corresponds to the different rate-limiting mechanisms in a heterogeneous catalyst, which include reaction kinetics, pore diffusion, and bulk mass transfer.

Loss of active sites, associated with catalyst sintering or selective poisoning, will affect the reaction kinetics. The activity will shift to higher temperatures, but the maximum

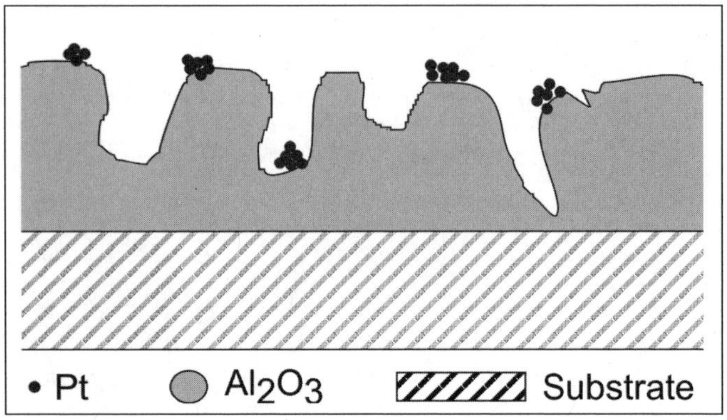

Figure 18.10 Catalyst sintering.

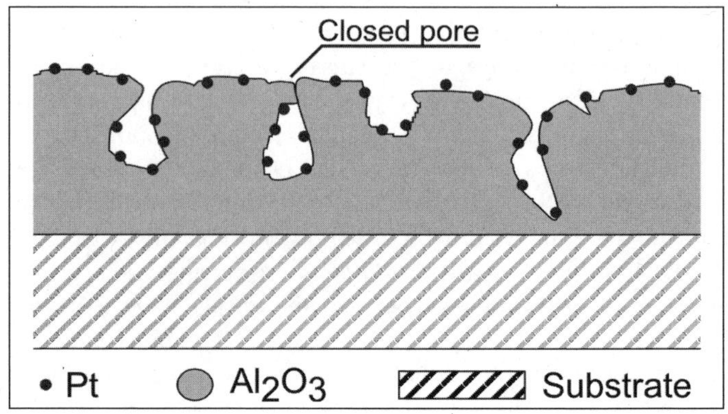

Figure 18.11 Washcoat sintering.

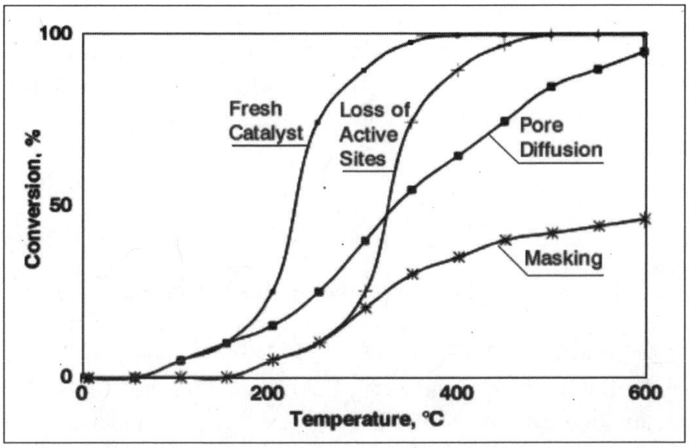

Figure 18.12 *Changes in conversion for various deactivation modes. (Reprinted from Ref. 5 with permission of John Wiley & Sons, Inc. ©1995)*

conversion may be unaffected. Catalyst deactivation associated with increased pore diffusion resistance manifests itself in lower conversions (change in the profile slope) in the pore diffusion control regime, which limits the process at medium temperatures. Catalyst performance at low temperature may be unaffected. Masking is a combination of both loss of surface sites and pore blocking. The conversion is shifted toward higher temperatures with a significant change in slope.

18.3 Cellular Monolith Substrates

18.3.1 Catalyst Substrates

Many of the early automotive catalytic converters in the 1970s utilized pellet or bead-shaped supports. A volume of spherical particles (pellets) made of gamma-alumina, 2.5 to 5 mm in diameter, was placed into a steel shell and contained between two screens to form the catalytic converter. The noble metal catalyst and stabilizers were incorporated into the pellets. That catalyst design originated from catalytic reactors used in the chemical processing industry. Pelleted catalysts had several disadvantages including a high-pressure drop and gradual catalyst loss due to attrition. Monolithic honeycombs, which were developed to address the shortcomings of pelleted supports, have become the standard substrate in today's automotive, diesel, and other emission control catalyst applications.

Monolithic catalyst supports are honeycomb structures with many small, parallel channels running axially through the part. They are sometimes called *flow-through* substrates. The cross section is typically circular or oval, but asymmetric contours are also manufactured. Exhaust gases flowing through the channels contact the catalyst, which is deposited on the channel walls. Major advantages of monolithic supports include high geometric surface area per unit volume (compactness), large open frontal area (low-pressure drop), and excellent attrition resistance.

Monolithic catalyst substrates (Figure 18.13) are made of ceramics or metal. Ceramic substrates (honeycombs) usually have square cells, whereas most metallic substrates have sinusoidal channels. Other channel cross sections are possible, including triangular, hexagonal, trapezoidal, and round. The number of cells can vary between 10 and more than 1000 cells per square inch (cpsi). Parts that were commercially used for internal combustion engine applications in the 1990s had cell densities between 200 and 600 cpsi, with 400 cpsi being the most common cell density in gasoline car applications. Since then, increasingly higher cell densities have been introduced in gasoline cars, with parts in excess of 1000 cpsi being available in both the ceramic [8] and metallic designs [9]. Most substrates for diesel applications remain in the 300- to 400-cpsi range.

Compared to materials used in catalysis, the walls of ceramic honeycombs have large pores and very low specific surface areas of about 0.3 m^2/g. Foils used for metal substrates have no porosity. Because high-surface-area carriers are required for catalysts, it is necessary to deposit a high-surface-area coating onto the channel walls. That coating, called the *washcoat*, is composed of porous, high-surface-area inorganic oxides such as gamma-Al_2O_3. The specific surface area of the catalyst washcoat materials is typically above 100 m^2/g. Noble metal catalysts, such as platinum, are deposited on the surface and within the pores of the washcoat. Exhaust gases in a catalytic converter diffuse through the washcoat pore structure to the catalytic sites where catalytic reactions occur.

The washcoat layer on a metallic foil and on a ceramic substrate is illustrated in Figure 18.14. The thickness of the washcoat layer is 20 to 40 μm. Much thicker washcoat deposits

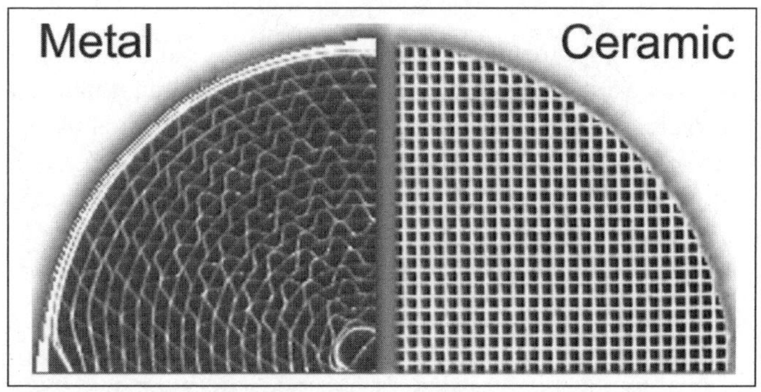

Figure 18.13 *Monolithic catalyst substrates.*

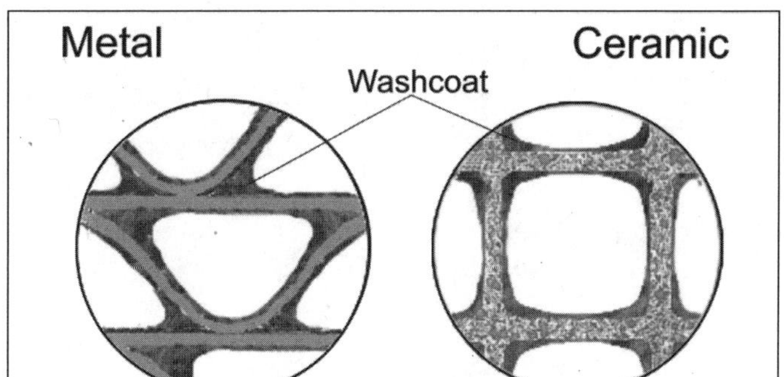

Figure 18.14 *Catalyst washcoat.*

(*fillets*) are formed in the cell corners, especially in the sinusoidal channels of metallic substrates. In some technologies, the metal foil is washcoated prior to forming the honeycomb. These technologies produce clean channels with no fillets in the corners but tend to compromise the mechanical strength and durability of the substrate.

Other, noncellular monolithic substrates have been also researched for catalyst applications. Popular designs include various rigid foams made of cordierite, silicon carbide [10], or metal [11]. Compared to the honeycomb monoliths, these products have low-geometric-surface areas and/or high-pressure drop. They found very limited commercial use as catalyst substrates.

18.3.2 Cellular Properties of Monoliths

A number of terms have been defined to describe the geometrical, mechanical, heat transfer, and hydraulics properties of cellular monoliths. All of these terms are related to the substrate cell geometry. A square cell matrix, typical for ceramic monoliths, is shown in Figure 18.15. Most of the cellular monolith properties are derived from and can be expressed as a function of the cell repeat distance, s, and the wall thickness, w.

Cell density (N) is defined as the number of channels per unit of cross-sectional area. In the case of a square cell geometry, it is a simple function of the cell repeated distance, s, as given by Eq. (18.24). Units of cell density are cells per square inch (cpsi) or cells per square centimeter ($1/cm^2$).

$$N = 1/s^2 \tag{18.24}$$

Specific geometric surface area (GSA), a_g, is the total channel surface area per unit of substrate volume. It is an important parameter of catalyst substrates that is always specified by the manufacturer. For square cells, it can be calculated from cell geometry using Eq. (18.25):

$$a_g = 4(s - w)/s^2 = 4(1 - w/s)N^{0.5} \tag{18.25}$$

Open frontal area (OFA), A_F, is the part of the total substrate cross-sectional area that is available for the flow of gas. The OFA, which is frequently expressed as a percentage of the total substrate cross section (and sometimes called the substrate void fraction), is another important parameter always included in the specifications of cellular products. For square cells, the ratio of the open area to the total cross-sectional channel area, (A_F/A), is given by Eq. (18.26):

$$A_F/A = (s - w)^2/s^2 = (1 - w/s)^2 \tag{18.26}$$

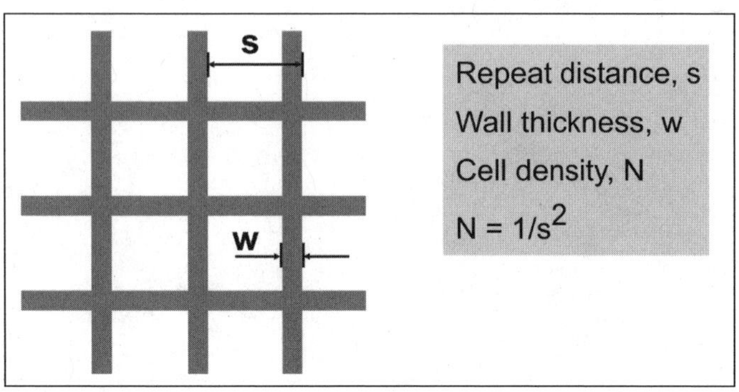

Figure 18.15 *Cell geometry.*

The hydraulic diameter of the substrate channels, d_h, is used as a measure of size for noncircular channels in all hydraulics, mass transfer, and heat transfer calculations. The definition of hydraulic diameter is given by Eq. (18.27). By that definition, the hydraulic diameter of a round cross-section channel is equal to its geometric diameter.

$$d_h = 4 \cdot A_{ch}/O_{ch} \tag{18.27}$$

where A_{ch} is the channel open area and O_{ch} is the channel wetted perimeter.

For cellular catalyst substrates, the preceding definition of hydraulic diameter can also be conveniently expressed by the ratio of the open frontal area to the specific geometric surface area:

$$d_h = 4 \cdot (A_F/A) / a_g \tag{18.28}$$

Equation (18.28) is valid for all cellular products with straight channels and can be used for all cell configurations including sinusoidal channels of metallic substrates.

The GSA, OFA, and the channel hydraulic diameter are different for uncoated and coated substrates. The catalyst coating changes the apparent substrate wall thickness and also, due to the washcoat fillets in the corners, the shape of the cell. Geometrical data that are available from substrate manufacturers usually refer to uncoated substrates. For best accuracy, the parameters of coated substrates should be estimated and used when modeling the catalyst performance.

Bulk density of a catalyst substrate, ρ_b, is expressed by Eq. (18.29).

$$\rho_b = \rho_m \cdot (1 - P) \cdot (1 - A_F/A) \tag{18.29}$$

where:

ρ_m = material density, kg/m^3
P = material porosity, dimensionless.

Total surface area, A_T, is a product of the catalyst volume, V, and its specific GSA:

$$A_T = V \cdot a_g \tag{18.30}$$

Heat capacity of the catalyst substrate, C, may be calculated from Eq. (18.31):

$$C = c_p \cdot \rho_b \cdot V \tag{18.31}$$

where c_p is the specific heat capacity of the substrate material.

18.3.3 Substrate Configuration and Performance

The substrate configuration and geometry has to be carefully selected for a given catalyst application. The monolith matrix has an effect on all aspects of the catalyst performance, including:

- Catalyst conversion efficiency
- Catalyst light-off
- Pressure drop
- Resistance to fouling.

Emission control catalysts operate in the mass transfer controlled regime at high exhaust temperatures. Therefore, at a constant volume of the substrate, any changes that enhance the mass transfer conditions or increase the mass transfer surface area will increase the catalyst conversion efficiency. In particular, increasing the cell density improves the catalyst performance. The practical limit on cell density is set by the substrate manufacturing technologies, the catalyst coating technologies, and by the catalyst pressure drop, which increases with smaller cells.

Catalyst substrate optimization for light-off performance has been conducted for gasoline engine applications. Even though diesels have lower cold start emissions than gasoline engines, light-off performance may be important in some applications, such as for diesel cars in Europe. The influence of the substrate configuration on catalyst light-off is related to the warm-up time of the substrate walls. During light-off, catalysts work in the reaction kinetics controlled regime. The rate of reaction on a given catalyst depends on the catalyst temperature. The catalyst is deposited on the substrate walls, which, at the moment of engine start-up, are cold. A certain time is required before the hot exhaust gases warm up the substrate walls to the temperature required for a catalytic reaction. Any changes to the substrate configuration that increase the heat transfer conditions or the heat transfer surface area will also shorten the catalyst warm-up time and improve the cold start performance. It was suggested that the influence of the catalyst washcoat, which increases the heat capacity of the substrate, is particularly important for cold start performance [12]. The washcoat can increase the catalyst heating time by as much as 40% compared to an uncoated support.

At a constant catalyst substrate volume, the exhaust gas pressure drop depends on the cell density, open frontal area, and the substrate cross section (or aspect ratio). The pressure drop increases with increasing cell density and with decreasing open frontal area. Therefore, substrates with thin walls, which yield high OFAs, are desirable to minimize catalyst pressure drop. The pressure drop can also be minimized by shortening the catalyst and increasing its cross-sectional surface area. This approach, however, is limited by its adverse effects on flow distribution, space constraints, and problems with canning of the monolith.

Fouling of the monolith channels by soot is another source of increased pressure drop in the diesel catalyst. Heavy-duty applications of high exhaust gas temperature are less likely to develop fouling problems. The particulates generated at high temperatures are "dry" and do not adhere to the substrate walls. At low temperatures, however, "wet" soot buildup may develop in the channels, which increases the pressure drop and decreases conversion efficiency. The risk of fouling increases at higher cell densities. Substrates of 300 and 400 cpsi have been successfully used on new, clean diesel engines. Lower cell density substrates of 200 or 300 cpsi are usually used for retrofit diesel applications to minimize the risk of plugging the substrate channels by soot.

18.3.4 Catalyst Pressure Drop

The total pressure drop across a catalyst substrate is a sum of three components: (1) pressure drop in the channels, ΔP_{ch}, (2) pressure loss at the channel entrance, ΔP_{in}, and (3) pressure loss at the channel exit, ΔP_{out}, as represented by Eq. (18.32).

$$\Delta P = \Delta P_{ch} + \Delta P_{in} + \Delta P_{out} \tag{18.32}$$

The channel inlet and exit losses are associated with the contraction and expansion of gas at the inlet and outlet catalyst face, respectively. They are different from pressure losses in the inlet and outlet headers of the catalytic converter, which depend on the exhaust pipe diameter and the header geometry. The pressure drop in headers is not discussed here. The channel pressure drop is the most significant component in the total catalyst substrate pressure drop. The inlet and outlet pressure drop constitutes in most cases less than 10% of the total pressure drop. The channel portion of substrate pressure drop is directly proportional to the gas flow rate. The inlet and outlet pressure drop contributions increase with the square of flow rate.

The frictional pressure drop of exhaust gases in substrate channels can be calculated using the Darcy equation:

$$\Delta P_{ch} = 4f \cdot (L/d_{ch}) \cdot (\rho v^2/2) \tag{18.33}$$

where:

ΔP = pressure drop, Pa
f = Fanning friction factor, dimensionless
L = substrate length, m
d_{ch} = channel diameter, m
ρ = gas density, kg/m^3
v = linear velocity of gas in channels, m/s.

The hydraulic diameter, d_h, is used instead of d_{ch} for noncircular channel cross sections. The gas velocity, v, can be calculated as follows:

$$v = W / (\rho \cdot A_F) \tag{18.34}$$

where:

W = total mass flow rate, kg/s
A_F = substrate open frontal area, m^2.

The dimensionless Reynolds number N_{Re} is used in fluid mechanics as a criterion for the flow regime. Laminar flow is observed at $N_{Re} < 2100$ for the flow of fluids in pipes and channels. In all practical cases, monolithic catalyst substrates operate in the laminar flow regime. The Reynolds number is defined by the following equation:

$$N_{Re} = v \cdot d_{ch} \cdot \rho/\mu \qquad (18.35)$$

where μ = dynamic gas viscosity, kg/(m·s).

The laminar flow Fanning friction factor can be calculated as follows:

$$f = K / N_{Re} \qquad (18.36)$$

where the magnitude of K depends on channel cross section (see Table 18.1). The friction factors given in Table 18.1 were compiled from the literature [13, 14]. Laboratory evaluation of ceramic monoliths with square cells and metal monoliths with sinusoidal cells showed very good correlation with the respective coefficients in Table 18.1 [15].

Because the substrate inlet and outlet pressure losses are relatively insignificant in the total catalyst pressure drop, few experimental data have published that relate specifically to catalyst substrates. The inlet and outlet pressure losses are likely to be influenced by the test equipment, thus, making the experimental studies more difficult. The exhaust gas flow, which is always laminar inside channels, can be turbulent or transitional when outside of the catalyst substrate.

It is convenient to express the inlet and outlet pressure drop by a formula with structure similar to that of Eq. (18.33):

$$\Delta P_{in/out} = K_{in/out} \cdot (\rho v^2/2) \qquad (18.37)$$

The friction coefficients K_{in} and K_{out} can be estimated using known methods for pressure losses due to gas contraction and expansion. For contraction K_{in} can be calculated from the substrate open frontal area as follows:

$$K_{in} = -0.415 \cdot A_F/A + 1.08 \qquad (18.38)$$

Equation (18.38) has been computed from literature data for laminar flow [16]. It is believed that in some situations the calculated pressure drop may be overestimated. Coefficients K_{in} for turbulent flow are approximately 50% lower than those of Eq. (18.38). Frictional losses for transitional Reynolds numbers range between those for the laminar and turbulent losses.

TABLE 18.1
FANNING FRICTION FACTOR

Channel Cross Section	$K = (f \cdot N_{Re})$
Round	16.00
Square	14.23
Equilateral triangle	13.33
Sinusoidal[a]	10.96

a. Duct open height to open width ratio = 0.425.

The friction coefficient for gas expansion with turbulent flow can be calculated from the simple Borda-Carnot equation:

$$K_{out} = (1 - A_F/A)^2 \qquad (18.39)$$

By combining Eq. (18.32) with the Darcy equation [Eq. (18.33)] and Eq. (18.37), the total catalyst substrate pressure drop can be expressed as follows:

$$\Delta P = [4f \cdot (L/d_{ch}) + K_{in} + K_{out}] \cdot (\rho v^2/2) \qquad (18.40)$$

18.4 Ceramic Catalyst Substrates

18.4.1 Introduction

The catalyst substrate is a crucial component that influences the performance, robustness, and durability of catalytic converter systems. Design targets for ceramic catalyst substrates include the following:

- High geometric surface area (GSA)
- Large open frontal area (OFA)
- Low thermal mass and heat capacity
- High use temperature
- Low coefficient of thermal expansion
- Good coatability
- Washcoat compatibility
- Strength
- Oxidation resistance.

Synthetic cordierite is a magnesium-alumino-silicate ceramic material that meets each of these requirements. Practically all of today's ceramic monolithic catalyst substrates are made of cordierite.

A ceramic catalyst substrate is shown in Figure 18.16. Ceramic monoliths are manufactured in an extrusion process. Raw materials such as alumina, talc, clay (kaolinite), and silica are reduced to powder and blended into a water-based paste. Additives that may be used in the process include lubricants (e.g., ethylene glycol), bonding agents (e.g., methylcellulose), and fluxes (alkaline hydroxides). Extrusions made from the paste are dried, cut to length, and calcined at temperatures above 1400°C.

The dimensions of substrates change slightly during calcination. Parts of small size, such as automotive, can be manufactured in a one-step process by predicting that change and extruding parts to dimensions that are slightly different from the specifications of the finished product. That procedure is not possible with large parts, such as those used for heavy-duty diesel trucks. These substrates are initially extruded and calcined to larger dimensions, after which they undergo further processing to match their dimensional specifications. They are then cut to

Figure 18.16 *Ceramic catalyst substrate.*

the final length, their contours machined, their outer skin applied using a ceramic paste, and the parts are then oven dried.

18.4.2 Material Characterization Parameters

Developers of ceramic catalyst supports characterize and compare their materials using a number of mechanical and thermal properties, such as various strength parameters, E-modulus, or fatigue constant. A comprehensive explanation of these properties can be found in texts on ceramics [17]. The following discussion covers definitions of selected material parameters that may be important to the performance of ceramic catalyst substrates [18].

18.4.2.1 Thermal Expansion and Thermal Strain

The coefficient of thermal expansion (CTE) of a solid material is a measure of its expansion when heated. The CTE is defined as follows:

$$CTE = \Delta L/(L\Delta T),\ 1/K \text{ or } 1/^{\circ}C \qquad (18.41)$$

where:

ΔL = length increase upon heating, m
L = original length of the body, m
ΔT = temperature difference, K or °C.

A large positive CTE means that the body expands significantly when heated; a negative CTE means that the body contracts when its temperature increases. When the material heating is not uniform, as may be caused by nonuniform gas velocity or catalyst exotherms, some parts of the catalyst support are expanding more than others, thus straining the material between the hotter and cooler regions. The strain for a given temperature difference (ΔT) is given by Eq. (18.42):

$$\varepsilon = CTE \cdot \Delta T,\ \text{dimensionless} \qquad (18.42)$$

High CTE coefficients in combination with high temperature gradients will cause high internal strain. Therefore, it is desirable for catalyst support materials to have low thermal expansion coefficients.

Cellular cordierite substrates typically have lower CTE values along the axial direction of the part. Higher CTE coefficients are measured along the tangential direction (i.e., perpendicular to the channels), which makes the tangential direction less resistant to thermal shock than the axial direction.

18.4.2.2 Elastic Modulus and Thermal Stress

The elastic modulus (E-modulus) is a measure of the stiffness or rigidity of the body. It can be measured in material samples in both the axial and tangential direction in the substrate using a nondestructive sonic technique (ASTM C 623). The E-modulus is important in determining the resistance to failure under an imposed thermal gradient. The amount of thermal stress (σ) that is generated within the substrate due to differences in thermal expansion of different regions of its body is related to the corresponding amount of strain by the E-modulus:

$$\sigma = E \cdot \varepsilon,\ \text{Pa} \qquad (18.43)$$

Materials of low E-modulus are desirable for increased durability and life span of the substrate, due to lower thermal stress. The E-modulus typically increases with temperature (as well as with washcoat loading), implying higher thermal stresses at higher temperatures.

18.4.2.3 Strength

Substrate strength is important for withstanding packaging loads and subsequent use in the vehicle exhaust stream with the related exposure to engine vibrations, road shock, and temperature gradients. High-strength substrates are desirable for robust catalytic converter systems. The strength of the ceramic material itself is controlled by the type of intra- and intercrystalline bonding, the porosity, pore size distribution, and flaw population. The strength of the cellular structure of the substrate is further determined by its dimensions, cross-sectional symmetry, and its cellular properties, such as cell density, channel geometry, and wall thickness. Substrate strength must exceed the stress to which the material is exposed during both packaging and operation. If the stress exceeds the strength, the substrate will crack.

A number of different strength tests and associated strength parameters, such as tensile strength, isostatic strength, or crush strength, can be determined along different substrate axes. Measurements are performed on specimens carefully obtained from the substrate. The specimens should provide a representative statistical distribution accounting for variations in the substrate pore network and cellular structure. The specimen size should be chosen to represent a sufficient number of cells, typically 200 to 400 cells across the specimen cross section.

Tensile Strength

The tensile strength of cordierite substrates is measured in a flexure test illustrated in Figure 18.17 (ASTM C 158). The specimens are broken in two directions, yielding the axial and tangential modulus of rupture (MOR), typically expressed in MPa or in psi. High MOR values are an indication of good thermal shock resistance of the substrate. The strength of cordierite substrates, as indicated by MOR measurements, increases with temperature, as well as with the washcoat loading.

Compressive Strength

The compressive strength represents the resistance of the substrate to stresses during the packaging process and to pressures from the packaging system during operation. Three basic methods are used to measure the compressive strength:

- Uniaxial crushing along the axial (A), tangential (B), and diagonal (C) directions (Figure 18.18)
- Isostatic pressurization of the whole substrate (Figure 18.19, 3-D test)
- Biaxial compressive strength of whole substrate (Figure 18.19, 2-D test).

The crush test is typically performed using 1″ × 1″ × 1″ cube specimens. Even though the uniaxial crush test does not provide the absolute compressive strength of the whole substrate, it is a simple test for assessing relative strengths along the substrate axes. Its results are very useful in designing of packaging systems for noncircular substrates where the mounting pressure is not uniform.

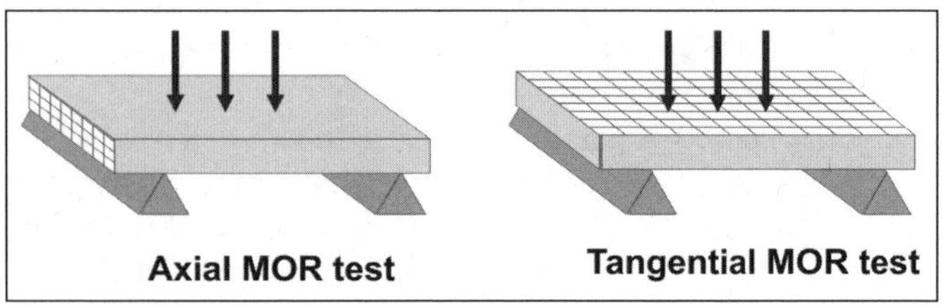

Figure 18.17 *Tensile strength measurement using the flexure test.*

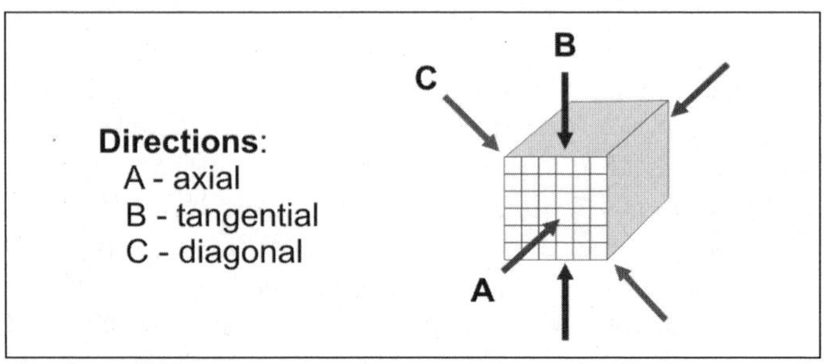

Figure 18.18 *Compressive strength measurement using the crush test.*

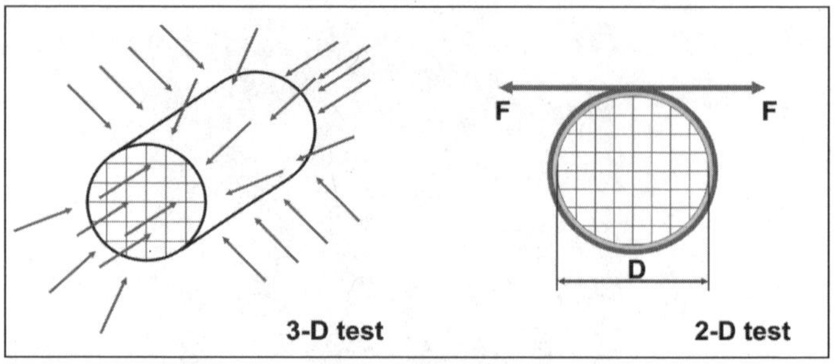

Figure 18.19 *Compressive strength measurement using isostatic tests.*

The 3-D isostatic test and the 2-D isostatic test (i.e., the biaxial compressive test) are more representative of the absolute compressive strength. These tests are well suited for round substrates with uniform mounting pressure, as typically encountered in diesel applications. Isostatic strength results should specify the size of substrate that was used for testing.

18.4.2.4 Fatigue

Ceramic materials are susceptible to loss of strength when stressed for long periods of time in a corrosive environment. Such a loss of strength, know as stress corrosion or fatigue, can be considered a stress-enhanced chemical activity of water vapor at the crack tip. Fatigue occurs under both dynamic and static loading conditions. Dynamic fatigue is a characteristic phenomenon in ceramics, whereby strength varies with the loading rate. The static fatigue, on the other hand, refers to the delayed failure of ceramic materials under constant load; that is, the higher the load is, the quicker the failure will occur.

Fatigue behavior is characterized by the fatigue constant (n). The fatigue constant can be obtained through a dynamic fatigue experiment, where axial MOR is measured at operating temperature and relative humidity as a function of the stress rate ($d\sigma/d\tau$). The n value is then determined as the slope of the MOR vs. stress rate plot. By combining equations describing dynamic and static fatigue, one arrives at the following power law formula expressing the safe allowable stress (σ_s) as a function of time (τ):

$$\sigma_s = \text{MOR}\, [\tau_0/(\tau(n + 1))]1^{/n} \qquad (18.44)$$

where τ_0 is the test duration, in seconds, for measuring MOR.

This equation can be used for either predicting a safe allowable stress for ensuring the specified substrate life (τ) or for estimating the life under a specified stress level. Published data for commercial cordierite formulations indicate that based on the power law fatigue model, the net tensile stress in the substrate must be kept below about 50% of the MOR value to ensure substrate durability of 2500 hours (i.e., 100,000 miles at 40 mi/h) [18].

18.4.2.5 Thermal Shock Parameter

One of the key durability requirements of ceramic catalyst supports is an adequate thermal shock resistance to survive temperature gradients. Due to heat losses from the catalytic converter, the center region of the support experiences higher temperatures than its periphery, which induces tensile stresses in the outer region (the center region may experience tensile stresses during cooldown). The magnitude of these stresses depends on the CTE, the E-modulus, and the radial temperature gradient ΔT. These stresses must be kept well below the modulus of rupture of the support to prevent material failure.

The thermal shock parameter (TSP), defined as the ratio of mechanical strain tolerance to thermal strain imposed by the radial temperature gradient [Eq. (18.45)], is often used as a relative measure of the material ability to withstand steep temperature gradients. The higher the TSP, the better the thermal shock capability of the material.

$$TSP = (MOR/E)/[\alpha_c(t_c - 25) - \alpha_p(t_p - 25)], \text{ dimensionless} \quad (18.45)$$

where:

- MOR = modulus of rupture, Pa
- E = elastic modulus, Pa
- α_c, α_p = coefficient of thermal expansion (CTE) at the center and peripheral region of the support, $°C^{-1}$
- t_c, t_p = temperature at the center and peripheral region of the support, °C.

In gasoline applications, the TSP is typically calculated for $t_c = 825°C$ and $t_p = 450°C$. If TSP values from different sources are to be compared, they should be computed using the same temperature gradient. The TSP may be computed for both the axial and tangential directions, by substituting the appropriate axial/tangential values for MOR, E, and α.

A simplified TSP formula is given by Eq. (18.46) [19]:

$$TSP = MOR/(E \cdot \alpha) \quad (18.46)$$

TSP values computed from Eqs. (18.45) and (18.46) have totally different numerical values [formally, Eq. (18.46) describes a TSP for the temperature difference of 1°C]. While Eq. (18.45) produces cordierite TSP of typically less than 1 (825 – 450°C), TSPs according to Eq. (18.46) range from hundreds to a few thousand. Care must be taken that TSPs for substrate comparisons are calculated using the same method.

18.4.3 Cordierite

Cordierite is an anisotropic, crystalline material of the formula $2MgO{\cdot}2Al_2O_3{\cdot}5SiO_2$. The major phase in commercial materials is cordierite (over 90%) with a few percent of spinel as the minor phase.

The key property of cordierite for the catalyst substrate application is its low thermal expansion. The orthorhombic cordierite crystal is shown in Figure 18.20. Coefficients of thermal expansion for the crystallographic dimensional axes a_1 and a_2 equal $39{\cdot}10^{-7}/°C$ and $26{\cdot}10^{-7}/°C$, respectively. The coefficient of thermal expansion along the c axis is negative, and equals $-13{\cdot}10^{-7}/°C$. All preceding values represent averages for 25 to 800°C. The resulting average coefficient of thermal expansion for random crystal orientation is $17{\cdot}10^{-7}/°C$. The substrate extrusion process, however, produces the preferred orientation in the raw materials, which upon firing causes alignment in the cordierite so that the lowest CTE axis of the orthorhombic crystal lies in the axial direction of the part (which is the same as the direction of extrusion and the direction of gas flow through the catalyst support). Nonaligned crystals generate localized stresses due to their expansion anisotropy, leading to microcracking. The resulting CTE values are determined by the combination of preferred orientation and microcracking. Typically, axial CTE coefficients of the extruded parts range from $1{\cdot}10^{-7}/°C$ to $10{\cdot}10^{-7}/°C$—much below the value for random crystal orientation [18].

Other important thermal properties of cordierite include specific heat (c_p) and thermal conductivity (k). Their average values are [18]:

$$c_p = 1.05 \text{ kJ/kg·K}$$
$$k_\varsigma = 0.209 \text{ W/m·K (axial direction)}$$
$$k_\theta = 0.419 \text{ W/m·K (tangential direction).}$$

Specific heat and thermal conductivity of cordierite are relatively insensitive to temperature and wall porosity.

The maximum continuous use temperature of cordierite is approximately 1200°C. Below this point, the strength of cordierite increases with increasing temperature, presumably due to healing of microcracks. Above about 1200°C, cordierite behaves like a viscoelastic material, that is, exhibits permanent deformation without fracture, and its strength gradually decreases. The melting point of cordierite is 1450°C [19].

18.4.4 Commercial Substrates

Commercial lines of ceramic catalyst substrates have been developed by several manufacturers, including Corning, NGK, and Denso. Corning substrates are presented as an example commercial product. Four cordierite compositions by Corning

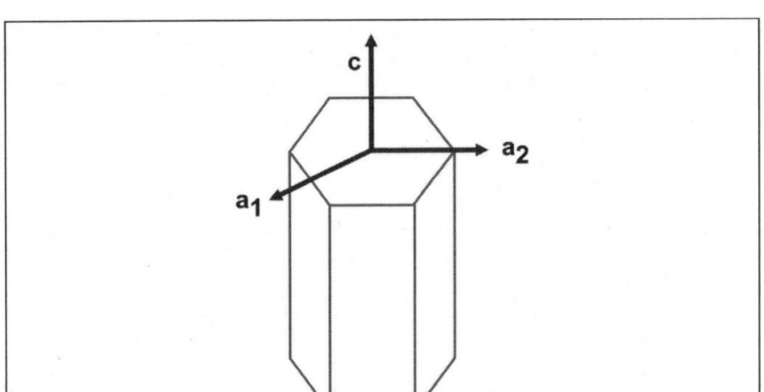

***Figure 18.20** Orthorhombic cordierite crystal.*

are listed in Table 18.2. The first three formulations are used for flow-through catalyst substrates for both gasoline and diesel applications. EX-20 is the standard and most commonly utilized composition. The fourth formulation listed, EX-80, has been developed and is used for wall-flow diesel filter application.

Diesel catalyst substrates usually have cell densities between 200 and 400 cpsi. OEM applications, including U.S. trucks and European diesel cars, use mainly 400-cpsi parts, but sometimes 300- and 350-cpsi substrates are also used. Cell densities for retrofitting applications, such as urban buses or underground mining, are typically 300 or 200 cpsi. Lower cell densities in retrofitting are dictated by the possibility that the channels will become plugged by diesel particulates when used on older, more polluting engines.

Progress in technology has allowed for the manufacture of ceramic cordierite substrates with decreasing wall thickness. The standard configuration for automotive applications, 400/6.5, has a cell density of 400 cpsi and walls of 6.5 mil (i.e., 0.0065″ or 0.17 mm). Substrates of thinner walls (5.5 to 4 mil) are already used for commercial applications. Diesel catalysts, because of their larger size, have thicker walls than their automotive counterparts. Material configurations used by Corning in the large frontal area (LFA) diesel substrates [20], developed for the North American truck application, are listed in Table 18.3. Production capability of the LFA substrates extends to diameters from 152 to 267 mm and lengths from 102 to 204 mm. Typical dimensional tolerances are ±2 mm. Standard LFA substrates manufactured by Corning [20] are listed in Table 18.4.

TABLE 18.2
CORNING CORDIERITE FORMULATIONS

Formulation	Mean Pore Size, μm	Wall Porosity, %	Water Absorption, %
EX-22	1.5	22	9
EX-20	3	35	18
EX-32	9	42	20
EX-80	9.5–14.5	50	23

18.5 Metallic Catalyst Substrates

18.5.1 Introduction

In their most common design, metallic catalyst substrates are made of thin metal foils, flat and corrugated, formed into a honeycomb structure that is placed inside a metal shell, as shown in Figure 18.21. The advantages of metal substrates are their high geometric surface area and low pressure drop associated with the thin walls. Advanced substrates where the foils are brazed/welded together also show good mechanical durability and resistance to thermal shock. The major disadvantage of high-quality metallic substrates is their high cost.

The major area of application of metallic substrates are close-coupled preconverters in gasoline cars. In this

TABLE 18.4
DIMENSIONS OF LFA DIESEL SUBSTRATES

Size	Cells/Wall	Volume, in.^{3}
5.039 × 7.992 × 8.66 in. 128 × 203 × 220 mm	400/7	296
5.764 × 9.767 × 6.0 in. 146.4 × 248.09 × 152.4 mm	400/7	286
5.764 × 9.767 × 7.68 in. 146.4 × 248.09 × 195 mm	400/7	366
9.5-in. diameter × 6.0 in. 241.3 × 152.4 mm	400/7 300/8 200/12	426
10.5-in. diameter × 6.0 in. 266.7 × 152.4 mm	400/7 300/8	520
7.5-in. diameter × 7 in. 190.5 × 177.8 mm	300/8 200/12	310

TABLE 18.3
CONFIGURATIONS OF LFA DIESEL SUBSTRATES

Configuration	200/12	300/8	400/7	236/11.5
Cell shape	Square	Square	Square	Triangular
Material composition	EX-20	EX-20	EX-20	EX-32
Cell density, cpsi	200	300	400	236
Wall thickness, in. (mm)	0.012 (0.30)	0.008 (0.21)	0.007 (0.18)	0.0115 (0.29)
Open frontal area, %	69	74	74	64
Geometric surface area, in.^{2}/in.^{3} (m^2/L)	47 (1.85)	60 (2.36)	69 (2.72)	56 (2.21)
Hydraulic diameter, in. (mm)	0.059 (1.50)	0.050 (1.27)	0.043 (1.09)	0.046 (1.17)
Compressive Strength, B-axis, psi	980	750	525	915
Axial MOR, psi	625	540	440	n/a
Isostatic strength (9.5″ × 6″), psi	1590	1100	1040	n/a
Density, g/in.^{3}	8.7	7.3	7.3	9.6

Figure 18.21 Metallic catalyst substrate (Emitec).

high-temperature application, metallic substrates allow for the elimination of the ceramic mounting mats, thus resulting in a robust catalytic converter system. In some cases, mostly in certain more expensive car models, metallic substrates have been also used in the main catalytic converter. Metallic substrates are uncommon in OEM diesel applications—light or heavy-duty alike—but have been utilized for many diesel aftermarket/retrofit converters.

Most metallic converters are cellular structures made of thin foils in various configurations, with channels formed through corrugation of the foil. Examples of other concepts can be given, such as metal foams [11] or assemblies of flat, perforated foils [21]. These alternative designs have not yet found commercial use.

18.5.2 Substrate Materials

Important material properties for metallic catalyst substrate application include sufficient oxidation resistance, reduced tendency for grain growth (high ductility), and low cost. Ferritic, iron-chromium-aluminum (Fe-Cr-Al) alloys can be tailored to meet the operating temperature requirements for a range of applications and are now in commercial use. They usually contain 15% to 25% Cr and about 5% Al. Sometimes nickel (Ni) containing materials are also used. Aluminum plays an important role in the high-temperature durability mechanism of these alloys. The chromium oxide layer that protects stainless steel surfaces decomposes and loses this ability at temperatures above 900°C. Aluminum oxide is capable of performing this protective function at significantly higher temperatures.

The oxidation behavior of high-temperature alloys is evaluated by a mass gain test. A sample of the alloy foil is exposed to an oxidizing atmosphere at constant, elevated temperature. Materials for automotive catalyst substrates are typically tested at 1100 to 1200°C. During the experiment, the weight of the sample increases due to metal oxide layer formation on the sample surface. That weight increase is a measure of the extent and rate of oxidation. The oxidation behavior of an aluminum-containing alloy is illustrated in Figure 18.22.

Parabolic oxidation is the first oxidation stage. As a protective surface layer of aluminum oxide is formed, a rapid initial increase in sample weight is observed. After the Al_2O_3 layer is developed, it acts as a barrier that isolates the depth of the alloy from oxygen. Cracks appearing in that layer are sealed by aluminum diffusing from the depth of the alloy to the surface. The thickness of the protective oxide layer slowly increases. There is a slow, linear sample weight increase during this stage of the process. A critical transition point occurs when the aluminum in the alloy is exhausted. As a result, cracks appearing in the protective Al_2O_3 layer cannot be sealed. Oxygen reacts freely with the alloy components (Fe, Cr), and the weight rapidly increases due to formation of the corresponding oxides. This stage is called *breakaway,* or *catastrophic, oxidation.*

The useful life of a material is represented by the parabolic and linear oxidation stages. The objective in material engineering is to extend the plateau part of the curve in Figure 18.22. Increases in aluminum content can extend durability, but are limited by deterioration of other alloy properties, such as increased brittleness. Addition of small quantities of rare earth metals also improves the oxidation behavior of these alloys. Yttrium improves the oxidation resistance at concentrations as low as 0.05% [22]. Other additives commonly present in materials used for catalyst supports include La, Ce, Zr, and Ti.

Exposures to high temperatures cause grain growth in Fe-Cr-Al alloys, which leads to lower ductility. Alloys of grain sizes of more than 110 μm lose their durability and become prone to fracture. Ductility is also improved by the addition of rare earths.

Because of their chromium content (more than 13%), these materials are classified as stainless (rust-proof) steels. The chromium content makes them fairly resistant against oxidizing acids. They are, however, sensitive to the attack of reducing acids, such as hydrochloric acid. Frequently, brownish rust spots may be observed on their surfaces. A claw-like crystal of iron chloride grown through the washcoat layer is shown in the microphotograph in Figure 18.23. To avoid corrosion problems,

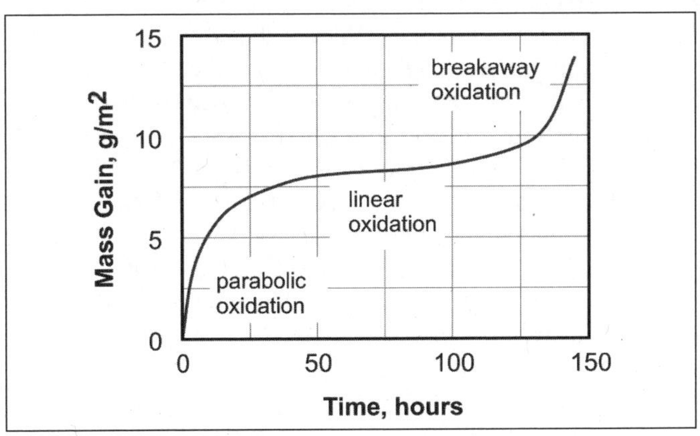

Figure 18.22 Oxidation of Fe-Cr-Al alloy material.

Figure 18.23 Chloride corrosion of Fe-Cr-Al alloy foil. (Reprinted with permission from Ref. 22.)

chlorine-containing compounds should not be used in the washcoating and noble metal impregnation of metallic substrates.

Materials for metallic catalyst supports are available from several steel manufacturers in Europe, Japan, and North America. Example materials from Krupp VDM are Aluchrom Y and Nicrofer 6025 HT [23]. Aluchrom Y is composed of Fe-20%Cr-5%Al + additions of Y, Zr, and Ti. Nicrofer is formulated of Ni-25%Cr-10%Fe-2.1%Al + Y, Zr, Ti. These materials maintain excellent properties at temperatures exceeding 1100°C.

Early metal substrates experienced problems with washcoat adhesion. Usually the metal foil requires surface treatment to produce roughness necessary for the adhesion of washcoat material. Today's metal substrates are typically preoxidized to form surface Al_2O_3, which is compatible with the washcoat alumina. It was also found that the presence of rare earths in the foil material enhances washcoat adhesion.

An alternative process for manufacturing Al-containing foils for catalytic converter applications was proposed by Texas Instruments [24, 25]. In this process, clad foil was fabricated through a rolled bonding technology. The clad foil was composed of a chromium stainless steel layer that was "sandwiched" between two pure aluminum layers. When diffusion annealed, the clad material yielded an alloy (TI DuraFoil) that was suitable for catalytic converter applications. The diffusion annealing process occurred at temperatures between 900 and 1150°C. It was meant to be performed at some point after the corrugation of foil (ideally during final assembly/brazing stage) in the catalyst support manufacturing process. Potential advantages of the clad metal technology include better handling during corrugation (less brittle before the annealing process), higher oxidation resistance, and lower cost.

18.5.3 Substrate Designs

18.5.3.1 Overview

Catalyst substrates are made of very thin metal foils. A foil thickness of 0.04 to 0.05 mm is typically used in substrates have 100- to 500-cpsi cell densities. Even thinner foils having a 0.03-mm thickness are used in high cell density substrates for gasoline applications. Commercial technologies use different combinations of corrugated and flat foils to assemble the substrate. One of the most common designs is to interchange flat and corrugated foil layers, as illustrated in Figure 18.24. The corrugation is adjusted to provide the desired cell density of the substrate. Usually sinusoidal corrugation is used, but other cell shapes are also possible.

Cellular honeycombs can be assembled using a number of methods, such as spiral winding of two foil layers—one flat and one corrugated—into round cross-sectional substrates, more complex nonspiral types of winding, or else stacking straight layers of corrugated/flat foils. The resulting substrate cross section can be round (especially compatible with winding methods), but also elliptic, rectangular, or hexagonal. Once assembled into the substrate matrix, the foils must be held together and attached to the shell to provide adequate mechanical integrity. This can be done by brazing or else by securing the foils mechanically using pins, retaining rings, and/or cross members at the substrate inlet and outlet face. Of these two methods, properly designed brazed substrates can offer superior mechanical and thermal shock durability. However, the substrate has to be skillfully designed, both in terms of the structure itself and the brazing (quality and quantity of the brazing material, areas of the matrix that are brazed, areas of brazing to the shell, etc.). Otherwise, the brazed design will face a number of failure risks, as follows [26]:

- The joint between the shell and the substrate is a critical area that is prone to separation due to different thermal expansion rates of the substrate matrix and the shell.
- Tensile stress in the foil occurs due to uneven temperature distributions in the substrate and simultaneous engine vibration and exhaust gas pulsation. Overly rigid brazed/welded

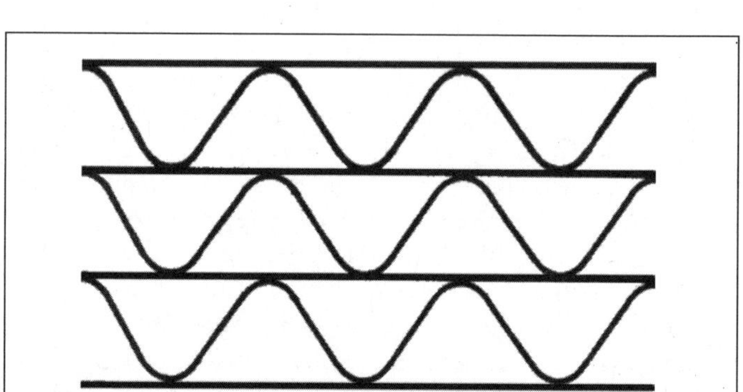

Figure 18.24 Flat and corrugated foils forming a sinusoidal cell pattern.

structures may not provide enough freedom in distributing these tensions through the part. As a result, cells can become distorted and, in extreme conditions, the foil can break.
- In high-temperature gasoline applications, the melting point of the brazing material (reported at 1160°C [26]) may be near the operating temperature.

Based on the catalyst application method, metallic substrates are divided into two types:

1. Uncoated foil substrates
2. Catalyst-coated foil substrates.

In the first type of technology, honeycombs are fabricated using uncoated foil. Finished honeycomb substrates are then washcoated in a process similar to that for ceramic catalyst substrates. In the other type of technology, a catalyst coating is applied onto the foils prior to their assembly into honeycombs. These products provide high-quality, uniform coating of good catalyst utilization, no channel blockage risk during coating, and no washcoat fillets in cell corners. Catalyst-coated foils, however, are much more challenging to braze or weld. As a result, coated foil substrates are typically held together using pins and retaining lips, often yielding a product of inferior mechanical and thermal shock properties in comparison to state-of-the-art brazed substrates.

18.5.3.2 Uncoated Foil Designs

The most common design among early metallic catalytic converters was spiral winding. Two layers of foil, one flat and one corrugated, were wound together into a spiral. The honeycomb matrix was fixed to the shell by either brazing/welding, or else by using retainer rings/lips. A disadvantage of the simple spiral-wound design is its tendency to "telescope." If brazed, spiral designs are also too rigid; when used at high temperature and under thermal shock conditions, cell deformation and detachment of the matrix from the shell may occur. As an answer to these problems, Emitec GmbH developed the so-called S-shaped design [27], which is compared with the spiral winding in Figure 18.25. In the S-shaped substrate, several flat and corrugated foil layers are stacked and twisted around two centers. The wound foils are then put into a steel shell (the mantle) and brazed together, and to the shell, in a high-temperature vacuum process. Compared with spiral, the S-substrate provides more favorable distribution of thermal stress throughout the part, resulting in extended length of service. This design has been successfully used in numerous automotive applications.

Following the introduction of the S-substrate, Emitec developed the SM-substrate design intended primarily for larger diameter parts. In the SM-substrate, three stacks of flat and corrugated foils are first folded and then wound together around the three centers, as illustrated in Figure 18.26.

The production program of Emitec brazed substrates ranged from 33 to 362 mm in the part diameter, and 25 to 174 mm in the part length (1994 data). Elliptical and racetrack-shaped parts were also available. Technical data on selected Emitec substrates are listed in Table 18.5. The data refer to the following foil parameters: thickness, 0.05 mm; density, 7150 kg/m^3; and specific heat, 0.5 kJ/kg·K. Coated substrate parameters assume a 0.05-mm coating thickness. More recently, parts of higher cell density were introduced for gasoline applications (prototypes of densities as high as 1600 cpsi have been reported [9]).

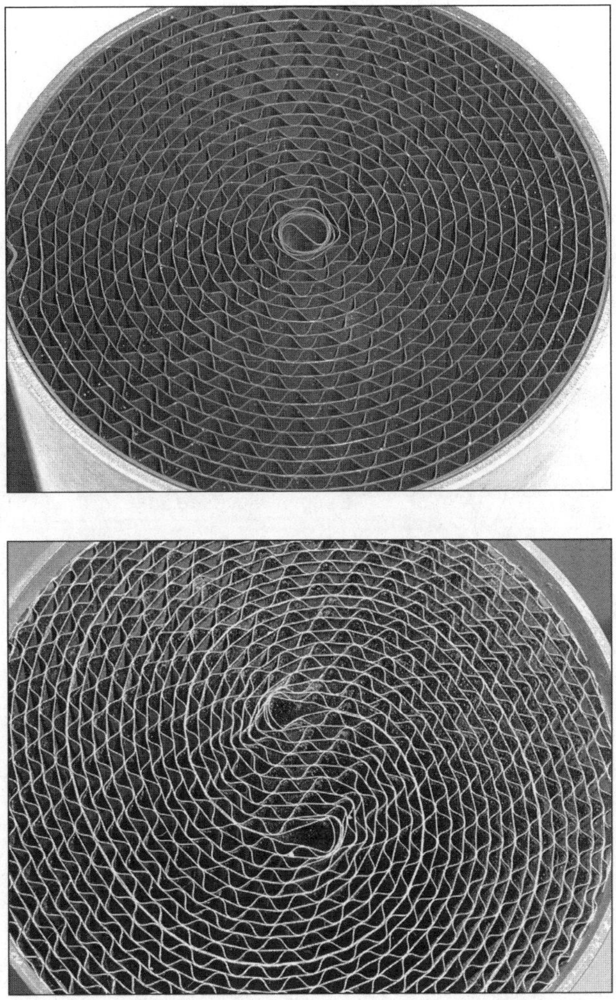

(a)

(b)

Figure 18.25 *(a) Spiral and (b) S-shaped designs.*

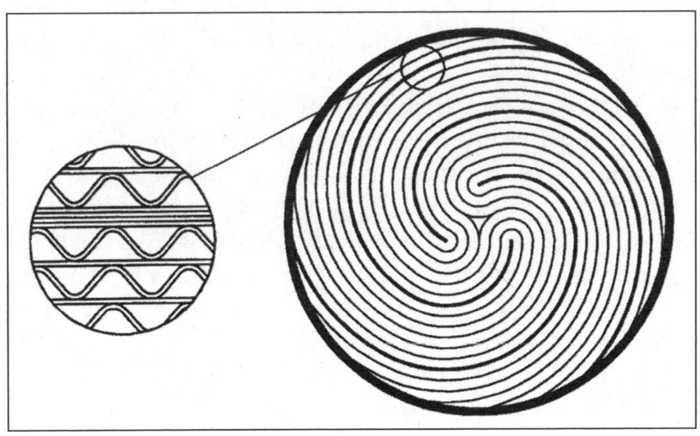

Figure 18.26 *Schematic of Emitec SM-substrate. (Courtesy of Emitec.)*

TABLE 18.5
METALLIC CATALYST SUBSTRATES (EMITEC)

Cell Density, cpsi	100	200	300	400	500
Matrix density, kg/m^3	340	550	630	740	—
Matrix specific heat, J/K·cm^3	0.17	0.28	0.32	0.37	—
Uncoated Substrate					
Hydraulic diameter, mm	2.192	1.419	1.211	0.997	0.914
Open area, %	96	94	93	92	91
Geometric surface area, m^2/L	1.75	2.65	3.07	3.67	3.97
Coated Substrate					
Hydraulic diameter, mm	2.35	1.60	1.34	1.08	0.998
Open area, %	89	83	81	78	76
Geometric surface area, m^2/L	1.52	2.07	2.41	2.89	3.05

Spiral-wound parts are still available from a number of suppliers. They are especially common in catalytic converters for large stationary engines and industrial (nonengine) emission control. Most of these large parts are not brazed or welded and require cross-member supports to prevent telescoping. Round metallic substrates for stationary emission control are available in diameters as large as 813 mm (32 in.) and more.

18.5.3.3 Catalyst-Coated Foil Designs

Application of a catalyst washcoat on the foil before it is assembled into a substrate allows for a better quality, uniform washcoat, as well as for more flexibility in designing complex, multilayered washcoat and catalyst systems [28]. Not only can multilayered washcoats be easily applied on both flat and corrugated foils, but a sequence of different washcoat/catalyst systems can also be applied in the axial direction of the substrate. The application of washcoats/catalyst on foils is performed in automated production lines, were the foil is continuously moving over a system of rollers and passing through a number of stations, such as washcoat application (spraying), drying, and high-temperature calcination. The width of the foil band corresponds to the substrate length. The finished, catalyst-coated foil is stored in rolls, ready for the substrate (catalytic converter) assembly.

Substrates can be assembled from precoated foils by spiral winding of flat and corrugated foil layers, just as was the case with noncoated foils. This type of catalytic converter for both OEM and aftermarket applications has been manufactured by Kemira. The foils are typically held in place by pins, which pierce the part in the radial direction.

In another design, straight stacks of flat and corrugated foils are cut to size and arranged into rectangular parts [29]. Ends of the foils are secured to the rectangular casing by welding. Larger size catalytic converters can be assembled using a number of such modular blocks. This type of modular approach is also common in large converters for stationary emission sources.

Some coated foil converters are designed using only corrugated foil (i.e., without the use of flat foil). In this case, the corrugation in adjacent foil layers must be arranged in a way that prevents interlocking and a collapse of channels. In one design the foil is corrugated in a special "herringbone" pattern, as shown in Figure 18.27. The converter is then manufactured from a single strip of foil, which is folded to fit any shape (e.g., the foil in Figure 18.27 is folded to fit a round profile). It should be noted that channels in this type of substrate are not straight, thus contributing to increased turbulence and better mass transfer at the expense of higher pressure drop. The folded foil package is placed into its shell and secured at both ends with rings or lips. Larger diameter parts may require strengthening with diagonally inserted metal strips. Converters of this design have been manufactured by Engelhard (former W. R. Grace) for use mainly in

Figure 18.27 *"Herringbone" corrugated, coated, and folded foil.*

stationary applications. Attempts to introduce brazing of the precoated foils in order to improve the mechanical and thermal shock durability of the substrate have been reported [30].

A somewhat similar design was proposed by Kemira in the so-called EcoXcell substrate [26]. The substrate was made by stacking foils of straight corrugation so that culmination lines were not parallel, but crossed each other at an angle. Also in this case the part was not composed of straight channels. Rather, the flow pattern resembled that in static mixers, resulting in enhanced mass transfer. Foils in this design could be fixed together at the crossover points by resistive welding and the resultant matrix was laser welded to the shell (welded substrates are made of uncoated foil and catalyzed after the welding).

18.5.3.4 Special Designs

Substrates made of metal foil offer a significant degree of flexibility in shaping the channels. A number of special designs have been developed to address various application needs. For instance, conical-shaped substrates were proposed to improve exhaust gas flow distribution through the converter [31]. A considerable amount of effort was devoted by Emitec to the development of electrically heated catalysts for California ultra low emissions vehicle (ULEV) gasoline cars—a concept without doubt compatible with metallic substrates [32]. Electrically heated catalysts were not commercialized, because the ULEV standards could be met without catalyst heating.

An interesting class of designs from the diesel perspective is the substrates optimized for diesel particle oxidation. It is generally agreed that carbon particulates are not oxidized in conventional flow-through substrate catalysts due to their large size, low diffusional mobility, and short residence time. Considering the laminar flow regime inside substrate channels, a large proportion of solid carbon particulates never contact the catalyst sites. Even if they do, carbon particles would require much longer residence time than the small gas molecules for the oxidation to take place. Substrates are being developed that increase the contact between particles and the catalyst through increased turbulence and also provide a degree of particle trapping to facilitate longer residence times.

One such substrate (the EcoCat substrate commercialized in the diesel "Particle Oxidation Catalyst," or POC, by Kemira) features protrusions in the channels that force the gas to make diversions, thus breaking the laminar flow pattern [33]. Catalysts utilizing this concept were reportedly able to remove 15% to 25% of the solid particulate mass. A similar flow principle was earlier proposed in the Emitec TS substrates to increase mass transfer in gasoline converters [34]. The diesel PM performance of the TS design has not been reported.

In a more recent Emitec substrate, specifically developed for diesel PM control, the flat foil is replaced by a porous sintered metal layer, as illustrated in Figure 18.28 [35]. The corrugated foil includes specially shaped openings, which direct the gas flow to impinge onto the sintered metal. This configuration acts as a "partial filter," holding some of the particulates in the sintered metal pores until they can be oxidized by nitrogen dioxide and/or oxygen.

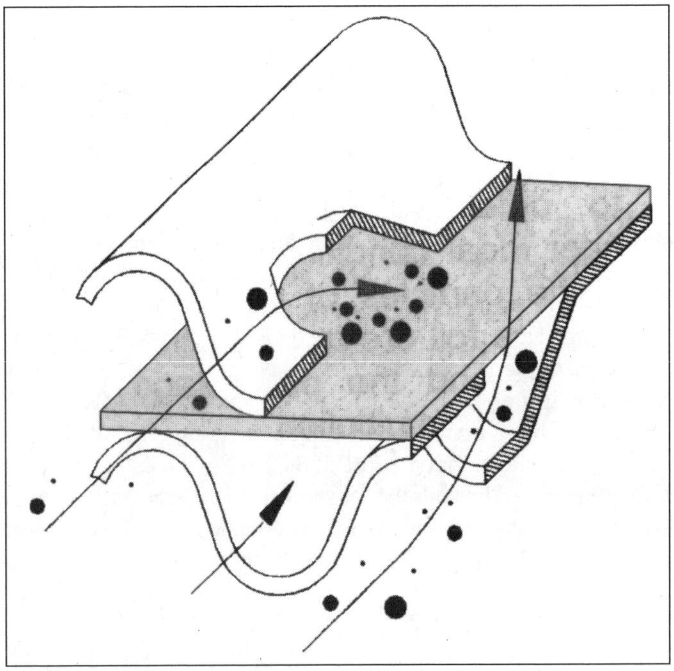

Figure 18.28 *Particle deposition mechanism in partial filter substrate. (Reprinted from Ref. 35.)*

18.6 Catalytic Coating and Materials

18.6.1 Catalyst Washcoats

18.6.1.1 Overview

Modern emission control catalysts utilize monolithic flow-through supports coated with high-surface-area (HSA) inorganic oxides and precious metals. The catalyst coating is usually done in two stages: (1) washcoating and (2) impregnation with noble metal(s).

The washcoat is a porous, HSA layer bonded to the surface of the support. Its exact role, which is certainly very complex, is not clearly understood or explained. The main function of the washcoat is to provide very high surface area, which is needed for the dispersion of catalytic metals. Additionally, the washcoat can physically separate and prevent undesired reactions between components of a complex catalytic system.

Washcoat materials include inorganic base metal oxides such as Al_2O_3 (aluminum oxide or alumina), SiO_2, TiO_2, CeO_2, ZrO_2, V_2O_5, La_2O_3, and zeolites. Some of them are used as catalyst carriers; others are added to the washcoat as promoters or stabilizers; still others exhibit catalytic activity of their own. Good washcoat materials are characterized by high specific surface area and thermal stability. The specific surface area is determined by a nitrogen adsorption measurement technique in conjunction with mathematical modeling known as the BET (Brunauer, Emmet, and Teller) method. Thermal stability is evaluated by exposing samples of given material to high temperatures in a controlled atmosphere, usually in the presence of oxygen and water vapor. The loss of BET surface area, which is remeasured at different time intervals during the test, indicates the degree of thermal deterioration of the tested material.

The washcoat is applied to the catalyst support from a water-based slurry. The wet washcoated parts are then dried and calcined at high temperatures.

A precious metal catalyst(s) is usually applied in a second step called impregnation. During impregnation, the washcoated monolith is exposed to a water-based solution containing catalytic precursors. The supported catalyst is then dried and calcined to its final form. During the calcination, the catalyst precursors decompose to form the final catalyst, usually a metal or a metal oxide. The most common catalysts are precious metals such as platinum (Pt), palladium (Pd), and rhodium (Rh).

Many base metal oxides, such as V_2O_5 or CeO_2, also exhibit catalytic activity. Some of the base metal catalysts (e.g., Ce) are applied through the washcoating process, while others (e.g., V) may be impregnated from aqueous solutions of their precursors.

18.6.1.2 Washcoating Process

The quality of the catalyst washcoat can significantly influence the performance and durability of the finished catalyst. Because the noble metal is subsequently applied to the washcoated parts by impregnation, that is, by "soaking" the washcoat porosity with the catalyst solution, the washcoat loading will determine the noble metal catalyst loading in the finished product. Therefore, it is extremely important for the washcoating process to produce a very repeatable and uniform washcoat layer. The details on the washcoating process and its parameters are guarded as trade secrets by all catalyst makers.

Catalyst washcoat is applied from an aqueous slurry. The alumina powder and/or other washcoat oxides are milled to the required particle size. The particle size distribution of the washcoat powder affects the mechanical strength of the finished washcoat and its adhesion to the substrate, as well as the rheological properties of the slurry during the washcoating process. Alumina, a very hard material, is usually milled using air-jet or ball mills.

In the next step the materials are dispersed in acidified water in a tank with a high-shear mixer. The solid content in the slurry is typically 30% to 50%. After prolonged mixing, the alumina suspension becomes a stable colloidal system.

The amount of washcoat deposited on the substrates depends on, and can be controlled by, the rheological properties (viscosity) of the slurry. The aluminum oxide slurry is a non-Newtonian fluid that changes its viscosity with time and with the amount of mechanical energy supplied to the system (shear rate). At any steady sheer rate, the viscosity of the slurry is a function of its pH. Therefore, the viscosity can be controlled by pH adjustment. Precise viscosity control, however, is probably the biggest challenge in the washcoating process due to the non-Newtonian character of alumina systems.

The washcoat slurry is applied to the substrates, either by dipping or by pouring over the parts, in a specialized coating machine. Excess slurry is cleared from channels with compressed air. The monoliths are then dried and calcined to bond the washcoat to the monolith walls. An electron microscope image of the washcoat surface of a commercial diesel oxidation catalyst is shown in Figure 18.29.

The washcoat can be applied in one, two, or even more layers. Each layer is dried and calcined before the processing of the next layer. Multilayer washcoats are applied for these reasons: (1) The catalyst design may require a different chemical formulation for each layer, and (2) coating/process equipment constraints exist, for example, an inability to handle the very viscous slurries that are needed to apply a thick washcoat in a one-pass operation.

The typical thickness of a washcoat layer is 20 to 40 μm. These numbers correspond to a washcoat loading of about 100 g/L on a 200-cpsi substrate, up to about 200 g/L on a 400-cpsi substrate. The specific surface area of catalyst washcoat materials is typically between 100 and 200 m^2/g.

18.6.1.3 Segregated Washcoat Technology

Noble metals in a complex catalyst system may react with each other, with washcoat components, or with the support material to produce undesired, catalytically inactive compounds. If such reactions occur in a given catalytic system, they are difficult to prevent in the conventional washcoat technology. Because the catalytic metals are impregnated onto the finished washcoat layer, the contact between reacting components cannot be avoided.

Segregated washcoat technologies have been developed to physically separate noble metals by fixing them on a particular base metal oxide of the washcoat before the washcoat is applied to the substrate (Figure 18.30). Through the use of washcoat layers with different oxides and/or noble metals, the components of a catalytic system can be separated. Additional benefits of this technology are that it provides precise control of the noble metal/base metal ratio and an improved noble metal dispersion.

Segregated washcoats were first applied to automotive three-way catalysts. An example of such a catalyst is a trimetal system that includes platinum, palladium, and rhodium [36].

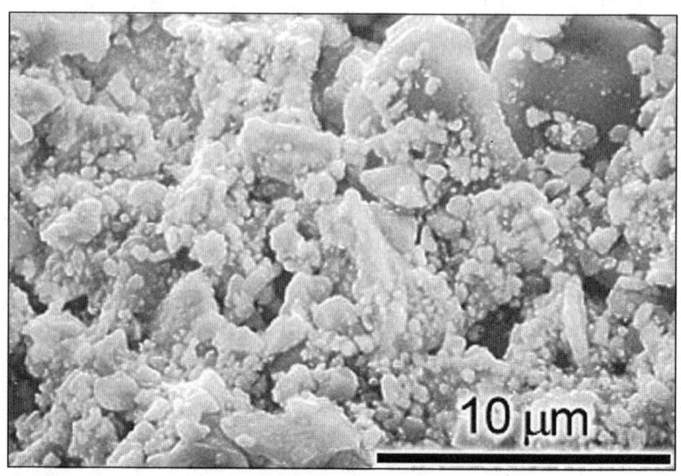

Figure 18.29 Catalyst washcoat.

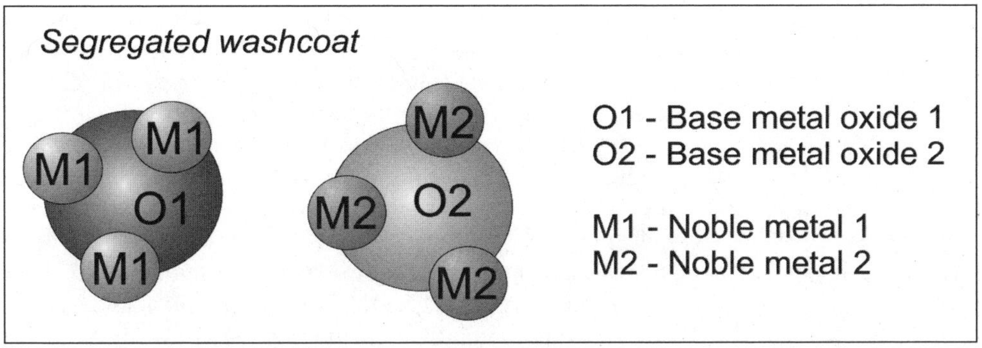

Figure 18.30 *Segregated washcoat schematic.*

The first layer of the catalyst is composed of Pd/Al_2O_3. The second (surface) layer is composed of Rh/Pt/Ce-Zr. That design prevents the formation of palladium-rhodium alloys that otherwise could cause catalyst deactivation.

18.6.2 Aluminum Oxide

Aluminum oxide or alumina is the basic material for an emission control catalyst washcoat; its physical properties are shown in Table 18.6. The HSA gamma crystalline structure (γ-Al_2O_3) is used for catalyst applications. It is characterized, as all catalytic materials, by very high purity. A disadvantage of alumina is its reactiveness with sulfur compounds, which has to be addressed during development of diesel catalysts.

A monohydrate ($Al_2O_3 \cdot H_2O$) alumina compound called bohemite is the starting point for preparation of γ-Al_2O_3. The bohemite precursor undergoes a phase transition during calcination at temperatures of 500 to 850°C. During this process, γ-Al_2O_3, the highest surface area form of alumina, is formed. A typical BET surface area for γ-Al_2O_3 is between 100 and 200 m^2/g. An example specification for a commercial γ-Al_2O_3 for catalyst washcoat application [37] is listed in Table 18.7. The cumulative pore volume distribution of the same alumina material is shown in Figure 18.31 as a function of its pore size.

Continuous heating of γ-Al_2O_3 causes loss of surface hydroxyl groups and sintering. At about 1150°C it converts to the lowest physical surface area structure called α-Al_2O_3. A typical pore diameter for α-Al_2O_3 is 1000 Å and the surface area 1 to 1.5 m^2/g.

TABLE 18.6
PHYSICAL PROPERITES OF ALUMINUM OXIDE

Name	Aluminum oxide
Formula	Al_2O_3
Formula weight	101.94
Appearance	White powder
Density	4 g/cm^3
Melting point, °C	1999–2032
Suppliers	LaRoche, Condea-Vista, Alcoa, W. R. Grace

TABLE 18.7
EXAMPLE SPECIFICATIONS OF COMMERCIAL γ-Al_2O_3

Test	Specification	Typical
Chemical Composition		
Al_2O_3 content,[a] wt%	97.5 min	98
Total ignition loss,[a] wt%	2.5 max	2
Residual bohemite, wt%	2.0 max	Nil
SiO_2, wt%	0.030 max	0.02
Fe_2O_3, wt%	0.015 max	0.01
Na_2O, wt%	0.015 max	0.01
Physical Properties		
Crystal structure	—	gamma
Surface Area, m^2/g	130–200	150
Crystallite size, Å	—	55
Loose bulk density, g/dm^3	600–800	650
Particle Size		
• below 45 μm, wt%	30 min	45
• above 90 μm, wt%	—	15
• average size, μm	—	60
Pore Size Distribution		
• below 65 Å, %	3–20	10
• below 80 Å, %	20–60	50
• below 100 Å, %	60–90	80
Total pore volume, cm^3/g	0.30 min	0.42

a. Predried at 250°C for 30 min.

The presence of certain elements in the Al_2O_3 can have a very significant influence on its thermal stability, both negative and positive. Small amounts of Na_2O present in Al_2O_3 act as a flux, enhancing the sintering of alumina. In contrast, several metal oxides, including La_2O_3, SiO_2, BaO, and CeO_2, have a stabilizing effect on the alumina surface area and reduce its sintering rate. Stabilized aluminas are available from some suppliers.

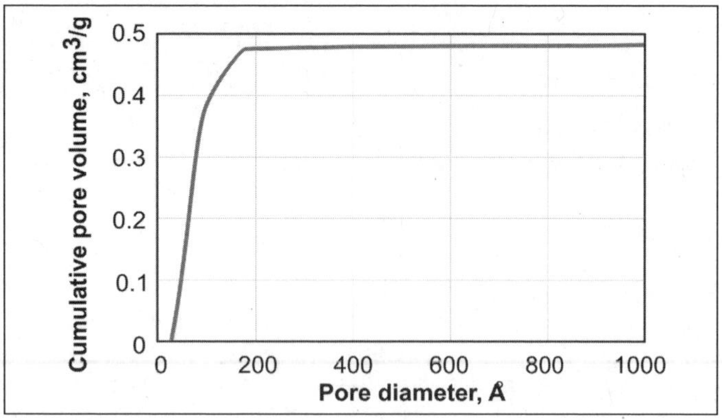

Figure 18.31 Pore distribution of a commercial gamma-alumina.

18.6.3 Silicon Oxide

Silicon oxide or silica exhibits low activity toward reacting with sulfur-bearing compounds in exhaust streams, which makes it a suitable catalyst carrier for some diesel catalyst applications. (Some of its properties are listed in Table 18.8.) High surface SiO_2 materials can provide BET surface areas as high as 300 to 400 m^2/g. They have a small amount of chemically held water, giving rise to some surface acidity.

Silica can be used either as a catalyst carrier or as a stabilizer of alumina. Pure silica, compared to alumina, exhibits an inferior thermal stability at higher temperatures, which disqualifies it as a carrier for high-temperature applications, such as a gasoline engine catalyst.

18.6.4 Cerium Dioxide

Cerium dioxide or ceria is a standard component of the automotive three-way catalyst washcoat, added in quantities up to 25%. Ceria is the most important promoter in the automotive emission control catalyst (see Table 18.9). The most important function of ceria in the three-way catalyst is to provide oxygen storage, which is possible through a cycling between Ce^{4+} and Ce^{3+}. Other effects attributed to ceria include stabilization of alumina, promotion of the steam re-forming reaction, promotion of noble metal dispersion, and promotion of noble metal reduction.

Certain diesel oxidation catalyst formulations include high loadings of ceria. The function of ceria is catalytic oxidation/cracking of the SOF fraction of diesel particulates (see Chapter 19).

HSA cerium oxide is produced by calcination of cerium compounds. The BET surface area of ceria can be as high as 270 m^2/g, but most automotive three-way applications use ceria of about 150 m^2/g surface area. High-temperature stabilized varieties, which are capable of withstanding 900 to 1000°C, have surface areas of 6 to 60 m^2/g.

Zirconium oxide is frequently added to ceria in order to increase its thermal stability.

TABLE 18.8
PHYSICAL PROPERTIES OF SILICON OXIDE

Name	Silicon dioxide
Formula	SiO_2
Formula weight	60.06
Appearance	White powder

TABLE 18.9
PHYSICAL PROPERTIES OF CERIUM DIOXIDE

Name	Cerium dioxide, ceric oxide
Formula	CeO_2
Formula weight	172.1
Appearance	Light-brown to pale-yellow powder
Density	7.28 g/cm^3
Melting point, °C	~2500
Suppliers	Rhone-Poulenc, Molycorp, W. R. Grace

18.6.5 Titanium Dioxide

Titanium dioxide is used in some diesel catalysts as an inert, nonsulfating carrier (Table 18.10). There are two important crystal structures of titanium dioxide: anatase and rutile. The anatase form is important for catalyst applications. It has the highest surface area of 50 to 120 m^2/g and is thermally stable up to 500°C. The rutile structure has a low surface area of below 10 m^2/g. A conversion of anatase into rutile, which takes place at about 550°C, leads to catalyst deactivation.

18.6.6 Zirconium Oxide

Zirconium oxide can be used as a thermal stabilizer and promoter of ceria in the automotive three-way catalyst and also as a nonsulfating component of diesel oxidation catalyst washcoats (Table 18.11). Zirconium oxide has a BET surface area of 100 of 150 m^2/g. It rapidly loses its surface area at 500 to 700°C. Better thermal stability can be achieved by the use of a wide range of dopants including La, Si, Ce and Y.

18.6.7 Zeolites

Zeolites are synthetic or naturally occurring alumina-silicate compounds with well-defined crystalline structures and pore

TABLE 18.10
PHYSICAL PROPERTIES OF TITANIUM DIOXIDE

Name	Titanium dioxide
Formula	TiO_2
Formula weight	79.9
Density	3.84 g/cm^3 (anatase) 4.26 g/cm^3 (rutile)

TABLE 18.11
PHYSICAL PROPERTIES OF ZIRCONIUM DIOXIDE

Name	Zirconium dioxide
Formula	ZrO_2
Formula weight	123.22
Suppliers	Magnesium Elektron, Rhone-Poulenc

TABLE 18.12
PHYSICAL PROPERTIES OF ZEOLITES

Name	Zeolites, molecular sieves
Formula	$xSiO_2 \cdot yAl_2O_3$
Suppliers	GSA, UOP, Degussa, Zeolyst, Zeochem

sizes (Table 18.12). The dimensions of zeolite pores are typically between 3 and 8 Å, which falls into the range of molecular sizes. Any molecule of a larger cross-sectional area is prevented from entering the channel of the zeolite cage. For this reason, zeolites are often referred to as molecular sieves. There are many types of zeolites, some commercial materials including ZSM-5, A, mordenite, Y, and beta.

An important characterization of zeolites is based on the number of silica ($SiO_{4/2}$) and alumina ($AlO_{4/2}$) tetrahedrons (T-units) used to build the pore framework. The Si/Al ratio reflects the chemical and physical properties of zeolites and is a measure of the electron charge and the polarity of the zeolite. The following is a commonly used structural classification of zeolites with respect to their pore size [38]:

- *Large pore zeolites:* 12 T-units form the pore opening (e.g., ZSM-12).
- *Medium pore zeolites:* 10 T-units form the pore opening (e.g., ZSM-5).
- *Small pore zeolites:* 8 T-units form the pore opening (e.g., ZSM-58).

Zeolites are characterized by high specific surface areas. For example, the ZSM-5 zeolite has a surface area of ~400 m^2/g, and zeolite mordenite is 400 to 500 m^2/g. Most zeolites are thermally stable up to 500°C.

Zeolites for some catalytic applications are ion exchanged with metal cations. The acid form of zeolite (HZ) is first treated with an aqueous solution containing NH_4^+ (NH_4NO_3) to form the ammonium exchanged zeolite ($NH_4^+Z^-$). This is then treated with a salt solution containing a catalytic cation forming the metal exchanged zeolite (MZ).

Zeolites, due to their repeatable and well-defined pore structure, are excellent adsorption materials. They have been used as adsorbents in numerous applications including drying, purification, and separation. Synthetic zeolites are also used as catalysts in petrochemical processing.

In recent years, zeolites have been increasingly used for diesel emission control, both as catalysts (SCR, lean NO_x catalyst) and adsorbers (hydrocarbon traps in diesel oxidation catalysts).

18.7 Catalytic Converters

18.7.1 Introduction

Catalyst substrates coated with the active catalyst washcoat are packaged in steel housings to form catalytic converters. The major catalytic converter design considerations for catalytic converters are as follows:

- Mechanical durability
- Heat losses
- Flow distribution
- Pressure drop.

Emission performance durability and mechanical durability are two key aspects of the overall durability of an emission control system. The emission durability depends on the quality of the catalyst coating and on the operating conditions such as temperature or levels of catalyst poisons in the exhaust gas. The catalytic converter design has to provide the required mechanical system durability.

Catalytic converters must provide adequate protection for the substrate under harsh operating conditions in the vehicle's exhaust system. Despite exposure to high temperatures and thermal shock, moisture and corrosive environments, as well as mechanical vibration [39] (Table 18.13), they endure hundreds of thousands of kilometers. The final mechanical durability of the emission control system is a combination of the substrate durability, durability of packaging materials, and packaging technology.

The durability of converter prototypes is usually evaluated by a hot vibration test. Most vehicle manufacturers have their internal specifications for such tests. Generally, the converter is placed in a vibration test stand under controlled vibration acceleration and frequency conditions. Accelerations as high as 75 g are sometimes used for testing. Hot gases of controlled temperature flow through the converter during the test. The converter integrity is evaluated after a prescribed test time. Tests might include various modifications, such as switching between hot and cold gas flow (thermal cycling) or periodic spraying of water onto the hot outside converter walls.

Because the diesel engine is more durable than its gasoline counterpart, the required life expectancy for diesel catalytic converters is also longer than that for gasoline converters. For

TABLE 18.13
OPERATING CONDITIONS OF CATALYTIC CONVERTERS

	Gasoline	Diesel
Temperature range, °C	300–1100	100–650
Temperature gradient, °C	100–300	100–200
Space velocity, 1/h	30,000–100,000[a]	60,000–150,000 [a]
Vibration acceleration, g	28	10–20

a. Higher space velocity may be used in aftermarket applications.

example, the current U.S. EPA durability requirement for emission control systems on heavy-duty diesel engines is 8 years or 290,000 miles (467,000 km), whichever occurs first. That requirement was extended to 10 years or 22,000 hours or 435,000 miles (700,000 km) in the year 2004.

In situations where thermal losses from the converter are important, they have to be modeled during the converter design. Double-walled designs with either air gaps or ceramic fiber insulation are commonly used on gasoline converters in the close-coupled location, which are optimized for cold start HC performance. Because diesel cold start emissions are much less critical, such designs have not been used for diesel converters. However, due to the low temperature of diesel exhaust gases, diesel converters should be placed close to the exhaust manifold or exhaust pipe insulation should be applied to ensure satisfactory catalyst performance. Low-temperature performance, including during a cold start, is becoming increasingly important for diesel catalytic converters, especially in light-duty applications.

The geometry of converter headers, especially that of the inlet header, can influence the exhaust gas flow distribution in the catalyst. It is believed that flow maldistribution negatively affects catalyst performance and/or durability. That opinion, although not supported by convincing experimental data, became a widely accepted consensus. Even if emission performance is not improved, use of a skillful design for the headers can certainly decrease the total catalytic converter pressure loss.

18.7.2 Packaging Materials

18.7.2.1 Overview

The following basic materials are used for the packaging of ceramic catalyst substrates:

- Ceramic mat or steel wire mesh for wrapping the core
- End seals to prevent exhaust gas leaks through the wire mesh or to protect the mat
- Steel for the converter shell.

18.7.2.2 Ceramic Mats

Ceramic monoliths are wrapped in a special packaging material that holds them securely in the steel housing, uniformly distributing pressure and preventing cracking. Ceramic fiber mats are most commonly used for packaging of catalytic converters for both gasoline and diesel applications. These packaging mats can be classified as intumescent (heat-expandable) mats or nonintumescent mats.

Intumescent mats are most commonly used for today's gasoline and diesel catalytic converters. They have the property of partly irreversible expansion once exposed to high temperatures. Thermal expansion curves for the mats are available from their manufacturers (3M, Unifrax). Once expanded, they increase the holding pressure on the substrate, thus providing a very secure mounting system. Thanks to their temperature expansion characteristics, the intumescent mats can actually increase their mounting pressures at high temperatures, compensating for the mounting pressure loss due to the thermal expansion of the steel housing.

The intumescent mats are made of alumina-silica ceramic fibers and contain vermiculite, which provides their thermal expansion capability. Typical compositions have 30% to 50% of alumina-silica fibers, 40% to 60% of vermiculite, and 4% to 9% of an organic binder. After the converter is assembled, the mat has to be exposed to temperatures in excess of 500°C to achieve the initial expansion. In gasoline applications, this is done on the vehicle during the initial hours of engine operation. Oven treatment of catalytic converters may be required for some diesel applications, where converter temperatures may never reach 500°C during regular vehicle operation. The organic mat binder, which decomposes at high temperatures, is responsible for the characteristic odor emitted when the mat is first heated.

The vermiculite component imposes relatively low maximum operating temperature limits for intumescent mats. The mats lose their holding pressure dramatically at temperatures of about 750°C. That temperature is usually defined as the average mat temperature. Therefore, mats can be used at higher exhaust temperatures provided there is a temperature gradient across the mat due to heat losses from the outside converter shell. The use of intumescent mats may be limited in hot isothermal applications where no heat loss occurs through the converter wall. Such situations include catalysts mounted inside mufflers, for example, for motorcycle applications.

Nonintumescent mats provide much higher temperature limits of about 1250°C. They have been developed for high-temperature applications and products that cannot withstand high mounting pressures, for example, ceramic substrates of very thin walls. Because they do not depend on vermiculite expansion, they do not require an oven treatment when used in low-temperature diesel applications.

The main component of nonintumescent mats is alumina fiber. The mats do not contain vermiculite but they still include organic binders. The mat mounting pressure is approximately constant over the entire temperature range. However, because the converter shell expands with temperature, a decrease in the effective converter mounting pressure is observed at higher temperatures. Therefore, the nonintumescent mats, quite opposite to the vermiculite mats, hold the catalyst substrate most securely at low temperatures. As the temperature increases, the substrate inside the converter is held with decreasing force.

The catalytic converter shell has to be designed in such a way as to provide the required mounting pressure for a given catalyst substrate and a given mat. The mounting pressure of the mat increases exponentially during the compression from its initial bulk density to the final target density. The mounting pressure exhibits viscoelastic relaxation; that is, the peak initial pressure that occurs at the canning decreases significantly during the first seconds or minutes thereafter due to the realignment of mat fibers. The mounting pressure loss of intumescent mats due to the relaxation varies, depending on the speed of canning, between 30% and 60% of the initial peak mounting pressure. Higher closing speeds result in higher initial peak pressures and more pressure relaxation. Low closing speeds result in lower peak pressures and less pressure relaxation. Further pressure losses are incurred due to the organic binder burnout when the mat is first heated. Finally, there is gradual mounting pressure loss during vehicle operation. That pressure loss is influenced by several operational factors. For example, a significant and irreversible loss of holding pressure is incurred when intumescent mats are soaked with water.

Due to the relaxation and pressure losses, the mounting pressure is not a convenient parameter for the canning process specification. Instead, the mounting density (also called the gap density) is commonly used for that purpose. Typical mat mounting densities along with their bulk (uncompressed) densities are listed in Table 18.14. The exact density for a given application should be determined in consultation with the mat manufacturer.

Another important property of catalytic converter mounting mats is their weight/area ratio, sometimes also called the *basis weight*. The weight/area ratio is expressed in g/m^2 or kg/m^2. (Because these are units of mass rather than weight, the customary term *weight/area* should be, strictly speaking, replaced by *mass/area*.) Available mats have their weight/area in the range from 1050 to 6200 g/m^2 and an uncompressed thickness between 1.5 and 10 mm. Intumescent mats of 3000 to 4000 g/m^2 are typically used for automotive converters. Higher mat weights, such as 6200 g/m^2 (6.2 kg/m^2), are recommended for more demanding applications or for large converters.

The gap thickness for a given target mat density and mat weight can be calculated from the following equation:

$$s = M/\rho_m \qquad (18.47)$$

where:

s = gap thickness,
M = mat weight/area, kg/m^2
ρ_m = target gap density, kg/m^3.

Packaging mats undergo erosion caused by the impingement of hot exhaust gases. Resistance to erosion is an important characteristic of the mat, which should be provided by its supplier. The resistance to erosion depends strongly on the gap density. An exponential relationship between these variables is illustrated in Figure 18.32 [40].

TABLE 18.14
MOUNTING DENSITIES OF CERAMIC MATS

	Bulk Density, kg/m³	Mounting Density, kg/m³	Approximate Mounting Pressure, kPa
Intumescent (vermiculite) mats	400–700	900–1100	200–1000
Nonintumescent mats	100–200	400–600	200–1000

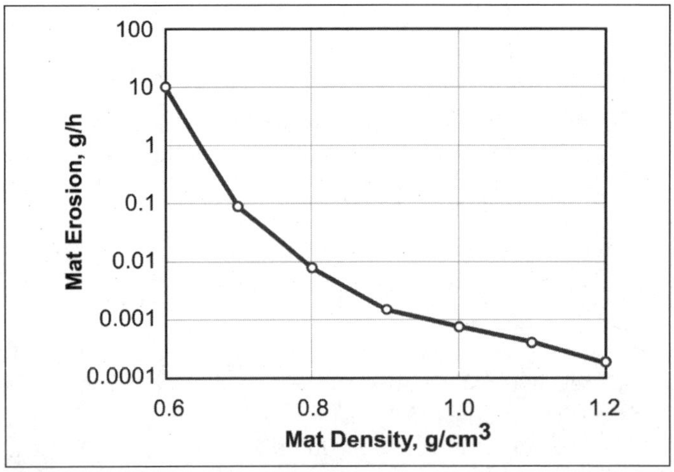

Figure 18.32 *Mat erosion as a function of gap density. (Data from Ref. 40.)*

Many other properties of the mounting mats are specified and/or tested and are available from mat manufacturers. The list of these properties includes thermal conductivity, gas sealing properties, friction coefficients, and more.

18.7.2.3 Wire Mesh

Knitted stainless steel wire mesh may be used instead of mats to package ceramic catalyst substrates. The wire mesh exhibits less favorable mounting pressure characteristics than mats and also requires end seals to prevent bypassing of the substrate by gas.

End seals, usually made of compressed wire mesh, may be used either on the inlet side or on both sides of the substrate, between the substrate and the shell. The end seals may serve two functions:

- To prevent exhaust gas from bypassing the catalyst substrate when used together with the wire mesh mount
- To form an erosion barrier when used with ceramic mat mount.

18.7.2.4 Steel

Steel used for a robust packaging system should have a high yield strength, good oxidation resistance, and low thermal expansion. The most common material employed for catalytic converters is the ferritic stainless steel type AISI 409. It contains 11% chromium and exhibits good resistance to atmospheric corrosion and to stress corrosion cracking. It exhibits the lowest thermal expansion coefficient among stainless steels. Its yield strength is excellent at low temperatures but deteriorates rapidly above 500°C.

An alternative material for very demanding, high-temperature applications is the type AISI 439 stainless steel. It is a titanium stabilized ferritic steel with 18% chromium alloy. The type 439 steel exhibits better yield strength and oxidation resistance than the 409 type.

18.7.3 Canning Methods

18.7.3.1 Overview

Ceramic monoliths are wrapped in the mat and canned into a steel housing. Four basic methods are used for canning and for the gap control, as follows:

- Clamshell
- Tourniquet
- Shoebox
- Stuffing.

A schematic illustration of these methods is presented in Figure 18.33.

In the clamshell method, the mat-wrapped monolith is placed between two stamped profiles. The profiles are then pressed together, until their flared edges meet, and welded. In the tourniquet technology, the substrate is wrapped in the mat and placed in a round, longitudinally split can. Next, a controlled force is applied to the can, which causes the metal walls to overlap. The overlapped edge is welded while still under pressure. The stuffing method utilizes a section of a tube. The monolith is enfolded in the mat and pushed under pressure into the tube. Finally, the shoebox method utilizes a can composed of two profiles. The overlapping edges of the profiles are closed and welded under a controlled force.

The last step in the converter manufacturing process is to weld cones (in the case of the clamshell method, cones are usually a part of the stamped profiles), flanges, or fittings to both ends of the can to facilitate installation in the exhaust system.

The preceding canning methods utilize two different gap control mechanisms: (1) fixed gap and (2) fixed canning force. From the welding process perspective, the methods produce converters with one or two seams. These classifications [40] are illustrated in Table 18.15.

Closing the can using a fixed force offers more accurate gap density control by eliminating the dimensional tolerance influence of the substrate, can, and the mat itself. Closing the can to a fixed gap has the advantage of producing a converter of fixed final dimensions, which simplifies the converter design, primarily with respect to welding of cones to the finished can.

TABLE 18.15
FOUR BASIC CANNING METHODS

	Fixed Gap	Fixed Force
Single seam	Stuffing	Tourniquet
Dual seam	Clamshell	Shoebox

The single-seam design is usually preferred for round or oval converters of low aspect ratios, where it can provide uniform gap density distribution. Single-seam shells also provide more manufacturing flexibility and require less expensive tooling. The dual-seam design is usually required for oval converters with high aspect ratios. In this case, reinforcing ribs are stamped in the shell to prevent its deformation and the resulting gap nonuniformity. The dual-seam shells are produced in stamping processes that require very expensive tooling and have to be justified by high production volumes.

18.7.3.2 Clamshell

The clamshell design is the most common design for automotive catalytic converters in North America. Construction of a clamshell catalytic converter is illustrated in Figure 18.34. The ceramic catalyst substrate(s) is wrapped in the mat and placed in the bottom part of the shell. Then, another symmetrical part of the shell (not shown in Figure 18.34) is placed on the top, pressed together, and the two pieces welded.

A tongue-and-groove design is used for the mounting mat to avoid bypassing of the substrate by exhaust gases. The converter illustrated in Figures 18.34 also includes end seals. The seals are used here to protect the mat against gas impingement and erosion, rather than to prevent leaks. Most converters that utilize mats do not have end seals. Whenever wire mesh mount is used instead of the mat, the end seals are required, at least at the inlet face of the substrate.

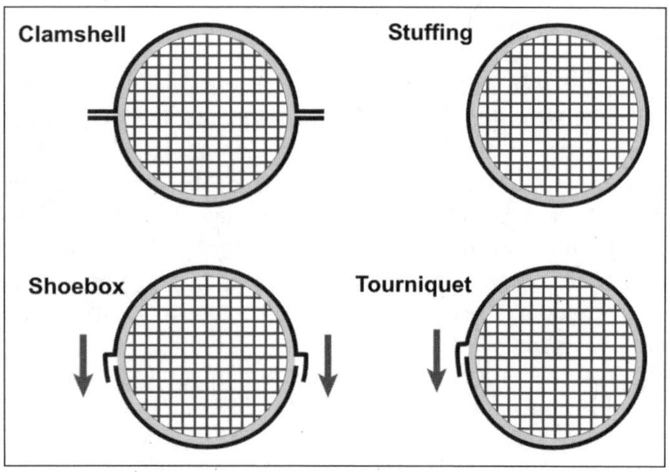

Figure 18.33 Catalytic converter canning methods.

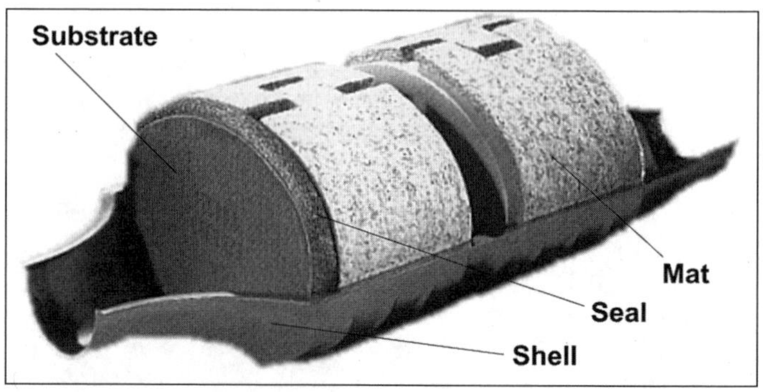

Figure 18.34 Clamshell catalytic converter.

Catalytic converters may be equipped with external heat shields to protect surrounding objects from high temperatures. Internally insulated designs were also developed, with the inside of clamshell stampings lined with an extra layer of thermal insulation.

Older designs of catalytic converters included support rings or deep pockets in the clamshell stampings to prevent axial movement of the substrate within the can. In a properly designed converter, which utilizes intumescent mats of high holding pressures, such measures are not required. Many automotive converters do not provide axial support for the substrate, but they still show an impressive durability record. However, axial support may be required for larger and heavier substrates or when nonintumescent mats of lower holding pressure are used. Another consideration is the erosion of the mat. The converter shell profiles or end cones should be designed in such a way as to shelter the mat from direct impingement of hot exhaust gases. Some converter manufacturers impregnate mat edges, which are exposed to the gas, with chemicals to improve their erosion resistance. Modern converters, however, which are characterized by high mat holding pressures, are also less susceptible to mat erosion compared to converters that have low holding pressures.

Dual-monolith converters are used in many automotive applications. Two, or more, monoliths may be used due to monolith length manufacturing constraints or to combine catalysts of different specifications in one converter. In most dual-monolith converters, the substrates are separated by space, which is maintained by forming separate pockets in the clamshell stamping. In some designs, the space between substrates is maintained by a metal or ceramic ring. A butted monolith position, with no space in between, is also possible. The butted design, which offers less pressure drop than the spaced design, has been used in some commercial converters for gasoline engines [41].

The converter shell geometry has to provide the required mat compression. For a given mat weight and target mat density, the gap between the substrate and the shell is calculated from Eq. (18.47). The clamshell profiles include stamped reinforcing ribs in order to provide the necessary stiffness and uniform pressure distribution. This is especially important for flat oval catalyst substrates. The clamshell method puts high requirements on the dimensional tolerances of the monoliths, as well as the clamshell stampings. Mat compression during clamshell canning proceeds until the half-shells close, producing a certain gap thickness. The gap thickness is determined by the dimensions of the monolith and of the shell. Therefore, any variations in the size of monoliths result in variations in the mat density and, consequently, in the canning pressure that may cause converter durability problems.

18.7.3.3 Tourniquet

The tourniquet method is the most common method that allows for direct control of the mounting pressure during the canning process. Because the tourniquet method is insensitive to dimensional differences that may occur among the substrate monoliths, it is capable of producing the most robust catalytic converters. In practice, the tourniquet method is limited to round or close-to-round catalyst substrate cross sections. Its suitability to oval or flat-oval automotive converters that are used in the underfloor position is very limited. The tourniquet shape is quite popular among European car makers but it is rarely used in North America. It may become more common as automotive converters migrate from the underfloor location to the engine close-coupled position. A tourniquet catalytic converter is shown in Figure 18.35.

In the tourniquet technique the substrate is first wrapped in a tongue-and-groove shaped mat. Then, the wrapped monolith is placed inside a longitudinally split can. The can is fabricated by rolling a rectangular piece of sheet metal. The part of the rectangle that goes underneath the overlap is usually tapered. In some designs the overlapping part of the can is formed, through an additional stamping operation, into a protruding lip to provide space for the can edge underneath the overlap. Such design prevents the inside edge of the can from cutting into the mat or creating a local pressure buildup that can damage the canned part, especially when thin mats are used. Next, the can with the wrapped monolith is placed in the tourniquet machine, which applies a controlled force to the assembly. The can is tack welded when still under pressure, removed from the machine, and seam welded. A push-out test is sometimes carried out as a quality assurance measure. Axial monolith displacement caused by applying a controlled force is measured in a special test apparatus. Finally, the converter headers or end cones, as well as flanges and/or ports, are welded to the converter body in a separate operation. The final assembly may be tested for weld quality by pressurizing with air while the unit is submerged in water.

The tourniquet machine includes a loop of a steel band that applies the force to canned parts. One end of the loop is attached rigidly to the machine while the other is pulled by a pneumatic or hydraulic actuator. In some machines vibration is applied during canning to minimize the closing force and to ensure a more uniform distribution of pressure.

Figure 18.35 *Tourniquet catalytic converter.*

The tourniquet closing force F (Figure 18.36) is a sum of the force needed for mat compression and a friction force. It can be estimated from the following formulas:

$$F = F_c + F_f \quad (18.48)$$

$$F_c = P_m DW / 2 \quad (18.49)$$

$$F_f = \mu P_m LW \quad (18.50)$$

where:

- F_c = compression force created by the mounting system (mat), N
- F_f = static friction force, N
- P_m = peak pressure created by the mounting system during canning, Pa
- μ = coefficient of friction for mounting system
- D = system diameter (= monolith diameter + 2 × mat thickness), m
- W = mounting system width, m
- L = mounting system length (= perimeter corresponding to diameter D), m.

The preceding equations assume static friction, while the actual condition would be, obviously, dynamic. Therefore, these equations are expected to produce conservative (i.e., too high) results that should be safe for designing tourniquet processes.

Coefficients of friction for catalytic converter mounting systems against stainless steel vary from about 0.15 for wire mesh mounts to more than 0.75 for ceramic fiber papers. The 3M intumescent mats, Interam 100, have friction coefficients of about 0.40.

The peak mounting pressure for a given mat is a function of mounting mat density and closing speed. Peak mounting pressures for intumescent mats compressed to 1 g/cm^3 are typically between 350 kPa for low closing speeds (<1.5 mm/min) and 700 kPa for fast closing speeds (>300 mm/min). Graphs showing the mounting pressures and their relaxation for different mat densities are available from mat manufacturers.

The actual required canning force to achieve a target mount density for a given mat should be determined through a series of tests. Several converters should be closed using different closing forces. Then, the gaps, s, should be measured and the resulting mat densities, ρ_m, should be calculated from Eq. (18.47). The canning force that produced the desired mat density should be selected. Due to mat pressure relaxation, it is important for the tourniquet machine to produce repeatable closing speed and time patterns. After the target closing force is reached, the machine should maintain the can at constant position, to allow for tack welding, rather than at constant force. Applying constant force to the can as the mat pressure relaxes would cause overcompression of the mat.

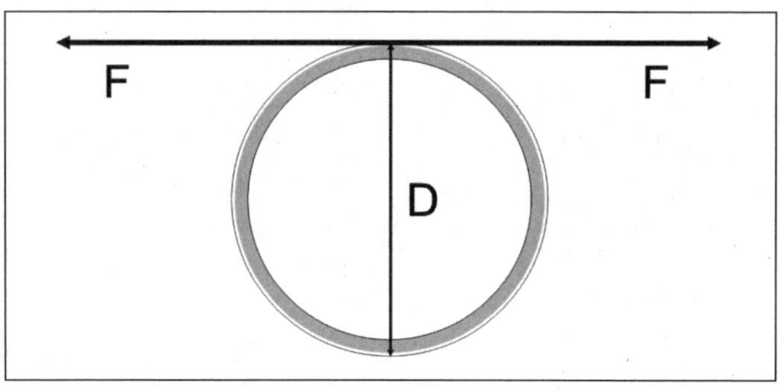

Figure 18.36 *Tourniquet canning force.*

18.7.3.4 Shoebox

The shoebox technique utilizes a split, two-part shell similar to the clamshell method. The shell, however, is closed under fixed force with the edges of one of the half-shells overlapping those of the other. Therefore, the shoebox offers the robust packaging benefits of the tourniquet technique with respect to its insensitivity to dimensional tolerances of the substrate.

Reinforcing ribs can be stamped on the shoebox shells. Thus, this technique allows for canning of flat-oval substrates in cases where the tourniquet style would be inadequate.

18.7.3.5 Stuffing

In the stuffing technique, the mat-wrapped monolith is pushed into a cylindrical can. The can is usually made of a section of tube but it can also be rolled of sheet metal and welded. Noncylindrical shapes (e.g., trapezoidal) are also possible. A stuffing cone is used to facilitate smooth insertion of the monolith. After completing the operation, the end cones are welded to each end of the cylinder to complete the can assembly.

Although stuffed converters are similar in appearance to the tourniquet assemblies, the actual mechanism used to hold the substrate is the same as in the clamshell design. In particular, the mat mounting pressure is determined by the geometrical dimensions of the shell and of the monolith. As a consequence, high repeatability of substrate diameters is required when stuffing technique is used.

18.7.4 Header Design

Catalytic converter headers provide the transition between the inlet and outlet pipes and the substrate cross section. Most converter headers are shaped like cones or funnels with axial gas flow. Other designs, such as truncated headers [42] or headers with a tangential gas inlet, are possible but uncommon. The function of the inlet header is to diffuse the inlet flow, that is, to decrease the gas velocity and increase its static pressure with as little loss in total pressure as possible. In practice, combined header losses can constitute from 10% to 50% of the overall converter pressure drop, depending on geometry and flow conditions. These pressure losses can be minimized by designing converter inlet headers that would provide more uniform flow distribution. There is also a notion that uniform flow distribution in a catalytic converter improves its emission performance and/or durability.

Figure 18.37 illustrates the geometry and flow distribution in a typical converter header. For most designs the ratio of the inlet header length, L, to the inlet pipe diameter, D_1, is between 1 and 5 and the divergence half-angle of the header to the major axis of the substrate is 30° to 60°. Headers of such geometry provide little gas diffusion. The flow separates at the interface of the pipe and diffuser (inlet cone) and proceeds as a constant diameter jet to impinge on the catalyst substrate [42]. As the inlet pipe diameter is decreased, the kinetic energy of the separated jet increases, increasing substrate flow maldistribution and pressure loss.

As flow proceeds through a diffuser, the velocity field adjusts in response to the changes in flow area and static pressure. Near the diffuser wall the axial component of the velocity decreases and eventually changes in sign. If the diffuser is too long and/or the divergence angle is too great, the backflow causes the flow to separate at the entrance of the diffuser and form a constant diameter jet. At this point the flow is insensitive to the contour of the diffuser.

Wendland *et al.* [43] tried to formulate stability criteria for catalytic converter inlet-cone diffusers. Based on literature data for two-dimensional diffusers, they suggested the following equation:

$$D_2/D_1 = 1 + 0.4887\,(L/D_1)^{0.456}R^{0.544} \qquad (18.51)$$

For any inlet diffuser diameter (D_1) and diffuser length (L), the maximum outlet diameter (D_2) can be calculated beyond which the diffuser becomes unstable and the boundary layer may start separating from the walls. Variable R is an arbitrary risk parameter, changing from 0 to 1. For R = 0 the diffuser becomes a piece of straight pipe; R = 1 represents a "line of first stall," where the diffuser becomes unstable.

A converter header design was then proposed with a short, shallow-angle diffuser modeled according to Eq. (18.51). Following the diffuser was a more abrupt expansion to the substrate cross section [44]. Several 2.8-L converter prototypes were tested with diffuser half-angles of 3.5° to 11° and lengths of 152 to 50 mm, respectively. All diffusers had a 63-mm inlet diameter, which matched the inlet pipe diameter, and an 83-mm outlet diameter, which expanded into a 161-cm^2 catalyst cross section. Total converter pressure loss reductions of 24% to 29% were reported [43].

If practical applications of diffuser headers are attempted, care must be taken to provide a precise and smooth transition from the inlet pipe to the diffuser. Weld intrusions, minor diameter mismatches, or split-pipe connections would likely "trip" the boundary layer, resulting in weaker performance than that observed with laboratory units.

18.7.5 Catalytic Mufflers

Space constraints existing in many diesel vehicle applications make it difficult to design a stand-alone catalytic converter. As a space-saving solution, the catalyzed substrate can be installed inside the vehicle's muffler. Such muffler design is referred to as the *catalytic converter muffler, catalytic muffler,* or *catalyst muffler.* Catalytic mufflers can also provide some pressure drop benefit, when compared with the two-piece catalytic converter-muffler configuration. Catalytic muffler designs were common in many U.S. diesel truck and bus applications in 1994 and later years.

The designer of a catalytic muffler has to observe all catalytic converter design considerations discussed to this point, such as robust, durable packaging or flow distribution issues. In addition, the catalyst muffler must provide the required noise attenuation performance. Typically, catalytic mufflers are composed of three functional sections, as shown in Figure 18.38:

- Flow distribution and acoustic section, including piping and baffles upstream of the catalyst
- Catalytic conversion section, including a catalyzed ceramic or metallic substrate

Acoustic section, typically including a system of tubes and baffles designed to provide noise attenuation.

Conflicting design targets have to be met in the catalytic muffler design, with the most important trade-off existing between the noise attenuation and pressure drop. In U.S. heavy-duty vehicle applications, mufflers are expected to provide noise attenuation of about 20 dB(A) to meet the EPA 50-foot drive-by limit of 80 dB(A). The maximum exhaust system pressure drop (or engine backpressure), always speci-

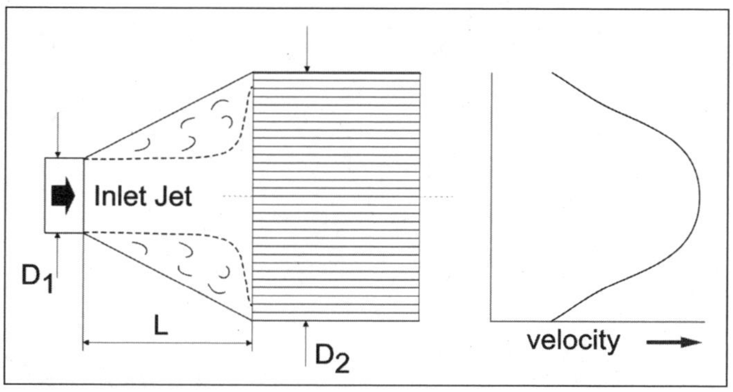

Figure 18.37 *Flow distribution in a catalytic converter.*

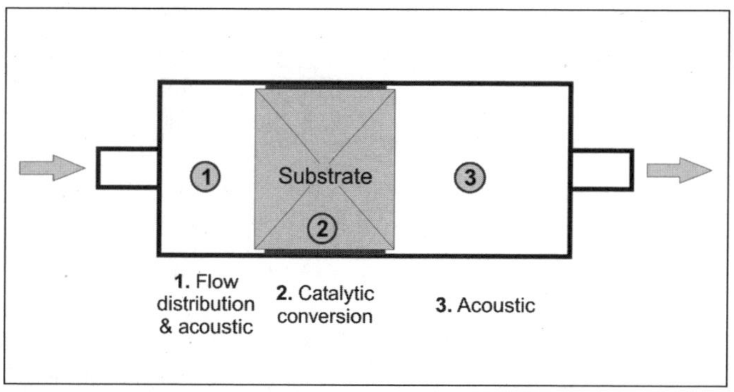

Figure 18.38 *Catalytic muffler schematic.*

fied by the engine manufacturer, includes pressure drop in the exhaust piping, in the catalytic muffler, and in the tailpipe. The acoustic/flow distribution sections in a muffler are the biggest pressure loss contributor, followed by the catalytic substrate. Some example exhaust pressure drop inventories [45] are listed in Table 18.16.

No significant noise attenuation performance has been found in the catalytic substrate. The substrate can produce a small "negative attenuation" at low frequencies; no attenuation benefit in the 200- to 500-Hz range, where most of the truck's engine noise occurs; and a small noise attenuation at higher frequencies [45]. In comparison to the noncatalyst design, an increase in the length of the catalytic muffler body may be necessary to compensate for the volume taken by the low acoustic performing substrate.

References

1. Summers, J.C., J.E. Sawyer, and A.C. Frost, "The 1990 Clean Air Act and Catalytic Emission Control for Stationary Sources," in *Catalytic Control of Air Pollution,* ACS Symposium Series 495, American Chemical Society, Washington, DC, 1992, pp. 98–114.
2. Hougen, O.A., and K.M. Watson, *Chemical Process Principles, Part Three: Kinetics and Catalysis,* 7th ed., John Wiley, New York, 1959.
3. Sabatier, P., *La catalyse en chimie organique,* rev. ed., Béranger, Paris, 1920.
4. Tauster, S., "High Temperature Three-Way Catalysts," Paper presented at SAE Advancements in Automotive Catalyst Technology TOPTEC, Dearborn, MI, 1997.
5. Heck, R.M., and R.J. Farrauto, *Catalytic Air Pollution Control: Commercial Technology,* Van Nostrand Reinhold, New York, 1995.
6. Bonacci, J.C., *et al.,* "Catalytic Incineration of Hazardous Wastes," in *Encyclopedia of Environmental Technology,* Vol. 1, Gulf Publishing, Houston, TX, 1989.
7. Reid R.C., J.M. Prausintz, and B.E. Poling, *The Properties of Gases and Liquids,* 4th ed., McGraw-Hill, New York, 1987.
8. Kikuchi, S., *et al.,* "High Cell Density and Thin Wall Substrate for Higher Conversion Ratio Catalyst," SAE Paper No. 1999-01-0268, SAE International, Warrendale, PA, 1999.
9. Marsh, P., *et al.,* "Application Guideline to Define a Catalyst Layout for Maximum Catalytic Efficiency," SAE Paper No. 2001-01-0929, SAE International, Warrendale, PA, 2001, http://www.emitec.com/pdf/SAE2001-01-0929.pdf.
10. Stankiewicz, E.P., *et al.,* "Properties and Performance of UltraCat Open-Cell Silicon Carbide Foam Catalyst Substrate," SAE Paper No. 980669, SAE International, Warrendale, PA, 1998.
11. Jatkar, A.D., "A New Catalyst Support Structure for Automotive Catalytic Converters," SAE Paper No. 971032, SAE International, Warrendale, PA, 1997.
12. Day, J.P., "Substrate Design for Cold Start," Paper presented at SAE Advancements in Automotive Catalyst Technology TOPTEC, Dearborn, MI, 1997.
13. Perry, R.H., *Perry's Chemical Engineers' Handbook,* 6th ed., McGraw-Hill, New York, 1984.
14. Rohsenow, W.M., *et al.* (Eds.), *Handbook of Heat Transfer Fundamentals,* 2nd ed., McGraw Hill, New York, 1993.
15. Day, J.P., "Substrate Effects on Light-Off—Part II: Cell Shape Contributions," SAE Paper No. 971024, SAE International, Warrendale, PA, 1997.
16. "VDI—Waermeatlas, 6. Aufl.," VDI-Verlag Duesseldorf, Germany, 1992.
17. Kingery, W.D., H.K. Bowen, and D.R. Uhlmann, *Introduction to Ceramics,*" 2nd ed., John Wiley, New York, 1975.
18. Gulati, S.T., "Ceramic Catalyst Supports for Gasoline Fuel," in *Structural Catalysts and Reactors,* A. Cybulski (Ed.), Marcel Dekker, New York, 1998, pp. 15–58.
19. Cutler, W.A., and G.A. Merkel, "A New High Temperature Ceramic Material for Diesel Particulate Filter Applications,"

TABLE 18.16
EXHAUST SYSTEM PRESSURE LOSS INVENTORY

	Medium Truck[a], kPa	Heavy truck[b], kPa
Exhaust pipe and bends	1.6	2.4
Tailpipe and bends	1.3	0.5
Catalyst substrate	2.7	2.7
Total ΔP limit	10.2 (3 in. Hg)	10.2 (3 in. Hg)
Remainder—acoustics and flow distribution	4.6	4.6

a. Medium-duty truck with ϕ10.2-cm pipe, horizontal muffler and vertical tailpipe, exhaust flow rate 40 m^3/min @ 500°C.

b. Heavy-duty truck with ϕ12.7-cm pipe and dual exhaust, vertical mufflers and vertical tailpipes, exhaust flow rate 62 m^3/min @ 400°C.

SAE Paper No. 2000-01-2844, SAE International, Warrendale, PA, 2000.

20. "Corning Diesel Products," Corning Inc., 1997.
21. Roychoudhury, S., *et al.,* "Development and Performance of Microlith Light-Off Preconverters of LEV/ULEV," SAE Paper No. 971023, SAE International, Warrendale, PA, 1997.
22. Brill, U., and U. Heubner, "Werkstoffe fuer Metalltraeger von Automobil Abgaskatalysatoren," *MTZ Motortechnische Zeitschrift* **49,** Jahrgang, Heft 9/1988.
23. Grosch, B., "Heating for a Cleaner Environment," Krupp VDM GmbH, *Metal Times* 13/95, 1995, p. 49.
24. Chang, C.S., *et al.,* "Aluminum Clad Ferritic Stainless Steel Foil for Metallic Catalytic Converter Substrate Applications," SAE Paper No. 960556, SAE International, Warrendale, PA, 1996.
25. Jha, S.C., and J.A. Forster, "Catalytic Converters—Metal Foil Material for Use Therein, and a Method of Making the Material," U.S. Patent 5,366,139, Texas Instruments, 1994.
26. Tuomola, H., R. Lylykangas, P. Matilainen, and A. Lievonen, "A New Metallic Catalyst," SAE Paper No. 2002-01-0357, SAE International, Warrendale, PA, 2002.
27. Pelters, S., *et al.,* "The Development and Application of a Metal Supported Catalyst for Porsche's 911 Carrera 4," SAE Paper No. 890488, SAE International, Warrendale, PA, 1989.
28. Harkonen, M., *et al.,* "Advanced Metallic Three-Way Catalysts with Optimized Washcoat Performance," SAE Paper No. 960560, SAE International, Warrendale, PA, 1996.
29. Bonnefoy, F., *et al.,* "High Flexibility in Converter Design by Using Modular Block Metallic Substrates," SAE Paper No. 2000-01-0497, SAE International, Warrendale, PA, 2000.
30. "A Catalyst for Change," Product Brochure, Metreon, 1996.
31. Maus, W., and R. Brueck, "The Conical Catalytic Converter and Its Potential for Future Close-Coupled Converter Concepts," SAE Paper No. 980414, SAE International, Warrendale, PA, 1998.
32. Kaiser, F., *et al.,* "Electrically-Heated Catalytic Converter System Calculations and Application," SAE Paper No. 930384, SAE International, Warrendale, PA, 193.
33. Lylykangas, R., and T. Maunula, "Particle Oxidation Catalyst for Heavy-Duty Diesel Engines," *AutoTechnology,* May 2002, pp. 57–59.
34. Held, W., *et al.,* "Improved Cell Design for Increased Catalytic Conversion," SAE Paper No. 940932, SAE International, Warrendale, PA, 1994.
35. Brueck, R., P. Hirth, M. Reizig, P. Treiber, and J. Breuer, "Metal Supported Flow-Through Particulate Trap; A Non-Blocking Solution," SAE Paper No. 2001-01-1950, SAE International, Warrendale, PA, 2001.
36. Steger, J., "Advances in Emission Control Technology," Paper presented at SAE Catalysts, Emission Control and Catalytic Combustion TOPTEC, Costa Mesa, CA, October 1995.
37. "HiQ®-7000 Series Alumina," Product Specification, Alcoa Industrial Chemicals, 1998.
38. Leyrer, J., *et al.,* "Advanced Studies on Diesel Aftertreatment Catalysts for Passenger Cars," SAE Paper No. 960133, SAE International, Warrendale, PA, 1996.
39. Gulati, S., "Design of LFA Substrates," Paper presented at SAE Diesel Exhaust Aftertreatment TOPTEC, Atlanta, GA, May 1995.
40. Rajadurai, S., *et al.,* "Single Seam Stuffed Converter Design for Thinwall Substrates," SAE Paper No. 1999-01-3628, SAE International, Warrendale, PA, 1999.
41. Kuisell, R.C., "Butting Monoliths in Catalytic Converters," SAE Paper No. 960555, SAE International, Warrendale, PA, 1996.
42. Wendland, D.W., *et al.,* "Effect of Header Truncation on Monolith Converter Emission Control Performance," SAE Paper No. 922340, SAE International, Warrendale, PA, 1992.
43. Wendland, D.W., *et al.,* "Reducing Catalytic Converter Pressure Loss with Enhanced Inlet-Header Diffusion," SAE Paper No. 952398, SAE International, Warrendale, PA, 1995.
44. Kreucher, J.E., *et al.,* "Integral Cast Diffuser for a Catalytic Converter," U.S. Patent 5,408,828, 1995.
45. Barris, M.A., "Development of Diesel Exhaust Catalytic Converter Mufflers," SAE Paper No. 920369, SAE International, Warrendale, PA, 1992.

Chapter 19

Diesel Catalysts

19.1 Overview

19.1.1 Catalyst Technologies

Three-way catalyst (TWC) technology introduced in the 1980s became an integral part of today's spark-ignited engines. The TWC, operating on the principle of nonselective catalytic reduction of NO_x by carbon monoxide (CO) and hydrocarbons (HCs), requires that the engine be operated at a nearly stoichiometric air-to-fuel (A/F) ratio [1]. Modern catalyst systems for gasoline or natural gas engines include an oxygen sensor in front of the catalyst and a closed-loop electronic control system. The electronic controller, based on a feedback system from the oxygen sensor, maintains the A/F ratio within a narrow range around the stoichiometric point, to ensure maximum catalyst efficiency. In the presence of oxygen, the three-way catalyst becomes ineffective in reducing NO_x.

For this reason, three-way catalysts cannot be employed for NO_x control on diesel applications, which, being lean-burn engines, contain high concentrations of oxygen in their exhaust gases at all operating conditions. Diesel emissions that can be controlled with high efficiency by today's catalyst technologies are CO and HCs, including such HC material as the organic fraction of diesel particulates (SOF) or polynuclear aromatic hydrocarbons (PAHs). Catalyst systems for the reduction of NO_x from diesel engines are either still under development or at early stages of commercialization, as summarized in Table 19.1.

The design target of early diesel oxidation catalyst (DOC) formulations was high removal efficiency of gas-phase diesel pollutants, including carbon monoxide and hydrocarbons. At the time of their introduction in the 1970s, carbon monoxide and hydrocarbon emissions from diesel engines were many times higher than they are in today's engines. The first diesel catalysts

TABLE 19.1
DIESEL CATALYST TECHNOLOGIES

Catalyst Technology	Reaction Type	Reduced Emissions	Commercial Status
Diesel oxidation catalyst	Oxidation	CO, HCs (incl. PAHs), PM (SOF), odor	Established commercial technology
Selective catalytic reduction (SCR)	SCR by ammonia/urea	NO_x	Commercial technology for stationary sources; some installations on marine engines; under development for truck and car engines
DeNOx (lean NO_x) catalyst	SCR by hydrocarbons	NO_x, CO, HCs, PM (SOF)	Some commercial oxidation catalysts incorporate small NO_x reduction activity (passive deNOx)
NO_x adsorbers	Adsorption (trapping) of NO_x from lean exhaust, followed by release and catalytic reduction under rich conditions	NO_x, CO, HCs	Technology under development, commercialized on lean burn gasoline engines and certain stationary sources

were used in confined space applications, such as in underground mining, where air quality was extremely important.

Introduction of fuels with a reduced sulfur content (<500 ppm S) made it possible to achieve small to moderate reductions of particulate matter (PM) emissions with the DOC. In highway engine applications of modern diesel catalysts, the PM reduction became an important—in some cases, the *only* important—catalyst function. Optimization of PM performance of the diesel oxidation catalyst was the main objective of catalyst research in the late 1980s and early 1990s, both in the United States and the European Union (EU). These diesel catalysts have been commercialized on some heavy-duty diesel engines in the United States and on light-duty diesel engines in the EU.

Future expectations focus on the NO_x reduction function of the diesel catalyst. Technologies for the reduction of NO_x from lean-burn engines, both diesel and gasoline, have been researched for many years by catalyst manufacturers and research institutes worldwide. This research has produced some promising catalyst concepts, some of which are near commercialization.

A number of future emission standards, such as the Euro 4 or the US2007, legislate NO_x emission levels that will require highly efficient NO_x emission aftertreatments. For example, the US2007 NO_x limit of 0.2 g/bhp-hr represents an approximate 90% reduction relative to the 2004 standard. The regulators expect that NO_x reduction catalysts will mature to provide this magnitude of NO_x control on time. Indeed, the very future of the diesel engine may depend on technological developments in the catalytic reduction of NO_x under lean conditions.

19.1.2 Diesel Oxidation Catalyst

The diesel oxidation catalyst, historically the first type of diesel catalyst, is still the only catalyst technology that has demonstrated the required robustness and durability and has been commercially established in a number of light- and heavy-duty applications.

DOCs promote a range of oxidation reactions utilizing oxygen, which is in ample supply in diesel exhaust at all engine operating conditions. These reactions can convert virtually all emission components of chemically reducing character, such as CO, HCs, and HC derivatives, into CO_2 and water vapor. The DOC chemistry is schematically shown in Figure 19.1.

At sufficiently high exhaust temperatures, diesel oxidation catalysts can provide very effective control of HC and CO emissions, with reduction efficiencies in excess of 90%. The DOC hydrocarbon activity extends to compounds such as PAHs or the SOF fraction of diesel particulates. The catalytic oxidation of SOF results in reduction of diesel PM emissions. Depending on the SOF contents of the particulates, the PM conversion efficiency typically ranges between 15% and 30%. DOCs may be also effective in controlling diesel odor.

High conversion efficiencies for CO and HC are realized through the use of highly active, noble metal (e.g., platinum) based catalysts. Many DOCs developed in the 1990s for highway applications were optimized for PM emission reduction, as opposed to the abatement of CO or HC. To suppress undesired

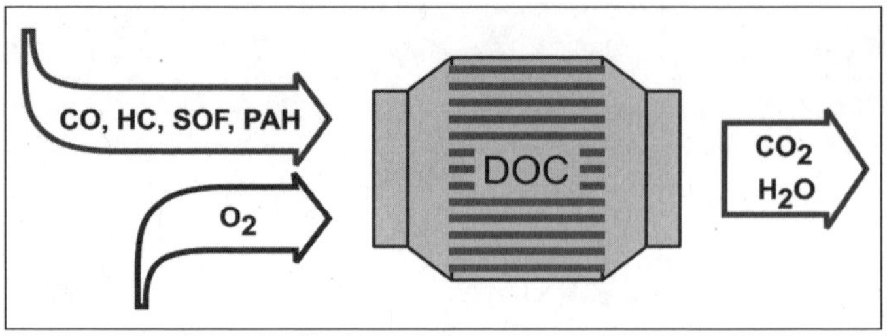

Figure 19.1 Conversion of pollutants in diesel oxidation catalyst.

oxidation of sulfur compounds, these catalysts generally utilized formulations of reduced catalytic activity (e.g., through the use of base metals instead of platinum or the use of very low platinum loadings). We should emphasize that CO and gas-phase HC conversion in these PM-optimized catalysts may be very low. Chemical reactions and DOC performance are discussed in more detail in Section 19.2.

The diesel oxidation catalyst was introduced in the 1970s to protect the ambient air quality in underground mines. The objective of the catalyst was to reduce carbon monoxide emissions and odors from diesel-powered mining equipment. The use of the DOC extended into other enclosed space applications including construction, tunneling, and materials handling. In most cases, tailpipe emission regulations are not considered in these applications. Rather, the use of catalysts is driven by occupational health and safety standards, which protect the ambient air quality at the workplace.

Early diesel catalysts, used throughout the 1970s and 1980s, were simple oxidation formulations, virtually the same as those developed for gasoline cars. Certain disadvantages of these catalysts, such as increased sulfate PM emissions, were gradually discovered by diesel research. As a result, the overall air quality benefit from using diesel oxidation catalysts in enclosed space applications was questioned [2]. The introduction of emission standards for highway diesel engines in the 1990s stimulated catalyst research, which led to the development of next-generation diesel oxidation catalysts. These new catalysts, when used with low-sulfur fuels (<500 ppm S), were able to control diesel particulate emissions in many diesel engine applications. Availability of the improved diesel catalyst technologies reinstated an endorsement for the use of DOCs in underground mining and other enclosed space applications [3].

The first automotive application of diesel oxidation catalysts was the diesel-fueled Volkswagen Golf "Umwelt" (German for "environment") in 1989. The catalyst was designed to reduce CO and HC emissions. The use of a catalyst was not necessary to meet German or European emission regulations at that time. Rather, it was introduced as a sign of an environmentally responsible attitude on the part of the car maker. Since the introduction of the Euro 2 emission standards in the mid-1990s, the diesel oxidation catalyst has become a standard component in diesel-fueled cars in Europe. The task of the catalyst is to reduce the emissions of particulates; it also provides some reduction in CO and HC emissions.

In the United States, the use of oxidation catalysts for highway heavy-duty engines was triggered in 1994, when the U.S. Environmental Protection Agency (EPA) introduced a PM emission standard of 0.1 g/bhp-hr for trucks and 0.05 g/bhp-hr for buses. Diesel oxidation catalysts were introduced on several truck and bus engines in order to meet that PM standard. However, the use of catalysts on U.S. heavy-duty engines had already been steadily decreasing, because the 0.1 g/bhp-hr PM standard could be achieved in more advanced, electronically controlled engines without the use of aftertreatment devices. The reduction of PM emissions was the only purpose of the DOC; no reductions of gaseous HC or CO were necessary under the 1994 regulations. DOCs are being introduced again on certain heavy-duty engines, such as on the Caterpillar ACERT engines, to meet the 2004 emission standards.

Oxidation catalysts have also been used in many diesel retrofit programs worldwide, mostly for heavy-duty engines. The U.S. Urban Bus Retrofit/Rebuild program, which became effective in 1995, triggered retrofitting of pre-1994 urban bus engines with emission control devices, mostly diesel oxidation catalysts. DOCs were also used in the initial stage of the Swedish Environmental Zones program, and in diesel retrofits in a number of cities including Hong Kong and Tokyo.

19.1.3 NO_x Reduction Catalysts

Catalyst research has been increasingly focusing on catalysts for the reduction of oxides of nitrogen from lean exhaust gases. NO_x emissions from both heavy- and light-duty diesel engines need to be reduced significantly below the levels currently being achieved. This is being done by means of engine design and calibration. Highly efficient NO_x reducing catalysts would also allow the diesel engine to be calibrated for maximum fuel economy, which is currently being sacrificed for the sake of in-cylinder NO_x control.

NO_x reducing catalysts are needed not only for diesels, but also for lean-burn SI engines. Because the three-way catalyst remains the only commercially available, highly efficient NO_x control technology for mobile applications, today's gasoline engines are designed to operate at a stoichiometric A/F mixture. A significant fuel economy (and greenhouse gas emission) penalty is associated with use of SI engines operating with a stoichiometric A/F ratio. An impressive improvement in fuel economy could be achieved by operating the gasoline engine on a lean mixture. Catalysts that could bring nitrogen oxide emissions from lean-burn gasoline engines to the levels regulated in Europe and in North America would open the door for such engine technology.

The following catalytic approaches have been investigated for the control of NO_x in lean exhaust gases:

- NO decomposition catalyst
- Selective catalytic reduction (SCR) with ammonia
- SCR with hydrocarbons (deNOx or lean NO_x catalyst)
- NO_x adsorber-catalyst systems.

Decomposition of nitric oxide to elements is described by the following equation:

$$2NO = N_2 + O_2 \qquad (19.1)$$

The decomposition of NO is thermodynamically favored under pressures and temperatures found in diesel exhaust. However, the rate of the spontaneous reaction is practically zero. Catalysts have been researched to accelerate that process but no significant progress has been achieved. One of the best catalysts suggested for NO decomposition was a copper exchanged zeolite, Cu/ZSM5. Copper exchanged zeolites were previously known to exhibit three-way catalyst activity for reducing NO_x in stoichiometric exhaust gas compositions [4]. In a series of papers published in the late 1980s, Iwamoto and coworkers [5] showed that Cu/ZSM5, as well as other transition metal-containing molecular sieves, were able to catalytically decompose NO. Although initially promising, catalytic decomposition of NO has proven difficult to realize. The decomposition of NO on Cu/ZSM5 is subject to inhibition by water, is very sensitive to poisoning by SO_2, is effective only at low space velocities, and the catalyst activity and selectivity are not satisfactory [5].

Selective catalytic reduction of NO_x can be realized if a reducing agent is injected into the gas upstream of the catalyst bed. SCR processes utilizing nitrogen-containing reductants such as ammonia or urea are commercially available for stationary diesel engines and for industrial sources. Figure 19.2 illustrates SCR performance using ammonia as the reductant.

The primary selective reaction in this system is expressed by Eq. (19.2). This reaction is promoted by the catalyst over the competitive reaction of NH_3 with oxygen, which is abundant in the system.

$$4NO + 4NH_3 + O_2 = 4N_2 + 6H_2O \qquad (19.2)$$

The main advantage of SCR, which contributed to the widespread use of SCR in industrial (nonengine) applications, is its high NO_x conversion efficiency—it can reach as high as 90%. Disadvantages of SCR include high capital and operating costs, space requirements, generation of ammonia emissions (ammonia slip), and fouling of equipment with ammonium sulfate, a hazardous waste.

SCR has been successfully adapted to control NO_x emissions from large, stationary diesel engines. It is used on diesel power generation and cogeneration systems, primarily in Europe and Japan. There is also ongoing interest in the application of SCR for mobile applications, including marine engines, trucks,

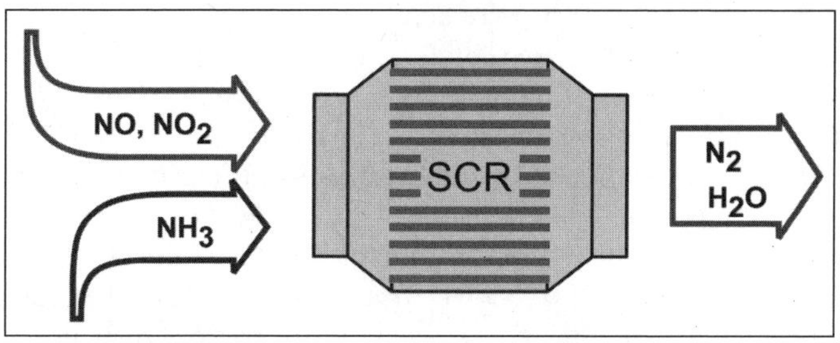

Figure 19.2 *NO_x reduction in an SCR catalyst.*

and diesel cars [6, 7]. Because mobile applications of SCR would require a sophisticated, active catalyst system and the availability of a reductant (urea) fueling infrastructure, that technology is not the most attractive option for diesel trucks and cars. However, in the absence of more elegant NO_x reduction catalyst systems, SCR is an option for meeting future emission standards, in particular the Euro 5 standards for heavy-duty engines. Demonstration programs showed that NO_x emissions from existing heavy-duty Euro 2 engines (7 g/kWh) can be reduced by SCR to levels below 2 g/kWh, that is, the Euro 5 standard [8]. SCR demonstration programs have been also conducted in the United States [9]. We discuss the SCR technology for stationary and mobile diesel engines in Section 19.3.

One can substitute ammonia as the reducing agent in Eq. (19.2) with hydrocarbons. By arbitrarily choosing propane as a representative HC, the reaction can be rewritten as follows:

$$10NO + C_3H_8 = 5N_2 + 3CO_2 + 4H_2O \qquad (19.3)$$

Catalysts that promote the preceding reaction over the competitive, nonselective oxidation of hydrocarbons by oxygen are commonly referred to as deNOx or lean NO_x catalysts (LNCs). This selective deNOx chemistry can be realized using either hydrocarbons emitted by the engine (so-called "passive deNOx" configuration) or else through the addition of extra hydrocarbons, such as diesel fuel, upstream of the catalyst ("active deNOx"). Active deNOx systems have higher NO_x conversion efficiencies at a cost of increased system complexity and a fuel economy penalty. Selective catalytic reduction of NO_x by hydrocarbons is discussed in Section 19.4.

We must emphasize that the use of such terms as the "SCR" (i.e., ammonia/urea based) or the "lean NO_x/deNOx" (i.e., HC-based) catalyst system is purely customary. Considering the nature of the chemical reactions, both processes represent selective catalytic reduction of NO_x. Both types of systems can also be described as lean NO_x catalysts, as determined by their capability to reduce nitrogen oxides under oxidizing, lean exhaust conditions. To complicate things even more, some authors use the term "deNOx" in reference to any NO_x reducing catalyst technology, not necessarily limited to selective reduction by hydrocarbons. These terms should be used with caution to avoid confusion.

A combination of the lean NO_x catalyst with the diesel oxidation catalyst is sometimes called a *four-way catalyst.* The four-way catalyst is capable of simultaneous reduction of four diesel exhaust pollutants: CO, HC, PM, and NO_x.

Two different classes of deNOx catalysts have been developed: (1) platinum-based catalysts and (2) base metal catalysts (Cu/ZSM5). Each of these catalysts has its specific drawbacks, related primarily to narrow temperature windows and insufficient thermal durability and/or sulfur tolerance. More importantly, however, peak NO_x conversions of known deNOx catalysts are in the range of 40% to 60%. Combined with relatively narrow temperature activity windows, these numbers translate to only about a 10% to 20% NO_x conversions of passive systems on regulatory test cycles, such as the ECE+EUDC. Therefore, deNOx catalysts are not currently seen as a technology capable of coping with the stringent future emission targets.

Ongoing research aimed at improving deNOx catalysts has resulted in commercialization of this technology, primarily in the form of limited, passive NO_x removal functionality in oxidation catalysts. Several commercial diesel catalysts developed for the light-duty Euro 3 standard (year 2000) incorporate a passive deNOx function and are capable of a 5% to 15% NO_x reduction in addition to their main oxidation activity with respect to PM, CO, and HCs [10].

NO_x storage materials were first incorporated into deNOx catalysts in an attempt to widen their temperature window. These compounds were supposed to adsorb NO_x during periods of low exhaust gas temperature, when the catalyst activity was low, and to release it at temperatures that would favor the selective reaction with hydrocarbons. This idea was then coupled with the three-way catalyst, producing nitrogen oxide adsorber-catalyst systems, also known as NO_x traps, which represent the newest concept in lean NO_x control research. First NO_x adsorbers have been commercialized on certain lean-burn, DI gasoline engines in Japan and the EU. A massive amount of development work is being conducted worldwide to adapt this technology to the diesel engine. Laboratory experiments have shown that NO_x adsorbers can achieve a 90% NO_x reduction efficiency over a relatively wide temperature window.

These systems cycle through two stages of operation, including (1) storage (chemisorption) of NO_x in the catalyst washcoat during lean operation and (2) regeneration of the trap through desorption and nonselective catalytic reduction of NO_x during periods of rich operation. These phases of operation are schematically represented by the top and bottom parts of Figure 19.3, respectively. NO_x is stored in the form of a metal nitrate, shown as $MeNO_3$ in the schematic.

Regeneration of NO_x adsorbers involves reaction mechanisms virtually identical to those found in the gasoline three-way catalyst. A reductant, presumably carbon monoxide, reacts with NO_x in the absence of oxygen to form elemental nitrogen. Also the catalyst systems, such as the most common platinum/rhodium formulation, are very similar to the three-way catalyst. Because the regeneration requires periodic short pulses (say, a couple of seconds every minute) of rich A/F mixture, this technology requires extremely tight integration with the engine control system and is probably feasible only in conjunction with electronic fuel injection systems, such as the common rail. It also involves certain fuel economy penalty, which depends both

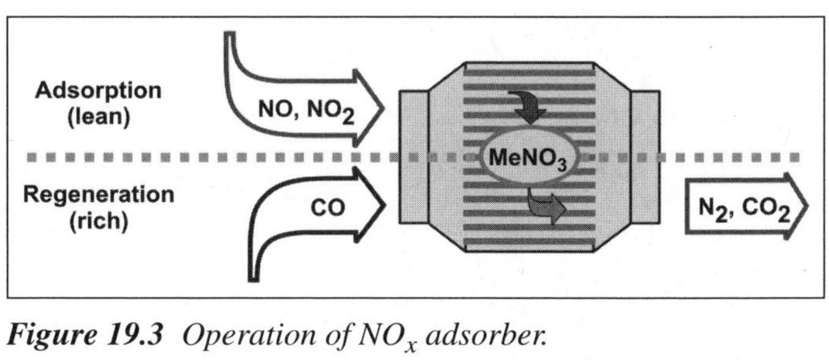

Figure 19.3 *Operation of NO_x adsorber.*

on the catalyst system and on the quality of the regeneration algorithm.

The major stumbling block in the development of diesel NO_x adsorbers is their susceptibility to sulfur poisoning. Ultra-low-sulfur diesel fuels, although a necessary condition, are not sufficient for the commercialization of NO_x adsorbers. Even if ultra-low-sulfur fuel is used, sulfur management strategies, including periodic desulfation, are required by this technology [11]. The application of NO_x traps to diesel engines is discussed in detail in Section 19.5.

A different NO_x adsorber concept has been termed selective NO_x recirculation (SNR) [12]. In the SNR technique, two NO_x adsorbers are installed in parallel in the engine exhaust system. Control valves allow the gas flow to be switched, so that each of the adsorbers alternates between the adsorption and desorption modes. While in the desorption mode, the NO_x carrying gas from the adsorber is recirculated to the engine air inlet. In this way, the desorbed NO_x can be reduced through in-cylinder reactions during regular combustion, through processes that appear to have something in common with the mechanisms of NO_x reduction via exhaust gas recirculation (EGR). The SNR technique eliminates the need for catalytic NO_x reduction, requiring a simpler catalyst system. Most likely, there is also less demand on the engine operating regime. However, SNR adsorber regeneration strategies still need to be demonstrated. In experiments involving feeding NO/NO_2 from bottles to the diesel engine air intake port (i.e., not accounting for the adsorber performance), the SNR achieved a NO_x reduction efficiency of 60%—much less than achieved by the adsorber-catalyst system.

It is common for commercial diesel catalysts to incorporate a number of technologies into one catalytic device. Examples include oxidation and lean NO_x catalysts that incorporate hydrocarbon adsorbers/traps (see Section 19.6). Catalysts are also combined with noncatalytic devices to enhance their performance, the most prominent example being the plasma-assisted deNOx catalyst (see Chapter 23).

19.2 Diesel Oxidation Catalyst

19.2.1 Reactions in Diesel Oxidation Catalyst

When passed over an oxidation catalyst, a number of exhaust gas components are oxidized by the oxygen present in ample quantities in diesel exhaust. The following diesel pollutants can be oxidized to harmless products, and thus can be controlled using a diesel oxidation catalyst:

- carbon monoxide
- gas-phase hydrocarbons
- organic fraction of diesel particulates.

Additional benefits of the DOC include oxidation of several nonregulated, HC-derived emissions, such as aldehydes or PAHs, as well as reduction or elimination of the odor of diesel exhaust.

The emission reductions in the DOC occur through chemical oxidation of pollutants occurring over the active catalytic sites. These processes can be described by the following chemical reactions:

$$\{HCs\} + O_2 = CO_2 + H_2O \tag{19.4}$$

$$CO + 1/2O_2 = CO_2 \tag{19.5}$$

Hydrocarbons are oxidized to form carbon dioxide and water vapor. In fact, the reaction of Eq. (19.4) represents two processes: the oxidation of gas-phase HCs and the oxidation of SOF compounds. Reaction (19.5) describes the oxidation of carbon monoxide to carbon dioxide. Because carbon dioxide and water vapor are considered harmless, the preceding reactions bring an obvious emission benefit.

However, an oxidation catalyst will promote oxidation of all compounds of a reducing character; some of the oxidation reactions can produce undesirable products and, in effect, be counterproductive to the catalyst purpose. Oxidation of sulfur dioxide to sulfur trioxide with the subsequent formation of sulfuric acid (H_2SO_4), described by reactions (19.6) and (19.7), is perhaps the most important of these processes.

$$2SO_2 + O_2 = 2SO_3 \tag{19.6}$$

$$SO_3 + H_2O = H_2SO_4 \tag{19.7}$$

When the exhaust gases are discharged from the tailpipe and mixed with air, either in the environment or in the dilution tunnel that is used for particulate matter sampling, their temperature decreases. Under such conditions the gaseous H_2SO_4 combines with water molecules and nucleates to form (liquid) particles composed of hydrated sulfuric acid. This material, called sulfate particulates, contributes to the total particulate matter emissions from the engine. Catalytic formation of sulfates, especially in conjunction with high sulfur content diesel fuel, can significantly increase the total PM emissions and, thus, become prohibitive for the catalyst application.

Oxidation of NO to NO_2 is another reaction that may be considered undesirable in some applications:

$$NO + 1/2O_2 = NO_2 \tag{19.8}$$

Concerns have been raised that the catalytic generation of NO_2, which is more toxic than NO, can create air quality problems in some underground mine applications [13]. Due to the thermodynamic equilibrium of Reaction (19.8), which is reached in the atmosphere after some time regardless of the original composition of NO_x, the oxidation of NO is probably of less concern in surface applications. Interestingly enough, nitrogen dioxide can be effectively used to facilitate the regeneration of diesel particulate filters. As diesel oxidation catalysts and catalytic particulate filters become more widespread, more research will be needed to clarify the health and environmental effects of increased NO_2 levels that may result from the use of these devices.

The reaction mechanism in diesel oxidation catalysts is explained by the presence of active catalytic sites that are deposited on the surface of the catalyst carrier and have the ability to adsorb oxygen. The catalytic reaction has three stages: (1) Oxygen is bonded to a catalytic site; (2) reactants, such as CO and hydrocarbons, diffuse to the surface and react with the bonded oxygen; and (3) reaction products, such as CO_2 and water vapor, desorb from the catalytic site and diffuse to the bulk of the exhaust gas.

19.2.2 Emissions Performance

19.2.2.1 Carbon Monoxide and Hydrocarbons

The diesel oxidation catalyst is an effective device to control carbon monoxide and hydrocarbon emissions from diesel engines, including the PAH and hydrocarbon derivatives such as aldehydes. Figure 19.4 illustrates catalyst performance as a function of temperature. The catalyst shows no activity at low exhaust gas temperatures. As the temperature increases, so does the oxidation rate of CO and HC. This is called *catalyst light-off.* At high temperatures the catalyst performance stabilizes to form the characteristic plateau on the light-off curve. For most catalyst systems, including the classic Pt/Al_2O_3 system, the conversion of carbon monoxide is higher than that of hydrocarbons at any given temperature. In a given catalyst there is also a competition between the oxidation of CO and HC species, causing the conversion rates in a mixture of gases, such as in engine exhaust, to be lower than laboratory conversions determined using pure CO gas in the absence of HC, or vice versa.

The high-temperature conversion rate for CO or HC, or the height of the plateau, depends on the mass transfer conditions in the catalyst. Consequently, it can be increased or decreased by changing factors that affect either the mass transfer coefficient or the mass transfer surface area. For example, the maximum conversion of CO or HC will increase when a larger catalyst or a higher cell density substrate is used. In fact, the high-temperature conversion efficiency can be brought as close to 100% as desired by enlarging the catalyst size. If needed, the conversion efficiency can be limited by using a smaller catalyst, or one of a lower cell density (larger cells).

The light-off temperature, on the other hand, depends mainly on the chemical reaction kinetics in the catalyst. Therefore, it is influenced primarily by the catalyst noble metal/washcoat system. The overall performance of a catalyst system, as measured on standard emission test cycles, is, thus, a combined result of the applied catalyst technology and the substrate geometry. Among the two common oxidation catalysts—platinum and palladium—platinum (Pt) is most active for the oxidation of CO and HC in diesel exhaust, as illustrated in Figure 19.5 [14]. The applied catalysts used 400-cpsi substrates having a 5.07-L volume. CO emissions with the Pt catalyst were less than a half of those with the palladium (Pd) catalyst. Also the hydrocarbon emissions were lower when the platinum catalyst was used, although the performance difference between Pt and Pd was smaller.

The lower activity of Pd in diesel exhaust—widely confirmed in the literature [15, 16]—is quite opposite that of its activity in gasoline exhaust. Palladium, due to its superior low-temperature hydrocarbon activity, is commonly used in close-coupled light-off catalysts in gasoline cars [1]. The difference in activity is explained by the different chemical composition of gasoline and diesel exhaust hydrocarbons. Gasoline HCs have short carbon chains and contain many unsaturated compounds, whereas diesel hydrocarbons are characterized by long carbon chains and mostly saturated bonds.

Observed heat effects associated with the oxidation of carbon monoxide and hydrocarbons are different in the gasoline and diesel oxidation catalyst. The oxidation of CO and HC involves exothermic reactions with significant release of heat. Of course, the heat of reaction for particular chemical compounds is the

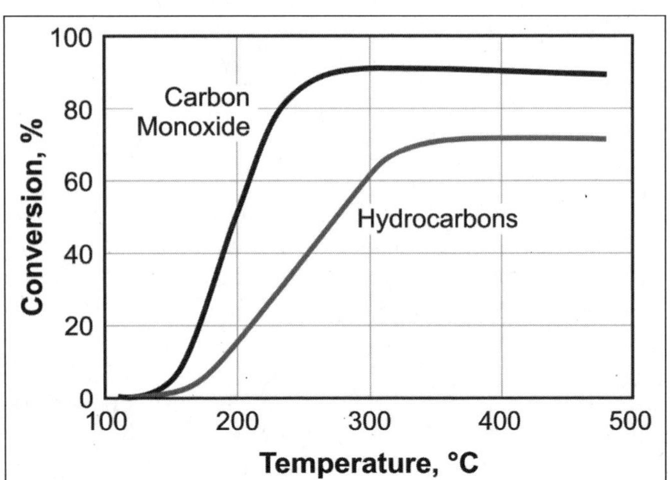

Figure 19.4 *Conversion of CO and HC over a diesel oxidation catalyst.*

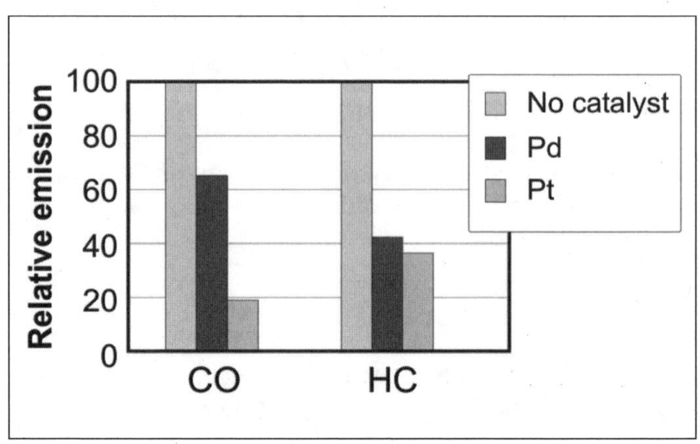

Figure 19.5 *Platinum and palladium diesel oxidation catalyst applied to a 6.925-L, DI, turbocharged, aftercooled diesel engine for the HD FTP Transient test. (Based on data from Ref. 14.)*

same whether they originate from diesel or gasoline exhaust. The difference is in the concentrations of CO and HC, which are many times higher in gasoline exhaust than in diesel exhaust. As a result, a temperature increase of a few hundred degrees Celsius is not uncommon in the gasoline catalytic converter, whereas the temperature rise across a diesel catalyst is rarely more than 10 to 20°C.

The removal of hydrocarbons and their derivatives in the DOC also results in reductions in the levels of odor in diesel exhaust [17]. The effectiveness of catalysts is difficult to quantify, because no universal methods exist to measure the intensity of odors. Authors who worked on developing such methods reported decreases of a threshold value of diesel odor between 3% and 17% on a heavy-duty engine with a 10 g/ft^3 Pt catalyst, and between 52% and 62% on a light-duty engine with a 90 g/ft^3 Pt catalyst [18].

19.2.2.2 Nitrogen Oxides

In most cases, the total nitrogen oxide emissions are not changed by the diesel oxidation catalyst. A considerable amount of research is being done do develop a catalyst technology capable of removing NO_x from lean exhaust. However, highly effective and durable lean NO_x catalyst systems are not yet available. Small NO_x reductions, usually not more than a few percent, are sometimes measured in diesel oxidation catalysts. There are two possible explanations for such activity:

- Lean NO_x performance
- Storage of nitrates in the catalyst washcoat.

Some catalyst systems, including the common Pt/Al_2O_3 formulation, exhibit lean NO_x performance through selective catalytic reduction of nitrogen oxides by hydrocarbons. Diesel oxidation formulations that contain such components may facilitate some NO_x reduction, especially at higher HC/NO_x ratios. In the case of platinum/alumina catalysts, maximum lean NO_x activity occurs at low temperatures of approximately 200 to 250°C. Note also that NO_x is reduced mainly to nitrous oxide (N_2O) over Pt catalysts. N_2O, although not a regulated emission, is not a desirable reaction product.

Storage of nitrates may occur on some catalysts, causing an apparent NO_x reduction effect. Catalyst materials capable of reacting with nitrogen oxides to form nitrates include barium, a common alumina washcoat stabilizer in oxidation catalysts. Because barium nitrate is not decomposing under typical conditions in diesel exhaust, repeated testing shows a declining NO_x activity in this type of catalyst. NO_x adsorbers —a different catalyst technology based on this type of chemistry—require alternating adsorption and regeneration cycles.

While the total NO_x remains essentially unchanged, nitrogen oxides do undergo chemical changes over the diesel oxidation catalyst. Engine-out NO_x emissions are composed mostly of nitric oxide with typically a few percent of nitrogen dioxide. An oxidation of NO to NO_2, following reaction Eq. (19.8), occurs over the diesel oxidation catalyst. As mentioned earlier, that process is considered undesirable is some applications.

The NO/NO_2 conversion depends on many factors such as the catalyst formulation and size. Example data are shown in Figure 19.6, where NO_2 concentration—in the raw exhaust and after the catalyst has been used—is plotted versus exhaust gas temperature. The catalytic generation of nitrogen dioxide was measured on a 7-L Caterpillar 3304 PCNA diesel engine rated at 100 hp @ 2200 rpm equipped with a 2.5-L diesel oxidation catalyst [13].

A large increase in NO_2 concentration was observed at a temperature of about 300°C. The nitrogen dioxide concentration increased from about 10 ppm in the raw exhaust to a maximum of 120 ppm after the catalyst. When temperatures approached 400°C, the NO_2 concentration decreased to reach practically zero at about 460°C. That decrease in NO_2 is related to the thermodynamic equilibrium of Reaction (19.8), which shifts to the left side of the equation at higher temperatures (compare with a CRT filter, Chapter 22). Because the concentration of nitric oxide during the experiment was between 800 and 900 ppm, the observed maximum conversion efficiency of NO to NO_2 was approximately 12%. Higher NO/NO_2 conversions of around 30% have been seen in various types of catalytic particulate filters [19, 20], which are typically larger than catalytic converters and incorporate more active oxidation catalyst formulations.

Literature data on catalyst formulations of suppressed NO/NO_2 activity are scarce. Some researchers believe that formulations developed for sulfate-suppressed catalysts, or at least some of them, are effective in preventing NO_2 formation. Published data support this notion in the case of vanadium, the addition of which suppressed both the sulfate make and the NO/NO_2 shift [21].

19.2.2.3 Particulate Matter

Transformation of PM in Oxidation Catalyst

The total diesel particulate matter (TPM) emission is composed of three major fractions including the carbonaceous particulates, the organic particulates (SOF), and sulfates (SO_4). Each of these fractions behaves differently over the

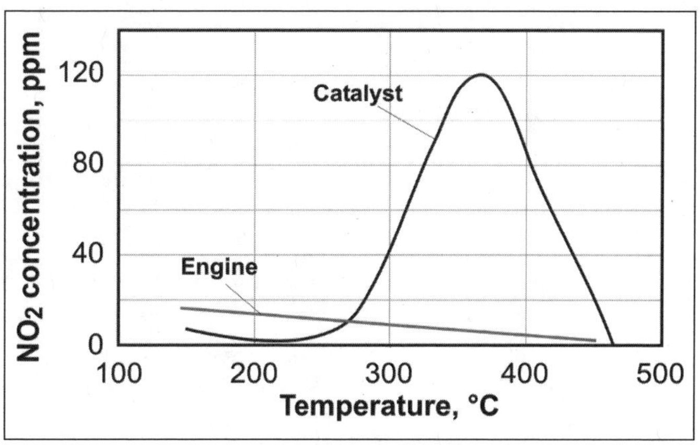

Figure 19.6 *Concentration of NO_2 with diesel oxidation catalyst. (Based on data from Ref. 13.)*

diesel oxidation catalyst. In general, the overall effect of the DOC on the total PM emission could be a decrease, as well as an increase. Typical transformations of the three fractions and the resulting total PM emissions are schematically illustrated in Figure 19.7. As is apparent from the graph, PM emissions can be reduced in the DOC through the removal of their organic fraction (SOF). Under certain conditions, however, the SOF decrease can be more than offset by an increase in sulfate PM, leading to an overall increase in TPM emission. (If high sulfur fuels are used, sulfate particulate emissions may be much higher than those shown in Figure 19.7.)

While literature reports exist that show both small decreases and small increases of solid, insoluble PM in the DOC, it is generally agreed that carbonaceous particulates pass virtually unchanged through oxidation catalysts [1]. Considering their large size in comparison to gas molecules, carbon particles are not likely to come into contact with the catalyst. Even if they do, both the residence time and the temperature in diesel exhaust are insufficient for any significant oxidation rates. Any carbon oxidation that may occur is likely attributed to NO_2, which is active at lower temperatures than oxygen [22]. Various catalyst substrates have been under development that increase both the particle residence time and their contact with the catalyst in an attempt to maximize catalytic oxidation of solid carbon (see Chapter 18). In these designs, which involve metallic substrates, flow turbulence is induced by special shaping of the channels, such as by introducing protrusions that deflect the laminar flow and create swirl motion. Removal of 15% to 25% of solid carbon particulates (in addition to the SOF removal) has been reported in a DOC using one such substrate design [23]. In another design, carbon particulates are impinged onto a porous metal layer, to allow for the oxidation by nitrogen dioxide to take place [24]. The performance of this type of system can be further improved by using an additional NO_2 producing catalyst in the upstream position [25].

Reports on changes in PM carbon fraction over DOCs must be always approached with careful scrutiny. The magnitude of change in many such reports is on the threshold of detection. Measurement error can be magnified due to the analytical methods that determine insolubles only indirectly, by subtracting sulfate and SOF from total PM. If sulfuric acid reacts with metals, such as with calcium from lube oil additives, the resulting sulfate salts may be accounted for as insoluble material.

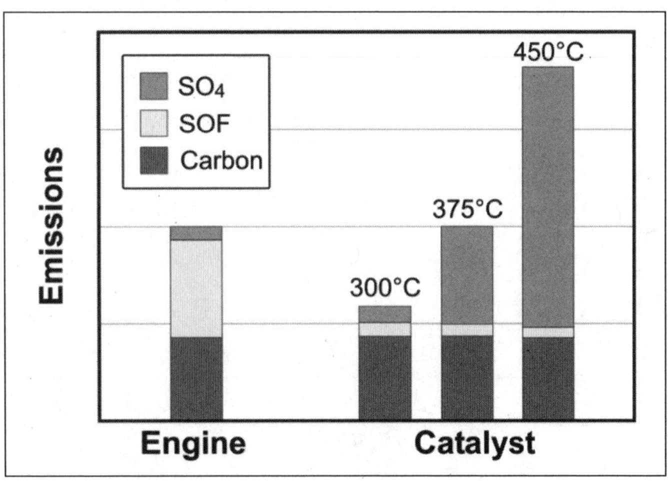

Figure 19.7 *Impact of diesel oxidation catalyst on PM emissions.*

The organic fraction of diesel particulates, composed of heavy hydrocarbons of high boiling temperatures, is very effectively oxidized in the catalyst. SOF removal is the major effect contributing to a decrease in the TPM emissions over the diesel oxidation catalyst. SOF oxidation is similar to the oxidation of gas-phase hydrocarbons, that is, a certain temperature has to be reached for the catalyst to light-off (Figure 19.4). After the light-off temperature is reached, the conversion of SOF shows little change with further temperature increases.

The sulfate fraction of diesel particulates (SO_4) is increased in the DOC due to the oxidation of SO_2 with subsequent formation of sulfuric acid [Eqs. (19.6) and (19.7)]. This is a counterproductive process causing an increase in TPM emissions. The intensity of the sulfate make increases with exhaust gas temperature and becomes critical at about 400°C. Special catalyst formulations are used to suppress that process and, thus, to make the diesel oxidation catalyst a viable PM reduction approach. Low-sulfur fuels minimize sulfate make and allow the use of more active diesel oxidation catalyst formulations. The phenomenon of catalytic formation of sulfates is of critical importance for diesel catalyst and particulate filter technologies alike; we will return to that topic later.

The net effect of the DOC on diesel particulate matter is a combination of the SOF and SO_4 reactivities, as shown earlier in Figure 19.7. (For the sake of simplicity, the dependence of engine-out PM composition on exhaust temperature has been neglected.) The catalytic oxidation of SOF contributes to the decrease in TPM, whereas the generation of sulfates increases particulate emissions. The desired SOF effect prevails at low temperatures. In practice, if sulfur is present in the fuel, there is always a temperature above which an increase of PM emissions will be recorded. Selective catalysts are formulated to suppress the sulfate generation in order to extend the catalyst PM control function into higher exhaust temperatures. A considerable amount of catalyst research produced diesel catalysts specifically optimized for PM control that can tolerate certain sulfur levels in the fuel and still control PM emissions. These selective formulations, however, compromise the catalyst activity, resulting in less effective gas-phase HC and CO control. If sulfur-free fuel is used, the conventional highly active platinum/alumina-based catalysts may be still the best choice for both the particulate and the gas-phase emission control.

PM Emission Reduction

The potential of catalytic PM emission reduction critically depends on the composition of engine-out particulates. Only particulates having a high SOF content, so-called "wet" particulates, can be effectively controlled by catalysts. In practice, different engine manufacturers design the combustion in different ways to meet emission standards. As a result, the PM portion that is susceptible to oxidation varies from engine to engine. In a system approach to the design of the engine-catalyst package, particulates can be purposely made "wet" by engine design measures to allow for more efficient catalytic control. In a given engine, SOF content depends on the operating conditions, changing widely from just about a few percent at high engine loads to well above 50% (in some cases approaching 90%) at low loads. Example data illustrating SOF content as a function of engine operating conditions were presented in Section 7.3. The average SOF content over an engine test cycle depends on its average load factor. Cold test cycles at low engine loads will produce wet particulates, which can be effectively controlled by the DOC (as long as the temperatures are sufficient for the catalyst to light-off). Hot, highly loaded test cycles, on the other hand, tend to produce dry particulates, which can be controlled to a small degree with DOC technology.

Wet particulates—collected over lightly loaded test cycles—can have SOF contents of more than 50%. Because the SOF fraction can be almost entirely removed by the catalyst, the potential for catalytic particulate control in such cases can be greater than 50%. In practice, PM reductions are additionally limited by the sulfur level in the fuel and by the catalyst temperature. Advanced catalyst formulations, in conjunction with low-sulfur fuels, can be very effective in controlling particulates from low-temperature applications. PM removal efficiencies of 50% and more have been recorded with diesel cars tested on the relatively cold ECE+EUDC cycle using fuels of less than 500 ppm sulfur [10]. In heavy-duty applications using the same 500 ppm sulfur fuel, the PM emission reduction potential is in most cases limited to 20% to 30%. In the U.S. EPA Urban Bus Retrofit Rebuild (UBRR) program, a number of catalysts were certified to achieve a 25% PM emission reduction over the (medium-load) FTP Transient cycle on two-stroke engines. Most of these catalysts were not able to achieve the same 25% PM emission reduction on four-stroke engines, which produce PM emissions that have a lower SOF content than the two-stroke technology. Even when tested on rather cold test cycles, such as the Japanese 13-mode, PM emission from heavy-duty engines could be reduced by only about 20%, due to high sulfate production during the high-temperature modes [26]. It could be reasonably expected that higher PM emission reductions would be possible with ultra-low-sulfur diesel fuels. On the other hand, oxidation catalyst technology has not been successful in controlling diesel particulates from very hot applications and test cycles, such as the European R-49, ESC, or ETC.

Finally, remember that diesel particulates are sometimes defined in different ways, which may have implications for catalytic control technologies. In some occupational health regulations, diesel particulates are determined as elemental carbon (EC). In such case, the DOC technology would be ineffective in controlling PM emissions. In certain other regulations, diesel particulates are defined as total carbon (TC), that is, including the SOF part, but excluding sulfates. In this case, catalysts can be a very effective control approach, as the sulfate make is no longer an issue.

Mechanism of SOF Removal

To analyze the mechanism of catalytic removal of SOF, one needs to recall again the definition of diesel particulate matter. Diesel PM is defined as everything that is collected from diluted diesel exhaust gas on an in-line filter at a temperature not higher than 52°C. Temperatures in the raw exhaust and in the catalyst are much higher, usually somewhere between 150 and 450°C. A significant proportion of the hydrocarbon species that contribute to the SOF material are in the vapor phase under these conditions and can undergo catalytic oxidation in exactly the same way as gas-phase HCs. After the catalyst, when exhaust gases mix with air in the dilution tunnel, heavy hydrocarbons condense and/or adsorb onto the carbon particles to form the SOF.

Although the preceding reasoning explains how SOF compounds can come into contact with the catalyst, closer examination of the control of PM by catalysts revealed facts that were not consistent with the oxidation theory, suggesting that other mechanisms of SOF removal must be also taking place in the DOC. Once past the light-off, emission control catalysts operate in the mass transfer control regime (compare Chapter 18). Reaction rates depend on mass transfer coefficients for the particular reactants, which in turn depend on their diffusion coefficients. If so, hydrocarbons of shorter carbon chains, characterized by higher diffusion coefficients, should have higher conversion rates over the catalyst in comparison to heavier, longer carbon chain hydrocarbons (provided all species are above their light-off; short-chain HCs are known to have high light-off temperatures). Experiments, however, were in disagreement with theoretical modeling based on oxidation alone, showing higher conversion rates for heavier hydrocarbons [27]. Higher catalytic conversion with increasing HC molecular weight has since been measured by other authors as well [28]. This behavior is likely explained by catalytic cracking of heavy HCs into lighter compounds, apparently catalyzed by washcoat oxides. Cracking appears to be a very important, maybe even the dominant, SOF removal mechanism over base metal catalysts, as reported in the case of cerium oxide catalysts [29] or zeolite catalysts (a rare earth metal exchanged zeolite oxidation catalyst, not optimized for HC storage) [30]. The cracked, short-chain HC products are then measured as gas-phase HCs, and not as part of SOF. This nonoxidation SOF removal path also seems to be confirmed by the observation that while catalytic oxidation of CO and gas-phase HC increases with noble metal (Pt) loading, the conversion of SOF is relatively independent from the Pt load [26, 29].

The explanation of the fate of SOF over the diesel oxidation catalyst by a combination of catalytic oxidation and catalytic cracking is still not entirely sufficient. Another important mechanism that influences DOC performance is HC storage and release in the washcoat. HC storage effects are seen mostly at low catalyst temperatures. Results of laboratory experiments on storage and release using feed gases with different HCs are illustrated in Figure 19.8 [31]. Adsorption of some gases in the washcoat, especially those of higher boiling point such as decane or toluene, resulted in apparent conversion efficiency of as much

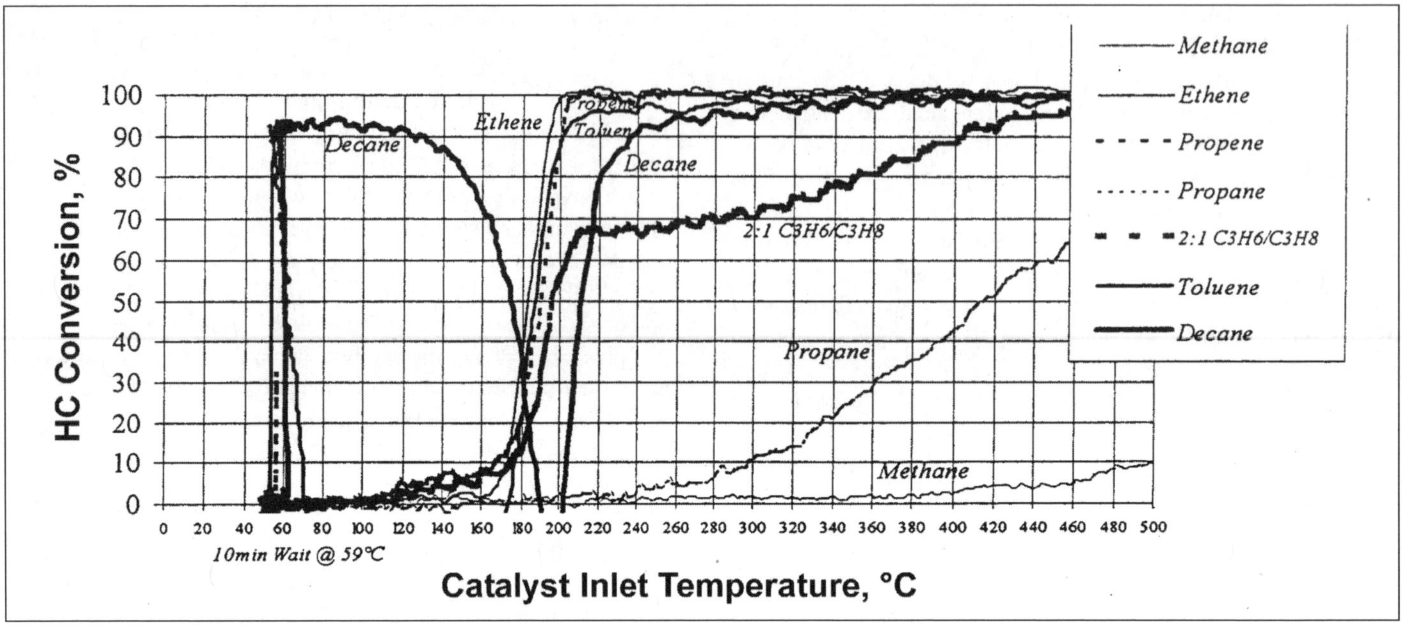

Figure 19.8 *HC storage in diesel oxidation catalyst. Temperature ramp experiment (10 min wait @ 59°C); feed gas contains 3% to 4% water; SV = 33,333 hr^{-1}. (Reprinted from Ref. 31.)*

as 90% at very low temperatures. As temperature increased, the HCs were desorbed, but the desorption temperatures varied among species. Toluene was almost entirely desorbed at 70°C, while the desorption of decane extended through 190°C. It was also observed that HCs and water adsorbed in competition.

Because the HCs in Figure 19.8 were released from the washcoat mostly below the light-off temperature, little or no emission benefit was realized due to the storage effect. Catalyst manufacturers have developed materials that are able to hold the adsorbed HCs at higher temperatures, say, 250°C, in order to catalytically oxidize them once released from the washcoat. These materials, which typically are various types of zeolites, are known as hydrocarbon traps. Hydrocarbon traps eventually reach their saturation, thus losing their effectiveness in a steady-state experiment, but may be very effective in enhancing PM conversion and improving the apparent HC light-off over transient test cycles. Many commercial diesel catalysts introduced since the late 1990s for both light- and heavy-duty applications include zeolites optimized for HC trapping [32, 33].

Impact on Particle Number Emission

Contradictory literature reports exist on the effect of diesel oxidation catalysts on particle number emissions. Because diesel particle number emissions can be attributed primarily to nuclei mode particles, which are composed mostly of hydrocarbon and sulfuric acid condensates, one can easily explain the performance of the DOC by analyzing its effect on nuclei mode particle precursors. If the catalyst removes hydrocarbons (gas phase and SOF), it prevents their subsequent nucleation, thus reducing the particle number emission. If, however, the catalyst produces sulfates—an effect more prominent with high-sulfur fuels and more active, noble metal catalysts—the particle numbers may be increased due to sulfuric acid nucleation.

Experiments that attempt to quantify the impact of DOCs on particle numbers must be very carefully designed. Catalysts can be a source of additional error, such as sample loss due to thermophoretic forces or sample additions due to solid particle blow-off and/or release of condensates from the washcoat.

19.2.2.4 Polynuclear Aromatic Hydrocarbons

PAH emissions are divided between the gas and particulate phases. The heaviest PAHs constitute part of the SOF fraction of diesel particulates, while the lighter species can be identified as part of the gas-phase HC emission. The HC and SOF effectiveness of the diesel oxidation catalysts extends on the PAH class of compounds and other HC derivatives. Table 19.2 lists PAH results from two DOCs tested on a heavy-duty diesel engine (test cycle: FTP Transient; engine: 1998 model 12.7-L DDC series 60 EUI engine, turbocharged and aftercooled, rated 400 hp @ 1800 rpm). The two catalysts provided average PAH emission reductions of 68% and 54%, respectively [34].

19.2.2.5 Mutagenic Activity

Mutagenic activity of diesel exhaust is often evaluated by the Ames test. In the Ames assay the number of bacteria that mutated after being exposed to the tested material are counted and given as the test result. That result is typically expressed as number of revertants per 1 μg of sample. There has been no agreement as to effect of the diesel oxidation catalyst on the mutagenicity of diesel exhaust. Some studies reported an increase of the mutagenicity when diesel catalysts were used

TABLE 19.2
CATALYTIC CONVERSION OF PAH EMISSIONS

Compound	Baseline, µg/bhp-hr	Cat B, µg/bhp-hr	Cat D, µg/bhp-hr	Red Cat B, %	Red Cat D, %
Napthalene	295	159	182	46.1	38.3
2-Methylnapthalene	635	278	277	56.2	56.4
Acenapthalene	40	13	13.6	67.5	66.0
Acenapthene	46	25	24.4	45.7	47.0
Fluorene	72	29	28.9	59.7	59.9
Phenanthrene	169	54	56	68.0	66.9
Anthracene	10	2.6	2.8	74.0	72.0
Fluoranthene	7.7	2.6	4.9	66.2	36.4
Pyrene	14	5	6.4	64.3	54.3
Benzo(*a*)anthracene	0.22	0.05	0.18	77.3	18.2
Chrysene	0.51	0.16	0.33	68.6	35.3
Benzo(*b*)fluoranthene	0.26	0.09	0.12	65.4	53.8
Benzo(*k*)fluoranthene	0.15	0.05	0.08	66.7	46.7
Benzo(*e*)pyrene	0.26	0.08	0.14	69.2	46.2
Perylene	0.01	0	0	100.0	100.0
Indeno(123-*cd*)pyrene	0.13	0.04	0.07	69.2	46.2
Dibenz(*ah*)anthracene	0.01	0	0	100.0	100.0
Benzo(*ghi*)perylene	0.32	0.1	0.22	68.8	31.3
Total	1290.57	568.77	597.14	55.9	53.7
Average reduction				68.5	54.1

Source: Data after Ref. 34.

[35]. In other studies mutagenicity of diesel exhaust extracts decreased with the use of diesel oxidation catalysts. McClure *et al.* [36] found that the mutagenicity of the SOF portion of diesel particulates decreased by 41% and the mutagenicity of the TPM sample decreased by 51% with a DOC.

It is quite possible that the biological activity of diesel exhaust varies with the catalyst formulation, engine, test cycle, and so on. Experiments with different diesel fuels revealed changes in the exhaust mutagenicity [37]. At present, not enough data are available to compare the effects of different catalysts on the biological activity of diesel exhaust.

19.2.3 Sulfate Formation

Catalyst selectivity was the focus of the DOC development throughout the 1990s. A selective catalyst would exhibit high activity toward the oxidation of hydrocarbons but, at the same time, low activity in the oxidation of sulfur dioxide and low sulfate particulate make. A lot of research has been conducted to understand the mechanisms of catalytic sulfate formation on different catalyst systems and with fuels having different sulfur contents. A number of commercial catalysts have been developed for the 500 ppm S fuel, which represents the certification fuel in the United States (1994), Europe (1996), and Japan (1996). Lower sulfur levels always yield better PM control, especially at higher temperatures. The selectivity of diesel catalysts and the sulfate emission problem will likely become less important in the future, when ultra-low-sulfur diesel fuels of 10 to 15 ppm S are commonly used.

The generation of sulfates in the diesel oxidation catalyst depends on the following factors:

- Sulfur content in the fuel
- Catalyst temperature
- Catalyst formulation.

Sulfur dioxide, the precursor of sulfate particulates, originates from the fuel sulfur. The lower the content of sulfur in the fuel, the less SO_2 present in the exhaust gas and the fewer the sulfates generated in the catalyst. If the catalytic conversion of SO_2 into SO_3 [Eq. (19.6)] and the brake-specific fuel consumption are known, one can easily estimate the sulfate particulate emissions. The following formulas can be used for the calculation. It is assumed that the sulfate particulates include all of the sulfuric acid and the hydration water. The water contribution depends on the assumed hydration ratio, as discussed in Chapter 7. A thermodynamically favored ratio during sulfate formation is 2.67, while a typical hydration ratio in samples collected and preconditioned for TPM determination is about 7. Equation (19.9) provides results in g/kWh;

Eq. (19.13) (later in this chapter) produces concentrations in kg/m^3. The mass and volume (**m**, **V**) are brake energy-specific magnitudes (per kWh):

$$\mathbf{m}_{sulfate} = K\,\mathbf{m}_F\,w_S\,M_{sulfate}/M_S \quad (19.9)$$

$$M_{sulfate} = M_{H2SO4} + xM_{H2O} \quad (19.10)$$

$$\mathbf{m}_{exh} = \mathbf{m}_F\,(1 + R_{A/F}) \quad (19.11)$$

$$\mathbf{V}_{exh} = \mathbf{m}_{exh}/\rho_{exh} \quad (19.12)$$

$$c_{sulfate} = \mathbf{m}_{sulfate}/\mathbf{V}_{exh} \quad (19.13)$$

where:

K = fractional conversion of SO_2 to SO_3
m = mass, kg/kWh
w_S = weight fraction of sulfur in fuel
M = molar mass, kg/kmol
V = volume, m3/kWh
$R_{A/F}$ = engine air-fuel ratio, kg/kg
ρ_{exh} = exhaust gas density, kg/m^3
ξ = hydration ratio of H2SO4 in sulfate particulates
c = concentration in the exhaust gas, kg/m^3

Index subscripts: F, fuel; S, sulfur; sulfate, sulfate particulates; exh, exhaust gas.

Very illustrative experimental results on the dependence of sulfate formation on fuel quality are shown in Figure 19.9, which compares the total particulate matter emissions with a platinum-based nonselective oxidation catalyst for three different diesel fuels containing 1500, 500, and 25 ppm of sulfur [38]. For the sake of clarity, the composition of PM (carbon, sulfate, SOF) is not shown in the graph. The particulates are composed mostly of SOF at the lowest (130°C) temperature. At higher temperatures, the particulates are composed primarily of sulfates. The carbon fraction is not more than 0.01 g/kWh at any catalyst temperature.

A substantial increase in the TPM emissions was observed with the 1500 ppm S fuel. Obviously, the nonselective catalyst is totally inappropriate for this application. Also the 500 ppm S fuel, known as the "low sulfur fuel," still contributed to a significant sulfate generation. When used with the 500 ppm S fuel, the nonselective catalyst can still increase the engine-out PM emissions several times, especially at higher temperatures, due to the sulfate generation. Only with the 25 ppm S fuel was the catalyst sulfate make almost insignificant over the entire temperature range. Hence, a nonselective oxidation catalyst can be used only in conjunction with fuels of very low sulfur levels (or with a sulfur-free fuel), without increasing the PM emissions.

The generation of sulfates increases with temperature up to about 450°C, where it exhibits a maximum (Figure 19.9). This maximum can be explained by the thermodynamic equilibrium of the oxidation of SO_2 to SO_3 [refer back to the reaction of Eq. (19.6)]. At high temperatures the reaction equilibrium shifts to the left side of the equation; that is, the higher the temperature the less SO_2 can be oxidized, and less SO_3 is produced. An equilibrium curve, computed for conditions typical for diesel exhaust, is presented in Figure 19.10 [39].

The equilibrium is expressed in terms of maximum (equilibrium) conversion of sulfur dioxide as a function of temperature. The declining equilibrium conversion of SO_2 is in agreement with experimental data, which typically show a gradual increase of PM emissions with temperature up to about 400°C. At these temperatures the sulfate formation increases due to the increased reaction kinetics at higher temperatures. After the temperature reaches 400 to 450°C, the reaction equilibrium limits the rate of sulfate formation and, despite very high reaction rates, the overall rate of sulfate formation decreases.

In addition to the steady-state oxidation of SO_2 with the following formation of sulfuric acid, as described by Eqs. (19.6) and (19.7), storage and release of sulfates from the washcoat can occur in the diesel oxidation catalyst. Sulfur can be stored either

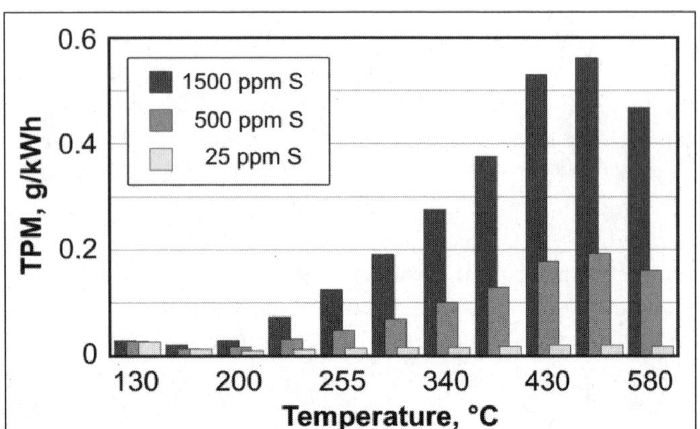

Figure 19.9 Particulate emissions with an oxidation catalyst. (Based on data from Ref. 38.)

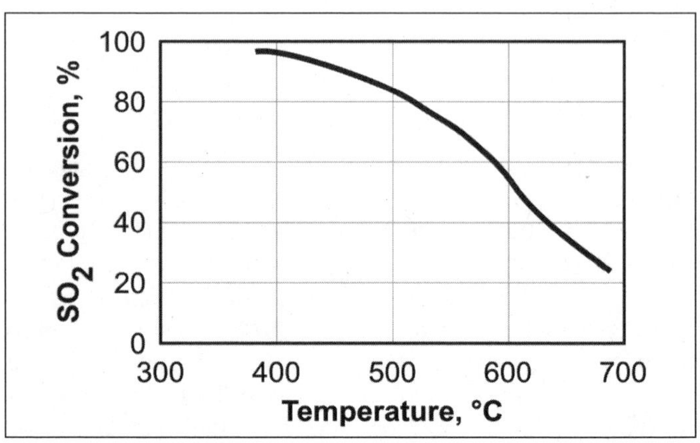

Figure 19.10 Thermodynamic equilibrium for oxidation of SO_2 to SO_3. (Based on data from Ref. 39.)

as SO_2 or as aluminum sulfate on alumina, or other materials, used as the catalyst washcoat. At a certain temperature, the stored material can be released and converted to sulfate. The sulfur release can be described by the following reactions:

$$Al_2(SO_4)_3 = Al_2O_3 + 3SO_3 \quad (19.14)$$

$$Al_2O_3{\cdot}SO_2 = Al_2O_3 + SO_2 \quad (19.15)$$

Some studies measured the quantity of stored sulfur at up to around 2% of the catalyst washcoat mass [39]. Considering typical catalyst sizes and washcoat loads, the sulfur storage can influence the sulfate response of the diesel catalyst.

Figure 19.11 shows the results of 12 emission test repetitions of a catalyst-equipped diesel engine over the U.S. heavy-duty transient cycle [14]. The sulfur content in the test fuel was 400 ppm by weight. An increase in TPM emissions was recorded over the subsequent tests. The particulate analysis indicated that the sulfate particulates were increasing with the test repetitions. In the first test the catalyst reduced the PM emissions by about 25%. In the last test, the PM emissions were increased by about 30%, relative to the engine baseline. Apparently, sulfates were stored in the catalyst washcoat but were not released under the test conditions. During the first tests, sulfur oxides were adsorbed in the washcoat and, thus, removed from the system. A net decrease of TPM emissions was observed. As the catalyst washcoat gradually saturated, more sulfate particulates appeared after the catalyst, eventually leading to an increase in the PM emissions. This example should be remembered whenever emission performance of fresh catalysts is tested. It is important to precondition or "degreen" fresh catalysts before testing. If the degreening is not properly done, the test results may be very misleading.

The sulfur storage phenomena were extensively studied in laboratories. A correlation was found between the catalyst sulfur storage capacity and the formation of sulfate particulates.

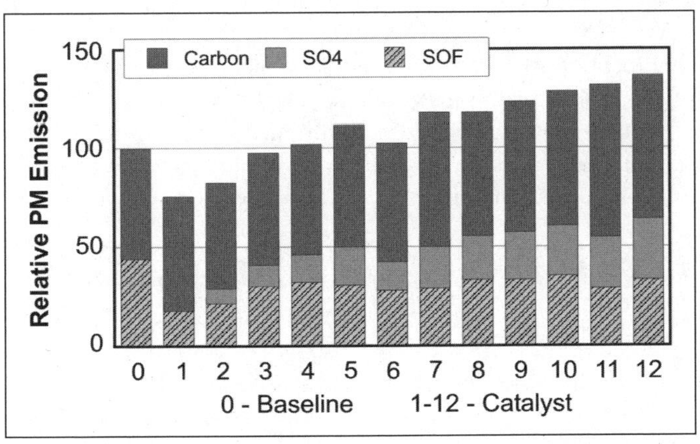

Figure 19.11 *Particulate emissions over repeated FTP cycle. (Based on data from Ref. 14.)*

The catalyst washcoat that stores the highest amount of sulfur is alumina. The Pt/alumina catalyst is known as the most active oxidation catalyst, very effective for oxidation of SOF but also for generating the highest sulfate particulate emissions. The sulfur retention in the Pt/alumina catalyst may be close to 2% of the washcoat mass. The alumina surface area is not sufficient for a monolayer adsorption of that quantity of sulfur. The storage mechanism must also involve some formation of bulk aluminum sulfates. Catalyst systems with lower sulfur storage capacity also exhibit lower rates of sulfate particulate formation. Examples of such systems are Pt on silica or Pd on silica with sulfur retention typically less than 0.05% of the washcoat mass.

In spite of the correlation between sulfur storage and the catalyst sulfate make, it is not clear if the sulfate release is an important contribution to sulfate emissions. It is difficult to set up an experiment that would convincingly separate the contribution of the steady-state SO_2 oxidation from that of the sulfate release from the washcoat. These two processes are also interconnected by the equilibrium of the SO_2 oxidation. If SO_3 is released from the washcoat at higher temperatures, it influences the equilibrium of Reaction (19.6), preventing more SO_2 from being oxidized.

Catalyst selectivity is the focus in the design of catalysts to control particulate emissions from diesel engines. There is always a trade-off between the catalyst activity and its selectivity. A catalyst with a suppressed sulfur dioxide activity will also show lower HC and CO conversions—and vice versa: The catalysts that are most active in the oxidation of CO and HC (and SOF) generate the most sulfate particulates. The objective in catalyst design is to minimize the CO/HC activity penalty for a given sulfate suppression. The following approaches have been used to design the selective diesel catalyst:

- Special catalyst preparation methods such as precalcination
- Selection and loading of the noble metal
- Selection of washcoat materials
- Sulfate suppressing washcoat additives.

Thermal treatment (calcination) of the finished Pt/Al_2O_3 catalyst improves its selectivity. During that process some sintering of platinum occurs. Apparently, the finer dispersions of Pt favor the oxidation of SO_2 to SO_3, while larger Pt grains make the catalyst more selective.

Platinum catalysts are much more active in the oxidation of sulfur dioxide than palladium-based systems. Some diesel oxidation catalysts utilize palladium rather than platinum, or Pt/Pd blends, as a means of sulfate suppression. The addition of rhodium was also found effective in controlling the sulfate formation at high temperatures. The activity of platinum toward SO_2 oxidation depends on the Pt loading to a much higher degree than other functions of the catalyst. Modern diesel oxidation catalysts use platinum at very low loadings of 0.5 to 2 g/ft^3.

Washcoat formulation can strongly influence catalyst performance. Several nonsulfating, chemically inert oxides, such as silica, titania, and zirconia, can be used to suppress the sulfate. Some washcoat oxides, such as ceria, exhibit catalytic properties of their own and can reduce SOF emissions even without the presence of noble metals. There is no sulfate problem in these base metal catalysts.

Various base metal oxides (V, Mo, Nb, etc.) can be added to the catalyst washcoat to inhibit the oxidation of SO_2. Vanadium oxide, which has been identified as one of the most effective sulfate suppressing additives [40], is used in some commercial diesel oxidation catalysts.

The addition of a sulfate suppressing agent may also have a negative impact on catalyst durability. The exact mechanism varies from system to system, but acceleration of support sintering, pore blockage, and formation of inactive surface compounds with the noble metal component are possible.

19.2.4 Catalysts and Configurations

The first diesel catalysts for mining engines and other nonregulated applications utilized pelleted or beaded supports, which were later replaced by monolithic supports. To prevent clogging of the catalyst by diesel particulates from old, dirty engines, large cells, typically having a 200-cpsi cell density, were used. Because neither the tailpipe emissions nor the catalyst durability was regulated, small catalyst volumes were used with space velocities up to 300,000 1/h. Monoliths having high diameter-to-length ratios were used to minimize the pressure drop at high exhaust gas velocities. Such catalyst configurations are still used today in many aftermarket products.

Diesel oxidation catalysts for regulated OEM applications are larger, with space velocities ranging from 50,000 to 200,000 1/h. Typically, the catalyst volume equals approximately the displacement volume of the engine, and exhaust gas space velocities are below 150,000 1/h. Due to much lower particulate emissions, smaller cells can be used to enhance catalyst performance. Cell densities of 400 cpsi are now commonly used in diesel applications [41]. In most cases ceramic cordierite monoliths are used but metallic substrates have been applied in some applications. The catalyzed washcoat loading is about 120 g/dm^3 (2 g/in.^{3}) and the washcoat thickness is typically less than about 40 to 80 μm in the thickest locations (corners) of the monolith cells.

Ceramic catalysts are contained in steel cans, with a mounting mat wrapped around their outside diameter to ensure mechanical integrity and resistance to vibration. Catalytic mufflers, with the catalyst installed inside a silencer, are also used. Catalyst substrates for diesel truck applications have usually about 180 to 250 mm (7 to 10 in.) diameters and 150 to 180 mm (6 to 7 in.) lengths. Smaller substrates are used for car applications.

The early diesel catalysts were generic, active oxidation formulations not optimized for diesel applications. The noble metal was platinum or a platinum/palladium blend of high metal loading, usually between 30 and 50 g/ft^3. Most catalysts utilized an alumina-based washcoat with rare earths or alkaline earths stabilizers (La, Ce, Ba, Sr, etc.).

Modern diesel oxidation catalysts use different strategies for sulfate suppression. Precise information on catalyst formulations is not always available. Many different types of catalysts have been reported in the literature. An Engelhard catalyst was composed of about 50% catalytically active cerium oxide and an equal amount of alumina with 0.5 to 2 g/ft^3 of Pt [29]. The catalyst was later modified by the addition of zeolite [33]. It is used for truck, bus, and selected passenger car applications in the United States and Europe. Johnson Matthey published results [40] using a variation of Pt and vanadia for selected applications. ICT (a joint venture of Degussa and Nippon Shokubai) patented the use of Pd (40 g/ft^3) on zirconia washcoat with rare earth promoters for truck applications [42]. A Degussa catalyst, primarily for use in European passenger car applications, contained 20 to 60 g/ft^3 of Pt on a TiO_2 washcoat with 3% to 10% V_2O_5 [43, 44]. Zeolites are increasingly used by different catalyst makers, especially for European car applications. For example, ASEC is using a Pt catalyst on a zeolite-alumina washcoat [10]. Formulations and performance of selected commercial diesel catalysts are discussed in more detail in Section 19.6.

19.3 Selective Catalytic Reduction

19.3.1 Introduction

Selective catalytic reduction (SCR) of NO_x by nitrogen compounds, such as ammonia or urea—commonly referred to as simply "SCR"—has been developed for and well proven in industrial stationary applications. It has been used since the 1970s in such applications as plant and refinery heaters and boilers in the chemical processing industry, gas turbines, and coal-fired cogeneration plants. The list of fuels used in these applications includes industrial gases, natural gas, crude oil, light or heavy oil, and pulverized coal [45].

The application of SCR to mobile diesel engines requires several problems to be overcome, as discussed later. Due to their active character, SCR emission control systems are likely to be complex and expensive. However, SCR is still the only catalyst technology capable of reducing diesel NO_x emissions to levels required by a number of future emission standards. Notably, SCR has been strengthening its position as the technology of choice for meeting the Euro V (2008) NO_x limit of 2 g/kWh for heavy-duty truck and bus engines. SCR systems for both heavy- and light-duty engines are currently being developed in Europe, as well as in the United States in the context of the 2007 NO_x limit of 0.2 g/bhp-hr. However, the U.S. clean air authorities have voiced concerns as to the enforceability of this technology. From the regulatory perspective, SCR poses enforcement problems, both in terms of ensuring that the reductant is available together with diesel fuel throughout the nationwide distribution network, and that it is always replenished in a timely manner by vehicle operators.

19.3.2 Reductants and Catalytic Reactions

19.3.2.1 Ammonia

Two forms of ammonia can be used in SCR systems: (1) pure anhydrous ammonia and (2) aqueous ammonia. Anhydrous ammonia is toxic, hazardous, and requires thick-shell, pressurized storage tanks and piping due to its high vapor pressure. Aqueous ammonia, $NH_3 \cdot H_2O$, is less hazardous and easier to handle. A typical industrial-grade ammonia, containing about 27% ammonia and 73% water by weight, has nearly atmospheric vapor pressure at normal temperatures and can be safely transported on highways in the United States and other countries.

Several chemical reactions that occur in the ammonia SCR system are expressed by Eqs. (19.16) to (19.20). All of these processes represent desirable reactions that reduce NO_x to elemental nitrogen. Equation (19.17) is the dominant reaction mechanism [46]. Reactions given by Eqs. (19.18) through (19.20) involve a nitrogen dioxide reactant. NO_2 concentrations in most flue gases, including diesel exhaust under normal conditions, are low. The importance of these reaction paths increases with feed gases containing increased levels of NO_2.

$$6NO + 4NH_3 = 5N_2 + 6H_2O \qquad (19.16)$$

$$4NO + 4NH_3 + O_2 = 4N_2 + 6H_2O \qquad (19.17)$$

$$6NO_2 + 8NH_3 = 7N_2 + 12H_2O \qquad (19.18)$$

$$2NO_2 + 4NH_3 + O_2 = 3N_2 + 6H_2O \qquad (19.19)$$

$$NO + NO_2 + 2NH_3 = 2N_2 + 3H_2O \qquad (19.20)$$

Undesirable processes occurring in SCR systems include several competitive, nonselective reactions with oxygen, which is abundant in the system. These reactions can either produce secondary emissions or, at best, unproductively consume ammonia. Partial oxidation of ammonia, given by Equations (19.21) and (19.22), may produce nitrous oxide (N_2O) or elemental nitrogen, respectively. Complete oxidation of ammonia, expressed by Eq. (19.23), generates nitric oxide (NO).

$$2NH_3 + 2O_2 = N_2O + 3H_2O \qquad (19.21)$$

$$4NH_3 + 3O_2 = 2N_2 + 6H_2O \qquad (19.22)$$

$$4NH_3 + 5O_2 = 4NO + 6H_2O \qquad (19.23)$$

At low temperatures (below about 100 to 200°C), ammonia can also react with NO_2 to produce explosive ammonium nitrate (NH_4NO_3):

$$2NH_3 + 2NO_2 + H_2O = NH_4NO_3 + NH_4NO_2 \qquad (19.24)$$

This reaction can be avoided by making sure that the temperature never falls below about 200°C. The tendency of NH_4NO_3 formation can also be minimized by supplying into the gas stream less than the precise amount of NH_3 necessary for the stoichiometric reaction with NO_x (in a 1-to-1 mole ratio).

When the flue gas contains sulfur, as is the case with diesel exhaust, SO_2 can be oxidized to SO_3 with the following formation of H_2SO_4 on reaction with H_2O. These reactions are the same as those occurring in the diesel oxidation catalyst. In another reaction, NH_3 combines with SO_3 to form $(NH_4)_2SO_4$ and NH_4HSO_4 [Eqs. (19.25) and (19.26)], which deposit on and foul the catalyst, as well as downstream piping and equipment. At low exhaust temperatures, generally below 250°C, the fouling by ammonium sulfate may lead to a deactivation of the SCR catalyst [47].

$$NH_3 + SO_3 + H_2O = NH_4HSO_4 \qquad (19.25)$$

$$2NH_3 + SO_3 + H_2O = (NH_4)_2SO_4 \qquad (19.26)$$

The SCR process requires precise control of the ammonia injection rate. An insufficient injection may result in unacceptably low NO_x conversions. An injection rate that is too high results in release of undesirable ammonia to the atmosphere. These ammonia emissions from SCR systems are known as ammonia slip. The ammonia slip increases at higher NH_3/NO_x ratios. According to the dominant SCR reaction of Eq. (19.17), the stoichiometric NH_3/NO_x ratio in the SCR system is about 1. Ratios higher than 1 significantly increase ammonia slip. In practice, ratios between 0.9 and 1 are used, which minimize the ammonia slip while still providing satisfactory NO_x conversions. Figure 19.12 presents an example relationship between the NH_3/NO_x ratio, NO_x conversion, temperature, and ammonia slip [48]. Note that ammonia slip decreases with increasing temperature, while the NO_x conversion in an SCR catalyst may either increase or decrease with temperature, depending on the particular temperature range and catalyst system, as will be discussed later.

In stationary applications, the maximum permitted NH_3 slip is always specified, with a typical specification at 5 to 10 vppm

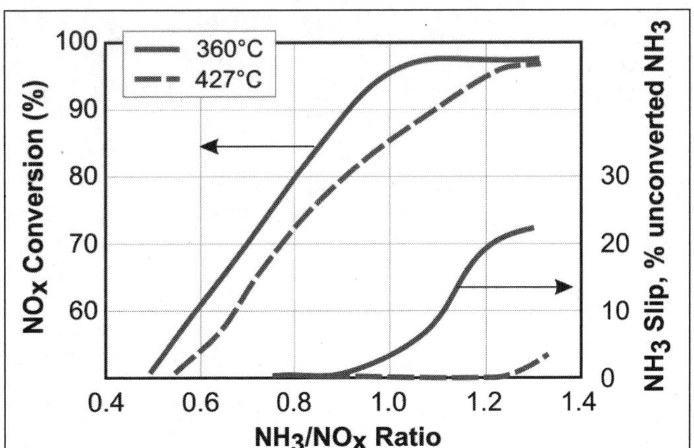

Figure 19.12 *NO_x conversion and ammonia slip for different NH_3/NO_x ratios. V_2O_5/TiO_2 SCR catalyst, 200 cpsi. (Based on data from Ref. 48.)*

NH_3. These concentrations of ammonia are generally undetectable by the human nose. Optionally, ammonia slip can be also controlled by a guard catalyst (oxidation catalyst) installed downstream of the SCR catalyst.

19.3.2.2 Urea

Due to the toxicity and handling problems of ammonia, more convenient SCR reductants are desirable. From the technical point of view, the alternative reductant has to easily and completely decompose to ammonia, producing no harmful by-products, under the conditions in the SCR reactor. From the commercial perspective, the perfect reductant would be nontoxic, easy to transport and handle, inexpensive, and commonly available. The most widely accepted reductant is urea, $CO(NH_2)_2$. Other alternatives that have been considered include carbamate salts, for example, ammonium carbamate, NH_2COONH_4, [49].

Urea meets the criteria of nontoxicity and safety in handling and transportation and is commercially available. For these reasons, urea is now practically the only reductant considered for use in mobile SCR applications. Water solutions appear to be the preferred form of urea. The use of solid urea has been suggested [50], but so far the idea seems to be gaining little acceptance.

Urea is a commodity produced on a large scale. Worldwide demand is estimated at 100 million ton/year, compared to a production capacity of 133 million ton/year [51]. It is used primarily in the food processing and fertilizer industries. Under normal conditions, urea is a solid substance. Even though the solubility of urea in water is better than 50% (108 g/100 mL H_2O at 20°C), aqueous urea solutions used in SCR systems typically have a concentration of 32.5 wt. %. At this concentration urea forms an eutectic solution characterized by the lowest crystallization point of –11°C. The use of eutectic solution provides an additional advantage of equal concentrations in the liquid and solid phases during crystallization. Even if partial freezing occurred in the urea tank, crystallization would not change the concentration of urea solution fed to the SCR system.

Currently, there is no industry standard for urea used in SCR applications. The specifications for aqueous urea used by Siemens in SCR demonstration programs with highway diesel engines [8, 9] are shown in Table 19.3.

In the SCR process, water solutions of urea are injected into the process gas stream. At temperatures above 160°C, urea starts to decompose and hydrolyze according to the following reaction equations:

$$CO(NH_2)_2 = 2{\cdot}NH_2 + CO \tag{19.27}$$

$$CO(NH_2)_2 + H_2O = 2NH_3 + CO_2 \tag{19.28}$$

The thermal decomposition of Eq. (19.27) is confirmed by an evidenced formation of CO during SCR processes with urea [6, 52]. The $\cdot NH_2$ radical can then react with NO as follows:

TABLE 19.3
SPECIFICATION OF AQUEOUS UREA FOR SCR APPLICATION

Property	Value
Name	Aqueous urea solution
CAS #	57-13-6
Chemical formula	$(NH_2)_2CO{\cdot}H_2O$
Molecular weight (anh. urea)	60.06 kg/kmol
Urea content	32.5 ±0.5%
Density at 15°C	1085 kg/m^3
pH	9–11
Appearance	Colorless
Point of crystallization	–11°C
Alkalinity as NH_3	Max. 0.4%
Carbon as CO_2	Max. 0.4%
Biuret	Max. 0.4%
Calcium	Max. 1 mg/dm^3

$$\cdot NH_2 + NO = N_2 + H_2O \tag{19.29}$$

If urea is fed into the system at lower temperatures, it may foul and deactivate the catalyst, presumably due to the production of polymeric species that mask the catalyst surface. This may be a serious consideration for low-temperature applications.

Some SCR systems proposed for mobile applications include a dedicated urea hydrolysis catalyst placed between the urea injection point and the SCR catalyst [53]. The hydrolysis catalysts selectively promotes urea hydrolysis according to Eq. (19.28).

Ammonia generated from the hydrolysis of urea reacts with NO_x according to the already discussed reactions of Eqs. (19.16) through (19.20). We can use Eq. (19.28) to calculate that 1 kg of urea is equivalent to 0.566 kg of ammonia reagent. This ratio represents the stoichiometric requirement for pure urea reactant. For 32.5% urea solution, 1 kg of the solution is equivalent to 0.184 kg of ammonia.

19.3.3 Catalysts

19.3.3.1 Types of SCR Catalysts

The selective catalytic reduction of NO_x with ammonia was first discovered over a platinum catalyst. The Pt technology can be used only at low temperatures (<250°C), due to its poor selectivity for NO_x reduction at higher temperatures. Base metal catalysts have higher SCR temperature windows. A comparison of the operating temperature ranges for various catalyst technologies available for SCR NO_x [54] is shown in Figure 19.13.

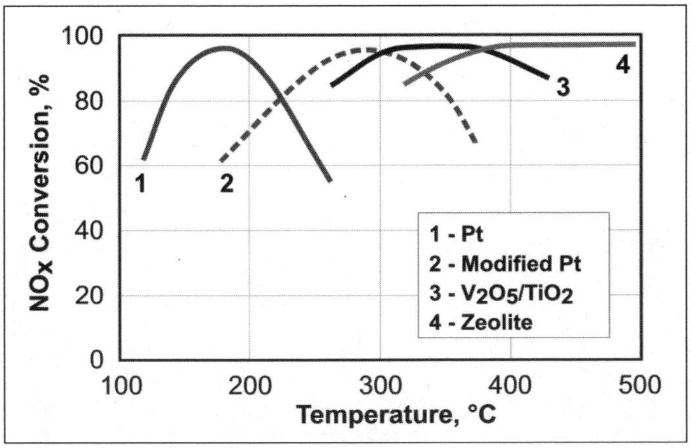

Figure 19.13 Operating temperature windows for different SCR catalysts.(Based on data from Ref. 48.)

Platinum catalysts lose their NO_x reduction activity above approximately 250°C. A V_2O_5/Al_2O_3 catalyst was used first for higher temperature applications. However, its use was limited to sulfur-free exhaust gases because the alumina reacted with SO_3 to form $Al_2(SO_4)_3$, resulting in catalyst deactivation. To solve this problem, a nonsulfating TiO_2 carrier was used for the V_2O_5, which then became the catalyst of choice. These catalysts functioned at higher temperature and over a broader temperature range than Pt. Other base metal oxides, such as tungsten trioxide (WO_3) and molybdenum trioxide (MoO_3), are often added to V_2O_5 as promoters to further decrease the SO_3 formation and to extend catalyst life. Finally, zeolite-based catalysts have been developed that function at high temperatures. The active catalytic components and the corresponding temperature ranges are classified in Table 19.4. These temperature ranges should be considered approximate. Catalysts are under development, especially for mobile SCR applications, which are characterized by increasingly wider temperature windows.

Pt Catalysts

At low temperatures, the SCR reactions of Eqs. (19.16) through (19.20) dominate over the Pt catalyst, so NO_x conversion increases with increasing temperature as shown in Figure 19.13.

TABLE 19.4
SCR CATALYST TECHNOLOGIES

Catalyst	Temperature Range, °C
Platinum (Pt)	175–250
Vanadium (V_2O_5)	300–450
Zeolite	350–600[a]

a. Ion-exchanged zeolites of greatly improved low-temperature activity have been developed for mobile applications.

At about 225 to 250°C, the oxidation of NH_3 to NO_x and H_2O [Eq. (19.23)] becomes dominant. As a result, the conversion versus temperature plot reaches a maximum and begins to fall. To utilize the Pt based catalyst, one must control the process gas temperature so that it remains above approximately 200°C to avoid NH_4NO_3 formation [see Eq. (19.24)], but does not exceed about 225°C, at which point the catalyst loses its selectivity toward the NO_x reduction reaction. This narrow window for temperature control adds expense and complexity to the overall process design. Consequently, this technology is not commonly used today.

Vanadia/Titania Catalysts

Medium-temperature V_2O_5-based catalysts operate best in the temperature range between 260 and 450°C. This has the obvious advantage of a wider temperature window than Pt. However, it also exhibits a maximum followed by a decline in NO_x conversion where the catalyst loses activity, as shown in Figure 19.13. NO_x conversion initiates at about 225°C, rises to a plateau at about 400°C, and then falls as the rate of ammonia oxidation begins to dominate. According to Figure 19.13, the selectivity is lost above about 425°C. The catalyst operating window of newer formulations (Figure 19.14) extends up to about 500°C.

If the exposure temperature of the V_2O_5/TiO_2 catalyst exceeds a certain level, the active, high-surface-area anatase phase of TiO_2 irreversibly converts to rutile with a surface area of less than 10 m^2/g. Normally, this conversion takes place at about 500 to 550°C, but catalysts may include stabilizers to increase their thermal durability. Tungsten trioxide (WO_3) is the most frequently used stabilizer for SCR vanadia/titania formulations. $V_2O_5/WO_3/TiO_2$ systems became a very common SCR catalyst for both stationary and mobile sources [53, 55].

Stabilized V_2O_5/TiO_2 catalysts were reported to be thermally stable up to 700°C. Example data for such a catalyst [53], showing a dramatic loss of activity after aging at 750°C, is presented in Figure 19.14.

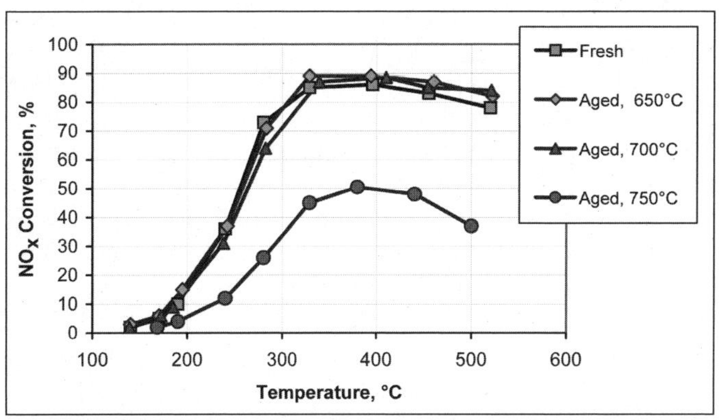

Figure 19.14 Thermal aging of $V_2O_5/WO_3/TiO_2$ catalyst for 100 hours of furnace aging in 10% H_2O, 20 ppm SO_2, air.

Zeolite Catalysts

The first zeolite identified as an active SCR catalyst was mordenite [48]. Common mordenites have a well-defined crystalline structure with a SiO_2:Al_2O_3 ratio of about 10. Manufacturers usually do not disclose the precise chemical composition of the zeolites, so it is not possible to describe them in detail. Zeolite SCR catalysts that are commercially available for stationary engines can operate at temperatures as high as 600°C. When NO_x is present, this catalyst does not oxidize ammonia to NO_x according to Eq. (19.23). Therefore, unlike the Pt and V_2O_5 catalysts, its selectivity toward NO_x conversion continually increases with temperature as shown in Figure 19.13.

Zeolite based catalysts may be prone to stability problems when exposed to high temperatures in the presence of water vapor. At exposure temperatures above 600°C, in a high water content process stream, zeolites tend to deactivate by de alumination, whereby the Al^{+3} ion in the SiO_2–Al_2O_3 framework migrates out of the structure. This leads to permanent deactivation and, in extreme cases, collapse of the crystalline structure.

A different type of a "low-temperature" zeolite catalyst has been developed for mobile engine applications [56, 57]. A starting point for this formulation was the Cu exchanged ZSM5 zeolite, well known from the extensive lean NO_x catalyst research conducted in the 1990s (compare Section 19.4). When used as an SCR catalyst, the Cu/ZSM5 was active in reducing NO_x within the temperature range of about 200 to 400°C, but its thermal durability was insufficient. New formulations were developed by modifying and ion exchanging the zeolite with various undisclosed transition metals. The final "low-temperature" zeolite catalyst was thermally stable up to 650°C. The normal NO_x reducing activity of this catalyst was low. The formulation has been specifically designed for operation in nitrogen dioxide containing gases, which significantly improved its NO_x conversion and extended the temperature window. In the presence of NO_2, the catalyst yielded better than 90% NO_x reduction over the 150 to 500°C temperature range.

Impact of NO_2 on SCR Catalysis

SCR catalysts used for mobile sources are derived from those used in stationary applications, with the V_2O_5/TiO_2 catalyst being the formulation of choice. Operating temperature windows of these catalysts of about 250 to 450°C provide the best match with temperatures found in the diesel exhaust. Their low-temperature activity, however, is insufficient for diesel engines operated at low loads. The low-temperature performance of the catalyst may be increased by using more catalyst volume or by increasing the NO_2 content in the exhaust gas.

Engine out NO_x emissions are composed in more than 90% of NO and only in a few percent of NO_2. It was discovered that increasing the NO_2 fraction in the feed gas can improve the low-temperature activity of the V_2O_5 [58], as illustrated in Figure 19.15. A similar performance enhancement was observed with the Cu/ZSM5 and other "low-temperature" zeolite catalysts [56].

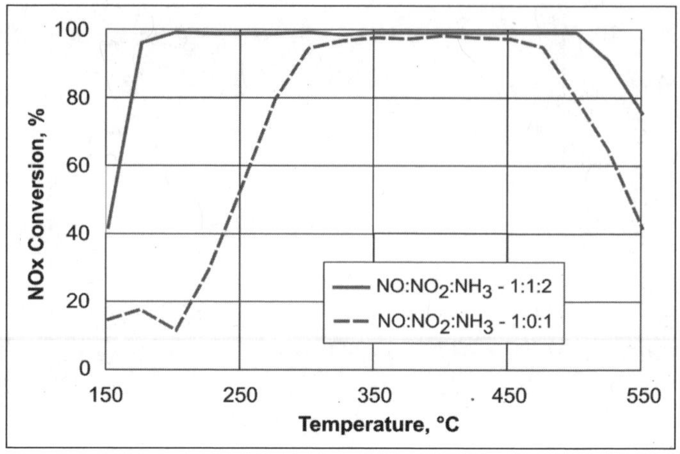

Figure 19.15 Effect of NO_2/NO ratio on NO_x conversion in the V_2O_5/TiO_2 catalyst.

The optimum NO_2 content was reported to be 50% [53]. At higher contents NO_x conversion efficiency decreases, due to the increased ammonia demand resulting from the substitution of NO by NO_2 [Eqs. (19.17) and (19.18)]. Increased NO_2 levels can be realized in practical installations by an oxidation catalyst that promotes oxidation of NO. The oxidation catalyst should be placed upstream of the urea injection point. With the use of the oxidation catalyst, the light off temperature of V_2O_5 catalysts can be decreased to as low as 150°C. Drawbacks of the preoxidation catalyst are (1) increased oxidation of sulfur dioxide and sulfate PM make with fuels of higher sulfur content and (2) increased make of nitrate PM [59].

NO_x conversion in the V_2O_5 SCR catalyst can be also impaired by hydrocarbons [53], as illustrated in Figure 19.16. In these results, generated with laboratory model gas, an increase in the concentration of n decane caused increased light off temperatures for NO_x. An oxidation catalyst placed upstream of the SCR system can also remove hydrocarbons from the exhaust gas, further improving system performance.

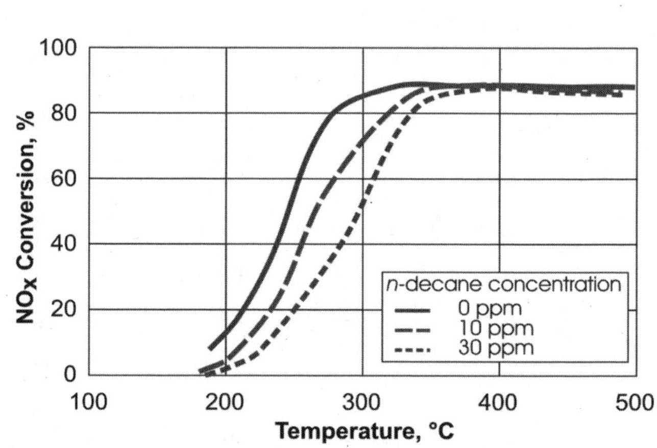

Figure 19.16 Effect of hydrocarbons on NO_x conversion in the V_2O_5/TiO_2 catalyst.

The preoxidation catalyst, similar to that used in the CRT particulate filter (Chapter 22), is optimized for high NO_2 generation through the use of high loadings of noble metals. In one engine study, a 2-L catalyst of 90 g/ft^3 Pt loading was used on a 4.0-L DI TCI engine [56].

19.3.3.2 Auxiliary Catalysts

As mentioned earlier, SCR systems may also include an oxidation catalyst downstream of the SCR catalyst to control the ammonia slip. It is a typical Pt/Al_2O_3 oxidation catalyst, similar in size to the preoxidation catalyst, but of low metal loading, which was reported by some studies at 10 g/ft^3 [56].

Some SCR systems use a dedicated urea hydrolysis catalyst, positioned upstream of the SCR catalyst. The urea hydrolysis catalyst is typically a base metal oxide formulation. For example, Eberhard and coworkers [50] disclosed a mixture of TiO_2 at 110 g/L, Al_2O_3 at 30 g/L, and SiO_2 at 10 g/L, coated on a 150-cpsi substrate and sized at a space velocity of 90,000 1/h.

19.3.4 Stationary SCR Systems

SCR catalysts for industrial applications can be prepared in a number of different geometric structures. Some are extruded into pellets or homogeneous monoliths, others are supported on parallel metal plates or ceramic honeycomb structures, and still others are fixed on a wire mesh. The appropriate structure depends on the particular application. SCR systems used for NO_x abatement from diesel engines typically utilize monolithic catalysts, which can be either homogeneous or supported on ceramic substrates. Square catalyst modules are usually assembled into layers (baskets), which in turn fit into the steel catalyst housing. The principal SCR suppliers and the catalyst types [45, 52] are identified in Table 19.5.

Cell densities in monolithic SCR catalysts vary from less than 10 cpsi to 200 cpsi. The smaller cells (larger cell densities) are used for "clean" gas applications and larger cells for "dirty" gas applications. Units with large cells are usually very big. The process gas space velocities can be as low as 3000 to 5000 1/h. Small cell substrates, such as 200 cpsi, are usually sized for space velocities of 10,000 to 20,000 1/h, but sometimes as high as 40,000 1/h.

Most stationary SCR systems use ammonia as the reducing agent. However, since the late 1980s, urea-based systems have become increasingly popular.

The traditional method of injecting ammonia into the exhaust gas stream utilizes pure anhydrous ammonia. Liquid ammonia is stored in a pressurized tank. From there it is piped to a heater (typically an electric heater), where the liquid vaporizes. The ammonia vapor is next routed into a mixing chamber,

TABLE 19.5
SUPPLIERS OF INDUSTRIAL SCR CATALYSTS

Supplier	Catalyst Type	Operating Temperature, °C
Japan		
Babcock Hitachi	Base metal/metallic substrate	250–416
Hitachi Zosen	Base metal/ceramic monolith or wire mesh	330–421
Ishikawajima-Harima Heavy Industries	Base metal/ceramic monolith	204–400
Kawasaki Heavy Industries	Base metal/ceramic monolith	300–400
Mitsubishi Heavy Industries	Base metal/ceramic monolith	204–400
UBE	Base metal/ceramic monolith	250–400
United States		
W. R. Grace	Noble metal/metallic substrate	225–275
Engelhard	Base metal/ceramic monolith	302–400
Johnson Matthey	Base metal/metallic substrate	343–427
Norton	Zeolite	221–521
Europe		
BASF	V_2O_5-TiO_2-WO_3/homogeneous monolith	—
Siemens	V_2O_5-TiO_2-WO_3/homogeneous monolith	—
Steuler	Zeolite	300–521
Haldor Topsoe	V_2O_5-TiO_2/homogeneous monolith	—

where it mixes with ambient air supplied by a fan or blower in a predetermined ratio. The ammonia/air mixture is then directed to the distribution grid system for subsequent injection into the flue gas stream at a location upstream of the SCR reactor. The NO_x concentration downstream of the reactor is used as a feedback to control the amount of injected ammonia. A precise ammonia injection control is essential to obtain high NO_x conversion and low ammonia slip.

An alternative approach for ammonia injection is the use of aqueous ammonia, $NH_3 \cdot H_2O$, which has lower vapor pressure, is easier to handle, and is less hazardous. A schematic of an aqueous ammonia injection system [46] is shown in Figure 19.17. It contains two trains of piping, one for aqueous ammonia and the other for carrier air, which feed into a vaporizer. Aqueous ammonia is stored at ordinary temperature in a tank. From there it is pumped, filtered, metered, and sprayed through atomizing nozzles into the vaporizer. In the vaporizer, ammonia is evaporated by mixing with hot air. Ambient air is drawn by a blower and heated, then enters the vaporizer through nozzles at the top of the vessel.

The air heater is usually electric, but heating coils placed directly in the process gas or heated by steam can also be used. The vaporizer is a column apparatus, usually packed with metallic rings or other packing to promote the mixing and evaporation processes. The mixture of ammonia, air, and water vapor is then injected into the flue gas stream. It is important to obtain a homogeneous NH_3/NO_x ratio in the entire process gas stream. The injection grid network consists of a number of pipes connected in parallel, each of which contains several orifice holes. The pipes and holes are sized to provide balanced ammonia flow distribution throughout the process gas. A static mixer may be placed in the exhaust duct to further improve mixing.

Stationary SCR systems often feature a closed-loop control system. The concentration of NO_x and/or NH_3 in the exhaust gas is continuously monitored and used as a feedback signal to control the amount of ammonia/urea to be injected.

19.3.5 SCR Systems for Mobile Engines

19.3.5.1 Overview

At this time, SCR remains the only proven catalyst technology that is capable of reducing diesel NO_x emissions to levels required by future emission regulations. The use of ammonia has been practically ruled out, due to safety concerns, and urea is the reductant of choice. Most development work is currently targeting the heavy-duty Euro V (2008) standard of 2 g/kWh NO_x. SCR is envisioned as one of the most promising technologies to simultaneously meet the Euro V limits for both NO_x and for PM [60]. It is possible to calibrate heavy-duty engines for very low particulate emissions, below the Euro V limit of 0.02 g/kWh. Such low PM engine calibration leads to higher NO_x emissions; a worst case baseline NO_x of 7 to 8 g/kWh may be increased to some 9 to 11 g/kWh. A urea-SCR aftertreatment system can then be used to bring down the NO_x emissions to less than 2 g/kWh. The required reducing capacity of the SCR system of about 80% to 85% is believed to be technically feasible. SCR systems of reduction efficiencies in this range have already been demonstrated [60]. The need for a separate diesel particulate filter is eliminated in this strategy, resulting in smaller size, complexity, and cost of the overall emission aftertreatment system.

SCR systems are also being developed for a number of other applications, including light-duty (both in the EU and the United States) and the US2010 heavy-duty NO_x standard of 0.2 g/bhp-hr. This standard is much more stringent compared to the Euro V. Relative to the 2002/2004 standards, which can be met by the cooled EGR technology, a more than 90% NO_x reduction over the transient FTP test is necessary; SCR systems of such performance have not yet been demonstrated [61].

Calibrating engines for low PM and high NO_x involves advanced injection timing, resulting in improved fuel economy. The "Euro V calibration" mentioned earlier brings an estimated 3% to 5% fuel savings over the regular engine calibration [60].

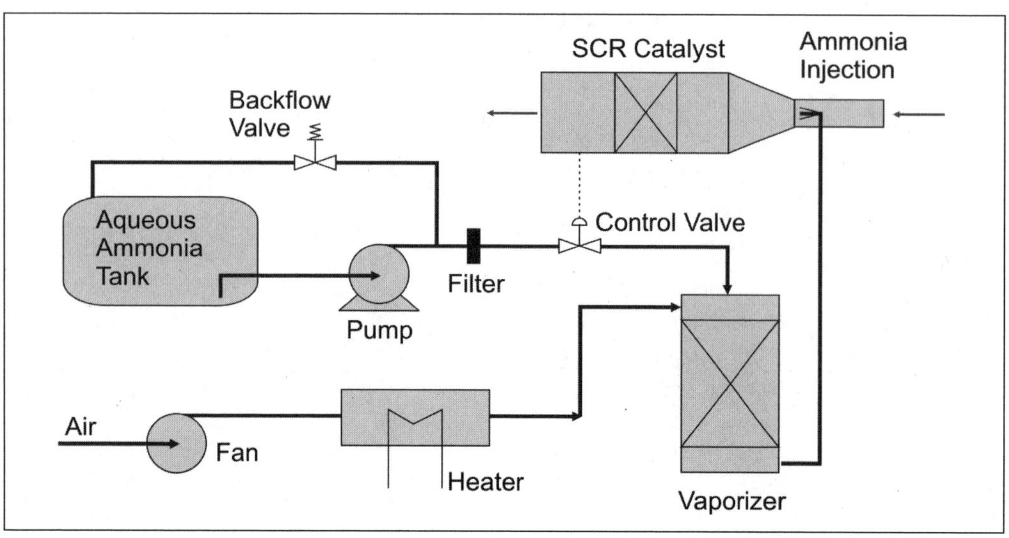

Figure 19.17 Aqueous ammonia injection system.

Some authors conclude that these fuel savings would provide a quick payback for the cost of the SCR system. One needs to be skeptical and carefully verify such claims. In general, the potential savings depend on the fuel costs and driving distances, which vary in different geographical areas. The fuel savings are also directly offset by the cost of urea, which is consumed at about 3% of the fuel volume. Finally, in applications where stringent NO_x standards are in place, the SCR technology may have little or no extra capacity to handle the additional engine-out NO_x levels that result from fuel-efficient engine calibration.

Depending on the particular application and NO_x reduction requirement, the application of SCR technology to mobile engines still remains a challenging task that requires solution of a number of technical, regulatory, and urea distribution-related problems. The following are the most important issues:

- *Catalyst selection and system configuration:* Most stationary applications, both industrial and diesel engines, are characterized by steady-state operating conditions. Mobile diesel engines, on the other hand, operate under transient duty cycles with great variation of mass flow, exhaust temperature, and NO_x concentration. The catalyst has to operate under a much wider temperature window than in stationary applications. In addition, the transient operation presents challenges in the development of a urea injection strategy, making it more difficult to control the formation of secondary emissions (NH_3, N_2O, NH_4NO_3, etc.), and to achieve required catalyst durability.
- *Size, complexity, and cost of SCR systems:* SCR catalysts are much larger than those used by other catalyst technologies. Most proposed systems also include auxiliary catalysts to enhance the performance of the SCR catalyst and/or to control ammonia slip. In addition, a complex reductant injection/control system and on-board storage capacity are needed. All this translates to expected high costs of SCR systems.
- *Regulatory compliance and urea distribution issues:* The diesel engine can run without the reductant. Because there is no incentive to the vehicle operator to replenish used reductant, regulatory authorities are concerned that vehicles may be operated without urea making SCR systems inactive. Other problems include the lack of distribution infrastructure and freezing of urea solutions in winter.

19.3.5.2 Experience

Not surprisingly, the first mobile SCR NO_x control systems were installed on marine applications. Large sizes and steady-state operation of marine units, similar to stationary diesel engines, made the adaptation of stationary SCR technology straightforward. The first SCR units were installed in 1989 and 1990 on two Korean 30,000 metric ton carriers [62]. The ship operator was seeking a permit from the Bay Area Air Quality Management District (BAAQMD) to allow the reduced-emission ships in its docks. Both ships were powered by MAN B&W two-stroke 8-MW diesel engines. The ships were equipped with an ammonia SCR system designed for 92% NO_x reduction. The SCR reactor included a bypass system. Exhaust gases were passed through the reactor only when the ships were sailing in waters subject to NO_x emission regulations.

Commercial SCR systems have been also installed on ferries. In 1992 the ferry *Aurora of Helsingborg* that shuttled between Sweden and Denmark was equipped with a urea SCR system [52]. The engine was a 2.4-MW Wartsila, type 6R32E. The reactor included three layers of monolithic extruded SCR catalysts and one layer of an oxidation catalyst. SCR technology has also been considered for NO_x control from locomotive diesel engines [63].

Numerous development projects aimed at adapting SCR technology to diesel-powered truck and car engines have been conducted in Europe. In a project sponsored by the Netherlands Ministry of Environment, a 12-L, heavy-duty diesel engine equipped with a urea SCR catalyst system was tested under both steady-state and transient conditions [7, 64]. In another study by FEV and Ford, a urea SCR system was tested on a 2.5-L, light-duty diesel engine [65]. TNO Automotive and Engelhard have developed an SCR system for heavy-duty engines, which was tested on DAF and Renault trucks [66].

Not surprisingly, mobile systems have been also developed by companies with traditional expertise in stationary installations, such as Haldor Topsoe [60] or Siemens with its automotive SCR system termed SiNOx. Reported SiNOx demonstration programs involved bench testing, chassis dynamometer tests, and vehicle road tests with a number of heavy-duty trucks that were operated by trucking companies in regular duty [8]. A prototype of the Siemens system has also been tested on a heavy-duty truck on U.S. highways [9].

19.3.5.3 System Configuration

A final configuration for an SCR system for mobile engines has not yet been established. There are a number of differences between systems proposed by different developers and systems evolve with time. The general layout shown in Figure 19.18 combines a number of features from different SCR systems.

Because suitable, fast-response NO_x or NH_3 sensors are not available, the outlet exhaust gas concentration cannot be used as feedback for the process control. Instead, the SCR systems for mobile engines work in an open-loop configuration in which a preprogrammed map of engine NO_x emissions is used to control the urea injection rate as a function of engine speed and load. This open-loop configuration is capable of about an 80% NO_x reduction, as required for heavy-duty Euro V applications. However, closed-loop systems are likely to be required for more demanding applications of 90%+ NO_x reduction targets. It was estimated that SCR systems to meet the US2007 standards would require NO_x sensors of 40 to 20 ppm NO_x sensitivity and little cross-sensitivity to NH_3 [61]. The use of closed-loop systems would also minimize the amount of engine calibration work that is required in the development of open-loop systems.

The urea solution is pumped from the urea tank and sprayed through an atomizing nozzle into the exhaust gas stream. Thorough mixing of urea with exhaust gases and uniform flow

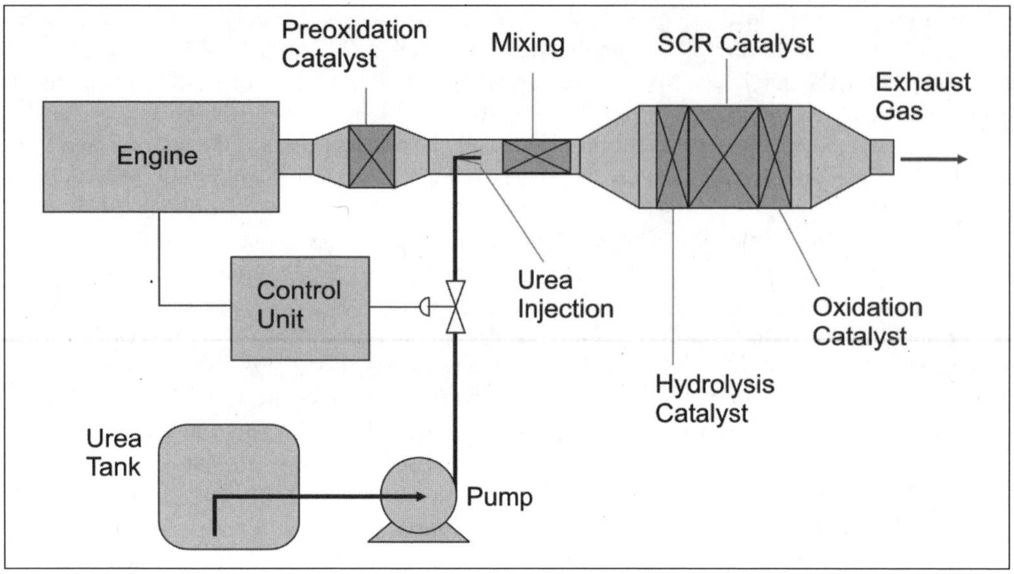

Figure 19.18 *Open-loop urea SCR system for mobile diesel engines.*

distribution are important in achieving high NO_x conversions. Flow maldistribution may be minimized by avoiding elbows and other flow disturbances upstream of the SCR catalyst [67]. In addition, static mixing devices [68] may be provided downstream, as well as upstream (not shown in Figure 19.18), of the urea injection point. In some systems used for marine engines, urea was first mixed with air, supplied by a dedicated fan. A similar injection system supplying urea aerosol has been designed for trucks that utilizes compressed air from the vehicle's suspension system [66]. Regardless of the injection method, we must emphasize that optimization of the urea injection process, in terms of accuracy and reproducibility, is one of the most important areas where SCR performance improvement can be achieved.

Once mixed with the hot exhaust gas, the urea undergoes hydrolysis and thermal decomposition to produce ammonia. In some systems the urea hydrolysis is additionally promoted by a dedicated hydrolysis catalyst installed upstream of the SCR catalyst [53]. In most systems, however, the SCR catalyst itself promotes both urea hydrolysis and the SCR reaction.

The mixture of exhaust gases and ammonia (decomposed urea) enters the SCR catalyst where nitrogen oxides are reduced to nitrogen. SCR catalysts for mobile applications have been using both coated metallic/ceramic substrates and homogeneous V_2O_5/TiO_2 extrusions, which were developed for stationary applications. Homogeneous catalysts of corrugated construction with cell densities as high as 300 cpsi have also been developed [60]. Reducing the catalyst size is one of the major objectives of ongoing development. The exhaust gas space velocities in the first mobile SCR catalysts were typically between 20,000 and 30,000 1/h. Lower space velocities provided higher NO_x reductions and lower ammonia slip, but resulted in large catalyst sizes. In one of the early SCR projects, a 12-L, heavy-duty diesel engine was equipped with 70 L of catalyst volume at a space velocity of 21,500 1/h [64].

Catalyst volume reductions can be achieved by optimization of its geometry, such as increasing the cell density. Coated SCR catalysts can have cell densities up to 400 cpsi [66], matching those used in other diesel catalyst technologies. Homogeneous catalysts became available with gradually increasing cell density. For example, an increase in the cell density of the SiNOx homogeneous extrusion from 100 to 200 cpsi allowed for reductions in the catalyst volume on a 12-L, heavy-duty engine from 63 to 45 L [9]. Further significant catalyst size reductions may be realized by installing a preoxidation catalyst, designed to generate NO_2 and remove hydrocarbons, upstream of the urea injection point. In one of the studies that used precatalysts, SCR catalysts volumes of 4.6 to 9.2 L were tested on a 4.0-L, heavy-duty engine [53]. The volume of the precatalyst was 2 L. In another study, also utilizing precatalysts, the volume of a homogeneous SCR catalyst was reduced to about 20 L on a 12-L, 400-hp engine by increasing the cell density to 300 cpsi [60].

Many systems include an oxidation catalyst positioned downstream of the SCR catalyst. The volume of this catalyst is comparable to that of the NO_2 precatalyst, but the noble metal loadings are much lower (one study reported 10 g/ft^3 Pt compared to 90 g/ft^3 Pt in the precatalyst [56]). The function of this "guard catalyst" is to oxidize ammonia slip that may appear due to the imperfect control strategy or in the case of an SCR catalyst failure. An added benefit of the oxidation catalyst is the control of hydrocarbons and carbon monoxide emissions, as well as the diesel odor. Disadvantages of the ammonia slip catalyst include (1) increased proportion of nitrous oxide, N_2O in the exhaust gases [59] and (2) increased NO, both generated though the oxidation of ammonia according to Eqs. (19.21) and (19.23), respectively. The oxidation of NH_3 to NO may cause an apparent decrease of NO_x conversion efficiency in SCR systems equipped with ammonia slip catalysts at higher exhaust temperatures [60].

Typically, the urea injection controller also monitors the exhaust gas temperature. When the temperature drops below a predetermined value, somewhere between 150 and 300°C depending on the catalyst type and configuration, the controller closes the urea supply to prevent catalyst deactivation (e.g., by ammonium sulfate or unhydrolyzed urea) and secondary emissions that may occur at low temperatures.

An alternative reductant supply approach is to use solid urea rather than water solutions. The advantages of solid urea include (1) less on-vehicle space requirement for storing the reductant (or achieving a higher driving range at a given storage volume), (2) avoiding the exhaust gas cooling effect associated with the evaporation of water, and (3) avoiding the freezing problems in cold climate areas. However, systems proposed for dosing solid urea appear to be more complex than those utilizing solutions. MAN has developed a system utilizing small-diameter urea pellets [50]. The system includes a urea pellets container, a motorized feeder, and a grinder. The urea powder produced in the grinder is carried by compressed air to the exhaust system and mixed with exhaust gases in a special mixing chamber upstream of the hydrolysis catalyst. Reported problems included (1) adsorption of atmospheric moisture, which made the urea pellets cluster, leading to clogging of the dosing system, and (2) corrosion problems resulting from melted urea, requiring either expensive steel grades or protective surface coatings.

19.3.5.4 Urea Consumption and Replenishment

Most proposed SCR systems require that aqueous urea solution be carried in an on-board storage tank that requires periodic replenishment. Reported SCR developments on light-duty engines indicated an average consumption of 2.5 cm^3 of urea solution per kilometer (0.25 l/100 km) on a 2.5-L diesel engine. At this consumption level, a 40-L on-board urea tank would have to be refilled every 15,000 km [65].

In an SCR demonstration project with a U.S. Class 8 highway truck, diesel fuel and urea consumption were found to be 5.2 mpg (45.2 L/100 km) and 94 mpg (2.5 L/100 km), respectively [9]. At this rate, a 15-gal (570-L) urea tank would allow for 1410 miles (2260 km) of truck operation between refills. In another heavy-duty engine project, a urea consumption of 3% of the fuel by volume (or 4% by mass) was reported [66].

Commercial introduction of SCR systems would require establishing a urea distribution infrastructure. The implementation of such a system poses a number of logistics and regulatory emission compliance challenges. Urea solutions would need to be distributed along with the diesel fuel throughout the fuel distribution network. The fueling system should prevent, or at least minimize, the possibility of operating the vehicle with an empty urea tank. "Intelligent," computerized fueling terminals have been proposed, which would communicate with the vehicle's electronic control unit and simultaneously dispense the required amount of fuel and urea through a special dual fueling nozzle [69, 70].

Maintaining the urea infrastructure also has serious economic implications. The infrastructure cost may, in fact, be the major component of the urea price at the pump, more than offsetting any fuel economy gains due to the use of SCR with fuel-efficient engine calibration. A comprehensive urea infrastructure cost analysis was carried out by the U.S. Department of Energy [51]. The costs of producing and distributing urea were estimated separately. Depending on the assumptions (demand, the number of retail points, and the level of product segregation), the distribution cost ranged from \$0.70 to \$35 per gallon (\$0.18 to \$9.25 per liter) of 32.5% SCR-urea solution. The lower end of the range assumed high-throughput truck stops, whereas the upper part represented light-duty retail outlets with low throughput of urea. Production costs were estimated to range between \$0.12 and \$0.30 per gallon (\$0.03 to \$0.08 per liter), with an estimated \$0.05 to \$0.10 per gallon (\$0.01 to \$0.03 per liter) to be added to the cost when the urea solution is sold.

Cold temperature properties of urea solutions would also need to be addressed in commercial SCR systems. Because urea solutions have freezing temperatures of –11°C, which is not acceptable for winter conditions in most of Europe or North America, the storage tanks and tubes would need to be heated. Alternatively, additives could be used to lower the crystallization temperature of aqueous urea solutions to acceptable levels below –30°C.

19.3.5.5 Emission Performance

Several studies have demonstrated that the SCR catalyst is capable of efficiently reducing diesel engine NO_x emissions. Reported NO_x conversion efficiencies in heavy-duty engines ranged from about 55% to 90%, depending on the application and test method [7–9]. Higher conversions are measured on EU steady-state cycles, such as the ESC. The U.S. FTP procedure has been more challenging due to both its transient character and the low exhaust temperatures and low conversions in the initial portion of the test.

It has been convincingly demonstrated that Euro II engines can meet the Euro V NO_x emission standard of 2.0 g/kWh when equipped with SCR systems. As an example, the study of the TNO/Engelhard system on DAF and Renault trucks showed NO_x conversions of 81% to 84% on the ETC test and 72% over the ETC test, with almost all results below 2 g/kWh NO_x [66]. These results were achieved through careful optimization of the urea injection. The catalyst system itself was rather simple, with a single SCR catalyst sized at 45,000 1/h space velocity (34 dm^3 catalyst on a 400-cpsi substrate on a 12-L engine) and no oxidation pre- or ammonia slip catalysts. U.S. 1994 engines also met the 2004 NO_x standard utilizing SCR. However, the ability to reduce NO_x emissions to 0.2 g/bhp-hr, the U.S. NO_x standard for the year 2010, has not yet been demonstrated. It may quite likely require closed-loop SCR controls to become feasible [61].

In an SCR study with a 2.5-L, light-duty diesel engine, NO_x conversions of up to 60% on the ECE+EUDC cycle and up to 83% on the FTP-75 cycle were reported [65].

In addition, a number of undesirable emissions might be created by SCR catalysts. SCR systems that do not include oxidation catalysts may produce emissions of (1) ammonia (ammonia slip) and (2) ammonium nitrate particulates. If the system also includes an oxidation catalyst, additional secondary emissions may be produced, such as (3) sulfate particulates (ammonium sulfate) and (4) nitrous oxide.

Progress made in the ammonia dosing equipment made it possible to lower the ammonia slip in mobile systems to levels almost equivalent to those seen in stationary SCR systems. A number of publications reported ammonia slips of less than 10 ppm. In the TNO/Engelhard system, which did not include an ammonia slip catalyst, an average ammonia slip of 6 ppm and a maximum slip of 37 ppm were measured [66].

The tendency of the diesel oxidation catalyst to form sulfate emissions is commonly known. It increases with temperature and with sulfur content in the fuel. A 40% increase in PM emissions due to sulfate formation was reported in an SCR/oxidation system on the hot ECE R49 cycle [7]. An increase in PM emissions in an SCR catalyst (in a system that also included an oxidation catalyst) was also reported when using ultra-low-sulfur fuels [58]. The most likely explanation was a combined effect of sulfate and ammonium nitrate particulate emissions generated in the SCR catalyst. Decreases of total PM emissions (by 4% to 23%) were measured in SCR systems that did not include oxidation catalysts [66].

Increased emissions of nitrous oxide (N_2O) reported in many publications are specifically related to the ammonia slip in the SCR catalyst. A high proportion of ammonia slip can be converted in the oxidation catalyst to N_2O, according to Eq. (19.21). One study found that proportion to be between 25% and 50% of the total ammonia emission [7]. Nitrous oxide levels of 11 to 20 ppm were measured after the oxidation catalyst. It was also found that the engine-out nitrous oxide, of about 3 ppm, was practically unchanged in the SCR catalyst. The N_2O emission could be avoided by the elimination of the oxidation catalyst. In theory this could be achieved through better urea injection control, preferably in the closed-loop configuration, resulting in acceptably low ammonia slip. However, because the oxidation catalyst downstream of the SCR catalyst allows a larger NH_3:NO_x window, its elimination in systems having high NO_x reduction requirements may be problematic.

19.4 Lean NO_x Catalyst

19.4.1 Selective Reduction of NO_x by Hydrocarbons

19.4.1.1 Chemical Reactions

After it became apparent that NO decomposition catalysts had too many shortcomings to produce a robust, commercial catalyst system [5], research turned toward selective reduction of NO_x by compounds of combustion gases. It was discovered that several catalysts promoted selective catalytic reduction of NO_x by hydrocarbons or other exhaust gas components (carbon monoxide, alcohols, etc.) [71]. Catalysts selectively promoting the reduction of NO_x by hydrocarbons have been termed *lean NO_x catalysts* (LNCs) or *deNOx catalysts.*

NO_x reduction by HCs was found to be less susceptible to sulfur poisoning than NO decomposition, and higher conversion efficiencies were demonstrated. In the case of the diesel application, diesel fuel was the obvious source of hydrocarbons necessary for the reaction. Because the enrichment of exhaust gases with diesel fuel seemed more straightforward than carrying on-vehicle ammonia vessels, catalyst research focused on that process and produced numerous technical publications. In fact, selective NO_x reduction by hydrocarbons became the mainstream in NO_x reduction catalyst research during the early to mid-1990s.

In the selective catalytic reduction, hydrocarbons react with NO_x, rather than with O_2, to form nitrogen, CO_2, and water:

$$\{HC\} + NO_x = N_2 + CO_2 + H_2O \qquad (19.30)$$

Assuming that a single hydrocarbon species of the formula C_mH_n reacts with nitric oxide, the preceding equation can be written as a properly balanced chemical reaction, as follows:

$$C_mH_n + (2m + \tfrac{1}{2}n)NO = (m + \tfrac{1}{4}n)N_2 + mCO_2 + \tfrac{1}{2}nH_2O \qquad (19.30a)$$

The competitive, nonselective reaction with oxygen is given by the general Eq. (19.31):

$$\{HC\} + O_2 = CO_2 + H_2O \qquad (19.31)$$

or, in the case of the hydrocarbon C_mH_n, by Reaction (19.31a):

$$C_mH_n + (m + \tfrac{1}{4}n)O_2 = mCO_2 + \tfrac{1}{2}nH_2O \qquad (19.31a)$$

DeNOx catalysts have to be optimized to selectively promote the desired reaction, Eq. (19.30), with hydrocarbons over the undesired reaction with oxygen, Eq. (19.31). The catalyst selectivity is determined by the catalyst formulation, but it also depends on a number of other factors, such as the hydrocarbon species used for the reaction, the temperature, the exhaust gas oxygen content, and the HC/NO_x ratio.

The selectivity of a reactant X for reduction of NO_x is defined as the ratio of the number of moles of X that react with NO_x to the total number of moles of X consumed. In the case of the hydrocarbon C_mH_n, its selectivity for Reaction (19.30a) is given by the following equation:

$$\text{Selectivity} = \{C_{NO} \cdot [NO]\} / \{(2m + \tfrac{1}{2}n) \cdot C_{CmHn} \cdot [C_mH_n]\} \qquad (19.32)$$

where:

[NO] and $[C_mH_n]$	=	inlet molar concentrations of NO and HC (actual, not as C_1)
C_{NO} and C_{CmHn}	=	fractional conversions of NO and hydrocarbons.

Equations (19.30) and (19.30a) present the ideal reaction path, with the formation of elemental nitrogen, N_2, as the only nitrogen-containing compound. In practice, selective reduction of NO_x by hydrocarbons produces a mixture of nitrogen and nitrous oxide, N_2O, which is an undesired product. The proportion of N_2O depends, among other parameters, on the catalyst formulation. Catalysts may be characterized by their selectivity toward the formation of nitrogen, as opposed to nitrous oxide. This selectivity, defined as the molar ratio of nitrogen to nitrous oxide that were produced in the reaction, is given by the following equation:

$$N_2 \text{ Selectivity} = 1 - 2[N_2O]_{out}/(C_{NOx} \cdot [NO_x]) \quad (19.33)$$

where:

$[N_2O]$ = outlet molar concentration of N_2O (assuming zero at the inlet)

$[NO_x]$ = inlet molar concentration of NO_x ($NO + NO_2$)

C_{NOx} = fractional conversions of NO_x ($NO + NO_2$).

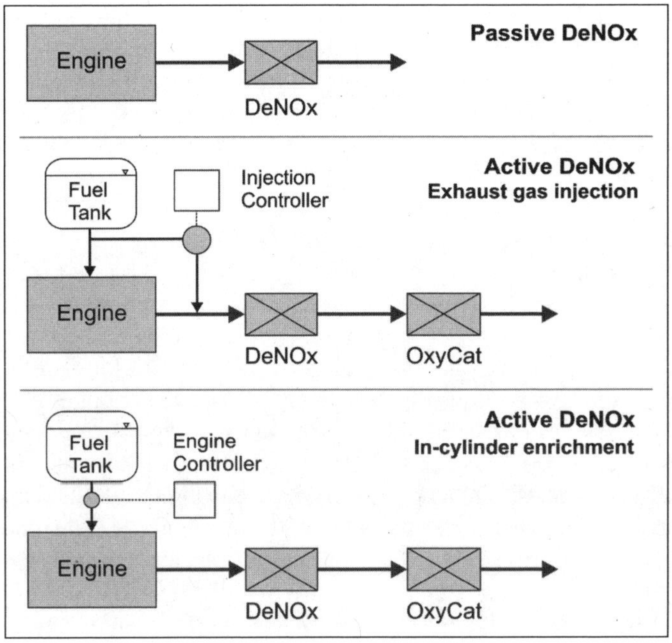

Figure 19.19 *DeNOx catalyst configurations.*

19.4.1.2 Passive and Active DeNOx

The most attractive source of hydrocarbons for reducing NO_x is the diesel exhaust itself. Systems utilizing exclusively the native diesel HCs are referred to as *passive deNOx* catalysts.

Passive systems, thanks to their simplicity, reliability, and lower cost, are always the preferred emission control option. However, due to the selective character of the reaction, deNOx catalysts show increasing NO_x conversion rates with increasing hydrocarbon concentrations. The limited supply of native diesel hydrocarbons (which is actually quite low in comparison to the NO_x emission levels) may constitute a barrier in achieving higher NO_x conversions, especially if the catalyst selectivity is low. Enrichment of the exhaust gases with additional HC material has been perceived as a solution to this problem. In general, such enrichment could be realized by two methods:

- Injection of hydrocarbons, preferably diesel fuel, into the exhaust system upstream of the catalyst, or
- Late in-cylinder injection in a common-rail fuel system (or merely late injection timing in conventional fuel systems).

Catalysts with HC enrichment have been termed *active deNOx* systems. The concept of passive and active deNOx catalysts is illustrated in Figure 19.19.

Systems for injecting hydrocarbons into the exhaust gas can be designed to reliably provide the required quantity of hydrocarbons to the deNOx catalyst. They are, however, quite complex and, thus, costly. By using common-rail post-injection, the cost of implementation of an active system can be fairly low. The biggest problem, especially in view of the continuous character of fuel enrichment that is required, is cylinder wall wetting by the fuel spray and its impact on the lube oil film and engine wear [72]. The application of simple late injection timing, which would have a less detrimental effect on engine durability, was reported to be less effective due to the propensity of hydrocarbons for partial combustion/cracking before reaching the catalyst [73].

Both types of HC enrichment also involve a certain fuel economy penalty, which depends on the quantity of injected fuel. An additional oxidation catalyst may be necessary in the active configurations to oxidize hydrocarbons that may pass through the deNOx catalyst.

19.4.2 Catalysts

19.4.2.1 Overview

Numerous catalyst systems, both precious and base metal based, were reported to exhibit lean NO_x reduction activity. Some of these formulations [74], grouped by washcoat technology, are listed in Table 19.6.

Some very optimistic results that were published gave the encouraging impression that the problem was almost solved. In reality, the multitude of systems under development proved that

TABLE 19.6
LEAN NO_X CATALYST SYSTEMS

Washcoat	Metal
Zeolite based (mainly ZSM5)	Cu, Pt, Mn, Fe, Co, Ni, Zn, Ag, Ce, Ga
Alumina based	Pt, Pd, Rh, Ir, Ru, Pt-Rh, Pt-W, V, Cr, Mn, Fe, Co, Ni, Cu, Zn, Ag
Silica-alumina	Co, Cu
Titania, sulfated titania	
Zirconia, sulfated zirconia	

the catalysts were far from being perfect. Some of the reported laboratory conversion efficiencies and their corresponding temperatures are listed in Table 19.7 [38]. Most of these catalyst systems were eventually disqualified as deNOx catalyst candidates. Eventually, a consensus had been reached that the following two groups of catalyst showed the most promise for further development: (1) copper exchanged zeolite (Cu/ZSM5) and (2) platinum on alumina(Pt/Al_2O_3).

Each of the catalysts listed in Table 19.7 exhibits a maximum NO_x conversion at a certain exhaust temperature. The Cu/ZSM5 is active at high temperatures, between 350 and 500°C, while the platinum catalysts are active at lower temperatures, approximately 200 to 300°C. Their maximum NO_x conversions measured in laboratory reactors typically amount to 50% to 60%, as illustrated in Figure 19.20. The combined Pt/Al_2O_3 and Cu/ZSM5 curves reflect the maximum NO_x reduction potential of that technology. We should emphasize that the presented performance is typical for lean exhaust gas condition of high O_2 concentration (say, 5% to 10%), as is the case in diesel exhaust. Lean NO_x catalysts can perform significantly better with low oxygen content (see Figure 19.23 in a later section and the following discussion).

TABLE 19.7
REPORTED LABORATORY LEAN NO_x PERFORMANCE

Catalyst	Peak NO_x Conversion (%)	Temperature (°)	Reducing Agent
Cu/ZSM5	60	400	C_2H_4
H-mordenite	62	400	C_3H_8
Ga-ferrite	100	600	C_3H_8
Ga/ZSM5	100	500	C_2H_6
SnO_2	67	400	C_2H_4
Ag/Al_2O_3	80	400	C_3H_6

Source: Data after Ref. 38.

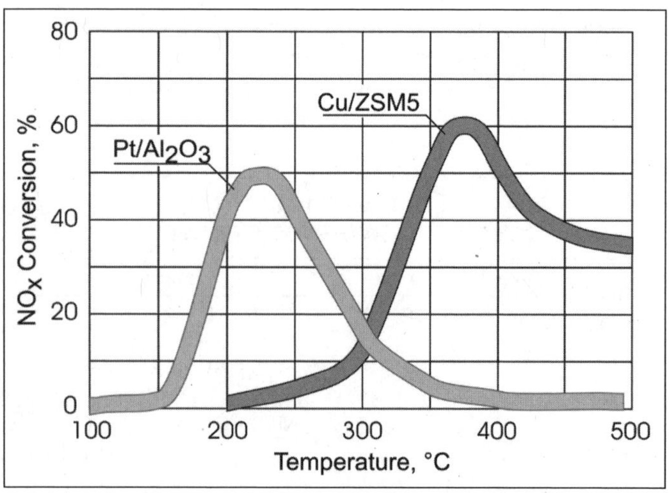

Figure 19.20 *Selective catalytic reduction of NO_x by hydrocarbons. Typical laboratory performance, HC1/NO_x ~ 6–12.*

No single catalyst would be able to cover the entire temperature range. This is a more serious limitation of the technology than it appears to be, because the operating temperature windows for the platinum and zeolite catalysts do not match the desired temperature range for most diesel engines. In many diesel engine applications, a significant portion of NO_x emissions can be generated at temperatures between 250 and 350°C, which is exactly in the area between the capabilities of the platinum and zeolite catalysts.

Both catalyst systems work most effectively using hydrocarbons as the reducing agent. Carbon monoxide and hydrogen are not effective reductants. The maximum selectivity of these catalysts is only about 20%; that is, 20% of HC reacts with NO_x and 80% with oxygen. Thus, relatively large amounts of HCs are necessary for the reduction of NO_x. HC1 to NO_x ratios between 3 and 12 were usually used in laboratory evaluations, with higher ratios resulting in better catalyst performance, while the ratio of HC to NO_x in native diesel exhaust is typically below 1. Therefore, higher NO_x conversions can be achieved in active deNOx systems, which feature exhaust gas hydrocarbon enrichment. In most diesel applications, even the performance of a 100% selective catalyst could be improved by hydrocarbon injection. If fuel was injected, injection rates of approximately 3% of the engine fuel consumption would be necessary to achieve HC1 to NO_x ratios greater than 3 [74].

19.4.2.2 Cu/ZSM5 Catalyst

The Cu/ZSM5 catalyst appears to be the best high-temperature deNOx catalyst. The typical catalyst performance, exhibiting a maximum conversion at a certain temperature, is explained by two competing reactions. The desired reduction of NO_x by hydrocarbons dominates at lower temperatures. Because it was found that, on the zeolite catalyst, the reaction also requires oxygen, Eq. (19.30) should be rewritten as follows:

$$\{HC\} + NO_x + O_2 = N_2 + CO_2 + H_2O \qquad (19.34)$$

At higher temperatures the catalyst selectivity decreases and the O_2 reduction [Eq. (19.31)] becomes dominant. Most hydrocarbons react with oxygen with a declining NO_x reduction efficiency.

The following hydrocarbon activity trend was found for NO_x reduction on the Cu/ZSM5 catalyst: $C_2H_4 > C_3H_6 > C_3H_8 >> CH_4$. The reactivity of different classes of compounds in deNOx catalysts was found to be aromatics > olefins > paraffins [75]. The type of hydrocarbon also influences the temperature at which maximum NO_x conversion occurs. Hydrocarbons of higher NO_x reduction activity move the catalyst window of operation toward lower temperatures. Cu/ZSM5 catalyst generates some carbon monoxide while reducing NO_x. The maximum CO concentration corresponds to the maximum NO_x conversion. A minimum HC1/NO_x mole ratio of about 3 is usually necessary to achieve NO_x conversion levels of practical importance [74], as illustrated in Figure 19.21.

A considerable amount of research has been published on reaction mechanisms of NO_x reduction on the Cu/ZSM5 catalyst

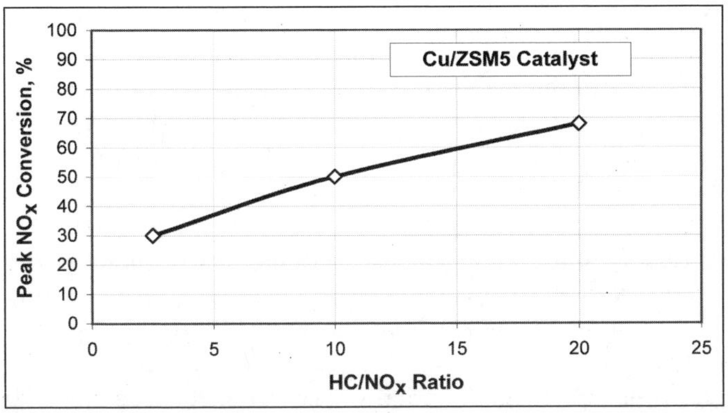

Figure 19.21 NO_x *conversion over Cu/ZSM5 catalyst.*

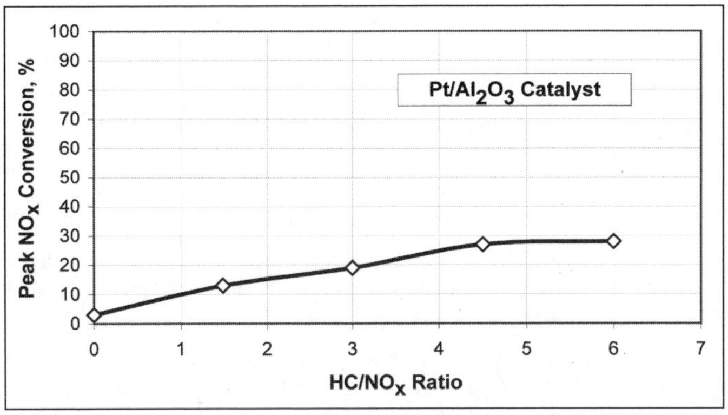

Figure 19.22 NO_x *conversion over a* Pt/Al_2O_3 *catalyst.*

[76]. According to one possible reaction path [74], the NO_x reduction on the Cu/ZSM5 catalyst may include two intermediate products. Hydrocarbons are believed to react on zeolite sites to form coke-like species. Cu-NO_2 species are formed on the copper sites by the oxidation of nitric oxide. The Cu-NO_2 reacts with the "coke" species to form nitrogen, a mixture of CO and CO_2, and water. The catalyst support acidity is important for hydrocarbon activation, while oxygen is necessary for the formation of the Cu-NO_2 species, but is not required for the activation of hydrocarbons.

Cu/ZSM5 catalysts have been formulated with about 5% Cu content. The catalysts were typically evaluated at space velocities ranging between 10,000 and 50,000 1/h.

The Cu/ZSM5 catalyst is highly selective toward the formation of N_2. Very little nitrous oxide, N_2O, can be found among the reaction products. The activity of Cu/ZSM5 is only slightly inhibited by SO_2 and H_2O.

Some of the most serious unsolved problems related to the Cu/ZSM5 system include unsatisfactory hydrothermal durability and susceptibility to poisoning. The zeolite in the presence of water vapor undergoes de-alumination at higher temperatures. In extreme cases the collapse of the zeolite crystalline structure may occur. In conjunction with these phenomena, further deactivation of the Cu/ZSM5 is also attributable to the sintering of the active copper species [likely to be ion-exchanged Cu(I) species] to CuO [77]. In diesel applications the Cu/ZSM5 is also prone to poisoning by sulfur and other catalyst poisons. Unless a solution is found, these problems disqualify the Cu/ZSM5 as a commercial catalyst system. There is ongoing research to modify this catalyst and improve its properties and durability. For example, a combination of copper and silver (Ag) with ZSM5 was shown to exhibit superior performance [78].

19.4.2.3 Pt/Al_2O_3 Catalyst

In comparison to the Cu/ZSM5 system, platinum/γ-alumina catalysts are active at lower temperatures and tend to exhibit a narrower temperature window. As a rule, Pt catalysts that are active at lower temperatures have higher peak NO_x conversions. Peak NO_x conversions over an example Pt-based catalyst are shown as a function of HC1/NO_x ratio [74] in Figure 19.22.

A possible mechanism of NO_x reduction in the Pt catalyst involves formation of an NO_2 species by the oxidation of NO on certain catalytic sites. On other sites, hydrocarbons react with oxygen to form reaction intermediates corresponding to the formula $C_xH_yO_{2z}$. The NO_2 and oxygenated HC intermediates react together to form CO_2, H_2O, and either N_2 or N_2O. In-depth discussion on various reaction models over the Pt catalyst can be found in the literature [79, 80].

Although the Pt/Al_2O_3 system is still the best deNOx catalyst option for low temperatures, it has serious drawbacks. The main disadvantage of platinum catalysts is generation of N_2O among the reaction products. The catalyst selectivity toward N_2 is only 10% to 40%, that is, between 60% and 90% of the reduced NO_x reacts to N_2O. Although N_2O is not a regulated emission, it has adverse environmental effects and its generation in the catalyst should be avoided. Another disadvantage of Pt catalysts is their SO_2 oxidation activity, contributing to increased sulfate particulate emissions at higher exhaust temperatures. The catalyst is also prone to deactivation by sulfur.

A number of modified Pt/Al_2O_3 catalysts have been proposed to address these drawbacks. For example, it was reported that the sulfur tolerance, as well as NO_x conversion, can be enhanced by incorporating 12-tungstophosphoric acid into the catalyst formulation [81].

19.4.3 Advanced Catalyst Systems

19.4.3.1 Issues

The lean NO_x catalysts, as discussed earlier, have never been developed to provide the high levels of NO_x reduction required by future emission standards. The major unsolved problems can be summarized as follows:

- Catalysts require some form of HC enrichment; HC levels in diesel exhaust, especially in modern engines, is too low for any significant NO_x reduction using passive deNOx catalysts.
- Even with HC enrichment, the maximum conversion efficiency at realistic HC/NO_x ratios is low (30% to 50%).
- The temperature window of known catalysts is narrow and does not always corresponds to the exhaust gas temperature range at which most NO_x is emitted from the engine.

- Catalyst durability needs to be improved, in terms of both resistance to poisons and, in the case of Cu/ZSM5, thermal durability.
- Catalysts need to be more selective toward reducing NO_x to nitrogen, as opposed to nitrous oxide.

Solutions to these problems can be classified into two groups: (1) optimization of catalyst formulation and (2) optimization of the emission control system. Better catalysts could provide such benefits as improved efficiencies, wider temperature window, or, last but not least, better durability. Even in the absence of new, breakthrough catalyst formulations, significant improvement can be achieved by system optimization. Reported ideas include using two catalysts, possibly of different formulations, optimized catalyst locations (close-coupled, underfloor), optimized HC enrichment strategies (including exhaust flow bypasses), and exhaust gas cooling.

A multitude of studies, especially during the 1990s, focused on improving the performance of deNOx catalysts. The following sections provide a brief overview of selected deNOx research. The selection of discussed studies is by no means complete. However, considering the limited degree of deNOx catalyst commercialization and the only moderate future potential, we hope that sufficient insight into the various directions of deNOx research is provided.

19.4.3.2 Catalyst Chemistry Optimization

Complex Catalysts

One approach to address the drawbacks of the "classic" deNOx catalysts is to combine different catalyst technologies. In particular, there have been numerous attempts to extend the catalyst window by designing complex catalysts that would incorporate a number of catalytic components, which would be active at different temperature ranges (e.g., to combine both ZSM5 and Pt catalysts or their modifications together).

An example of such formulation combines three components, as follows [82]. A first catalytic component, cobalt (Co) on a γ-alumina washcoat, is active at high exhaust temperatures. A second component, Pt/Rh/Co alloy dispersed on BaO-stabilized δ-alumina, provides medium-temperature activity. Finally, a third component, Pt/Rh on α-alumina, is active at low temperature. These three components may be placed in the exhaust system as three separate catalysts, arranged in sequence, or may be coated on a single substrate as three catalyst layers (in this configuration, the third component forms the first catalyst layer, that is, is deposited on the substrate first). The NO_x reduction activity of this catalyst was claimed to span the 200 to 800°C temperature range.

Combined catalyst systems have been also developed to enhance the CO utilization in the deNOx catalyst. In other words, catalyst systems were optimized to facilitate selective reduction of NO_x not only by hydrocarbons, but also by carbon monoxide. A two-stage catalyst system proposed to achieve this objective included an upstream Cu/ZSM5 catalyst operating through reactions with hydrocarbons, and a downstream Au-Pt/TiO_2 catalyst that enhanced the reactions with CO [83]. That catalyst system was also reported to provide enhanced sulfur poisoning resistance.

Iridium Catalyst

Iridium was first used as a lean NO_x catalyst in combination with platinum. A two-layer catalyst, combining (1) Pt/Ir on γ-alumina/ceria with (2) Cu/Fe mordenite, was reported to show higher temperature activity and a wider temperature window relative to the base Pt/Al_2O_3 formulation [84].

Figure 19.23 shows the performance of a pure Ir catalyst (i.e., not containing Pt) on a ZSM5 zeolite carrier. A de-aluminized zeolite is used, with a SiO_2 to Al_2O_3 ratio of above 100, in order to improve its thermal stability [85]. It is interesting to see how the NO_x conversion changes with

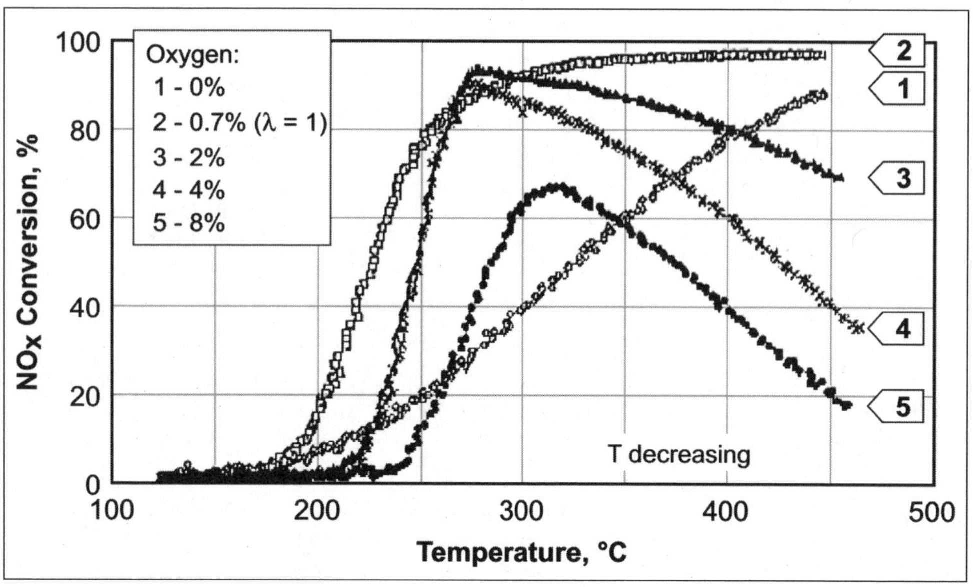

Figure 19.23 NO_x conversion over Ir catalyst. Synthetic gas: H_2O, 10%; NO, 270 ppm; C_3H_6, 1650 ppm; space velocity, 51,000 1/h. (Reprinted from Ref. 85.)

the oxygen content. The highest NO_x conversion, in excess of 90%, was measured at 0.7% O_2, which corresponded to the stoichiometric A/F ratio. From this point, NO_x reduction decreased for both richer and leaner mixtures. The lean mixture curves (3, 4, and 5 in Figure 19.23) also show gradually narrower temperature windows, due to the increasing reaction rate between oxygen and hydrocarbons at higher temperatures and higher O_2 concentrations.

19.4.3.3 System Optimization

Dual Catalyst Systems

An important area of deNOx system optimization is related to widening the catalyst temperature window. By splitting the catalyst into two pieces and placing them at different locations, one can extend the deNOx temperature range. In light-duty applications, the first catalyst would be placed in the close-coupled position, exposed to higher temperatures, while the second catalyst would be in the colder, underfloor position.

This concept was developed by Daimler and Degussa for a light-duty diesel vehicle [72]. An interesting internal exhaust gas bypass was designed to prevent depleting the gas from hydrocarbons in the upstream catalyst. The bypass involved leaving the perimeter of the front, close-coupled deNOx catalyst substrate uncoated. A portion of the exhaust that passed through these uncoated channels provided a source of HC for the downstream (underfloor) deNOx catalyst. Because of the differences in temperature of the two converters, the effective operating window of the deNOx system was broadened. An active deNOx catalyst system using this internal bypass concept and in-cylinder HC enrichment was commercialized over a brief period of time on C Class Mercedes cars.

Further widening of the temperature window, in particular over the European ECE+EUDC cycle, is possible through introduction of exhaust gas cooling during hot driving segments [86]. The system shown in Figure 19.24 was proposed based on a comprehensive optimization study involving different catalyst formulations, configurations, and HC injection strategies on both light- and heavy-duty engines [87]. For light-duty applications, NO_x reductions up to 36% over the ECE+EUDC cycle were measured when using a heat exchanger and HC injection in front of each of the two catalysts. For heavy-duty applications (ESC test cycle), NO_x reduction of about 20% was measured at a 3% fuel economy penalty, and about 40% at a 9% fuel economy penalty.

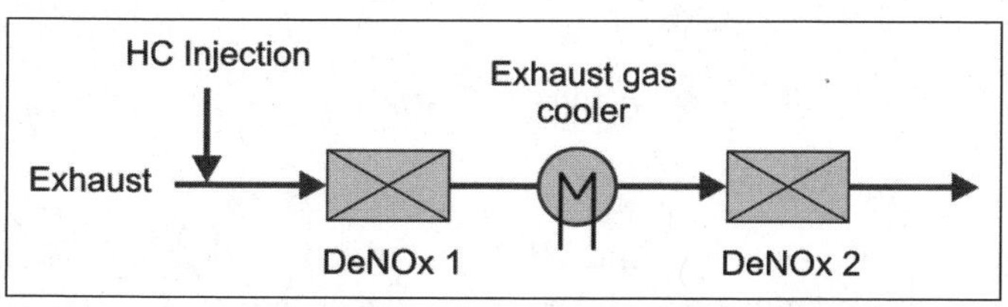

Figure 19.24 Two-catalyst DeNOx layout with exhaust gas cooling. DeNOx 1, close coupled with bypass; DeNOx 2, underfloor.

The system in Figure 19.24, proposed for light-duty vehicles, includes two deNOx catalysts, each of them having a volume of 0.8 to 1.2 of the engine displacement. The first catalyst, installed in the close-coupled position, includes a bypass. A heat exchanger in front of the second, underfloor catalyst cools the exhaust gas to between 200 and 280°C at the catalyst inlet. The first catalyst reduces NO_x during the ECE part of the European driving cycle, while the second catalyst is active during the hotter, EUDC part of the cycle.

Catalyst-Adsorber Combinations

NO_x conversion efficiencies of deNOx catalysts can be improved by storing HC, NO_x, or both compounds in the catalyst washcoat and releasing them at conditions that are more favorable for the catalytic reactions. Such storage is realized by incorporating HC or/and NO_x adsorbers, also known as HC and NO_x traps, into the catalyst formulation. Through this approach, the deNOx efficiency can be increased without the need for an active system (i.e., without hydrocarbon injection) and the associated fuel economy penalty.

HC traps, usually in the form of zeolites, can adsorb HCs at low temperatures when the catalyst is still inactive [88, 89]. At higher temperatures, the adsorbed HC material is released to increase the local HC/NO_x ratio at the catalytic sites and, thus, to enhance the NO_x reduction performance of the catalyst.

The concept of NO_x adsorbers was originally considered in an attempt to widen the temperature window of the Cu/ZSM5 catalyst. The idea was to incorporate NO_x trapping compounds into the catalyst washcoat, which would adsorb NO_x during periods of low exhaust gas temperature. Then, at higher temperatures, the stored NO_x would be released and reduced in the catalyst. This idea has evolved into an independent NO_x adsorber technology, discussed in detail in Section 19.5, where deNOx catalysts were replaced with three-way catalysts reducing stored NO_x during periodic enrichment of the air-to-fuel mixture.

Some researchers have also proposed incorporating both HC and NO_x traps in one catalyst system [90]. The catalyst, in its first layer, incorporates an alkaline earth oxide based NO_x adsorber. A second layer, in direct contact with exhaust gases, incorporates a noble metal catalyst and a zeolite HC trap. The macroscopic desorption temperatures for HC and NO_x are 200 and 400°C, respectively. However, due to microscopic-scale processes underlying the dynamics of NO_x adsorption equilibrium, NO_x is being reduced by hydrocarbons significantly below its desorption temperature. The catalyst NO_x conversion efficiency measured on a light-duty diesel vehicle on the ECE+EUDC cycle was 25%, up from 13% in reference systems without NO_x storage. (Data on long-term durability and sulfur poisoning of the NO_x storage function were not presented.)

19.4.4 Diesel Engine Experience

19.4.4.1 Demonstration Programs

Several deNOx catalyst formulations (ICT-Degussa), including precious metal and base metal (Cu/ZSM5) catalysts, were tested on a Caterpillar 3116, 6.6-L, four-cylinder, intercooled and turbocharged diesel engine [91]. Emission tests were measured under steady-state conditions as well as on the U.S. FTP Transient cycle. Diesel fuel was sprayed into the exhaust stream as a source of secondary hydrocarbons. Diesel fuel of 0.03 wt. % S, a cetane number of 45.2 and 35% aromatics were used for the tests.

Steady-state tests were conducted at 1400 rpm. The catalyst size was 5.66 in. diameter × 6.0 in., resulting in a space velocity of 55,000 1/h. Secondary fuel was injected into the catalyst at a rate of 3% of the engine fuel consumption. It was found that precious metal catalysts could reduce NO_x but greatly increased PM emissions due to the generation of sulfate particulates. On the other hand, the base metal catalysts of high light-off temperature were not able to oxidize the added secondary fuel at low temperatures and, as a result, increased CO and HC emissions.

A larger catalyst size utilizing two round substrates 7.5 in. × 7 in. each was tested on the FTP cycle with optimized fuel injection patterns. The best combined NO_x/PM results were achieved using a base metal catalyst and 3% secondary fuel injection. A simultaneous 12% NO_x and 25% PM reduction was demonstrated.

Another engine dynamometer evaluation was performed on the same Caterpillar 3116 engine model [92]. Nine "black box" catalysts supplied by different manufacturers were tested under steady-state and transient conditions. The space velocities during the steady-state tests varied between 10,000 and 40,000 1/h. Catalyst volumes from 7 to 28 L were used for the U.S. FTP Transient test evaluations. Diesel fuel or ethanol was used as a supplemental reductant.

NO_x reductions from 17% to 44% were seen during the steady-state tests. However, due to the narrow catalyst temperature windows, much lower NO_x conversions were recorded on the transient test. Several of the transient tests produced marginal NO_x improvements of not more than 5%. The best FTP cycle NO_x reduction, achieved using a 28-L catalyst, amounted to 24% but it was accompanied by a 26% increase in PM emission.

The best NO_x performance was observed when the HC1/NO_x ratio was higher than 6. It was also found that, depending on the tested catalyst, 35% to 75% of NO_x was converted to N_2O rather than to nitrogen. A linear correlation was found between the generation of N_2O and reduction of NO_x.

19.4.4.2 Commercial Status

Unless more selective catalysts are found, the lean NO_x technology represents only small to moderate NO_x reduction potential. Even though peak NO_x reductions measured in laboratories are frequently in excess of 60% and more, performance of these catalysts over regulatory test cycles is usually limited to 10% to 20% for the following reasons:

- Temperature windows of deNOx catalysts tend to be narrow and to not always correlate well with the temperatures seen in real duty operation or over the regulatory emission test cycles.
- Passive systems are further limited by insufficient hydrocarbon supply in the native diesel exhaust.

Other deNOx issues include durability and sulfur tolerance concerns, as well as secondary emissions. Because most deNOx catalysts that have been commercialized are based on the platinum formulation, it can be reasonably conjectured that a significant portion of NO_x is reduced to nitrous oxide (N_2O), rather than to N_2, over these devices. Because N_2O is not a regulated emission, it is not measured during certification testing (no data are published by the catalyst manufacturers). However, despite its unregulated status, N_2O is not a desired product of catalytic reactions.

For all of these reasons, commercial deNOx applications usually take the form of a small, passive NO_x reduction functionality incorporated into other technologies, primarily into diesel oxidation catalysts. Iridium-based catalysts in the passive configuration, discussed earlier, were introduced by Degussa to some models of diesel fueled cars in Europe in the mid-1990s as (probably) the first commercial lean NO_x catalyst application.

Since the late 1990s, a number of diesel car models in the EU are using Pt-based oxidation catalysts incorporating zeolite HC traps and optimized for passive deNOx activity. In one EURO III application, the overall NO_x emission reduction of such a catalyst over the ECE+EUDC cycle, as reported by ASEC, amounted to 19% [10]. (See Section 19.6 for more details on that technology.)

A Pt-based deNOx catalyst was a part of the "QuadCAT" kit that was being developed by California-based Ceryx Inc. for retrofitting heavy-duty diesel engines [93]. The QuadCAT electronic control unit could be programmed to deliver different amounts of diesel fuel upstream of the deNOx catalyst, to serve as hydrocarbon enrichment. NO_x reductions were reported to be strongly dependent on the amount of supplementary diesel fuel added. For example, approximately 2 g fuel per 1 g of NO_x were needed to achieve a 25% NO_x emission reduction over the AVL eight-mode, heavy-duty test cycle. Assuming an engine of 6 g/bhp-hr NO_x and 300 g/bhp-hr BSFC, the associated fuel economy penalty would be 4%. More than twice as much fuel was needed to achieve a 40% NO_x reduction.

19.5 NO_x Adsorbers

19.5.1 NO_x Adsorber Concept

19.5.1.1 Overview

The concept of the *NO_x adsorber/catalyst* has been developed based on acid–base washcoat chemistry. It involves adsorption and storage of NO_x in the catalyst washcoat during lean driving conditions and release of it under rich operation. The released NO_x is catalytically converted to nitrogen just as it happens over the three-way catalyst (TWC), which is widely used in stoichiometric gasoline engines. Normally, three-way catalysts are inactive in converting NO_x under lean exhaust conditions, when oxygen is present in the exhaust gas. Through the lean storage and rich release and conversion, the applicability of the three-way catalyst is being extended to lean-burn engines, such as gasoline direct-injected (GDI) engines and compression ignition engines. NO_x adsorbers, which are being developed by catalyst manufacturers with notable progress, are perceived by many as the diesel NO_x control technology of the future.

Different authors use different terms when discussing NO_x adsorbers, such as lean NO_x traps (LNT), NO_x storage/reduction catalysts (NSRC), or NO_x adsorber catalysts (NAC). All of these names are synonyms, describing the same emission control technology. The term *lean NO_x catalyst*, on the other hand, refers to the selective catalytic reduction of NO_x by hydrocarbons—an entirely different technology that should not be confused with NO_x adsorbers.

Adsorption and desorption of NO_x in the NO_x adsorber is primarily related to lean and rich cycling during vehicle operation. Desorption with accompanying conversion, also referred to as *regeneration,* occurs during rich air-to-fuel mixture excursions. Quite clearly, this concept is very compatible with "partial lean-burn" GDI engines, where the engine controller can easily provide mixture enrichment when needed. With modern, electronically controlled fuel systems, it is also possible to periodically enrich the A/F mixture in the diesel engine in order to facilitate the regeneration of diesel NO_x adsorbers. However, unlike more conventional diesel catalysts, this technology requires an unparalleled degree of integration with the engine management system. Therefore, while NO_x adsorbers are a promising technology for new engine/vehicle applications, they are not particularly suitable for retrofit emission control systems.

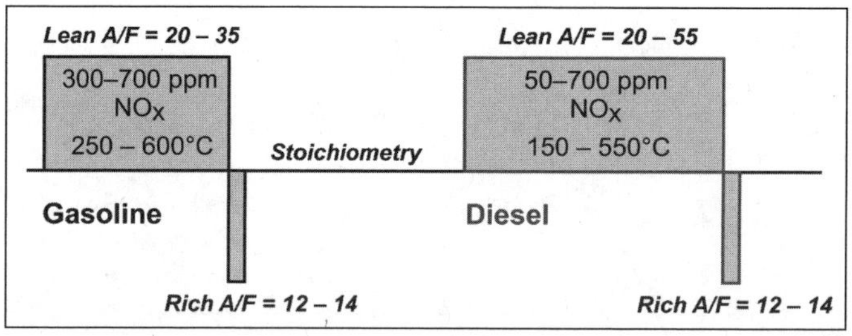

Figure 19.25 *Operation of NO_x adsorbers on gasoline and diesel engines. (Reprinted from Ref. 94.)*

The discussion of adsorber/catalyst systems in this section is based on the developments for both lean-burn gasoline and diesel engine applications. Although there are many similarities between catalysts used in these applications, there are also important differences that affect the catalyst system design, as listed in Table 19.8 and illustrated graphically in Figure 19.25 [94]. Depending on the fuel quality, diesel adsorbers may also require higher sulfur tolerance than gasoline adsorbers. However, considering the great sensitivity of this technology to sulfur, NO_x adsorbers will likely require ultra-low-sulfur levels in both gasoline and diesel.

19.5.1.2 Chemical Reactions

The NO_x adsorption/reduction mechanism is illustrated in Figure 19.26. The catalyst washcoat combines three active components: (1) an oxidation catalyst, for example, Pt; (2) an adsorbant, for example, barium oxide (BaO); and (3) a reduction catalyst, for example, Rh.

TABLE 19.8
NO_x ADSORBER CATALYSTS FOR GASOLINE AND DIESEL ENGINES

	GDI	Diesel	Comment
A/F ratio	Partially lean (lean A/F 20–35)	Continuous lean (A/F 20–50)	GDI is operated stoichiometric at high-speed, high-load conditions.
Catalyst functionality	NO_x adsorber + three-way catalyst	NO_x adsorber only	Due to the partial stoichiometric operation, gasoline adsorbers must be also good three-way catalysts with a wide A/F catalyst window, high HC conversion, etc.
Rich regeneration	Easy	Difficult	Rich operation of diesel is problematic, may cause high PM emission.
Fuel economy penalty	Lower	Higher	
Operating temperature	250–600°C	150–550°C	Diesel adsorbers must be optimized for low-temperature performance, which may require different formulations and higher adsorber loadings [95].
Reductants	CO, HC, H_2	HC, CO, H_2	
Durability required	800–850°C	650–700°C	

Source: Data after Ref. 94.

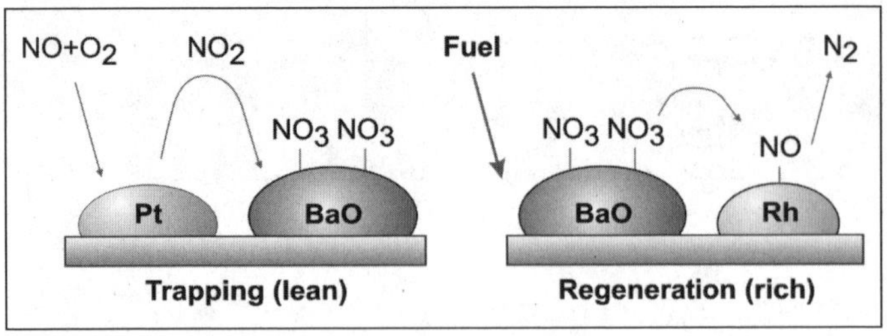

Figure 19.26 *NO_x adsorption and reduction mechanism.*

The adsorption process involves two steps, as represented by Eqs. (19.35) and (19.36), which occur during lean engine operation. NO_x emissions from the diesel engines are typically composed of 90% to 95% nitric oxide. In the first step, the NO reacts with oxygen on active oxidation catalyst sites (e.g., platinum, Pt) to form NO_2. The second step involves adsorption of the NO_2 by the storage material in the form of an inorganic nitrate:

$$NO + \tfrac{1}{2}O_2 = NO_2 \qquad (19.35)$$

$$BaO + NO_2 + \tfrac{1}{2}O_2 = Ba(NO_3)_2 \qquad (19.36)$$

When the engine runs under excessive fuel conditions or at elevated temperatures, the nitrate species become thermodynamically unstable and decompose, producing NO or NO_2, according to Eqs. (19.37, 19.37a) [96, 97]. Under rich conditions, these nitrogen oxides are subsequently reduced by carbon monoxide, hydrogen, and hydrocarbons to N_2 over the reduction catalyst, in a conventional three-way catalyst process. One of the possible reduction paths is shown by Eq. (19.38).

$$Ba(NO_3)_2 = BaO + 2NO + 1\tfrac{1}{2}O_2 \qquad (19.37)$$

$$Ba(NO_3)_2 = BaO + 2NO_2 + \tfrac{1}{2}O_2 \qquad (19.37a)$$

$$NO + CO = \tfrac{1}{2}N_2 + CO_2 \qquad (19.38)$$

This simplified set of reactions allows for an understanding of the basic NO_x adsorber chemistry. A more detailed analysis should also include other reaction paths and species, for example, carbonates that may be formed in reactions between barium nitrate and carbon dioxide [98].

NO_x adsorbers also show some undesired reactivity, primarily with regard to sulfur compounds that are present in exhaust gases from both diesel and gasoline engines. Reactions of sulfur are basically equivalent to the reactions of NO_x [Eqs. (19.35) through (19.37)] and can be written as follows:

$$SO_2 + \tfrac{1}{2}O_2 = SO_3 \qquad (19.39)$$

$$BaO + SO_3 = BaSO_4 \qquad (19.40)$$

First, sulfur dioxide is oxidized to sulfur trioxide over platinum in Eq. (19.39). Then, the SO_3 reacts with BaO to form barium sulfate in Eq. (19.40). This causes gradual saturation of the barium sites with sulfur and loss of activity toward the adsorption of NO_2. $BaSO_4$ can be thermally decomposed in a process equivalent to the reaction of Eqs. (19.37, 19.37a). However, sulfates of barium and other NO_x adsorber compounds are more stable than their nitrates and require much higher temperatures to desulfate. For this reason, sulfur deactivation is the major problem in the development of NO_x adsorber systems.

There are a number of other possible reactions in NO_x adsorbers that can produce secondary, unregulated emissions. These are mainly reduction processes, also known to occur in the three-way catalyst, which involve NO/NO_2, as well as SO_2, and generate products other than nitrogen. The main secondary emissions include nitrous oxide (N_2O), and hydrogen sulfide (H_2S).

19.5.1.3 Catalyst Systems

NO_x adsorber systems generally consist of a conventional three-way catalyst that additionally incorporates NO_x storage components. These storage components in NO_x adsorbers are typically compounds of the following elements [99]:

- *Alkaline earths:* barium (Ba), calcium (Ca), strontium (Sr), magnesium (Mg)
- *Alkali metals:* potassium (K), sodium (Na), lithium (Li), cesium (Cs)
- *Rare earth metals:* lanthanum (La), yttrium (Y).

While most catalysts incorporate adsorbers in the form of oxides, a number of defined systems, such as perovskites or metal substituted zeolites, have also been tested [100].

Different adsorbing elements, their chemical compounds, and/or mixtures may yield catalysts of different properties, such as different NO_x storage capacity, thermal stability of the nitrate and its desorption temperature, susceptibility to sulfur poisoning, or sulfur desorption (desulfation) temperature. These properties are important in designing the catalyst system and its regeneration strategy, as will be discussed later. Compounds of more basic chemical character can provide higher NO_x storage capacity. Importantly, alkali metals, such as potassium, exhibit superior NO_x adsorption performance at high temperatures relative to barium, as illustrated in Figure 19.27 [101]. If sodium, potassium, and/or cesium are incorporated into a NO_x adsorber washcoat containing barium, they significantly increase NO_x conversion in the temperature range of about 350 to 600°C. Alkali metal oxide adsorbers (up to 50% Na in alumina washcoat) were also reported to show superior resistance to sulfur poisoning [102].

As research has continued, it has become apparent that catalyst formulations can provide the most desired combination of properties. Most current NO_x adsorbers are based on one of the following systems:

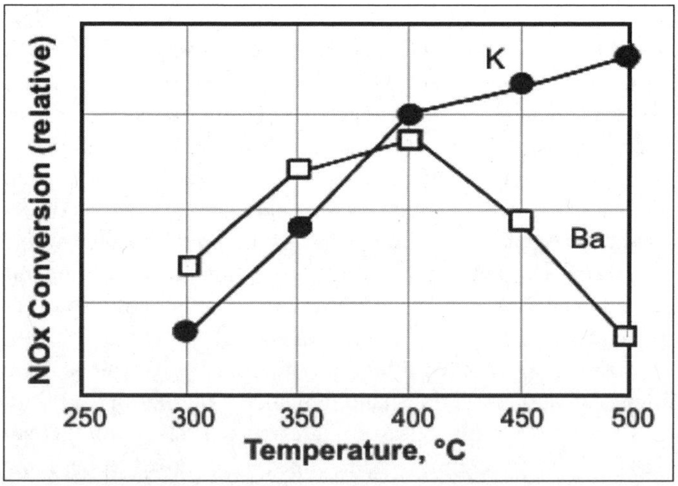

Figure 19.27 Temperature activity of storage elements. (Based on data from Ref. 101.)

- Alkaline earth, primarily Ba-based adsorbers
- Combined, Ba + alkali metal (K, Na, etc.) adsorbers.

Ba + alkali metal adsorbers provide superior NO_x conversion, both fresh and aged, at temperatures greater than 250 to 300°C, with only small NO_x conversion penalty at lower temperatures [103]. Ba + alkali metal adsorbers offer also better nitrate stability. Finally, Ba + alkali metal adsorbers have better sulfur poisoning resistance than Ba-only adsorbers, with higher NO_x conversion after repeated poisoning and desulfation cycles. Their disadvantage is poor hydrocarbon conversion, which is lower in comparison to Ba adsorbers at all temperatures. Among all alkali metals, Cs is the most effective at enhancing NO_x conversion, followed by K, Na, and Li, as shown in Figure 19.28 [103].

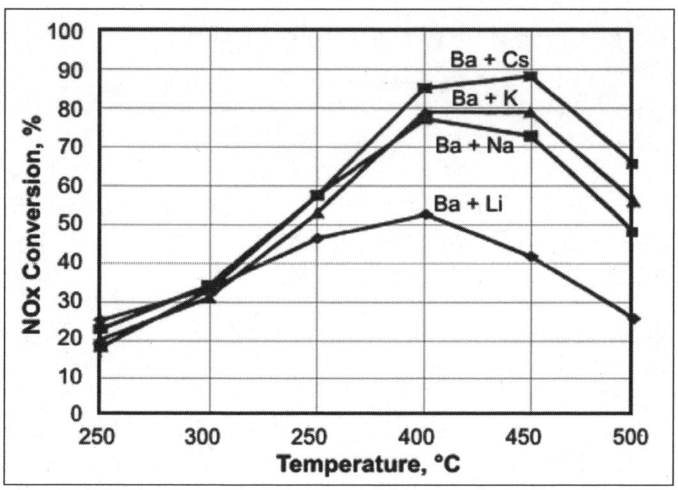

Figure 19.28 NO_x conversion with different alkali metals for a 5.0-L, V8 gasoline engine; modulation A/F=21.5 for 30 s; A/F=12.5 for 2 s; space velocity =50,000 1/h; inlet NO_x = 500 ppm; after 50 h of aging @ 800°C. (Based on data from Ref. 103.)

Alkali metal catalysts also have a number of disadvantages, such as a potential negative impact on the durability of platinum catalysts by promoting noble metal sintering, or the already mentioned poor HC and low-temperature performance. Possibly the most serious issue is related to the high mobility of water-soluble alkali metal compounds, such as nitrates. In catalysts, alkali metals migrate with time away from the adsorber washcoat. Due to their strong affinity to silicon oxide, a component of cordierite, alkali metals accumulate within the ceramic catalyst substrate , causing loss of its physical strength [104]. Cs was found to have the highest migration intensity, followed by K and Na, as illustrated in Figures 19.29a–c [103].

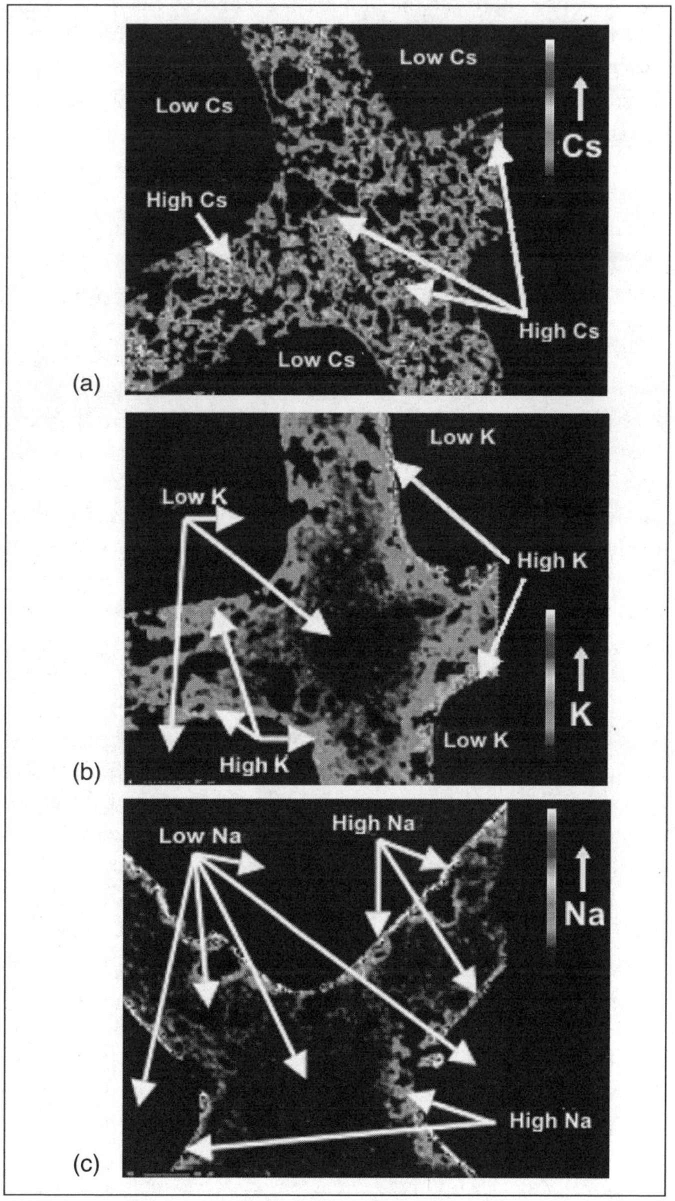

Figure 19.29 Maps of (a) Cs, (b) K, and (c) Na in aged NO_x adsorbers coated on cordierite substrates; after aging at 900°C for 16 h under air/H_2O. (Reprinted from Ref. 103. Courtesy of Delphi and Corning.)

The alkali metal maps shown in the figure were obtained by analyzing (scanning electron microscopy) aged catalyst samples of the same molar amount of metals. The Cs map shows that Cs migrated throughout both the cordierite substrate cell corner and cell wall. Lower cesium concentration was detected on the washcoat after aging. A significant amount of potassium migrated away from the washcoat and accumulated at the interface of the washcoat and cordierite substrate. Little K was detected at the center region of the cell corner. In contrast to Cs, Na is concentrated at the interface of cordierite substrate and washcoat, with little Na found in the middle of the cell wall and the center of the cell corner. After aging at 900°C, Cs and K containing NO_x adsorbers showed skin cracking while the Na containing adsorber was free of any visible cracking.

Because alkali metals that are more mobile have also better NO_x conversion effect, catalyst designers face the difficult task of optimizing their formulations to achieve the required NO_x performance, while minimizing the chemical attack on the cordierite substrate. Advanced catalysts and substrates with barrier Si coatings have been proposed to minimize the impact of alkali metals [105]. NO_x adsorbers coated on metallic substrates have been also tested [95, 106]. However, because material durability issues were not included in these studies, it is not certain whether metallic substrates would be sufficiently resistant to attacks from alkali metals in NO_x adsorbers.

Due to the importance of high-temperature performance in spark-ignited engines, the Ba + alkali metal system has become the NO_x adsorber of choice in gasoline applications [101, 103]. Adsorbers for light-duty diesel engines, which operate at lower temperatures, are likely to be able to avoid the necessity of using alkali metals. Ba + alkali metal formulations, however, may be needed in heavy-duty diesel applications, especially those that require emission certification on hot test cycles such as the U.S. steady-state Supplemental Emission Test (SET), which is based on the European ESC [107].

Most NO_x adsorber washcoats are based on γ-alumina, Al_2O_3, which is also the main washcoat material in gasoline three-way catalysts. Three-way catalysts for SI engines also include high loadings of zirconia-stabilized cerium oxide (ceria), which provides the oxygen storage capacity (OSC). By storing oxygen in a lean condition and releasing it at a rich condition, the OSC component is able to extend the three-way catalyst window. The OSC function is less important in diesel NO_x adsorbers, but it is important for lean-burn gasoline applications that spend significant operating time at stoichiometry. Another important function of the ceria is to promote the reaction between NO_x and CO. Because of the high level of instantaneous NO_x conversion that is required for rapid, efficient adsorber regeneration, this feature becomes extremely important, in both the diesel and gasoline applications. Unfortunately, the incorporation of ceria also has a negative effect on fuel economy due to the oxygen storage. Some NO_x adsorber systems include ceria in the catalyst washcoat, while others do not.

Other common washcoat materials used in NO_x adsorbers include titania (TiO_2) and zirconia (ZrO_2). Acidic oxides, such as TiO_2, are unable to store sulfur oxides as effectively as high-surface-area alumina, thereby enhancing sulfur resistance of NO_x adsorbers. For this reason, titania was incorporated in NO_x adsorbers for the Toyota D-4 GDI engine in Japan in 1996 [108]. Further addition of Rh/ZrO_2 to BaO-based adsorbers was found to catalyze the water shift reaction with an increased generation of hydrogen and, as a result, improved desorption of sulfur [108]. The use of titania and zirconia was further investigated with NO_x adsorbers utilizing a Ba + alkali metal system (Ba/K) [101]. While the acidic TiO_2 was confirmed to enhance sulfur resistance and to promote sulfur desorption at lower temperatures, it was also found to reduce the high-temperature stability of KNO_3, thus defying the function of alkali metals and causing poor NO_x performance at higher temperatures. The more basic oxide ZrO_2, on the other hand, enhanced the thermal stability of KNO_3, but its sulfur poisoning resistance was far lower than that of TiO_2. A complex oxide catalyst, designed to combine the sulfur resistance of titania systems with the alkali adsorber stability of zirconia systems, was developed and commercialized in 2001 on the D-4 GDI engine [101].

Noble metal systems used in most NO_x adsorbers include platinum and rhodium, the classic three-way catalyst components. Platinum is the most active catalyst for NO oxidation [Eq. (19.35)]. Because this step is critical for the NO_x storage efficiency, platinum plays an important role in adsorber catalysts. Rhodium is used as the NO_x reducing catalyst in most adsorbers. Trimetal formulations (Pt/Pd/Rh), which are common in stoichiometric three-way catalysts, can also be used in NO_x adsorbers, as long as their platinum content remains relatively high. For instance, Delphi reported two trimetal formulations with precious metal loadings of 80 g/ft^3 (60:15:5) and 100 g/ft^3 (70:20:10) [103]. The catalysts were designed for GDI engines, where they may be required to perform as typical three-way catalysts (i.e., under stoichiometric conditions) over prolonged periods of time; it is not clear whether palladium would offer a benefit in diesel adsorbers.

19.5.2 Performance and Operation

19.5.2.1 Conversion Efficiency

NO_x adsorber catalysts exhibit high NO_x conversion efficiencies, in excess of 80% to 90%, as illustrated by example data in Figure 19.30 [96]. The activity of NO_x adsorbers covers a fairly wide catalyst temperature window, extending from about 200°C to 450 to 500°C. The lower end of the temperature window, T_1, is determined by the catalyst activity with regard to the oxidation of NO to NO_2, as well as NO_x release and reduction (three-way function). The upper temperature, T_2, is related to the thermodynamic stability of nitrates, which undergo thermal decomposition at higher exhaust temperatures, even under lean conditions.

The NO_x adsorber temperature window correlates fairly well with the diesel engine loads and exhaust temperatures at which most diesel NO_x generation occurs. It is this favorable position of the temperature window and the high NO_x conversion efficiency that makes NO_x adsorbers an extremely attractive diesel NO_x control technology.

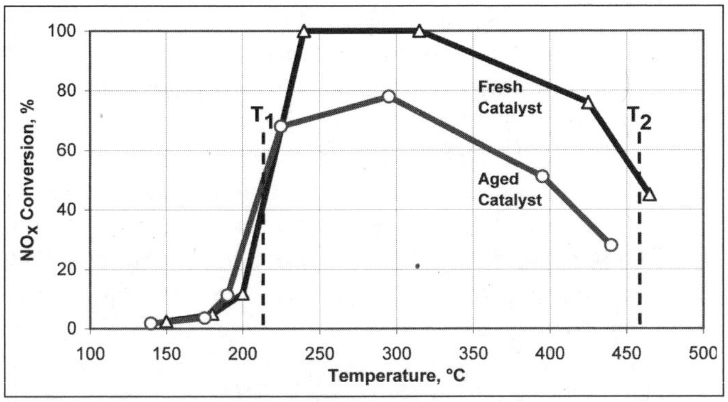

***Figure 19.30** NO_x adsorber temperature window; 2.47- L adsorber on a 1.8-L SI engine cycled 59 s lean/1 s rich. (Based on data from Ref. 96.)*

In gasoline engine applications, the NO_X adsorber performance was reported to be enhanced by a three-way catalyst positioned upstream, in a configuration typical for many GDI studies [96, 97, 109]. In diesel applications, NO_X conversion efficiency can be improved by positioning an oxidation catalyst upstream of the adsorber [94].

19.5.2.2 Regeneration

During the adsorption cycle, the adsorber is gradually converted into its nitrate form (e.g., barium nitrate) and the adsorption capacity becomes saturated. At this time the stored NO_X needs to be released and catalytically reduced in a process called *regeneration.* At lean exhaust conditions, barium nitrate decomposes at temperatures greater than 450 to 500°C. The regeneration occurs at much lower temperatures if a short pulse of fuel-rich mixture is provided. NO_X adsorbers can fully regenerate at 250°C, with the onset of partial regeneration at temperatures as low as 150°C, if the air-to-fuel equivalence ratio is maintained at $\lambda < 1$ [110]. Therefore, the operation of the adsorber catalyst system involves continuous cycling through lean and rich fuel conditions.

The designer of an adsorber system has to analyze very carefully all pertinent operation temperatures, including exhaust gas temperatures during real-life duty cycles and during emission certification tests, the NO_X adsorber temperature window, the rich regeneration temperature, and the lean decomposition temperature. During lean operation, it is important for adsorber temperatures to be below the thermal decomposition temperature of the stored nitrate (e.g., barium nitrate). Otherwise, NO_X may be released at lean conditions, leading to a decrease in the average conversion efficiency. Although diesel adsorbers are not likely to see temperatures of 450 to 500°C, especially if installed away from the exhaust manifold, this may be a problem with gasoline engines, which experience higher combustion temperatures.

An example of lean/rich cycling is illustrated in Figure 19.31, which shows concentrations of NO_X upstream and downstream of the adsorber, and the exhaust gas temperature [111]. The data were generated on a light-duty DI diesel engine. Rich spikes were achieved by simple throttling of the intake air, resulting in an oxygen concentration decrease to less than 0.4%. The 60-s storage/regeneration pattern used during the test is visible in the peaks in NO_X concentration. Engine-out NO_X shows minimums, which are caused by lower pressures/temperatures in the combustion chamber during intake air throttling. The tailpipe NO_X, on the other hand, shows maximums that represent the portion of released NO_X that has not been reduced over the rhodium catalyst. The engine was tested at a constant speed, but different, steady-state torque levels, resulting in stepwise changes in the exhaust temperature.

At the highest tested temperature of about 330°C, the adsorber catalyst achieved 95% NO_X conversion efficiency. As the exhaust temperatures were lowered, the conversion

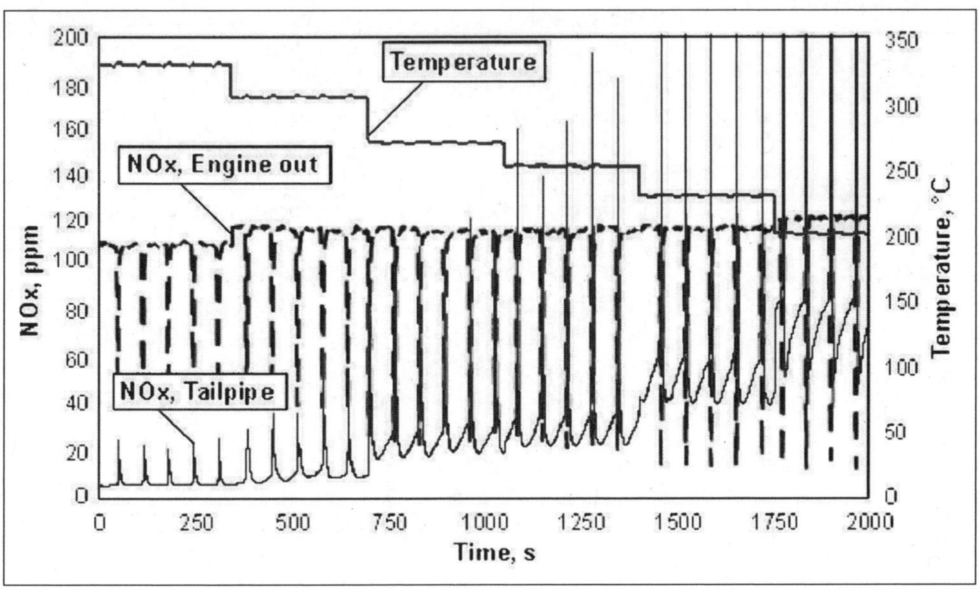

***Figure 19.31** NO_x adsorber regeneration for a 1.7-L adsorber (4.66 in. × 6 in.) on a 1.9-L, DI diesel engine, 1200 rpm/15–50 Nm, space velocity =20,000–60,000 1/h, fuel sulfur 2 ppm, cycled 58 s lean/2 s rich. (Courtesy of Johnson Matthey.)*

gradually decreased to reach about 50% at about 200°C. The declining catalyst efficiency manifests itself by increasingly higher tailpipe NO_x peaks during regeneration, as can be seen in Figure 19.31.

The author reported that HC breakthrough was observed during rich spikes at all temperatures, whereas no CO breakthrough was seen. This result is an indication that the NO_x reduction mechanism may involve reactions with CO rather than with HC.

To maximize NO_x conversion efficiency, the storage capacity and frequency of regeneration must be optimized during the design of the adsorber system. The typical capacity of barium adsorbers in the fresh state amounts to around 2 g NO_x per liter of catalyst volume [96]. Depending on the engine emissions, catalyst size and condition, and the desired NO_x reduction, regeneration must be performed every 30 to 120 s. The duration of NO_x adsorber regeneration is short, between one and a few seconds.

19.5.2.3 Inhibition by Sulfur

Diesel exhaust contains certain quantities of sulfur, primarily as sulfur dioxide, derived from diesel fuel and engine lubricating oil. In the presence of an oxidation catalyst, these compounds form stable sulfates with the NO_x storage materials [see Eqs. (19.39) and (19.40)]. The adsorption of sulfur is preferential over the adsorption of NO_x. As a result, the catalyst performance gradually declines as fewer sites are available for NO_x adsorption.

Higher levels of sulfur in fuel result in faster and more severe deactivation, as shown in Figure 19.32 [112]. Note, however, that even sulfur levels of less than 10 ppm eventually lead to NO_x adsorber poisoning [11], not to mention sulfur contribution from the engine lube oil. Ultra-low-sulfur fuels are the necessary condition for implementation of this technology, but even if such fuels are available, NO_x adsorbers are still likely to require some form of desulfation mechanism.

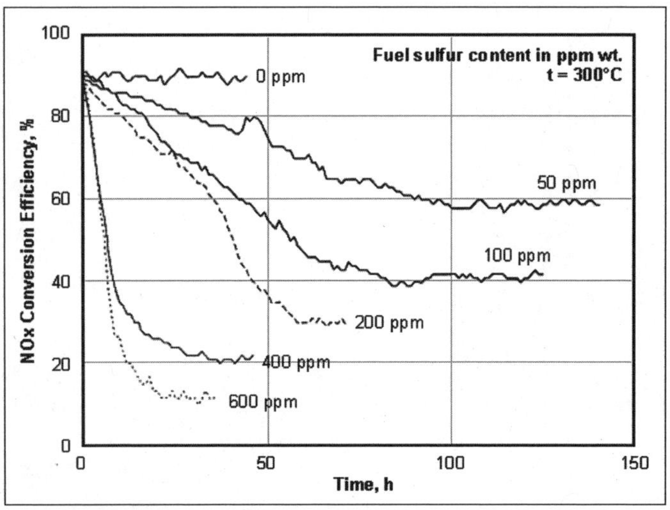

Figure 19.32 NO_x adsorber efficiency at different fuel sulfur levels.

Sulfur poisoning begins on the surface of the catalyst inlet and progresses deeper into the washcoat and in the axial direction [97, 113]. Sulfates derived from the known NO_x storage materials are more thermally stable than the corresponding nitrates. They do not decompose at conditions that are usually encountered during adsorber operation, including both the adsorption and NO_x regeneration cycles. The problem of sulfur deactivation affects equally the diesel and GDI engine applications.

19.5.2.4 Desulfation

In general, the sulfur poisoning is reversible and site activity can be restored by a desulfation process involving decomposition of the sulfate species. The desulfation of NO_x adsorbers requires temperatures between 500 and 700°C, accompanied by mixture enrichment. It was reported that optimum desulfation of barium NO_x adsorbers is achieved at 650°C and $\lambda = 0.98$ [114]. Studies on gasoline engines have found that adsorber desulfation was most effective in the presence of H_2 and CO, with H_2 being more effective. However, H_2 desulfation produced a mixture of H_2S and SO_2, while only SO_2 was detected among desulfation products when CO was used [97].

In theory, the desulfation of NO_x adsorbers can restore their full adsorption capacity. In practice, a permanent and irreversible poisoning of some barium sites has been reported [113]. Another source of permanent performance loss after repeated desulfation is thermal degradation of washcoat and catalyst materials due to high temperatures during the desulfation process [115]. The desulfation strategy is a critical function in NO_x adsorber design and is still far from being solved, especially for diesel engines. If sulfates are left in the catalyst for prolonged periods of time, the NO_x conversion efficiency is compromised. Frequent desulfation, on the other hand, may involve significant fuel economy penalties and accelerated thermal deterioration of the catalyst.

Various methods of desulfation through the increase of exhaust gas temperatures have been under development. For diesel engines, exhaust temperatures can be increased by post-injections of fuel. Algorithms have been developed that facilitated catalyst desulfation through system integration with the engine control unit, the common-rail fuel injection system, and the on-board diagnostics [116]. Desulfation strategies for gasoline engines can also include a rapid, large-amplitude modulation of the A/F ratio to create exothermal reactions increasing the catalyst temperature and minimizing H_2S release [109]. A desulfation strategy involving short A/F ratio pulses has been also proposed for the diesel engine [95]. Desulfation through a series of short rich pulses, rather than a single, continuous rich period, allowed both H_2S emissions and the fuel economy penalty (reported at 1% when using 10 ppm S fuel) to be minimized. The pulses of $\lambda = 0.95$ and 30-s duration were repeated every 250 s over a nearly 1-h-long desulfation process.

A NO_x adsorber desulfation strategy was also developed and demonstrated on a diesel engine by the Diesel Emissions

Control—Sulfur Effects (DECSE) program of the U.S. Department of Energy [117, 118]. The strategy, developed for a single point on the engine map for an exhaust temperature of 400°C, involved common-rail post-injection and a close-coupled, warm-up catalyst. Hydrocarbons generated through the post-injection process were oxidized in the warm-up catalyst to create an exotherm increasing the NO_x adsorber inlet temperature to 700°C (Figure 19.33).

DECSE experiments to develop the strategy were conducted by operating the engine on commercial, 380 ppm S fuel for approximately 2 h, until the initial NO_x conversion efficiency of 80% dropped to 60% due to sulfur poisoning. At that moment, the desulfation event was triggered. The adsorber inlet temperature of 700°C was achieved 90 to 180 s after initiating the desulfation, depending on the post-injection quantity and the EGR rate (a quantity of 30 mm³/stroke and an EGR of 26.6% were eventually selected). Restoring the NO_x efficiency to the original 80% required a duration of the desulfation event of about 5 to 6 min. The procedure was applied to a number of catalysts of different poisoning histories, all of which were restored to more than 85% NO_x reduction efficiencies over the catalyst inlet operating temperature window of 300 to 450°C. This performance level was achieved while staying within the 4% fuel economy penalty target defined for the adsorber regeneration. Variations of the engine torque due to post-injections during the desulfation event were controlled to within 1%.

19.5.2.5 Sulfur Traps

Obstacles encountered in the development of adsorber desulfation strategies stimulated the development of alternate methods of protection from sulfur, leading to the idea of sulfur traps. A sulfur trap, also called the SO_x trap, is another adsorber catalyst, specifically designed to store sulfur. It is placed in the exhaust system upstream of the NO_x adsorber, to reduce its rate of poisoning by sulfur. Due to the high-temperature requirements for sulfur trap regeneration, its preferred position is close to the exhaust manifold, as illustrated in Figure 19.34. The NO_x adsorber, which operates at lower temperatures, can be positioned away from the engine, such as in the underfloor position on passenger cars. Although sulfur traps were developed that can effectively protect NO_x adsorbers, the trap regeneration (desulfation) management remained an open issue; unless an exhaust gas bypass is used, the sulfur released from the trap is readsorbed while passing through the downstream adsorber, thus making the application of sulfur traps problematic.

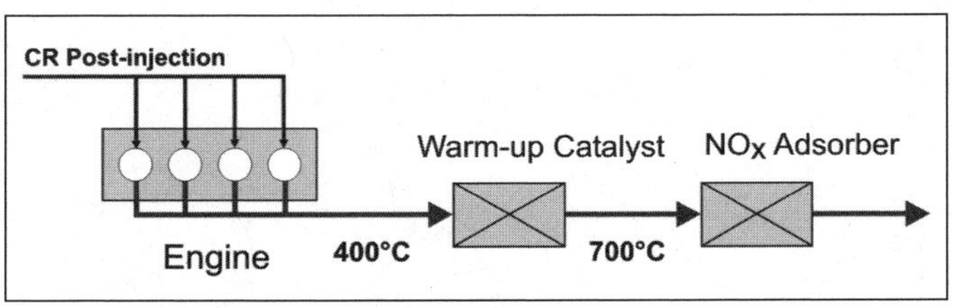

Figure 19.33 Schematic of DECSE desulfation strategy. Engine: 1.9-L HSDI, 1943 cc, 81 kW @ 4200 rpm, turbocharged, common-rail, EGR. Warm-up catalyst: 2.5-L (5.66 in. diam. × 6 in.), 400 cpsi/6.5 mil, Pt, 70 g/ft³. NO_x adsorber catalyst: 2.5-L (5.66 in. diam. × 6 in.), 400 cpsi/6.5 mil, 10:3.9:1, 164 g/ft³.

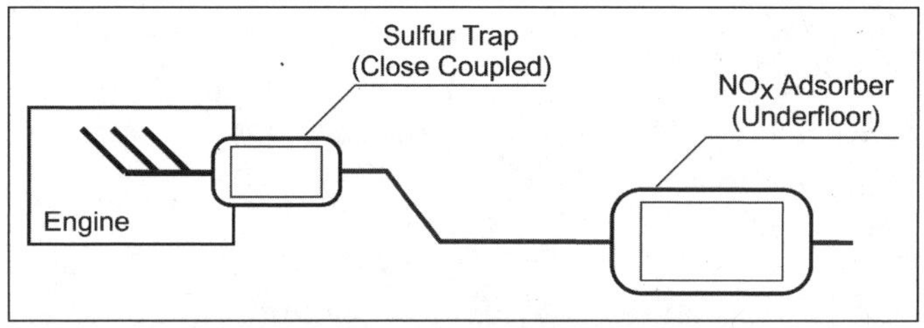

Figure 19.34 Sulfur trap configuration.

The sulfur storage mechanism in NO_x adsorbers and in many sulfur traps is similar. Because sulfur is stored in the form of a metal sulfate, the sulfur dioxide has to first be oxidized to SO_3 per Eq. (19.39). Therefore, sulfur traps must include an oxidation catalyst (e.g., platinum) to facilitate this reaction. Also the storage systems are similar to those used in NO_x adsorbers, but specifically optimized for the following features:

- High adsorption capacity for sulfur
- Optimal desulfation temperatures for a given system
- Fast and complete release of sulfur (regeneration) after reaching the desulfation temperature
- Selectivity toward sulfur adsorption
- No secondary emissions
- Thermal durability.

Typical sulfur storage systems are based on alkaline earth or alkali metal oxides and their mixtures on an alumina (titania, zirconia) washcoat. A number of other materials have also been tested, including zinc, nickel, chromium, copper, and silver. These metals or their oxides are used either as stand-alone scavengers or as promoters to modify trap performance, strengthen desired reactions, and influence the adsorption/desorption temperatures [99]. Silver, for example, has been reported to enhance the sulfur trap selectivity toward adsorption of sulfur oxides over nitrogen oxides and to lower the regeneration temperature [119].

After its storage capacity is saturated, the sulfur trap has to be regenerated. Unless a complex piping system involving valves and bypasses is used [120], sulfur released from the trap must pass through the downstream NO_x adsorber. Depending on the temperatures in both devices, a fraction of sulfur released from the trap will be readsorbed in the NO_x adsorber. This presents one of the major obstacles in implementing sulfur traps.

Thermal release of sulfur under lean conditions is possible, but would require temperatures beyond the thermal stability limits of existing NO_x adsorber designs. On the other hand, materials have been developed that allow desulfation of sulfur traps under rich conditions at temperatures as low as 300 to 350°C [11, 119]. For this reason, most studies have focused on rich regeneration in a manner parallel to that of the NO_x

adsorbers. There are two possible maintenance strategies for the sulfur trap:

- *Continuous maintenance strategy,* in which sulfur scavengers having a low regeneration temperature are regenerated every time the NO_x adsorber is regenerated
- *Periodic regeneration strategy*, where the sulfur trap is regenerated less frequently than the NO_x adsorber.

In the periodic strategy, the sulfur trap is regenerated at higher temperatures, thus minimizing the readsorption of sulfur in the NO_x adsorber. A certain fuel economy penalty would be associated with increasing the temperature of the sulfur trap. The continuous regeneration approach, while minimizing the fuel economy penalty, presents a larger material development challenge to develop scavengers that would regenerate at low temperatures and over short periods of time. Irrespective of the strategy, periodic desulfation of NO_x adsorbers is likely to be required, even in the presence of sulfur traps.

The desulfation of sulfur scavengers involves release of secondary emissions, including hydrogen sulfide (H_2S) and carbonyl sulfide (COS) [99]. The ratio of H_2S/SO_2 released during regeneration increases with decreasing A/F ratio. If low A/F ratios are used for regeneration, which typically reduce the duration of regeneration, most of sulfur may be released as H_2S. Even though the tendency to release sulfur as SO_2 can be maximized by the selection of sulfur scavengers [11], the SO_2 may be subsequently reduced to H_2S in the downstream NO_x adsorber under rich conditions. H_2S production increases with the gas residence time in the catalyst; that is, larger sulfur traps and NO_x adsorbers produce a higher proportion of hydrogen sulfide. In the system shown in Figure 19.35, more than 90% of the sulfur is released from the adsorber as H_2S [99].

Considering the obnoxious rotten egg odor of hydrogen sulfide, it is desirable that sulfur be released from the vehicle tailpipe in the form of SO_2 rather than H_2S. If satisfactory H_2S-suppressed sulfur scavengers are not developed, it may be necessary to provide an additional catalyst, downstream of the NO_x adsorber, to store the hydrogen sulfide during rich regeneration periods and oxidize and release it as SO_2 during lean operation. Suitable H_2S scavengers, based on nickel, manganese, zinc, or iron, have been developed for the gasoline three-way catalyst [99, 121].

Formation of another secondary emission, ammonia (NH_3), was reported over a silver-based sulfur trap at lean condition with a resulting decrease in NO_x adsorber efficiency [119]. An ammonia decomposition catalyst was installed between the sulfur trap and the NO_x adsorber to alleviate the problem.

19.5.3 Diesel Engine Systems

19.5.3.1 Mixture Enrichment Strategies

To minimize torque disturbances, mixture enrichment ($\lambda <$ 1) on lean-burn gasoline engines may require some coordination of the spark advance with fuel injection, but otherwise seems to be fairly straightforward. Generation of rich A/F mixtures on the diesel engine, which normally operates at $\lambda > 1.3$ and leaner, is certainly more challenging.

Theoretically, the methods of diesel mixture enrichment can be grouped into two categories: (1) exhaust gas enrichment and (2) in-cylinder enrichment. Exhaust system enrichment may be realized through injection of diesel fuel upstream of the catalyst, in a manner similar to that suggested for active deNOx catalysts, which work through selective NO_x reduction by hydrocarbons. However, efficient NO_x adsorber regeneration requires not only the presence of reductants, but also the absence of oxygen. In the case of exhaust enrichment, the existing oxygen needs to be combusted to CO_2 and H_2O before NO_x conversion can proceed. This can be realized in the NO_x adsorber itself, or in an upstream oxidation catalyst. Disadvantages of this process include exotherms, which can be damaging to the catalyst, and an additional fuel economy penalty.

Theoretically, reducing gases other than hydrocarbons derived from the diesel fuel can also be used for the regeneration of NO_x adsorbers. Regeneration of a NO_x adsorber through the injection of syngas (CO + H_2) into the exhaust system was

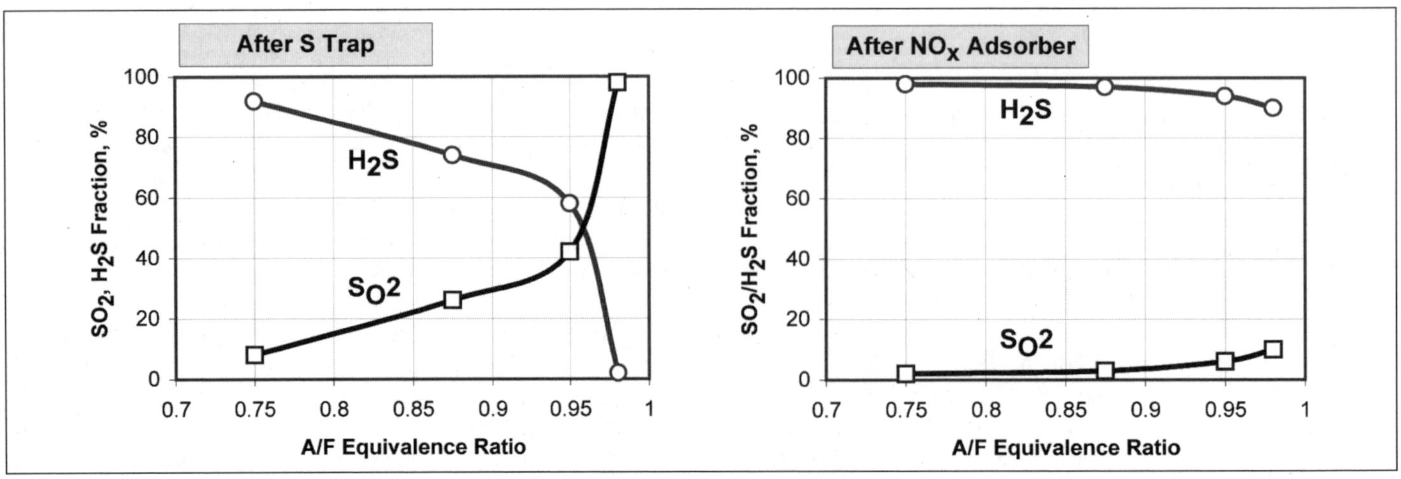

Figure 19.35 *SO_2 and H_2S fraction during regeneration with an 0.8-L sulfur trap on an SI passenger car engine, regenerated at 640°C.*

demonstrated on a light-duty vehicle [122]. In practical applications, the reductants would have to be generated on board by reforming diesel fuel. It was proposed that hydrogen-rich gas ($CO + H_2$) be produced from diesel fuel by electric discharge (plasma) continuously applied to a flowing fuel/air mixture in a device termed the *diesel plasmatron reformer* [123].

In-cylinder enrichment involves such methods as altering the fuel injection timing and rate, throttling, and using EGR [124]. Advanced, electronically controlled diesel fuel injection systems, such as the common rail, can now offer much more flexibility for practical in-cylinder enrichment strategies. The common rail is the first diesel injection system in which both the injection rate and timing can be controlled totally independent of the engine speed and load condition. In particular, it is possible to provide post-injections of fuel, as may be required to increase the exhaust gas temperature.

Common-rail injection has already been used for increasing diesel exhaust gas temperature as a means of assisting the regeneration of a particulate filter [125]. The task was accomplished with three injections and required a turbocharger adjustment in order to maintain torque. The filter application, however, differs from that of the NO_x adsorber in the fact that quantities of oxygen remain in the exhaust gas. Systems specifically developed for regeneration of NO_x adsorbers are likely to require a combination of methods, for example, implementing intake air throttling in addition to multiple injections [126].

A new diesel combustion process, called *low-temperature combustion,* has been developed by Toyota for the regeneration of NO_x adsorbers at low-load and -speed conditions in light-duty engines [127]. The process involves massive EGR, intake air throttling, and injection timing designed for smokeless combustion despite a rich A/F ratio.

19.5.3.2 Integration of an Engine-Catalyst System

One of the most important aspects of utilizing NO_x adsorber technology is to establish engine operating conditions that will facilitate a satisfactory level of NO_x conversion through proper adsorber regeneration and desulfation, while minimizing the associated fuel economy penalty. This optimization is achieved by defining a lean/rich modulation strategy, while paying close attention to resulting NO_x, CO, and HC concentrations, as well as to exhaust temperatures. More fuel-rich modulations typically result in faster and more complete regeneration of the adsorber and, thus, in higher average NO_x conversion efficiencies.

The DECSE program concluded that a more than 80% peak NO_x conversion efficiency can be achieved at a fuel economy penalty of less than 4% [124]. While the 4% figure provides a useful reference number in terms of the anticipated fuel cost of this technology, there is still room for optimization of the engine-catalyst system. This is best illustrated in Figure 19.36, taken from the same DECSE study. The two NO_x conversion curves shown in the graph were generated using two different engine calibrations. In both cases, the fuel penalty was kept below 4% and identical rich/lean timing was used, but different NO_x conversions were seen. It is clear that the regeneration cycle modulation has to be very carefully tuned to match the catalyst requirement; otherwise, quantities of fuel may be wasted through unproductive mixture enrichment.

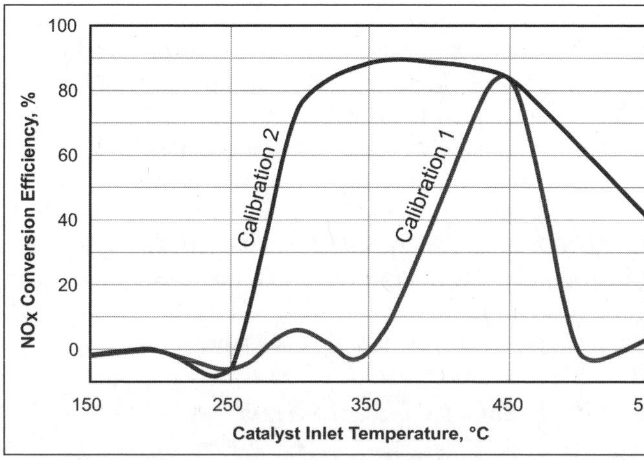

Figure 19.36 *Influence of engine calibration on NO_x conversion efficiency; 2.5-L fresh catalyst on a 1.9-L HSDI engine rated 81 kW @ 4200 rpm, 3 ppm S fuel. (Based on data from Ref. 124.)*

Whereas most of the early laboratory studies were performed at steady-state conditions, real-life NO_x adsorber systems must provide efficient NO_x reductions at all operating conditions, including engine transients. The regeneration and desulfation cycles in high-efficiency systems require closed-loop control, based on the concentrations of NO_x and oxygen, temperature, and other parameters. The feedback signals must be provided by sensors, including NO_x sensors and A/F ratio sensors. The control system, integrated with the engine control module, has to determine the regeneration and desulfation parameters (timing, duration, A/F ratio, etc.) and the enrichment strategy depending on the process variables and the engine operating conditions. The optimization of the regeneration/desulfation control and the integration of the engine-catalyst system remain perhaps the most challenging tasks on the road toward commercial introduction of NO_x adsorbers.

19.5.3.3 Diesel Engine System Concepts

Since the late 1990s, notable progress has been achieved in the integration of NO_x adsorbers with diesel-fueled vehicles, especially in light-duty applications for the European and Japanese markets. As research progresses, more attempts have been made to demonstrate NO_x adsorber systems on diesel engines, including heavy-duty applications targeting the US2007 standards. Because NO_x adsorbers can only be effective in controlling NO_x, CO, HC, and SOF emissions, particulate filters will be required for PM reductions needed to meet emission standards (Euro5, US2007, Tier 2). For this reason, any serious demonstration programs need to integrate NO_x adsorbers with diesel particulate filters (DPFs). There are at least two such integrated emission control systems, one for heavy- and one for light-duty engines, that deserve more detailed discussion:

- An EPA 2007 "Proof-of-Concept" system has been developed as a laboratory demonstration of the technical feasibility of the US2007 emission standards. While not optimized and far from mature, the system became a benchmark and a starting point for NO_x adsorber-based emission control system development by U.S. heavy-duty engine manufacturers.
- The Toyota DPNR system is perhaps the most advanced NO_x adsorber system, which is nearing commercial deployment on diesel cars. The DPNR features a very elegant, compact integration of a DPF and the NO_x adsorber, with the latter being simply coated on the filter substrate.

EPA 2007 Proof-of-Concept System

One of the first NO_x adsorber demonstrations on a heavy-duty engine was presented by the U.S. EPA as a "proof-of-concept" for the US2007 heavy-duty diesel emission standards (NO_x = 0.2 g/bhp-hr; PM = 0.01 g/bhp-hr) [128]. The system, installed on a production Cummins ISB engine (5.88-L, six-cylinder, turbocharged-aftercooled, DI, four valves per cylinder, 194 kW/260 hp @ 2500 rpm), included two catalyzed DPF-NO_x adsorber banks in parallel, as shown in Figures 19.37 and 19.38. The exhaust flow was periodically switched through a valving system to direct most of the flow to the adsorber in operation, and only a small portion of the gas stream to the adsorber in regeneration. A secondary fuel injection was applied during regeneration upstream of the DPF. The injected fuel was oxidized over the DPF catalyst, depleting oxygen and increasing the temperature. An oxidation catalyst, shared by both DPF/adsorber banks, was included at the end of the system to control HC emissions that may have broken through the regenerating adsorber. The system also included a number of zirconia NO_x sensors (before DOC, after adsorbers, in the engine exhaust manifold) and oxygen sensors (after adsorbers) to provide feedback for the control system.

The EPA system provided impressively high conversion levels of 98% reduction in NO_x (down to 0.25 g/bhp-hr) and 93% reduction in PM (0.002 g/bhp-hr), as measured over the FTP Transient cycle. The engine was operated on a 5 ppm S fuel. A fuel economy penalty of 2.3% was attributed to the regeneration of the NO_x adsorber. (This figure did not include any penalties that may have resulted from the system pressure drop.) An oversized catalyst system, including 70 L of combined catalyst/DPF volume, was used in the study. No adsorber desulfation tests were conducted; adsorber deactivation by sulfur would eventually occur in real life, even when using diesel fuel of only 5 ppm S.

In the second stage of the study, the engine was modified by adding a high-pressure common-rail injection system with a Bosch/ETAS engine management system, and a high-pressure loop EGR [129]. In this configuration, the adsorber system yielded an average NO_x emission of 0.13 g/bhp-hr (down from a 2.67 g/bhp-hr baseline) over a hot start FTP test at a reductant fuel economy penalty of 1.49%.

The dual adsorber/DPF bank approach allows the fuel economy penalty to be minimized by regenerating the adsorber at a low exhaust gas flow; however, it results in a very large total catalyst volume. Reduction in the system size may be realized by decreasing the catalyst volume, as well as by optimizing the split between the catalyst volume in the operation and regeneration modes. The system shown in Figure 19.38 includes

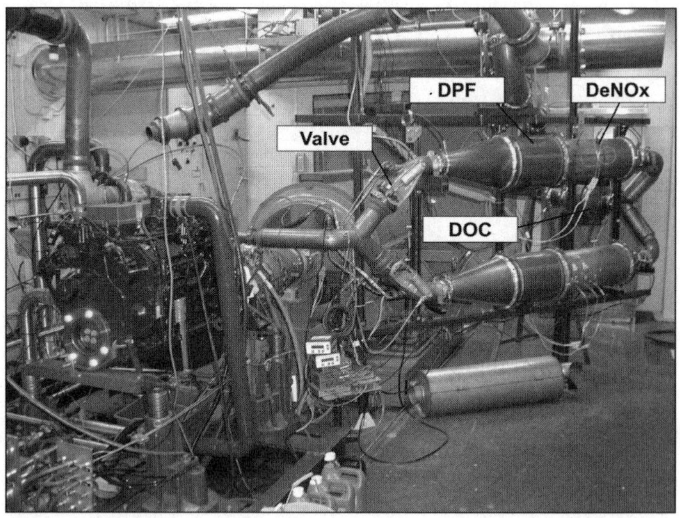

Figure 19.38 EPA 2007 "Proof-of-Concept" system in the EPA emission laboratory.

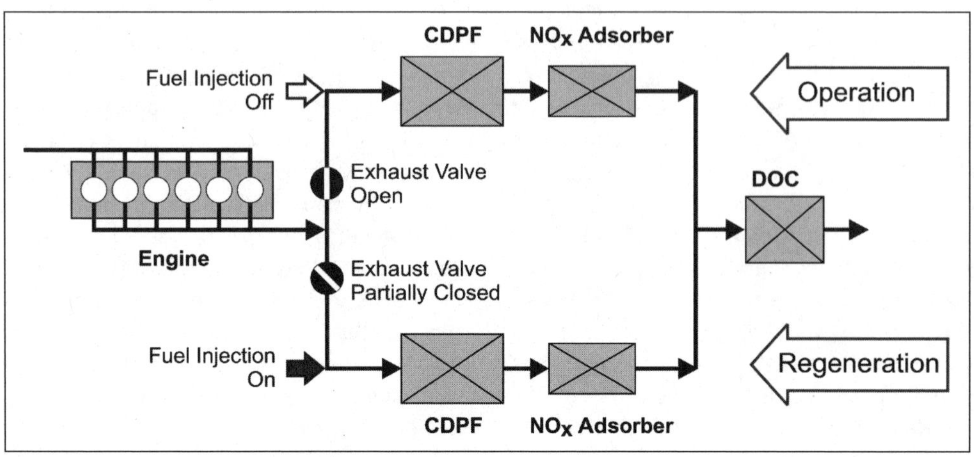

Figure 19.37 NO_x adsorbers in the EPA 2007 "Proof-of-Concept" system.

equal volumes in each mode. However, the regeneration time that is required for the NO_X adsorber is shorter than the time needed to saturate the adsorber with NO_X. This results in a part of the system volume "idling" after the regeneration cycle is completed. The EPA system size would be further reduced if the regeneration was conducted using a smaller proportion of the total catalyst volume. The EPA planned on splitting the system into four sections, three of which would be in the operating mode, while one would be regenerated. Exhaust gas flow would be directed to the proper section(s) using a rotary valve system upstream of the filter/catalyst body. The standard truck muffler visible on the floor underneath the system in Figure 19.38 represents the target volume for the entire system utilizing that concept.

The EPA system configuration, based on a production diesel engine, is an example of a retrofit NO_X adsorber system. In future engines, the adsorber management strategy will be integrated with that of the engine, thus allowing for the optimization of the regeneration process and, preferably, allowing for use of a single-bank configuration having a much smaller catalyst volume.

Toyota's DPNR System

A very elegant and compact system, combining a NO_X adsorber and a DPF on one substrate, was developed by Toyota and termed the *Diesel Particulate-NO_X Reduction*, or DPNR, system [130–132]. The DPNR utilizes a cordierite wall-flow particulate filter substrate, with channels that are alternatively plugged at the ends to force the gas through the porous walls, where the particulates are mechanically trapped. The NO_X adsorber catalyst is coated on the filter substrate, over the channel walls and inside the pores. The system adsorbs and reduces NO_X over a lean/rich modulation cycle typical for NO_X adsorbers. Particulate matter collected on the filter is continuously regenerated over the Pt-containing catalyst, just as is done in catalyzed particulate filters. Interestingly, a synergistic effect was found between the NO_X and PM control functions in the DPNR, with the rich pulsing enhancing the filter regeneration process.

A schematic of the DPNR system integrated with a prototype diesel engine is shown in Figure 19.39. The in-line, four-cylinder, 2.0-L, water-cooled, TDI engine is equipped with a common-rail fuel system and cooled, high-pressure loop EGR. The NO_X adsorber/DPF unit is installed downstream of the turbocharger. The reported volume of the DPNR substrate varies between 1 and 1.5 times the engine displacement; it was 2.8 L in the system shown in Figure 19.39 [132]. A substrate geometry of 300 cpsi, a 0.3-mm wall thickness, 55% porosity, and a relatively large mean pore size of 25 μm to enable catalyst coating were reported [132]. The DPNR substrate is followed by an oxidation catalyst, which removes HC emissions that may pass through the NO_X adsorber during regeneration. The engine is also equipped with a fuel injector in the exhaust manifold. A number of sensors are needed to provide feedback to the engine controller, including gas temperature, DPNR

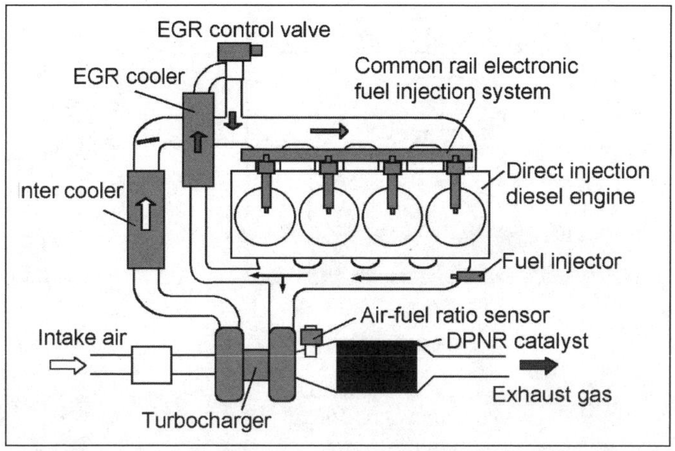

Figure 19.39 Schematic of DPNR system. (Reprinted from Ref. 132.)

pressure drop, intake air flow, and A/F ratio sensors, as shown in the schematic.

Parameters of the NO_X adsorber operation varied in particular tests, but have been generally reported around 60-s lean operation, followed by a 1- to 3-s rich pulse at A/F = 11.5 to 12.5 [132]. Most testing has been performed with 30 ppm S fuel. Adsorber desulfation is performed at around a stoichiometric condition over a period of about 50 s and a catalyst bed temperature of about 600°C. H_2S emissions during desulfation are minimized by lean/rich switching. The PM regeneration in the filter is controlled through the A/F ratio, which impacts the catalyst temperature through both the exhaust gas temperature and the temperature increase in the catalyst due to HC oxidation. The control of the A/F ratio involves three mechanisms, as follows:

- "Low-temperature combustion" (high EGR + injection timing + intake air throttle)
- Post-injection of fuel to the engine cylinders
- Exhaust manifold injection of fuel.

The combination of control mechanisms depends on the engine operating condition, as well as on the state of the DPNR system. For instance, increased temperatures are needed to force the PM oxidation if the filter becomes overloaded with soot. The DPNR control strategies are illustrated in Figure 19.40 [133].

It was found that the rich pulsing enhances the oxidation of PM, thus making it possible to reliably control filter regeneration at widely ranging operating conditions. A switch-flow version, in which the direction of flow through the filter could be reversed by a valve, was considered during early development but was later abandoned. The PM regeneration enhancement relative to conventional catalyzed filters is not entirely understood, but it may be attributed to a combination of several mechanisms, as follows:

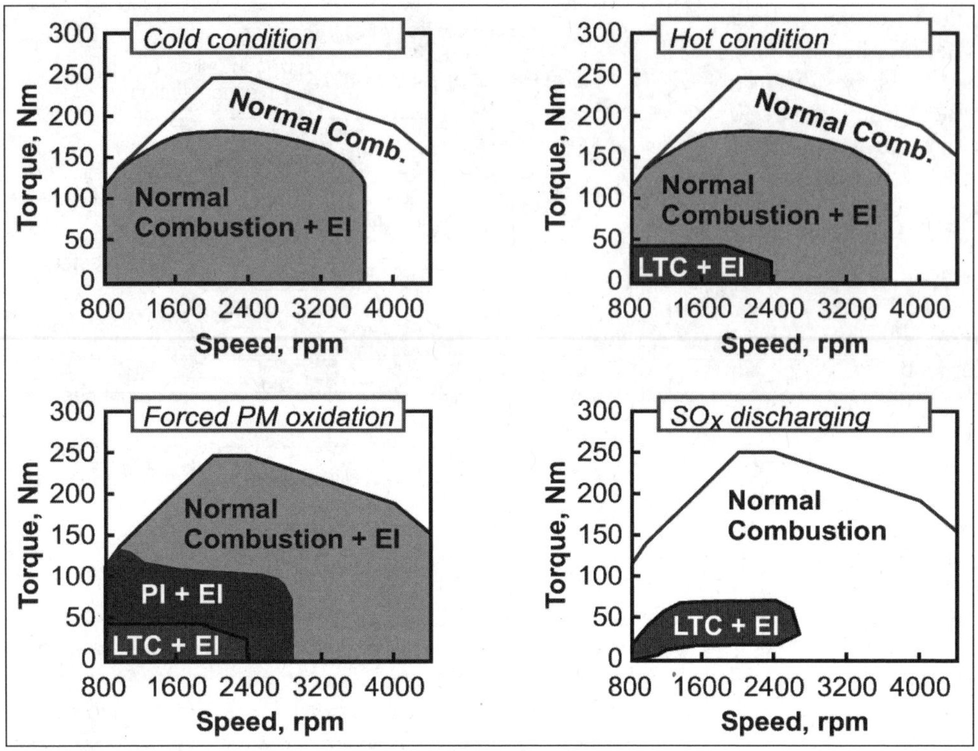

Figure 19.40 *DPNR system control strategies. EI, exhaust manifold injection; PI, common-rail post-injection; LTC, low-temperature combustion. (Based on data from Ref. 133.)*

- Nitrogen dioxide, which is an intermediate product in the process of lean NO_x adsorption [Eq. (19.35)] is well known to enhance DPF regeneration [22].
- Toyota researchers found that active oxygen species that are formed during both NO_x adsorption and rich regeneration are very effective in soot oxidation [132].
- The PM oxidation rates were found to be higher with "fresh" soot, as is the case with the continuously regenerating DPNR. Soot that may accumulate to higher loadings in conventional filters exhibits particle growth and collapsing of its micropore structure, both effects leading to slower oxidation rates. Maintaining low soot loadings also improves the contact between carbon particles and the catalyst [132].

According to Toyota, a fresh DPNR catalyst can reduce both NO_x and PM emissions by more than 80%. An aged catalyst achieved 0.13 g/km NO_x and 0.005 g/km PM over the ECE+EUDC cycle on a 1400-kg car [133]. Introduction of the DPNR system on production vehicles is expected in the near future.

19.5.3.4 Issues

Before adsorber-catalysts can become a widespread commercial NO_x control technology, more progress has to be made in the NO_x/SO_x adsorber technology itself and, very importantly, in many areas relevant to the system integration with the diesel engine. The following is a summary of issues that need to be addressed:

- Methods of mixture enrichment need to be developed. Potential problems include torque/drivability issues and high PM and HC emissions during the enrichment process.
- NO_x adsorber regeneration and desulfation strategies have to be developed. They will require closed-loop control in tight integration with the engine management and other vehicle systems, such as the on-board diagnostics.
- Optimized NO_x adsorbers in terms of adsorption capacity, rich regeneration temperatures, lean thermal stability, and so on, are necessary for efficient application of the technology.
- All of the preceding issues affect the fuel economy penalty associated with adsorber regeneration, which needs to be minimized.
- Secondary unregulated emissions, including H_2S and N_2O, from adsorber systems, need to be minimized. Depending on the regeneration/desulfation strategies, adsorber systems may also produce off-cycle emissions of regulated pollutants.
- Thermal durability of catalysts and resistance to poisons has to be demonstrated. The impact of phosphorus and zinc, known poisons of the three-way catalyst, on NO_x adsorbers is still unknown.
- NO_x adsorbers will increase the cost of diesel emission control systems, especially in view of the rising cost of noble metals. Diesel systems, especially heavy-duty systems, are quite large, which will create additional usage demand and potentially a further increase in cost.
- Furthermore, the commercialization of this technology is only possible in conjunction with legislation and widespread availability of ultra-low-sulfur diesel fuels.

19.6 Commercial Catalyst Technologies

19.6.1 Introduction

Most modern commercial diesel catalysts are diesel oxidation catalysts (DOCs) optimized for the reduction of diesel particulate matter emissions through the removal of their heavy organic fraction known as SOF. Certain emission regulations—for instance, EU standards for diesel passenger cars—also establish stringent limits on gaseous emissions of HC and CO, which must be reduced in the catalyst as well, sometimes with high efficiencies. Other regulations—for instance U.S. standards for heavy-duty engines—set CO and HC limits that are easily met without a catalyst. Control of NO_x emissions would be an extremely desirable function of the diesel catalyst. While existing technologies cannot yet provide significant degrees of NO_x reduction, small NO_x control functionality has been incorporated into commercial diesel oxidation catalysts. Such catalysts, able to control all four regulated pollutants, are sometimes referred to as *four-way catalysts*. Commercial diesel catalysts also reduce several unregulated emissions, such as aldehydes and PAHs, as well as the diesel odor.

Mechanisms of catalytic PM emission control were explained when discussing the DOC technology (Section 19.2) . In short, SOF can be removed in the DOC with very high efficiency through catalytic oxidation and/or cracking. Storage and release of hydrocarbons in the catalyst washcoat may also play an important role and influence the apparent PM conversion efficiency. On the other hand, the carbonaceous fraction of diesel particulates—or dry soot—is practically not oxidized in the catalyst. Thus, the potential for TPM reduction depends on what fraction of TPM is composed of SOF. For example, if SOF forms 40% of the total particulate matter emission, the maximum possible conversion of TPM in the catalyst is limited to about 40%.

Oxidizing the SO_2 to SO_3 is a counterproductive process that occurs in the diesel oxidation catalyst, which increases TPM emissions by the generation of sulfate particulates. The fundamental problem in the development of catalysts for the removal of diesel particulates has been to design a selective catalyst, that is, one that will remove SOF without oxidizing SO_2. Catalyst performance and its effectiveness in reducing PM emissions is strongly affected by the following parameters:

- Sulfur content in the fuel (and in the lube oil in the case of ultra-low-sulfur fuels)
- SOF content of diesel particulates (and, thus, the engine test cycle)
- Exhaust gas temperature at the catalyst inlet (and, thus, the engine test cycle and catalyst location)
- Catalyst washcoat and precious metal system
- Catalyst volume and substrate configuration (e.g., cell density).

As indicated, two of the parameters listed—SOF content and exhaust temperature—critically depend on the engine/vehicle test cycle. Modern DOCs are tailored for their applications and, most importantly, for the test cycle to be applied during the emission certification or type approval process. Generally, the catalytic oxidation rates of SO_2 increase with temperature. Thus, the catalyst sulfate make becomes difficult to control over cycles characterized by heavier load factors, which produce high exhaust temperatures. The SOF content follows the exactly opposite tendency; the highest SOF content in the TPM is measured at low engine loads and exhaust temperatures. Therefore, oxidation catalysts are most effective in controlling diesel PM emissions over lightly loaded, low- to medium-temperature test cycles. The catalyst light-off temperature presents the low-temperature limit for catalyst effectiveness. As will be discussed, many modern DOCs include zeolite-based HC traps, which can extend the low-temperature SOF activity by storing hydrocarbons in the catalyst washcoat. Nevertheless, a low light-off temperature is a very important and desirable property of the diesel oxidation catalyst.

Diesel car emissions in the EU are tested over the ECE+EUDC cycle, which is characterized by low exhaust gas temperatures, typically below 300 to 350°C. Under these conditions PM can be controlled effectively by the DOC, due to both high SOF content in the particulates and relatively straightforward control of sulfate formation in the catalyst. Catalytic PM emission reduction of as much as 60% to 70% may be possible over the European driving cycle. Temperatures during the heavy-duty FTP Transient cycle, which is used for the certification of truck and bus engines in the United States, average around 350°C. Catalytic particulate matter control might be difficult, especially on hot running engines, and advanced catalyst formulations are necessary. The possible degree of PM control over the FTP Transient test is in most cases limited to some 25%. Test cycles used for type approvals of heavy-duty diesel engines in Europe—including the ESC, ETC, as well as the pre-Euro 3 ECE R-49—are all characterized by heavy engine loads and high temperatures, often in excess of 500°C. Catalytic PM control over these cycles is very difficult or impossible due to both sulfate formation and low SOF content.

Performance of three catalyst formulations of increasing selectivity in three different test cycles is illustrated in Figure 19.41 [38]. The catalyst formulations utilize the Pt/Al_2O_3

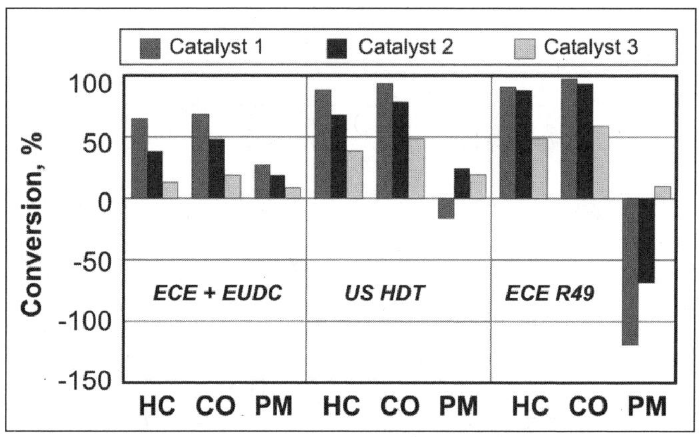

Figure 19.41 *DOC performance in different test cycles. (Reprinted from Ref. 38.)*

system with a base metal oxide additive to suppress sulfate formation. Catalyst 1 of the lowest selectivity can effectively control PM emissions under the ECE+EUDC cycle, but generates high quantities of sulfate particulates when tested in the ECE R-49 test cycle (negative conversions in the graph indicate an increase in TPM emissions). Even the most selective catalyst, catalyst 3, shows only a small TPM emission reduction over the ECE R-49.

Improvements in catalyst selectivity are always accompanied by a deterioration in their activity. The decreasing performance for the oxidation of hydrocarbons and carbon monoxide with increasing catalyst selectivity is well pronounced in Figure 19.41. The same three catalyst formulations are compared in Figure 19.42, which illustrates the SO_2 activity as a function of the CO light-off temperature. Catalyst 3 exhibits an excellent CO light-off of 180°C, but oxidizes about 90% of the sulfur dioxide. The selective catalyst 3 oxidizes less than 20% of SO_2, but its CO light-off is 260°C.

Introduction of diesel fuels of decreased sulfur content makes the sulfate issue less prominent, thus allowing for use of more active catalyst formulations. The sulfate make, however, always needs to be evaluated in the context of existing PM emission standards. Assuming 10 ppm S fuel plus an equivalent sulfur amount from the lube oil, an active DOC can easily produce sulfates on the order of 0.01 g/bhp-hr (see Section 19.2). This figure may be insignificant when compared with the U.S. 1994 PM standard of 0.1 g/bhp-hr, but is clearly prohibitive in view of the US2007 standard of 0.01 g/bhp-hr.

19.6.2 Light-Duty Engines

The main area of DOC application in light-duty vehicles is EU cars, which are emission tested for the type approval over the ECE 15+EUDC driving schedule. This test involves four repetitions of the slow, urban driving ECE segment, followed by one repetition of a faster, highway driving EUDC segment. Originally, the test was started after a 40-s warm-up period, which was eliminated at the Euro 3 stage (2000). The new, cold start version of the test is sometimes referred to as the New European Driving Cycle (NEDC). In the United States, light-duty vehicles are emission certified over the FTP-75 test, which is characterized by somewhat higher engine load factors than the European cycle (but still produces low exhaust temperatures in comparison with tests for heavy-duty engines).

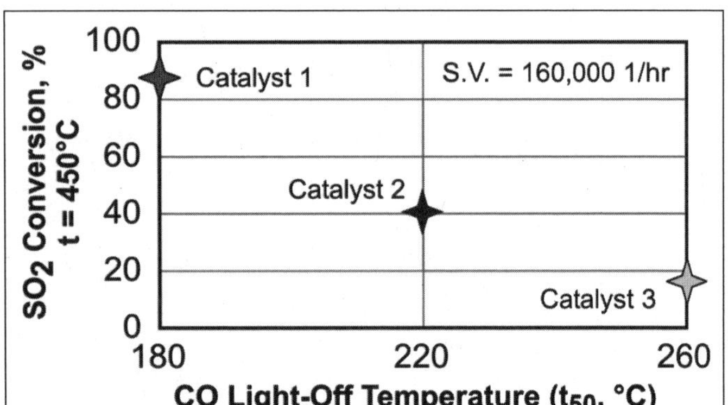

Figure 19.42 *Activity vs. selectivity of DOC.*

Exhaust gas temperatures from a number of diesel cars and light trucks (vehicles C and D) [134] are shown in Figure 19.43. Most of the data were collected over the cold start ECE+EUDC test (NEDC), as indicated in the chart. The T_{10} values, shown as the left end of each bar, are defined as temperatures above which the exhaust gas temperature is 90% of the time. The T_{90} values, shown as the right end of each bar, are defined as temperatures below which the exhaust gas temperature is 90% of the time.

While there is a certain variation between vehicles and engines, exhaust temperatures are generally low, in some cases below 300°C. These low exhaust gas temperatures result in relatively high SOF content in the TPM. At the same time, the sulfate make in the catalyst becomes less critical than in high-temperature applications, providing favorable conditions for catalytic PM control. The major challenge facing catalyst developers is to develop formulations of low light-off temperatures.

In the European Union, diesel oxidation catalysts were introduced on a number of Euro 2 (1996) and practically all Euro 3 (2000) diesel passenger car models. Commercial use of catalysts was made feasible by the introduction of the 500 ppm sulfur limit in diesel fuel in 1996. The catalyst, in addition to the particulate matter removal, has to be effective in the oxidation of gas-phase hydrocarbons and carbon monoxide. The cool driving conditions contribute to increased formation of odors from unburned hydrocarbons and aldehydes. Although odors are not regulated, their control is also a desirable feature of the diesel car catalyst. Furthermore, some NO_x reduction activity may be required in Euro 3 applications.

The configuration of the first diesel catalyst systems in light-duty vehicles featured a single catalytic converter in the underfloor position. Elimination of the warm-up period in the European test cycle at the Euro 3 stage created more challenges related to cold start emissions. As a result, catalysts were moved closer to the exhaust manifold to expose them to the highest possible exhaust temperatures. In some Euro 3 cars, catalyst systems include two catalytic converters: a smaller close-coupled precatalyst, followed by the main converter in the underfloor position [135]. This configuration—resembling catalyst systems used in many gasoline applications—was introduced mainly in large size engines, where catalyst light-off and cold start issues are especially challenging. It was also proposed (but not commercialized) that the precatalyst be moved upstream of the turbocharger to expose it to even higher temperatures and maximize its performance [25, 136].

Advanced commercial catalysts for diesel cars can incorporate a number of technologies, such as oxidation catalysts, lean NO_x catalysts, and HC adsorbers (traps).

Considering the need for simultaneous PM, CO, and HC control, commercial catalyst technologies for EU diesel cars are typically active, medium- to high-loading Pt formulations with sulfate inhibitors (V_2O_5) or with nonsulfating washcoat materials (TiO_2, SiO_2, ZrO_2). These diesel oxidation catalysts

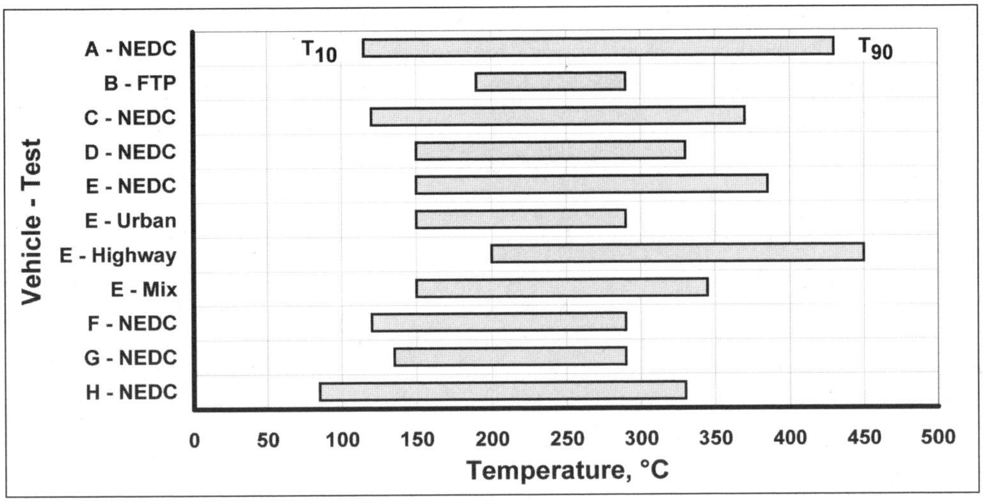

Figure 19.43 *Exhaust temperature in light-duty vehicles. A: 1.9-L IDI; B: 1.9-L TDI; C: 2.5-L DI; D: 2.5-L TDI; E: 1.9-L TDI; F: 1.9-L TDI; G: 1.9-L TDI; H: 3-L IDI.*

are optimized for good HC light-off and sulfate suppression. Some catalysts also incorporate a limited NO_x reduction activity (on the order of 15%). These formulations typically utilize lean NO_x activity of the Pt/Al_2O_3 system. Hydrocarbon traps can be considered another spin-off concept from the lean NO_x catalyst research that has been incorporated into the diesel oxidation catalyst technology. Zeolites, extensively tested in the past in many lean NO_x formulations, became a standard component in modern diesel oxidation catalysts. Zeolites are used to enhance the low-temperature hydrocarbon performance through storage of HCs in the catalyst washcoat [89]. The general idea is to adsorb (store) the hydrocarbons that are produced during low-temperature operation and release them when a certain adsorber temperature is reached. It is essential for the adsorbing material to be capable of dragging the stored hydrocarbons into a temperature range where the true catalytic oxidation of the HCs starts. Otherwise the hydrocarbons would desorb at temperatures before they can be destroyed at the active sites of the catalyst without any advantage in the overall hydrocarbon removal.

An example catalyst incorporating all of the above functionality is a Pt/zeolite/alumina formulation developed by ASEC (Delphi) [32]. The catalyst washcoat is composed of zeolites and alumina. The zeolites must be capable of trapping diesel HCs at temperatures below about 250°C and releasing them at above 250°C. The required temperature durability is 750 to 850°C. The preferred types of zeolite are beta and Y, with a Si/Al ratio of 5 to 50 and a pore size of about 0.6 nm. A mixture of two zeolites can be used, one having a pore size above 0.6 nm to trap longer chain hydrocarbons, the other with smaller pores to trap shorter chain HCs. Alumina is added to the washcoat to improve adhesion of the zeolite to substrate, as well as to aid in the oxidation of CO. Usually gamma alumina is used. Typical washcoat loadings are 90 to 100 g/dm^3 zeolite and about 15 g/dm^3 alumina. The noble metal is in most cases pure Pt, typically used at 10 g/ft^3. If higher CO/HC conversions are needed, Pt loadings may be increased. Some commercial catalysts are believed to have Pt loadings up to 28 g/ft^3 (1 g/dm^3).

In applications where sulfate m-ake is a problêm, vanadium oxide is used as a sulfate suppressant. Typical vanadium loading is 5 g/ft^3 (as V metal). This is equivalent to a V/Pt ratio of about 2 by mole or 0.5 by mass. Loadings of both Pt and V may be varied. In some catalysts described by ASEC, Pt loadings are decreased to 5 g/ft^3 and V loadings are increased to 60 g/ft^3.

The combined adsorption/catalytic oxidation effect on the apparent HC conversion in the Pt/zeolite/alumina catalyst [10] is illustrated in Figure 19.44. Three oxidation catalysts incorporating zeolite HC traps were tested on a quasi-steady-state engine dynamometer test. The exhaust temperature was increased stepwise during the test. Diesel fuel was injected into the exhaust gas to maintain an $HC1/NO_x$ ratio of 10.

The HC conversion observed below approximately 200°C is attributed to adsorption into the micropore volume of the zeolite. The declining HC conversions are likely caused by gradual saturation of the trap adsorption capacity. Catalytic oxidation is the HC removal mechanism at higher temperatures.

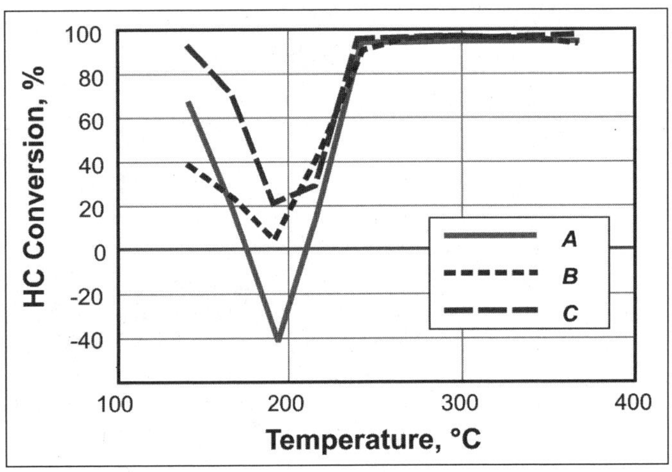

Figure 19.44 *Performance of HC trap/catalyst systems.*

In catalyst A, desorption apparently preceded catalyst light-off, resulting in a negative HC conversion around 190°C. Catalysts B and C showed increasingly better HC adsorption/desorption characteristics.

The zeolite HC trap technology is compatible with the lean NO_X activity of Pt/Al_2O_3 catalysts. These catalysts promote selective reactions between HC and NO_X. Therefore, their NO_X conversion increases with increasing HC/NO_X ratio in the exhaust gas. Zeolites that would desorb hydrocarbons at temperatures of maximum NO_X reduction activity can enhance the lean NO_X function of the catalyst. This catalyst strategy has been adopted in several systems for Euro 3 cars. An example of the four-way performance of the Pt/zeolite/alumina catalyst is shown in Figure 19.45. The overall emission reduction of the catalyst over the ECE+EUDC cycle was HC, 92%; CO, 90%; NO_X, 19%; and PM, 71% [10].

The maximum NO_X reduction activity of the Pt catalyst occurs at about 200 to 250°C. For this reason the catalyst would "lose" its lean NO_X activity when tested on a hotter test cycle, such as those used for heavy-duty engines. However, the low-temperature HC performance benefit due to the zeolite can be easily applied in catalysts for heavy-duty engines. In fact, very similar catalyst technology is believed to be offered in certain aftermarket catalysts for occupational health and safety applications (mining, engines operated indoors, etc.). These catalysts can offer superior HC and diesel odor control at low temperatures, for example, during operation with frequent engine idling. Note also that the zeolite adsorber catalysts have a limited HC adsorption capacity and are truly effective only under transient engine conditions, such as those in diesel-powered cars during city driving. Under low-temperature, steady-state operation or testing, the HC emissions would be removed only by a fresh catalyst and for a limited time, until the zeolites are saturated.

Because the catalyst light-off and cold start performance remain a challenge, catalyst manufacturers continuously develop formulations for improved low-temperature activity. To close this section, we should mention one such more recent catalyst, disclosed by Degussa (OMG), which uses platinum on an alumina/silica washcoat with lead (Pb) as a promoter to enhance low-temperature performance [137].

19.6.3 Heavy-Duty Engines

The main area of diesel oxidation catalyst use on new heavy-duty engines are U.S. truck and bus applications, which are emission certified over the heavy-duty FTP Transient test. At the medium to high temperatures observed during that cycle—as illustrated in Figure 19.46—particulate matter emissions can be controlled by a catalyst with suppressed sulfate make. In 1994, when the 0.1 g/bhp-hr PM emission standard was introduced by the EPA (in coordination with 500 ppm sulfur diesel fuel), several models of truck and urban bus diesel engines in the United States were equipped with catalytic converters. The use of catalysts then gradually declined, as the 0.1 g PM standard could be met on advanced, electronically controlled engines without catalysts. DOCs, however, continued to be used on urban bus engines, which have a more stringent PM standard of 0.05 g/bhp-hr. Starting in October 2002, DOCs were introduced again on certain 2004-compliant engines, such as the Caterpillar ACERT engines.

The only required function of the catalyst on U.S. truck and bus engines has been the control of PM emissions; no CO or HC activity has been needed as the respective emission standards are fairly relaxed and are easily met by diesel engines without aftertreatment. Due to the higher exhaust temperatures over the FTP test (as compared to the light-duty applications), highly selective catalysts with good sulfate make control are required. These selective catalysts have reduced activity, resulting in a certain performance penalty. Many of these catalysts, especially the base metal formulations, can have very low CO and HC activity. Considering also the decreased SOF content (relative to LD engines), the degree of PM emission reduction over the FTP test is limited to about 25%, with many commercial catalysts providing as little as about 10% of PM benefit.

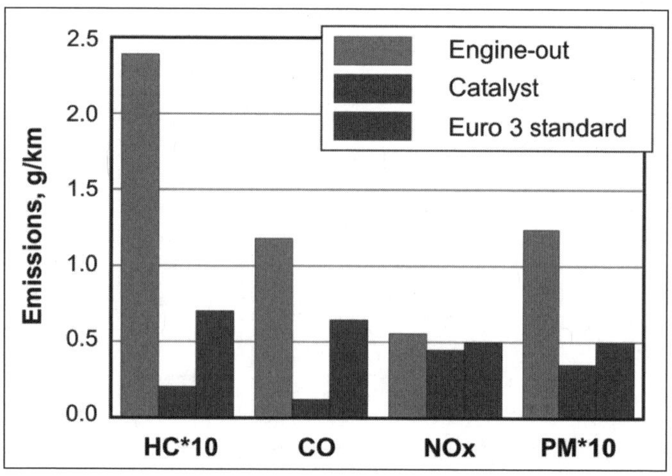

Figure 19.45 *Four-way activity of a Pt/zeolite/alumina catalyst.*

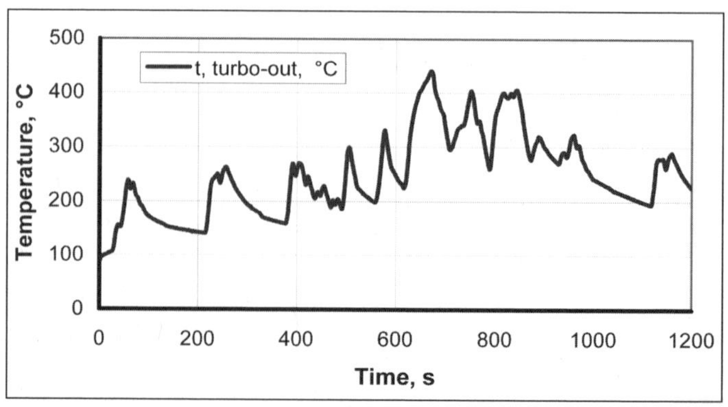

Figure 19.46 *Exhaust gas temperature over FTP Transient cycle; model year 1999, heavy-duty, DI, turbocharged, intercooled diesel engine.*

Active high Pt loading catalysts generate too much sulfate over the FTP test. Several alternative catalyst strategies or their combinations have been pursued for the U.S. heavy-duty truck engines:

- Base metal catalyst
- Platinum catalyst with low precious metal content and sulfate suppressants
- Palladium catalyst.

Platinum catalyst activity toward the oxidation of SO_2 increases with the precious metal loading. Maintaining low Pt loading, such as between 0.1 and 5 g/ft^3, produces catalysts with suppressed sulfate formation. Pt catalysts of medium metal loadings, such as 10 g/ft^3, are still viable on the FTP test, provided a sulfate suppressant is added (such as vanadium). Palladium catalysts were also used of Pd loadings in the range of 40 to 60 g/ft^3. Washcoat formulations are optimized to include various base metal oxides that inhibit sulfate make or exhibit catalytic properties toward the reduction of SOF emissions.

A low Pt catalyst with a catalytically active base metal oxide has been developed and commercialized by Engelhard [29]. In its first generation, the catalyst washcoat was composed of high-surface-area, stabilized alumina and ceria in approximately equal loadings. Alumina and ceria may be mixed together and form one layer, or may be applied as two separate washcoat layers. The particulate matter performance of the catalyst is explained by the catalytic activity of ceria. The SOF fraction of diesel particulates is removed over the cerium oxide catalyst through a mixed cracking and oxidation mechanism. Platinum content is apparently not important (or has little importance) for the PM removal function of the catalyst but is necessary for the conversion of gas-phase HC and CO.

The performance of the ceria catalyst was evaluated by steady-state testing on a Cummins 6BT DI turbocharged diesel engine having a 5.9-L displacement and rated at 190 hp [29]. Three catalyst samples, S4, S5, and S6, contained 0.0, 0.5, and 2.0 g/ft^3 Pt, respectively, and were coated on 400-cpsi cordierite substrates of 229 mm (9 in.) in diameter and 152 mm (6 in.) long (6.25 L). A catalyst washcoat comprised of a γ-alumina undercoat at a loading of 61 g/dm^3 (1.0 g/in^3) was coated with a top layer comprised of 64 g/dm^3 (1.05 g/in^3) γ-alumina plus 55 g/dm^3 (0.90 g/in^3) alumina-stabilized ceria (2.5% Al_2O_3). The catalyst samples were subjected to about 25 h of engine aging prior to testing. Selected test results are shown in Figures 19.47 and 19.48.

Figure 19.47 presents the effect of Pt loading on the TPM conversion. It is apparent that catalysts with different metal loadings show a very similar PM removal effectiveness indicating that the role of platinum is negligible in the SOF removal mechanism. Negative TPM conversions at temperatures greater than 500°C indicate increased particulate emissions due to the formation of sulfates in the catalysts. All catalyst samples, including the Pt-free formulation (S4), make sulfates at very high temperatures.

There is a well-pronounced relationship between Pt loading and gaseous HC conversion, as shown in Figure 19.48. The Pt-free formulation S4 is relatively inactive with maximum HC conversion of about 40%. The 2 g/ft^3 Pt formulation S6 exhibits a substantial improvement with HC efficiency around 80%. Additional testing demonstrated very little performance improvement when Pt loading was further increased beyond 2 g/ft^3. Conversion of CO showed trends that were very similar to those recorded with hydrocarbons. These results indicate that the function of platinum in the ceria catalyst is to enhance the gas-phase (HC + CO) catalyst performance.

The ceria catalyst showed SOF performance even at low exhaust temperatures (<200°C), presumably due to storage of hydrocarbons in the catalyst washcoat. A 1000-h engine testing demonstrated satisfactory durability of the ceria catalyst, despite its low platinum content. However, in many real-life applications, Pt loadings higher than 2 g/ft^3 are used, such as 5 g/ft^3 or higher. The reason for higher Pt loadings are (1) increased durability and (2) increased conversion efficiency, especially for CO and gaseous HC.

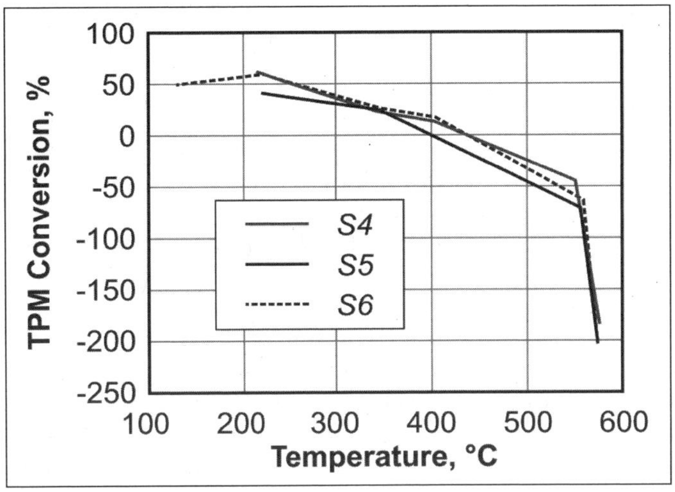

Figure 19.47 *Effect of Pt loading on TPM removal.*

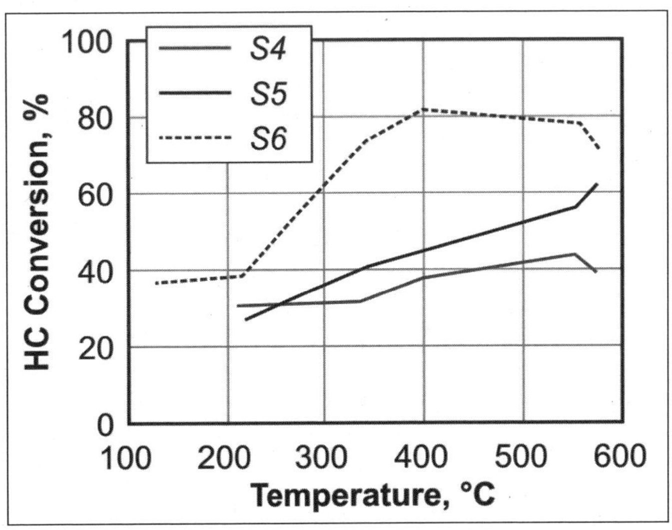

Figure 19.48 *Effect of Pt loading on gas-phase HC removal.*

A second generation of this ceria catalyst additionally incorporates beta zeolite, which further enhances the HC storage function for improved low-temperature performance [33]. The washcoat is composed of equal amounts of the zeolite, alumina, and alumina-stabilized ceria. The preferred Pt loading according to the examples in the patent appears to be 2.5 g/ft^3.

Diesel catalyst technologies have not been used to control PM emissions at high exhaust gas temperatures, such as those encountered over the EU test cycles for HD engines. Sulfates produced over the catalyst during the high engine load modes would more than offset any SOF removal benefit. A system with a temperature-controlled catalyst that was once proposed for ECE R-49 applications is shown in Figure 19.49 [138].

Exhaust gas temperatures upstream (T_1) and downstream (T_2) of the catalyst would be continuously monitored by an electronic control unit (ECU). When temperatures exceeded 380°C, the exhaust gases would bypass the catalyst, preventing excessive formation of sulfates. Systems of this type were never commercialized, however, because emission standards for heavy-duty engines were met without aftertreatment. Considering the low SOF content on high-temperature cycles, the PM emission reduction benefit of the catalyst would likely be very limited.

19.7 Deactivation of Diesel Catalyst

19.7.1 Catalyst Durability

19.7.1.1 Durability Requirements

Deactivation of the diesel catalyst causes gradual deterioration of its performance, resulting in increased emission levels from the vehicle. Most emission standards specify "emission durability" requirements, defined as the minimum mileage and/or time period after which the engine is still expected to conform with applicable emission limits. New emission standards typically increase the required emission durability targets, putting higher demands on emission catalysts and emission control systems in general.

Most U.S. emission standards adopt a two-tier approach to emission durability, which involves (1) emission warranty and (2) full life durability demonstration. Engine manufacturers warrant to the purchaser that the engine is designed and built to conform with applicable emissions standards. The following are the warranty periods for different classes of vehicles through the model year 2003:

- 5 years or 100,000 miles for heavy-duty diesel engines
- 5 years or 100,000 miles for medium-duty diesel engines
- 5 years or 50,000 miles for light-duty diesel engines.

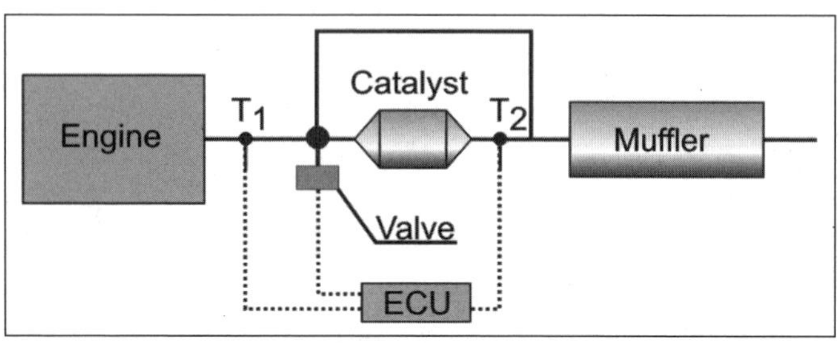

Figure 19.49 *Temperature-controlled catalyst. (Based on data from Ref. 138.)*

As part of the engine emission certification process, manufacturers are required to demonstrate compliance with emission standards over the "full useful life" of the engine. Useful engine lives are defined as follows:

- 8 years or 290,000 miles for heavy-duty diesel engines
- 8 years or 185,000 miles for medium-duty diesel engines
- 8 years or 110,000 miles for light-duty diesel engines.

Beginning in model year 2004, the EPA established increased useful life periods for all pollutants. The revised requirements for the heavy-duty diesel engine service class are 435,000 miles, 22,000 hours, or 10 years, whichever occurs first. Very durable catalyst technologies will be needed to meet these increased requirements.

19.7.1.2 Determination of Catalyst Durability

Catalyst deactivation is usually a gradual process. The determination of catalyst durability involves testing of a fresh catalyst, followed by the test of a catalyst that was aged over a given period of time. Based on the measured performance, emission deterioration factors can be calculated for each of the pollutants by dividing the emission result with the aged catalyst by that with the fresh catalyst. The knowledge of emission deterioration factors, which can be determined not only for catalysts but also for the entire engine-aftertreatment system, is necessary to provide the required engineering margins when designing emission control systems.

Many catalysts exhibit a certain activity loss in the short, initial period of their operation. After this initial activity loss, the catalyst performance remains stable over a long period of time. For this reason, new catalysts should be "degreened" before emission tests. The catalyst degreening involves a short period of aging performed usually on an engine dynamometer bench. Emission test results of "fresh catalysts" are typically measured with degreened units. The duration and conditions of degreening vary greatly between manufacturers and catalyst types. A quick degreening process may involve a few hours of operation under a steady-state engine condition. More typically, the degreening lasts 10 to 20 h (a 20-h green aging at a steady-state condition resulting in 375°C at the catalyst inlet was reported in a diesel oxidation catalyst study [39]). Some automotive catalysts are degreened over transient tests that are designed to represent the equivalent of as much as 4000 miles of real driving.

There are three methods of catalyst aging: (1) field aging, (2) engine bench aging, and (3) oven aging. In the field aging, the catalyst is installed on a vehicle, such as a car, which is then operated over the desired time period. Field aging is an expensive and time-consuming method. A 50,000-mile catalyst aging on a car requires several months of time and usually involves high costs for the car purchase/lease, its operation (fuel, maintenance, etc.), and the driver expense. To minimize the time and cost, accelerated engine dynamometer aging procedures have been developed. The catalyst is installed on an engine operated through a transient test, which can simulate hundreds of miles of driving in 1 h of testing. The test cycle must be designed to provide a high amount of "stress" to the catalyst, such as high exhaust temperatures, transients, or, in the case of spark-ignited engines, lean/rich cycling. The most challenging task in designing accelerated aging tests is the determination of the equivalency between the real-life mileage and the duration of the test. Ideally, the equivalency is verified by comparing the performance loss of identical catalysts aged in the field and in the bench test. Accelerated aging tests are very common in the development of automotive catalysts [139].

For the purpose of catalyst development and comparison of thermal stability between catalyst formulations, oven aging is commonly used by catalyst manufacturers. Oven aging of diesel formulation is typically performed over a number hours at elevated temperature—typically between 600 and 900°C—in a controlled atmosphere containing oxygen and usually about 10% water vapor. As an example, oven aging at 650°C (20 ppm SO_2, 10% H_2O, 10% O_2, balance N_2) and at 800°C was used in the development of diesel SCR catalysts, with aging periods ranging from 48 to 100 h [53, 56]. Oven aging methods are fast and simple, but in most cases cannot be directly related to the emission durability in real catalyst operation.

19.7.1.3 Sources of Deactivation

In general, there are several sources of catalyst deactivation, as discussed in Chapter 18. In the following sections we discuss the two most important modes: thermal degradation and catalyst poisoning, as applied to diesel catalysts.

19.7.2 Thermal Degradation

Thermal degradation is the most common mode of the loss of activity of SI engine catalysts, which may be exposed to temperatures in excess of 1000°C. In the diesel engine, both the exhaust temperature and the temperature rise (exotherm) in the catalyst due to catalytic reactions are lower than those in gasoline engines. The peak temperatures in the diesel catalyst rarely exceed 600°C. Because most of the catalytic materials that were used in early diesel catalysts had been developed for high-temperature gasoline applications, thermal degradation of diesel catalysts has not been considered a major problem in the past. This notion has changed in view of the following considerations:

- New diesel-specific materials have been introduced to catalyst formulations, which may require thermal stabilization.
- Catalysts used in low-temperature applications have extremely demanding light-off temperature requirements and cannot tolerate thermal deterioration.

Examples of materials introduced more recently for diesel catalysts include various zeolites or titania, which are used in both diesel oxidation catalysts and NO_x reduction catalysts such as SCR. It is well known that the anatase form of titanium dioxide used in catalysts is thermally stable only up to about 500°C. In the case of titania-based SCR catalysts, this temperature limit can be increased by stabilizing titania using tungsten trioxide (WO_3), as discussed in Section 19.3.

Another source of demand for good thermal durability of diesel catalysts comes from low-temperature passenger car applications in the European Union. This application became even more challenging after the year 2000, when the initial warm-up period was eliminated from the ECE+EUDC test cycle used for emission measurement. This test cycle is characterized by low engine loads, resulting in very low exhaust temperatures over a long portion of the cycle, especially in the case of large passenger car engines of more than four cylinders. The catalyst must provide reductions of gaseous emissions, including carbon monoxide, with required CO light-off temperatures (t_{50}) as low as 150°C. An important source of light-off temperature deterioration (increase) in catalysts is loss of active sites due to sintering. Catalysts that have to meet challenging light-off temperature demands, therefore, must be designed for excellent thermal stability.

19.7.3 Poisoning

19.7.3.1 Common Diesel Catalyst Poisons

A big challenge in the development of durable diesel catalysts is poisoning. There are two major sources of catalyst poisons in diesel exhaust: (1) lube oil additives, including phosphorus (P), zinc (Zn), calcium (Ca), and sulfur (S); and (2) additional quantities of sulfur from the diesel fuel.

Because diesel engines burn larger quantities of oil than their gasoline counterparts, the diesel catalyst must be more resistant to the oil and its additives. Under the cooler modes of operation, the unburnt oils and their additives deposit within the catalyst washcoat. Unlike the organic portion of the oil, the additives remain after the oil is catalytically oxidized. Such substances as phosphorus, zinc, and calcium oxide accumulate on the surface or within the catalyst. These oil-derived poisons typically result in irreversible catalyst deactivation.

Traditionally, diesel fuels contained significant quantities of sulfur, several times higher than those present in gasoline. Those high-sulfur fuels may be another source of catalyst deactivation. Sulfur poisoning is frequently reversible by high temperatures, under which the sulfur compounds decompose and are released from the catalyst washcoat. However, due to the low diesel exhaust temperatures, in many diesel engine applications the conditions needed for catalyst regeneration may never be reached.

According to a study conducted by Volkswagen/Degussa [44], sulfur and the lube oil additives (phosphorus) can produce cumulative poisoning effects of comparable magnitudes. The loss of catalyst activity can be represented by its loss of carbon monoxide conversion efficiency (Table 19.9). Among three engine-aged samples of the same catalyst, the highest loss of activity was observed when both high sulfur fuel and high phosphorous lubrication oil were used. Aging with low S fuel

TABLE 19.9
INFLUENCE OF POISONS ON CATALYST ACTIVITY

Catalyst Sample	S in Fuel, %	P in Oil, %	CO Conversion, %
Fresh	0.15	0.1	94
Aged 32 h	0.15	0.1	80
Aged 32 h	0.04	0.1	89
Aged 32 h	0.15	0.05	89

Source: Data after Ref. 44.

(0.04% S) and high P oil induced the same loss of activity as aging with low P oil (0.05% P) and high S fuel.

Certain poisons can be also derived from vehicle components. An important example is silicon (Si), contained in sealants, washers, and so forth, which can enter the combustion chamber and/or the exhaust stream being carried by the fuel or through outgassing. Si was reported to be an effective poison of gasoline catalysts [140]. It can be reasonably expected that diesel catalysts, if exposed, would also be deactivated by Si. This kind of poisoning is also known in engines operated with certain types of biogas fuels that contain silicon compounds.

Contamination with silicon is apparent as crystalline deposits of SiO_2, which in extreme cases may plug the inlet face of the monolith. Silicon deactivation was reported to be reversible through washing in hydrofluoric acid and ultrasonic treatment in soapy water [1].

19.7.3.2 Distribution of Poisons

Diesel catalyst makers developed special test methods and conducted several studies to determine how different poisons are deposited within the catalyst. Knowledge of the poison locations and their interactions with the active catalytic sites allows the catalyst manufacturer to design more poison tolerant catalysts.

Figure 19.50 shows microprobe traces for S and P, which were scanned across a 75-μm-thick washcoat fillet [48]. Sulfur was found to accumulate throughout the entire depth of the washcoat, while the phosphorous compounds were concentrated in the outer 15 to 20 μm. The washcoat interface with the monolith is at about 75 μm, as is evident by the drop in the S profile to zero. Zn- and Ca-containing compounds, not shown in Figure 19.50, were concentrated practically entirely at the washcoat surface, within the outer 10- to 15-μm layer.

The distribution of poisons across the washcoat thickness, as well as along the catalyst length, was also investigated in the previously mentioned work by Volkswagen/Degussa [44]. Adsorption of sulfur, carbon, phosphorus, and zinc were measured on a Pt-based diesel catalyst. The findings, which are in general agreement with the data from Figure 19.50, can be summarized as follows:

- Sulfur compounds, which included sulfites, sulfates, and thiosulfates, were uniformly distributed over the catalyst length and the washcoat depth. The amount of S in the catalyst depended both on the washcoat type and on the noble metal loading.
- Carbon formed a separate layer at the washcoat surface. The amount of adsorbed carbon decreased at higher noble metal loading but was independent of the washcoat technology.
- Phosphorus was selectively adsorbed at the inlet side of the catalyst and in the outer layer of the washcoat. The adsorption was independent of the washcoat technology or of the platinum loading.
- Zinc was found only within the carbon layer and not within the washcoat.

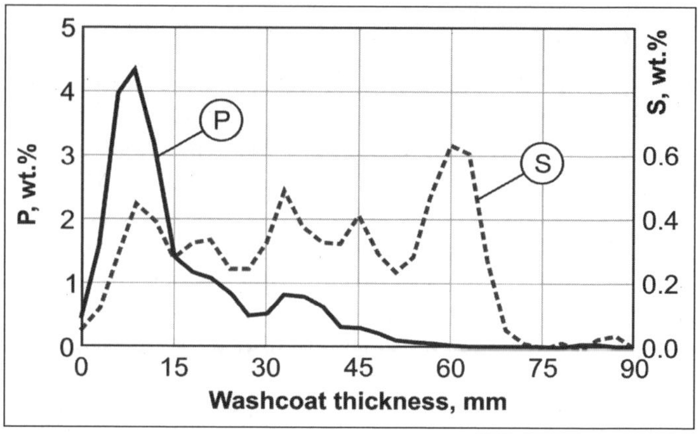

Figure 19.50 Phosphorus and sulfur distribution in the washcoat of an aged diesel catalyst. (Based on data from Ref. 48.)

Considering the distribution of oil-derived poisons in the catalyst, poison-tolerant catalysts can be designed by locating the active catalytic species in protected depths within the washcoat. The washcoat must also be designed to be chemically inert to the poisons. Furthermore, the washcoat pore structure should be designed to tolerate large amounts of the fouling materials to prevent pore blockage and the subsequent pore diffusion limitations that result.

The main catalyst can be protected (at least to some degree) from poisons that adsorb selectively at the inlet side by an upstream guard catalyst. Guard catalysts, commonly known in the chemical processing industry and also used in stationary engine applications, are "blank," nonprecious metal-containing catalysts designed to provide adsorption capacity for poisons. Replacement of a poison-saturated guard catalyst is much more cost effective than replacing a poisoned precious metal catalyst. Guard catalysts can be coated with alumina washcoat, or else with coatings optimized for the adsorption of particular poisons.

19.7.3.3 Restoring Catalyst Activity

In some cases the activity of a poisoned catalyst can be restored, at least in part. As mentioned before, certain poisons (e.g., sulfur) may be driven off the catalyst washcoat through exposure to high temperatures, either on the engine or in an oven. Many cases of catalyst fouling or masking can be remedied by cleaning of the catalyst. Catalyst cleaning, not usual with small catalysts in mobile engine applications, is a commonplace procedure in stationary applications using large and expensive catalysts [141].

There are three groups of catalyst cleaning methods:

- Air cleaning
- Washing (in water or solvents)
- Chemical washing.

Fouling and ash buildup can sometimes be cleared using either compressed air or a vacuum. The preferred method in the case of ash accumulation is to simply vacuum the catalyst face. Washing using water or other solvents can restore catalyst activity by removing poisons, masking, fouling, or inhibiting agents, provided such agents are soluble in the solvent. The most simple washing procedure is to rinse the catalyst using water or, better, deionized water. More effective cleaning procedures use various chemicals to remove poisons. An alkaline (high-pH) wash may be very effective in removing organic resins, residues, and some deposits resulting from lube oil contamination. For many inorganic substances, an acidic (low pH) wash will be more effective. Solutions of EDTA (ethylenediaminetetraacetic acid), a commonly known chelating agent, are sometimes used as acidic catalyst wash. Washing should be performed in consultation with the catalyst manufacturer, because chemicals may cause damage to the catalyst or its substrate.

References

1. Eastwood, P., *Critical Topics in Exhaust Gas Aftertreatment,* Research Studies Press, Baldock, Hertfordshire, England, 2000.
2. Mogan, J.P., *et al.,* 1987. "Diesel Emission Control Catalyst: Friend or Foe?," U.S. Bureau of Mines Information Circular IC 9141, Minneapolis, MN, 1987, p. 140.
3. "Practical Ways to Reduce Exposure to Diesel Exhaust in Mining—A Toolbox," U.S. Department of Labor, Mine Safety and Health Administration, 1998; retrieved from www.msha.gov/S&HINFO/TOOLBOX/Tbcover.htm.
4. Ritscher, J.S., and M.R. Sandner, "Three-Way Catalytic Process for Gaseous Streams," U.S. Patent 4,297,328, 1981.
5. Iwamoto, M., *et al.,* "Removal of Nitrogen Monoxide through a Novel Catalytic Process," *Journal of Physical Chemistry* **95,** 1991, pp. 3727–3730.
6. Held, W., *et al.,* "Catalytic NOx Reduction in Net Oxidizing Exhaust Gas," SAE Paper No. 900496, SAE International, Warrendale, PA, 1990.
7. Havenith, C., and R.P. Verbeek, "Transient Performance of a Urea DeNOx Catalyst for Low Emissions Heavy-Duty Diesel Engines," SAE Paper No. 970185, SAE International, Warrendale, PA, 1997.
8. Fritz, N., *et al.,* "On-Road Demonstration of NO_x Emission Control for Diesel Trucks with SiNOx Urea SCR System," SAE Paper No. 1999-01-0111, SAE International, Warrendale, PA, 1999.
9. Miller W.R., *et al.,* "The Development of Urea-SCR Technology for U.S. Heavy Duty Trucks," SAE Paper No. 2000-01-0190, SAE International, Warrendale, PA, 2000.
10. Kharas, K.C., O.H. Bailey, and J. Vuichard, "Improvements in Intimately Coupled Diesel Hydrocarbon Adsorber/Lean NO_x Catalysis Leading to Durable Euro 3 Performance," SAE Paper No. 982603, SAE International, Warrendale, PA, 1998.
11. Bailey O.H., D. Dou, and M. Molinier, "Sulfur Traps for NOx Adsorbers: Materials Development and Maintenance Strategies for Their Application," SAE Paper No. 2000-01-1205, SAE International, Warrendale, PA, 2000.
12. Krutzsch, B., *et al.,* "Reduction of NOx in Lean Exhaust by Selective NO_x-Recirculation (SNR-Technique). Part I: System and Decomposition Process," SAE Paper No. 982592, SAE International, Warrendale, PA, 1998.
13. Ambs, J.L., and B.T. McClure, "The Influence of Oxidation Catalysts on NO_2 in Diesel Exhaust," SAE Paper No. 932494, , SAE International, Warrendale, PA, 1993.
14. Harayama, N., *et al.,* "Effects of Sulfate Adsorption on Performance of Diesel Oxidation Catalysts," SAE Paper No. 920852, SAE International, Warrendale, PA, 1992.
15. Kawanami, M., M. Horiuchi, H. Klein, and M. Jenkins, "Development of Oxidation and de-NOx Catalyst for High Temperature Exhaust Diesel Trucks," SAE Paper No. 981196, SAE International, Warrendale, PA, 1998.
16. Majewski, W.A., J.L. Ambs, and K. Bickel, "Nitrogen Oxides Reactions in Diesel Oxidation Catalyst," SAE Paper No. 950374, , SAE International, Warrendale, PA, 1995.
17. Zelenka, P., *et al.,* "Diesel Oxidation Catalyst Strategies with Special Emphasis on Odour Reduction," SAE Paper No. 942066, SAE International, Warrendale, PA, 1994.
18. Hamm, E., G. Hohenberg, U. Standt, and P. Zelenka, "The Effect of Fuel Specifications and Different Aftertreatment Systems on Exhaust Gas Odour and NonRegulated Emissions at Steady-State and Dynamic Operation of DI-Diesel Engines," SAE Paper No. 1999-01-3559, SAE International, Warrendale, PA, 1999.
19. LeTavac, C., J. Uihlein, and K. Vertin, "Year-Long Evaluation of Trucks and Buses Equipped with Passive Diesel Particle Filters," SAE Paper No. 2002-01-0433, SAE International, Warrendale, PA, 2002.
20. Ayala, A., N. Kado, and R. Okamoto, "ARB Study of Emissions from Late-Model Diesel and CNG Heavy-Duty Transit Buses," California Air Resources Board, 2001.
21. Layrer, J., *et al.,* "Oxidative Diesel Control Catalyst," U.S. Patent 5,371,056 (Degussa), 1994.
22. Cooper, B.J., and J.E. Thoss, "Role of NO in Diesel Particulate Emission Control," SAE Paper No. 890404, SAE International, Warrendale, PA, 1989.
23. Lylykangas, R., and T. Maunula, "Particle Oxidation Catalyst for Heavy-Duty Diesel Engines," *AutoTechnology,* May 2002, pp. 57–59.

24. Brueck, R., P. Hirth, M. Reizig, P. Treiber, and J. Breuer, "Metal Supported Flow-Through Particulate Trap; A NonBlocking Solution," SAE Paper No. 2001-01-1950, SAE International, Warrendale, PA, 2001.
25. Saroglia, G., G. Basso, M. Presti, M. Reizig, and H. Stock, "Application of New Diesel Aftertreatment Strategies on a Production 1.9 L Common Rail Turbocharged Engine," SAE Paper No. 2002-01-1313, SAE International, Warrendale, PA, 2002.
26. Mogi, H., K. Tajima, M. Hosoya, and M. Shimoda, "The Reduction of Diesel Engine Emissions by Using the Oxidation Catalysts of Japan Diesel 13 Mode Cycle," SAE Paper No. 1999-01-0471, SAE International, Warrendale, PA, 1999.
27. Johnson, J.E., and D.B. Kittelson, "Physical Factors Affecting Hydrocarbon Oxidation in a Diesel Oxidation Catalyst," SAE Paper No. 941771, SAE International, Warrendale, PA, 1994.
28. Voss, K., *et al.,* "Zirconia Based Ceramic, In-Cylinder Coatings and Aftertreatment Oxidation Catalysts for Reduction of Emissions from Heavy Duty Diesel Engines," SAE Paper No. 970469, SAE International, Warrendale, PA, 1997.
29. Farrauto, R.J., *et al.,* "Ceria-Alumina Oxidation Catalyst," U.S. Patent 5627124 (Engelhard), 1997.
30. Standt, U., and A. Koenig, "Performance of Zeolite-Based Diesel Catalysts," SAE Paper No. 950749, SAE International, Warrendale, PA, 1995.
31. Adams, K.M., *et al.,* "Laboratory Screening of Diesel Oxidation Catalysts and Validation with Vehicle Testing: The Importance of Hydrocarbon Storage," SAE Paper No. 962049, SAE International, Warrendale, PA, 1996.
32. Sawyer, J.E., *et al.,* "Treatment of Diesel Exhaust Using Zeolite Catalyst," U.S. Patent 5,849,255 (ASEC), 1998.
33. Yavuz, B.O., *et al.,* "Zeolite-Containing Oxidation Catalyst and Method of Use," U.S. Patent 6,274,107 (Engelhard), 2001.
34. "Demonstration of Advanced Emission Control Technologies Enabling Diesel-Powered Heavy-Duty Engines to Achieve Low Emission Levels," Manufacturers of Emission Controls Association, June 1999, retrieved from www.meca.org/SwRIHDE.PDF.
35. Hunter, G., *et al.,* "The Effects of Fuels on Diesel Oxidation Catalyst Performance and the Physical, Chemical, and Biological Character of Diesel Particulate Emissions," *SAE Trans.* **90,** 3679.3700, SAE International, Warrendale, PA, 1990.
36. McClure, B.T., *et al.,* "The Influence of an Oxidation Catalytic Converter and Fuel Composition on the Chemical and Biological Characteristics of Diesel Exhaust Emissions," SAE Paper No. 920854, SAE International, Warrendale, PA, 1992.
37. Rasmussen, R.E., "Effect of Fuel Properties on Mutagenic Activity in Extracts of Heavy Duty Diesel Exhaust Particulate," *Journal of Air Waste Management Association* **40,** 1990, p. 1391.
38. Smedler, G., *et al.,* "Diesel Catalysts for Europe beyond 1996," *Automotive Engineering,* October 1995, p. 41.
39. Henk, M.G., W.B. Williamson, and R.G. Silver, "Diesel Catalysts for Low Particulate and Low Sulfate Emissions," SAE Paper No. 920368, SAE International, Warrendale, PA, 1992.
40. Wyatt, M., *et al.,* "The Design of Flow-Through Diesel Oxidation Catalysts," SAE Paper No. 930130, SAE International, Warrendale, PA, 1993.
41. "Corning Diesel Products," Corning Incorporated, 1997.
42. Horiuchi, M., *et al.,* "Exhaust Gas Purification Catalyst," U.S. Patent 5,000,929, 1991.
43. Domesle, R., *et al.,* "Catalyst for the Purification of Exhaust Gases of Diesel Engines and Method of Use," U.S. Patent 5,157,007, 1992.
44. Beckmann, R., *et al.,* "A New Generation of Diesel Oxidation Catalysts," SAE Paper No. 922330, SAE International, Warrendale, PA, 1992.
45. Cobb, D., *et al.,* "Application of Selective Catalytic Reduction (SCR) Technology for NOx Reduction From Refinery Combustion Sources," *Environmental Progress* **10,** 1991, p. 49.
46. Cho, S.M., "Properly Apply Selective Catalytic Reduction for NOx Removal," *Chem. Eng. Prog.,* January 1994, pp. 39–45.
47. Lepperhoff, G., and J. Schommers, "Verhalten von SCR-Katalysatoren im dieselmotorischen Abgas," *MTZ* **49,** 1988, p. 1.
48. Heck, R.M., and R.J. Farrauto, *Catalytic Air Pollution Control: Commercial Technology,* Van Nostrand Reinhold, New York, 1995.
49. Hoffman, J., "Process for the Selective Catalytic Reduction of Nitrogen Oxides," International Patent Application, WO 96/06674 (Nalco), 1996.
50. Eberhard, J., and J. Kreutmair, "Verfahren und Vorrichtung zur selectiven katalytischen Reduktion von NOx in sauerstoffhaltigen Gasen," European Patent, EP 0 615 777 A1, 1994.
51. Fable, S., F. Kamakaté, and S. Venkatesh, "Selective Catalytic Reduction Urea Infrastructure Study," Report NREL/SR-540-32689, National Renewable Energy Laboratory, Golden, CO, July 2002, retrieved from www.ott.doe.gov/pdfs/adl_urea.pdf.
52. Hug, H.T., *et al.,* "Off-Highway Exhaust Gas After-Treatment: Combining Urea-SCR, Oxidation Catalysis and Traps," SAE Paper No. 930363, SAE International, Warrendale, PA, 1993.
53. Gieshoff, J., *et al.,* "Improved SCR Systems for Heavy Duty Applications," SAE Paper No. 2000-01-0189, SAE International, Warrendale, PA, 2000.
54. Heck, R.M., *et al.,* "Operating Characteristics and Commercial Operating Experience with High Temperature SCR NOx Catalyst," *Environmental Progress* **13,** 4, 1994, pp. 221–225.
55. Janssen, J.J., "Environmental Catalysis—Stationary Sources," in *Handbook of Heterogeneous Catalysis,*

G. Ertl *et al.* (Eds.), John Wiley, New York, 1997, pp. 1636–1644.
56. Gieshoff, J., *et al.,* "Advanced Urea SCR Catalysts for Automotive Applications," SAE Paper No. 2001-01-0514, SAE International, Warrendale, PA, 2001.
57. Spurk, P.C., M. Pfeifer, J. Gieshoff, and E. Lox, "Ein SCR-Katalysator auch fuer den Einsatz im Pkw," 10. Aachener Kolloquium Fahrzeug- und Motorentechnik, 2001.
58. Chandler G.R., *et al.,* "An Integrated SCR and Continuously Regenerating Trap System to Meet Future NOx and PM Legislation," SAE Paper No. 2000-01-0188, SAE International, Warrendale, PA, 2000.
59. Koebel, M., M. Elsener, and G. Madia, "Recent Advances in the Development of Urea-SCR for Automotive Applications," SAE Paper No. 2001-01-3625, SAE International, Warrendale, PA, 2001.
60. Gekas, I., *et al.,* "Urea-SCR Catalyst System Selection for Fuel and PM Optimised Engines and a Demonstration of a Novel Urea Injection System," SAE Paper No. 2002-01-0289, SAE International, Warrendale, PA, 2002.
61. Song, Q., and G. Zhu, "Model-Based Closed-Loop Control of Urea SCR Exhaust Aftertreatment System for Diesel Engine," SAE Paper No. 2002-01-0287, SAE International, Warrendale, PA, 2002.
62. Gibson, J., and O. Groene, "Selective Catalytic Reduction on Marine Diesel Engines," *Automotive Engineering,* October 1991, pp. 18–22.
63. Weaver, C.S., and D.B. McGreger, "Emission Control Technology for Locomotive Engines," SAE Paper No. 940453, SAE International, Warrendale, PA, 1994.
64. Havenith, C., *et al.,* "Development of a Urea DeNOx Catalyst Concept for European Ultra-Low Emission Heavy-Duty Diesel Engines," SAE Paper No. 952652, SAE International, Warrendale, PA, 1995.
65. Lueders, H., *et al.,* "An Urea Lean NOx Catalyst System for Light Duty Diesel Vehicles," SAE Paper No. 952493, SAE International, Warrendale, PA, 1995.
66. van Helden, R., *et al.,* "Engine Dynamometer and Vehicle Performance of a Urea SCR-System for Heavy-Duty Truck Engines," SAE Paper No. 2002-01-0286, SAE International, Warrendale, PA, 2002.
67. Mathes, W., F. Witzel, and S. Schnapp, "Exhaust Gas Control System for Diesel Engine Exhaust Gases," International Patent Application, WO 99/05402, 1999.
68. Berner, G., *et al.,* "Static Mixer," International Patent Application, WO 93/00990, 1993.
69. Baker R.E., and B. Bates, "Automotive Vehicle Fueling System," U.S. Patent 6,032,703 (Ford), March 7, 2000.
70. Levin, M., and R. Baker, "Co-Fueling of Urea for Diesel Cars and Trucks," SAE Paper No. 2002-01-0290, SAE International, Warrendale, PA, 2002.
71. Shelef, M., "Selective Catalytic Reduction of NOx with N-Free Reductants," *Chem. Rev.* **95,** 1995, pp. 209–225.
72. Peters, A., *et al.,* "Catalytic NOx Reduction on a Passenger Car Diesel Common Rail Engine," SAE Paper No. 980191, SAE International, Warrendale, PA, 1998.
73. Kharas K.C., M.J. Miller, and J.Y. Yan, "Analyses of System Factors Affecting Performance in Lean NOx Catalysis. 2. The Deleterious Role of Parasitic Homogeneous Hydrocarbon Oxidation on the Performance of High Temperature Lean NOx Catalysts," SAE Paper No. 982604, SAE International, Warrendale, PA, 1998.
74. Truex, T.J., "Lean NOx Catalysts," Paper presented at SAE Catalysts and Emission Control TOPTEC, Dearborn, MI, September 1994.
75. Engler, B.H., *et al.,* "Catalytic Reduction of NOx with Hydrocarbons under Lean Diesel Exhaust Gas Conditions," SAE Paper No. 930735, SAE International, Warrendale, PA, 1993.
76. Kharas, K.C., D.J. Liu, and H.J. Robota, "Structure-Function Properties in Cu-ZSM-5 NO Decomposition and NO SCR Catalysts," *Catalysis Today* **26,** 1995, pp. 129–145.
77. Kharas, K.C., H.J. Robota, and A. Datye, "Deactivation of Cu-ZSM-5: Probable Irrelevance of Modest Zeolite Degradation Compared to Cupric Sintering to CuO," Chap. 4 in *ACS Symposium Series No. 552: Environmental Catalysis,* John N. Armor (Ed.), American Chemical Society, 1994.
78. Kharas, K.C., "Copper-Silver Zeolite Catalysts," U.S. Patent 5,908,806 (ASEC), 1999.
79. Masel, R.I., "An Experimental Test of Various Models of the Active Site for Nitric Oxide Reduction on Platinum," *Catal. Rev. Sci. Eng.* **28,** 2–3, 1986, pp. 335–369.
80. Burch, R., P.J. Millington, and A.P. Walker, "Mechanism of the Selective Reduction of Nitrogen Monoxide on Platinum-Based Catalyst in the Presence of Excess Oxygen," *Applied Catalysis B: Environmental* **4,** 1994, pp. 65–94.
81. Hoost, T.E., R.J. Kudla, and M.S. Chattha, "Sulphur Resistant Lean-NOx Catalyst for Treating Diesel Emissions," European Patent Application, EP 0 864 354 A1 (Ford), 1998.
82. Kharas, K.C., H. Robota, and M.G. Henk, "Reduction of Nitrogen Oxides in the Exhaust Gases from Internal Combustion Engines," International Patent Application, WO 00/00272 (ASEC), 2000.
83. Arakawa, K., S. Matsuda, and H. Kinoshita, "Progress in Sulfur Poisoning Resistance of Lean NOx Catalysts," SAE Paper No. 980930, SAE International, Warrendale, PA, 1998.
84. Leyrer, J., *et al.,* "NOx Vermindering im mageren Abgas von Kraftfahrzeugmotoren," European Patent, EP 0 559 021 A2 (Degussa), 1993.
85. Van den Tillaart, H., *et al.,* "Katalysator zur Reduktion von Stickoxiden in oxidierender und reduzierender Atmosphaere," European Patent Application, EP 0 968 754 A1 (Degussa), 2000.
86. Hartick, J., "A Compact Temperature Control System for Lean Exhaust Gas to Improve Conversion of NOx Aftertreatment," SAE Paper No. 1999-01-3497, SAE International, Warrendale, PA, 1999.

87. Klein, H., S. Lopp, E. Lox, M. Kawanami, and M. Horiuchi, "Hydrocarbon DeNOx Catalysis—System Development for Diesel Passenger Cars and Trucks," SAE Paper No. 1999-01-0109, SAE International, Warrendale, PA, 1999.
88. Deeba, M., *et al.*, "Catalytic Abatement of NOx from Diesel Engines: Development of Four Way Catalyst," SAE Paper No. 952491, SAE International, Warrendale, PA, 1995.
89. Leyrer, J., *et al.*, "Advanced Studies on Diesel Aftertreatment Catalysts for Passenger Cars," SAE Paper No. 960133, SAE International, Warrendale, PA, 1996.
90. Klein, H., *et al.*, "Katalysator fuer die Reinigung der Abgase eines Dieselmotors," European Patent Application, EP 1004 347 A2 (Degussa-Huels), 2000.
91. Kawanami, M., *et al.*, "Catalytic NOx Reduction for On-Highway Diesels," *Truck Engineering*, November 1995, p. 3.
92. Heimrich, M.J., "Demonstration of Lean NOx Catalytic Converter Technology on a Heavy-Duty Diesel Engine," SAE Paper No. 970755, SAE International, Warrendale, PA, 1997.
93. Page, D.L., *et al.*, "The QuadCAT Four-Way Catalytic Converter: An Integrated Aftertreatment System for Diesel Engines," SAE Paper No. 1999-01-2924, SAE International, Warrendale, PA, 1999.
94. Dou, D., *et al.*, "A Systematic Investigation of Parameters Affecting Diesel NOx Adsorber Catalyst Performance," Paper presented at 8th Diesel Emissions Reduction Conference (DEER), San Diego, CA, August 2002, retrieved from www.orau.gov/deer2002/Session11/dou.pdf.
95. Klein, H., *et al.*, "NOx- Nachbehandlung fuer Diesel PKW geloest? Entschwefelung von NOx-Speicher-Katalyzatoren," 22nd Internationales Wiener Motorsymposium, Fortschritt-Berichte VDI (VDI Verlag, Duesseldorf), Reihe 12, Nr. 455, Band 2, 2001, pp. 192–215.
96. Brogan, M., A.D. Clark, and R.J. Brisley, "Recent Progress in NOx Trap Technology," SAE Paper No. 980933, SAE International, Warrendale, PA, 1998.
97. Erkfeld, S., *et al.*, "Sulphur Poisoning and Regeneration of NOx Trap Catalyst for Direct Injected Gasoline Engines," SAE Paper No. 1999-01-3504, SAE International, Warrendale, PA, 1999.
98. Kobayashi, T., *et al.*, "Study of NOx Trap Reaction by Thermodynamic Calculation," SAE Paper No. 970745, SAE International, Warrendale, PA, 1997.
99. Strehlau, W., *et al.*, "Verfahren zum Betrieben einer Abgasreinigungsanlage ehthaltend eine Schwefelfalle und einen Stickoxid-Speicherkatalysator," European Patent Application, EP 0 945 608 A2 (Degussa), 1999.
100. Chaize, E., *et al.*, "Reduction of NOx in Lean Exhaust by Selective NOx-Recirculation (SNR-Technique). Part II: NOx Storage Materials," SAE Paper No. 982593, SAE International, Warrendale, PA, 1998.
101. Hachisuka, I., T. Yoshida, H. Ueno, N. Takahashi, A. Suda, and M. Sugiura, "Improvement of NOx Storage-Reduction Catalyst," SAE Paper No. 2002-01-0732, SAE International, Warrendale, PA, 2002.
102. Hepburn, J.S., and W. Watkins, "Sulphur Resistant Lean-Burn NOx Catalyst for Treating Diesel Emissions," European Patent Application, EP 0 857 510 A1 (Ford), 1998.
103. Dou, D., and J. Balland, "Impact of Alkali Metals on the Performance and Mechanical Properties of NOx Adsorber Catalysts," SAE Paper No. 2002-01-0734, SAE International, Warrendale, PA, 2002.
104. Cutler, W.A., and J.P. Day, "Mechanical Durability of Cordierite-Based NOx Adsorber/Catalyst Systems for Lean Burn Gasoline Applications," SAE Paper No. 1999-01-3500, SAE International, Warrendale, PA, 1999.
105. Iwachido, K., H. Tanada, T. Watanabe, N. Yamada, O. Nakayama, *et al.*, "Development of the NOx Adsorber Catalyst for Use with High-Temperature Conditions," SAE Paper No. 2001-01-1298, SAE International, Warrendale, PA, 2001.
106. Bergmann, A., R. Brück, S. Brandt, and M. Deeba, "Design Criteria of Catalyst Substrates for NOx Adsorber Function," SAE Paper No. 2000-01-0504, 2000, retrieved from www.emitec.com/pdf/sae2000-01-0504-gb.pdf.
107. McDonald, J., "EPA's Perspective on the 2007 Heavy-Duty Truck Diesel Engine Emissions Compliance," Paper presented at the 8th Diesel Emissions Reduction Conference (DEER), San Diego, CA, August 2002, retrieved from www.orau.gov/deer2002/Session3/Joseph%20McDonald%202002%20DEER%20Conference.pdf.
108. Hachsuka, I., *et al.*, "Deactivation Mechanism of NOx Storage-Reduction Catalyst and Improvement of Its Performance," SAE Paper No. 2000-01-1196, SAE International, Warrendale, PA, 2000.
109. Asik, J.R., G.M. Meyer, and D. Dobson, "Lean NOx Trap Desulfation through Rapid Air Fuel Modulation," SAE Paper No. 2000-01-1200, SAE International, Warrendale, PA, 2000.
110. Pott, E., "Verfahren und Vorrichtung zur Regeneration einer Schwefelfalle," European Patent Application, EP 0 891 806 A2 (Volkswagen), 1999.
111. Brogan, M., "Catalyst Developments for Direct Injection Gasoline Engines," Paper presented at SAE Advancements in Automotive Catalyst Technology TOPTEC, Dearborn, MI, 1997.
112. "Comments of Ford Motor Company on the Diesel Fuel Quality ANPRM," U.S. EPA Air Docket A-99-06-II-D-68, Ford Motor Company, 1999.
113. Dou, D., and O.H. Bailey, "Investigation of NOx Adsorber Catalyst Deactivation," SAE Paper No. 982594, SAE International, Warrendale, PA, 1998.
114. Strehlau, W., *et al.*, "Engine and the Environment," in *Proceedings AVL Conference*, Graz, Austria, 1997, pp. 15–30.
115. Theis, J.R., J.J. Li, R.G. Hurley, and J.A. Ura, "The Desulfation Characteristics of Lean NOx Traps," SAE Paper

No. 2002-01-0733, SAE International, Warrendale, PA, 2002.
116. Pott, E., and G. Splisteser, "Verfahren and Vorrichtung zur de-sulfatierung eines NOx-Speicherkatalysators," International Patent Application, WO 00/23702 (Volkswagen), 1999.
117. "Phase II Summary Report: NOx Adsorber Catalysts," Diesel Emissions Control—Sulfur Effects Program, U.S. Department of Energy, Washington, DC, October 2000, retrieved from www.ott.doe.gov/decse/pdfs/decse2final.pdf.
118. Tomazic, D., *et al.,* "Development of a Desulfurization Strategy for a NOx Adsorber Catalyst System," SAE Paper No. 2001-01-0510, SAE International, Warrendale, PA, 2001.
119. Nakatsuji, T., *et al.,* "Highly Durable NOx Reduction Systems and Catalysts for NOx Storage Reduction System," SAE Paper No. 980932, SAE International, Warrendale, PA, 1998.
120. Hiromi, T., "Exhaust Emission Control Device for Internal Combustion Engine," Japanese Patent, JP11159322 (Mitsubishi), 1999.
121. Kim, G., *et al.,* "Catalysts for Controlling Auto Exhaust Emissions Including Hydrocarbon, Carbon Monoxide, Nitrogen Oxides and Hydrogen Sulfide and Method of Making the Catalyst," U.S. Patent 4,780,447 (W.R. Grace), October 25, 1988.
122. West, B.H., and C.S. Sluder, "NOx Adsorber Performance in a Light-Duty Diesel Vehicle," SAE Paper No. 2000-01-2912, SAE International, Warrendale, PA, 2000.
123. Bromberg, L., D.R. Cohn, J. Heywood, and A. Rabinovich, "Onboard Plasmatron Generation of Hydrogen Rich Gas for Diesel Aftertreatment and Other Applications," Paper presented at 8th Diesel Emissions Reduction Conference (DEER), San Diego, CA, August 2002, retrieved from www.orau.gov/deer2002/Session10/bromberg.pdf.
124. "Diesel Emission Control Sulfur Effects Program, Phase I Interim Data Report No. 2: NOx Adsorber Catalysts," Diesel Emissions Control—Sulfur Effects Program, U.S. Department of Energy, Washington, DC, October 1999, retrieved from www.ott.doe.gov/decse/pdfs/nox_report.pdf.
125. Salvat, O., P. Marez, and G. Belot, "Passenger Car Serial Application of a Particulate Filter System on a Common Rail Direct Injection Diesel Engine," SAE Paper No. 2000-01-0473, SAE International, Warrendale, PA, 2000.
126. Terazawa, Y., *et al.,* "Exhaust Gas Purifying System for diesel Engine," European Patent Application, EP 0 9907880A2 (Mazda), 2000.
127. Sasaki, S., T. Ito, and S. Iguchi, "Smoke-less Rich Combustion by Low Temperature Oxidation in Diesel Engines," *Aachener Kolloquium Fahrzeug- und Motorentechnik* **9,** 2000.
128. Schenk, C.R., J.F. McDonald, and B.A. Olson, "High-Efficiency NOx and PM Exhaust Emission Control for Heavy-Duty On-Highway Diesel Engines," SAE Paper No. 2001-01-1351, SAE International, Warrendale, PA, 2001.
129. Schenk, C.R., J. McDonald, and C. Laroo, "High-Efficiency NOx and PM Exhaust Emission Control for Heavy-Duty On-Highway Diesel Engines—Part Two," SAE Paper No. 2001-01-3619, SAE International, Warrendale, PA, 2001.
130. Yamaguchi, J., "Toyota Diesel Catalytic Converter," *Automotive Engineering*, February 2001, pp. 81–84.
131. Tanaka, T., "Simultaneous Reduction of PM and NOx—A New After-Treatment System," 22nd Internationales Wiener Motorsymposium, Fortschritt-Berichte VDI (VDI Verlag, Duesseldorf), Reihe 12, Nr. 455, Band 2, 2001, pp. 216–228.
132. Nakatani, K., *et al.,* "Simultaneous PM and NOx Reduction System for Diesel Engines," SAE Paper No. 2002-01-0957, SAE International, Warrendale, PA, 2002.
133. Fujimura, T., S. Matsushita, T. Tanaka, and K. Kojima, "Development Towards Serial Production of a Diesel Passenger Car with Simultaneous Reduction System of NOx and PM for the European Market," in *Proceedings of 23rd International Vienna Motor Symposium,* April 25–26, 2002, VDI Verlag.
134. Lueders, H., P. Stommel, and R. Backes, "Applications for the Regeneration of Diesel Particulate Traps by Combining Different Regeneration Systems," SAE Paper No. 970470, SAE International, Warrendale, PA, 1997.
135. "The Laguna II Turbodiesel Range Is Extended with a New 150bhp 2.2 dCi Version," Press Release, Renault, December 10, 2001.
136. Reizig, M., R. Brueck, R. Konieczny, and P. Treiber, "New Approaches to Catalyst Substrate Application for Diesel Engines," SAE Paper No. 2001-01-0189, , SAE International, Warrendale, PA, 2001, retrieved from www.emitec.com/pdf/paper_2001_01_0189.pdf.
137. Klein, H., *et al.,* "Oxidation Catalyst for Internal Combustion Engines," U.S. Patent 6,077,489 (Degussa), 2000.
138. Graf, U., and P. Zelenka, "Oxidationskatalysatorstrategien fuer Dieselmotoren," in *Proceedings of 15th Internationales Wiener Motorensymposium,* April 28–29, 1994, VDI Verlag.
139. Kisenyi, J.M., *et al.,* "Correlation between Dynamometer- and Vehicle-Aged Catalysts for Automotive Lean-Burn Applications," SAE Paper No. 980934, SAE International, Warrendale, PA, 1998.
140. Gandhi, H.S., *et al.,* "Silicon Contamination of Automotive Catalysts," SAE Paper No. 860565, SAE International, Warrendale, PA, 1986.
141. "Catalyst Cleaning," Miratech Corporation, retrieved November 2002 from www.miratechcorp.com/support/faqs.asp.

Chapter 20

Diesel Particulate Filters

20.1 Introduction

Diesel particulate traps are devices that physically capture diesel particulates to prevent their release to the atmosphere. Some of the diesel filter materials that have been developed show quite impressive filtration efficiencies, frequently in excess of 90%, as well as acceptable mechanical and thermal durability. In fact, diesel traps are the most effective control technology for the reduction of particulate emissions with high efficiencies. More precisely, due to the particle deposition mechanisms utilized in these devices, traps are effective in controlling the solid fraction of diesel particulates, including elemental carbon (soot) and the related black smoke emission. Remember that traps may have limited effectiveness, or be totally ineffective, in controlling the nonsolid fractions of particulate matter (PM), such as the soluble organic fraction (SOF) or sulfate particulates. For this reason, trap systems designed to control the total PM (TPM) emissions are likely to incorporate additional functional components targeting the SOF emission (e.g., oxidation catalysts) and sulfate particulates (e.g., ultra-low-sulfur fuels).

All diesel particulate traps of practical importance are *diesel particulate filters* (DPFs). Several particle deposition mechanisms other than filtration are commonly employed in industrial dust separation equipment. These mechanisms include gravity settling, centrifugal separation, and electrostatic trapping. Gravity settling and centrifugal separation (cyclones) are ineffective in capturing diesel particulates due to their small, submicron size and low density. Diesel aerosol tends to behave in cyclones like a gas rather than a gas/solids mixture. A combination of electrostatic coagulation of fine diesel particles (upstream) with a cyclone (downstream) can theoretically increase the recovery efficiency. The electrostatic traps, however, appear to be impractical for diesel engine application.

Even though the term *trap* covers a wider range of devices, it is often used as a synonym for "diesel filter." Use of the term *diesel trap* in reference to filter devices was more common in older literature. In recent years, the trend has been toward replacing it with the more precise term *diesel particulate filter.*

Due to the low bulk density of diesel particulates, which is typically below 0.1 g/cm^3 (the density depends on the degree of compactness; as an example, a number of 0.056 g/cm^3 was reported by Wade [1]), diesel particulate filters can quickly accumulate considerable volumes of soot. Several liters of soot per day may be collected from a heavy-duty truck or bus engine. The collected particulates would eventually cause excessively high exhaust gas pressure drop in the filter, which would negatively affect engine operation. Therefore, diesel filter systems have to provide a way of removing particulates from the filter to restore its soot collection capacity. This removal of particulates, known as *filter regeneration,* can be performed either continuously, during regular operation of the filter, or periodically, after a predetermined quantity of soot has been accumulated. In either case, the regeneration of filter systems should be "invisible" to the vehicle driver/operator and should be performed without intervention. In most cases, thermal regeneration of diesel filters is employed, in which the collected particulates are removed from the trap by oxidation to gaseous products, primarily to carbon dioxide (CO_2). The thermal regeneration, schematically represented in Figure 20.1, is undoubtedly the cleanest and most attractive method of operating diesel filters.

To ensure that particulates are oxidized at a sufficient rate, the filter must operate at a sufficient temperature and oxidizing gases, such as oxygen or nitrogen dioxide (or its precursors), must be supplied to the filter. In some filter systems, the source of heat (as well as of the oxidizing gases) is the exhaust gas stream itself. In this case, referred to as a *passive filter,* the filter regenerates continuously during regular operation of the engine. Passive filters usually incorporate some form of a catalyst, which lowers the soot oxidation temperature to a level that can be reached by exhaust gases during the operation of the vehicle. Another approach that may be needed to facilitate reliable regeneration involves a number of active strategies for increasing the filter temperature (engine management, electric heaters, etc.). Regeneration by such systems, known as *active filters,* is usually performed periodically, as determined by the

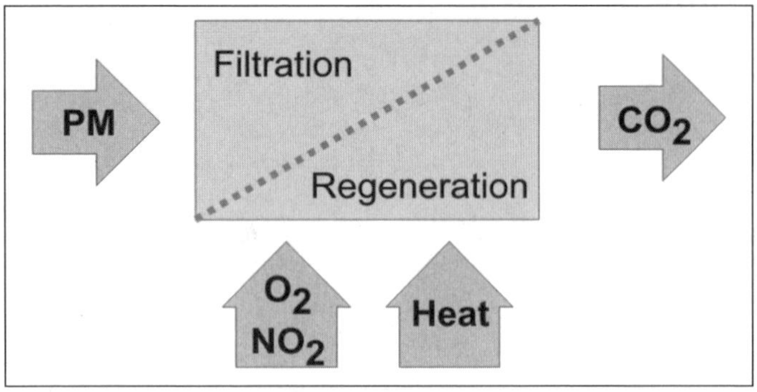

Figure 20.1 *Particulate filter with thermal regeneration.*

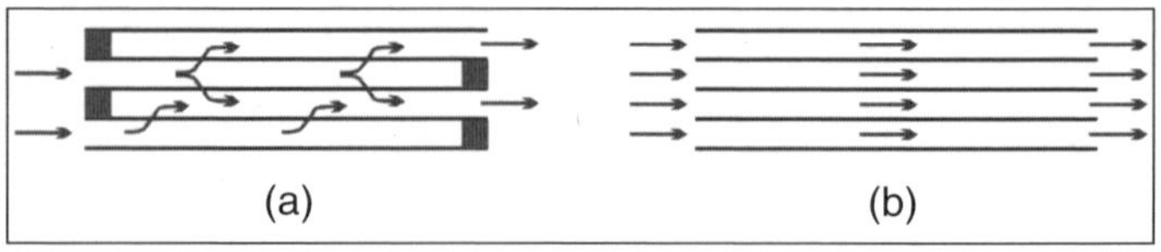

Figure 20.2 *Flow pattern of (a) wall-flow and (b) flow-through substrates.*

vehicle's control system. The classification of particulate filter systems is presented in Chapter 22.

An alternative strategy involves the use of disposable filter cartridges, which are replaced with new units once filled with soot. Particulate traps of this kind are used in some occupational health environments. Such maintenance-intensive filter systems are clearly not acceptable in mobile, highway vehicle applications.

20.2 Principle of Operation

20.2.1 Filter Configuration

The most common design for a diesel particulate filter is the *wall-flow monolith.* This is an extruded, usually cylindrical ceramic structure with many small, parallel channels running in the axial direction. Those familiar with catalyst technologies realize that this description, so far, matches that of the catalytic converter substrate (Chapter 18). Indeed, the design of diesel filter monoliths has been derived from automotive, "flow-through" catalyst substrates. There are, however, two important differences between these structures: (1) The wall-flow monoliths are made of ceramics of higher and more precisely controlled porosity, and (2) adjacent channels in the wall-flow filters are alternatively plugged at each end, thus forcing the gas to flow through the porous walls, which act as a filter medium. The flow pattern difference between the flow-through and the wall-flow substrate is illustrated in Figure 20.2.

The wall-flow filter substrates are easily distinguished from catalyst substrates by the characteristic checkerboard pattern created by the open and plugged cells at their inlet (as well as outlet) face, as illustrated in Figure 20.3.

A number of other, alternative filter designs and materials have been developed. Examples include cartridges made of ceramic fibers, various types of foams, or sintered metal structures. An overview of these technologies is presented in Chapter 21.

Regardless of the type of material or design, the filter media is ultimately packaged into a steel container that is installed in the exhaust system of the vehicle. The DPF can be packaged as a stand-alone unit resembling a catalytic converter but usually of bigger size. In an alternative design, the DPF is placed inside a muffler. However, due to the good noise attenuation properties of many filter media, the filter unit can frequently replace the muffler without any additional provisions for noise attenuation. An example design of a diesel particulate filter is shown in Figure 20.4.

20.2.2 Filtration Mechanisms

The particulate collection by any type of diesel filter is based on separation of the gas-borne particles from the gas stream by deposition on a collecting surface. This separation involves passage of the gas through a porous barrier that retains the particulates. Filters, depending on the type of the barrier, may be divided into (1) deep-bed filters and (2) surface-type filters. In the deep-bed filters, the mean pore size of the filter media is bigger than the mean diameter of collected particles. The particles are deposited on the media through a combination of depth filtration mechanisms driven by various force fields. As an example, the force fields may be related to velocity or

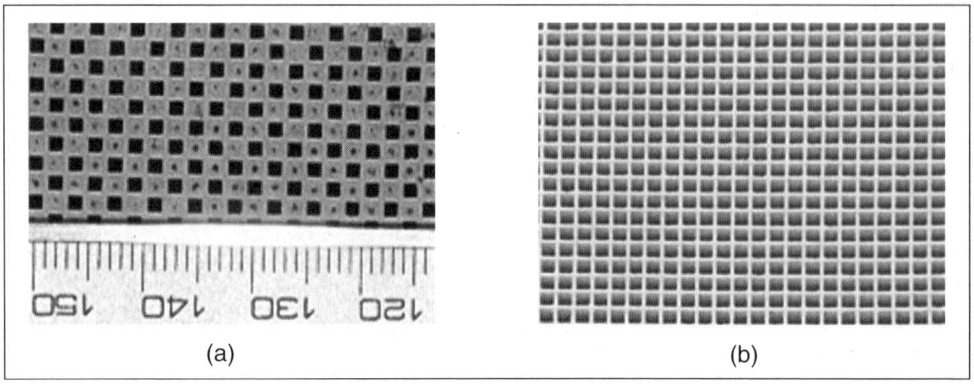

Figure 20.3 *Inlet cell pattern for (a) wall-flow and (b) flow-through substrates.*

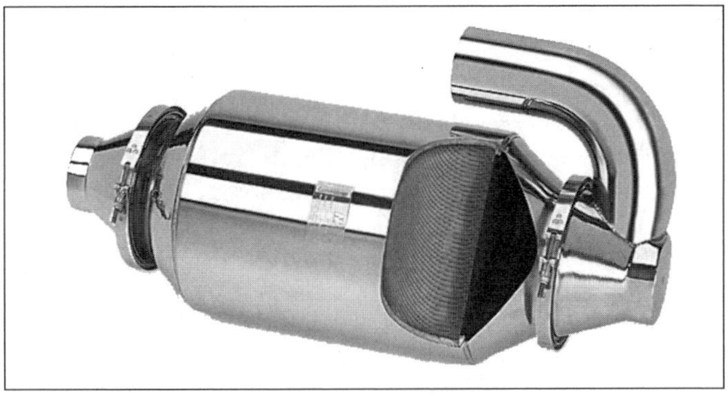

Figure 20.4 Cutout of catalyzed DPF for urban bus and nonroad applications. (Courtesy of Nett Technologies.)

concentration gradients in the gas. In the surface-type filters the pore diameter is less than the particle diameter. The particles are deposited on the media through sieving. These two types of filtration are shown in Figure 20.5. Depth filtration is the main filtration mechanism in most dust collection devices, as well as in the majority of diesel filters. For example, engine intake-air cleaners work predominantly through the depth filtration.

The layer of collected dust, or diesel particulates, is itself the principal filter medium in the surface-type filters. That layer is commonly referred to as *filtration cake* and the process is called *cake filtration.* Cake filtration is very common in liquid filtration. Pure cake filtration is not observed in diesel filters or, generally, in dust collecting filters. However, some diesel filters, such as ceramic wall-flow monoliths, may work through a combination of depth and surface filtration. A filtration cake develops at higher soot loads, when the depth filtration capacity is saturated and a particulate layer starts covering the filtration surface. A visible filtration cake would also eventually develop at the pleated surface of an engine intake-air cleaner (in that case, however, it would indicate that the air cleaner service has been long overdue).

The process of depth filtration relies on three mechanisms of aerosol deposition [2]:

- Diffusional deposition
- Inertial deposition
- Flow-line interception.

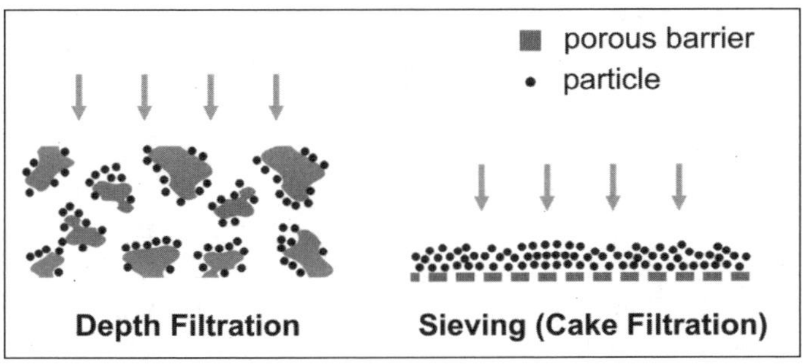

Figure 20.5 *Types of filtration.*

These mechanisms are illustrated in Figures 20.6, 20.7, and 20.8, respectively. The large circle in the middle represents a collecting body in the filter media, say, a fiber in a ceramic fiber filter. Diesel exhaust gas flows around the filter media, as indicated by the streamlines. The small spheres represent diesel exhaust particles traveling with the gas stream.

Diffusional deposition (Figure 20.6) depends on the Brownian movement exhibited by smaller particulates, particularly those below 0.3 μm in diameter. Those particulates do not move uniformly along the gas streamlines. Rather, they diffuse from the gas to the surface of the collecting body and are collected.

Inertial deposition (Figure 20.7), also called inertial interception, becomes more important with increasing particle size (mass). On approaching the collecting body, particles carried along by the gas stream tend to follow the stream but may strike the obstruction because of their inertia.

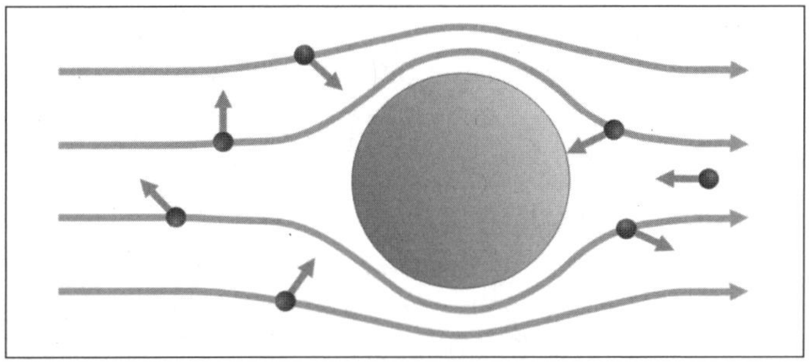

Figure 20.6 *Diffusional deposition.*

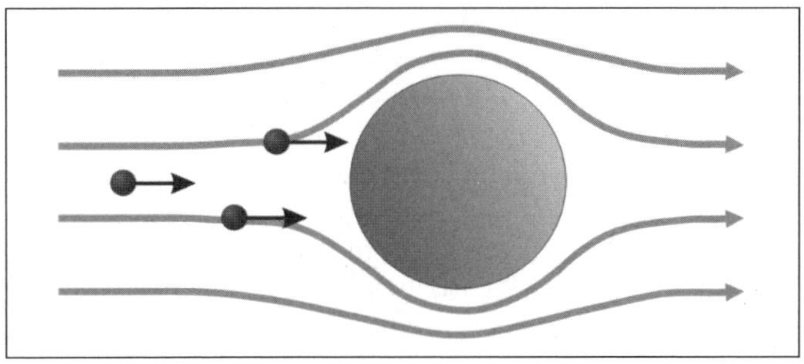

Figure 20.7 *Inertial deposition.*

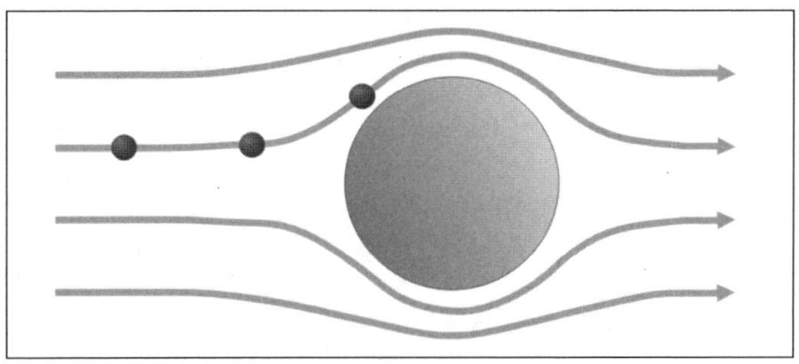

Figure 20.8 *Flow-line interception.*

Flow-line interception may occur when a fluid streamline passes within one particle radius of the collecting body. Then, a particle traveling along the streamline will touch the body and may be collected without the influence of Brownian diffusion or inertia.

Performance of diesel particulate filters may be expressed as the filtration efficiency (E) or penetration (1 – E). The filtration efficiency is the mass ratio of the particulate matter collected on the filter to the particulate matter entering the filter. The penetration is the mass ratio of the particulates escaping the filter to the particulates entering the filter. Both filtration efficiency and penetration may be expressed as percentages.

Depth filtration is characterized by somewhat lower filtration efficiency and lower pressure drop than cake filtration. Particulates that were already deposited on a deep-bed filter may be also reentrained by the gas, causing a decrease in the observed filtration efficiency. This phenomenon is termed *blow-off* of the collected particulate matter. Blow-offs occur at high exhaust gas flow rates or at rapid flow accelerations.

Cake filtration is characterized by higher filtration efficiency and relatively high pressure drop. The pressure drop steadily increases with the increase of filtration cake thickness.

20.2.3 Regeneration

All filter materials are designed to hold a certain quantity of soot. If the filter becomes overloaded, the particulates create an obstruction to the gas flow, which manifests itself in increased pressure drop and may lead to clogging of the filter. Therefore, the filter system has to provide reliable regeneration mechanisms to ensure problem-free operation. The soot holding capacity in most filter systems is sufficient only for some hours or, at most, a few of days of engine operation. Some DPF systems regenerate continuously. Others, which regenerate periodically, must perform this operation fairly frequently, usually at least once per day.

The exhaust gas and/or filter temperature is the most important parameter influencing filter regeneration. Oxidation of carbon particulates can be considered a kinetic process, the rate of which can be described by the laws of chemical reaction kinetics. Again, certain analogies can be drawn between the oxidation of gases in the catalytic converter and the oxidation of soot in diesel filters, especially those which operate in the continuous regeneration regime. Filters do provide longer residence times (relative to catalytic converters), which are required to oxidize the large size particles (relative to gas molecules). Similar to gases, the soot oxidation rates increase at higher temperatures and decrease, possibly slowing down to zero, at low temperatures. Contrary to the catalytic converter, however, material that is not oxidized cannot be released from the filter. Instead, unoxidized particles are accumulated until conditions change (e.g., as a result of increased temperature) and oxidation rates increase. It is an important function of the filter system to provide mechanisms that prevent accumulation of soot that would exceed the filter holding capacity.

Failures of filter systems usually result in more serious consequences to the vehicle driver than failures of catalytic converters. A deactivated catalyst in the converter results in excessive emissions, but in most cases does not create adverse effects to the engine or vehicle. On the other hand, a diesel particulate filter that fails to regenerate is likely to become clogged, effectively stalling the engine and immobilizing the vehicle.

Particulate filters also capture inorganic ash particles contained in the PM emissions. These incombustible particles cannot be removed from the filter through thermal regeneration. Even though the contribution of ashes in the TPM is much less than that of inorganic carbon, the accumulation of ash causes a gradual increase in the pressure drop in the filter over its life span. Filters can be cleaned from ashes only through dedicated maintenance procedures, such as washing with water or reversing the direction of flow through the filter. The frequency of cleaning is typically not more than once per year.

20.3 DPF Performance

20.3.1 Particle Mass

Due to the physical particle collection mechanisms, as discussed earlier, filters are effective in collecting the solid fraction of diesel particulate matter (DPM), including inorganic carbon and ash. The effectiveness of diesel filters in controlling the organic fraction of DPM (SOF) depends on the type of filter and on its operating conditions. Furthermore, diesel particulate filters are not only ineffective in controlling sulfate particulates, but, in the case of catalytic filters, may significantly increase this emission through catalytic oxidation of sulfur dioxide.

The impact of particulate filters on PM composition is illustrated in Figure 20.9 [3]. The filter, a two-stage system incorporating an oxidation catalyst and a wall-flow DPF, was installed on a Euro 1 engine and tested on the R-49 schedule, which is dominated by the full-speed/full-load mode, resulting in high average exhaust gas temperatures. It is clear from the graph that the DPF is extremely effective in removing elemental

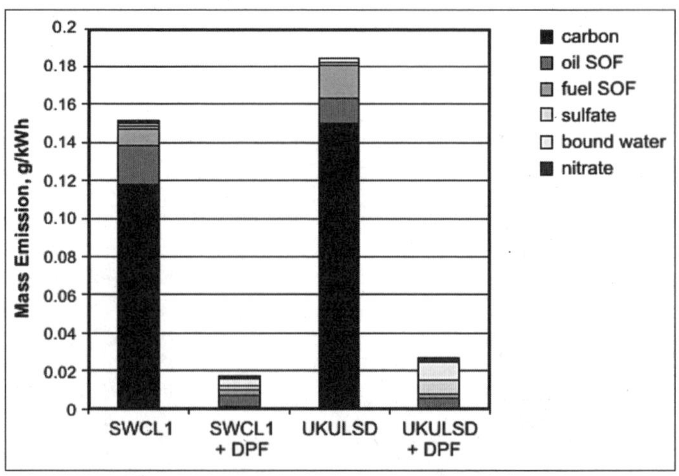

Figure 20.9 *Effect of particulate filter on PM composition for a Euro 1 engine on the R-49 cycle; SWCL1 (Swedish Class 1) fuel, <10 ppm S; UKULSD (U.K. ultra-low-sulfur diesel) fuel, 50 ppm S. (Based on data from Ref. 3.)*

carbon particulates (black, bottom portion of each bar). SOF fractions from fuel and oil are somewhat reduced by the catalyst portion of the filter system. Despite the use of ultra-low-sulfur fuels, sulfate particulates are increased (especially so in the test with the UKULSD fuel, which has a higher sulfur content than the other fuels shown).

To understand the performance capabilities of diesel filters, one has to be familiar with the definition and composition of diesel particulates. Diesel particulate matter is composed of three main fractions: (1) elemental carbon and ash (solid), (2) an organic fraction (SOF, liquid), and (3) sulfate particulates (liquid). The described particle deposition mechanisms are effective in collecting solid and, to a degree, liquid particles. However, a significant portion of the liquid particulates, both SOF and sulfates, is in the gas phase under the high-temperature conditions experienced by the filter. The formation of liquid particles (through nucleation and condensation) occurs only after the exhaust gases are cooled down, typically in the particulate sampling equipment downstream of the particulate filter, as illustrated in Figure 20.10. Obviously, particle formation downstream of the trap will reduce the observed filtration efficiency.

The higher the filter temperature, the more SOF particulates that pass through as vapors. Therefore, the SOF efficiency of an uncatalyzed diesel trap will decrease with increasing temperature. Most passive diesel filters incorporate catalysts that lower the soot oxidation temperature to facilitate regeneration. Because these catalysts also oxidize hydrocarbons, catalyzed filters are more effective in the removal of SOF particulates.

However, the catalyst may also generate sulfate particulates, especially at elevated temperatures and in conjunction with high-sulfur diesel fuel, as discussed in Section 19.2. Sulfate particulates are composed primarily of sulfuric acid, which is formed by heteromolecular nucleation of water vapor and catalyst-generated SO_3 molecules. The nucleation occurs only at low exhaust temperatures. Thus, a high-temperature diesel filter is ineffective in capturing the sulfate particulates. For this reason, catalytic diesel filters may show declining filtration efficiencies at high temperatures due to the generation of sulfate particulates. Filters utilizing oxidation catalysts of high activity must be operated on ultra-low-sulfur fuels (say, below 50 ppm S) to control sulfate particulate emission. If fuels of higher sulfur levels are used, the catalyst-generated sulfates may more than offset the benefit realized in capturing the solid particulates, resulting in a negative filtration efficiency for TPM emissions.

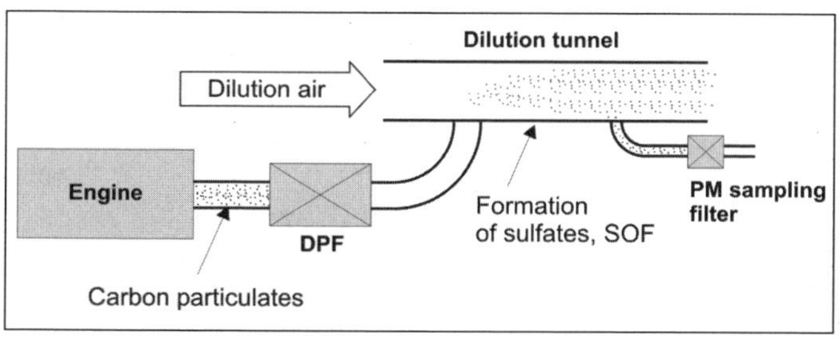

Figure 20.10 Formation of liquid particulates downstream of the DPF.

The performance of several types of diesel filters is discussed in the following sections. The general trends are summarized in Table 20.1. The performance, especially that related to sulfates, depends very strongly on the fuel sulfur content and the emission test cycle (temperature). The efficiencies listed in this table are based on the assumption that sulfates are included in the definition of PM, as is the case in public health regulations. However, sulfates are not considered diesel particulates in several occupational health regulations (Germany [4], U.S. noncoal mines [5]). According to the letter of such regulations, the generation of sulfuric acid in the filter is not considered a counterproductive process, making it possible to operate catalytic filters with high-sulfur fuels (unless separate exposure limits for SO_3/sulfuric acid that may also be in place are exceeded when operating the DPF). If sulfates are neglected, the total efficiency of most filter systems will be in the 85% to 95% range and above.

20.3.2 Particle Numbers

While most PM mass emissions are composed of solid matter (or solid particles with adsorbed vapors), liquid material constitutes a very significant part of diesel nanoparticles, which are the main contributor to particulate number emissions (compare Section 7.4). If the liquid material, including sulfates and SOF, is formed in the PM sampling system, that is, downstream of the filter, the DPF will be ineffective in reducing nanoparticle and particle number emissions. Even worse, by retaining carbon particulates, the DPF removes the material that otherwise acts as a "sponge" for condensates formed in the sampling system. Therefore, particulate filters tend to increase the formation of nanoparticles through nucleation. In effect, DPFs reduce the numbers of mostly solid agglomeration-mode particles, replacing them by mostly liquid nuclei mode nanoparticles.

TABLE 20.1
TYPICAL FILTRATION EFFICIENCY OF DIESEL PARTICULATE FILTERS (PM MASS)

Fraction of Diesel PM	Efficiency, %	Comments
Elemental carbon	95–99.9	Physical filtration.
Organic fraction (SOF)	50–90	Conversion primarily by catalytic oxidation, depends on temperature. Lower conversion in noncatalytic filters.
Sulfate particulates	No filtration	Catalytic oxidation of SO_2 to SO_3 can increase SO_4, causing negative efficiency when high-sulfur fuel is used.
TPM	70–95	

As a result, several studies measured increased particle numbers with particulate filters, as illustrated in Figure 20.11 [3]. The data for the figure were generated on the hot ECE R-49 test cycle with a catalytic filter system. High particle number emissions are clearly related to sulfates, which are generated at high exhaust temperatures in the catalyst, as indicated by the higher particle numbers in the tests with higher sulfur contents in the fuel.

In the general case, however, the impact of DPFs on particulate number emissions has to be described as inconclusive; the particle numbers may be either increased or decreased. The main parameters influencing the DPF performance or the performance assessment can be listed as follows:

- *Presence of nanoparticle precursors:* This includes SO_3/sulfuric acid, which can be catalytically generated from sulfur precursors in the fuel. Hence, a lower fuel sulfur content will result in fewer nanoparticle emissions. Hydrocarbons are another nanoparticle precursor. If HCs are removed from the system, for example, by using a catalyst, lower nanoparticle emissions may be expected. Such phenomena as adsorption/desorption of HCs on accumulated soot in the exhaust system may also contribute to the overall particle number emission.
- *Exhaust gas temperature:* Sulfate particulates are formed at high-temperature conditions, such as at full engine load. Therefore, in catalytic systems, higher particle numbers will be measured over hot engine test cycles.
- *Particulate sampling parameters:* In the absence of standard measuring methods of particle number emissions, laboratories are free to choose any measuring setup and parameters. As discussed in Section 7.4, the choice of parameters, such as dilution ratios or dilution tunnel residence times, can critically influence the measurement. Measurements are particularly unstable and irreproducible in systems with high rates of particulate nucleation and condensation (as opposed to testing configurations that attempt to measure exclusively solid nanoparticles).

The preceding discussion is based on a quiet assumption that liquid condensates are counted as particles. As standards are being developed to quantify diesel particle number emissions, a controversy exists as to the inclusion of liquid condensates in particle number measurements [6]. Several laboratories use sample conditioning techniques that "dry" the gas sample by removing volatile nanoparticle precursors before particle number measurements are taken. If particulate filters are evaluated using this technique, they show consistently good particle number reduction performance, as shown in the example plot in Figure 20.12 [7]. This is an indication that solid nanoparticles are retained with high efficiency in a variety of DPF media.

20.4 Commercial Status

The most important issue with diesel particulate filters is reliable filter regeneration. Generally speaking, today's diesel filter systems still have insufficient dependability and may experience excessively high pressure drop or clogging. Furthermore, excessive heat released during regeneration may lead to a loss of mechanical integrity and catastrophic failure of the filter material. Although these issues are being addressed through intensive development efforts in the areas of filter materials and regeneration strategies, the technology has achieved a sufficient level of maturity to allow for the first commercial applications on new light-duty (LD) and heavy-duty (HD) diesel engines.

Clean air regulatory authorities in the European Union (EU) and in the United States have adopted diesel emission standards that are believed to require the use of particulate filters to meet the PM limits. These standards, including the HD Euro 4, HD US2007, LD Euro 4, and US LD Tier 2, represent

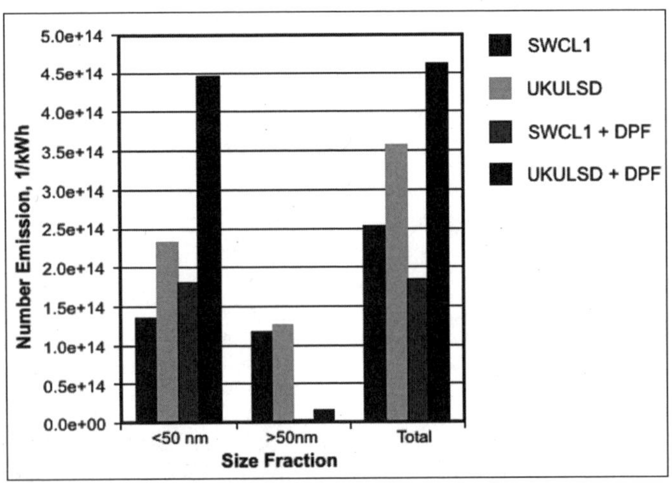

Figure 20.11 Effect of particulate filter on particle number emissions for a Euro 1 engine on the R-49 cycle; SWCL1 fuel, <10 ppm S; UKULSD fuel, 50 ppm S. (Based on data from Ref. 3.)

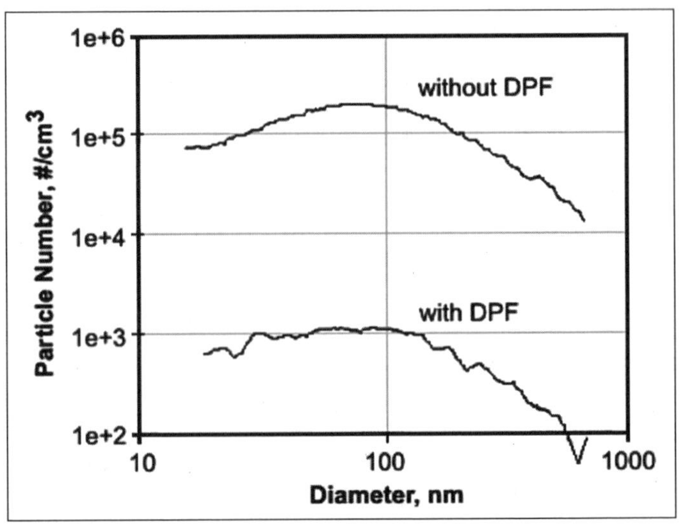

Figure 20.12 Particle size distribution with and without DPF for solid particulates, using a ceramic monolith filter on a Caterpillar construction engine. (Based on data from Ref. 7.)

tightening of diesel PM emission limits from 50% to 90%, relative to their respective predecessors. The European standards, both effective in 2005, introduce PM limits of 0.025 g/km for light-duty vehicles and 0.02 g/kWh for heavy-duty engines. The U.S. standards become effective later than those in the EU (Tier 2: phase-in during 2004–2009), but introduce even more stringent PM limits: no more than 0.02 g/mi for LD vehicles and 0.01 g/bhp-hr for HD engines. If these standards are upheld by the authorities, a wide-scale launch of particulate filters on new diesel engines for highway vehicles can be expected between 2005 and 2007.

Historically, the first diesel particulate filters were introduced in 1985 on Mercedes cars sold in California [8]. The traps, which utilized passive catalyzed filter technology, experienced a number of technical problems and were soon taken off the market. After this experience the technology disappeared from new vehicles, but was quickly adapted for retrofitting of selected off-road applications, mainly in the underground mining, tunneling, and materials handling industries. Most mining applications took place in Australia and the United States, while diesel forklifts and tunneling machinery was retrofitted mainly in Europe (Germany, Switzerland, etc.). In the late 1990s, once more retrofit experience had been collected, several programs were launched that targeted heavy-duty highway vehicles, such as urban bus fleets. The first program of this type, known as the Environmental Zones, was started in 1996 in three major Swedish cities. Retrofit programs for diesel buses and other vehicles were subsequently adopted in other European cities (Paris, Berlin, Munich, cities in the United Kingdom), in the Far East (Taipei, Tokyo), and in the United States (New York City, California).

In 2000, diesel particulate filters made their return to the automobile application. The first "new era" filter application on a diesel car was the PSA Peugeot 607 HDi 2.2 (model year 2000), which came equipped with an active particulate filter system regenerated through a sophisticated strategy involving the use of fuel additive and common-rail post-injection of fuel [9]. In 2001 this filter system was introduced on the Peugeot model 406 HDi 2.2, and then on a number of other cars powered by Peugeot engines including the Peugeot 307, 807, and some Citroen C5s and C8s. By the end of 2002, the PSA Group had sold nearly 400,000 particulate filter–equipped vehicles [10].

The first heavy-duty engine with a diesel particulate filter was introduced by Navistar in 2001. The engine, a 530-in.^{3} displacement, 275-hp unit, was certified by the U.S. Environmental Protection Agency (EPA) and the California Air Resources Board (ARB) to the 0.01 g/bhp-hr PM standard (US2007). Equipped with a passive catalyzed diesel filter, the engine was developed for school bus applications, primarily in California.

20.5 Diesel Filter Regeneration

20.5.1 Oxidation of Diesel Soot

Thermal regeneration of diesel particulate filters involves the oxidation of solid particulates to gaseous products. The rate of thermal regeneration is limited by the kinetics of soot oxidation. The oxidation of soot, as in any other chemical process, has a reaction rate that depends on the temperature and the concentration of reactants, and it can be accelerated by catalysts. The rate of soot oxidation is significantly lower than that of the oxidation of gaseous exhaust pollutants, such as carbon monoxide. This is one of the reasons why carbonaceous particulates are not oxidized in flow-through catalysts that provide relatively short particle residence times. The oxidation of soot requires diesel traps that extend the residence time by physically capturing the particulates.

Laboratory-measured oxidation rates of diesel soot are shown in Figure 20.13 [11]. The mass fraction of soot that was oxidized is plotted against the oxidation time for different temperatures. At low temperatures (360°C, 400°C), the oxidation rates are very slow and regeneration is incomplete. Only at temperatures of about 600°C was the soot oxidized quickly and completely. At any given temperature, the oxidation rate changes with the amount of soot in the sample. Fast oxidation rates (i.e., steep curves in Figure 20.13) are observed at the beginning of each experiment. As the mass of soot in the sample decreases, so does its oxidation rate. This behavior is frequently seen in diesel filters, which tend to regenerate faster when the soot loading (and the pressure drop) is higher.

The temperature at which soot starts oxidizing at an appreciable rate is sometimes referred to as the *soot ignition temperature.* The soot ignition temperature is not a physical property of soot or carbon in the sense as, for example, melting and boiling temperatures are physical properties of liquids. Rather, it depends both on the properties of the soot and on the particular apparatus used for the measurement, as is the case of, for example, flash points of combustible liquids. Therefore, soot ignition temperatures measured using different methods or different equipment are not necessarily comparable.

Two laboratory procedures may be used to determine the soot ignition temperature: (1) thermal gravimetric analysis (TGA) or (2) differential thermal analysis (DTA). In both

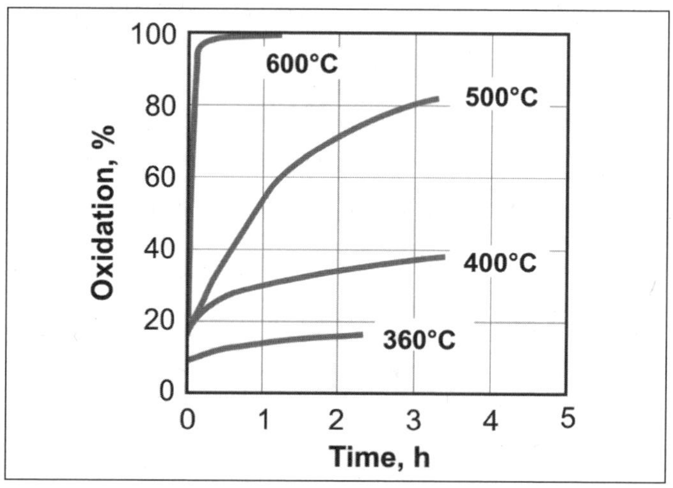

***Figure 20.13** Diesel soot oxidation. (Based on data from Ref. 11.)*

methods an oven of controlled atmosphere and temperature is used. The diesel soot or graphite sample to be analyzed, pure or mixed with a catalyst, is placed in the oven. Then, the oven temperature is slowly increased. In TGA, the sample is placed on a balance that is continuously recording its weight. In DTA, a thermocouple placed inside the sample monitors the temperature difference between the oven atmosphere and the sample. The oxidation rate of soot increases during the experiment until the sample completely burns. The TGA analyzer shows a drop in the sample mass. The DTA recording indicates a peak in the sample temperature due to the heat effect of combustion. The temperature corresponding to the TGA mass drop or the DTA temperature peak is the soot ignition temperature.

The soot ignition temperature is an excellent gauge to compare, on a relative scale, the performance of different catalysts. It may, however, have little relevance to the regeneration temperature of an actual diesel particulate filter.

20.5.2 Regeneration as an Equilibrium Process

The regeneration of a diesel particulate filter can be analyzed as a continuous, equilibrium process, the mass balance of which is given by Eq. (20.1) [12]. The first term in the equation, $dM/d\tau$, represents the accumulation of soot in the filter. The two remaining terms in Eq. (20.1) describe the oxidation of soot and the deposition of fresh soot on the filter, respectively.

$$dM/d\tau + M \cdot k(T) - e_F \cdot m = 0 \qquad (20.1)$$

where:

M = particulate mass on the filter, kg
τ = time, s
$k(T)$ = reaction rate constant for particulate oxidation, 1/s
T = temperature
e_F = filter collection efficiency, dimensionless
m = particulate mass flow from the engine, kg/s.

Depending on the sign of the accumulation term, the filter may be in one of the following operation modes, as illustrated in Figure 20.14:

1. Accumulation ($dM/d\tau > 0$)
2. Equilibrium ($dM/d\tau = 0$)
3. Regeneration ($dM/d\tau < 0$).

The plot of Figure 20.14 represents a hypothetical engine dynamometer experiment, during which the filter is exposed to stepwise temperature (T) changes at a constant mass flow of the exhaust gas. The exhaust gas pressure drop across the filter (ΔP), which is plotted versus the time of experiment, can be considered a measure of the filter soot load at any given moment.

The experiment starts with a new clean filter having a low pressure drop. The filter is first exposed to a relatively low exhaust temperature. During that phase (mode 1 in Figure 20.14), the oxidation rate of soot is very low and the trap accumulates particulates. As a result, a steadily increasing pressure drop is recorded.

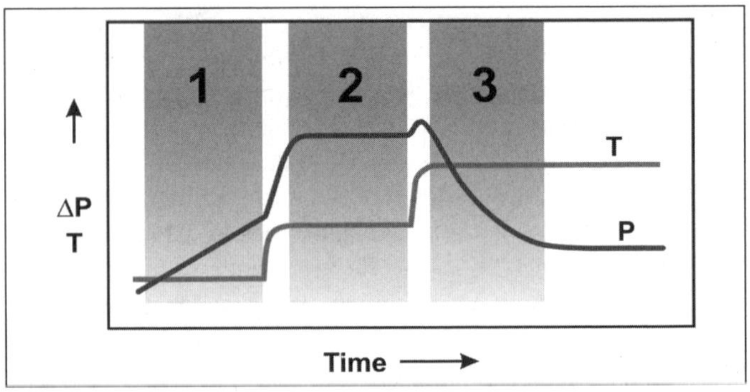

Figure 20.14 Particulate filter operation modes.

If that temperature were to be maintained for a prolonged time, the trap could become clogged with soot.

When the exhaust temperature is raised, the pressure drop increases in response to the increased volumetric gas flow, as well as to further accumulation of particulates. However, the rate of soot oxidation is now much faster. Eventually, the filter reaches a mass equilibrium that is reflected by flattening of the pressure drop profile. At the equilibrium, the mass of soot captured on the filter per unit of time is equal to the mass of soot oxidized in the filter, as described by Eq. (20.2). The filter operates with a constant soot loading and at a constant pressure drop. The temperature at which the filter reaches mass equilibrium is called the balance temperature:

$$M \cdot k(T) = e_F \cdot m \qquad (20.2)$$

With a further increase of exhaust temperature the filter enters the regeneration mode. The temperature is so high that the amount of soot oxidized is higher than that captured in the filter per unit of time. As a result, the filter soot loading and the pressure drop decrease.

The filter soot loading will not drop to zero during the regeneration phase, unless the filter is exposed to extremely high temperatures. Due to the decreasing soot mass in the filter, the soot oxidation rate decreases during the regeneration. Finally, another mass equilibrium condition will be reached at higher temperature, lower pressure drop, and lower soot loading than those at the starting point.

According to this model, as well as in agreement with both common logic and practical experience, the exhaust gas temperature is a key variable influencing particulate filter regeneration. However, the temperature is not the only important variable. As shown in Eq. (20.1), the rate of soot oxidation also strongly depends on the particulate mass collected in the filter (M). Even though the reaction rate constant for particulate oxidation, k(T), has been reported to decrease with increasing M [13], the rate of soot oxidation in kilograms per second expressed by the product $M \cdot k(T)$ increases with increasing M. As a consequence, particulate filters can operate at a number of balance temperatures corresponding to different soot load levels, as illustrated in Figure 20.15 [14]. If the

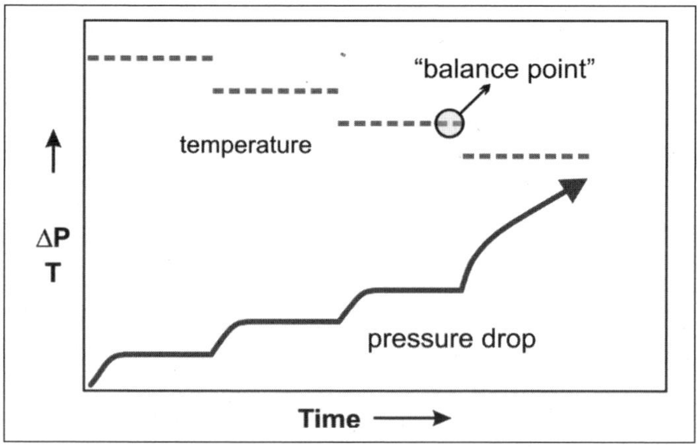

Figure 20.15 *Particulate filter balance conditions.*

exhaust temperature decreases, the filter will still regenerate, but its soot load will increase to maintain the required soot oxidation rate. The increased soot loads manifest themselves through increased pressure drop levels. In other words, particulate filters operated at higher temperatures will be holding less soot and experience less pressure drop compared to filters operated at lower temperatures.

The operation at a balance condition, or at a number of balance points of fluctuating temperatures and corresponding pressure drop levels, is seen in continuously regenerating filters. The continuous regeneration is characterized by constant steady-state pressure drop readings. In transient operation, pressure drop fluctuates with the engine speed and exhaust gas temperature, but the maximum ΔP peaks reach an approximately constant level, as shown in Figure 20.16. The temperature and pressure traces shown in the plot (both measured upstream of the particulate filter) correspond to the repeated duty cycle of a heavy-duty truck in an underground mine.

Such a continuous-operation mode requires a certain minimum exhaust temperature. If the temperature is too low, the soot oxidation rate will become too slow to sustain an appreciable regeneration rate and the filter will accumulate quantities of soot, as indicated by the increasing pressure drop line in Figure 20.15. This condition could lead to either (1) clogging of the filter with soot or (2) a stochastic regeneration. Clogging occurs when the pressure drop buildup in a soot-overloaded filter makes it impossible to operate the engine. If no exhaust gas bypass mechanism is provided in the filter system, this condition will effectively immobilize the vehicle.

The stochastic regeneration involving a rapid oxidation, or burn-out, of the accumulated soot can be considered the opposite of the continuous regeneration. The oxidation of large soot loads involves the release of a significant quantity of heat, which makes the stochastic regeneration self-sustaining even if the temperature of exhaust gas entering the filter is low. Stochastic regenerations can be induced under laboratory conditions by loading filters to high soot loads, as shown in Figure 20.17 [15]. A fuel additive was used in this experiment, but the exhaust gas temperature was maintained at a low level that was insufficient to support continuous regeneration. Rather, the filter was experiencing a series of soot accumulation periods followed by very fast stochastic regenerations, resulting in the sawtooth pattern.

The term *runaway* or *uncontrolled* regeneration is frequently used when discussing particulate filter systems. Because the same physical phenomena underlie both runaway and stochastic regeneration, these terms can be considered synonyms. A research worker trying to theoretically model filter regeneration would use the term *stochastic regeneration* to emphasize its random, unpredictable character. A filter system designer would use the term *runaway regeneration* to express the difficulties he is experiencing in controlling the process.

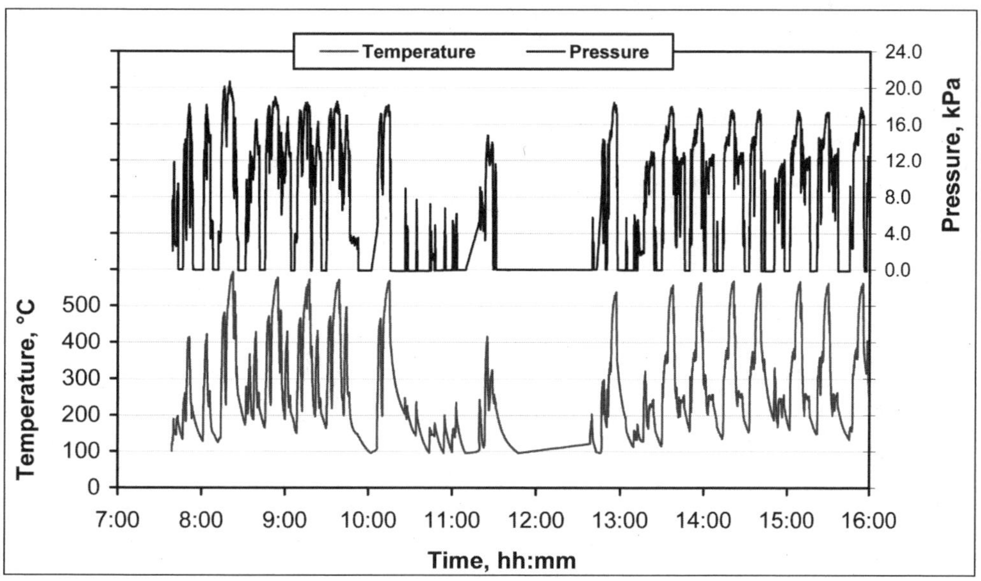

Figure 20.16 *Continuously regenerating DPF—exhaust pressure and temperature during a field test.*

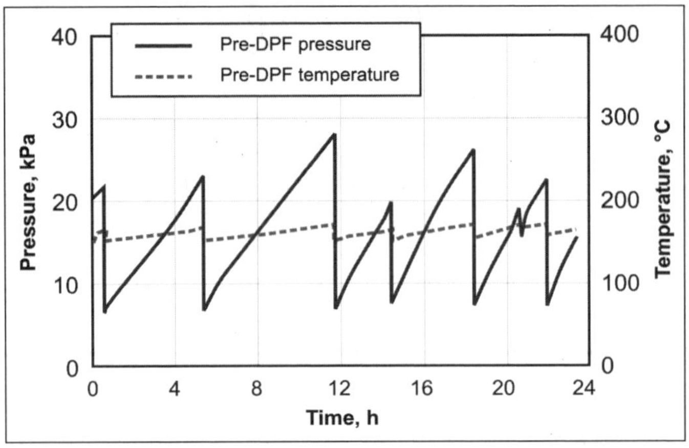

Figure 20.17 *DPF operation with periodic stochastic regenerations; steady-state engine operation, fuel-borne catalyst.*

In summary, the regeneration of diesel traps strongly depends on (1) the temperature and (2) the soot mass in the filter. At higher temperatures the rate of soot oxidation increases, and the filter holds fewer particulates and exhibits lower pressure drop. At lower temperatures the filter needs higher soot loadings to reach mass equilibrium. There is a certain trade-off between the exhaust temperature and the filter soot loading. Therefore, traps installed on hot engines work with lower pressure losses, whereas colder exhaust gases cause an increase in soot loading and pressure drop. Operating the filter at low temperatures and high soot loads increases the chances for stochastic regenerations, involving rapid burn-out of the accumulated particulates sustained by the released heat of reaction.

The discussion of continuous regeneration of diesel particulate filters would not be complete without mentioning the time aspect of reaching a mass equilibrium state. As shown in Figures 20.14 and 20.15, if the exhaust temperature changes, the filter responds by reaching a new mass balance. In particular, if the temperature is increased, the filter will lower its soot load and experience lower pressure drop. However, reaching the new condition is not instantaneous and requires a certain period of time. Because the changes of soot mass in the filter depend on the kinetics of soot oxidation [Eq. (20.1)] with reaction rates increasing with temperature, increased temperatures will shorten the time period needed to reach equilibrium.

Figure 20.18 compares the time needed for a "complete" regeneration of filters preloaded with soot and exposed to a constant exhaust temperature of 370°C or 450°C [16]. The times are theoretical extrapolations calculated based on laboratory measurements of the duration of incomplete regeneration [in a regeneration experiment the soot mass in the filter, M in Eq. (20.2), can never reach zero unless the temperature is infinitely high]. As seen in the graph, the filters regenerated quickly at 450°C, while changes in the soot mass were very slow, especially when starting with higher soot loads, at 370°C. In practical terms, if a passively regenerated filter becomes overloaded with soot (e.g., due to an extended engine idling incident), it will quickly return to its normal operating conditions and lower the pressure drop in a high-temperature application. On cold engines, on the other hand, it may take a very long time (e.g., days of operation) before the filter regains its normal pressure drop level.

Regeneration of diesel filters is also influenced by many other factors, including the following:

- *Engine-out PM emissions:* Higher particulate emissions [m in Eq. (20.2)] require more soot to be oxidized per unit of time, and higher temperatures for continuous regeneration. Very likely, this is the most important reason why filters regenerate at lower temperatures when installed on newer technology diesel engines having low PM emissions.
- *Engine-out PM composition:* Particulates of different SOF content may exhibit different rates of oxidation and different ignition temperatures. There are contradicting reports as to what effect is seen as a result of increasing SOF levels on the balance temperature. A significant portion of SOF passes through the filter as gas-phase vapors, especially at higher temperatures. It is logical to assume that the SOF that is bound to the solid particulates in the filter, whether as a liquid or as adsorbed vapor, would have more impact on the filter balance temperature than the gas-phase SOF. Even

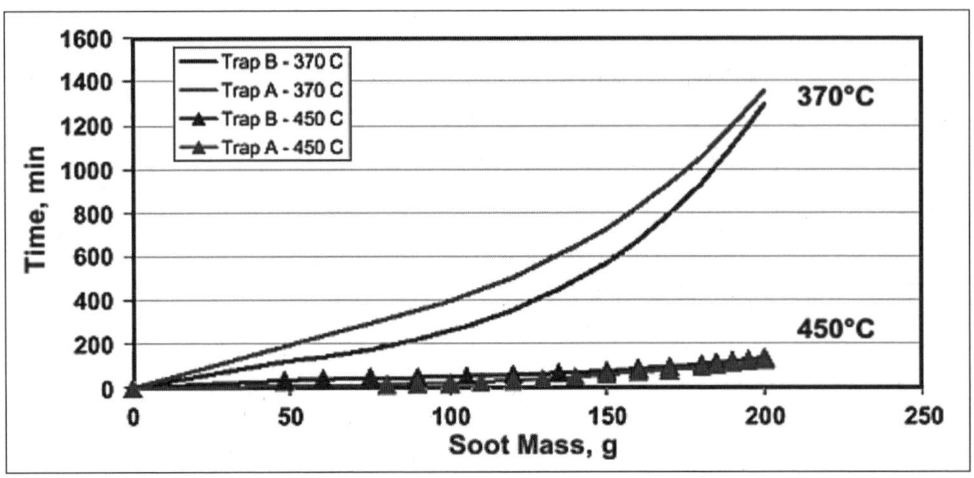

Figure 20.18 *Filter regeneration time at different soot loads and exhaust temperatures. Engine: Cat 3126B, HEUI, 190 hp. (Reprinted from Ref. 16.)*

though it is difficult to know what fraction of the total SOF is bound to solid particulates in the filter, the issue has to be taken into consideration. Many literature reports fail to do so. It is also possible that the impact of SOF depends on the type of catalyst or additive used to facilitate regeneration.

- *Fuel sulfur content:* In the case of catalytic filters, lower regeneration temperatures are observed with fuels having a lower sulfur content.
- *Exhaust gas flow rate:* The exhaust gas flow rate has an influence on temperatures due to its cooling or heating effect on the soot deposit and the filter body.
- *Oxygen content in the exhaust gas:* Shortages of oxygen may slow down the soot oxidation rate and negatively affect the regeneration process. However, there is no evidence that the variation of oxygen levels that is typical for the diesel engine, with the minimum O_2 concentration of no less than 3%, can impair regeneration.

Theoretical models are being developed to simulate the regeneration of diesel particulate filters. Some models are quite sophisticated and describe a number of parameters and processes, such as pressure drop on clean and loaded filters [17, 18], catalytic and NO_2 assisted regeneration, transient filter loading, aging effects of ash [19], or spatial nonuniformities in filter regeneration [20]. If theoretical models are developed to reach practical applicability, they may become very valuable tools for performing system-level optimizations and the integration of particulate filters with powertrain management.

20.5.3 Determination of Balance Temperature

Because the balance temperature depends on engine emissions and a number of operational parameters, it is not a universal property of a given particulate filter. Balance temperatures determined on different engines or using different measuring procedures are in general not comparable. Nevertheless, a procedure is needed that will allow for the determination of relative tendencies between balance temperatures of particulate filters. Such determination is a common task in DPF catalyst development, catalyst durability work, or in research aimed at comparisons among different filter technologies.

Repeatability of results is the major challenge in developing such a procedure. For example, because filter regeneration depends on the amount of accumulated soot, the balance temperature may change depending on the "history" of the tested filter unit. To minimize this variability, balance temperature procedures must include a filter preconditioning protocol, which will ensure that the filter is preloaded to exactly the same soot loading at the beginning of the test. Another common source of variability is the test engine itself. If, for any reason, the engine's PM emissions change from test to test, the determined balance temperatures will not be repeatable.

The simplest version of the balance temperature test involves installing the filter in the engine exhaust line and ramping the engine load. The engine torque is increased slowly, either continuously or in small (e.g., 1-min) steps. The exhaust gas temperature at the filter inlet and the pressure differential across the filter are recorded during the ramp. The exact timing (torque increase rate) must be selected experimentally for a given engine and filter size, so as to provide an easy-to-interpret pressure drop curve. Some tests are designed for a fairly long duration of 4 to 6 h or more. The results are plotted in the coordinates of pressure drop versus temperature, as illustrated in Figure 20.19. At first, the ΔP line increases with temperature, indicating accumulation of soot in the filter. At a certain time/temperature, the pressure drop stabilizes, and then gradually decreases. The filter inlet temperature corresponding to the maximum of the ΔP(T) function is considered the filter balance temperature.

To achieve repeatable results, the filter starting condition must always be the same. Some authors prefer to start the test with a clean filter, after completely regenerating the unit through a sufficiently long exposure to high temperature [21]. This can be achieved by preconditioning the filter at a high engine load condition. Others develop a standard preloading procedure and start the ramp with a filter having a relatively high, but always the same, soot load [22]. Preloading allows for a shorter ramp duration, but it takes more effort to accurately reproduce the filter soot load than does the clean filter condition.

A more precise balance point procedure was developed by the U.S. Department of Energy's Diesel Emissions Control—Sulfur Effects (DECSE) program [23]. The test involves recording of the filter pressure drop while exposed to several (five to seven) stepwise changed filter inlet temperature levels, as shown in Figure 20.20. The filter is preloaded to a predetermined soot loading (pressure drop) before the test.

The filter balance temperature is determined as follows (for a five-mode test):

1. Following stabilization of ΔP at each temperature, the least-square slopes of the ΔP versus time curve are calculated for each of the five segments. Data are first reduced by selecting the linear portion of each curve segment (generally, the 10-min period beginning 5 min after each change in exhaust temperature).
2. The established slopes provide five points of the relationship between the ΔP slope (kPa/min) and the filter inlet temperature, $\Delta P_{slope} = f(t)$.

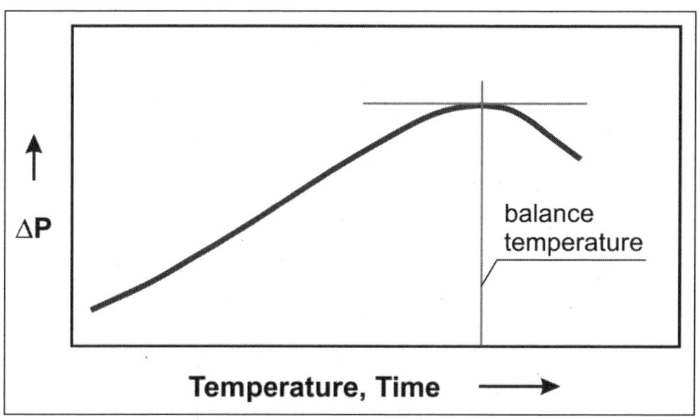

Figure 20.19 Laboratory determination of filter balance temperature.

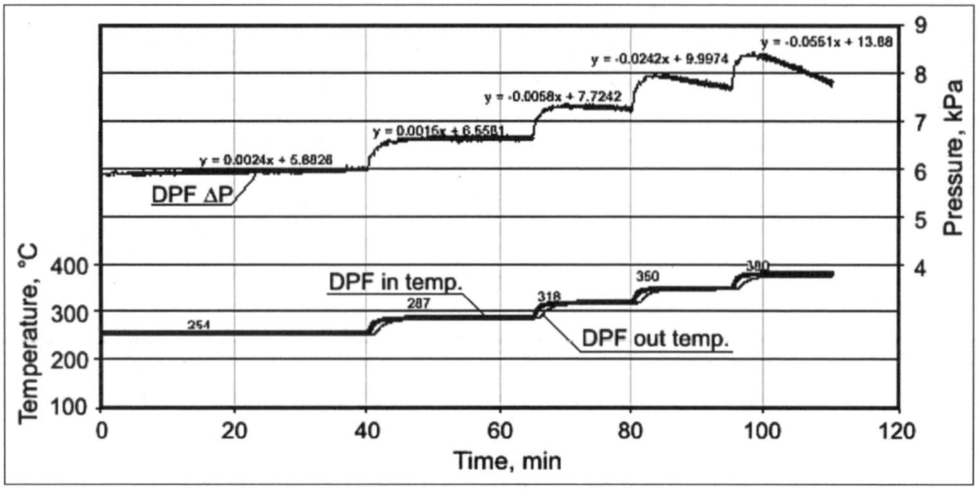

Figure 20.20 DECSE five-mode balance point test—example run data. (Reprinted from Ref. 23.)

3. Based on the five ΔP_{slope} points, a regression model is fit to establish a functional relationship in the form $\Delta P_{slope} = a + bt + ct^2$.
4. The balance point is determined as the filter inlet temperature at which the slope of ΔP equals zero.

The test was repeated and statistical analyses were performed to analyze run-to-run variability. It was also found that the best repeatability was achieved when the same fixed test temperatures were used for all measurements. Changing the temperature points for different measurements resulted in different "histories" for the filters and, thus, in differences in the determined balance points.

20.5.4 Regeneration Methods

Thermal regeneration of diesel particulate filters requires temperatures of 550 to 650°C. Such exhaust temperatures are not encountered in too many practical applications of the diesel engine. Most heavy-duty diesel applications have exhaust temperatures in the range of 300 to 450°C. Light-duty diesel applications may be as cold as 200 to 300°C. The following three approaches are used, either separately or in combination, to facilitate diesel filter regeneration:

1. Increasing the exhaust gas temperature
2. Decreasing the soot ignition temperature by use of catalysts
3. Decreasing the soot ignition temperature by use of reactive oxidants.

20.5.4.1 Increased Exhaust Temperature

The exhaust gas temperature may be increased by two groups of methods: (1) exhaust gas heating or (2) use of an higher combustion temperature (in-cylinder). The following methods of heating the exhaust gases have been investigated:

- Electric heaters
- Fuel burners
- Microwave heating
- Injection of combustibles into the exhaust system, alone or in conjunction with warm-up catalysts.

Systems utilizing any of these methods are classified as active filter systems or filters with assisted regeneration. Systems that do not utilize catalysts and rely solely on the exhaust gas temperature to regenerate have to be able to increase exhaust temperatures to 600 to 700°C. To save energy, regeneration is usually performed periodically. Regenerations are triggered by an electronic controller after a predetermined amount of soot is collected in the trap. Therefore, active filters cycle through long accumulation periods alternated with short high-temperature regenerations. The regenerations clean the trap almost completely, lowering the pressure drop to a level comparable with that of a new filter.

A moderate increase in exhaust temperature may be needed in systems that utilize catalysts or fuel additives to lower the soot ignition temperature. Target exhaust gas temperatures in these system may be anywhere between 350 and 450°C. That temperature level may be reached through several engine management measures, all of which increase the combustion and the exhaust temperature [24]:

- Exhaust gas recirculation (EGR), which is a very effective solution to minimize NO_x emissions in diesel engines, can be used (in its uncooled configuration) as an effective measure to increase the exhaust gas temperature.
- Post-injection of fuel into the combustion chamber is possible in electronically controlled injection systems such as the common rail.
- Increases in the exhaust gas pressure can be controlled with a throttle valve installed in the exhaust system.
- Retarded injection timing is another method of controlling NO_x emissions that also increases exhaust temperatures.

- Increasing the intake air temperature is a simple method for turbocharged and intercooled diesel engines. It can be implemented by bypassing the intercooler.
- Decreasing the boost pressure can be attempted with turbocharged engines by opening the wastegate.
- Increasing the intake air vacuum by means of a throttle valve may be applied to naturally aspirated engines.

All of these methods cause some adverse effects on engine operation, including increased fuel consumption, increased particulate and hydrocarbon emissions, and a lower exhaust oxygen concentration, which may negatively influence filter regeneration. Some of the methods, for example, EGR or fuel post-injection, may also cause premature engine wear problems.

20.5.4.2 Catalysts and Fuel Additives

Application of a catalyst allows for decreases in the soot ignition temperature to a point where the filter would self-regenerate under the regular exhaust temperatures experienced in a given application. The mechanism of catalytic oxidation of soot is similar to the catalytic processes in the gas phase. The particulates have to physically contact active catalyst sites, where they are oxidized by oxygen atoms, which are adsorbed on the catalyst.

Diesel filter systems that utilize catalysts are in most cases classified as passive or self-regenerating traps. These filters operate for most of the time under conditions close to mass equilibrium . The amount of soot entering the trap equals approximately the amount being oxidized. Properly operating filters exhibit relatively small fluctuations of soot load and pressure drop over time. It is very important for problem-free operation that the exhaust gas reach a certain minimum temperature for a given fraction of the total operating time. If the minimum temperature conditions are not met, the filter will exhibit increasing pressure drop and may eventually overload and clog with soot. Overloaded filters may be prone to rapid uncontrolled regeneration should a high-temperature spike occur and ignite the soot loading.

Two methods are used to introduce the catalyst to the system: (1) catalyzing the filter media and (2) introducing the catalyst to the fuel. *Catalyzed diesel particulate filters* (CDPFs) utilize catalysts deposited directly on the filter media. Both precious and base metal catalysts are used, such as platinum, silver, copper, vanadium, iron, molybdenum, manganese, chromium, and nickel. Depending on the type of filter, the catalyst can be impregnated directly into the media (e.g., on cordierite filters) or an intermediate washcoat layer can be used (e.g., on wire-mesh filters). Catalyzed filters require exhaust temperatures of about 325 to 420°C for regeneration, depending on engine technology (PM emissions) and fuel quality (sulfur content).

The other method of introducing catalyst utilizes *fuel additives*, also known as fuel-soluble catalysts. Because the catalyst is continuously supplied with the fuel, a bare uncatalyzed trap can be used. The soot oxidation chemistry with fuel additives is actually very similar to that in the catalyzed filter. However, when fuel additives are used, the catalyst is physically incorporated into the particulates. It is introduced to the fuel, proceeds through the combustion chamber, and leaves the engine as part of the PM emission. The physical contact between soot and the catalyst is much better than in the case of a catalyzed filter. Therefore, filters with fuel additives can typically regenerate at somewhat lower exhaust temperatures than catalyzed traps. Depending on the additive and its concentration in fuel, additive-based systems require temperatures of 300 to 400°C for regeneration.

20.5.4.3 Oxidation with Nitrogen Dioxide

The rate of soot oxidation can be increased, and the balance temperature decreased, if an oxidant more active than oxygen could be introduced to the particulate filter. It was found that nitrogen dioxide can oxidize diesel particulate at higher rates than oxygen [22]. NO_2 is present in the native diesel exhaust in very small quantities, but its concentration can be increased by using an oxidation catalyst or other methods (e.g., by using plasma; see Chapter 23).

A filter system that uses NO_2 for regeneration has been introduced by Johnson Matthey under the trade name "CRT." NO_2 is generated in the CRT in an oxidation catalyst, which is placed upstream of an uncatalyzed filter. The CRT filter is very sensitive to the fuel sulfur content. With fuels having a very low sulfur content (3 ppm S), the filter can regenerate at very low exhaust temperatures of about 275°C [25]. The system cannot be used with fuels of more than 50 ppm S. It also requires a certain minimum NO_x/PM ratio for proper operation.

References

1. Wade, W.R., *et al.,* "Diesel Particulate Trap Regeneration Techniques," SAE Paper No. 810118, SAE International, Warrendale, PA, 1981.
2. Perry, R.H., *Perry's Chemical Engineers' Handbook,* 6th ed., McGraw-Hill, New York, 1984.
3. Andersson, J., and B. Wedekind, "DETR/SMMT/CONCAWE Particulate Research Programme," Summary Report DP01/0515, Ricardo Consulting Engineers, May 2001, retrieved from www.ricardo.com/downloads/SummaryReport.pdf.
4. "TRGS 554—Dieselmotoremissionen (DME)," BArbBl. (German Ministry of Labor), March 2001, p. 112.
5. "Diesel Particulate Matter Exposure of Underground Metal and Nonmetal Miners, Final Rule (30 CFR Part 57)," *Federal Register* **66,** 13, January 19, 2001, p.. 5706, retrieved from www.msha.gov/REGS/FEDREG/FINAL/2001finl/01-996.pdf.
6. Burtscher, H., "Literature Study on Tailpipe Particulate Emission Measurement for Diesel Engines," Report for Particle Measurement Programme, BUWAL/GRPE, March 2001, retrieved from www.akpf.org/pub/burtscher_bericht.pdf.
7. Mayer, A., "Particulate Traps for Heavy Duty Vehicles," Environmental Documentation No. 130,Swiss Agency for

the Environment, Forests and Landscape, 2000, retrieved from www.akpf.org/pub/bericht_um130_en.pdf.

8. Abthoff, J., *et al.,* "The Regenerable Trap Oxidizer—An Emission Control Technique for Diesel Engines," SAE Paper No. 850015, SAE International, Warrendale, PA, 1985.
9. Salvat O., P. Marez, and G. Belot, "Passenger Car Serial Application of a Particulate Filter System on a Common Rail Direct Injection Diesel Engine," SAE Paper No. 2000-01-0473, SAE International, Warrendale, PA, 2000.
10. "PSA Peugeot Citroen Exceeds 2002 Sales Target," Press Release, PSA Group, January 7, 2003.
11. Goldenberg, E., "Depollution des gaz d'echappement des moteurs Diesel au moyen de pots catalytiques," *Revue Inst. Franc. du Petrole* **38,** 1983, pp. 793–805.
12. Widdershoven, J., *et al.,* "Possibilities of Particle Reduction for Diesel Engines," SAE Paper No. 860013, SAE International, Warrendale, PA, 1986.
13. Tanaka, T., "Simultaneous Reduction of PM and NOx—A New After-Treatment System," in *22nd Internationales Wiener Motorsymposium,* Fortschritt-Berichte VDI (VDI Verlag, Duesseldorf), Reihe 12, Nr. 455, Band 2, 2001, pp. 216–228.
14. Jelles, S.J., M. Makkee, J.A. Moulijn, G.J.K. Ackres, and J.D. Peter-Hoblyn, "Diesel Particulate Control: Application of an Activated Particulate Trap in Combination with fuel Additives at an Ultra Low Dose Rate," SAE Paper No. 1999-01-0113, SAE International, Warrendale, PA, 1999.
15. Richards, P., B. Terry, M.W. Vincent, and S.L. Cook, "Assessment of the Performance of Diesel Particulate Filter Systems with Fuel Additives for Enhanced Regeneration Characteristics," SAE Paper No. 1999-01-0112, SAE International, Warrendale, PA, 1999.
16. Opris, C.N., "Catalyzed Diesel Particulate Trap: Technology Assessment," Paper presented at 7th Diesel Emissions Reduction Workshop (DEER), Portsmouth, VA, August 2001, retrieved from www.osti.gov/hvt/deer2001/opris.pdf.
17. Masoudi, M., *et al.,* "Validation of a Model and Development of a Simulator for Predicting the Pressure Drop of Diesel Particulate Filters," SAE Paper No. 2001-01-0911, SAE International, Warrendale, PA, 2001.
18. Suresh, A., A. Khan, and J.H. Johnson, "An Experimental and Modeling Study of Cordierite Traps—Pressure Drop and Permeability of Clean and Particulate Loaded Traps," SAE Paper No. 2000-01-0476, SAE International, Warrendale, PA, 2000.
19. Konstandopoulos, A., *et al.,* "Fundamental Studies of Diesel Particulate Filters: Transient Loading, Regeneration, and Ageing," SAE Paper No. 2000-01-1016, SAE International, Warrendale, PA, 2000.
20. Konstandopoulos, A., *et al.,* "Spatial Non-Uniformities in Diesel Particulate Trap Regeneration," SAE Paper No. 2001-01-0908, SAE International, Warrendale, PA, 2001.
21. Bickel, K., and W.A. Majewski, "Evaluation of a Catalyzed Ceramic Diesel Particulate Filter and Catalytic Converter on an Underground Mine Vehicle," SAE Paper No. 932493, SAE International, Warrendale, PA, 1993.
22. Cooper, B.J., and J.E. Thoss, "Role of NO in Diesel Particulate Emission Control," SAE Paper No. 890404, SAE International, Warrendale, PA, 1989.
23. "Diesel Emission Control—Sulfur Effects Program, Phase I Interim Report No. 1," Diesel Emission Control—Sulfur Effects Program, U.S. Department of Energy, Washington, DC, August 1999, retrieved from www.ott.doe.gov/decse/pdfs/interim.pdf.
24. Lemaire, J., W. Mustel, and P. Zelenka, "Fuel Additive Supported Particulate Trap Regeneration Possibilities by Engine Management System Measures," SAE Paper No. 942069, SAE International, Warrendale, PA, 1994.
25. "Phase I Interim Data Report No. 4: Diesel Particulate Filters," Diesel Emission Control—Sulfur Effects Program, U.S. Department of Energy, Washington, DC, January 2000, retrieved from www.ott.doe.gov/decse/pdfs/interim4.pdf.

Chapter 21

Diesel Filter Materials

21.1 Overview

21.1.1 Material Requirements

The filter substrate is the key component of the diesel filter system, affecting both its performance and durability. Its task is to physically capture solid particulates and hold them until they can be removed in the regeneration process (periodic or continuous). Design targets for diesel particulate filter materials include the following:

- High filtration efficiency: particle mass, particle number
- High maximum use temperature
- High soot holding capacity
- Thermal shock resistance
- Strength and mechanical integrity
- Low pressure drop: empty, loaded with particulates and ashes, low scatter of pressure drop (i.e., repeatable ΔP at a given gas flow rate and soot load)
- Chemical stability: resistance to exhaust gas components including sulfur, low reactivity with ash compounds, oxidation resistance
- Compatibility with regeneration methods: coatability with a catalyst, compatibility with fuel additives, other methods
- No generation of secondary emissions (such as fiber loss in fibrous media)
- Small size and low weight
- Long life and high durability
- Low cost.

Depending on the regeneration method, diesel filter materials may be exposed to very high temperatures (significantly in excess of 1000°C) and also rapid temperature changes. These thermal conditions are caused by the release of heat during rapid oxidation of soot accumulated in the filter. Because the distribution of soot is not necessarily uniform throughout the filter, the thermal stress frequently has a local character. Thermal phenomena, both high temperature and thermal stress, are responsible for most instances of filter failure, such as melting or cracking.

A very important aspect of filter durability, often underestimated in the older literature, is the reactivity of filter material with ash. Various metal oxide ashes originating from lube oil additives and other sources are always present in diesel particulates. Much higher quantities of ash will be present in filters that utilize metal-based additives to facilitate regeneration. These ashes can either chemically attack and corrode the filter media through solid-state reactions (e.g., creating pinholes) or they can form eutectic-type systems, which may result in dramatic deterioration of the material melting point. Ash-related problems were reported with all common filter media, including wall-flow monoliths [1] and ceramic fiber cartridges [2]. In many cases reactions with ash, rather than the material melting temperature, determine the maximum operating temperature of a given filter material.

Typical reactions of ash with a filter material [3] are illustrated in Figure 21.1. Phenomena occurring at lower temperatures, such as ash sintering, depend primarily on the properties

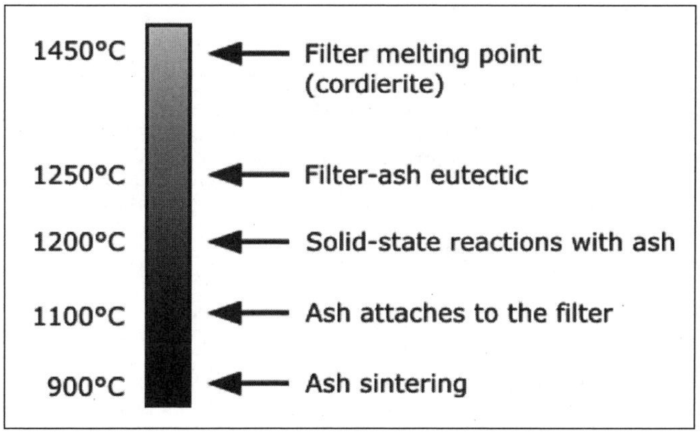

Figure 21.1 *Ash reactions in particulate filter.*

of the particular ash, rather than the filter media. Phenomena in the midtemperature range, such as solid-state corrosion or eutectic melting, depend on the interaction between ash and the particular material. In the case of cordierite, a common wall-flow filter material with a melting point of 1450°C, the maximum operating temperature is limited to about 1200°C, primarily due to ash reactions.

All of the existing filter materials have certain advantages and shortcomings. Development of new materials is ongoing, as discussed in the following section. The estimated share of particular filter materials in the global particulate filter market is shown in Table 21.1 [4].

21.1.2 Types of Materials

21.1.2.1 Wall-Flow Monoliths

Ceramic wall-flow monoliths (discussed later in Section 21.2) are by far the most common type of diesel filter. They have been utilized in the majority of commercial diesel filter systems, as well as in numerous research and demonstration projects. Monolithic filters, derived from flow-through supports used for automotive catalytic converters, were developed by Corning in cooperation with partners from the automotive industry [4]. A patent for that design—issued in 1981—was owned by General Motors [5]. The structure of a ceramic wall-flow monolith is shown in Figure 21.2. The channels are alternately plugged at each end in order to force the diesel aerosol through the porous walls. Thus, the walls of the structure act as a filter.

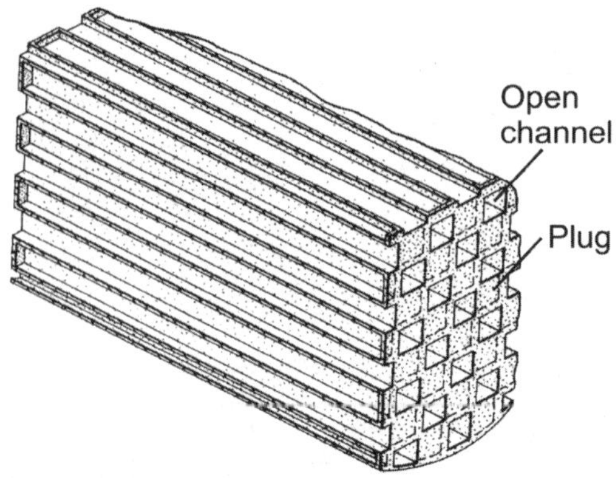

Figure 21.2 *Wall-flow monolith structure. (Reprinted from Ref. 5.)*

Monolithic filters are made of cordierite—a synthetic ceramic composition of the formula $2MgO\text{-}2Al_2O_3\text{-}5SiO_2$ that has been developed for automotive catalytic converters. Cordierite is characterized by a very low thermal expansion coefficient, which makes the material resistant to extreme thermal cycling. It also exhibits high temperature resistance (~1200°C) and good mechanical strength. Wall-flow monoliths are also made of silicon carbide (SiC). Silicon carbide is characterized by better temperature resistance (~1800°C) than cordierite. Its drawback is a higher thermal expansion coefficient and higher cost.

The filtration mechanism in ceramic monoliths is usually a combination of depth and surface (cake) filtration. In the initial phase, diesel particulates are collected in the pores within the walls through inertial and diffusional deposition. With increasing filter load, however, a layer of soot may build up at the surface of the inlet monolith channels. After that layer develops, it acts as a filtration cake.

TABLE 21.1
ESTIMATED UTILIZATION OF FILTER MATERIALS

Material	Utilization (%)
Ceramic wall-flow monolith	70
Ceramic fiber (wound and knitted)	25
Sintered metal	4
All others	1

Source: Data after Ref. 4.

Wall-flow monoliths are characterized by very good filtration efficiencies, frequently higher than 90%. The cellular design results in the compact size of the filter (although a typical monolithic filter would be still much larger than a catalytic converter for the same engine). Both cordierite and SiC monoliths can be coated with catalysts for use in passive filter systems. Drawbacks of monolith filters include a relatively high pressure drop and a fast pressure drop increase with increasing soot load, which may lead to complete clogging. For long filter durability, the regeneration process of wall-flow monoliths has to be designed so as to eliminate high temperature peaks due to the exothermic soot combustion. Ceramic monoliths may suffer premature thermal damage (melting, cracking) in filter systems where such conditions are not prevented.

21.1.2.2 Ceramic Fibers

Overview

Various types of high-temperature ceramic fibers have been used in numerous designs of diesel filtration media. In one group of designs, fiber filter cartridges are made by winding or otherwise packaging continuous fibers around steel support members. Another method involves making thin ceramic fiber sheets that resemble paper or fabric. Various configurations of filter cartridges are then made using this ceramic fiber paper.

Common regeneration methods for fiber filters include fuel additives, fuel burners, and electric heaters. Due to the difficulties associated with applying catalyst coatings to ceramic fibers, fiber cartridges usually cannot be used as catalyzed filters. In view of the advantages of passive or "passive-active" filter systems, the incompatibility with catalysts may present a serious drawback of fiber filters. Fiber filters have been commercialized to a much lesser degree than wall-flow monoliths.

The possibility of emissions of fibers exists with all types of fiber filters. Fiber fragments may be emitted during regular operation, during the initial operating period for a new filter, during filter failure, or under other circumstances. Depending on their geometry and other properties, the emitted fibers may be a health issue with this type of filters.

Wound and Knitted Fibers

One typical fiber filter cartridge design involves a layer of continuous fibers wound around a perforated metal tube. A number of cartridges in a diesel filter system are arranged in such a way that the exhaust gas is forced to flow inside the tubes and through the yarn layer, where diesel particulates are deposited on the fibers. An example of a wound fiber cartridge is a design that utilizes Nextel fibers developed by 3M [6]. A number of other designs have been developed that utilize knitted fibers [7] or support fiber layers on wire mesh, formed to provide a spiral flow path [8]. These designs are discussed in detail in Section 21.3.

Regardless of the cartridge design, ceramic fiber filters are deep-bed filters, working through practically pure depth filtration mechanisms. As such, they tend to show a somewhat lower gravimetric filtration efficiency than the wall-flow monoliths, but may still provide high reductions of particle numbers [9]. If used with fuel additives, fiber filters may be more tolerant to ash deposits than monoliths. The ash, however, may cause fiber durability problems at higher temperatures.

Fiber Papers and Fabrics

In an alternative fiber filter design, the fibers are first formed into thin sheets. Most commonly, ceramic fiber "paper" is made in a process generally similar to that used in paper making, often followed by high-temperature firing of the media. Many different designs for filter cartridges made of ceramic fiber sheets (papers or fabrics) have been developed. A pleated media diesel filter design, very similar to the typical engine intake-air cleaner element, is shown in Figure 21.3 [10]. This and other ceramic fiber paper filters operate mainly as deep-bed filters, but as the amount of captured soot increases, a surface filtration mode is also possible.

High-temperature pleated media filters have been also proposed for electrically regenerated traps [11]. Ceramic silicon carbide fibers were chosen as the filter material. The filter media included two layers—coarse and fine—of the SiC "paper" and a wire cloth, which acted as an electric heater to facilitate regeneration. The clean media showed a relatively low filtration efficiency of about 65%, which increased to more than 90% once the soot load increased. Prototype systems tested on urban buses accumulated 40,000 km.

Composite SiC-coated alumina fiber media and a wall-flow filter element design have been developed by Fleetguard [12]. The media is based on varying lengths of alumina fibers having diameters on the order of 3 μm. An aqueous slurry is prepared containing the alumina fibers, organic fibers, and a water-soluble thermoplastic binder. The slurry is formed into paper-like sheets and dried. These flexible sheets are then formed into cellular, honeycomb-type filter elements, as shown in Figure 21.4. Opposite ends of alternate channels of the honeycomb are sealed with high-temperature cement, thus forcing the gas flow through the walls. The channels have a triangular cross section with the length of a side of the triangle between 3 and 5 mm and a channel length in the range of 70 to 350 mm [13].

These "green" filter elements are finished by coating with a phenol-formaldehyde resin binder, drying, and firing at about 1000°C, followed by the application of a final silicon carbide coating by means of a chemical vapor deposition process. The resulting filter element structure was reported to be stable at temperatures of 650 to 700°C.

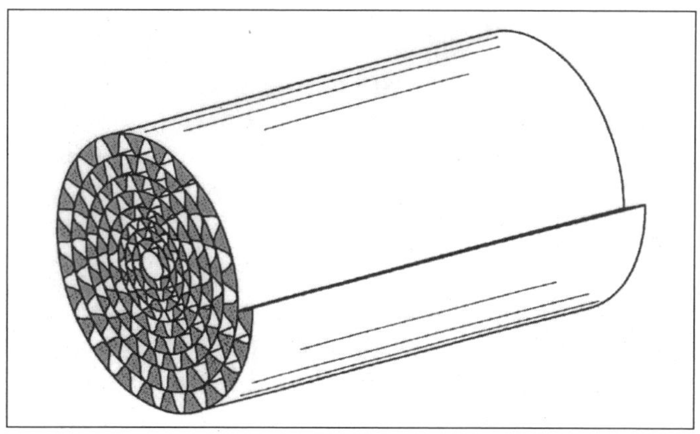

Figure 21.4 *Cellular wall-flow filter substrate made of ceramic fiber media. (Reprinted from Ref. 12.)*

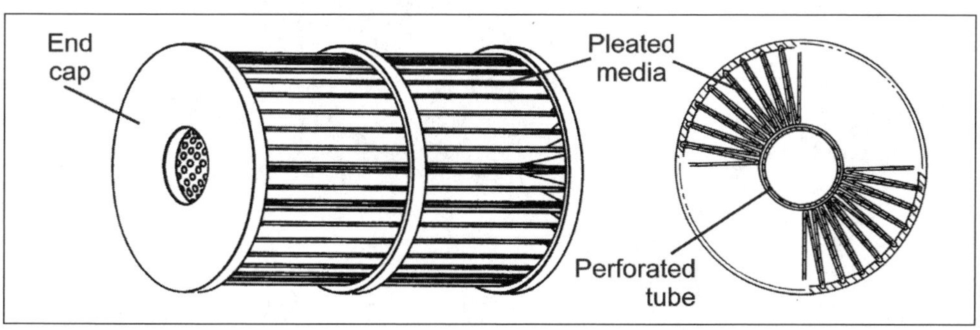

Figure 21.3 *Pleated filter made of ceramic fiber media. (Reprinted from Ref. 10.)*

21.1.2.3 Ceramic Foams

Rigid ceramic foam structures were once widely researched as diesel particulate filters. A schematic of a ceramic foam filter substrate is shown in Figure 21.5 [14]. Ceramic foam filters attract less attention in the more recent literature.

Ceramic foam filters have large, mostly open, circular pores of 250 to 500 μm in diameter and pore densities of 20 to 30 pores/cm^2 [15]. The foam is manufactured by impregnating a polyurethane foam matrix with ceramic paste. The parts are then calcined to decompose the polyurethane and obtain the rigid ceramic structure. The filters can be made, in a manner quite similar to that of the wall-flow monoliths, of either cordierite or silicon carbide. Ceramic foam filters can be also catalyzed [16].

Inertial deposition is the major filtration mechanism in ceramic foams. Typical filtration efficiencies are relatively low, usually between 60% and 70%. Due to low soot holding capacity, large filter volumes are needed with ceramic foams.

21.1.2.4 Metal Fiber Filters

Traps made of packages of wire mesh, metal wool, or metal fleece are another example of typical deep-bed filters. The attractiveness of wire mesh filters has been related to their potential compatibility with a variety of regeneration methods, including catalyzing with alumina-based washcoat or electrical regeneration. A disadvantage of many metal fiber substrates—often characterized by void volumes as high as 95%—is a tendency for blow-off of the accumulated soot.

In some designs, the metal was coated with alumina in order to enhance the collection efficiency due to the alumina whiskers on the filter surface. This type of coated steel wool filter was developed by Texaco [17]. The PM mass collection efficiency in the Texaco filter ranged between 50% and 70% [18]. The collection efficiency initially increased as the particle mass was accumulated, reached a peak, and then decreased. The decrease in efficiency was explained by reentrainment of agglomerates of collected particles; the coarse particle mass fraction leaving the filter increased markedly with time.

Metal fleece for diesel filter applications is being developed by Bekaert [19]. The particle trapping medium is sintered fleece consisting of drawn metal filaments having a medium diameter of 21 μm. The fleece is made of a Fe-Cr alloy with the addition of Y (Fecralloy A by AERE Harwell). Sheets of the fleece are formed into pleated cartridges. An electrically regenerated design was also developed, in which the fleece itself is used as a resistive heater [20].

21.1.2.5 Sintered Metal

Sintered metal powder materials have been developed that show high filtration efficiencies for diesel particulates. The most common design is based on a stainless steel wire mesh sheet, which is coated with metal powder mixed with additives. Once the coated sheet is sintered, it can be formed into various filter assemblies through standard sheet metal processing techniques, such as cutting, stamping, seaming, and welding.

In filters manufactured by Purem, sintered metal sheets are typically cut into annular plates, which are alternately joined along their inside and outside perimeters to form filter structures [21], as shown in Figure 21.6. Stamped protrusions ensure that there is always space for gas flow between adjacent sheets. Selected properties of Purem sintered metal are listed in Table 21.2 [21]. The mechanical properties are dominated by the wire mesh; the porosity is given through the metal powder and its sintering process.

Sintered metal filters cannot be catalyzed. They have been tested primarily together with fuel additives [22] or else with an upstream NO_2 catalyst (CRT configuration). According to some opinions, sintered metal substrates are compatible with additives, showing more resistance against plugging with additive ash than other types of substrates.

Sintered metals have a number of advantages related to a metal material. They can be flexibly fabricated in a variety of shapes to accommodate space constraints. On the negative side, they are characterized by their relatively heavy weight.

21.1.2.6 Other Metal Filters

A number of other metal filter designs have been proposed, including a porous metal substrate made of Ni-Cr-Al alloy [23]. This porous metal media is made of thin sheets (50 to 100 μm) of metallic foam with pores of 150 to 400 μm in diameter. Sheets

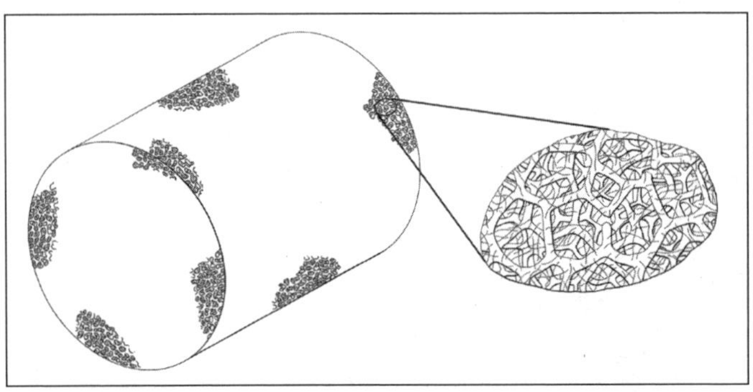

Figure 21.5 Ceramic foam filter. (Reprinted from Ref. 14.)

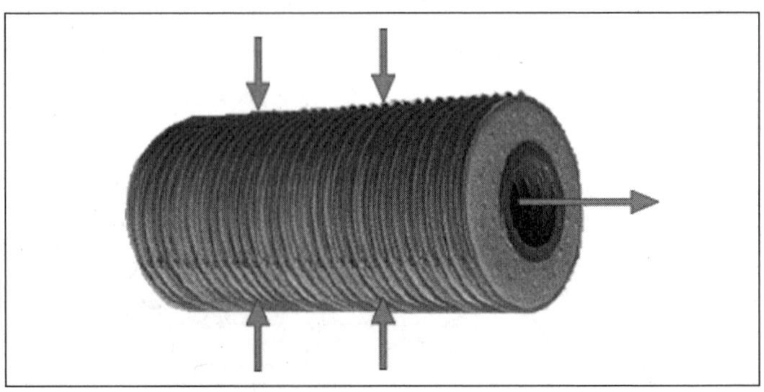

Figure 21.6 Flow direction in sintered metal filter assembly. (Courtesy of Purem.)

TABLE 21.2
PROPERTIES OF PUREM SINTERED METAL SHEETS

Property	Value
Material name	Sintered Metal Light
Material type	1.4845
Weight	2200 g/m^2
Thickness	0.55 mm
Porosity	50%
Mean pore size	10 μm
Permeability (Darcy coefficient)	1.5[a] μm^2
Heat conductivity (20°C)	14 W/m K
Specific heat	500 J/kg K
Thermal expansion coefficient (20–1000°C)	18 × 10^{-7} K^{-1}
Tensile strength	60 MPa
Young's modulus	14–26[a] GPa
Elongation at break	24[a] %

a. Intrinsic material property.

of porous metal media were alternated with electric heater sheets in a particulate filter prototype.

Another filter design, utilizing spirally wound flat sheets of metallic fabric alternated with corrugated sheets of metal foil, is shown in Figure 21.7 [24]. The inlet to the corrugated area is alternately sealed at the opposite ends, making this structure a wall-flow type of filter. As the exhaust gases flow through the metallic fabric, particulates are trapped on the metal filaments and the alumina whiskers.

21.1.2.7 Pleated Paper Filters

Pleated paper cartridges, which resemble regular engine intake-air cleaners, can be used as disposable diesel filters. Paper filters exhibit excellent, close to 100%, filtration efficiency. However, such systems require the exhaust gas to be cooled. There are two temperature considerations with paper filters: (1) the mechanical durability of the filter media and (2) the possibility of starting a fire as a result of soot accumulated in the filter. The maximum temperature limit of regular filter media varies from 80 to 120°C [25]. Application of such media requires that substantial quantities of heat be dissipated from the exhaust gases, for example, through the installation of a dedicated exhaust gas cooler [26]. Synthetic papers of increased maximum operating temperature offer only limited advantage, due to the possibility of fire in hot filters, which is a critical safety issue. Such high-temperature paper filters had been under development [27], but the idea was later abandoned. Other issues with paper filters include generation of solid waste and maintenance intensity.

Paper diesel filters found limited use in underground coal mining in the United States [28]. Such filters are used on some permissible vehicles which operate in explosive atmospheres. The maximum temperature of exhaust gases discharged from permissible vehicles is limited by safety regulations. On-vehicle exhaust coolers (water scrubbers or heat exchangers) that are used to meet exhaust temperature requirements make these vehicles compatible with paper diesel filters. It is important for the exhaust gas coolers to be equipped with fail-safe controls. Otherwise, fire hazard exists whenever hot exhaust gases enter the filter cartridge.

21.2 Wall-Flow Monoliths

21.2.1 Monolith Construction

Ceramic wall-flow monoliths, which are derived from the flow-through cellular supports used for catalytic converters, are the most common type of diesel filter substrate. They are distinguished, among other diesel filter designs, by high surface area per unit volume and by high filtration efficiencies. Monolithic diesel filters consist of many small parallel channels, typically of square cross section, running axially through the part. Diesel

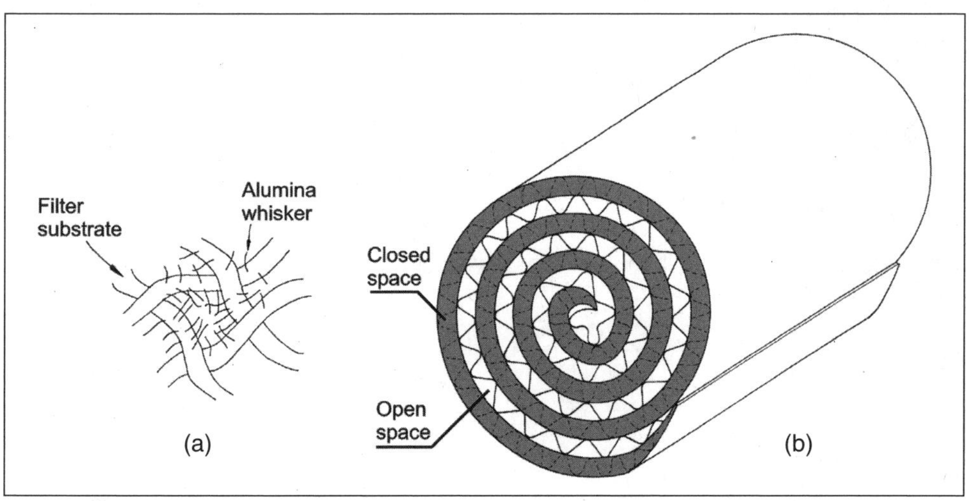

Figure 21.7 *Metal wall-flow monolith: (a) metal fiber fabric detail and (b) wall-flow substrate. (Reprinted from Ref. 24.)*

filter monoliths are obtained from the flow-through monoliths by plugging channels as shown in Figure 21.8. Adjacent channels are alternatively plugged at each end in order to force the diesel aerosol through the porous substrate walls, which act as a mechanical filter. To reflect this flow pattern, the substrates are referred to as wall-flow monoliths.

Wall-flow monoliths are most commonly available in cylindrical shapes, as shown in Figure 21.9, although oval cross-section parts are also possible for space-constrained applications.

Wall-flow filter walls have a distribution of fine pores whose production has to be carefully controlled during the manufacturing process. The total material porosity is typically between 45% and 50%, while the medium pore sizes usually range from 10 to 15 μm. Older materials, which are obsolete today, had pores up to about 35 μm in diameter. The filtration mechanism on monolith wall-flow filters is a combination of cake and depth filtration. Depth filtration is the dominant mechanism on a clean filter as the particulates are deposited in the inside of pores. As the soot load increases, a particulate layer develops at the inlet channel walls and cake filtration becomes the prevailing mechanism. Typically, monolith filters have filtration efficiencies between about 70% and 95% of total particulate matter (TPM). Higher efficiencies are observed for solid particulate matter (PM) fractions, such as elemental carbon and metal ash [as already discussed, particulate filters may be ineffective in controlling soluble organic fraction (SOF) and sulfate particulates].

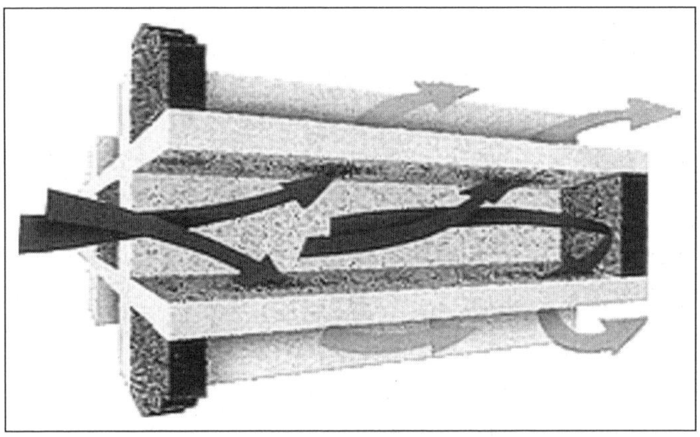

Figure 21.8 *Gas flow in a monolith filter. (Illustration copyright Corning Incorporated. All rights reserved. Reprinted by permission.)*

Figure 21.9 *Cylindrical wall-flow monoliths: (left front) silicon carbide and (center and right) cordierite. (Courtesy NGK Automotive Ceramics USA, Inc.)*

Wall-flow monoliths are typically extrusions made from porous ceramic materials. Two types of filter materials have been commercialized, cordierite and silicon carbide (Table 21.3). Cordierite is a synthetic ceramic developed for flow-through catalyst substrates and subsequently adapted for filter applications. Silicon carbide has been used for a long time in a number of industries for such applications as semiconductors, abrasives, or high-temperature/molten metal contact materials. More recently, it was successfully introduced as a diesel filter material.

21.2.2 Monolith Characterization Parameters

21.2.2.1 Cellular Properties

A number of cellular properties were introduced in the discussion of cellular monolith substrates for flow-through catalytic converters in Section 18.3. In many cases definitions of these properties apply to wall-flow filters as well. In some properties, however, changes need to be introduced to account for the plugged channel ends and different flow pattern. In such cases, the corresponding properties in flow-through and in wall-flow substrates should not be confused.

Cell geometry in the wall-flow monolith with its characteristic checkerboard cell plugging pattern is shown in Figure 21.10. A number of monolith properties can be expressed in terms of the wall thickness (w) and the cell repeat distance (s), also known as the cell spacing [29].

The substrate cell density (N), defined as the number of channels per unit of cross-sectional area, is given for square cell geometry by Eq. (21.1). The metric unit of cell density is $1/cm^2$, but the U.S. customary unit of cells per square inch (cpsi) remains widely used in the industry.

TABLE 21.3
COMMERCIAL MONOLITH MATERIALS

Material	Formula	Monolith Suppliers
Cordierite	$2MgO\text{-}2Al_2O_3\text{-}5SiO_2$	Corning, NGK, Denso
Silicon carbide	SiC	Ibiden, NGK, NoTox/Stobbe DPF

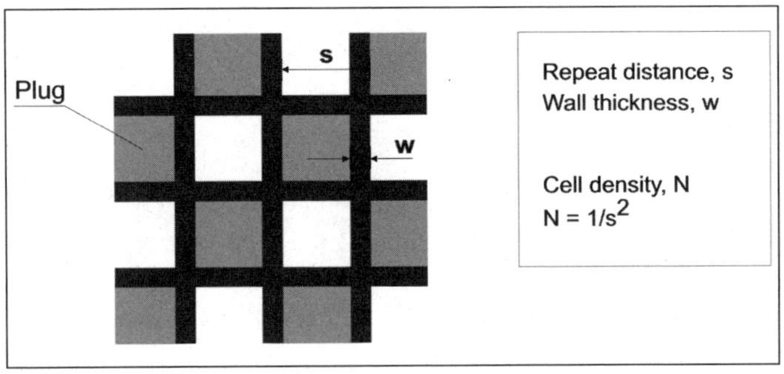

Figure 21.10 *Cell geometry.*

$$N = 1/s^2 \quad (21.1)$$

Cell hydraulic diameter (d_h) is a measure of channel size in hydraulics, mass transfer, and heat transfer calculations. In the case of square cells, it can be simply expressed as the difference between repeat distance and wall thickness, as follows:

$$d_h = s - w \quad (21.20)$$

Open frontal area (OFA) is defined as the part of the total substrate cross-sectional area that is available for the flow of gas (i.e., the cross-sectional area of the filter inlet channels). It is typically expressed relative to the total substrate cross section, as given by Eq. (21.3):

$$OFA = \tfrac{1}{2}\,[(s - w)/s]^2 = \tfrac{1}{2}\,(1 - w/s)^2 \quad (21.3)$$

Specific filtration area (SFA) is the total channel surface area per unit of substrate volume. For square cells, it can be calculated from Eq. (21.4). It should be noted that the SFA increases with increasing cell density (N).

$$SFA = 2\,(s - w)/s^2 = 2\,(s - w)\,N \quad (21.4)$$

Total filtration area (TFA) can be calculated as the product of SFA and the filter volume (V_f), as follows:

$$TFA = SFA \cdot V_f \quad (21.5)$$

$$V_f = \tfrac{1}{4}\pi D^2 L \quad (21.6)$$

where D is the filter diameter and L is the filter length. (Depending on the required precision, L may be corrected for the "inactive" channel length taken by the end plugs.)

The filter cell configuration also influences its mechanical integrity. Cell configurations of higher mechanical integrity have a higher mechanical integrity factor (MIF). For a given wall porosity, the MIF is defined by Eq. (21.7):

$$MIF = w^2/[s\,(s - w)] \quad (21.7)$$

Selected cellular properties calculated for a number of monolith configurations (some commercial, others under development) are listed in Table 21.4.

Configuration 200/12 from Table 21.4 has been designed to provide the same OFA and, thus, the same flow velocity in channels as the basic configuration 100/17. Commercial 200-cell substrates, however, use thicker walls than 0.012 in. For example, Ibiden commercialized the 200/14 geometry with a 0.014-in. wall. The Corning RC-200/19 provides an even more sturdy structure. In this case, the thick, 0.019-in. wall was introduced to provide certain thermal properties (to be discussed later), rather than just mechanical strength.

TABLE 21.4
PROPERTIES OF MONOLITHS WITH DIFFERENT CELL CONFIGURATIONS

Cell Configuration	100/17	200/12	200/14	200/19	300/12
Commercial monolith	Corning EX-80	—	Ibiden	Corning RC-200/19	—
Cell density, N, cpsi	100	200	200	200	300
• 1/cm^2	15.5	31	31	31	46.5
Repeat distance, s, mm	2.54	1.80	1.80	1.80	1.47
• inch	0.100	0.071	0.071	0.071	0.057
Wall thickness, w, mm	0.43	0.30	0.36	0.48	0.30
• inch	0.017	0.012	0.014	0.019	0.012
Open frontal area, OFA	0.344	0.345	0.322	0.268	0.312
Specific filtration area, SFA, 1/cm	6.5	9.2	8.9	8.1	10.9
• in.^{2}/in.^{3}	16.6	23.4	22.6	20.6	27.7
MIF	0.035	0.034	0.048	0.098	0.056

21.2.2.2 Physical and Thermal Properties

Filter monolith materials are characterized using basically the same set of properties that is used for the description of ceramic flow-through catalyst supports (Section 18.4). Certain groups of properties, however, play a more important role in filter monoliths than in the flow-through substrates. Due to the possibility of the release of high quantities of heat during filter regeneration, filter materials are required to demonstrate excellent thermal properties in terms of resistance to both high temperatures and high temperature gradients. Insufficient temperature tolerance may result in melting of the material, while insufficient thermal shock resistance causes cracking.

Contrary to flow-through substrates in which support material is separated from gases by a thick layer of washcoat, filter material comes in much more direct contact with the exhaust gas. Material pore characteristics (size, percent porosity, pore connectivity, open vs. closed pores, etc.) influence the physical interaction between gas and filter material and affect such properties as filtration efficiency and pressure drop. Furthermore, filter durability depends on the material resistance to chemical attack by exhaust gas components. In particular, materials need to be resistant to corrosion by the metal ash that may be part of diesel particulates.

Thermal Expansion and Strength

The coefficient of thermal expansion (CTE) is a measure of how much a given body expands (or contracts) on heating. When the filter monolith is exposed to thermal gradients, for example, as a result of localized heat release during regeneration, the hotter parts expand relative to the colder regions straining the material. This strain creates thermal stress in the material, which increases with increasing thermal gradient (or the resulting temperature difference, ΔT), increasing CTE, and increasing elastic modulus of the material (E). If the thermal stress exceeds the tensile strength of the material, the monolith will crack. In practical applications, tensile stress in the substrate must be kept well below its strength value to provide sufficient margin for the material fatigue (i.e., a loss of strength when stressed for a long time). Therefore, filter materials of low CTE and low E are preferred.

While the tensile strength is important for thermal stress durability during filter operation, its compressive strength determines the maximum allowable stress from the canning system, during both packaging and operation. In general, the strength parameters can be classified as follows:

Tensile Strength

- Modulus of rupture (MOR) in axial and tangential direction of the part

Compressive Strength

- Uniaxial crush strength along the axial (A), tangential (B), and diagonal (C) directions
- 3-D isostatic strength, determined by pressurization of the whole substrate
- 2-D isostatic strength, determined by biaxial compressive crushing of the whole substrate

Thermal expansion and strength parameters were discussed in more detail in Section 18.4.

Thermal Shock Resistance

The most common measure of thermal shock resistance is the thermal shock parameter (TSP), which represents the ratio of mechanical strain tolerance to thermal strain imposed by temperature gradient. The higher the TSP, the better the thermal shock capability of the material. The TSP values presented throughout this section are calculated from the following formula:

$$TSP = MOR/(E \cdot CTE) \tag{21.8}$$

where:

MOR = modulus of rupture, Pa
E = elastic modulus, Pa
CTE = coefficient of thermal expansion, 1/°C

TSP is a dimensionless factor; for the sake of simplicity, temperature difference (°C) has been omitted in Eq. (21.8).

Another method of quantifying the thermal shock resistance is the quenching or water immersion test [30]. Hot filter samples are immersed in water and evaluated for cracks. The quench test rating is determined as the temperature difference between the part and the water at which cracks appeared in the part (i.e., the higher the quench test temperature, the better the thermal shock resistance).

Melting Point and Maximum Operating Temperature

The melting point of the monolith material must be sufficiently high to withstand the normal use temperature, as well as temperature spikes occurring during controlled and uncontrolled regeneration. Uncontrolled regenerations can produce very high temperatures. In several applications the melting point of cordierite of 1450°C has been proven to be insufficient, thus driving the development of improved filter materials [31].

A quiet assumption was made in the older literature that the maximum operating temperature of a diesel filter was equal to its melting temperature (or a temperature where the material lost its properties, if a significant property loss occurred below the melting point). Newer studies show that the maximum practical use temperature of a diesel particulate filter may, in fact, be limited by the interaction of the material with ash, rather than by the melting point of the pure material [1]. These considerations are certainly valid with regard to long-term filter durability. Diesel PM ashes start sintering at about 900°C. If the temperature is further increased, the ash will adhere to the filter surface, thus making it impossible to periodically remove ashes from the filter, resulting in irreversible increase of pressure drop.